Vegetable Crops Science

Vegetable Crops Science

Edited by
Dr. M.K. Rana
Professor
Department of Vegetable Science
CCS Haryana Agricultural University
Hisar, Haryana, India

CRC Press
Taylor & Francis Group
Boca Raton London New York

CRC Press is an imprint of the
Taylor & Francis Group, an **informa** business

CRC Press
Taylor & Francis Group
6000 Broken Sound Parkway NW, Suite 300
Boca Raton, FL 33487-2742

First issued in paperback 2020

© 2018 by Taylor & Francis Group, LLC

CRC Press is an imprint of Taylor & Francis Group, an Informa business

No claim to original U.S. Government works

ISBN 13: 978-0-367-57294-5 (pbk)
ISBN 13: 978-1-138-03521-8 (hbk)

Visit the Taylor & Francis Web site at
"http://www.taylorandfrancis.com" http://www.taylorandfrancis.com

and the CRC Press Web site at
"http://www.crcpress.com" http://www.crcpress.com

Contents

Preface

Vegetables, the cheapest and nutrient-rich food sources within the economic reach of the poor, play a vital role in the human diet since they provide carbohydrates, protein, fat, minerals, vitamins, fibres, and phytochemicals (nonnutrient bioactive compounds), which are essential for making the body's immune system strong, detoxifying carcinogens, reducing muscular degeneration, and protecting the body from infectious ailments. The importance of eating vegetables for the maintenance of good health is now well recognized all over the world. Vegetables, besides having medicinal values, also play an important role in nutritional security and national economy of India since they are quick growing and short duration, and their yield per unit is 5–10 times higher than the cereal crops.

In the last few decades, vegetable production has accomplished great significance in the world trade, and with high-income suppleness, the demand for vegetables has escalated, and simultaneously the production of vegetables has increased enormously. India has emerged as the second largest producer in the world next to China due to the sincere efforts made by the Indian Council of Agricultural Research, State Agricultural Universities, State Agricultural Departments, Private Seed Agencies, and growers in this direction. The fast increase in technology generation in the field of vegetable science has helped in achieving this level of vegetable production in India. The country today is almost self-sufficient in the production of vegetables, and even in a position to export some of its surplus vegetables; however, the self-sufficiency in a true sense can only be realized when each individual in India is assured with the recommended quantity of vegetables, which is 300 g per head per day, including 120 g leafy, 90 g roots and 90 g other vegetables. However, in India per capita per day, the consumption of vegetables is much lower than the standard requirement; hence, to overcome the problems of under- and malnutrition, the intake of vegetables shall have to be still increased substantially.

India is endowed with varied climatic conditions and a tremendous wealth of biodiversity with respect to genetic material of vegetable crops. After independence, tremendous progress has been made in the development of high-yielding varieties and hybrids, covering more than 80% of the area. During the recent past, progress has been made in the evolution of off-season varieties of certain vegetable crops, which have made the cultivation of those crops round the year. Similarly, valuable progress has been made in the development, standardization and management of technology in respect to the judicious use of organic manure, mineral nutrients, bio-fertilizers, weedicides, irrigation, plant growth regulators, plant protection measures including bio-control, and the vegetable-based cropping system for higher and quality production of vegetable crops. Likewise, the technology for quality seed production is progressing to meet the future needs and challenges. Production technology is incessantly changing; hence, based on the evolution of new varieties, the new insect-pests, diseases and other inputs and ever-arising needs the technology is simultaneously refined and tuned.

Teaching courses on production technology of vegetable crops to B.Sc. Agriculture, B.Sc. Home Science, and MSc Vegetable Science students aims to provide essential knowledge concerning basic principles for growing vegetable crops at home or on commercial scale. This book, containing very concise and precise information on production aspects of vegetable crops, has been written in a very simple language, which can be explicable to the undergraduate students. This book has also been prepared to provide every aspect of the production technology of vegetable crops along with the knowledge regarding origin, introduction, composition, uses, medicinal uses, improved varieties/hybrids, sowing time, seed rate, requirement of climate, soil, manure, fertilizers, irrigation, intercultural operations, weedicides, harvesting, yield, postharvest management, insect-pests and diseases along with their control measures. It also contains the information on best possible solutions for problems faced by the students during rural agriculture work experience (RAWE). Moreover, vegetable growers would find this book very useful and valuable for raising vegetable crops successfully and economically. The information given in this book is truly based on scientific records of scientists working on vegetables in various organizations.

Considering the importance of vegetable production in view and making the students familiar about technical aspects of vegetable cultivation, the author was obliged to prepare a book that may sequentially presents vegetables in alphabetical order by family. This would help them grasp a basic knowledge of vegetable production. Though very few books on production aspects of vegetable crops are available in the library, this book on production technology of vegetable crops compiled for the students of undergraduate and postgraduate programs is one such attempt to make them learn and understand the subject of vegetable production more precisely and motivate them to improve their knowledge in the field of vegetable science to meet the future needs and serve their best to the vegetable growing community. In addition, this book may be user-friendly to others who have the concern to expand their basic knowledge in the field of vegetable production and wish to grow vegetable crops profitably at their farms. Earning scientific knowledge will undoubtedly be rewarding to its users and finally to the nation.

– M.K. Rana

Editor

Dr. M.K. Rana, MSc and PhD in vegetable science and presently working as a professor in the Department of Vegetable Science, CCS Haryana Agricultural University, Hisar, Haryana, India, has a brilliant academic career standing first in all the examinations. Professor Rana also served as a lecturer of agriculture in NCERT Regional College of Education, Ajmer, Rajasthan, India. He is specialized in the field of abiotic stress, vegetable physiology, and postharvest technology and processing of vegetables.

He has an excellent experience of 28 years in the field of teaching in CCS Haryana Agricultural University, Hisar, and as a teacher, he has taught more than 20 different courses to the students of undergraduate, postgraduate, and doctorate classes, delivered several lectures in the trainings and given talks on radio as well as on television on different topics related to vegetable science. As a proficient writer, Dr. Rana has published 164 book chapters, 12 research reviews, 90 research papers, 37 research abstracts, and 112 scientific articles in sundry books, journals, and magazines of scientific repute, has published 14 books and is publishing 2 books within the year, 12 teaching manuals, 2 technical bulletins on spices and edited 3 symposia abstracts, 2 symposia proceedings, 2 issues of Haryana Journal of Horticultural Sciences, 1 compendium, and has guided 15 MSc and 3 PhD students.

He served as an executive councilor and editor of the Haryana Journal of Horticultural Sciences and as a joint secretary of the National Symposium organized by the Horticultural Society of Haryana. He has also worked as paper reviewer in several national societies. Dr. Rana is a life member of 11 professional bodies and societies and has been an examiner in different universities. Dr. Rana has been awarded the ICAR Team Research Award for the year 1993–1994 in recognition of his outstanding research contribution in the field of abiotic stress and University Best Teacher Award for the year 2004–2005 in recognition of his dedicated service rendered in the university. He had the opportunity to visit Michigan State University for 3 months training in the field of postharvest handling and processing of fruits and vegetables. He has been the member secretary of the Board of Subject Matter Area Committee (ICAR).

Contributors

Baseerat Afroza
Division of Vegetable Science
Sher-e-Kashmir University of Agriculture
    Science and Technology
Srinagar, Jammu and Kashmir, India

Tarique Aslam
Department of Vegetable Science
CCS Haryana Agricultural University
Hisar, Haryana, India

Mahesh Badiger
Division of Vegetable Science
Indian Agricultural Research Institute
Pusa, New Delhi, India

Shashi Kumar Bairwa
Krishi Vigyan Kendra
S.K.N. Agriculture University
Jobner, Rajasthan, India

M.V. Bharathkumar
Department of Vegetable Science
CCS Haryana Agricultural University
Hisar, Haryana, India

Archana Brar
Department of Vegetable Science
CCS Haryana Agricultural University
Hisar, Haryana, India

Navjot Singh Brar
Department of Vegetable Science
CCS Haryana Agricultural University
Hisar, Haryana, India

Sachin S. Chikkeri
Department of Vegetable Science
CCS Haryana Agricultural University
Hisar, Haryana, India

B.R. Choudhary
Central Institute for Arid Horticulture
Bikaner, Rajasthan, India

J. Debata
Department of Fruit Science
College of Horticulture
Orissa University of Agriculture and Technology
Sambalpur, Odisha, India

Sugani Devi
Department of Horticulture
Swami Keshwanand Rajasthan Agricultural
    University
Bikaner, Rajasthan, India

S.K. Dhankhar
Department of Vegetable Science
CCS Haryana Agricultural University
Hisar, Haryana, India

D.K. Dora
Dean Postgraduate Studies
Orissa University of Agriculture and
    Technology
Bhubaneswar, Odisha, India

D.S. Duhan
Department of Vegetable Science
CCS Haryana Agricultural University
Hisar, Haryana, India

Shashikant Evoor
Department of Vegetable Science
KRCCH, Arabhavi, University of Horticultural
    Sciences
Bagalkot, Karnataka, India

Nivedita Gupta
Department of Vegetable Science &
    Floriculture
CSK Himachal Pradesh Agricultural University
Palampur, Himachal Pradesh, India

R.K. Gupta
Vegetable Science and Floriculture
Sher-e-Kashmir University of Agricultural
    Sciences and Technology of Jammu
Srinagar, Jammu and Kashmir, India

Surbhi Gupta
Agriculture Officer, Himachal Pradesh
Crop Diversification Promotion Project, JICA
    ODA, DPMU
Palampur, Himachal Pradesh, India

S.M. Haldhar
Central Institute for Arid Horticulture
Bikaner, Rajasthan, India

V.S. Hooda
Department of Agronomy
CCS Haryana Agricultural University
Hisar, Haryana, India

K.B. Jagtap
Department of Vegtable science
College of Horticulture, Pune
Mahatma Phule Krishi Vidyapeeth
Rahuri, Maharashtra, India

P. Jansirani
Horticultural College and Research Institute
Tamil Nadu Agricultural University
Coimbatore, Tamil Nadu, India

Pradeep Kumar Jatav
Department of Vegetable Science
CSK Himachal Pradesh Agricultural
    University
Hisar, Haryana, India

Naval Kishor Kamboj
Department of Vegetable Science
CCS Haryana Agricultural University
Hisar, Haryana, India

Shiwani Kamboj
Department of Vegetable Science
CCS Haryana Agricultural University
Hisar, Haryana, India

Prashant Kaushik
Department of Vegetable Science
Instituto Universitario de Conservación y
    Mejora de la Agrodiversidad Valenciana
Universidad Politécnica de Valencia (UPV)
Valencia, Spain

D.S. Khurana
Department of Vegetable Science
Punjab Agricultural University
Ludhiana, Punjab, India

Raj Kumar
Division of Vegetable Science
Indian Agricultural Research Institute
Pusa, New Delhi, India

Sanjay Kumar
Department of Vegetable Science
CCS Haryana Agricultural University
Hisar, Haryana, India

Sunny Kumar
Department of Vegetable Science
CCS Haryana Agricultural University
Hisar, Haryana, India

Suresh Kumar
Central Institute for Arid Horticulture
Bikaner, Rajasthan, India

N.C. Mamatha
Department of Vegetable Science
CCS Haryana Agricultural University
Hisar, Haryana, India

N. Meenakshi
Horticultural College and Research Institute
Tamil Nadu Agricultural University
Coimbatore, Tamil Nadu, India

Harshita Mor
Department of Vegetable Science
CCS Haryana Agricultural University
Hisar, Haryana, India

Ravindra Mulge
Department of Vegetable Science
KRCCH, Arabhavi, University of Horticultural
    Sciences
Bagalkot, Karnataka, India

Manoj Kumar Nalla
Division of Vegetable Science
Indian Agricultural Research Institute
Pusa, New Delhi, India

A.K. Pandav
Department of Vegetable Science
CCS Haryana Agricultural University
Hisar, Haryana, India

Asish Kumar Panda
Department of Horticulture
Tamil Nadu Agricultural University
Coimbatore, Tamil Nadu, India

V.P.S. Panghal
Department of Vegetable Science
CCS Haryana Agricultural University
Hisar, Haryana, India

K.L. Patel
Department of Horticulture
Indira Gandhi Agricultural University
Raipur, Chhattisgarh, India

P.V. Patil
Department of Horticulture
Mahatma Phule Krishi Vidyapeeth
Rahuri, Maharashtra, India

L. Pugalendhi
Department of Vegetable Crops
Horticultural College and Research Institute
Tamil Nadu Agricultural University
Coimbatore, Tamil Nadu, India

Monalisha Pradhan
Department of Vegetable Science
College of Agriculture
Orissa University of Agriculture &
    Technology
Bhubaneswar, Odisha, India

Chanchal Rana
Department of Vegetable Science and
    Floriculture, College of Agriculture
CSK Himachal Pradesh Agricultural University
Palampur, Himachal Pradesh, India

M.K. Rana
Department of Vegetable Science
CCS Haryana Agricultural University
Hisar, Haryana, India

Pooja Rani
Department of Vegetable Science
CCS Haryana Agricultural University
Hisar, Haryana, India

S.A. Ranpise
Department of Horticulture
Mahatma Phule Krishi Vidyapeeth
Rahuri, Maharashtra, India

P. Karthik Reddy
Department of Horticulture
S.V. Agriculture College
Tirupathi Acharya N.G. Ranga Agricultural
    University
Hyderabad, Andhra Pradesh, India

G.K. Sadananda
Department of Post-harvest Technology
University of Horticultural Sciences Bagalkot
Bagalkot, Karnataka, India

A.T. Sadashiva
Division of Vegetable Crops
Indian Institute of Horticultural Research
Bangalore, Karnataka, India

Kiran Sagwal
Department of Vegetable Science
CCS Haryana Agricultural University
Hisar, Haryana, India

Dipika Sahoo
Department of Vegetable Science
Orissa University of Agriculture and
    Technology
Sambalpur, Odisha, India

T. Saraswathi
Horticultural College and Research Institute
Tamil Nadu Agricultural University
Coimbatore, Tamil Nadu, India

Akhilesh Sharma
Department of Vegetable Science and
    Floriculture
CSK Himachal Pradesh Agricultural University
Palampur, Himachal Pradesh, India

Bunty Sharma
Department of Genetics and Plant Breeding
    (Oilseeds Section)
CCS Haryana Agricultural University
Hisar, Haryana, India

G.L. Sharma
Department of Horticulture
Indira Gandhi Agricultural University
Raipur, Chhattisgarh, India

Pardeep Kumar Sharma
Department of Vegetable Science &
    Floriculture
CSK Himachal Pradesh Agricultural University
Palampur, Himachal Pradesh, India

M.K. Shivaprasad
Department of Vegetable Science
CCS Haryana Agricultural University
Hisar, Haryana, India

Hira Singh
Department of Vegetable Science
Punjab Agricultural University
Ludhiana, Punjab, India

Ravinder Singh
CSK Himachal Pradesh Agricultural University
Palampur, Himachal Pradesh, India

Ruchi Sood
Agriculture Expert, Himachal Pradesh Crop
Diversification Promotion Project JICA ODA,
    DPMU
Palampur, Himachal Pradesh, India

Sonia Sood
Department of Vegetable Science &
    Floriculture
CSK Himachal Pradesh Agricultural University
Palampur, Himachal Pradesh, India

R. Sridevi
Department of Seed Science and Technology
Tamil Nadu Agricultural University
Coimbatore, Tamil Nadu, India

Mekala Srikanth
Department of Vegetable Science
CCS Haryana Agricultural University
Hisar, Haryana, India

S.C. Swain
Department of Horticulture
Orissa University of Agriculture and
    Technology
Bhubaneswar, Odisha, India

R.K. Tarai
Department of Fruit Science
Orissa University of Agriculture and
    Technology
Sambalpur, Odisha, India

Bhupender Singh Thakur
Regional Horticultural Research and Training,
    Centre
YS Parmar University of Horticulture and
    Forestry
Solan, Himachal Pradesh, India

K. Shoba Thingalmaniyan
Department of Vegetable Crops
Horticultural college and Research
    Institute
Tamil Nadu Agricultural University
Coimbatore, Tamil Nadu, India

Pradyumna Tripathy
Department of Vegetable Science
Orissa University of Agriculture &
    Technology
Bhubaneswar, Odisha, India

B. Varalakshmi
Division of Vegetable Crops
Indian Institute of Horticultural Research
Bangalore, Karnataka, India

Deven Verma
Division of Vegetable Crops
Indian Institute of Horticultural Research
Bangalore, Karnataka, India

A. Vijayakumar
Department of Seed Science and Technology
Tamil Nadu Agricultural University
Coimbatore, Tamil Nadu, India

Chandanshive Aniket Vilas
Department of Vegetable Science
CCS Haryana Agricultural University
Hisar, Haryana, India

Flemine Xavier
Division of Vegetable Crops
Indian Agricultural Research Institute
Pusa, New Delhi, India

Neha Yadav
Department of Vegetable Science
CCS Haryana Agricultural University
Hisar, Haryana, India

1 New Zealand Spinach

S.K. Dhankhar and Chandanshive Aniket Vilas

Botanical Name   : *Tetragonia tetragoniodes L.*

Family   : Aizoaceae

Chromosome Number : 2n = 32

ORIGIN AND DISTRIBUTION

New Zealand spinach is native to the islands of New Zealand, as many coastal parts of New Zealand have been the homes to this creeping plant prior to the colonial period. For more than 200 years, it was the only exported vegetable native to New Zealand or Australia. This herb is found growing wild along the margins of salt marshy lands, in protected sites all along the coast, in all parts of New South Wales, Chile, Japan, across the Australia, Argentina and surrounding areas. It can be found as an invasive plant in North and South America and has been cultivated along the East Asian rim. After exploring the coasts of New Zealand, Captain Cook recognized it to be a nutritious vegetable and introduced it to the world in the 17th century. The sailors on the *Endeavour* prized this plant because they desperately needed vitamin C. The native Maori who inhabit the island country referred it to as *kokihi* or *rengamtu*. The species was rarely used by the indigenous people as a leafy vegetable. Sooner or later, it was taken back to England, where it was introduced in 1772 by Sir Joseph Banks, an explorer and botanist. It spread during the last half of the 18th century towards many parts of the world. It was first listed in a seed catalogue by Fearing Burr in 1863.

INTRODUCTION

New Zealand spinach, also known as Botany Cook's cabbage, bay spinach, sea spinach, everbearing spinach, everlasting spinach, perpetual spinach, Warringal greens, tetragon and Della Nuova Zelanda, is a leafy groundcover vegetable and is an evergreen perennial plant in cold climate but widely grown as an annual herbaceous tender plant. The native people of New Zealand also call it Maori spinach. Because of its succulent leaves, it is sporadically referred to as ice plant. In spite of having similarity in name, it is entirely different from common spinach but used in same way and is more resistant to bolting during the hot summer months as compared to regular spinach. It can be grown in tropical, subtropical, Mediterranean, or temperate climatic regions. It is a warm season perennial often used as an excellent alternative to regular spinach since it thrives well in hot and dry conditions. It can be grown until frost in temperate areas and year-round in subtropical areas. Its flavour is very similar to common spinach with the same sharp taste but it does not get as bitter a taste.

COMPOSITION AND USES

COMPOSITION

New Zealand spinach is nutritionally almost similar to common spinach. It is an excellent source of minerals like iron, potassium, sodium and magnesium and vitamin A, vitamin B$_1$, vitamin B$_2$, vitamin C and vitamin K. It is low in fat, fibre and calorie. It has a balance of calcium to phosphorus levels that make it ideal for calcium absorption in the body. The nutritional composition of New Zealand spinach is given in Table 1.1.

USES

The leaves and young shoot tips of New Zealand spinach are eaten raw or cooked as a green vegetable. However, it is a delicious curry when it is eaten after stir-frying. Its edible leaves also used as salad or as a bed for meats, cooking fish, combined with cheese or pork and pickled. It is used in casserole, soups, sautéed, steamed, braised, or in lasagna. Owing to its sprawling nature and attractive foliage, it is a very good plant for borders and beds used ornamentally as edible landscaping. It is also used as an ornamental plant for ground cover.

MEDICINAL USE

New Zealand spinach is effective in fighting against scurvy disease. It reduces the risk of heart disease, cancer, Alzheimer's, sleep disorders, macular degeneration and several other degenerative diseases. Its leaves are rich in carotenoids, which reduce the damage caused by super oxides to the body. Its consumption regulates body metabolism, aids in digestion, prevents people suffering from constipation and itching, ulcers and dryness of eyes due to vitamin A deficiency and keeps the hair moisturized through increased sebum production. Its intake not only improves the immune system but can also create and maintain collagen, an essential protein found in hair and skin. Its vitamin K improves calcium absorption, reduces urinary excretion of calcium and acts as a modifier of bone matrix proteins.

BOTANY

New Zealand spinach is a perennial vegetable but grown as an annual crop. A low-growing, weak, fragile vine-like stem produces brittle small abundant green fleshy leaves and young stem tips. The

TABLE 1.1
Nutritional Composition of New Zealand Spinach (per 100 g Edible Portion)

Constituents	Contents	Constituents	Contents
Water (g)	84	Manganese (mg)	0.53
Carbohydrates (g)	2.13	Zinc (mg)	0.31
Protein (g)	1.3	Vitamin A (IU)	3622
Fat (g)	0.17	Thiamine (mg)	0.03
Dietary fibre (g)	1.4	Riboflavin (mg)	0.11
Calcium (mg)	48	Niacin (mg)	0.39
Phosphorus (mg)	22	Pantothenic acid (mg)	0.26
Potassium (mg)	102	Pyridoxine (mg)	0.24
Sodium (mg)	107	Vitamin C (mg)	16
Magnesium (mg)	32	Vitamin K (µg)	292
Iron (mg)	0.66	Energy (kcal)	12

stems and leaves of this spreading herb are covered with small blisters filled with shiny liquid. Foliage colour is medium green. Lower portion of the plant is large and hardy. It grows to a foot in height and spreads 30–90 cm. The leaves are fleshy, succulent, almost triangular to oval shaped, smaller and fuzzier than regular spinach and are pale to dark green and grow from 5 to 10 cm long. Its small yellow flowers are pollinated by wind. Fruits are small, hard, conical capsules covered with small horns, containing several seeds, which are covered with wooly wings which help in dispersion by wind or water.

CLIMATIC REQUIREMENT

New Zealand spinach thrives best in frost-free cool weather coupled with full sunlight and keeps producing quality leaves throughout the summer in temperate regions. Like common spinach, it is not frost hardy. The optimum temperature for getting higher yield ranges from 15°C to 25°C. It prefers comparatively moist weather for its growth. Long days coupled with warm weather favour the development of seed stalk but low-temperature treatment alters its photoperiodic response since it is a biennial crop. Plants exposure to 10–15°C for a month can induce flowering even under short-day conditions. Long and hot summer days induce bolting, leading to poor leaf quality. For its cultivation, the best soil temperature ranges from 10°C to 25°C. Partial shade in afternoon in the warmer regions is beneficial for the formation of leaves.

SOIL REQUIREMENT

New Zealand spinach, being a halophyte, grows well in coastal saline soils and can withstand conditions of drought. However, well-drained sandy loam soils rich in organic matter having good moisture retentive capacity and pH ranging from 6.8 to 7 are most suitable for its cultivation.

CULTIVATED VARIETIES

Maori is the most commonly grown variety. Use bolt resistant varieties for later plantings. Spinach may be of two types, *i.e.*, (i) the cool season or broadleaf type and (ii) the warm season or New Zealand spinach, which is taxonomically different from broadleaf spinach but both are similar in taste and appearance. Another variety, which is widely used for cultivation, is Winter Bloomsdale.

PLANTING TIME

In temperate regions, New Zealand spinach is sown in early spring; however, in tropical and subtropical regions, it is sown in early autumn. For a steady supply of spinach, successive sowings are done at 2- to 3-week intervals.

SEED RATE

For sowing a hectare land area of New Zealand spinach, 10–15 kg seed is often used. Seed rate should be reduced when New Zealand spinach is to be grown at high temperatures.

PLANTING METHOD

New Zealand spinach is propagated by seeds, which are actually the fruits containing multi-germ seeds similar to beetroot and produce several seedlings in clusters. The seeds are sown directly in the field at a depth of 1–2 cm keeping a row spacing of 60–90 cm and plant spacing of 30–45 cm since its seedlings sprawl on the ground. A single fully developed plant covers a ground area of one square meter. The seeds germinate slowly and irregularly. Hence, at the time of sowing, the field must have plenty of soil moisture for better seed germination. Before sowing, the seeds should be soaked overnight in ordinary water or 3 hours in warm water to fasten the germination process. If the germination is high and the seedlings are in clusters, the extra seedlings are removed as and when they attain a height of 3–4 cm after emergence, however, the thinning operation is not necessary if the entire plants are to be harvested.

NUTRITIONAL REQUIREMENT

The crop responds well to the application of manure and fertilizers, thus should be supplied to obtain rapid growth of the plants. New Zealand spinach requires more manure and fertilizers as compared to the common spinach. The quantity of farmyard manure to be applied depends on soil texture, structure and crop grown in previous season. Also, nitrogen is applied 100 kg, phosphorus 25 kg and potash 150 kg/ha. However, application of the excessive dose of nitrogenous fertilizer should be avoided. The supplemental nitrogen should be applied only if the leaves are pale green. Farmyard manure should be incorporated into the soil at the time of land preparation. Phosphorus and potash are applied as basal dose at the time of planting, while the nitrogen is applied as top dressing in split doses, as it stimulates regrowth after harvesting. If the soil pH is high, lime is added to normalize the soil pH level.

IRRIGATION REQUIREMENT

New Zealand spinach is a shallow rooted crop, thus requires consistent moisture for better plant growth as water is a critical component for growth and development of the plant. Providing plants with adequate moisture by frequent irrigation when necessary prevents bolting. However, excessive irrigation should be avoided, as it is harmful to the crop. The plant produces best flavour when watered consistently. On the other hand, drying of soil affects the leaf quality, hence, light irrigation at regular intervals is recommended to keep the plants healthy and to produce the best quality leaves. In coastal areas, it grows well in sandy soils too as these soils supply consistent soil moisture to produce leaves with the best flavour.

INTERCULTURAL OPERATIONS

The New Zealand spinach field should be kept weed free to reduce competition with crop plants and to get higher yield of best quality leaves. The weeds are controlled by hand pulling or shallow hoeing close to the plant.

Pinching of top shoot and new buds is an essential operation in New Zealand spinach to encourage side shoots and also to lengthen the harvest period. It is a weak-stemmed vegetable and trails across the garden, thus, pinching is of great significance.

Covering soil surface with mulch after sowing is helpful in conserving soil moisture, suppressing weeds and increasing economic yield of the crop. Floating row covers can also be used against insect-pest damage.

HARVESTING

New Zealand spinach is usually harvested 40–80 days after sowing, depending upon the variety and growing conditions. First harvesting of new shoots or 15–20 cm tops may be done when the main branch has grown to a length of 30 cm or as the plant covers the soil surface. For best flavour produce, the young and tender leaves or shoot tips are picked at weekly intervals over a period of several months. Harvesting is usually done either by cutting the plants with a sharp knife just below the soil surface or by handpicking the outer leaves and retaining younger leaves for future pickings. However, this practice is commonly followed in kitchen gardens, whereas, the crop at commercial scale is completely removed after some months when the yield and quality become poor. Harvesting once or twice a week is recommended to promote the shoot growth. Picking of older leaves or shoots with fruits for consumption should be avoided as they become bitter and fibrous. It is advisable to harvest the crop at the edible stage before the deterioration of leaves or the development of seed stalks. After the development of seed-stalks, the entire plant is uprooted.

YIELD

New Zealand spinach on an average gives a yield of 10 quintals per hectare at first harvest and a total yield of 30 quintals per hectare with continuous harvesting over a period of 3–4 months.

POST-HARVEST MANAGEMENT

New Zealand spinach for fresh market should be harvested in evening hours, not in morning hours, to prevent undue breakage of leaves since the leaves in morning hours have high turgor, which will cause breaking of leaves during harvesting. All yellow, injured, insect damaged and diseased leaves should be removed before packaging. Slightly wilted spinach will be less subject to breakage during packaging. Usually, spinach is packed in baskets topped with ice crystals. New Zealand spinach cannot be stored for more than a day at ambient room temperature, but in the refrigerator, it can be stored for about a week, thus, it is advisable to consume it soon after harvest. It can also be canned, frozen, or dehydrated.

INSECT-PESTS

Lygus Bug (*Lygus* spp.)

Lygus bugs can cause different kinds of damage at different growth stages, *e.g.*, blackheart, blasting on flower tissues, collapsing and decreased leaf yield. The bug also causes damage in common spinach seed crop by feeding on the developing flowers and seed. At feeding injury site, a black sticky substance is often produced.

Control
- Follow long crop rotation with non-host crops.
- Remove alternate hosts plants.
- Use natural bio-control agents of Mymaridae family.
- Use biological agents such as big-eyed bugs and damsel bugs.
- Spray the crop with acephate 0.1%, aldicarb 0.05%, or dimethoate 0.03% at 10- to 15 days intervals.

Spinach Leaf Miner (*Pegomya hyoscyami*)

Leaf miner makes tunnels within leaf beneath the epidermis. When the attack is severe, it produces small black specks within leaf mine, and its larvae feeding within leaf tunnels produce numerous black faces. Epidermis of the leaf pulled back from a leaf mine to reveal two larvae feeding within the leaf. Its adult causes scarring symptoms on the leaves.

Control
- Follow long crop rotation with non-host crops.
- Grow resistant varieties, if available.
- Spray the crop with *neem* oil 0.4%, dimethoate 0.03%, DDVP 0.05%, Fenitrothion 0.05%, or quinalphos 0.05% at 15 days intervals.

SPIDER MITE (*Tetranychus* spp.)

Tetranychus spp., including two-spotted spider mite (*Tetranychus urticae*), strawberry spider mite (*Tetranychus turkestani*) and Pacific spider mite (*Tetranychus pacificus*) affect the crop. They generally live on lower surface of leaves in colonies covered by white silky webs. Their adults and nymphs suck cell sap from lower surface of the older leaves, resulting in yellow or white spots on upper surface of the infested leaflets. Heavy infestation results in bronzing of leaves followed by defoliation.

Control
- Remove and destroy the affected leaves.
- Spray *neem* oil 0.4%, methyl parathion 0.01%, phosalone 0.25%, acaricides 0.25%, abamectin or dicofol 0.1%, wettable sulfur 80 WP @ 3 g/litre, or fenazaquin 10 EC 1 ml/litre of water on lower surface of leaves at 10- to 12-day intervals.

DISEASES

DAMPING-OFF (*Aphanomyces, Fusarium, Pythium* and *Rhizoctonia* spp.)

Generally, post-emergence damping-off disease damages the seedlings more. The roots of damped-off seedlings show brown and blackened roots. *Damping-off* is a term often used for sudden-death seedlings. The attack of these fungi on collar region causes toppling of seedlings over the ground under wet conditions.

Control
- Disinfect the soil either through solarization or by drenching with 1% formaldehyde.
- Treat the seeds with Dithane M-45, copper oxychloride, phenyl mercury, or captan 3 g/kg of seed.
- Drench the field soil after sowing with 0.05% copper sulfate, captan, or Aretan, 0.1% Bassicol, or 0.3% copper oxychloride.
- Spray the crop with 0.3% copper oxychloride, or 0.25% Diethane M-45 or captan 50 WP at 10- to 12-day intervals.

FUSARIUM WILT (*Fusarium oxysporum* f. sp. *spinaciae*)

A soil-borne pathogen enters the plant body through roots. It is a most serious disease in poorly drained soils. The initial symptoms appear as yellowing of lower leaves, including leaf blades and petioles and continue to successive younger leaves. The leaves of infected plants hang around the pseudostem and wither. In pseudostem of diseased plant, yellowish to reddish streaks are noticed with colour intensification towards roots. Warm soil temperature, poor drainage, light soils and high soil moisture are congenial for the spread of this disease.

Control
- Follow deep summer ploughing to expose the soil to atmospheric heat.
- Follow long crop rotation with non-host crops.
- Grow resistant varieties, if available.
- Uproot and destroy the infected plants from the field.
- Apply bioagents such as *Trichoderma viride* or *Pseudomonas fluorescence* in soil.
- Treat the seed with carbendazim, Benomyl, or thiram @ 3 g/kg of seed.
- Drench the field with 0.3% Lanacol or Brassicol.

Verticillium Wilt (*Verticillium dahliae*)

The fungus is soil born. Symptoms appear as interveinal chlorosis of lower leaves, which progresses to interveinal necrosis. Vascular browning of entire plant takes place immediately after infection. As a result, drying and wilting of entire plant takes place. In case wilting occurs late, defoliation of lower leaves and wilting of upper leaves are the most characteristic symptoms. The root system is destroyed and the resulted plants topple over the ground surface.

Control

- Follow deep summer ploughing to destroy the inoculums.
- Follow long crop rotation with non-host crops.
- Grow resistant varieties, if available.
- Use disease-free planting material.
- Treat the seed with Thiram @ 3 g/kg of seeds.
- Maintain good sanitation in the field.
- Collect and destroy the infected plant debris.
- Drench the soil with 0.3% Lanacol or Brassicol.

Anthracnose (*Colletotrichum dematium*)

The fungus attacks the plant leaves and stems. Water soaked lesions or necrotic spots appear on leaves, resulting in thin and papery leaves and produce dull salmon pinkish slimy mass under a condition of high humidity. Similar lesions appear on cotyledons and young seedlings. The symptoms appear on leaves as dark brown sunken spots with reddish margins.

Control

- Use disease-free planting material.
- Keep the field weed free.
- Follow clean cultivation.
- Dip the seeds in 0.15% solution of Cerasan or thiram for 30 minutes.
- Rogue out the diseased plants from the field.
- Spray the crop with Bavistin or Benlate 0.1% at 10-day intervals.

Cladosporium Leaf Spot (*Cladosporium variabile*)

The disease is characterized by round tan leaf spots, which rarely exceed 6 mm in diameter. Initially, the symptoms appear as necrotic lesions on leaves, and later, dark green spores and mycelium develop in the center of these spots. The presence of dark green sporulation distinguishes *Cladosporium* leaf spot from *Anthracnose* and *Stemphylium* leaf spot diseases, both of which also form circular lesions.

Control

- Grow resistant variety, if available.
- Use healthy certified seeds for planting.
- Maintain proper sanitation in the field.
- Follow proper cultural practices.
- Spray the crop with Dithane M-45 or copper hydroxide 0.1% twice at 10 to 12 days intervals.

Alternaria Blight (*Alternaria brassicae* and *alternaria brassicicola*)

The disease appears as small, dark brown-blackish' sooty and velvety spots with round concentric rings on stem and leaves. Centre of spots soon dries and drops out. When the infestation is severe, the leaves and upper parts of stem wither. Infection at seed stage results in loss of seed yield and germination power. The pathogen remains viable in soil debris.

Control

- Plough the field in summer to reduce soil inoculums.
- Grow resistant variety, if available.
- Follow long crop rotation and field sanitation.
- Use disease-free healthy and certified seeds.
- Collect and destroy the infected plant debris.
- Treat the seeds with metalaxyl @ 6.0 g/kg, or thiram or carbendazim @ 2 g/kg seed.
- Spray the crop with aqueous extract of garlic bulb or *Eucalyptus* leaf extract 2.0% w/v basis, or oxadixyl or mancozeb 75 WP @ 0.25% at 15-day intervals.

DOWNY MILDEW (*Peronospora farinosa* f. sp. *spinaciae*)

This disease alone or in combination with white rust causes severe yield loss. Symptoms appear on all above ground plant parts, particularly on leaves and inflorescence. Chlorotic lesions appear on upper leaf surface infected with downy mildew. Purplish-brown spots appear on underside of the leaves. Fungus stimulates hypertrophy and hyperplasia. The most conspicuous symptom is deformation of various floral organs. The affected inflorescence does not bear seeds or produce abnormal seeds.

Control

- Maintain good sanitation in the field.
- Remove and destroy the affected plants.
- Spray the crop with 0.2% zineb, maneb, or copper fungicide prior to the onset of rainfall.

WHITE RUST (*Albugo candida*)

The fungus causes severe infection in New Zealand spinach. The symptom appears as light brown water soaked lesion covered with white growth of mycelium on all aerial parts except root, while prominent white or creamy yellow pustules appear on lower surface of leaves. The small pustules may also appear on upper leaf surface. The infected parts, usually the top leaves, become deformed and the resulted seeds may become smaller and shriveled.

Control

- Plough the field in summer to destroy the soil inoculums.
- Plant the crop at right time.
- Follow long crop rotations with non-host crops.
- Maintain good sanitation in the field.
- Remove and destruct the affected plant parts.
- Spray the crop with garlic bulb extract or *Eucalyptus* leaf extract 2.0%, or mancozeb 75 WP @ 0.2% at 50 and 70 days after sowing.

■■■

2 Chinese Chive

Dipika Sahoo and Manoj Kumar Nalla

Botanical Name : *Allium tuberosum* Rottl. Ex Spreng

Family : Amaryllidaceae

Chromosome Number : 2n = 4x = 32

ORIGIN AND DISTRIBUTION

Chinese chive is native to Asia temperate and widely cultivated crop in temperate and subtropical regions of Asia, Australia, Europe and America. It was introduced mainly by the Chinese into other Asiatic countries, including North India. It has been changed little from wild and was domesticated in the Orient. Domestication must have happened in North China where it is still occurring as wild. Especially in China, good variability with late and summer flowering population varying in flower characters are known to occur, which is scarcely known elsewhere. In India, this is a less known cultivated vegetable species to occur in the Western Himalaya (altitude 1,500–3,000 m) but it occurs as semi-wild population in Spiti and Kinnaur regions (1,600–3,300 m) of Himachal Pradesh. It is also naturalized in Northeastern regions mainly in Manipur, Sikkim, Khasi hills of Meghalaya, Darjeeling (West Bengal) and in the hills of Western Himalaya. In Delhi, Agra and Uttar Pradesh, it is cultivated as ornamental plant under the name American grass and in Japan as Green Nira Grass. The crop is still under experimental cultivation in Punjab and Tamil Nadu.

INTRODUCTION

Since ancient times, many *Allium* species such as onion, garlic, leek and chives have been used as food, spices and herbal remedies in widespread areas of the world, especially in the northern hemisphere. The genus *Allium* with about 881 species is a rich source of steroidal saponins, alkaloids and sulfur-containing compounds. Chinese chive, commonly known as garlic chives, Oriental garlic, Chinese leek, Indian leek, *nira*, green *nira* grass, etc., is among the less known cultivated species grown as home garden or backyard cultigen in tribal dominated tracts of Indian Himalayas. For domestic consumption, it can be grown in pots or small beds as perennials in hills and plains. In off-season from September to February when onion prices are very high, Chinese chive is used in place of onion. It is one of the daily edible green vegetables for Chinese since it is low in calorie (30 kcal/ 100 g). It is distributed all over mainland China and used not only as food but also as medicine. The freshly harvested leaves as well as dehydrated leaves have great potential for export purposes. Evergreen look of the field throughout the year and star-shaped white showy fragrant flowers make this species an ideal plant for barren land and landscaping for home gardens as an ornamental use, especially during rainy season. Chinese chive with wider adaptability and year-round growth can be a potential commercial vegetable that can supplement the requirement of bulbous crops.

COMPOSITION AND USES

COMPOSITION

Chinese chive is a valuable source of minerals and vitamins. It is known to be low in fat and calorie but relatively high in carbohydrates, protein and dietary fibre. It is an excellent source of calcium and potassium. This truly gifted vegetable is also potent in beta-carotene, thymine, riboflavin and vitamin C. Its seeds contain many new and known steroidal saponins, alkaloids and amides. The nutritive composition of Chinese chive is given in Table 2.1.

USES

The leaves of Chinese chive are consumed as raw, cooked, pickled, or in processed form. The sundried petioles, leaves and flower parts are used as a spice for flavouring food items during off-season. Like onion and garlic, its leaves, flower stalks and immature unopened flower buds are used as a seasoning ingredient. The edible part consists of 5–10 mm wide green leaves from the middle part. In Khasi region of India, the bulbs are consumed raw or cooked. In Nepal, vegetable dish of Chinese chive and potatoes is known as *dunduko sag*. It is a common ingredient in Chinese, Japanese and Korean cuisine. In Vietnam, its leaves are chopped into small pieces and used as the only vegetable in a broth with sliced pork kidneys.

MEDICINAL USES

Chinese chive carries many health benefits, as it is rich in beta-carotene and vitamin C. Another thing that makes it a popular ingredient in Chinese medicine is that it is an excellent source of calcium. The plant has great medicinal value. It is locally used in lowering blood cholesterol. Seeds possess an aphrodisiac property. It is one of the 76 plant species used in traditional hair care lotion locally known as *Chenghi* by women folk of Meitei Community of Manipur, India. Its seeds are reputedly used as a traditional Chinese medicine for treating both impotence and nocturnal emissions. The *n*-BuOH extract preparation of its seeds possessing aphrodisiac property is used as a traditional remedy. The latest findings reveal that allyl sulfide is preventive against blood clots and cancer of skin, lung, liver and large intestine. With all its benefits, Chinese chive has been introduced into daily diet as a healthy vegetable with stamina enhancing properties.

TAXONOMY

Chinese chive belonging to Butomissa section is generally regarded as a cultivated species with tetraploid chromosome number (2n) 32, although wild population and population with double haploids

TABLE 2.1

Nutritional Composition of Chinese Chive (per 100 g Edible Fresh Weight)

Constituents	Contents	Constituents	Contents
Water (g)	93.1	Iron (mg)	10.6
Protein (g)	2.1	Vitamin A (IU)	1800
Fat (g)	0.1	Thiamine (µg)	60
Sugar (g)	2.8	Riboflavin (µg)	190
Crude Fibre (g)	0.1	Niacin (µg)	600
Calcium (mg)	50	Vitamin C (mg)	25
Phosphorus (mg)	32	Energy (kcal)	19

have also been reported. This species differs from its close relative *Allium ramosum* in flat leaves and perianth segments usually with a green or yellowish green mid-vein, which has perianth segments with a pale red mid-vein.

A cultivar of Chinese chives locally known as *Zimu* was developed in 1969 at College of Agriculture, Solan (Himachal Pradesh), using *Allium tuberosum* from local flora of mid hills of Himalaya. Due to characteristic odour closer to onion and morphology of leaf like that of the garlic, *Zimu* was wrongly identified as an inter-specific hybrid between onion and garlic. Later, this species was taxonomically identified as *Allium tuberosum*. In Uttarakhand, *Allium tuberosum* is locally called *doona* though this local name was also assigned to other *Allium* species, including *Allium ascalonicun* and *Allium cepa* var. aggregatum, which are used in same way. However, its plant has a distinctive growth habit with flat leaf like garlic and mild flavour more like onion. It can be grown by expanding perennial rhizome as well as by seed.

BOTANY

Chinese chive has a distinctive growth habit with strap-shaped leaves unlike either onion or garlic and straight thin white-flowering stalks that are much taller than the leaves. The plant has clustered cylindrical bulbs arising from horizontal rhizome. Leaves are linear, shorter than scape, flat and solid with smooth margin, which is 25–60 cm, covered with leaf sheaths at base. Spathe is persistent with 2–3 valves. The plant usually blossoms from August to October with 20–40 small white flowers, developing a hemispheroidal umbelliforum. The umbels are loose and hemispheric to sub-globose, containing many small white star-shaped flowers with brown-striped tepals. The seeds are produced in autumn, but in tropical lowlands, flowering is sparse and seeds do not always develop. Pedicels are sub-equal, 22–24 in number and are as long as the perianth. Perianth with lobes are white and segments usually with green or yellowish green mid vein, outer ones oblong-ovate to oblong-lanceolate, 42–47 × 1.8–2.3 mm and inner ones oblong-obovate, 4–7 × 2.1–3.5 mm. Filaments are narrowly triangular, equal, as long as the perianth segments, connate at base and adnate to perianth segments and inner ones slightly wider than outer ones at base. Ovary is obconical-globose and minutely tuberculate without concave nectaries at base. Papery capsules are filled with hard rhomboidal shaped black seeds.

CLIMATIC REQUIREMENT

Chinese chive grows well in cool climate but high humidity and high rainfall are the main limiting factors for its growing. The best temperature for its cultivation is around 20°C. In warmer areas, it may remain green round the year, but in cold climate, aerial parts die back completely to the ground and the roots or rhizomes overwinter and re-sprout in spring. It can be grown in full sun to partially shaded areas.

SOIL REQUIREMENT

Chinese chive being a hardy crop grows well on a wide range of soils provided they are well drained, fertile and rich in organic matter but sandy loam or loam soils are better for higher yield. The optimum pH range is between 5.8 and 6.5. The land is prepared by giving 5–6 ploughings followed by planking.

CULTIVATED VARIETIES

Chinese chive can be recognized as (i) the large leaf types and (ii) the smaller leaf types. The large leaf types are very popular in China and Taiwan, especially when grown as a blanched crop. Chinese chives can also be divided into two groups, *i.e.*, (i) Winter Dormant and (ii) Winter Non-dormant types. Winter Dormant types need low temperature about 2 months for dormancy to plant next year. These chives can survive very low temperature in the winter and sprout in next spring when temperature rises, therefore, they are recommended for outdoor cultivation, especially in cold areas. Non-dormant chives are suitable for greenhouse production and tropical areas. They do not require dormancy and they keep growing when temperature is appropriate.

SOWING TIME

Chinese chive can be grown round the year, making it multipurpose plant in substituting the onion in various food items, curries, etc. It is grown as a perennial crop ranging from 7 years to 30 years. The best time for the planting of its rhizomes is October-November for commercial or domestic use.

SOWING METHOD

Chinese chive grows from slowly expanding perennial clumps as well as from seeds, thus, the plant is multiplied through seed as well as rhizome of old plants, which produce numerous pseudostems having 4–5 leaves. However, the old rhizomes are preferred more for the production of true to type plants. Seedlings are either raised from seed or by division of the clumps, consisting of tuberous rhizomes, which are separated into individual plantlets. Seedlings or rhizomes are planted on raised beds in a square spacing of 20 × 20 or 25 × 25 cm, depending on vigour of the clone. When planting material is produced through seed, the seedlings are planted in groups of 3–4 approximately at a spacing of 20 cm each way.

NUTRITIONAL REQUIREMENT

The nutritional requirement of this crop depends upon the type of soil, spacing, cultivar, irrigation facilities and growing conditions. Over fertilization can be detrimental to the plants of Chinese chive. Farmyard manure 15–20 t/ha, nitrogen 50–60 kg, phosphorus 40–50 kg and potash 20 kg/ha should be applied at the time of field preparation to obtain a good yield. Like many herbs, slower and more compact growth leads to healthier plants with stronger flavour. A soil rich in organic matter can provide sufficient nutrients; however, if there is a continuous and intermittent harvesting each year, NPK fertiliser dose of 5-10-5 kg/ha may be applied once at spring to boost up the vigour of its plants. Top-dressing of nitrogen is very beneficial to crop harvested frequently.

IRRIGATION REQUIREMENT

Chinese chive requires regular watering throughout the growing season for best production. Soil moisture is maintained near field capacity. It can survive during the period of drought stress but growth as well as productivity is decreased. Long dry period should usually be avoided, as it may result in splitting and untimely maturity. The total water requirement varies with the soil type, climatic conditions and crop duration. Overwatering or too much soil moisture is detrimental to its growth, as it may favour the development of diseases.

INTERCULTURAL OPERATIONS

Weeds should be controlled through regular hoeing but it should be done shallow to avoid damage to the plant roots. Controlling weeds is important particularly during the first 2 months when plants have slow growth. Chemical weed control along with one hand weeding is indispensable. Fluchloralin 0.5–1.0 litre a.i. per hectare as pre-emergence is found effective in controlling weeds. Mulching can also be done to control weed growth. Organic mulch is better than inorganic mulch since it conserves moisture, supplies extra nutrients and controls weeds.

HARVESTING

Both, the young leaves and flowers are used for culinary purposes; thus, harvesting of chives should be done when the leaves grow large enough to be used. First, the leaves are gathered into a bunch and then the leaves are cut with a pair of sharp and clean scissors. However, the leaves should not be cut too close to the ground, as it affects regrowth of plants adversely.

YIELD

During early growing season, the cutting of leaves is done at an interval of 15–20 days. Although the Chinese chive can be grown throughout the year but it gives lesser yield in winter due to poor growth caused by low temperature. During winter, growing Chinese chive in protected structures is profitable, as it reduces the crop duration. The scape in August–September produces flowers and seeds, which consequently decline both the yield and quality as the vegetative growth of plants is reduced in these months.

POST-HARVEST MANAGEMENT

The freshness of Chinese chive is lost within 2–3 days after harvest, thus, it has very short shelf life under normal conditions but it can be stored well at 0–2°C temperature and 90–95% relative humidity. For prolonged use, it may be subjected to sun drying until the produce becomes dry and crispy.

PESTS AND DISEASE

No serious insect-pest or disease problem occurs in this crop due to the release of volatile compounds from root exudates and fallen leaves of Chinese chive. As a result, it is used as a rotating or intercrop of banana for avoiding the occurrence of Panama disease. Chinese chive intercropped with tomato hampers the occurrence of bacterial wilt in tomato, as the root exudates of Chinese chive prevent *Psuedomonas solanacearum* from infecting tomato plants but some of the pests and diseases reported in Chinese chives are discussed in the following sections.

INSECT-PEST

THRIPS (*Thrips tabaci*)

Thrips are small winged insects that attack many types of plant and thrive in hot and dry conditions (like polyhouse). They can multiply quickly, as the female lays eggs up to 300 in a life span of 45 days without any mating. Most of the plant damage is caused by the young larvae, which feed on tender leaf and flower tissue, which leads to distorted growth, injured flower petals and premature flower drop. They can also transmit many harmful plant viruses.

Control
- Keep the weeds under control in Chinese chive field.
- Use screens on polyhouse windows to keep the thrips out.
- Fix yellow sticky cards on different locations of the field to trap the thrips.
- Remove or discard the infested plants.
- Take advantage of natural enemies such as predatory mites.
- Give a good steady shower of water to wash them off the plant.
- Spray the crop with 0.1% profenofos, or 0.5% carbosulfan.

ONION FLY (*Delia antique*)

The pest is a grey-coloured fly that lays long whitish colour eggs on under surface of bulb near soil. Its maggots eat into seedlings between the leaves and small and older bulbs at the bottom of bulbs. This pest eventually causes the plant leaves pale, wilt and then die off. The growth of bulbs is hampered, tissues become decayed and bulbs become unfit for consumption and marketing.

Control

- Adopt long crop rotation with non-host crops.
- Treat the seed with 5% malathion dust.
- Remove and destroy the affected plants immediately.
- Keep the weeds under control in and surrounding the field.
- Avoid use of undecomposed organic matter.
- Avoid planting in rows to stop maggots moving from plant to plant.
- Spray the crop with malathion 1 liter per hectare.

DISEASES

PURPLE BLOTCH (*Alternari porri*)

The disease symptoms appear as small sunken whitish flecks with deep purplish centres on leaves and flower stalks, however, these spots are very prominent on flower stalks. Further, large purple area develops forming dead patches. In advanced stages, these lesions may girdle leaves and stalks, which drop after 2–3 weeks.

Control

- Follow long crop rotation of 2–3 years with non-host crops.
- Use disease-free healthy planting material.
- Follow clean cultivation and proper drainage.
- Use drip irrigation system to avoid the disease occurrence.
- Spray the crop with mancozeb 0.25%, tricyclazole 0.1% and hexaconazole 0.1% with surfactant.

STEMPHYLIUM LEAF BLIGHT (*Stemphylium vesicarium*)

The disease appears as small light yellow to brown and water soaked lesions on leaves, which later grow into elongated spots, which frequently coalesce, resulting in blighted leaves. Lesions usually turn light brown to tan at the center and later dark olive brown to black as the spores of this pathogen develop. The fungus normally invades dead and dying tissues, such as leaf tips, purple blotch and downy mildew lesions and injured and senescent tissues. Infection usually remains restricted to leaves but never extends to the bulb scales. Lesions generally occur on side of the leaf facing the prevailing wind. Long period of warm and wet conditions encourages the disease development.

Control

- Follow long crop rotation of 2–3 years with non-host crops.
- Grow resistant varieties.
- Remove and destroy the infested plant debris of previous season.
- Keep the planting rows parallel to the prevailing air current to allow adequate airflow.
- Avoid long period of leaf wetness of plants.
- Spray the crop with mancozeb 0.25%, Dithane M-45 0.1%, or hexaconazole 0.1%.

Downy Mildew (*Peronospora destructor*)

The symptoms appear as violet fungal growth on surface of the leaves and flower stalks under humid conditions. These later become pale-green, yellow and finally collapsed. The bulbs fail to attain normal size and remain soft and immature. The disease affected bulbs in storage become soft and shrivelled with amber outer fleshy scales. The fungus overwinters in crop residues and persists through mycelium in infected bulbs during storage.

Control

- Plough the field repeatedly in hot summer months.
- Follow clean cultivation and maintain proper field sanitation.
- Avoid application of excessive irrigation.
- Remove infected plant debris from the field.
- Avoid overcrowding of plants by adopting proper plant spacing.
- Plough the field soon after harvesting the crop.
- Treat the seed with Metalaxyl 6 g/kg of seed.
- Spray the crop with Difolatan 0.2% or Ridomil MZ 72 WP 0.2% at 10- to 12-day intervals.

Rust (*Puccinia porri*)

The symptoms appear as small white flecks on underside of lower leaves and stems, which develop into circular or elongated brown pustules. They slowly spread on all the leaves and green parts. In case of severe infestations, it can cause the leaves to yellow and die. Most rusts are host specific and overwinter on leaves, stems and exhausted flower debris. This fungus is favoured by moist weather but low rainfall and the spores can spread over long distances by wind, splashing water or rain.

Control

- Grow resistant only varieties especially in infested areas.
- Use well drained field for the planting of Chinese chive.
- Follow proper crop sanitation by removing and destroying all plant debris and infected material.
- Avoid watering through overhead irrigation system.
- Spray the crop with zineb 80 WP 2 g/liter or Mancozeb 75 WP 2 g/liter.

Botrytis Leaf Blight (*Botrytis cineraria and Botrytis squamosa*)

The pathogen causes fusiform brown or black spots and patches that give blighted appearance, which may be either ragged or circular with a water soaked or yellow-edged appearance. Insects, rain and dirty garden tools and implements spread the disease.

Control

- Use disease-free planting material.
- Remove diseased and decayed plants and debris.
- Collect and destroy older leaves around the plant base.
- Avoid excess application of nitrogenous fertilizers.
- Avoid watering through overhead irrigation system.
- Treat the seed with benomyl + thiram 1 g each per kg of seed.
- Spray the crop with 0.1% benomyl or 0.2% mancozeb before harvesting the crop.

Fusarium Bulb Rot (*Fusarium oxysporum*)

Fusarium bulb rot, which attacks the growing plant and stored bulbs, can sometimes be a serious problem in Chinese chive. The fungus usually introduced in chive field by infected bulbs, soil, or even tools and enters the plant through abrasion to the plant tissue. It is a serious problem in warm climates where temperature rarely drops to a freezing point, thus, the fungus persists in soil for several years.

Control

- Follow long crop rotation of at least 4 to 5 years with non-host plants.
- Avoid the use of infected bulbs or introduction of infected soil and implements.
- Follow good drainage practices.
- Plough the field repeatedly in hot summer months.
- Avoid injuries to the bulbs and roots during hoeing and weeding.
- Adopt mixed cropping with tobacco and sorghum.
- Treat the seed with *Trichoderma viride* and *Pseudomonas*.
- Dip the sets and seedling roots in 0.2% carbendazim solution.
- Spray the crop with 0.1% carbendazim before harvest to avoid storage loss.

Bacterial Soft Rot (*Erwinia carotovora*)

The infected bulb becomes pale-brown, soft and watery. It appears healthy from outside but when cut open, some of the inner scales are brown, wet and have a cooked appearance. The neck of infected bulb becomes soft and when pressed emits a foul smell. Onion maggot is an important vector for carrying the bacteria from one plant to another and causing wounds for the infection to enter the bulb. The disease often occurs during storage. Improperly stored bulbs or bulbs that are too wet in their dormant stage are more susceptible to bacterial soft rot disease.

Control

- Follow long crop rotation with non-host crops.
- Avoid planting bulbs in poorly drained soils.
- Control nematodes and other insect pests.
- Avoid injury to the plants during hoeing and weeding.
- Remove plant residues from the field after harvest.
- Separate the infected plants immediately from the lot.
- Store the bulbs properly after harvest.

Iris Yellow Spot Virus–IYSV (*Tospovirus*)

The symptoms vary among bulbs and seed crops but often appear as straw coloured diamond shaped lesions on leaves and scapes with twisting or banding flower bearing stalks. Some lesions have distinct green centre with yellow or tan borders, while other lesions appear as concentric rings of alternating green and yellow or tan tissue. Infected plants become scattered. Large necrotic region may develop on scapes and cause a collapse of escape. Diseased plants may be scattered across the field but the highest incidence of disease is often found on the field edges.

Control

- Follow long crop rotation with non-host crops.
- Control thrips in chive field to avoid the spread of virus from unhealthy to healthy plants.
- Keep the weeds in and around the chive field under control.
- Use organic mulch for keeping the thrips population low.
- Spray the crop with the insecticides formulations like 0.4% *neem* seed kernel extract or spinosad.

■■■

3 Asparagus

M.K. Rana and Shiwani Kamboj

Botanical Name : *Asparagus officinalis* L.

Family : Liliaceae

Chromosome Number : 2n = 20

ORIGIN AND DISTRIBUTION

The exact centre of origin of asparagus is not known but the garden asparagus is believed to be originated from Eastern Mediterranean countries and traces of wild varieties have been discovered from Africa. The name asparagus is believed to be derived from the ancient Greeks, who referred the word *asparagus* to all tender shoots. The Romans and Greeks cultivated this plant over 2000 years ago, and it can still be found growing wild in Britain, Europe and as far as central Asia. There are 300 varieties known, but out of them, only 20 varieties are found to be edible.

INTRODUCTION

Asparagus is herbaceous perennial plant with an erect branched stem growing to 1–3 m height. It is also known as *satavar* or *shatamuli* in Hindi. It was being grown in English gardens in 1534. In sixteenth century, asparagus was served in the royal courts of Europe, but in the eighteenth century, asparagus made its appearance in the local market and in numerous culinary works. Once asparagus starts producing, it continues to give economic yield for 10–15 years. The countries like United States of America, China, Germany, Spain and France have large area under this crop especially green varieties, while the white varieties are cultivated mainly in Europe. It has good potential for export as fresh vegetable. In India, the area and production of asparagus are negligible. It is also known as the *Queen of Herbs* in the *Ayurvedic* system of medicine since it promotes love and loyalty.

COMPOSITION AND USES

COMPOSITION

Asparagus shoots provide a good amount of calcium, phosphorus and vitamin C and it is a good source of energy. It contains an appreciable amount of potassium too. The nutritional composition of asparagus is given in Table 3.1.

USES

The tender shoots or spears are used for making juices and cooked as vegetable. Mustard sauce is also prepared with asparagus spears. It may be eaten raw as salad and the starchy tuberous roots are used for the preparation of soups. It is canned and frozen in large quantity as a processed product. The juice obtained from spears contains a white crystalline substance called *aspergine*, which has diuretic properties and used especially in cardiac dropsy and chronic gout.

TABLE 3.1

Nutritional Composition of Asparagus (per 100 g Edible Portion)

Constituents	Contents	Constituents	Contents
Water (g)	91.7	Potassium (mg)	278
Carbohydrates (g)	5.0	Iron (mg)	1.0
Protein (g)	2.5	Zinc (mg)	0.54
Fat (g)	0.2	Thiamine (mg)	0.18
Dietary fibre (g)	2.1	Riboflavin (mg)	0.2
Calcium (mg)	22.0	Vitamin C (mg)	33
Phosphorus (mg)	62	Energy (kcal)	20

MEDICINAL USES

Asparagus is also known for its numerous medicinal properties as it is anti-inflammatory, diaphoretic and demulcent. It stimulates and strengthens kidney function and provides an excellent diuretic action to human body. It revitalizes bladder function, helps in cases of edema and swollen joints, and protects small blood vessels from rupturing. It can also be used as an aphrodisiac. It is an excellent remedy for those suffering from retention of liquids. Its stalks have high antioxidant properties, and its high dietary fibre content makes it a good laxative. According to the reports published in different journals, it can be used in the treatment and prevention of various types of tumor. It can also be used to balance the levels of hormones in females, promote fertility and help pregnant women against neural tube defects in babies.

BOTANY

Asparagus is an herbaceous perennial plant growing to 100–150 cm tall. The root system is adventitious and the root type is fasciculate. It has succulent, edible and stout stems with branched feathery foliage. The leaves are glabrous, needlelike cladodes (modified stems), which are reduced to small scales on main stem. They are 6–32 mm long, 1 mm broad and clustered 4–15 leaves together. The flowers are produced either singly or in clusters of two or three in junctions of branchlets. The flowers are bell-shaped, greenish-white to yellow and solitary at nodes. They are 4.5–6.5 mm long with six tepals partially fused together at the base. It is usually dioecious with male and female flowers on separate plants, but sometimes, it bears hermaphrodite flowers. It is propagated by crown, which is made up of roots and rhizomes. The aerial stem (spear) arises from a bud on the rhizomes. The fruit is a small red berry with 6–10 mm diameter, which is known to be poisonous to humans.

CLIMATIC REQUIREMENT

Asparagus is a high-value crop and the earliest producing spring vegetable. It is a temperate crop and requires a temperature range of 16–24°C in most of the growing seasons for good spear production but low temperature when plants are dormant is necessary for 60–90 days to produce good-quality spears. Temperature plays an important role in influencing flowering and can tolerate great temperature variations. It needs moist temperate regions that have long growing season and sufficient light for maximum photosynthesis. The optimum temperature for the germination of seeds is 25–30°C. It can tolerate freezing to some extent but continuous freezing is also dangerous and affects the crop adversely.

SOIL REQUIREMENT

Asparagus can be grown on a wide range of soils but it thrives best in sandy, silt loam or alluvial soils with a high content of organic matter in soil. It is known for its tolerance to soil salinity but the pH 6.0–6.7 is suitable for its higher production with superior quality spears. The soil should be deep with good drainage facilities so that the spears may emerge from the soil without any damage.

CULTIVATED VARIETIES

Green asparagus or green spears are more popular and produced mainly for fresh market, while the light green or white spears are mainly used for processing. Several varieties and hybrids of asparagus are described below in the following subsections.

PERFECTION

An early variety with uniform maturity has large, succulent green spears having high copper content.

MARY WASHINGTON

A high-yielding variety produces spears with purplish tips if not exposed to light, which turn dark green if exposed to sunlight.

CONOVER'S COLOSSAL AND MAMMOTH WHITE

These are high-yielding varieties commonly grown for marketing purposes and both produce light green to whitish-green spears.

GIJNLIM

A very early hybrid produces medium thick and straight spears of excellent quality.

WHITE GERMAN

A variety selected from Germany grown widely by Germens produces round tipped good quality spears.

ANETO AND DESTA

The inter-clone hybrids evolved through *in vitro* tissue culture are high yielding with less heterogeneity.

The other cultivars grown widely with lesser or no description are UC-12, UC-66, UC-157, SL-831, Violet Dutch, Faribo, Paradise, Jersey Giant, Jersey Queen, Brock Imperial 84, Schewetzinger, Lara and Mira.

SOWING TIME

The sowing time varies considerably according to the type of variety and climatic conditions of that region where crop is to be raised. Sowing of seeds of a particular variety at its right time is very essential to get higher yield of better quality spears since unfavourable environmental conditions have adverse effect on these components. The best time for sowing asparagus is March to May in hills and July to November in plain areas. Under average field conditions, asparagus seed is slow to germinate and seed takes 3–6 weeks to germinate.

SEED RATE

The germination percentage of seed chiefly affects the seed rate. It may also vary due to growing season, method of sowing and soil properties. A seed rate of 3–4 kg is enough for sowing a hectare area.

PLANTING METHOD

Seeds are soaked in water for 24 hours before sowing in well-prepared nursery beds in rows 30 cm apart with 4–6 cm between seeds at a depth of 3–4 cm. The seedlings are allowed to grow in nursery beds for about 7–12 months and then transplanted in field on ridges about 75–90 cm apart and 50–60 cm between plants in rows. Seedling emergence may become more rapid if seeds are primed with -0.6 MPa polyethylene glycol 8000 at 20°C than water-soaked seeds.

PROPAGATION

The asparagus crop can also be propagated with crowns taken from a year old plants. The crowns should be dug out carefully with less injury to the roots and consequently with less interruption to growth are used for propagation. Well-grown crowns of good size containing large buds are selected, separated from mother plants before the buds sprout and planted within few hours. They are planted 20–25 cm below the ground level and 30 cm apart, and thereafter, the soil around the crown is pressed firmly. The distance between rows is kept 1.8 m for the production of green spears and 2–2.5 m for white spears.

NUTRITIONAL REQUIREMENT

Asparagus can be grown successfully with well rotten farmyard manure, which is incorporated @ 20 t/ha into the soil during field preparation. Application of nitrogen 50–60 kg, phosphorus 25 kg and potassium 40 kg/ha is required every year for the production of better quality spears. The entire dose of phosphorus and potash along with half dose of nitrogenous fertilizer is applied before planting and rest half of nitrogen is top dressed 30–35 days after planting.

IRRIGATION REQUIREMENT

The first irrigation is applied immediate after planting the crowns and second irrigation is applied before the cracks develop in soil around the crowns for their better establishment. Further irrigations are given depending on soil and environmental factors of the growing region. Keeping the soil moist during cutting season is essential for higher yield of good quality spears. In humid regions, irrigation is not essential to apply.

INTERCULTURAL OPERATIONS

HOEING AND WEEDING

For getting higher yield from asparagus crop, the field should be kept weed free in early stages of crop growth since the plants may not compete with weeds for nutrients, light, moisture and space due to their poor growth rate in starting. For keeping the weeds under control, 2–3 shallow hoeing are done between rows without injuring the spears. Hoeing is also essential during cuttings of spears. It is advisable to avoid the intercultural operations during morning hours as the spears are turgid, brittle and liable to break. Pre-emergence application of metribuzin @ 1 kg/ha controls the broad-leaved weeds in asparagus field.

BLANCHING

Mounting up of soil around the plants up to a height of 25–30 cm is essential to obtain whitish spears particularly for canning purpose, as it is a common practice to blanch the asparagus for canning. This operation is done a week before harvesting the spears for fresh market. Blanching may also be done by wrapping the spears with newspaper.

HARVESTING

The spears become ready for harvesting in the last week of March to the first week of April. The spears of 10–15 cm are generally harvested. The first cutting is done approximately 2 years or 13–20 months after planting. Well-grown spears are to be cut with a knife known as *asparagus knife*. While cutting the spears, care should be taken to avoid any damage to the young spears. Crooked, curved and damaged spears are rejected. Male plants give higher total yield, while the female plants produce larger individual spears.

YIELD

The yield varies with crop variety, soil and environmental factors, irrigation facilities and the cultural practices adopted by the growers. On an average, asparagus gives a spear yield of 2.5–4.0 kg/10 m² or 25–40 q/ha.

POST-HARVEST MANAGEMENT

Immediately after harvesting, the spears are shifted to the packinghouse station where the post-harvest operations like sorting, grading, trimming, washing, bunching and packing are done and thereafter, the spears are sent to the market. If spears are stored at room temperature, the tissues become tough and unfit for consumption. The spears can safely be stored for 3–4 weeks at 0°C temperature and 90% relative humidity.

INSECT-PESTS

ASPARAGUS BEETLE (*Corioceris duodecimpunctata* L.)

Both, adult and larvae harm the asparagus plant. The larvae feed on tender portion of the plant and beetle damages the shoots, leaves, flowers and seeds, which results in weakening of roots and causes harm to the crowns, thus, it causes huge loss to the growers.

Control
- The damage may be reduced by cutting all the spears after every 3–5 days.
- Spray the crop with 0.1% malathion or Nuvan.

GARDEN CENTIPEDE (*Scutigerella immaculata*)

This pest also causes severe damage to the spears by making round and small holes in them. Its attack makes the spears unfit for consumption because tissues around holes become tough.

Control
Flood the field if infestation is scattered in the whole field.

DISEASES

FUSARIUM WILT (*Fusarium* sp.)

The *Fusarium* species attack the roots and collar region of the plant. The affected spears show brown discoloration and become wilted and stunted. The disease is not so common, but sometimes, it causes huge loss. The disease is soilborne.

Control
- Avoid asparagus planting on infected soils.
- Use healthy planting material for propagation.

- Avoid excessive irrigation.
- Treat the crowns with 0.2% captan solution.
- Drench the field with captan 3 kg and Bavistin 1.5 kg/ha.

ASPARAGUS RUST (*Puccinia asparagi*)

This is the most serious disease of asparagus, which causes a great economic loss to the farmers. The small reddish-yellow spots on the main stem and branches are the main symptoms of this disease. These spots later coalesce and develop into a large lesion. It causes great reduction in yield as firstly it weakens and ultimately kills the plant.

Control

- Grow rust-resistant varieties like Jersey Giant and UC-157 to overcome the problem.

■■■

4 Chive

M.K. Rana and Navjot Singh Brar

Botanical Name	: *Allium schoenoprasum* L.
Family	: Alliaceae
Chromosome Number	: 2n = 16

ORIGIN AND DISTRIBUTION

Chive (*Allium schoenoprasum* L.) is the only species of *Allium* group, which is native to both the Old and New World. Though little is known about the early history of chive as a cultivated species, it has been established that the plant was known by the Chinese 5,000 years ago. It was not known until the 16th century, but thereafter, it became popular in European gardens. Chive at that time was used mostly for food, even though the English botanist John Gerard wrote a medical guide in 1597, in which, he described chive as an herbal remedy. It was introduced to Germany by the Italians in the 17th century. During the 1600s, chive became a peasant food in England, and since then, it is used periodically as a potherb. The plant was featured in the list of American esculents in 1806.

The different species of chive that are found cultivated in different parts of the world are mentioned as under:

- Giant Siberian (*Allium ledebourianum*)
- Chinese or garlic chive (*Allium tuberosum*)
- Siberian garlic chive (*Allium nutans*)
- European or common chive (*Allium schoenoprasum*)

In India, both European and Chinese chives are minor vegetable and grown on large areas. The European chive is found wild in western Himalayas from Kashmir to Kumaon hills (Uttrakhand) up to 3000 m and also in hills of south India, and the species *Allium tuberosum* has sporadic distribution in western Himalayas, extending to northeastern hills, and the Chinese chive is grown in West Bengal for flavouring the foodstuffs.

INTRODUCTION

Chive (*Allium schoenoprasum* L.) is the smallest species of *Allium* genus and family Alliaceae, which includes other crops such as shallot, onion, garlic, leek and bunching onion. The crop is perennial in growing habit but it can also be cultivated as an annual where the entire plant is harvested after a 1- to 2-month growing cycle. Compared to onion and other alliums, chive has a strong tillering habit, forming dense clumps without well-formed bulbs. The herb has culinary and medicinal properties. The fresh leaves are used for making herbal butter, vinegar and a flavouring agent in different dishes. Its English name, chive, has been derived from a French word *cive*.

COMPOSITION AND USES

COMPOSITION

Chive is a nutrients rich food, meaning that while it is low in calorie, it is high in beneficial nutrients like minerals, vitamins and antioxidants too. It is an excellent substitute of salt and a perfect aid for those who are on a low-fat and salt-restricted diet. It is also a good source of folic acid and dietary fibre. Chive surprisingly comprises more vitamin A as compared to any other member of *Allium* family. A 100 g of fresh chive leaves contain 4353 IU of vitamin A. The nutritive analysis of raw chive (per 100 g of edible portion) has been given in Table 4.1.

USES

Chive is mainly used as garnish, and its fresh chopped leaves are stir-fried and used as flavoring in dumplings, soups, stews, pasta, salads, etc. The leaves stalk, which gives out a light *oniony* taste, is used in cooking. The herb is also used in muffins, scones, quiche, pizza, omelets and biscuits.

MEDICINAL USES

Chive is rich in flavonoids like organosulfides, which have anti-cancer and anti-inflammatory effects. It is a good source of β-carotene, which helps in improving eyesight and clear acne. The presence of different minerals helps in improving metabolic activity of the body. The choline in chives is an important nutrient that helps in sleeping, muscle movement, learning and memory. Choline also helps in maintaining the structure of cellular membranes, aids in the transmission of nerve impulses, assists in the absorption of fat and reduces chronic inflammation. Folic acid also found in chive may help in relieving depression by preventing an excess of homocysteine.

BOTANY

Chive is a monocot, small perennial herb growing in clumps. Plants can be grown from seed or divisions of 2 to 3 bulbs. The leaves of European chive are hollow and tubular. The plant height is 15–30 cm. The flower stems, which rise directly from the base, grow slightly taller than leaves and bear small clusters of purple or pink flower heads. The inflorescence is an umbel, containing 10–30 flowers together. The flowers are pale purple or pink in colour and star-shaped with six petals.

TABLE 4.1

Nutritional Composition of Chive (per 100 g Edible Portion)

Constituents	Contents	Constituents	Contents
Water (g)	92.0	Manganese (mg)	0.37
Carbohydrates (g)	4.35	Vitamin A (µg)	218
Sugars (g)	1.85	Thiamine (mg)	0.07
Protein (g)	3.27	Riboflavin (mg)	0.11
Fat (g)	0.73	Niacin (mg)	0.64
Dietary fibre (g)	2.50	Pantothenic acid (mg)	0.32
Calcium (mg)	92.0	Vitamin C (mg)	58.1
Phosphorus (mg)	58.0	Vitamin E (mg)	0.21
Potassium (mg)	296.0	Vitamin K (µg)	217.7
Magnesium (mg)	42.0	Energy (kcal)	30.0
Iron (mg)	1.60		–

The petal length is 7 to 15 mm. Flowering depends on chilling temperature coupled with short days. Opening of flowers starts first from top of the umbel. It flowers in June–July, and the seeds mature in cold and moist conditions though rarely allowed to do so under cultivation. Fruit is a capsule with length of 4 mm.

Chive is a diploid cross-pollinating, self-compatible and insect-pollinated perennial herb. The European chive has chromosomes number $2n = 16$ and Chinese chive has $2n = 32$. The leaves have an onion-like flavour. The small bulbs produced by the plants are not eaten as food.

CLIMATIC REQUIREMENT

Chive is a cool season crop. Its plant is perennial, cold-hardy and tolerant of high temperature and grows well under mild climate without extremes of heat and cold. Optimum temperature of its successful cultivation is 17°C to 25°C. They tolerate light shade but 6–8 hours of direct light is the best for its cultivation.

SOIL REQUIREMENT

Chive prefers a well-drained, loose and friable soil rich in organic matter with a pH of 6.0–7.0. It grows well in moist habitat but not in waterlogged soils. It is a shallow-rooted crop (25–30 cm), so land preparation should not be deep. For field preparation, two or three light ploughings followed by planking are required to make the soil loose, friable and porous. For continuously cultivated lands, the field is forked and raked before planting.

CULTIVATED VARIETIES

There is no distinct variety or type available in India. Most of the varieties of chive are exotic, which are dwarf or taller types, although they may not be readily available.

CURLY MAUVE

This variety has prostrate foliage that curls in different directions and has gray-lavender flowers held erect above the short curled foliage.

The other important varieties are Forescate, Marsha, Pink Giant, Profusion, Snowcap and Album.

PLANTING TIME

The right time for the planting of divisions of clumps is August to November in the north Indian plains. In hilly tracts, planting is done in mid February to March, depending on altitude.

Chive can also be grown indoors easily in a bright sunny location. The pots must have holes in the bottom for draining out the excess water, and soil should be well drained, not the heavy one. A good houseplant potting mix is used to fill the pots. During winter, when light is poor, the plant will not grow much and may even die back a bit but it grows with the return of brighter sun in spring.

PROPAGATION

Chive is propagated by seeds or from the divisions of clumps. The propagating chive vegetatively by division of existing clumps of bulbs is a common practice. The easiest and most successful means of propagating chive is planting rooted clumps. Initially, they are planted in multiple rows at very densities. The established plants usually need to be divided every 3–4 years to prevent overcrowding. Divisions should include at least five bulblets.

A spacing of 10 cm within rows and 10 cm across rows is maintained. Increasing plant density results in depressed yield, while decreasing plant density results in underutilization of land resources, thereby decreasing productivity. Although chive can be propagated by seeds, this method is not suitable for commercial cultivation because of the long crop cycle.

For planting in hilly tracts, the divisions of clumps of previous crop have a dormancy period of 4 to 6 weeks in winter season. This dormancy can be broken by soaking the clumps in warm water for 8–16 hours at 40°C temperature followed by normal cooling in water. The treatment possibly leaches growth inhibitors and increase the metabolic rate.

NUTRITIONAL REQUIREMENT

Nutritional requirement generally depends on soil type and fertility status of the soil. Therefore, it is essential to conduct a soil analysis so that a precise recommended dose of fertilizers can be applied. Generally, 15–20 t/ha of well-decomposed farmyard manure should be applied at the time of field preparation. In addition, nitrogen is applied @ 50–60 kg, phosphorus 40–50 kg and potash 20 kg/ha 6 weeks after planting to obtain a good yield.

IRRIGATION REQUIREMENT

The frequency of irrigation depends on soil type and climatic conditions. Chive responds well to regular irrigation, especially in dry season. When rain is infrequent, the field should be irrigated to make sure that the soil does not dry out around the root zone. Irrigation should be applied in the evening, not in the morning. In general, crop is irrigated at an interval of 10–15 days.

INTERCULTURAL OPERATIONS

Chive is poor competitor with weeds for water, space, nutrients and light due to its shallow adventitious root system and the lack of an aerial canopy to shade out other vegetation. Therefore, two weedings should be done frequently in order to keep the weeds under control. A light mulch of dried ground leaves, compost, or grass clippings will also help to control weed and to retain soil moisture.

HARVESTING

Chive requires 75–90 days to be mature after sowing. Generally, the chive becomes ready to harvest when it attains a height of 7.5–10 cm. Multiple hand-cuttings of chive leaves are made about 2 to 3 cm above the ground surface. In first year, usually, two or three cuttings are taken. After the plants have built up their food reserve, cuttings may be taken more frequently. Harvesting should be done early in the morning to avoid the harming effect of heat as the leaves should be crispy, clean and relatively free from discoloration and this may be possible when harvesting is done in morning hours. All disease-infected or insect-damaged and yellow leaves are removed from the harvested plants. The roots are thoroughly washed to remove all the adhered soil particles. The produce is then tied into small bunches as per the requirement of traders and consumers in the market.

YIELD

Depending on fertility of the soil and the management practices adopted, a yield of 70–80 q/ha can be obtained from six to seven cuttings each year. Only leaves of chive are eaten since its bulbs are too small.

POST-HARVEST HANDLING

Chive is a highly perishable crop and therefore should reach the market within 24 hours of harvesting. The leaves at the time of harvesting should be young, uniform, firm, fresh and deep green with mild flavour. They are tied in bunches for sale. The larger leaves tend to have strong onion-like pungent flavour. Fresh leaves of chive can be packed in polyethylene bag and kept in the refrigerator for 48 hours. The dried leaves may be placed in an airtight container and stored in a cool and dark place. Storing at 0°C to 1°C temperature and 95–100% relative humidity can extend the shelf life of chives for 7 to 14 days. In markets, fresh as well as dried (dehydrated) leaves are available. Chives can be made available year-round if it is grown as potherb.

PLANT PROTECTION MEASURES

Chive is almost completely free from insect-pests and diseases. However, occasionally, it may suffer from onion thrips. It is usually because the chive is usually planted near an onion crop which is attacked by thrips. The solution is not to plant chive near onion crops.

INSECT-PESTS

ONION THRIPS (*Thrips tabaci*)

Minute pale insects feed on foliage and produce whitish dot-like spots followed by curling.

Control

- Follow long crop rotation.
- Spray the crop with malathion (0.1%) or methyl demeton (0.025%) at 15-day intervals.

■■■

5 Leek

M.K. Rana and Pradeep Kumar Jatav

Botanical Name : *Allium porum* L.

Syn. Allium ampeloprasum L.

Syn. Allium ampeloprasum var.

porrum L.

Family : Alliaceae

Chromosome Number : 2n = 32 (x = 8)

ORIGIN AND DISTRIBUTION

Leek has originated from the east Mediterranean region, where it was cultivated by the Egyptians, Greek and Romans since ancient times, and later, it was spread to middle Europe. It has also been suggested that leek is a native to Europe and central Asia. The Greek philosopher Aristotle created clear voice with the dieting of leeks, while the Roman emperor Nero used to eat leek every day to make his voice stronger. The Romans introduced leek to the United Kingdom, where it adapted to cool weather. Now, leek has become an important crop in many countries like Belgium, Denmark, France, Greece and Germany. However, more recently, it is becoming more popular in United States of America. In India, it is not grown commercially so far but its cultivation has been noticed in temperate regions of north India and the Niligeris hills of South India.

INTRODUCTION

Leek is an important non-bulb-forming crop of Alliaceae family. It is robust herbaceous biennial cultivated species but has not been found anywhere in wild form. The economic parts for which it is mainly cultivated are blanched stem and leaves. It has flattened leaf blades rather than the radial, and it is bigger than onion. As compared to onion, it has milder and more delicate flavour. Being hardy biennial winter crop, it is grown in temperate regions of the county. In India, leek is not under commercial cultivation though it may be best suited for kitchen gardening.

COMPOSITION AND USES

COMPOSITION

Leek contains more protein and minerals than the onion based on fresh and dry weight. Its energy value per 100 g of edible portion is also higher than the onion. Its major storage tissue is leaf sheath. It provides a good amount of minerals like phosphorus, iron and vitamins like vitamin A and C, carbohydrates, protein and fibre. Its greens are also good source of calcium, iron and vitamin A, C and thiamine (Table 5.1).

TABLE 5.1

Nutritional Composition of Leek (per 100 g Edible Portion)

Constituents	Contents	Constituents	Contents
Water (g)	78.9	Iron (mg)	2.3
Carbohydrates (g)	5.0	Vitamin A (IU)	30
Protein (g)	2.0	Thiamine (mg)	0.23
Fat (g)	0.3	Riboflavin (mg)	0.04
Dietary fibre (g)	1.3	Vitamin C (mg)	23
Phosphorus (mg)	70.0	Energy (kcal)	77

USES

The blanched stem and green leaves are used as salad and cooked, boiled and fried in butter with other vegetables. The elongated blanched stems and its leaves are also used for making soup.

MEDICINAL USES

The small-sized bulbs of leek are used in the medicine as a heart stimulant.

BOTANY

Leek belongs to monocotyledonous species, and it has white narrow bulbs with broad leaves and etiolated pseudostem formed by the leaf sheath. Leek grown in summer has long pseudostems, whereas, the crop during the winter season gives a short stem growth. It does not produce bulbs, but after flowering stage, it forms bulbils in leaf axil at the base of flower stalk. The inflorescence is umbel, which is similar as in onion. The umbel contains numerous small flowers of light pink to dark purple colour. Although leek is a highly cross-pollinated crop due to the presence of male sterility and protogynous type of dichogamy, self-pollination is also reported extending from 10% to 30% in the individual flower of open pollinated variety. The cross pollination is mainly carried out by honeybees.

The plant has solid, flat and longitudinal leaf blades arranged in cross section, forming a V-shape. The erect leaves are arranged in opposite rows and they rise progressively above the other leaves. The height of leek plant may vary from 40 to 80 cm. The plant has a disk like small stem. The tube shaped leaves with 5 to 55 cm length develop from apical portion of the stem. Due to continuous growth, development and elongation, leaf is modified as a solid leaf blade above the sheath portion and young leaves developed from the previous leaf sheath. Many leaf sheaths are combined to form the pseudostem.

CLIMATIC REQUIREMENT

Leek, well adapted to wider climatic conditions, has no rest period and not much effect of photoperiod, which is the reason for its wider adoption. In spite of a cool season crop, it can withstand heat and cold better than onion. Germination of seed is found to be good within the temperature range of 11–23°C and reduced above 27°C. The best temperature for rapid vegetative growth is 21–24°C.

Leek does not form bulbs under the most cool conditions, however, bulb formation increases under long day conditions with a temperature range of 15°C to 18°C as the accumulation of dry matter is noticed more at this temperature. Fresh as well as dry matter of plant is increased with the rising in temperature up to 21°C and day length. The number of adventitious roots is found more in increased temperature and day length.

Temperature also influences the characteristics of pseudostems, which vary from cultivar to cultivar and sometimes with cultural practices. The cultivars grown in winter season produces short, thick and small pseudostem, while the summer crop produces thin and long pseudostem. Short pseudostem growth is associated with winter hardiness. Leek requires vernalization or low chilling temperature for flower induction though some cultivars may bolt even at 21°C. Short days along with low temperature meet the requirement of vernalization. Sometimes, the lateral bulbs, which can be utilized for propagation, are produced by the plants in leaf axils during long day conditions.

Leek is more resistant to cold than the onion but bulbs are not formed. Leek is not having any rest period and its growth is not influenced by the photoperiod but temperature is more important than the day length in the development of flower stalk.

SOIL REQUIREMENT

Leek can be grown on a wide range of soils. However, a rapid plant growth is noticed in sandy loam soils with high content of humus and good moisture holding capacity. Soil should be loose-textured and well-drained, rich in organic matter and initial plant nutrients. The clayey and very sandy soils should be avoided for it cultivation. This crop can be grown successfully in soils with neutral to slightly acidic in soil reaction, having the pH range from 6 to 6.8.

CULTIVATED VARIETIES

Many high-yielding and promising varieties have been developed by following different breeding methods. The varieties that are under commercial cultivations are London Flag, American Flag, Elephant, Gennivillies, St. George, Copenhagen Market, Bluevetia, Borde Pearl, Belgium Winter Giant, Large Musselburgh, etc.

RENOVA: A variety developed through a selection from the variety Regions produces marker leaves, longer stalk and fewer tendency to form bulbs and is more uniform in size and shape.

MINER: An autumn or early winter variety derived from St. George produces long stem and high greens yield.

SPLENDID: This variety is suitable for cultivation throughout the season.

LONDON FLAG: It is most generally cultivated variety with tall and broad leaves.

MUSSELBURGH: It is very long and hardy variety. Its leaves are broad and tall and stem long and thick.

LARGE ROVEN: Its leaves are very dark green, broad and thick and stem rather short but remarkably thick. It grows as thick as a man's arm. It is found to be the best kind of forcing, as it requires a sufficient thickness of stem sooner than any other varieties.

LINCOLN: It is early Bulgarian giant type leek, which gives uniform long and thick stems. It is suitable for *Rabi* season planting and matures in 80 days after transplanting.

CARLTON: This hybrid variety is very fast in growth and leaves are erect and green in colour. Its long shaft nodes have excellent uniformity.

ROXTON: This variety possesses long shaft length with excellent uniformity. The growth of this hybrid is very fast with green to dark green erect leaf colour.

PARTON: Medium growth of this hybrid produces medium shaft length with excellent uniformity. Leaves are dark green in colour.

SOWING TIME

In India, seeds are sown in nursery during August to October and the seedlings are transplanted in field 2–3 months after seed sowing in spring season.

SEED RATE

Depending on seed viability, soil fertility and climatic conditions of the growing region, about 5 to 7 kg seed is sufficient to raise seedlings enough for planting a hectare land area.

SOWING METHOD

Leek is grown by either direct seeding or transplanting. However, the usual propagation of leek is by seed produced transplants because of its longer growing period of about 12 weeks and slow germination of seed. The overall development is promoted by seed priming and germination of seeds prior to sowing. Seeds are sown shallow at a depth of <1 cm. At the time of planting, the seedlings should attain a height of 15 cm.

When the stem attains a diameter of 5–6 mm, the seedlings are planted about 10–15 cm deeper into a shallow trench or furrow in order to assist subsequent blanching of base of the developing pseudostem since the length of blanched portion of pseudostem influences its market value. This is achieved by deep planting so that more pseudostems may develop below the soil surface or by hilling soil against the plant as it grows. Bulbils can also be used for propagation.

NUTRITIONAL REQUIREMENT

Leek requires soil having plenty of organic matter since it is larger than the onion. The soil should be prepared with green manure plough down or farmyard manure to enhance organic matter content in soil and to provide sufficient nutrients and extra moisture holding ability for the crop.

The diameter and the length of pseudostem increase with the application of fertilizers. Therefore, nitrogen is applied @ 200–250 kg, phosphorus 50–100 kg and potash 150–200 kg/ha. One-third dose of nitrogen along with full dose of phosphorus and potash is applied at the time of planting. The remaining two-thirds is applied in two split 1 and 2 months after planting as side dressing and top dressing, respectively. As compared to nitrogen, the phosphorus and potash requirement of the crop is low.

IRRIGATION REQUIREMENT

The seeds for direct seeded crop are sown in dry soil, and thereafter, light irrigation is applied. Irrigation should be provided at frequent interval until the seedlings emerge out and develop a good root system. The top soil should not be allowed to dry up until the seedlings are well established. The seedlings are best planted in moist soil followed by irrigation. Traditionally, leek is cultivated under overhead sprinkler systems. However, drip irrigation is equally good. Overhead sprinkler irrigation gives very good results in dry years and good results in average years. Light sprinkling during dry spells is economical even in wet years.

INTERCULTURAL OPERATIONS

HOEING AND WEEDING

Leeks do not shade out the weeds that emerge in between the rows. Field infested with nut sedge and other hard seeded legume weeds should be avoided for the cultivation of leek. Since the crop is long duration, controlling weeds is essential to reduce competition with crop plants for light, nutrients, moisture and space. The problem of weeds can be reduced largely by preparing the land well ahead of planting and using paraquat herbicide in seedbeds as pre-emergence. Pre-emergence and early post-emergence application of herbicides may control many weeds for 4–6 weeks. When the leek is intercropped with the celery, it considerably shortens the critical period of weed control in the plot as compared to leek pure stand. Since leek is shallow rooted crop, care must be taken not to damage the root system while doing hoeing. Hoeing is occasionally recommended for the large broadleaf weeds since labour is very costly.

BLANCHING

Blanching should be carried out 3 weeks after planting when plants have attained a height of 15–20 cm and the diameter of about 2.5–5.0 cm. To improve the quality, blanching is done by placing the soil around the stem to a certain height. The blanching should be done at an interval of 15 days. It may also be done by wrapping brown paper around the plant keeping the top leaves unwrapped.

BOLTING

Development of seed stalk from pseudostem is undesirable in leek crop except when it is raised for seed production. The tendency of plants to bolting depends on growing temperature and type of variety. Bolting is not a photoperiodic but only a low temperature response, therefore, temperature is more important to influence the physiology of biennial plants rather than photoperiod. Exposure of plants to temperature below 7°C always tends to accelerate flower induction. The chilling temperature required for flowering in leek varies from cultivar to cultivar. In general, lower the atmospheric temperature, shorter is the period of exposure to chilling temperature. A temperature range of 2°C to 8°C is required during chilling period of 1 to 2 months for the induction of flowering. However, the plant has to complete the basic vegetative phase before it responds to low temperature. The chilling requiring species or varieties have definite inductive stage, at which, they are subjected to cold stimulus. The plants at juvenile stage do not respond to cold stimulus as the sensitivity of plants to vernalization varies with root size. In leek, generally, normal flowering takes place in the second year after vernalization.

HARVESTING

A growth period of 120–150 days after transplanting is usually required to obtain high yield. The crop becomes ready for harvesting when the blanched portion of leaves attains at least a diameter of 1.25 cm.

The machinery used has made the harvesting process less difficult. Harvesting is done by uprooting the plants, which require considerable efforts because of more vigorous and extensive root system. Thereafter, the roots are cut and senescent leaves are removed. The remaining leaves are shortened and the plants thereafter are packed in boxes.

In addition to uprooting, washing and trimming is tedious. Washing is important to remove the soil particles present on blanched portion of the peudostem. The blanched section of the stem should be about 10 to 15 cm long. Roots are trimmed closer to the base of pseudostem, outer dried and damaged leaves are removed, and the remaining leaves are cut back to a length suitable for better marketing.

YIELD

The best yield can be obtained with plants sown in mid March and planted in the first half of June. Its yield increases when older and thicker seedlings are transplanted. The peudostem typically represents 40% of the total plant weight. The yield varies from 250 to 500 q/ha, depending on crop variety, soil type, environmental conditions and the cultural practices adopted by the growers.

POST-HARVEST MANAGEMENT

Generally, leek can have a post-harvest life of about 2 months when they are stored at 0°C temperature and 95% relative humidity. Storing leek in controlled atmospheric (10% CO_2 + 1% O_2) can further extend the storage period. The continuous storage of leek in 10% carbon dioxide reduces the losses by 44–48%. The keeping quality may be improved by keeping the leeks in moist peat or sand.

Spraying the crop with growth regulators like GA_3 (100 ppm), benzimidazole (500 ppm), or Benomyl (500 ppm) extends the shelf life by 9–10 days.

INSECT-PESTS

THRIPS (*Thrips tabaci*)

These are 2 mm long insects hidden between the inner leaf blades, where they feed on leaf fluids. Both adults and immature thrips cause damage to this crop. The green leaves lose their colour as the empty surface cells form thousands of highly visible grey spots. The economic damage due to quality loss is serious.

Control
- Migration can be monitored by blue sticky traps.
- Mulch with coloured plastic is effective.

EUROPEAN CORN BORER (*Ostrinia nubialis*)

The European corn borer that feeds only on leeks in late summer or fall is a rare pest of leek. Initially, its larvae feed on outer cuticle of the leaves but continue feeding inside the leaf.

LEEK MOTH (*Acrolepiopsis assectella*)

A serious pest in some parts of Europe bores into the folded leaves towards centre of the plant, causing series of pinholes on inner leaves.

Control
- Follow long crop rotation strictly.
- Transplant the seedlings bit late.
- Collect and destroy the pupae or larvae.
- Remove old and infected leaves.

DISEASES

LEEK RUST (*Puccinia porri*)

The insect causes a serious damage to European leeks. The disease first appears as small, circular, and white to tan lesions along the leaf veins. If disease pressure is severe, the leaves turn yellow and die prematurely.

Control
- Follow long crop rotation.
- Keep proper isolation between leek and onion field.
- Spray the crop with protectant fungicides like maneb or zineb at 0.2%.

WHITE TIP DISEASE (*Phytopthora porri*)

This fungal disease becomes prevalent after heavy rainfall in late summer. Water soaked areas appear at leaf margins near the tip. Older plants when slightly affected with pathogen wilt rapidly after harvesting.

Control

- Spray the crop with maneb 0.2% during September and October.
- As and when the first symptoms appear, spray the crop with more systemic products such as benalaxyl or metalaxyl at 0.2%.

WHITE ROT (*Sclerotium cepivorum*)

This is a soilborne fungal disease, which can be devastating if pathogen presents in farm soils. The first signs are yellowing and dying back of leaves starting from the tip and progressing downwards. Young plants wilt and collapse and are easily dislodged. Cool wet growing season favours the development of white rot.

Control

- Follow long crop rotation.
- Provide proper drainage.
- Spray metalaxyl at 0.1% or azoxystrobin at 1%.

■■■

6 Shallot

M.K. Rana and Harshita Mor

Botanical Name	: *Allium cepa* var. aggregatum
	Syn. *A. ascalonicum*
Family	: Alliaceae
Chromosome Number	: 2n = 16

ORIGIN AND DISTRIBUTION

Shallot is thought to have originated in China around 2000 B.C., and travelled from there to India and eastern Mediterranean region. It was first introduced to Europeans during the 12th century. Crusaders brought it home as *valuable treasure* from the ancient city of Ascalon in Israel. Originally, it was called *Allium ascalonicum* after Ascalon and was considered a sacred plant by Persians and Egyptians. It is grown largely in southern and temperate areas of Russia, Estonia, Georgia, Ukraine and Armenia. In Africa, shallots are cultivated in the area around Anloga in southeastern Ghana.

INTRODUCTION

Shallot, also known as golden shallot, French shallot, *khan kho, scalogna, schalotte, eschalot, kaanda, gundhun, cheriya ulli, praan, chyapi, bawang mera, hom,* or *brambang,* is a type of small onion, specifically a botanical variety of the species *Allium cepa*. However, it was formerly classified as a separate species, namely *Allium ascalonicum*. It looks more like garlic and its mild flavour is reminiscent of a sweet onion with a hint of garlic. It contains more flavonoids and phenols than other members of the *Allium* genus.

COMPOSITION AND USES

COMPOSITION

Shallot has more impressive nutritional profile than onions. Its bulb contains more sugars than other species of *Allium*. It is an excellent source of potassium, phosphorus and vitamin A, good source of calcium, magnesium, iron and folic acid and a fair source of zinc, manganese, copper and B group vitamins. The nutritional composition of shallot is given below in Table 6.1.

USES

Shallots can be eaten raw, boiled, pickled, baked, or fried and its young outer leaves can be used like chives. The finely sliced and deep-fried shallots are used as a condiment in Asian cuisine and served with porridge. In Nepal, it is used as one of the ingredients for making *momos*, while in Kashmir, in preparation of *wazwan kashmiri* cuisine. In south India, these are widely used in curries and different types of *sambar*. In Iran, shallot is grated and mixed with dense yogurt, which is served with grills or kebabs. Shallot is also used to make different types of *torshi*, a sour Iranian side dish

TABLE 6.1

Nutritional Composition of Shallot (per 100 g Edible Portion)

Constituents	Contents	Constituents	Contents
Water (g)	89.0	Zinc (mg)	0.4
Carbohydrates (g)	16.8	Manganese (mg)	0.292
Protein (g)	2.5	Copper (mg)	0.088
Fat (g)	0.1	Vitamin A (IU)	1160
Sugars (g)	7.87	Thiamine (mg)	0.06
Dietary fibre (g)	3.2	Riboflavin (mg)	0.02
Calcium (mg)	37	Folic acid (µg)	34
Phosphorus (mg)	60	Panthothenic acid (mg)	0.29
Potassium (mg)	334	Niacin (mg)	0.2
Magnesium (mg)	21	Pyridoxine (mg)	0.345
Selenium (µg)	1.2	Vitamin C (mg)	8
Iron (mg)	1.2	Energy (kcal)	72

consisting of a variety of vegetables in vinegar and eaten with main dishes in small quantity. It is also pickled along with other vegetables called *shour* in Persian and served as *torshi*. It enhances the flavour of many Southeast Asian dishes, such as fried rice variants, and its crispy chips are also used in southern Chinese cuisine. In Indonesia, it is sometimes made into pickles that are added to several traditional foodstuffs. The shallot based products such as shallot pulp, frozen shallots or chopped shallots and peeled shallots are also available. These are also used to flavour meats, stews, soups and sauces. When used raw for salads and vinaigrettes, it provides a subtle yet distinct flavour with only a slight amount of heat. Its leaves are used as potherb.

MEDICINAL USES

Shallot is rich source of antioxidants such as quercetin, kemferfol, etc. Further, they contain sulfur compounds such as diallyl disulfide, diallyl trisulfide and allyl propyl disulfide, which through enzymatic action are converted into allicin. Therefore, it reduces the cholesterol production, blood pressure, blood vessel stiffness and overall risk of Coronary Artery Disease (CAD), Peripheral Vascular Disease (PVD) and stroke. It is also found to have antibacterial, antiviral and antifungal activities. It is used in medicine for eyes and stomach diseases.

BOTANY

Shallot is a hardy biennial, but for bulb production, it is grown as an annual crop. It forms a cluster of offsets with a head composed of multiple cloves like garlic. Its plant grows to a height of 20–25 cm in a clump with narrow green leaves and roots that look like small onion bulbs, which are about 1–2 cm in diameter at maturity. The plant shows intense branching habit and has fibrous root system. Its bulbs and leaves differ according to the variety. During the first year, delicate medium-sized, 25–45 cm long, light green, tubular and subulated leaves, and small bulbs are formed, while during the second year, the leaves appear on thin peduncle, which is 40–80 cm long. The plant forms up to 30 bulbs with average total weight of 500 g. The bulbs are compact and ovate or pear-shaped with diameter 0.5–2 cm, length 1.5 cm and weight up to 100 g. The skin colour varies from golden brown to grey or rose red, and their off-white flesh is usually tinged with green or magenta colour. Shallots do not have the concentric whorls like onion bulbs though they are sectioned into cloves like garlic. Inflorescence is spherical multiflorous umbel with small light bluish-pink or lilac-coloured flowers, sometimes with light reddish veins. Instead of seeds, small bulbils are formed from flowers.

CLIMATIC REQUIREMENT

Shallot can be grown over a wide range of climatic conditions, as it does not have the specific day length or temperature requirement. It is well adapted to cool mild to mild tropical climate. The seeds germinate at a temperature range of 10–30°C with optimum germination temperature 18–24°C. The ideal shallot growing temperature is 13–24°C. However, the temperature exceeding 27°C should be avoided in daytime by planting it timely. It is more vulnerable and less frost-resistant as compared to common onion. Though shallot can be cultivated in full sun, it can tolerate partial shade also. For the initiation of bulbs, it needs specific day length; however, if a sufficient number of days exceeds its minimum requirement for flowering, it starts bolting.

SOIL REQUIREMENT

Shallot for giving higher yield requires well-drained friable loam soil rich in organic matter and initial plant nutrients. Very sandy and clayey soils should be avoided for its cultivation, as both types of soil affect the bulb quality adversely. Shallots can lose their flavour if grown in clay soil. Incorporation of too much manure in soil should be avoided since very high nitrogen content reduces the keeping quality of its bulbs. Shallot responds well to soil amendments particularly in light-textured soils with low cation exchange capacity. It prefers a soil pH of 6.5–7.5. Liming is necessary if the soil pH is below 5.5 since shallots are sensitive to low pH soil conditions. The soil should thoroughly be prepared few months before planting.

CULTIVATED VARIETIES

AMBITION F₁ AND PRISMA: Both the cultivars produce globe-shaped bulbs.

BANANA: This cultivar has been rated best seed sown and produces long shallots with red-blush skin and thick neck, which is difficult to dry.

GOLDEN GOURMET: It produces large, flattened globe-shaped golden bulbs.

JERMOR: Its firm long bulbs are rated best among the autumn-planted sets.

LONGOR: It produces best-quality long and firm golden bulbs with good keeping quality.

MIKOR: It produces round to long pinkish bulbs.

PIKANT: It produces small bulbs with robust flavour and more resistance to bolting.

RED SUN: It produces red-skinned, squat, globe-shaped firm bulbs.

TOPPER: It produces small and slightly flattened round bulbs.

The other cultivars are Atlantic, Atlas, Dutch Yellow, French Shallots, Frog Leg Shallots, Giant Red, Grey Shallot, Odetta's White Shallot, Pink Shallots and Success.

SEED RATE

The germination percentage of seed chiefly affects the seed rate. The seed rate also varies due to growing season, method of sowing and soil properties. The seeding rate is around 8–10 kg to plant a hectare land area.

SOWING TIME

The sowing time varies considerably in relation to the type of variety and climatic conditions of the growing region. Sowing of seeds of a particular variety at its right time is very essential to get higher yield of better quality bulbs since unfavourable conditions have adverse effect on these components. In north plain areas, the sets are planted in autumn, and at high altitude, they are planted in spring months. Research evidences in literature indicate that the spring-planted sets at high altitude generally produce bigger bulbs than autumn-planted sets in plain areas, though the autumn-planted sets become ready for harvesting about 2 weeks earlier than the sets planted in spring months. Shallot seeds are sown in flat seedbeds at a depth of 1–2 cm from the last week of October to the first week of November.

SOWING METHOD

The seedlings after attaining a height of 15 cm are planted in a well-prepared field at a row distance of 30 cm and plant distance of 15 cm. It is advisable to transplant the seedlings at closer spacing to avoid the splitting of bulbs, as the single bulb stores better.

The healthy shallot sets are planted at a distance of 30 × 20 cm keeping the growing tip just below the surface of well-prepared ground. If large-sized sets are planted, they may develop the flower stalks during the growing season; however, these flower stalks should be removed at initiation stage to improve the size of bulbs. The bulbs that have produced a flower stalk should be consumed first, as they are unlikely to store well.

NUTRITIONAL REQUIREMENT

Shallots generally have nutritional requirement similar to other alliums. The quantity of manure and fertilizers depend on climate, growing season, soil type and location. About 20–25 tonnes of well-decomposed farmyard manure is applied at the time of land preparation. Besides, nitrogen is applied at 130 kg, phosphorus 50 kg and potassium 60 kg/ha. The entire quantity of phosphorus and potash and half of the nitrogen are given as basal dose, and the remaining half of the nitrogen is top dressed in two equal halves at 30 and 45 days after planting.

IRRIGATION REQUIREMENT

The first light irrigation should be applied immediately after transplanting of seedlings or planting of sets if planting is done in dry soil. The topsoil should not be allowed to dry out until the seedlings are well established. Frequent light irrigations are required until the plants emerge and develop a good root system. The frequency of irrigations depends on the prevailing weather conditions of the growing region. Usually, overhead sprinkler system is used to apply irrigation in shallot crop. However, drip irrigation system is also suitable.

INTERCULTURAL OPERATIONS

HOEING AND WEEDING

Removing weeds is essential to harvest a good yield since shallot is a poor competitor of weeds, which compete with crop plants for nutrients, moisture, light and space. Being closely planted and shallow rooted crop, any root injury may reduce the bulb growth. Hence, shallow hoeing once or twice in early stages of plant growth will provide better environment to roots for their proper growth and will keep the weeds down. If manual hoeing and weeding are not possible, Stomp can be applied @ 3.35 litres/ha immediately after transplanting.

EARTHING UP

In shallot, earthing up is essential to provide better soil atmosphere to the developing bulbs and to protect the bulbs from excessive heating and greening effect of sunrays on exposed portions. This is done in such a way that an adequate amount of soil may be placed around base of the plants.

HARVESTING

Shallots become ready for harvesting 10 to 12 weeks after planting but some of the hybrids may mature a little earlier. Its bulbs at the time of harvesting should ideally be more than 25 cm long and 8–10 mm thick at stem end. Shallots are pulled out by hand, washed and bunched after trimming roots. The yellow, damaged, or diseased leaves should also be trimmed off. The tops are cut to a consistent length. Thereafter, they are tied with rubber bands in bunches of 10–12, depending on size, and after packaging in approved vegetable cartons, they are sent to the market for sale. The packed shallots in cartons are forced air cooled to maintain freshness and shelf life.

YIELD

The yield of shallot depends on the variety, soil type and fertility, growing season, growing conditions and package of practices adopted by the growers. On an average shallot gives a bulb yield of 50–70 q/ha.

POST-HARVEST MANAGEMENT

Fresh shallots can be stored for 6 months or longer in cool and dry area at 0–5°C temperature and 60–70% relative humidity. Low relative humidity and low temperature are important factors to keep the shallots dormant and free from sprouting and root growth. A condition of humidity >70% and temperature 5–8°C favours the development of sprouts and roots and decaying of shallots. With good airflow and humidity control, shallots can be stored well for about 8–10 months. Its bulbs are resistant to frost, thus, they have the good storage potential.

INSECT-PESTS

ONION THRIPS (*Thrips tabaci*)

Thrips are elongated and creamy yellow tiny (1–2 mm long) insect. Both, its nymphs and adults suck cell sap from the stalk and young leaves. The central leaves become curled, while the outer leaves turn brown at the tips. The severely infected plants may show a silvery appearance. A long dry spell favours its multiplication.

Control
- Strictly follow at least a 2–3 years crop rotation with non-host crops.
- Use overhead irrigation system to reduce thrips population.
- Remove and destroy the weeds and alternate hosts surrounding the shallot field.
- Spray the crop with Nuvacron 0.1%, Thiodan 0.2%, or Rogor 1 litre/ha at 10- to 12-day intervals.

DISEASES

PURPLE BLOTCH (*Alternaria porri*)

The disease incidence symptoms appear as water soaked areas on leaves and seed stalk, and soon after, they turn brown with purplish center. When the disease incidence is severe, the leaves and seed stalks bend downwards. Moist weather favours the spread of disease.

Control
- Strictly follow long crop rotation with cereal crops.
- Provide good drainage facility to drain out the excess water from the field.

- Treat the seed with captan, thiram, captafol, foltaf, or captaf 2 g/kg of seed.
- Spray the crop with Dithane M-45 or copper oxychloride 0.2%.

Downy Mildew (*Peronospora destructor*)

The disease incidence may appear at any stage of plant growth, even at flowering stage. Purplish or yellowish-brown spots appear on leaf surface followed by fluffy downy growth. The affected leaves drop down from the point of infection and die. Moisture and temperature both are important factors for the multiplication and further spread of the disease. The incidence is very severe under humid conditions. The disease is soil- as well as seedborne.

Control
- Follow long crop rotation with non-host crops.
- Use disease-free healthy seed.
- Destroy the infected plant debris from the field.
- Treat the seed in hot water at 50°C for 30 minutes to kill the pathogens present on seed surface.
- Spray the crop with Dithane M-45 or Dithane Z-78 0.2%.

White Rot (*Sclerotium cepivorum*)

Under successive or repeated cropping systems, this is one of the most damaging diseases of alliums, causing heavy losses since the fungus is soilborne. Unbalanced nutrition and heavy rains favour the development of the disease. The symptoms include stunted plant growth, yellowing and wilting of leaves and fluffy white growth on base of the bulb.

Control
- Strictly follow long crop rotation with non-allium crops.
- Treat the seeds with hot water before planting.
- Avoid transferring of soil or plant material from infected to healthy sites.
- Remove and destroy the infected plants debris from the field.
- Follow prophylactic sprays of streptocycline 0.1% or achromycin (tetracyclin hydrochloride) 0.04% at 7-day intervals.

Pink Root (*Pyrenochaeta terrestis*)

Generally, the incidence is noticed when crop reaches maturity. Though the infection took place much earlier to roots, the infected roots become pink and then turn to red, purple, brown and black.

Control
- Follow long crop rotation with non-allium crops.
- Disinfect the soil with methyl iodide or chloropicrin.

■■■

7 Welsh Onion

M.K. Rana and Harshita Mor

Botanical Name : *Allium fistulosum* L.

Family : Alliaceae

Chromosome Number : 2n = 16

ORIGIN AND DISTRIBUTION

Allium fistulosum, known only as cultivated species but not as wild, is originated in Northwestern China, probably in the region of Mongolia, Siberia, or China. Its nearest wild relative appears to be *Allium altaicum*, which is widespread in parts of Mongolia and Siberia, where it is occasionally collected as a vegetable for local use as well as for export to China. Cultivation of welsh onion dates back to at least 200 BC in China. It reached Japan before 500 AD and spread further to Southeast Asia and Europe. The crop is grown throughout the world but the main area of its cultivation remains Eastern Asia, *i.e.*, from Siberia to Indonesia, and elsewhere, it is mainly a crop of kitchen gardens. In Africa, it is reported from Sierra Leone, Ghana, Cameroon, Congo, DR Congo, Sudan, Kenya, Zambia and Zimbabwe. A leaf onion reported from Nigeria also belongs to the *Allium fistulosum* species.

INTRODUCTION

Welsh onion, also known as Japanese bunching onion, bunching onion, *cibol, cebolinha* and *cozida*, is a species of perennial onion. It is cultivated in vegetable and flower gardens as it takes very little space. The common name welsh onion is rather a misnomer though it is probably related to the old German *welsche*, meaning foreign. *Allium fistulosum* is the most important *Allium* species fulfilling the culinary role of common onion in China and leek in Europe. In Japan, it is now second in importance to the bulb onion (*Allium cepa* L.). The species is very similar in taste and odour to the bulb onion. Welsh onion does not develop bulbs (or only small bulbs) and possesses hollow leaves (*fistulosum* meaning hollow) and scapes. Large varieties of welsh onion resemble the leek (*Allium ampeloprasum*), while small varieties resemble the chives (*Allium schoenoprasum*). Many varieties of welsh onion multiply by forming perennial evergreen clumps.

COMPOSITION AND USES

COMPOSITION

The nutritional composition of green tops and blanched pseudostem differs, green tops being more nutritious than its blanched pseudostem. It is an excellent source of potassium, calcium, phosphorus and vitamin A; a good source of iron, magnesium, vitamin C and K; and a fair source of zinc, manganese, and vitamin E and B group vitamins. The nutritional composition of welsh onion is given in Table 7.1.

TABLE 7.1

Nutritional Composition of Welsh Onion (per 100 g Edible Portion)

Constituents	Contents	Constituents	Contents
Water (g)	90.5	Manganese (mg)	0.137
Carbohydrates (g)	6.5	Vitamin A (IU)	1160
Protein (g)	1.9	Thiamine (mg)	0.05
Fat (g)	0.4	Riboflavin (mg)	0.09
Dietary fibre (g)	2.4	Niacin (mg)	0.4
Calcium (mg)	52	Panthothenic acid (mg)	0.169
Phosphorus (mg)	49	Folic acid (µg)	16
Potassium (mg)	212	Vitamin C (mg)	27
Magnesium (mg)	23	Vitamin E (mg)	0.51
Sodium (mg)	17	Vitamin K (µg)	193.4
Iron (mg)	1.22	Energy (kcal)	34
Zinc (mg)	0.17	–	–

USES

Japanese bunching onion pseudostems are eaten as a potherb, *e.g.*, in sukiyaki and chicken dishes, whereas, welsh onion green leaves are used in salads or as an herb to flavour soups and other dishes. In the Brazzaville-Kinshasa area, the whole plants are eaten as a boiled vegetable. In Southeast Asia, the plants are eaten whole either steamed or after heating for a short time, and in Japan, small seedlings are used in special dishes. Welsh onion has not yet been used in a processed form until a dehydration industry started. Its products are used mainly as an additive to preprocessed food such as instant noodles. Sometimes, the young inflorescence is deep-fried and eaten as a snack.

The plant is said to prevent termite infestation in gardens. In China, its diluted extract is used against aphids.

MEDICINAL USES

Welsh onion have many therapeutic properties. In Chinese medicine, it is used to improve the functioning of internal organs and the metabolism to prevent cardiovascular disorders and to prolong life. It has also been reported to improve eyesight and to enhance recovery from common cold, headache, wounds and festering sores.

BOTANY

Welsh onion is a perennial glabrous herb but usually grown as an annual or biennial plant, which is 50–100 cm tall with indistinct, ovoid to oblongoid bulb up to 10 cm long, few to several or virtually absent lateral bulbs and is tunic white to pale reddish brown. Leaves are 4–12 in number, distichously alternate, glaucous with tubular sheath, hollow cylindrical blade with acute apex and somewhat round in cross-section. Inflorescence is spherical umbel, which is 3–7 cm in diameter placed on a long, erect and hollow 50–100 cm long scape having 2.5 cm diameter. The umbel is composed either of flowers or of bulbils only.

Ostensibly, *Allium fistulosum* has strong resemblance to *Allium cepa* but does not form bulb, although some slight thickening of pseudostem base may occur. The difference in flowers is more prominent. Flowering in welsh onion progresses from top of the umbel downwards. Its flowers are larger and they lack bracteoles, while those of *Allium cepa* are smaller, stellate and have bracteoles. The flowers of welsh onion are bisexual, narrowly campanulate to urceolate having slender pedicel, up to 3 cm long. Tepals are six presenting in two whorls, free, ovate-oblong to oblong-lanceolate, 6–10 mm long and white with greenish mid vein. The stamens are also six in number, exceeding tepals, connate at base and adnate to tepals. Ovary is 3-celled and superior. Style is also slender, exceeding tepals. Fruit is a globular capsule, 5 mm in diameter, splitting loculicidally and few-seeded. Seeds are black in black.

In Europe, welsh onion has been classified into two cultivar groups, *i.e.*, (i) Japanese bunching group with single-stemmed cultivars grown for their thickened pseudostem in Eastern Asia and (ii) Welsh onion group with multi-stemmed cultivars grown for their leaves. In Japan, it has been classified into 3 groups, *i.e.*, (i) Kaga Group grown in coolest parts of Japan for its pseudostems, (ii) Kujyo Group grown in warmest parts of Japan mainly for its leaves and (iii) Senju Group, which is intermediate in use and ecological requirements.

Allium fistulosum hybridizes easily with *Allium altaicum*. The hybrids have high pollen fertility and develop enough seeds. Its hybridization is also possible with *Allium cepa*. The resulted hybrids are perennial, forming small bulbs. Several old hybrids between *Allium fistulosum* as female parent and tropical shallots (*Allium cepa* Aggregatum Group) as male parent still exist, *e.g.*, (i) the Indonesian shallot *Sumenep*, which is well adapted to tropical conditions and never flowers and (ii) the top onion, in which, the inflorescence is composed of bulbils. At vegetative stage, these plants resemble *Allium fistulosum*. The commercial hybrids between *Allium fistulosum* as female parent and *Allium cepa* as male parent grown mainly in the United States and Europe for their green tops include *Beltsville Bunching*, a seed-producing tetraploid; *Louisiana Evergreen*, a sterile diploid; and *Delta Giant*, a sterile triploid originating from a backcross of the hybrid with *Allium cepa*. Another group of hybrids grown in Japan is Yakura Negi Group, the plants of which tiller abundantly in spring and summer and become dormant in winter. Their inflorescence produces no flowers but produces only bulbils, thus, they are propagated by division or by bulbils.

CLIMATIC REQUIREMENT

Welsh onion is amazingly adapted to a wide range of climates. It is highly tolerant to cold weather and can overwinter even in Siberia. It is also tolerant to hot and humid conditions occurring near Brazzaville and Kinshasa in Central Africa. It can be grown well above an altitude of 200 m but it enjoys the climate more above 500 m mean sea level. Thus, most of the local cultivars of welsh onion are well adapted to variations in rainfall and more tolerant to heavy rainfall than other *Allium* species. The plants after establishment are tolerant to moisture stress and are very less susceptible to drought conditions. Usually, its seeds need 15–20°C temperature to germinate.

Flower induction is governed by temperature and day length. Exposure to low temperature coupled with short days is essential to induce flowering but requirement varies strongly with cultivars. Flowering is generally induced by temperature below 13°C when seedlings have completed the basic vegetative (juvenile) phase, *i.e.*, formation of 6–7 leaves and pseudostem of certain thickness. The tropical conditions favour only the vegetative phase, not the reproductive phase, hence, only the well-adapted cultivars can flower under such conditions.

SOIL REQUIREMENT

Welsh onion can be grown on almost all types of soil but loamy soils rich in organic matter are preferred more. Sandy soils requires frequent irrigation and favour early maturity, while heavy soils lead to development of misshaped bulbs and make the harvesting of bulbs difficult. The most desirable soil for welsh onion is that which retains moisture for longer period and allows the proper development of bulbs. It is very susceptible to water-logging conditions, which promptly kill the active roots, thus, the soil should be well drained. Its good growth is possible even at a soil pH of 8–10 but a soil with neutral pH is required for its optimal growth. Its plant growth is generally poor in acidic soils.

CULTIVATED VARIETIES

Numerous cultivars including some with red skin, *e.g.*, Santa Clause and Red Beard, some with shorter and thicker stalks, *e.g.*, Shimonita, some with larger bulbs, *e.g.*, Yoshima and some with exceptional cold-hardiness such as White Lisbon, Evergreen White Bunching and Winter Over are available. It is not unusual to harvest these under snowy conditions. Bunching onions have also been hybridized with other *Allium* species, especially with common bulbing onion. However, the resulted hybrids are sterile and can be propagated by division of side shoots. Beltsville Bunching is one such hybrid, which is noteworthy for its tolerance to hot and dry conditions.

SEED RATE

The germination percentage of seed chiefly affects the seed rate. It may also vary due to growing season, method of sowing and soil properties. A seed quantity of 8–16 kg/ha is needed for direct seeding crop and 2–4 kg/ha for transplanting crop.

SOWING TIME

The sowing time varies considerably in relation to the type of variety and climatic conditions of growing region. Sowing of seeds of a particular variety at its right time is very essential to harvest economic yield. Its seeds are sown in the first week of November and seedlings are transplanted in the last week of December to the first week of January. In cold areas, the crop sown directly in autumn takes 46 weeks to produce large bulbs; however, the crop sown in February under cloches or directly in March–April takes 22 weeks to harvest large bulbs. In cold areas, the seeds can be sown in greenhouse in January and the seedlings can be transplanted in outdoor fields in April. Under mild climatic conditions, it can be grown throughout the year.

SOWING METHOD

In tropics, welsh onion is propagated mainly by basal tillers. Seedlings are rarely raised from seed since it is time-consuming and difficult under tropical conditions. In temperate areas where seed production is more successful, propagation of welsh onion is mainly done by seed, which is sown directly in the field or first in nurseries and then seedlings are transplanted in field. In nursery beds, seeds are sown indoors in lines at 5–7.5 cm distance using normal potting compost. The seedlings will appear a week to 10 days later, and till then, the nursery soil is kept moist. When the seedlings are 15–20 cm tall and pencil like thick, they are transplanted in the field at a distance of 20 cm between lines and 15 cm between plants.

Welsh onion is very much similar to other onions, which have a bulbous root and green leaves. The bulbs that multiply quickly over the years make its propagation easiest. The clumps of bulbs are dug out in March or October and carefully separated into individual bulbs, which are used to replant keeping their tips at the level of soil surface. Growing by this method, welsh onion thrives best since it alleviates congestion in the bulb clumps.

For green leaf or blanched pseudostem production, the field is deeply cultivated and tillers or seedlings are transplanted on raised beds or ridges that are alternated with 10–20 cm deep furrows made for irrigation and drainage by throwing soil to one side forming a ridge, which will support the young plants and facilitate earthing up later. Planting is done at a row distance of 20 cm and plant distance of 25 cm, accommodating 200,000 plants per hectare. About one-third top portion of the tillers is usually trimmed to reduce moisture loss through transpiration.

NUTRITIONAL REQUIREMENT

The quantity of manure and fertilizers depend on climate, growing season, soil type and location. About 20–25 tonnes of farmyard manure is applied at the time of land preparation. Besides, nitrogen is applied @ 100 kg, phosphorus 50 kg and potassium 50 kg/ha. The entire quantity of phosphorus and potash and half dose of nitrogen are given as basal dose and remaining half of the nitrogen is top dressed in two equal halves at 30 and 45 days after planting.

IRRIGATION REQUIREMENT

Welsh onion crop needs plenty of water. Sometimes, the seeds do not germinate due to lack of moisture. Therefore, the soil must have plenty of moisture for better seed germination and subsequent growth of the plants. A light irrigation is appreciable just after transplanting. Subsequent irrigation should be supplied as per the crop need. The quantity of water and its frequency depend on crop stage, soil type and planting season. Frequent irrigation is necessary during dry season.

INTERCULTURAL OPERATIONS

HOEING AND WEEDING

Removing of weeds is essential to harvest good yield since welsh onion is a poor competitor of weeds, which competes with crop plants for nutrients, moisture, light and space. Being closely planted and shallow rooted crop, any root injury may reduce bulb growth. Hence, shallow hoeing once or twice in early stages of plant growth will provide better environment to roots for their proper growth and will keep the weeds down. If hoeing and weeding manually are not possible, Stomp can be applied @ 3.35 litres/ha immediately after transplanting.

THINNING

Thinning in direct sown welsh onion is an essential operation to maintain optimum spacing between plants to reduce competition among plants for space, light, nutrients and moisture and to provide better conditions to the plants for their growth and development. Thinning is done as and when the plants are large enough to handle. After completion of thinning operation, a light irrigation should be applied to establish the disturbed seedlings and to fill the soil gaps developed due to the uprooting of seedlings.

EARTHING UP

Earthing up is also an essential operation in welsh onion to blanch and keep the pseudostem soft. The soil is mounted up around the pseudostems and bulb clumps to avoid their exposure to sunlight, which may cause greening of pseudostems and bulbs. Earthing up also improves aeration around the roots. However, this operation should be done gradually, not too early, as it may check plant growth.

HARVESTING

In tropics, welsh onion can be harvested year-round. However, in the Brazzaville-Kinshasa area, it is mainly harvested during rainy season. For pseudostems, the plants are pulled out about 2.5 months after planting the tillers. The part used as planting material for the next crop is left in the field until it is needed. Harvesting is a labour-intensive operation, especially for pseudostems, which have to be dug up, cleaned and bundled. The equipments for mechanical harvesting have been developed in Japan.

YIELD

The average yield of welsh onion in Japan and Korea is about 25 t/ha, in Taiwan 10–15 t/ha. In Indonesia, the yield is considerably lower, averaging 7 or at the most 15 t/ha; however, the growing period is only 2.5–3 months compared to 9 months in East Asian countries.

POST-HARVEST MANAGEMENT

Method of harvesting is very important factor in shelf life of welsh onion. After harvesting, the leaves and pseudostems are cleaned, dried, or the damaged leaves are removed, and thereafter, the pseudostems are bunched and packed in boxes or baskets for sending to the market.

INSECT-PESTS

ONION THRIPS (*Thrips tabaci*)

The thrips are very tiny (1–2 mm long), elongated and creamy yellow insect. Both, its nymphs and adults suck cell sap from the stalk and young leaves. The central leaves become curled, while the outer leaves turn brown at the tips. Severely infected plants may have a silvery appearance. A long spell of dry weather is favourable for its multiplication.

Control

- Follow long crop rotation with non-host crops.
- Use overhead irrigation system to reduce thrips population.
- Remove the grasses and weeds from the field to destroy the alternate hosts.
- Spray the crop with Nuvacron 0.1%, Thiodan 0.2%, or Rogor @ 1 litre/ha 2–3 times at 10-day intervals.

BEET ARMYWORM (*Spodoptera exigua*)

These are greenish-brown, soft, bulging caterpillars with dark longitudinal stripes. The adult is brownish or grey-coloured moth with 2–3 cm wingspan. The larvae feed on both foliage and fruit. Young larvae feed gregariously and skeletonize the foliage. Smaller larvae eat the parenchyma of leaves greedily, leaving only thin epidermis and veins. As the larvae mature, they become solitary and make large irregular holes on leaves. They attack shoot buds and new growth, preventing new buds to sprout and flowers to open. As the smaller larvae move about, they leave strands of silk behind, netting the leaves with a silvery film.

Control

- Follow clean cultivation.
- Remove broad leaf weeds through weeding.
- Spray the crop with malathion 0.05% or dichlorvos 0.05%.

AMERICAN BOLLWORM (*Helicoverpa armigera*)

Moths are variable in size and colour, and larvae are greenish or brown in colour. Their body has longitudinal dark and pale bands. Its adults with stout body, light brown colour and wingspan of 32–40 mm are nocturnal.

Control
- Destroy the pupae in the soil by ploughing the field deep in hot summer months.
- Remove the weeds in and around the field.
- Spray the crop with quinalphos 3 ml, chloripyriphos 2.5 ml, fenvaralate 2 ml, or deltame-thrin 1.6 ml/litre of water.

DISEASES

PURPLE BLOTCH (*Alternaria porri*)

Incidence starts with the appearance of water-soaked areas on leaves and seed stalk and soon they become brown in colour with a purplish center. Severity of disease causes bending of leaves and seed stalks. Moist weather is favourable for the spread of disease.

Control
- Strictly follow long crop rotation with non-host crops.
- Provide better drainage facility to drain out excess water.
- Treat the seed with fungicide like captan, thiram, captafol, or ziram 2 g/kg of seed.
- Spray the crop with Dithane M-45 @ 0.2% or copper oxychloride @ 0.2%.

DOWNY MILDEW (*Peronospora destructor*)

This disease may attack the plant at any stage of growth, even at flowering stage. On mature plant parts, purplish or yellowish-brown spots appear on leaf surface followed by fluffy downy growth. The infected leaves turn yellow and later die. Moisture and temperature both are important factors for the multiplication of pathogen and further spread of this disease. The incidence is very severe under very wet weather conditions. The disease is soil- as well as seedborne.

Control
- Strictly follow long crop rotation with cereal crops.
- Avoid planting of infected sets.
- Remove and destroy infected plant debris.
- Treat the seed with hot water at 50°C for 30 minutes to kill the pathogens present on seed.
- Follow prophylactic sprays of Dithane M-45 or Dithane Z-78 0.2%.

WHITE ROT (*Sclerotium cepivorum*)

This is one of the most damaging diseases of bulb crops worldwide, causing serious losses under successive or repeated cropping since the pathogen persists in soil. The fungus can survive in soil for 20 years. Unbalanced nutrition and heavy rains stimulate its development. The symptoms include stunted plant growth, yellowing of older leaves, death of all leaves and fluffy white growth on base of the bulb.

Control
- Avoid including bulbous crops in crop rotation.
- Treat the seeds with hot water prior to planting.
- Avoid transferring of soil or plant material from infected to healthy sites.
- Remove and destroy the affected plants and crop debris.
- Spray the crop with streptocycline 0.1% or achromycin (tetracyclin hydrochloride) 0.04% at an interval of 7 days.

Welsh Onion Yellow Stripe Virus (WoYSV)

The symptoms show yellow leaf striping, chlorotic mottling and streaking, stunting and distorted flattening of the leaves. Aphid acts as virus vector for the spread of this disease.

Control
- Uproot and destroy the affected plants to check further spread of the disease.
- Spray the crop with insecticides to control aphids responsible for the spread of disease.

■■■

8 Amaranth

Dipika Sahoo and Pradyumna Tripathy

Botanical Name : *Amaranthus* spp.

Family : Amranthaceae

Chromosome Number : 2n = 32 and 34

ORIGIN AND DISTRIBUTION

Amaranth is one of the oldest food crops in the world, with evidence of its cultivation reaching well back to 6700 BC. The genus *Amaranthus* consists of nearly about 60 species, which include leafy vegetables, vegetables for grains, ornamental plants as well as weeds. The centres of diversity for vegetable amaranth are Central and South America, India and Southeast Asia with secondary centres of diversity in West and East Africa. Most probably, it is native to India from where it spread to neighbouring countries like Nepal, Tibet and Afghanistan. Numerous different amaranth varieties are grown throughout the country but the Himalaya region is known as the *centre for diversity* of amaranth due to the existence of a number of species in this region. Today, the amaranth is treated as a staple crop in the hilly regions of Northwestern India, where up to half of the non-irrigated land is covered with this crop. Extensive variations in growth habit, disease resistance, taste and quality are reported to exist within species, which offer considerable scope for future breeding programmes. The crop is cultivated in most parts of tropical and subtropical regions of the world successfully.

INTRODUCTION

The word *amaranth* has actually been derived from the Greek word *amarantos*, which means *unwithering* and the people who used it symbolize immortality. The Hindi term for amaranth, *Ramdana*, means *God's own grain*. *Amaranthus tricolour* L. is the most common species grown as leafy vegetable during summer and rainy season in India but it can be grown throughout the year under mild conditions. Its leaves and succulent stems are good sources of iron (38.5 mg/100 g), calcium (350–400 mg/100 g), carotene (pro-vitamin A) and vitamin C, and at maturity, it has high fibre content. It contains about four times as much calcium and twice as much iron and magnesium as wheat. In Southeast Asian countries, especially Malaysia, Indonesia, Southern China as well as hot and humid regions of Africa, the amaranth is widely grown as a leafy vegetable. It fits well in different crop rotations because of its very short duration, quick response to manure and fertilizers, high yield, easiness in cultivation, suitability to different soil and agro-climatic conditions and large yield of edible matter per unit area. It is usually grown in kitchen and market gardens. Several species of the genus *Amaranthus*, which are cultivated for grain, vegetable and ornamental purposes, have the unique features.

Amaranth is more widely used as a potherb. It is generally a very common leafy vegetable grown during both summer and rainy season. In some instances, it supplies a substantial part of protein, minerals and vitamins in the diet, which makes it an easy crop to cultivate and domesticate. Due to very short maturity period, the crop is very much suited in multiple cropping systems.

COMPOSITION AND USES

COMPOSITION

The leaves and tender stems of vegetable amaranth are rich in carbohydrates, protein, minerals like calcium, phosphorus, potassium, magnesium, sodium, sulfur, iron, zinc, copper and manganese as well as vitamins such as vitamin A, thiamine, riboflavin, niacin, folic acid and vitamin C. On a dry weight basis, the crude protein content in its leaves ranges from 20% to 32%. A loss of 80.3% ascorbic acid from parboiling for 5 minutes and up to 91.5% after final cooking for 5 minutes has been reported. High oxalate (1–2%) and nitrate (1.8–8.8 g/kg dry matter) levels are reported from leaves of various amaranth species. Cooking of fresh *Amaranthus hybridus* leaves results in a 35% loss of ascorbate and a reduction of the chemical score from 71 to 58. The nutritive value of leafy amaranth has been shown in Table 8.1.

USES

The fresh tender leaves and stem of amaranth taste mellow with a hint of sweetness, possesses a tender texture and are delicious when cooked like other fresh leafy vegetables. The mature leaves, on the other hand, are much tougher and bitter in taste. To reduce their astringency, mature leaves are often boiled, steamed, or sautéed. Its grains are used as human food in a number of ways but the most common conventional usage is to grind its grains into flour for mixing in breads, noodles, pancakes, cereals, granola, cookies, or other flour-based products. The grain can be popped like popcorn or flaked like oatmeal. More than 40 products of amaranth are currently available in the markets.

MEDICINAL USES

Amaranth, which is quite nutritious and rich in calcium and beta-carotene, keeps bone strong and reduces risk of osteoporosis. However, the absorption of beta-carotene, calcium and iron by human body is relatively lower, which primarily depends on quality of the fresh products, the mode of preparation and other physical conditions of the consumers. Calcium in amaranth assists muscle regeneration and stabilizes blood pressure in human body. Its high vitamin C content helps the body to produce collagen, which aids in joint health and keeps the skin glowing. The folic acid in amaranth leaves decreases cardiovascular damage and keeps the memory strong. The soup of amaranth leaves soothes inflammation and sore throat. Besides, the amaranth contains protein, which is rich in sulfur containing amino acids, particularly methionine, lysine, cystene, etc., making it a good combination with cereals – the most ideal combination among the Indian diets. Daily consumption of 50–100 g of amaranth leaves reduces incidence of blindness in malnourished children. It has a

TABLE 8.1
Nutritional Composition of Amaranth Leaves (per 100 g Edible Portion)

Constituents	Contents	Constituents	Contents
Water (g)	85.7	Sodium (mg)	**230.0**
Carbohydrates (g)	6.3	Sulfur (mg)	**61.0**
Protein (g)	4.0	Iron (mg)	25.5
Fat (g)	0.5	Vitamin A (IU)	9200
Dietary fibre (g)	1.0	Thiamine (mg)	0.03
Minerals (g)	2.7	Riboflavin (mg)	0.10
Calcium (mg)	397.0	Nicotinic acid (mg)	1.00
Phosphorus (mg)	83.0	Vitamin C (mg)	99
Potassium (mg)	341.0	Energy (kcal)	36
Magnesium (mg)	247.0	–	–

great potential for combating under- and malnutrition problems. Tribals of Tamil Nadu hills, use the juice of leaves and stems of *Amaranthus spinosus* in the treatment of kidney stones. Another species of *Amaranthus viridis* is used in treating snakebites. In spite of several advantages of amaranth as leafy vegetable, both its stems and leaves also contain the anti-nutritional factors such as nitrate and oxalate, which can be removed by cooking with ample water. In general, the presence of oxalates causes the kidney stone, while that of nitrates cause methaemoglobin in blood. Therefore, it is recommended not to consume leaves of more than 200 g/day/head.

CYTOLOGY

The chromosome number of amaranth varies with different species. Cytological studies reveals that with the exception of one tetraploid species *Amaranthus dubius* (x = 16, n = 32, 2n = 64) belonging to *Amaranthus* section, all the species examined so far have chromosome number of either 32 (n = 16) or 34 (n = 17). The genus is dibasic with x = 16 and 17, almost equally distributed in section *Amaranthus* and *Blitopsis*. The later chromosome counts indicated *Amaranthus tricolour* to be 2n = 2x = 34, *Amaranthus cruentus* to be 2n = 32 and *Amaranthus dubius* to be 2n = 64.

TAXONOMY

Amaranth belongs to the genus *Amaranthus* and family Amaranthaceae, which comprises of 65 genera and 850 species. The genus *Amaranthus* includes 50–60 species, the leaves of which are edible. These are the most important leafy vegetables of the tropical countries in South Asia, Southeast Asia, East Africa, Central Africa, West Africa, Ethiopia, the Pacific and Far East. The important species of leafy amaranth are *Amaranthus tricolour* L., *Amaranthus dubius* Mart. Ex Thell., *Amaranthus lividius*, *Amaranthus blitum*, *Amaranthus tristis* L., *Amaranthus spinosus* L., *Amaranthus viridis* L. and *Amaranthus graecizans* L., and the most popular grain amaranth species are *Amaranthus hypochondriacus* L., *Amaranthus cruentus* L. and *Amaranthus caudatus* L. Inter-specific and inter-varietal hybridization in nature have caused a wide variation differing in pigmentation of plant parts as well as in the inflorescence. Taxonomists recognize two sections in amaranth *viz.*, *Amaranthus* and *Blitopsis*, with equal number of species in each. *Blitopsis* has been described as a section, where all species under this have flower clusters in axils, while species under section *Amaranthus* have only terminal flower clusters. Section *Amaranthus* includes important grain types, while section *Blitopsis* consists of the green types. The cytological taxonomy of 11 *Amaranthus* species is presented in Table 8.2.

TABLE 8.2
Cytological Studies in *Amaranthus*

	Haploid Chromosome Number	
Investigated Cytological Taxonomy	Presently Worked Out	Previous Report
Amaranthus caudatus L. (both red and yellow form)	n = 16	2n = 32, 34
Amaranthus hyrbidus L. var. paniculatus L	n = 16	2n = 32, 34
Amaranthus hybridus L. var. frumentaceous Roxb.	n = 16	–
Amaranthus giganteus Koenig	n = 32	2n = 34
Amaranthus caturtus Heyne.	n = 32	–
Amaranthus tricolour L.	n = 17	2n = 34
Amaranthus tricolour L. var. tristis Haines	n = 17	–
Amaranthus blitum L.	n = 17	2n = 34
Amaranthus mangostanus L.	n = 17	2n = 32, 34
Amaranthus gracilis Desf.	n = 17	n = 17 & 2n = 34

Several types of amaranth grown in India are given as under:

- *Amaranthus caudatus*: This species is also known as *Love Lies Bleeding* for its red beautiful overhanging inflorescences.
- *Amaranthus cruentus*: A green species that occasionally appears in beautiful shades of purple, red, ginger, or cinnamon.
- *Amaranthus hypochondriacus*: This ornamental species, usually known as *Prince of Wales Feather*, has tall, thin and spiky inflorescences. Though this type originates in Mexico, it is seen throughout the Himalayas.
- *Amaranthus paniculatus*: This species is also known as *Autumn palette*, which is named as such for its gorgeous colours of gold, rust and delicate orange.
- *Amaranthus spinosus*: The species is best described by its other monikers, *i.e.*, *Prickly Amaranthus* and *Spiky Amaranthus*.
- *Amaranthus tricolour*: This species is named as such because of its tricolour leaves, *i.e.*, flaming yellow, green and red.
- *Amaranthus vridis*: This species is also called *green amaranth* and *slender amaranth*. It is one of the most common varieties sold in the country.

BOTANY

Amaranth is an annual dicot herb with erect or spreading growth, scarce to profuse branching habit and shallow to deep taproot system. Stem is glabrous, succulent and green, purple, or mixed shades of these two. Leaf is simple, alternate, exstipulate with obviate to lanceolate shape and acute tip. Leaf colour is green, red, or with different shades of above. Sometimes, purple colouration, which is prominent in young leaves, fades away at maturity. Flowers are borne terminally and in axils of leaves in clusters. The basic unit of inflorescence is called as glomerule. Flowers are small, unisexual, pentamerous, bracteate and bracteolate, which are scaly, perianth 4–5, imbricate, membranous and often persistent. Most of cultivated types are monoecious. Proportion of male and female flowers varies in an inflorescence. Each glomerule consists of a staminate flower and a number of pistillate flowers. In all species except *Amaranthus spinosus*, the first flower of each glomerule is staminate and the rest are pistillate. In *Amaranthus spinosus*, each glomerule bears flowers of same sex with pistillate flowers in the axils of branches and at the base of terminal inflorescence, while the staminate flowers are terminal on the main axis and lateral branches.

The extent of cross-pollination is governed by proportion of male and female flowers in an inflorescence and position of inflorescence in plant. Stamens are 2–5 placed opposite to perianth parts, often some reduced to staminodes, filaments either free or united below, anthers one- or two-celled and carpels are two to three, which are syncarpous. Ovary is superior with commonly one campylotropous ovule. Style is single to two- or three-branched. Percentage of male flowers in a glomerule is 0.5 in grain types and 10.25 in leaf types. Leaf amaranths are predominantly self-pollinated due to the presence of a large number of male flowers per glomerule, terminal inflorescence and development of axillary glomerules, while the grain types are cross-pollinated. Stigma of pistillate flower becomes receptive several days prior to opening of staminate flowers in an inflorescence. Wind helps in the transfer of pollen grains from male to female flowers of a glomerule, inflorescence, or plant to another glomerule, inflorescence, or plant but the grain type species with colourful inflorescence are occasionally visited by bees. Fruit is utricle, indehiscent and enclosed by the persistent perianth parts. In leaf-type amaranths, the seed colour is black, while in grain-type amaranths, the seed colour is sandal yellow. The number of tiny lens-shaped seeds per gram may be up to 3000.

CLIMATIC REQUIREMENT

Amaranth is widely distributed in both tropical and subtropical regions of the world. Leaf type amaranth is a warm season crop adapted to hot and humid climatic. It is grown throughout the year in tropics and in autumn, spring and summer season in temperate regions. Most of the leaf types are day neutral in habit but differ in their day length requirements and respond differently to changes in photo- and thermo-periodism. Red amaranth requires bright sunlight for colour development. Grain types like *A. caudatus*, *A. cruentus* and *A. edulis* are short day species, while *A. hypochondriacus* is day neutral species. However, all the species have early and late cultivars. Leafy ones are grown best in the plains. Amaranth seeds need soil temperature between 18°C and 25°C to germinate and an air temperature above 25°C for optimum growth. The growth is ceased at temperature below 18°C. The number of growing day degrees during the growing season is a major determinant of amaranth plant growth. Lower temperature and shorter days will induce flowering with a subsequent reduction in leaf yield. Frost plays an important role in the harvesting of crop. Amaranth being an annual crop does not mature completely in areas with a short growing season, thus, frost is necessary to terminate the crop's growth.

SOIL REQUIREMENT

Amaranth is adapted to a variety of soil types, including marginal soils but it performs best on well-drained deep fertile soils. However, sandy-loam soil slightly acidic in nature is preferred more. The ideal pH range is between 5.5 and 7.5. However, some strains of amaranth can be grown in soils with pH as high as 10. Loose and friable soils with high organic matter content are ideal for an early and higher yield. Minimisation of crusting in seedbeds with low clay content can help in ensuring good crop stands. Amaranth seeds require good seed-soil contact for rapid germination and emergence, and adequate soil moisture must be maintained at seeding depth throughout initial establishment. Though amaranth can be grown on a wide range of soils, heavy clayey soils with very poor drainage or sandy soils with very poor water-holding capacity are unsuitable for its cultivation.

The seedlings are fragile, thus, it is important to prepare fine seedbeds after incorporation of cattle, chicken, or compost manure at a rate of one bucket per square meter or one to two teaspoons of mineral compound fertilizers per metre row. Since the crop is directly sown and the seed size is small, the field is prepared to a fine tilth and free from clods by ploughing with a disc or spike-tooth harrow repeatedly followed by cult packing and planting, preferably using a planter with press wheels.

CULTIVATED VARIETIES

ARKA SUGUNA

A multi-cut green variety, belonging to the species *Amaranthus tricolour*, developed at Indian Institute of Horticultural Research (Hesserghatta) Bangalore through pure line selection from an exotic collection from Taiwan (IIHR-13560) produces light green broad leaves with succulent stem, giving first cutting 25–30 days after sowing and total cuttings 5–6 in crop duration of 90 days. It is moderately resistant to white rust under field conditions. Its average yield is 25–30 t/ha.

ARKA ARUNIMA

A multi-cut purple variety, belonging to the species *Amaranthus tricolour*, developed at Indian Institute of Horticultural Research (Hesserghatta) Bangalore through pure line selection from a local collection IIHR 49 produces broad dark purple colour leaves. It becomes ready for first cutting 30 days after sowing and gives two subsequent cuttings at 10- to 12-day intervals. Leaves are rich in calcium and iron. Its average yield is 27 t/ha.

ARKA SAMRAKSHA

A high-yielding amaranth variety developed at Indian Institute of Horticultural Research (Hesserghatta) Bangalore produces leaves with high antioxidant activity of 499 mg (AEAC units) and minimum nitrate content of 27.3 mg and oxalates of 1.34 g/100 g fresh weight of leaves. It is a pulling type amaranth variety with green leaves and stem. Its yield potential is 10.9 t/ha in 30–35 days crop duration.

ARKA VARNA

A high-yielding amaranth variety developed at Indian Institute of Horticultural Research (Hesserghatta) Bangalore produces leaves with high antioxidant activity of 417 mg (AEAC units), nitrate content of 37.6 mg and oxalates of 1.4 g/100 g fresh weight of leaves. It is also a pulling type amaranth variety with green leaves and pink stem. Its yield potential is 10.6 t/ha in 30–35 days duration.

CO 1

A cultivar tolerant to insect-pests and diseases developed at Tamil Nadu Agricultural University, Coimbatore in 1968 through a selection from a local type belongs to the species *Amaranthus dubius*. It is mainly grown for its tender greens as well as immature stems, which are thick and fleshy. The leaves are broad, thick and dark green in colour. The crop duration is 25 days for tender greens and 50–60 days for immature stems. It produces 7–8 quintals of green matter from a hectare area. Seeds are small and black in colour.

CO 2

A variety tolerant to insect-pests and diseases developed at Tamil Nadu Agricultural University, Coimbatore through a selection from local type (*Amaranthus tricolour*) and released in the year 1976 is especially suitable for early harvest. The stems remain tender and succulent, thus, both leaves and stems are used as vegetable. Plants medium tall with 125 cm height, leaves green, lanceolate, slightly elongated, stem green and succulent, spikes terminal and axillary, seeds bolder and black and exhibit early germination with vigorous growth. The crude fibre content is less (1.3%) with 19.0 mg of iron and 20.0 mg of calcium per 100 g edible portion. The crop duration is 25 days for tender greens and 35–50 days for immature stems. It produces green matter of 10–11 t/ha.

CO 3

A clipping-type variety developed at Tamil Nadu Agricultural University in 1981 belongs to *Amaranthus tristis* (Synonym *Amaranthus tricolour* var. amaranthus). The leaf to stem ratio is high (2.0) and this enhances the palatability of cooked greens largely. The greens of this cultivar are nutritious with 25.2 mg of ascorbic acid per 100 g of fresh matter and 2.48% calcium, 0.8% iron and 1.74% crude fibre. As a seed crop, this cultivar flowers in 35–40 days after sowing and matures in 85–90 days. The cultivar lends itself for 10 clippings commencing from 25 days after sowing and provides a continuous supply of luscious tender green for a period of 3 months at weekly intervals. The total green matter yield would be 31 t/ha.

CO 4

A green-cum-grain type variety developed at Tamil Nadu Agricultural University, Coimbatore through a selection belongs to the species *Amaranthus hypochondriacus*. The plants that make rapid vegetative growth within a period of 20–25 days are dwarf in growing habit, for which, they are thinned at a spacing of 30 × 30 cm. By this way of thinning, a greens yield of 7–8 t/ha can be obtained. The remaining plants are allowed to produce the inflorescence and set seeds. The grains are separated after attaining proper maturity. The grain yield is 2–2.5 t/ha in 80–90 days. The seeds are rich in protein (15.95%) and essential amino acids like lysine (7.5 mg/100 g), phenylalanine (5 mg/100 g), leucine (1.2 mg/100 g) and isoleucine (1.8 mg/100 g). The grain can be substituted for minor millets like *Ragi* and *Thenai*. It is amenable for various food preparations like any other grains.

CO 5

A variety developed at Tamil Nadu Agricultural University, Coimbatore through a single plant selection (A166-1) from an open-pollinated plant introduction has double colour (green and pink) large obovate leaves. The plants are medium tall with high biomass and nutritive value. It gives a rosette growth in early stages. The stem and petiole, which are free of fibre, are pinkish red in colour. It is also ideal for pot or container cultivation. The crop can be harvested at 30 days for tender greens (10 t/ha) and at 50 days for immature tender stem (30 t/ha).

Pusa Chhoti Chaulai

A variety developed at Indian Agricultural Research Institute, New Delhi belongs to the species *Amaranthus blitum* and responds well to cuttings, thus, it is suitable for its leafy shoots. Plants are erect in growth habit and slightly dwarf with thin stem. Its leaves are small and green in colour. Flowers are borne in clusters in leaf axils. It is best suited for cultivation in early summer but can be grown in rainy season too.

Pusa Badi Chaulai

Another variety developed at Indian Agricultural Research Institute, New Delhi belongs to the species *Amaranthus tricolour* and responds well to cuttings since the economic part of this cultivar is leafy shoot. Stem is thick and green and leaves large and green. This variety is distinguishable by its much longer growing period. It is best suited for cultivation in summer season.

Pusa Kirti

A variety developed at Indian Agricultural Research Institute, New Delhi through a selection from a heterozygous collection from Tamil Nadu Agricultural University, Coimbatore belongs to the species *Amaranthus tricolour*. The leaves are green with broad ovate lamina and stem is green. The leaves will be ready for harvest in 30–35 days after sowing and the harvest extends up to 70–85 days. The variety is specifically suited for cultivation in summer season. On an average, it produces green matter yield of 55 t/ha.

Pusa Kiran

Another variety developed at Indian Agricultural Research Institute, New Delhi through a selection (S 44-6) from a natural cross between *Amaranthus tricolour* and *Amaranthus tristis* is a combination of good characters of both the species but predominantly of *A. tricolour*. The leaves are glossy green with broad ovate lamina. The stem is glossy green, and the stem to leaf ratio is 1: 4.6. The leaves will be ready for harvest in 21–25 days after sowing, and the duration of harvest extents up to 70–75 days. It takes 90–100 days from sowing to flowering. It is best suited for growing in plains of North India during *Kharif* season. The average green matter yield is 35 t/ha.

Pusa Lal Chaulai

A vegetable amaranth variety developed at Indian Agricultural Research Institute, New Delhi belongs to the species *Amaranthus tricolour*. The upper surface of the leaf is deep red or magenta, whereas, the lower surface is purplish red. It provides first flush of leaves in 24–33 days after sowing.

Subsequently, four cuttings can be taken at an interval of 8–10 days. On an average, its green matter yield is 45–40 t/ha. It can be grown in *Kharif* as well as in spring season.

SIRUKEERAI

An amaranth variety belongs to the species *Amaranthus blitum*. The plants are short with alternate leaves on the main stem. The leaves are small when compared with *Amaranthus dubius*, obovate with blunt bifurcated tip and have long petioles. Juncture of the stem base and root is dark pink in colour. A miniature branch, which has very small alternate leaves, starts from each leaf axil. Plants are ready for cutting in 25 days after sowing. The leaves and tender stems are used and 2–3 cuttings can be taken.

KONKAN DURANGI

An early maturing variety is suitable for cultivation in warm and humid climate of Konkan region of Maharashtra. Plants are erect and leaves are broad with prominent venation. Upper surface of leaves is attractive green with red tinge toward petioles. Branching habit is spreading type. Days required from sowing to flowering are 45 (low bolting).

KANNARA LOCAL

A multi-cut amaranth variety developed at Kerala Agriculture University produces dark red season bound leaves. The plants flower in November–December, thus, its planting is adjusted accordingly.

ARUN

A maroon, red-leaved high-yielding amaranth variety is resistant to leaf spot diseases. It is photoinsensitive variety, which is suitable for multi-cut and once over harvests. Its average yield is 20 t/ha.

KRISHNASREE

A multi-cut red amaranth variety developed at Kerala Agriculture University through selection has fleshy stem with high nutritive value, low anti-nutritional factors and yield potential of 14.8 t/ha.

MOHINI

An attractive green leaved amaranth variety is photoinsensitive in nature and high yielding. It is suitable for planting in spring–summer.

RENUSREE

A green leaved variety has purple fleshy stem with low anti-nutritional factors. The average anthocyanin content in stem is 5.33 mg/100 g of fresh sample with average yield potential of 12.5 t/ha.

SOWING TIME

Under south Indian conditions, amaranth is sown almost throughout the year, whereas, in north India, it is sown in February–March as spring–summer crop and in June–July as rainy season crop.

SEED RATE

The seed rate of amaranth is 2 kg/ha for direct sowing and 1 kg/ha for crop grown through transplanting mostly in South India. A seed rate of 2.5 kg/ha is recommended for sowing especially in Tamil Nadu state.

SOWING METHODS

Amaranth is usually propagated by direct seeding but the crop can also be raised by transplanting the seedlings. The methods of sowing and transplanting are given below:

- **Direct sowing:** For direct sowing, the field after ploughing is divided into small plots of about 3.0–3.6 m length and 1.5–1.8 m width with irrigation channels running between every two plots. Flat seedbeds are better than ridges or raised beds since they have a high infiltration rate and conserve water for longer period in summer months, while raised beds are better for raising the crop in rainy season to avoid waterlogging conditions. The seeds are sown directly in the field by either broadcasting or drilling in lines 20–30 cm apart, according to variety and kind of the crop. Amaranth seeds, being small, are mixed with fine sand in the ratio of 1:10 and sown uniformly by broadcasting at a shallow depth of 1–1.5 cm, depending on soil texture and surface moisture. The seeds are covered either by raking up soil or with a thin layer of sand or soil. The field at the time of sowing should have sufficient moisture for normal germination of seeds. If the field is deficient in moisture, sowing should be followed by a light irrigation and kept moist until the completion of germination by applying light irrigation at frequent intervals. Seedlings' emergence can easily be hampered by a thin crust on soil surface, which generally forms after rain. Crusting can sometimes be a serious problem but no solution has been researched. Rotary hoeing may be a little bit helpful in breaking the hard curst.

- **Transplanting:** In south India, there is also a practice of transplanting amaranth with a seed rate of 1 kg/ha, especially for the variety Badi Chauli grown either as a pure crop or along the border of beds of other vegetables. Seeds can also be sown in protrays and after 3–4 weeks when the seedlings are about 15 cm tall transplanted in the field. Shallow trenches of 30–35 cm width are made 30 cm apart and well-decomposed farmyard is mixed with soil in the trenches. In rainy season, planting shall be done on raised beds. During transplanting, the water is poured into the trenches or holes, into which, the plants are to be planted and the water is allowed to seep into the soil. The small seedlings along with their roots are planted in the holes, which are covered thereafter with soil. Fertilizer mixture is not placed in the same hole rather it is placed about 10 cm away to avoid scorching of roots. Kerala Agricultural University has recommended transplanting 20–30 days old seedlings in trenches at a distance of 20 cm. Transplanting seedlings at 10 × 10 cm spacing is advantageous for obtaining higher yield per unit area; however, seedlings transplanted at a spacing of 20 × 10 cm exhibit better growth and more number of harvests since the plants transplanted closely flower faster than the widely spaced ones.

NUTRITIONAL REQUIREMENT

Amaranth, being high-yielding, is a nutrient-exhaustive crop. It responds very favourably to organic matter and fertilizers. Well decomposed farmyard manure 20–25 t/ha is incorporated in the soil at the time of land preparation, and nitrogen is applied 50 kg, phosphorus 25 kg and potash 20 kg/ha. Full dose of phosphorus and potash along with half dose of nitrogen is applied at the time of last ploughing or just before sowing, and the remaining half of the nitrogen is applied in split doses after sowing since split application of nitrogen is preferred for maximum yield. However, it is advisable to avoid excess application of nitrogen, as it leads to accumulation of more nitrates – the anti-nutritional factor for human health. Nitrogen is the most limiting nutrient in most of the soils, thus, it should also be top dressed @ 20 kg/ha after each cutting or clipping. For clipping type varieties like CO 3, it is advisable to apply nitrogen 75 kg, phosphorus 25 kg and potash 20 kg/ha since a high level of nitrogen is essential for the regrowth of leaves after harvesting. To promote better regrowth, a top dressing of limestone ammonium nitrate 28 kg can be given at monthly intervals. Foliar spray of 1% urea or diluted cow urine at every harvest is good for promoting further growth and higher yield of amaranth leaves.

IRRIGATION REQUIREMENT

Although amaranth is drought tolerant but it performs optimally under irrigation conditions. It is always better to sow the seeds in field containing plenty of moisture; however, if moisture in the field at the time of sowing is not sufficient, the first irrigation is given immediately after sowing for proper germination of seeds under field conditions. Subsequently, the field should be irrigated based on seed germination, crop growth, soil moisture status and prevailing weather conditions. In sandy soils, an irrigation frequency of 4–5 days is maintained in summer months, while in the rainy season, the irrigation frequency is based on soil moisture level. However, drainage system is more important for rainy season crop than irrigation. After sowing, the first irrigation should be supplied slowly and steadily since heavy flooding and rapid water flow may wash off the seeds to one end of the bed, resulting in an uneven stand of the crop. The crop cannot withstand water logging, as it has a relatively low capacity for water consumption. Though the crop is regarded as drought tolerant, the precise mechanism involved is not well understood.

INTERCULTURAL OPERATIONS

Amaranth is a short duration and shallow rooted crop. Hence, the field should be kept weed free, especially during initial stages as it grows slowly during first few weeks. Weeds – *viz.* lamb's-quarter, redroot pigweed, kochia, cheat grass and various other grasses – are the biggest pest in amaranth production. Since the crop is sown at a closer spacing, weeds do not pose much problem in this crop, removal of weeds periodically in early stages of crop growth is essential to reduce competition with crop for nutrients, moisture, space and light, resulting in better growth of crop. Usually, hoeing is done after each cutting with a hand hoe in a transplanted or line sown crop. Thus, hoeing in the early stage not only removes the weeds but also creates better aeration into the soil and mixes the fertilizers that are top-dressed in standing crop.

In case of leafy amaranth, 10–15 days after sowing, thinning is done, leaving one or two plants per hill at a spacing of 20 cm between plants. Alternate foliar application of water soluble NPK 19:19:19 at fortnight intervals improves the quality and yield of the foliage.

BOLTING

Premature flowering or bolting is a serious problem in amaranth cultivation. The exposure of plant to severe drought induces early flowering and halts the production of leaves. Yield and quality are deteriorated after flowering. Bolting is usually associated with planting of short day varieties during November–December, deficiency of nitrogen, extreme high temperature and poor soil aeration. Practices like raising of crop at ideal time depending on locality, frequent application of nitrogenous fertilizer and manure and keeping the soil loose by light hoeing may avoid premature flowering in amaranth crop.

HARVESTING

Amaranth greens become ready for cutting or clipping when they are 8–10 cm tall at about 3–4 weeks of sowing, depending on species, variety, soil fertility, growing season, climatic conditions and cultural practices adopted during cultivation of the crop. Amaranth is harvested early in the morning by hand only. In vegetable types, pull out the entire plant with the roots on the 25th day after sowing and at the 40th day after sowing for immature stems. Plants may be cut back to 15 cm to encourage lateral growth for successive harvesting. However, in multi-cut varieties like CO 3, first clipping is done 3 weeks after sowing and subsequent clippings are done at weekly intervals up to 3 months. Normally, cuttings are made until the flowering begins and suitable vegetable material is no longer available. When the plants are harvested at regular intervals, picking of leaves should be started 8 weeks after sowing or 4 weeks after transplanting. Small quantity of leaves can be harvested on a daily basis. Leaf production can be sustained by the removal of flowers. The concentration of nutrients is observed highest when harvesting is done about 20 days after transplanting.

YIELD

A healthy crop with good management practices gives a greens yield of about 20–30 t/ha, depending on the variety, soil conditions, growing conditions and cultural practices adopted by the growers. Spring sown crop gives higher yield as compared to June sown crop. The leaf type varieties give a yield of tender greens 10 t/ha and immature stems 15–30 t/ha, while an average greens yield of clipping type variety is 30 t/ha in 6–8 cuttings.

POST-HARVEST MANAGEMENT

Leafy amaranth does not stand storage for more than a few hours under ambient room conditions, as they lose water rapidly, resulting in rapid wilting and decrease in chlorophyll, ascorbic acid and soluble protein content with an increase in amino acids content. It can be stored for 6 days at 24–28°C temperature. Vacuum cooling method and water spray vacuum cooling are used primarily for pre-cooling of leafy vegetables like amaranth. Among different methods of storage, refrigerated storage is better than other methods of storage.

The storage of leaves for different duration does not affect the nutrient content except ascorbic acid but the loss of nutrients during cooking is found to be lesser during steaming, while it is higher with respect to ascorbic acid in leaves cooked by boiling. The iron content is found to increase when cooked in iron cauldron. Loss of ascorbic acid is found to be significantly higher when cooked in open copper and iron vessels.

INSECT-PESTS

A wide range of insects attacks the amaranth crop. Amaranth stem weevil, amaranth caterpillar or Webber and leaf Webber are regarded as the potentially significant insect-pests of amaranth. Generally, tender leaves and stems are the economic parts, thus, normally no insecticide is used for the control of insect-pests.

AMARANTH STEM WEEVIL *(Hypolixus truncatulus)*

The adult amaranth stem weevil feeds on tender leaves, makes circular holes in stems, branches and mid-ribs but the larval stage is more damaging because the larvae bore into the central tissue of roots and occasionally to stems, causing rotting and potential lodging. Grubs bite into stems, feed on pith region making irregular zigzag tunnels filled with excreta. Stems split longitudinally. Plants dry completely. Full-fed grubs form a greyish-brown hard, compact gall-like chamber and pupate therein. They remain inside the stem for 5–6 days, then cut epidermal membrane and emerge. Its attack causes stunting of plants, twisting and swelling of branches and stem and suppression of shoot and leaf production.

Control
- Collect and destroy wild amaranth hosts in the vicinity of cultivated crop.
- Collect and destroy the affected plant parts along with grubs and adults.
- Spray the crop with malathion 50 EC 500 ml or dichlorvos 375 ml in 500 litres of water per hectare after harvesting the crop.

Amaranth Caterpillar *(Hymenia recurvalis)*

Larvae scrape the epidermal and palisade tissues of leaves along with webbing the leaves with silken threads, resulting in drying of webbed leaves and the entire plant. In the Indian subcontinent, it is found year-round but it is more active during warmer, rainy and early winter months. Its adult is a dark brownish black moth with white wavy markings on wings. The adult female lays spherical snow-white eggs singly or in batches of 2–5 in grooves of leaf veins. Caterpillars are greenish in colour with white lines and black crescents on thorax below lateral line. Fully fed, caterpillars drop down and pupate in soil.

Control

- Collect and destroy the affected plant parts along with caterpillars.
- Use light traps @ 1–2/ha to attract and kill adults.
- Spray the crop with malathion 50 EC 500 ml or dichlorvos 375 ml in 500 litres of water per hectare after the harvest of leaves and stems.

Leaf Webber *(Eretmocera impactella and Psara basalis)*

It is a sporadic pest, which is widely distributed in the Indian subcontinent. The adult female lays eggs on leaves or on top shoots. Caterpillars web the leaves with white silken threads and remain hidden in folds, feeding from inside leaf. Full-grown caterpillars are cylindrical, brownish-yellow to brownish-grey in colour with a broad sub-median dark stripe and black tubercles, bearing several divergent longitudinal hairs. Long brown pupae in white silken cocoons remain attached to leaves. In severe cases, the leaves dry up.

Control

- Spray the crop with Rogor or malathion 0.15%.
- Apply 2% plant extract of *Thevatia neriifolia* followed by 4% of *Azadirachta indica*, 4% of *Clerodendron infortunatum* and *Eupatorium odoratum* against leaf webber *Psara basalis*.
- Spray 4% extract of *Azadirachta indica*, 2% of *Thevatia neriifolia* and 4% of *Clerodendron infortunatum* against *Atractomorpha crenulata* and *Rotoita basalis*. The seed oil emulsion of *neem* or *samanea* at 10% is also found effective for the control of leaf webber.

Other Insect-Pests

Snout beetles, flea beetles, tortoise beetle, stinkbugs, grasshoppers, blowflies, ants and termites are the other burrowing insects but they are less damaging.

Control

- Apply lindane 13% dust @ 10 kg/ha for the control of ants, termites and other burrowing insects.
- Spray the crop with malathion 50 EC 500 ml or dichlorvos 375 ml in 500 litres of water per hectare after the harvest of leaves and stems.

DISEASES

No significant disease problem has been identified for amaranth crop. However, there are instances of leaf spot, leaf blight and white rust, which affect the amaranth plants.

Leaf Spot *(Cercospora sp.)*

This disease causes damage mostly during rainy season where infection occurs within 15 days of planting and results in the highest percentage of disease severity. In the beginning, it is characterized by the presence of numerous small brown circular spots with concentric rings on the leaves, but later on, these spots increase in size, and sometimes, they coalesce. Low temperature, high relative humidity and high rainfall favour the development of disease.

Control
- Avoid cultivation of red leaved varieties, as they are highly susceptible.
- Spray the crop with copper oxychloride 0.3% three times at an interval of 15 days.

LEAF BLIGHT (*Rhizoctonia solani*)

Leaf blight is the most severe disease during rainy season under warm and humid conditions. Symptoms include appearance of white and irregular spots on leaf lamina, wilting and yellowing of the foliage followed by death of tissues on the older leaves. Petioles remain attached to the crown after they die. The base of the petioles will have dark brown to black lesions, and the infected root tissues are dark brown to black, making the produce unmarketable. The organic debris contains the *Rhizoctonia solani* that survives as mycelium or sclerotia. The disease is more prevalent when soil temperature reaches 25–33°C and is favoured by poor soil structure and high soil moisture.

Control
- During raining season, grow resistant green amaranth variety CO 1 in combination with red types that are susceptible to get a mixed stand.
- Avoid splash irrigation and overwatering.
- Apply microbial antagonists *Trichoderma longibrachiatum* in the soil followed by foliar spray at 15-day intervals starting from 25 days after sowing.
- Spray the crop with mancozeb 4 g/litre of water. Cover the plants thoroughly so that spray solution may reach under surface of leaves at weekly intervals.

WHITE RUST (*Albugo bliti*)

The disease is characterized by appearance of typically yellow lesions on upper leaf surface and white blisters like circular or irregular pustules on underside of the leaf. Under severe infection, the leaves turn brown and die. The pathogen is spread by wind, water and insects.

Control
- Follow long crop rotation with non-host crops.
- Destroy the diseased plants and plant parts by deep ploughing or disking.
- Remove host weeds from the field and follow other phytosanitary measures.
- Apply Dithane M-45 or any other copper-based fungicide 0.2% at 10- to 12-day intervals to the soil and subsequently to the foliage.

VIRAL DISEASES

The virus is found readily sap transmissible but the insects *Bemisia tabaci*, *Aphis craccivora* and the caterpillars of *Hymenia fascialis* fail to transmit the virus. Mosaic virus of *Amaranthus viridis* is transmitted by sap inoculation and insect vectors such as *Aphis gossipii*. Mosaic virus of *Amaranthus viridis* infects *Amaranthus caudatus*, *A. gangeticus gangeticus*, *Celosia cristata* and *Gomphrena globosa*, causing mosaic mottling on leaves and severe yellowing of veins. The disease is characterized by the appearance of irregular chlorotic patches alternating with dark green areas over the entire leaf lamina.

Control

- Rogue out and burn the infected plants as early as possible from the field.
- Control the vectors like aphids by spraying acephate 75 SP 1 g/litre, methyl demeton 25 EC @ 2 ml/liter, or phosalone 35 EC @ 2 ml/liter 2–3 times at weekly intervals.

■■■

9 Sea Beet

M.K. Rana and Kiran Sagwal

Botanical Name   : *Beta vulgaris* var. maritima

Family   : Amranthaceae

Chromosome Number : 2n = 18

ORIGIN AND DISTRIBUTION

The genus *Beta* was first originated in Mediterranean Europe but later on diversified and spread north and eastward. Its secondary centre of biodiversity is around Near East. Carl Linnaeus divided the species into wild and cultivated type, giving the name *Beta maritima* to the wild taxon, which is native to the coasts of Europe, northern Africa and southern Asia, and grown mostly in southwestern Europe. Sea beet occurs along the seacoast of all Mediterranean Sea countries, most of the Middle East countries and along the Atlantic coast of northern Europe and British island. It is also found in some inland sites in the Mediterranean region, Iran and Azerbaijan. The distribution of plants within the section *Beta* is centered on the eastern Mediterranean region.

INTRODUCTION

The genus *Beta* includes annual, biennial and perennial plants. *Beta maritima*, also known as wild spinach, is believed to be the common ancestor of all cultivated beets *viz.*, beet leaf, sugar beet, and Swiss chard, which have been developed by means of selective breeding. It is an herbaceous short statured plant, which grows as perennial plant in Mediterranean region and in the coastal regions of Northern Europe. In Germany, it is known to exist in Helgoland and on the Baltic coast, predominantly on the island of Fehmarn. Sea beet currently seems to be spreading along the German Baltic coast. It grows well on stony and sandy beaches, rocky cliff sides, coastal grasslands and salt marshes, being very dependent upon a silty environment. Now a days, sea beet is even cultivated as a silty vegetable. It is used as an important source of resistance against Cercospora leaf spot and cyst nematode in breeding of beet leaf, sugar beet and Swiss chard.

COMPOSITION AND USES

COMPOSITION

Sea beet is a prominent source of magnesium and sodium, good source of vitamin A and C, which are important antioxidants and a fare source of carbohydrates, protein, calcium, iron and vitamin B_6. Its leaves are very low in calorie. The nutritional composition of sea beet is given in Table 9.1.

TABLE 9.1

Nutritional Composition of Sea Beet Leaves (per 100 g Edible Portion)

Constituents	Contents	Constituents	Contents
Water (g)	93	Sodium (mg)	226
Carbohydrates (g)	4.3	Iron (mg)	14
Protein (g)	2.2	Vitamin A (mg)	126
Fat (g)	0.1	Pyridoxine (mg)	5
Calcium (mg)	11	Vitamin C (mg)	50
Potassium (mg)	762	Energy (kcal)	22
Magnesium (mg)	17	–	–

USES

Every part of sea beet is edible, including succulent leaves, roots and flowers. Its flowers are perhaps the sweetest part of the plant but the leaves are consumed after boiling, blanching, steaming, or eaten raw. Its leaves giving taste very much like spinach are very good for cooking as greens since they are superior in flavour, texture and nutritional value to any of their progeny. Its flavour is stronger as compared to cultivated spinach. The older leaves may sometimes be quite bitter if eaten raw. However, this bitterness is customarily vanished with cooking, thus, can be used in any recipe as a substitute for spinach. Leaves are pretty yummy and used as a vegetable year yound. It can be added in soups, tarts and salads. It makes a very good combination with other vegetables like a potato. Sea beet makes a good adjunct to any kind of meat, fish or shellfish.

MEDICINAL USES

Sea beet is widely used in modern herbal medicinal preparations. It is used in the treatment of tumors. A decoction prepared from its seeds has been used in the treatment of intestinal and genital tumors. Its plant juice helps in the treatment of anemia, yellow jaundice, ulcers, leukemia and other forms of cancer *viz.*, breast, esophagus, glands, head, intestines, leg, lip, lung, prostate, rectum, spleen, stomach and uterus cancer. Its decoction is used as a purgative by those individuals who are suffering from hemorrhoids. Leaves in association with roots are used as an emmenagogue. The plant has been found effective in the treatment of feline ascariasis. In the ancient time, beet juice was put into the nostrils to purge the head, clear ringing ears and alleviate toothache. Its juice is said to clear the liver and spleen obstructions and with vinegar it makes the scalp dandruff free and prevents hair loss. It is also used to cure headache, vertigo and other brain afflictions.

BOTANY

Sea beet is a leafy annual or perennial growing up to a height of 1.2 m. In first year, it forms a rosette of leaves and in second year a large inflorescence. Its 20- to 40-cm-long leaves are green in colour occasionally with purple dots, and in some cases, it may be either purple or red. Sea beet has long, wedge-shaped, thicker and fleshier leaves in a rosette arrangement with a smaller tuberous root as compared with cultivated beet. The plant parts are often red. Although it bears very small hermaphrodite flowers of green or yellow colour having both male and female organs but they are cross-pollinated in nature. The pollination is performed by wind and honeybees. Its pollens remain viable only for 24 hours. Seeds are black usually bound in a sort of square brown and woody structure containing several seeds. Such type of structure helps the seeds to move with wind.

The glomerules occur in groups of three and they are crowded together on the inflorescence. Through its smaller glomerules and flowers that are less flat, it is further differentiated from subspecies *adanensis*. It is not surprising since sea beet is an ancestor of cultivated beet leaf and beetroot. Indeed, sea beet is also known as wild spinach, which produces shiny dark green oval or diamond shaped leaves in rosettes year yound. It flowers in spring–summer. Its stems and leaf stalks sometimes may have a purplish red colour, which shows its close relationship with beetroot. Fresh leaves give a nutty taste. Its wild forms are the source of resistance for various insect-pests, diseases and three economically important root knot nematodes, *i.e.*, *Meloidogyne incognita* Race 1, 2 and 4, *Meloidogyne javanica, Meloidogyne arenaria* Races 1 and 2, and *Meloidogyne hapla* .

CLIMATIC REQUIREMENT

In southern Europe, sea beet is an annual, while it is biennial or perennial in northern regions. It can be grown successfully in subtropical climate from September to January without any bolting and in temperate climate from March to May. The locations with well-distributed rainfall of 230 to 315 cm across the growing period are suitable for its cultivation. Such conditions favour vegetative growth and provide a base for leaf expansion. The crop needs an optimum temperature of 5.0°C to 26.6°C for its better growth and development. The crop also requires abundant sunshine during its growth. Shady conditions are inappropriate for its cultivation, as shade reduces the nutritive quality of the plant. The plant is not too frost-hardy and can survive only in sparsely populated biotopes.

SOIL REQUIREMENT

Sea beet can be grown well on a wide range of soils from sandy loam to clayey loam textured soil but it prefers deep, friable, weed-free, well-drained moist soil with porous structure. The organic matter rich soils are more suitable for its cultivation. Being a halophyte plant, it is well adapted to soils with high salt concentrations, *i.e.*, alkaline and saline soils, but the optimum pH range for its normal growth and development is 4.5 to 8.5. All beets are notably tolerant to manganese toxicity.

CULTIVATED VARIETIES

Sea beet is an underexploited wild plant of Amaranthace family, thus, improvement work has not yet been done in India to develop varieties or hybrids. However, several local selections without proper nomenclature are available. Farmers usually grow the locally available varieties.

SOWING TIME

Sowing time varies with the type of variety and climatic conditions of the growing region. Sowing of a particular variety seeds at its right time is very essential to get higher yield with better quality leaves since unfavorable weather conditions produce adverse effect on these components. Sea beet is mainly sown in frost-free autumn–winter season in subtropical climate and during spring–summer (March to May) in temperate climate.

SEED RATE

The crop grown for fresh market is hardly ever thinned since the operation is too expensive, thus, to obtain optimum plant population per unit area, the seed rate is ascertained according to germination percentage, seed vigour, seed size, soil type and soil fertility. The optimum plant density can be obtained by sowing disease-free healthy seeds. About 37–45 kg seed is enough for sowing a hectare of land.

SOWING METHOD

Sea beet is propagated by seeds, which are sown directly in the main field. Its method of sowing is almost similar to beet leaf crop. However, spacing between sea beet plants influences the number of leaves and their size. High plant density contributes to early development of seed stalk (bolting). For fresh market production, the plant density should be maintained about 60 plants per m². Lower plant density facilitates hand harvesting for bunching of leaves, and on the contrary, high plant density results in more upright leaf growth, which is suitable for mechanical harvesting.

The seeds are sown either by broadcasting or in lines. However, sowing seeds in lines is always better over broadcasting method, as it makes the intercultural operations and harvesting easy. Based on soil type and moisture availability, the seeds of sea beet are sown at a depth of 2–5 cm in lines 10–20 cm apart. However, sowing seeds at a spacing of 18–22 cm encourages the plant growth. Depending on drainage facilities, the crop can be grown either in raised or flat beds. Soaking seeds before sowing in tap water overnight is helpful in improving germination, as soaking in water results in leaching of inhibitors and softening of seed coat. The dormancy in sea beet seeds persists for over 10 years. Usually, the seeds after sowing take 7–20 days for germination.

NUTRITIONAL REQUIREMENT

The nutritional requirement of sea beet is more or less same as that of beet leaf crop. Application of nitrogenous fertilizer normally promotes fresh green leaves yield and total crop dry matter of sea beet. The crop is fertilized well to meet the plants requirement for their rapid growth in a shortest period. Since the deficiency of nitrogen in soil results in yellowing of leaves and the deficiency of phosphorus and potash leads to drying of leaf tops and leaf necrosis, respectively. About two third of the total crop biomass is produced by the crop during last third of the crop duration, thus, to meet out the nutrients demand of quick growing crop plants, correct scheduling of fertilizers application is necessary. Well-decomposed farmyard manure should be applied @ 10–15 t/ha, nitrogen 70–75 kg/ha, phosphorus 55–60 kg/ha and potash 25–30 kg/ha depending on soil type and irrigation facilities.

Farmyard manure is applied at the time of land preparation and mixed thoroughly into the soil by repeated ploughing. Full dose of phosphorus and potassium along with one-third dose of nitrogen should be applied at the time of seed sowing and remaining two-thirds dose of nitrogen should be applied in two equal splits after first and second cuttings. Besides, urea may be sprayed at 1.5% concentration at 15 days after germination and after every cutting to hasten the growth.

IRRIGATION REQUIREMENT

Irrigation requirement of sea beet depends on soil type, growing season and climatic conditions of growing region. Sea beet in general needs plenty of water for its rapid growth. Pre-sowing irrigation is essential, as plenty of soil moisture is essential for proper germination. As usual, first irrigation is crucial for early establishment of the crop. In light-textured soils, irrigation is applied once in 6–8 days, and in heavy textured soils, it is applied once in 10–12 days. Applying light and frequent irrigation is always beneficial for maintaining optimum soil moisture. Sea beet is undoubtedly a moisture-loving crop but water stagnation is detrimental to leaf yield and quality.

INTERCULTURAL OPERATIONS

HOEING AND WEEDING

The intercultural operations to be adopted for sea beet are similar to spinach beet. Keeping the field weed free is very essential especially for the crop grown for processing since weeds and sea beet plants are similar in appearance and difficult to separate, therefore, the removal of such weeds is most indispensable. Regular hoeing after each cutting can keep the weed population low.

Sea beet faces severe competition from weeds in early stages of crop growth. Therefore, the crop should be kept weed free for the first 45–60 days.

HARVESTING

Sea beet is harvested at very young stage when the leaves are glossy, juicy and succulent. It gives several cuttings if sown in early autumn, but in the summer, it gives only a single cutting, as the plants start developing seed stalk. Harvesting in summer should be done at an earlier stage since increase in temperature develops bitterness in leaves and reduces their market value. The stage of harvesting usually depends on the purpose of use. For salad purposes, young tender leaves should be harvested, and for cooking purposes, slightly mature leaves are harvested by cutting the plant at soil surface just below the lowest leaves. However, for processing purposes, harvesting is done by cutting the plants about 2–3 cm above the soil surface. Cutting is usually done with a sharp knife to avoid of tissues at cut end. In advanced countries, sea beet is harvested by machines. It is advisable not to harvest the crop early in the morning since the leaves are very turgid and break easily when wet; thus, sea beet should be harvested a bit late in the day when the leaves are less turgid since a slight wilting may prevent breaking of leaves.

YIELD

First cutting is taken 35–40 days after sowing of the crop. Sea beet on an average gives a yield of fresh green biomass of about 80–100 q/ha in 4–6 cuttings.

INSECT-PESTS

APHIDS (*Myzus persicae, Brrevicoryne brassicae* and *Liphaphis erysimi*)

Both nymphs and adults of this tiny insect suck cell sap from under surface of the tender leaves, reducing plant vigour. The leaves curl up and ultimately wilt. The aphids excrete honeydew, on which, black sooty mould develops which hampers the photosynthetic activity. Besides, these aphids act as vector for the transmission of virus. The severe infestation affects the yield and quality of roots adversely.

Control

- Avoid excessive use of lime.
- Spray the crop with simple water with strong jet pressure to wash the aphids.
- Remove and destroy the portion of the plant attacked by the aphids.
- Spray the crop with malathion 0.05%, monocrotophos 0.04%, methyl demeton 0.05%, dimethoate 0.03%, or phosphomidon 0.03% at an interval of 10–15 days.

SLUGS (*Arion hortensis* agg.)

Slugs, the shell less terrestrial gastropod mollusk, live on soil surface and in salt water. These are nocturnal animals, as during daytime they hide themselves beneath flowerpots, stones, leaves, etc. and feed on leaves during night and on cloudy or foggy days. They feed on a variety of living plants and on decaying plant refuse. They make irregular holes with smooth edges by feeding on leaves. As a result, the plant growth is completely ceased.

Control

- Avoid excessive use of lime in soil.
- Spray the crop with simple water with strong jet pressure.

- Avoid excessive supply of irrigation and allow soil surface to dry out between irrigations.
- Use copper borders around the field.
- Use iron phosphate pellet containing baits to attract the slugs.
- Use seaweeds around the plant base and perimeter of bed as mulch to repel slugs.

PRODENIA (*Spodoptera litura*)

Prodenia, also known as armyworm, is one of the destructive pests of sea beet. The insect at larval stage is very damaging. Larvae feed on the foliage leaving small irregular holes on leaves, leading to complete stripping of the plants.

Control

- Follow clean cultivation.
- Avoid excessive use of lime.
- Spray the crop with simple water with strong jet pressure.
- Use *Nicotiana tobacum* powder to control the insect.
- Spray the crop with *Neekunchi* [*neem + aloe vera + chi* (tobacco snuff decoction)] @ 100–150 ml of extract diluted in 15 liters of water at 10-day intervals.
- Apply *neem* seed-kernel extract 5% during early stages of crop growth.
- Apply *Pongamia glabra* (*pongamia* oil or *Karanj*) oil on plant.

DISEASES

DAMPING-OFF (*Rhizoctonia solani* and *Pythium* spp.)

This is very common soil borne fungal disease in early stages of crop growth. The fungus has a wide host range. The pathogen may cause damage to the seeds during germination (pre-emergence damping-off) or seedling after emergence (post-emergence damping-off). In damping-off, the fungus primarily attacks the seeds below the soil surface and causes death of the seeds. In post-emergence, the fungus attacks the seedlings at collar region near the ground line. The seedlings as a result happen to discolour, wilt suddenly and die. High humidity and moderate temperature favour the development of this disease.

Control

- Follow long crop rotation with non-host crops.
- Sterilize the soil with formaldehyde (1: 100 parts) about 2 weeks before seed sowing.
- Use less quantity of seeds per unit area to avoid overcrowding of seedlings.
- Use disease-free healthy seeds.
- Minimize soil compaction and increase air circulation.
- Avoid excessive supply of irrigation.
- Treat the seed with captan, thiram, foltaf, or captafol 2 g/kg of seed.
- Drench the seedbeds with captan at 0.1% solution.

WHITE RUST (*Albugo bliti*)

The fungus develops white circular or irregular blister on lower surface of the leaves, and opposite each blister, chlorotic lesions and sometimes leaf galls are developed on upper surface of the leaves. More than 50% yield is reduced due to the incidence of this disease. Plants can be stunted with distorted leaves and flowers. Infection and disease development occur during cool and moist conditions. The optimal temperature for infection is between 12°C and 25°C. Leaves must remain wet for a minimum of 2 to 3 hours for the spores to germinate.

Control

- Follow long crop rotation with non-host crops.
- Keep the field weed free.
- Destroy the infected plant debris immediately after harvesting of crop.
- Spray the crop with fungicides containing mefenoxam or fosetyl-aluminum.

ANTHRACNOSE (*Colletotrichum dematium* f. spinaciae)

The fungus attacks all plant parts at any growth stage. The affected leaves show water-soaked small and irregular yellow, brown, dark-brown, or black spots. Later, these spots coalesce and cover the completely affected area and cause drying of the elongated grayish spots on seed. The colour of the infected part darkens as it ages. The disease can also produce cankers on petioles and stem, which cause severe defoliation. As it ages, the center of older spots becomes blackish and emits gelatinous pink spore masses.

Control

- Follow long crop rotation with non-host crops.
- Use disease-free healthy seeds.
- Remove and destroy the crop debris.
- Treat the seed with captan, thiram, or Bavistin @ 2 g/kg of seed.
- Spray the crop with Dithane M-45, Dithane Z-78, Blitox, or Bavistin 0.2% at 15- to 20-day intervals.

DOWNY MILDEW (*Peronospora schachtii* and *Peronospora farinosa* f.sp. *betae*)

The disease can appear at any stage of plant growth but it causes significant damage at seedling stage. The infected cotyledons become discoloured and distorted. The light yellow irregularly shaped lesions without distinct margins develop on upper surface and grayish to dark brown fungal growth mainly on lower surface opposite to the lesions, spreading to the upper surface in wet conditions. As the disease progresses, the leaves begin to curl and wilt. Whitish gray mycelium grows on lower surface in high humidity, while the mycelium is generally found absent in low humid atmosphere. The disease develops more in cool and moist weather. The infection of downy mildew occurs primarily in the field but it becomes prominent after harvesting of the crop, especially at 14–24°C and above 85% relative humidity.

Control

- Follow long crop rotation with non-host crops.
- Use only disease-free healthy seeds.
- Remove and destroy crop residues or disease debris.
- Spray the crop with 0.2% Dithane M-45 or Dithane Z-78 at 15-day intervals.

BACTERIAL SOFT ROT (*Erwinia carotovora* subsp. *betavasculorum*)

This is a post-harvest bacterial disease causing severe losses to sea beet during transportation. The bacteria usually enter only where the host skin has injured in some way or through fungal lesions. There is decaying of packed leaves due to the lack of proper aeration and cleanliness. The infected leaves first show water-soaked lesions, which rapidly grow wider and deeper as the tissue turns gray green in colour, become soft and give a mushy look. The decay progresses at a faster rate at high humidity and temperature above 7°C.

Control

- Remove all the broken leaves before packing.
- Clean the harvested leaves before packing.
- Avoid injuries to the plant tissue as much as possible.
- Pack the leaves loosely and use clean packing material.
- Maintain proper sanitation in packinghouse.
- Control nematodes and other insect-pests that serve as vectors (carriers) of the bacteria.
- Send the harvested produce to the market as soon as possible.
- Spray all the storage walls and floors with formaldehyde.
- Store the produce at perfectly cleaned place before marketing.

■■■

10 Elephant Foot Yam

M.K. Rana and V.P.S. Panghal

Botanical Name : *Amorphophallus campanulatus*
(Roxb.) Blume

Family : Araceae

Chromosome Number : 2n = 28

ORIGIN AND DISTRIBUTION

Elephant foot yam probably originated from India or Sri Lanka. It is grown in Bangladesh, India, Southeast Asia and Africa. It is also found growing wild in the Philippines, Malaysia, Indonesia and Southeastern Asian countries.

INTRODUCTION

Elephant foot yam also known as *Suran* and *Zamikand* is a potential tropical tuber crop. Though the crop is perennial due to underground stem, it is treated as an annual with duration of 8–9 months. It is widely grown for its edible tubers, which are important source of carbohydrates in India and Indonesia and a valued secondary crop throughout tropical Asia. It is also cultivated as an ornamental for its striking compound foliage and unusual and dramatic flowering and fruiting structures. In India, it has attained the status of a cash crop and the area under its cultivation is increasing fast. It is a remunerative and profitable stem tuber crop, which is gaining popularity due to its shade tolerance, easiness in cultivation, high productivity, less incidence of insect-pests and diseases, steady demand and reasonably good price. In India, this crop is traditionally cultivated in Andhra Pradesh, Gujarat, Maharashtra and Kerala states. The Orissa climate is highly suitable for its cultivation, where it can be grown under rain-fed conditions with protective irrigation. Its wild forms are also found throughout Orissa. The tubers of wild plants are highly acrid, causing irritation in throat and mouth due to excessive calcium oxalate content present in its tubers. Acridity is usually removed by boiling fairly for a long time. However, through research and development, high-yielding non-acrid varieties have been released by different organizations.

COMPOSITION AND USES

COMPOSITION

The tubers are rich in carbohydrates, starch, calcium, phosphorus, potassium and vitamin C. The nutritional composition of elephant foot yam is given in Table 10.1.

USES

Tubers of elephant foot yam are mainly used as vegetable after thorough cooking. Its tubers, rich in starch, are used for making chips. Tender stem and leaves are also used as vegetables. Its tubers are also used for making pickles. It is also widely used as fodder for cattle.

TABLE 10.1

Nutritional Composition of Elephant Foot Yam (per 100 g Edible Portion)

Constituents	Contents	Constituents	Contents
Water (g)	69.9	Potassium (mg)	816
Carbohydrates (g)	27.9	Iron (mg)	0.54
Starch (g)	18	Vitamin A (µg)	7
Protein (g)	1.5	Thiamine (mg)	0.112
Fat (g)	0.17	Riboflavin (mg)	0.032
Calcium (mg)	17	Vitamin C (mg)	17.1
Phosphorus (mg)	55	Energy (kcal)	118

MEDICINAL USES

The tubers of elephant foot yam have the medicinal properties, thus, they are used in many *Ayurvedic* preparations. The tubers are considered to have pain-relieving, anti-inflammatory, anti-flatulence, digestive, aphrodisiac, rejuvenating and tonic properties. Traditionally, its tubers are used in the treatment of a wide range of health problems, including parasitic worms, flatulence, coughs, inflammation, constipation, anaemia, haemorrhoids and fatigue.

BOTANY

The cultivated species of elephant foot yam is a robust herbaceous plant, having 1.0–1.5 m height. Large dissected tripartite leaves constitute the luxuriant outspreading crown-like foliage, borne on a thick single upright stem. The aerial pseudostem, which is round with characteristic irregular blotches, is botanically a leaf petiole. Tuber, a thickened part of underground stem, is dark brown in colour and flattened globe-shaped up to 50 × 30 cm with prominent root scars, weighing up to about 15 kg. It flowers once in 1–3 years. Inflorescence, which is borne on a very short stalk, consists of a bell-shaped spathe surrounding a central yellow spadix. It appears almost at ground level. Although the wild species flowers set seeds profusely but the cultivated species fail to set seeds under normal conditions due to extreme protogyny condition coupled with delay in opening of spathe. The spadix (flower spike) is up to 70 cm long. The lowermost portion of the spadix is female and is covered with pistils (female parts). Each pistil consists of a pale green or maroon ovary with a maroon stalk (style) and two- or three-lobed yellow head (stigma). The next floral zone is male and contains tightly packed yellowish stamens. Spadix at the tip is a bulbous, dark maroon, rounded to deeply wrinkled appendix. It emits an odour reminiscent of rotting flesh, which attracts pollinators such as carrion flies and beetles. Spathe (bract surrounding spadix) is bell-shaped, broader than length, up to 45 × 60 cm and pale green to dark brown with paler blotches on exterior. The spadix opens outwards to form a frilled, glossy maroon and collar-like structure around. The basal portion of interior is pale green-yellow. Fruit is about 2 × 1 cm and bright red when ripe. The fruit borne on a spike up to 50 cm long and 8 cm in diameter and the fruiting part held aloft on a peduncle (stalk) of 20–100 cm length.

CLIMATIC REQUIREMENT

Elephant foot yam grows well in hot and humid climate. For vigorous growth, it requires 25°C–30°C temperature and 80–90% relative humidity at initial stages of crop growth, whereas, dry climate facilitates tuber bulking at later stage. Well-distributed rainfall of 1000–1500 mm is helpful for good growth and tuber yield.

SOIL REQUIREMENT

The crop can be grown in any types of soil. However, well-drained, fertile, sandy loam soil rich in humus is ideal for the cultivation of elephant foot yam successfully. It does not thrive on wet, poorly drained, alkaline and saline soils. Field must be well leveled, and for getting good quality tubers, enough arrangement for drainage should be made. To make the soil loose, friable and free from clods, four to five ploughings followed by planking should be done.

CULTIVATED VARIETIES

GAJENDRA

A variety developed at Andhra Pradesh Agricultural University, Hyderabad through a local selection from Kovvur area of Andhra Pradesh yields 500–600 q/ha. Its tubers are non-acrid, well shaped and generally devoid of cormels or propagules.

SREE PADMA

A variety developed at Central Tuber Crops Research Institute, Trivandrum has a yield potential of 400 quintals per hectare. The plant generally produces one mother corm and a few cormels or propagules with non-acrid tubers.

SREE ATHIRA

A first genetically improved variety developed at Central Tuber Crops Research Institute, Trivandrum produces tubers with very good cooking quality.

KUSUM

A variety developed by Vidhan Chandra Krishi Viswavidyalaya, West Bengal has yield potential and other features similar to variety Gajendra.

SEED RATE

Initial size of planting material plays most significant role in determining final size of the harvested tubers. The whole tubers of 400–500 g size are more suitable for raising a commercial crop. Tubers of 3–4 kg can be harvested after 6 to 7 months. This size is most suitable from marketing and transport point of view. About 12,000 cut pieces weighing about 12 tonnes are required for planting a hectare area. Most of the seed tubers will sprout within a month after planting.

SOWING TIME

Elephant foot yam is a long duration crop and generally attains maturity in 6–7 months. Under irrigated conditions, it is planted in March and harvested in November. Under rain-fed conditions, the crop is planted at the onset of monsoon, preferably in June. In southern parts of the country, the crop has the sustainability to grow at any time of the year, provided the temperature is congenial and available soil moisture is adequate.

PLANTING METHOD

Cut tubers of 50–100 g size are produced from the planting material of 500–1000 g size. Although the cut tubers can be used as planting material but the use of whole tuber is significantly superior over cut tubers in terms of sprouting percentage and overall yield. When cut tubers are used for planting, certain precautions and treatments are needed, as the cut tubers are prone to decay after planting due to possible presence of numerous soilborne pathogens.

The cut pieces of 50–100 g are treated with thick cow dung slurry containing mancozeb 0.2% and Monocrotophos 0.05% for 5–10 minutes, followed by drying in shade for 24 hours. For organic production, the cut pieces are treated with slurry containing cow dung, *neem* cake and *Trichoderma harzianum* (5 g/kg of seed) and dried under shade before planting. At the time of planting, *Trichoderma harzianum* mixed with farmyard manure is applied in pits @ 3 kg/pit. *Trichoderma* can be multiplied in farmyard manure alone but it will take 15 days to form sufficient inoculums as compared to 7–8 days if *neem* cake is also used along with farmyard manure, as it will be effective against collar rot caused by *Sclerotium rolfsii*.

The field should thoroughly be ploughed, levelled and tilled before planting. The planting should be done at 90 × 90 cm spacing for commercial crop. The pits of 60 × 60 × 60 cm size are dug out and refilled with a mixture of same soil and well-decomposed manure before planting for facilitating tuber bulking. At the time of planting, *neem* cake may also be applied in pits @ 80–85 g/pit.

INTERCROPPING

During initial period of 2–3 months after planting, the crops like leafy vegetables, green gram, black gram, cowpea, cucumber, etc. can be grown as intercrop. Intercropping of elephant foot yam in banana, coconut and other newly planted orchards gives additional income to the farmers. Cowpea may also be sown as intercrop between elephant foot yam pits for green manuring and incorporated in pits after 45–60 days of sowing. Ash may also be applied @ 250 g/pit at the time of incorporation of green manure in pits.

NUTRITIONAL REQUIREMENT

Elephant foot yam has high nutrients requirement. The pits are filled with a mixture of soil and well-decomposed farmyard manure (20–25 t/ha). Although the fertilizer dose depends on the soil fertility and nutrients status but a fertilizer dose of 100 kg nitrogen, 80 kg phosphorus and 100 kg potash per hectare has been found to be optimum. A cowpea crop (seed rate @ 20 kg/ha) can be raised as green manure prior to elephant foot yam and the green matter is incorporated into the soil 45–60 days after sowing.

IRRIGATION REQUIREMENT

For summer crop, a light irrigation should be provided immediately after planting. Depending on the soil moisture availability, irrigation should be given at regular intervals until the arrival of monsoon. However, care should be taken to prevent water stagnation at every stage of crop growth. Irrigation should be withheld during later stage of crop growth after 5–6 months of planting to allow the crop to be mature.

INTERCULTURAL OPERATIONS

HOEING AND WEEDING

Hoeing and weeding may be done as and when required. If no weedicide has been applied and the weeds have come up, then weeding may be done at about 30 days after planting followed by earthing up when rest dose of nitrogen is applied. Weeds may be controlled by applying Lasso @ 5–6 litre or Stomp 30% @ 4–5 litres within 10 days of planting in 625 litres of water per hectare. Enough moisture in soil at the time of weedicide spraying is essential for effective control of weeds.

MULCHING

Mulching with organic waste, *e.g.*, compost, sawdust, paddy straw, dry grasses, sugarcane tresses, and aluminium foil or back polyethylene sheet, helps in reducing the weed growth and conserving soil moisture.

HARVESTING

Underground corms are harvested with pickaxe or by digging when the tops are completely withered and fallen over the ground. Crop will be ready for harvesting in 8–9 months after planting. However, the tubers can be harvested 5–7 months after planting when price in the market is high. Normally, it is harvested when the tops become yellow and wither.

YIELD

The yield varies with soil type, variety, growing season, growing purpose, harvesting stage, weather conditions, irrigation facilities and incidence of insect-pests and diseases. On an average, elephant foot yam gives a yield of 300–400 q/ha.

POST-HARVEST MANAGEMENT

Fully mature, graded and cured tubers are stored as planting material. The places where the elephant foot yam is to be stored should be well ventilated and cool. The tubers should be stored in a single layer, but if the storage place is insufficient, the tubers can be stored in two layers. However, storing the tubers in a heap should be avoided.

INSECT-PESTS

MEALY BUGS (*Plancoceus citriculus* and *Rhizoceus* spp.)

This insect usually feed on roots and tubers, rarely on stem and leaves. The infected tubers are shriveled and do not sprout when they are planted. The infestation is generally found when tubers are harvested.

Control
- Use pest-free seed corms for planting.
- Treat the infected seed corms with monocrotophos 0.05% for 10 minutes.

ROOT KNOT NEMATODE (*Meloidogyne incognita*)

It attacks the roots and corms of elephant foot yam. At early stage of crop growth, yellowing, stunting and premature death of the crop are noticed. In roots, it causes typical swelling or knotting, and on the corms, irregular swelling on the surface is observed.

Control
- Use nematode-free planting material.
- Apply castor or *neem* cake @ 10 q/ha.
- Apply carbofuran @ 5 g/plant.

DISEASES

COLLAR ROT (*Sclerotium rolfsii*)

It is a destructive disease, which is widespread in rainy period following warm dry season. Symptoms appear when the crop is 2 to 3 months old. Brownish lesions first appear on collar regions, which later spread to the entire pseudostem and cause complete yellowing of the plant. In severe cases, the plant collapses leading to complete crop loss. Water logging, poor drainage and mechanical injury at collar region favour the disease incidence.

Control

- Use disease-free planting material.
- Remove the infected plants from the field.
- Maintain proper sanitation in the field.
- Provide good drainage facilities.
- Apply *Neem* cake in soil @ 250 g/pit.
- Use biocontrol agents like *Trichoderma* @ 2.5 kg/ha mixed with 50 kg of farmyard manure.
- Drench the beds with captan 0.2%.

LEAF SPOT (*Corynespora cassiicola*)

The symptoms are initiated as reddish-brown spots on the leaves. Later, the spots increase in size up to 2 cm in diameter, become irregular and develop into light brown necrotic areas encircled by reddish brown to black margin. In severe cases, the affected plants dry and collapse.

Control

- Spray the crop with captan 0.2% or mancozeb 0.25% at 15-day intervals.

MOSAIC DISEASE

Primary spread of the disease is through planting material and secondary spread is through insect vectors such as *Myzus persicae* Sulz., *Aphis gossypii* Glover, *Aphis craccivora* Koch. and *Pentalonia nigronervosa* Coq. The disease symptoms include mosaic mottling of leaves and distortion of leaf lamina. Corms produced by the mottled plants are much smaller than the corm produced by plants without mottled leaves.

Control

- Use virus-free planting material.
- Rogue-out infected plants from field.
- Spray the crop with 0.05% Rogor or Metasystox insecticide at fortnightly intervals.

■■■

11 Tannia

K.B. Jagtap and Chandanshive Aniket Vilas

Botanical Name : *Xanthosoma sagittifolium* L.

Family : Araceae

Chromosome Number : 2n = 26

ORIGIN AND DISTRIBUTION

Tannia originated in tropical America, but later, it spread towards Southeast Asia, the Pacific islands and Africa. The spread of this crop to the South Pacific and Africa occurred in recent times, probably in the 19th century.

INTRODUCTION

Tannia produces large corms with coarse starch grains and is often acrid in nature. However, its cormels are quite popular for edible purpose. It is quite popular in many parts of the country like Maharashtra, North Eastern states, Orissa, Uttar Pradesh, Bihar, West Bengal and the Andhra Pradesh. Tannia is grown especially for its corms and cormels. The crop is also considered ideal as an intercrop in perennial and plantation crops as it is tolerant to shade.

COMPOSITION AND USES

COMPOSITION

Tannia corms are a good source of the carbohydrates, calcium, phosphorus, vitamin C and a fair source of protein, iron, carotene, thiamine, riboflavin and niacin. Its corms also provide a good amount of energy. The nutritional composition of tannia is given in Table 11.1.

TABLE 11.1
Nutritional Composition of Tannia (per 100 g Edible Portion)

Constituents	Contents	Constituents	Contents
Water (g)	65	Iron (mg)	1
Carbohydrates (g)	31	Carotene (mg)	2
Protein (g)	2.0	Thiamine (mg)	1.1
Fat (g)	0.3	Riboflavin (mg)	0.3
Dietary fibre (g)	1.0	Niacin (mg)	1
Calcium (mg)	20	Vitamin C (mg)	90
Phosphorus (mg)	47	Energy (kcal)	133

USES

Tannia is usually grown for its corms as well as leaves. The fresh leaves are consumed as a cooked, baked, or fried vegetable. Corms are used for making flour from dried and peeled corms. In southern states of India, it is used to prepare a fermented product named *fufu*. Its leaves and petioles are also used as vegetable in some of the places, and leaves are used for the preparation of rolls.

MEDICINAL USE

The juice extracted from corms and cormels is beneficial in healing wounds. Infestation of diarrhea may be controlled by herbal tea prepared from tannia leaves. Its decoction is used to reduce the body temperature of feverish patients.

BOTANY

Plant height is around 1.5 to 2 m and leaves are sagittate arrow-shaped with sharp tips and deep with wider basal lobes. The leaves have prominent marginal veins of about 50–75 cm length. Tannia seldom bears flowers but some cultivars can bear flowers. Inflorescence usually appears 6 months after planting. Large inflorescence arises in leaf axils in clusters of two to three, depending on the species. The flowers may be unisexual or bisexual depending on its species. The flowers are sessile and form on a spadix enclosed in large spathe. In unisexual inflorescence, the pistillate flowers borne at the base of spadix, staminate flowers at the top and in between are a group of abortive sterile flowers. Insects pollinate the flowers but the setting of fruits and seeds is rare. Botanically, its fruits are the berries, each containing 2–5 small (1 mm), oval, hard and difficult-to-germinate seeds. It produces enlarged corms that may or may not be edible. Underground club-shaped cormels are also produced during later growth stages.

CLIMATIC REQUIREMENT

Tannia prefers the warm humid climate prevalent in tropical regions of the world. For sustained growth and development, it requires an evenly distributed rainfall of about 1800–2000 mm. It grows better in dry areas as compared to the other aroids but a long dry spell, cool climate and frost are not good for its cultivation. It can successfully be cultivated in areas where the mean annual temperature is 24–30°C with maximum variations ranging from 13°C to 29°C.

SOIL REQUIREMENT

The crop thrives best in well-drained loamy soil with slightly acidic reaction of 5.5 to 6.5 pH. It grows well as an upland rain fed crop on soils where rainfall is evenly distributed throughout the year. Tannia grows best when the soil is turned 2 to 3 months before planting since ploughing the field few months earlier helps in decomposing of previous crop residues and reducing the population of nematodes and disease-causing pathogens. In general, the fields that have not been under tannia crop in the last 2 to 3 years are preferred to grow tannia. The fields that have very high nematode populations are usually avoided for raising tannia crop.

CULTIVATED VARIETIES

No specific improved variety of tannia is available in the country. Only local selections with low acridity, good texture and suitability of leaves for making deep-fried rolls are available. Where corms and cormels yield is preferred, local collections with high yield and less acridity are desired. For making rolls, the cultivars with high leaf yield and less acridity are preferred.

Kokan Harit Parni

A variety recently developed at Balasaheb Sawant Konkan Krishi Vidyapeeth, Dapoli is suitable for cultivation in Konkan region of the Maharashtra state. Its foliage is erect with dark green leaves and corms are small and brown-colored. Its average yield is 3–5 tonnes per hectare in crop duration of 6 months from planting to harvest.

Another variety of tannia is South Dade White, which produces white-fleshed cormels.

PLANTING TIME

The planting of tannia depends on climatic conditions of the growing region. In north Indian plains under rain fed conditions, it is planted in the month of June-July. In south India under irrigated conditions, it can be grown year yound. However, the main planting seasons for tannia planting are April–May with the onset of southwest monsoon and September–October with the onset of northeast monsoon. Planting can also be done during February–April, provided sufficient soil moisture is made available through irrigation. For maximum corm production, April–May planting is preferred since the crop can utilize both the monsoons effectively, and the second best season is September–October.

SEED RATE

The quantity of planting material required varies with the size of corms, soil fertility, growing conditions and availability of irrigation facilities. In general, the recommended seed rate is for planting a hectare area is 5–10 quintal corms.

PLANTING METHOD

Tannia is although a perennial plant but cultivated as an annual crop. Corms and cormels are used as propagating material. Disease-free healthy cormels of bigger size and 20–25 cm long are commonly used. The sets from top portion of the main corm with a thickness of 5–10 cm containing apical bud may also be used for propagation. In Maharashtra, it is usually grown as a ratoon crop in kitchen gardens. The cormels are separated annually from the main corm, which is covered with soil after removing the cormels. In some parts of Maharashtra and Gujarat where only leaves are harvested, the main corms and cormels that are left in the field continue to act as new planting material.

The field is prepared by ploughing repeatedly in summer before the onset of monsoon followed by the establishment of corms in rows on flat land and formation of ridges 90 cm apart, which provide a place for the development of storage organ, improve drainage and facilitate harvesting. Corms or cormels are planted on ridges keeping a spacing of 60 to 90 cm. For leaf purposes, it is planted on both sides of the ridge.

In southern parts of the country, tannia is intercropped with plantation crops such as cocoa, rubber, banana and coconut.

NUTRITIONAL REQUIREMENT

Well rotten farmyard manure @ 12–15 t/ha is incorporated into the soil at the time of land preparation. Nitrogen, phosphorus and potassium are applied at the rate of 80, 60 and 80 kg/ha, respectively. Traditionally, full dose of phosphorus and potash along with half dose of nitrogen is applied as a basal dose at the time of ridge formation and the remaining half of nitrogen is applied a month after planting at the time of earthing up. In general, tannia requires nitrogen medium to high, phosphorus medium to low and potash medium doses.

IRRIGATION REQUIREMENT

Tannia prefers upland rain-fed conditions rather than low land water logging conditions. However, it needs sufficient irrigation for optimum vegetative growth and leaf production, especially during the period of carbohydrate storage. Under rain-fed conditions, a short-term drought may often accelerate crop maturation.

The soil at the time of planting should have sufficient moisture. If planting is done in dry soil, irrigation is applied immediately after planting and one week later for uniform sprouting of corms. Subsequent irrigations may be given at an interval of 12–15 days, depending on moisture retention capacity of the soil. The irrigation is stopped 3–4 weeks before harvesting the crop. Tannia requires 9–12 irrigations from planting to harvesting. In rain fed crop, supplementary irrigation is applied to provide sufficient moisture under conditions of prolonged dry period after planting. Proper water management is essential for obtaining optimum size of corm and cormels.

INTERCULTURAL OPERATIONS

Weeding is an essential operation, which should be done 1 and 2 months after planting to keep the field weed free and to reduce competition between crop plants and weeds. Weeding during early growth phase is beneficial since the crop may not compete with weeds at early stages due to slow growth rate of plants. Earthing up before the initiation of corm formation is also essential for proper development of corms and cormels. If labour is not available for hoeing and weeding, Diuron herbicide may be applied at the rate of 1.6 kg/ha as pre-emergence to control the weeds.

Application of mulch soon after planting is beneficial to the crop as it conserves soil moisture, regulates soil temperature around the sets and also suppresses the weed emergence.

HARVESTING

Tannia becomes ready for harvesting 8–10 months after planting. The crop is usually harvested by digging out mature corms with the help of spade. If the crop is grown for leaf purpose, the leaves can be harvested as and when they are fully expended and mature. A leaf size of 20–30 cm x 20–30 cm is ideal for making rolls. A clump of plants gives about 40–50 leaves during its entire growth period. The leaves to be ready for harvesting take 40–50 days after emergence in the field. The leaves are harvested along with their petioles. Petioles are usually used as a cooked vegetable. If the crop is intended for taking corms and cormels, it can be harvested when older leaves started turning yellow. Its corms and cormels do not deteriorate if left in the field, and in this way, harvesting can be done from 6 to 9 months. Care should be taken to avoid injury to the corms and cormels during harvesting.

If tannia is to be grown as a ratoon crop, the soil around the clump should be dug to remove the mature cormels. The main corm is retained along with roots and covered with soil. This can be practiced for 4–5 years.

YIELD

The yield of tannia corms depends on variety, soil type and fertility, irrigation facilities available, growing season, climatic conditions of the growing region and package of practices adopted by the growers. On an average, the yield of tannia corms and cormels is 200 to 250 quintals per hectare.

POST-HARVEST MANAGEMENT

Curing of corms is done for removing excessive moisture and soil particles adhered with their surface, healing wounds and enhancing shelf and storage life. The corms and cormels after harvesting are heaped under shade for 4–5 days for their curing. However, before curing, the damaged and decayed corms are sorted out from the lot.

The corms after proper curing can be stored conventionally in dry soil pits or in sand for a period of 5–6 months. The corms can also be stored for a period of 4–5 months under ventilated, dry and semi-dark conditions. Under artificial storage conditions, the corms can be stored for about 2 months at 7°C temperature coupled with 80% relative humidity. While harvesting, care must be taken to avoid bruising to the corms and cormels otherwise they are liable to develop serious rots during storage.

INSECT-PESTS

Aphids (*Aphis gossypii*)

Being sucking pest, it sucks cell sap from tender leaves, on which it secretes honeydew, a sticky sugary substance, which is covered by black sooty mould, affecting the photosynthesis of plants. It transmits a virus, which causes yellowing and curling of the leaves. It attacks the plants mostly at the early stage of crop. High humidity and high temperature favour the occurrence of pest.

Control

- Grow resistant variety.
- Spray the crop with malathion 0.05% at interval of 10–12 days.

Thrips (*Heliothrips indicus*)

It is commonly observed in the months of February–May. It sucks the cell sap and cause white specks with black dots. The affected leaves get brown and distorted. Dry weather is favourable for the rapid multiplication of thrips.

Control

- Grow resistant variety.
- Follow rotation with non-host crops.
- Collect and destroy the infected leaves.
- Apply phorate 2 kg a.i. per hectare in soil.

Root Weevil (*Diaprepes abbreviatus l*)

It is the most destructive pest of tannia. Both its grubs and adults cause damage by burrowing and making tunnels into corms and feeding on soft tissues. The pest causes damage both in field and stores. Larval feeding causes physical damage to the cormels, causing them to be placed in a lower grade. However, there is no documentation of economic loss due to this pest.

Control

- Keep the field fallow.
- Follow long crop rotation with non-host crops.
- Use weevil-free corms and cormels for planting.
- Mount up soil on both sides of ridges 30 and 60 days after planting.
- Install sex pheromone trap in the field.
- Spray fenthion or fenitrothion 0.05% at monthly intervals.
- Apply phorate 2 kg a.i. at 2, 6 and 10 weeks of planting.

ROOT-KNOT NEMATODES (*Meloidoginy* spp.)

The root knot nematodes are polyphagous and widely distributed in the country. Nematode is the most serious problem in light sandy soils. The infected roots are not able to utilize soil nutrients and moisture efficiently. Stunted plant growth, leaves chlorosis, wilting and death of plants are the major symptoms if the infection is very severe. Nematode infection causes rotting of corms in storage and reduces storability of corms.

Control

- Follow long crop rotation with trap crop.
- Grow nematode-resistant varieties.
- Avoid planting of nematode infested corms.
- Apply high doses of potash fertilizer.
- Turn the soil two to three times at 10-day intervals in summer.
- Sterilize the nursery beds through soil solarization in hot summer months.
- Fumigate the field with Nemagon before planting the sets.
- Add organic amendments in soil, *i.e.*, groundnut, *neem*, *mahua*, or *karanj* cake.

TANNIA BEETLE (*Ligrus ebenus*)

It is the most destructive pest of tannia that feeds on leaves. It is a polyphagus in habit and problematic in the states of Kerala, Maharashtra and Andhra Pradesh.

Control

- Spray Ridomil at 0.05% at intervals of 60 days after the planting.
- Store only healthy corms in the sand or sawdust.

DISEASES

LEAF BLIGHT (*Phytophthora colocasiae*)

Blight is the most destructive fungal disease, which appears as pale green to brown round spots on leaves when the crop is at maturity stage in monsoon season. Later, the spots spread and coalesce rapidly over the whole leaf and form large lesions. High humidity and high temperature in the months of August to October favour the occurrence of this disease.

Control

- Grow resistance variety.
- Follow field sanitation.
- Destroy infected plant debris after harvesting the crop.
- Spray the crop with 0.05% Ridomil, 0.25% maneb or zineb formulations.
- Apply *Tricoderma viridae* as a bio control agent.

SOFT ROTS (*Pythium ultimum*)

The pathogen enters the plant through wounds and damages the tissues. Symptoms appear on leaves as small water-soaked lesions, which later enlarge and turn brown. The infected tissues become soft and shiny with foul smell. High humidity favours the disease occurrence. The loss due to this disease is not too much but it causes rotting of corms in storage.

Control

- Strictly follow a crop rotation of at least 3 years.
- Use disease-free planting material.
- Avoid bruising during harvesting, handling and transportation.

MOSAIC DISEASE

The disease appears as mosaic mottling and blistering of leaves. Leaves become small and mis-shapen and plants stunted. The infected plants bear a much lower yield of small and deformed corms. The disease is generally transmitted by white fly and aphids.

Control

- Use disease-free healthy planting material.
- Keep the field weed free.
- Follow proper field sanitation.
- Apply Furadon 1.5 kg a. i. per hectare in soil.
- Spray the crop with Nuvacron 0.05% at 10-day interval.

12 Taro

M.K. Rana and P. Karthik Reddy

Botanical Name : *Colocasia esculenta* (L.) Schott

Family : Araceae

Chromosome Number : Like other tuber crops, polyploidy and diploidization in taro have also occurred during evolution, thus, the chromosome number (2n) varies with cultivars and also within a cultivar from 22 to 48. In India, the chromosome number in different cultivars is 24, 28, 42 and 48, and triploids with chromosomal number (3n) 42 are also available. However, diploids are predominant in south India, while the triploids are found in north and higher altitude regions.

ORIGIN AND DISTRIBUTION

Taro has originated in Northeastern India and Indo-Malayan region of the Southeast Asia but it was originally cultivated in India or in Malaysia. *Colocasia* species are distributed in Southeast Asia (Indonesia and Malaysia), Bangladesh, South America, which includes Guatemala and Peru. In India, it is distributed from foothills of Himalayas to Maharashtra and Meghalaya.

INTRODUCTION

Taro, also known as arum, *Colocasia*, *Ghuiyan* and *Arvi*, is an ancient crop grown throughout the humid tropics for its corms, stalks and leaves. It is a most commonly grown vegetable in the tropical countries. The greatest intensity of its cultivation and its highest percentage contribution to diet occurs in the Pacific Islands. However, the largest area of cultivation is in West Africa, which therefore accounts for the greatest quantity of production. A significant quantity of taro is also grown in the Caribbean and virtually in all humid or sub-humid parts of Asia. It is widely cultivated in Nigeria, Malaysia, Indonesia, Philippines, West Indies, Brazil, Egypt, Bangladesh and India. In India, it is extensively grown in Madhya Pradesh, Maharashtra (Konkan region), Gujarat, West Bengal, Bihar, Uttar Pradesh, Assam, Orissa, Kerala, Andhra Pradesh, Karnataka and Tamil Nadu.

Among the cultivated species, there are two types of taro, namely, the *eddoe* type (*Colocasia esculenta* var. antiquorum) and the *dasheen* type (*Colocasia esculenta* var. esculenta). Eddoe types have the cormels bigger than the mother corms, while the dasheen types have the mother corms bigger than the cormels. The species *Colocasia esculenta* (L.) Schott. is found growing wild in sub-Himalayan tracts, peninsular region and Northeastern regions, mostly waterlogged humid tropical areas and its large variability is in eastern region. The crop has evolved with the cultures of the people in the Asia-Pacific region. It has acquired considerable socio-cultural importance for the people. Among food crops in Oceania, the adulation and prestige attached to taro is equal to yam in certain localities.

COMPOSITION AND USES

COMPOSITION

The corms are rich in carbohydrates and also contain a good amount of protein, calcium, phosphorus and β-carotene. Taro contains about 7% protein on dry weight basis, which is more than yam, cassava and sweet potato. The protein fraction is low in histidine, lysine, isoleucine, tryptophan and methionine but otherwise rich in all other essential amino acids. The protein content of corm is higher towards the corm's periphery than towards its center. The taro leaf, like most of the higher plant leaves, is rich in protein, as it contains about 23% protein on a dry-weight basis. It is also a rich source of thiamine, riboflavin, niacin and vitamin C, calcium, phosphorus and iron, which are important constituents of a human diet. The fresh taro lamina has about 20% dry matter, while the fresh petiole has only about 6% dry matter. The nutritional composition of corms, petioles and leaves is given in Table 12.1.

USES

All parts of taro plant, *i.e.*, leaves, petioles, stalks, corms and cormels can be consumed as vegetable or in other forms. The corms and cormels are boiled, baked, roasted, or fried and consumed in conjunction with fish, coconut preparations, etc. A favorite and peculiarly Pacific way to prepare taro is to roast it on hot stones (*mumu* or *umu*) in dugout earth ovens. This is quite common when taro is used in feasts and ceremonies. The leaves are usually boiled or prepared in various ways mixed with other condiments. The high protein content of leaves favorably complements the high carbohydrates content of corm, which goes with it. In northern India, the fresh tender leaves are used for the preparation of *pakora*. Processed or storable forms of taro are not common in the Asia-Pacific region. Geographically, the most widespread use of taro for human consumption is in the form of chips and noodles. Taro flour is available in some places and used as a thickener in soups, infant formulations and other preparations. Another processed and packaged form of taro is *poi*, a fermented sour paste made from boiled taro. In the Hawaiian Islands, its leaves are used for the preparation of *fafa* and *palusami* dishes. Taro is especially useful to persons allergic to cereals and it can be consumed by the children who are sensitive to milk. Its acridity factors can be separated by gravitational means and the material can be stabilized by dehydration.

TABLE 12.1

Nutritional Composition of Taro Corms, Petioles and Leaves (per 100 g Edible Portion)

Constituents	Leaves	Petioles	Corms
Water (g)	78.8	94.0	73.1
Carbohydrates (g)	8.1	3.6	21.1
Protein (g)	6.8	0.3	3.0
Fat (g)	2.0	0.3	0.1
Dietary fibre (g)	1.8	0.6	1.0
Minerals (g)	2.5	1.2	1.7
Calcium (mg)	460	60	40
Phosphorus (mg)	125	20	140
Iron (mg)	0.98	0.5	0.42
β-carotene (μg)	120	104	24
Thiamine (mg)	0.06	0.07	0.09
Riboflavin (mg)	0.45	0.07	0.03
Niacin (mg)	1.9	0.10	0.4
Folic acid (μg)	–	–	5.4
Vitamin C (mg)	63	3	–
Energy (kcal)	77	18	97

MEDICINAL USES

Fermented sour paste made from boiled taro, *i.e.*, *poi*, is an excellent food against food allergies, and consumption of cooked corms helps in relieving pain from morphia. It reduces ulcers caused due to tuberculosis, and it expedites healing of fungal abscess in the animals. Corm juice is used as poultice and the petiole juice is used as stypic. Its consumption as vegetable curry gives relief in hemorrhoids.

BOTANY

Taro is an herbaceous plant, which grows to a height of 1–2 m. The plant consists of a central corm (lying just below the soil surface), from which, the leaves grow upwards and roots grow downwards, while cormels, daughter corms and runners (stolons) grow laterally. The root system is fibrous and lies mainly in top one meter soil. The corm is a short upright underground stem, which is encircled with rings from where the leaves arise. These are seen as dark scaly or papery sheaths. The flesh is purple, white, yellow, or pinkish in color depending on the variety and somewhat slimy in consistency. The bulbous corms are cylindrical and vary in size, normally 30 to 40 cm long. In dasheen types of taro, the corms are cylindrical, large up to 30 cm long and 15 cm in diameter and constitute the main edible part of plant. In eddoe types, the corms are small, globoid and surrounded by several cormels (stem tubers) and daughter corms. The cormels and the daughter corms together constitute a significant proportion of the edible harvest in eddoe taro. Daughter corms usually give rise to subsidiary shoots even while the main plant is still growing but the cormels tend to remain dormant and give rise only to new shoots if left in the ground after the death of the main plant. Each cormel or daughter corm has a terminal bud at its tip and axillary buds in the axils of numerous scale leaves all over its body.

Corms, cormels and daughter corms are quite similar in their internal structure. The outermost layer is a thick brownish periderm, within this, the starch-filled ground parenchyma lies. Vascular bundles and laticifers ramify throughout the ground parenchyma. Idioblasts (cells that contain raphides or bundles of calcium oxalate crystals) also occur in the ground tissue and in nearly all other parts of the taro plant. The raphides are associated with acridity or itchiness of taro, a factor that will be taken up in greater detail when the utilization of taro is discussed. The density and woodiness of corm increase with age. In eddoe and dasheen types of taro, the central corm represents the main stem structure of the plant. The surface of each corm is marked with rings showing the points of attachment of scale leaves or senesced leaves. Axillary buds are present at nodal positions on the corm. The apex of the corm represents the growing point of the plant and is usually located close to the ground level. The actively growing leaves arise in a whorl from the corm apex. These leaves effectively constitute the only part of the plant which is visible above ground. They determine height of the plant in the field. All parts of taro plant are acrid due to the presence of calcium oxalate crystals, which exist in two forms, *i.e.*, (i) *druses* (80–95% of the total) and (ii) *raphides*. The density of crystals in corms may be as high as 1,20,000 cm^{-1}. In leaves, the number of crystals is bit higher. The acrid principles causing irritation in the mouth and esophagus can be destroyed by cooking or fermentation.

CLIMATIC REQUIREMENT

Taro is a crop of tropical and subtropical regions, thus, warm humid climate is ideal for its normal growth. Usually, taro crop requires a rainfall of 1500–2000 mm for its optimum yield and it thrives best under very wet or flooded conditions. Dry conditions result in reduced corm yield. Corms produced under dry conditions also tend to have a dumb-bell shape (the constrictions) and reflect the period of reduced growth during drought. The crop performs better under a temperature range between 21 and 27°C as it is well adapted to high temperature and high humidity. It can thrive well in a frost-free subtropical environment with an average temperature of 20°C. It can be cultivated up to an elevation of 1500 m.

SOIL REQUIREMENT

Taro can be grown on a wide range of soils from light to heavy soils. Loam and sandy loam soils are ideal for its successful cultivation. The soil should be well drained, fertile, loose friable, rich in organic matter and should have good water holding capacity, which may favor the uniform development of tubers. Dasheen types perform well in heavy soils with good water-holding capacity, while the eddoe types prefer well-drained loamy soils with a high water table. The optimum pH for its successful cultivation is 5.5–7.0. The land prepared and brought to a fine tilth by ploughing or digging to a depth of 20–25 cm. Permanently moist soils appear to be the most desirable to maximize the plant growth and yield. Moisture stress can be detrimental to growth, thus, supplementary irrigation is necessary during dry periods.

CULTIVATED VARIETIES

Several local varieties of taro like Sar Kachu, Satamukhi, Sahasramukhi, Topi, Jhankri, Astamukhi, Panchamukhi, Bansi, Faizabadi, Bengali Banda, Desi Banda and Kovvur are available. Breeding work with respect to taro crop is being carried out at the Central Tuber Crops Research Institute, Thiruvananthapuram, Kerala. The characteristic descriptions of some selected improved cultivars developed by different centers are as follows:

CO 1

A variety developed at Tamil Nadu Agricultural University, Coimbatore is high yielding with a yield potential of 243 quintals per hectare. The corms containing 22.5% starch and 2.4% protein content on fresh weight basis are mildly acrid with good cooking quality.

KOVVUR

A variety developed through a selection from the local material obtained from Andhra Pradesh matures in 180 days after planting. Its plants are about 70 cm tall, and leaves are narrow with purplish green petiole. Its corm yield is 200 to 230 q/ha. It is recommended for cultivation in Andhra Pradesh, Maharashtra, Kerala, Karnataka, West Bengal and Bihar.

MUKTAKESHI

A variety developed at the Regional Centre of CTCRI is resistant to taro leaf-blight disease. Its corm with excellent cooking quality has 16–18% starch and average tuber yield is 250–300 q/ha in crop duration of 150–180 days.

PANISARU 1

This variety has been released for cultivation in Orissa under wet land ecosystems. The corms have a brown-colored flesh fibre with 12–13% starch content. It takes 180–210 days after planting to be ready for harvesting and yields about 160 q/ha.

PANISARU 2

This variety has been released for cultivation in Orissa under wet land ecosystems. The corms have a brown-colored flesh fibre with 16–18% starch content. Its average yield is 130 q/ha in crop duration of 180–210 days.

PANCHAMUKHI

A cultivar developed through a selection from local material of the Northeastern regions of India produces broad leaves on green petiole. Its plants are erect in growth habit with medium height of 70 cm. It matures in 180–195 days and yields ranging from 120–150 q/ha.

PUNJAB ARVI 1

A first variety of taro developed at Punjab Agricultural University, Ludhiana has been recommended for cultivation all over Punjab except the district of Bathinda. Its plants are tall with dark green, large and obliquely erect leaves. Its petioles are long and sheathering at the base. The corms are brown, long of medium thickness with creamy inner flesh. Its average yield is 225 q/ha in crop duration of 176 days.

SAHASRAMUKHI

The plants of this cultivar are erect in growth habit with plant height up to 70 cm. It takes 5.5 to 6.0 months to be mature after planting. The yield ranges from 120 to 150 q/ha.

TAMARAKANNAN

It is a local collection from Kerala, which matures in 165 to 180 days after planting. The characteristic plant exhibits erect growth with a height of 70 cm. The leaves are medium sized with green petiole. It is a high yielding variety with a yield potential ranging from 120 to 150 q/ha and the corms have low acridity.

SREE KIRAN

The first hybrid variety of taro in India developed and released by Central Tuber Crop Research Institute, Thiruvananthapuram, Kerala during 2004 gives an average yield of 180 q/ha in crop duration of 210 days. The corms contain 16–18% starch. The higher cormel yield, good cooking quality and larger storability of tubers are the desirable attributes of this hybrid.

SREE PALLAVI

A variety developed at Central Tuber Crop Research Institute, Thiruvananthapuram, Kerala is high yielding with a yield potential of 150–180 q/ha in crop duration of 205–210 days. Its plants are erect and 1–1.5 m tall. Its leaves are green and glaucous on purple petiole. Its tillering rate is low. The corms containing 16–17% starch and 2–3% protein on fresh weight basis are small but more in number (20–25 per plant).

SREE RASHMI

A variety developed at Central Tuber Crop Research Institute, Thiruvananthapuram, Kerala is high-yielding with a yield potential of 150–200 q/ha in crop duration of 7–8 months. The plants are erect and 100–150 cm tall. The petiole and the leaf are green in color and the lower surface of the leaf is light green. All parts of the plant are free from acrid substance. Its corms containing 15% starch and 2.5% protein on fresh weight basis have good cooking quality and excellent taste.

VALLABH NIKKI

A cultivar developed by selection from gemplasm maintained at the Department of Horticulture, Sardar Vallabh Bhai Patel University of Agriculture and Technology, Meerut is resistant to leaf blight. Its plant height is 38–42 cm, leaf length 2 9–30 cm and width 18-20 cm, number of leaves 4-6 and number of corms 24-26 per plant. Its yield potential is as high as 281–305 q/ha. It is grown successfully in plains as well as in hills. It is superior to other varieties in yield and well accepted by farmers due to other desirable characters as well.

WHITE GUARIYA

The plant is 30 cm tall with medium leaves on green petiole. It takes 180 to 195 days after planting to be ready for harvesting and yields about 120–150 q/ha. The corms of this variety are free from acridity and all parts of the plant are edible.

PLANTING TIME

Taro is planted in April-June in Kerala, February in Andhra Pradesh, May in Tamil Nadu, February and June in Bihar and Eastern Uttar Pradesh and in April in Assam. In southern parts of the country, the crop under rain fed conditions is planted in May–June to October–November in rainy season, but under irrigated conditions, the planting can be done throughout the year as per the availability of moisture, especially in February to March; however, rainy season is the most commonly adopted planting time.

SEED RATE

Taro is commercially propagated by cormels of ideally 20–25 g weight. Mother tubers of 50–75 g size, suckers that are produced as a result of lateral proliferation of the main plant in the previous crop and *huli*, *i.e.*, 1–2 cm apical portion of the corm with basal 15–20 cm of the petioles attached, can also be used for planting. Microsetts of 2 g give higher cormel yield, while apical and large sized microsetts are suitable for getting better sprouting and yield. A pre-planting soaking of corms in 1000 ppm solution of morphactin or chlorflurenol and in 100 and 1000 ppm solution of coumarin significantly delays sprouting and reduces the percent sprouting of corm setts, respectively. However, soaking in 1–10 ppm solution of GA_3 promotes early sprouting of corms.

About 37,000 vegetative propagules are required for planting a hectare of area at a spacing of 60 cm line to line and 45 cm plant to plant within a line.

PLANTING METHOD

The preparation of land differs from place to place. Hawaii is the only country where taro is cultivated with scientific methods. The land is puddled before planting in Hawaii, while it is planted on ridges and furrows in parts of South India and in beds in Northeastern regions of the country. The soil should be ploughed well up to a depth of 20–35 cm and made into a fine tilth before planting. The depth of planting varies between 5 and 10 cm on the ridges, which are prepared 60 cm apart, and the plant to plant distance is maintained 45 cm on these ridges. The spacing also varies with the growing region. The recommended spacing for planting in Kerala (India) is 60 × 45 cm, whereas, in Cuba, it has been suggested to plant at a distance of 90 × 30 cm from line to line and plant to plant. In Japan, it is planting at a spacing of 60 × 100 cm for getting higher yield.

NUTRITIONAL REQUIREMENT

Like other tuber crops, taro is responsive to the application of fertilizers. The requirement of crop depends upon the soil type, soil fertility, climate, spacing, cultivar and cultural practices adopted.

Nutrients play an important role in plant growth and vigor at the initial stages of crop growth and provide protection against insect pests and diseases. The plants suffering from nitrogen deficiency shows stunted growth and short petioles with yellowing of the plants. Phosphatic fertilizers have shown positive effect on increasing the dry matter content of corms, promoting root growth and hastening the corm maturity, whereas, the deficiency of phosphorus leads to defective root system. It requires more potassium than nitrogen and phosphorus as it increases the starch, protein, sugar and carbohydrates content of corms but it has negative effect on dry matter content of the corms. Potassium deficiency causes marginal chlorosis of leaves.

Under conditions of Kerala, nitrogen has been recommended 103.9 kg, phosphorus 74.1 kg and potash 135.7 kg/ha. In general taro crop removes nitrogen 125 kg, phosphorus 19.2 kg, potassium 135 kg, calcium 30.4 kg, magnesium 15.87 kg, sulfur 11.80 kg, iron 10.25 kg, manganese 1.69 kg, zinc 647 g and copper 93 g from the soil with a production of 170 quintals of edible corms per hectare, and the peak period of removal of NPK is 3–5 months after planting, calcium, iron and manganese 3rd month after planting and zinc and copper at maturity stage.

Taro suffers from *Metsubure* disorder when it is grown under calcium-deficient soils. Its deficiency leads to interveinal chlorosis and necrosis, failure of leaf blades to unfurl, collapse of petioles, dieback of roots and death of growing point. Use of lime can solve the problem of calcium deficiency. In calcareous soils, the zinc deficiency is characterized by interveinal chlorosis, stunting, narrow leaves and cupping of leaves.

IRRIGATION REQUIREMENT

Water is one of the essential components required for growth and development of the crop. The total water requirement varies with the soil type, climate and crop duration. Taro is one among a few crop species that can adapt to a wide range of moisture regimes. Overwatering is not harmful to its growth but supplementary irrigation is essential during dry season. Growth is restricted and suckering and/or reduced quality may occur if plants are water stressed. Corm yield increases with increasing levels of water because of extensive rooting system. It requires plenty of water during its growth since the plant is having shallow root system. Optimum irrigation should be made available to the crop for uniform sprouting. Irrigation immediately after planting and 1 month after planting is crucial. Dasheen types thrive well even under flooded lowland conditions but cannot tolerate drought. The eddoe types are mostly grown in upland conditions. In general, dry conditions cause lower corm yield, deformed and cracked corms, which finally lead to rotting of corms in storage or in the field itself. Subsequent irrigations may be given at fortnightly intervals. Moisture stress may be detrimental to growth, thus, supplementary irrigation has to be given during dry period. Under moisture stress, plant growth becomes stunted and yield of corms is reduced drastically. During summer, irrigation is given at 10-day intervals.

INTERCULTURAL OPERATIONS

HOEING AND WEEDING

Weeding becomes natural under flooding conditions. Keeping the field weed free during the initial stage of the crop growth helps in better growth of the plant. Three hand weedings, two coinciding with the two splits of fertilizer application and earthing up, and one at a later stage, are recommended to get higher corm yield. Herbicides like propanil, prometon and nitrofen are effective in wet lands, while in uplands, ametryn is recommended. Earthing up of soil around base of the plants is advisable during weeding so that the developing corms may be protected from exposure to light. Weeding and earthing up at 45 to 60 days after planting are usually considered the best for getting higher yield from taro crop.

Removing of Suckers

Suckers inhibit the growth of corms as they compete for nutrients, space, sunlight and water. Therefore, desuckering is very important operation in the cultivation of taro crop. It is done by removing the side sprouts a month after emergence and only three suckers per plant are retained to obtain economic corm yield. Corm size, yield and quality can be increased by desuckering unhealthy suckers and retaining the healthy ones.

Mulching

Mulching is considered a good agronomic practice to increase the productivity of taro as it helps in controlling weeds and conserving soil moisture. The organic or inorganic mulch can be used for the purpose. Organic mulch includes crop residues, green leaf mulch, straw, hay, etc., which conserves soil moisture and controls weeds and at the same time adds organic matter to the soil, which improves the tops and root growth and yield along with moisture conservation and weed control. Among the various mulches tried, paddy-husk gives the maximum cormels yield. Plastic mulch does the same but does not add organic matter to the soil. Mulching with dry leaves and other organic materials provides nutrients along with controlling the weeds. Polythene mulch can also be used for controlling weeds.

HARVESTING

The crop duration depends on the type of cultivar, soil type, soil fertility, climatic conditions of the growing region and the cultural practices followed during crop cultivation. There are reports saying that taro matures in 3 months in Ceylon, while it takes 12–15 months in Hawaii. In India, the crop matures in 5–7 months after planting. Dasheen types mature in 8–10 months and eddoe types in 5–6 months. Harvesting is signalled by decline in height of the plant and general yellowing of the leaves. At maturity, the main corms begin to push out of the soil surface. This is usually an indication that the crop is ready for harvesting. Irrigation should be stopped 3 to 4 weeks before harvesting. Harvesting is most commonly done by means of hand tools. The soil around the corms is loosened, and the corms are pulled out by grabbing base of the petioles. For flooded taro, harvesting is more tedious because of the need to sever the living roots that still anchor the corm to the soil. Even in mechanized production systems, harvesting is still done by hand, thereby increasing the labor and cost of production. After harvesting of corms, these are then washed to remove the roots and soil before grading and packing.

YIELD

The yield of taro also varies with cultivar, soil type, soil fertility, climate and the cultural practices adopted during cultivation. In Kerala, the reported yield varies from 160 to 260 q/ha, while in neighboring state of Tamil Nadu, the recorded yield is 130 q/ha. Its average yield in general ranges from 120 to 150 q/ha.

POST-HARVEST MANAGEMENT

The corms after harvest are washed with clean water to remove adhered soil and other particles soon after harvesting. Thereafter, they are sorted and graded according to size or weight depending on the market demand. During sorting and grading, the deformed, cracked and decayed corms are rejected. Studies on post-harvest management of taro are meager, though in general, taro cannot be stored for long time, thus, the corms should be consumed within 2 weeks of harvesting. Refrigeration appears to extend storage life. Corms are stored at room temperature only for few days to cure bruises and then at 7–10°C temperature with good ventilation for long-term storage. Taro corms harvested at full mature stage can be stored up to 30 days in sawdust and sand media. High temperature and relative humidity greater than 85% cause deterioration, and a very low temperature results in death of corms within two months. The storage can be extended up to 4–5 months if care is taken during harvest and handling. Storage in pits is also practiced in many countries.

Taro leaves with folded tops that are packed in low-density polyethylene bags and stored at 10°C temperature can last for 2 weeks. Unpackaged leaves turn brown-yellow after 8 days and lose more weight due to lower humidity. Chilling injury has also been observed as browning discoloration after 12 days at 3°C temperature.

INSECT-PESTS

APHIDS (*Aphis gossypii*)

This is a cosmopolitan pest. In India, it is prevalent in humid tropical regions. Adults and nymphs damage the crop by sucking sap from tender leaves, causing the leaf curl and appearing sick. Aphids secrete a honeydew-like substance, which attracts the ants, and on which the black sooty mold starts growing. The population build up is high during early stages of the crop with peak infestation from August to October. The maximum activity of the aphid population is observed during cloudy days when humidity is high and rainfall is low.

Control

- Two parasitoids, *viz. Aphelinus mali* and *Aphidencyrtus aphidivorous*, and two aphidophagous predators, *viz. Menochilus sexmaculata* and *Ischiodom scuttelaris*, have been found very effective natural enemies of aphids. Out of them, *M. sexmaculata* has a high predation rate of 15–28 aphids per day.
- Spray the crop with malathion 50 EC 2 ml/litre of water or dimethoate 30 EC 1 ml/litre of water.

LEAF-EATING CATERPILLAR (*Spodoptera litura*)

It is commonly known as tobacco caterpillar and one of the key pests of aroids, which is polyphagous. The young larvae are gregarious, scrapping the leaf tissues, and later on, segregate and become solitary. It feeds voraciously on leaves and completes its life cycle in 30–40 days.

Control

- *Charops hersei*, which feeds on the leaf-eating caterpillar, is an endoparasite, and the parasitization ranges from 6% to 10%. Parasitoids are at the peak during the months of September–October.
- Spray the crop with malathion 0.5% to control these larvae.

THRIPS (*Caliothrips indicus*)

The nymphs and adults cause damage by sucking cell sap from the tender leaves. It is a seasonal pest attacking during summer months from January onwards and the peak period is from February to May. The damage is characterized by silvery-white specks with black dots on the leaves due to continuous sucking of sap by the nymphs and adults.

Control

- Irrigate the field at frequent intervals to keep the soil moist.
- Spray the crop with dimethoate @ 5 ml/litre of water for effective control of thrips.

Taro Leafhopper (*Tarophagus proserpina*)

A reddish crust of sap exudates is produced when the hopper punctures the plant during feeding. The adults feed on the plants and cause wilting and death. It is visible on the outside of the leaf stalks near the ground. Its eggs are deposited in pairs in small holes made by the female of taro leafhopper in the midrib on the underside of the leaf and in base of the leaf stalks or petioles. Inspect leaves particularly on under surface that are starting to unfold and at base of the younger petioles. When a threshold of 1 nymph per 5 plants has been reached, chemical control is warranted.

Control

- Smoking taro plants with smoldering bundles of coconut leaves will be effective in reducing the pest incidence.
- Follow long crop rotation with non-host crops. If possible, relocate the crop to a new land area on which dasheen crop has not previously grown but with an unrelated crop, *e.g.*, maize, sorghum, pearl millets, etc.
- The egg-sucking bug, *Cyrtorhinus fulvus*, has given satisfactory biological control in Hawaii and the Philippines and is effective in controlling *T. proserpina*.
- Spraying imidacloprid (Confidor®) or thiamethoxam (Actara®) is beneficial for controlling the leafhoppers.

Red Spider Mite (*Tetranychus* spp.)

The red spider mite attack is found in Andhra Pradesh, Kerala and Tamil Nadu. The nymphs and adults of red spider mites are polyphagus, which suck cell sap from leaves, resulting in yellowish-white specks and chlorotic patches on lower leaf. It reduces plant growth, leaf chlorophyll and photosynthesis.

Control

- Among predators, thrips (*Scolothrips indicus*) and coccinellid beetle (*Stethorus gibviferous*) prey upon mites. Thrips prey on 29–45 eggs or 22–48 nymphs per day.
- Spraying dimethoate 30 EC at 3 ml/l or quinalphos or phosalone at 5 ml/litre of water is found effective for controlling thrips.

Taro Beetle (*Papuana* spp.)

The adult beetles fly from the breeding sites to the taro field and tunnel into the soil just at base of the taro corm. They then proceed to feed on the growing corm, leaving large holes that degrade the eventual market quality of corm. Also the wounds that they create while feeding promote the attack of rot-causing organisms. Both, larvae and adults bore into the corms at growing points of the roots, resulting in wilting of plant and even death of the affected plants. The damaged corms with holes become unfit for marketing. This pest appears in irrigated taro during dry weather.

Control

- Certain pathogens of the beetle have been identified. These include a fungus (*Metarhizium anisopliae*), a bacterium (*Bacillus popilliae*) and the protozoa *Vavraia*.

Root Knot Nematode (*Meloidogyne javanica* and *M. incognita*)

The foliage of the nematode infested plant at the outset turns yellow, followed by brown colour and ultimately dies. The adult female is pear-shaped, 0.4–1.3 mm long and 0.27–0.75 mm wide and the male root-knot nematodes are worm-like 1.2–1.5 mm long and 30–36 mm in diameter. The second stage larva of *Meloidogyne* species enters a taro root and begins feeding on cells around its head by inserting a sharp mouthpart called a stylet. The larva injects saliva, which stimulates cell enlargement, forming large nurse cells, from which the larva continues to feed. The nematode goes through three molts and in fifth stage, the female swells to a visible size and it lays as many as 500 eggs in a gelatinous coating just outside the root. Larvae go through one molt inside the eggs, and then, enter the soil as second stage juveniles. It may complete its life cycle in 25 days at 27°C temperature. The corms become deformed with galls or blister-like swellings. Nematode-infected corms usually rot during storage.

Control

- Healthy nematode-free corms should be used as planting material.
- The planting material should be given hot water (50°C) for 40 minutes before planting.
- Apply fumigants and nematicides (DBCP 3.6 l/ha) to control the nematodes.

DISEASES

More than 100 species of fungi cause damage to Aroids, out of which, 21 species have been reported from India. Their detailed descriptions are given below:

LEAF BLIGHT (*Phytophthora colocasiae*)

It is the most serious disease caused by *Phytophthora* sp. in India and commonly occurs under wet land conditions. It is widespread in North and Eastern regions and is severe problem in Orissa, Bihar, West Bengal and parts of Uttar Pradesh. The disease is favoured by high relative humidity, night temperature of 20–22°C and day temperature of 25–28°C. The disease appears on the foliage, first as purple to brownish circular water-soaked lesions 1–2 cm in diameter, usually at the tip, base and margins of the leaves. These lesions gradually enlarge by coalesce small spots and the entire leaf dies. As the disease progresses, the spots coalesce and form characteristic rings of yellow or brown colour. Eventually, the whole leaf may be affected and die. The losses caused due to this disease range from 25% to 50%.

Control

- A number of approaches has been used to control the taro leaf blight. Agronomic methods that have given partial success include careful choice of planting material, planting at high density, intercropping taro with other crops rather than growing it as a sole crop and long crop rotation with non-host crops.
- Planting of early maturing varieties during May for rain-fed crop and during February/March for irrigated crop found useful to escape from the disease infection.
- Field sanitation by removal of infected leaves has been found useful in controlling the disease but it is extremely laborious.
- Drying seed-corms for 2 to 3 weeks before planting kills most of the spores of *Phytophthora colocasiae.*
- When young plants are affected by this disease the infected leaves are rogued out to control the further spread of infection. The rogued leaves should be burnt and care should be taken to avoid spreading of spores from diseased leaves to healthy ones.
- In severe cases, the disease can effectively be controlled by spraying copper fungicides, particularly copper oxychloride or blue copper 0.3%, folpet 0.2%, Dithane M-45 0.2%, Benlate 0.1%, or Dithane Z-78 0.2% at an interval of 21 days for about 4 months starting from 6 weeks after planting.
- Use tolerant varieties like Muktakesi, Jankhari, Nadia local or Topi.

BROWN SPOT (*Cladosporium colocasiae*)

Brown spot or ghost spot is a fungal disease of older leaves. It is called ghost spot as the lesions are often less evident on opposite surface of the leaf. The disease is characterized by appearance of numerous reddish-brown spots on leaves, which are more prevalent under humid conditions. High humidity helps in the spread of disease.

Control

- Cultural operations like field sanitation and crop rotation can be followed to control the disease.
- Spray copper fungicides like blue copper or copper oxychloride 0.3% at an interval of 21 days for about 4 months starting from 6 weeks after planting.
- Similar chemical treatments, which are used to control blight, can also be used to control brown spots.

Soft Rot (*Pythium* spp.)

Several species of *Pythium*, which are soil borne, infect the roots and corms. Soft rot is also called corm rot or root rot. When the plant is infected at early stages it shows chlorosis of the leaves and the affected plants gradually wilt. The rotting of corm starts at the base and affects the whole corm. The tissues become soft, dark in colour and emit a foul odour. The growth of fungi is encouraged by warm humid conditions coupled with poor soil conditions. Physical damage to the plants during intercultural operations or by any other means is favourable for the spread of disease. Over-mature corms left in the field help in the proliferation of fungi. The fungi can also survive on disease debris.

Control

- Field sanitation, crop rotation and use of disease-free planting material are some of the cultural practices to control the disease and its further spread.
- Dipping the corms in copper fungicides solution @ 2 g/litre for half an hour before planting is beneficial. Broadcasting captan 10 kg/ha in soil one week before planting or soil drench with captan 2 g/litre controls the disease significantly.

Sclerotium Rot (*Corticium rolfsii*)

Infected plants gradually withered, a white mycelial mat appeared and numerous sclerotia developed on the surface of petioles near the soil line. The heavily infected petioles start rotting, and eventually, the entire plant dies. The rotten tissues turn soft, brown and stringy, which makes the corms unfit for market.

Control

- Cultural practices like crop rotation with non-host crops for long duration, timely harvesting of crop and field sanitation by destroying the infected plants can control the disease to a certain extent.
- Dipping the corms in carboxin solution (50 mg/litre of water) for 30 minutes and drenching the soil with Dicloran 50 mg/litre of water effectively control the disease.

Dasheen Mosaic Virus

Dasheen mosaic virus is prevalent in most of the taro-producing countries in the Asia-Pacific region. It is caused by a stylet-borne, flexuous, rod-shaped virus, which is spread by aphids (*Aphis gossypii*). Chlorotic and feathery mosaic patterns on leaf, distortion of leaves and stunted plant growth are characteristic symptoms of the disease. The disease is not lethal but yield is depressed.

Control

- Use of virus-free healthy planting materials.
- Roguing and burning of infected plants are the primary steps to control the disease.
- Apply dimethoate @ 5 ml/litre of water to control the virus vectors, *i.e.*, aphids.

13 Giant Taro

M.K. Rana and Archana Brar

Botanical Name	: *Alocasia macrorrhizos*
	Alocasia indica
	Alocasia cucullata
Family	: Araceae
Chromosome Number	: 2n = 28

ORIGIN AND DISTRIBUTION

Giant taro is native to rain forests from Malaysia to Queensland and has long been grown on many Pacific islands and elsewhere in the tropics. Giant taro is commonly known as elephant ear taro, and, in different Polynesian languages, it is known as both kape and ape, while in Australia, it is known as *cunjevoi*. *Alocasia* species are commonly found in Samoa, Tonga and other parts of Polynesia. It is thought to have originated in Southern or Southeast Asia and to have been spread to Oceania through the Island of New Guinea many centuries ago. It has evolved with the ethnicity of the Asiatic and Pacific region. It has acquired substantial socio-cultural importance for the people of these regions. Among the vegetable crops in Oceania, the reverence and status appended to giant taro is equaled only by yam in certain regions.

INTRODUCTION

Giant taro is a colossal plant with characteristic leaves and is valued as an ornamental plant. Its swollen underground stems are used as food and animal fodder. It is grown as an ornamental plant, for its edible corms and a conventional starch staple in tropical regions. The corms are edible if cooked properly but its sap is skin irritant due to the present of calcium oxalate crystals or rap hides, which are needle like structures. The leaves and roots of some wild types have been reported to contain oxalate crystals, which cause irritation in the mouth. The farmers use its leaves and roots as animal feed. Giant taro growing naturally in moist and partially shaded areas gives a high yield of above-ground biomass. The heart-shaped leaves make umbrellas in tropical downpours. The new leaves are produced at 2- to 3-week intervals, eventually dying and falling to the ground after 15 days. By far, the largest quantity of giant taro produced in the Asia and Pacific region is utilized starting from the fresh corms or cormels. Its major producers in Asia are China, Japan, Philippines and Thailand, while in Oceania, Papua New Guinea, Samoa, Solomon Islands, Tonga and Fiji are the countries dominating in it production.

The root and leaves have a finicky flavour, which gives an acrid taste due to the presence of calcium oxalate crystals; however, cooking destroys their acridity. However, the cultivated cultivars are almost free from calcium oxalate crystals. Its main economic parts are the corms, cormels and leaves. The size of its starch particles is small, thus, it is easily digestible. This characteristic makes it a suitable food for infants allergic to wheat starch and people suffering from stomach disorders; however, this feature of giant taro starch makes it less suitable as a source of industrial starch. The base of the corm contains more starch content as compared to its apex, while the protein content is higher towards the corm's periphery than towards its centre, thus, care should be taken while peeling the corms; otherwise, a great amount of protein is lost in the peel. The high protein content of the leaves favourably complements the high carbohydrate content of the corm.

COMPOSITION AND USES

COMPOSITION

Giant taro is a rich source of calcium, phosphorus, iron, thiamine, riboflavin, niacin and vitamin C, which are important constituents of human diet. The fresh giant taro lamina dry matter is about 20%, while the fresh petiole dry matter is only 6%. Its corms contain a range of minerals and vitamins, which are listed in Table 13.1.

USES

Leaves, young shoots and corms of giant taro are eaten as food. Its corms and cormels are boiled, baked, roasted, fried, barbecued, or cooked in curries and consumed in conjunction with fish, coconut preparations, etc. Leaves are usually not eaten but some of the ethnic people cook them in various ways mixed with condiments. In Colombia, its roots after chopping and cooking are usually fed to chickens and pigs, while some of the farmers use its leaves as chicken and fish feed. Giant taro peels and wastes are fed to domestic livestock. In Hawaii, efforts have been made to prepare silage from its tops, which are left after the harvesting of corms.

The processed forms of giant taro are not common in the Asia and Pacific region but the most widespread forms of giant taro are the chips prepared by peeling the corm, washing, slicing into thin pieces and blanching. The pieces are then fried in vegetable oil, allowed to cool and then packed. Although giant taro chips are made in most of the countries but their availability is sporadic since they are prepared in small quantity. Another processed form of giant taro is poi, a sour paste, which is made after boiling; however, its production and utilization are quite restricted, mainly in the Hawaiian Islands. In some places, its flour is used as a thickener for soups and other preparations.

TABLE 13.1
Nutritional Composition of Giant Taro (per 100 g Edible Portion)

Constituents	Contents	Constituents	Contents
Water (g)	70–81	Iron (mg)	0.8–1.2
Carbohydrates (g)	17–27	Vitamin A (SI)	0.1
Protein (g)	0.6–3.3	Thiamine (mg)	0.09–0.18
Fat (g)	0.1–0.2	Riboflavin (mg)	0.02–0.04
Dietary fibre (g)	0.60–1.18	Niacin (mg)	0.5–0.9
Calcium (mg)	46–153	Ascorbic acid (mg)	15–17
Phosphorus (mg)	45–72	Ash (g)	1.2–1.5
Calcium oxalate in skin (mg)	451	Energy (kcal)	102–105.12
Calcium oxalate in centre (mg)	84–182		

MEDICINAL USES

The giant taro roots and leaves are used medicinally in some of the countries. In Hawaii, it has been used to treat digestive complaints and as a topical dressing for burns. In the Philippines, the leaf stalks are used to relieve toothache, and in Java, the roots and leaves are used to relieve pain and redness. The rhizomes contain an anti-fungal protein called *alocasin*.

BOTANY

Giant taro is an herbaceous plant, which grows to a height of 1–2 m. The plant consists of a central corm, lying just below the soil surface, from which the leaves grow upwards and roots downwards, while the corms, cormels, daughter corms and runners grow laterally. Its root system is adventitious lying mainly in top 1 m soil. Corms, cormels and daughter corms are quite similar in their internal structure. The outmost thick brownish layer is periderm, within which, the starch-filled ground parenchyma lies. Vascular bundles and laticifers ramify all through the ground parenchyma. The idioblast cells containing bundles of calcium oxalate crystals (raphides) also occur in the ground tissue and almost in all other plant parts. Raphides are linked with acridity or mouth irritation. The density and hardness of the corm increase with age.

Leaf is made up of an erect 0.5- to 2-m-long petiole and a 20- to 50-cm-long lamina. The petiole is flared out at its base, where it is attached with corm, so that it may be well griped around the corm apex. The petiole at its base is thickest and thinner towards its attachment to the lamina. From inside, the petiole is spongy in texture and has abundant airspaces, which facilitate gaseous exchange when the plant is specifically grown under swampy or flooded conditions. In most of giant taros, the petiole attachment to the lamina is peltate, meaning that the petiole is attached, not at lamina edge but at some point in the middle. This peltate leaf attachment makes the giant taro different from tannia, which has a hastate leaf, *i.e.*, the petiole is attached at the lamina edge. The leaf lamina is oblong-ovate with rounded basal lobes. It is entire, not the serrated, glabrous and thick. The main three veins radiate from the point of petiole attachment, one going to the apex and one to each of the two basal lamina lobes. Some important veins arise from these three main veins but the overall leaf venation is reticulate.

Giant taro flowers occasionally but flowering can be induced artificially by applying gibberellic acid. The inflorescence arises either from the centre of unexpanded leaves or from the leaf axils. Each plant bears more than one inflorescence, which is made up of a short peduncle, a spadix and spathe. Botanically, spadix is a spike with fleshy central axis, with which, the small sessile flowers are attached. The spadix is 6–14 cm long with female flowers at the base, male flowers towards the tip and sterile flowers in between in the region compressed by the spathe neck. The spadix tip produces no flower, thus, called the sterile appendage. The spathe is a 20-cm-long yellowish bract, which sheathes the spadix. The lower part of the spathe tightly wraps the spadix and completely occludes the female flowers. The top portion of the spadix is rolled inward at the apex but is open on one side to reveal the male flowers on the spadix. The top and bottom portions of spadix are separated by a narrow neck region, corresponding to the region of sterile flowers on the spadix.

Pollination in giant taro is probably performed by flies. Under natural conditions, it produces fruits and seeds rarely; if it produces, only the lower portion of the spadix may produce fruits. Each fruit is a berry measuring 3–5 mm diameter and containing numerous seeds. In addition to the embryo, each seed contains a hard testa and endosperm.

CLIMATIC REQUIREMENT

Giant taro thrives best under very wet or flooded conditions. Generally, a rainfall or irrigation of 1,500–2,000 mm is required for obtaining economic yield since the dry conditions result in reduced yield and dumbbell-shaped corms. It requires a daily mean temperature above 21°C for its normal production. It cannot tolerate frost. Because of somewhat temperature sensitivity, it is essentially a lowland crop. At high altitudes, it tends to give very less yield. It seems shade-tolerant more than most of the vegetable crops, meaning that its reasonable yield can be obtained even in shade where other crops completely fail to give yield. However, its maximum yield can be obtained under full sunlight. This peculiar characteristic enables giant taro to fit into exclusive intercropping systems even in orchards. Its growth and development are also influenced by the day length. The induction of corms and cormels is stimulated by short days, while flowering is stimulated by long days.

SOIL REQUIREMENT

Giant taro can successfully be cultivated on soils, in which, flooding and water logging can occur since its plants are able to transport oxygen down to the roots through aerial parts under flooding soil conditions, which enable the roots to respire and grow normally in flooded and oxygen deficient soils. However, the flooded giant taro fields in practice must be aired sporadically in order to avoid iron and manganese toxicity under such soil conditions. The soil pH 5.5–6.5 is optimum for its cultivation. The most advantageous characteristic of this plant is that some of its cultivars can successfully be cultivated in saline soils.

After removing weeds, the field is prepared by ploughing repeatedly followed by planking to ensure a friable soil texture. If the soil is heavy textured, the field is ripped deeply to improve root penetration. Thereafter, the mounds are prepared in wet coastal tropical regions and in areas where water logging occurs.

CULTIVARS

Over 200 cultivars of giant taro are available but only a few of the cultivars are in vogue. The cultivars may be distinguished by plant height, corm size, sucker production, leaf shape and colour of leaf and stem. Cultivars also differ in hard or softness of corms and duration required for crop maturity. However, the growers understand the basic difference in appearance, texture and flavour of the corms of diverse cultivars, as these characteristics of corms influence the marketing success. Another grouping of cultivars is based on corm size whether it is *jap* giant taro or small corm type. Some of the important cultivars are as follows:

SATOIMO: This is the small corm-type cultivar.

ISHIKAWA WASSE: This is a leading variety in Japan, the corm of which retains firm texture even after cooking. This type of cultivar is generally preferred by the people of Pacific and Torres Strait origin.

TOAKULA: This is the cultivar imported from Fiji and grown only on small areas.

PAPUA: This is New Guinea cultivar grown by Torres Strait Islanders but mainly consumed locally.

SAMOAN PINK: The corms of this cultivar are pink in colour.

PLANTING TIME

Giant taro, being a crop of the tropical regions, can be planted at any time of the year. However, with a definite alternation of dry and wet seasons, it is generally planted at or just before the beginning of wet season.

PROPAGATION

Since flowering is rare in giant taro, it is propagated vegetatively by setts, which consist of the lower 30–40 cm of the leaf stalk together with top 1–3 cm of sucker corms or full corms. Healthy and vigorous medium to large setts from the mother plant usually give higher yield as compared to smaller setts. The larger setts also grow faster, and hence, they smother the weeds better. The setts to be used for planting must be clean without any discoloration or blackening and be free from mould and soft spots.

PLANTING MATERIAL

Actually, four types of planting material are essentially used for the planting of giant taro:

- **Side suckers:** These are produced because of lateral proliferation of main plant in the previous crop.
- **Small corms:** These are unmarketable corms resulting from the main plant in the previous crop.
- **Huli:** This is 1–2 cm apical portion of the corm with 15–20 cm basal portion of the petiole attached. Using *huli* is advantageous, as it is not the edible material. Besides, *huli* establishes very quickly and results in vigorous plant. Nevertheless, *huli* is best adapted to situations where planting is done soon after harvesting since extended storage of *huli* is not at all advisable.
- **Corm pieces:** These are the pieces made from large corms by cutting. However, it is advisable to pre-sprout the pieces in a nursery before they are planted in the field, which enables the sprouts to appear on the pieces before they are taken to the field. Small corms and side suckers may also be retained in the nursery to develop good sprouts, particularly when the time between the previous harvest and the next planting is long.

Availability of planting material is a major problem in the production of giant taro, especially where occasional drought reduces the quantity of planting material. The following methods can be used for the rapid multiplication of planting material:

- The first method is to use a mini-sett technique, in which small corm pieces of 30–50 g are protected with seed dressing. The pieces are first allowed to sprout in the nursery and then planted in the field. The resulting small corms and suckers are used as subsequent planting material.
- The second rapid method of generating planting material is tissue culture, as thousands of plantlets can be generated in a few months through meristem culture but it has not yet become a commercial method for the generation of giant taro planting material.
- The third method is to use true seed of giant taro for planting but the infrequent nature of flowering is a big problem in giant taro. Although one successful crossing in giant taro can produce hundreds of seeds but the smallness of the resulting seedlings is likely to be a problem due to segregation in subsequent generations.

PLANTING METHOD

Giant taro can be planted in single rows, twin rows, or in plots. The holes for planting are made slightly larger than the corm size, usually 10–20 cm, depending on corm size. The other options are to prepare ridges and furrows, or flat seedbeds for planting corms. Planting the corms shallow results in developing corms above the ground and the exposed corms are more likely to be damaged by insect-pests and rodents. The corms are generally planted @ 13 corms/m^2 with a planting density of 10,000 to 45,000 plants per hectare. Higher-density plantation produces more but smaller corms and average yield per plant is reduced. In single row spacing method, the spacing is kept 60–160 cm between rows and 50–60 cm between plants. In twin rows method, the spacing is maintained 80–260 cm between rows (centre to centre) with 30–60 cm between the twin rows and 20–50 cm between plants. However, a single row spacing of 90 × 60 cm is common and a slightly wider spacing can be maintained where cloud cover is frequent and growing conditions are not optimum. Very wide spacing is not desirable, as it increases the weed intensity.

PRODUCTION SYSTEMS

Usually, two production systems, *i.e.*, (i) wetland cultivation and (ii) dry land cultivation, are used for the cultivation of giant taro, as follows:

- **Wetland cultivation:** This method is used where abundant irrigation water sources like rivers and stream are available and the level of water can be controlled or the land is swampy in nature. Wetland cultivation is restricted only to certain locations where the economics of production and water availability permit the system to thrive. The soil must be deeply clay and properly leveled so that it may allow the impounding of water without much loss through percolation. The field is puddle to retain water and flooded just before or after planting. The water level in starting is kept low but it is raised gradually as the season progresses so that the base of the plant may remain in water constantly. Occasionally, water from the field is drained out to apply fertilizers but re-flooded after 2–3 days. Giant taro is one of the few crops that can be grown successfully under flooded conditions. The large airspaces present in the petiole allow the flooded parts to maintain gaseous exchange. The water, in which, giant taro is growing should be cool and flowing incessantly so that it can have a maximum of dissolved oxygen since warm stagnant water results in low oxygen content and causes decaying of giant taro basal portion. Growing giant taro under controlled flooding is more advantageous than dry land production, as the corm yield is much higher in wetland system of cultivation, weed infestation is less and off-season production is possible, resulting in very attractive prices. However, wetland production needs longer duration to be mature and involves higher infrastructure investment and operational cost.

- **Dry land cultivation:** This method of cultivation is used where irrigation water is not available abundantly. In the Asia and Pacific region, giant taro is largely grown in dry lands, essentially depending on rains. Sprinkler or furrow irrigation is used as substitute of rains to keep the soil moist, not to get the field flooded. However, in rain-fed giant taro cultivation, time of planting is critical. The crop will usually be planted only at the onset of rainy season, and the rainy season must last long enough, *i.e.*, 6–9 months, to permit the giant taro crop to be mature. After applying farmyard manure, the field is prepared as usual by ploughing repetitively with harrow followed by planking. If the soil is deep and loose, the crop can be grown in flatbeds otherwise, the crop is planted on ridges. Ridges are usually made 70–100 cm apart and plant spacing is maintained 50–90 cm. Although dry land giant taro is frequently grown as intercrop but it is also grown as sole crop. Giant taro in dry land is planted by opening up the soil with spade, inserting the planting pieces and covering them with soil. Giant taro in dry lands matures earlier than flooded giant taro but the yield is comparatively low.

NUTRITIONAL REQUIREMENT

Giant taro responds well to the application of manure and fertilizers, showing improved corm size and number. The crop needs higher doses of potassium, as it is essential for the synthesis of starch. Liming is necessary in acid soils to increase the availability of nutrients. Nitrogenous fertilizers should be applied after developing full canopy, as it affects the eating quality and palatability of corms' however, their excessive use reduces the quality and storability of corms. The types of fertilizer and their quantity vary widely with the growing region. Usually, it is better to apply nitrogenous fertilizers in splits. The first dose is applied at the time of last ploughing, which will promote early plant establishment and leaf expansion, while the second dose is applied 3–4 months later when the corms start enlarging. The plant is able to develop beneficial relationship with vesicular arbuscular mycorrhizae, which facilitates nutrients absorption, especially phosphorus. In flooded fields, it is useful to culture *Azolla* in standing water, which increases the supply of nitrogen to giant taro plants, which are sensitive to certain nutrients deficiency and develop deficiency symptoms. Potassium deficiency causes death of the roots and chlorosis of the leaf margins and zinc deficiency results in interveinal chlorosis, while the phosphorus content in leaf petiole below 0.23% causes the older leaves to turn blue but not yellow.

IRRIGATION REQUIREMENT

Giant taro is one of the few crops that can adapt to a range of moisture regimes. Overwatering is not harmful to its growth; however, supplementary irrigation is necessary during dry period, as the giant taro plant growth is restricted and corms quality is reduced if plants are cultivated under water stress. Water can be applied to giant taro by sprinkler, furrow, or drip irrigation.

INTERCULTURAL OPERATIONS

HOEING AND WEEDING

In flooded giant taro, weed infestation is not a problem but some aquatic weeds may grow, thus, pulling out such weeds manually is essential to reduce completion with crop plants for nutrients, moisture, space and light. The herbicide nitrofen 3–6 kg/ha may be added to the irrigation water. In dry-land giant taro, weed control is necessary only during first 3 months, if the crop spacing has been kept close enough. Thereafter, the crop canopy becomes so large that further weed control is not necessary. In the last 2 months, the average plant height becomes less and spaces between plants open up again. Thus, the weeds may reappear but their potential for economic damage is very low. Hoeing with spade is the most widespread practice in dry land giant taro. Giant taro roots are very shallow and can be damaged very easily by deep hoeing. Hence, hoeing in giant taro should be done shallow to avoid damage to the roots. In dry land giant taro, chemical weed control is very effective. The herbicides to be applied are Promtryne 1.2 kg, Dalapon 3 kg, Diuron 3.4 kg, or Atrazine 3.4 kg/ha.

MULCHING

Covering soil surface with organic mulch, especially in dry land taro field, is useful operation, provided the soil is not already too high in organic matter. As the organic mulch breaks down, it improves growth rates by providing nutrients, improving soil structure and conserving soil moisture.

EARTHING UP

Earthing up of soil around the plants base is advisable during hoeing to provide more space to the growing corms for better development and to protect them from sun exposure, which may deteriorate the quality of the corms.

HARVESTING

The maturity in dry land giant taro is indicated by a decline in plant height and a general yellowing of leaves. Similar indication is observed in flooded giant taro but leaf senescence is only partial since the root system of flooded giant taro remains alive and active due to continuous and abundant water supply. In dry land, the crop needs 5–12 months, and in wetlands, 12–15 months, depending on cultivar and the prevailing conditions during the growing season. Harvesting is largely done by manual means. The soil around the corms is made loose with the help of spade and the corm is pulled up by clutching the petioles base. In wetlands, harvesting is more tiresome since it is difficult to cut the living corm roots, which remain affixed with soil. Even in mechanized systems of farming, harvesting is mostly done by hand, thereby increasing the labour cost.

YIELD

In general, the wetland giant taro yields comparatively more than the dry land giant taro. The average yield of giant taro varies from 62 to 126 q/ha, depending on variety, growing system, climatic conditions and package of practices adopted by the growers during crop cultivation.

POST-HARVEST MANAGEMENT

The corms are graded according to size, depending on the consumers' preference in the market. The corms of 1–1.5 kg weight are preferred more, depending on variety. The cut flesh should look and smell fresh and juicy. Only quality corms with no rot or other blemishes should be packed and sent to the market. The corms of giant taro cannot be stored well for more than a few days at ambient room temperature, as they have high water content. The corms of giant taro should be consumed within 2 weeks. If not, the post-harvest losses in giant taro will be too high. Preparation of value-added durable products could be a partial solution. Its life may be extended by harvesting the corms with 5 cm green tops and putting them in cold storage.

INSECT-PESTS

GIANT TARO BEETLE (*Papuana woodlarkiana, Papuana biroi, Papuana huebneri* and *Papuana trinodosa*)

The adult beetle is black, shiny and 15–20 mm long. Its many species have a horn on the head. The adult beetles fly from breeding sites to the giant taro field, make tunnels into the soil just at the base of the corm and feed on the growing corm, leaving large holes, which degrade the market quality of corm. The wounds are later attacked by decay-causing organisms. Its feeding can cause wilting and even death of the affected plants. After feeding for about 2 months, the female beetle shifts to neighbouring bushes to lay eggs 5–15 cm beneath the soil close to a host plant. After hatching, the larvae feed on plant roots and leave a dead organic matter at the base of the host plants.

Control

- Follow crop rotation with cereal crops.
- Use clean planting material (free from soil, grubs and beetles).
- Destroy breeding sites surrounding the crop.
- Use polythene, coconut husk or grass mulch to cover the soil surface to avoid its entry.
- Use biological control, *i.e.*, a fungus (*Metarhizium anisopliae*), a bacterium (*Bacillus popilliae*) and the protozoa (*Vavraia*).
- Spray the crop with imidachloprid (Confidor) or cypermerthrin (Mustang) 0.3 ml/litre of water at 10 days interval.

APHIDS (*Aphis gossypii*)

Aphids are tiny pear-shaped insects with soft fragile bodies. They pierce leaves to suck cell sap. If aphids are present in high number and rainfall is low, leaves senesce faster than normal. In severe cases, the plants wilt and may become stunted. Indirect damage is caused by the accumulation of honeydew produced by the aphids. Honeydew serves as a substrate for sooty moulds. The older leaves become distorted and covered in honeydew, which encourages the growth of sooty mould, which blacken the leaves, reducing photosynthesis and plant vigour. The presence of ants often indicates that aphids are also present. Aphids are vectors of Dasheen mosaic potyvirus.

Control

- Avoid planting new crops downwind, as aphids are not strong fliers.
- Destroy the heavily infested leaves.
- Introduce predators such as ladybird beetles, syrphids and lacewings, in particular.
- Spray the crop with horticultural oils and insecticidal soaps as alternatives to synthetic pesticides.
- Spray the crop with 0.05% Fernos or Metasystox, 0.075% dimethoate, or 0.05% quinalphos at 10- to 12-day intervals.

TARO ROOT APHID (*Patchiella reaumuri*)

The giant taro root aphid infests the roots, and when population is high, it infests lower parts of the petioles. The taro root aphid feeds on the roots of taro, causing them to rot. Damage to the roots produces stunted plants with yellowish small leaves. It effect is greater on young plants than on mature plants, and worse on all stages during periods of drought. Infestations appear as a white mould on the fibrous taro roots. Since it is on the roots, it may not be detected until aboveground symptoms occur. The base of the petioles, corm and roots should be inspected to see if masses of white, cottony, waxy threads are present (these cover the insects). The presence of ants on the soil and lower parts of taro can be an indication of its presence.

Control

- Follow quarantine measures strictly in order to prevent further spread.
- Follow long crop rotation with non-host crops.
- Plough the field deep and leave it fallow for a year.
- Destroy the crop if infestation is heavy.
- Spray insecticidal soaps (1% active ingredient) to control its infestation.
- Spray the crop with 0.05% chlorpyriphos or Metasystox, 0.06% Nuvacron, or 0.15% Carbetox at 10- to 12-day intervals.

ARMYWORM (*Spodoptera litura*)

Armyworm is the main giant taro insect pest. Outbreaks are sporadic and vary in location and intensity. A female moth lays 200–300 eggs in a cluster on a leaf. Eggs hatch in 4–8 days. Larvae feed alongside on the leaf, and within a few days, only the skeleton-like leaf veins may remain.

Control

- Practice long crop rotation with non-host crops.
- Use tolerant varieties, if available.
- Use grass barriers around field.
- Preserve and encourage the natural enemies, *e.g.*, spiders, lacewings, ladybugs, wasps.
- Grow few lines of *Coleus blumei* around the giant taro field, as this plant emits an odour that repels the adult moth.
- Use polythene, coconut husk, or grass mulch to cover the soil surface to avoid its entry.
- Handpick and destroy when infestation is light.
- Remove and burn the heavily infested material.
- Spray the crop with chlorpyriphos, monocrotophous or azinphos-ethyl 0.05%, or methamidophos 0.06% at 10 to 12 days intervals.

Taro Plant Hopper (*Tarophagus proserpina*)

The female plant hopper cuts slits in giant taro stems and lays two eggs in each slit. There are two adult forms, short- and long-winged. The latter is most common during cool weather. Long-distance spread is by the long-winged form. Two serious giant taro virus diseases are transmitted by the plant hopper.

Control

- Grow few lines of *Coleus blumei* around the giant taro field to discourage plant hoppers.
- Plant the crop in clean fields to delay the onset of an attack. Some farmers torch the insects
- Introduce mired bug (*Cyrtorhinus fulvis*) to controg plant hoppers.
- Spray the crop with 0.4% *neem* oil, malathion, fenitrothion, or fenthoate 0.05% at 10- to 12-day intervals.

Spider Mites (*Tetranychus spp.*)

Spider mites are difficult to see with the naked eye. Often, their infestation can be detected by the presence of webbing but this can be seen with the naked eye only when mites are present in large numbers. Their presence is indicated by tiny spots on the upper leaf surface. They are reddish-brown to yellowish-green, depending on the species. Spider mites suck sap from the leaf tissue, resulting in characteristic white to pale yellow speckling, usually between the main veins. When infestations are severe, the speckling is seen all over the leaf. Webs occur on the underside of the leaves. As the infestation advances, the leaves turn yellow and die prematurely. Spider mite outbreaks commonly occur on water-stressed plants.

Control

- Preserve the natural enemies.
- Use the predatory mite *Phytoseiulus persimilis* as bio-agent in the ratio of one predator to four pests.
- Irrigate the crop regularly to suppress mite populations.
- Spray the crop with miticides hexythiazox, Sanmite, spiromesifen, or bifenazate formulation 22.6% a.i. 0.3 g/litre of water a single time.

Spiralling whitefly (*Aleurodicus disperses*)

The damage is caused by piercing the leaf and sucking cell sap, which leads to premature death of the plant when infestation is high. Indirect damage is caused by the accumulation of the honeydew and the waxy, white, fluffy, woolly material produced by the whiteflies. Honeydew serves as a substrate for sooty moulds, which blacken the leaf, retarding photosynthesis and reducing plant health. It lays its eggs in distinctive spiral patterns on underside of the leaves, as its name suggests.

Control

- Follow effective quarantine measures to prevent its spread.
- Use parasitic wasp (*Encarsia haitiensis*) to control spiralling whitefly efficiently.
- Spray insecticidal soaps (1% active ingredient) to control its infestation.
- Dip the corms in 0.1% solution of imidachloprid.
- Incorporate Furadon @ 1.5 kg a.i./ha in the nursery 15 days before uprooting the seedlings.
- Use predator *Chrysoperla carnea* (25,000 larvae/ha) + spray Econeem 0.3% (0.5 litre/ha) three times at 15 days interval starting from 45 days after planting.
- Spray the crop with dimethoate or Monocrotophos 0.02%, Metasystox 0.15%, or Confidor 0.3 ml/litre of water at 10 days interval during vegetative stage

Tobacco Whitefly (*Bemisia tabaci*)

The tobacco whitefly is commonly seen on giant taro but it is not considered a major pest, as it does not transmit viruses to giant taro. Its attack is rarely large enough to cause feeding damage and symptoms rarely occur on giant taro since the number of whitefly is usually insufficient to cause direct damage, although they may increase wilting in times of drought. The adults and nymphs suck the sap from leaves, which can cause yellowing, wilting and early plant death. However, the buildup of sooty moulds on honeydew deposits is not frequently seen. Whiteflies are usually detected by examining the undersides of leaves and searching for the tiny yellow-cream, scale-like instars or nymphs. The nymphs also occur occasionally on the upper surface of the leaves and can range from being scattered widely to forming dense clusters. Shaking the plant may disturb the small white adults, which fly out and quickly resettle.

Control
- Adjust sowing time to avoid the attack of whitefly.
- Limit the irrigation frequency to reduce the population of whitefly.
- Treat the corms by dipping in 0.1% imidachloprid solution.
- Apply Furadon @ 1.5 kg a.i./ha in the nursery 15 days before uprooting the seedlings.
- Use predator *Chrysoperla carnea* (25,000 larvae/ha) + spray Econeem 0.3% (0.5 litre/ha) three times at 15 days interval starting from 45 days after planting.
- Spray the crop with dimethoate or Monocrotophos 0.02%, Metasystox 0.15%, or Confidor 0.3 ml/litre of water at 10 days interval during vegetative stage

Lesion nematode (*Pratylenchus coffeae* and *Meloidogyne incognita*)

The root lesion nematode is commonly found associated with the roots of giant taro. It has consistently been found associated with root decay, reduced number of cormels and stunting and death in taro. The problem appears to be exacerbated by continuous cropping. It is not clear whether this nematode damages taro in Pacific island countries sufficiently to cause above-ground symptoms. It is reported to cause localised necrosis (darkened areas of dead tissue) on roots and corms.

Control
- Follow quarantine measures strictly in order to prevent further spread.
- Follow long crop rotations with non-host crops.
- Use planting material free from soil and roots.
- Dip the planting material in hot water (50°C) for 40 minutes before planting.
- Grow cover crops such as marigolds produce chemicals that are toxic to nematodes.
- Remove the necrotic tissues from the tops (approximately 40 cm of petiole and 1–2 cm of corm) and suckers, which should then be washed with running water and allowed to dry before planting.
- Apply nematicides, *e.g.*, dazomet, oxamyl and fenamiphos, before planting to reduce nematode.

DISEASES

Corm rot (*Athelia rolfsii*)

This is a soil borne fungus, which infects giant taro at soil level, causing the corms and roots to decay and leaves to wilt. Infection starts from base of the petioles. The disease is characterised by the presence of more dead leaves than normal. Pale cream to reddish-brown 1- to 2-mm-size sclerotia is usually present at the site of infection. The fungus also causes a post-harvest pinkish corm rot, infecting corms through wounds made when suckers are detached from the plant. In giant taro plants with wilted leaves, the base of the petioles at soil level should be inspected for white mycelia and sclerotia.

Control

- Strictly follow crop rotation with cereal crops to reduce numbers of infective propagules in a field.
- Plough the field deep in hot summer months to expose the propagules to the atmosphere.
- Cover the moist soil with clear polyethylene film and allow the sun to heat the trapped air.
- Add lime into the soil to raise the pH.
- Remove and destroy the infected plants by burning.
- Apply *Trichoderma* spp. to inhibit the fungal growth.
- Dip the planting material in 0.2% copper fungicides solution for half an hour before planting.
- Broadcast captan 5 kg/ha in soil 1 week before planting.
- Drench the field with captan 2 g/litre of water.

BROWN LEAF SPOT OR GHOST SPOT (*Cladosporium colocasiae*)

This is a fungal disease of older leaves. Lesions are often less evident on opposite leaf surface thus called ghost spot. The pathogen causes reddish-brown circular or irregular diffuse spots or blotches on either side of the leaf, sometimes with dark, diffuse centres. However, the spots are usually less evident on the opposite surface of the leaf. Sometimes, the spots are surrounded by a yellow halo or have a dark brown, diffuse border. Spots can be up to 15 mm in diameter but are usually much smaller when there are many spots on a single leaf. These become darker with the onset of sporulation.

Control

- Follow long crop rotation with non-host crops.
- Use disease-free healthy planting material.
- Follow quarantine measures strictly in order to prevent further spread.
- Maintain phyto-sanitary conditions in the field.
- Remove and destruct the infected leaves and plants by burning to reduce the inoculums.
- Spray the crop with blue copper or copper oxychloride 0.3% at 10- to 12-day intervals.

GIANT TARO LEAF BLIGHT (*Phytophthora colocasiae*)

The disease appears as purple-brown water-soaked lesions on leaves, which ooze a clear yellow liquid. These lesions later enlarge, coalesce together and eventually destroy the entire lamina in 10–20 days. Water droplets on older leaves, high temperature and high humidity are conducive to the onset and spread of the disease and germination of the spores. The disease spreads from plant to plant through wind and splashing rain. Spores stay alive in planting material for 3 or more weeks, thus, the infected planting material is one of the common means of spreading the disease over long distances and from season to season.

Control

- Follow long crop rotation with non-host crops.
- Use disease-free healthy planting material.
- Dip the corms in bleach (1% sodium hypochlorite) and store them in polyethylene bags.
- Plant the crop at wider spacing to maintain high plant density.
- Intercrop giant taro with other crops rather than growing it as a sole crop.

- Collect and destroy the infected leaves.
- Spray the crop with systemic fungicide blue copper 0.3%, folpet 0.2%, Ridomil MZ 72, Benlate 0.1%, Dithane Z-78 0.2%, or metalaxyl or mancozeb 0.2% at 10- to 12-day intervals.

Bacterial Soft Rot (*Erwinia chrysanthemi*)

This is a bacterial disease, which causes a soft rot of corms in the field as well as in storage. In field, infection causes a foul-smelling, creamy-white corm soft rot and plants wilt suddenly. A similar rot occurs in harvested corms stored at high temperature and high humidity. Collapsing of mature leaves is often indicative of bacterial soft rot of the corm. At this stage, corms are usually so decayed that plants can topple over in the wind. In storage, in soil-pits or plastic bags, the bacterium can be detected by the presence of soft rot with a strong, unpleasant smell.

Control

- Follow long crop rotation with non-host crops.
- Always use the disease-free healthy propagating material.
- Maintain low temperature and low humidity in storage.
- Treat the corms with bleach (1% sodium hypochlorite) before storage.
- Spray the crop with copper-based bactericides initiated 2 weeks before corm initiation and continued at 5- to 10-day intervals.

Alomae Virus Disease

The disease is caused by a complex of two or more viruses, *i.e.*, (i) giant taro large bacilliform virus, which is transmitted by grasshopper (*Tarophagus proserpina*), and (ii) giant taro small bacilliform virus, which is transmitted by mealy bug (*Planococcus citri*). These viruses are neither transmissible by mechanical means nor their host range, which is limited to aroids only. The disease occurs when these two viruses attack together. The presence of TLBV alone results in bobone, a milder form of the disease. Alomae appears as a feathery mosaic on leaves. Lamina and veins become thick. The young leaves are crinkly and do not unfold normally. The petiole becomes short and manifests irregular outgrowths on its surface. The entire plant grows to be stunted and ultimately dies. The symptoms of bobone are alike but the leaves are more stunted and the lamina is curled up and twisted. Bobone does not cause complete death of the entire plant. Severe infection of alomae can cause total crop loss, while bobone can result in yield loss up to 25%, as the bobone-infected plants may recover.

Control

- Grow some tolerant cultivars bred through recurrent selection.
- Use disease-free healthy planting material.
- Pull out and destroy the diseased plants from the field.
- Introduce bio-agents like flower bugs, ladybird beetles, praying mantis, green lacewing, long horned grasshoppers, dragonflies, spiders, etc.
- Spray the crop with *neem* seed-kernel extract 4%, phosphamidon 0.03%, malathion 0.05%, or dimethoate 0.03% to control the aphids, grasshopper and mealy bugs.

Dasheen Mosaic Virus

The problem of dasheen mosaic virus is worldwide. It is caused by a stylet-borne, flexuous, rod-shaped virus, which is spread by aphids. The disease is characterized by chlorotic and feathery mosaic patterns on the leaf, distortion of leaves and stunting plant growth. The disease is not lethal but its infection reduces the yield too much.

Control

- Use disease-free healthy planting material.
- Maintain field sanitation by removing the dry leaves.
- Follow quarantine measures.
- Apply *neem* seed-kernel extract 4%, Rogor 0.5% or malathion 0.1% to control the aphids.

■■■

14 Giant Swamp Taro

M.K. Rana and Sachin S. Chikkeri

Botanical Name	: *Cyrtosperma chamissionis* (Schott) Merr.
	Syn. *Cyrtosperma merkusii* (Hassk.) Schott.
Family	: Araceae
Chromosome Number	: 2n = 26

ORIGIN AND DISTRIBUTION

The genus *Cyrtosperma* has 12 species, which, except *Cyrtosperma chamissionis*, are confined to New Guinea and its associated islands, Philippines, Borneo, Sumatra, Java and Oceania. Giant swamp taro, the largest among the taros, is terrestrial, usually grown in atoll pit gardens and in freshwater swamp or marsh area. It is probably originated in Southeast Asia, perhaps in Indonesia. It spread throughout Malayan Archipelago and Pacific territories. It is minor crop in Melanesia and parts of Polynesia but a major food crop in Micronesia.

INTRODUCTION

Giant swamp taro is mainly cultivated for its underground corms grown along stream banks in the tropical and subtropical forests. The crop is cultivated and used to very marginal extent in India wherein, the crop is harvested by the tribal people of northeast in patches, *i.e.*, the lowlands of Assam, Meghalaya and Manipur. The giant swamp taro is introduced from Indo-Malaysian to the Pacific region. This species is most widely distributed and cultivated in Micronesia and the Western Pacific region, in particular, on atoll islands where it is either the first or the second most important cultivated aroid. It is a major crop in most atolls and low islands of the Pacific Ocean but is a minor crop today in the high islands of the Pacific Ocean. It can easily be field stored in the ground for very long period (up to 30 years or more) and has traditionally been an important emergency crop at the time of natural disaster. Its strong root system and thick dense foliage made it to withstand strong winds and hurricanes, thus providing food security.

COMPOSITION AND USES

COMPOSITION

Giant swamp taro is very rich in nutrients like calcium, iron and zinc. The yellow corm varieties are generally higher in β-carotene as compared to other coloured varieties. The nutritional composition of giant swamp taro is given in Table 14.1.

USES

Giant swamp taro is the dominant tuber vegetable in coral islands of the Pacific. The primary produce of this crop is the underground corm, which varies in characteristics with cultivar and age.

TABLE 14.1

Nutritional Composition of Giant Swamp Taro (per 100 g Edible Portion)

Constituents	Contents	Constituents	Contents
Water (g)	60–70	Iron (mg)	0.9–1.4
Carbohydrates (g)	28–36	β-carotene (μg)	28
Protein (g)	0.5–1.4	Thiamine (mg)	0.03–0.06
Fat (g)	0.1–0.5	Riboflavin (mg)	0.08–0.11
Calcium (mg)	301–598	Vitamin C (mg)	1.0
Phosphorus (mg)	28–79	Energy (kcal)	72

It is a staple food in many subsistence communities, particularly in the Pacific islands. Its corms can be roasted, boiled, or baked whole, or mashed or grated and combined with other starches for eating. Apart from tubers, the other plant parts like young leaves and inflorescences are eaten as vegetable, and the petioles yield a fibre for weaving. The leaf is used as a food wrapper and cover for the earth oven. The plant has been used in traditional medicine in many of the high and low islands of Micronesia. The yellow mould from sliced and sun-dried corms is also used to treat skin infections.

Medicinal Uses

Giant swamp taro is used in traditional medicines in many of the low and high islands of Micronesia. The yellow mould from sun-dried corms sliced also used to treat skin infections. The consumption of yellow-fleshed varieties that are very rich in β-carotene, containing 1900–4450 mg/100 g of edible portion, cures vitamin A deficiency disorders such as night blindness, softening of cornea, soreness of eyes, roughness of skin, respiratory tract infections and other degenerative diseases and improves body immunity. The decoction prepared from its flower clusters acts as emmenaggogue and ecbolic.

BOTANY

Giant swamp taro is the rosette (with no stem) herbaceous perennial plant, usually 1.8–2.4 m high, even more up to a height of 5 m and is not only the largest in the Aroideae family but also the largest plant in the world that yields an edible corm. In some varieties, the corm weight could be as much as 100–120 kg if left to grow for a number of years. The 6–8 large arrow-shaped leaves with relatively short leaf sheath arranged in a spiral manner are about 1.5 m long on thick and long petioles. The leaf blades have sharply pointed basal lobes. The petiole has a diameter of about 10 cm, and in some cultivars, the lower part is covered with spines. There is a considerable variation among cultivars in leaf shapes, colour and degree of spines on petioles. The corm develops by thickening at base of the stem and almost cylindrical in shape and externally resembles a banana sucker. It bears inflorescences in the leaf axils or from the centre of cluster of unexpanded leaves, beginning around the second year of growth and thereafter continuing for a couple of years. Naturally, the flowering occurs only sporadically but flowering can be promoted artificially with the application of gibberellic acid. Each plant may bear more than one inflorescence, which is made up of a short peduncle, a spadix and spathe. The spadix is botanically a spike with a fleshy central axis, with which, the small sessile flowers are attached. The inflorescence is large with 25–65 cm open spathe. The young purplish spadix is about 20 cm long, tubular to cylindrical and covered with a bract, gradually tapering towards the top. It bears both male and female flowers. The fruit is berry, and seeds of cultivated forms are fertile.

CLIMATIC REQUIREMENT

The giant swamp taro grows in the elevation range from sea level to 600 m (as in Papua New Guinea) but the crop is not suitable for growing in uplands under rain fed conditions. It needs a constant supply of fresh water and can be grown in slightly brackish water too. However, it not only thrives in freshwater swamps but also thrives in swiftly flowing rivers and streams. As the crop indicated by its name, giant swamp taro has adapted to growth within fresh water and coastal swamps. This species is well adapted to moist tropical climate. The temperature range of 23 to 31°C is optimum for better crop growth. It also does well in warm and seasonally moist climate that has a short dry season and variable precipitation. The main factor for its growth is a continuous supply of water, although it can tolerate short period of dryness and maximum temperature of 35–38°C. The crop is also found to withstand a minimum monthly mean temperature of about 15.5°C. Probably, it cannot survive where the annual rainfall is unable to support a more or less constant fresh water supply as the amount of rainfall is not relevant, thus, the species must grow in marshy or swampy land where the water table is near the surface. It is also commonly grown in purpose-built swamp pits in low-lying coral islands.

SOIL REQUIREMENT

Giant swamp taro is a water-loving plant (hydrophyte), which is well adapted to fresh to brackish water conditions in coastal marshes, natural and manmade swamps and pit depressions. Deep soils are preferable, as the crop grows both upward and downward in contrast to arum, tannia and taro, which grow upward only. It can be grown in a wide range of soil reaction, *i.e.*, from 4.5 to 7.3 pH.

CULTIVATED VARIETIES

The cultivation of giant swamp taro is limited to the swampy lowland regions of the world like a few countries of Southeast Asia and the coral islands of the Pacific Ocean. As a result, no such commercial variety has been identified or notified but a few varieties are classified based on colour and β-carotene content.

SEED RATE

Giant swamp taro is a vegetatively propagated plant. Seed rate largely depends on factors like size of planting material, soil fertility and spacing, type of cultivar and availability of water. About a tonne of planting material is enough for planting a hectare land area following a spacing of 50 × 50 cm, while higher planting density (30 × 30 cm) requires about 1.5 to 2 t/ha. The plant population of around 40,000 to 80,000 per hectare could be maintained.

SOWING TIME

Generally, the time of planting in giant swamp taro is not season bound but the usual practice is to grow plant types that mature in 1 or 2 years with harvest usually made at the time of flowering. While intercropping with other plants of its kind, it is normally planted during rainy season but it can be grown anytime if sufficient water is available.

SOWING METHOD

In giant swamp taro, suckers (sprouting cormels) are the commonest planting material, although sometimes, the tops of the corm of harvested plant (setts) are also used as a planting material. In each case, one or two of the youngest leaves in the shoot are retained. Planting methods and techniques vary greatly depending on the habitat. The simplest systems can be found in the freshwater marshes of high islands where the marsh is cleared of its vegetation (mainly a reed) and then planted with setts from previously harvested giant swamp taro. It may be planted either as a single plant, short rows of 6–10 plants, or intercropped in arum, tannia, or taro crop.

In the pit planting method, spacing between plants is kept 40 to 100 cm of larger-type varieties and 30 × 30 cm or 50 × 50 cm of smaller type cultivars. In wet land cultivation, the giant swamp taro is often intercropped with arum, which is planted at a distance of 1–1.5 m. The arum may be replanted for three annual crops, but subsequently, the giant swamp taro is allowed to grow alone for years.

In atolls where there is subterranean fresh water lenses, pits are dug deep enough to reach the fresh water layer, which may be 0.5–3 m below the surface. The pits may extend to 10 × 20 m across, and once the fresh water is reached, the individual holes are dug for each plant and filled with organic material, *e.g.*, chopped leaves, covered with sand, and the suckers or setts are planted in sand so that its upper roots are at the water level.

The *bottomless basket* technique is mostly used to cultivate the crop. The basket is woven *in situ* from screw pine or coconut leaves, and the enclosed area is filled with a mixture of chopped leaves, soil and compost. This type of basket cultivation gives the largest corms but the growth rate is slow. In some areas, planting methods similar to those employed for wet land cultivation of taro (*Colocasia*) are used, but more emphasis is given for mulching and shading until the plants reach a height of 1–2 m. However, in case of non-puddled or compact soils, deep planting holes, or furrows are prepared (15–100 cm deep), and after the setts or suckers are placed in position, the furrows are partially filled with a mixture of soil and compost up to 10–15 cm above the base of the sucker or setts.

NUTRITIONAL REQUIREMENT

Unlike other aroid crops, no standard recommendation for manure and fertilizers application has been made, as it is grown only in swampy lowland environments, whereas, in case of constructed pits, various kinds of organic matter and fertilizer are usually added to the pit to improve the soil fertility, physical properties and water-holding capacity. However, the pits may be filled with chopped leaves of *Guettarda speciosa* and *Tournefortia argentea* plants and then covered with a layer of black humic sand. In starting, well-decomposed farmyard manure is incorporated in pit soil, and later, the extra organic matter is added as each new leaf emerged. Each plant is supplied with compost at least four times a year until the corms are harvested.

IRRIGATION REQUIREMENT

In giant swamp taro, irrigation has least influence on sucker initiation but moisture stress adversely affects its vegetative growth, which largely reduces the dry matter production and tuber yield. However, the corm yield is increased with increasing water level because of very extensive rooting beyond 9 months under flooded conditions. Irrigation throughout the life cycle of giant swamp taro results in higher total yield of corms and cormels as compared to unirrigated plots.

In the Pacific coral islands where it is mainly cultivated, three horticultural water control methods are practiced. In first method, the crop is planted in pits dug in the marshy lowlands in such a way that roots could tap lens of water. In second kind of water control method, the crop is planted in so-called *raised bed* or *garden island* system, which is applied on land having too much permanent surface water. It is consisted of a network of ditches dug throughout the marsh and the crop is planted above water on the mounds prepared by excavating the soil. In third method, *pondified irrigation* involves the transfer of water from a more or less continuous source, *e.g.*, a stream or spring, to a site that is not watered enough by rain for cultivation.

INTERCULTURAL OPERATIONS

The intercultural operations like thinning and earthing up are not at all required to be practiced in giant swamp taro, as its cultivation is restricted to the freshwater and costal swampy plots in coral islands of Oceania; however, mulching and weeding operations are very important in the cultivation of this crop.

HOEING AND WEEDING

Since giant swamp taro is grown in hot and humid climate, weeds pose problem and reduce its yield drastically. Therefore, removing weeds in giant swamp taro field is essential to reduce competition with crop plants for nutrients, moisture, space and light. Hand hoeing in the crop is one option, but not only it is the time consuming but also labour intensive. Giant swamp taro is cultivated under flooded conditions, thus, weeds are naturally controlled. For better growth of the crop plants and higher tuber yield, it is essential to keep the field weed free during initial growth phase.

MULCHING

Mulching with dry leaves, straw, grasses and other organic material is found to be beneficial for getting higher yield as mulching with organic material provides additional nutrients, conserves soil moisture, ensures quick sprouting, regulates soil temperature and reduces weed growth, resulting in better tuberization and corm production.

HARVESTING

The duration to reach maturity varies depending upon the type, cultivar and climate. It is a slow growing crop, which can take from couple of months to 15 years for maturation. The tubers are dug by hand as and when required. The crop is cultivated for subsistence rather than for sale, thus, continuous harvesting and replanting is the normal procedure adopted by a family patch. Storage is not usually practiced, but sometimes, the tubers are buried in a damp place where they may be kept for up to 6 months.

YIELD

Yield of giant swamp taro is highly variable and dependent on cultivar and age of the planting material. An individual plant can give a corm yield of 25–90 kg. On an average, the yield of giant swamp taro is 75–100 q/ha in crop duration of 18–24 months. The weight of individual corm is widely variable.

POST-HARVEST MANAGEMENT

The corms are eaten within 2–3 days of harvesting though the corms can be stored in moist ground for up to 6 months. A corm if not cooked instantly after harvesting is said to form bitter spots, which are removed during peeling prior to cooking. In some areas, the corms are cut into small pieces, sundried for a week and stored for half a year or longer. However, the frozen corms may have a higher content of carotenes as compared to the dehydrated corms. Hence, dehydration of corms is a poor method to adopt for preserving the quality characteristics of fresh corms.

INSECT-PESTS

TARO BEETLES (*Papuana armicollis* and *Papuana huebneri*)

Adult beetles, which bore and tunnel into the corms to feed, are the damaging stage. In severely damaged plants, the tunnels run together to form large cavities and secondary rots often develop. Their infestation destroys the corms and even very low infestation reduces marketability. The moist and organic matter rich loam or silt loam soils are most suitable to the attack of this insect. The insect multiplication takes place in any habitat that is moist and protected from natural enemies.

Control
- Restrict movement of taros between countries.
- Allow imports of only sterile material.
- Multiply planting material from pathogen-tested plantlets through tissue culture.
- Apply imidachloprid (Confidor) 5 g per plant in soil near the plants.
- Dip the corms in 0.1% solution of cyperm erthrin (Mustang).

APHIDS (*Aphis gossypii*)

Aphids are one of the important sucking pests of this crop. They cause both direct and indirect damage to the giant swamp taro plants. Direct damage to plants occurs from feeding activity of the aphid nymphs and adults, which suck cell sap from tender parts of the plant, causing mottling, reduced vigour, yellowing, speckling and curling of leaves and sometimes they may cause wilting followed by the death of the plant. Aphids attack both sides of the leaves and petioles. Its maximum activity on aroids is observed during cloudy days with high humidity and high temperature. It is a potent vector of *dasheen mosaic disease.*

Control
- Install yellow sticky traps to monitor aphid movement into fields.
- Remove and destroy the plant portion attacked by aphids.
- Use natural enemies like ladybugs and lacewing to keep the aphid population low.
- Spray fungi such as *Beauveria bassiana, Metarhizium anisopliae, Verticillium lecanii* and *Neozygites fresenii.*
- Spray the crop with malathion 0.05%, monocrotophos 0.04%, dimethoate 0.03%, phosphomidon 0.03%, or methyl demeton 0.05%,

TOBACCO CATERPILLAR OR COMMON CUTWORM (*Spodoptera litura*)

The tobacco caterpillar or common cutworm is most destructive pest of giant swamp taro, arum and elephant foot yam. The caterpillars feed on leaves by scraping gregariously for few days, and later on, they disperse and move to other leaves to feed voraciously, causing significant leaf damage. They hide during the daytime in crevices or among plant residues and become active during dusk to dawn.

Control
- Mechanically destroy all egg masses laid on the lower surface of leaves and gregarious larvae.
- Install pheromone traps, 5–10 traps per hectare.
- Spray the crop with *neem* formulations 0.5%, Polytrin C 0.1%, Lannate 0.3–0.5%, or Ekalux 0.15%.

MEALY BUG (*Pseudococcus citriculus* and *Rhizoecus* sp.)

These species of mealy bugs are found together. Their nymphs and adults suck cell sap from tender parts of the plant. The mealy bugs multiply faster during high temperature and high humidity. They cover the tuber surface along with dirty white powdery mealy substance. The severely infested tubers are shriveled, affecting their quality and marketability adversely.

Control

- Dislodge the mealy bugs with a steady stream of water.
- Dip the seed corms in 0.05% solution of monocrotophos before planting.
- Spray the plants with soaps and dish detergents to control the mealy bugs to some extent.
- Use beneficial insects such as beetles (*Cryptolaemus montrouzieri*) and parasitic wasp (*Leptomastix dactylopii*).

DISEASES

PYTHIUM ROT (*Pythium aphanidermatum* and *Pythium ultimum*)

This is the most destructive disease, which attacks the plant in field but mainly in storage. High soil moisture favours the infection of *Pythium*. A number of species of *Pythium*, which are soil borne, are able to infect the roots and corms. The primary symptoms appear as chlorosis of leaves. The affected plants gradually wilt and their corms become soft and start rotting.

Control

- Strictly follow long crop rotation with non-host crops.
- Use disease-free healthy planting materials.
- Maintain proper field sanitation.
- Provide good drainage facilities.
- Drench the crop field with captan 0.2% or Bavistin 0.1%.

LEAF BLIGHT (*Phytopthora colocaciae*)

This is a serious menace in the cultivation of giant swamp taro. The disease is wide spread in tropical and subtropical regions. The symptoms first appear on leaves as typical round shaped lesions and rapidly spread over whole leaf lamina. Petiole infection leads to collapse of entire plant. The disease may appear at any stage of crop growth. It causes a yield reduction of 25–50%.

Control

- Strictly follow field sanitation.
- Apply *Trichoderma viridae* formulations.
- Spray the crop with 4% *neem* seed kernel extract, Ridomil 0.2%, or mancozeb 0.25% at 10- to 12-day intervals.

DASHEEN MOSAIC VIRUS

The disease spread through virus vectors like aphids. The disease is not lethal but its chief effect is to retard plant growth and reduce corm yield. The virus infection leads to mottling, yellowing, speckling and curling of leaves. Chlorotic patches appear on leaves due to the attack of this disease.

Control

- Uproot and destroy the affected plants from the field to check further spread of the disease.
- Remove other plants of this family, as they serve as an alternate host.
- Spray systemic insecticides to control aphids.

15 Yam

M.K. Rana and Tarique Aslam

Botanical Name   : *Dioscorea alata* (Greater yam)

                          Dioscorea esculenta (Lesser yam)

                          Dioscorea rotundata (White yam)

Family   : Dioscoraceae

Chromosome Number : 2n = 40

ORIGIN AND DISTRIBUTION

Most of the yams are believed to originate from Southeast Asia and West Africa. Some taxonomists have suggested that there was an early separation in the *Discorea* development into old and new world species, which were reflected in cytological differences. The old world species (Africa and South Asia) have the basic chromosome number 10 and the new world species 9. Most of the species are tetraploid and hexaploid. In Asia, *Dioscorea alata* seems to have developed from the wild species, *i.e.*, *Dioscorea hamiltonii* and *Dioacorea persimilis* as a result of human selection. Among the species, the species cultivated for food are *D. cyenensis* (yellow yam), *D. batatus* (Chinese yam), *D. bulbifera* (air potato) grown for its aerial tubers, *D. trifida* (cush-cush) producing small sized tubers of superior quality and *D. alata* (greater yam) producing very large-sized tubers.

INTRODUCTION

The genus *Dioscorea* is the largest genus in the family Dioscoreaceae, including about more than 500 species. Most of the yams are monocots but some of their wild primitives exhibit the characteristic of both monocots and dicots as that of earlier angiosperm. Yam is an underutilized tuber crop cultivated only in the tropics and a very important starchy staple food crop of West Africa, West Indies and tropical America and Southeast Asia. Yams are usually herbaceous climbers with rhizome or tubers, which are an important component of diet in Asia and Africa. Its plant is perennial and deciduous in nature, and its vines may be winged or round with or without prickles or hairs. It is essentially a crop of subsistence agriculture.

COMPOSITION AND USES

COMPOSITION

Edible yams are rich in carbohydrates and are a good source of protein, more than other tuber crops. Its tubers contain almost all the essential nutrients, including minerals, vitamins and amino acids. It has been reported that diosgenin, a steroid saponin, is commonly used as a raw material, from which, a number of steroid hormones including active birth control *pills* are manufactured. The other important sapogenins found in yam are yamogenin, botogenin and kryptogenin, and minor sapogenins like pannogenin and tigogenin in certain of its species. The species that contain appreciable amount of sapogenin are *D. deltoidea*, *D. floribunda*, *D. composite*, *D. friedrichshali*,

TABLE 15.1

Nutritional Composition of Yam (per 100 g Edible Portion)

Constituents	Contents	Constituents	Contents
Water (g)	65–73	Manganese (mg)	0.39
Carbohydrates (g)	27.9	zinc(mg)	0.24
Protein (g)	1.50	Vitamin A (µg)	7.01
Fat (g)	0.17	Thiamine (mg)	0.11
Dietary fibre (g)	4.10	Riboflavin (mg)	0.03
Calcium (mg)	17.0	Vitamin C (mg)	17.1
Phosphorus (mg)	55.0	Vitamin E (mg)	0.35
Potassium (mg)	816	Vitamin K (µg)	2.30
Magnesium (mg)	21.0	Ash (g)	0.67
Iron (mg)	0.54	Energy (kcal)	118

D. spiculiflora and *D. prazeri* and the saponin bearing species are *Dioscorea floribunda*, *D. deltoidea* and *D. composite* that yield about 0.6% to 10.3% of diosgenin on dry weight basis. The nutritional composition of yam is given in Table 15.1.

Uses

Yam is still processed by conventional method though there is a need to modernize processing and utilization of yam. The conventional processing methods of yam are roasting, boiling and frying with other vegetables. In West Africa, yam is mostly consumed as *fufu*: stiff glutinous dough. It is also used for making chips, flakes and flour. Yam flour is also used for human consumption as *kokonte* in northern zones of West Africa. Biscuits made from *Dioscorea alata* tubers are found to have good acceptability by the consumers.

Medicinal Uses

Yam tubers are used as a folk medicine in China, Korea and Japan as it has antispasmodic properties (capable of relieving spasm). It helps in curing the nausea of pregnant women. The mucilaginous tuber milk contains allantoin, which helps in healing process when applied externally to ulcers, boils and abscesses. The decoction of tubers is helpful in stimulating appetite and relieving bronchial irritation, cough, etc. It is also used in neuralgic affection, painful cholera with cramp, spasmodic asthma and spasmodic hiccough. It supports healthy menstrual cycles by attenuating pain and cramps. Diosgenin in wild yam is a precursor in the synthesis of progesterone, a hormone whose functions include uterine health and warding off premenstrual syndrome. Some of the herbalists believe that it triggers the production of progesterone in women and balances against high levels of estrogen. Wild yam gives relief to women from vaginal dryness, insomnia, headaches, migraines, flashes and cramps, and its progesterone creams are applied to thin areas of skin.

BOTANY

Yam is a deciduous perennial plant. Yam plants produce tubers or rhizomes, from which, the vines and roots emerge annually during the growing season. The roots developed from lower part of tuber are thick and unbranched at the beginning, which become fibrous and branched later. Yam tuber is neither a root nor a stem but it develops from hypocotyl structure. The annual stems are weak, thus, they are trained with the help of twine. However, in some species like *D. rotundata*, it remains erect to a height of about 1 m. In some species, stems have the spines, which protect the plant from wild animals and provide support to the stem on host plant. The leaves are simple having many veins from base of lamina to reticulate subsequently. However, *D. spida* and *D. dumentorum* have compound leaves usually with three or more leaflets. The size of leaf lamina may vary from 50 to 200 cm2. Yams are dioecious, where male and female flowers are borne on separate plants. Generally, female flowers are bigger in size than the male flowers.

The individual size of male flowers is small and inconspicuous and varies from 2 to 4 mm in diameter. The colour of male flower is white to creamy white and borne on axillary spike or panicle drooping downwards. Male flowers have six sepals and six stamens in two whorls. Pollen grains are usually small and sticky. Female flowers have an inferior ovary, trilocular with two ovules in each locule. The pistil is single and occupies the centre of the flower. Fruit is dehiscent trilocular capsule, which differs in shape and size from one species to another. Seeds are small, light in weight, flattened and winged all around the seeds, which helps in dispersal by winds. The seeds of many species have the dormancy of 3 to 4 months but the seeds of *D. alata* species have no seed dormancy.

CLIMATIC REQUIREMENT

Yams are mainly grown in tropics and subtropics of the world. However, the major growing areas are in tropical regions. The optimum temperature for its growth and development is 25–30°C. When the temperature falls below 20°C, growth is affected adversely. Yam cannot tolerate frost. High rainfall is essentially required for maximum production. Yams require a well distributed rainy days for 7–10 months. Low moisture at early stage may kill the young shoots, which have exhausted reserve food of the old set. Day length more than 20 hours during early stage promotes the vine growth, while short day length during later stage of the growing period favours satisfactory tuber production. Yams are relatively tolerant to dry conditions but thrive best when supplied with ample moisture. White yam (*D. rotundifolia*) requires optimum rainfall of 100–150 cm distributed over 6–7 months for optimum yield, whereas, lesser yam (*D. esculenta*) grows well in areas having a rainfall of 90–100 cm.

SOIL REQUIREMENT

The development of tubers is mainly dependant on texture and preparation of soil. The tubers of some species go deep into the soil and of other species spread on upper surface of the soil. For proper development of tubers, yam requires a deep friable and well-drained sandy loam soils rich in organic matter. Soils should have good water retention capacity as well as good porosity. In sandy soil, the plant declines, and in stiff clay soils, deformed tubers with poor quality are formed. The yam can be grown on wide range of soil pH ranging from 5 to 7.

CULTIVATED VARIETIES

GREATER YAM (*Discorea alata*, 2n = 20, 30 and 40–80)

Sree Keerthi

A variety developed at Central Tuber Crops Research Institute, Kerala through a selection made from germplasm collection was released in 1987. Plants are climber with 4–5 m long vine twining in clockwise direction on support and emerging leaves are light brown in colour. Tubers are conical in shape with white flesh and good cooking quality. Starch content ranges from 20% to 22%. The average tuber yield is 300 q/ha in a crop duration of 9–10 months.

Sree Roopa

A variety developed at Central Tuber Crops Research Institute, Kerala through a selection from germplasm collection has a yield potential of 250–300 quintals per hectare. Plant is climber with

4–5 m long vine twining in the clockwise direction on support. Emerging leaves are grey in colour and stem is purple. Tubers, which are suitable for culinary purpose, are digitate with black skin and white flesh. The tubers contain 17–19% starch and 1–2% protein.

Sree Shilpa

A variety developed at Central Tuber Crops Research Institute, Kerala has a yield potential of 280 quintals per hectare in crop duration of 8 months. Tubers are swollen, oval and smooth with black skin and white flesh.

White Yam (*Dioscorea rotundifolia*)

Sree Shubra

A variety developed at Central Tuber Crops Research Institute, Kerala through a selection from open pollinated progeny of the exotic cultivars and introduction from Africa produces 1–2 cylindrical tubers with brown skin. The tubers with white flesh possess 21–23% starch and good keeping quality. The average yield is 350–400 quintals per hectare in crop duration of 9–10 months.

Sree Priya

A variety developed at Central Tuber Crops Research Institute, Kerala through a selection from a population derived from open pollinated seed of variety Umidika received from Africa has a yield potential of 350–400 quintals per hectare in crop duration of 9–10 months. Plant is climber vine twining to the right direction on the support. The vines are spiny. It has dark glossy leaves, unifoliate with wavy margin. The plant produces 2–3 long tubers with good cooking quality. The starch content is 20–21% and protein content is 2–3%.

Sree Dhanya

A variety developed at Central Tuber Crops Research Institute, Kerala through a selection from seedlings progeny is dwarf bush type. The main advantage of variety is that it does not require staking, as the plants do not grow beyond 50 cm height. Its tubers yield is about 200–250 quintals per hectare.

Lesser Yam (*Discorea esculenta* , 2n = 40, 60, 90 and 100)

Sree Lata

A variety developed at Central Tuber Crops Research Institute, Kerala through a selection from D. esculenta germplasm has the yield potential of more than 250 quintals per hectare in crop duration of 7.8–8 months. The plants are climbing in nature with a vine length of 2.5–3 m. Vine is green with prickles and leaves are green with petioles brownish green and swollen at both the ends. Tuber shape is oblong to fusiform, skin grayish brown and flesh is creamy white with a starch content of 18.4%. Its tubers' shelf life is 2.5–3 months.

Sree Kala

A variety developed at Central Tuber Crops Research Institute, Kerala through a selection and introduction from Puerto Rico has the yield potential of 200 quintals per hectare in crop duration of 8 months. The tubers are smooth and possess excellent cooking quality.

PLANTING TIME

Time of planting may vary from place to place but the best time for planting yam is June to July, *i.e.*, beginning of the rainy season.

PLANTING METHOD

Yams are mainly propagated by means of vegetative propagation and the material mainly used for planting is tuber piece or small whole tubers. It is possible to propagate yams through vine cuttings although the tuber production by this method is low. However, it is better to use cutting for successful production of *D. floribunda*, *D. dumetorum* and *D. rotundata*. For propagation, the whole tuber is cut into small seed pieces consisting of tops (proximal), middle portion and the bottom end (distal end). The seed tubers from top portion (proximal) are preferred more for planting because of their earliness and uniform sprouting.

Before planting, the tuber pieces are dipped in cow dung slurry and allowed to dry under shade. The size and weight of tuber greatly influence the yield of the crop. It has been reported that large sized seed tubers give higher yield than the small size tubers. A seed tuber having a weight of 200–250 g is ideal for optimum production of *D. alata*. Conversely, for optimum production of *D. esculenta*, the seed tuber weight should be 100–125 g. The planting of tubers may be done on flatbeds, raised beds, mounds, ridges, or in trenches.

Flatbed method of planting is adopted in light soils; however, in heavy soils, planting is done on mounds or ridges. In this method, ridges are prepared by drawing the top soil up to 40–50 cm height, keeping the base wide. Then, the tubers are planted deeply on ridges. In trench planting method, the trenches are opened and filled with a mixture of soil and compost, and thereafter, the tubers are planted in trenches. A distance of 150 cm is kept between the rows and 50 cm between the plants within a row. By following this spacing, about 13,300 tubers are needed to plant a hectare area.

SEED DORMANCY

Yam tubers generally have a dormancy period of about 2–2½ months. During storage, early sprouting causes considerable damage to the seed tubers, thus, the sprouts should be removed frequently. If freshly harvested tubers are to be used for planting, the dormancy should be broken artificially by dipping seed tubers in 4–8% solution of ethylene chlorohydrin followed by drying in shade.

NUTRITIONAL REQUIREMENT

Yams require plenty of organic manure for successful production. Well decomposed farmyard manure of 10–15 t/ha is incorporated into the soil during land preparation. Integrated application of farmyard manure 12.5 t/ha as a basal dose and NPK 80:60:80 kg/ha are usually recommended for *D.alata* and *D. esculenta*. Half dose of nitrogen and potassium along with full dose of phosphorus should be applied at the time of planting, and the remaining half dose of nitrogen and potassium should be applied 1 month later at the time of interculturing.

Use of some biofertilizers, namely *Azospirillum* and *Phosphobacterium* 2 kg/ha, along with compost, can reduce the requirement of nitrogen and phosphorus to 50% of the recommended dose and maintain the soil nutrients status.

IRRIGATION REQUIREMENT

Yams are usually raised as a rain-fed crop, but in parts of nontraditional areas, it is cultivated under irrigation conditions. First irrigation is given on the day of planting followed by two irrigations at an interval of 3 to 5 days until the plants are established. It is needed to supply regular water through rains or artificial irrigation since irregular water supply affects the tuber development adversely. The plants are also irrigated after fertilizer application if the soil moisture is not optimum.

INTERCULTURAL OPERATIONS

HOEING AND WEEDING

In yam crop, hoeing is done shallow to provide better aeration for plant growth, development and for controlling the weeds. Light racking of soil followed by earthing up twice in 2 months after planting is essential. Earthing up helps in providing support to the vines and prevents water stagnation near the base of the vine. Yams are sensitive to competition of weeds during the early stage of their growth. The critical stage of weeds in yam synchronizes with the plants when they are at leaf development and tuber bulking stage. For effective control of weeds in yams, simazine or altazine is applied @ 3 kg/ha as pre-emergence.

MULCHING

Mulching immediately after planting is beneficial for increasing the tuber yield. It provides protection from excessive temperature, conserves soil moisture, ensures quick and uniform sprouting of tubers and suppresses the weeds.

TRAILING OF VINES

It is an important operation to expose the plant to sunlight. Yams must be staked for getting a good yield since staked plants give higher yield than the non-staked plants, but staking is costly and laborious operation. The support should be provided to emerging shoots to avoid any injury to the tender shoots. Trailing should be done within 15 days of sprouting by using coir rope attached to artificial support or trees growing in that area. The most effective system is bower system, which produces a yield of 220 q/ha, while the plants without trailing gives a yield of only 147 q/ha. Yams are also stacked on living, dead trees, or other quick-growing plants. In Kerala, vines are usually trailed on ropes tide with coconut and other tall trees.

HARVESTING

Harvesting can be done as and when the leaves turn yellow and the vines completely dry up. The crop is ready for lifting about 8–9 months after planting. *Dioscorea esculenta* matures early as compared to other species. In yams, there is a practice of double harvesting by removing the mother tubers after 2 months of growth and allowing the plants for the production of side tubers. However, it is not economical as compared to single harvesting. For convenient harvesting, the soil is dug around the plants with the help of a spade. However, care should be taken to avoid injury to the tubers at the time of harvesting. Tubers are removed from the soil carefully. To produce seed tubers, harvesting is done by cutting the tubers from crown, which can be planted.

YIELD

The yield of tubers depends upon the species and cultural practices adopted by the growers. A tuber yield of about 200–350 q/ha may be obtained if good management practices are adopted by the farmers.

POST-HARVEST MANAGEMENT

Among tuber crops, yams can easily be stored for a longer time, as they possess an excellent keeping quality. The bulkiness of tubers and high moisture content make the post-harvest handling of tubers most difficult. Being bulky, space requirement for its storage is large and shipping and storage cost is also considerably high. Curing has been suggested to develop a few layer of cork cell around it and to protect the tubers from water loss. Curing can be done for 4 days at 29–30°C temperature and 90–95% relative humidity. The loss of weight during storage may be as much as 20% due to various factors. In tropical countries, yams are stored at a temperature of 30°C. Chilling injury may occur when tubers are stored below 12°C. Many methods of storage are followed in different countries; however, the most commonly practiced method is to leave the yam in the field and harvest the tubers whenever required. In this method, the quality of tubers deteriorates after the onset of rain. In some cases, tubers may be stored by (i) tying them on vertically shaded area, (ii) heaping the tubers in pyramid shape in the well-ventilated room or under shade, or (iii) placing the tubers in a layer or in racks under shade. There is a considerable loss in weight during storage due to evaporation of water from the tubers and respiration.

INSECT-PESTS

SCALE INSECT (*Aspidiella hartii* and *Aspiditous destructor*)

Both nymphs and adult of *A. hartii* are whitish in colour. They are usually found in group on tubers causing shriveling of tubers. Occasionally, they also feed on foliage. The nymphs and adults of *Aspidiella destructor* suck sap from ventral surface of the leaves and tender shoots,; hence, the plant growth is checked.

Control

- Use healthy tubers for planting.
- Avoid the infested field for the planting of crop.
- Collect the affected plant parts and destroy them.
- Remove the insects manually from the tubers by scrapping individual tubers with brush.
- Dip the tubers in 0.03% Rogor solution just before planting.
- Spray Rogor @ 0.03% on standing crop as and when the insects appear and repeat the spray depending on the intensity of infestation.
- Fumigate the storage with carbon disulfide (27 g/m^3).

WHITE GRUB (*Leucopholis coneophora*)

A polyphagous insect attacks a number of plants, including yam. The grubs attack the tuber and form shallow round pits.

Control

- Application of 20–25 kg/ha of Sevidol found effective.

YAM BEETLE (*Galerucida bicolor*)

A red and black beetle appears with the onset of monsoon. The yam beetle is a pest that starts as a curled grub, which changes into a beetle. Pupation takes place in the soil. The grubs skeletonize the leaves and later attack on the petiole and sometimes on the main stem. Yam beetles may spoil the tubers by making holes into them, and then the tuber rots.

Control

- Follow long crop rotation with non-host crops.
- Wash the planting pieces in lead arsenate and Bordeaux mixture.
- *Cymbopogun citrates* L. and *Ocimum viride* L. were superior as repellant or antifeedant botanicals to the beetle.

DISEASES

LEAF BLIGHT (*Cercospora* spp.)

This fungal disease favoured by high relative humidity. The symptoms appear as brown to black leaf spot in the beginning of rainy season. The spots gradually coalesce and blight the leaf. The disease causes defoliation and even the death of plants if the infection is severe.

Control

- Follow long crop rotation with non-host crops.
- Maintain field sanitation.
- Avoid excess irrigation and provide proper drainage.
- Obtain seed tubers from healthy seed lot.
- Collect the defoliated leaves and burn to reduce the inoculums at source.
- Spray the crop with Bavistin @ 0.1% or NF44 (thiophanate-methyl) 3–4 times at an interval of 10–15 days.

ANTHRACNOSE (*Colletotrichum gloeosporiodes*)

This disease appears as brownish spots with concentric rings on leaf and stem. The spots gradually coalesce. Tissues of necrotic spots on leaves collapse, become thin and papery, and produce abundant pinkish slimy mass in humid weather. Infected stem crack and rot. The incidence of disease takes place in July due to wet weather.

Control

- Follow long crop rotation with non-host crops.
- Maintain field sanitation.
- Obtain seed tubers from healthy seed lot.
- Spray the crop with mancozeb @ 0.25% or carbendazim @ 0.1% twice at 10- to 15-day intervals.

YAM MOSAIC VIRUS

Chlorosis is the main symptom of this disease. The virus-infected plants do not grow properly and remain very stunted. The leaves due to the attack of this disease are reduced in size. This virus ultimately affects the tuber adversely and deteriorates its quality.

Control

- Uproot affected plants from the field and destroy them to check further spread of the disease.
- Use virus-free planting materials.
- Control vectors by applying dimethoate @ 5 ml/litre of water.

16 Cardoon

M.K. Rana and Harshita Mor

Botanical Name : *Cynara cardunculus* L.

Syn. *Cynara cardunculus* var. altilis DC

Cynara cardunculus var. scolymus L.

Family : Asteraceae

Chromosome Number : 2n = 34

ORIGIN AND DISTRIBUTION

Cardoon is thought to be the native of the Western and Central Mediterranean regions, where it was domesticated in ancient times. The cultivated cardoon is grown on small scale in northern Italy, southern France and in Spain and widely distributed across the Mediterranean region from Cyprus and Black sea to Atlantic Spain, Portugal and Canary Islands. It is naturalized mainly in southeastern parts of Australia and Victoria and has a scattered distribution in sub-coastal regions of New South Wales. It is also found growing in southwestern Australia, southeastern Queensland and possibly naturalized in Tasmania, overseas in western USA (California, Oregon and Washington), Mexico, South America (Argentina, Brazil, Chile and Uruguay), United Kingdom and New Zealand. This species is principally a weed of temperate regions but it is also found occasionally in subtropical. Semiarid environments are its natural habitat. It is also a weed of crops, grasslands, open woodlands, roadsides, gardens, waste areas and disturbed sites.

INTRODUCTION

Cardoon, also known as the artichoke thistle, *cardone, cardoni, carduni*, or *cardi*, is although very similar to globe artichoke in appearance but it tends to form taller plants. It is well adapted to the xerothermic conditions of southern Europe and typical conditions of arid and semiarid areas of the Mediterranean environment. Like other close relatives, its flower buds are consumed but they are comparatively smaller and very tricky to handle. Its edible leaves are more valuable than its flower buds. It is a multipurpose crop, which can be utilized as a raw material in paper pulp industry, as forage in wintertime but most importantly as solid and/or liquid biofuel in bioenergy sector. As compared to globe artichoke, it has been grown on small scale and appreciated less by the Greeks but has been popular among the Romans. Sometimes, it is usually cultivated around the Mediterranean regions and is the most popular vegetable in France and Italy. In the United Kingdom, it is gaining popularity as an unusual vegetable. It is also grown in some parts of United States and Australia. In warmer climate, its seeds in pastoral locations can readily be dispersed by winds and animals, and pose a potential weed problem in California and Southern Australia.

COMPOSITION AND USES

COMPOSITION

Cardoon contains unique health benefiting plant nutrients such as antioxidants, minerals and vitamins as it is rich in folic acid, magnesium, calcium, potassium, phosphorus and sodium. It has a good level of dietary fibre, which includes inulin. It is a fair source of vitamin C, thiamine and riboflavin and also provides a good amount of carbohydrates, dietary fibre and energy. Like celery, cardoon is also one of very low-calorie leafy vegetable, carrying just 17 calories per 100 g. The nutritional composition of cardoon per 100 g edible portion is given in Table 16.1.

USES

Chiefly the flower buds are eaten much as small artichokes, but more often, the stems are eaten after cooking. In southern Italy and Sicily, the flower buds widely collected from wild cardoons are still used as vegetable. In Spain and Portugal, its flower buds and their pistils are also employed in making cheese. It is also used as a vegetarian source of enzymes for the preparation of cheese. In Portugal, it is traditionally used to prepare curd. The leaf stalks, looking like large celery stalks and having artichoke-like flavour, are eaten after steaming or cooking. Only the innermost white leaf stalks are harvested in winter and spring just before flowering. These stalks in Spain particularly in northern regions, where they are grown on commercial scale are considered a delicate dish. These stalks combined with artichokes are steamed to soften them and eaten after adding almond sauce. In markets, it is often sold as preserved vegetable in brine so that it can be made available round the year. In Spain, it is used as an ingredient in one of the national dishes such as *cocido madrilène*, a slow-cooking vegetable dish simmered in broth.

The selected cultivars of cardoon with very bright silvery-grey foliage and large flowers are also grown as an ornamental plant because of their architectural appearance. Recently, it has become a possible source of biodiesel. The oil extracted from its seeds is similar to safflower and sunflower oil in composition and use. It has been the biological material for the first bio-refinery in the world, converting the petrochemical plants and oils for the building blocks of bio-plastics.

TABLE 16.1

Nutritional Composition of Cardoon (per 100 g Edible Portion)

Constituents	Contents	Constituents	Contents
Water (g)	10.6	Zinc (mg)	0.17
Carbohydrates (g)	4.07	Manganese (mg)	0.256
Protein (g)	0.7	Thiamine (mg)	0.02
Fat (g)	0.1	Riboflavin (mg)	0.03
Dietary fibre (g)	1.6	Niacin (mg)	0.3
Calcium (mg)	70	Pantothenic acid (mg)	0.338
Phosphorus (mg)	23.0	Pyridoxine (mg)	0.116
Potassium (mg)	400	Folic acid (µg)	68
Magnesium (mg)	42	Vitamin C (IU)	2
Sodium (mg)	170	Energy (kcal)	17
Iron (mg)	0.7		

MEDICINAL USE

Cardoon has become important as a medicinal herb in recent years. The pharmacological active compounds, *i.e.*, inulin in roots and cynarin and silymarin, extracted from cardoon, are also of a potential use. The latter two bitter-tasting compounds found in its leaves detoxify the body, improve liver and gall bladder function and stimulate the secretion of digestive juices, especially bile, and it lowers the cholesterol levels in blood stream through inhibiting its synthesis and increasing its excretion in the bile. The leaves are anticholesterolemic, antirheumatic, cholagogue, digestive, diuretic, hypoglycaemic and lithontriptic. It is used internally in the treatment of chronic liver and gall bladder diseases, jaundice, hepatitis, arteriosclerosis and the early stages of late-onset diabetes. Cardoon contains phytonutrients such as luteolin, silymarin, caffeic acid and ferulic acid and dicaffeoylquinic acid, which protect cellular proteins, membrane lipids and DNA from oxidative damage caused by free radicals, which can cause cancer. Scientific studies have shown that adequate levels of folic acid in the diet during pre-conception period and during early pregnancy can help in protecting from neural tube defects in the newborn baby.

BOTANY

Cardoon is an upright perennial herbaceous plant, which is usually grown to a height of 75–150 cm, occasionally reaching up to a height of 2 m. It is a rosette plant with large stalked leaves, which along the upright stems are significantly smaller and arranged alternately. The topmost leaves are sessile (without petiole). Its leaves are deeply serrated with lobes and upper surface greyish-green covered with slight hairs and lower surface whitish woolly in nature due to dense covering of hairs. Their tips and margins are armed with yellowish-orange spines along the leaf stalks. Its upright thick and rigid stems, which are usually branched in upper parts, are longitudinally ribbed and covered with cottony pubescence. The plant bears very large sub-globular flower heads singly at the tips of branches, consisting of many blue, pink, or purple florets enclosed within several layers of large spine-tipped bracts. These flower heads, containing numerous large seeds of light grey, brown, or black colour topped with large feathery hairs, readily fall off.

CLIMATIC REQUIREMENT

Cardoon thrives best in regions with warm and humid climate. It shows great adaptation to Mediterranean climate, which is characterized by moist and mild cold winters and dry and hot summers. Seedlings are frost sensitive but the young cardoon plants at rosette stage with 4-6 well developed leaves endure low winter temperature as low as −5°C, that is why, early sowing two to 3 months before frost is done. Under hot climatic conditions, the leaves become bitter in taste. Although its roots can usually tolerate slight frost but severe frost kills its roots and less severe frost tends to cause the tops to die back. It prefers full sunshine but can tolerate partial shade too. Like any other crop grown under rain-fed conditions, the productivity of cardoon is closely related to the amount of water available. It performs well when the precipitation accumulated in the period comprised between sprouting and blossom is about 450 mm. Lower precipitation may result in reduction of biomass production.

SOIL REQUIREMENT

Cardoon can be grown on a wide range of soils but prefers a deep loamy soil fairly rich in nutrients and organic matter. It is a deep rooted plant, thus, should be planted on soils that afford adequate area for root development. Its growth is poor in clay soils due to restricted penetration of roots, thus, its cultivation in clay soil should be avoided. It cannot tolerate water-logging conditions. It grows best in slightly acidic soils with 5.5 to 6.5 pH.

CULTIVATED VARIETIES

Spineless and spiny cultivars exist in its natural habitat. Generally, the spineless varieties are preferred. Not too much breeding work has been done on this crop to develop its improved varieties, thus, only the locally available cultivars *viz.,* Tender Heart, Gigante, Bianco Ameliore, Large Smooth, Large Smooth Spanish Ivory, White Smooth, White Improved and Italian Dwarf are grown.

SOWING TIME

Cardoon is a perennial plant but grown as an annual crop. Its seeds need 24°C temperature to germinate properly, thus, it is sown in last week of February to first week of March in unprotected areas. However, for taking an early crop of cardoon, the seeds are sown individually in portrays in the month of January to raise seedlings under protected conditions (greenhouse), and later, the 6 weeks old well-grown healthy seedlings are shifted in the garden 3–4 weeks after the last frost date in spring since cardoon is sensitive to frost.

SEED RATE

The seed rate of cardoon depends on growing season, soil type, soil fertility, irrigation facilities and package of practices followed during its cultivation. In general, approximately 3–5 kg seed is needed to sow a hectare land area.

SOWING/PLANTING METHOD

Cardoon is often propagated by seed but its production is best when sections of young shoots with some roots taken from old plants and set out in field in spring. Seeds are sown in 20–25 cm deep trenches at a depth of about 2.5 cm. Its seeds germinate within few days of sowing. As the plants grow, additional soil is added until the trenches are filled. Seeds are sown individually in cells of portrays to raise seedlings before planting in field when the risk of frost has passed.

The sections of young shoots or suckers along with some roots are separated gently from the crown of already established plants in spring and planted in pots or in field at a line spacing of 90–120 cm and plant spacing of 45–60 cm. Its tuberous roots in late autumn are divided and planted in beds or pots over the winter. In spring, the established tuberous roots in the field start forming new leaves and roots. Generally, the plant once established does not require stacking. The plants are rejuvenated by division after every 2–3 years.

NUTRITIONAL REQUIREMENT

A balanced use of manure and fertilizers is very important for successful crop production. Well-decomposed farmyard manure is incorporated into the soil @ 20–30 t/ha at the time of field preparation. Besides, nitrogen is applied @ 30–110 kg, phosphorus @ 110–225 kg, potash @ 225 kg and sulfur @ 15–25 kg/ha. The whole dose of phosphorus, potash and sulfur along with one third dose of nitrogen is applied at the time of sowing or planting and rest of the nitrogen is applied in two equal splits 1 and 2 months after planting.

IRRIGATION REQUIREMENT

Cardoon requires a continuous supply of moisture since its leaves and stalks become pithy and unmarketable when subjected to water stress. Irrigation frequency and amount of water usually depend on the type of soil, growing season, relative humidity and wind velocity. It necessarily requires irrigation water about 25–30 cm, which should be distributed uniformly throughout its growing period. In light soils, irrigation is applied more frequently at short intervals, while in heavy soils it is applied at long intervals.

INTERCULTURAL OPERATIONS

HOEING AND WEEDING

Controlling weeds at early stages of crop growth is important for successful production of cardoon since initially the crop growth is poor, thus, the weeds compete with crop plants for nutrients, moisture, light and space. This crop-weed competition during first 4 weeks of crop in field can result in 30–40% reduction in yield and quality of cardoon. Hoeing is an effective method of controlling weeds in early stages. However, care must be taken to avoid damage to crop roots as the feeding roots are present in the top few centimetres of soil. It has been shown in literature that the quality of cardoon is reduced if mechanical cultivation or hand-hoeing is not done properly.

BLANCHING

Cardoon is normally blanched to improve its flavour and to keep its petioles white and tender. Blanching is done by tying the leaves together in a bunch and wrapping newspaper around the stems up to a height of about 45 cm, or mounting up of soil around the stems about 3 to 4 weeks before harvest, when the plant has attained a height of 90 cm.

HARVESTING

Cardoons are usually harvested during winter months and often treated as annuals if they are grown for culinary purposes. In areas with mild winters, cardoon can be harvested from November to February, and then, starts new crop in early spring. Its harvesting commences 120 days after planting and continues up to 150 days of planting. It is harvested by cutting the blanched parts with the help of sharp knife just below the crown. The extra loose leaves are trimmed off, leaving only the blanched heart, which is 45–60 cm in length and 5–7.5 cm in diameter.

YIELD

Number of factors such as variety, growing season, soil type, soil fertility, irrigation facilities and the cultural practices adopted during the course of crop cultivation affect the yield of cardoon. The above ground biomass average yield of cultivated cardoons is 190–200 q/ha per year with maximum yield of 260 q/ha per year.

POST-HARVEST MANAGEMENT

Cardoon hearts are seldom stored, but for temporary holding, it is kept at 0°C temperature and 95–100% relative humidity to avoid wilting or drying. A good-quality cardoon may be kept in good condition for 2–3 weeks without any decay or freezing injury at 0°C temperature. Its stalks wrapped in paper or polyethylene film bags can be kept in refrigerator for 1 to 2 weeks. Like celery, cardoon can also be frozen, canned, or dehydrated.

INSECT-PESTS

ARTICHOKE APHID (*Capitophorous elaeagni*)

The tiny insects having pale green to yellowish green colour suck cell sap from tender leaves of the plant. As a result, the affected leaves curl and turn yellow, plant growth becomes stunted, buds small and deformed and stalks cannot support the weight of buds, thus, drop down. These insects secrete honeydew like substance, on which, the black sooty mould grows, which affects the photosynthetic activities of the leaves adversely. The insect becomes more problematic at high temperature coupled with high humidity.

Control

- Remove and destroy the infested portion of the plant to prevent further spread of the insects.
- Wash aphids from the plants with a strong stream of water.
- Spray the crop with 0.05% malathion or 4% *neem* seed kernel extract.

ARTICHOKE PLUME MOTH (*Platyptilia carduidactyla*)

The artichoke plume moth varies in colour from buff to brownish buff, with a wing expanse of 2–3 cm. Both fore and hind wings are divided into lobes, giving an appearance of several pairs of wings. The hind wings are fringed. This insect makes holes in leaves and stems, which are discoloured black and filled with frass (insect excreta). The insect is more problematic when the crop is taken for several years on same piece of land. If infested crowns are used for vegetative propagation, the pest will establish itself quickly in the newly planted field. Larvae may also bore into the crown below the soil surface.

Control

- Pick all infested buds at harvest and destroy.
- Cut the infested plant stems above ground and bury into the soil.
- Spray *Bacillus thuringiensis* at 1.0–1.5 g/l twice or thrice at 10-day intervals.
- Spray the crop with spinosad 0.02%, permethrin 0.4%, or cypermethrin 0.1%.

ARMYWORMS (*Spodoptera exigua* and *Spodoptera ornithogalli*)

Both the species of armyworm make closely grouped circular to irregularly shaped holes on foliage. Heavy feeding by young larvae leads to skeletonized leaves and shallow dry wounds on fruit. Feeding by its larvae on young seedlings soon after transplanting results in stand loss. Its damage is hardly serious enough to consider replanting. Once the seedlings are established and start new growth, the crop can tolerate its damage. Armyworm causes negligible bud damage in annual crop.

Control

- Spray *Bacillus thuringiensis* at 1.0–1.5 g/l twice or thrice at 10-day intervals.
- Spray the crop with spinosad 0.02%, permethrin 0.4%, or cypermethrin 0.1%.

LOOPERS (*Trichoplusia ni* and *Autographa californica*)

Caterpillars walk in a *looping* manner. They move by holding on with their front legs, arching the middle portion of their body to bring the hind legs forward, and then extending the front of the body holding on with the hind legs. The caterpillars cause an extensive damage to the plant by making large or small holes on leaves. The insect survives as pupae in plant debris in soil. Its adult is a dark-colour moth, which has a wide range of hosts.

Control

- Pick the larvae from the plant manually.
- Eliminate weeds of Brassicaceae family such as wild mustard, peppergrass and shepherd's purse, as they are alternate hosts for this insect.
- Spray *Bacillus thuringiensis* at 1.0–1.5 g/l twice or thrice at 10-day intervals.
- Spray the crop with permethrin 0.4%, or cypermethrin 0.1%.

FLEA BEETLE (*Systena blanda*)

The dark colour small (1.5–3.0 mm) beetle, which is often shiny in appearance, makes small holes or pits in leaves, which give the foliage a characteristic *shot hole* appearance. As a result, the plant growth becomes stunted. Its severe infestation may kill the plant. Younger plants are more susceptible to flea beetle damage than the older ones. In winters, these beetles survive on weed species, in plant debris, or in the soil.

Control

- Use floating row covers to provide a physical barrier against beetles to protect young plants.
- Sow the seeds early to establish the seedlings before the beetles become a problem since mature plants are less susceptible to beetles damage.
- Grow few border rows of trap crops (cruciferous plants).
- Apply a thick layer of mulch on soil to prevent the beetles reaching the surface.
- Apply *neem* seed kernel extract 4%, spinosad 0.02% or permethrin 0.4%.

SPIDER MITES (*Tetranychus urticae*)

The spider mite is a cosmopolitan agricultural pest with an extensive host plant range and an extreme record of pesticide resistance. The leaves with yellow spots become speckled due to web covering. Mites may be visible as tiny moving dots on webs or underside of the leaves. Consequently, the leaves turn yellow and drop down from the plant. Spider mites thrive in dusty conditions. The water-stressed plants are more susceptible to its attack.

Control

- In kitchen gardens, spray the plants with a strong jet of water to reduce the population of spider mite.
- Apply low doses of nitrogenous fertilizer.
- Irrigate the field at frequent intervals to keep the mite population low.
- Spray the crop with 4% *neem* seed kernel extract, dicofol 18.5 EC 0.05%, Rogor 30 EC 0.06%, or Metasystox 25 EC 0.02%.

DISEASES

BACTERIAL CROWN ROT (*Erwinia chrysanthemi*)

Stunted plant growth, wilting of leaves at high temperature, collapsing of plant and narrowing, browning and drying of new leaves are the major symptoms of this disease. The crown tissues become soft, discoloured and rotten when stem is cross-sectioned. The disease may spread through cutting or digging tools. Splitting of crowns may cause new plantings to become infected.

Control

- Avoid the use of infected crowns as planting material.
- Propagate the crop by seeds or disease-free transplants.
- Spray the crop with streptocycline or plantomycin @ 500 ppm in combination with copper oxychloride 2000 ppm at 15-day intervals.

BOTRYTIS ROT OR GREY MOULD (*Botrytis cinerea*)

The crown of plant becomes slimy and emits foul smell, on which, fuzzy white to grey mould grows. The disease is more prevalent under a condition of high rainfall. This fungus survives on decaying organic matter in and around cardoon fields. Senescent leaves may also support the growth of this fungus. This disease is also of a post-harvest concern since the damaged or improperly handled cardoon can develop symptoms in storage.

Control

- Grow the crop on well-drained light soils.
- Use precise quantity of seed to avoid overcrowding of plants.
- Avoid too deep sowing of seeds.
- Follow clean cultivation and maintain phyto-sanitary conditions.
- Apply light irrigation to avoid wetting of plants base.
- Avoid overhead irrigation and keep the irrigation period as short as possible.
- Remove crop debris from soil after harvest.
- Spray the crop with benomyl 0.2% or apply vinclozolin or iprodione @ 850–1100 g/ha twice.

ARTICHOKE CURLY DWARF VIRUS (ACDV)

Reduced plant growth, lack of plant vigour, distorted leaves with dark necrotic spots or patches and deformed buds are the prominent symptoms of artichoke curly dwarf virus. Infected plants are less productive, having up to 40% less yield than healthy plants. Buds that are produced are deformed and hence unmarketable. Severely affected plant may die.

Control

- Use only disease-free healthy planting material.
- Remove and destroy the infected plants to limit the further spread of disease.

■■■

17 Chicory

M.K. Rana and Shiwani Kamboj

Botanical Name	: *Cichorium intybus* var. sativum (Witloof chicory)
	Cichorium intybus var. foliosum (Radicchio chicory)
Family	: Asteraceae
Chromosome Number	: 2n = 18

ORIGIN AND DISTRIBUTION

The exact center of origin is not known though Mediterranean region and Europe have been suggested as center of chicory origin. It is known that it is cultivated in ancient Egypt, but now, it is being cultivated all over the world. A slight confusion between chicory and endive is that these two crops are the same; however, both are entirely different from each other. The leaf chicory, a Mediterranean plant, is widely grown in Europe, Western Asia, Egypt and North America. In India, this plant is found wild in Punjab, Kashmir and area of southwest India. The chicory has successfully been cultivated in Coimbatore since 1918 and subsequently in the Nilgiris of Tamil Nadu, Nadiad, Amalsad and Jamnagar of Gujarat, Munnar and Vettavoda in Kerala and in parts of Karnataka, Maharashtra, Himachal Pradesh and Assam.

INTRODUCTION

Chicory, a perennial plant, is an important salad or leafy vegetable grown throughout the northern plains and also southwest area of India since long time. The other names of chicory are *Kasni*, *Hinduba*, *succory*, etc. The major chicory growing states in India are Jammu and Kashmir, Punjab, Karnataka, Maharashtra, Himachal Pradesh and Assam. The plant is successfully cultivated in Kinnaur district of Himachal Pradesh and meets major demand of the country. It is mainly grown for its leaves and roots. The edible portion is young enlarged compact and etiolated terminal bud, which is composed of young leaves and partially suppressed but enlarging floral stem known as *Chicons*, and the roots are used as coffee supplement. Chicory comes in many types, *e.g.*, the two important garden types are Witloof and Radicchio. Witloof is grown for its root, which is forced in winter to produce a tight white set of leaves, *e.g.*, Chicons, and Radicchio is grown for its loose leaves and tight head. Nowadays, it is also gaining popularity as a cover crop in vineyards. It is also important as forage for livestock since it provides a good balance between crude protein, energy and minerals to animals but it is also used as vegetable in many parts of the world.

COMPOSITION AND USES

COMPOSITION

Chicory leaves are a fair source of carbohydrates, proteins, energy, fibres and provide appreciable amount of vitamin B, vitamin C, phosphorus, iron, etc. The nutritional composition of chicory greens is given in Table 17.1.

USES

Chicory is especially grown for its delicate leaves and roots, which are used in several ways. The roasted pieces of roots are used for blending and flavoring coffee or as coffee substitute. Young tender roots are also boiled and eaten and may be used to produce blue dye. The leaves are delicious when cooked and served like spinach and the young tender leaves may be used in salad. Chicory is used in fresh salads, whereas, the chicons are used as a cooked food in stuffed witloof, potato puree, witloof salad and many other cooked dishes, as well as raw in salad.

The leaves left after collection of roots may serve as fodder for livestock, as it provides good amount of protein, energy and crude fibre to them. However, chicory leaves, if fed alone, have some laxative and diuretic effect, and thus, they may be fed with feeds like cottonseed or cake, mango seed kernels, etc.

MEDICINAL USES

Chicory has several medicinal properties. The plants are used in indigenous medicines for the treatment of liver disorder, gallstones and inflammation of urinary track. It is used little in modern herbalism, though it is often used as part of the diet. The root and the leaves are appetizer, cholagogue, depurative, digestive, diuretic, hypoglycaemic, laxative and tonic. It is also used in the treatment of insulin independent diabetes and type II diabetes. The latex of stem is applied to warts in order to destroy them. In Indian traditional medicines, *Ayurveda* tonic prepared from this is used to treat fever, diarrhea and enlargement of spleen. Its leaves have been effective in the treatment of jaundice, liver enlargement, gout and rheumatism. Its roots are a good stomachic and diuretic. The wild form is considered as tonic and alexiteric.

Beside, the roots have the potential to produce biomass for industrial use. They are rich in starch inulin, which can easily be converted to alcohol. A blue dye is obtained from its leaves. The flowers are an alternative ingredient of QR herbal compost activator. This is a dried and powdered mixture of several herbs, which can be added to a compost heap in order to speed up bacterial activity and thus shorten the time needed to make the compost.

TABLE 17.1
Nutritional Composition of Chicory Greens (per 100 g Edible Portion)

Constituents	Contents	Constituents	Contents
Water (g)	92.0	Iron (mg)	0.9
Carbohydrates (g)	1.1	Phosphorus (mg)	47.0
Protein (g)	1.7	Thiamine (mg)	0.06
Fat (g)	0.3	Riboflavin (mg)	0.1
Dietary fibre (g)	3.1	Vitamin C (mg)	24.0
Calcium (mg)	100	Energy (kcal)	17.0

BOTANY

Chicory, the herbaceous perennial, produces a continuous rosette of leaves on a shortened stem. Plant has alternate leaves with lobed or toothed margins. The plant forms a distinctive deep fleshy taproot. Its roots are dirty brownish or yellow from outside and white from inside. The inflorescence is a receme within 9–18 separate clusters of 1–9 sub-sessile heads alternating along branches, branchlets and upper stems. The flowering cycle develops an elongated hairy stem, forming flowering heads-axillary bearing in cluster of sessile perfect, blue, purplish and occasionally white cross-pollinating flowers. The flower heads are bright blue, solitary terminal, often clustered and the axillary flowers are arranged along stout, erect, green, nearly leafless spreading stem. The flowers are 2.5–4 cm across the ray florets.

The bracts are green, outer lanceolate and spreading and the inner one longer and erect. The flowers are hermaphrodite (having male and female organs) and are pollinated by honeybees or self. The plant is self-fertile. Fruits are achene, single seeded and dark brown in colour.

CLIMATIC REQUIREMENT

Being cool season crop, it thrives best under mild climatic conditions and requires a hot and humid climate if raised for roots, mostly cultivated in southern states, but for raising seed crop, it is grown in regions with dry temperate climate. For better germination, it needs a minimum temperature of 21°C, and for better plant growth, a uniform temperature of 18–24°C is favourable. Being a hardy plant, it can tolerate mild frost, drought and an extreme temperature during its vegetative and reproductive phase, but at very high temperature, the leaves may become fibrous. It needs chilling temperature for about 90 days to enter reproductive from vegetative phase. Like lettuce, chicory responds to day length and high temperature relative to bolting.

The chicory plant can tolerate a wide range of precipitation but an annual precipitation of 300 to 400 cm is better during its vegetative phase. It requires a well distributed rainfall but can also be cultivated with careful irrigation. Moderate rainfall and plenty of sunshine favour its vegetative growth but rains hinder harvesting and curing of crop.

SOIL REQUIREMENT

The chicory crop can be grown on a wide range of soils. However, it prefers fertile, especially top 20 cm, well-drained soil rich in organic matter for best vegetative growth. The crop performs well on light soils but fertile loam with slightly clay content gives higher yield with good-quality roots. Heavy soils are not preferred because of difficulty in securing a good crop stand and smooth root production. In hilly tracts, alluvial soils are most preferred for its production, while in plains, light or somewhat sandy soils are considered best for obtaining large, thick and fleshy roots. Although it can tolerate a soil pH of 4.5–8.3, but for getting higher yield, the soil pH should be 5.5–7.0.

CULTIVATED VARIETIES

The description of varieties of both Radicchio and Witloof are given as follows:

Radicchio Type Chicory

A chicory variety resembles a small red head of lettuce or cabbage. It can range in size from a large radish to a large grapefruit. The leaves range in color from bronzy green to wine red or magenta with white veins. The important varieties of this group are discussed as follows:

CHIOGGIA (ROSA DI CHIOGGIA)

The most common Radicchio resembles a compact head of cabbage with dark wine coloured leaves and white ribs with a bittersweet to bitter taste.

VERONA RED (ROSA DI VERONA)

A variety resembles an elongated butter head lettuce with tender but firm leaves that are slightly bitter. It produces a bright red head with prominent midribs and veins.

TREVISO (ROSSA DI TREVISO)

A first chicory variety developed during the 16th century is green and bitter during hot weather. Its second growth produces red cone-shaped head with pure white central ribbing and leaf veins.

EARLY TREVISO

A variety resembles a small reddish purplish lettuce with a slender and tapered head of reddish-purple leaves and white ribs.

CESTELFRANCO

A variety developed probably through a cross between a radicchio and escarole, produces crinkled lettuce-like leaves, eggshell coloured blotched with wine red spots, and with mild flavor and tender texture.

WITLOOF TYPE CHICORY

The witloof type chicory varieties are Witloof de Brussels (forcing type), Zoom, Sugarhat and San Pasquale.

SOWING TIME

The sowing time of chicory varies considerably in relation to the type of variety and climatic conditions of that region where crop is to be raised. Sowing of seeds of a particular variety at its right time is very essential for getting higher greens yield and better quality roots since unfavourable weather conditions produce adverse effect on these components.

Radicchio is a hardy cool-season perennial chicory grown best in spring and early summer in cold winter province and in fall and winter in warm-winter province. Chicory seeds are sown in gardens 2 to 3 weeks before the average date of last frost in spring. The sweetest chicory is produced in the cool season. Radicchio heads are most often grown indoors as the second stage of plant growth.

For forcing chicory, seeds are directly sown in the field in the last week of April to the beginning of May. For better roots and head, the crop may be sown in June–July so that they can be harvested in the months of October to November.

SEED RATE

The germination percentage of chicory seed chiefly affects the seed rate. It may also vary due to growing season, method of sowing and soil properties. Chicory seed is slow growing, hence, 3 to 5 kg seed is enough for sowing a hectare area. Higher seed rate per unit land area decreases the colour development in roots.

SOWING METHOD

Chicory is propagated by seed. Sowing can be done by broadcasting the seeds; however, line sowing gives a yield higher than the broadcasting method and the seed wastage can be reduced by sowing in lines. Chicory seeds are sown 6 mm deep at a spacing of 60 to 80 cm between rows and 2.5 to 5 cm between plants within row.

NUTRITIONAL REQUIREMENT

Chicory can successfully be grown with 20–25 t/ha well rotten farmyard manure supplemented with nitrogen 60–70 kg, phosphorus 75–80 kg and potassium 100 kg/ha, while for seed crop, nitrogen 80 kg, phosphorus 120 kg and potassium 160 kg/ha is needed. The farmyard manure should be incorporated into the soil during field preparation. Half of the dose of nitrogen and the entire dose of phosphorus and potash fertilizers is applied before sowing the seeds and the remaining half of nitrogen is applied in the form of top dressing at the time of second or third irrigation. A moderate quantity of nitrogen should be applied, as application of its higher doses increases nitrates content in leaves and makes the plants susceptible to rotting.

IRRIGATION REQUIREMENT

For better seed germination and subsequent growth of plants, pre-sowing irrigation is indispensable as the seed germination takes place better where soil is properly moist. If there is lack of moisture, the seed may not germinate. Therefore, in field where chicory seeds are to be sown, the soil should have plenty of moisture at time of sowing.

The water requirement of chicory depends on soil type and temperature. Usually, it needs water frequently to maintain uniform soil moisture. In first week, water is applied on alternate days, but once the plants are established, irrigation can be reduced, applying at 10-day intervals. However, irregular supply of moisture may cause discoloration of younger leaves edges. Chicory may not tolerate waterlogging conditions.

INTERCULTURAL OPERATIONS

HOEING AND WEEDING

Chicory field should be kept free from weeds, as the plants cannot compete with weeds, thus, weed control is necessary during early stage of crop growth. At least two shallow hoeings are necessary for proper soil aeration and keeping the weeds down at seedling establishment stage. Weeding should always be done shallow to ensure better root development and uninterrupted plant growth. Close spacing may also suppress weed growth and establishment to certain extent.

THINNING

Thinning is an important operation to maintain optimum spacing among the plants and to provide favourable conditions to the plants for their growth and root development. When the seedlings are large enough to handle (10 cm tall), the plants are thinned out keeping 15 to 35 cm apart since it is essential to reduce competition among plants for space, light, nutrients and moisture.

FORCING

While the leaves of witloof chicory can be eaten during summer, most force new growth during winter. This new growth is called as *chicon*. It is an important practice to produce chicons. For forcing chicory, the tops are cut off within 2.5 cm of the crown and the roots placed closely together in deep pots of moist soil. Another pot is to be put on it to exclude light and a drainage hole should be plugged.

The chicory can be forced in pits, cold frames, or houses. Pits are dug 18–20 cm deep. The roots trimmed about 1 cm above shoulder to the depth of pits are planted closely in upright position and covered with moist soil, sand, peat, or combination of all three. A layer of straw mulch can also be used. The pit or bed is covered with opaque material allowing at least 40 cm space over the bed for shoot growth. Proper moisture and humidity should be maintained throughout growth. If light reaches the plant, bitter flavour and green colour will develop.

HARVESTING

Chicory is harvested when heads are compact, brightly coloured and attain appropriate size. Witloof roots used for forcing chicons should be harvested just before the first frost date. The leaves are cut 1 cm above the crown and the roots are lifted from the soil and trimmed to uniform size. The trimmed roots should be stored for 3–7 weeks before forcing. Forced witloof heads should be harvested 3–5 weeks after the roots are transferred to pots.

Radicchio should be harvested when the leaves or heads reach full size. The delayed harvesting may cause bitter flavour and toughness of Radicchio. The stem is cut just below the leaves to keep the head together and diseased or damaged heads are trimmed out. The heads are then washed and cooled by either vacuum cooling or hydro cooling at 0°C. The heads are then packed in wire bound crates, cartons, or baskets.

YIELD

The yield may vary depending on crop variety, soil fertility, environmental factors and cultural practices adopted by the growers. On an average, the crop gives a wet root yield of 200 q/ha. For forcing chicory, each root provides one chicon.

POST-HARVEST MANAGEMENT

The shelf life and the storage requirements of chicory are similar to other cool season leafy vegetables.

INSECT-PESTS

Aphids (*Myzus persicae, Brrevicoryne brassicae* and *Liphaphis erysimi*)

Both nymphs and adults of this tiny insect suck cell sap from under surface of the tender leaves reducing plant vigour. The leaves curl up and ultimately wilt. The aphids excrete honeydew on which black sooty mould develops, which hampers the photosynthetic activity. Besides, these aphids act as vector for the transmission of virus. The severe infestation affects the yield and quality of roots adversely.

Control
- Avoid excessive use of lime.
- Spray the crop with simple water with pressure.
- Remove and destroy the plant portion attacked by the aphids.
- Spray the crop with endosulfan 0.07%, malathion 0.05%, monocrotophos 0.04%, methyl demeton 0.05%, dimethoate 0.03%, or phosphomidon 0.03%.

DISEASES

ALTERNARIA LEAF SPOT *(Alternaria raphani* and *A. dauci)*

This disease is quite wide spread, and when the foliage of chicory is infected, the roots may become infected and show symptoms. The spots are nearly circular often zonate and show various shades of brown or sometimes black colour.

Control

- Follow long crop rotation strictly.
- Always use disease-free seed to avoid the incidence of this disease.
- Treat the seed with hot water at 50°C for 30 minutes to kill the pathogen present on or below the seed coat.
- Spray the crop with Bavistin, mancozeb, Blitox, or Blitane at 0.2% to control this disease.

POWDERY MILDEW (*Erysiphe* sp.)

The infected leaves, stems and other succulent plant parts show symptoms of grayish-white powdery patches of various sizes. The white floury coating, which appears on leaves, causes yellowing of leaves, and subsequently, rapid defoliation of leaves. Dry conditions are always conducive for the spread of this disease. It is a seed-borne disease.

Control

- Irrigate the crop frequently.
- Remove the infected leaves and destroy.
- Treat the seed with hot water at 50°C for 30 minutes.
- Spray the crop with Karathane 0.5%, calaxin 0.1%, Cosan 0.2%, sulfex 0.2%, moroside or topaz 0.1%, Benlate 0.15%, Bavistin 0.15%, or Dithane M-45 0.2% 3–4 times at an interval of 10 days just after the appearance of disease.

DOWNY MILDEW (*Peronospora parasitica* pers.)

This disease may attack the plant at any stage of growth, even at flowering stage. However, it is not as serious as the powdery mildew. In young seedlings, discolouration occurs, and in severe cases, the whole plant dies. In mature plants, purplish or yellowish-brown spots appear on upper surface of the leaf, while fluffy downy growth is observed on the lower leaf surface. The infected leaves turn yellow and later die. During flowering stage, blackish patches appear on the seed stalks as a result the seed production is affected adversely. Moisture and temperature both are important factors for reproduction of the pathogen and further spread of this disease. The incidence is very severe under very wet weather conditions. The disease is soil- as well as seed-borne.

Control

- Follow long crop rotation.
- Remove infected leaves and destroy them.
- Follow prophylactic sprays of mencozeb 0.3%, Dithane M-45 0.2%, or Dithane Z-78 0.2%.
- Treat the seed with hot water at 50°C for 30 minutes to kill the pathogens present on seed.

18 Dandelion

M.K. Rana and D.S. Duhan

Botanical Name : *Taraxacum officinale* Weber

Family : Asteraceae (Compositae)

Chromosome Number : 2n = 8

ORIGIN AND DISTRIBUTION

Dandelion, a very common weed, grows wild almost everywhere. Probably a native of Eurasia, it is now widespread in North America and over 60 countries throughout the world. In Ohio, it is the most prevalent flowering plant. It is thought that the genus *Taraxacum* has been derived from a Persian word *tarashaquq*, which means bitter potherb and refers to the plant of culinary use. It was recognized and used by the Persian pharmacists around 900 A.D. The species *officinale* has been derived from a Latin word *officina*, meaning an office, storeroom, or pharmacy. In India, it is found throughout the temperate Himalaya from Kashmir to Khasi hills of Meghalaya and Mishmi hills of Arunachal Pradesh. In Himachal Pradesh, the plant is found widely distributed in pastures. The Russian dandelion (*Taraxacum kok-saghyz* Rodin) is a native of Uzbekistan and Kazakhstan and was first brought to the United States in 1942 and used as an alternative source of rubber production. During World War II, dandelion was grown by several Eastern European nations for rubber production since tropical sources were unavailable.

There are two species of dandelion, described as follows:

- **Common dandelion (*Taraxacum officinale* ssp. officinale):** It is an introduced weed in North America. In the mid 1600s, European settlers brought it to Eastern America and cultivated it in their gardens for food and medicine. Since then it has spread across the continent as a weed.

- **Native dandelion (*Taraxacum ceratophorum* ssp. ceratophorum):** Native dandelion is the species that occurs naturally in an area. There are several native dandelions in North America but only one of these, a small alpine species called northern dandelion, grows in Alberta. In Alaska, common dandelion can be distinguished from native dandelion by the absence of horns on its involucral bracts and by its substantially larger flower heads. Native dandelion species are found primarily in undisturbed herbaceous alpine pastures.

INTRODUCTION

The dandelion belongs to Asteraceae family, containing over 22,000 species. Sunflower, aster, artichoke, chrysanthemum, marigold and lettuce are its close relatives. The name 'dandelion' has been derived from the French word, *dents de lion*, meaning teeth of the lion, and refers to the jagged leaves edges. In Alberta, it is listed as a nuisance or noxious weed, which must be checked as it can become a problem in gardens and lawns if allowed to grow unchecked. It is an early herbaceous perennial, meaning that it consists solely of soft tissue and an indefinite lifespan and it is commonly found on disturbed sites in nature, *i.e.*, lawns, pastures, roadsides, waste places and meadows. No lawn or garden can escape dandelion seeds that blow in wind. Its seeds are never fully dormant even though it grows very little in winter. It becomes rapidly a troublesome weed in agronomic fields under reduced tillage regimes. In spite of its status here as a noxious weed, the plant has both culinary and medicinal uses. People hate and love common dandelions. Children love to pick them, make chains from them and blow the seeds into the air.

COMPOSITION AND USES

COMPOSITION

Nutritionally, dandelion has remarkable nutritive value, as it contains almost as much iron as spinach and four times vitamin A content. It has higher potassium content than banana and more vitamin A than carrots. An analysis of dandelion shows it to consist of carbohydrates, protein and fat. It provides calcium, phosphorus, iron, magnesium, sodium, vitamin A and C. It contains a bitter crystalline principle taraxacin and a crystalline substance taraxacerin. The leaves also contain phytosterols: taraxasterol and humotaraxasterol. The nutritional composition of dandelion herb is given in Table 18.1.

USES

Dandelion tender leaves and crowns are used as a tasty salad or cooked vegetable. To retain pungency, the leaves before use should be torn to pieces rather than cut. Thereafter, the pieces are cooked in a little water or in combination with beet leaf. A tasty soup can be prepared from chopped leaves and its flowers are fermented to make wine. The dried leaves are used in tea and as an ingredient in diet drinks. The roasted and ground roots are often used to give stronger taste to tea or coffee, which is a natural harmless conventional beverage. In the era when minerals and vitamins were not available in the form of tablets, millions of people died due to their deficiencies. *Scurvy* disease in that era was as dreaded as AIDS in today's era. Dandelion being powerhouse of vitamin

TABLE 18.1
Nutritional Composition of Dandelion Herb (per 100 g Edible Portion)

Constituents	Contents	Constituents	Contents
Water (g)	83	α-Carotene (µg)	363
Carbohydrates (g)	9.20	β-Carotene (µg)	5854
Protein (g)	2.70	Cryptoxanthin (µg)	121
Fat (g)	0.70	Lutein-zeaxanthin (µg)	13610
Cholesterol (µg)	0	Vitamin A (IU)	10161
Dietary fibre (g)	3.50	Thiamine (mg)	0.190
Calcium (mg)	187	Riboflavin (mg)	0.260
Phosphorus (mg)	66	Niacin (mg)	0.806
Potassium (mg)	397	Pantothenic acid (mg)	0.084
Sodium (mg)	76	Pyridoxine (mg)	0.251
Magnesium (mg)	36	Folic acid (µg)	27
Iron (mg)	3.10	Vitamin C (mg)	35
Manganese (mg)	0.342	Vitamin E (mg)	3.44
Selenium (mg)	0.5	Vitamin K (µg)	778.4
Zinc (mg)	0.41	Energy (kcal)	45

A, vitamin C, iron, calcium and potassium, perhaps helps in alleviating many deficiencies and ailments. Dandelion is commonly eaten by mouse, bears, sharp-tailed grouse, pocket gophers, deer, elk and big horn sheep. Common dandelion is an important source of nectar and pollens for bees in Alaska. Therefore, its presence may alter the pollination ecologies of co-occurring plants. Its roots bear milky latex, which is processed into rubber.

In the 1930s, Russia produced nearly 30% of their domestic rubber from 67,000 hectares of this plant. The roots are harvested prior to the first fall frost and they can be stored fresh or dried for longer-term storage prior to the extraction of latex. Its roots also contain substantial amount of starch inulin, which can be fermented to produce fuel ethanol.

MEDICINAL USES

Dandelion has been recognized for a number of medicinal properties. Its leaves are used as a diuretic. It has been known to treat a wide variety of ailments from acne to heart stroke, liver ailments and night blindness. It helps in making the liver and the gall bladder normal by handling fats within the body and aiding the detoxifying role of liver and exercises a beneficial effect upon nervous system. It acts as a general body stimulant but chiefly acts on liver and kidneys. It is used as a bitter tonic in atonic dyspepsia, mild laxative, appetizer and digestion promoting. Its leaves juice mixed with carrot and turnip leaves juice containing readily available magnesium is valuable for all bone disorders. The entire plant is used in many forms to cure urinary disorders and helpful in cases of slow start of passing urine. Its plant extract shows strong activity against HIV-1 and retrovirus replication. Its roasted roots are also used in dyspepsia, rheumatism and gout. Its milk is smeared twice or thrice a day to cure warts. The leaves are effective against baldness, dandruff, toothache, sores, fevers, rotting gums, weakness, lethargy and depression. Spring roots are higher in taraxacin, which stimulates bile production.

SAFETY MEASURES

The use of herb may have side effects and interact with other herbs, supplements, or medications, thus, herb should be taken with care under the supervision of a health care provider. Though dandelion is considered safe, its touching may cause allergy or mouth sores in some people. If an individual is allergic to ragweed, chrysanthemums, marigold, chamomile, yarrow, daisies, or iodine, should avoid the use of dandelion in any form. In some people, it can cause hyperacidity, heartburn and skin irritation. People suffering from kidney, gall bladder, or gallstone problems should ask their health care provider before its use. Pregnant or breast-feeding women should avoid its consumption. Diabetics should be aware that its consumption could drop sugar in blood.

BOTANY

Dandelion, a member of Asteraceae, composite, or daisy family, is a perennial plant, which grows to a height of nearly 30 cm and live for as long as 10 years. Its dark green leaves measuring 5–25 cm length and 1.2–4 cm width are divided into tooth-like lobes that point back toward base with sparsely hairs. Leaves are alternate, which are difficult to distinguish as the stems are so compressed that the nodes are generally at or below the soil surface. The shape of leaves may vary with cultivar. They grow directly from the crown and form a rosette of leaves at soil surface. It develops a strong taproot, which may extend to a depth of 1.8 m. Old roots divide at the crown into several heads. The roots are perennial and tapering, simple or more or less branched, fleshy and brittle, externally of a dark brown, internally white and abounding in an inodorous milky juice of bitter but not disagreeable taste.

Flowering begins in spring, when the day length is <12 hours or in summer if there is sufficient shade. Most plants have several flower heads, each carried separately on a long hollow stem, which releases a thick white juice when broken. Each flower head with 100–300 petals is a bright yellow tuft that appears as a single complete flower, measuring 2–4 cm across. A compact mass of many petal-like flowers is called a composite. Although the bees find its flowers a favourite source of food but they do not pollinate dandelion flowers. Dandelion is apomictic and seeds are produced asexually without fertilization since most of the pollen grains are sterile. Seeds produced are genetically similar to the parent plant. The plants regenerate from its root fragments. The flower bud develops on a short stem near the ground for about a week, and thereafter, the stem elongates swiftly, lifting the flower bud above the leaves. The flower opens early in the morning and closes the same evening. The stem droops and the closed head rests near the ground for several days until the seeds mature. The stem straightens again and the flower reopens as familiar fluffy white ball of parachutes, containing individual seeds, each with an attached cluster of hairs called pappus, which breaks off easily and with the seed still attached is carried away by wind.

TYPES OF DANDELION

WHITE FLOWERED JAPANESE DANDELION (*Taraxacum albidum*)

A plant native to southern Japan is perennial type that can be found in eastern areas of Eurasia. It has deeply lobed leaves and hollow flower stalks. It is a hermaphrodite and self-fertilized plant. Once the flower pollinates it closes and will later open like the other dandelion species.

JAPANESE DANDELION (*Taraxacum japonicum*)

A second species of dandelion, which is native to Japan, can also be found in eastern Eurasia. In the United States, it is considered an invasive weed. It is a smaller dandelion species than the common species.

CALIFORNIA DANDELION (*Taraxacum californicum*)

California dandelion found in the San Bernardino Mountains is listed as an endangered species. The flower, which is very similar to common dandelion, is a small wildflower. The yellow-flower heads yield brown and white fruits when in bloom. There are fewer than 20 known occurrences of the California dandelion with recreational development, cattle grazing and trampling.

RUSSIAN DANDELION (*Taraxacum kok-saghyz*)

Russian dandelion, found in the Soviet Union, is used to produce high-quality rubber. Appearance wise, it is very similar to other dandelion species. However, it has greyish-green leaves, which are smaller and horn-like around the bud. It has also been cultivated in the United States, Germany, Spain and the United Kingdom. During World War II, it was used as a source of rubber.

COMMON DANDELION (*Taraxacum officinale*)

The common dandelion can be found around the world. It grows on lawns, roadsides and any other temperate regions with good soil and moisture. Though many people consider it a weed, it has many different uses from medicine to salads and even wine. In summers, it is common to see children picking dandelions and blowing the silver tufts into the wind.

CLIMATIC REQUIREMENT

The ideal conditions for its cultivation are moderate rainfall and moderate temperature ranging from 23°C to 28°C. It requires a long growing season and develops best at low temperature. Although dandelion requires full sunlight but it can also do well in partial shade. The plant can tolerate hot summers and cold winters. Cold conditions closer to freezing temperature are not suitable for growing dandelion. This species can withstand temperature up to −3.8°C. However, such conditions on south Indian hills lead to majority of the plants flourishing in large number. It requires 100 frost-free days to grow and reproduce successfully.

SOIL REQUIREMENT

Being a hardiest plant, dandelion can grow everywhere, regardless of soil conditions but it thrives best in moderately fertile loam soil having good water retention capacity. It does not grow well in poorly drained soil, as in such soils, it grows slowly with a poor root system. Any well-drained garden soil is good for its cultivation. It is well adapted to all types of soil with pH from 4.8 to 6.5, but optimally, it grows in soils with 7.5 or a slightly alkaline reaction. Sulfur-based products or sawdust, wood chips and peat moss can be used to lower down the soil pH.

VARIETIES

Broad Leaved: It is a thick- or cabbage-leaved variety harvested 95 days after planting. Its leaves are large, broad, dark-green, thick, tender and more deeply lobed along the leaf axis than those of the wild dandelion. In rich soil, its plant spreads 40–60 cm across.

Some other varieties of dandelion are Vert de Montmagny Dandelion (or French Dandelion), Amélioré à Coeur Plein Dandelion, Pissenlit Coeur Plein Ameliore Dandelion, Improved Broad Leaved Dandelion, Arlington Dandelion and Dandelion Ameliore.

SOWING TIME

Dandelion can be planted in the months of August–September when lettuce is planted in north plain areas of India. In southern parts of the country, the seed may be sown during autumn, and the plant will be ready for use in spring. In temperate climate, the crop is sown in spring months.

SEED RATE

Dandelion is easy to propagate by seed. About 2–3 kg seed is required for sowing a hectare area if sown by drills, although in Europe, the seed rate is as high as 4 kg/ha. In dandelion, no seed treatment is required but stratification for 1 week may increase germination rate to 90%. The seed maintains viability only for 1 year or less, thus, always, the fresh seed should be used for sowing.

SOWING METHOD

The field is divided into small seedbeds and the seeds are evenly sown in these beds at 0.5–1.0 cm depth keeping 5–6 cm distance between plants. As and when the seedlings attain a height of 5 cm, the row and plant spacing is adjusted to 20 cm. Being positive photoblastic, its seeds require light to germinate, therefore, the seeds should not be covered with soil. Thereafter, irrigation is applied in such a way that the top few millimetres of soil may be kept fairly moist until the seeds germinate. Its seedlings grow too quickly when well watered, thus, it is advisable to control the amount of water so that the seedlings may not grow too much. The seedlings are thinned out after forming a first pair of true leaves. The weak and unhealthy seedlings are pinched off keeping the larger seedlings at a spacing of 25–30 cm. The plants are allowed to grow for several weeks until they begin to produce broad green leaves.

For winter sowing, the seeds are sown in cold frames just on soil surface, which should not be dried out. The seeds as usual take about 2 weeks to germinate but 2 weeks of stratification may speed up the germination process. Mature seeds are capable of germinating almost immediately. After attaining optimum height, the seedlings are replanted in field or pots. If baby greens (leaves harvested at relatively small stage) are to be produced rather than full-grown greens, the seeds are relatively sown densely, otherwise, sown about 5–6 cm apart. If dandelion greens are to be grown indoors where there is no risk of wind blowing the seeds away or birds eating them, the seeds are not covered with soil since the dandelion seeds germinate best when exposed to light. In colder parts of the country, it may be desirable to mulch slightly during winters, using leaves, straw, or compost.

PLANTING

Dandelion root fragments have the capacity to regenerate new plant, especially the upper thicker root sections regenerate most rapidly and prolifically. The root segments measuring 1.5–4 mm diameter and 2–10 mm length are found to regenerate. Burying the roots turning upside down or horizontally reduces survival and increases regeneration time. Burying the roots to a depth of 10 cm is essential for better regeneration of plant. The roots taken from the plants at flowering stage have the meagre regeneration capacity. The root fragments can economically be collected from a ploughed field. The collected roots after washing and drying are graded into different grades. The largest (pencil size) roots are cut into straight 5–7 cm long pieces after removing smaller side roots. Deep taproots are difficult to uproot manually since a large taproot can reach a depth of 2 meters. A polyethylene tunnel can be placed over the row to force growth for late winter or early spring cutting. Dandelion emits ethylene gas, which is detrimental to health of other plants, thus, this plant should be grown in isolation.

NUTRITIONAL REQUIREMENT

Dandelion requires high fertility for obtaining economic yield. It is also very responsive to nitrogen fertilization. Phosphorus and potash applied at the time of sowing should be in moderate to optimum range. Assuming the soil moderately fertile, nitrogen, phosphorus and potash are applied @ 50, 25 and 25 kg/ha, respectively. Full dose of phosphorus and potash along with half of the nitrogen is applied at the time of planting to stimulate dandelion establishment and the remaining half of nitrogen is applied at the time of flowering. If the crop is grown without phosphorus and potash, nitrogen may be applied @ 60–80 kg/ha in two splits, half at the time of planting and remaining half in early spring when it becomes green. If dandelion is planted after legume crop, annual application of nitrogen is restricted to three-fourths of the recommended dose.

IRRIGATION REQUIREMENT

Dandelion plant has taproot system. The crop prefers a uniform and frequent supply of water for its fast and tender growth. Irrigation requirement depends upon the growing season, type of variety being grown and seasonal variation. Depending on the season and availability of soil moisture, irrigation is usually applied at an interval of 7–10 days. It is advisable not to irrigate the field too frequently though care should be taken that the field may not dry too quickly and become compact too much. Otherwise, the root development is affected adversely. Soil type does not affect the amount of total water needed though it affects the frequency of water application. An irregular supply of moisture may cause discolouration of young leaves margins. The plant is sensitive to waterlogging conditions. Where the rainfall is <1000 mm, supplying irrigation is very essential.

INTERCULTURAL OPERATIONS

HOEING AND WEEDING

Common dandelion competes with native plants for nutrients, moisture, light and space. Thus, weeding and hoeing are done regularly to retain moisture and soil porosity. In early stages, the crop should be kept weed free since the crop plants in early stages may not compete with weeds due to their slow growth rate. At least two hoeing are necessary to keep the weeds down and provide good soil environment to the roots for their development. Hand weeding within the rows is also necessary. No herbicide is registered for controlling weeds in dandelion crop. Organic practices such as mulching may be considered important in controlling weeds and conserving soil moisture.

BLANCHING

Dandelion quality can greatly be improved if blanched by covering the leaves with paper. Blanching not only keeps the leaves tender but also checks the synthesis of alkaloid responsible for their bitter taste. A week before harvesting, the leaves are tied up in order to blanch them and to moderate their bitterness. Blanching can also be done by covering the individual plant with a pot or container to exclude light. There are cultivars that are self-blanched, especially when planted densely.

HARVESTING

Dandelion leaves are harvested by cutting them just below the crown with a sharp knife. The leaves are ready for harvesting when their length is less than 25 cm or before they become bitter in taste or the plant flowers. Some of the plants are retained in the field without harvesting for the next year's generation. In winters, its plants are covered with a cold frame to provide protection against damage caused by low or freezing temperature.

Dandelion is harvested mechanically since digging of roots by hand is too cumbersome and labour intensive. Only large and vigorous plants are selected to harvest the roots since the small and spindly plants are supposed to have small roots, which are uneconomical. The desired roots are those, which are large, fleshy and single-branched. Its roots are harvested, when the plant is dormant and has stored up full energy in its root. For digging the roots satisfactorily, a carrot digger may be used. After collecting roots from the field, they are separated from other roots and sorted based on their size and quality. Before digging the roots, the soil is made moist by applying very light irrigation, which facilitates the removal of roots if digging is done manually. Before uprooting, the tops are removed closer to the root. All leaves traces are removed from the roots and washing is done properly. Simultaneously, the diseased roots are discarded from the lot.

Harvesting the roots before winters increases inulin (insoluble fibre) levels and lowers the fructose levels. The freezing of winter converts the inulin to fructose, which makes spring roots more palatable for eating, as they will be less bitter and chewy.

After flowering, the upper half portion of dandelion is removed with a cutter and the lower half portion is left for growing continuously. Thereafter, fertilizers and irrigation are applied to bloom the plant again. The upper half portion is harvested again and this operation is done time and again. The planting is done in such a way that the crop may be harvested several times.

YIELD

Under favourable conditions, 10–15 q/ha dry root yield can be obtained from 2-year-old plants. In New Zealand, the 2-year-old plants give a fresh root yield of 100 quintals per hectare and the 6-month-old plants 60–70 q/ha. The difference in plant density affects the size of individual root, which can vary from 3 to 25 g/root. During drying, the roots shrink too much since its roots contain about 83% water.

The price of thick roots in the market is higher as compared to thin roots. The smaller roots are usually trimmed off and used for making dandelion coffee. Every part of the root is thus used for one or the other purposes. Before drying, the leaf-bases and other traces of roots are removed as their presence lessens the root value.

The *clocks* (heads) take around 12 days for the seed to ripen. Before they become ready to blow away in the wind, they are gathered. These fresh seeds may be sown immediately as the germination rate of such seeds is high. Under optimum conditions, a large dandelion plant can produce 48–146 flowering heads with an average of about 250 seeds per head.

POST-HARVEST MANAGEMENT

For cooking and medicine purposes, the fresh dandelion roots are used, but for long-term storage, the roots are dried to a safe moisture level. Roots are scrubbed before cutting them into pieces. Thick roots are sliced lengthwise into strips of uniform thickness to shorten the drying period and encourage uniform drying. However, for medicinal use, drying of intact roots is preferred since oozing of sap from the roots after slicing reduces their quality. Slicing large roots lengthwise into 8–15 cm pieces facilitates drying. Electric ovens are used to dehydrate the roots at 35–40°C temperature until they become brittle. Alternatively, the root pieces are spread on a mesh screen in a cool and dry place having good airflow. The root pieces under such conditions will take 3 to 14 days to be brittle and dry, depending on atmospheric conditions. The roots packed in metal containers are stored in a dry and pest-free place as the dandelion roots attract the larvae of moths. The roots to be used for medicine are not stored for more than 1 year, as prolonged storage reduces the medicinal value of roots.

INSECT-PESTS AND DISEASES

Although dandelion is usually not attacked by insect-pests and diseases but few insect-pests and diseases may attack the dandelion plant. Their descriptions are given as under:

INSECT-PESTS

Aphids (*Myzus perisicae*)

Aphids are soft body tiny insects, which tend to have an oval body. They usually attack almost all tender parts of the plant in groups and suck cell sap from under surface of newly emerged leaves. Heavy infestation distorts the plant with necrotic spots on leaves. Some species of aphid transmit leaf curl virus. Feeding aphids excrete a honeydew-like sticky sugary substance, on which, the black sooty mould grows, which inhibits the photosynthesis of plant.

Control

- Spray simple water with pressure using strong jet sprayer.
- Use floating row covers to keep out the pest during early establishment.

- Use reflective plastic mulch to deter the aphids from feeding on plants.
- Expose the plants to natural enemies such as lady beetle, flower fly larvae, lacewing larvae and parasitic wasps.
- Spray the crop with *neem* seed kernel extract 4%, malathion 0.05%, phosphamidon 0.04%, o r dimethoate 0.045%.

THRIPS (*Frankliniella occidentalis* and *Thrips tabaci*)

Straw-coloured, yellowish brown and elongated adults measuring 1 mm in length suck cell sap from young leaves and shoots and lacerate the tissues, resulting in silvery or brown necrotic pin dot-like spots. The infested plants grow slow and the leaves become wrinkled, curl upwards and distorted with white shiny patches. Rusty appearance in patches develops on undersurface of the leaves. The infested crop presents rusty appearance from a distance. Heavy infestation during growth phase results in late bud formation. Once acquired, the insect retains the ability to transmit virus for the remainder of its life.

Control

- Avoid the use of preferable hosts in crop rotation.
- Bury the plant residues deep into the soil.
- Destroy the natural habitat surrounding the dandelion field.
- Use shiny aluminium foil as mulch around base of the plants.
- Apply row covers before the crop emerges.
- Avoid excessive applications of nitrogenous fertilizer.
- Irrigate the field at frequent intervals to keep the soil moist.
- Use yellow or blue sticky cards to trap the adult thrips (one trap per 100 m^2).
- Use natural predators such as mites, green lacewing larvae, minute pirate bugs, damsel bugs and certain parasitic wasps.
- Spray the crop with *neem* seed kernel extract 5%, acetamiprid 20 SP 0.2 g/l, fipronil 5 SC 1 ml/l imidacloprid 200 SL 0.5 ml/l, or thiamethoxam 25 WG 1–1.5 g/l of water.

CUTWORM (*Agrotis* and *Euxoa* spp.)

The cutworms are greasy blackish brown in colour. They cut the plants at ground level and feed on tender parts only. They are nocturnal in habit as they feed only at night, and during daytime, they hid underground. They also attack root and make holes. Most of these cause damage in late June and early July by feeding on roots until cool weather arrives. They spend winter 10 to 20 cm underground well below the frost line. In spring, they come on surface, feed again and then burrow back into the earth where they pupate, emerging as adults after several weeks. The dandelion is damaged by the larvae in fall season.

Control

- Plough the field deep in hot summer months, as the beetles prefer a moist environment for their survival.
- Use only well-rotten farmyard manure.
- Discourage the adults to lay eggs.
- Deep and infrequent watering can reduce the grub population.
- Pick up the beetles by hand and destroy them.
- Apply phorate 10 G at 10 kg/ha in soil around the plant and rake the soil thereafter.
- Spray the crop with Dursban 20 EC at 2.5 ml/l of water or drench the soil around the root system where the damage is noticed.

DISEASES

ANTHRACNOSE (*Colletotrichum* spp.)

Tan-coloured sunken spots or lesions with raised dark brown margins appear on affected foliage. The spots become dark with age, and in moist conditions, pink spores are followed by black fruiting bodies. The fungus also causes a major post-harvest problem and remains dormant in the tissues for many months. The disease is seed-borne and carried over on crop residue in the soil. It spreads through water droplets and becomes worse in warm humid weather. The disease is most severe during wet weather.

Control

- Follow long crop rotation with non-host crops.
- Treat the seed with hot water (52°C) for 30 minutes.
- Provide good drainage facilities.
- Avoid excessive use of nitrogenous fertilizer.
- Spray the crop with 0.2% Dithane M 45 or Dithane Z 78.

DOWNY MILDEW (*Peronospora* spp.)

The disease appears as yellow to white patches on upper surface of the older leaves, and on undersides, these areas are covered with bluish cottony fungal growth, which is often noticed after rain or heavy dew and disappears soon after sunny weather resumes. As the disease progresses, the leaves may eventually turn crisp and brown and fall off. The disease occurs mostly in early spring or late fall season and overwinters on plant debris and in the soil, thus, this is soil as well as seed borne disease. Fungal spores spread through insects, wind, rain, or garden tools.

Control

- Strictly follow long crop rotation with non-host crops.
- Treat the seed with hot water (52°C) for 30 minutes.
- Remove and destroy the infected plants from the field.
- Follow phytosanitary conditions during fall and winter to prevent the spread of disease.
- Spray the crop with 0.2% Dithane M 45 or Dithane Z 78.

POWDERY MILDEW (*Erysiphe* and *Sphaerotheca* spp.)

The disease most often appears in mid to late summer. Unlike most fungi, it does not require a lot of moisture. It appears as small, white, round, powdery spots on top sides of the older leaves but quickly covers whole leaves. In severe cases, the leaves turn brown and die. It is more problematic in shady and humid areas with poor air circulation.

Control

- Provide good air circulation.
- Use low doses of nitrogenous fertilizer.
- Remove infected leaves to reduce the spread of disease.
- Destroy the disease debris in the fall season to reduce overwintering fungal structures.
- Dust the crop with sulfur at 3 kg/ha.
- Spray the crop with Karathane (dinocap), Sulfex, or Thiovit 0.2%, Calixin or Bavistin 0.1% at 10-day intervals.

19 Globe Artichoke

M.K. Rana and Tarique Aslam

Botanical Name : *Cynara scolymus* L.

Family : Asteraceae (Compositae)

Chromosome Number : 2n = 34

ORIGIN AND DISTRIBUTION

Globe artichoke has been originated from North Africa and other parts of Mediterranean region. Out of 12 species, *Cynara scolymus* L. is cultivated commercially. Its wild relatives are *Cynara cardunculus* L., *C. syriaca*, *C. cornigera* (syn. *C. sibthorpiana*), *C. algarbiensis*, *C. baetica*, *C. humilis* and *C. cyrenaica*, etc., which are distributed in Spain, Morocco, Algeria, Tunisia, Naples, Sicily and Palestine. Three species, *i.e.*, *Cynara scolymus*, *C. cardunculus* and *C. syriaca*, are fully cross-compatible with one another and their F$_1$ hybrids are fertile. It was introduced into France in 1548 and was first mentioned in the United States in 1806. In the United States, it was introduced primarily through French immigrants in Louisiana and Italian immigrants in California. It is supposed to have been introduced into Europe by the Moors of Spain, where only young blanched leaves were used. In Italy, the first record of artichoke cultivated for the receptacle of flowers was at Naples around 1400 and they were used by Romans as food 2,000 to 2,500 years ago.

INTRODUCTION

The globe artichoke, commonly known as artichoke and *Hathichoke* in Hindi, is an important vegetable crop in France and other European countries, North Africa (Egypt, Morocco, Algeria and Tunisia), South America (Argentina, Chile and Peru) and the United States (mainly in California). Its cultivation is now spreading in China also. A significant amount of artichoke is grown on a limited scale in some parts of the country as well. The crop is chiefly raised for its immature flower buds or flower heads and the edible bracts, which are harvested before they bloom. The name 'globe artichoke' has been derived from two Arabic words, *i.e.*, *ard*, meaning earth, and *shoke*, meaning thorny or barbs of the earth. It belongs to the family Asteraceae, which also includes lettuce, endive, aster, sunflower, thistles and other cultivated and weed species.

COMPOSITION AND USES

COMPOSITION

Nutritionally, artichoke is not a valuable crop. Though it is low in protein, the levels of minerals and vitamins rank the artichoke seventh among the top vegetables. It contains carbohydrate, protein, dietary fibre and minerals like calcium, phosphorus, potassium, sodium, iron and many of the vitamins essentially required in our diet. The nutritional composition of artichoke is given in Table 19.1.

TABLE 19.1

Nutritional Composition of Artichoke (per 100 g Edible Portion)

Constituents	Contents	Constituents	Contents
Water (g)	83.7	Sodium (mg)	43
Carbohydrates (g)	10.7	Iron (mg)	1.3
Protein (g)	2.9	Vitamin A (IU)	200
Dietary fibre (g)	5.4	Riboflavin (mg)	0.5
Calcium (mg)	51	Vitamin C (mg)	12
Phosphorus (mg)	82	Energy (kcal)	60

USES

Globe artichoke is especially grown for its young immature heads, enlarged receptacle, fleshy inner bracts and choke, *i.e.*, immature flower buds, pappus and bristles. The small heads are cooked as vegetable either alone or mixed with other vegetables and can be used raw in salads, while large heads are consumed only after cooking. Its fleshy petioles and offshoots can also be consumed raw or after cooking. It has delicate nutty flavour, which makes it a prized dish. The crop is also processed by both freezing and canning, in which thick receptacles or *hearts* are used.

MEDICINAL USES

In addition to use as food, it has been reported to have medicinal benefits, including reduction of blood clotting and capillary resistance, neutralization of some toxic substances and amelioration of gastrointestinal process. It is very useful for the patients suffering from diabetes as the flower buds contain insulin.

BOTANY

The cultivated forms of globe artichoke are dicotyledonary herbaceous thistle-like perennial plant whose edible part is fleshy base of immature flower bud. Its root system is mostly fibrous, but during initial year of growth, the largest roots thicken and become more of fleshy taproot. Plants reach a height of 1–1.5 m. Each year the plant develops from the shoots produced at the base, a few at first to as many as a dozen in older plants. Each shoots form a cluster or large basal rosette of leaves covered with dense white hairs from which the stem grows.

In the first year of vegetative growth, the plant forms a rosette of large divided or lobed pubescent greyish green wooly leaves of 5–8 cm length, which are attached with a compressed stem. The inflorescence is a capitulum, and flowers are borne terminally on the main stem and the laterals. Flowers are white, blue, or violet in colour and highly cross-pollinated mainly due to their protandrous nature. Capitulum or head size ranges from 2.5 to 10 cm in diameter and composed of many florets, which are fertile. These heads consists of thick bracts, which are edible along with the large fleshy receptacles.

CLIMATIC REQUIREMENT

Globe artichoke is essentially a cool season crop and can be grown where the winters are mild and summers are relatively cool. It cannot withstand freezing temperature but can tolerate a light sort duration frost. Air temperature and moisture are the primary concerns in determining the productivity of artichokes. Temperature below −1°C may kill the buds and excessively warm days will impair quality as bracts of flower buds open and become tough. The temperature of 12–18°C is most favourable for its growth and development. Along with cool temperature, high relative humidity enhances the bud quality. A light frost may not destroy the edible quality of bud but causes blistered appearance, which will affect the marketability.

SOIL REQUIREMENT

Globe artichoke can be grown on a wide range of soils. However, the best-quality buds can be produced on sandy loam soils with good moisture holding capacity. For getting higher yield, it should be grown in deep, well-drained, friable loam soils that are rich in organic matter and initial plant nutrients. The clayey and very sandy soils should be avoided for its cultivation. This crop grows well in soils with neutral to slightly alkaline reaction having pH range from 6 to 7.5.

CULTIVATED VARIETIES

GREEN GLOBE

This is a high-yielding and very popular commercial variety in coastal California, accounting for about 90% of production. It has large green capitula and thick fleshy scales. It is propagated by seeds.

PURPLE GLOBE

This is also a high-yielding variety producing purple capitula and thick fleshy scales. It can be propagated through seeds.

IMPERIAL STAR

It is another popular seed propagated cultivar grown mainly in dry areas of California and Arizona in the United States.

TEROM

A variety developed in Italy through clonal selection from the progeny of an open-pollinated plant of Violetta di Toscana is propagated vegetatively.

CAMERYS AND CARIBOU

These are early varieties, which are developed through clonal selection from the progeny of hybrid between Camus de Bretagne and Romanesco. They produce green globular heads.

SALANQUATE AND CACIQUE

These violet heads producing varieties are high-yielding clones selected from vegetative propagation of two hybrids between inbreed lines.

JAJA

This variety has been developed through a clonal selection from commercial cultivar Precoce di Jesi.

Fresh market cultivars: Violet de Provence, C. Atanese and Spinosa Sarda.

Processing varieties: Bianco, Tarantino, Precoce di Jesi, Bull, Tudella, E15, Brindisino and Romanesco.

Some other clonal selections are Balca de Espana in Spain, Violetta di Sicilia in Italy and INRA VP 45 in France.

PROPAGATION

Globe artichoke can be propagated by seed as well as by using sucker or offshoots. Since the plants propagated by seeds exhibit undesirable variability, propagation by using suckers or offshoots is preferable. The plant parts used for vegetative propagation are the divisions (rooted basal stem portion), suckers (rooted offshoots) and ovoli (enlarged detached axillary buds), which are obtained from established plantings. Dividing the old crown into pieces with a stem and a piece of crown has the advantage of providing new plant with more reserve food than the offshoot would contain. In globe artichoke, the plants are spaced at a distance of 1 × 1.2 m or equidistant at 1 m, accommodating 15,000 plants per hectare.

SOWING/PLANTING TIME

The sowing time varies considerably in relation to the type of variety and climatic conditions of the growing region where crop is to be raised. In plains, the artichoke is sown or planted in August–October and in hilly tracts in March–May.

NUTRITIONAL REQUIREMENT

Globe artichoke can successfully be grown with farmyard manure 25 t/ha supplemented with nitrogen 60–100 kg, phosphorus 50–60 kg and potassium 150–200 kg/ha. The farmyard manure should be incorporated into the soil at the time of field preparation. The entire dose of phosphorus and potash and half dose of nitrogenous fertilizer should be applied at the time a new planting or when the plants have been cut back at the end of previous cutting period. Rest half of nitrogen should be supplied in the form of top dressing in furrows or through sprinkler irrigation water.

IRRIGATION REQUIREMENT

For better establishment and subsequent growth of the plants, the soil at the time of sowing or planting must have ample moisture, and for this purpose, a pre-sowing or pre-planting irrigation is applied. Light and frequent irrigation may be given to obtain early and higher yield. Lack of moisture will cause development of loose flower buds of inferior quality. Therefore, it requires a relatively constant supply of water. In flat beds, furrow irrigation is commonly practiced. Although artichoke requires constant water but it can not tolerate flooding. Irrigation just after sowing is advantageous in light soils but it hinders germination in heavy soils where the soil surface after drying forms a hard crust before the sprouts come out of the soil. Depending on the season and availability of soil moisture, the crop needs irrigation at an interval of 10–15 days. It is advisable not to irrigate the field very frequently; however, care should always be taken that the field does not dry and become compact too much, otherwise the plant development is affected adversely.

INTERCULTURAL OPERATIONS

HOEING AND WEEDING

The field should be kept weed free in early stage of crop growth since the plants may not compete with weeds due to poor growth in early stages. Thus, at least two shallow hoeing are beneficial and necessary for keeping the weeds down and providing good soil environment to the plant roots for their development. Hoeing is necessary after irrigation to control the weeds. Application of trifluralin 1.2 kg/ha plus Diuron 1.0 kg/ha is applied before planting to control the weeds successfully.

HARVESTING

The plants that are propagated by seed become ready for harvest in about 8 months, whereas, in vegetatively propagated crop, the first harvesting commences 6–7 months after planting. In north India, the flower heads emerge from February to May. Harvesting of globe artichoke for marketing depends on size and type of variety. Flower heads become ready to harvest as and when they attain full size prior to opening of involucral bracts. If harvesting is delayed, the buds become loose and fibrous. Based on compactness, size and age, the buds are selected for harvest and terminal buds are more desirable for harvesting. Harvesting is done by cutting the stem 2.5–3.0 cm from base of the bud. Harvesting should be carried out at an interval of 4–10 days.

YIELD

The yield of globe artichoke depends on crop variety, soil, environmental factors and cultural practices adopted by the growers. Generally, the average yield varies from 100 to 200 q/ha.

POST-HARVEST MANAGEMENT

The method of harvesting is very important factor in shelf life of globe artichoke. The produce should be hydro cooled to 4.4°C temperature. It can be stored for a few weeks at 0°C temperature and 95% relative humidity. At low humidity, the artichoke buds shrivel badly and are more likely to decay than if kept in a moist atmosphere.

INSECT-PESTS

Aphids *(Myzus persica, Brrevicoryne brassicaec* and *Liphaphis erysimi)*

Both nymphs and adults of this tiny insect suck cell sap from under surface of the tender leaves, reducing plant vigour. The leaves curl up and ultimately wilt. The aphids excrete honeydew, on which, black sooty mould develops, which hampers the photosynthetic activity. Besides, these aphids act as vector for the transmission of virus. The severe infestation affects the yield and quality of buds adversely.

Control
- Avoid excessive use of lime.
- Spray the crop with simple water with pressure.
- Remove and destroy the portion of the plant attacked by the aphids.
- Spray the crop with malathion 0.05%, monocrotophos 0.04%, methyl demeton 0.05%, dimethoate 0.03%, or phosphomidon 0.03%.

Plume Moth *(Platypilia cardidactyla)*

The larvae feed on all part of the plants including bud and cause economic loss (2–50%) and make them unmarketable.

Control
- Follow clean cultivation.
- Follow long crop rotation.

- Collect and destroy the caterpillars manually from the infected parts.
- Spray the crop regularly with contact insecticides to reduce the insect population.
- Spray of fenvalerate 50 g, cypermethrin 30 g, or deltamethrin 10 g a.i/ha controls these pests effectively.

SLUGS

These pests feed on leaves and tender parts of the crops and damage the growing tips of plants. Slugs leave a slimy shiny trail though not as pronounced as that of snails.

Control

- Use the poison bait of metaldehyde and bran (1: 25) in 12 litres of water to trap the insects.
- Alum solution @ 2% also acts as repellant to these pests.

Note. These chemicals should not be used very close to harvest though the sprays should be stopped about 15 days before harvesting the crop.

MITES

Both nymphs and adults damage the crop but the nymphs are more harmful. Colonies of mite protected under fine silken webs can be seen on ventral surface of thick leaves of infested plants, sucking the vital sap. The damage caused by de-sapping and covering leaves with thick webs on which soil particle collect during windy weather results in leaf drop, which in turn affects the growth and flowering adversely.

Control

- Clean cultivation and avoid the dust deposition on the plants.
- Dust the crop with sulfur or spray 0.1% wettable sulfur 80WP.
- Spray the crop with carbaryl 0.2% as and when the mites appear.

DISEASES

POWDERY MILDEW (*Leveillula taurica, Erysiphe cichoracearum*)

The disease first appear on the leaves and then on other parts of the plant. White floury patches appear on both sides of leaves as well as floral parts and stems. These patches originate as minute discoloured specks from which a powdery mass radiates on all sides. When the attack has advances, large areas on the aerial parts of the host may be covered with these white floury patches. The superficial mass consists of the mycelium and spores of the fungus.

Control

- Remove plant debris and wild hosts from the field.
- Follow long crop rotation with non-host crops.
- Spray a systemic fungicide like tridemorph @ 0.1% at 10-day intervals.
- Spray 0.2% wettable sulfur at appearance of first symptom and repeat at 20 days.
- Follow 3–4 sprays of dinocap 100 ml in 150–200 litre of water or 0.05% chinomethionat (Morestan) at weekly intervals.

GREY MOULD (*Botrytis cinerea*)

The symptom appears as greyish or brownish indefinite areas on upper surface of the leaf. It sporulates, readily forming grey/silver to brown mould over affected areas. The stem and floral parts are also affected under favourable condition.

Control

- Maintain field sanitation.
- Avoid excess irrigation and provide proper drainage.
- Obtain planting materials from healthy plants.
- Collect the affected debris and burnt to reduce the inoculums at source.

■■■

20 Jerusalem Artichoke

M.K. Rana and M.K. Shivaprasad

Botanical Name	: *Helianthus tuberosus*
	Syn. *Helianthus tuberosus* var. fusiformis
Family	: Asteraceae (Compositae)
Chromosome Number	: 2n = 102

ORIGIN AND DISTRIBUTION

Jerusalem artichoke is primarily originated in North America. In the early 17th century, it was introduced in Europe and widely distributed in temperate and tropical regions. It is cultivated to a lesser extent in high-altitude places like the Caribbean, Malaysia and East and West Africa. In India, it is cultivated in kitchen gardens and hill stations at an elevation of 300–800 m above mean sea level. It is also introduced in some parts of India like Assam, West Bengal, Uttar Pradesh, Maharashtra, Andra Pradesh and Himachal Pradesh.

INTRODUCTION

Jerusalem artichoke is a perennial erect hardy plant belonging to the Asteraceae (Compositae) family, which also includes lettuce, sunflower and globe artichoke. It is called Jerusalem artichoke but it is not actually originated in Jerusalem and is an Italian word meaning *girasole*, wrongly pronounced as Jerusalem. It is cultivated as an annual plant, arising from fleshy rootstock with a height of 2–3 m and width of 0.6 m, bearing oblong tubers of 7.5–10 cm length with many colours like white to yellow and red to blue. Its tubers are knobby and contain inulin carbohydrate that gives fructose after hydrolysis, which is useful for diabetic patients. The tubers are harvested, at which, the tops die off in early winter. The plant, similar to sunflower, produces edible tubers where the globe artichoke produces immature flower. It is commercially soled under several names like *sunchoke* and *lambchoke*. Its tubers resemble a potato but the taste is just like a water chestnut.

COMPOSITION AND USES

COMPOSITION

Jerusalem artichoke green tops provide a good amount of protein, fibre and mineral matters and tubers provide carbohydrates in the form of inulin. The calcium and phosphorus content is also high in tubers as compared to green tops. The carbohydrate inulin is converted into fructose, and as a result, the tubers develop a much sweeter taste when stored in the ground or a refrigerator. The nutritional composition of Jerusalem artichoke green tops and tubers is given in Table 20.1.

USES

Jerusalem artichoke is mainly grown for consumption as human food, alcohol and fructose production and livestock fodder purposes. Its tubers are used in ways similar to potato. Besides, the tubers

TABLE 20.1

Nutritional Composition of Jerusalem Artichoke Green Tops and Tubers (per 100 g Edible Portion)

Green tops		Tubers	
Constituents	Contents	Constituents	Contents
Dry matter (%)	72.8	Nutritive ratio	21.6
Protein (%)	1.4	Water (g)	79.0
Fat (%)	0.3	Carbohydrates (g)	17.0
Fibre (%)	4.9	Protein (g)	2.0
Mineral matters (%)	2.4	Fat (g)	0.1
Calcium (%)	0.4	Dietary fibre (g)	1.2
Phosphorus (%)	0.03	Calcium (mg)	32
Potassium (%)	0.37	Phosphorus (mg)	88
Digestible protein (%)	0.8	Iron (mg)	0.4
Digestible nutrients (%)	18.1	Calories (kcal)	67

are eaten as raw, made into flour or pickled. Tubers used as boiled or baked and very fresh grated raw into salads should be scrubbed, not peeled. The plant is used as summer windbreak for the vegetable garden.

In France, artichoke has been used for the production of wine and beer for many years. It contains ethanol and butanol that are fuel-grade alcohols. However, the cost of producing ethanol currently is not economical as compared to gasoline. Therefore, the extraction of ethanol is limited.

The artichoke tops are suitable for livestock fodder since the tops are considered as maintenance feed. Compared to alfalfa, the crude and digestible protein concentrations are low. The total digestible nutrients content is high in artichoke compared to perennial fodder crops but the total digestible nutrients in artichoke are lower than maize silage. Protein level will be at its maximum in crop harvested in mid September and have optimal forage quality with reduced tuber yield. For getting greater tuber yield, the tops are harvested after a severe frost. Roots, tubers and tops are fed as a fodder in combination. Tops can be fed fresh or ensiled, although the forage does not ensile well because of its high concentration of soluble sugars and high moisture content.

MEDICINAL USES

The tubers contain carbohydrates in the form of inculin and laevulin, which are especially suitable for diabetic patients since these are readily metabolized as the natural sugar laevulose.

BOTANY

Jerusalem artichoke belongs to the genus *Hellianthus* and species *tuberous* of family Asteraceae with diploid chromosome number (2n) 34. The genus includes 75 species and grows to a height of 2–4 meters. It is perennial herb having stem covered with hirsute and often grooved. Leaves 4–8 cm long opposite orientation in below and alternate above with finely pubescent beneath. Flowers are in large capitula, 4–8 cm in diameter, flat receptacle, ray florets normally, yellow in colour and sterile, and the disc florets are hermaphrodite with yellow, purple and brown colour. Tubers produced from underground stolons with oval or irregular shape and vary with skin colour: red, purple, or white contain carbohydrate inculin which gives rise to fructose after hydrolysis.

CLIMATIC REQUIREMENT

Jerusalem artichoke is a warm season plant. It requires 140 frost-free days for harvesting a good crop. It performs well in temperate climate too but it is more productive in warmer and more humid areas. Growing in frost improves the flavour of mature tubers in ground when the foliage has died down. Tuberization initiates with decreasing day length concurrent with translocation. High temperature above 27°C may reduce the yield. More than 500 m elevation is good for getting satisfactory yield of this crop.

SOIL REQUIREMENT

It is best to choose a permanent location for planting Jerusalem artichoke. Being hardy herbaceous, the plant is well adapted to a wide range of soil types and pH levels but slightly alkaline soil reaction favours the yield of artichoke. The plant grows poorly on heavy clay soils, particularly if there is a condition of waterlogging. Plants at tuber development stage are sensitive to poorly drained soils. Hence, it prefers well-drained light soils rich in organic matter. Sandy to sandy-loam soils are best as compared to heavy soils since heavier soils adhere to the tubers at the time of harvesting or digging. The field should be tilled well and free from perennial weeds and grasses although minimal soil preparation may also be enough for its planting.

CULTIVATED VARIETIES

Mammoth White French and Sutton's White are the two cultivated varieties of Jerusalem artichoke. Based on locations, the varieties are selected for cultivation.

PLANTING TIME

Usually, Jerusalem artichoke is cultivated in spring-summer in north plain areas where planting is done in last week of February to first week of March. In tropical zones, it can be grown successfully year-round by regular replanting. However, the crop gives higher yield when it is planted between September and November in tropical areas. Planting should usually be delayed until normal time; otherwise, the frost in winter months may cause severe damage to the young plants.

SEED RATE

Jerusalem artichoke is vegetatively propagated through tubers. The size of seed tubers is related directly to its yield. Tubers used for planting are also called sets. The sets of 56 g having 2–3 buds are most suitable for planting purpose. About 3–5 quintals tubers or sets are needed to plant a hectare area.

PLANTING METHOD

Jerusalem artichoke is replanted from stored tubers. Being vegetatively propagated, the crop is usually planted by individual tubers or sets of 60 g containing two to three buds. The sets are planted on well-prepared raised beds at 2.5–5 cm depth keeping a distance of 45–75 cm between lines and 30–45 cm between plants. Mulching and earthing up are done after planting and tuber development stage, respectively.

MANURIAL REQUIREMENT

Not much information on the fertilizer requirement of this crop is available in the literature. Before planting, incorporation of farmyard manure in soil may be beneficial for the crop grown on sandy soils, which will supply organic matter and help in retaining moisture in the soil for a longer time.

Nitrogen is applied through urea @ 60 kg/ha and phosphorus is applied at the time of planting in the form of double superphosphate @ 500 to 700 kg/ha. It is better not to use phosphorus in the form of ordinary superphosphate as it contains higher level of cadmium, which is a toxic heavy metal. Potassium is considered the important nutrient required for better growth of Jerusalem artichoke so muriate of potash is applied @ 60 kg/ha as side-dressings to light sandy soils. Magnesium sulfate is applied @ 100 kg/ha at 1 and 2 months after planting. It has been recommended that soil nutrients and irrigation water analysis should be made before planting the tubers, as well as one to two analyses of the youngest mature leaves after planting.

IRRIGATION REQUIREMENT

The water requirement of crop in general is not so high, but for obtaining optimum yield from Jerusalem artichoke, total 4 to 5 irrigations are needed. Although it can tolerate dry periods, applying irrigation regularly gives the best tuber production. If soil moisture is less than 30% of field capacity at the time of tuber formation (early September to November), the yield of tops and tubers is limited. The crop is affected adversely by heavy rainfall particularly at tuber development stage as it leads to rotting of tubers. The water requirement of crop is less in areas where the rainfall is more than 1500 mm per annum.

INTERCULTURAL OPERATIONS

For planting Jerusalem artichoke, the soil must be free of perennial weeds seeds and grasses. Since weeds are very hardy and very tolerant of soil conditions, it is difficult to eliminate them once they are well established, thus, it is essential to keep the artichoke field weed free throughout the growing period. Its plants though are extremely vigorous and easily compete with weeds. Once the plants have started growing, they compete with weeds very successfully. The artichoke plants grow very well under mulch with dry hay or straw.

Early season cultivation reduces the emergence of weeds, with a subsequent tillage operation. When the plants are young, a light hoeing is done between rows to control weeds and the last hoeing should be done before the plants become 1 m tall since the tubers begin to form at this stage around base of the plant; thus, they must not be disturbed. The volunteer Jerusalem artichokes can be a serious problem in the following crop when the crop is retained in overwinter. The possible herbicide to eliminate the volunteer artichoke plants is Roundup (glyphosate).

HARVESTING

The crop is harvested 4 to 6 weeks after flowering. If the flower buds are pinched off as soon as they appear, the yield will be higher. In well-drained soils, it is better to dig up the tubers in the cooler period of the day. It is advisable to dig up the entire crop at one time since the tubers may rot if left in the ground for a longer time.

Harvesting is done as the potato is harvested, with few exceptions, as in case of potato, the vines are weak and usually senesced before harvest, which is in contrast to the continued growth of strong stems in artichoke. In potato, tubers from the plant separate easily, while the artichoke tubers are strongly attached and intertwined with the roots.

POST-HARVEST MANAGEMENT

The tops, roots and tubers are stored immediately after harvesting or they can be dried and then sorted. Generally, its tubers do not store well out of the ground; however, they can be stored well in damp sawdust or sand in a dark place or stored in bottom of the fridge with the use of a perforated plastic bag. After harvesting, if the tubers are kept at room temperature, they rapidly shrink and deteriorate. If the tubers are healthy and disease-free, they can successfully be stored for 90–150 days at 0°C temperature and 90–95% relative humidity. Tubers can also be stored for long period without deterioration of protein and sugar in dehydrated and ground form.

YIELD

Jerusalem artichoke plant produces a good number of edible tubers. With better management practices, in India, the average yield is 120–250 q/ha and 300–400 q/ha in other countries.

INSECT-PESTS

Insect-pests are not a serious problem in Jerusalem artichoke. However, if the crop is grown in a large number of hectares, the attack of stalk borer is observed but it usually causes limited damage. No insecticide is currently registered and recommended for use on the crop to control the stalk borer.

APHIDS (*Aphis* spp.)

Both nymphs and adults of this tiny insect suck cell sap from under surface of the tender leaves, reducing plant vigour. The leaves curl up and ultimately wilt. The aphids excrete honeydew, on which black sooty mould develops, which hampers the photosynthetic activity. The severe infestation affects the yield and quality of green tops.

Control

- Spray the crop with simple water with pressure.
- Remove and destroy the portion of the plant attacked by the aphids.
- Spray the crop with malathion 0.05%, monocrotophos 0.04%, methyl demeton 0.05%, dimethoate 0.03%, or phosphomidon 0.03%.

ROOT KNOT NEMATODE (*Meloidogyne* spp.)

This is the most widely occurring pest in all most all the crops including Jerusalem artichoke. The affected plant shows stunted or retarded growth with chlorotic leaves, which reduce the total tuber yield of Jerusalem artichoke.

Control

- Plough the field deep repeatedly during summer months to expose the soil to high temperature.
- Grow a trap crop like marigold and wheat as rotation crop.
- Incorporate *neem* or castor in the field.
- Apply granular insecticide like Phorate 10G @ 25 kg/ha or Cabofuran 3G @ 35 kg/ha at the time of planting.

DISEASES

The pathogen *Sclerotinia*, which causes stalk rot and deterioration of tubers, is reported in Jerusalem artichoke. Downy mildew, rust and southern stem blight have also been reported but do not cause economic loss. Currently, there is no proper recommendation of fungicides for the control of these diseases in Jerusalem artichoke.

Stalk Rot (*Sclerotium rolfisii*)

The outer skin of infected plants remains firm and hard but the stalk becomes soft. Cuts and injuries are the venue for the entry of pathogens, which cause rotting of tissues in tubers and plant stalks.

Control

- If possible, follow long crop rotation with pseudocereal crops or maize.
- Avoid excessive hoeing to avoid injury to the tissues.
- Remove the infected plants along with tubers and destroy.
- Dip the tubers in 1% solution of ziram for 30 minutes before planting them in the field.
- Spray the crop with streptocycline 0.1% or Achromycin (tetracycline hydrochloride) 0.04% at an interval of 7 days.

■■■

21 Lettuce

Baseerat Afroza and M.K. Rana

Botanical Name : *Lactuca sativa* L.

Syn. *Lactuca scariola* var. sativa (Moris)

Syn. *L. scariola* var. integrata (Gren. and Godr.)

Syn. *L. scariola* var. integrifolia (G. Beck)

Family : Asteraceae (Aster or Sunflower family)

Chromosome Number : 2n = 18

ORIGIN AND DISTRIBUTION

The species was first described in 1753 by Carl Linnaeus in the second volume of his *Species Plantarum*. Lettuce is native to the Mediterranean region, and it was first cultivated by the ancient Egyptians in around 4,500 B.C., who turned it from a weed, whose seeds were used to produce oil, into a plant grown for its leaves. Cultivated lettuce was probably derived from the so-called wild or prickly lettuce (*Lactuca sierriola*). Lettuce spread to the Greeks and Romans, the latter of whom gave it the name *lactuca*, from which the English *lettuce* was ultimately derived. Salad lettuce was popular with the Ancient Greeks and Romans. By the 50 A.D., multiple types were described and lettuce appeared often in medieval writings, including several herbals. The crop was also used in China by the 7th century A.D. The primitive forms of lettuce were loose and leafy. Firm-heading forms became well developed in Europe by the 16th Century. The 16th through 18th centuries saw the development of many varieties in Europe like oak-leaved and curled-leaf types of various colours, and by the mid 18th century, the cultivars were described that can still be found in gardens. In India, lettuce was introduced by Portuguese or British during 16th century.

Europe and North America originally dominated the market for lettuce, but by the late 1900s, the consumption of lettuce spread throughout the world. At present, though China is the top world producer of lettuce, the majority of crop is consumed domestically. Spain is the world's largest exporter of lettuce, with the US ranking second. Its popularity is increasing in Asia and Africa. In India, lettuce is widely grown not only in kitchen gardens but also on commercial lines to meet increasing demand of continental hotels and restaurants.

INTRODUCTION

Lettuce is one of the most important vegetable crops of temperate regions and is also widely cultivated in tropical as well as subtropical regions of the world. Lettuce is an annual plant. It is most often grown as a leafy vegetable, but sometimes, it is also grown for its stem and seeds. It is now one of the most important salad crops.

COMPOSITION AND USES

COMPOSITION

Lettuce is a good source of vitamin A, vitamin K and potassium, as well as a minor source for several other minerals and vitamins. The content of nutrients is highest in the dark green outer leaves but it is low in calorie. It also provides some dietary fibre (concentrated in the spine and ribs), carbohydrates, protein and a small amount of fat. Lettuce naturally absorbs and concentrates lithium. The nutritional value of butter head lettuce is given in Table 21.1.

USES

Lettuce is used mainly in salads but the leaves may even be boiled like spinach. It is commonly used in sandwiches. It is also used in the preparation of soups and wraps and can also be grilled. In addition to its main use as a leafy green, it has also gathered religious and medicinal significance over centuries of human consumption. One variety, the *Woju* or asparagus lettuce, is grown for its stems, which are eaten either raw or cooked. In some parts of the world, the leaves are used to make a cigarette that does not contain nicotine. Seeds of a primitive form found in Egypt are used to manufacture some edible oil.

MEDICINAL USES

A sleep-inducing medicine is manufactured from latex found in *Lactuca virosa* L. due to the presence of two sesquiterpene lactones called lactucarium or lettuce opium. Lettuce extracts are sometimes used in skin creams and lotions for treating sunburn and rough skin. Folk medicine has also claimed it as a treatment for pain, rheumatism, tension and nervousness, coughs and insanity. Despite its beneficial properties, lettuce when contaminated is often a source of bacterial, viral and parasitic outbreaks in humans, including *Escherichia coli* and *Salmonella*. Lettuce contains several defensive compounds, including sesquiterpene lactones and other natural phenolics such as flavonol and glycosides, which help to protect it against pests.

TABLE 21.1
Nutritional Value of Butter Head Lettuce (per 100 g Edible Portion)

Constituents	Contents	Constituents	Contents
Water (g)	95.63	Zinc (mg)	0.2
Carbohydrates (g)	2.23	β-carotene (µg)	1987
Sugars (g)	0.94	Lutein and zeaxanthin (µg)	1223
Protein (g)	1.35	Vitamin A (µg)	166
Fat (g)	0.22	Thiamine (mg)	0.057
Dietary fibre (g)	1.1	Riboflavin (mg)	0.062
Calcium (mg)	35	Pantothenic acid (mg)	0.15
Phosphorus (mg)	33	Pyridoxine (mg)	0.082
Potassium (mg)	238	Folic acid (µg)	73
Magnesium (mg)	13	Vitamin C (mg)	3.7
Sodium (mg)	5	Vitamin E (mg)	0.18
Iron (mg)	1.24	Vitamin K (µg)	102.3
Manganese (mg)	0.179	Energy (kcal)	13

BOTANY

Lettuce plant is a lactiferous herb. Plants generally have a height of 15 to 30 cm. The leaves are colourful, mainly in green and red colour spectrums, with some variegated varieties. There are also a few varieties with yellow, gold, or blue-teal leaves. Lettuce has a wide range of shapes and textures, from dense heads of iceberg type to the notched, scalloped, frilly, or ruffly leaves of leaf type varieties. Lettuce plant has a root system that includes a main taproot and smaller secondary roots. Depending on the variety and time of year, lettuce generally lives 65–130 days from planting to harvesting. After flowering (bolting), lettuce becomes bitter and unusable. Hence, the plants grown for consumption are rarely allowed to bolt. Once the plants move past the edible stage, they develop flower stalks with small yellow blossoms. Lettuce inflorescence (flower heads or capitula) is composed of multiple florets (18 to 20), each with a modified calyx called a pappus (which becomes the feathery *parachute* of fruit), a corolla of five petals fused into a ligule or strap and the reproductive parts. These include fused anthers that form a tube, which surrounds a style and bipartite stigma. Flowers are protandrous but autogamy is predominant due to elongation and growth of pistil through the anther tube. The ovaries form compressed, obovate (teardrop-shaped) dry fruits measuring 3 to 4 mm long that do not open at maturity. The fruits have 5–7 ribs on each side and they are tipped by two rows of small white hairs. The pappus remains at the top of each fruit as a dispersal structure. Fruit is a typical achene and the seed is exalbuminous. Each fruit contains one seed and each plant produces approximately 1,500 seeds, which can be white, yellow, gray, or brown, depending on the variety of lettuce.

The cultivated lettuce is never found in wild stage but *L. serriola* is the closest relative and probable ancestor of cultivated lettuce. Two other closely related species, *i.e.*, *L. drageana* and *L. altaica*, can be considered conspecific to *L. serriola*. There are three other species forming part of the *sativa* gene pool, which is called the *L. sativa* group:

- *L. serriola*: Containing small seeds, narrow toothed leaves with spines, pediceled flowers.
- *L. saligna*: It is similar to *L. serriola* but with narrower leaves and sessile flowers.
- *L. virosa*: It is morphologically distinct to the two species discussed earlier, flat seeds, strong rosette.
- *L. serriola*: It is closest to lettuce and crosses freely in both directions. The most distant species in the group is *L. saligna*.
- There are four horticultural varieties:
- *Lactuca sativa* var. capitata (head lettuce) having a structure similar to cabbage. It includes iceberg type (crisp head) and butter head, which has less compact head than crisp head, which is flattened and has a more delicate taste.
- *Lactuca sativa* var. longifolia (cos or romaine lettuce) having crispy and elongated leaves overlapping loosely with each other but without forming a head.
- *Lactuca sativa* var. crispa (curled lettuce) having loose leaf forming a rosette.
- *Lactuca sativa* var. asparagina (asparagus lettuce or celtuce), which is a non-heading type grown for its thick edible stems.

CLIMATIC REQUIREMENTS

Lettuce is a cool season crop that grows best within a temperature range of 12–20°C. The optimum temperature range for seed germination is 16 to 20°C. At soil temperature over 27°C, the germination of seed is poor. The temperature above 27°C also affects head development and edible quality and promotes development of seed stalk at premature stage. High temperature can cause a high incidence of tip burn. It does not suffer from light frost and winter cold except near maturity. Hardened seedlings are tolerant (−5 to −7°C) to frost but mature plants are more sensitive to frost (−1°C) depending on cultivar. With light frosts, only the wrapper leaves are injured. Severe frost before harvest can scorch the leaves and heads.

Warm and dry conditions promote flowering and development of seed stalk, known as bolting, which usually occurs where temperature over 20°C is maintained day and night. High-temperature conditions also cause bitter flavour. Cool nights are essential for the production of quality lettuce. Considerable differences in heat tolerance exist among the lettuce varieties. These differences are the primary reasons that some lettuce varieties can be grown in warmer climates too.

SOIL REQUIREMENTS

The plant grows well on a wide variety of soils ranging from light sand to heavy clay. However, the best results can be obtained on fertile loam soils rich in organic matter. A pH between 6.0 and 6.8 is optimum for its successful cultivation. Lettuce should be grown on soils with a high water-holding capacity and proper drainage for good root growth and plant performance. Warm sandy soils are preferred for the early harvest and loam to clay loam or peat for late but bumper production. A regular supply of water is essential but high humidity and excess water close to the time of harvest can be destructive to yield and quality of the crop.

CULTIVARS

There are several types of lettuce but three (leaf, head and cos or romaine) are the most common. Seven main cultivar groups of lettuce, each including many varieties, are discussed as follows:

- **Leaf type:** It is also known as loose leaf, cutting, or bunching lettuce. This group does not form heads though characterized by soft loosely bunched leaves and is the most widely planted. It is used mainly for salad. It grows well both in open field and under protection, and it can be shipped over longer distances. It is represented by cultivars such as Grand Rapids for greenhouse cultivation and Black-seeded Simpson, Prize Head, Australian and Salad Bowl for outdoor cultivation. This type of lettuce is also grown in India.
- **Romaine/Cos type:** This type of lettuce forms long upright heads, which are used mainly for salads and sandwiches. The cos lettuce has a loose head with narrow soft leaves. The outer leaves are dark green, coarse and have heavy ribs, while the inner foliage is lighter. It is more resistant to cold than the other groups of lettuce. This type of lettuce is usually grown in open fields. The cultivars in this group include Eiffel Tower, Paris Island, Paris White and Dark Green Valentine.
- **Crisp head type:** Better known as *iceberg* lettuce, which is the most popular lettuce in the United States of America. This type of lettuce is characterized by firm heads with crisp and curly leaves. The outer leaves are dark green, while the inner ones are pale and lack chlorophyll. This type of lettuce is very heat-sensitive and was originally adapted for growth in the northern United States. It ships well but is low in flavour and nutritional content, being composed of more water than other lettuce types. The cultivars in this group include Great Lakes, Del Rio, Del Oro, Frosty, Winter Crisp and Winter Supreme, which are cold tolerant. Aviram, Commander, Tropical Emperor, Empire 2000, Summer Gold and Victory are the cultivars within this group that are better adapted to warmer conditions. The cultivars are also well adapted for field growing and for long-distance shipments.
- **Butter head type:** Also known as Boston or bibb lettuce, this type of lettuce forms a relatively small loose head, which is somehow similar to that of cabbage in shape and has soft, waxy, flexible leaves with loose arrangement. It is known for its sweet flavour and tender texture. It is sensitive to hot weather. Cultivars in this group include Kragramer Sommer,

Big Boston, White Boston, May King and All Year Round. Butter head lettuce is more popular in Europe and is less adapted to field growing or long-distance shipments.

- **Summer crisp type:** Also called Batavian or French Crisp, this type of lettuce is midway between the crisp head and leaf type. These lettuces tend to be larger, bolt-resistant and well flavoured.

- **Stem type:** Also called *celery lettuce*, this type of lettuce is grown for its seed stalk or thick stem rather than its leaves and is used in Asian cooking, primarily Chinese as well as stewed and creamed dishes. Celtuce is a lettuce variety grown primarily in Asia for its stems.

- **Oilseed type:** This type of lettuce is grown for its seeds, which are pressed to extract oil mainly used for cooking. It has few leaves, bolts quickly and produces seeds around 50% larger than other types of lettuce.

The *butter head* and *crisp head* types are sometimes together known as *cabbage* lettuce since their heads are shorter, flatter and more cabbage-like than romaine lettuce, which has requirements similar to head lettuce, except it can withstand more heat. Butter head and leaf types can withstand even more heat and have a longer season of production.

Lettuce varieties are crossable with each other. Keeping isolation distance of 1.5 to 6.1 m between varieties is necessary to prevent contamination while producing seeds. Lettuce is also crossed with *Lactuca serriola* (wild lettuce), with the resulting seeds often producing a plant with tough bitter leaves. Celtuce is crossed easily with lettuces grown for their leaves. However, this property for crossing has led to breeding programs using closely related species in *Lactuca* such as *L. serriola*, *L. saligna* and *L. virosa*, to broaden the available gene pool.

In India, various cultivars like Great Lakes (crisp head), Chinese Yellow (leaf type), Slobolt (leaf type), White Boston (butter head type), Dark Green (cos type), Imperial 859 (crisp head type), Black seeded Simpson, Golden Ball, Little Gem, Iceberg, Paris Cos, Wonderful, etc. have been recommended for cultivation.

GREAT LAKES

A crisp head type variety recommended by Indian Agricultural Research Institute, New Delhi, for cultivation in India is resistant to tip burn. It produces large and firm heads with inner green and outer-blistered leaves.

PUNJAB LETTUCE NO. 1

A heading type variety developed at Punjab Agricultural University, Ludhiana, it is cultivated for its loose, light green, shinning and crisp leaves.

SEED RATE

Approximately 400–500 g of seed will provide seedlings sufficient for planting a hectare area. To raise seedlings for planting a hectare area, a nursery area of about 60 to 70 m² is required. Direct seeding lettuce requires seeds 1 to 2 kg/ha.

SOWING TIME

In plains of India, lettuce is sown during September to November. For raising seedlings, its seed is usually sown in the greenhouse from March to July, *i.e.*, 6 to 8 weeks before field planting. In hills, it is sown from March to June.

SOWING METHOD

Lettuce crop can be raised either by transplanting or by direct seeding. The sowing method depends mostly on the availability of seedlings and the season, in which, the crop is to be grown. Springhead lettuce often fails because it is planted too late, and for this reason, the transplants should be considered important. Fall crop lettuce is most often started when the climatic conditions are hot and dry. In this period, direct seeding may be a good choice provided irrigation is constantly available until the plants are well established.

Lettuce seed germinates best at relatively cool temperature. The seeds should be pre-germinated in cool rooms during summer, as the conditions are generally too hot for good germination even if greenhouses are well ventilated. Seed should be kept at cool temperature and not exposed to high relative humidity in order to prevent seed deterioration, which is otherwise a naturally occurring phenomenon.

Lettuce seed is very small and as such requires soils that are not prone to crusting. The soil should be prepared to a fine tilth, clods should be broken and it should be leveled in order to ensure a more uniform emergence. Raised beds are ideal for lettuce production as they help in preventing damage from soil compaction and flooding. They also improve airflow around plants, resulting in reduced disease incidence. Seedlings for transplanting may also be raised in plastic trays or seedbeds and transplanted about 5 weeks after sowing.

Seedlings should be hardened before transplanting. Hardening is done by withholding water for about a week. Lettuce is regularly sown directly in the field to a depth of 1.0 to 1.5 cm. However, weed control is usually a problem with direct seeded fields. The seedlings are transplanted at a line spacing of 30 cm. Plants within a row should be spaced 25 to 40 cm apart for head lettuce and 20 to 30 cm apart for leaf and Bibb lettuce. For getting economic yield, plant population is maintained from 60,000 to 100,000 per hectare.

NUTRITIONAL REQUIREMENT

Lettuce with its limited (small and shallow) root system is considered a poor nutrients consuming crop. It requires moderate amounts of soil fertility. Manure is generally not recommended due to containing weed seeds and the lack of good selective herbicides registered on this crop. However, 10–15 tonnes of well-rotten farmyard manure may be incorporated at the time of field preparation. The NPK doses may be recommended only after soil analysis. However, nitrogen 50 kg, phosphorus 90 kg and potash 50 kg/ha may generally be applied to meet nutrient requirement of lettuce. Half of the nitrogen and full dose of phosphorus and potash may be broadcast before planting and mixed into the soil. The remaining dose of nitrogen is side dressed 3 weeks after transplanting the crop or after thinning of the seeded crop. Over application of nitrogen can result in rapid growth and development of tip burn disorder. Lime should be applied to maintain a soil pH in the range of 6.5 to 6.8. Under hot conditions, lettuce may benefit from foliar sprays of calcium to prevent tip burn. Sulfur is recommended for soils low in organic matter, which is being intensively cropped. Generally, these are sandy soils. Application of boron, copper and molybdenum is needed in organic or peat soils. Application of boron may be beneficial for crop grown on some mineral soils. Minor elements are most conveniently applied mixed with fertilizers but may be sprayed on the soil and incorporated before planting or applied as a foliar spray to the crop. Manganese deficiency may show up in soils having high pH. Apply foliar sprays starting after establishment with manganese sulfate at 1 to 2 kg of manganese per hectare in 3,000 litres of water.

IRRIGATION REQUIREMENT

Lettuce has a relatively high water requirement and not more than 50% of the available water in the root zone should be depleted before irrigation. Lettuce also has a shallow root system and as such requires frequent but lighter irrigations. First irrigation is given soon after transplanting and subsequent irrigations are given at an interval of 4–5 days in light soils and 8–10 days in heavy soils. The roots penetrate the soil to a depth of only 300 mm. Water should be applied throughout the growing period and reduced when the heads attain full size. A water shortage tends to stunting growth, affect head quality and promote bolting. Irrigation greatly reduces risk of crop failure. Usually, irrigation is made to soften the soil prior to thinning and cultivation. Afterwards, adequate moisture should be maintained throughout the plant development. During harvesting, irrigation may be necessary between cuttings if multiple harvests are taken.

INTERCULTURAL OPERATIONS

HOEING AND WEEDING

In lettuce, shallow hoeing is done because of its shallow root system. Weeds are controlled mechanically, manually, or chemically. Lettuce should be planted to land free of perennial weeds, where the annual weed seed population has been reduced by cultural practices such as crop rotation, fallowing, or stale seedbed. Hand weeding is a common practice to control weeds. Weeds are removed by hand hoeing or pulling between plants in the rows. Three to four hand weedings are adequate for entire cropping season. Mechanical weed control can be practiced only before planting because of close spacing. Chemical weed control can be achieved through the application of propyzamide shortly after sowing, which can last 12 months and longer in the soil. However, herbicides recommended for use in lettuce will not provide total and season-long weed control. Use of black plastic mulches controls weeds.

MULCHING

Lettuce crop may be mulched to serve specific purposes. Mulching during winter months raises soil temperature and thereby stimulates plant growth. If rains coincide with cropping season, mulching helps in preventing soil erosion. Mulching also prevents leaves from touching the soil that in turn prevents them from decaying. Different types of mulch may be used but availability and economics should be a consideration. As such, paddy straw is a good choice.

THINNING

Lettuce is sometimes directly seeded in the main field that may lead to non-uniform growth of plants, increased competition between plants for space, moisture, nutrients and sunlight and enhanced weed problems due to difficulty in weeding operations. Hence, the lettuce seedlings are thinned out at 2 to 3 leaf stage to the desired spacing. The thinned-out plants may sometimes be used for transplanting at other places in the field.

HARVESTING

In most instances, the head lettuce will be ready for harvesting in 70 to 80 days after seeding or 60 to 70 days after transplanting. Most leaf types become ready in 50 to 60 days after seeding and 30 to 45 days after transplanting. The heading types are harvested when the heads are grown fully and firm. The loose-leaf types are picked when the leaves have reached the required size. Lettuce is harvested by hand by cutting off the plant just above the soil surface to keep most of the outer leaves around the head. Harvesting should be done very early in the morning, and if possible, kept cool until marketed because lettuce wilts rapidly.

YIELD

Average yield of lettuce ranges from 10 to 14 tonnes per hectare. A good head yield for lettuce is about 400 to 500 crates per acre and for leaf types 800 to 1,000 crates per acre.

POST-HARVEST MANAGEMENT

Leaves that are loose, discoloured, damaged, soiled and diseased are removed. The butt ends are cut cleanly for packing. Leaf, butter head and cos types are cut, trimmed and tied into compact bundles before being placed in cartons. It is graded according to head size. Good quality lettuce is that, which is free from wilting, shriveling, or bitter taste and is firm, fresh, clean and crispy.

Most of the lettuce is field packed, meaning that the produce is harvested, packaged in the field, and shipped to market with no further processing. Since most lettuce undergoes little processing, great emphasis is given to produce a high-quality lettuce. It is essential that the produce is free from pest damage and contamination at harvest. Field packaged lettuce may be packed without plastic wrapper, film-wrapped in perforated or non-perforated cellophane, or bagged in perforated plastic bags. Perforated cellophane is used to prevent moisture accumulation and extend shelf life. Bagged lettuce is packaged in perforated plastic bags, with usually six heads per bag. A wire-bound or waxed fibre carton designed to hold 20 to 24 heads is also used for its packing. Lettuce should be packaged *flat pack* (non-bulge) to avoid crushing of heads. There are two layers of heads, *i.e.*, the bottom layer is packed stem-down and the top layer stem-up. This keeps the milky latex from the stems from smearing the heads. All heads in a crate should be of similar size and weight.

Lettuce should be transported in refrigerated vehicles. A controlled atmosphere of 2% carbon dioxide and 3% oxygen is recommended if lettuce is to be shipped to long-distance markets for a month. It is said that the reduction in decay achieved by 2% carbon dioxide outweighs the danger of damage.

Lettuce may sometimes be transported to a cooling shed and distribution centre, where it is stored at 2–3°C. Though the lettuce storage life under these conditions is 16 to 20 days, almost all lettuce is shipped within 48 hours. The high water content of lettuce (94.9%) creates problems when attempting to preserve firmness as it cannot be successfully frozen, canned, or dried and must be eaten fresh.

Unless the crop is to be sold locally, the lettuce must be cooled before shipment. Vacuum cooling is the primary method for cooling in the major lettuce growing areas. The most practical method is to stack the cartons in a cold room with high humidity and force cold air through the stack with high-velocity fans. Internal head temperature should be lowered to 2–3°C for best shipping and holding conditions. Head lettuce can be held for 2 to 3 weeks at 0°C temperature and 95% relative humidity. Leaf and butter head types can be kept for 1 to 2 weeks in cold store. Lettuce should not be stored with produce such as, apples, pears, or cantaloupes that give off ethylene as ethylene increases russet spotting. The crisp head and cos lettuce have a longer shelf life than the butter head and the loose-leaf types.

Lettuce seeds (relatively short lived) keep best when stored in cool conditions, and unless stored cryogenically, remain viable the longest when stored at −20°C. At room temperature, lettuce seeds remain viable for only a few months. However, when newly harvested lettuce seed is stored cryogenically, its life increases to a half-life of 500 years for vaporized nitrogen and 3,400 years for liquid nitrogen. This advantage is lost if seeds are not frozen promptly after harvesting.

Bulk harvesting of lettuce has become more popular in recent years because of the popularity of prepackaged salads. Lettuce, after harvest, is transported to salad plants where it is sorted, washed with a dilute chlorine solution or fumigated with ozone and then chopped for prepackaged or ready-made salad in sealed plastic bags. This type of processing is known as *value-added* packaging. Similar to non-processed lettuce, value-added packaged lettuce is stored prior to transport at 2–3°C. Value-added lettuce has a shelf life of 12 to 14 days, and hence, it is shipped from the salad plant within 1 or 2 days.

PHYSIOLOGICAL DISORDERS

Tip Burn

Tip burn, which is visible on lateral margins of the inner leaves of mature head, is more common in greenhouse than in open cultivated crop. Tip burn may be caused by high temperature, high light intensity and duration, calcium and boron deficiency, high manganese content, high soil moisture content, high endogenous level of indole acetic acid (IAA), or ontogenic age of plant.

Control
- Raise crop under ideal conditions of temperature, light, moisture and nutrition.

Pink Rib

It first appears as a pink discoloration at the base of mid veins of lettuce leaves. This discoloration extends throughout the veins of outer leaves and then extends to the younger leaves. The cause is unknown but high temperature may be involved since it shows up most during summer harvests. This problem can cause a high loss of marketable yield.

Control
- Avoid July to mid August harvest schedules.
- Cultivate romaine and leaf lettuce, which seem to be less prone to this disorder.

Nutrient Deficiency

Nutrients deficiency, including lack of boron, phosphorus, calcium, lithium, molybdenum, or copper, can cause a variety of plant problems that range from malformed plants to a lack of head growth.

Control
- Manage macro- and micronutrients after proper soil analysis.

Slime

Slime develops due to a disease in hot and humid weather, which is often aided by bacteria. It is the greatest single danger to mature lettuce. It produces a wet slimy decay on lettuce in the field, transit, or market. Usually, the large internal leaves are affected first.

Control
- Aim for a sequence of harvests at optimal maturity by successive sowings.
- Avoid overcrowding of plants and excessive watering.
- Harvest the heads as soon as they mature, pre-cool them to 1°C and keep them cool.

INSECT-PESTS

Cutworms (*Agrotis* spp.)

Cutworms may be problematic at seedling stage. They are usually found 2–5 cm below the soil surface near cut-off plants. They become active during night and cut the stems just above or below the soil surface.

Control
- Use poison baits to collect the insect.
- Dust the crop with lindane 2% @ 35 kg/ha.

APHIDS (*Myzus persicae, Myzus ascalonicus* and *Macrosiphum euphorbiae*)

Lettuce aphids are small, green, or pinkish and feed on the inner leaves and within the heads. Their presence contaminates the heads, thereby affecting their appearance. Heavy population of aphids can result in stunted growth of young plants. Other aphids feed on undersides of the leaves and curl or stunt them. Lettuce root aphids will also feed on the roots, thereby stunting the plants especially under dry conditions. Aphids act as vectors of some viral diseases.

Control
- Irrigate the field regularly for controlling the root aphids partly.
- Fumigate the stores with acetaldehyde.
- Avoid growing second crop of lettuce unless insecticide is banded.
- Spray the crop with Metasystox @ 1.5 ml/litre or 0.2% malathion or thiodon.

ASTER (SIX-SPOTTED) LEAFHOPPER (*Macrosteles fascifrons* stal)

The small, slender, wedge-shaped, greenish-yellow leafhoppers can carry aster yellows from plant to plant, as they feed. Once a leafhopper feeds on an infected plant, it can spread it to other plants also as it feeds on. They feed on many other crops such as potato, tomato, celery, spinach, lettuce, onion and squash.

Control
- Apply Metasystox-R 2 EC @ 1.5–2 pints per acre.
- Spray the crop and field boundaries with appropriate insecticides from the time of transplanting or emergence of seedlings.
- Bury the crop residues deep immediately after harvest.

CABBAGE LOOPER (*Trichoplusia ni* hubner)

Cabbage looper gets its name from the way it forms a loop as it walks. It is a smooth green larva with two white stripes along the back and two along the sides. It prefers lettuce as alternate host for oviposition. The cabbage looper, which is capable of causing significant damage to lettuce, tends to be more problematic during late summer.

Control
- Monitor moth numbers with traps.
- Spray the crop with malathion or thiodon at 0.2%.

TARNISHED PLANT BUG (*Lygaeus lineolaris* palisot de beauvois)

Tarnished plant bug is a small 6 mm long mottled white and yellow insect with some black marking on the wings, somewhat oval shaped. It flies in nearby hay fields or weedy areas along the margins of fields. It feeds along the main veins of leaf and causes browning along the lesions they make while feeding.

Control
- Spray the crop with 500 ml of malathion 50 EC dissolved in 500 litres of water per hectare or with any other appropriate insecticide as needed, *e.g.*, Fenvalerate or organochlorine, organophosphate, carbamate and pyrethroid insecticides.

SLUGS (*Elysia crispata*)

Slugs eat leaves of the crop and cover it with unsightly slime tracks. Irregularly shaped holes in leaves and stems are formed. Flowers and fruits may also be damaged. If infestation is severe, leaves may be shredded.

Control

- Practice good field sanitation.
- Spread wood ashes or eggshells around plants.
- Handpick the slugs at night to decrease their population.
- Avoid growing lettuce in areas, which are cool, moist and high in organic matter.
- Use baits containing metaldehyde, a specific poison for slugs.
- Use ferrous phosphate for organic gardens and baits containing metaldehyde (*e.g.*, Bug-geta) and carbaryl (*e.g.*, Sevin bait) for non-organic growers.
- Since slugs can over-winter easily, cultural practices aimed at controlling them should begin at least 1 year before the susceptible crop is put in.

WIREWORM (*Limonius agonus*)

The yellow, white, or darker brown wireworms of all size and age are present in soil throughout the year as there is always an overlapping of generations. Wireworms are often numerous in lands those have been in sod for several years. They are also more abundant in heavy poorly drained soils.

Control

- Avoid planting crops highly susceptible to wireworms in a field that has been recently in sod.

AMERICAN BOLLWORM (*Helicoverpa armigera* hubner)

The larvae of American bollworm penetrate at the bottom of leaves and eat their way into the heads.

Control

- Spray the crop with fenvalerate 20 EC @ 250 ml dissolved in 625 litre of water when the pest is noticed.
- Spray the crop with malathion 5% dust @ 25 kg/ha.

DISEASES

DAMPING-OFF (*Pythium* spp., *Rhizoctonia* spp. and Other Fungi)

This is a frequent problem in greenhouse and field and lettuce is extremely susceptible. Seedling emergence may be poor due to the attack of above fungi, causing rotting of seeds. There may be sudden collapsing and death of seedlings. Cool and damp conditions favour the multiplication of these fungi, causing this disease. Soilless medium may also contain high levels of *pythium* unless it is properly treated before use.

Control

- Use raised beds or well-drained soils for early seeding or transplanting.
- Treat the seeds with Thiram, Captaf, Ceresan, or Agrosan G.N. @ 2 g/kg of seed and Thiram or Captaf @ 2 g/litre for soil drenching in greenhouse.
- Properly condition the greenhouse soils before seeding and maintain favourable conditions for seedlings growth.

Septoria Leafspot (*Septoria lactucae*)

The disease is widespread and damaging in hot weather. Infected plants have small, yellowish spots on the outer leaves. These spots grow and become large, irregular and brown. The centre of the spot is pale creamy brown with many black dots. The disease is characterized by spots or patches of white to grayish powder-like growth. Tiny pinhead sized spherical structures that are first white, later yellow-brown and finally black may be present singly or in groups.

Control
- Use disease-free seeds.
- Provide hot water treatment to seeds prior to sowing.
- Grow lettuce in areas where Septoria is uncommon and rainfall is low.
- Spray the crop with Maneb 80 WP @ 1.5 to 2 lbs/acre or Mancozeb 75 WP @ 0.3% at 7- to 10-day intervals.

Downy Mildew (*Bremia lactucae*)

Downy mildew affecting seedlings or mature plants occurs frequently during cool and moist weather in spring or early autumn and in cool and humid areas in summer. Light green or yellow lesions on upper surface of the leaves are first noticed on older leaves. The lesions later become necrotic, limited by veins and angular. A white downy mould is also noticed on lower surface of the leaves. The fungus overwinters on crop residue. Spores are spread by wind. Spore production is greatest at temperature below 19°C and at relative humidity approaching 100%.

Control
- Grow only resistant cultivars in areas prone to downy mildew.
- Use disease-free healthy seeds.
- Carry out at least a 2 to 3 years' rotation.
- At first sign of disease, spray the crop with Ridomil MZ 72 @ 0.025% followed by 0.2% Dithane Z-78 at 7- to 10-day intervals.

Powdery Mildew (*Erysiphe cichoracearum*)

The disease occurs frequently and reduces the quality of crop. The disease is characterized by spots or patches of white to greyish coloured powder-like growth. Tiny pinhead-sized, spherical structures that are first white, later yellow-brown and finally black may be present singly or in groups.

Control
- Follow long crop rotation, especially with non-host crops.
- Use disease-free healthy seeds.
- Always try to grow disease-resistant cultivars.
- Apply sulfur at first sign of symptoms, as long as the temperature is high enough.

Lettuce Drop or Sclerotinia Rot (*Sclerotinia minor* and *Sclerotinia sclerotiorum*)

Symptoms begin on the stem near the soil surface. A severe wet rot develops rapidly and spreads downward to roots and upward through the head. Once the base of the leaf is rotten, the leaf wilts, withers and dies. Symptoms successively develop from outer to inner leaves. The head becomes a wet slimy mass. During wet conditions, a white cottony mould develops on decayed plant parts. Hard irregular black sclerotia (pea-sized bodies) may occur in the mould.

Control

- Avoid growing lettuce in land where clover, soybean, snap, or dry beans, lettuce, pea, tomato, carrot, cucurbits, or cabbage have been grown recently.
- Practice a 3 years' rotation with non-susceptible crops (grasses, corn, cereals, onions, or beets).
- Disc the field promptly after harvest to destroy the production of sclerotinia.
- Use row and plant spacing to encourage good air movement or plant on raised beds.
- Plough the field deeply in hot summer months.
- Apply fungicides like carbendazim @ 0.2%, Ridomil MZ 72 @ 0.025% immediately after thinning plants or apply DCNA 2–5 kg/ha at various stages especially after thinning.

ANTHRACNOSE OR SHOT HOLE (*Microdochium panattonianum*)

Small water soaked tan spots appear on outer leaves, which may expand and turn straw-coloured. Centres fall out of mature lesions, giving the plant a shot-hole appearance. Fungus survives on crop debris in soil and disease spreads by splashing water.

Control

- Carry out a long crop rotation.
- Bury the disease debris into soil.
- Control wild lettuce population around main plantation.
- Avoid overhead irrigation.

BACTERIAL ROT COMPLEX (*Pseudomonas cichorii*)

The disease is caused by a combination of bacteria that occur on the leaf surface without causing damage. The bacteria get into the leaves and cause rotting when the plant is stressed or damaged. Symptoms are leaf spots, which start off under water-soaked leaves and later turn brown, before developing soft head rot and dying.

Control

- Use disease-free healthy seed.
- Grow cultivars resistant to this disease.
- Discard infested plants and remove these from the field.
- Always keep the field weed free.

GRAY MOULD (*Botrytis* sp.)

Gray mould can appear on plants at all stages although initial infection is often on seedlings in the greenhouse. Seedlings look like they have damping-off, while older plants rot at the stem or on lower leaves in contact of soil. A slimy rot spreads upwards into the head. A dense fuzzy grey mould appears on infected areas and dark hard sclerotia (fruiting bodies) may develop. Disease spreads mostly under moist conditions.

Control

- Follow a 3 or 4 years' crop rotation.
- Follow field sanitation.
- Use sterile seedbeds and flats.
- Treat the seed with suitable fungicide like Thiram or Captaf @ 2 g per kg of seed.
- Avoid thick stands of seedlings.
- Orient rows in direction of prevailing winds.
- Spray the greenhouse seedlings with Bavistin 0.2% at weekly intervals and once or twice later in the field at 10-day intervals.

Bottom Rot (*Rhizoctonia solani*)

Infection occurs on lower leaves touching the ground. The disease progresses up into the head causing a dark brown slimy decay. Later, the head may dry out, leaving a dry mummified plant. The pathogen lives indefinitely in the soil.

Control

- Avoid rotating potatoes and other susceptible crops with lettuce.
- Grow lettuce on 8 to 15 cm high ridges.
- Destroy the crop residues immediately after harvest.
- Combine cultural control with fungicide application.
- Avoid irrigation near harvest.
- Grow varieties with an erect growth habit to reduce leaf contact with soil.
- Apply appropriate foliar fungicides like Mancozeb 75 DF @ 1 to 1.5 lb/100 gallon of water.

Lettuce Mosaic

The disease is seed-borne and is transmitted by the green peach aphid. Vein clearing followed by mottling, recurving of the leaves and increased marginal frilliness results from early infections. Infected mature plants are yellow and stunted and cannot be harvested.

Control

- Control aphids using systemic insecticides.
- Use disease-free healthy seeds.
- Grow disease-resistant cultivars.
- Always keep the field weed free.

Big Vein (*Mirafiori lettuce big-vein virus or MiLBVV*)

Veins enlarge, clear and show yellow discolouration. Leaves become puckered or ruffled and ultimately thicken. The outer leaves become upright. Virus is soil borne and is introduced to plants *via Olpidium brassicae* fungus. Disease is more prevalent during cool weather.

Control

- Grow disease-resistant cultivars.
- Use disease-free healthy seeds.
- Annually fumigate soil with methyl iodide to control the virus vectors.
- Spray the crop with insecticides like Metasystox @ 1.5 ml/litre or 0.2% malathion or thiodon to control virus vectors.

Aster Yellows (*Mycoplasma*)

Yellowing and curling occur on the youngest leaves. At the time of heading, the leaves become dwarf and curled and heads remain soft. The mycoplasma, which can overwinter in many perennial weeds, is spread to lettuce by leafhoppers during their feeding.

Control

- Eliminate weed hosts from the field headlands and ditch banks.
- Avoid growing lettuce adjacent to earlier lettuce plantings that contain infected plants.
- Avoid planting grain, grass, carrot, or celery crops close to lettuce.
- Apply insecticide to control leafhopper.

22 Salsify

M.K. Rana and V.P.S. Panghal

Botanical Name	: *Tragopogon porrifolius*
Family	: Asteraceae
Chromosome Number	: 2n = 12

ORIGIN AND DISTRIBUTION

Salsify is native to eastern Mediterranean region and Southern Europe, where it has been in cultivation for at least 2,000 years. In the 13th century, it was first mentioned in cookbooks, and its cultivation widely began in the 16th century in France and Italy. Presently, its roots are cultivated and eaten most frequently in France, Germany, Italy and Russia. In the United Kingdom, it was initially grown for its flower and later it became a mildly popular vegetable in the 18th century but thereafter declined in popularity. It is widely naturalized in southern and eastern Australia and in temperate regions of the world, including many parts of Canada and the United States of America. In Victoria, South Australia, Tasmania and Western Australia, it is regarded as an environmental weed.

The genus *Tragopogon* has several species, *e.g.*, *T. orientalis* L., *T. dubius* L. and *T. pratensis* L., which are native to the Czech flora, but the species *T. porrifolius* L. is not native to the Czech flora. It was cultivated as a vegetable in the middle Ages and only a few records document its rare escapes from gardens in the last two centuries. Escaped plants survived only for a short period in human-made habitats close to the place of their cultivation.

INTRODUCTION

Salsify (*Tragopogon porrifolius* L.) is a cool-weather crop, roots of which are of great nutritional importance. It is seldom cultivated though it deserves larger popularization and utilization in human diet. Its swollen roots and young leaves are suitable for human consumption. Like parsnip, the flavour of its roots is improved after hard frosts resulting in sweetening of roots. Some say salsify has a delicate sweet flavour like oysters which is why it is sometimes called an oyster plant or vegetable oyster. Some also say salsify has a flavour like asparagus or artichokes and others say that it has no flavour. In cold climates, it flowers from June to September, and in warm climates, it blooms from April onward and the seeds ripen from July to September.

COMPOSITION AND USES

COMPOSITION

Salsify roots contain carbohydrates, proteins, fats, minerals *viz.*, calcium, phosphorus, iron and small amounts of vitamins such as vitamin A, B_1, B_2 and C. However, the most valuable component is inulin-fructan with pre-biotic properties having a positive influence on digestive tract functions. The nutritional value of salsify is attributed to its monounsaturated and essential fatty acids, vitamins, polyphenols and fructo-oligosaccharides. The nutritional composition of salsify is given in Table 22.1.

Uses

All parts of salsify plant, *i.e.*, roots, leafy shoots and open flowers, are edible and consumed as cooked and raw. In Lebanon, the shoots are more frequently consumed than the roots. However, in other parts of the world, salsify is grown mainly for its edible roots. Its roots and sometimes its young shoots are used as a vegetable. After peeling, its roots are usually consumed like carrot and parsnip. It is used as parboiled and baked in a casserole with herbs and butter, pureed, battered and fried, or eaten after cooking. Its young roots are used in salads and older roots are better cooked and they are usually used in soups or stews. The latex derived from its roots is used as a chewing gum. The flowering shoots can be used like asparagus either raw or cooked and the flowers can be added to salad, while the sprouted seeds can be used in salads or sandwiches. Its brown seedlings or the long clumps of grass-like green leaves are also edible and can be added to mixed salads.

The roots before cooking are first washed properly and then the outer most skin is removed by peeling. While cutting into pieces, the roots exude a sticky milky white liquid. Hence, the roots are quickly put into cool water containing lemon juice or vinegar to prevent discoloration, and thereafter, they are fried in cooking oil. However, sautéing or steaming is the simplest method of cooking salsify roots.

Medicinal Uses

The plant has been used by the herbalists since classical times. It is a cleansing food with a beneficial effect on liver and gallbladder. The roots are diuretic, antibilious, slightly aperient, deobstruent and sometimes used as a remedy for upset stomach and other gastric disorders. It is specific in the treatment of obstructions of gallbladder and jaundice and is used in the treatment of arteriosclerosis and high blood pressure. Its colour preserved in acidulated water is a useful antioxidant. However, for medicinal purposes, it should be taken in small quantities since large quantities may have adverse effects, depending on the individual characteristics of those who take it.

Its methanolic extract, rich in flavonoids, quercetin and luteolin, has antioxidant capacity comparable to that of ascorbic acid and beneficial effects on cardiovascular diseases, cancer and oxidative stress inducing carcinogenesis. Besides, its flavonoids are also known to produce antitumor activity through inhibition of proliferation, metastasis and invasive effects, induction of apoptosis, suppres-

TABLE 22.1

Nutritional Composition of Salsify (per 100 g Edible Portion)

Constituents	Contents	Constituents	Contents
Moisture (g)	–	Iron (mg)	0.7
Carbohydrates (g)	18.6	Zinc (mg)	0.38
Protein (g)	3.3	Thiamine (mg)	0.08
Fat (g)	0.2	Riboflavin (mg)	0.22
Dietary fibre (g)	3.3	Niacin (mg)	0.5
Calcium (mg)	60	Pantothenic acid (mg)	0.37
Phosphorus (mg)	75	Pyridoxine (mg)	0.28
Potassium (mg)	380	Folic acid (µg)	26
Magnesium (mg)	23	Vitamin C (mg)	8
Manganese (mg)	0.27	Vitamin K (µg)	0
Sodium (mg)	20	Energy (kcal)	82

sion of tyrosine kinase activity and antiangiogenesis. Phenolic substances in salsify have also been reported to possess anticancer effects and quercetin inhibits the proliferation of colon and breast cancer cells.

BOTANY

Salsify is a hardy biennial vegetable but grown as an annual. Its carrot-like roots are tender when harvested young, long, tapered and white-skinned with shining white flesh. The plant commonly grows to a height of 60 cm and diameter 5 cm. Its stem is largely unbranched, and the leaves are somewhat grass-like. It exudes a milky juice from the stems. Its roots are usually studded with root-lets and have shoot-like leaves reaching from its top. Being a biennial crop, it completes its vegetative phase in first season and flowers in its second season. The flower head is about 5 cm across, and each is surrounded by green bracts, which are longer than the petals. Its delicate flowers are larger about 30–50 mm across and dull purple in colour. The flowers are hermaphrodite, having both male and female organs, but they are pollinated by insects.

CLIMATIC REQUIREMENT

Being a temperate crop, salsify grows best in cool season. It is very winter hardy and can withstand freezing weather without any damage to its roots. Indeed, freezing has been reported to improve the oyster-like flavour of its roots. It can successfully be cultivated where it may receive full sunlight. Salsify essentially requires a cool climate during seed germination, seedling growth and root development, and warm and sunny days during flowering to seed maturity. High humidity accompanied by cloudy weather and rainfall at the time of flowering results in incidence of insect pests and diseases and poor seed set.

SOIL REQUIREMENT

Although it can be cultivated on all types of soil from light to heavy soils, it prefers loose, friable, well-drained light soils rich in organic matter and fair in water-holding capacity. If it is planted in heavy clay soil, it is essential to incorporate plenty of sand and compost to make it loose and friable. It cannot tolerate acid and alkaline soils. The ideal soil reaction for its successful cultivation is 6.0 to 6.8 pH. The plant can stand against strong winds but can not withstand waterlogging conditions.

The field is prepared to a fine tilth by ploughing 20 to 30 cm deep with turning plough. Like other root crops, it is better not to grow salsify in fields containing fresh manure or stones as it may cause the roots to *fork*. The stones and previous crop residues are removed before preparing the seedbeds since any obstacle in downward movement of roots in soil can cause forking and splitting, thus, only the well-decomposed farmyard manure should be incorporated into the soil to avoid forking of roots.

CULTIVATED VARIETIES

Mammoth Sandwich Island

This variety with oyster-like flavour is very easy to grow. Its roots are best when harvested after a frost. It matures in 120 days after sowing and has excellent keeping quality.

Belstar Super

A variety producing black roots with white flesh takes 90 days to be mature.

The other varieties are Giant Russian, Scorzonera, French Blue Flowered, Improved Mammoth, etc.

SOWING TIME

Salsify is a hardy cool-weather root crop. In temperate climate, its seeds are sown in spring (March–April) about 2 weeks before the last expected frost when the soil temperature has reached about 4°C to 5°C. In spring for a fall harvest, it is better to sow the crop as early as possible. However, in regions with mild winter, it is sown in early autumn (September–October) for a winter harvest. In regions with hot summer and mild winters, planting in autumn is considered the best.

SEED RATE

Salsify has large, narrow, brown seeds. Like parsnips, always fresh seeds are used as they lose their viability quickly. Like other root crops, it may not be raised well by transplanting; thus, the seeds are usually sown directly into the field. The seed rate of salsify is not being given since the quantity of seed required for sowing a unit area has nowhere been mentioned in the literature.

SOWING METHOD

The salsify seeds are sown at a depth of 2.0–2.5 cm on ridges made 45 to 60 cm apart since sowing on ridges provides good conditions to the roots for their proper growth and development. After germination, when the seedlings are large enough to handle, thinning is done keeping healthy seedlings at 8–10 cm spacing. It will reduce competition among the plants for nutrients, moisture, light and space. The seeds are sown in lines over a deeply loosened and composted raised bed and covered with a thin layer of topsoil or fine compost. Like parsnips, always its fresh seeds are sown as they lose their viability quickly.

NUTRITIONAL REQUIREMENT

The nutritional requirement of salsify depends on soil type, variety, growing season and crop grown in previous season. Adequate supply of nitrogen, phosphorus and potash is important for optimal development of salsify. Nitrogen is applied @ 100 kg/ha, out of which, 60 kg/ha is supplied at the time of sowing and 40 kg/ha as a side dressing. However, the quantity of nitrogen applied to rich soils is reduced to 60 kg/ha. Nitrogen fertilizers should be applied carefully since its excessive use may result in luxury growth, which makes the plants susceptible to disease causing organisms. If manure is added into the soil or the field in previous season was occupied by legume crop, the nitrogen dose should be reduced. The farmyard manure is applied @ 10–12 t/ha, which will not only supply the essential nutrients but also keep the soil moisture conserved. The doses of phosphorus and potash should be decided based on soil analysis test. The manure to be applied should be well decomposed since the incorporation of fresh or undecomposed manure may cause the roots to *fork* or *split*.

IRRIGATION REQUIREMENT

The soil is kept moist in order to avoid hardening of roots and to produce the quality roots. Irrigation of 25–40 mm is provided at an interval of 7 to 10 days or when demanded by the crop. Soluble fertilisers may be applied through fertigation. The roots should not be allowed to dry out but watering should be reduced after the development of first leaf to encourage the roots to grow deeper. Thereafter, the soil is kept moist, especially in dry periods to produce sweet and silky roots by mid autumn.

INTERCULTURAL OPERATIONS

Salsify is a slow-growing crop in its young stage; thus, shallow hoeing is necessary to keep the field weed free. Otherwise, the weeds will compete with crop plants for nutrients, moisture, light, space, etc. Use of straw or black polyethylene film in beds is also useful to check the weed growth, especially when harvesting is planned after the onset of freezing weather. Hoeing should be done shallow to avoid damage to its developing feeding roots.

THINNING

Thinning is an essential operation where a bit large quantity of seed is used to ensure optimum plant density per unit area, which is essential to harvest an economic yield. The plant density is an important factor, which affects the root size at harvest. Overcrowding of seedlings may create a problem of competition among the crop plants for nutrients, moisture, light and space. As and when the seedlings attain a height of 8–10 cm, the extra seedlings are pooled out, maintaining 8–10 cm spacing between plants. This operation must be followed by a light irrigation to fill gaps created by uprooting the seedlings and to establish the disturbed roots.

MULCHING

Mulch is often used to moderate soil temperature, control weeds and erosion, conserve moisture and to protect the plants from insect-pests and diseases. Saw dust, wheat or paddy straw, dry grasses, sugarcane tresses, aluminium foil, or black polyethylene films can be used as soil mulch. In addition to regulate soil temperature, opaque mulches control weed growth. However, in summer, black polyethylene films become hot, causing burning of plants touching them during the daytime. Aluminium foils used as mulch disrupts aphid flights and decreases the chances of aphid-transmitted diseases.

HARVESTING

Salsify is a slow-growing crop, thus, it requires 120 to 150 days to be mature. As the roots attain a length of 20–30 cm and diameter 2.5 cm at crown region, they become ready for harvest. Its roots can withstand freezing temperature in soil, thus, they can be left in the field until early spring. With late harvesting unique oyster flavour develops in roots. The roots left in the ground undoubtedly remain fresh until the following spring, but generally, the longer they are left in the ground the less they are said to have the characteristic subtle oyster flavour. If the winters are mild enough without freezing the ground, digging of fresh roots may be kept continued all over the winter instead of storing in cold store. As the temperature rises above 30°C, the roots become stringy and fibrous. The roots are pulled out either by hand or with the help of spading fork. A light irrigation should be applied before harvesting the crop to avoid breaking of roots since any breakage causes discolouration and spoilage of roots quickly. Harvesting commences from late autumn and continues up to early spring. The roots are harvested in small quantity as needed. The mature salsify roots left in the ground over the winter sprout new leaves in spring and produce white tender shoots known as salsify *chards*, if covered in 15–20 cm high mound after sprouting, and in subsequent season, they produce flowers and seeds. A few plants can be left in the field without harvesting to get seeds for future planting. They continue to grow all through the next year until they bloom and produce seeds. Like dandelion seeds, its seeds also have feather-like hair, hence, the seed should be collected in time before the winds take them away.

YIELD

The yield varies with soil types, varieties used, growing season and growing purpose, stage of harvesting, weather conditions, irrigation facilities and incidence of insect-pests and diseases. In general, on an average, salsify gives a root yield of 180 to 200 q/ha.

POST-HARVEST MANAGEMENT

Salsify roots without tops packed in low density polyethylene film bags can be kept in the refrigerator for 3 to 4 weeks. However, in cold stores, its roots can safely be stored for 2 to 4 months at near 0°C temperature but above freezing and high humidity. In the absence of storage facilities, the salsify roots can be stored in the field itself by mulching the salsify rows with straw, and the roots can be harvested as per need throughout the winter.

INSECT-PESTS

Salsify is seldom attacked by insect pests but some of the insect pests attacking the salsify crop are given as follows:

CARROT RUST FLY (*Psila rosae*)

The carrot rust fly lays its eggs near base of the plant, and the newly hatched larvae or maggots start eating the plant root and make burrow into roots. This pest also damages carrot, celery, parsley and parsnip plants by chewing roots and leaving reddish brown excreta in their tunnels. The plant growth becomes stunted, and the infestation leads the way for soft-rot bacteria.

Control

- Follow the long crop rotation with cereal crops.
- Use yellow sticky traps @ 10 per acre for the sticking of adult flies.
- Follow clean cultivation by destroying the plant refuse.
- Grow crop on raised beds.
- Apply rock phosphate or pulverized wormwood around base of the plants to repel the fly.
- Grow few lines of leek, onion, pennyroyal, rosemary, sage, black salsify, or coriander to repel this insect.
- Soil drench with insecticides diazinon or cypermethrin in early spring may be beneficial.

WIREWORMS (*Selatosomus destructor* and *Ctenicera pruinina*)

Wireworms are the larvae of click beetle and look like reddish-brown worms. Click beetles usually lay eggs in grass. After hatching, the larvae shift to salsify crop growing near the lawn. Wireworms may remain in larval stage for 2–5 years moving up and down in the soil depending on temperature and moisture. The larvae of click beetles spread out quickly in areas of both conventional and organic farming. Its adults feed little but larvae feed on underground roots, doing their worst when they bore straight in roots. Wireworm tunneling also creates an entry point for plant pathogens, eventually leading to rotting of roots. During their 3–5 years of development, the larvae can cause great damage to underground parts of plant. Root feeding causes wilting, stunting and distortion of seedlings and ultimately kills the plant. They are usually most damaging in poorly drained areas on upland soils.

Control

- Rotation with non-host crops such as onions, lettuce, alfalfa, sunflower and buckwheat reduces wireworm population.
- Plough the field deep in hot months to destroy pupae as well as larvae.
- Alternating periods of flooding and drying can increase wireworm mortality.
- High rates of compost may reduce wireworm damage.
- Application of organophosphate chemicals like ethoprop (Mocap), phorate (Thimet) and diazinon (Diazinon) as a pre-plant broadcast or in furrow at planting provides the best and most consistent control of wireworms.
- Apply imidacloprid 350 ml/ha as soil drench.

Root Knot Nematode (*Meloidogyne incognita, M. javanica, M. hapla* and *M. arenaria*)

Root knot nematodes can cause substantial damage to salsify roots. Typical symptoms of nematode injury can involve both aboveground and belowground plant parts. The aboveground symptoms generally involve stunting and general unhealthiness, premature wilting and leaf chlorosis (yellowing). Their parasitic activity can damage the growing root tip, which results in forking, distortion, stunting and deformation of taproot. Besides, they may cause reduction of plant stand and yield. Typically, this occurs within the first few weeks after seed germination. These nematodes induce characteristic galls on feeding roots. The stubby root nematode causes short roots with a stubby appearance, and feeding by needle nematode causes cessation of root growth, forking, root branching, swelling of root tips and sometimes root tip necrosis.

Control

- Grow resistant varieties.
- Prevent movement of infested soil by using clean machinery and equipments.
- Sanitation aids in reducing plant parasitic nematode populations.
- Soil solarization during hottest period of the year can temporarily reduce many soil-borne diseases (including those caused by plant-parasitic nematodes) and weeds.
- Add peat, manure and composts to the soil to reduce the effect of nematodes on crop plants.
- Avoid the use of pond water for irrigation.
- Irrigate the crop frequently since nematodes damage water-stressed plants more readily.
- Rotation with marigolds will suppress the nematode population.

DISEASE

Bacterial Soft Rot (*Bacillus carotovorus* Jones)

The symptoms appear as a soft, watery and slimy root decay, which consumes the root core rapidly, leaving the epidermis intact. Roots lose firmness and become soft. The symptoms also include yellowing, wilting and collapsing of foliage as the roots break down. The soft rot lesions are invaded by secondary fungi or bacteria. A watery or slimy consistency increases as the rot progresses. The rotten tissues keep their natural colour until they decay completely and emit a foul odour. The bacteria can be active at 0°C –32°C temperature but the ideal temperature for its activity is 21°C –27°C. Post-harvest storage and transportation are difficult in tropical and warm environments when the air is not properly ventilated. Higher temperature and high humidity favour the multiplication of bacteria.

Control

- Follow long crop rotation with non-host crops like soybean, forage legumes and small grains.
- Plough the field deep in hot summer months.
- Follow phyto-sanitary conditions by destroying the infected plant debris.
- Avoid wounds and injuries to plant roots during hoeing and harvesting.
- Grow the crop on ridges or raised beds.
- Avoid excessive supply of irrigation.
- Store the roots at 0°C temperature and 85–90% relative humidity.
- Wash the roots immediately harvest with clean chlorinated water.
- Control nematodes and other insect-pests that are carriers of bacteria to invade the plant tissues.

White Blister (*Albugo* species)

White blister is a foliar disease, which is closely related to the organisms causing downy mildew diseases, and sometimes, both occur together. Like downy mildew, infection and spread of white blister are favoured by wet weather. The disease appears as pale leaf spots, which eventually develop into blister-like white pustules releasing number of tiny powdery spores. The pustules usually appear on the underside of the leaf and may develop singly or in concentric rings. The affected spots are often surrounded by purple pigmentation. The organism also attacks the stems, flowers and pods, which often become distorted, resulting in bulge or sunken spots on leaves, curved stems and abnormally shaped inflorescence or pods. Plant vigour is reduced and severely affected plant parts can shrivel and die.

Control

- Follow long crop rotation to reduce the incidence of disease.
- Remove and destroy the severely infected plant or part of the plant.
- Avoid close spacing of plants, which encourages high humidity and infection.
- Remove weed hosts of the disease surrounding the field.
- In organic field, spay 4% *neem* seed kernel extract.
- Spray the crop with 0.2% Dithane Z-78, Ridomil MZ-72, maneb, zineb, Blitox, captofol, captan, or wettable sulfur.

■■■

23 Malabar Spinach

M.K. Rana, Ravindra Mulge and Shashikant Evoor

Botanical Name : *Basella alba* L.

                            Basella rubra L.

Family : Basellaceae

Chromosome Number : *Basella alba* 2n = 48 and *Basella rubra* 2n = 44

ORIGIN AND DISTRIBUTION

Basella alba is native to the Indian subcontinent, Southeast Asia and New Guinea. It is reportedly naturalized in the China, Czech Republic, tropical Africa, Brazil, Colombia, the West Indies, Fiji and French Polynesia, whereas *Basella rubra* L. is probably native to India or Indonesia (Malabar is a coastal region in southwestern India). The plant is used in traditional cuisines as far westward from its point of origin as Japan and eastward as Africa. It has also been introduced to South America and the Caribbean.

INTRODUCTION

Malabar spinach, which is an excellent substitute of spinach in hot weather, is an edible fast-growing soft-stemmed vine. It is also known by many names such as Indian spinach, Ceylon spinach, Surinam spinach, Chinese spinach, Malabar nightshade, or Vietnamese spinach as well as flowing water vegetable. The other common names are vine spinach, red vine spinach, climbing spinach, creeping spinach and buffalo spinach in South Asia. It is also known as *pui shak* in Bengali *poi ni bhaji* in Gujarati, *basale soppu* in Kannada, *valchi bhaji* or *vauchi bhaji* in Konkani, *vallicheera* in Malayalam, *mayalu* in Marathi, *poi saaga* in Oriya, *kodip pasali* in Tamil and *bachhali* in Telugu. Malabar spinach or *Basella alba* is not related to spinach, though it can be used as a substitute for spinach in dishes and salads. Malabar Spinach or Indian spinach is used most widely in the tropics of India and Asia. *Basella alba* is known as *huang ti cai* in China, which translates into *Emperor's vegetable.*

Although it is well recognized by its nutritional value but its cultivation is restricted only to kitchen garden, thus, its cultivation practices have not been standardized thoroughly. It is found growing in tropical Asia and Africa where it is widely used as leafy vegetable. Malabar spinach or vine spinach is a popular tropical leafy green vegetable commonly grown as backyard plant in homestead gardens, and often it is grown as an ornamental plant. In a true sense, it is different from English spinach (*Spinacea oleracea*) in that the plant is a creeping vine, and its leaves are glossy, broad, deep green, thick and mucilaginous. When it is raw, Malabar spinach has very fleshy, thick leaves that are juicy and crisp with tastes of citrus and pepper. Commonly found in backyard gardens of many south Asian families, it is gaining popularity in some of the tropical and temperate climates of America, Australia and Europe for its succulent nutritious greens and tender stems. In its dormant state during cooler temperature, it will still grow but much slower. Malabar spinach is grown in Australia and in the Continental United States as an annual, but in Puerto Rico and Hawaii, it is grown as a perennial.

Being quick-growing in a growth habit, it can be best fitted in a kitchen garden. Tender shoots and leaves are used as vegetable in many parts of the world. It is available year yound due to its tropical growing environment.

COMPOSITION AND USES

COMPOSITION

Malabar spinach, a typical leafy vegetable, holds an incredibly good amount of carbohydrates, starch, iron, calcium, potassium, magnesium, vitamin A, carotene, thiamine, riboflavin, ascorbic acid and vitamin E. Dark green leafy vegetables, like spinach, contain lutein and zeaxanthin, both carotenoids. It is low in calorie but high in protein per calorie. The succulent mucilage is a particularly rich source of soluble fibre (2.37%). It is one of the versatile leafy green vegetables revered in some East Asian cultures for its wholesome phytonutrients profile. It has been shown to contain certain phenolic phytochemicals that act as antioxidant. Its thick and fleshy leaves are a good source of non-starchy polysaccharide, *i.e.*, mucilage. The nutritional composition of Malabar spinach is given in Table 23.1.

USES

Malabar spinach leaves, with a pleasant mild spinach flavour, can be used as spinach. The mucilaginous qualities of the plant make it an excellent thickening agent in soups and stews. Like spinach, it can be used to complement other foods. The leaves can be prepared in stews, gratins and soufflés. It can also be used as a filling for quiche, omelettes, savoury turnovers and potpies. An infusion of the leaves is used as a tea substitute. Raw leaves can be added to green salad and stir-fries with garlic and hot peppers, steamed with tofu and ginger and used as a substitute of okra. If they are overcooked, they become slimy. It can be sautéed as a vegetable or eaten raw. Since red-stemmed Malabar spinach can lose a lot of its red colour when cooked, perhaps it is best utilized in raw dishes. The juice from the berries is so intensely purple that it puts beet juice to shame. Hence, it is used as a natural food colourant for agar (vegetable gelatin) dishes, sweets and pastries.

In the Philippines, its leaves are one of the main ingredients in all vegetable dishes called *Utan*, which is served with rice. In Karnataka cuisine (Karavali and Malnad regions), the leaves and stems are used to make *basale soppu saaru* (a curry especially in combination with jackfruit seeds). In Bengali cuisine, it is widely used both in vegetable dish, cooked with red pumpkin, and in a non-

---

TABLE 23.1

Nutritional Composition of Malabar Spinach (per 100 g Edible Portion)

Constituents	Contents	Constituents	Contents
Water (g)	93	Selenium (μg)	0.8
Carbohydrates (g)	3.40	Zinc (mg)	0.43
Protein (g)	1.80	Vitamin A (IU)	8000
Fat (g)	0.30	Thiamine (mg)	0.050
Calcium (mg)	109	Riboflavin (mg)	0.155
Potassium (mg)	510	Niacin (mg)	0.500
Magnesium (mg)	65	Pantothenic acid (mg)	0.053
Sodium (mg)	24	Pyridoxine (mg)	0.240
Iron (mg)	1.20	Folic acid (μg)	140
Copper (mg)	0.107	Vitamin C (mg)	102
Manganese (mg)	0.735	Energy (kcal)	19
		Soluble fibre (g)	0.9

vegetarian dish, it is cooked with the bones of Ilish fish. In Andhra Pradesh, a curry of Malabar spinach and yam is made popularly known as *Kanda Bachali Koora* (yam and Malabar spinach curry). It is also used to make the snack item, *i.e.*, *bachali koora bajji*. In Odisha, it is used to make curries and saag (a type of dish made from green leafy vegetables). In the Western Ghats in Maharashtra and Karnataka, it is used to make *bhaji* and often used in soups and stir-fries. In Vietnam, particularly in North Vietnam, it is cooked with crabmeat, *luffa* and jute to make soup. In Africa, the mucilaginous cooked shoots are most commonly used. It can be found at many Chinese, Vietnamese, Korean, Indian, etc. grocery stores, as well as farmers' markets. It is a popular green in Asian, Indian and even African cuisine although in Africa they tend to eat the tender shoots more than the leaves themselves.

MEDICINAL USES

Fresh leaves, particularly of *Basella rubra*, are rich source of several vital carotenoids such as β-carotene, lutein and zeaxanthin, which act as protective scavengers against oxygen-derived free radicals and reactive oxygen species (ROS) that play a healing role in ageing and different diseases, preventing macular degeneration and the formation of cataracts. In addition to regular fibre (roughage) found in stem and leaves, mucilage facilitates in smooth digestion, brings reduction in cholesterol absorption and helps in preventing bowel movement problems. Being rich in vitamin A, it helps in maintaining healthy mucus membranes, skin and eyesight, and its consumption offers protection from the lung and oral cavity cancers and keeps the cardiovascular system healthy, thereby reducing the risk of strokes and heart attacks. Being a good source of vitamin C, a powerful antioxidant, it helps the body to develop resistance against infectious diseases and scavenge harmful free radicals. Folic acid is essential for the production of red blood cells and for normal growth, and may reduce the risk of certain cancers. It is particularly important for pregnant women. Hence, anticipating and pregnant women are advised to include it in their diet to prevent neural tube defects caused by folate deficiency in offspring since it contains good amount of many B-group vitamins such as folate, riboflavin and pyridoxine. Regular consumption of Malabar spinach helps in preventing osteoporosis (weakness of bones) and iron-deficiency-causing anaemia. Besides, it is believed to protect the body from cardiovascular diseases and cancers of colon.

The astringent cooked roots are used in the treatment of diarrhoea. The cooked leaves and stems are used as a laxative and flowers as an antidote to poisons. The plant juice being febrifuge is a safe aperient for pregnant women and a decoction is used to alleviate labour pain. The leaves juice is a demulcent and used in dysentery. The leaf juice is used in Nepal to treat catarrh, inflammation of the nose and throat. A paste of root or leaves is externally applied to swellings and boils. Studies done by the National Institutes of Health have found that the total body vitamin A in Bangladeshi men is increased when Malabar spinach is consumed on a daily basis.

Malabar spinach also contains antinutritional substance phytate and dietary fibre in its leaves, which may interfere with the bioavailability of iron, calcium and magnesium. Like spinach, Malabar spinach also contains oxalic acid, a naturally occurring substance found in some vegetables, which may crystallize as oxalate stones in the urinary tract of some people. People with stones in their urinary tract are advised not to eat this vegetable. However, adequate intake of water maintains normal urine output.

BOTANY

The cultivated forms of Malabar spinach with thick, green semi-succulent, heart-shaped leaves and mucilaginous texture are dicotyledonary perennial vines. It is a fast-growing vine reaching several metres in one season and bears white or white-pink colour tiny hermaphrodite flowers born in clusters on nodes at the end of short stems, depending upon the species. Upon fertilization, the flowers develop into small, highly ornamental, single-seeded purple berries. Its stems are purplish or green. Ovate or heart-shaped leaves with mild flavour and mucilaginous texture are thick, broad, fleshy, 5–12 cm long, stalked and tapering to a pointed tip with a cordate base. Red venation in the leaves adds another level of colour contrast. Spikes are axillary, solitary and 5–29 cm long. Ovoid or spherical fruit without stalk is fleshy, 5–6 mm long and purple when mature. *Basella rubra* features pink or purplish stems and pink colour veins running in the leaves. The stem of *Basella alba* is green and that of *Basella rubra* is reddish purple.

CLIMATIC REQUIREMENT

Malabar spinach, with a climbing growth habit, is a frost-tender perennial but grown in cold areas as an annual cool-season vegetable. One of the most valuable virtues of Malabar spinach seems to be that it grows well in conditions that other greens find unfavourable. This particular plant is tolerant of high temperature, which makes it easy to grow during hot weather when other plants struggle to flourish. As a result, Malabar spinach is a good choice for warm-weather gardens that cannot sustain other harvestable crops. It is best suited in subtropical to tropical conditions where the rain and heat of summer suit it perfectly but will also grow with a shorter season in cooler climates. Being native to tropical Asia, it grows well under full sunshine in hot and humid climates in areas lower than 500 meters above mean sea level. Partial shade increases its leaf size but not the nutritive quality. Its growth is slow at low temperature, resulting in poor yield. The day length shorter than 13 hours results in flowering. It is extremely tolerant to heat and high rainfall but intolerant to chilling temperature.

SOIL REQUIREMENT

Malabar spinach can be grown on a wide range of soils but it grows best in well-drained moisture-retentive and fertile sandy loam soils rich in organic matter. It tolerates poor soils too but does much better in rich soils. It can tolerate damp soil with pH ranging from 4.3 to 7.0 but the ideal pH range for its successful cultivation is 6.5 to 6.8.

CULTIVATED VARIETIES

Malabar spinach has two botanical varieties, *i.e.*, (i) *Basella alba* with green stems and deep-green leaves and (ii) *Basella rubra* with purplish stems and deep-green leaves with pink veins.

MALABAR SPINACH

This vine-type warm-season variety with large meaty leaves is also known as *Lo Kui* and *Indian Spinach*. It starts indoors in north and takes 110 days to be mature.

GREEN MALABAR

This vine-type warm-season variety with glossy thick Savoy-type leaves is also known as *Malabar Green Stem* and *Green Vine*. It becomes mature in 60–85 days after planting in summer.

RED MALABAR

This vine-type Malabar spinach warm-season variety is also known as *Malabar Red Stem* and *Red Vine*. Its leaves are darker green than Green Malabar leaves and the stems are deep purple-red. It takes 60–85 days to be mature.

The green leaved varieties are popular in Uttar Pradesh and Punjab, and Green- and Red-leaved varieties are very common in West Bengal, Assam and the states of South India.

SOWING/PLANTING TIME

The seeds can be sown in March–April when the danger of frost has passed and night temperature has become above 16°C, though a minimum of 18–21°C temperature is required for seed germination, which takes 10 to 21 days at 20°C temperature. However, pre-soaking the seed for 24 hours in warm water shortens the germination time. In northern and eastern plains of India, the seeds are sown from March to May, while is southern parts, it is sown twice, once in June and again in October to November. Early summer is the best time for sowing on hilly tracts.

SEED RATE

Depending on seed viability, growing season, fertility of the soil and irrigation facilities, 8 to 10 kg seeds are required for sowing a hectare area.

SOWING/PLANTING METHOD

With the help of a plough or mechanized bed former, the beds of 15 cm height during dry season and 20–25 cm height during wet season are prepared keeping 90 cm width with 50 cm space between two beds and convenient length. Although it is commercially propagated by seeds but can also be propagated by cuttings of 20–25 cm with 3 to 4 internodes. The seeds before sowing are scarified to speed up germination, if not it may take 3 or more weeks. The scarified seeds are soaked in lukewarm water for 24 hours to shorten the germination time. For direct seeded crop, the seeds are sown in beds 1.5–2.5 cm deep and 5 cm apart in rows with spacing of 15 cm between rows. Usually, thick cuttings of about 20 cm length are preferred for easy propagation and fast growth. Being a vine crop, the plant requires trellising for its spread at a faster rate. As soon as the seedlings become large enough to handle, they are pricked out into individual pots containing compost rich mixture. In field, the emerged seedlings are thinned out to 30 cm apart.

NUTRITIONAL REQUIREMENT

The nutrients requirement has not yet been studied for this crop. It has been noticed that amending soils with biogas plant residues enhances the vegetative growth of Indian spinach. Among the rates of tested biogas plant residues, about 40 t/ha has been proved best in enhancing the vegetative growth of Indian spinach. Application of poultry manure @ 20 tonnes combined with 6,000 litre of water per hectare recorded the highest agronomic parameters as well as nutritional quality of Indian spinach. Besides, nitrogen of 60–80 kg, phosphorus 60 kg and potash 40 kg/ha can be applied before sowing or transplanting.

IRRIGATION REQUIREMENT

The crop requires consistent moisture to stay devoid of flowering and of bitterness of leaves. Thus, depending upon soil type and prevailing weather conditions, the soil should be kept moist for its better growth. In rainy season, the irrigation depends on rain frequency and intensity. It will thrive in moist, fertile and well-drained soils. If irrigation is not given at frequent intervals and the soil conditions are too dry, it tends to develop tough leaves or bolt to seed.

INTERCULTURAL OPERATIONS

HOEING AND WEEDING

The Indian spinach field should be kept weed free in the early stage of crop growth because the seeds take more time to germinate and plants may not recover from weed infestation. One or two hoeings and weedings are necessary to suppress the weeds. Hoeing not only removes the weeds but also provides good aeration into the soil, which can help the roots for better growth.

TRAINING

Indian spinach is a fast-growing vine crop. Hence, it produces best when trellised. If plants are not harvested in early stage for consumption, then staking has to be provided. Plant requires trellis or other support for twining the vine. The vines can be trailed on bower system or poles to keep the plants erect so that the tender shoots can be clipped, leaving a few leaves for further growth of plant for multiple harvests. The plants may also be allowed to crawl on the ground if support is not possible to provide. Under normal conditions, it is a very attractive plant, quickly growing up trellises and other plants for most of the warm season.

MULCHING

The soil may be covered with organic mulch after transplanting when seedlings reach a height of 10 to 15 cm. It will keep the soil moisture conserved for a long period, which essential for fast growth and a higher yield.

HARVESTING

Malabar spinach normally becomes ready for harvesting in 45 to 60 days after planting. The leaves and terminal tender 20–30 cm stems are usually harvested about 35 to 45 days after planting (about 50 days after seeding). Entire plant can be lifted for once over harvest or leaves and tender shoots are clipped to have many harvests. Shoots of 15–25 cm long and leaves are cut, washed and tied in bundles for sale. With multiple harvests, young shoots and leaves can be harvested at weekly intervals. Frequent harvesting can delay flowering and stimulate growth of side shoots. The top 15–20 cm shoot tips are harvested 55–70 days after seeding. Further harvesting of new growing stems can be made throughout the season.

The leaves of Indian spinach have a large surface area, thus, they lose their water and turgidity soon after harvest. To reduce water loss, harvesting has to be carried out during cooler hours of the day like early morning or late in the afternoon and then the produce is kept in a cool shaded place. The shoots are tied in bundles to avoid transpiration loss and to give an attractive look to the harvested produce in the market.

YIELD

Depending on variety, soil fertility, irrigation facilities and cultural practices followed during crop cultivation, its average yield is 150 to 200 q/ha in a growing period of 120 to 150 days.

INSECT-PESTS AND DISEASES

Leaf miners and root knot nematodes are sometimes serious pests. *Neem* cake @ 5–8 q/ha or Furadan 4 g/m^2 may be applied to beds before sowing or planting, and the crop may be sprayed with *neem* oil at 3 ml/litre of water to control the insect pests. Since it is leafy vegetable, the chemical pesticides should be applied at least 10 days before harvesting the crop.

The disease-like leaf spots caused by *Cereospora*, *Alternaria*, or *Colletotrichum* are observed sporadically. Cultural practices like crop rotation, filed sanitation, wider spacing, etc. can be followed to control the diseases. Indian spinach is a leafy vegetable, thus, use of chemical fungicides should be the last resort.

■■■

24 Ulluco

Raj Kumar and Flemine Xavier

Botanical Name : *Ullucus tuberosus*

Family : Basellaceae

Chromosome Number : 2n = 24

ORIGIN AND DISTRIBUTION

Ulluco originated in the Andes Mountains in Peru, Bolivia and Northern Argentina. Primarily, it was found in Peru, Bolivia, Ecuador and then to a lesser extent in Colombia, Venezuela, Argentina and Chile. It is cultivated in scattered pockets from Venezuela to Northern Argentina, usually between 30,000 and 40,000 m above mean sea level and in lower valleys of Peru between 1000 and 2000 m above mean sea level. It is still a popular crop in its native region but has not spread to the outside world due to the difficulties in its cultivation and availability of potato as its substitute in Europe. Its use is associated with socio-cultural aspects of the Andean people and their traditional production systems. The attempts to introduce it to Europe at the time of potato famine in the mid 19th century had been unsuccessful. In the early 20th century, it was introduced to Sri Lanka but its cultivation is still nearly confined to the Andes. It has considerably resistant to frost, thus, it is well suited to the Andean altiplano or high valley conditions.

INTRODUCTION

Ulluco, commonly known as *papa lisa*, *olluco*, or *melloco*, is very similar to the potato but not a close relative. All parts of the plant are edible at all stages of growth but tuber is the primary edible part used as food. The leaves, giving taste similar to spinach, can also be eaten raw or after cooking. Its tubers can be cooked like other root vegetables and are nearly interchangeable with potatoes, although they do not mash or fry as well. Being a tuber crop, it is grown very much like a potato but requires more specific growing conditions as compared to the potato. The plant is similar in size but more trailing than the potato. After cooking, the flavour is similar to a beet but the texture is intense and somewhat similar to a boiled peanut. It remains firm even after a long cooking. The tubers become green when exposed to light but greening does not alter their flavour and they do not become toxic as the potatoes do. The tubers are small as compared to a potato, mostly having the diameter of 2–5 cm and typically cooked whole with the skin. It was used by the Incas prior to arrival of Europeans in South America.

COMPOSITION AND USES

COMPOSITION

Ulluco tubers are known to contain high levels of protein, carotene and calcium. The high level of ascorbic acid is also noteworthy. Its leaves are rich source of carotene and calcium. The nutritional composition of ulluco tubers is presented in Table 24.1.

TABLE 24.1

Nutritional Composition of Ulluco Tubers (per 100 g Edible Portion)

Constituents	Contents	Constituents	Contents
Moisture (g)	85.9	Iron (mg)	0.8
Carbohydrates (g)	12.5	Thiamine (mg)	0.04
Protein (g)	1	Riboflavin (mg)	0.02
Dietary fibre (g)	0.6	Niacin (mg)	0.3
Calcium (mg)	3	Vitamin C (mg)	23
Phosphorus (mg)	35	Energy (kcal))	214

Uses

Ulluco is very popular tuberous vegetable and used in traditional preparations such as *mellocos* soup, *olluquito con charqui* and *ullucu*. It is also used with potato, meat, egg, cheese stew and *ajíde papalisas* as ullucu pepper in Bolivia, Peru and Ecuador. It can be eaten raw as salad but is very mucilaginous. Its tubers are eaten like a potato after boiling. The cooked tubers are used in several dishes, soups and stews.

Medicinal Use

Ulluco is believed to be eaten in the Andes for ease of childbirth.

BOTANY

Ulluco is grown as an annual crop, though it is a perennial herb with a small number of 20–30 cm high, erect stems and fibrous roots, some of which are thicken at the end and produce tubers. Stolons arise in the leaf axils and trail over the ground, rooting and producing small tubers at the nodes. Its tubers are borne aerially, often on the stolons that do not reach at the ground. The alternate leaves are broadly oval to cordate, 5–20 cm long and 5–12 cm broad, somewhat fleshy and variable in colour according to cultivar, ranging from dull green to bright green with red spots and purplish yellow borders. The skin colour of the tubers also varies with cultivar, being white, yellow, red, or red-spotted and the flesh is normally yellow. The tubers are small, rather longer than broad, measuring 3–7 cm in length.

CLIMATIC REQUIREMENT

The plant grows best in a cool and moist climate. It also grows well in mild climate with short day lengths of about 12 hours but the production of stolons and stolon-borne tubers is stimulated by shorter days of 10 hours. It gives higher yield in full sun where summer temperature is cool enough to allow it to grow. It prefers shade where temperature frequently exceeds 21°C. When it is cultivated at elevations between 1500 and 4000 m, it gives a yield higher than the Andean cultivars of potato. The plant is sensitive to drought, heat and frost.

SOIL REQUIREMENT

Ulluco thrives best even in relatively poor soils but harvesting is easy in loose and light soils. Moderately acidic soils are ideal for its cultivation. The soil having good amount of compost rather than fertilizer gives better tuber yield.

CULTIVATED VARIETIES

More than 70 cultivars have been recognised for its cultivation worldwide, but generally, local and traditional cultivars are used by the farmers to grow. The high-yielding varieties, which are widely used for its cultivation, are Pica de Pulga and Purple.

PLANTING TIME

Ulluco is planted in spring months after risk of frost has passed, as frost kills its sprouts. However, in areas of mild winter, the tubers are planted in September–October.

SEED RATE

The crop is normally propagated by small tubers, stem cuttings, or pieces of tubers containing nodes. About 10–12 quintals disease-free healthy tubers of 20 g are enough for planting a hectare land area.

PLANTING METHOD

Ulluco is planted almost exclusively from seed tubers. Usually small tubers, weighing about 20 g are used. Cutting of tuber is usually not recommended for its planting, as it is difficult to identify eyes before cutting to ensure a minimum of two eyes in pieces. If large tubers are available, it is advisable to make pieces after the tubers have sprouted. The tubers should be allowed to sprout in light, not in darkness since the sprouts become spindly and fragile. Tubers are planted in barely damp soil at a depth roughly equal to the diameter of the tuber with the tip of the sprout exposed outside the soil. The highest yield of usable tubers can be obtained when planting is done keeping 80–90 cm distance between rows for vine-type and 40 cm for bush-type varieties. The plant to plant spacing is kept 30 cm between plants for both types of variety. The tubers of vine and bush type varieties are planted in mounds and furrows, respectively.

NUTRITIONAL REQUIREMENT

Generally, ulluco is grown without fertilizers, thus, it requires fewer amount of the fertilizers for its growth and development. It grows best in soil containing a moderate amount of well-decomposed compost or farmyard manure, hence, it requires good amount of compost. It is better to add compost than fertilizer as overly rich fertilizer may harm the roots.

IRRIGATION REQUIREMENT

Ulluco prefers cool and moist soils for good vegetative growth and tuber development. Therefore, light irrigation at frequent intervals is preferred for its growth, especially during initial period. Heavy irrigation at long intervals especially during tuber development should be avoided, as it affects tuber quality adversely, leading decaying of the tubers. Drip irrigation is preferred over other methods of irrigation to avoid insect-pest and disease infestation.

INTERCULTURAL OPERATIONS

Hand weeding and shallow hoeing close to plants are important in initial stages of crop growth to reduce crop weed competition for nutrients, moisture, space and light. Ulluco when planted densely grows as a thick ground cover and is able to suppress weeds very effectively. The plants are earthed up two or three times without disturbing stolons during their growing period to provide more space for the developing tubers. Floating row covers can also be used against insect-pest damage.

HARVESTING

The growing duration may be as short as 150 days but the duration of 180–240 days is more common, and at high elevations, the duration of 270 days or more is required. For obtaining good yield, ulluco crop needs at least 7 months. Harvesting is usually done manually with the help of spades. After harvesting, the tubers are categorized into five grades, *i.e.*, table >10 g, unblemished large >10 g, blemished medium 5–10 g, small 0.5–5 g and micro <0.5 g.

YIELD

Yield depends on several factors such as variety, soil type, soil fertility, growing region and package of practices followed by the farmers. It gives an average tuber yield of 50–110 q/ha.

POST-HARVEST MANAGEMENT

Compared to other Andean roots, the ulluco tubers have better shelf life, as they can be stored up to a year at ambient room temperature in cooler areas of the Andes, however, it can be stored for 2 years under storage conditions of 1–3°C temperature and 90% relative humidity. They must be stored in complete dark, as their skin turns green in sunlight. The small tubers should also be kept in a cool and dark place to protect them from dehydration. Even so, the badly desiccated ulluco tubers may usually survive when introduced to some damp soil. Ulluco tubers store very well in cool conditions as long as the air is not too dry, which may cause them to dehydrate. Ancient Andean people have made attempts to store the surplus tubers by freezing, drying and processing. Ulluco tubers can also be canned nicely.

INSECT-PESTS AND DISEASES

Among all the Andean tuber crops, ulluco is probably the most vulnerable to both insect-pests and diseases.

ANDEAN WEEVIL (*Premnotrypes solani*)

The weevils burrow into the tubers where they can lie dormant and then spread further when the tubers are used as planting material, thus, their infestation affects the crop stand, yield and quality of tubers adversely.

Control
- Use weevil-free healthy planting material.
- Rogue out the volunteer plants.
- Use chickens as predators.
- Harvest the crop bit early.
- Apply ash at the base of the plants to prevent weevil infestation.
- Use parasitic fungus *Beauveria brongniartii* at the time of first earthing up and at storage.
- Destroy the pupae after harvesting by ploughing the field.

NEMATODES (*Globodera* spp.)

Globodera species can cause severe economic loss in cultivation causing in retarded growth, severe damage to root system, leading to early senescence if present in high densities.

Control

- Follow soil solarization to destroy eggs.
- Use disease-free planting material.
- Follow clean cultivation.
- Maintain proper field sanitation.
- Follow crop rotation with non-host crops.
- Use plant extracts like *neem* oil, *neem* extracts and tobacco extracts for soil drenching.

SLUGS AND SNAIL (*Cornu aspersum / Deroceras reticulatum*)

Slugs are primarily a threat at initial stage of crop although they can damage the established crop too when they are present in large numbers. They make irregular holes in leaves and remain active at night. During the daytime, they hide under decomposing plant parts and other dark areas.

Control

- Follow soil solarization to destroy eggs.
- Follow clean cultivation.
- Install drip irrigation system to reduce humidity and moist surfaces.
- Handpick and destroy the snails and slugs.
- Spray household ammonia solution 5%–10% on collected slugs.
- Use metaldehyde baits and spray 1% Bordeaux mixture or copper sulfate.
- Use copper barriers with the slime secreted by slugs, causing a disruption in their nervous system.

DISEASES

Most of the common fungal, bacterial and viral infections of ulluco are not terribly serious but somewhat reduce the yield. Bacterial and fungal diseases can be controlled simply by ensuring that ulluco is frequently watered at ground level by using a drip line. Damp soil and dry leaves are good combination for ulluco.

POTATO DEFORMING RUST (*Aecidium cantense*)

Symptoms include orange spores on leaf surface and urediospores are seen under favourable conditions.

Control

- Cut and burn the infected leaves or plants.
- Avoid sprinkling and splashing of water to prevent spread of spores.
- Treat the seed with Vitavax or thiram 2 g–3 g or tebuconazole 1 g/kg of seed.
- Spray the crop with maneb or zineb 0.25%, or propiconazole 25 EC 0.1% at the appearance of rust.

VERTICILLIUM WILT (*Verticillium alboatrum* and *V. dahliae*)

Verticillium wilt is the most problematic disease of ulluco at low altitudes. The symptoms first appear on lower parts and gradually progress to upper parts of the plant. Chronic symptoms include stunted, chlorotic and deformed foliage, leaf scorch, slow growth, abnormally heavy seed crops and dieback of shoots. Midday wilting followed by evening recovery is common for a time. Vascular discoloration or streaking, consisting of dark-colour elongated necrotic tissue, can be seen in a cross section of infected stems. In a cool climate, the plant may be asymptomatic, but as the temperature increases, the symptoms become more severe. Fast wilting at temperatures above 24°C is a likely indication of verticillium infection.

Control

- Follow long crop rotation of 2–3 years with non-host crops.
- Follow deep summer ploughing to destroy the inoculums.
- Grow resistant varieties, if available.
- Use healthy disease-free tubers for planting.
- Collect and destroy the infected plant debris.
- Maintain good sanitation in the field.
- Apply bioagents such as *Trichoderma viride* or *Pseudomonas fluorescens* in the soil.
- Fumigate the soil with a mixture of methyl iodide or chloropicrin.
- Dip the tubers with 0.1% solution of Bavistin before planting.

BLACK SCURF AND STEM CANKER (*Rhizoctonia solani*)

Black scurf is the most noticeable sign of *Rhizoctonia*. The fungus attacks underground sprouts before they emerge from the soil. Stolons that grow later in the season can also be attacked. Severe lesions interfere with the normal functioning of stems and stolons in translocating starch from leaves to the storage tubers. If the fungal lesion expands quickly, the stolon or stem can be killed. Damage is most severe at cold temperature, when emergence and growth of stems and stolons from the tuber are slow. Wet soils also contribute to damage because they warm up more slowly than dry soils and excessive soil moisture slows plant development and favours fungal growth.

Control

- Follow long crop rotation with non-host plants.
- Use sclerotia-free seed tubers.
- Provide proper drainage facilities to reduce waterlogging.
- Harvest the crop as soon as possible.

VIRAL DISEASES

Viral diseases are common in tropical and subtropical climatic regions, where conditions are favourable for the perpetuation of viruses and their vectors. A number of viruses have been found in virtually all ulluco growing areas; these include the ulluco strain of tobacco mosaic virus (TMV/U), papaya mosaic virus (PMV/U), potato latent virus, potato leaf roll virus and ulluco viruses A, B and C. Apart from slight leaf mottling, these viruses appear to be symptomless but yield is considerably reduced by their presence. Some of the viruses can transmit from potato crop.

Control

- Use of resistant varieties, if available.
- Use disease-free healthy tubers for planting.
- Maintain proper field sanitation by destroying discarded tubers to prevent sprouting, roguing out.
- Maintain proper isolation distance between ulluco and potato plants.
- Rogue out the infected plants along with their tubers as soon as they appear.
- Dip the tubers with 0.1% solution of Bavistin before planting.
- Spray the crop with methyl dematon @ 0.075% at 10- to 15-day intervals.

■■■

25 Cactus

M.K. Rana and Bunty Sharma

Botanical Name	: *Opuntia ficus-indica*
	Opuntia megacantha
	Opuntia amychlea
Family	: Cactaceae
Chromosome Number	: 2n = 11 and 33

ORIGIN AND DISTRIBUTION

Cactus is almost exclusively New World plant, meaning that it is native only to North America, South America and the West Indies. In Mexico, the United States of America, Spain, Italy and northern Africa, cactus form an important part of the people's dietary requirement. Pitaya (*Stenocereus griseus*) is a cactus fruit used as a complementary part of diet since ancient times in Southern Mexico. In this region, approximately 30 types of Pitayas like Amarilla, Jarra, Melon, Crema and Olla have been identified. However, Olla and Jarra showed the best quality attributes for the fresh and processed market. Therefore, drought-resistant plants are increasing in popularity.

Many cacti are grown as wild plants in arid and semiarid regions of India. In the 7th century, the British introduced cactus to India for cochineal dye production but these plantations gradually disappeared due to pests and flooding of the areas. Some countries, such as Australia, have water restrictions in many cities.

Among the remains of the Aztec civilization, cacti can be found repeatedly in pictorial representations, sculpture and drawings, principally *Echinocactus grusonii*. This cactus, also known as *Mother-in-law's Cushion*, has great ritual significance. Human sacrifices were carried out on these cacti. Economic exploitation of the cactus can also be traced back to the Aztecs. The North American Indians exploited the alkaloid content of many cacti for ritual purposes. Particularly in South America, dead pillar cacti were used as valuable wood for construction. From the moment of their discovery by early European explorers, cacti have aroused much interest. Christopher Columbus brought the first *Melocactus* to Europe. Scientific interest in them began in the 17th century. By 1737, 24 species were known, which Linnaeus grouped together as the genus *Cactus*. The word *cactus* is ultimately derived from the Greek word *Kaktoς kaktos*, used in classical Greek for a species of spiny thistle, possibly the cardoon, and used as a generic name, *Cactus*, by Linnaeus in 1753. With the passage of time, cacti enjoyed increasing popularity.

INTRODUCTION

Cactus (plural: cacti, cactuses, or cactus) is a member of succulent plant family Cactaceae. It is often grown as ornamental plants but many of the cacti are cultivated as crop plants. In India, an important segment of the population is settled in rain-fed dry areas, which need perennial vegetation to protect them from erosion using drought-hardy and economically viable plants. Cactus seems to be an option to sustain livelihood, reduce poverty and generate employment opportunities. It is drought tolerant due to its carbon dioxide fixation capacity. It is well suited to dry areas, where it can be used as an alternative food and fodder, as well as a live fence to protect agricultural fields.

The high water use efficiency of cactus is attributed to Crassulacean Acid Metabolism (CAM), which is present in rapidly growing cactus species such as *Opuntia ficus-indica*, *Opuntia megacantha* and *Opuntia amychlea* that produce forage for animals, vegetables and fruits with 14% glucose. Cacti are distinctive and unusual plants, which are adapted to extremely arid and hot environments, showing a wide range of anatomical and physiological features, which conserve water. Their stems have expanded into green succulent structures containing the chlorophyll necessary for life and growth, while the leaves have become the spine for which cacti are so well known.

COMPOSITION AND USES

COMPOSITION

Fruits are rich in sugars, vitamins, minerals and amino acids. Fruit weight of cactus pear varies from 80 to 140 g and average edible portion is 54.18%. The pulp yield of cactus pear fruits is approximately 62% and the total and reducing sugar content approximately 9% and 8%, respectively. Ascorbic acid and β-carotene are found present in moderate levels, *i.e.*, 14.7 mg/100 g and 334.0 μg/100 g, respectively. The total pectin content of cactus pear ranges from 5.32 to 14.19%, while the mucilage content varies between 3.78 and 8.5%. Cactus fruits are considered rich source of betalain pigments. The cactus pear cultivar *Gialla* contains 13 kinds of betaxanthins. In *Opuntia undulata* and *Opuntia ficus-indica*, both betacyanins and betaxanthins have been identified, while in *Opuntia stricta* fruits, only betacyanins (betanin and isobetanin) have been detected. *Opuntia stricta* fruits show highest betacyanin content, *i.e.*, 80 mg/100 g fresh fruit. The nutritive value of cactus is comparable with those of other popular fruits as in Table 25.1.

USES

As Food

With excellent quality and flavour of fresh fruits, its young leaves serve as a nutritious vegetable and salad dish and the immature fruits for making mock-gherkins. Enzyme activity levels are negligible in green immature fruits and increased with the fruit development and during storage, concomitant with the timing of linalool accumulation in fruits. Jotilla, a fruit of cactus tree *Escontria chiotilla*, is a non-climacteric fruit with sour-sweet flavour. Its pH is 4.2 with 10–12° Brix, which makes it tasty as a natural dessert. A proximal analysis of fruit determines its suitability for marmalades and

TABLE 25.1
Nutritive Value of *Opuntia* Fruits (per 100 g Edible Portion)

Constituents	Contents	Constituents	Contents
Water (%)	85.6	Pectin (%)	0.19
TSS (°Brix)	12–15	Calcium (mg)	27.60
Sucrose (%)	0.2	Phosphorus (mg)	15.40
Glucose (%)	7.0	Potassium (mg)	161.00
Fructose (%)	4.8	Magnesium (mg)	27.20
Crude protein (%)	0.21	Sodium (mg)	0.80
Crude fat (%)	0.12	Iron (mg)	1.50
Crude fibre (%)	0.02	Vitamin C (mg)	22
Acid (%)	0.02	Calorific (kcal)	47.30
pH	5.8	Ash (%)	0.44

jams as well as dressing products: the product is minimally processed and frozen with sugar. Purple cactus fruits are a source of betalains, which are a potential colorant similar to the pigment obtained from red beet, which is widely used in the food industry.

The fresh juice is clarified by ultra-filtration, and then, it is clarified by osmotic distillation up to Total Soluble Solids (TSS) content of 61° Brix. Combination with pledged filtration effectively reduces the loss in polysaccharide content. Homogenization with higher pressure and shorter pulping time effectively retains the polysaccharide content. The peeling process has little effect on polysaccharide content, suggesting that the peel should be kept to preserve the nutrients in cactus.

Different parts of cacti are used as vegetable and also for salad purposes. The young tender spring vegetative growth of wild cactus (nopalitos) has been extensively consumed by Hispanics during Lent, with only recently plantations of spineless cacti being established in Texas for nopalito production. It is also used for the preparations of mock-gherkins, jams and syrups, soap from its leaves, alcoholic drinks, and honey and cheese production from its seeds. Coloured cactus fruit concentrates are used in yoghurt and ice cream.

Medicinal Uses

Though cactus is traditionally used as a valuable health-supporting nutritious food, the vegetative parts of *Opuntia* spp. are scarcely used in modern nutrition and medicine. It also has applications in pharmaceutical industries. Peyote (*Lophophora williamsii*) extract has been associated with stimulating the central nervous system and regulating the blood pressure, sleep, hunger and thirst. Also, one of the most known cacti since remote times by its use as hallucinogen is peyote, which is a well-known psychoactive agent used by Native Americans in the Southwest of the United States of America. Some species of *Echinopsis* (previously *Trichocereus*) also have psychoactive properties. *Opuntia streptacantha dialysate* could be considered as a new approach in treating Non-Insulin-Dependent Diabetes Mellitus (NIDDM).

Prickly pear is widely used as folk medicine for burn wounds, oedema and indigestion, and it is found that the effect of fruit extract is better than those of stem extract. Cactus pear fruit contains vitamin C and the radical scavenging properties. Cactus pear is a food of nutraceutical and functional importance. Purple fruits are a potential source of antioxidants. Consumption of cactus pear fruit positively affects the body redox balance, decreases oxidative damage to lipids and improves antioxidant status in healthy humans. Supplementation of vitamin C at a comparable dosage enhances overall antioxidant defence but does not significantly affect body oxidative stress.

A study of antioxidant compounds in cactus pear fruits reveals that kaempferol is found in green-skinned, purple-skinned and red-skinned varieties and isorhamnetin in green-skinned and purple-skinned varieties. The red-skinned fruit contains the most ascorbic acid and the yellow-skinned fruits the most carotenoids. Thermal treatments increase the extractability of these pigments and the antioxidant activity is related to carotenoid concentration. Total phenolic content decreases after the thermal treatments. However, this result had little effect on the antioxidant activity. Both antioxidative and DNA damage-reduction activities are increased with increasing cactus pear fruits extract. Cactus pear fruits extract constituents' antioxidative and DNA damage-reduction efficacies are a potential source of raw material for pharmaceutical and functional food industries. Arizona cactus pear extracts effectively inhibit cell growth in several different immortalized and cancer cell cultures, suppress tumour growth in nude mice and modulate expression of tumour-related genes.

The seeds of cordon cactus (*Pachycereus pringlei*) are edible and highly nutritious and plants have been used in traditional medicine. Supplementation with cactus pear oil or cactus pear seeds is useful in reducing the serum cholesterol level, and thus, reducing the atherogenic risk factors in rats.

A novel food product (NeOpuntia) made from dehydrated leaves of the cactus *Opuntia ficus-indica* is found to have hypolipaemic properties, and hence, useful for patients with lipid metabolism disorders. The stems of *Selenicereus grandiflorus* and other species contain a glucoside, which in extract or tincture form constitutes a diuretic and also has cardiac properties. The stems of buckhorn cholla are burnt, and the ash is applied to cuts and burns to aid in the healing process.

USE AS FODDER

Prickly pears can be a good alternative forage crop on land, which is presently deemed marginal for other crops. The spines can be burnt before feeding the cattle. In Mexico and southwest Texas, propane torches known as pear burners have been used to burn off cactus spines. Although cactus leaves are low in proteins, they can serve as a very good source of roughages, especially in rain-fed and severe drought-prone areas. *Nopalea cochenillifera* (*Opuntia cochenillifera*) is the species widely used in semiarid regions as a forage crop. The material not sold in vegetable market is being used as dairy cattle fodder. It imparts better flavour and quality to milk and also enhances colour of the butter. Although milk production and composition are not affected by replacing corn and Tifton hay with forage cactus (*Opuntia ficus-indica*), significant changes are found in the milk profile of medium- and long-chain fatty acids by increasing the proportion of forage cactus in the diet. In Texas, the primary use of cactus has been in times of drought when the spines have been burnt off the cactus to feed cattle. A daily ration of 40 kg of cactus, 0.5 kg of mineral salts and 0.5 kg of protein supplement is sufficient to permit excellent live weight gain, reproduction and lactation from nursing cattle. The fruits of two columnar cacti, cardon dato (*Stenocereus griseus*) and cardon lefaria (*Cereus repandus*), have good potential as food and feed in arid zones.

OTHER USES

Cactus pears from *Opuntia stricta* are considered as a potential source of natural red colourants. The use of prickly pear cactus mucilage as an edible coating leads to increased strawberry shelf life. Opuntia genus is widely known for its mucilage production. Mucilage, a complex carbohydrate with a great capacity to absorb water, should be considered a potential source of industrial hydrocolloid. Mucilage contains varying proportions of *L*-arabinose, *D*-galactose, *L*-rhamnose and *D*-xylose, as well as galacturonic acid. The mucilage content found in the cactus cladodes is influenced not only by the management of crop but also by the temperature, irrigation and the rain. In some countries, small farmers use cactus mucilage to purify drinking water. Another traditional use is for improving house paint. Recently, a cactus cladode extract has been tested to improve water infiltration in soil. The cladodes, dehydrated and ground into powder, are a source of dietary fibre, which may be used as natural ingredient in different foods to enhance their beneficial properties. Cactus mucilage can be used as natural thickener. All these components could be used as natural ingredients in foods to enhance their healthy properties. The new functions of some compounds open new possibilities for adding values to the cactus pear, a new crop for semiarid regions of the world. Cacti are commonly used for fencing material where there is a lack of either natural resources or financial means to construct a permanent fence.

BOTANICAL CLASSIFICATION

Cactus belongs to the kingdom Plantae, division Magnoliophyta, class Magnoliopsida, order Caryophyllales and family Cactaceae. It is commonly known as prickly pear, having about 130 genera and 1,500 species. *Cacti* exist in a wide range of shapes and sizes. The tallest is *Pachycereus pringlei* with a maximum height of 19.2 m and the smallest is *Blossfeldia liliputiana*, only about 1 cm diameter at maturity. Cactus flowers are large and, like the spines and branches, arise from areoles. Many cactus species are night-blooming, as they are pollinated by nocturnal insects or small animals, principally moths and bats. Numerous species have entered widespread cultivation, including members of *Echinopsis*, *Mammillaria* and *Cereus* among others. *Opuntia ficus-indica* is the most important cactus species in agriculture today and the first fruit crop especially adapted to semi-arid and non-irrigated lands. It has usefulness as food, fodder, dye, source of energy and has a role in ecosystem remediation.

SPECIES OF CACTUS

Some of the useful species of cactus are as follows:

- *Carnegiea gigantean* (Saguaro): The fruits are also considered of excellent quality. Fruit pulp is processed into jelly and wine. It is part of Papago Indians' diet. Seeds are also ground and eaten.
- *Cereus peruvianus* (Koubo): It is a commercially grown columnar cactus that produces an apple-size, berry-like, edible fruit. The unique aroma of this fruit is largely due to *S*-linalool and linalool derivatives.
- *Echinocactus* sp. (Barrel cactus): The spines of this genus were fashioned into phonograph needles and fishhooks.
- *Echinocereus enneacanthus* (Strawberry hedgehog): Edible fruit tastes similar to strawberry.
- *Echinocereus stramineus* (Straw-coloured hedgehog): Edible fruit tastes similar to strawberry.
- *Echinopsis chiloensis* (Quiska): Chilean cactus is used in the manufacture of rain sticks.
- *Epithelantha bokei* (Button cactus): Edible fruits are fed to cattle.
- *Escontria chiotilla* (Jiotilla): This species produces edible fruits known as jiotilla.
- *Ferocactus hamatacanthus* (Texas barrel cactus): Juicy brown fruit is used as lemons and limes.
- *Ferocactus wislizenii* (Candy barrel): Animals eat the fruit and stems, and fruits of this species are used to make cactus candy.
- *Hylocereus undatus* (Pitaya, dragon fruit, or strawberry pear): It bears edible bright red or pink fruit with green scales. It is eaten raw or made into wine and other drinks.
- *Lophocereus schottii* (Senita): Stem is processed into drugs to fight cancer and diabetes.
- *Lophophora williamsii* (Peyote or mescal buttons): Plant contains mescaline, a hallucinogenic drug capable of inducing visions.
- *Myrtillocactus geometrizans* (Blue myrtle or whortleberry cactus): Blue fruit resembling a blueberry is edible.
- *Nopalea cochenillifera* (Nopal Cactus): Plant is used as a host for the female cochineal insect *(Dactylopius coccus)*. Cochineal, a crimson dye, is processed from the body of this insect.
- *Opuntia acanthocarpa* (Buckhorn cholla): Pima Indians steam and eat flower buds.
- *Opuntia ficus-indica* (Indian fig): The edible fruit of this cactus, commonly known as a tuna, has a sweet taste similar to watermelon. Fruits are also processed into jams and jellies.
- *Opuntia spinosior* (Cane cholla): Skeleton of dead plants is used for making furniture.
- *Pachycereus pectin-aboriginum* (Hairbrush cactus): Indians used the bur-like fruit of this cactus as a hairbrush.
- *Peniocereus greggii* (Queen of night): Edible root and fruits are eaten by the Indians. Poultice is reportedly used for respiratory ills.

- *Pereskia aculeata* (Barbados gooseberry): The small yellow fruit is used in jellies and preserves. Fruit is juicy and slightly acidic.

- *Schlumbergera truncates* (Christmas cactus): It is perhaps the most commercially grown cactus.

- *Selenicereus grandiflorus* (Night-blooming cereus): Stems and flowers are processed into homeopathic medicine for urinary tract infections and angina. It has been reported to have a digitalis-like effect on the heart.

- *Stenocereus gummosus* (Pitahaya agria): Stems of this cactus are crushed and thrown into water by natives. Substances in the cactus act as a fish poison and stun fish.

- *Stenocereus thurberi* (Organ pipe cactus): Its fruits are edible.

- *Trichocereus pachanoi* (San Pedro cactus): Plant contains mescaline, a hallucinogenic drug capable of inducing visions.

The other species of *Opuntia* like *Opuntia tuna*, *Opuntia streptacantha* and *Opuntia cardona* are also usually cultivated with same purpose. The bluish berries of *Myrtillocactus geometrizans* are sold in Mexican markets with the name of *garambullos*.

CULTIVARS

An improved Nopalito cultivar with greater year-round production, lack of spines, lack of glochids (nearly microscopic spines) and low mucilage is available as *Texas A&M 1308*. In contrast, spineless *Opuntia* plantations for Nopalito production have been common in central Mexico (Milpa Alta) for many years. Vegetable clone *1308* can yield biomass production of 80–90 tonnes per hectare per year. In some of the American and African countries, its produce is used as vegetable.

CLIMATIC REQUIREMENT

Since cacti are basically drought resistant, they are well suited to areas having scanty rainfall and poor irrigation facilities. The areas having high rainfall are not much suitable for its cultivation as the crop becomes more liable to insect-pests and diseases. Optimum conditions for its growth are available in summer rainfall regions having average rainfall of 300–600 mm. Hot sunny days and cool dry winter where temperature does not fall below −5°C are most suitable for cactus production. Therefore, a large part of India is ideal for its cultivation.

SOIL REQUIREMENT

Cactus thrives best on sandy and sandy loam soils. However, it still performs better on heavy soils with adequate drainage. Gravelly or stony lands especially the foothill slopes are also suitable for its cultivation. Further, it performs best on slightly alkaline soils rich in calcium and potassium. Any type of soil which is not suitable for other crops can be planted with cactus provided that it is not subjected to prolonged waterlogging.

All cacti require fast-draining soils. Special *cactus mixes* are available in the market. These mixes should be used for potted plants as poorly drained soils will predispose the plants to attack by soil-borne root-decay-causing pathogens, which may result in death of the plant. Careful irrigation management is essential in heavy clay soils. In open fields, a copious amount of sand should be incorporated if the soil is clay or clay loam to improve drainage and soil structure. Growing cactus in sandy soils should be avoided since sandy soil may not conserve sufficient moisture and nutrients. They should be amended by incorporating a large amount of well-decomposed compost. Partially decomposed or fresh manure should not be incorporated in soil as soil amendment since the high salt content found in fresh manure may be detrimental to root development

PLANTING

Cacti are both self- and cross-pollinated. They can be raised from both seed and vegetative parts. However, the most common method of propagating prickly pears is by 1-year-old leaves. Upright method of planting is ideal. In this method, the leaves are planted upright in the soil with cut end in the ground, keeping one-third of the leaf below the ground. Before planting, the leaves should be allowed to wilt for 4–6 weeks in semi-shade, and then, the leaves should be dipped in some fungicide solution to check root rot. Spacing depends on type of cactus and purpose for which it is grown. In general, prickly pears give best production when planted at 4 × 3 m spacing.

Proper planting depth is essential for the survival of many cacti. In some of the cacti, the green portion of the stem should not be kept below the ground level since it leads to poor establishment and death of the plant. To keep the plants identical in height, planting should not be done with some deeper than others. However, the young prickly pear (*Opuntia* spp.) should be planted deeply enough so that they remain upright after planting.

NUTRITIONAL REQUIREMENT

Cacti have relatively low nutrient requirements. They need fertilizer only once or twice a year during the late spring or summer when they are actively growing. Use a houseplant food, which is higher in phosphorus than nitrogen, diluted to half the recommended rate. Other succulents may be fertilized in the same manner three or four times during the brighter months. Some people use earthworm castings for extra nutrients and bonemeal some of the time. Incorporating nitrogen in soil encourages good top growth and it helps to build protoplasm, protein and other components of plant cells, phosphorus encourages good root growth and aids in ripening of fruit and seed germination (seed viability), and potassium encourages good flowers and fruits. Too much nitrogen reduces the fruit firmness and increases fungal decay, and no negative effects of nitrogen deficiency on fruit quality have been detected, so use a low-nitrogen fertilizer at about one-half to one-fourth the recommended rate.

IRRIGATION REQUIREMENT

While growing cactus, one of the most serious abiotic problems is over-watering, which combined with poorly drained soils causes failure of plant. Irrigations should be applied when the soil is moderately dry, depending on the plant type and species. On the other hand, during drought, supplemental irrigation will minimize the problems associated with heat and sun damage. As a general rule, irrigation in well-drained soils at 10- to 14-day intervals is sufficient to insure plant health and growth during summer. In heavy soils such as clay soils, irrigation interval may be decreased. In light soils such as sandy or very rapidly draining soils, irrigation frequency may be increased. If the soil in root zone (5–7.5 cm below the surface) is even slightly damp, wait for irrigation until it has dried. A thorough irrigation followed by a 2–14 days' dry period will give the soil a chance to dry off completely, which in turn will reduce the incidence of soil borne pathogens. As the daylight decreases in winter, irrigation frequencies may be reduced. When the night temperature drops below 16°C, irrigation is discontinued. Most of these plants can survive on natural rainfall through late fall, winter and early spring.

If inadequate water is supplied at prevailing high temperature, cactus plants stop growing and may shed their leaves and the apical tip of stems may die, which is followed by die back. Cacti may shrink back to the potting mixture and possibly take on a reddish or purplish colour due to the synthesis of stress pigments. Shrinkage of a cactus in some cases during drought produces irreversible folds in the plant body, which never fill out again. However, careful watering usually reverses the effects of drought. To water-starved plants, only a small amount of water should be applied at first since some of the roots have been lost.

Excessive watering is probably the single most common cause of failure of cactus plants to thrive. The roots may suffer in wet soil and start to decay invisibly but the plant still looks well since the few remaining roots are able to absorb sufficient water. As the roots continue to die in water-stagnant soil, a point is reached at which they are unable to supply sufficient water and the plant appears to be suffering from lack of water. If more water is supplied, the situation gets worse and the rot may spread upwards into the basal stems. Ultimately, the plant is observed to be yellow or greyish. At this stage, it is difficult to save it. The other reason for loss of roots may be pest damage. Watering a cactus plant at the wrong stage, especially when it is dormant, can cause decaying as effectively as can also happen if the roots have been eaten by insect-pests.

USE OF GROWTH SUBSTANCES

Application of ethaphon at a concentration of 500 and 250 ppm induces the beginning of fruit ripening about 9 and 5 days before the natural ripening, respectively.

HARVESTING

Production of fruits in cactus starts 3 years after planting. Normally, fruits appear on 1- to 2-year-old cladodes. Unlike other crops, fruit is fully formed when flowers first appear; after pollination and fertilization all that happens is that the cavity is filled with seed and the edible portion. A well-established crop with a spacing of 4×2 m can give a fruit yield of 10–40 tonnes per hectare per year with proper irrigation and even two crops per year can be taken.

POST-HARVEST MANAGEMENT

Out of season, the fruits are regular in size and per cent flesh with only a slight reduction in total soluble solids content. Seedless fruits are obtained by emasculating flowers 2 days before bloom and by spraying indole butyric acid (IAA) and gibberellic acid (GA). Ethaphon is used for the ripening of cactus pear fruits. Cactus plantations in semiarid lands produce large quantities of cactus fruits with short shelf life in sparsely populated areas where production rapidly surpasses demand. Improvements in shelf life are needed for following reasons:

- To permit international shipments by refrigerated ship of approximately 3 weeks' duration.
- To be compatible with 2°C fruit fly quarantine treatments.
- To function without the use of fungicide, as none is currently registered and as registration for additional minor crops is not likely.

Unfortunately, the low acidity and the highly soluble solids content make the fruit very attractive for growth of microorganisms, requiring a thermal treatment (>115.5°C) to obtain good control of the microbial invasion. Cold storage maintains fruit firmness and reduced water loss and fungal decay. Hot water treatment is highly beneficial in reducing chilling injury, fungal development and visual quality, and the temperature of 2°C approved for the control of fruit fly will not cause chilling injury. The respiration rate and tissue permeance decreased as the relative humidity is increased. Relative humidity in the range of 65% to 90% has a marked effect on the rate of gas exchange, especially for carbon dioxide and consequently on the gas equilibrium inside Modified Atmosphere Package (MAP). Fresh prickly pear cactus stems can be stored up to 32 days in modified atmosphere package with carbon dioxide concentration of 20 kPa without significant increase in microbial population.

The shelf life of fresh-cut cactus pear, packed in bidirectional polypropylene bags and stored at 4°C, can be extended up to 20 days without affecting the quality, whereas storage at 2°C is optimum for preserving the minimally processed cactus fruits up to 12 days. Minimal processing of cactus fruits is important and has a great potential because it offers the opportunity to differentiate and add value to cactus pears and nepalitos as long as nutritive quality, sensorial quality and safety are guaranteed. The minimally processed cactus cladodes have a shelf life of 1–2 days at room temperature, which can be extended to 7 days at 5°C. The drawbacks that shorten the shelf life of cactus cladodes are browning and mucilage secretion. Controlled atmosphere storage of minimally processed cactus pear fruits at 2°C in 10% carbon dioxide showed high values of visual quality, soluble sugar content and sugar content, as well as a reduced tendency toward browning. Under these conditions, the quality of minimally processed fruits was preserved up to 20 days.

PHYSIOLOGICAL DISORDERS

Improper cultural conditions are a major cause of poor growth. The solo major cultural problem is over-watering, with the roots left wet for excessively long periods, resulting in decaying of roots. The other growth problems are related to insufficient light and too low or too high temperatures. Most of the cacti are expected to flower when they reach the mature size, or even before, and failure to flower may indicate substandard growing conditions.

POOR LIGHT

Undoubtedly, natural sunlight is the best way of providing light to the plants, but sometimes in arid hillsides under blazing sunlight, the plants get scorched. Cactus plants kept in insufficient light grow with pale or yellow leaves, sometimes stunted leaves and elongated relatively thin stems, which is known as *etiolation*. This condition can generally be reversed by keeping the plant under stronger light. These plants can often be pruned to restore their shape. Cactus plants will not usually become etiolated in dark conditions if kept cool and absolutely dry.

HEAT DAMAGE

Scorch can affect cactus if there is a sudden period of blazing sun after the dark winter days, or even after a prolonged cloudy period during summer. The sun-exposed portion develops sunken brown or white patches and the green chlorophyll is destroyed. Sometimes, a greenhouse plant loses its green pigment due to exposure to excessive heat alone, albeit the plant is not exposed to direct sunlight. This scorching can be avoided by providing shade to the greenhouse, improved ventilation and air circulation within the growing area. When plants are to be moved to direct sunlight, harden them off gradually in diffuse sunlight or put them under mesh shading for a few days to acclimatise.

COLD DAMAGE

Although many cacti are surprisingly cold-hardy if kept absolutely dry during winter, some species from perpetually tropical climates (*e.g.*, Madagascar) can suffer damage to the soft tissues at their growing points, and scarring and collapse of their stems leading to fungal attack and death of the tissues. The only solution is to maintain higher temperature for susceptible plants. Some species such as *Echincactus grusonii*, which are easy to grow, can develop hideous brown marks, which spoil the plant, when temperature is too low.

FREEZE DAMAGE

The cactus plants suffer freeze damage, if temperature for several hours remains below freezing point. The symptoms become evident as blackening of the exposed parts, but after a few weeks, the blackened area gets dried and becomes crisp, which cannot be repaired. If temperature is not significantly below freezing and the freeze lasts only for a few hours or less, only the epidermis may be damaged and the plant will outgrow the damage in several years. When the freeze is of long duration and extreme, the damage may be severe.

To prevent freeze damage, often, the protective measures should be taken to assure survival. Cold-sensitive plants should be grown in areas that receive radiated heat at night from a courtyard or wall where the air temperature is slightly warmer. The plant may be covered with a cotton sheet during nights. Another strategy is to place a 60 watt light bulb under a cotton sheet, taking care that the light bulb does not touch the plant or the cotton sheet. If the plant is covered with a plastic sheet, it will transfer cold directly to the plant wherever it touches the plant. The plants in container can be moved indoors until freezing weather has passed.

SUNSCALD

Sunscald is a common problem in greenhouses and is also problematic in cacti grown in open field or in nurseries when the seedlings are transplanted in open field. Those parts that have not been acclimatised to direct sun will burn easily. The sunburnt plants turn yellow, and as the damage progresses, the epidermis turns straw-yellow and dies. Barrel and columnar cacti are very vulnerable to sunburn. The entire plant usually does not die but it is permanently scarred. The apex of the leaf or the area where the leaf naturally curves will become yellow. In severe cases, leaves may die, but if damage is not severe, they may recover at cooler temperatures. Sunburn can be minimized by placing shade (30%) or cheesecloth over newly acquired plants. If possible, purchase those plants that have been grown in full sun. Using inorganic mulch such as gravel or decomposed granite around new plants also reduces sunburn. Application of mulch about 5 to 10 cm deep helps in conserving soil moisture and reducing reflected light and heat but the mulch should not cover the base of the plant. Many species will benefit from partial shade. The plant may also be protected from summer sun and winter freezes by planting cactus near deciduous trees,

HAIL DAMAGE

In some areas, hailstorms cause considerable damage to young cacti. Hail damage on cactus results in wounds at the point of impact, leaving scars that can vary from very small spots to larger lesions. In some species, leaves may be perforated by the impact of hail. When hailstorms are accompanied by strong winds, damage to the plant may be on one side of the plant only. This distribution of damage is helpful in recognizing hail damage. It is difficult to provide protection from hail unless the plants are covered before a hailstorm begins. If the area is small it may be covered with a heavy blanket to minimize hail damage.

INSECT-PESTS

A number of insect-pests can potentially damage the cactus plant but most of the insects do not require chemical treatment for their control. A healthy plant can withstand the occasional insect-pest better than a plant stressed by planting in poorly drained soil, improper planting location or general neglect. The most harmful insects are described as follows:

MEALY BUGS (*Planococcus citri* , *Pseudococcus longispinus* and *P. calceolariae*)

Probably, this is the most common pest of cacti. White or grey mealy bugs about 3 mm in length suck cell sap from cactus and reproduce rapidly by laying their eggs underneath a cotton-like elliptical covering. Mealy bugs often accumulate to feed on the tender tissues at or near the growing point. Very often, when nesting, they hide around the base of the succulent plants, just below the soil. A plant infested with mealy bugs is covered with these white spots and sticky honeydew, which can encourage black mould. No part of the plant is left without attack but most often it attacks the stem. Infested plants stop growing, look sickly and deformed in appearance and start to shrivel. Mealy bugs may also attack the roots of cactus, causing the plant to appear sickly.

Control
- Take cuttings from a healthy plant.
- Apply systemic and contact insecticide to the cuttings.
- Remove the mealy bugs mechanically or spraying water with pressure.
- Use biological controls like predatory ladybugs and lacewings.
- Introduce predator *Cryptolaemus montrouzeri*, which requires a temperature about 21°C.
- Spray the plant with malathion and dimethoate after adding surfactant at weekly intervals.

RED SPIDER MITE (*Tetranychus urticae*)

Mites are not insects but are closely related to spiders. Mites are very small (less than 0.2 mm long) and can be observed only with a magnifying lens or microscope. Red-colour insects are often found in their whitish webs where they suck cell sap on cactus sap. Pale yellow/white spots, which later turn rusty brown, appear on plants infested with red spider mites. Damage occurs mainly at the top of the plant. Weakened plants are also become susceptible to viral, bacterial, or fungal diseases.

Control
- Dry indoor air lessens the problem.
- Keep the plants healthy by providing proper water.
- Maintain high humidity around the plant using periodic misting or washing.
- Spraying water with pressure may remove the red spider mites. As the plant permits, a strong stream of water should be used when watering.
- Use *Phytoseiulus persimilis* predator, but it requires a temperature of about 21°C.
- Spray the crop with a miticide.

SCALES *(Dactylopius coccus* or *Coccus cacti)*

After mealy bugs, scale is probably the most commonly encountered pest of cactus. Pinhead-size insects present on plant as raised brown spots, which resemble marine limpets, are actually hard shell coverings that protect the insect or eggs underneath. The insect secrets a sugary honeydew-like substance, which encourages black sooty mould. The insect under the protective scale feeds on the plant sap and can transmit virus diseases between plants. These insects are like little armoured tanks that clamp themselves to a leaf's surface and do their damage. The infected plants appear yellow and become weak. The infested bud and shoot drop down.

Control

- Wash the infested plant with a weak detergent solution.
- Remove the scales from the infested plant mechanically with a toothpick.
- Spray the crop with malathion at 0.05%.

THRIPS (*Scopaeothrips bicolor* and *Rankliniella occidentalis*)

This minute insect sucks cell sap from the tender cladodes and leaves behind telltale yellow or white spots on plant leaves.

Control

- Spray the plants with a solution of nicotine sulfate or miticide.

APHIDS (*Aphis gossypii* and *Myzus persicae*)

These minute green-colour insects often inhabit soft parts of the cactus. Majority of aphids are female and produce a rapid succession of live young. They often target the blooms and blooming parts of the cactus. A colony of aphids appears as tiny green spots. Aphids produce a sugary honeydew-like substance, which in turn encourages black mould.

Control

- Spray the plants with malathion at 0.05% at 10-day intervals.

WHITEFLY (*Trialeurodes vaporariorum*)

These are a minute (2 mm) white flying insect but not very common on most succulents. A cloud of these insects rises if the foliage is disturbed. They suck cell sap from tender parts of the plant and make the plant weak. They lacerate the tissues and secrete a sugary honeydew-like substance, which encourages black sooty mould on plants, which in turn affect photosynthesis of the plant.

Control

- Use parasitic wasp (*Encarsia formosa*) to control whitefly in greenhouse.
- Spray the plant with Confidor™ (imidacloprid).

NEMATODES (*Cactodera cacti*)

These are microscopic worm-like animals living in the soil. Nematodes burrow into plant roots, causing root galls. The infected plants become stunted or pale yellow.

Control

- Take out the plant from the soil and remove all soil from roots. Carefully cut off all the nematode-infested roots and let the plant or cutting dry for several days. Replant the plant or cutting in sterile soil.
- Spray the infested plant with systemic insecticides.

ANTS (*Brachymyrmex patagonicus*)

Ants often make their nest in soil near the plant and dine on young stems, grains of seeds from ripening fruits and stamens of flowers. They are often an indicator of an aphid problem. These ants often eat the honeydew excreted by aphids.

Control
- Wash the plant with a strong lukewarm stream of water.

DISEASES

A number of fungal and bacterial diseases affect cacti plants, some of which can collapse and die swiftly due to the attack of diseases. The atmosphere is full of fungal spores, which are opportunists, waiting for the proper environment for their germination. Generally, fungi do not affect cactus since they are cultivated under relatively dry conditions. Damp atmosphere is a universal requirement for activation of fungal spores. Cactus seedlings are vulnerable to fungal attack of the lower stem, which causes damping-off. Once the seedling has wilted, it is usually too late to save it, thus, preventative measures are a better option. Some of the destructive diseases are discussed as follows:

ALOE RUST (*Phakopsora pachyrhizi*)

It is a fungus that causes round brown or black spots on leaves of cactus. The black colour is caused by oxidation of phenolic substances in the sap, which seals of the affected area. Once formed, the black spots are permanent and can be unsightly but do not usually spread.

Control
- Spray the plants with a systemic fungicide but prevention is the best option.
- Avoid leaving water to lie on the leaves for long time.
- Avoid excess damp in cool weather.
- Arrange for plenty of air circulation and sunlight.

BLACK OR SOOTY MOULD (*Cladosporium, Aureobasidium* and *Antennariella* spp.)

The fungus is ubiquitous, which is often seen on plants covered with honeydew excreted by aphids, whitefly, mealy bugs, etc. or on plants with nectar-producing glands such as certain *Ferocacti*. Generally, sooty mould is more gruesome than harmful on otherwise healthy plants. However, it will attack seedlings following mechanical damage or excessively wet conditions and other weak or damaged plants.

Control
- Use *neem* oil formulations to control sucking insects' population that excretes honeydew.
- Use systemic insecticides such as Orthene, malathion, or diazinon.
- Spray insecticidal soap, dish soap, or detergent dissolved in water on plants.
- Prune the infested plant parts judiciously.
- Also, keep the ants away from honeydew-producing insects by applying a sticky compound around the trunk and trimming limbs touching buildings or other access points.
- Wiping affected leaves and other plant parts with water is sometimes enough to remove the sooty mould growth.
- Lukewarm water can be more effective.

BASAL STEM ROT AND DAMPING-OFF (*Fusarium oxysporum, Phytophthora parasitica* and *Pythium aphanidermatum*)

Excessive cold or damp may lead to rotting of stem, often just around the soil level where damp soil may be in prolonged contact with plant stems. In many cases, fungal attack and poor culture are linked. The rotten tissues may become black or reddish brown depending on plant and organism attacking it. Numerous brown or grey spots on leaves and corky brown marks on stems are undoubtedly due to fungal attack following damage or prolonged contact with drops of water. Others may reflect poor cultural conditions or the natural development of corky or woody stems as the plant matures. Growers of Asclepiads will be familiar with black spots developing on the stem which spread and develop into sunken patches of dead tissues. This fungal infection can spread to the whole plant unless the affected part is removed promptly or treated with fungicide. Usually, this happens perhaps there where water droplets fail to evaporate because of cold conditions.

Control

- If the stem is cut well above the rotten part, it may be possible to re-root or graft the healthy tissues and save the plant.
- Improve ventilation in greenhouse.
- Control temperature, watering and application of fertiliser.
- Avoid injury to the plants at the soil line.
- Apply a fungicide to protect healthy plants.
- Nurelle (pyroxychlor) provides good control of both the pathogens.
- Truban (ethazole) provides control of *P. aphanidermatum* equal to that achieved with Nurelle.
- Spray the potting mix with a systemic fungicide such as Benlate or copper sulfate.
- Remove the infected seedlings promptly before the spores are produced, and spray the remaining seedlings with fungicide.
- Allow the surface moisture to evaporate to keep the soil surface dry.

PHYLLOSTICTA PAD SPOT (*Phyllosticta* spp.)

The disease is found throughout the desert. Lesions are almost completely black because of the presence of small black reproductive structures called pycnidia produced on the surface of infected plant tissue. The spores produced within the reproductive structures are disseminated by wind-blown rain or dripping water and they can infect new sites on nearby pads. Pads on lower part of the plants are often most heavily infected since the humidity is higher and moisture often persists after rain. Once pads dry, the fungus becomes inactive and the lesions may fall out.

Control

- Remove severely infected pads or entire plants from the field or pot to prevent spread of the fungus.
- Spray the crop with Copral 50% WP (copper oxychloride) and Cabriotop 38% WG (pyra-clostrobin + metiram).

ANTHRACNOSE (*Colletotrichum agaves*)

This can be a problem during moist conditions or occasionally when cactus is grown in shade and irrigated by overhead method. Infections cause lesions on the leaves. When the fungus is active, it produces a red to orange spore mass within the lesions. The spores are distributed by splashing water and windborne rain.

Control

- Remove leaves with active lesions.
- Avoid overhead watering in areas where disease has occurred.
- Spray the crop with fungicides such as thiophanate-methyl during wet weather.

BACTERIAL NECROSIS (*Erwinia cacticida*)

The initial symptom is a small, light-coloured spot with a water-soaked margin on surface of the trunk or branches that may easily go unnoticed. The tissue under the infection site soon becomes brown or black. As the disease progresses, the tissue may crack and exude a dark brown liquid. As the infected tissue breaks down, the woody skeleton is exposed. Its pathogen survives in soil or plant tissue for a long period. The bacterium is spread by insects and in infested soils. Infection takes place at wound sites on roots, trunks and branches, especially those made by insects, weather-related events or rodents. Dead and dying plants serve as reservoirs of the bacterium and as sources of bacterial inoculum for new infections.

Control

- Carefully remove the infected tissue along with a small margin of healthy tissue using a clean sharp knife when very small to avoid spread of infection to new sites.
- Allow the wound to heal on its own.
- Remove the plants with advancing decay to prevent infection of nearby plants.
- Remove all the infested plant material completely from the site.

STEM AND LEAF LESIONS (*Pseudocercospora opuntiae*)

Stem and leaf lesions on cacti have been associated with other fungi as well. They usually occur during prolonged periods of cool and wet weather. The infection usually develops during spring rains and is more common on plants that are shaded. Infections rarely kill the plant but cause irreversible cosmetic damage. The pathogen produces airborne spores abundantly, which are easily disseminated in splashing water and windborne rain. They require free water for germination and penetration into the host. The fungus is found on pad cacti occasionally. The first symptom is dark, sunken areas that are soft and water-soaked. Sunken soft-rotten areas appear on the pad surface.

Control

- Manage watering to prevent this disease.
- Remove heavily infected plants, place in plastic bags and dispose of in the trash to reduce the spread of fungi to nearby susceptible plants.
- Apply fungicides such as thiophanate-methyl.

OPUNTIA SAMMONS' VIRUS

The only known native host is the species of *Opuntia*. Opuntia Sammons' virus, which causes light yellow rings with some mosaic pattern in the pads and which is often called a ring spot virus, is common on Engelmann prickly pear (*Opuntia englemannii*). However, no harm to the plant is noticed due to this virus. Opuntia Sammons' virus has no known insect vector but it is sap transmissible. It is a member of the genus *Tobamovirus* and is closely related to viruses such as tobacco mosaic virus. It is typically spread by propagating cuttings from infected plants.

Control

- Do not use infected plants as a source of propagation material.
- No chemical control is recommended.

26 Chenopod

M.K. Rana and A.K. Pandav

Botanical Name	: *Chenopodium album* L.
	Syn. *Chenopodium reticulatum* Aell.
	Anserina candidans (Lam.) Mont.
Family	: Chenopodiaceae
Chromosome Number	: 2n = 18, 32, 36, or 54

ORIGIN AND DISTRIBUTION

Chenopod is thought to be a native of Western Asia. However, the exact location of its origin is unknown. It is found throughout the world from sea level to 3,600 m in elevation and from 70°N to 50°S latitude except in extreme desert climates. According to the book *Food in China*, chenopod has been a food of several old civilizations. It was perhaps cultivated in Neolithic Europe from 7,000 to 1700 BC and was found in China around the 5th century AD. According to most botanists, it has indeed originated in Europe, and evidence supports the claim that hunter-gatherers ate *bathua* throughout the Bronze Age and Iron Age. The edible species *Chenopodium gigantium* D. Don (Syn. *C. amaranticolor*) is thought to be its nearest relative and probably derived through its cultivation. This species is commonly known as purple goosefoot or tree spinach. In Madagascar and Zambia, *Chenopodium gigantium* is considered an excellent cooked vegetable and it occurs widely in India. The genus *Chenopodium* is large, having about 100–150 species mainly occurring in temperate zones throughout the world. It consists of leafy vegetables and pseudocereals, which are abundantly consumed in the Andenian region of South America as well as in Africa and some parts of Asia and Europe.

INTRODUCTION

Chenopod, also known as pigweed, lamb's quarters, white goosefoot, muck weed, dung weed, fat hen, or wild spinach, belongs to chenopodiaceae or *goosefoot* family. It is commonly known as *bathua* in Hindi, *parippukeerai* in Tamil, *chandanbethu* in Bengali and *vastuccira* in Malayalam. The name *Chenopodium album* L. has been derived from Latin word *cheno*, meaning goose, and *podos*, meaning footed, and the species *album* is in reference to its white powdery bloom. It is considered a popular wild vegetable in southern Africa and grown as a traditional leafy vegetable in many parts of India. This is a polymorphic plant and more commonly found in north India from Sikkim to Kashmir but it is also available in southern parts of the country. Usually, it is not grown commercially, but in parts of Kulu Valley and Rajasthan, it is being cultivated on a small scale. The crop is chiefly raised for its young, tender leaves and stems. Being quick-growing in habit, it can be fitted well in multiple cropping systems. In Himachal Pradesh, it is cultivated as an intercrop with finger millet, potato, maize, rice, amaranth, foxtail millet, sesame, soy bean, taro, cowpea or common bean. Its leaves and seeds are important food for livestock, poultry and other birds, especially in Europe and some parts of Asia. However, it is also used as vegetable in many parts of the world and sometimes as a famine food.

COMPOSITION AND USES

COMPOSITION

Chenopod leaves are an inexpensive source of nutrients as they provide a good amount of calcium, phosphorus, sodium and vitamin C, and they are also good source of carbohydrates and protein and contain an appreciable amount of vitamin B too. The mature leaves are higher in magnesium, calcium and sodium, whereas the young shoots are higher in copper and iron. The nutritional composition of chenopod leaves and seeds is given in Tables 26.1 and 26.2.

USES

Chenopod is grown for its succulent and tender leaves, and to some extent, for its edible shoots. It is a great alternative when other greens like amaranth and fenugreek are out of season. The young tender leaves of cultivated varieties are used in salads but the older leaves are primarily used as either cooked or steamed vegetable for the preparation of various dishes in different parts of the world like spinach as a leafy vegetable. The seeds being rich in vitamin A, protein, calcium, phosphorus and potassium are ground into flour and used for making pancakes, bread and sometimes along with dhal. In Morocco, the seeds are eaten as a famine food. In Shimla, a large portion of the seed is used for making the fermented product *Sura* and an appetizing drink as well as *ghanti*, a local alcoholic drink.

MEDICINAL USES

Chenopod has some medicinal properties and is useful in curing anorexia, cough, dysentery, diarrhoea, oedema, piles, heart and blood diseases, and it kills small worms. It is also manages biliousness, hepatic disorders, throat troubles, spleen enlargement, burns, ulcers, abdominal pain and eye

TABLE 26.1
Nutritional Composition of Chenopod Leaves (per 100 g Edible Portion)

Constituents	Content	Constituents	Content
Water (g)	85.7	Iron (mg)	6.6
Carbohydrates (g)	7.3	Zinc (mg)	0.3
Protein (g)	4.3	Vitamin A (IU)	7817
Fat (g)	0.8	Thiamine (mg)	0.15
Calcium (mg)	258	Riboflavin (mg)	0.4
Potassium (mg)	288	Niacin (mg)	1.3
Sodium (mg)	265	Vitamin C (mg)	90.0
Magnesium (mg)	23	Energy (kcal)	44

TABLE 26.2
Nutritional Composition of Chenopod Seeds (per 100 g Edible Portion)

Constituents	Contents	Constituents	Contents
Carbohydrates (g)	66.0	Fat (g)	7.0
Protein (g)	16.0	Energy (kcal)	395

diseases. Its seeds are traditionally used to improve appetite and as an anthelmintic, laxative, aphrodisiac and tonic. Ethanol extract of its fruits has shown antinociceptive and antipruritic properties. In India, the plants were used since ancient times as it is evidenced in *Rig Veda* and *Atharva Veda* that it cures all diseases, acts as a laxative and is beneficial in piles and clearing worms. The leaves are a source of ascaridole, which is used to treat infestations of roundworms and hookworms. It is mentioned in *Charaka Samhita* that it enhances digestive power, while it is reported in *Sushruta Samhita* that it enhances memory, appetite, digestive power, body strength and destroys all worms. As per the book *Indian Medicinal Plants*, chenopod acts as a laxative, anthelmintic and blood purifier. Its leaves have antibacterial properties against human pathogenic bacteria, including *Salmonella typhimurium*, *Staphylococcus aureus* and *Pseudomonas aueruginosa*. The plant produces toxic substance of oxalates and saponins, usually in very small quantity, which may be harmful to human beings and livestock when consumed in large amount in a short period under certain conditions. These toxic substances are absorbed poorly by the intestine but they are denatured by prolonged cooking.

BOTANY

The cultivated forms of chenopod are dicotyledonary herbaceous annuals for the purpose of tender leaves and succulent shoot production. It is a polymorphic, mealy white, erect annual herb, which may extent to a height of 1.0 to 2.5 m, depending upon fertility and moisture in soil. The entire plant is covered with varying amount of a waxy substance giving the plant a light green appearance. The stems are soft, varying in colour from reddish light bluish-green to striped with purple and green, and single or may have a few rigid angled branches. Young stems have mealy white covering with tiny white hairs, whereas, older stems are more smooth, stout, grooved, or ribbed.

Chenopod produces simple, alternate, pinnately lobed and oblong to triangular leaves of variable shape and size, varying from 2.0 to 6.0 cm long and 0.5 to 2.0 cm broad, which have thin stalks that are about half as long as the leaf blade. The leaf blades are usually ovate to lanceolate. The leaves are dull green with a pale pink centre with a waxy coating and wavy margin. The leaves appear to be opposite to one another along the stem and are almost equal or somewhat larger than the cotyledons. The lower leaves usually have three main veins that extend from the base, which is usually less than 1–1.5 times the width.

The inflorescence is standing terminal and lateral, and flowers are dimorphic, bisexual, or pistillate, leafy panicle, grey to green in colour and may be cross or self-pollinated. Grain chenopod is predominantly self-pollinating but cross-pollination up to 8–10% may also take place mainly by wind. The 10–40 cm long inflorescence consists of spicately or paniculately arranged in dense heads of small rounded clusters or *glomerules* and having five stamens with yellow anthers tips. The ovary is superior (hypogynous) in nature, depressed globose, short styled and one-celled with two stigmas. Flowers are symmetric, regular bearer, pentamerous, without petals and have sepals slight to sharply ridged, nearly covering the mature fruit, which is a nut having a thin-walled indehiscent single seed.

Chenopod produces two types of seed, *i.e.*, (i) smaller black and (ii) larger brown, which are disc-shaped with a notch, glossy and 1.2–1.6 mm diameter. It has a thin papery covering, which often persists, giving them a dull appearance. The five-segmented perianth may also be attached with the seed, which may detach during collection and subsequent handling. The black seeds are found abundantly and hard coated with almost smooth to finely striate, rugulose, or pitted. A smaller percentage of seeds (5%) are relatively large and brown with thin smooth seed coats.

CLIMATIC REQUIREMENT

Chenopod is a fast-growing, frost susceptible, annual weed, which is very commonly found in temperate regions. Being long-day plant, it flowers usually from March to April in northern plain areas.

The plants under long photoperiod (16–17 h) produce a higher seed than plants grown under eight hours photoperiod. It is readily available during winter months at an elevation up to 4700 meters and may grow as a summer crop in irrigated areas. Its wide distribution as a weed shows a great tolerance to temperature with average temperature ranging from 5 to 30°C though the optimum temperature for higher germination rate ranges from 15°C to 25°C. Seed germination is stimulated by red or white light and inhibited by far red light. It is also promoted by stratification and alternating temperature and light interaction.

SOIL REQUIREMENT

Chenopod is found growing wild in pastures, agronomic fields and vegetable croplands, gardens, orchards, vineyards, landscape areas, roadsides and other disturbed areas. It thrives on all types of soil from sandy and heavily clay soil and over a wide range of soil reaction. However, the best quality greens can be produced in fertile soils rich in organic matter and good in moisture holding capacity. For getting higher yield with broader leaves, the chenopod should be cultivated in well-drained, friable sandy-loam soils rich in organic matter and initial plant nutrients.

CULTIVATED VARIETIES

PUSA BATHUA NO.1

A variety developed at Indian Agricultural Research Institute, New Delhi through pure line selection from local collection. It is a first improved variety, which contains more vitamin C and beta-carotene about 60% and 10%, respectively in its edible parts than the wildly grown chenopod. Its purplish-green leaves are 10.5 cm long and 3.0 cm broad. The plants may extent to a height of 2.25 m and thickness of clipped stem is 0.45 cm with internodal length 2.5 cm. Its stem is purplish-green and tender with fine texture. It takes 175–180 days to flower but its tender leaves may be picked up 45 days after sowing and harvesting continues for 150 days. It is suitable for sowing from October to March and the average yield is 300 q/ha.

OOTY 1

A dual-purpose dwarf variety developed at Tamil Nadu Agricultural University, Coimbatore may have a plant height of 38–40 cm and produces unfolded dark green to pinkish leaves. It is suitable for leaf and grain production. This variety is resistant to *Cercospora* leaf spot, *Colletotrichum* leaf spot and *Macrophomina* foot rot. It gives an average greens yield of 289 and 170 q/ha in crop duration of 55 days in hills and plains, respectively. Its grain yield is 12 q/ha.

Some other genotypes are Sel-2, Local-1 and Local-2.

SEED RATE

Chenopod is generally not grown commercially as it is found growing naturally at disturbed sites. However, it can be propagated through seeds. The brown seeds germinate earlier with higher germination rates than the black seeds, which persist longer in soil. Seed is very small with test weight about 1.4 g. A seed rate of about 6–10 kg for line sowing and about 20 kg for broadcasting is required for sowing a hectare area.

SOWING TIME

In plain areas, chenopod can be cultivated year yound due to its wider adaptability to climate except in extreme winter, as it is susceptible to frost. In temperate regions, the seedlings that emerge in the autumn rarely survive in winters. For getting higher yield with best quality greens, it is sown from October to March in northern plain areas.

SOWING METHOD

Chenopod plant has a taproot system so may not be transplanted. The seeds, though tiny, can be sown directly in the field either by broadcasting or in lines at spacing of 20–30 cm row to row and 10–15 cm plant to plant. The emergence of seedlings depends upon the frequency of soil cultivation. Spacing significantly influences the plant growth. If seeds are sown at wider spacing, plants grow vigorously. For grain purposes, the seeds are sown at a depth of 1–2 cm in rows 25–50 cm apart, depending on soil moisture content.

NUTRITIONAL REQUIREMENT

Chenopod does not require too many nutrients since it is very efficient in extracting nutrients from the soil. However, application of nitrogen, phosphorus and magnesium in moderate amounts is essential for harvesting the best quality greens. It grows well in soils rich in magnesium so it is used as an indicator of magnesium status in the soil. Nitrogen and phosphorus are the useful fertilizers for vigorous growth of the plants and reducing the percentage of underdeveloped seeds.

IRRIGATION REQUIREMENT

For better seed germination and subsequent growth of plants, the soil at the time of sowing must have ample moisture. No pre-sowing irrigation is given for preparing seedbeds if the crop is sown on conserved soil moisture. Sometimes, the seeds do not germinate due to lack of moisture. Under such situation, a light irrigation may be given to obtain early and higher percentage of germination and for subsequent seedlings' growth. Depending on the season and availability of soil moisture, the crop needs irrigation at an interval of 10–15 days.

INTERCULTURAL OPERATIONS

Chenopod grows naturally as a weed in growing fields of potato, tomato, pepper, beans, lettuce, egg-plant, radish, carrot, turnip, beetroot, cauliflower, cabbage, broccoli, knoll kohl, etc. However, if it is grown commercially, the field should be kept weed free in early stages of crop growth since the weeds, especially the *Chenopodium murrel*, which is morphologically similar, compete with the main crop for nutrients, moisture, light and space. Thus, at least two shallow hoeings are beneficial and necessary for keeping the weeds under control and providing good soil environment to the roots for their development.

HARVESTING

Harvesting may be done by either uprooting the whole plants or clipping the leaves. The first picking may be done 40–45 days after sowing and further if is harvested for 100–150 days, depending upon variety. The shoots along with leaves are cut at ground level and when the plants attain a height of 10–15 cm and when they are tender and crisp.

Harvesting of chenopod for grains is almost similar to cereal crops. The crop is harvested when the seeds are ripe and most of the leaves have turned yellow. The plants after cutting are bundled and left on the threshing floor for few days to dry them. When the plants are fully dried, the seeds are separated by either rubbing the flower spikes onto a tarp or beating them lightly with a stick, and thereafter, the seed is cleaned by winnowing. The seeds at harvest contain moisture around 20%, therefore, they are artificially dried to a safe moisture level (12–14%) so that they may be stored safely for a longer period.

YIELD

The average yield of chenopod varies from 250 to 300 q/ha in hills and 150 to 200 q/ha in plains, depending on crop variety, soil and environmental factors and cultural practices adopted by the growers. The average seed (grain) yield varies from 8 to 12 q/ha.

PHYSIOLOGICAL DISORDER

No specific physiological disorder is observed in cultivated chenopod.

INSECT-PESTS

LEAF STICKER (*Eurysacca melanocompta*)

This pest attacks the plants at all stages of growth particularly at maturation of grains. It has a longitudinal dark band on the wings and two dark spots surrounded by light scales. A female lays 80–300 eggs into the leaf tissues. The egg-laying range mainly depends on temperature. The larvae feed on foliage, with a preference for inflorescence. The loss due to the attack of this pest is considered severe and an average yield loss ranges from 20% to 50%.

Control

- Spray the crop with simple water with pressure.
- Remove and destroy the portion of the plant heavily attacked by leaf sticker.
- Spray the crop with Metasystox 0.1%, malathion 0.05%, monocrotophos, dimethoate 0.03%, or phosphomidon 0.03%.

APHIDS (*Myzus persicae*)

Both nymphs and adults of this tiny insect suck cell sap from under surface of the tender leaves, reducing plant vigour. The leaves curl up and ultimately wilt. The aphids excrete honeydew-like substance, on which black sooty mould develops, which hampers the photosynthetic activity. The severe infestation affects the yield and quality of leaves and inflorescence adversely.

Control

- Spray the crop with simple water with pressure.
- Remove and destroy the portion of the plant infested with aphids.
- Spray the crop with malathion 0.05%, monocrotophos 0.04%, methyl demeton 0.05%, dimethoate 0.03%, or phosphomidon 0.03%.

DISEASES

DOWNY MILDEW (*Peronospora farinosa* f.sp. *chenopodii* or *Peronospora chenopodii*)

A most destructive disease causes much damage in chenopod growing areas all over the world. The disease may attack the plant at any stage, even at flowering stage. The leaves affected by this fungus show light yellow irregularly shaped areas without distinct margins on their surface. Whitish-grey mycelium grows on leaf surface at high humidity, while the mycelium is generally found absent in low humid atmosphere. The disease develops in warm and moist weather. Its infection occurs primarily in the field but it becomes prominent after harvesting, especially at 25–28°C temperature and >85% relative humidity.

Control

- Follow long crop rotation with non-host crops.
- Always use disease-free seed to avoid the incidence of this disease.
- Remove the infected leaves and destroy.
- Spray the crop with Indofil M-45 or Indofil Z-78 at 0.2% at 10- to 15-day intervals.

LEAF BLIGHT (*Alternaria alternate*)

The initial infection is noticed as small brown necrotic spots with concentric rings, which rapidly increase in number and size, and later on, these spots coalesce to form large irregular blotches; as a result the leaves wilt, and ultimately, die and fall down. In advanced stages, the burning effect and blight symptoms are also seen after coalescing of adjoining spots. The disease comes in the middle of February and remains throughout summer. The crop shows about 10–60% mortality in different areas, which is higher in the rainy season.

Control

- Irrigate the crop frequently.
- Use disease-free high-quality seed.
- Treat the seed with hot water at 52°C for 30 minutes to kill the pathogen present on or below the seed coat.
- Spray the crop with Bavistin, mancozeb, Blitox, or Blitane at 0.3% 3–4 times at an interval of 10 days just after the appearance of disease.

CERCOSPORA LEAF SPOT (*Cercospora chenopodii*)

The disease is characterized by the presence of numerous small brown circular spots on the leaves. In the beginning, the spots are small round with concentric rings, but later on, these spots increase in size, and sometimes, they coalesce. The infected leaves dry and fall down.

Control

- Follow field sanitation.
- Grow resistant varieties like Ooty 1.
- Spray the crop with Bordeaux mixture (5:5:50) or Blitox 0.3% thrice at an interval of 15 days.

COLLETOTRICHUM LEAF SPOT (*Colletotrichum* spp.)

The disease is caused by fungus, which affects all above ground parts of the plant and may infect at any stage of plant growth. The affected leaves shows round to angular reddish brown spots on older leaves. Spots may later dry, turn almost black and tear out, giving the leaf an irregular appearance. Severity of infection depends almost entirely on environmental conditions. A combination of high temperature and high humidity is very conducive for the spread of this disease.

Control

- Use disease-free healthy seed.
- Adopt field sanitation by burning crop debris.
- Treat the seeds with 0.25% carbendazim @ 2.5 g/kg of seed before sowing.
- Spray chlorothalonil 0.2% for good control of this disease.
- Grow resistant varieties like Ooty 1.

27 Spinach

Bhupinder Singh Thakur

Botanical Name : *Spinacia oleracea* L.

Family : Chenopodiaceae

Chromosome Number : 2n = 12

ORIGIN AND DISTRIBUTION

The earliest record of spinach is available in Chinese, whereby its introduction from India to China through Nepal has been stated probably occurring in 647 AD, although to date, it is not clear who introduced this crop to India. Iran and neighbouring countries of Arabian- Mediterranean regions are believed to be the centre of origin of spinach as 10th-century work on medicine by Rhazes and two separate works by Ibn Washiyah and Qustus al-Rumi has given clearly written evidence for spinach cultivation in this region. During the later part of the 12th century, spinach was introduced to Spain from where it spread to other parts of the world, probably by Moores. In Germany, the prickly seeded cultivars were known by the 13th century, whereas smooth seeded cultivars were not described until 1552. The first English cookbook *The Forme of Cury* (1390) has referred to spinach as *spinnedge* or *spynoches*.

INTRODUCTION

The word 'spinach' has been derived from a Spanish word *Hispania*, however, the Latin meaning of spinach is spiny fruit and *oleracea* means herbaceous garden herb. *Spinacia tetrandra* and *Spinacia turkistanica*, which have been described in Northwest India and Nepal, are the two diploid relatives of *Spinacia oleracea*. Although it is the most important potherb, its cultivation is restricted to smaller areas in the hills and parts of north as well as south India. It is also a well-accepted vegetable in the United States of America, Canada and Europe. France, Germany, Italy, the United States and the Netherlands are the leading countries where it is grown on a commercial scale for shipment to distant markets and for canning and freezing purposes.

Spinach is usually classified as a dioecious plant though that is not exactly true since varying sexual types of plant, *i.e.*, male, female, or both male and female, and infrequently, hermaphrodite flowers, occur in a spinach field. Based on flowering or sex expression, four types of plant and their characteristics are given as follows:

- **Extreme males:** Such plants are short in stature (smaller) with minimal foliage, earliest to flower and die after flowering. They produce only staminate (male) flowers and act useful pollen source for pollination.
- **Vegetative males:** Such plants are giant type (tall and larger) with more foliage and flower slightly later than extreme males. They also bear only staminate (male) flowers.

- **Monoecious plants:** Such plants have well-developed foliage and are slow in flowering. They bear both staminate (male) and pistillate (female) flowers on the same plant but separately at different positions and are able to produce seeds.

- **Female plants:** Such plants have well-developed foliage and are very late in flowering. They bear only pistillate (female) flowers and produce seeds, provided, they are grown with a few rows of male plants, which supply pollens for the pollination.

Though the male plants are inherently having short stature (smaller) with minimal foliage it is very tricky to ascertain the predominating sex expression of spinach plants prior to flowering. From economic point of view, female and vegetative male plants are preferred since they have well-developed (larger) foliage and high yielding capacity and are slow in bolting.

COMPOSITION AND USES

COMPOSITION

Spinach has a high nutritional value as it is a rich source of vitamin A, C, K and minerals like magnesium and manganese. It also provides enough quantity of essential dietary fibre to the human body. It has moderate calcium content but is among the least bio-available food calcium sources. The human body can absorb only around 5% of the calcium from spinach as it is found to be less soluble than that from skim milk. Even cooking spinach in water for 20 minutes had no effect on the availability of calcium. The nutritional value and chemical composition per 100 g edible portion of spinach are given in Table 27.1.

According to the United States Department of Agriculture, it is rich in iron, as a 180 g serving of boiled spinach provides 6.43 mg of iron, though spinach contains iron-absorption-inhibiting substances, including high levels of oxalate, which can bind to iron to form ferrous oxalate and render much of the iron in spinach unusable by the human body. Spinach contains appreciable quantity of ascorbic acid and riboflavin. Ascorbic acid content increases with plant age. Phylloquinone (vitamin K_1) at a level of 299.5 µg/l00 g is reported in spinach, and its level increases with plant maturity. Its leaves also contain histamine.

TABLE 27.1
Nutritional Composition of Spinach (per 100 g of Edible Portion)

Constituents	Contents	Constituents	Contents
Water (g)	91.4	Zinc (mg)	0.53
Carbohydrates (g)	3.6	Vitamin A (IU)	9,377
Protein (g)	2.9	Thiamine (mg)	0.078
Fat (g)	0.4	Riboflavin (mg)	0.189
Dietary fibre (g)	2.2	Niacin (mg)	0.724
Mineral (g)	1.7	Pantothenate (mg)	0.27
Calcium (mg)	99	Pyridoxine (mg)	0.195
Phosphorus (mg)	49	Folic acid (µg)	194
Potassium (mg)	558	Vitamin C (mg)	28
Magnesium (mg)	79	Vitamin E (mg)	2.0
Sodium (mg)	79	Vitamin K (µg)	483
Sulfur (mg)	30	Oxalic acid (mg)	658
Iron (mg)	2.71	Nitrate (ppm)	1703
Copper (mg)	0.01	Nitrite (ppm)	4.7
Chloride (mg)	54	Energy (kcal)	26
Manganese (mg)	0.897		

USES

Like other leafy vegetables, spinach is widely consumed as cooked vegetable either alone or mixed with other vegetables and can be consumed raw in salads. Spinach leaves along with their tender petioles are cooked, steamed, or boiled for the preparation of dishes like curries, pastries, soufflés and soups. Spinach is also processed by freezing and canning.

MEDICINAL USES

The significance of dark green leafy vegetables in human nutrition is world over well recognized. The inclusion of dark green leafy vegetables that are innately rich in β-carotene (pro-vitamin A) is strongly recommended by nutritionists and dieticians to cure vitamin A deficiency. Being rich in fibres, spinach supplies roughage essentially required in the daily diet for normal movement of bowel. A daily dose of spinach extract prevents some loss of long-term memory and learning ability. It is most potent in protecting different types of nerve cells in various parts of the brain against the effect of ageing. With age, body cells become sluggish in response to chemical stimulation but spinach has been scored best among the fortified diets in a test of nerve cells in the cerebellum, a part of the brain that maintains balance and coordination. Spinach, because of having specific phytonutrients, is considered more effective than strawberries in protecting cerebellar physiology.

Spinach also contains many anti-nutritional components like oxalic acid and nitrate in abundance. In some individuals who are predisposed to stone formation, its frequent consumption in large quantity is very deleterious and causes serious health hazards, especially to infants. It contains oxalic acid, which in combination with calcium forms insoluble calcium oxalate, reducing dietary iron and magnesium availability. Protein in spinach has been revealed to have hypocholesterolemic properties. Excessive accumulation of nitrate in spinach has been found to be associated with the disease methemoglobinemia in humans. In spinach, excessive nitrate reduced to nitrite either in saliva during chewing or in the upper gastrointestinal tract not only causes methemoglobinemia in infants but also amplifies the jeopardy of carcinogenic nitrosamines formation. The plants that accumulate higher nitrates are also known to contain sizeable amounts of nitrite. Post-harvest storage of such vegetables also leads to the accumulation of nitrite due to reduction of nitrate through either enzymes present in plants or activities of microbes, however, exposure of spinach plants to light before harvesting decreases the nitrate levels, thus, harvesting should be done in the afternoon since low light and low temperature favour the accumulation of nitrate. Conversely, storage of spinach leaves for 24 hours is reported to increase the nitrate, therefore, it is suggested that spinach should be stored for a minimum time. High nitrate content may also be due to processing problems like can corrosion. Variations in nitrate levels of spinach tissues may be due to cultivar, application of higher doses of nitrogenous fertilizers, atmospheric temperature, or activity of nitrate-reductase enzyme in leaf blades, hence, developing cultivars having a tendency to accumulate fewer nitrates would be rather advantageous to the consumers.

The content of oxalate in tissues of spinach intensifies with growth, light intensity and an increase in atmospheric temperature, though it is influenced rather less by the type of fertilizer used during cultivation. Synthesis of oxalic acid is greater at lower rather at higher temperature. It is assumed that L-ascorbic acid present in oxalate-accumulating plants, including spinach, might be metabolized into oxalic acid. Glycolate is also supposed to be an efficient precursor of oxalic acid in spinach leaves.

BOTANY

Spinach is an annual herbaceous plant for leaf production and biennial for completing its life cycle, *i.e.*, reproduction or seed production. It has a shallow taproot system, which gives rise to many fibrous roots. The plant produces rosette leaves from a much shortened stem near the ground surface during vegetative phase. The leaves vary in shape and size and the margins may be smooth or wavy. The leaf surface is smooth, savoyed, or heavily savoyed. Leaf colour varies from light green to dark red. Its thin petioles are habitually as long as the leaf blades and often become hollow when the leaves are fully expanded. On the onset of the reproductive phase, the stem elongates and forms flower stalks, which bear 6 to 12 flowers per cluster. The flowers are axillary, small and greenish in colour and borne on both large stems and small branches. Flowers may be staminate, pistillate, or hermaphrodite. The flowers remain receptive for 1 week or longer. Pollination is helped by wind only and the fertilized ovary develops into a single seeded fruit (*utricle*), which bears seeds. Seed may have either a smooth round or an irregularly prickly shape. Seeds in their outer layer contain inhibitor.

CLIMATIC REQUIREMENT

Spinach is an absolutely cool season crop, thus, it grows best in an average temperature range of 16 to 18°C. The seeds germinate better at 5–10°C than at 25°C temperature. They can germinate even at 0°C but emergence is tremendously slow. The optimum temperature for proper growth and development of leaves is 18–20°C. The growth becomes slow below 10°C temperature. Low temperature tends to increase leaf thickness but decreases leaf size and smoothness. Thermo-quiescence stage usually occurs above 30°C temperature. The spinach plants can withstand frost better than other winter vegetable crop plants, and after acclimatization, it can tolerate freezing temperature as low as −10°C. The tolerance in spinach to freezing temperature is attributed to the presence of cold-regulated proteins in its leaves.

An increase in light intensity and atmospheric temperature increases the production of biomass and the contents of carotene and ascorbic acid. It produces higher yield under short days and mild temperature, hence, it is more popular in the hills under temperate climatic conditions. Since it is a long-day plant for flowering, spinach tends to develop seed stalk under long-day and warm weather conditions. Therefore, it is not possible to grow spinach during spring season in plains. The spinach for processing purposes is normally grown under a condition of short days so that more than one cutting can be taken before the plants start bolting. The critical photoperiod to develop seed stalk ranges between 12.5 and 15 hours. Although photoperiod is the key stimulus to induce flowering, high temperature exposure after flower initiation causes rapid stem elongation, which almost inhibits vegetative growth, and as a result, the new leaves formed are small, narrow and pointed. Higher quality of spinach could be obtained from a night-cooling greenhouse in summer.

SOIL REQUIREMENT

Spinach can be grown on any good soil provided it has high moisture-holding capacity and a proper drainage system but sandy loam soil rich in organic matter is the best. Proper land preparation and maintenance of adequate soil moisture are must to ensure satisfactory seed germination and further plant growth. It can be grown successfully on soils having a pH range of 6.0 to 6.8. However, it can be grown at a pH up to 10.5. It is sensitive to soil acidity and may exhibit low germination, leaf tip yellowing or browning, root burn and overall slow growth if soil pH is too low. On soils with high pH, spinach leaves may become chlorotic. If the soil pH is less than 5.5, liming is essentially required.

CULTIVATED VARIETIES/HYBRIDS

In a spinach breeding programme, resistance breeding and breeding for high-yielding F_1 hybrids are most important objectives to improve the spinach crop. The different sex types in spinach are highly suitable for F_1 hybrid production as utilizing female lines in combination with a known pollen (male) parent is very easy. Monoecious plants on self-segregate into females, monoecious and male plants in varying proportion. Selection of a high degree of female plants (80–90%) and crossing them with the pollen parent results in F_1 hybrid. Spinach cultivars differ in seed and leaf morphology, thus, the cultivars can broadly be grouped according to their seed and leaf morphology:

A. Based on seed morphology

According to seed morphology, the spinach cultivars can be categorized into two groups, given as follows:

Prickly-seeded: Its seeds have a jagged or toothed surface. Prickly-seeded varieties perform better in hilly tracts than the smooth-seeded varieties.

Round- or smooth-seeded: Most of the commercial cultivars are smooth-seeded, which makes them easier to sow and plant accurately. Round-seeded varieties perform better in plain areas than do prickly-seeded varieties.

B. Based on leaf morphology

Based on leaf morphology, the spinach cultivars can be classified into three groups given as follows:

Savoy (wrinkled leaf): This type of spinach has a highly wrinkled leaf surface and is very productive and tolerates cold better than most other types of spinach. The lower temperature causes deep wrinkles on their leaves and a low growth habit that makes cleaning leaves a chore, especially in silty soils where grit splashes onto the leaves and makes them look dirty and unpleasing. These types of spinach are preferred for fresh market only.

Semi-savoy (semi-wrinkled leaf): These varieties have a more upright growth habit that makes mud splash less likely, and the leaves are not as crinkly as the savoy type varieties. They also tend to have better disease and bolt resistance, thus, they may usually be the best choice when growing spinach in kitchen gardens. These are preferred for both fresh marketing as well as processing purposes.

Smooth or flat leafed: The leaves of this group spinach are flat and smooth surfaced and used for processing purpose.

The characteristic features of some important improved varieties and F_1 hybrids are described as follows:

SAVOY TYPES

Virginia Savoy

A savoy-leaf type cultivar resistant to cucumber mosaic virus-1 has been developed through a cross between domestic savoy type and an exotic introduction from Manchuria. Its leaves are blistered large, thick and dark green with round tips. The plants are vigorous and upright in growth habit. It has smooth seeds, which are easy in threshing and grading. It is a late-bolting cultivar suitable for sowing in September–October and gives 5–6 cuttings at 15- to 20-day intervals.

Regiment (F_1 hybrid)

A F_1 hybrid resistant to powdery mildew produces a high yield of deep green leaves that stay tender even when large enough. It becomes ready for harvest in 37 days after sowing.

Bloomsdale

This is typically thick leaved, succulent savoy spinach that tolerates cold better than most of the other varieties. It produces large yield in early summer but has limited bolt resistance.

Vitro Flay

The leaves of this variety plants are slightly savoy type in cool weather.

SEMI-SAVOY (semi-wrinkled leaf)

SH 1 (F₁ Hybrid)

A F_1 hybrid developed at IARI Temperate Vegetable Research Station, Katrain (Kullu Valley) Himachal Pradesh through a cross between Virginia savoy and Japanese varieties. Its leaves are green semi-savoy with blunt apex. This hybrid gives yield 20% higher than the yield of its parents.

Old Domination

This is a variety with medium-savoyed leaves.

Tyee

A F_1 hybrid resistant to downy mildew and bolting becomes ready for harvest in 45 days after sowing. It is a hybrid with vigorous upright growth habit and dark green leaves. It can be grown year yound in mild winter areas.

Catalina

A F_1 hybrid that becomes ready for harvesting after 48 days of sowing has thick, succulent, spear-shaped leaves and moderate bolting resistance.

Teton

A F_1 hybrid with deep green oval leaves on upright plants is suitable for cultivation throughout the year and very slow in bolting. It becomes ready for harvesting in 40–45 days after sowing. The hybrid also has resistance against downy mildew.

SMOOTH-LEAVED SPINACH

Space

A F_1 hybrid that becomes ready for harvesting in 45 days after sowing is resistant to downy mildew and slower in bolting than most of the smooth-leaved varieties.

Red Cardinal

A variety that matures in 25–30 days after sowing has deep red stems and red veins on leaves. It is harvested as baby greens, which make a beautiful addition to a salad but it bolts faster than any green-leaved spinach, thus, it is harvested at young stage.

Spinach Early

An early variety developed by Tropica Green Company Limited has plants with upright growth habit and deep green leaves. It is a medium bolter but can be harvested year yound in tropical high lands. It is resistant to downy mildew races 1 and 3, heat and cold.

Early Smooth Leaf

A smooth round seeded cultivar produces small, light green, thin and smooth leaves with a pointed apex.

Sporter

A high yielding F_1 hybrid developed by Bejo Sheetal Seeds produces broad leaves without fibres, which become very soft on cooking. It matures in 45 days after sowing and is suitable for cultivation in both *Kharif* and *Rabi* season. It is tolerant to downy mildew races 1, 2 and 3.

Dongxin 1

A broad-leaved late-bolting new spinach F_1 hybrid from China is suitable for protected and open field cultivation. The indices of resistance to downy mildew (*Peronospora farinosa*) and virus are lower (by 28.8% and 25.1%, respectively). It bolts 15 days later than the Japanese broad-leaved spinach.

Amsterdam Giant

The variety with prickly seeds is sown bit late.

Dark Green Bloomsdale

Being medium-early variety, it is also sown bit late.

Hollandia

This is a variety with no smooth leaves and grown in cool weather.

Juliana

The plants of this variety have blue-green leaves.

King of Denmark

This variety also has plants with blue-green leaves.

Long-Standing Bloomsdale

The plants of this variety have dark green leaves.

Nobel

This variety can be distinguished by its medium green leaves.

The choice of a cultivar depends on its leaf characteristics, critical day length relative to the growing latitude and seasonal temperature since the cultivars of spinach are sensitive to both temperature and photoperiod. Usually, slow-bolting cultivars are preferred when planting is scheduled to get the crop mature during long days and warm temperature. Earliness of a variety is associated with its growth rate since fast-growing varieties are harvested earlier. Spinach growers prefer to grow varieties that are fast-growing, slow bolters, high-yielding and suitable to the growing conditions. In greenhouse only the cultivars having rapid foliage growth under a condition of short days and low temperature are chosen for cultivation. For processing purposes, smooth or semi-savoy cultivars are preferred more since they are fast-growing, high-yielding and washed more easily, whereas the leaf surfaces of savoy cultivars may not easily be made soil-particle free by washing. Cultivars with savoy surface texture are ideal for fresh market since such leaves resist compression during packaging, and thus, allow better aeration, which helps in fast cooling and prolonging the post-harvest life.

SOWING TIME

Sowing time of spinach varies from region to region as follows:

The crop sown in spring–summer in Indian plains usually has a greater tendency to bolt. The prickly seeded cultivars are best suited for autumn–winter crop in hills, whereas, smooth seeded cultivars are suitable for spring–summer in the hills and autumn sowing in the plains. The growth and yield of spinach crop are influenced by the date of sowing. Early sowing gives higher yield with poor ascorbic acid and oxalic acid content. In northern Europe, spinach is grown in greenhouse on a small scale during winter and early spring when outdoor cultivation is not possible due to adverse temperature conditions.

SEED RATE

The seed rate generally varies with the method of sowing and the intended use of crop. Spinach requires relatively higher seed rate than beet leaf since due to its dioecious nature about 50% of the plants may be the male with poor growth, which affects the yield. It is impossible to recognize and remove the male plants unless the vegetative phase of plants shifts to reproductive phase. Secondly, the germination of seeds, which is poor due to the presence of endogenous inhibitors, is also affected by the sowing depth, environment and soil temperature. Therefore, about 37–45 kg of seed is required to sow a hectare land area. However, its germination can be improved by treating the seeds with sulfuric acid or bichromate mixture (sodium + potassium), which destroys the endogenous inhibitors present below the outer seed layer.

Region	Sowing time
In plains	September to October
In lower hills	July to November
In higher hills	March to April

METHOD OF SOWING

Like beet leaf, spinach is also propagated by seeds but high economic efficiency in terms of growth rate, yield and quality can be obtained with transplanting compared to direct seeding. The method of sowing is almost similar to beet leaf crop, however, spacing between spinach plants influences the number of leaves and their size, and high plant density contributes to early bolting. The plant density for fresh market production averages about 60 plants per square meter area, whereas, plant density is doubled when the crop is raised for processing purposes. Lower plant density facilitates hand harvesting for bunching of leaves or intact plants. On the other hand, high plant density results in more upright leaf growth, which is desirable for mechanical harvesting.

The seeds are sown either by broadcasting or in shallow furrows. However, sowing seeds in lines is better over the broadcasting method as it facilitates easy cultural operations and harvesting of crop. Depending on soil type and moisture availability, the seeds are sown at a depth of 3–4 cm in lines 10 cm apart. However, sowing seeds at a row spacing of 20–22 cm encourages optimum plant growth. Either the crop can be raised in narrow raised or flat beds depending on drainage facilities. Soaking seeds before sowing in tap water or commercial acid can improve germination since soaking in water or acid removes water-soluble inhibitors and softens the pericarp.

NUTRITIONAL REQUIREMENT

Spinach is fertilized well to meet the requirement of plants for their fast growth that arises in a short period since lack of nitrogen results in yellow leaves, while lack of phosphorus and potash results in dry leaf tops and leaf necrosis, respectively. About two-thirds of the total crop biomass is produced by the crop during the last of the growth duration, thus, to meet out the nutrients demand of quick-growing crop plants, appropriate scheduling of fertilizer application is essential. The crop needs farmyard manure 10 t/ha, nitrogen 75 kg, phosphorus 60 kg and potash 30 kg/ha. The crop requires more nitrogen in the form of ammonium sulfate along with phosphorus and potassium for increasing the fresh green leaf yield and total crop dry matter, besides the micronutrients, i.e., zinc, manganese and boron, which are also important for increasing the yield as well as carotene and ascorbic acid content.

Farmyard manure should be applied at the time of land preparation and mixed thoroughly into the soil by repeated ploughings. Full dose of phosphorus and potassium along with one-third dose of nitrogen should be applied before sowing at the time of last ploughing and remaining two-thirds dose of the nitrogen should be applied in standing crop in two equal splits after first and second cuttings. Besides, foliar spray of urea 1.5% may be done 15 days after germination and repeated after every cutting to increase the growth rate.

Spinach has a limited root system and is not efficient in pulling nutrients from the soil. If spinach is stressed by a lack of nutrients, vegetative growth is slowed and the plants are more prone to bolting. Nitrogen deficiency may be confused with magnesium deficiency in spinach since the symptoms can be similar. If older leaves are yellow or the plant has a general pale green colour, soil for magnesium deficiency should be analyzed before applying additional nitrogen. Excess nitrogen can cause nitrates to build up in the plants. Low soil magnesium levels should be made correct by adding high magnesium lime to the soil, adding magnesium to the fertilizer, or by making a foliar application of 4.5–6.5 kg magnesium sulfate (Epsom salts) in 350 to 400 litres of water. Spinach responds quickly to foliar applications when magnesium is lacking. Slower growth of the spinach plants and bluish green colour of the older leaves indicate towards phosphorus deficiency. Browning of leaf tips is the result of poor potassium availability to the plants.

IRRIGATION REQUIREMENT

The irrigation requirement of spinach crop is usually not too high because of less loss of water from the plants through a process of transpiration since its cultivation is done in cool season, thus, on an average, 250 mm of irrigation water is normally enough for this crop. Yet, the plants can easily be stressed from inadequate moisture due to its shallow root system. Waterlogging conditions are also detrimental to the crop. The first light irrigation is given just after sowing if soil moisture in the field at the time of sowing is not sufficient. In general, irrigation is applied at an interval of 4–6 days in summer and 10–12 days in winter for providing soil moisture adequate for normal growth of plants. The spinach plants grow markedly better when irrigation is applied through a sprinkler system than through a strip irrigation system.

INTERCULTURAL OPERATIONS

Keeping the field weed free is very essential in spinach. Regular hoeing after each cutting and use of effective weedicides can keep the weed population low. Pre-emergence weedicides are highly effective in checking the early weed growth, and hence, competition of any kind with the crop is reduced. Weed control also reduces chances of contamination with the diseases and insect-pests. All the plants are nicely exposed to good sunlight, and hence, high growth accompanied by accumulation of nutrients is ensured in the plant leaves. Weed control is very essential in the crop grown for processing since weeds especially that are similar to spinach in appearance and difficult to separate are contaminants in crop. Therefore, the removal of such weeds is most indispensable.

The crop grown for fresh market is hardly ever thinned since the operation is too labour-intensive, hence, to obtain optimal plant density, the seed rate is ascertained according to germination percentage, seed vigour, seed size and field conditions. Higher yield of spinach can be obtained by growing high-performance hybrids having a high degree of femaleness.

HARVESTING

Spinach crop becomes ready for harvest when the plants attain 5–6 leaves, which are usually attained 40–50 days after sowing, depending on temperature and other climatic components, and harvesting continues over winter until just before the seed stalk develops. Higher yield from spinach crop can be obtained when the crop is harvested after attaining the leaves of marketable size. Delaying in harvesting increases the leaves weight, however, leaf quality is affected adversely. For the fresh market, it is harvested by cutting the plant at soil surface just below the lowest leaves. However, for processing purposes, harvesting is done by cutting the plants about 2–3 cm above the soil surface.

Harvesting is usually done by using a sharp knife to avoid the maceration of tissues at the cut end. In advanced countries for the fresh market, spinach is harvested mechanically, and the leaves are gathered and bulk handled rather than bunched. Separate machines have been manufactured to cut the whole plants for fresh market and to cut only leaves for processing, though *clip-topped* spinach is preferred more than the crown-cut produce since the yield of clip-topped spinach is more and labour required for packaging is less.

The harvesting machine is so adjusted that it may cut the plants 10–15 cm above the growing point to reduce the undesirable portion of petioles and also to allow the plants to grow again for a possible second harvest 3–4 weeks later since a high leaf blade to petiole ratio, which increases the value of processed product, is preferred more by the processing industries. It is advisable not to harvest the crop immediately after rain or heavy dew since the leaves are very turgid and break easily when wet, however, a slight wilting may prevent this breaking so as to minimize leaf damage, hence, spinach should be harvested later in the day when the leaves are less turgid.

YIELD

A spinach plant in its growth duration may produce about 25 leaves, and on average, it gives a yield of fresh green biomass about 50–60 q/ha in 3–4 cuttings. Processing crop yield ranges from 50 q/ha from single cut to as much as 200 q/ha from multiple cuts, however, fresh-market yield of spinach is normally much less.

POST-HARVEST MANAGEMENT

All yellow, injured and diseased leaves are trimmed off during preparation of produce for the market, and the produce is handled carefully to prevent bruising or breaking of leaves and stem. Only the intact plants are trimmed off, not all hand-cut leaves. Washing of produce should generally be avoided since washing usually results in leaf injury. Spinach should be harvested only during cooler parts of the day to minimize wilting, a major disorder of spinach, which is known to hasten the loss of ascorbic acid in spinach leaves but is of much less importance than unfavourable temperature. Spinach loses water more swiftly under warm weather conditions than under cool conditions even at high relative humidity. Spinach has a very high rate of respiration, hence, must be cooled rapidly to prevent weight loss and decay since overheating swiftly destroys its quality. The temperature of produce may be lowered by vacuum cooling adopted by large operations or hydro cooling adopted by small operators. Hydro cooling is very effective to maintain firmness of the leaves but it causes decaying of leaves, however, spinach could be package-iced, hydro-cooled, or vacuum-cooled equally well. If a hydro-cooling system is used, excess water is removed by centrifuging about 3 minutes to retard decaying under higher temperature conditions. Hydro-vacuum cooling, where moisture is added during vacuum cooling, is useful in limiting wilting. Using chlorinated water (100 ppm) for washing or cooling the spinach prevents build-up of bacteria in water but has a slight effect on its decaying.

Spinach is highly perishable due to its succulent large leaf surface area and high rate of respiration, thus, should be disposed of immediately after harvesting. The several harvested intact plants before marketing are tied together in small bunches with twine or string for local market, however, for shipment, the harvested leaves after pre-cooling are packed in containers including round bushel baskets, wire bound crates, or hampers. Crushed ice is placed on top of the produce packed in bushel baskets to keep the leaf tissues cool during shipment when refrigerator vehicles are used for distant transportation. As a rule, the quantity of ice placed in the container is about equal to the weight of spinach. For distant markets, it is packed in polyethylene bags. In retail stores, spinach is sold in prepackaged transparent polyethylene film bags, which allow the exchange of metabolic gases and maintain high humidity by reducing evaporation.

Being highly perishable vegetable like beet leaf, spinach cannot be stored for long periods. High temperature, low humidity and rapid air movement often result in rapid wilting of green leafy vegetables, making the produce less attractive. Wilting is known to hasten the loss of ascorbic acid in spinach but is of much lesser importance than unfavourable temperature. Spinach can be stored for 10 days at 0°C temperature and 95% relative humidity. The storage in Controlled Atmosphere (CA) prolongs the shelf life of spinach by about 1 week at 5°C temperature. Carbon dioxide–enriched (10% CO_2) atmosphere retards yellowing of spinach and maintains the produce in good quality for 3 weeks at 5°C temperature.

INSECT-PESTS

This crop is generally less susceptible to insect-pests since it is harvested at very short intervals. However, its major pests are aphids and leaf-eating caterpillars.

APHIDS (*Myzus personae* sulzer)

Aphid, especially the green peach aphid, can be a major pest of spinach from seedling stage to harvesting. Green peach aphids are approximately 2 mm long and vary in colour. Both winged and wingless forms damage the crop. Their larvae and adults suck cell sap from tender leaves and reduce plant vigour if its population is high enough. Its population on leaves can build up to a high number, and thus, most of the buyers reject spinach contaminated with insect excreta. Its population peaks up during late summer season, hence, it needs monitoring from the beginning of the season for the production of good-quality leaves.

Control

- Grow the crop at recommended spacing for full flow of winds in the field, which disperses the aphids from the field.
- Eliminate weeds giving shelter to aphids.
- Harvest the crop as early as possible to minimize vulnerability to late-season aphid colonization and virus infection.
- Spray the crop with malathion 50 EC 0.1%, imidacloprid (Provado 1.6F) 0.04%, or pymetrozine 50% (Fulfill) 0.025%.

LEAF-EATING CATERPILLARS (*Spodoptera exigua*)

The caterpillars feed on crown portion of the spinach plant and can severely stunt or kill the seedlings. The potential for damage and contamination continue until harvest. Its population in fall season is much more damaging than population in spring. Its cutworms feed at or below ground level. Therefore, the field is monitored before the emergence of seedlings for eggs and young larvae in surrounding weeds as the young larvae are more sensitive to the insecticides to be used for sprays.

Control

- Plough the field immediately after harvest to kill larvae and pupae.
- Destroy the weeds along the field borders.
- Spray the crop with permethrin 0.05% , diazinon 0.1%, cypermethrin 25 EC 0.04%, or malathion 50 EC 0.1%.

DISEASES

DAMPING-OFF (*Rhizoctonia, Pythium, Fusarium* and *Phytophthora* spp.)

These pathogenic fungi are found present in most of the soils and primarily cause a pre-emergence decaying of seeds and seedlings. If the soil moisture is very high, these fungi attack the seedlings at collar region near the ground. Due to their attack, the seedlings start rotting and collapsing. Many of the attacked seedlings die and fall over. High humidity and moderate temperature favour the development of this disease. These are the damaging pathogens, especially when debris from the previous crop is not adequately decomposed before planting.

Control

- Follow long crop rotation with non-host crops.
- Sterilize the soil with formaldehyde (1: 100 parts) about 2 weeks before seed sowing.
- Use less quantity of seeds per unit area.
- Avoid excessive irrigation application.
- Treat the seed with captan, Foltaf, captafol, or Agrosan GN 2 g/kg of seed.
- Drench the field with captan or thiram at 0.1% solution.

LEAF SPOT (*Phyllosticta chenopodii* Sacc. and *Cladosporium variabile* cke)

Leaf spot, which shows light yellow specks, is caused by many fungi but the main leaf spot is caused by *Phyllosticta chenopodii* fungus. Later, the affected leaves are shriveled and dried up. The fungus *Cladosporium variabile* restricts to prickly-seeded cultivars and causes circular chlorotic spots on older leaves with brown or purple margins, which enlarge and multiply until they cover most of the leaf surface, which turn yellow, wither and die. There is greenish black mould on both sides of leaf, made up of large olive conidia. On severely infected foliage, lesions coalesce to produce relatively large necrotic regions. Leaf spot of spinach sometimes becomes severe in cold and wet weather.

Control

- Grow disease tolerant/resistant cultivars of prickly-seeded type.
- Grow the crop on well-drained soil.
- Spray the crop with Blitox (copper oxychloride) 0.2%.

WHITE RUST (*Albugo occidentallis* wilson)

This fungus causes white blister-like circular or irregular pustules, which appear on the lower surface of the leaves, and opposite each blister on the upper surface a yellow patch is developed. As the disease develops, pustules release spores that create secondary infections in other plants if conditions are favourable for spore germination. More than 80% yield is reduced due to this disease. The disease is considered sporadic but it can cause heavy losses to the crop due to deterioration of the leaf quality. The disease is favoured by warm sunny days followed by cool nights with dew. Spores are more viable when they experience a period of drying but will not germinate until leaves are wet. Oospores may survive 1 year or more in soil and infested crop debris, leading to primary infection of leaves closest to the soil.

Control

- Keep the field free from weeds.
- Follow long crop rotation with non-host crops.
- Grow flat or smooth leaf cultivars, which are usually more resistant than the savoy or crinkled leaf cultivars.

DOWNY MILDEW (*Peronospora effusa*)

The leaves affected by this fungus show light yellow irregularly shaped areas without distinct margins on their surface. Whitish-grey mycelium grows on leaf surface at high humidity, while the mycelium is generally absent in low humid atmosphere. The pathogen persists only for 2–3 years in the soil as oospores. The disease develops in cool and moist weather. It appears at any stage of the plant growth but the attack is more prominent at seedling stage. The cotyledons are systemically infected, becoming discoloured and distorted. Yellow lesions appear on the older leaves. The infection of downy mildew occurs primarily in the field but it becomes prominent after harvesting of crop, especially at 14°C–24°C temperature and above 85% relative humidity.

Control
- Grow resistant varieties, if available.
- Use only disease-free healthy seeds.
- Follow long crop rotation with non-host crops.
- Remove and destroy crop residues (debris).
- Spray the crop with Dithane M-45 or Dithane Z-78 0.2% at 15-day intervals.

ANTHRACNOSE (*Colletotrichum spinaciae* ellis and Halsted)

This disease is more prevalent in the crops planted in rainy and autumn season as lower temperature along with wet conditions is favourable for the development of pathogen. This disease is caused by fungus, in which the affected leaves show water-soaked spots. Later, these spots coalesce and cause drying of the elongated greyish spots. Tiny black fruiting bodies on diseased tissue distinguish this pathogen from other leaf spot pathogens. The disease often infects leaves that are already infected with other pathogens, especially white rust. In case of seed crop of spinach, black spots on seeds are also formed.

Control
- Use disease-free healthy seeds.
- Remove and destroy the crop debris.
- Follow long crop rotation with non-host crops.
- Treat the seed with Captan, thiram, or Bavistin @ 2 g/kg of seed or dip the seed in 0.125% solution for half an hour.
- Spray the crop with Dithane M-45, Dithane Z-78, Blitox, or Bavistin 0.2% at 15- to 20- day intervals.

BACTERIAL SOFT ROT (*Erwinia carotonova*)

The soft-rot-causing bacteria damage the succulent plant parts by degrading pectate molecules that bind plant cells together, causing the plant structure to eventually fall apart. It is a post-harvest bacterial disease causing severe losses during transportation. The bacteria generally enter through wounds or via fungal lesions. Decaying occurs in the packed leaves due to the lack of aeration and cleanliness. Initially, the bacterial soft rots cause water-soaked spots. These spots enlarge over time and become sunken and soft. Interior tissues beneath the spots become mushy and discoloured, with the discolouration ranging anywhere from cream to black. Soft rot is known for a strong disagreeable odour that accompanies the breakdown of plant tissue. The infected leaves become grey-green and give a mushy look. Water-soaked areas are developed on leaves. The decay progresses at a faster rate at high humidity and temperature above 7°C, with the worst decay between 21 and 26°C.

Control
- Clean the harvested leaves before packing.
- Remove all the broken leaves before packing.
- Pack the leaves loosely.
- Use clean packing material.
- Send the harvested produce immediately to the market.
- Store the produce at a perfectly cleaned place before marketing.

28 Swiss Chard

M.K. Rana and Pooja Rani

Botanical Name

: *Beta vulgaris* L. var. cicla

B. vugaris L. var. flavescens
(Rainbow Swiss chard)

B. vulgaris L. var. maritime
(wild)

B. vulgaris L. var. macrocarpa
(wild)

B. vulgaris L. var. adanensis
(wild)

Family

: Chenopodiaceae

Chromosome Number : 2n = 18

ORIGIN AND DISTRIBUTION

Swiss chard has been cultivated since 300 B.C. and the roots of wild chard were used as medicine. The wild form of Swiss chard is found in the Canary Island, Mediterranean region and East to South Asia. It is thought to have originated in the Mediterranean and has been gradually selected for its edible leaves with their prominent petioles and wide leaf blades. At the same time, it spread to many other regions from where it was adopted as an important leafy vegetable, *e.g.*, in warmer parts of Nepal. In most African countries, Swiss chard is far more important than the garden beet, which is primarily consumed by Europeans. Fordhook Giant with a green petiole and prominent white lamina, originally bred in United States of America, is now widely grown in tropics and subtropics.

INTRODUCTION

Swiss chard is a leafy vegetable, which is scientifically classified as *Beta vulgaris* var. cicla. Swiss chard is actually a common name, but in fact, it is simply called *vegetable chard*. Yet, for common purposes, people continue to refer to it as Swiss chard. It is closely related to beet and actually has the same scientific classification but beet is usually cultivated for roots and Swiss chard for its leaves. It is grown most commonly and easily in the Northern Hemisphere and is very popular as an ingredient for Mediterranean cooks. The first documented use of chard in cooking was in Sicily. It is usually eaten raw in salads, cooked, or sautéed. The bitterness of raw leaves dissipates when cooked, leaving a soft delicious flavour, more subtle than spinach.

COMPOSITION AND USES

COMPOSITION

Chard is known to be a nutritional powerhouse vegetable packed with minerals, vitamins and health benefits. It is a rich source of vitamin A, B_2, B_6, C, K and E. In term of minerals, it has a wealth of magnesium, manganese, potassium, iron, sodium and copper. All parts of chard plant contain oxalic acid. Furthermore, in addition to dietary fibre, it has a significant amount of polyphenolic antioxidants, phytonutrients and enzymes that are unique and highly beneficial to human health. The nutritional composition of Swiss chard is given in Table 28.1.

USES

Fresh young chard can be used raw in salads and added in omelettes, soups and stews. Mature chard leaves and stalk are cooked or sautéed. Leaves of any size can be steamed, stir-fried and mixed with other greens. In Egyptian cuisine, chard is normally cooked with coriander in a light broth. In Africa, the whole leaf blades are usually prepared with the mid-rib as one dish. Chard can be blanched and frozen for later use just like spinach.

MEDICINAL USES

Swiss chard has anti-cancerous properties due to the huge amount of antioxidants found in it. It contains significant amounts of vitamin E, vitamin C, lutien, beta-carotene, zinc and quercetin. Many of these have been connected to preventing a variety of cancers, specifically colon cancer. The most important benefit of Swiss chard is its ability to regulate the blood sugar level. Syringic acid is one of the unique flavonoids found in Swiss chard and prevents diabetes. Vitamin A and beta-carotene reduce the risk of cataracts by as much as 39%. It has an ability to improve digestion, boost the immune system, reduce fever, lower the blood pressure, prevent heart disease, increase bone strength and development, detoxify the body and strengthen the functioning of the brain.

BOTANY

Swiss chard is a biennial plant but it is grown as an annual. It is dicotyledonous and usually herbaceous in nature. The plant has a moderately deep root system. The roots are hardy and woody. The stalks resemble those of spinach but are fleshy and white or red in colour. Thick and ruffled leaves have a stiff mid-rib and heavily veined leaves. The leaves grow from a crown at the base of the plant. In early stages of plant growth, it produces succulent tender edible rosette leaves on a small thick stem. The large, glossy, crispy and dark green leaves can grow to 37 cm long and 25 cm wide. The flower produces abundant, small and light pollen grains, which are carried by wind, leading to wind pollination. Hence, the crop is highly cross-pollinated. In order to induce flower stalk formation, the chard roots must be exposed to a chilling temperature. Chard has a highly vegetative phase. As the first-formed or outer ring of petioles is harvested, new growth occurs and the outer petioles mature successfully for constant harvesting.

Inflorescence is racemose with bracts of cymose type and emerges either in the axil of the leaf or from a terminal bud. Swiss chard, being biennial, flowers after a chilling period of 6–12 weeks at

TABLE 28.1
Nutritional Composition of Swiss Chard (per 100 g Edible Portion)

Constituents	Contents	Constituents	Contents
Water (g)	92.65	Magnesium (mg)	86
Carbohydrates (g)	4.3	Iron (mg)	2.26
Protein (g)	1.88	Vitamin A (IU)	6124
Fat (g)	0.08	Thiamine (mg)	0.034
Dietary fibre (g)	2.1	Riboflavin (mg)	0.086
Calcium (mg)	58.0	Vitamin C (mg)	18
Phosphorus (mg)	33.0	Vitamin K (mg)	3.27
Potassium (mg)	549.0	Energy (kcal)	20

10°C or less. The flowers are small, bisexual, perfect, bracteates (leafy) and are borne at the top of the stalk. The seed is actually a fruit containing anything from 2 to 8 fine seeds. The anthesis takes place between 7:00 a.m. and 5:00 p.m. with peak period between 11:00 a.m. and 1:00 p.m. The flowers open mostly during midday. In an individual flower, anthesis is completed within 2 hours and hastened by high temperature and low humidity. Anther-dehiscence takes place between 8:30 a.m. and 6:30 p.m. with a peak period from 12:30 to 2:30 p.m., depending on temperature and humidity. Stigma receptivity begins 8 hours before anthesis, reaches the maximum just after anthesis and continues for 10 hours.

CLIMATIC REQUIREMENT

Swiss chard is a cool season crop that grows best at a temperature ranging from 7°C to 24°C. Being half-hardy, it can withstand light frost. Prolonged exposure to temperature less than 5°C will induce bolting, usually in spring season. Bolting induced by long days (14 h) followed by cold temperature. Leaves remain small and are of inferior quality under hot weather conditions. In late summer, particularly the foliage of plants is subjected to a fungal leaf spot disease, which may be a limiting factor for the cultivation of Swiss chard. The plant has a moderate root system, but like other leafy vegetable crops, it should not be allowed to suffer moisture stress, thus, it requires fairly frequent irrigation to ensure that the soil does not dry out to less than 50% available water. For better growth, the moisture in field should not be limiting. Swiss chard requires full sun to partial shade for higher yield.

SOIL REQUIREMENT

Swiss chard can be grown on any type of soil having sufficient fertility and proper drainage system but sandy loam or loam soils are ideal for its cultivation. The crop is highly susceptible to acidic soil but it can tolerate slightly alkaline soil. For higher yield, the optimum soil pH should range from 6 to 7. The crop is susceptible to boron deficiency, which can be corrected by adding borax (sodium tetraborate) @ 2 g/m². Soil temperature should be around 20°C for its optimum growth, yield and quality. For raising a crop, the soil is prepared by ploughing the field 3 to 4 times followed by planking so that soil may be well pulverized and properly levelled since well-prepared soil improves the quality of the crop.

CULTIVATED VARIETIES

There are mainly three basic types of chard, as follows:

Types of chard	Description	Varieties
White-stemmed	The varieties of this group have broad white stems, often flattened with glossy green leaves and tend to be more productive and tolerant to both cold and heat than coloured varieties.	Fordhook Giant, Lucullus and Silerado
Coloured	The varieties of this group are excellent ornamentals. They are tall, pink, or yellow in colour. The red-stemmed varieties such as Pink Passion and Burgundy often grow quite tall and tend to be highly productive. Red varieties may bolt if planted too early during spring season.	Pink Passion, Burgundy, Orange Fantasia, Golden Sunrise, Bright Light, Rhubarb, Ruby Red and Rainbow chard
Perpetual	The varieties of this group usually popular in Europe have the capacity to regenerate new leaves quickly after harvesting of outer leaves. The green-leafed perpetual chard bears mild-flavoured greens over a very long season.	Perpetual and Verde Da Taglio

SOWING TIME

In frost-free regions, Swiss chard is usually sown from February to August, and in very cold regions, it is sown from August to September and even until February. In most other parts of the country, it is sown twice, once in spring-summer from January to April and second in rainy season from July to September.

SEED RATE

Generally, direct seeding is practiced and later the extra plants are thinned to optimum plant stand. For direct seeding crop, the seed rate varies from 7 to 9 kg/ha, depending on soil type, growing season and the irrigation facilities available.

SOWING METHOD

In the broadcasting method, seeds are broadcast in well-levelled beds and mixed in soil thoroughly. However, this method is not so desirable. Line sowing is a more suitable method, which facilitates weeding, hoeing and cutting of leaves. If the soil moisture and temperature are at optimum level, the seeds take 10 days to germinate. In line sowing method, the seeds are sown at a depth of 2 cm keeping a distance 45 to 60 cm between rows and 20 to 30 cm between plants.

NUTRITIONAL REQUIREMENT

The crop responds well to the application of organic manure. Farmyard manure should be incorporated in the soil @ 35–40 t/ha during field preparation. Besides, nitrogen, phosphorus and potash are applied @125, 20–25 and 125 kg/ha, respectively at the time of sowing and 25–35 kg/ha nitrogen as top dressing after each cutting for quick growth of tender succulent leaves. The rates of inorganic fertilizers needed are usually determined by soil analysis. Swiss chard also responds to periodic side dressing of nitrogen. Being a leafy vegetable, it requires a bit more nitrogen for leaf growth. The crop is tolerant to magnesium deficiency.

IRRIGATION REQUIREMENT

The irrigation frequency depends on the soil conditions and growing season. This crop has a moderately deep root system, but like other leafy vegetable crops, it should not be allowed to suffer moisture stress, thus, irrigation is applied a bit frequently for quick growth of the crop. If ample soil moisture is not available at the time of sowing, a light irrigation is applied before sowing to ensure good germination. Chard prefers moist soil to encourage production of new leaves. First irrigation is applied just after sowing if optimum moisture is not available in the field at the time of sowing and second irrigation is applied after first cutting and then according to necessity. In light soils, irrigation is applied more frequently than in heavy soils. Moisture fluctuation causes the leaves to become tough and develop slowly. During dry conditions, irrigation at 14- to 21-day intervals ensures continuous yield and quality chard.

INTERCULTURAL OPERATIONS

HOEING AND WEEDING

To keep the field weed free and to loosen the soil for proper aeration, 2–3 hoeings-cum-weedings are needed. Weeds can be controlled both mechanically and chemically. Chemical weed control can be achieved by applying pyrazone @ 2.4–2.8 kg/ha as pre-emergence to control weeds. The herbicide chloriazon, sold as Pyramin in the market, is registered for use in Swiss chard to control annual broad-leaf weeds.

MULCHING

Mulching conserves the soil moisture. The soil is covered with partially decomposed compost to prevent weeds, keep the soil evenly moist and provide disease protection.

THINNING

In Swiss chard, thinning is an important operation to maintain optimum spacing between plants and to provide better conditions to the plants for their growth and root development. It is important to remove excessive seedlings, keeping 10 cm spacing between plants, not later than 3 weeks after the seedlings have emerged. This operation is essential to reduce competition among plants for space, light, nutrients and moisture. The removed seedlings may be used for transplanting to fill any gap in the field. The total population should be 60,000 to 80,000 plants per hectare. Sometimes, the seedlings are raised in nursery beds or protrays for later transplanting.

HARVESTING

Usually, chard is harvested manually by removing only the outer leaves along with large petioles approximately 50–60 days after planting when the leaves have reached a length of 10–12.5 cm. Care should be taken not to damage the growing tip. Yellow and decayed foliage is removed when first seen. Leaves are harvested by cutting with a sharp knife about 5 cm above the ground level. The plants should not be overharvested at any picking to avoid weakening and affecting the later pickings as well as reducing yield. Leaves can be harvested from the same plant for several months, often until the leaf spots become too severe or plant starts bolting. When the day temperature becomes 30°C, harvest season ends.

YIELD

The average yield of Swiss chard varies from 200 to 300 q/ha, depending upon cultivar, soil type, environmental conditions and cultural practices adopted by the growers. Yield up to 400 and more per hectare can also be obtained under good management practices.

POST-HARVEST MANAGEMENT

After harvesting, the leaves are generally trimmed, washed and tied into bunches. The leaves must be free of insect injury, worm, mould, decay or other serious injury that affect its appearance. One or two dozen Swiss chard are packed in wax cardboard boxes. Constant top icing is necessary to maintain its shelf life. It may be stored for 10–14 days at 0°C temperature and 90–95% relative humidity.

INSECT-PESTS

LEAF MINER (*Liriomyza sativa*)

Larvae can produce discoloured patches on leaves. As the larvae feed on mesophyll tissue, they create extensive tunnelling within the leaves. The width of these tunnels increases as the larvae grow. This discolouration of leaves renders the Swiss chard unmarketable. Leaf miner will pupate between the leaves, contaminating the plant. These bodies may also die and rot, providing a medium for pathogen growth.

Control

- Remove and destroy the affected leaves earlier to prevent further spreading.
- Avoid planting near cotton, alfalfa and other host field to avoid migration from these fields.
- Apply registered chemicals like dimethoate and permethrin.

APHIDS (*Myzus persicae, Brrevicoryne brassicae* and *Liphaphis erysimi*)

Both nymphs and adults of this tiny insect suck cell sap from the tender leaves, reducing plant vigour. The leaves curl up and ultimately wilt. The aphids excrete a honeydew-like substance, on which black sooty mould develops, which hampers the photosynthetic activity. Besides, these aphids act as vector for the transmission of virus. Severe infestation affects the yield and quality of roots adversely.

Control

- Avoid excessive use of lime.
- Spray the crop with simple water with pressure.
- Remove and destroy the portion of plant attacked by the aphids.
- Spray the crop with malathion 0.05%, monocrotophos 0.04%, methyl demeton 0.05%, dimethoate 0.03%, or phosphomidon 0.03%.

LEAF-EATING CATERPILLAR (*Laphygma exiqua*)

Caterpillars feed on seedling and skeletonize the older plants. The larvae often feed on leaves and make holes. If population is high, they can destroy an entire seedling stand.

Control

- Avoid planting near cotton, alfalfa and other host fields to avoid migration from these fields.
- Spray the crop with malathion 50 E.C. @ 1 litre per hectare dissolved in 625 litres of water.
- A ditch of water containing oil or detergent that surrounds the perimeter of the field can also be used as a barrier.

DISEASES

CERCOSPORA LEAF SPOT (*00*)

Brown dead circular spots appear on leaves, turning grey with reddish purple border at a later stage. Infection affects the appearance of leaves and such leaves become poor in quality. The optimum temperature range for fast development of leaf spots is 25°C–30°C. Severe infestation of disease may be from mid-summer to early autumn.

Control

- Always use disease-free seed to avoid the incidence of this disease.
- Treat the seed with hot water at 50°C for 30 minutes to kill the pathogen present on or below the seed coat.
- Spray the crop with Bavistin, mancozeb, Blitox, or Blitane at 0.2%.

CURLY TOP

Infection is characterized by clearing of leaves veins and leaf curling with sharp protuberances from the veins on the leaf undersides. Leaves may be thickened and somewhat brittle.

Control

- Rotate the crop from one location to another in subsequent years.
- Use good cultural practices such as proper watering and fertilizing.
- Grow resistant varieties.

29 Beetroot

Sonia Sood and Nivedita Gupta

Botanical Name	: *Beta vulgaris* L. subsp. vulgaris
	Syn. *Beta vulgaris* L. var. crassa (Alef.) J. Helm.
Family	: Chenopodiaceae
Chromosome Number	: 2n = 18

ORIGIN AND DISTRIBUTION

The name *Beta* came from Romans. It has its origin in the Mediterranean region and North Africa. This species is thought to be derived from its ancestor *Beta vulgaris* ssp. *maritima*, known as sea beet by hybridization with *Beta patula*. It was taken to Northern Europe by the invading armies and widely used throughout Europe. Wild forms of *Beta vulgaris* occur along the shores of the Mediterranean, extending eastwards as far as Indonesia and westwards along the coasts of Atlantic up to southern Norway. Beetroot was considered for cultivation in the eastern Mediterranean or the Middle East and first mentioned in the literature in Mesopotamia in the 9th century B.C. Originally, beets were grown mainly for their leaves. The first use of beetroot was the edible leaves and roots for medicinal purposes by the Romans during the second and third centuries. Towards the end of the Middle Ages, the garden beet, with its thick cylindrical or globular root, had become an important vegetable in Central Europe. In Germany, it was first described in 1557 as Roman beet. It has been used for fodder in Europe since the 16th century. Beetroot is an important crop in Europe for the production of sugar. In 1800, the beetroot was introduced in the United States of America.

INTRODUCTION

Beetroot, also known as table beet, garden beet and sugar beet, are thick fleshy roots which store large amounts of reserve food. It is not as commonly grown as radish and carrot in the country. Based on the edible part, beets are generally classified into two major groups. The first is the *Crassa* or garden beet group, including yellow beets, common table beet, or red beet, mainly grown for its fleshy roots; the second is the *Cicla* group, including leafy beets and Swiss chard, mainly grown as a leafy vegetable. Table beet is an important root crop in Eastern and Central Europe, mainly Poland, Russia and the Middle East. It is a popular root vegetable grown mainly for its fleshy enlarged roots, in both home and market gardens in northern and southern parts of India.

COMPOSITION AND USES

COMPOSITION

Beetroot is rich in protein, carbohydrates, calcium, phosphorus, iron, vitamin A and vitamin C. The tops of garden beet and the foliage of Swiss chard used as greens are excellent source of vitamins

A and B_2 and are a good source of vitamin B_1. The nutritive value of beetroot per 100 g of edible portion is given in Table 29.1.

USES

Beetroot is consumed in the form of fresh salad, vegetable and processed products as pickled and canned. After sugarcane, it is the second most important crop for the preparation of sugar. Beet leaves probably were used in prehistoric times as potherb (green). By the 16th century, it was widely used for feeding animals in Rome, particularly during winters. The large-sized beetroots are used for canning. The tender leaves may be used much like spinach, as leafy greens for salad or boiled.

MEDICINAL USES

The roots and leaves of beet have been used by the ancient Romans in folk medicines to treat a wide variety of ailments, *e.g.*, in fevers, constipation, liver disorders and in reducing cholesterol level in the bloodstream. They also considered beetroot juice as an aphrodisiac. Hippocrates advocated the use of beet leaves for healing wounds. From the Middle Ages, beetroot was used as a treatment of blood- and digestion-related problems, disorders of the respiratory tract and liver, fevers and infections. Betanin pigment present in beetroot has been used in European countries to protect against oxidative stress. However, the formation of kidney stones has been implicated due to high oxalic acid content in its roots.

TAXONOMY

Botanically, beetroot can be classified based on binomial nomenclature as well as horticultural classification system. Binomial nomenclature is the system of classification in which the organisms are assigned to class, order, family, genus and species. This system was devised by the Swedish botanist Carl Linnaeus (1707–1778). Binomial nomenclature can be extended below the species level (infra-specific classification) by including the ranks of subspecies (subsp.) and variety (var.). The rules of binomial nomenclature have most recently been laid out in the International Code of Botanical Nomenclature (ICBN), which was adopted by the International Biological Congress in 1981 and published in 1983.

TABLE 29.1

Nutritional Composition of Beetroot (per 100 g Edible Portion)

Constituents	Root	Leaves	Constituents	Root	Leaves
Water (g)	87.7	92.3	Iron (mg)	0.80	16.2
Carbohydrates (g)	9.56	6.5	Zinc (mg)	0.35	–
Sugars (g)	6.61	–	Chromium (mg)	0.012	–
Protein (g)	1.7	3.4	Copper (mg)	0.29	–
Fat (g)	0.1	0.8	Chlorine (mg)	24	–
Dietary fibre (g)	2.80	1.70	Carotene (mg)	0.0	5,862
Calcium (mg)	18.3	380	Thiamine (mg)	0.04	0.26
Phosphorus (mg)	55.0	30	Riboflavin (mg)	0.09	0.56
Potassium (mg)	325	–	Niacin (mg)	0.4	3.3
Sodium (mg)	78.0	–	Choline (mg)	242	–
Magnesium (mg)	23	–	Vitamin C (mg)	10	70
Sulfur (mg)	14	–	Energy (kcal)	43	46
Manganese (mg)	0.32	–	Oxalic acid (mg/100 g)	40	–

Linnaeus in 1753 first described three varieties within species of *Beta vulgaris*, *i.e.*, wild ances-tral beet (*Beta vulgaris* var. *perennis*), leaf beet (*Beta vulgaris* var. *cicla*) and garden beet (*Beta vulgaris* var. *rubra*), but in 1763, he changed the ancestral name *Beta vulgaris* var. *perennis* to *Beta vulgaris* subsp. *maritima* (L.). However, all cultivated forms of *Beta vulgaris* belong to one subspe-cies, *i.e.*, *Beta vulgaris* subsp. *vulgaris*.

In taxonomic schemes, variety (var.) is the level below subspecies. According to horticultural classification system, *Beta vulgaris* subsp. *vulgaris* is distinguished at variety level as *Beta vulgaris* var. *cicla* (chard), *Beta vulgaris* var. *conditiva* (table beet), *Beta vulgaris* var. *alba* (fodder beet) and *Beta vulgaris* var. *altissima* (sugar beet).

BOTANY

The herb produces thick and fleshy roots, which store a large amount of reserve food. Morphologically, the upper portion of root develops from the hypocotyl and lower portion from the taproot, from which the secondary roots arise. During the development of fleshy roots, vascular tissues and stor-age parenchyma tissues are formed alternately, which result in concentric rings and are visible in a median cross section of mature beetroot. The exterior colour of the root varies from orange-red to dark purple-red, depending on the cultivar, while the interior colour is influenced by cultivar, temperature, season, soil, nutrition, etc. The colour of root is due to the presence of red-violet pig-ment β-cyanins and yellow pigment β-xanthins. Leaves are often ovate and cordate, dark-green, or reddish. The rosette of leaves develops in a spiral with the oldest ones on the outside. The size, shape and colour of foliage are also influenced by temperature, season, spacing and soil-moisture. Inflorescence is a large spike, which normally develops in the second year. The flowers are sessile, arising in clusters of 3–4 in the axils of bracts of inflorescence axis and its secondary branches. The flowers are small, inconspicuous and without corolla, but with green calyx, which becomes thicker towards the base as the fruit ripens. Flowers have five stamens and stigma with three segments. Fruits are mostly aggregate formed by the cohesion of two or more fruits and held together by swol-len perianth (calyx) base and thus form an irregular dry cork-like body known as seed ball or seed. The true seed is small, about 3 mm long, kidney-shaped and shiny to reddish-brown in colour. The seed ball contains more than one seed, which remains viable for 5–6 years under normal storage conditions.

CLIMATIC REQUIREMENT

Beetroot is a cool season crop but it can be grown under warm climate also. The optimum tem-perature for proper growth, development and good quality root production is 18.3°C –21.1°C. Temperature below 10°C causes bolting before attaining marketable size of roots, which are unde-sirable. At 30°C or above, the accumulation of sugar in the roots is ceased. The roots grown at high temperature have less intense red colour and distinct zoning inside flesh. It requires abundant sunshine for the development of roots. Excessively hot weather causes the appearance of alternating light and dark red concentric circles in the root. The roots develop good colour and texture under cool weather conditions. However, fluctuating temperature causes zoning, which is marked by the appearance of alternating light- and dark-red concentric circles in the root.

SOIL REQUIREMENT

Beetroot can be grown on a wide range of soils, however, deep and well-drained loam or sandy loam soil is considered best. The crop grown in heavy soils produces roots of asymmetrical shape and small, forked, or branched. It can be grown on acidic as well as alkaline soil up to pH as high as 9–10. The ideal soil pH is 6–7 and acidic soils with pH below 5.8 affect the yield. In highly alkaline soils, scab may develop.

CULTIVATED VARIETIES

The varieties below are grouped according to their shape, *i.e.*, flat, short top, deep oblate to round, globular or oval, half-long and long. Recently a few F₁ hybrids have been introduced from abroad by private Indian seed companies for commercial cultivation like Solo (British-bred monogerm), Alto, Super King, Red Cloud, Majesty, Fancon, Cardinal and Super Queen. The characteristics of individual groups are as follows:

Flat shaped	The modern type varieties produce thicker but not so flat roots as the older types, *e.g.*, Flat Egyptian.
Short top shaped	The roots are flattened at top and bottom with rounded sides and conical or tapered base, *e.g.*, Crosby Egyptian, Light Red Crosby, Strawberry Crosby, Ferry Crosby and Vermilion Crosby.
Deep oblate to round	Roots are round or globular in shape, *e.g.*, Early Wonder, Green Top Early Wonder, Asgrow Wonder, Morse Detroit, Ohio Canner, and Perfected Detroit.
Globular to oval	Roots are globular to oval in shape, *e.g.*, Detroit Dark Red, Good for All, Crimson Globe, Green Top Bunching, Asgrow Canner and Crimson King.
Half long	Length is shorter than long types, *e.g.*, Half Long Blood and Early Blood Turnip.
Long	Roots are long, may grow as much as 40 cm and they are quite popular in Europe, *e.g.*, Long Dark Blood, Long Smooth Blood, Lutz Green Leaf, or Winter Keeper.

CRIMSON GLOBE

Tops are medium to tall, leaves bright green with maroon shade and flesh dark-red without zoning. The roots are globular to flattened-round and medium red. They become ready for harvesting in 60–65 days for salad purposes. It takes 80–90 days when grown for seed production. It is suitable for salad, canning and making pickles.

DETROIT DARK RED

Tops are small, leaves dark green tinged with maroon, roots round with deep red skin and flesh deep red without zoning. It is suitable for salad and becomes ready for harvesting in 80–100 days.

CROSBY EGYPTIAN

The roots are flat globe with a small top root and a smooth exterior. The top is medium tall and green with red veins. The internal colour is dark purplish red with indistinct zones when cut. Roots take 55–60 days to be mature. When grown in warm weather, roots show pronounced white zoning.

EARLY WONDER

The roots are flattened globular with rounded shoulders and a smooth, dark red skin. Flesh is dark red with some lighter-red zoning. This cultivar takes 55–60 days after sowing to reach harvest maturity.

OOTY 1

A selection from the local type is suitable for growing in south Indian hills. The skin is thin and the flesh colour is blood red. It is a high-yielding variety with a potential of 310–450 q/ha. The crop duration from sowing to harvest is 120–130 days.

PANT S 10

A variety tolerant to *Cercospora* leaf spot and *Sclerotium* root rot developed at G.B. Pant University of Agriculture and Technology, Pantnagar. It is a high-yielding variety with a potential of 725–825 q/ha. It is rich in sugar (sucrose about 14.5–15%). The variety is recommended for cultivation in Uttar Pradesh, Rajasthan, Punjab and West Bengal.

SEED RATE

Beetroot is propagated directly by seed. The seed ball usually contains one or more seeds, thus, it is called *seed ball*. The seed rate of beetroot varies according to variety, seed viability, sowing time, growing region, growing conditions, soil type, soil fertility, irrigation facility, etc. In general, a seed rate of 7–9 kg is enough for sowing a hectare area.

SOWING TIME

In the northern plains of India, the seeds are usually sown from September to November, while in the southern plains, the sowing of beetroot may extend from July to November. In hills, the seed is sown from the first week of March to the end of July. Several successive sowings at an interval of 2 to 4 weeks ensure a continuous supply of marketable roots. Seeds sown early may produce roots with coarse and woody flesh and dull colour due to high temperature.

SOWING METHOD

Seeds are sown directly on ridges or in flat seedbeds. On commercial scale, a seed drill may also be used for this purpose. The plant population per unit area depends upon spacing, variety used and agro-ecological conditions. A population of 3,00,000 plants per hectare can be achieved by keeping the spacing 50–60 cm between rows and 8–10 cm between plants. The seeds are sown at a depth of 1.5–2.5 cm to ensure good germination. The cultivars with small tops are sown comparatively at closer spacing than those with heavy tops.

Seed treatment with thiram or captan at the rate of 2.5 g/kg of seed gives better seedling emergence and controls pre- and post-emergence damping-off disease. The seeds are soaked in water for about 12 hours before sowing to facilitate better germination in the field. The seedbeds should not be too wet or too dry to facilitate uniform and early germination.

NUTRITIONAL REQUIREMENT

Well-decomposed farmyard manure 20–25 tonnes, nitrogen 100 kg, phosphorus 50 kg and potash 70 kg/ha should be applied to harvest a bumper crop. Farmyard manure is incorporated into soil at the time of field preparation. Full dose of phosphorus and potash along with half dose of nitrogen is applied at the time of last ploughing or ridge formation. The remaining half of the nitrogen is top-dressed 30 days after sowing. In boron-deficient soils, borax 20–25 kg/ha may also be applied at the time of last ploughing. Beetroot has a relatively high boron requirement for quality root production since its deficiency causes internal breakdown as black spot or dry rot. Nitrogen application increases the size of roots but reduces the total soluble solids and red pigment contents in roots. Single or combined application of micronutrients such as copper, zinc, manganese, boron and molybdenum is essential for seed production of red beet.

IRRIGATION REQUIREMENT

Sufficient soil moisture facilitates the germination, thus, a regular water supply is essential for both seed germination and high yield of good-quality roots. The water requirement has been reported to be 300 mm supplied in five to seven irrigations, however, when there is winter rains, only about three irrigations are sufficient for harvesting a good yield. The crop is irrigated at an interval of 10–12 days in winter and 4–6 days during summer. The roots crack or become hard and stringy when the soil lacks sufficient moisture but stagnation of water is as harmful as the scarcity of moisture.

INTERCULTURAL OPERATIONS

EARTHING UP AND WEED CONTROL

One or two earthings up is essential to prevent the exposure of roots to sunlight, which causes greening and lowers the quality of produce. In early stages, the crop growth is very slow and the crop may not compete with weeds for nutrients, space, light and moisture, hence, it is essential to remove the weeds for reducing their competition with the crop. Light weeding and hoeing at least twice during the early stages of crop growth are required to keep the weeds under control.

In India, the chemical weedicides have not been experimented, however, pre- and post-emergence sprays of propachlor (2.5 kg/ha) control the weeds for 40 to 50 days and increase the crop yield. Several chemicals, such as Pyramin (pyrazon) 3.74–4.0 kg/ha applied during pre-emergence and Venzar (lenacil) 0.75–1.0 kg/ha applied either before sowing by mixing with the top 2–3 cm of soil or pre-emergence at 1.5 kg/ha, give good weed control and have no adverse effect on yield and quality of roots.

THINNING

In beetroot, each seed ball produces 2–6 seedlings, thus, thinning is an essential operation. Removing extra seedlings is necessary to maintain a distance of 10–15 cm between plants, which is important to have roots of uniform shape and size. Hand-thinning is done as and when the seedlings are large enough to handle. The roots grown for processing are normally not thinned because of high labour cost and preference for small beetroots in the markets. Two thinning can be followed to maintain the optimum plant spacing within the row. Thinning the seedlings to 3 cm produces roots of best quality. Gap filling has to be taken up for maintaining optimum plant population in poorly germinated crops.

HARVESTING

The roots are harvested by pulling out the plant after about 8–10 weeks of sowing, depending upon the variety and location. The marketable maturity is judged depending on the size ranging from 3 to 5 cm diameter of root. Roots are washed and graded according to size before they are sent for marketing. Oversized roots are not in demand because they are tough and woody and crack on the surface. Hence, the roots should be harvested as and when they attain a marketable size. In developed countries, mature roots are mechanically harvested, detopped, washed, graded and packed in polythene bags.

YIELD

Depending upon the season and variety, the average yield varies from 250 to 300 quintals per hectare having a sucrose content of 15–17%.

POST-HARVEST MANAGEMENT

Edible mature roots should be harvested for storage and fresh market. Immature roots may store poorly, sprout soon after storage and develop an undesirable flavour. Removal of tops and packing in polythene bags increase the shelf life by reducing water loss during transit and storage.

Beetroots are waxed to avoid wilting in a dry atmosphere and to retain good quality and appearance in marketing. The consumers in some markets demand waxed-beetroot to the exclusion of untreated roots. The finished product should be covered evenly with wax, the surface should be bright and the wax should be so transparent that the root colour may be seen clearly.

Roots of uniform size with 4 cm or less in diameter should be canned without cutting. Besides, the roots can be peeled and processed in light syrup. In slicing, it is not essential to have small beets, though they should not exceed 6 cm in diameter since large roots tend to be coarse and poorly coloured. The roots are cut into cubes about 1.25 cm on a side by a slicing machine. Slicing requires a greater tonnage of beets than whole canning because more sliced cubes or squares can be placed in a can. When whole beets are canned, the number of roots used per can varies with the size of the can, and usually, 8–10 roots are adjusted in a can.

The roots can be stored up to 2–3 days at room temperature. However, they can be stored well for 10–14 days at 0°C temperature and 95% relative humidity. The usable period for well-stored beets extends to late winter or early spring. After digging out, the roots are topped and stored immediately. All injured roots should be sorted out, as they encourage rots. Crown removal increases the loss of sucrose because of highly accelerated rate of wound respiration caused due to injury. The large exposed surface of the crowned root also provides storage pathogens with easy access to the most susceptible tissue of the root. Crowned roots, which are infected with these pathogens, decay much faster in storage than uncrowned infected roots. It has been reported that cracked and bruised roots have an initial respiration rate higher than undamaged beets, and therefore, they lose more sucrose. To prevent shriveling and decaying of roots, large piles should be avoided. Certain chemicals may prove very useful in delaying sprouting and keeping the roots firm.

PHYSIOLOGICAL DISORDERS

INTERNAL BLACK SPOT, BROWN HEART, HEART ROT, OR CROWN ROT

This disorder is caused due to boron deficiency. The deficient plants usually remain dwarf and stunted. The leaves are smaller than normal. The young unfolding leaves fail to develop normally and eventually turn brown or black and die. The leaves may assume a variegated appearance due to the development of yellow and purplish red blotches over parts or the whole, while the stalk of such leaves shows longitudinal splitting. Frequently, the affected plant has twisted leaves and exhibits a slight shortening and distorting of its leaf stalk in the center of its crown. The growing point may die and decay. The roots do not grow to full size, and under severe deficiency, they remain very small and distorted and have a rough unhealthy and greyish appearance instead of being clean and smooth. Their surface is often wrinkled and cracked. Within the fleshy roots, hard or corky spots are found scattered throughout the roots but more numerous on light-coloured zones or cambium layers. Boron deficiency characteristically affects the young cells and tissues first.

Control

- Avoid sowing in acidic soils.
- Avoid drought conditions by supplying regular irrigation.
- Apply borax from 5 to 50 kg/ha, depending upon the nature of soil, soil moisture and soil reaction, to manage the disorder.

DISTINCT ZONING

The beetroots have alternating circular bands of conducting and storage tissues. The bands of storage tissues are broader and darker red in colour than the conducting tissues that are narrow and light in colour. The contrast colour between alternating bands is known as zoning, which varies greatly between cultivars and within a cultivar. The colour development in beetroots is due to the presence of betaine, which may be suppressed at high temperatures, resulting in a marked increase in zoning, particularly during hot weather.

Speckled Yellows

This disorder is due to the deficiency of manganese. The leaves of affected plants show yellowish-green chlorotic mottled areas, which become necrotic, resulting in breaking of lamina. The leaf margins roll upward and turn into an arrow-shaped outline, which remains upright. Generally, the deficiency of manganese is observed in very sandy and very alkaline soils.

Control

- Avoid planting on very sandy and alkaline soils.
- Apply manganese sulfate 5–10 kg/ha in soil or spray manganese sulfate 0.25% two or three times.

INSECTS-PESTS

Beet Leaf Miner or Spinach Leaf Miner (*Pegomyia hyocyami*)

White larvae attack the young seedlings, thus, the damage is more severe in early stages than at later stages. It eats tissues between upper and lower surfaces of the leaves and causes blister-like areas, and thus, the plants remain stunted. The leaves become blistered and they wither and die.

Control

- Destroy all the infested leaves.
- Apply either carbofuran or phorate 0.5 kg a.i./ha at sowing followed by another application of 0.5 kg a.i./ha 40 days after germination for effective control of leaf miner.
- Spray or drench the crop with chloropyrifos or malathion 0.1%.

Beet Webworm (*Hymenia fascialis* cramer and *Loxostega* sp.)

The larvae feed on leaves by either spinning small webs or feeding on the underside protected by a small web. Larvae can also affect flowers and pods. They consume large amounts of foliage by skeletonizing the leaves and can completely defoliate the plants in a very short period of time. Webworms spin a web on leaves together or folding individual leaf to form a tube, in which they hide when disturbed.

Control

- Remove weeds that are preferred by adults for laying eggs.
- Turn the soil during the pupal stage to kill some pupae.
- Set up light traps @ 12/ha to attract and kill the adult moths.
- Set up pheromone traps @ 12/ha to attract the male moths of webworm.
- Spray *Bacillus thuringiensis* at 2 g/litre, dimethoate 0.03%, or quinalphos 0.25% at 10-day intervals.

Cutworms (*Agrotis ipsilon*)

The larvae are dark brown with a red head. Larger larvae may notch the seedlings stems just below the soil surface, which can cause the plants to wilt and die. They may completely cut through stalks, which can result in severe stand reductions. Black cutworms usually feed at night or during overcast days. They sometimes drag cut plants under dirt clods or into small holes in the soil to continue their feeding during the daylight hours.

Control

- Keep the weeds under control in field and surroundings.
- Remove the entire crop residue to lessen the possibility of cutworm outbreak.
- Drench the field with chlorpyriphos 0.1% on the appearance of cutworms.
- Apply phorate or Thimet 4–6 kg/ha before sowing of seeds.

Leaf-Eating Caterpillar (*Spodoptera litura*)

Leaf-eating caterpillars make holes in the leaves and sometimes in flower buds. As the caterpillars grow, the size of the holes they make also increases. Caterpillars eat the buds of young tender plants and often cover the feeding surface with webs. The green caterpillars feed on the foliage and damage the green foliage badly by eating from the margins inwards, leaving intact the main leaves. The larvae are greenish yellow with a black head and black dot on the dorsum.

Control

- Grow at least four rows of mustard to trap the caterpillers.
- Handpick the caterpillars and egg masses and crush or drop them in soapy water.
- Dust the crop with 5% malathion or 2% lindane.
- Spray the crop with *neem* seed-kernel extract 4% in early stage of larvae and cypermethrin 0.01%, quinalphos 0.2%, or fenvalerate 0.05% at 10–15 days to control large larvae.

Semi-looper (*Plusia nigrisigna*)

The caterpillars feed on the foliage and affect the crop seriously. They make holes of varying size. The larvae are pale green, plump and move by forming semi-loops. They are in abundance during rainy season. The adult moths hide at base of the plants in rice fields or in grassy areas during daytime and remain active at night.

Control

- Dust the crop with 5% malathion powder @ 25 kg/ha.
- Introduce natural biological control agents such as small trichogrammatid wasps and spiders.
- Apply *neem* seed kernel extract 5% + *neem* oil 0.4%.
- Spray the crop with malathion 50 EC 0.0l%, or dimethoate 0.03%.

Root Knot Nematodes (*Heterodera schachtii* and *Globodera rostochinensis*)

The root knot nematodes are polyphagous and are widely distributed in the country. Nematode problem is most serious in light sandy soils. It is almost impossible to eliminate the nematode from the soil once it is infected. The infected plants with roots damaged by the nematodes are not able to utilize the nutrients and moisture available in the soil efficiently. As a result, the plant growth becomes stunted, and the leaves of infected plant show chlorosis and wilting symptoms. When its infection is severe, it causes death of the plants. Soon after invasion by *G. rostochinensis*, the roots begin to branch underground, making clamps of taproot.

Control

- Rotate the crop with non-host crops like wheat, maize, sorghum, pearl millet, groundnut, cotton and pigeon pea up to 4–5 years.
- Maintain the soil pH near 6.0.

- Plough the field 20 cm deep with turning plough, two to three times at 10-day intervals in hot summer months to reduce the viable cyst in the soil.
- Grow marigold as trap crop or as intercrop to reduce nematode population.
- Apply *neem* cake @ 20 q/ha after green manuring.
- Fumigate the soil with nematicides like Nemagon @ 25 litre/ha with irrigation or Furadan 3 G @ 40 kg/ha or carbofuran 3 G @ 25 kg/ha at the time of planting in the furrows.

DISEASES

Sclerotium Root Rot (*Sclerotium rolfsii*)

This is one of the most serious diseases of beetroot. The affected roots start decaying and leaves turn yellow and wilt. The symptoms appear from February to April when the weather warms up. Fungal growth and sclerotia appear in soil around the roots also. Numerous black small-sized sclerotia are visible on or embedded in the infected plant parts. The leaves become light green followed by yellowing and fall off prematurely. The infection on the stem spreads downwards to the roots.

Control

- Follow phytosanitary measures to reduce the inoculums load.
- Apply nitrogen @ 160 kg/ha through calcium ammonium nitrate or ammonium sulfate to reduce the disease incidence considerably.
- Treat the seed with Bavistin, captan, or thiram @ 2–3 g/kg of seed.
- Drench the soil with copper oxychloride 0.03% solution to a depth of 15 cm.

Cercospora Leaf Spot (*Cercospora beticola*)

The symptoms appear as yellowing of leaves and premature defoliation. Leaves near ground are infected first. Spots are nearly circular, measuring 2–5 mm diameter, tan, pale brown, grey or whitish and sometimes with a narrow brown or reddish purple border. Stomata are often visible as minute black dots scattered in the centre of the spots. Under humid conditions, abundant conidia are produced and spots appear velvety. In some plants, spots are very small (1–2 mm) without any border. The disease spots also appear on pods.

Control

- Follow crop rotation of at least 3 years with non-host crops.
- Use disease-free healthy seed for sowing.
- Plough the field deep to bury crop refuse after harvesting.
- Destroy the diseased plant debris in the field.
- Treat the seed with Bavistin 2 g/kg along with seedling dip and foliar spray.
- Spray the crop with 0.3% copper oxychloride, mancozeb 0.25%, or hexaconozole 0.05% alternatively at an interval of 12–15 days.

Phoma Blight and Heart Rot (*Pleospora bjorlingii*)

The pathogen causes distinct brown leaf spots, sometimes, with concentric rings. On mature spots, the fungus forms numerous glistening black pycnidia, which may be visible without magnification. Roots become weak followed by gradually weakening and yellowing of older leaves. Water-soaked areas appear, which turn brown quickly and finally become coal black. On seed stalks, brown to black necrotic streaks occur with a greyish center, on which the black pycnidia develops

Control

- Follow long crop rotation with non-host crops.
- Grow beetroot in well-fertilized friable soil.
- Use disease-free healthy seed for cultivation.
- Treat the seed with hot water at 58.8°C for 8 minutes followed by air-drying for 24 h and then treat the seed with thiram 2.5 g/kg.
- Avoid injury to the roots.
- Add a sufficient amount of sodium borate before sowing seeds.
- Use organo mercurials to eradicate the seed-borne inoculums effectively.

POWDERY MILDEW (*Erysiphe polygoni*)

The disease symptoms appear as yellowish spots of 2.5 cm diameter on the upper surface of older leaves and white powdery growth on lower surface below the yellow spots. The affected leaves become dry but remain attached to the stem. The fungus survives on other plant species or weeds also. White powdery patches appear on leaves and other green parts, which later become dully coloured. These patches gradually increase in size and become circular, covering the lower surface too. When the infection is severe, both the surfaces of leaves are completely covered by whitish powdery growth. Severely affected parts get shriveled and distorted. In severe infections, foliage becomes yellow, causing premature defoliation. The pathogen develops a delicate hyaline and cob-web-like growth of mycelium on the host surface on either leaves or twigs. If infection occurs early, the roots remain small and produce little sugar.

Control

- Maintain optimum soil moisture for the crop.
- Dust the crop with wettable sulfur 25 kg/ha.
- Spray the crop with Karathane 0.1%, Rubigan 0.025%, Calixin 0.03%, carbendazim 0.1%, or sulfur 0.2% for effective control of the disease.

DOWNY MILDEW (*Peronospora farinose* sp. *betae*)

The fungus may affect the plant at any stage of growth. Initially, discolouration and in severe cases death of seedling occur. Later, at adult stage, the infected plants show fluffy white downy fungus growth on lower surface of the leaf, while the corresponding upper surface has purplish or yellowish brown spots. Both the surfaces are covered with fruiting fungus, and later, they curl downward sharply. Similarly, the lesions formed on leaves, branches, bracts, flowers and on seed stalks cause distortion of these parts. The diseases appear when the humidity is high with light rains, prolonged dew or heavy fog in cool weather.

Control

- Follow long crop rotation with non-host crops.
- Avoid application of excessive irrigation.
- Avoid the use of sprinkler irrigation system.
- Harvest the seed only from healthy plants.
- Use Sandovit 0.1% with fungicide to avoid runoff of droplets.
- Spray the crop with 0.3% Dithane Z-78 or 0.25% mancozeb at the onset of disease along with one spray of metalaxyl + mancozeb 0.2%.

Rhizoctonia Root Rot (*Rhizoctonia solani*)

It is a most serious disease of beet. The fungus causes damping-off and wire stem of seedlings in seedbed, bottom rot and heart rot of older plants in the field and storage, and root rot of beetroot, turnip, radish and rutabaga. The base of the leaf petiole turns black and the leaves collapse and die. The fungus often penetrates the root crown, causing the flesh to become brown and the skin to crack. Affected roots start decaying and leaves turn yellow and wilt. Loss in crop occurs as reduced stands and lowered yield and quality.

Control

- Follow long crop rotation and phytosanitary measures.
- Apply *neem* cake @ 20 q/ha after green manuring.
- Treat the seed with captan or thiram @ 2 g/kg to reduce the inoculum.
- Drench the soil with copper oxychloride 0.03% and carbendazim 0.1%.

Bacterial Blight (*Pseudomonas syringe* pv. *aptata*)

Numerous water-soaked and irregular dark brown spots appear on leaves. In beginning, the spots are brown and isolated or as wedge-shaped areas near the margin, which later enlarge and turn black in colour. Petioles become soft in severe infections. The disease is seed-borne and destructive in rainy season.

Control

- Use disease-free healthy planting material.
- Treat the seed with hot water at 56°C for 10 minutes.
- Treat the seed with streptocycline solution 1000 ppm for 2 hours.
- Drench the field with streptomycin sulfate 2.5 kg/ha.

Beet Yellow Mosaic

This disease is caused by two viruses, *i.e.*, beet yellow virus (BYV) and beet mild yellowing virus (BMYV). Transmission of virus occurs through aphids, *i.e.*, *Myzus persicae* and *Aphis fabae*. Scattered small yellowish patches on leaves, vein clearing and reduction of leaf lamina are the major symptoms. Leaves appear puckered and mottled with light and dark areas. Plant remains stunted. Yellowing begins at leaf tips and upper margins and spreads downwards between the veins. As the infected leaves mature, the older invaded portions turn brown and die. If infection occurs early, the roots remain small and produce little sugar.

Control

- Remove the overwintering-virus-infected weeds.
- Rogue out the infected plants and destroy them.
- Grow root crops in widely separated areas.
- Spray the crop just after germination with mineral oil like Krishi oil 1.5% or Confidor 0.03%, Metasystox 0.1%, Nuvacron 0.05%, or Dimecron 0.05% to reduce the population of vectors and spread of disease.

Curly Top

Curly top disease is caused by beet curly top virus (BCTV), which is transmitted from plant to plant by feeding leafhoppers. The symptoms of curly top disease include warty leaf veins and rolled brittle and twisted foliage. The leaves of affected plants remain under sized and curled. The veins become transparent. The root system also gets reduced.

Control

- Follow long crop rotation with non-host crops.
- Use disease-free healthy seed.
- Uproot and burn the infected plants.
- Isolate the field from infected plants.
- Uproot the affected plants and burn them.
- Spray the crop with malathion 0.1% or dimethoate 0.03%.

Purple Leaf of Beet (A strain of tobacco mosaic virus)

The disease is transmitted by aphids and mechanical sap inoculation. It causes flecking and mosaic patterns on beet leaves and eventually causes major leaf abnormalities. The infected plants become stunted. Leaves stand erect and closely spaced. Infected leaves become deep purple in colour, especially the young emerging leaves. In some areas, minute necrotic lesions develop on entire lamina.

Control

- Rogue out and burn the infected plants.
- Remove all the weed hosts surrounding the crop.
- Fix yellow sticky traps @ 12/ha to trap adults.
- Install pheromone traps @ 12/ha in field to reduce beetle population.
- Apply phorate @ 1.0 kg/ha in soil followed by irrigation.
- Spray the crop with monocrotophos 0.05% or malathion 0.1% at 10- to 15-day intervals.

■■■

30 Water Spinach

M.K. Rana and Navjot Singh Brar

Botanical Name : *Ipomea aquatic* Forssk.

Family : Convolvulaceae

Chromosome Number : 2n = 30

ORIGIN AND DISTRIBUTION

Water spinach is considered a native to Africa, Asia and southwestern Pacific Islands. It has been a medicinal herb in southern Asia since A.D. 300 and 200 B.C. People still collect its plants from wild habitat and cultivate them for vegetable purpose. In the late 1400s with the arrival of European in these regions, they became aware of this medicinal herb, took it with them for cultural uses and began spreading it around the world. The herb was introduced into new areas with the migration of people from Asian countries to other parts of the world. Still, the doubt persists on its domestication but the evidence available on food and medicinal uses and regions of cultivation in literature suggests that it was first cultivated in southeastern Asia. It has also been domesticated in China and India but its cultivation is escaped, as it may become ecologically a hostile weed.

INTRODUCTION

Water spinach, also known as river spinach, water morning glory, water convolvulus, Chinese convolvulus, Chinese spinach, Chinese water cress, swamp cabbage and *kangkong* in Southeast Asia, is an herbaceous aquatic or semi-aquatic perennial plant of tropical and subtropical regions, though it is not known where it originated. In India, it is commonly known as *paani palak*, *kalmua*, *kalmi*, or *kalmisaag*. It has no association with common spinach though it is closely associated with sweet potato (*Ipomea batatas*). Although it has a creeping growth habit, it grows erect in water. Like other leafy vegetables, water spinach is very nutritious. The Chinese consider the white-stemmed varieties better flavoured and "more tender" than the green stemmed varieties.

COMPOSITION AND USES

COMPOSITION

Based on proximate composition, water spinach is rich in carbohydrates, minerals, vitamins, crude lipid and crude fibre. Its leaves are found to have high moisture and ash but low in crude protein. The nutritional composition of water spinach is given in Table 30.1.

TABLE 30.1

Nutrition Composition of Water Spinach (per 150 g Edible Portion)

Constituents	Contents	Constituents	Contents
Moisture (%)	72.83	Manganese (mg)	1.61
Carbohydrates (g)	4.65	Vitamin A (µg)	540
Protein (g)	3.3	Thiamine (mg)	0.15
Fat (g)	0.15	Riboflavin (mg)	0.3
Dietary fibre (g)	4.65	Niacin (mg)	1.5
Calcium (mg)	111	Pyridoxine (mg)	0.17
Phosphorus (mg)	66	Folic acid (µg)	180
Potassium (mg)	570	Vitamin C (mg)	28.5
Magnesium (mg)	42	Vitamin E (mg)	3.3
Iron (mg)	2.25	Vitamin K (µg)	375
Zinc (mg)	0.75	Energy (kcal)	26
Copper (mg)	0.3		

Uses

Practically, almost every part of the young plant is edible but its tender shoots and younger leaves are preferred more. It is consumed differently in Western and Chinese cuisines. The leaves are consumed either uncut or after cutting into small pieces. Like ordinary spinach, the stems, being hard, require slightly longer time for cooking than the leaves. It is better to use water spinach exclusively after stir-frying. The hard stems and leaves are often used for cattle fodder.

Medicinal Uses

The pharmacological effects of water spinach include laxative, anti-inflammatory, antitoxin, diuretic, homeostatic and sedative. It has a soothing effect in the intestine, as it is purgative because of its high fibre content. Its juice is widely used as an alternative remedy for emetic, while its buds are used to treat ringworm. It prevents diabetes as it acts like insulin to control blood sugar. In Philippines, it has been acknowledged to cure constipation, haemorrhoids and is a good food for those who want to lose weight. In *Ayurveda*, the leaves' extract is used to cure jaundice and for the treatment of poison and nervous, liver and eye disorders. Being rich in antioxidants, it neutralizes the super oxides synthesized by the body, which cause degenerative diseases especially cancers.

BOTANY

Water spinach is an herbaceous aquatic or semi-aquatic perennial plant with an adventitious root system. It has long hollow stems, which allow the vines to float on water or creep across muddy ground. The stem exuding a milky juice on breaking may be white or green in colour, depending on the varieties. It generates roots at nodes, when they come in contact of water or moist soil. Depending on the variety, its flat leaves vary in shape from heart-shaped to long, narrow and arrow-shaped, measuring 20–30 cm long and 1–2.5 cm wide, while the broad leaves are up to 15–25 cm long and 5 cm wide. The leaves have a very pleasant, mild, sweet flavour and a slightly slippery texture. Like convolvulus or bindweed flowers, its large flowers are usually white, sometimes with a pinkish centre, attractive, trumpet-shaped and typically opened; however, the wild species may have purple or violet flowers.

CLIMATIC REQUIREMENT

Water spinach is perennial in warm climates and annual in cool climates. In summer, it can be grown in open but cool areas; it can be grown well in heated greenhouse. In tropical climate, it can be grown round the year. It cannot tolerant climates with mean temperatures below 10°C, as the plant growth is ceased below 10°C temperature. The optimal temperature for its successful cultivation is 20–30°C. It can tolerate very high rainfall but it is very sensitive to cold and can easily be killed by even a light frost. Flowering commences from mid-summer onwards under short-day conditions. It prefers full sunshine, but where temperature in summer is very high, it is sometimes grown as a ground cover beneath climbing plants. The crop needs shelter against strong winds for it successful cultivation.

SOIL REQUIREMENT

Although water spinach can be grown on wide range of soils, it performs well in fertile loam or light clay soils rich in organic matter. The most suitable soil pH ranges from 5.5 to 7.0.

CULTIVATED VARIETIES

Water spinach is an underutilized crop at world level, which is why no research work has yet been done for its genetic improvement, but now the crop is attracting the attention of research scientists for systemic and concerted research work. Presently, only two varieties are widely adopted by the growers for cultivation in different regions.

CHING QUAT

The variety, also known as *Green Stem*, has narrow pointed leaves and white flowers and is well adapted to moist soils, thus, it can be grown in seedbeds, having always plenty of moisture.

PAK QUAT

The variety, also known as *White Stem* or *Water Ipomea*, has broad arrow-shaped leaves and pink flowers. Like paddy crop, it is well adapted to aquatic conditions.

SOWING TIME

If the seedlings are to be raised in portrays under protected conditions, the seeds can be sown 1 month before the last frost. However, if the seeds are to be sown directly in the seedbeds, the seeds are sown after the last frost is over. In tropical climate, the seeds can be sown round the year.

PLANTING METHOD

In moist soil, the seeds are sown directly or seedlings are transplanted on 60–100 cm wide raised seedbeds. The seeds should not be more than 2 years old, as they are poor in viability and vigour. Soaking of seeds in ordinary tap water for 24 hours before sowing improves germination. The soil temperature at the time of sowing should be at least 20°C for normal germination of seeds.

To produce healthy and stout seedlings, the seed should be sown at 5–10 mm depth in trays with potting mixture deep enough to allow the plants to develop a good root system. As and when the seedlings attain a height of 10–15 cm or four true leaves, they are transplanted in the seedbeds at a spacing of 15 × 15 cm for getting higher crop yield.

Usually, water spinach is propagated by seeds but it can also be propagated by stem cuttings taken from the young growth just below a node. About 30–40 cm long cuttings with seven to eight nodes are planted 15 cm deep in nursery to prepare plants. To ensure earliness, the growers in China sometimes pick up the roots at the end of growing season, store them carefully in winter and plant the shoots from them in spring. Like paddy planting, cuttings taken from the broadleaved cultivars are directly planted 15–20 cm deep in puddled field at a row spacing of 30–40 cm and plant spacing of 20 cm.

NUTRITIONAL REQUIREMENT

Water spinach responds well to nitrogenous fertilizer application, thus, sufficient quantity of nitrogen fertilizer must be applied to the soil before planting to raise healthy crop. After proper establishment of the plants, nitrogen in ammonium form should be applied at the rate of 40–50 kg/ha, and thereafter, the water level is raised to a depth of 15–20 cm. However, overfeeding of nitrogen should be avoided since high nitrate concentration in leaves and stems is undesirable. Regular applications of a small quantity of organic liquid fertilizers should be applied regularly after every 2 weeks for getting best results in terms of biomass. Liquid fertilizers should usually be diluted before their application.

IRRIGATION REQUIREMENT

For aquatic culture, the field after planting is flooded to a depth of 3–5 cm, keeping the water flowing continuously. In moist soil culture, irrigation is applied after every 1–2 days for high-quality shoots if rainfall is low. However, overwatering should be avoided as it may leach out the readily available nutrients, which will affect the yield adversely. Therefore, slow-releasing fertilizers are used to avoid the loss of nutrients.

INTERCULTURAL OPERATIONS

Water spinach is similar to American lotus (*Nelumbo lutea*), white water lily (*Nymphaea odorata*) and floating bladderwort (*Utricularia radiata*) plants, which grow in fields of each other as weed and cover the field completely, therefore, hand weeding is the only possible solution for the control of these plants in the fields of each other. To manage these plants in the field of water spinach, the weeding should be done at 10- to 15-day intervals.

HARVESTING

Water spinach is usually harvested before the emergence of flowering when the plants have attained a size of about few inches. In semi-aquatic type, the crop becomes ready for harvesting 50–60 days after sowing. For its harvesting, the entire plants are pulled out along with their roots, washed properly, tied in bundles and sent to the market for sale. Several harvestings can also be made if the shoots are cut a few centimetres above ground level, allowing the secondary shoots to grow from nodes below the cut.

In aquatic type, first harvesting can be done after attaining a good growth. The top shoot tips are harvested leaving few inches lower stem intact since pinching of the main shoot stimulates growth of the secondary shoots, which can be harvested at 4- to 6-week intervals, depending on plant vigour and growing conditions. The shoots are harvested at regular interval before they become tough and useless. The interval of harvesting depends on growth rate of the crop. Since consumers in the market prefer a long-stemmed plant, harvesting is done bit late. The harvested shoots are tied in bundles of 8–10 shoots and sent to the markets for sale.

YIELD

The yield varies with several factors such as variety, soil type and its fertility, growing season and cultural practices adopted during the course of crop cultivation. However, on an average, water spinach gives a yield of about 90 to 120 tonnes per hectare per year.

POST-HARVEST MANAGEMENT

The shoot tips and leaves of water spinach deteriorate rapidly, thus, they should be handled swiftly and carefully just after harvesting to avoid quality loss. It is better to consume it fresh soon after harvesting. It can safely be stored at 10°C temperature and high relative humidity with good venti-lation. However, under ordinary conditions, the leaves after bunching are finely sprayed with cold water and kept away from the wind in a cool moist place.

PHYSIOLOGICAL DISORDERS

OEDEMA

Blisters or water-soaked swellings develop on under surface of the leaves, especially when soils are wet and warm and night is cool and saturated. Initially, these blisters are about 1–2 mm in diameter, but later, they become lager. The water-soaked areas later turn tan or brown and become corky. Under severe conditions, the leaves turn yellow and abscise from the plant.

Control

- Ensure optimum watering and temperature.
- Avoid excessive irrigation when the variation in day and night temperature is too much.
- Grow crop at wider spacing so that the plants may get sufficient light.
- In greenhouse, reduce humidity by heating and venting in the evening and early in the morning.
- Use horizontal airflow fans to keep better air circulation in greenhouse.

COLD INJURY

Since water spinach is a plant of tropical and subtropical climate, cold weather, particularly frost, causes the water in plant cells to freeze, which leads to death of the cell inclusions. Frost-injured plants defrost too quickly, killing leaves and stems. The temperature below 10°C results in chilling injury to the plant.

Control

- Provide shelter to the plants against strong cold winds.
- Protect the seedlings and transplants using row covers or plastic tunnels.
- In frost-prone regions, grow the crop in greenhouse.

INSECT-PESTS

A number of insect-pests and diseases attacks water spinach crop during the growing season. Some of the most destructive insect-pests and diseases are given as follows, along with their control measures:

TARNISHED PLANT BUG (*Lygus rugulipennis*)

The adults of this insect feed on succulent parts of the plant, particularly on the new developing leaves. A black dead area at growing tip of the plant is evidence of their feeding. Its feeding develops holes on the leaves, which with the expansion of leaf often enlarge in a somewhat symmetrical pattern on two halves of the leaf and reduce marketability of the greens.

Control
- Eliminate the natural habitat (weeds) surrounding the production areas.
- Use row covers or plastic tunnels to exclude the tarnished plant bugs.
- Install green cross-vane funnel trap in a water spinach growing filed.
- Spray the crop with methyl demeton 0.0025% or phosphamidon 0.05% for effective control.

Golden Tortoise Beetle (*Metriona bicolor*)

Both adult and larvae inhabit the lower surface but eat the foliage entirely. Its adults can change their colour from shiny gold to dull rust and its larvae have the habit of piling frass on their backs. Their feeding creates numerous small to medium-sized irregular holes on foliage.

Control
- Plough the field deep in hot summer months to expose the beetle to natural predators.
- Spray 0.01% *neem* seed kernels extract and 0.4% *neem* oil to control the beetles effectively.
- Spray the crop twice or thrice with Rogor 0.1% near base of the plants.

Cucumber Beetle (*Diabrotica undecimpunctata*)

The adults of cucumber beetle feed on plant leaves and its larvae feed on small plants' roots at seedling stage. However, the large plants with fully developed root system are damaged reasonably less. The adult beetles rarely reach numbers high enough to cause significant damage.

Control
- Plough the field repeatedly in hot summer months.
- Follow phytosanitary measures.
- Grow cucurbits a bit early as trap crop to attract adults.
- Spray 0.01% *neem* seed kernels extract and 0.4% *neem* oil to control the beetles effectively.
- Spray the crop with dimethoate 0.03% at 10- to 15-day intervals.

Aphid (*Mysus persicae*)

The nymph and adult both suck sap from tender parts of the plant. The affected plants remain stunted, leaves curled and plant vigour is reduced. They excrete honeydew-like sugary substance, on which the black sooty mould develops, which checks photosynthesis, and ultimately, it reduces the crop yield.

Control
- Follow clean cultivation.
- Set up yellow pan traps in the field.
- Introduce natural parasites and predators such as lady beetles and their larvae, lacewing larvae, and syrphid fly larvae to control aphids.
- Spray the crop with dimethoate or methyl demeton 0.03% at 10-day intervals.

Wireworms (*Aeolus* spp., *Anchastus* spp., *Melanotus* spp. and *Limonius* spp.)

Adults cause no damage but its larvae damage the young roots during plant establishment. They girdle the stems, which causes death of the seedlings, resulting in reduced plant stand. Its larvae are yellow-brown thin worms with shiny skin and found in soil when the soil is dug around the stem. Its infestation is generally not uniformly distributed throughout the field, thus, the damage is seen in patches, not in the entire field.

Control

- Plough the field deep in hot summer.
- Keep the land fallow to reduce wireworm population.
- Rotate the crop every third or fourth year.
- Avoid inclusion of wireworm-susceptible crops like potato in rotation.
- Apply parathion or ethoprophos granules 4 kg a.i./ha in soil at the time of sowing.

DISEASES

BLACK ROT (*Ceratocystis fimbriata*)

In the beginning, symptoms appear as stunted growth and wilting of plants. When the infection is severe, the leaves become yellow and shrivelled, and ultimately, it causes death of the plant. The disease is primarily seed-borne.

Control

- Follow a crop rotation of at least 3–4 years.
- Use only disease-free healthy seed.
- Collect seed material from disease-free plants.
- Treat the seed with captan or Bavistine fungicide 2 g/kg of seed prior to sowing.
- Treat the cuttings with thiabendizole (Mertect 340-F) as a pre-plant dip to kill the fungus spores contaminating the root surface.

STEM ROT (*Fusarium oxysporum*)

Water spinach plants are more prone to this fungus in aquatic cultivation. The initial symptoms are small irregular black lesions on leaves near water level, which expand with the advancement of the disease. Infected stems start rotting. Numerous tiny white or black sclerotia along with mycelium can be seen inside the infected leaves.

Control

- Rotate the crop every third or fourth year.
- Select the disease-free propagating material carefully.
- Burn the straw, stubble and any other crop residues after harvesting.
- Drain out the field water to reduce sclerotia.
- Use clean land or fumigate the soil with proper fumigates like formalin or methyl iodide.

■■■

31 Arugula

M.K. Rana and Sunny Kumar

Botanical Name	: *Eruca sativa* (Miller) Thell.
	Eruca vesicaria
Family	: Cruciferae (Brassicaceae)
Chromosomes Number	: 2n = 22

ORIGIN AND DISTRIBUTION

Arugula is native to Israel. However, N.I. Vavilov considered the northwest regions of India, Tadzhikistan, Uzbekistan, West Tien Shan province and Southwest Asia the centres of origin and temperate regions around the world, including Northern Europe and North America the the secondary centre of origin. The species *Eruca vesicaria* synonym *Brassica vesicaria* L. is considered a native to Western Mediterranean region, Morocco, Algeria, Spain and Portugal and closely related to *Eruca sativa*. The subspecies *vesicaria* and *pinnatifida* (Desfontaines) are endemic to Spain and North Africa, which are rarely cultivated in Europe. Some of the botanists considered *Eruca sativa* as a subspecies of *E. Vesicaria*, but later, *Eruca sativa* (having persistent sepals) was distinguished from *E. vesicaria* because of having early deciduous sepals. However, others still do not consider any difference between these two species.

INTRODUCTION

Arugula, also known as garden rocket, rocket salad, *taramira*, *roquette*, *rugula*, or *rucola*, is one of the most nutritious green-leafy vegetables. It is an aromatic, pepper-flavoured salad greens, which is popular in Italian cuisine and has become ubiquitous in the United States of America. Its pungent leafy greens resemble lettuce. Besides leaves, its flowers, young pods and mature seeds are also edible. The young tender leaves are sweet-flavoured and less peppery in contrast to mature greens, which are of strong spicy flavour. It has been grown as an edible herb in the Mediterranean area since Roman times. In north plain areas of India, arugula flowers in December–January. Its mature seeds in India are known as *gargeer*, which in West Asia and Northern India are pressed to expel *taramira* oil. In India, arugula is mainly grown in Rajasthan, Haryana, Punjab, Madhya Pradesh and Uttar Pradesh. Rajasthan has the highest area under arugula cultivation. Arugula is also found in Western Himalayas up to 3000 m.

COMPOSITION AND USES

COMPOSITION

Arugula is a very low-calorie vegetable, but it provides immensely health-benefiting phytochemicals, antioxidants and vitamins such as vitamin A, thiamine, riboflavin, niacin, pantothenic acid, pyridoxine, vitamin C and vitamin K. Its fresh leaves contain adequate amount of minerals, especially copper, iron and small amount of some other essential minerals and electrolytes such as calcium, potassium, manganese and phosphorus. Like other leafy greens, arugula contains very high nitrate levels (more than 250 mg/100 g). The nutritional composition of arugula leaves is given in Table 31.1.

TABLE 31.1

Nutritional Composition of Arugula Leaves (per 100 g Edible Portion)

Constituents	Contents	Constituents	Contents
Water (g)	91.8	Copper (mg)	0.076
Carbohydrates (g)	37	β-carotene (µg)	1424
Protein (g)	27	Vitamin A (IU)	2373
Fat (g)	0.2	Thiamine (mg)	0.044
Dietary fibre (g)	0.90	Folic acid (µg)	97
Calcium (mg)	160	Pyridoxine(mg)	0.073
Phosphorus (mg)	46	Choline (mg)	19
Magnesium (mg)	45	Vitamin C (mg)	15
Potassium (mg)	7.4	Vitamin K (mcg)	102
Sodium (mg)	0.5	Ash (g)	1.6
Iron (mg)	0.8	Energy (kcal)	21

USES

Arugula fresh and tender leaves often together with flowers, young pods and mature seeds are frequently consumed in salad. It is also used as raw with pasta or meats in northern Italy and in western Slovenia. In Italy, raw arugula is often added to pizzas just before the baking ends or immediately afterwards so that it may not wilt in the heat. Large amounts of coarsely chopped arugula leaves are added to pasta seasoned with a homemade tomato sauce and pecorino as well as in many unpretentious recipes. Its seeds are used as a condiment for cold meats, fish and seafood dishes. The mature arugula seeds are pressed to extract oil, which is used in pickles and as cooking oil. The seed cake is used as animal feed.

MEDICINAL USES

The phytochemicals such as *indoles*, *thiocyanates*, *sulforaphane* and *isothiocyanates* present in arugula help in countering the carcinogenic effects of oestrogens and thus help in protecting against prostate, breast, cervical, colon and ovarian cancers by virtue of their cancer-cell growth inhibition and cytotoxic effects on cancer cells. Being rich in vitamin C and vitamin K, its leaves help protect the human body from scurvy disease, develop resistance against infectious agents by boosting immunity and scavenge harmful and pro-inflammatory free radicals from the body. Its chlorophyll has been shown effective at blocking the carcinogenic effects of heterocyclic amines generated when grilling food at high temperatures. Leafy greens containing alpha-lipoic acid have been shown to lower glucose levels, increase insulin sensitivity and prevent oxidative stress-induced changes in patients with diabetes.

BOTANY

Arugula is a fast-growing annual herb with 70–75 cm height. Its branched stems are usually glabrous, hirsute or hispid. Leaves are variable in size like biseriate, sessile, or alternate and the petiole 0.2–0.4 cm long with membranous and very short stalks. The leaves are deeply and pinnately lobed with 4 to 10 small lateral lobes and a large terminal lobe. Basal leaves occur in a rosette and a lyrate-pinnatifid, and the cauline leaves are lobulate and dentate. Inflorescence is a terminal raceme without bracts. The flowers are bisexual, 2–4 cm in diameter and arranged in a corymb with creamy white petals and yellow stamens. The sepals are plunged down soon after the flower opens. Flowers having genetic male sterility are allogamous with a complex system of self-incompatibility (gametophytic). The fruit is a siliqua (pod) up to 4 cm long with an apical beak, containing several edible seeds. Seed colour is pale or greyish brown.

CLIMATIC REQUIREMENT

Arugula is a cool season crop, which performs best in autumn–winter under subtropical conditions, while in temperate regions, it is grown in late fall or early spring. Too much drought and heat may cause the leaves to be smaller and more peppery. Temperature beyond 30°C is conducive for bolting. Growing arugula under mild frost conditions hinders the growth of the plant, as well as turns the green leaves to red. The seed germination takes about 3–4 days at 25°C and 4–5 days at lower temperature. The crop needs sunny location for its cultivation.

SOIL REQUIREMENT

Arugula although can be grown on a wide range of soils but sandy-loam to loam soils are best suited for its cultivation, while a high concentration of soluble salts in soil has adverse effects on growth and development of the plant. Its growth and yield are more when it is cultivated in well-drained fertile soils rich in organic matter and good in water-retention capacity. It may not grow vigorously in poor or depleted soils unless a considerable quantity of organic matter and other nutrients is incorporated. The ideal soil pH for its cultivation is 6 to 8.

CULTIVATED VARIETIES

Arugula is very less-preferred minor crop of the country and cultivated on a small area. Thus, very little research work on the improvement of this crop has been done and only a few cultivars have been developed for cultivation in some potential regions of the northern plains.

TMLC-2

A variety developed at Punjab Agricultural University, Ludhiana has longer main shoot length, more number of pods on main shoots and more seeds per pod. It is recommended for cultivation in the areas of Bathinda, Sangrur, Ferozpur, Hoshiarpur and Ropar districts of Punjab. It gives a seed yield of 5–6 q/ha in crop duration of 150 days.

Other varieties of arugula are Karan Tara (RTM 314), RTM 1212, RTM 1301, RTM 603, RTM 673, RTM 1107, RTM 1356, RTM 1354, TMC 1, ITSA, T 27, T 42 and Jawahar Toria 1.

SEED RATE

The seed rate undoubtedly varies with soil fertility, germination percentage, irrigation facilities and climatic conditions of the growing region. Generally, about 3–5 kg of seed is enough for rain-fed areas to sow a hectare land area and 5–6 kg for irrigated areas.

SOWING TIME

For vegetable purposes, arugula is generally sown in the months of September to November in northern regions of the country, and in a temperate climate, it is sown in late winter or early spring, depending on altitude. In Rajasthan, it is sown in the months of November–December, in Haryana, October–November, and in Punjab, it is sown in October.

SOWING METHOD

Arugula is propagated by seed since raising crop by seed is much easier and more common. The seeds for greens are sown 0.5–1.0 cm deep in rows 20–30 cm apart. Within the row, the plant spacing is about 5–7 cm. A good yield can be obtained by line sowing over broadcasting method of sowing. The soil at the time of sowing must have plenty of moisture for better germination of seeds. Soaking of seeds in water for 6 h and sowing in the evening improve seed germination.

NUTRITIONAL REQUIREMENT

The crop does not demand much quantity of fertilizers for its growth and yield. However, the application of small amount of farmyard manure along with nitrogen, phosphorus and potash fertilizers undoubtedly improves the yield and quality of the crop. The farmyard manure should be applied @ 4 t/ha at the time of land preparation. Although there is no such standard recommendation of fertilizers for this crop, application of nitrogen 60 kg, phosphorus 40 kg, potassium 40 kg and sulfur 20 kg/ha gives higher greens yield. The entire amount of phosphorus and potash along with one-third dose of nitrogen is applied as basal dose at the time of sowing and rest of the nitrogen is applied as top dressing after taking first and second cuttings.

IRRIGATION REQUIREMENT

Although arugula is a drought-resistant crop but irrigation during warmer days increases the number of cuttings. After every clipping, it is essential to apply irrigation regularly during rain-free periods so that the new growth may emerge out. Arugula requires 250–400 mm water to its entire growing period but the number of irrigations varies with the number of cuttings. First irrigation should be applied at branching stage, *i.e.*, 30 days after sowing, and second at 60 days after sowing, as it helps the plants to branch well, which results in higher greens yield, profuse flowering and fruiting. At pre-flowering stage, a single irrigation is sufficient for optimum crop performance.

INTERCULTURAL OPERATIONS

Arugula possesses medicinal properties, hence, weeds do not pose much problem in this crop, even though, removal of weeds periodically in early stages of crop growth is essential to reduce competition with crop for nutrients, moisture, light and space, resulting in better growth of the crop. The most common weeds of arugula are *Chenopodium album* (*bathua*), *Lathyrus* spp. (*chatrimatri*), *Melilotus indica* (*senji*) and *Cyprus rotundus* (*motha*). Initial 45 to 60 days are critical for crop weed competition. Weeds act as alternate host for many insect-pests and diseases. If weeds are not controlled in this period, they may cause 20–70% yield loss. The weeds in arugula are usually removed manually by hoeing. Under rain-fed conditions, hoeing is done 25–30 days after sowing, and under irrigated conditions, hoeing 25 and 45 days after sowing is necessary for controlling weeds effectively.

The chemicals can be applied as pre-sowing, pre-emergence and post-emergence. Pre-sowing application of fluchloralin 1.0 kg/ha or pre-emergence application of pendimethalin 1.0 kg/ha controls the weeds effectively in arugula. If weeds emerge after sowing, isoproturon 0.75 kg/ha can be applied 30 days after sowing to control the weeds effectively.

HARVESTING

First harvesting of leaves can be done 50 days after sowing by pinching off the lower leaves with fingers or snipping them with a pair of scissors. The centre of the plant is left undisturbed to promote new growth. To avoid high nitrate content, the leaves should not be allowed to get too old and harvesting must be done in evening hours.

For seed purpose, the crop is considered mature and ready for harvesting when the siliquae start turning their colour from green to brownish yellow and the seeds inside the pod become slightly dry. In general, the crop matures in about 140 to 150 days and is harvested manually using sickles. Harvesting should always be done in morning hours to avoid shattering. The harvested material is brought to threshing floor without any delay to reduce seed moisture content to 8% and for threshing. After winnowing, cleaning and grading, the seed is stored in a cool and dry place until the seed is processed for the extraction of oil.

YIELD

The greens yield of arugula depends on several factors such as variety, soil type, soil fertility, growing season, irrigation facilities and crop management. The average greens yield of arugula varies from 3.5 to 4 kg/m^2 and 300 to 500 q/ha. A single plant gives a yield of fresh green biomass about 1–3 kg per plant per year. Its average seed yield is 5–6 q/ha.

POST-HARVEST MANAGEMENT

The simplest and easiest way to clean the leaves is to place them in large bowl under tap water and wash them properly with hands. This will remove sand or dirt particles from the arugula leaves. This process is repeated until no sand or dirt particle remains in the water. Before washing, the diseased, misshaped and yellow leaves are discarded from the lot. After washing and proper drying the arugula leaves, small bundles of one-half to 1 kg weight are made for sale in the market.

INSECT-PESTS

Arugula is not much seriously damaged by insect-pests but some insects like the aphid, mustard sawfly, painted bug, flea beetle and Bihar hairy caterpillar sometimes do considerable damage to the crop. The description of these insect-pests along with their management is given as follows:

APHIDS (*Aphis malvae*)

These are tiny sucking-type green-colour insects attacking the tender growing leaves. The damage is caused by direct as well as indirect methods. The nymphs and adults attached with growing shoots and leaves suck cell sap and thereby reducing vigour and growth of the plant. The leaves become poor in quality and unfit for consumption. They also act as a virus vector, resulting in curling of leaves, reduction of leaf area and stem size. Heavily infested plants remain stunted and bear no pod. The aphids secrete honeydew, which invites fungi to grow on it. Its severe infestation causes wilting and death of the plant.

Control

- Grow early bulking and maturing variety.
- Adjust sowing time to avoid the attack of aphids.
- Spray simple water with strong jet pressure to wash the aphids.
- Remove and destroy all the infested leaves during early stage of attack.
- Spray *neem* oil or *neem* seed kernel extract 5% before egg laying.
- Spray the crop with malathion 50 EC 0.1%, Metasystox 25 EC 0.025%, monocrotophos 0.04%, or methyl demeton 0.05%.

MUSTARD SAWFLY (*Athalia proxima*)

This is most destructive pest of arugula particularly at seedling stage. The larvae feed on the leaves, make holes in the leaves and eat up the entire leaf lamina leaving behind the midrib. When the infestation is severe, the plant is completely killed. This pest appears in early stages of the crop growth, *i.e.*, October–November. The adults are orange yellow wasps with smoky wings, black head and legs. Its larvae are light green to dark green in colour. The size of full-grown larvae is 1 to 1.5 cm.

Control

- Plant the crop timely.
- Avoid the excessive use of nitrogenous fertilizer.
- Spray *neem* oil or *neem* seed kernel extract 5%.
- Irrigate the crop regularly, which helps in killing the larvae through sinking.
- Spray the crop with malathion 0.1% or dichlorvos 0.05%.

Painted Bug (*Bagrada cruciferum*)

This insect suck cell sap from the leaves and tender stems at the time of germination, resulting in pale yellow colour of the leaves, poor growth and ultimately death of the plant. Sometime, it results in complete failure of the crop in case of severe infestation. The painted bug appears during October–November and March–April when the maximum temperature is about 30°C. Its adults are oval-shaped and flat-bodied with pale and orange markings. The nymphs are small and wingless. Adults and nymphs suck cell sap from tender parts of the plant. Infested leaves become bushy with white blotchy spots. Its attack in later stage results in loss of oil.

Control

- Avoid the sowing of crop too early when the day temperature is above 30°C.
- Apply first irrigation 20–30 days after sowing.
- Collect and burn the infected plant parts.
- Treat the seed with imidacloprid 70 WS @ 5.0 g/kg of seed.
- Spray the crop with malathion 50 EC 0.1%.

Flea Beetle (*Phyllotreta* sp.)

The flea beetle attacks the leaves of arugula. The affected leaves become peppered with small holes and the damaged areas turn brown. This pest damages the vegetative parts of the plant by making small circular holes in them and it becomes very active in sunny days during spring months. When its population is high, it causes defoliation and death of the plant. Adults are small, shiny, dark brown or black beetles with large hind legs that allow them to jump when disturbed. Some species may have white or yellow stripes on their wings. Larvae are small cream-colour worms. They live underground and feed on the roots and underground parts of young plants as well as on germinating seeds.

Control

- Follow phyto-sanitary measures.
- Pick the beetles manually by hand and drop them into soapy water.
- Remove the alternate hosts from the surrounding of crop.
- Grow trap crops like marigold (125 plants/ha).
- Use pheromone traps (10 traps/ha) to trap the male moths.
- Set light traps (one light trap/2 ha) in the field to kill the moths.
- Release predators like coccinellids, *Chrysoperla carnea* @ 1 larva/head.
- Release parasitoids, *i.e.*, *Trichogramma* spp., *Bracon* spp., or *Campoletis* spp. (50,000/ha).

- Spray *neem* oil or *neem* seed kernel extract 5% before egg laying.
- Spray the crop with malathion 0.2% or monocrotophos 0.04%.

Bihar Hairy Caterpillars (*Spilosoma oblique*)

The caterpillars feed voraciously on leaves and make holes in the leaves, which deteriorate their quality. The infested leaves become papery, devoid of chlorophyll and almost transparent. The adult moth is dull coloured with red abdomen. The body of the larvae is covered with long hairs, which make their preying difficult by predators. The size of full-grown larva is 4–5 cm.

Control

- Remove the alternate hosts from the surroundings.
- Remove the infested leaves containing larvae and dip them in insecticidal solution.
- Avoid the use of undecomposed manure.
- Intercrop with pigeon pea at a row ratio of 2:1 to reduce the insect attack.
- Spray the crop with malathion 0.1% or phosalone 35 EC 1 litre/ha.

DISEASES

White Rust (*Albugo candida*)

The disease occurs in all cruciferous growing areas of the India and the extent of damage varies from 17% to 55%. The pathogen survives through oospores on crop residues, in soil and as a contaminant of seed. High moisture and cool temperature encourage the growth and spread of the disease. Pustules of 1–2 mm diameter having white to creamy yellow colour appear on leaves, stems, inflorescence and floral parts. These pustules coalesce to form large patches and liberate a white powdery mass on bursting. Swelling and distortion of the stems and floral parts result in hypertrophy and hyperplasia, commonly known as stag head.

Control

- Follow long crop rotation with non-cruciferous crops.
- Practice deep ploughing during summer to destroy the fungus.
- Always use disease-free seed to avoid the incidence of this disease.
- Sow seeds in the field timely.
- Apply a balanced dose of fertilizers.
- Avoid over-irrigation and drain out excess water.
- Remove and destroy the infected leaves.
- Treat the seed with metalaxyl (Apron 35 SD) 6 g/kg seed or thiram 2.5 g/kg seed.
- Spray the crop with 0.25% Ridomil MZ-72 followed by a spray of copper oxychloride 0.2%.

Downy Mildew (*Peronospora parasitica*)

This is a most destructive disease, which causes much damage in chenopod growing areas all over the world. The pathogen may attack the plant at any stage of growth, even at flowering stage. The leaves affected by this fungus show light yellow irregularly shaped areas without distinct margins on their surface. Whitish-grey mycelium grows on leaf surface under humid conditions, while the mycelium is found absent in low-humidity atmospheres. The warm and moist weather is most congenial for the development of the disease. The infection of downy mildew occurs primarily in the field but it also becomes prominent after harvesting of crop, especially at 25–28°C temperature and relative humidity >85%.

Control

- Follow long crop rotation with cereal crops.
- Plough the field deep during hot summer to destroy the fungus.
- Always use disease-free seed to avoid the incidence of this disease.
- Plant the crop at wider spacing to avoid overcrowding of the plants.
- Spray the crop with Dithane M-45 or Dithane Z-78 0.2% at 10- to 15-day intervals.

POWDERY MILDEW (*Erysiphe cruciferarum*)

The disease first appears on lower leaves of the plant near the soil surface through *Cleistothesia* present in soil. Dirty white circular floury patches develop on leaves, stems and siliquae. As the disease advances, the whole plant looks to be dusted with white talcum-like powder. Minute spherical black fruiting bodies called *Cleistothesia* can be seen on the affected parts. The disease usually appears late in the season when temperature begins to rise, as high temperature (15–28%) and low relative humidity (60%) favour the development of this disease.

Control

- Follow long crop rotation with non-cruciferous crops.
- Plough the field repeatedly during hot summer to destroy the fungus.
- Always use disease-free seed to avoid the incidence of this disease.
- Spray wettable sulfur 0.2 %, dinocap (Kerathane) @ 0.1 %, or carbendazim (Bavistin) @ 0.05% at the initiation of disease.

BACTERIAL LEAF SPOT (*Pseudomonas syringae* and *Xanthomonas axonopodis*)

Small water-soaked irregular spots are formed on leaves, which later increase in number and turn brown under favourable conditions. These spots in congenial atmosphere coalesce and make large lesions. In later stages, the prominent symptoms are vascular blackening, stunted growth, wilting and stem rot. As the pathogen proceeds from leaf margins to the veins, water stress and chlorotic symptoms appear due to the constriction of water-conducting vessels. Ultimately, defoliation occurs. The optimum temperature for bacterial growth is 25 to 30°C. Temperature below 5°C and above 35°C restricts growth of the bacteria. However, the infected hosts do not show any symptom below 18°C.

Control

- Follow at least 3–4 years' crop rotation with non-host crops.
- Practice deep ploughing during summer season to destroy the fungus.
- Use only certified healthy and disease-free seeds.
- Treat the seed with hot water at 52°C for 15 minutes.
- Treat the seed with sodium hypochlorite, hydrogen peroxide, hot cupric acetate, or zinc sulfate.
- Remove cruciferous weeds from the arugula field and its surroundings.
- Avoid excessive supply of irrigation and provide good drain facilities.
- Uproot and destroy the infected plants immediately to prevent spreading of the disease.

■■■

32 Sprouting Broccoli

Sonia Sood and M.K. Rana

Botanical Name : *Brassica oleracea* var. italica

Family : Cruciferae (Brassicaceae)

Chromosome Number : 2n = 18

ORIGIN AND DISTRIBUTION

Broccoli, commonly known as green cauliflower in India, was introduced from the United States of America and Europe. Botanically, it is known as *Brassica oleracea* var. italica, belonging to the family Brassicaceae with chromosome number 2n = 18. It is one of the important members of the cole group of vegetables, having a common progenitor, *i.e.*, wild cabbage (*Brassica oleracea* var. oleracea or sylvestris), which is found growing wild in southern Europe and along the Eastern Mediterranean Sea. In India, it was introduced during the second half of the 20th century.

INTRODUCTION

Broccoli also known as Calabrese, Italian Green, or Spear Cauliflower, is a nutritive green vegetable, and morphologically, it is quite similar to cauliflower but the plant forms heads rather than curd, consisting of green, purple, or white buds and a thick fleshy floral stalk. Besides the main head, long slender smaller heads, which are called spears, develop in the axils of leaves. It is of two types, *i.e.*, (i) green heading broccoli and (ii) purple sprouting broccoli. However, in India, the sprouting type broccoli is more popular. Heading broccoli consists of immature differentiated flower buds (winter cauliflower), while sprouting broccoli contains closely packed mature differentiated flower buds and a thick, edible, sturdy, fleshy flower stalk forming the head. In India, the cultivation of sprouting broccoli has not yet been exploited commercially and is limited to kitchen and market gardens located near the big cities and tourist areas. However, its popularity is increasing gradually because of awareness among the educated masses about its high nutritive value, cancer prevention properties, its potential in the developing tourism industry and also as a relatively new crop.

In India, broccoli production stands at fourth place after cauliflower, cabbage and kohlrabi, though no figure for area utilization and volume of the crop is available. It is mostly cultivated in hilly tracts of Himachal Pradesh, Uttar Pradesh, Jammu and Kashmir, Nilgiri hills and northern plains of India. The United States of America is the largest producer, though Italy, Northern Europe and cooler regions of Far East are also major producers of broccoli.

COMPOSITION AND USES

COMPOSITION

Sprouting broccoli is one of the low-calorie vegetables and the richest source of nutrients among the cole crops because of its relatively higher quantity of minerals, vitamins and proteins, especially minerals like phosphorus, potassium, calcium, sodium, iron and other trace elements and vitamin

A, B and C. In addition, it is low in carbohydrates and fat. Broccoli leaves are an excellent source of carotenoids and vitamin A, several times greater than that of their flower heads. Fresh heads are an excellent source of B-group vitamins like riboflavin (vitamin B_2), niacin (vitamin B_3), pantothenic acid (vitamin B_5), pyridoxine (vitamin B_6) and folate (vitamin B_9) and vitamin K. The flower heads also have some amount of omega-3 fatty acids. The nutritional composition of broccoli is mentioned in Table 32.1.

USES

Sprouting broccoli is consumed fresh or processed (frozen). Both its heads along with fleshy stems and spears are cooked and consumed like cauliflower. It is consumed as salad, soups, or cooked vegetable singly or mixed with other vegetables. Because of their high nutritional value, the leaves can also be used for human nutrition as a substitute of kale and the stems for the feeding of animals.

MEDICINAL VALUE

Broccoli is a storehouse of many phyto-nutrients such as thiocyanates, indoles, sulforaphane, iso-thiocyanates and flavonoids like beta-carotene, cryptoxanthin, lutein and zeaxanthin. The American Cancer Society also has suggested inclusion of broccoli in diet to reduce the risk of developing cancer. It also contains a high concentration of carotenoids, which are believed to be chemo-preventive and associated with decreased risk of various human cancers (prostate, colon, urinary bladder, pancreatic and breast cancers), diabetes, heart disease, osteoporosis and high blood pressure. Broccoli may also play a role in reducing the levels of serum cholesterol. It is rich in glucosinolates (40–80 mg/100 g of edible portion). The dominant glucosinolates in broccoli are glucoraphanin, glucorapin and glucobrassicin. At high intake levels, certain glucosinolates are associated with toxic effects, especially goitre development.

BOTANY

Sprouting broccoli plant grows erect about 60 cm high with large flower heads, which are arranged in a tree-like structure on branching sprout from a thick edible stalk. Pods are linear, cylindrical, or conical, which dehisce longitudinally. Flowers are yellow and seeds seriate. The inflorescence is a raceme. Flowers are actinomorphic and bisexual (complete). Calyxes are gamosepalous with four sepals and the corolla gamopetalous with four petals and cruciform. Pollination is entomophilous, which takes place through honeybees. It is highly cross-pollinated crop due to sporophytic self-incompatibility, which is determined by up to 10 alleles at S locus. The S alleles, which remain

TABLE 32.1
Composition of Sprouting Broccoli (per 100 g Edible Portion)

Constituents	Contents	Constituents	Contents
Water (g)	89.2	Sodium (mg)	23.0
Carbohydrates (g)	5.7	Iron (mg)	1.2
Proteins (g)	3.6	Vitamin A (IU)	3150
Fat (g)	0.35	Thiamine (mg)	0.10
Dietary fibre (g)	1.4	Riboflavin (mg)	0.23
Calcium (mg)	101	Niacin (mg)	1.0
Phosphorus (mg)	77	Vitamin C (mg)	109
Potassium (mg)	389	Energy (kcal)	32

inactive in bud stage, *i.e.*, 2–4 days before anthesis, become active in stigma at flowering stage. Anthesis takes place early in the morning between 7:30 and 8:00 a.m. and remains continued up to 11:30 a.m., depending upon prevailing weather conditions. Pollen fertility is maximum on the day of anthesis. Stigma is receptive 2–3 days before the day of anthesis due to protogyny condition of the flowers and remains receptive even up to 4 days after anthesis. The glycoproteins present in the stigma hinder the pollen germination and penetration of pollen tubes through the styler tissue. Cytoplasmic male sterility system in which male sterility is manifested due to interaction of cytoplasm, and nuclear gene is generally utilized in hybrid seed production of broccoli. The cytoplasm of Ogura male sterile radish has successfully been utilized in developing cytoplasmic male sterility in broccoli.

CLIMATIC REQUIREMENT

The climatic requirement of sprouting broccoli differs from cabbage and other Cole crops. Even different cultivars of same crop require different climatic conditions, especially temperature, for the development of their edible portion and conversion of vegetative to reproductive phase. Broccoli is relatively tolerant to environmental stress. Being cool hardy, it is grown as a winter crop in the plains, low and mid hills of north (up to 1500 m above mean sea level) and during spring–summer months in higher hills. It thrives well in a cool humid climate and can withstand light frost. Flavour (fresh and sweet taste) improves with cooler temperature because plant cells convert starch to sugar to protect the plants against cold. The low temperature injury to the developing heads causes considerable damage, resulting in freezing, browning and ultimately rotting of bud clusters. High temperature is responsible for the loosening of heads, premature opening of buds and the yellowing of petals exposed over the head surface. A temperature between 12°C and 18°C is suitable for proper head formation, comparable with mid–late group of Indian cauliflower (maturity group 3). Sprouting broccoli is an annual crop, which does not require vernalization (chilling temperature) for the initiation of bolting (development of flower stalk) to enter generative phase, whereas, heading broccoli is biennial and requires vernalization (low temperature below 10°C) for heading and to enter into reproductive phase.

SOIL REQUIREMENT

Broccoli grows in nearly all types of soil, provided adequate water supply is available. Well-drained loamy to sandy loam soils rich in organic matter are ideal for its cultivation. The plant is slightly tolerant to acidic soil, but the optimum soil pH for its cultivation is 6.0–6.5. Less fertile soils should be supplemented with manure and fertilizers so that the crop may be able to make rapid and uniform development immediately after transplanting. Dry soil conditions result in increasing fibre content in the shoots. Its plants in acidic soils with pH less than 5.5 demand relatively higher molybdenum, which can be supplied through foliar spray of sodium molybdate or liming the soil. The crop raised in alkaline soil requires addition of boron (addition of borax or foliar sprays of boron). Stagnation of water causes injury to the root system of the plant.

The field is prepared well in advance with one deep ploughing and 2–3 light ploughings followed by planking as essential to obtain a fine tilth. All weeds and stubbles are removed from the field and well-rotten farmyard manure is incorporated in the field thoroughly.

CULTIVATED VARIETIES

The varieties of sprouting broccoli are grouped according to their maturity, *i.e.*, early, medium and late maturing types. The early maturing cultivars are annual, requiring no vernalization to initiate inflorescence and flowering, while the late maturing cultivars are biennial type requiring vernalization to produce inflorescence. Both hybrids and open pollinated varieties are available in our country. Hybrids more commonly grown in agriculturally advanced countries have gained popularity because of their higher yield, uniformity in size and maturity and better quality produce. Some Japanese and European seed companies are also marketing hybrid cultivars in India. The open pollinated varieties released from the public-funded institutes are as follows:

PUSA BROCCOLI KTS-1

A variety developed at Indian Agricultural Research Institute (IARI), Vegetable Research Station, Katrain, through selection from a segregating exotic germplasm has medium tall (40–50 cm) plants. Its foliage is waxy and dark green with slightly wavy margins. Heads are solid green with small beads slightly raised in the centre. Its average main head weight is 350–450 g. It takes 90–105 days after transplanting to be mature in temperate climate, which may be 5–10 days earlier in the tropical climate. Its average yield is 160 q/ha.

PALAM SAMRIDHI

A variety developed at Himachal Pradesh Krishi Vishva Vidyalaya, Palampur (Himachal Pradesh), through selection from an exotic material produces compact green heads free from yellow eyes and bracts. The average large terminal head weight is 300–400 g. Plants are branched, bearing sprouts in the axil of leaves, which develop into small heads. Spears are also produced. The average yield varies from 150 to 200 q/ha. It is ready to harvest in 85–90 days after transplanting.

PALAM HARITIKA

A green sprouting broccoli with delayed maturity and dark green upright leaves with purple reddish tinge produces tender and crisp heads, which are full of flavour and suitable for salad as well as cooking. It attains marketable maturity in 145–150 days. Its average yield potential is 175–225 q/ha.

PALAM VICHITRA

A variety with open dark leaves developed at Himachal Pradesh Krishi Vishva Vidyalaya, Palampur (Himachal Pradesh) through recurrent breeding produces purple heads. Purple tinge is prominent on stem at seedling as well as on leaf margins at young plant stage. The head is medium sized, purple coloured, compact and attractive, attaining marketable maturity in 115–120 days after transplanting with an average yield potential of 225–250 q/ha.

PALAM KANCHAN

A heading type variety developed at Himachal Pradesh Krishi Vishva Vidyalaya, Palampur (Himachal Pradesh) has long, broad, bluish green upright leaves with prominent white midrib and veins. The heads are compact, attractive and yellowish green, attaining marketable size in 140–145 days after transplanting with an average yield potential of 250–275 q/ha.

PUNJAB BROCCOLI-1

A variety developed at Punjab Agricultural University, Ludhiana through selection is recommended for growing in Punjab. Foliage is dark green with smooth leaf surface and wavy margin with a bluish tinge. Main head and spears in the leaf axils are dark green with bluish tinge. It takes 65–70 days after transplanting to harvest and its average yield is 70 quintals per hectare.

The hybrids are also being sold in India by the private seed companies *viz.*, Fiesta and Lucky by Bejo Sheetal Seeds Private Limited, Green Beauty by Doctor Seeds Private Limited, CBH-1 and CBH-2 by Century Seeds Limited, Puspha and Prema by Seminis Vegetable Seeds India Limited, Premier, Premium Crop and Titleist by Pahuja Takii Seed Limited, Tahoe and Monfalis by Rizwan Seeds Company and Royal Queen by Rajendra Hybrid Semences Private Limited.

SEED RATE

The seed rate depends on the variety, plant spacing and real value of the seed. About 300–400 g seed is required to raise seedlings sufficient for transplanting a hectare area. A quantity of one gram contains 350 seeds. For direct sowing, the seed rate is about 1.2–1.6 kg/ha.

RAISING SEEDLINGS

Broccoli is propagated by seed, which is sown on well-drained finely prepared raised nursery beds, which are of 3.0 × 1.0 × 0.15 m size. Ten nursery beds are required to produce seedlings sufficient for planting a hectare area. The land selected for nursery should have sandy loam soil and plenty of sunlight and should be ploughed and prepared thoroughly. Fine and fully decomposed farmyard manure or compost @ 20–25 kg and single super phosphate @ 200–250 g per bed are incorporated in top 5 cm soil at the time of bed preparation. Alternatively, 50 g of (19:19:19) complex fertilizer, 200 g powdered *neem* cake and 10 g Furadan 3G per bed are added and mixed thoroughly in soil. The nursery bed should be disinfected with 0.3% solution of captan, thiram, mancozeb, or Bavistin as 2% solution @ 5 litre/m^2. Chlorpyrifos @ 2.5 ml/litre of water may be used to protect the nursery from soil-borne insect-pests.

Seeds are sown in rows 5.0 cm apart at a depth of about 1.0 cm. The beds after sowing are covered with sieved mixture of well-decomposed manure and soil in equal proportion, and thereafter, the nursery beds are covered with straw or long dry grass until seed germination. Polyethylene sheet can also be used to cover the beds except when protection from sun heat is required. Water is applied every day with fine rose can for first few days, and thereafter, light irrigation is provided to ensure uniform germination and good growth. Grass or polyethylene mulch must be removed as and when the seedlings start emerging. Light hoeing and weeding are needed. The nursery beds may be drenched with mancozeb, Ridomil MZ 72, or captan @ 0.25–0.3% at weekly interval. The nursery beds are protected from hot sunshine and heavy rains. Hardening of seedlings before transplanting is essential for enduring better the transplanting shock and setting in the field. Hardening is done by withholding water 4–5 days before uprooting the seedlings or adding 4000 ppm sodium chloride, or spraying 2000 ppm Cycocel (CCC). The nursery beds are made thoroughly moist 2–3 hours before lifting the seedlings to avoid root damage.

NUTRITIONAL REQUIREMENT

Fertilizer dose recommended for cauliflower is commonly applied to broccoli crop also. Besides the application of farmyard manure @ 20–25 t/ha, nitrogen is applied @ 150 kg, phosphorus 100 kg and potassium 50 kg/ha, depending upon soil fertility status and cultivars. Full dose of phosphorus and potash fertilizers along with one-third of nitrogenous fertilizer is applied as basal application. The remaining two-thirds dose of nitrogenous fertilizer is top-dressed in two equal splits within 4–6 weeks of transplanting and just before the initiation of heading. Weekly spray of 2% urea 20 days after transplanting increases the biomass of whole plants, marketable yield and total antioxidant activity.

Like cauliflower, broccoli is also sensitive to the deficiency of micronutrients, especially molybdenum, in highly acidic soils and boron in alkaline soils. Application of 1.0 kg sodium or ammonium molybdate and 10–15 kg borax is recommended in such soils. The highly acidic soils (pH <5.5) may be corrected by liming the soil to increase the availability of molybdenum as soils with pH more than 7 limit the availability of boron.

SOWING TIME

The time of sowing depends on variety, growing season and location. In north Indian plains, the optimum time for sowing seeds in nursery beds is from mid-August to mid-September. In hills where snowfall is rare or mild, seed sowing is carried out during July–August. In higher hills where snowfall is severe, nursery sowing is done in March–April. Mid-September at Bajaura (Kullu valley) and second fortnight of October to the first week of November at Dhaulakuan (low hills) have been reported to be the best period for transplanting broccoli in Himachal Pradesh. Depending upon the prevailing weather conditions, the seedlings become ready for transplanting after 4–6 weeks of sowing. In agriculturally advanced countries, direct seeding is done.

TRANSPLANTING

The seedlings are transplanted on flat or raised beds, preferably during afternoon hours. The yield and size of broccoli heads are affected by plant spacing. The yield is more but closer spacing reduces the size of main head, less production of spears and delays the horticultural maturity. The optimum spacing for planting broccoli in field is 45 × 45 cm. Spacing in the light soils can be reduced to 45 × 30 cm. At Indian Agriculture Research Institute (IARI), New Delhi, 50 × 40 cm spacing has been found best for the variety Pusa Broccoli Kt Sel-I. Depending upon the spacing, about 33,000 to 45,000 plants can be adjusted in a hectare area.

IRRIGATION REQUIREMENT

Being a shallow-rooted crop, broccoli requires light irrigation frequently. An even supply of moisture is needed for broccoli seedlings to be established. The crop grown in soil having longer water-retention capacity gives optimum yield. Dry spells during vegetative growth and head development stage result in small-sized heads and reduced yield. Depending upon weather conditions and soil types, irrigation may be given at an interval of 7–10 days to ensure optimum moisture for proper vegetative growth and head formation.

INTERCULTURAL OPERATIONS

HOEING AND WEEDING

Weeding, hoeing and earthing up are very important as weeds create serious problem in the cultivation of broccoli crop due to wider spacing, higher fertility and frequent irrigations. It is a shallow-rooted crop. Hence, deep hoeing and weeding should be avoided to keep the roots free from damage. About 2–3 shallow hoeings should be done at initial stage of crop growth to remove weeds and facilitate good aeration into the soil around the root system for proper growth of the plants. Two earthing-up operations are sufficient for optimum growth. Manual weeding and hoeing with hoe are done after 2 weeks of transplanting. Top-dressing with nitrogenous fertilizer followed by earthing up is carried out after 4–5 weeks of transplanting, and thereafter, spot weeding can be done if required.

Pre-plant soil incorporation of weedicides like fluchloralin (Basalin 48 EC) 1.0 kg/ha and nitralin 0.75–1.0 kg/ha or pre-plant surface application of pendimethalin (Stomp 30 EC) 1.0 kg/ha and oxadiazon 1.0 kg/ha or post-plant application of alachlor (Lasso) 1–2 kg/ha controls the weeds effectively.

HARVESTING

Harvesting is done when the heads have attained proper size. Central head constitutes the major portion of the yield, thus, should be removed along with 15–20 cm fleshy stem, when green buds are small and tightly closed. Delay in harvesting causes opening of buds or emergence of yellow petals and loosening of heads, thereby making them unfit for consumption. After the harvest of main head, the development of spears depends upon prevailing temperature. Overmature spears turn woody in texture and are unmarketable.

YIELD

Yield depends upon cultivars, maturing period and method of harvest. The total yield of central head constitutes about 60–70%, but under optimum environmental conditions, the spears may constitute up to 50% of the total yield. Its yield varies from 7 to 20 t/ha.

POST-HARVEST MANAGEMENT

Broccoli is highly fragile as the respiration rate of freshly harvested heads or spears is extremely high. The shelf life of broccoli heads and spears is significantly affected by temperature. The harvested heads or spears are required to be hydro-cooled. At room temperature, the harvested produce can be kept for 2–3 days only, but at 0°C temperature and 95% relative humidity, it remains good for 10–15 days. Packaging in low-density polyethylene (LDPE) bags or films helps in keeping the broccoli heads more fresh and green under low temperature conditions. It should never be stored along with climacteric fruits, which can cause yellowing of green buds due to the production of large quantity of ethylene. Broccoli heads or spears are tied in bunches of about 250–500 g and packed in waxed boxes or plastic crates. Deterioration occurs quickly in non-wrapped heads due to yellowing, weight loss, stem hardening and chlorophyll degradation.

PHYSIOLOGICAL DISORDERS

The physiological disorders have the symptoms similar to certain diseases but no pathogen or casual organism is involved in these disorders. These may be due to either deficiency of essential elements or climatic factors.

HOLLOW STEM

This is a nutritional disorder resulting from the deficiency of boron and excessive use of nitrogen. It occurs more at wider spacing. Hollowness caused by boron deficiency is recognized as water-soaked areas, while the one caused due to excessive use of nitrogenous of fertilizers is distinguished by clear white stem with no sign of disintegration. In later stages and in seriously affected plants, the stem becomes hollow with water-soaked tissue surrounding the walls of the cavity. In more advanced stages, pinkish or rusty brown areas are developed on surface of the head, because of which, it is known as brown or red rot. The affected heads develop a bitter taste.

Control
- If deficiency is noticed during growth of the plant, spray the crop with borax 0.25–0.50%.

INTERNAL TIP BURNING

The disorder may be caused due to the prevalence of high temperature, calcium deficiency and imbalance of nutrients in soil along with certain weather conditions (high humidity, low soil moisture, high potash and high nitrogen aggravate calcium availability), high manganese content and high endogenous level of indole acetic acid (IAA). Broccoli requires low temperature for its normal growth but if it is grown at slightly higher temperature, the water requirement of crop is increased, which is not fulfilled due to poor water movement within the plant, as a result, the leaf margins starts burning.

Control

*   Spray the crop with 0.5% calcium chloride.

Heat Injury

In many parts of the world where temperature is higher than the normal requirement, the broccoli produces rough heads with elongated peduncles and uneven head size. This disorder occurs only in heat sensitive varieties. Hence, only heat resistant varieties should be grown in areas with high temperature.

Bolting/Loose Heads

Due to delayed planting, the young plants are exposed to low temperature, leading to bolting without forming heads or producing loose heads.

Control

*   Grow bolting resistant varieties.
*   Supply adequate amount of nutrients.
*   Follow proper plant protection measures.

Blindness

The blind plants fail to produce any head. In such cases, the terminal growing point of the plants is damaged by mechanical injury, insects, low temperature, or frost in early stages of plant growth. Such plants produce many shoots at ground level.

Floret (Bead) Yellowing

The florets are the most perishable part of the broccoli head. Yellowing may be due to over maturity at harvest, high storage temperature after harvest and exposure to ethylene. It is also called as marginal yellowing.

Leafy Heads

Small leaves develop and protrude through the head due to high temperature, drastic fluctuations in day and night temperature or improper nitrogen balance.

Broccoli Buttoning

Buttoning is the premature head formation of 2.5 to 10 cm diameter. Buttoning can occur anytime between seeding and almost mature plant stage but usually occurs shortly after transplanting into the field. Generally, foliar growth slows after buttoning, thereby resulting in too few nutrients to nourish the head to a marketable size.

INSECT-PESTS

Cutworms (*Agrotis* spp.)

Cutworms are greyish, fleshy caterpillars up to 5 cm long, which curl up when disturbed. They cut the stem of newly set seedlings at ground level and bite the foliage, which leads to uneven crop stand and too much gap filling.

Control

- Use well-rotten farmyard manure.
- Add chlorpyrifos 20 EC @ 2.0 litre/ha after mixing in 25 kg of sand before transplanting the seedlings.

Cabbage Caterpillar (*Pieris brassicae*)

The female adult lays eggs in clusters on upper surface of the leaves. The young caterpillars scrape the leaf surface and older caterpillars damage the leaves by feeding from leaf margins, and they proceed to the centre and skeletonize them. The larvae are greenish yellow with black head and black dots on the dorsum. Infested plants do not develop heads.

Control

- Follow mustard trap cropping.
- Pick the infested leaves and destroy them.
- Spray the crop with malathion 0.1%, dimethoate 0.05%, monocrotophos 0.05%, or Nuvan 0.05%.

Diamondback Moth (*Plutella* spp.)

The adult female of this insect lays eggs on undersurface of the leaves. The greyish green caterpillars feed on the leaves, thereby leaving intact the parchment-like epidermis and holes in the leaves. Leaves turn brown and fall down. The growth of young plants is greatly inhibited and such plants do not form heads.

Control

- Follow mustard trap cropping.
- Pick the infested leaves manually and destroy them.
- Spray *neem* seed kernel extract (5%) at regular intervals.
- Spray the crop with malathion 0.1%, Nuvan 0.05%, fenvalerate 0.01%, or quinalphos 0.2% at 10-day intervals.

Aphids (*Brevicoryne brassica* and *Myzus persicae*)

Aphids get shelter within the dense bud clusters and leaves and suck cell sap from tender portion. The black sooty mould develops on the honeydew excreted by aphids, which hinders photosynthetic activity of the leaves. This results in poor growth of young plants, yield and quality of the produce.

Control

- Set yellow pan traps in the field.
- Wash the crop with plain water at high pressure.
- Remove and destroy the affected plant portion in mild infestation.
- Spray malathion 0.1%, monocrotophos 0.03%, or Nuvan 0.05%.

DISEASES

DAMPING-OFF (*Pythium* spp., *Phytophthora* spp., *Rhizoctonia solani* and *Fusarium* spp.)

It causes pre- and post-emergence death of plants. Pre-emergence attack results in inhibition of seed germination and allows the pathogen to grow, whereas, post-emergence damping-off occurs above the ground on the young seedlings, in which the stem tissue becomes soft and water-soaked, and the plant collapses at collar region near the soil surface. The roots may also get affected, resulting in death of the seedlings. Excessive soil moisture or waterlogging, high temperature and poor aeration are favourable for the disease development and it spreads from the undecomposed infected plant debris present in the soil.

Control

- Adopt long crop rotation with non-host crops.
- Avoid waterlogging and provide good drainage.
- Treat seed with Bavistin, thiram, captan, mancozeb, or Apron 2.5 g/kg of seed.
- Drench (5 litre of water/m² area) the soil with fungicide captan or thiram 2.5 g/l of water.

DOWNY MILDEW (*Peronospora parasitica*)

The disease is characterized by yellowish, irregular, or angular lesions on lower leaf surface, which is soon covered with downy growth of fungus. The corresponding upper surface becomes chlorotic and leaves may drop off prematurely. The severely infected heads become rotten. The disease appears when the humidity is high with light rains, prolonged dews, or heavy fogs in cool weather.

Control

- Follow long crop rotation with non-host crops.
- Eradicate cruciferous weeds from the field.
- Avoid overhead irrigation and excess watering.
- Treat the seeds with hot water at 50°C for 30 minutes to kill pathogens present on seeds.
- Spray Indofil M-45 0.2%, Indofil Z-78 0.2%, or Ridomil MZ 0.25% at 7-day intervals.

STALK ROT OR WHITE ROT OR WATERY SOFT ROT (*Sclerotinia sclerotiorum*)

The lower leaves touching the ground turn light green followed by yellowing and falling off prematurely. The infection on stem spreads downward to the roots. On leaf petioles and stems, small irregular dark brown to black necrotic lesions are seen, which are covered with white cottony mycelial fungus growth. Stem, twigs and inflorescence become straw-like and dry in March. Pith of the infected plants is filled with black and hard sclerotia of fungus.

Control

- Adopt long crop rotation with non-host crops.
- Maintain good sanitation and good drainage facilities.
- Keep the field weed free.
- Flood the field for longer periods during hot summer months to destroy the sclerotia.
- Treat the seed by soaking in suspension of 50 ppm aureofungin and 50 ppm streptocycline for 30 minutes and air-dried before sowing in the nursery.
- Dip the seedlings in 0.25% Benlate suspension for 5–8 minutes before transplanting.
- Spray the crop with carbendazim 0.05% at an interval of 10 days.

BLACK LEAF SPOT (*Alternaria brassicae* or *Alternaria brassicicola*)

Small concentric dark spots up to 1.0 cm diameter on lower leaves are observed. The head is also infected and subsequently it rots. The fungus also causes damage to pods and seeds with brown to black spots at seed production stage and harvested seeds are contaminated resulting in poor germination. The disease spreads through infected seeds, crop residues, cruciferous weeds, infected seedlings and spores (conidia) transmitted by wind and water. Disease is found more commonly in seed crops.

Control

- Adopt crop rotation with non-cruciferous crops.
- Always use disease-free seed.
- Maintain good sanitation in the field.
- Remove and destroy all affected leaves and heads from the field.
- Treat the seed with captan, thiram, or ceresan at 2.5 g/kg of seed.
- Spray the crop with mancozeb or Ridomil MZ-72 at 10–15 days interval.

CLUB ROOT (*Plasmodiophora brassicae*)

It is caused by a root parasite. There are many races of the organism. It is a soil-borne fungus, which enters through root hairs. Yellowing of foliage, flagging and wilting of plants are the typical symptoms. Small spindle-like spherical club-shaped swellings develop on roots and rootlets. The disease has higher prevalence on acidic soils having pH below 7.0 and less serious on heavy soils and those containing low organic matter. The pathogen spreads through seedlings raised in infested seedbeds when transplanted in the field. It also spreads by irrigation water or infested soil.

Control

- Follow long crop rotation with non-cruciferous crops.
- Grow resistant cultivars.
- Raise soil pH above 7.0 by liming.
- Maintain good sanitation and eradicate weeds.
- Dip the roots in 0.05% solution of PCNB (pentachloronitrobenzene) at transplanting.

BLACK ROT (*Xanthomonas campestris*)

Initially, chlorotic lesions appear along the leaf margins, which progress towards forming V-shaped blotches followed by blackening of veins and death of the whole plant. The head formation is restricted if the attack is early and severe.

Control

- Follow long crop rotation with non-cruciferous crops.
- Grow resistant varieties.
- Seed beds to be located as far as possible from cruciferous crops.
- Give hot water treatment to seeds at 50±2°C for 30 minutes followed by 30 minutes' dip in streptocycline 100 ppm.
- Spray the crop with streptocycline 0.01% or 100 ppm at monthly intervals.

BROCCOLI HEAD ROT (*Pseudomonas* spp.*)

It is caused by a soil-borne bacterium. The symptoms appear when the heads remain wet for several days and bacteria are splashed up from soil to the head. Water-soaked areas appear on heads because biosurfactant is released by the bacteria in contrast to unaffected areas where the waxy surface of the florets causes the water to form beads. Small black lesions may develop in these water-soaked florets. During long periods of wetness, the decay spreads rapidly, resulting in a sunken area on head. Head rot develops most rapidly at high temperature (28°C). Frost injury and infection by downy mildew may also increase the disease prevalence.

Control

- Grow resistant cultivars.
- Avoid using high doses of nitrogenous fertilizers.
- Avoid applying pesticides during head formation.
- Transplant the seedlings at wider spacing to increase air movement through the crop.

■■■

33 Chinese Broccoli / Chinese Kale

M.K. Rana and P. Karthik Reddy

Botanical Name : *Brassica oleracea* var. alboglabra

Syn. *Brassica alboglabra*

Family : Cruciferae

Chromosome Number : 2n = 18

ORIGIN AND DISTRIBUTION

Brassica oleracea var. alboglabra is native to Central and South China. This cool season green traces its ancestry to *Brassica oleracea*, a fleshy-leaved short-lived perennial from coastal areas of western and southern Europe. It is now mostly cultivated in East and Southeast Asia, and in other parts of the world where there are large number of Chinese immigrants. The recent DNA evidence suggests that Chinese broccoli is closely related to Portuguese cabbage and kale, while some studies reported that early Portuguese explorers brought cabbage to Asia, and through generations of selections, it has developed into Chinese broccoli.

INTRODUCTION

Chinese broccoli, also known as Chinese kale, *kai-lan*, flowering kale, or *gai-lan*, is a leafy vegetable, featuring thick, flat, glossy bluish-green leaves with thick stems and a small number of tiny flower curds similar to those of broccoli. Chinese broccoli is an annual vegetable, which is popular in Asia and increasingly grown in Western gardens. It is grown for its loosely clustered green flower buds and pungent leaves, which are harvested and eaten before the flowers open. The flowering stalks, young leaves and buds possess excellent eating quality. The florets are smaller and less solid as compared to western broccoli. They are hardy, heat tolerant and can be grown easily in summer when cabbage, broccoli and cauliflower cannot be grown due to heat. Broccoli and *kai-lan* belong to the same species *Brassica oleraceae* but *kai-lan* belongs to the group *alboglabra* (Latin *albus* + *glabrus* = white and hairless). Its flavour is very similar to that of broccoli but slightly bitterer and also noticeably stronger.

COMPOSITION AND USES

COMPOSITION

Chinese broccoli is a rich source of vitamin A and C. It also contains minerals like potassium, calcium, sodium, magnesium and iron and is a good source of carbohydrates, protein and dietary fibre. The nutritional composition of Chinese broccoli is given in Table 33.1.

TABLE 33.1

Nutritional Composition of Chinese Broccoli (per 100 g Edible Portion)

Constituents	Contents	Constituents	Contents
Water (g)	89.2	Calcium (mg)	88.0
Carbohydrates (g)	3.8	Sodium (mg)	6.2
Sugars (g)	0.8	Magnesium (mg)	0.4
Protein (g)	1.1	Iron (mg)	0.5
Total fat (g)	0.7	Vitamin A (IU)	1441
Saturated fat (g)	0.1	Pyridoxine (mg)	0.5
Polyunsaturated fat (g)	0.3	Vitamin C (mg)	24.8
Dietary fibre (g)	2.5	Energy (kcal)	22
Potassium (mg)	261		

USES

Young flowering shoots and small leaves are eaten raw or cooked. *Kai-lan* is eaten widely in Chinese cuisine. The common preparations include *kai-lan* stir-fries with ginger and garlic and boiled or steamed. It is very delicious vegetable when young shoots are used since they become tough with age. Older stems should be peeled before they are consumed. All parts of the growing plant are used, including the developing inflorescence.

MEDICINAL USES

The potential health benefits of *gai-lan* broccoli include protection against cancer, cardiovascular benefits, asthma relief and improved eye health. *Gai-lan* does not provide as much isothiocyanates as some strong-tasting brassica vegetables such as broccoli, cabbage, or Chinese mustard, even though, it is still a fairly good source of these cancer-fighting compounds. Hot poultice of leaves is applied on arthritis, gout, rheumatism, pains. Pounded leaves are applied on boils, cuts, haemorrhoids, inflammations, swellings, tumours and wounds. In fact, according to a study, *gai-lan* provides more than three times as much isothiocyanates as *bok choy* and significantly more isothiocyanates than cauliflower or Chinese flowering cabbage. According to nutritional data of United States Department of Agriculture, *gai-lan* greens and stalks are packed with vitamin K, with one ounce of cooked *gai-lan* providing almost one-third of the daily vitamin K requirement. Recent scientific studies suggest that consumption of foodstuffs rich in vitamin K may offer significant cardiovascular benefits due the ability of this powerful vitamin to direct calcium into the bone tissue rather than the arteries but vitamin K is the only cardio-protective vitamin found in *gai-lan*. This variety of Chinese broccoli also provides tonnes of vitamin C. The quinine in Chinese broccoli plays a role in relieving summer heat and fever. It contains an organic base, which can stimulate the taste nerves and increase appetite.

BOTANY

Most representatives of the *Brassica* genus are annual herbs, although there are also a few perennial species. Its plant grows to a height of 30–45 cm. The leaves are generally alternate on the stem, estipulate (without stipules) and glabrous (without hairs). Leaf shape is lanceolate. Inflorescence is generally either racemose or comprised of perfect flowers. Furthermore, each flower has four distinct sepals and white-coloured petals, the latter of which are arranged in order to form a cross, giving the family its alternate name, Cruciferae. The ovary is superior and is comprised of two carpels. The stamens are six in number and are somewhat unusual in that four of them possess long filaments and two have much shorter filaments (tetradynamous). The fruit of Chinese broccoli is a siliqua, and

Chinese broccoli planting in field	Chinese broccoli crop in field
Flowering in Chinese broccoli	Chinese broccoli flower
For Marketing Chinese broccoli	Chinese broccoli for cooking

the seeds that are produced are generally quite small, weighing 1–5 mg, and may range in colour from yellow to black. Furthermore, the seed coat is composed of four distinct layers, *i.e.*, epidermal, sub-epidermal, palisade and parenchymatous. Unlike many other genera of the Brassiceae family, the seeds of *Brassica* genus do not produce mucilage when the seed coat is moistened.

CLIMATIC REQUIREMENT

Chinese broccoli is a cool season crop and resistant to mild frost but may be grown round the year in both tropics and temperate regions. The optimum temperature for germination is 25–30°C and the optimum temperature for growth is 18–28°C, though the lower temperature promotes early flowering and is necessary for complete floral development. Its plants enjoy daytime temperature around 10–20°C and certainly not higher than 28°C. Though it can tolerate frost (6.5°C), the growth is better above 10°C temperature. It grows well in no shade or semi-shade (light woodland). When the plant is young, uniform conditions are favourable, *i.e.*, not too wet, dry, shady or windy. Heaviest yield is obtained from mid to late summer sowings that mature in late summer and autumn. In cool areas, delay sowing until mid-summer, as earlier sowings may bolt prematurely.

SOIL REQUIREMENT

Chinese broccoli can be grown on light (sandy), medium (loamy) and heavy (clay) soils but it prefers well-drained heavy clay soil since it prefers bit high soil moisture, which is possible only in clay soils. The crop can successfully be cultivated on acid, neutral and alkaline soils but the soil with pH over 6.8 is necessary to manage club root disease, hence, addition of gypsum is beneficial. The plant can even tolerate marine exposure.

Soil should be prepared well before sowing. To obtain the required tilth, field is prepared by ploughing 2–3 times with harrow followed by planking.

CULTIVATED VARIETIES

Chinese broccoli varieties are of two types, *i.e.*, (i) one with white flowers and (ii) another with yellow flowers. The white flower variety is popular and grows up to 48 cm high. The yellow flower plant grows only to a height of 20 cml. Both the varieties are heat resistant and will grow through the winter in most areas. The important varieties of Chinese broccoli recommended for cultivation in Taiwan and Australia are Pugong, Helgelan, Gelanya, Kailaan White, Thainan, Dai Sum Kailaan, Sak Sum Kailaan, Green Lance, Mandy, Happy Rich, Hon Tsai Ta, Suiho, Summer Jean and Trong.

BROCCOLINI

It is a hybrid of standard broccoli (*Brassica oleracea* var. *italica*) with Chinese broccoli (*Brassica oleracea* var. alboglabra), which resembles long slender broccoli side shoots that are nearly ready to flower. The flavour has been described as similar to broccoli but sweeter and less pungent or as resembling asparagus. This hybrid developed by Sakata Seed Company is grown and marketed exclusively in the United States by Mann Packing Company.

SOWING TIME

The sowing time of Chinese broccoli varies with climatic conditions of the growing region, but in general, the seed sowing is done from April to September in temperate regions and September to October in subtropical areas.

SEED RATE

The seed rate may vary with variety, soil fertility, growing season, irrigation resources, climatic of the growing region, seed viability and sowing method. A seed rate of 3–4 kg/ha is required for direct seeding crop and 300–400 g/ha for transplanted crop. Seed should be purchased from reputable sources, which ensure high quality of seed and free from pathogens.

SOWING METHOD

The crop can be raised by sowing seeds directly in the field. Seeds, which take 10–20 days for germination, are sown at a depth of 1–1.5 cm. Its sowing is done on flat beds or on ridges in a well-prepared field.

Usually, Chinese broccoli is sown at a density of 108,000–220,000 plants/ha and then thinned after 3 weeks of growth to a spacing of 15–22 cm. Thinned plants can be sold as the first harvest. A high density slows down the maturation process and produces a more desirable product.

Chinese broccoli can also be transplanted 3–4 weeks after sowing. However, proper care is taken if the seedlings are raised in portrays since the growth rate is often too fast, particularly in warm weather, which causes early flowering and minimal vegetative production.

The seeds before sowing should be treated with Captaf or Thiram @ 2.5 g/kg of seed, taking it as a preventive measure. The seeds after sowing should be covered up to 0.5 to 1.0 cm layer with a mixture of sieved soil, farmyard manure and sand in 1:1:1 ratio.

NUTRITIONAL REQUIREMENT

Because of the influence of soil type, climatic conditions and other cultural practices, the Chinese broccoli crop responses from fertilizer may not always be accurately predicted. Soil test results, field experience and knowledge of specific crop requirements help in determining the nutrients needed and the rate of application. Use of optimum doses of fertilizers is important for its proper growth since both rapid and slow growth is undesirable. The bud clusters become loose and a hollow stem results from rapid growth, however, slow growth affects the yield adversely. Generally, application of 15–20 tonnes of farmyard manure, nitrogen 60–80 kg/ha and phosphorus and potash each 100 kg/ha is recommended. The doses differ from place to place depending upon fertility status of the soil. The full dose of phosphorus and potash along with one-third dose of nitrogen are applied at the time of land preparation. The remaining dose of nitrogen should be top dressed in two equal split doses. The first is applied 30 days after transplanting and second at 45 days after planting.

IRRIGATION REQUIREMENT

Chinese broccoli needs sufficient moisture in the soil for uniform and continuous growth of plants. Therefore, frequent irrigation is given at 10 to 15 days intervals, depending upon weather conditions, since dry conditions adversely affect the quality and yield of shoots, which become more fibrous. However, waterlogging conditions depress the plant growth. The early sown crop requires irrigation at an interval of 5 to 7 days, while mid and late sown crop require irrigation at 10- to 15-day intervals. Generally, furrow system of irrigation is practiced to avoid water stagnation conditions.

INTERCULTURAL OPERATIONS

The Chinese broccoli crop field should be kept weed free. Hoeing is done for breaking the surface crust to facilitate better aeration and water absorption. Since it is a shallow-rooted crop, hoeing should not be done beyond the depth of 5–6 cm close to the plant to avoid injuries to the roots. A light earthing-up at final hoeing is also beneficial. Besides, pre- or post-emergence herbicides can be used to control the weeds. Pre-sowing application of fluchloralin 0.75–1.5 kg/ha or pre-emergence application of oxadiazon 0.5 kg/ha, alachlor 1.5 kg/ha, or pendimethalin 2 kg/ha effectively controls the weeds in Chinese broccoli field.

HARVESTING

Chinese broccoli should usually be harvested 60 to 70 days after sowing when the leaves are large enough with maximum flavour and texture. Young flowering stems with compact florets and small leaves are selected for harvesting and cut with a sharp knife at 15 to 20 cm length. Delay in harvesting will results in tougher leaves that develop a more bitter taste. Either the whole plant can be harvested or if a further harvest is required, just the terminal shoot, which encourages the development of lateral shoots, is harvested. The white flower buds should be developed but not open. Harvest Chinese broccoli frequently to prevent bolting and toughening, particularly in summer. About three harvestings can be obtained from one stem, and the main stalk should be cut relatively short to enhance further growth. Each plant can be harvested several times. A liquid feed after harvest may benefit regrowth.

YIELD

The yield of Chinese broccoli varies with variety, growing region, sowing time, irrigation facilities, field conditions, harvesting frequency and package of practices followed during the cultivation of crop. The average yield in a season with two to three harvests is 60–110 q/ha.

POST-HARVEST MANAGEMENT

Stems are best harvested early in the morning to minimize water stress. After gentle washing, the 5–7 plants are tied in bunches and secured with rubber bands or string. The main stem should be 10–15 cm long and 1.5–2.0 cm wide at the base. Common post-harvest defects include open or deteriorating flowers and yellowed or decayed leaves. The Chinese broccoli can be stored for 27–30 days at 0°C, more than 21 days at 2–5°C and 10–14 days at 5–10°C temperature if the relative humidity is maintained 90–95%. Chinese broccoli is ethylene sensitive so should not be stored with ethylene-producing produce.

INSECT-PESTS AND DISEASES

A wide range of insect-pests and diseases, common to *Brassica* crops, attack the Chinese broccoli during the growing season. Their description along with control measures is given below:

INSECT-PESTS

Diamondback Moth (*Plutella xylostella*)

The caterpillars of diamondback moth are 1 cm long and light green or brownish green in colour. The larvae feed on all parts of the plant but prefer places around the bud of a young plant and undersides of the wrapper leaves. The caterpillars feed on lower side of the leaves producing typical whitish patches. Later on, the caterpillars bore small holes in the leaves, giving a shot hole appearance all over the leaves. Diamondback moth infestation is most serious when it damages the growing point of the young plant. The attack of diamondback moth is serious during August to September.

Control
- Follow long crop rotation with beans, peas, tomato and cucurbits.
- Remove and destroy all debris and stubbles after harvesting the crop.
- Spray the crop with *Bacillus thuringiensis* formulations at early infestation.
- Introduce natural parasites like ichneumonid wasp and *Diadegma insularis*.
- Use predators such as ground beetles, true bugs, syrphid fly larvae and spiders.
- Spray the crop with 4% *neem* seed kernel extract, 0.01% fenvalerate, or 0.03% Thiodan and repeat at 10-day intervals.

Aphid (*Lipaphis erysimi*)

Aphids are the greenish white small insects attacking most of the cole crops when the weather is cloudy. The nymphs and adults both attack the young plants and suck cell sap from growing tips and underside of the leaves. Leaves of infected plants will crinkle and the plant growth becomes stunted. In severe infestations, the plants wilt and die. The badly infested plants are covered with honeydew like substance secreted by aphids, which promotes the growth of sooty mould, inhibiting photosynthesis and reducing yield of the crop. Aphids are more troublesome during cool and dry weather.

Control

- Wash the plant with plain water using strong jet pressure to reduce aphid population.
- Remove infested culls and weed species around fields that harbour the aphids.
- Introduce predators and parasites like *Lysiphlebus testaceipes*, *Aphidius matricariae*, *Aphelinus semiflavus* and *Diaeretiella rapae* that attack the aphids.
- Apply carbofuran granules 1.0 kg a.i./ha in soil.
- Spray the crop with 4% *neem* seed kernel extract, 0.03% dimethoate, or 0.05% phosphamidon.

Painted Bug (*Bagrada cruciferarum* kirk)

Painted bug is a black and orange bug, newly hatched nymphs of which are bright orange in colour. Eggs are laid on leaves or in loose soil. Both adults and nymphs suck the cell sap from the plants and retard their normal growth.

Control

- Follow clean cultivation.
- Remove alternate host surrounding the field.
- Spray the crop with 5% malathion dust, 0.25% chloradane, or fresh oil resin soap.

Flea Beetle (*Phyllotreta striolata*)

The pest responsible for damage to the crop is a small (1.5–3.0 mm) dark coloured, shiny, hard beetle with enlarged hind legs that allow them to jump like fleas. Flea beetles feed on undersides of the leaves, creating small pits or irregularly shaped holes. The adults of beetle causes damage by making small holes or pits on leaves, which give the foliage a characteristic shot hole appearance. The young plants and seedlings are more susceptible to its attack. The affected plants become stunted, but under heavy infestation, the plant may be killed.

Control

- Use floating row covers prior to seedlings emergence to provide a physical barrier against beetles.
- Remove weeds along the Chinese broccoli field margins.
- Deeply disk the plant residues in infested fields after harvest.
- Spray the crop with 4% *neem* seed kernel extract, 0.03% dimethoate, or 0.05% phosphamidon.

Cutworms (*Agrotis ipsilon*)

The moths of cutworm are large with a wingspan of 1.5 to slightly over 5 cm. The front wings are dark brown with a lighter band near the end of each wing. The hind wings are whitish to grey. Larvae feed on young plants by cutting off leaves and later on the entire plants. The caterpillars cause damage by biting the foliage and cutting down the young seedlings just above the ground level. Cutworms feed at night, and during the daytime, they hide just below the soil surface.

Control

- Plough the field at least 10 days before planting to destroy larvae, food sources and egg-laying sites.
- Remove weeds from Chinese broccoli field margins.
- Follow clean cultivation of the crop.
- Apply chlordane @ 20 kg/ha, or chlorfenvinphos @ 2 kg a.i./ha in soil.
- Spray the crop with 4% *neem* seed kernel extract.

DISEASES

ALTERNARIA LEAF SPOT (*Alternaria brassicae* and *A. brassicicola*)

On leaves, the disease symptoms appear as minute dark brown to black spots, which may enlarge by forming concentric rings surrounded by a yellow halo. At seedling stage, the symptoms appear as small dark spots on stems, which eventually cause the seedling to topple over the ground. The disease is also a problem in storage of cole crops. In humid weather, fungal spores may appear as a bluish growth in the centre of spots.

Control

- Follow long crop rotation with non-host crops.
- Use disease-free healthy seed.
- Treat the seed with captan or thiram 2–3 g/kg of seed.
- Collect and destroy the infected plant debris.
- Spray the crop with mancozeb 0.25% or copper oxychloride 0.3% and repeat at 10- to 14-day intervals.

DOWNY MILDEW (*Peronospora parasitica*)

The symptoms usually appear as a purplish or yellow-brown spots on upper surface of the leaves, and white to grey fluffy downy mildew growth is observed on lower surface of the leaves. The infected areas enlarge and their centres turn tan to light brown with papery texture. During flowering stage, blackish patches appear on the seed stalks, thus, seed yield is adversely affected. When the disease is severe, the entire leaves turn yellow and die. A temperature of 10–20°C and relative humidity more than 90% favour the occurrence of disease.

Control

- Grow resistant varieties if available.
- Use disease-free healthy seed.
- Follow long crop rotation with cereal crops.
- Maintain field sanitation by burning crop debris to reduce the inoculums.
- Treat the seed in hot water at 50°C for 30 minutes.
- Spray the crop with 0.3% Dithane M-45 or Dithane Z-78.

POWDERY MILDEW (*Erysiphe cruciferarum*)

In initial stage, the symptoms appear as yellow spots on upper surface of the leaves, which later develop the powdery growth, and the disease then spread to underside of the leaves and stem too. The affected leaves turn completely yellow, curl, die and fall down. Powdery mildew is favoured by warm dry days and cool moist nights.

Control

- Follow long crop rotation with cereal crops.
- Grow resistant varieties if available.
- Spray 0.5% sulfex or other formulations of sulfur.
- Spray the crop with systemic fungicide like Karathane, Bavistin, or Benlate 0.2% at 10- to 15-day intervals.

Fusarium Yellows (*Fusarium oxysporum* f.sp. *conglutinans*)

The disease symptoms appear as yellowing of the lower leaves, which are often on one side of the plant. These leaves later turn brown and drop off. With time, the entire plant may yellow, wilt, and collapse. The affected plants are stunted, lopsided, yellowed and lose most of their lower leaves and have a brown to black discolouration in the veins.

Control

- Follow long crop rotation with cereal crops.
- Use disease-free healthy seed.
- Treat the seeds with thiram or captan @ 3 g/kg of seed.
- Apply less nitrogenous fertilizers.
- Drench the soil with 0.2% Brassicol.
- Spray the crop with 0.3% Bavistin or Dithane Z-78.

Black Leg (*Phoma lingam*)

The most serious symptoms occur on stems near the soil line where elongated, sunken, brown lesions form. These lesions may girdle the stem, resulting in stunting, wilting and poor growth of the plant. If the lesions enlarge, the stem may break, causing the plant to fall over. Lesions usually contain minute, spherical and dark structures that are fruiting bodies of the pathogen. Yellow spots with grey centres appear on the foliage. The severely infected plants usually topple over the ground, as the pathogen destroys the supportive stem tissue.

Control

- Follow long crop rotation with cereal crops.
- Grow resistant varieties if available.
- Collect and destroy the infected plant debris.
- Treat the seed in hot water for 30 minutes at 50°C.
- Treat the seeds with thiram or captan @ 3 g/kg of seed.
- Spray the crop with 0.2% Bavistin or captan at 10- to 15-day intervals.

Black Rot (*Xanthomonas campestris*)

The bacteria affect almost all the cole crops both in nursery and field. Due to this disease, the margin of leaves turns yellow, veins become dark and vascular discolouration takes place in main stem. It is a seed-borne disease.

Control

- Take no cruciferous crop in the rotation for 3 to 4 years.
- Treat the seed before sowing in hot water at 50°C for 30 minutes, 1% mercuric chloride solution for half an hour, or antibiotics like streptomycin or areomycin solution (1:1000).

CLUB ROOT (*Plasmodiophora brassica*)

This causal organism lives in the soil and enters the roots by which they enlarge due to characteristic swellings. These enlarged structures are called clubs. Secondary invasion by soft rot bacteria follows, forming material toxic to plants. As the disease advances, the deformed roots become incapable of supplying the plants with sufficient moisture and nourishment. This results in temporary flagging of leaves on bright days. The affected plants become dwarf and light in colour.

Control

- Avoid growing crop in infested soils
- Properly lime the soil to raise the pH.
- Treat the seed with mercuric chloride (1:500).

■■■

34 Chinese Cabbage

R.K. Gupta, D.S. Khurana and Hira Singh

Botanical Name : *Brassica rapa* L. subsp. *pikeensis* (Petsai [Heading type])

Brassica rapa L. subsp. *chinensis* (Pak-choi [Non heading type])

Brassica rapa L. subvar. *pe-tsai* (L. Bailey) Kitam.

Brassica rapa L. *pekinensis* Group

Brassica chinensis L. var. *pekinensis* (Rupr.) Sun.

Brassica rapa L. var. *amplexicaulis* Yoshio Tanaka & Ono.

Brassica pe-tsai L. Bailey

Brassica pekinensis (Lour.) Rupr.

Sinapis pekinensis Lour

Family : Brassicaceae (Crucifereae)

Chromosome Number : 2n = 2x = 20

ORIGIN AND DISTRIBUTION

Brassica is one of the most important genus containing around 37 different species. *Brassica campestris* emerging in Near East reached China through several pathways about 4000 years ago, and then, turnip in Northern area and *pak-choi* in Southern area of China developed separately. The earliest known records suggest that Chinese cabbage arose from a cross between *Pak-choi* (*Brassica rapa* var. chinensis) from South China and turnip (*B. rapa* var. rapifera) from North China. Artificial crosses between these subspecies also support this hypothesis. The earlier version was ancestral non-heading Chinese cabbage and it finally became heading Chinese cabbage in Qing of North China, where the basic types of cylindrical, oval and flat-headed were established. Later, the tropical types having extra-early maturity were derived from flat-headed types.

Until recently, seven groups of *Brassica rapa* types, commonly known as var. campestris, var. pekinensis, var. chinensis, var. parachinensis, var. narinosa, var. japonica and var. rapa, were considered as separate subspecies because of the wide range of variability. They represent their evolution in isolation from each other. It appears that *B. rapa* var. chinensis as a leafy vegetable differentiated from oilseed rape types of middle China. *Brassica rapa* var. parachinensis is a derivative of *B. rapa* var. chinensis. However, *Brassica rapa* var. campestris is considered the most primitive leafy vegetable. *Brassica rapa* var. narinosa is similar to var. chinensis in its adaptation. The *Brassica rapa* var. japonica is a leaf vegetable of Japan. Headed Chinese cabbage (*Brassica rapa* var. pekinensis) has its centre of diversity in Northern China and has some relationship with oilseed type, which is grown there. It is well adapted to somewhat cooler climate.

Pe-tsai cabbage, the heading type (*Brassica rapa* var. pekinensis or *B. pekinensis*), and pak-choi cabbage, the leafy type (*Brassica rapa* var. chinensis or *B. chinensis*), are now considered two different groups. Heading type cultivars can be further grouped into (i) cylindrical head forms called chichili (*B. rapa* var. pekinensis or cylindrica), (ii) round and compact heading types called chefoo (*B. rapa* var. pekinensis or cephalata*)* and (iii) open-headed form (*B. rapa* var. pekinensis or laxa).

INTRODUCTION

Chinese cabbage is closely related to turnip and swede than cabbage. It is grown for its edible leaves on large scale in Southeast Asia. It is quite popular vegetable in China, Taiwan, Korea, Japan and Mongolia, and recently, it has become popular in India. It includes both non-heading type (*Brassica rapa* L. var. chinensis) and heading type (*Brassica rapa* L. var. pekinensis). The non-heading group includes pak-choi, Chinese mustard, celery mustard and chongee, and its plant is characterized by several thick white petioles and glossy green leaf blades that form a celery-like bunch. The heading group includes *petsai*, celery cabbage, Chinese white cabbage, Peking cabbage, *Won Bok*, *Napa*, *Hakusai*, *Pao* and *Bow Sum*. Within heading group, there are chihili types that grow erect and develop cylindrical heads and *chefoo* types that form compact round heads of green-bladed and white petiole leaves. These are two distinctly different groups often used as leaf vegetables. Both have many variations in name, spellings and scientific classification. It is known as *Chinese Leaf* in the United Kingdom, *Wong Bok* or *Won Bok* in New Zealand and *Wombok* in Australia and the Philippines. Now, its cultivation has spread in many countries because of its high tonnage capacity, nutritional value and low oxalate content.

Chinese cabbage is generally grown as an annual crop. The leafy type Chinese cabbage has thick white leaf stalks (petioles) and smooth, glossy, dark green and almost round leaf blades. The headed types have cylindrical or barrel-shaped head with a broad central midrib. The head is firm but not as firm as cabbage at maturity. The outer foliage and wrapper leaves are pale green, while the inner leaves are blanched to a creamy white in colour. The crop takes about 55–100 days from sowing to maturity depending on genotype. Of two major types, the leafy types (loose-headed forms) are usually faster in growth than the headed forms. The loose-headed forms are usually suitable for the areas with hotter summers while headed forms for the cooler areas. The loose-headed forms are less prone to bolting and relatively tolerant to diseases. The small flowers are borne on top of the erect flower stalks varying in colour from yellow to purple depending on the variety.

COMPOSITION AND USES

COMPOSITION

Chinese cabbage is rich source of vitamins containing vitamin C content higher than lettuce. It also contains citric acid and anticancer compound glucosinolates. The total glucosinolates content varies from 0.097 to 0.337 and 0.39 to 0.704 g/kg of fresh edible portion in heading type and leaf types, respectively. It is also rich in calcium, potassium and magnesium. The nutritional composition of Chinese cabbage is given in Table 34.1.

USES

Like common cabbage, Chinese cabbage is used as cooked vegetable, as salad, in soups, other Chinese dishes and also processed as a brined product or pickles such as kimchi. The white tender leaves midribs are sliced or coarsely shredded in salads. Being low in oxalate, the patients suffering from kidney or gall bladder stone can consume Chinese cabbage *saag* otherwise those patients are advised not to consume either common *sarson saag* or one from raya because of rich oxalate content. Moreover, the leaves being succulent are easily cooked with minimum time and fuel. Its leaves are also dehydrated for off season use.

TABLE 34.1

Nutritional Composition of Fresh Pak-Choi Chinese Cabbage (per 100 g Edible Portion)

Constituents	Contents	Constituents	Contents
Water (g)	95	Sodium (mg)	65.0
Carbohydrates (g)	2.2	Iron (mg)	0.8
Protein (g)	1.5	Vitamin A (IU)	4468
Fat (g)	0.2	Thiamine (mg)	0.05
Calcium (mg)	105	Riboflavin (mg)	0.04
Phosphorus (mg)	40	Niacin (mg)	0.6
Potassium (mg)	250	Vitamin C (mg)	300
Magnesium (mg)	19.0	Energy (kcal)	13

MEDICINAL USES

Chinese cabbage containing high calcium and potassium and low sodium content regulates blood pressure and blood sugar, supports cardiovascular system and healthy brain function and relieves hypertension. Being high in iron it fights anaemia and fatigue and keeps the haemoglobin level high in blood. Chinese cabbage contains glucosynolates, which prevent cancer. Its large amount makes Chinese cabbage a rich source of flavonoids, vitamin A and C, which act as antioxidants and protect the body from free radicals, degeneration, cataracts and loss of vision. The body require Chinese cabbage to produce collagen, a protein, which keeps the skin youthful and elastic. It boosts and maintains a healthy body immune system. Chinese cabbage is antirheumatic, antiarthritic, antiscorbutic and resolvent.

BOTANY

Chinese cabbage is a leafy green or purple biennial plant, grown as an annual vegetable crop for its dense leaved heads. It is a multilayered vegetable. The plant grows to a height of 25 to 50 cm. Like common cabbage its head is not round though it is tall like a romaine lettuce head with a broad central midrib. The cylindrical head of Chinese cabbage is firm but not as firm as cabbage at maturity. The outer foliage and wrapper leaves are characteristically pale green, whereas the inner leaves are launched to a creamy white colour. The head length is about 20–25 cm and diameter is 15–20 cm. It comes in flowering from May to August. The flowers are borne in racemes rows of simple flowers arranged on an elongated axis that grows from the main stem and lateral branches. The flowers have four bright-yellow petals (12–25 mm long) that are broad on the distal end and appear in the form of a cross, hence, the former family name Cruciferae (cross bearing). There are six stamens, all of which face the style but two of the stamens are usually shorter and lean away from the style. The other four stamens are longer than the style. The single pistil contains a two-chambered ovary that contains many ovules. The pistil develops into a long, slender silique (pod), which projects upward at an angle. The flowers are hermaphrodite (have both male and female organs) but pollinated by honeybees due to self incompatibility. The open architecture of the flower allows access to many kinds of insect.

CLIMATIC REQUIREMENT

Chinese cabbage can be grown under moderate to cool climate with temperature range of 15–22°C. It is grown as cool season crop in tropics and subtropics and in temperate areas as warm season crop. The optimum temperature for growth during the initial vegetative phase is about 20°C, while during head initiation stage, the lower temperature around 15°C is considered desirable. High temperature during head formation reduces the compactness of the heads. In northern plains of India, the rise in temperature in spring season transits the plants into reproductive phase and flower formation occurs.

SOIL REQUIREMENT

Chinese cabbage can although be cultivated on a variety of soils from light to quite heavy but it prefers well-drained moisture-retentive soils with pH ranging from 6.5 to 7.5 and EC around 1.5 dS cm^{-1} since it is susceptible to club root and fungal diseases. If the pH is below 5.5 the soil should be limed to raise the pH since at low pH, the calcium and other major nutrients become unavailable to the plants. Liming also reduces the incidence of club root disease. The seedling emergence is markedly reduced in soils having an EC value of 2.5 dS cm^{-1} or above and plants show phytotoxicity symptoms characterized by leaves stunting, chlorosis and necrosis of leaf margins and root tip dieback. Well-prepared raised beds can help in preventing soil compaction, avoiding waterlogging conditions and improving air circulation near plant root zone, which will lessen the incidence and severity of diseases.

VARIETIES

Usually, two types of Chinese cabbage cultivars *viz.*, Pet-sai cabbage heading type (*Brassica rapa* var. pekinensis or *B. pekinensis*) and pak-choi cabbage leafy type (*Brassica rapa* var. chinensis or *B. chinensis*) are cultivated in India. In heading type, the heads are formed like cabbage, while in leaf type, the loose leaves are formed with open centre.

Leafy Types

Chinese Sarson No. 1

A leafy type variety with semi-erect growth habit developed at Punjab Agricultural University, Ludhiana, possesses field resistance to *Alternaria* leaf spot and bears light green, broad and puckered leaves. Its midrib is white, succulent and tender. It takes about 30 days for the first cutting and is quite rejuvenating, giving 6–8 cuttings with average yield of about 400 quintals per hectare.

Saag Sarson

A leafy type variety developed at Punjab Agricultural University, Ludhiana, was released in 2014. The leaves are large with prominent midrib. This is most suitable for processing and preparation of good quality *saag*, which is better than Raya and Gobhi Sarson. It is rich in ascorbic acid and minerals but low in oxalates. This is also approved by Markfed Canaries Jalandhar, Punjab for processing. The *saag* prepared from this cultivar is suitable for export. On an average, it yields about 500 quintals per hectare.

Palampur Green (PUCH-3)

A late-bolting leafy type variety developed at CSK Himachal Pradesh Krishi Vishvavidyalaya, Palampur produces tender green leaves with creamy stem. It takes about 25–30 days for first cutting and gives total 5–6 cuttings each at 15-day intervals.

Pusa Saag

A leafy type variety was developed at Indian Agricultural Research Institute, New Delhi, through a cross between Wong Bok (Suttons) and turnip. Its taste is like that of Local Sarson.

Solan Selection

A leafy type variety developed at Dr. Y.S. Parmar University of Horticulture and Forestry, Solan (Himachal Pradesh) produces light green smooth and tender leaves with fleshy petioles. Its average yield is about 150–190 q/ha.

Hybrid Shan Tung

A leafy type hybrid is relatively tolerant to heat and adapted well to subtropical and mild climate. Light green leaves and white thin petioles are very tender and delicious. This hybrid is fast-growing and matures in 3–4 weeks. Its young leaves can be picked for salad purpose. The leaves are suitable for stir frying and soups.

Green Seoul

A leafy type hybrid widely grown in Korea primarily for fresh market is fast growth, attaining maturity in about 4–5 weeks. Its leaves that are very tender and well flavoured and are used for cooking as well as for pickling.

Hybrid Taiwan Express

A leaf-type hybrid suitable for densely growing year yound in subtropical regions produces light green leaves with frilled edges, which can be harvested within 30 days after sowing. It can be grown very well in mild and warm climate.

Chin Sun

A leafy type Taiwanese hybrid suitable for salad, stir frying and soup preparation produces very tender green leaves and petioles with excellent texture and flavour. Its plants can be harvested at any growing stage.

Some of the other leafy type hybrids grown in Southeast Asia are Canton Pak-choi, Pak choy Green, Pak choy White, Green Boy, Lei Choi, Hung Chin, Shanghai Pak choi and Joi choi.

HEADING TYPES

Solan Band Sarson

A heading type variety developed at Dr. Y.S. Parmar University of Horticulture and Forestry, Solan (Himachal Pradesh) produces long barrel shaped light green semi-compact heads with 6–9 outer leaves. The head length is 20–25 cm, diameter 10–15 cm and average head weight about 700–1100 g. It becomes ready for harvesting in 120 days and its average yield is about 350–400 q/ha.

Optike

An early heading type hybrid resistant to *Fusarium* wilt developed by Bejo Sheetal Seed Private Limited produces barrel-shaped short and compact heads with vigorous growing habit. The head is of light green colour and good in taste.

The other head type varieties popular in Southeast Asia are Chefoo, Wong Bok, Spring Giant, Tokyo Giant, Tropical Pride, Tropical Delight, Early Top, Tip Top, China King, Winter Giant, Oriental King, Chihili and Michihili.

SOWING TIME

In North Indian plains and low hills, its seeds are sown in raised nursery beds in mid September and transplanting is done in mid October. In mid hills, the ideal sowing time is September to October and transplanting time is October–November. In high hills and dry cold desert areas, the crop is sown or transplanted from April to June.

SEED RATE

Approximately 500 g seed is required for raising seedlings sufficient for planting a hectare land area and about 2.5 kg seed per hectare is sufficient for direct seeding crop. For sowing in portrays, around 200–300 g seed is adequate for producing seedlings enough for planting a hectare land area.

RAISING OF SEEDLINGS

Seeds are usually sown in raised nursery beds that are approximately 15 cm high, 3.0 m long and 1.0 m wide. Nursery beds are usually covered with fine nylon mesh screen net so as to protect the nursery from heavy rain, blazing sun and virus transmitting insects. The upper 5–7 cm soil layer of bed is prepared by mixing garden soil, well-decomposed farmyard manure and sand in the ratio of 2:1:1. In well-drained soils, calcium ammonium nitrate is incorporated 40 g, superphosphate 50 g and muriate of potash 30 g/m². The seeds are sown 5 mm deep in rows spaced at 6 cm apart, and thereafter, the beds are covered with a thin layer of compost followed by a covering with paddy straw or dry grass and then irrigated with rose can. The mulch cover is removed after the completion of seed germination, *i.e.*, 4–5 days after sowing. The seedlings usually become ready for transplanting in 3–4 weeks after sowing or when they attain 4–5 leaves. Water supply is restricted 3–5 days before transplanting to harden the seedlings.

The seedlings can also be raised in protrays, which usually have individual cell-pack with a diameter of 15 mm and depth of 10 mm. In protrays, the seed requirement is almost half. The seedlings in protrays are usually raised under protected conditions. About 2–3 seeds per plug are sown at a depth of 0.5–1.0 cm. The seeds take 3–5 days to emerge and thinning is done at 2–3 leaf stage, keeping one seedling per plug. After thinning, water-soluble fertilizer is applied. Plug seedlings usually become ready for transplanting in field 3–4 weeks after sowing when they are 15 cm tall.

TRANSPLANTING

Direct seeding and transplanting are done in case of leafy type varieties, while heading types are usually transplanted. The distance between rows is maintained 30–45 cm and between plants 30 cm in heading types, while the leafy types are spaced at a distance of about 25–30 cm between lines and 25 cm between plants, however, in leafy type, the best results are obtained when transplanting is done at 45 × 45 cm spacing. In direct sown leafy type crop, thinning is normally done at 5–6 leaf stage, which is attained about 3 weeks after sowing.

NUTRITIONAL REQUIREMENT

The crop requires incorporation of well-decomposed farmyard manure about 30–40 t/ha at least 2–3 weeks before sowing/transplanting at the time of land preparation. Besides, nitrogen is applied 125 kg, phosphorus 75 kg and potash 75 kg/ha but the specific dose of fertilizers depends on soil fertility and agro-climatic conditions of the growing region. A full dose of phosphorus and potash along with one-third dose of nitrogen is given as basal application at the time of sowing or transplanting and remaining two-thirds dose of nitrogen is applied in two equal splits as top-dressing after second and fourth cuttings in leafy types and 30 and 45 days after transplanting in heading type. Excessive application of nitrogen can lead to increased susceptibility to insect-pests and diseases.

For heading types or chefoo types, some scientists recommend the use of 10 tonnes of well-decomposed farmyard manure along with nitrogen 75 kg, phosphorus 30 kg and potash 30 kg in a hectare land area. The compost is mixed thoroughly in top 10 cm soil while preparing the field. The whole quantity of farmyard manure, phosphorus and potash along with half dose of nitrogen is applied at the time of land preparation and remaining half of the nitrogen 1 month after transplanting.

IRRIGATION REQUIREMENT

Chinese cabbage is a shallow-rooted crop. Most of the roots remain in top 30 cm soil layer, while some penetrate to the depth of 60–70 cm. Its water requirement usually varies from 400 to 600 mm.

The crop needs a continuous supply of water with first irrigation given just after transplanting and subsequent at 10- to 15-day intervals as per need of the crop. In heading types, it is important to maintain adequate moisture during head formation since inadequate watering can delay maturity, reduce head size and thus yield. Excess irrigation during head formation can cause formation of poor-quality heads. Heavy irrigation after a dry spell especially at maturity can cause bursting of heads.

INTERCULTURAL OPERATIONS

The intercultural operations are needed to control weeds and provide aeration into the soil around roots. Crop weed competition has been reported between the second and seventh week after direct seeding or transplanting, therefore, the weed intensity should be low during this period. The weeds can be controlled manually by hand hoeing or in combination with pre-emergence application of herbicides. Two to three weedings are enough to keep the weeds under control. The time of weeding depends on the weed emergence behaviour in crop. Mounting up of soil around base of the plants especially in case of early crop favours better growth of the plants. The recommended herbicides should be used to control the weeds especially in transplanted crop in situations where labour is expensive. Pre-plant application of Stomp 30EC (pendimethalin) @ 2.5 l/ha followed by one hand hoeing has been found to be effective.

HARVESTING

High quality *Saag* can be prepared from leafy type Chinese cabbage due to its tender and succulent leaves. First cutting is done in mid November. Fully developed leaves are harvested near the base without injuring the central whorl. In heading types, the heads are usually harvested manually when they are well developed and firm but before the development of flower stalk. The heads should be handled carefully as they are very easily injured. Harvesting should always be done in cool hours of the day with the outer leaves trimmed off. The heading type Chinese cabbage is usually harvested 65–80 days after transplanting, while in leaf types, first picking can be done 30–45 days after sowing or transplanting.

YIELD

The yield of Chinese cabbage is influenced by several factors such as variety, soil type, soil fertility, growing conditions and the cultural practices followed during the course of cultivation. The average yield varies from 20 to 50 t/ha, however, the higher yield up to 100 t/ha can also be obtained using better varieties and improved production and protection techniques.

POST-HARVEST MANAGEMENT

The post-harvest handling and storage losses include mechanical damage, weight loss, leaf yellowing and decaying in storage. Mechanical damage in leafy type is caused by rough handling, washing, or loading on to the trucks and transport to retail markets. The weight loss occurs during transport to retail markets and home storage, while leaf yellowing is usually caused by high temperature. In head types, the mechanical damage is caused by rough handling at all stages in collection centres and during loading. Storage decaying is caused due to poor ventilation and freezing during storage. Immediately after harvesting, the heads should rapidly be cooled to as close to 0°C as possible. In leafy types, the fully developed leaves are harvested near the base without injuring the central growing point.

PHYSIOLOGICAL DISORDERS

The major physiological disorders developed in Chinese cabbage due to the environmental and nutritional factors are tip burn, pepper spot, bolting and petiole freckles.

TIP BURN

The tip burn disorder caused due to calcium deficiency in young leaves and a condition of high temperature and low relative humidity is characterized as brown to black necrotic areas on leaf margins. The disorder is also strongly associated with soil salinity, nitrogen fertilization and climatic conditions. To avoid the development this disorder, avoid using excessive amount of basal nitrogen and spray the crop with calcium citrate (25 g/100 litres of water) twice a week from 4 weeks onward after transplanting at the rate of 1500 litres of solution per hectare. Spraying daminozide at 1–5 g/litre of water at seedling stage also reduces the incidence of tip burn in Chinese cabbage.

PEPPER SPOT

Another major physiological disorder is pepper spot. However, still it is not certain whether the disorder is caused by a virus, high carbon dioxide within the head, genetics, environmental factors, or something else but high rates of nitrogen fertilizer make the disorder worse, especially when ammonium nitrate is used since the presence of nitrite in midribs has been detected before the black spots appear. Also, the soil pH near 8.0 may cause an increase in pepper spot. Pepper spot has also been associated with tissue levels that contain high levels of copper and low levels of boron. The other conditions that may influence the occurrence of pepper spot include air temperature, low light levels, soil nutrients levels and harvesting conditions. The disorder is recognized as small, dark, circular or elongated spots that first appear on white midribs of the outer leaves and then spread to the middle leaves. The spots develop on both outer and inner surface of the leaves. Midribs look much as if ground black pepper was sprinkled on them, hence, the name of the disorder. Symptoms can occur in the field during growth but cool storage followed by warm temperature seems to make pepper spot worse. The exact cause and remedy of this disorder are still not known, but elevated carbon dioxide of 10% in the storage atmosphere reduces pepper spot severity. The leaves and heads affected with pepper spot are often marketable, if the spots are not too much. At present, pepper spot incidence is best controlled by using tolerant cultivars and following good cultural and post-harvest practices. Excessive use of fertilizers, particularly ammonium nitrate, should be avoided and soil pH should be maintained 6.0 to 6.5.

BOLTING

Bolting is a process of developing seed stalk. Chinese cabbage when grown in late spring under long day conditions develops a seed stalk and thus undesirable poor quality heads result. Late sown crop normally tends to bolt as compared to fall season sown crop. In Chinese cabbage, bolting response is under the control of genetics. Hence, some of the cultivars bolt more swiftly than the other cultivars. If the sowing is to be done in spring months, slow-bolting cultivars should be selected for growing since these cultivars are tolerant to warm conditions and produce compact heads. With respect to flowering, this crop is not only sensitive to photoperiod but also sensitive to temperature. The evidences in literature indicate that exposure of plants to 4.4°C temperature for a week or 10°C for two weeks, or prolonged temperature below 13°C at seedling stage induces bolting. Growing the crop under short days and warmer temperature conditions keeps the plant in vegetative phase. Lastly, the factors that check growth of the plants such as nutrients deficiencies or water stress also induce bolting.

Petiole freckles

Boron deficiency causes brown streaks and corking of midribs. To overcome this problem, borax is applied 10–20 kg/ha at the time of sowing, depending on soil type. Acidic soils that have been heavily limed require double the rates.

INSECT-PESTS AND DISEASES

Chinese cabbage is susceptible to a wide range of insect-pests and diseases, which affect other crucifers, but damage is relatively low in early stages in leafy types that are commonly grown in Northern plains and low hills. In leafy types, losses have been reported due to the attack of aphids in late stages and in the seed crop. The major insect-pests include diamondback moth, cabbage maggot, aphids and beetles, the descriptions of which are given as follows:

INSECTS-PESTS

Diamondback Moth (*Plutella xylostella*)

The diamondback moth, sometimes called cabbage moth, is capable of migrating long distances. It is the most destructive pest of *Brassica* crops worldwide. Although the larvae are very small they can be quite numerous, resulting in complete removal of foliar tissue except the leaf veins. This is particularly damaging to seedlings and may disrupt head formation. Larvae usually eat nearly every plant part at all stages. In young plants, the growing tips are eaten and seedlings appear stunted. The amount of damage varies greatly depending on plant growth stage, larval density and larval size.

Control

- Cover the seedlings with nylon mesh net or plastic sheet to reduce the moth population.
- Keep the field weeds free.
- Collect and destroy plant debris from the field.
- Spray fenvalerate (1 ml/l) and repeat at 10-day intervals, if required.
- Spray the crop with 4.0% *neem* seed kernel extract twice or *Bacillus thuringiensis* formulation @ 500 g/ha at weekly interval.
- Use parasites *Trichogrammatoidea bactrae, Diadegma semiclausum, Diadromus collaris* and *Cotesia plutellae.*

Aphid (*Myzus persicae* and *Lipaphis erysimi*)

The small soft-bodied insect sucks cell sap from tender leaves and pods. Leaves infested with aphids become curled or twisted and when its population is high, the entire plants become wilted, distorted, or yellowish. Small plants can be killed or severely stunted. Its infestation becomes most serious when the temperature starts rising in the months of January–February. Aphids also transmit viruses and emit sticky honeydew-like substance, which supports the growth of a black sooty mould. With heavy infestation, sooty mould becomes thick enough to block sunlight, which reduces photosynthetic activity of the plant and affects the crop yield adversely.

Control

- Follow long-term rotation with non-host crops.
- Remove infested weed species around the field.
- Destroy the crop residue soon after harvesting the crop.

- Wash the plants with plain water using strong jet pressure to reduce aphid population.
- Spray the crop with *neem* seed kernels extract 5% or malathion 0.05% and repeat at 10-day intervals.

STRIPED FLEA BEETLE (*Phyllotreta striolata*)

Though the larvae of the beetle live in the soil, feeding on roots of the host plants, they are not significant pests. In early spring, beetles attack the seedlings and young plants. Small shiny black adult beetles cause primary damage by feeding on undersides of the leaves, creating numerous round *shot holes*. This type of injury is capable of killing young plants. The seedlings may also be killed if severe damage occurs. Besides, the beetles may act as vector of plant diseases. The damage is more severe in dry season. In dry summers, larvae attack the roots of plants and cause irreversible wilting.

Control

- Keep the field weed free.
- Protect the seedlings with fine mesh netting.
- Plough the field after harvesting the crop to expose the larvae to predators.
- Spray the crop with methyl parathion 8 EC 0.05 to 0.1%.

DISEASES

DAMPING-OFF (*Rhizoctonia solani, Phytophthora* spp., *Fusarium* spp. and *Pythium* spp.)

Damping-off is a very common fungal disease of seedlings in nursery as well as in field. The above fungi attack the seedlings at collar region and cause rotting, consequently, the seedlings fall over to the ground. Dense sowing of seeds, high temperature, excessive soil moisture or water logging and poor soil aeration are the major factors responsible for the occurrence of damping-off disease.

Control

- Follow long crop rotation with cereal crops.
- Plough the field repeatedly during hot summer months.
- Use disease-free healthy seed for sowing.
- Always sow the seed thinly to avoid overcrowding of seedlings.
- Provide good drainage facilities.
- Treat the seed with Bavistin 2 g or captan 3 g per kg of seed.
- Drench the soil with 0.2% captan to disinfect the soil.
- Avoid excessive irrigation and provide good soil drainage.

CLUB ROOT (*Plasmodiophora brassicae*)

The club root disease is caused by soil-borne fungus that enters the plant through root hairs in young plants or wounds on roots or stem. It causes the roots to become thick and distorted, which affects the absorption of water and minerals from the soil but the disease is not so serious in leafy type Chinese cabbage in northern plains. In heading types, the plants become stunted and they fail to produce marketable heads. The disease spreads readily through infected seedlings, water and soil attached to the equipment or legs of workers. It is very hardy organism and can survive and remain infectious for years even in the absence of a suitable host. Under a condition of low soil pH, the disease infestation can be minimized by liming to bring pH around 7.0 or above.

Control

- Grow resistant varieties.
- Follow long crop rotation with non-host crops.

- Apply good sanitation practices in Chinese cabbage field.
- Fumigate the field with metam sodium.
- Drench the field with benomyl 0.1 g a.i./200 ml per plant or mancozeb 3.0 kg a.i./ha.

DOWNY MILDEW (*Peronospora parasitica*)

The symptoms appear as leaf spots that enlarge to form yellow areas on upper side and mildew on lower side of the leaves. Irregular yellow patches, which later turn light brown in colour, appear on leaves. A fluffy white growth appears on lower surface of the leaves. Spots may appear on leaves, stems, siliquae and flowers. Disease develops most rapidly when night temperature is between 10°C and 15°C. It spreads quickly under cool and humid conditions.

Control
- Select well-drained and well-aerated sites for planting.
- Follow long crop rotation with non-brassica crops.
- Remove all plant refuse after harvesting the crop.
- Destroy the plant refuse of previously grown crop.
- Treat the seed before sowing with captan or thiram 3 g per kg of seed.
- Spray the crop with Dithane M-45 0.2%, if necessary.

BLACK LEG (*Leptosphaeria maculans*)

Black leg is not a serious problem in India, but sometimes, it develops in Chinese cabbage growing areas. Black pycnidia are visible as small dots on hypocotyl and cotyledons of the young plants, resulting in their death. In advanced stages, round or irregularly shaped grey necrotic lesions with dark purple or black margins appear on stem, which may extend to entire surface and damage the cortex of root. Later, these lesions may be covered with pink masses under favourable weather conditions. The disease appears under warm and wet conditions quickly but more visible symptoms of the disease develop at higher temperatures.

Control
- Plough the field deep in hot summer months.
- Use disease-free seed as a preventive measure.
- Strictly follow a crop rotation of at least 3 years with non-cruciferous crops.
- Use raised beds for the planting of crop.
- Provide hot water treatment at 50°C for 30 minutes.
- Collect and destroy crop debris after harvesting.
- Treat the seed with benomyl slurry combined with thiram to eradicate black leg disease.

SOFT ROT (*Erwinia carotovora*)

This bacterial disease is not serious in Northern plains and low hill regions. The symptoms usually appear on underside of the lower leaves near the soil surface as water-soaked spots, followed by brown or black discoloration of the vascular tissue. Subsequently, the entire plant rapidly rots with a creamy white colour and a strong smell. The soft rot is favoured by rainfall and high temperature. The infection occurs through wounds caused by leaf scars, insects, or mechanical injury.

Control
- Adopt long crop rotation with cereal or other non-susceptible crops.
- Use raised beds for the planting of crop.
- Cover the soil surface with straw mulch.

- Destroy the crop refuse after harvesting.
- Drench the beds with copper oxychloride (Blitox) 0.2% solution.
- Inoculate *Bacillus subtilis* in soil against the bacterium.

TURNIP MOSAIC VIRUS (TuMV)

Viral diseases in Chinese cabbage spread through aphids. The symptoms include mosaic mottling, black speckling, or stippling of heads, ring spot, malformation, discoloration, chlorotic lesions and puckering of leaves. Later, it causes lumpy or warty growths on veins of leaf undersurface and vein clearing. Infected plants become stunted with coarsely mottled and distorted leaves. Black spots develop on leaves, which prematurely drop. Early infection can reduce the crop yield by 75% but the late infection has little or no effect on yield. It also reduces seed yield. In stored Chinese cabbage, black sunken spots develop on leaves all over the head. The spots are much larger than those caused by cauliflower mosaic virus. However, cauliflower mosaic virus has a restricted host range, as it is infectious only to members of the *Brassicaceae* family plants.

Control

- Grow resistant varieties.
- Use reflective mulch to deter the aphids.
- Use biological predators to suppress the population of aphids.
- Rogue out the infected plants from the field.
- Protect the seedlings from aphids in nursery.
- Spray the crop with malathion 50 EC 0.1%, imidacloprid 0.04%, or pymetrozine 50% 0.025%.

■■■

35 Romanesco Broccoli

S.K. Dhankhar and Chandanshive Aniket Vilas

Botanical Name   : *Brassica oleracea* var. romanesco

Family   : Cruciferae (Brassicaceae)

Chromosome Number : 2n = 18

ORIGIN AND DISTRIBUTION

Romanesco broccoli, a native to Italy or eastern Mediterranean region, has been documented and cultivated in European regions since the 16th century. During the last 400–500 years, it spread from Italy to central and northern Europe, which became a secondary centre of diversity. Romanesco broccoli was probably evolved from wild or primitive cultivated forms in Roman times. It was cultivated in the area of Rome where it was discovered and improved by Dutch gardeners.

INTRODUCTION

Romanesco broccoli, also known as calabrese romanesco, turret cauliflower, coral broccoli and minaret broccoli, belongs to the *Botrytis* group of *Brassica oleracea*, which is a breed of cauliflower and is a highly decorative heirloom vegetable variety with pale green florets in a complex bewitching design, an amazing example of phyllotaxis that form an attractive fractal-like conical spiral pattern, which makes this vegetable attractive. Its most well-known common name actually makes it a type of cauliflower and its name romanesco broccoli is due to its pointy turret-like floret structure. Romanesco, an Old Italian vegetable variety that has been rediscovered, bears little resemblance to other types of broccoli but its flavour differs from broccoli. It is sometimes called broccoflower, which is also applied to green-curded cauliflower cultivars. It is considered one of the most easily digestible vegetables.

COMPOSITION AND USES

COMPOSITION

Romanesco broccoli is a good source of carbohydrates, protein, calcium, phosphorus, potassium and magnesium. It is rich in vitamin A and vitamin C but poor in fibre content. It also contains glucosinolates, which are responsible for its characteristic flavour and taste. The nutritional composition of *Romanesco* is given in Table 35.1.

USES

Romanesco broccoli in its very young stage is eaten raw or after cooking. Similar to kale, the leaves are also edible but very bitter in taste. They are used in pasta and stews, and sometimes, they are cut into small pieces and used raw in mixed salads or pickles. Romanesco broccoli is often used in Italian recipes. It is popular as a quick-frozen vegetable, particularly in Europe and the United States of America.

TABLE 35.1

Nutritional Composition of Romanesco Broccoli (per 100 g Edible Portion)

Constituents	Contents	Constituents	Contents
Water (g)	94	Sodium (mg)	19
Carbohydrates (g)	4.99	Iron (mg)	0.9
Protein (g)	2.42	Vitamin A (IU)	589
Fat (g)	0.25	Thiamine (mg)	0.1
Dietary fibre (g)	2.6	Riboflavin (mg)	0.1
Calcium (mg)	20	Niacin (mg)	0.1
Phosphorus (mg)	48	Vitamin C (mg)	49
Potassium (mg)	326	Vitamin E (mg)	0.2
Magnesium (mg)	17	Energy (kcal)	25

MEDICINAL USE

Romanesco broccoli has an immense status in medicinal use, as it helps in improving and protecting eyesight by supporting the vision. It has antiviral, anti-inflammatory, antibacterial and anti-microbial properties. It is best source of antioxidants like flavonoids called *kaempferol* that protects the body from free radicals and lowers the risk of bladder, breast, colon, prostate and ovary cancers and prevents the multiplication of new cells. It helps in healing wounds and protecting from flu and cold infections. It improves health of red blood cells, lungs, bones and muscle health, brain functions, etc. It boosts overall immune function, strengthens both the cardiovascular and respiratory tissues, detoxifies the blood and safeguards the central nervous system. It reduces blood pressure, depression, constipations, cholesterol levels and blood sugar level. It contains glucosinolates and thiocyanates like sulforaphane and isothiocyanates, which boost the immune system and the liver's ability to neutralize potentially toxic substances. Its high carotene content improves skin health, skin appearance and skin quality. It helps in curing anaemia and infertility. The fibre found in romanesco aids in digestion and helps in maintaining a healthy body weight.

BOTANY

The plant stem is slightly shorter and its thickness is almost similar to that of cauliflower. The plant is erect glabrous annual or biennial herb growing to a height of 80 cm at vegetative and 150 cm at flowering stage, with unbranched stem thickening upwards but root system is strongly branched. The leaves are large, usually oblong, the younger ones being almost sessile but closely arranged and more or less erect, forming a rosette surrounding the young inflorescence. Leaf blades are ovate to oblong up to 80 × 40 cm, coated with a wax layer and bluish-green with whitish veins. Unlike other cole crops, its buds do not usually arise in the leaf axils. The peduncles that are composed of thin-walled parenchyma and vascular bundles do not become woody. At initial stage, the young curd remains completely covered with foliage, however, on becoming visible, it may have 5 cm diameter, and after some time, the flower stems elongate and a number of apices develop into normal flowers. Its pedicel is 2 cm long, tetramerous and ascending. Flowers, which are bisexual with four oblong sepals, four 1.5–2.5 cm long erect petals, six stamens and two carpels, are borne in racemes on main stem and its branches. The carpels form a superior ovary with false septum and two rows of campylotropous ovules. The androecium is tetradynamous, containing two short and four long stamens. Fruit is a linear siliqua of 5–10 cm × 0.5 mm with a 5–15 mm long tapering beak, and 30 seeds, which are 2–4 mm in diameter, finely reticulate, globose and brown.

CLIMATIC REQUIREMENT

Romanesco broccoli requires temperature and moisture for optimum growth and development similar to other broccoli varieties but it is less adapted to extreme heat or cold. It grows well in cool and humid areas but not in very cold areas. The average minimum temperature for its successful cultivation is about 7°C. Its plants at young stage can tolerate light frost too, however, at mature head stage they can easily be damaged by frost, especially during sudden cold periods. Hot weather affects yield and quality of its curds, hence, those, which mature in summer months, will often be poorer in quality. It grows best under partially shaded conditions and in subtropical areas too.

Good curd induction requires a relatively low night temperature of 10–15°C for temperate and 18–22°C for tropical cultivars. Temperature higher than optimum during curding will result in delayed curd formation and development of disorders such as fuzziness, riceyness and leafiness. When the temperature is below optimum, premature flower bud formation causing a rough granular appearance can occur.

SOIL REQUIREMENT

Romanesco broccoli prefers well-drained, deep and fertile soils. Soils with high organic matter content and pH range of 6.5–7.5 are most suitable for the production of better quality curds. It can tolerate slightly alkaline soil up to pH 8.0 but cannot tolerate acidic soils. The field is prepared to a fine tilth by repeated ploughing and planking. The seedbeds of appropriate size depending on field level are made to transplant the seedlings.

CULTIVATED VARIETIES

Caesar is a unique romanesco hybrid with a high yield potential and medium maturity period. Its plants are medium in size, uniform and vigorous, and curds are light green, conical and compact. Best curd colour is obtained when the plants are grown with modest stress. Its curds can be harvested at slightly larger stage than the curds of open pollinated varieties. It is suitable for cultivation in cool season production.

PLANTING TIME

In subtropical regions, sowing of romanesco broccoli starts from September to November, and in temperate regions, its sowing is done in April–May. Romanesco flourishes in cool weather, thus, the growers in cooler areas are able to take advantage of good production during summer when there is a demand for quality curds.

SEED RATE

Several factors such as variety, soil fertility, growing season and region and seed viability influence the seed rate. About 100–200 g seed is required for raising seedlings enough for planting a hectare land area. Seed before sowing in nursery is treated with organo-mercurial chemical like thiram or captan 3 g/kg of seed before planting. Its seed test weight is 3–5 g.

PLANTING METHOD

The freshly harvested seeds of romanesco broccoli are in dormancy, which can be broken by soaking and rinsing in water overnight. Seeds are sown in well-prepared nursery beds 6 mm deep and 5–7.5 cm apart. The seedlings in young stage should be protected from blazing sun by covering with green shading net and from frost by covering with polyethylene sheet or low plastic tunnels.

Hardening-off is necessary, especially when the seedlings are to be planted in warmer months. Before uprooting the seedlings, the nursery beds should be moistened by sprinkling water with rose can, and for better survival of the seedlings, there must be minimum gap between uprooting and transplanting of seedlings.

In subtropical and temperate regions, nearly 4- and 8-week-old seedlings are ideal for transplanting, respectively. Healthy and stout seedlings free from disease and lankiness are best for transplanting in the field. Straight and stocky seedlings with three to four true leaves and 10–13 cm height are ideal for transplanting. The seedlings with terminal bud damaged should not be used for planting, as these seedlings will be unproductive. Vegetative propagation is also possible through transplanting of lateral shoots. Like other broccoli varieties, the seedlings of romanesco broccoli are planted at a spacing of 50 cm between rows and plants, accommodating around 15,000–35,000 plants per hectare.

While planting, the seedling roots point should be kept straight down and not be bent during the process of planting, otherwise the plant growth will be stunted and will not produce curds. In wet areas, planting on raised beds or ridges is advisable to reduce the risk of waterlogging and rotting of stem or roots. The width of beds is usually kept about 1 m and length according to convenience. The beds are usually raised about 15 cm above the ground and there must be a small furrow between two beds, which will also enable drainage.

NUTRITIONAL REQUIREMENT

Romanesco broccoli is a nutrients-exhaustive crop. Well-decomposed farmyard manure should be applied @ 20–25 t/ha at the time of field preparation. Other organic matter such as compost, wood ashes, bark, rabbit, poultry and barnyard manure are also good for its growth and development. Its high-yielding varieties need nitrogen 180–220 kg, phosphorus 25–40 kg and potash 200–300 kg/ha, depending on soil fertility. Full dose of phosphorus and potash along with one-third of nitrogen should be applied at the time of planting, and remaining two-thirds of the nitrogen dose is top-dressed in two equal halves 1 month after planting and at the time of curd formation. Nitrogen deficiency at early growth stage will cause buttoning, which shows stunting of plants with reduced leaf and curd development. The romanesco broccoli also needs boron and molybdenum, as the crop is sensitive to boron and molybdenum deficiency, hence, incorporation of borax 15–20 kg and ammonium molybdate 1–2 kg/ha into the soil at planting time is necessary to prevent browning and whiptail physiological disorders, respectively. Their deficiencies occur more commonly in alkaline soils, thus, the crop should be cultivated on fertile soils neutral in reaction.

IRRIGATION REQUIREMENT

A light irrigation is given just after transplanting for the establishment of seedlings. It requires plenty of regular moisture supply with high humidity conditions. However, over-irrigation should be avoided, as continuous wet conditions favour the decaying of seedling roots. Mid to late summer is a critical period for applying irrigation in romanesco broccoli, as the temperature is high and soil moisture is low during this period. Such conditions affect the growth and development of the plants adversely.

INTERCULTURAL OPERATIONS

Keeping the crop field weed free at initial stages is always beneficial. Weeding is an essential operation, which should be done frequently after planting to suppress the weeds and to reduce crop-weed competition for nutrients, moisture, space and light. Mulching with rice straw is beneficial to plant growth, as it retains soil moisture, keeps the soil temperature favourable and suppresses weeds.

CROP ROTATION

Following cropping pattern is important in reducing soil-borne pathogens and pests that survive in infected plant residues and have a specific host range. Rotation is often designed to include a green-manure crop in order to increase organic matter content in soil. The crops belonging to *Brassicaceae* family should not be planted in the same field more than once in every 3 years though the unrelated crops should be included in a rotational system. The cruciferous weeds must be controlled meticulously during the period when brassica crops are not grown otherwise much of the crop rotation benefits are lost. Legumes, which are the most suitable green maturing crops for brassicas, should be ploughed in the field when the plants are still tender and at least 8 weeks before planting.

HARVESTING

The crop usually becomes ready for harvesting about 75 days after planting. The curds are cut at the base when they have attained marketable size but before the opening of florets. The curds are cut close to the ground level with a sharp edge sickle. Harvesting of open pollinated varieties is done regularly over a period of 7–15 days though harvesting of F_1 hybrids can be done only in 2–3 cuttings. The leaves are trimmed, keeping a few leaves attached to protect the curds during packing and transport. The curd should have at least a 10- to 15-cm-long stem.

YIELD

The yield depends on variety, soil fertility, irrigation sources, growing season, climatic conditions of the growing region and package of practices adopted by the growers. Its average marketable yield is 160–250 q/ha.

POST-HARVEST MANAGEMENT

The unwanted lower leaves are trimmed off, leaving only a few jacket leaves projecting 2–3 cm above the curd to protect it from mechanical injury during handling and transportation. However, more jacket leaves are retained when romanesco is transported loose rather than packed in crates. Its curds deteriorate quickly unless pre-cooled soon after harvesting. Its storage life is usually much shorter unless the curds are wrapped in polythene film to prevent rapid desiccation and yellowing. However, at 1°C temperature and 95% relative humidity, the curds can be stored for about 3 weeks.

INSECT-PESTS

DIAMONDBACK MOTH (*Plutella xylostella*)

Diamondback moth, which is one of the serious pests of cruciferous vegetable crops, also attacks romanesco broccoli. Its mature moths are small greenish yellow caterpillars with black hairs. The body is slender and pointed at both the ends, with a distinctive V formed by two pro-legs at rear end. The feeding by larvae causes small bite holes on leaves mostly on the growing part, which usually results in excessive defoliation. The growing points of younger plants are chewed as a result, the plants become stunted. Some of the larvae may bore into curds, causing infection. When the attack is severe, plants never form curds and the yield loss may be up to 80–90%, especially during hot periods.

Control
- Grow resistant variety, if available.
- Grow mustard as a trap crop.

- Destroy plant debris and mustard-type weeds before planting.
- Use *Bacillus thuringiensis* formulations.
- Spray the crop with 4% *neem* seed kernel extract, malathion or Thiodon 0.2%, pyrethrins 0.1%, quinalphos, or chlorpyrifos 0.05% at 10- to 15-day intervals starting from 15 to 20 days after planting.

Aphids (*Brevicornea brassicae*)

Aphids are small soft-bodied insects with long slender mouthparts that they use to pierce leaves and other tender plant parts and suck cell sap, which turn the leaves yellow as a consequence, the plant becomes stunted. The insect produces large quantity of sticky exudates known as honeydew. The leaves often turns black due to the growth of sooty mould fungus on sticky sugary substance, affecting the photosynthetic rate of the plants. It attacks the plants mostly at seedling stage. High humidity and high temperature favour the occurrence of pest. It also causes the leaves to curl and further distorts the plant growth.

Control

- Grow resistant variety, if available.
- Use mustard as trap crop to attract aphids.
- Use yellow sticky traps in the field.
- Treat the seed with thiamethoxam 30FS 5 g/kg of seed.
- Spray the crop with 0.5% *neem* oil or 4% *neem* seed kernel extract.
- Spray the crop with malathion 0.05%, acephate 0.1%, dimethoate 0.03%, imidacloprid 0.1%, or oxydemeton methyl 0.05% at 10- to 12-day intervals.

Cutworms (*Agrotis ipsilon*)

Cutworms are dull brown caterpillars having a length of 2.5–5 cm when fully grown. When they are disturbed, they curl into a C-shape. Normally, they live just below the soil surface or on lower parts of the plants and become active at night. They are smooth-skinned and have markings that blend well with the soil. Cutworms feed on blossoms and leaves of many ornamental plants and attack most garden crops. They clip off seedlings near or just below the soil line. A few species make holes on leaves or bore into curds of cole crops.

Control

- Grow resistant variety.
- Destroy crop residues and keep the field weed free.
- Hand-pick the cutworms at night with a flashlight and destroy them.
- Clip and dispose of the infested foliage and blossoms.
- Protect the seedlings from cutworms with sticky collars, screen, or protective cloth.
- Spray *Bacillus thuringiensis* formulations on crop.
- Treat the stems or the foliage of the plant with permethrin, or cyfluthrin.

Stem Borer (*Hellula undalis*)

Larva looks brownish with black head and four longitudinal strips on the body. Moth appears with wavy lines and central elliptical spots on forewings. The pupation takes place inside burrow of the stem as well as in soil. Larvae web the leaves and bore into the main stem immediately after transplanting and prevent head formation, which results in production of multiple heads. Larvae also feed on stalk or leaf veins. When the attack is severe, it makes the produce unfit for consumption.

Control

- Keep the field fallow.
- Follow long crop rotation with non-host crops.
- Grow a few lines of mustard as a trap crop.
- Install sex pheromone trap in the field.
- Spray fenthion or fenitrothion 0.05% at monthly intervals.

Root-knot Nematodes (*Meloidoginy* spp.)

The nematode infection causes swelling of roots and formation of knots or galls, which make the roots unable to utilize soil nutrients and moisture efficiently since the infected roots are retarded in growth and lack fine feeder roots. In late season, the roots may rot due to the occurrence of several other infections along the same area. Under a condition of severe infection, the symptoms appear as stunted plant growth, leaf chlorosis, wilting, drastic reduction in yield and simultaneous death of plants.

Control

- Grow nematode-resistant varieties, if available.
- Deep ploughing of field during hot summer.
- Use suppressive plants like *Tagetes* species.
- Apply organic amendments in soil, *i.e.*, groundnut, *neem*, *mahua*, or *karanj* cake.
- Irrigate the field with frequent irrigation to reduce damage of nematode.
- Turn the soil two to three times at 10-day intervals in summer.
- Sterilize the nursery beds through soil solarization in hot summer months.
- Fumigate the field with Nemagon before planting.

DISEASES

Damping-off (*Pythium, Phytopthora, Fusarium, Sclerotium, Botrytis, Sclerotinia* and *Rhizoctonia*)

This is a serious fungal disease of nursery, which may be pre- and post-emergence. In pre-emergence damping-off, the pathogens attack the seed and cause decaying of seeds, but in case of post-emergence damping-off, pathogens attack the collar region of the young seedlings, which as a result topple over to the ground in a few days. High soil moisture high temperature and high humidity favour the incidence of disease.

Control

- Avoid overcrowding of seedlings to improve aeration.
- Follow proper planting depth to sow seeds.
- Avoid excessive irrigation.
- Treat the seed with thiram or captan 3 g/kg of seed.
- Drench the nursery beds with thiram, captan, or Bavistin 2 g/litre of water.

Black-leg (*Phoma lingam*)

The fungus often kills seedlings, which topple over to the ground due to the distortion of vascular bundles and root system. It produces black sunken cankers at the stem base, which cause stunting of surviving plants. Yellow to brown spots with grey centres appear on leaves. Discolouration of bundles can be seen when the stem of affected plant is split open vertically. The presence of tiny black bodies (pycnidia) on cankers is a characteristic symptom of black-leg disease. The black bodies contain fungal spores, which ooze out during wet weather.

Control

- Follow long crop rotation with non-cruciferous crops.
- Dip seeds in hot water at 50°C for 30 minutes.
- Treat the seeds with fungicides like thiram or captan 3 g/kg of seed.
- Drench the field with thiram, captan, or Bavistin 2 g/litre of water.

CLUB-ROOT (*Plasmodiophora brassicae*)

A soil-borne fungus enters roots through wounds and multiplies rapidly. It causes abnormal swelling or enlargement of roots and rootlets like a shape of a club. It retards growth and imparts a pale green colour to the plants. These roots often start decaying before the crop matures, releasing many latent spores, which can survive in soil for a decade in the absence of a susceptible host plant. Temperatures 17°C –23°C favour the growth of club-root disease.

Control

- Follow deep ploughing in hot summer months.
- Follow long crop rotation with non-cruciferous crops.
- Grow resistant variety, if available.
- Eradicate and destruct the host weed plants.
- Add lime in soil to raise the pH level.
- Treat the soil with 0.1% Brassicol to reduce the infestation.
- Dip the seedling roots in 4% calomel before transplanting.

FUSARIUM YELLOWS (*Fusarium oxysporum f.sp. conglutinans*)

The fungus-affected plants become stunted. Yellowing starts from lower leaves, resulting in brown to black veins discolouration. However, this yellowing may be distinguished from black rot yellowing because leaf dieback progresses from petiole or midrib outwards. The affected leaves are usually curved laterally, leaf margins may have reddish-purple discolouration and pockets of dark discolouration are not associated with the vascular system when crucifers are infected with yellows fungus.

Control

- Add organic matter to increase microbial activity to reduce fungus growth.
- Apply less quantity of nitrogenous fertilizers.
- Treat the seeds with 0.2% ziram solution at 35°C for 30 minutes.
- Spray the crop with 0.2% Bavistin or Benlate at 10-day intervals.

SCLEROTINIA BLIGHT (*Scierotinia sclerotiorum*)

The disease spreads through wind-blown spores or directly by fungal strands arising from hard black fungal bodies called sclerotia. Water-soaked spots appear on any part of the plant, usually on leaves close to the ground, or on curds. Affected tissue often turns grey, in wet weather giving rise to a fluffy white mould, which sooner or later is dotted with black sclerotia. Pale grey-bleached spots, stem-rot, stunting and premature death of the plant may occur during flowering, especially in densely populated plantings.

Control

- Follow deep summer ploughings.
- Follow crop rotation of at least 3 years with non-host crops.
- Grow resistance variety, if available.
- Use disease-free healthy seeds for cultivation.

- Treat the soil with a mixture of Brassicol and captan at the rate of 10 kg/ha in 1:1 ratio.
- Soon after harvesting, the diseased plants should be destroyed by burning.

Downy Mildew (*Peronospora parasitica*)

This is a serious disease of romanesco broccoli. Primarily, discolouration of young seedlings occurs. When the infection is severe, the whole plant may die. Small yellowish brown spots appear on lower leaves or greyish discolouration on curds, which eventually expand and develop greyish black lace-like markings. Under moist weather conditions, bluish-white fluffy downy mould growth appears on lower leaf surface. The disease develops rapidly at 8°C–16°C temperature and 98% relative humidity.

Control

- Grow resistant varieties, if available.
- Follow a crop rotation of at least 3 years with non-cruciferous crops.
- Maintain optimum plant density.
- Treat the seeds with hot water at 50°C for 30 minutes.
- Spray the crop with Ridomil MZ 72 @ 0.5 g /litre of water, nitrophenol 0.05%, mancozeb 0.2%, or chlorothalonil 2 g/litre of water at 10-day intervals.

Sclerotinia Stalk Rot (*Sclerotinia sclerotiorum*)

The infection begins as water-soaked circular areas, which soon become covered by white cottony fungal growth. The affected tissue becomes soft and watery as the disease progresses. The fungus eventually colonizes the entire curd and produces large black seed-like structures called sclerotia on infected tissue. The disease causes serious loss in field, storage and during transportation. Damp weather favours the occurrence of the disease. Infection may occur on stem at ground level, on leaf base, or on foliage that comes in contact with soil.

Control

- Use disease-free healthy planting material.
- Follow deep summer ploughing.
- Maintain proper aeration in the field.
- Avoid mechanical injuries to curds during harvesting and transportation.
- Treat the seed with captan 3 g/kg of seed.
- Spray the crop with 0.25% Dithane Z-78 or Dithane M-45 at 10- to 12-day intervals.

Black Rot (*Xanthomonas campestris* pv. *campestris*)

Infected seedlings show the symptom of yellowing, dropping of lower leaves, and finally, the plant may die. Only one side of leaves of a seedling may be affected. The symptoms are caused by local infection, which results when bacteria enter leaves through natural openings at leaf margins. The infected tissue turns pale green and then turns brown and dies. Affected areas that are usually V-shaped enlarge as the disease progresses and the severely affected leaves may drop off. The veins of infected leaves, stems and roots sometimes become black. The infected plants develop small and poor-quality curds, which are unfit for consumption. The disease is common in areas having a warm and wet climate. Plants can be infected at any stage and the symptoms resemble nutritional deficiencies.

Control

- Use resistant varieties, if available.
- Use disease-free healthy planting material.

- Follow long crop rotation with non-cruciferous crops.
- Apply recommended boron doses to reduce the disease.
- Treat the seeds with Agrimycin 100 or streptocycline at 100 ppm.
- Treat the seeds with hot water at 52°C for 30 minutes followed by 0.01% streptomycin dip.
- Spray the crop with streptomycin or Agrimycin 100 at 100–200 ppm at transplanting and curd initiation stage.

■■■

36 Brussels Sprout

Sonia Sood and M.K. Rana

Botanical Name : *Brassica oleracea* var. gemmifera

Family : Cruciferae (Brassicaceae)

Chromosome Number : 2n = 18

ORIGIN AND DISTRIBUTION

Brussels sprout has evolved from wild cabbage (*Brassica oleracea* var. oleracea, synonym sylvestris), which is found growing wild in cool northern Europe and Western Mediterranean and the plant in the present form appeared in Brussels (Belgium) in 1750 and reached England and France in 1800, however, the crop became popular in Europe and California in late eighteenth and early nineteenth century. In India, it was introduced during the second half of the twentieth century.

INTRODUCTION

Brussels sprout is an important cole crop in most of the European countries. It is a hardy, slow growing, long season and cold climate winter vegetable, which gained importance in the 19th century and named after the capital city of Belgium since it has been grown around Brussels for hundreds of years. From Europe, the crop was taken to other parts of the world, and gradually, it has become popular especially in temperate regions. It is mainly cultivated in United Kingdom and northwest Europe (the Netherlands, Belgium and France). It is also grown on small scale in the United States of America (California), Eastern Europe and Australia. In India, the cultivation of Brussels sprout has not been exploited commercially and is limited to kitchen gardens, places of tourist importance and near big cities. Being a nutritious vegetable, it is gaining importance by the high living people. It is cultivated in Himachal Pradesh, Jammu and Kashmir, hilly regions of Uttar Pradesh, Kodaikanal and Nilgiris of Tamil Nadu and Maharashtra. Among cole crops, it ranks at fifth place in India, though no figures for area utilization and volume of the crop are available.

COMPOSITION AND USES

COMPOSITION

Nutritionally, it is quite comparable with broccoli for vitamins and minerals except for vitamin A and calcium. It has more protein and carbohydrates content in comparison to other cole crops. It is rich in many valuable nutrients such as carbohydrates, sugar, soluble and insoluble fibre, sodium, minerals, sulfur, fatty acids and amino acids. It is one of the excellent vegetable sources for vitamin K (177 mg/100 g edible portion). The nutritional composition of Brussels sprout is given in Table 36.1.

TABLE 36.1

Nutritional Composition of Brussels Sprout (per 100 g Edible Portion)

Constituents	Contents	Constituents	Contents
Water (g)	85.5	Zinc (mg)	0.42
Carbohydrates (g)	7.1	Copper (mg)	0.07
Proteins (g)	4.7	Vitamin A (IU)	520
Fat (g)	0.5	Thiamine (mg)	0.05
Dietary fibre (g)	1.2	Riboflavin (mg)	0.16
Calcium (mg)	61	Niacin (mg)	0.40
Phosphorus (mg)	80	Pantothenic acid (mg)	0.09
Potassium (mg)	404	Pyridoxine (mg)	0.31
Magnesium (mg)	23	Folic acid (mcg)	61
Sodium (mg)	22	Vitamin C (mg)	72
Iron (mg)	1.5	Vitamin E (mg)	0.88
Manganese (mcg)	0.42	Energy (kcal)	52

Uses

Brussels sprout is a nutritive green vegetable and grown for fresh consumption as well as for processing (freezing and canning). Its taste is much like cabbage but has a slightly milder flavour and a denser texture. Sprouts are cooked and served in the same way as the cabbage heads. These are stir-fried, boiled, steamed, or cooked as single or mixed with other vegetables. Sprouts are eaten raw as salad and liked in soups. The leaves can also be used as a substitute of kale. The frozen product is popular in western countries like the United States of America and the United Kingdom. It is also used as a canned product. Sprouts in general, if harvested at the proper stage, have a longer shelf life, and a large number of hybrid varieties suitable for freezing are also now available; thus, there is a good potentiality for its export from cultivated areas to hotter places within and also outside the country as a fresh vegetable. There is also a good potential to export this vegetable in frozen form since a large number of hybrid cultivars suitable for freezing are available.

Medicinal Uses

Brussels sprout is fairly high source of glucosinolates (160–250 mg/100 g edible portion) predominantly sinigrin and glucobrassicin, which are known to fight cancer, including bladder, breast, colon, lung, prostrate and ovarian cancer. It prevents constipation, maintains blood sugar and checks overeating. Sulforaphane also protects the stomach lining by obstructing the overgrowth of *Heliobacter pylori*, a bacterium that can lead to peptic ulcer. Brussels sprout has been used to determine the potential importance of crucifers on thyroid function and provides health benefits without putting the thyroid gland at risk. Vitamin K has a potential role in bone health by promoting osteotrophic (bone formation and strengthening) activity. Adequate vitamin K levels in the diet helps in limiting neuronal damage in the brain and thereby preventing or at least delaying the onset of Alzheimer's disease. Beta-carotene of Brussels sprout acts as a potential antioxidant. It is a suitable food for those who are suffering from water retention, obesity, arthritis and constipation. It helps in healing wounds.

BOTANY

Brussels sprout (*Brassica oleracea* var. *gemmifera*) is a form of cabbage, belonging to the family Brassicaceae, which is widely grown for its edible buds or lateral small heads. The varietal name *gemmifera* means that the plant reproduces by gemmae of leaf buds, which when detached from the mother plant are capable of rooting and developing into a new plant. The plant looks like a non-heading cabbage with an unbranched upright-growing stem, which produces miniature cabbage heads (sprouts) in the axils of petiolated leaves from the bottom upwards. It is mainly of two types *viz.*, tall ones (0.6–1.2 m) preferred in areas with long growing season and short ones (<0.6 m). The edible part of Brussels sprout is a swollen auxiliary bud, which is about 5–8 cm in diameter and is composed of tightly packed leaves resembling a miniature cabbage head. It is also known by several other names such as sprouts, buttons, or mini cabbage heads. The sprouts are round and white, green, or purple.

Brussels sprouts are commonly grown as annuals, whereas, for seed production, they are biennial. Its inflorescence is a raceme and its flowers are yellow, actinomorphic and bisexual (complete). The flower heads arranged in a tree-like structure on branching sprout from a thick stalk. Calyxes are gamosepalous with four sepals and the corolla gamopetalous with four petals and cruciform. Pods are linear and dehiscent, beak of pods cylindrical or conical and seeds seriate. Pollination is entomophilous, which takes place through honeybees and butterflies. It is highly cross-pollinated crop due to sporophytic self-incompatibility, which remains inactive in bud stage, *i.e.*, 2–4 days before anthesis. It becomes active in stigma at flowering stage. Anthesis takes place early in the morning between 7:30 and 8:00 a.m. and continues up to 11:30 a.m., depending upon prevailing weather conditions. Pollen fertility is observed at its maximum on the day of anthesis. Stigma becomes receptive 2–3 days before the day of anthesis due to protogyny condition of the flowers. The glycoproteins present in the stigma hinder the pollen germination and penetration of pollen tubes through the styler tissue. Seeds are borne in siliqua. The plant for seed production needs exposure to low temperature (4–10°C) during winter to transit to generative phase. However, a certain amount of vegetative growth is necessary for the conversion of juvenile phase to reproductive phase. In Brussels sprout, 30 petiolated leaves are formed first, the apex increases in size and becomes globular in shape due to the formation of sessile leaves and the stem becomes thickened due to accumulation of carbohydrates. The plant at this stage becomes receptive to receive cold stimulus from the atmosphere. The flower primordia on exposure to low temperature at this stage are laid down and the leaf primordia become progressively smaller and finally only the floral primordial develops. As soon as the temperature start rising in spring, these plants bolt and bear flowers. Day length does not have any effect on these processes. A temperature of 7°C for 28 days is sufficient to induce flower initiation, whereas, the induction period is 35 days at 4–10°C. The plants at 14–23°C do not transform to generative phase and remain vegetative.

CULTIVATED VARIETIES/HYBRIDS

The productivity, harvest period, shape, size, colour intensity and compactness of the buds are important attributes of Brussels sprout. Its cultivars may be dwarf or tall. Dwarf varieties have stem less than 50 cm in length and sprouts are small and crowded on the stem. These are early and suitable for areas where growing season is short. The common cultivars are Improved Long Island, Early Morn, Dwarf Improved, Frontier Zwerg and Kvik. The tall varieties have plant stem more than 50 cm in length and are late in maturity. These are suitable for the areas where growing season is long, especially in England, other parts of Europe and north Indian hills. Late and hardier cultivars are Hilds Ideal, Red Vein and Amarger. However, the open pollinated varieties in both early and late group have been replaced by hybrids because of their wider adaptability, uniform sprout shape and maturity, which are suitable for mechanical harvesting. The hybrids marketed by different companies of Japan and Europe are Jade Cross, Pearl, Crystal, Dorman, Predora, Rovoka, Sonara, Oliver, Valiant, Rogor, Rider, Boxer, Richard, Royal Marvel, Asgard, Kundry, Rasmunda and Stephen. In India, the crop improvement work has been carried out only at Temperate Vegetable Research Station of Indian Agricultural Research Institute, Katrain, and the open pollinated variety Hilds Ideal has been found suitable for cultivation in hills as well as in Maharashtra. Early Dwarf, Early Morn and Bowler are the most promising genotypes exhibiting high sprout and seed yield at

Kalpa, Himachal Pradesh, India. The description of some important varieties cultivated in India is given as follows:

HILDS IDEAL

It is an open pollinated cultivar with plant height of 55–60 cm and 45–50 sprouts per plant. Each sprout on an average weighs 60–70 g and diameter 7–8 cm. Sprouts are light green, globular and solid and take 115 days for first picking after transplanting and give four to five pickings at about 10-day intervals. Its average yield is 160 q/ha.

JADE CROSS

A Japanese F_1 hybrid cultivar has become popular for its short stemmed, early, high-yielding and single-harvest quality. It belongs to a dwarf group of varieties.

RUBINE

It is a high-yielding commercially grown variety that bears red sprouts.

BRUSSELS DWARF

Plants are dwarf (50 cm in height), early and suitable for areas of short growing season. It is a high-yielding variety and its sprouts are medium sized.

CLIMATIC REQUIREMENTS

Brussels sprout requires a long cool and humid climate and grows well in regions with moderately severe winters. It can withstand frost and extreme cold weather. For good growth and development of sprouts, an average temperature of 16–20°C is favourable. Sunny days with light frost during the night result in the production of the best quality sprouts, which are solid, tender and sweet in taste. Puffy and loose sprouts are produced at temperatures above 27°C. Its late cultivars are frost resistant and can withstand temperature as low as −10°C. It can continue to grow even at 0°C. Being a strong biennial plant, it does not bolt and flower unless exposed to chilling temperature (<10°C) for 6–8 weeks.

SOIL REQUIREMENT

Brussels sprout is a nutrients exhaustive crop, which demands highly fertile soil with a pH between 6.0 and 7.0. In acidic soil, light dusting in the planting hole with lime reduces soil-borne disease called club root and increases the yield of Brussels sprout. It grows into a hard top-heavy plant with thin base, which is easily damaged by strong winds. If the plant falls over the ground will continue to grow but will not produce nearly as well as the plant that stays upright. Light sandy soils cannot adequately anchor the Brussels sprouts plant, even when the plant is supported with stakes or the soil is mounded up around base of the plant. Therefore, the ideal soil texture for its cultivation is dense clay loam with a few rocks clay since tight clay loam soil helps in holding the roots firm when the large plant is blown by high-speed winds.

NURSERY MANAGEMENT

The seedlings of Brussels sprout are raised from seeds sown on well-drained and finely prepared raised nursery beds of 3.0 × 1.0 × 0.15 m size. About ten nursery beds are required to produce seedlings sufficient for planting a hectare land area. Nursery management practices are similar to those followed in broccoli.

SOWING/PLANTING TIME

The time of sowing depends on variety, growing season and location. In north Indian plains and low hills, the optimum time for sowing seeds in the nursery beds is early September. The optimum time for nursery sowing in the hills up to 1500 m average mean sea level with mild or rare snowfall is early July, whereas, it is March–April in very high hills experiencing heavy snowfall. Nursery sowing during October has been found appropriate in the Mahableshbwar area of Maharashtra. The planting time should be adjusted in such a way that at the time of sprout formation, the temperature is mild.

SEED RATE

The seed rate depends on the variety, spacing and real value of the seed. About 300–400 g seed is required to raise seedlings sufficient for transplanting a hectare of land area. A quantity of one gram contains 350 seeds of Brussels sprout and seed size is comparable with that of cabbage. For direct seeding, the seed requirement is four times higher, *i.e.*, 1.2–1.6 kg/ha.

PLANTING METHOD

The seedlings become ready for transplanting in 4–5 weeks after sowing. The seedlings are transplanted in the evening either on flat or raised beds. In agriculturally advanced countries, direct sowing of seeds is done for commercial cultivation and thinning is done to maintain the recommended spacing.

The spacing depends upon the sowing time, variety used and method of harvesting. For single mechanical harvesting, close spacing between the plants is followed. Depending upon the variety and length of the growing period, inconsistent spacing is followed. Spacing of 60 × 45 or 60 × 60 cm is followed for dwarf varieties, whereas, wider spacing of 75 × 45, 75 × 60, or 90 × 90 cm is kept for tall varieties, especially where the growing period is relatively longer. For Hilds Ideal variety, a spacing of 60 × 45 cm, which accommodates about 33,000 plants in a hectare area, has been found appropriate for north Indian hills and Peninsular India. The sprout size increases as the distance between rows is increased.

NUTRITIONAL REQUIREMENT

Brussels sprout needs a heavy intake of nutrients. Besides the application of farmyard manure 25 t/ha, nitrogen is applied 200 kg, phosphorus 90 kg and potassium 80 kg/ha, depending upon fertility status of the soil and cultivars. The entire quantity of farmyard manure, phosphorus and potassium along with one-third dose of nitrogenous fertilizer is applied as basal dose prior to transplanting. The remaining two-thirds dose of nitrogen is top-dressed twice in two equal halves *viz.*, about 1 and 2 months after transplanting.

Increase in nitrogen levels results in the reduction of total sugar, dry matter content and ascorbic acid. Excessive application of nitrogen results in quick growth of plant and development of loose sprouts, thus, it should be avoided. Similarly, excessive application of potash imparts a bitter taste to the sprouts.

IRRIGATION REQUIREMENTS

An even moisture supply is needed to ensure continuous and optimum plant growth for getting higher yield. Surface irrigation or repeated spot-watering manually is carried out until the seedlings are established, and thereafter, the irrigation is given depending upon the soil type and prevailing weather conditions. Optimum moisture level at the time of head formation is very essential. In general, Brussels sprout field is irrigated at 8- to 10-day intervals. Every week, it needs 2–3 cm irrigation water. Fertilizer could be applied through an irrigation system.

INTERCULTURAL OPERATIONS

HOEING AND WEEDING

Brussels sprout is a shallow-rooted crop. Therefore, deep hoeing and weeding should be avoided to minimize injuries to its roots. Weeds create a serious problem in the cultivation of crop due to wider spacing, higher fertility and frequent irrigations. Hoeing within 2–3 weeks of transplanting is done to remove weeds and to facilitate good aeration into the soil. Top-dressing with nitrogenous fertilizers followed by earthing-up is carried out after 6–7 weeks of transplanting.

Weedicides recommended for cabbage and other cole crops prove effective in Brussels sprout as well. Application of weedicides like trifluralin 1 kg/ha, fluchloralin 0.5 kg/ha, nitrofen 2 kg/ha, oxyfluorfen 1 kg/ha, alachlor 0.2 kg/ha, or butachlor 2 kg/ha 2–3 days before transplanting helps in checking the growth of broad leaved weeds.

Removal of terminal bud of the plant about 3 weeks before harvest is done to ensure uniform sizing at maturity of sprouts in areas having short growing period, *e.g.*, north Indian plains. Pruning of lower leaves before the start of harvest of sprouts leads to yield reduction because of reduction in photosynthetic area.

HARVESTING

Harvesting is done when lower sprouts on the stem attain proper size, firmness and before they start loosening and bursting. Sprouts should be solid, upper leaves folded closely, attractive dark green in colour and uniform in size. Harvesting is done at least three to five times at 7- to 10-day intervals. The harvest period varies according to the agro-climatic zone. In north Indian plains, harvesting period is January–February, whereas, in hills and very high hills of northern India, the harvesting period is November–March and July–September, respectively. Generally, the leaves are removed from the plants before harvesting, which starts from bottom upwards and continues for several weeks. In agriculturally advanced countries, a single harvest with a machine is carried out for industrial processing. The entire crop is harvested after about 4–6 weeks of topping (removal of apical or terminal vegetative bud). Delay in harvesting results in poor-quality sprouts.

YIELD

The sprouts yield varies according to the variety, maturity period, cultural practices adopted during crop cultivation and the length of growing season. In India, a yield of 150 q/ha has been reported. Single harvested crop generally gives a yield of 30–50 q/ha.

POST-HARVEST MANAGEMENT

Brussels sprout is highly delicate as the respiration rate of freshly harvested sprouts is very high. After removing the leaves from the cut stem, the harvested Brussels sprout is required to be hydro-cooled. After grading, the fresh sprouts are packed in cardboard boxes with proper aeration and sent to the market at the earliest possibility. The entire plant can also be stored in a cool underground room and sprouts are removed as per the requirement of the family, especially in high hills. At room temperature, Brussels sprouts can be kept for 2–3 days only. At low temperature (0°C) and 95% relative humidity, it can be kept in good condition for 6–8 weeks. Sprouts packed in polythene ceramic film can be stored well for 16 weeks at 0°C temperature and only for 11 days at 20°C temperature. Storage of sprouts in sealed polyethylene packs at 0–10°C in modified atmosphere containing 2–5% CO_2 and 14.5–16.8% O_2 reduces the losses and improves their quality in comparison to open storage at the same temperature. Longer storage may result in black speckling of leaves, loss of bright green colour, decay and discolouration of the surface.

PHYSIOLOGICAL DISORDERS

The physiological disorders have the symptoms similar to certain diseases but no pathogen or casual organism is involved in these disorders. These may be due to either deficiency of essential elements or climatic factors.

LOOSE SPROUTS

Affected sprouts are large, leafy, open and with poorly formed heads. This may be due to several causes: loosely formed sprouts in Brussels sprout can be due to any stress that slows growth, making them small or open. Fluctuating temperatures and moisture will also cause less compact growth. In contrast, excess vegetative growth caused by excessive use of nitrogenous fertilizer can also be the cause of loose sprout in Brussels sprout. Premature flowering and open heads in Brussels sprout can be brought on by high temperatures.

TIP BURN OF BRUSSELS SPROUTS

Tip burn is a breakdown of plant tissue inside the head of individual sprouts. It is a physiological disorder, which is associated with an inadequate supply of calcium in the affected leaves, causing a collapse of tissue and death of cells. Calcium deficiency may occur where the soil calcium is low or where there is an imbalance of nutrients in the soil along with certain weather conditions, *i.e.*, high humidity, low soil moisture, high potash and high nitrogen aggravate calcium availability. Secondary rot caused by bacteria can follow tip burn of sprout.

This problem can cause severe economic loss. Internal leaves turn brown, especially on edges or tips, and inside sprouts, the central tissues often break down. It is a physiological disorder, which is associated with an inadequate supply of calcium to the young actively growing inner leaves. High humidity, low soil moisture, high potassium, high nitrogen and low soil calcium all influence calcium availability.

INTERNAL BROWNING

In Brussels sprouts, the symptoms are yellowing of leaves at distal ends within the sprout, generally on a line through the central region of the sprout and midway between its growing point and exterior surface. Internal browning is especially troublesome because it usually cannot be detected from external examination of the sprouts, yet quite a small number of affected sprouts may taint a large sample, especially when they are blanched for quick-freezing. The factors that have contributed to internal browning have done so by bringing about a temporary localized calcium deficiency in the sprouts. Rapid growth, low soil moisture and high relative humidity, which inhibit transpiration. The actual soil calcium levels have very little influence unless they are extremely low. Ensuring the soil moisture close to field capacity can reduce the risk of internal browning. Excessive nitrogen fertilizer should be avoided. Foliar sprays of calcium nitrate may prevent the onset of internal browning if applied before the onset of adverse growing conditions.

BLINDNESS

This disorder occurs not only in Brussels sprouts but also in cole crops. It results in no or multiple small heads formation. There are several causes, which include pre- and post-transplanting factors, such as damage to the terminal growing point due to low temperature, cutworm damage, or rough handling of transplants.

INSECT-PESTS

APHIDS (*Brevicoryne brassica* and *Myzus persicae*)

Aphids are tiny grey-green colour insects with soft bodies covered with a white waxy coating. They get shelter within the dense bud clusters and leaves and suck cell sap from tender parts of the plant. Its large population can cause stunting of growth or even death of the plant. As a result, the leaves become curly in appearance, and its nymphs and adults excrete a sticky honeydew-like substance, on which the black sooty mould develops, which hinders photosynthetic activity of the plant. This results in poor growth of young plants, yield and quality of the produce.

Control
- Grow tolerant varieties if available.
- Set yellow pan traps in the field.
- Wash the crop with plain water with high pressure.
- Remove and destroy the affected plant parts in mild infestations.
- Spray the crop with malathion 0.1%, monocrotophos 0.03%, or Nuvan 0.05%.

FLEA BEETLE (*Systena blanda*)

The pest responsible for the damage is a small (1.5–3.0 mm) dark-coloured beetle, which jumps when distributed. The beetles are often shiny in appearance. The beetles feed on leaves, which results in small holes or pits in leaves, giving the foliage a characteristic shot-hole appearance. The young plants and seedlings are more susceptible to its attack. The plant growth, due to its attack, is reduced. If damage is severe, the plant may be killed.

Control
- Application of a thick layer of mulch may help in preventing the beetles' reaching the surface.
- Spray the crop with spinosad 0.02%, bifenthrin, or permethrin 0.4% to control the beetles.

CUTWORMS (*Agrotis* spp.)

Cutworms are greyish, fleshy caterpillars up to a length of 5 cm, which curl up when disturbed. The young caterpillars feed on leaves and later on stems. Most of the damage is caused by mature caterpillars. They are capable of eating or destroying the entire plant. Damage is worse where cutworms are present in large numbers before planting. They girdle and cut off young seedlings at ground level during the night, dragging them into the tunnel in the soil and feeding on them during the day. Cutworms tend to be more frequent in soil containing plenty of decaying organic material. Cutworms often reoccur in the same field, coming with crop residues and dense stands of weeds.

Control

- Use well-rotten farmyard manure.
- Plough the field to expose the caterpillars to predators.
- Pick the caterpillars manually at night by torch or very early morning.
- Flood the field for a few days before sowing or transplanting.
- Pheromone traps can be used to monitor cutworm adults.
- Spray the crop with aqueous *neem* seed and *neem* leaf extracts three times at weekly intervals.
- Add chlorpyrifos 20 EC @ 2.0 litres/ha after mixing in 25 kg of sand before transplanting the seedlings.

CABBAGE CATERPILLAR (*Pieris brassicae*)

Cabbage caterpillars are the larval stage of various butterflies and moths. The larvae are greenish yellow with a black head and black dots on the dorsum. Brussels sprout is the most susceptible crop to this pest. The attack is rather localized and can lead to 100% crop loss in a certain area. The caterpillar can eat twice its weight per day. The female adult lays eggs in clusters on upper surface of the leaves. The young caterpillars scrape the leaf surface and older caterpillars damage the leaves by feeding from leaf margins, and they proceed to the centre and skeletonize the the leaves. Such plants do not develop heads.

Control

- Grow a few lines of mustard around the field as trap crop.
- Pick the infested leaves and destroy them.
- Spray profenofos 500 EC or dimethoate 40% EC @ 1 litre/ha.
- Spray the crop with malathion 0.1%, dimethoate 0.05%, monocrotophos 0.05%, or Nuvan 0.05%.

DIAMONDBACK MOTH (*Plutella xylostella*)

The adult female of this insect lays eggs on undersurface of the leaves. The greyish green caterpillars feed on leaves, thereby leaving the parchment-like epidermis intact and holes in the leaves. Consequently, the leaves turn brown and fall down. The growth of young plants is greatly inhibited and such plants do not form heads. Damage is small incomplete holes caused by young larvae and larger complete holes caused by mature larvae. Under heavy infestations, the entire plants may become riddled with holes. Brussels sprout develops deformed heads due to its attack, which later encourages soft rot.

Control

- Plough the field in hot summer months to expose the caterpillars to predators.
- Follow long crop rotation with non-crucifer crops.
- Grow few rows of mustard or marigold around the Brussels sprout field as trap crop.
- Pick the infested leaves manually and destroy them.
- Encourage the natural enemies by maintaining natural surroundings with plenty of breeding places like trees and shrubs.
- Remove and destroy all the plant debris, alternate hosts and weed hosts from the field.
- Apply irrigation frequently to reduce the mating of moths and wash off the caterpillars and pupae from plant leaves.
- Spray *neem* seed kernel extract 5% at regular intervals.
- Spray the crop with malathion 0.1%, Nuvan 0.05%, fenvalerate 0.01%, or quinalphos 0.2% with repetition at 10-day intervals.

DISEASES

DAMPING-OFF (*Pythium* spp., *Phytophthora* spp., *Rhizoctonia solani* and *Fusarium* spp.)

Damping-off is the pre- and post-emergence death of plants. Pre-emergence attack of these fungi results in death of the seeds before germination, whereas, post-emergence damping-off occurs above the ground on the young seedlings, in which the stem tissue at the collar region becomes soft and water-soaked, and the seedling collapses at the collar region near the soil surface. The roots may also get affected, resulting in death of the seedlings. Excessive soil moisture or waterlogging, high temperature and poor aeration favour the development of disease, and it spreads from the unde-composed infected plant debris present in the soil.

Control

- Plough the field repeatedly in hot summer months.
- Adopt long crop rotation with non-host crops.
- Avoid waterlogging and provide good drainage facilities.
- Treat the seed with Bavistin, thiram, captan, captafol, foltaf, or Apron 2.5 g/kg of seed.
- Drench the beds with captan or thiram 2.5 g/l of water.

FUSARIUM WILT (*Fusarium oxysporum*)

The disease appears as dull yellow green discolouration of leaves and drooping of the lower leaves. These symptoms often appear first on one side of a plant or on one shoot. Successive leaves become yellow, wilt and die, often before the plant reaches maturity. Older leaves usually die first in infected plants, commonly followed by death of the entire plant. Plants infected at young stage often die. Cross-sections of main stem may reveal brown streaks running lengthwise through the stem. This discolouration often extends far up the stem and is especially noticeable in a petiole scar. Masses of spore-bearing stalks are sometimes visible on dead tissue and may look like small pink cushions.

Control

- Plough the field two to three times in summer months.
- Grow only resistant varieties, if available.
- Grow seedlings on raised beds.
- Avoid excessive use of nitrogenous fertilizer.
- Avoid overwatering and provide good drainage facilities.
- Provide proper sanitation and cultural care to reduce plant susceptibility to infection.
- Most of the other control measures are ineffective once the disease has emerged.

DOWNY MILDEW (*Peronospora parasitica*)

The disease is characterized by small light greenish yellow lesions on upper leaf surface and yellow-ish, irregular, or angular lesions on lower leaf surface, which are soon covered with downy growth of fungus during the period of high humidity. Later, the leaf may become papery and die. The corresponding upper surface becomes chlorotic and leaves may drop off prematurely. The heads develop sunken black spots, which may be minute or larger than 2–3 mm in diameter. The severely infected heads become rotten. The disease appears when the humidity is high with light rains, prolonged dew, or heavy fog in cool weather. In seed crops, downy mildew on an inflorescence can lead to stag-head formation, abnormal shape and development of the inflorescence due to downy mildew or white rust. White rust and downy mildew are commonly found together in leaf colonies on crucifers.

Control

- Follow long crop rotation with non-host crops.
- Eradicate cruciferous weeds from the field.
- Avoid overhead irrigation and excess watering.
- Treat the seeds with hot water at 50°C for 30 minutes to kill pathogens present on seeds.
- Spray Dithane M-45 0.2%, Dithane Z-78 0.2%, or Ridomil MZ 0.25% at 7-day intervals.

BLACK LEAF SPOT (*Alternaria brassicae* or *Alternaria brassicicola*)

Small concentric dark spots up to 1.0 cm diameter appear on lower leaves. The head is also infected and subsequently it rots. The fungus also causes damage to pods and seeds with brown to black spots at seed production stage and harvested seeds are contaminated resulting in poor germination. The disease spreads through infected seeds, crop residues, cruciferous weeds, infected seedlings and spores (conidia) transmitted by wind and water. Disease is found more commonly in seed crop.

Control

- Adopt crop rotation with non-cruciferous crops.
- Always use disease-free healthy seed.
- Maintain good sanitation in the field.
- Remove and destroy all affected leaves and heads from the field.
- Treat the seed with captan, thiram, or Ceresan at 2.5 g/kg of seed.
- Spray the crop with mancozeb or Ridomil MZ-72 at 10- to 15-day intervals.

POWDERY MILDEW (*Erysiphe polygoni*)

The disease affects all parts of the plant and first appears as white powdery patches that may form on both sides of the leaves, which gradually turn completely yellow, curl, dry up and fall off. These patches gradually spread over a large area. Severely infected plants may have reduced yield. Dry weather, moderate temperature (15–25°C) and shady conditions are generally favourable for the development of this disease. Its spores are carried to new hosts by wind. Spores and fungal growth are sensitive to extreme heat (above 32°C) and direct sunlight.

Control

- Follow long crop rotation with non-host crops.
- Provide good air circulation and sunlight.
- Avoid excessive use of nitrogenous fertilizer.
- Avoid overhead sprinklers for irrigation.
- Spray the crop with Bavistin 0.1%, wettable sulfur 0.3–0.5%, Sulfex 0.2%, or Karathane 0.5%.

RING SPOT (*Mycosphaerella brassicicola*)

This is a seed-borne disease. The pathogen can attack the crop at all stages of growth from seedling to mature plant. However, it is normally most prevalent on older crops. Circular light brown to black spots appear on leaves. A yellow halo may surround the ring spot. In cases of severe infection, the entire leaf becomes yellow and leads to early senescence and defoliation. Gradually, these spots develop concentric rings much like a target pattern. Small spherical fruiting structures may also be observed within the leaf spots. In severe cases, the diseased leaves fall off prematurely. This pathogen may also infect the sprouts, causing dark lesions on outer leaves.

Control

- Follow long crop rotation with cereal crops.
- Use disease-free healthy seed.
- Avoid growing crop adjacent to brassica fields.
- Soak seeds in hot water (50°C) containing 0.2% captan or thiram for 20 minutes.
- Spray chlorothalonil 0.005% or a mixture of penconazole and mancozeb 0.2%.

Bacterial Black Rot (*Xanthomonas compestries* pv. *campestris*)

The symptoms vary considerably depending on host, cultivar, plant age and environmental conditions. The bacteria can enter plants through natural openings and wounds caused by mechanical injury on roots and leaves. The seed-borne bacteria infect the emerging seedlings through pores on margin of the cotyledons and then spread systemically through the seedlings. The disease often appears along the margins of leaves as chlorotic lesions, which progress towards forming V-shaped blotches followed by blackening of veins and vesicular bundles and ultimately death of the whole plant. The plants die at a later stage. The head formation is restricted if the attack is early and severe.

Control

- Grow resistant variety, if available.
- Follow long crop rotation with non-host crops.
- Follow sanitation to eliminate the initial source of disease.
- Eliminate refuse piles and eradicate the alternative hosts.
- Give hot water treatment to seeds at 50±2°C for 30 minutes followed by 30 minutes' dip in streptocycline 100 ppm.
- Spray the crop with streptocycline 0.01%, or Agri-Mycin or copper oxychloride 0.3% + streptomycin sulfate 100 ppm at transplanting.

Club Root (*Plasmodiophora brassicae*)

Club root is a disease caused by a root parasite with many races. It is a soil-borne fungus, which enters through root hairs. Yellowing of foliage, flagging and wilting of plants are the typical symptoms. Small spindle-like spherical club-shaped swellings develop on roots and rootlets. The disease has higher prevalence in acidic soils having pH below 7.0 and less serious on heavy soils and those containing low organic matter. The pathogen spreads through seedlings raised in infested seedbeds when transplanted in the field. It is also spread by irrigation water or infested soil.

Control

- Follow long crop rotation with non-cruciferous crops.
- Grow resistant cultivars, if available.
- Raise soil pH above 7.0 by adding lime.
- Maintain good sanitation and eradicate weeds.
- Dip the seedling roots in 0.05% solution of PCNB (Pentachloronitrobenzene) before transplanting.

Stalk Rot or White Rot or Watery Soft Rot (*Sclerotinia sclerotiorum*)

The lower leaves touching the ground turn light green followed by yellowing and falling off prematurely. The infection on stem spreads downward to the roots. On leaf petioles and stems, small irregular dark brown to black necrotic lesions are seen, which are covered with white cottony mycelial fungal growth. Stem, twigs and inflorescence become straw-like and dry in March. Pith of the infected plants is filled with black and hard sclerotia of fungus.

Control

- Adopt long crop rotation with non-host crops.
- Maintain good sanitation and good drainage facilities.
- Keep the field weed free.
- Flood the field for longer periods during hot summer months to destroy the sclerotia.
- Treat the seed by soaking in suspension of 50 ppm aureofungin and 50 ppm streptocycline for 30 minutes and dry them in air before sowing in the nursery.
- Dip the seedlings in 0.25% Benlate suspension for 5–8 minutes before transplanting.
- Spray the crop with carbendazim 0.05% at an interval of 10 days.

HEAD ROT (*Pseudomonas* spp.)

The disease is caused by a soil-borne bacterium. The symptoms appear when the heads remain wet for several days and bacteria are splashed up from the soil to the head. Water-soaked areas appear on heads because biosurfactant is released by the bacteria in contrast to unaffected areas where waxy surface of the florets causes the water to form beads. Small black lesions may develop in these water-soaked florets. During long periods of wetness, the decay spreads rapidly, resulting in a sunken area on heads. Head rot develops most rapidly at high temperature (28°C). Frost injury and infection by downy mildew may also increase the disease prevalence. A slimy soft rot develops on heads and an unpleasant odour is often associated with affected heads.

Control

- Grow resistant cultivars, if available.
- Use disease-free healthy seed.
- Avoid using high doses of nitrogenous fertilizers.
- Avoid sprinkler irrigation in the seedbed once the seedlings have emerged.
- Avoid excessive irrigation and waterlogging conditions.
- Avoid applying pesticides during head formation.
- Destroy the crop debris as soon as possible to remove the source of inoculum.
- Transplant the seedlings at wider spacing to increase air movement through the crop.
- Spray chlorothalonil 0.005% or a mixture of penconazole and mancozeb 0.2%.
- Store the harvested heads just above freezing point.

37 Collard

Akhilesh Sharma and Chanchal Rana

Botanical Name   : *Brassica oleracea* var. acephala

Family   : Cruciferae (Brassicaceae)

Chromosome Number : 2n = 18

ORIGIN AND DISTRIBUTION

Collard originated in the eastern Mediterranean or Asia Minor. It has been cultivated since the time of ancient Greek and Roman civilizations and spread into Europe around 600 B.C. through its introduction by a group of Celtic wanderers. It might have been introduced into the United States earlier through settlers from the British Isles probably in the late 1600s, though its first mention as collard greens dates back to the late 17th century. Leafy collard is the earliest vegetable cultivated by man, and it is thought that it has been consumed as food since prehistoric time.

INTRODUCTION

Collard (*Brassica oleracea* L. var. acephala DC) is one of the most primitive members of the cabbage family. Like other cole crops, it is an offspring of the wild cabbage (*Brassica oleracea* var. sylvestris) and closely related to kale and cabbage and can be described as a non-heading open leafy type cabbage, consumed in different parts of the world. The plants are cultivated for their edible large and broad, dark blue-green, tender, smooth-textured leaves, which are formed in clusters at top of the stalk. The leaves lack the frilled edges, which make them different from kale. In Kashmir, it is known as *haak* and can be found in almost every meal in Kashmiri households, where both leaves and roots are consumed. It is also grown as garden ornamental and considered an important traditional vegetable in the southeastern United States. It is mainly cultivated in Brazil, Southern United States, Portugal, different parts of Africa, Bosnia and Herzegovinia, Southern Croatia, Northern Spain and India.

COMPOSITION AND USES

COMPOSITION

Collard is a rich source of minerals, vitamins and proteins along with soluble fibres but low in calories. It provides conventional antioxidants such as vitamin A, vitamin C and vitamin E. However, collard greens carry antioxidant properties far beyond these conventional nutrients into the realm of phytonutrients by providing key antioxidant phytonutrients such as caffeic acid, ferulic acid, quercetin and kaempferol. It also contains a significant amount of phytonutrients called glucosinolates such as glucobrassicin, glucoraphanin, gluconasturtiin, and glucotropaeolin, which help in activating detoxification of enzymes by regulating their activity. Glucobrassicin is readily converted into an isothiocyanate called indole-3-carbinol (I3C), which prevents the initiation of inflammatory responses at a very early stage by operating at genetic level. The nutritional composition of collard greens is given in Table 37.1.

TABLE 37.1
Nutritional Composition of Collard Greens (per 100 g Edible Portion)

Component	Contents	Component	Contents
Water (g)	85.3	Magnesium (mg)	37
Carbohydrates (g)	7.5	Vitamin A (IU)	9300
Protein (g)	4.8	Thiamine (mg)	0.16
Fat ()	0.8	Riboflavin (mg)	0.31
Calcium (mg)	250	Niacin (mg)	1.7
Phosphorus (mg)	82	Pantothenic acid (mg)	3.8
Potassium (mg)	450	Pyridoxine (mg)	2.6
Sodium (mg)	40	Vitamin C (mg)	152
Iron (mg)	1.5	Energy (kcal)	46

USES

Both the roots and the leaves, which may be cooked together or separately, are consumed. *Haak-rus*, which is a soup prepared by cooking whole leaves in water, salt and oil, is usually consumed with rice in Kashmir. The leaves are also cooked by mixing them with meat, fish, or cheese. Collard leaves and roots are fermented in winter in Kashmir to form a very popular pickle called *haak-e-aachaar*.

MEDICINAL USES

Collard greens provide two hallmark anti-inflammatory nutrients, namely vitamin K and omega-3 fatty acids in the form of alpha-linolenic acid (ALA). It contains multiple nutrients with effective anticancer properties such as 3,3'-diindolylmethane and sulforaphane. Diindolylmethane is a strong modulator of essential immune response system with powerful antiviral, antibacterial and anticancer activity. Among different types of cancer, consumption of collard greens helps in preventing lung, breast, bladder, prostate, colon and ovarian cancer. Consuming raw or cooked collard greens also helps in lowering cholesterol. It contains a high amount of folate (a critical vitamin B), supporting cardiovascular health, which is 50% higher than broccoli, 100% higher than Brussels sprouts, three times higher than cabbage and seven times higher than kale. The fibre in collard (about 4 g per 100 g edible portion) is a natural choice for a healthy digestive system. However, it contains a high content of oxalate so should be carefully consumed by individuals prone to kidney stones.

BOTANY

Collard is dicotyledonous herbaceous plant, which is biennial or potentially perennial in nature under certain conditions, but commercially, it is cultivated as annual. It does not form a head though has a rosette form of vegetative growth on stem. Collard has large, spoon-shaped smooth and succulent leaves. It can produce a hard stalk up to a height of 120 cm, bearing a loose crown of leaves. In addition, collard, being self-incompatible, is a cross-pollinated plant, which produces viable seeds through insect pollination. Its inflorescence is terminal raceme. Flowers are hermaphrodite and bright yellow. Anthesis and dehiscence take place in the morning, and the pollen grains remain available until early afternoon hours, depending upon the prevailing temperature. The fruit dehisces longitudinally at maturity into distinct equal parts. Seeds are small, globular and smooth, having a uniform brown colour.

CLIMATIC REQUIREMENT

Collard is a cool season crop, which is grown during early spring or fall. The optimum soil temperature for seed germination is 10–29°C, while the optimum air temperature of 15–18°C is good for vegetative growth. It can tolerate severe cold weather in comparison to other members of the cole group. The mature plants can tolerate frost and light to medium freezing temperature. For higher yield, collard requires plenty of sunshine and warm weather. It is believed that collard greens are sweeter when harvested after a light frost, which improves its flavour.

SOIL REQUIREMENT

Collard may be grown in a variety of soils. Well-drained sandy loam soils are good for early spring crop, while higher yield can be obtained from heavy loam soils, which are rich in organic matter, and for its successful cultivation, the ideal soil pH ranges from 6.0 to 6.5.

CULTIVATED VARIETIES

Collard is not very popular in India and the varieties grown are exclusively exotics. It is a popular crop of the United States of America. The important varieties grown there are as follows:

OPEN POLLINATED VARIETIES

Georgia

An oldest open pollinated variety, also known as Georgia LS and Georgia Southern, produces blue-green and slightly savoyed leaves. Plant type is not very uniform having low to moderate yield potential depending upon growing conditions. It tolerates cool weather effectively but has a severe problem of bolting among the open pollinated varieties.

Morris Heading

An old open pollinated variety, also known as cabbage collards, has medium green, slightly savoyed, broad, waxy leaves. Many plants form loose and leafy heads late in the growing season. It is an old favourite variety with good flavour and nutrition. Frost makes its leaves sweeter. Its yielding ability varies from low to medium.

Vates

A non-heading variety with compact growing habit bears dark green, glossy, flat to wrinkled leaves on 75 cm plants. The variety with long standing is fairly uniform with moderate yield potential and has resistance against bolting and frost. This has been the standard variety for many years, especially for processing. It matures in 75–80 days.

Champion

A selection from Vates does not bolt readily in spring. It is very uniform open pollinated variety with smooth, deep green, flat to slightly savoyed leaves and short internodes. Plant habit is low-growing and has moderate yield potential. It takes 78 days from planting to harvesting.

Georgia Blue Stem

A darker blue-green variety with good yield matures in 60 days after planting.

HYBRIDS

Bulldog

A variety with upright growth habit bears dark green and slightly savoyed leaves.

Hi-Crop

A variety previously sold in the US market as Blue Max has very attractive, blue green, semi-savoyed leaves and requires 75 to 85 days from planting to harvesting. It has small intermodal distance, which makes it little more difficult to bunch than other varieties. It is prone to bolting.

Flash

A variety similar to Vates has uniform, smooth, dark green, glossy, flat to wrinkled leaves. It has excellent yield potential and does not bolt readily in spring. Unlike Vates, it has upright growth habit, thus, is good for bunching.

Tiger

A F_1 hybrid from private seed company has an upright plant growth habit like Georgia, having slightly savoyed, blue-green leaves. It is best suited for fresh market use.

Top Bunch

A Georgia-type variety with a uniform growth bears slightly Savoy type medium bluish-green leaves. It has tall and semi-erect growth habit and lends easily to bunching.

SEED RATE

Seed is the most critical and costly input, so optimum seed quantity should be used to raise the crop. The seed rate varies with method of planting, which may be direct seeding or planting seedlings. For direct-seeding crop, the seed rate varies from 1.5 to 2.5 kg/ha, and about 350–400 g of seed is required for raising seedlings enough for planting a hectare area.

SOWING TIME

Optimum sowing time is most important to exploit the yield potential of any vegetable crop. The planting time varies from region to region according to prevailing climatic conditions. It is a critical factor in determining the environmental conditions required at planting and plant development stages to raise successful crop of collard. The seeds can be sown in September–October in the northern Indian plains, August–September in hills and April–May in higher mountainous hills.

SOWING METHOD

Collard may be planted on flat or raised beds. It is grown by either sowing the seed directly in the field or raising the seedlings first in the nursery and then transplanting in the field on flatbeds. Direct seeding can be done manually but it requires more seed quantity followed by thinning of plants to avoid overcrowding. Seed should be placed in shallow and moist furrows at a depth of 1.5 cm but it should not be deeper than 2.5 cm. For harvesting young collard plants like mustard greens, direct sowing of seed is done 5–10 cm apart with inter-row spacing of 30–45 cm. The seeds are scattered lightly in the furrow and then covered with loose soil or compost. The seeds may germinate in 6–12 days. In the United States of America, several good precision seeders are available in the market, which reduce seed use by 40–70%, resulting in a much more uniform crop stand that requires little thinning.

The spacing for transplanting seedlings depends upon the stage at which the crop is to be harvested and which method is used for the harvesting of crop. For harvesting at early growth stage, i.e., half-grown plants may be spaced 25–40 cm apart, depending upon the variety. On the other hand, plants are raised at a spacing of 90 cm between rows and 40–45 cm between plants to carry harvesting at full-grown stage. Generally, 4- to 6-week-old seedlings are transplanted in the field preferably in the evening hours. Irrigation is applied immediately after transplanting to ensure good contact between the soil and seedling roots. After transplanting, it is important to maintain optimum soil moisture, especially within first 2 weeks so that the plants may become well established.

NUTRITIONAL REQUIREMENT

For obtaining optimum growth and yield, it is essential to apply only the required quantity of chemical fertilizers. Collard needs ample nitrogen for good green colour and tender leaves along with continuous growth for best quality. Under Indian conditions, well-decomposed farmyard manure is applied @ 20–25 t/ha at the time of field preparation. In addition, the crop needs nitrogen 100–150 kg, phosphorus 40–75 kg and potassium 50–100 kg/ha, depending upon fertility status of the soil. Full dose of phosphorus and potassium and one-third of nitrogen should be applied at the time of transplanting, and remaining two-thirds of nitrogen should be top-dressed in two splits at an interval of 1 month each. The leaf enlargement phase is the most critical stage for the application of irrigation.

IRRIGATION REQUIREMENT

Collard is a cool season crop and needs comparatively less water for its better growth and development. However, optimum irrigation will significantly increase its yield and quality. In direct seeding crop, pre-sowing irrigation is required to have optimum moisture at sowing time to get better seed germination. Depending upon the weather conditions and availability of soil moisture, irrigation should be applied at an interval of 10–15 days. It is advisable not to irrigate the field very frequently but it is essential to maintain proper moisture in the soil as it is a shallow-rooted crop and can use the moisture available in the shallow root zone very quickly. Frequent irrigation is important to obtain good plant stand during hot weather.

INTERCULTURAL OPERATIONS

HOEING AND WEEDING

Collard is a shallow-rooted crop, hence, it is essential to do shallow hoeing to remove weeds and to avoid injury to the roots. Regular hoeing operations keep the crop weed free and provide aeration to the root system. Critical period for crop-weed competition is between 30 and 50 days after planting. Therefore, it is essential to keep the field weed free in initial stages of crop growth. Herbicide is used in initial stages followed by hand weeding in later stages of plant growth along with fertilizer top dressings. Application of alachlor (Lasso) at 2 kg a.i./ha before transplanting or immediately after sowing is beneficial for controlling annual and broad leaved weeds. Pendimethalin (Stomp) at 1.2 kg a.i./ha or oxyflurofen (Goal) at 600 ml/ha can also be used if there is problem of only annual weeds. Trifluralin is also widely used at 0.60–1.0 kg a.i./ha as a pre-plant incorporation (PPI) to control many of the weeds such as grass, pigweed and purslane. Use of herbicides is a critical issue with many leafy greens as they are very sensitive to many soil active herbicides and great care must be taken to avoid such possibilities.

THINNING

Thinning, which is done to maintain optimum spacing and to reduce competition among plants, is an important operation in direct seeded crop. In the direct seeded crop, plants are left to grow to a height of 10–15 cm after the emergence or till there is crowding within the row. Then, thinning of plants is done gradually by maintaining a distance of 30–45 cm between plants. Crowding results in smaller and light green leaves.

EARTHING-UP

Being a shallow-rooted crop, it needs earthing-up essentially to cover the roots, which are exposed due to the application of irrigation water. This helps in preventing the plants from lodging and provides better soil atmosphere to the developing roots and thereby to the plants. Earthing up should be done 30–40 days after planting and may be repeated depending upon the prevailing conditions.

FLOWERING

Collard is sensitive to environmental conditions, which cause prolonged cessation of vegetative growth during early stages of plant development and leads to the onset of bolting, *i.e.*, development of flower stalks when the plant is still immature. Bolting is influenced by age of the plant and cold temperature. Flower stalks can emerge when plants are grown below 10°C and are exposed to a period of cold weather (2°C to 10°C) for more than 10 days. Old plants are more susceptible to bolting than the young plants. Susceptibility to bolting also depends upon genetic make-up of the variety.

HARVESTING

Collard becomes ready for harvest when a large rosette of leaves is developed at crown of the plant. It is appropriate to harvest collard greens, which have firm, tender vividly deep green leaves with no sign of wilting, yellowing, or browning. The plants may be harvested by machine or by cutting with hands. Collard is harvested by different ways depending upon the method of planting, which are given as follows:

- Collard is harvested at very young age, when the stalks are tender, clean and free of discolouration and wilted leaves.
- The whole plant is cut at very young stage just like mustard greens followed by successive cutting.
- The entire plant is cut when it is about half grown and tied in bunches of one to three plants with a rubber band or string.
- The entire plant is cut when it is full-grown.
- The tender leaves are harvested from full-grown plants and marketed by tying in one-half or one kg bunches.

It is necessary to trim the plants after picking to remove any yellow, brownish, or damaged leaves before they are tied into bunches and are sent to the market. It is essential to make bunches of collard leaves or plants having a uniform size. Coarse or tough stems of plants should not be packed.

The US standards for collard greens provide one grade, *i.e.*, US No.1, which consists of greens of similar varietal characteristics that are fresh, fairly tender, fairly clean, well trimmed and of characteristic colour of the variety or type and free from decay or damage caused by coarse stems and seed stems, discoloration, freezing, foreign material, disease, insects, or equipment.

YIELD

The average yield of collard green varies from 100 to 250 q/ha, depending upon crop variety, soil fertility, environmental conditions and cultural practices adopted by the growers.

POST-HARVEST MANAGEMENT

Collard greens should be cleaned before marketing. Quality is best achieved if collard greens are pre-cooled by hydro-, vacuum, or forced air-cooling before transportation as rapid air movement is needed to remove field heat from leaves. Collards are top iced at 0°C for sale, which is necessary for maintaining quality, as they are perishable. Plants are either bulk loaded into trucks or should be packed into bushel baskets, wire-bound boxes, crates, or cartons for distant shipment and supermarket sales. Collard greens, in general, are sold in bunches of two to three plants, depending on market demand.

Greens are not usually stored but will remain in good condition for 10–14 days at 0°C temperature and at least 95% relative humidity. In addition to cold storage, greens should be top-iced to retain crispness. These greens are also ethylene sensitive, thus, should not be stored, or shipped with ripening tomatoes, musk melon, etc.

PHYSIOLOGICAL DISORDER

BOLTING

It is characterized by the development of flower stalks at immature stage. It occurs when the crop is subjected to 2–10°C temperature for more than 10 days. The degree of temperature-induced bolting response depends upon the type of variety.

INSECT-PESTS

DIAMONDBACK MOTH (*Plutell xylostella*)

Spindle-shaped pale yellowish green larvae feed on lower side of the leaves, affecting the produce yield as well as quality. First instars mine leaf tissue, thereafter, they feed on the undersurface of the leaf. The larvae are very small but they are quite numerous and chew on leaves, resulting in holes and severe defoliation. The presence of larvae on the leaves leads to complete rejection of produce. Collard is preferred for egg-laying by the adults.

Control

- Grow Indian mustard as a trap crop for suppressing the incidence of diamondback moth and cabbage aphid.
- Release *Trichogrammatoidea bactrae* @ 0.5–0.75 lakh eggs per hectare at weekly intervals.
- The larval parasitoid *Diadegma insulare*, is also effective.
- Pheromone traps can be used to monitor adult populations.
- Spray the crop with malathion 0.05%, deltamethrin 0.028%, cypermethrin 0.01%, or lambda-cyhalothrin 0.004%.

CUTWORMS (*Agrotis* sp.)

Cutworms are dull brown caterpillars with 2.5–5 cm length at full-grown stage. They are, in general, found on or just below the soil surface or on lower parts of plants. These are quite active during night hours. They damage the crop at seedling stage early in the season. They cause major damage by chewing the stems of young plants at or slightly above or below the soil surface. Large population of some cutworms, *e.g.*, black, bronzed and army cutworms, cause severe damage as they cut the whole newly planted crop overnight. Other species like glassy cutworms remain inside the soil and feed upon roots and underground parts of the plant.

Control

- Use well-decomposed manure.
- Remove weeds and plant residue to check egg-laying sites.
- Plough the soil before planting to expose and kill the overwintering larvae.
- Collect and destroy the larvae after flooding the filed/beds.
- Drench the beds with chloropyrifos 0.04%.
- Spray the crop with cypermethrin 0.01% on foliage and surface.

Aphids (Cabbage aphid – *Brevicoryne brassicae*; Green peach aphid – *Myzus persicae*; Turnip aphid – *Lipaphis erysimi*)

Both, nymphs and adults of this tiny insect suck cell sap from the tender leaves. The aphids excrete honey-dew-like substance, which develops black sooty mould and hampers photosynthetic activity of the plants. They also act as vector for the transmission of virus. The severe infestation affects the yield and quality adversely. *Myzus persicae* has a very wide host range and is important vector of virus transmission.

Control

- Natural enemies include tiny wasps that parasitize the aphids, while lady birds, their larvae, lacewing and hoverfly larvae eat the aphids.
- Spray malathion 0.05% with the appearance of pest at 15-day intervals.
- Spray oxydemeton methyl 0.025%, dimethoate 0.03%, or phosphamidon 0.03% as soon as the aphid population is above 50 aphids per plant.
- In seed crop, apply phorate granules @ 1.5 kg a.i./ha as side dressing during mid February to early March.

Cabbage Butterfly (*Pieris brassicae*)

The butterfly is cream to yellowish with black-tipped forewings and central spots. They lay yellow eggs on undersurface of the leaves in clusters, which on hatching develop into caterpillars and cause damage. The larvae feed voraciously by chewing the lamina, leaving large holes in the leaves, and pollute the plants with their excreta. The younger caterpillars feed on undersides of these leaves and move towards centre of the plant as they grow. Older caterpillars usually appear on upper surface of the central leaves.

Control

- Collect and destroy yellow egg masses and early stage larvae of cabbage butterfly.
- Use *Cotesia glomeratus* and *Cotesia rubecula* which parasitizes the larvae.
- Spray malathion 0.05%, deltamethrin 0.0028%, cypermethrin 0.01%, or dichlorvos 0.05%.

Snails (Garden brown snail – *Cornu aspersum*; White garden snail – *Theba pisana*) and Slugs (Grey garden slug – *Deroceras reticulatum*; Banded slug – *Lehmannia poirieri*; Three-band garden slug – *Lehmannia valentiana*; Tawny slug – *Limacus flavus*; Greenhouse slug – *Milax gagates*)

They chew succulent leaves leaving irregular holes with smooth edges behind. They leave characteristic silvery mucous trails, which distinguish them from other chewing insects. The mucous trails deteriorate the quality of the produce. They become problematic under humid conditions, or if the crop is irrigated with sewer water.

Control

- Pick and destroy manually on regular basis.
- Traps can effectively be used with baits.
- Baiting with metaldehyde and bran (1: 25 in 12 litres of water) is effective for their control.
- As a repellant, alum may be sprayed @ 2% solution.

DISEASES

Damping-off (*Pythium* spp., *Rhizoctonia* spp. or *Fusarium* spp.)

This disease commonly affects seeds and young seedlings. Young seedlings are severely affected and show soft water soaked spots on the skin just above the ground. The cotyledons wither and the plant eventually falls over and perishes. High temperature and high humidity favour the infection.

Control

- Burn the trash or sterilize the soil with formaldehyde (1:50 water) drenching.
- Drench the nursery beds with 0.2% captan 15 days before sowing.
- Treat the seed with hot water at 52°C for 30 minutes.
- Spray mancozeb 0.25% or carbendazim 0.1% on seedlings.
- Drench the soil with 0.2–0.3% Brassicol solution.
- Spray the crop with Dithane-M-45, Difolatan 80, or Daconil @ 0.2% at 15-day intervals.

BLACK LEG (*Phoma lingam*)

It occurs in areas with continuous rainfall during the growing period. It is a seed borne disease and hence infests crop plants at early stages. Stem of the affected plant when split vertically shows severe black discolouration of sap stream. Whole root system decays from bottom upwards. Often, the affected plants collapse in the field.

Control

- Use disease-free seed.
- Treat the seed with hot water at 52°C for 30 minutes.
- Spray the seed crop with copper oxychloride 0.2%.

DOWNY MILDEW (*Peronospora parasitica*)

It causes serious damage at all stages of plant growth. Discolouration occurs in the young seedlings, and in severe cases, whole plant perishes. In adult plants, purplish leaf spots or yellow brown spots appear on upper surface of the leaves, while fluffy downy fungus growth is found on lower surface. During bolting, the seed stalk show blackish patches, and in severe cases, the whole curd is spoiled.

Control

- Avoid overhead watering.
- Remove weed hosts and crop debris.
- Treat the seed with metalaxyl @ 0.3–0.6 g a.i./kg of seed.
- Spray 0.2% Ridomil MZ 72 at 10 to 15 days intervals.

WIRE STEM ROT (*Rhizoctonia solani*)

Plants infected with wire stem shows reddish brown discoloration of stem near the soil surface, which is often severely constricted. The outer skin of the stem tissues is worn out, leaving a dark, wiry and woody inner stem. Plants may be twisted but do not split, and hence the name wire stem. Plants have stunted and weak growth and some of them may die.

Control

- Discard the infected seedlings at the time of transplanting.
- Follow crop rotation with non-cruciferous crop.
- Drench the soil with 0.2–0.3% Brassicol solution.
- Spray the crop with Dithane-M-45, Difolatan 80, or Daconil @ 0.2% at 15-day intervals.

BLACK SPOT (*Alternaria* sp.)

Initial symptoms are small light brown spots with concentric rings on older leaves. A prominent yellow halo around the lesion edge may be seen, which eventually give a shot hole appearance. The infected leaves become yellow and fall down when mature. The fungus is seed-borne. Young plants affected by the disease resemble those resulting from *Rhizoctinia solani*.

Control

- Treat the seeds with thiram @ 2.5 g/kg seed.
- Treat the seed with hot water at 52°C for 30 minutes.
- Spray the crop with 0.2% Dithane M 45 at 15-day intervals if needed.

WATERY SOFT ROT (*Sclerotinia sclerotiorum*)

Disease first appears as wet soft lesions on leaf. Later, the lesion enlarges into a watery rotten mass of tissues covered by white silvery appearance. At the point of attack, the seed stalks can break, wither and eventually die. Initially, there is formation of white sclerotia with often excessive tufts of mould, which later turn black, usually several millimetres in diameter. Damp weather favours the occurrence of this disease.

Control

- Spray the crop with carbendazim 0.1% or mancozeb 0.2% after appearance of disease at an interval of 15 days.

BLACK ROT (*Xanthomonas campestris*)

The tissue at leaf margin becomes yellow. Chlorosis progresses towards leaf centre and it develops a V-shaped area at the mid rib. These spots are associated with a typical black discolouration of the veins. In severe infestation, the whole leaf shows discolouration and eventually falls off.

Control

- Use disease-free certified seed.
- Treat the seed with hot water at 52°C for 30 minutes.
- Spray streptocycline (100 ppm) or Blitox (0.1%) of water after transplanting.
- High soil moisture helps reduce inoculum.

COLLARD GREEN MOSAIC VIRUS

The disease is transmitted by aphids and can be recognized by poor growth of plants with a mosaic pattern of light green on the normal green colour of leaves, crinkling and distortion of leaves.

Control

- Destroy the overwintering reservoirs of viruses by destroying wild and volunteer plants.
- Kill the overwintering aphids.

■■■

38 Garden Cress

M.K. Rana and Naval Kishor Kamboj

Botanical Name : *Lepidium sativum*

Family : Cruciferae

Chromosome Number : 2n = 16, 24 and 36

ORIGIN AND DISTRIBUTION

Garden cress is native to the Southwest Asia, which later spread to various parts of Europe, Britain, Italy, France, Germany, Syria, Greece, India and Egypt. Persian culativated this plant as early as 400 B.C. and used it even before bread. It was introduced to the United States from China. In India, it is cultivated in southern states at commercial scale.

INTRODUCTION

Garden cress, also known as garden pepper cress, pepper grass, or pepperwort, is a fast-growing edible herb, which is botanically related to water cress and mustard. In India, it is known as Asalio or chandrasoor. Its roots, leaves and seeds are economically important, however, the crop is commercially cultivated as an important medicinal crop for the production of seeds. It is a perennial plant but grown as an annual. It is an important green vegetable consumed by human beings most usually as a garnish or as a leafy vegetable in some places of the world. The leaves and seeds having a peppery taste are consumed in relatively small amount since their peppery taste can be overwhelming.

COMPOSITION AND USES

COMPOSITION

Garden cress contains significant amount of calcium, potassium, phosphorus, magnesium, iron as well as vitamin A and C, folic acid, thiamine, riboflavin and niacin. Its seeds are rich source of protein, dietary fibre, fatty acid and phytochemicals. The nutritional composition of garden cress is given in Table 38.1.

USES

Garden cress is usually cultivated for its leaves and seeds. Its leaves are used in salad, baby greens and as a garnish. Due to its health-boosting properties, its cut shoots are an excellent addition to sandwiches, steamed potatoes and soups for its tangy flavour. It is also eaten with cottage cheese, boiled eggs, or mayonnaise and in the form of chatuni along with bread in various parts of the world. Its sprouted seeds are used for the decoration of salad. In Arab countries, its boiled seeds are consumed with drinks and their ground powder with honey or hot milk. The seed oil is used for illumination and soap making.

TABLE 38.1

Nutritional Composition of Garden Cress (per 100 g Edible Portion)

Constituents	Contents	Constituents	Contents
Water (g)	89.40	Vitamin A (IU)	6917
Carbohydrates (g)	5.5	Thiamine (mg)	0.08
Sugars (g)	4.4	Riboflavin (mg)	0.26
Protein (g)	2.6	Niacin (mg)	1.0
Fat (g)	0.70	Pantothenic acid (mg)	0.247
Dietary fibre (g)	1.1	Pyridoxine (mg)	0.247
Calcium (mg)	81.0	Folic acid (µg)	80.0
Phosphorus (mg)	76.0	Vitamin C (mg)	69.0
Potassium (mg)	606.0	Vitamin E (mg)	0.7
Magnesium (mg)	38.0	Vitamin K (µg)	541.9
Manganese (mg)	0.553	Energy (kcal)	32.0
Iron (mg)	1.3		

MEDICINAL USES

In traditional system of Indian medicine, various parts of plant are used to treat many human ailments such as leprosy, skin diseases, dysentery, diarrhea, dyspepsia, eye diseases, leucorrhoea, scurvy, asthma, cough, cold and seminal weakness. The seeds are bitter and considered aborfacient, diuretic, expectorant, aphrodisiac, anti-diabetic, anti-bacterial, anti-spasmodic, anti-cancerous, anti-ashthamatic, hypocholesterolemic, analgesic, hepatoprotective, coagulant, gastrointestinal stimulant, gastro protective, laxative thermogenic, depurative, galactagogue, emmenagogue and ophthalmic. The paste of its seeds is useful in rheumatic joints to relieve pain and swelling. Ethanolic extract of seeds is effective in treating inflammatory bowel disease. Its leaves are stimulant, diuretic and anti-bacterial and useful in scurvy and hepatopathy. A preparation made of seeds, butter and sugar is a common household remedy for general weakness.

BOTANY

Garden cress is an erect, small glabrous annual herb with slender branches, attaining a height up to 50 cm with dark grey, black or reddish rough bark. Its leaves are opposite, ovate or ovate-lanceolate, glabrous, 1.5 to 8.0 cm long and 1.5 to 3.0 cm thin. The flowers borne in terminal and axillary paniculate cymes are brownish purple, bisexual, regular and tetramerous. Flower pedicel is 1.5 to 4.5 mm long, sepals ovate and 1–2 mm long, and petals spatulate with short claw, up to 3 mm long, white or pale pink. Stamens are six and anthers usually purplish. Ovary is superior, flattened, apex emarginate, style up to 0.5 mm long and stigma capitate. Fruit is round or ovate, flattened silique 4–6 mm × 3–5.5 mm, pale green to yellow, margins wing-like, apex emarginate, dehiscing by 2 valves and usually 2-seeded. Seeds are small, oval-shaped, pointed and triangular at one end, smooth, about 3 to 4 mm long, 1 to 2 mm wide and reddish brown in colour. When soaking in water, the seed coat swells and is covered with transparent, colorless, mucilage with mucilaginous taste. The transverse section of seed shows the presence of testa, tegmen, alleurone layer, endosperm and embryo. Testa is thick, single- or double-layered and appears yellowish brown, whereas, the tegmen layer is attached to inner side of testa as a single layer. Endosperm is composed of thick-walled polygonal cells. Embryo appeared as innermost structure surrounded by endospermic cells. The cells of embryo are small sized and polygonal in shape. Seed germination is epigeal, seedling cotyledons are tri-foliolate and leaflets are spatulate.

CLIMATIC REQUIREMENT

Being a cool season crop, garden cress requires cool and humid climate. It can successfully be cultivated in areas with full sunshine as well as partial shade. The ideal mean temperature for its cultivation is 12–22°C. The seed germination is slow and poor, when the soil temperature is less than 8°C. Under long day and warm temperature conditions, the plants start bolting, which spoils the quality of the produce.

SOIL REQUIREMENT

Garden cress can be grown in a variety of soils but it thrives best in well-drained fertile soils rich in organic matter and having a good moisture holding capacity. It prefers slightly acidic to neutral soils with pH 4.9 to 7.0 for its successful cultivation. The soil is prepared well to a fine tilth by repeated ploughing with harrow followed by planking to level the field properly.

CULTIVATED VARIETIES

DADAS

A newly registered variety for cultivation in the Eastern part of Turkey produces curly leaves.

BAHAR

A variety for cultivation in Turkey produces plain leaves.

WRINKLED

A variety with extremely slow bolting developed through a cross between curled and Persian cress is recommended for cultivation in United States of America. Its leaves with ruffled edges look like a curly parsley.

CURLED

A variety with dark green and curly leaves and slow bolting nature is recommended for cultivation in United States of America.

Some other important varieties of garden cress recommended for cultivation in United State of America are Crinkled, Crumpled and Persian.

SOWING TIME

The sowing time of garden cress varies with climatic conditions of the growing region, but in general, the seed sowing is done in last week of September to first week of October in subtropical regions and in March–April after winter frosts in temperate regions.

SEED RATE

Seed rate usually varies with viability, sowing time and climatic conditions of the growing region, but as usual, a seed rate of about 3.0–4.0 kg is sufficient to sow a hectare land area.

PLANTING METHOD

Garden cress is propagated by seed. Sowing is done either by broadcasting or in lines at 15–20 cm spacing and 0.5–1.0 cm depth. The seeds after sowing are covered lightly with fine soil or compost.

The seeds germinate within 4 to 6 days after sowing, depending on the climatic conditions. The crop bolts quickly in hot weather, thus, sowing is done after every 10 days to ensure a steady supply of young leaves throughout the season. It can also be grown year yound in pots, bowls, boxes, or flat plates, putting or hanging them in a windowsill.

NUTRITIONAL REQUIREMENT

Garden cress requires very low amount of fertilizer due to its very short growing period. However, to harvest a good yield, it is necessary to fertilize the crop periodically with soluble liquid fertilizers.

IRRIGATION REQUIREMENT

Garden cress is a shallow rooted crop, thus, it performs best if moisture in soil is kept at field capacity. In summer, irrigation is applied at 4- to 6-day intervals, and in winter, at 10- to 12-day intervals to keep the soil moist. The field is kept moist during dry periods since water deficit stress during growth period reduces the yield greatly.

INTERCULTURAL OPERATIONS

HOEING AND WEEDING

Hoeing and weeding are very important operation particularly at initial stage of crop growth, when the plants are establishing since the crop plants at this stage compete poorly with weeds. Hoeing should be done to avoid damage to its shallow roots, which slows down the plant growth.

MULCHING

The crop needs continuous moisture supply, which is impossible to provide all the time, however, the moisture can easily be conserved by covering the soil with straw, shredded newspaper, dry grass or polyethylene sheet. The weeds, which deplete field soil moisture, can also be checked with the use of mulch.

HARVESTING

Garden cress is a fast growing crop and it is harvested in the same month of sowing, *i.e.*, 2 or 3 weeks after emergence. The leaves as and when attain a height of 5 to 8 cm and turn green are harvested with the help of scissors. Only the older leaves without touching the younger ones are harvested for immediate use so that the plants may not stop growing.

YIELD

The yield depends on several factors such as variety, climatic conditions of the growing region and package of practices followed during the cultivation of crop. The average yield of garden cress leaves is 60 q/ha.

POST-HARVEST MANAGEMENT

Garden cress may not be stored longer, thus, it should be used soon after harvesting, as it loses its freshness and flavour very quickly.

INSECT-PESTS AND DISEASES

Garden cress is usually not attacked by much insect-pests and diseases since the crop is harvested within shortest period of growth. However, the plants are susceptible to some of the insects and diseases. Their description along with control measures is given as follows:

INSECT-PESTS

Aphid (*Lipaphis erysimi*)

Aphid is serious pest of garden cress. The nymphs and adults both attack the young plants and suck cell sap from growing tips and later from whole plant. It causes curling of leaves, reduces plant vigor, yield and quality of the produce. Aphids secrete a honeydew-like substance, which promotes the growth of sooty mould, which inhibits photosynthesis and ultimately reduces the yield of the crop.

Control
- Wash the plant with plain water using strong jet pressure to reduce aphid population.
- Spray the crop with 5% *neem* seed kernel extract, 0.03% dimethoate or 0.05% phosphamidon.

Flea Beetle (*Phyllotreta* spp.)

The pest responsible for damage to the garden cress plants is a small (1.5–3.0 mm) dark-coloured beetle, which is often shiny in appearance. The beetle causes damage by making small holes or pits on leaves, which give the foliage a characteristic shothole appearance. The young plants and seedlings are more susceptible. The affected plants become stunted, but under heavy infestation, the plant may be killed.

Control
- Use floating row covers prior to the emergence of seedlings to provide a physical barrier against beetles.
- Spray the crop with 5% *neem* seed kernel extract, 0.03% dimethoate or 0.05% phosphamidon.

DISEASES

Damping-off (*Pythium* sp.)

The first symptom of damping-off disease is the failure of some seedlings to emerge. If the seeds are attacked before their germination, they become soft and mushy, turn dark brown and later they decay. If seedlings are attacked after their emergence, stem tissues near the soil line become weak and decayed, usually causing the plants to topple over the ground and die.

Control
- Follow long duration crop rotation with cereal crops.
- Provide good drainage facilities in the field.
- Use disease-free healthy seed.
- Treat the seeds with captan, Foltaf, captafol, or thiram @ 2.5 g/kg of seed.
- Drench the beds with captan 0.2%, or Bavistin 0.1%.

Powdery Mildew (*Erysiphe cruciferarum*)

The fungus of powdery mildew attacks all parts of the plant. Initially, the symptoms appear as yellow spots on upper surface of the leaves, and these spots later develop the characteristic powdery growth and spread to under surface of the leaves and stem. The affected leaves turn completely yellow, curl, die and fall down. Powdery mildew is favoured by warm dry days and cool moist nights.

Control

- Follow long duration crop rotation with cereal crops.
- Grow resistant varieties if available.
- Spray 0.5% Sulfex or other formulations of sulfur.
- Spray the crop with systemic fungicide like Karathane, Bavistin, or Benlate 0.2% at 10- to 15-day intervals.

Downy Mildew (*Peronospora parasitica*)

The disease pathogen attacks the plants at any stage of development. The symptoms appear as a purplish or yellow-brown spots on upper surface of the leaves and white to grey fluffy downy mildew found on underside of the leaves. The infected areas enlarge and their centers turn tan to light brown with papery texture. When the disease is severe, the entire leaves turn yellow and die. A temperature of 10°C to 20°C and relative humidity more than 90% favour the occurrence of disease.

Control

- Grow resistant varieties if available.
- Follow long duration crop rotation with cereal crops.
- Give hot water treatment at 50°C for 30 minutes before sowing the seeds.
- Spray the crop with Dithane M-45 or Dithane Z-78 0.3%.

■■■

39 Upland Cress

M.K. Rana and Kiran Sagwal

Botanical Name : *Barbarea verna*

Barbarea praecox

Lepidum nativum

Family : Cruciferae (Brassicaceae)

Chromosome Number : 2n = 16

ORIGIN AND DISTRIBUTION

Upland cress is native to the parts of Central, Southeastern and Southwestern Europe, as it has been grown there as a leafy vegetable since the 17th century. Thereafter, this species has been widely introduced to rest of the Europe, Africa, America and Asia. The species is widespread in France, Italy, Spain and Belgium, however, in Malaysia, it is cultivated on a small scale.

INTRODUCTION

Cress is a general name for plants eaten as a sharp-tasting salad, garnish or potherb. Upland cress (*Barbarea verna*), locally known as American cress, dry land cress, cassabully and creasy salad, is one of three major cresses. The rest two are garden cress (*Lepidium sativum*) and watercress (*Nasturtium officinale*). In form and flavour, upland cress resembles watercress. The word *Lepidium* has been derived from a Greek word *Lepidion*, which means *a little scale*, as the pod shape is like scales, and the species *verna* simply means *of spring*. The two very similar but slightly dissimilar species of upland cress are winter rocket (*Brassica vulgaris*) and Belleisle cress (*Brassica praecox*). Normally, it is cultivated in hot areas of the world. Upland cress is occasionally found in the gardens of Florida. Being extremely hardy, it is grown during the coolest months of the year.

COMPOSITION AND USES

COMPOSITION

Upland cress is rich in minerals such as calcium and iron as well as in vitamins, chiefly vitamins A, B, C, E and K. It is also high in phytonutrients, having twice the amount of vitamin A than broccoli and twice the amount of vitamin C than oranges. It also contains lots of lutein, which is important for preventing macular degeneration. Its seeds contain high levels of protein and calories. The nutritional composition of upland cress is given in Table 39.1.

Upland cress calcium is comparatively less soluble than that from skim milk and deprives the availability of calcium from spinach. Even cooking upland cress in water for 20 minutes had no effect on the availability of calcium.

TABLE 39.1

Nutritional Composition of Upland Cress (per 100 g of Edible Portion)

Constituents	Content	Constituents	Content
Water (g)	89.4	Manganese (g)	0.56
Carbohydrates (g)	6.1	Copper (mg)	0.16
Protein (g)	2.4	Vitamin A (IU)	6915
Fat (g)	0.6	Thiamine (mg)	0.08
Dietary fibre (g)	1.1	Riboflavin (mg)	0.25
Calcium (mg)	80.5	Folate (mcg)	81
Phosphorus (mg)	77	Pantothenate (mg)	0.24
Potassium (mg)	602	Vitamin C (mg)	70
Magnesium (mg)	37	Vitamin E (mg)	0.7
Iron (mg)	1.3	Vitamin K (mcg)	540
Zinc (mg)	0.23	Calories	30

USES

The leaves of upland cress having a radish-like sharp spicy flavour are cooked as grilled vegetable and a garnish in fish dishes. It is also used raw in salads, sandwiches and in any recipe in which watercress is used. Its freshly harvested leaves, shoots or very young flower buds are used to add vigour especially to soups and egg preparations. Besides, its leaves may also be used as a substitute of parsley and young sprouts may be added together with mustard sprouts to give a spicy taste. The attractive leaf petioles may be used in arranging both fresh and dried flowers for craft purposes.

MEDICINAL USES

Upland cress, being a member of the brassica family, has the ability to fight cancer and other diseases. In the old world, it was used for the healing of wounds. During colonial times, it was consumed in the winter to ward off scurvy disease. The addition of dark green leafy vegetables that are innately rich in beta-carotene (pro-vitamin A) is strongly recommended by nutritionists and dieticians to cure vitamin A deficiency. The seeds are used as an expectorant and for treating asthma, indigestion, constipation, sore throat, cough and headache. Being fat and cholesterol free it is useful for weight loss and good for cardiovascular health. It is mild diuretic, helping to get rid of excess water from the body. Its lutein is important for preventing macular degeneration.

BOTANY

Upland cress is a biennial herb but it is cultivated as annual for leaf purpose. It has a diverging taproot system. The plant is low (30 cm) and spreading (30 cm), having an extremely small stem, bearing a cluster of crowded leaves that radiate away from the stem in a circle, which is more or less flat. The foliage of the plant is not so dense. The leaves are dark green, very short, stocky and almost square-shaped and with a zesty tenderness. The leaf petiole is 15 to 20 cm long and margins are slightly notching, resembling curled parsley. In the first season, it forms a rosette of leaves, and in the second season, it develops inflorescence. Its small yellow flowers are hermaphrodite, having both male and female organs, but they are cross-pollinated in nature. The plant is self-fertile. Pollination is performed by flies, bees and beetles. Flowering takes place from December to January and seeds mature from March to April. The flower stalk grows up to a height of 30–90 cm. The seeds remain viable for 3–5 years.

CLIMATIC REQUIREMENT

Upland cress, being a cool season crop, grows best in cool and moist climates with an average temperature range of 16°C to 25°C. The seed germination is better at lower (12–23°C) than at higher (25°C) temperature. The optimum temperature for seed germination is 15°C and for growth 18–20°C, however, the growth is slow at 5–7°C. Low temperature tends to increase leaf thickness but decreases leaf size and smoothness. It thrives in the open conditions as it can withstand frost although it needs protection in colder months at night and during very heavy frost, which can be provided using cloches with sacks. If it is grown at high temperature and as the plant matures, its flavour becomes very spicy but young leaves' taste is just fine. Under a condition of dry soil and high temperature, its leaves and stems become unpleasant and bitter in taste. Upland cress can be grown in a site with full sunshine as well as partial shade. The crop in summers can successfully be grown by providing some shade, while the winter crop succeeds in sunny locations, needing 6–8 hours of light for vegetative growth but increasing light intensity and atmospheric temperature increases the production of biomass since environmental conditions influence the leaf number and size. It can be grown almost round the year but its production is highest under short day and mild temperature conditions. Being a long-day plant for flowering, upland cress under long day and warm weather conditions has the tendency to develop seed stalks, therefore, it is not possible to grow it during spring–summer in plains, especially under long day conditions.

SOIL REQUIREMENT

Upland cress can be grown in any type of soil from light to heavy provided the soil has proper drainage facility and good water-holding capacity. However, sandy loam soils rich in organic matter are far better than heavier soils since the crop is susceptible to waterlogging conditions. The ideal soil for its cultivation is fertile, moist and properly drained with adequately decomposed compost. Although it can be grown at a soil pH 4.5–7.5 the soil pH 6.0 to 6.8 keeps the plants healthy and nourish the plant better. Proper land preparation is must to ensure satisfactory seed germination and for further plant growth.

CULTIVATED VARIETIES/HYBRIDS

Upland cress is an underexploited wild plant of the brassica family, thus, improvement work has not yet been done in India to develop varieties or hybrids. However, several local selections without proper nomenclature are available. Farmers usually grow the locally available varieties.

SOWING TIME

Sowing time of upland cress varies from region to region. In subtropical climate, the crop is sown from last week of August to first week of September, and in temperate climate, first sowing can be done in early spring after the danger of severe frost and continue during late summer for a winter salad crop, but under cloches, it can be grown any time of the year without any difficulty except January and February. In mild climates, it can be grown during any month for succession supply of the leaves, and for a steady supply, the sowings are made after every 3 to 4 weeks, however, during August–September, the plants are protected from extreme blazing sun and heat using shading net.

SEED RATE

The seed rate of upland cress for obtaining optimum plant population per unit area is usually ascertained according to seed size, seed vigour, germination percentage, soil type, soil fertility, sowing method and the intended use of crop. For fresh market production, relatively a slightly greater quantity of seed is used per unit area. A seed quantity of 1.7–2.25 kg is enough for sowing a hectare area.

SOWING METHOD

Upland cress can be propagated by root divisions or cuttings taken during spring months. However, propagation by vegetative parts is less adopted. Therefore, it is usually propagated by seeds, which are sown directly in the field by either broadcasting or in lines. However, sowing seeds in lines is always better over broadcasting method, as it facilitates intercultural operations and harvesting. A spacing of 15–20 cm is kept between lines and 8–10 cm between plants. Spacing between plants influences the number of leaves and their size. Lower plant density facilitates hand harvesting for bunching of leaves, and on the contrary, high plant density results in more upright leaf growth. The optimum plant density can be obtained by sowing disease-free healthy seeds. Based on soil type and moisture availability, the seeds are placed at a depth of 0.6–1 cm. The sowing depth is maintained a little greater in light soils and a little lesser in heavy soils. The seeds after sowing are covered with a light soil. Its seeds take 2–3 weeks to germinate.

NUTRITIONAL REQUIREMENT

The nutritional requirement of upland cress depends on variety, soil type, soil fertility, growing conditions and crop grown in previous season. Farmyard manure @ 8–10 t/ha should be applied at the time of land preparation and mixed thoroughly into the soil by repeated ploughings. Upland cress, being a leafy vegetable crop, needs more nitrogen as compared to other vegetable crops, as insufficiency of nitrogen in soil results in poor growth and yellowing of leaves. Lack of phosphorus and potash results in dry leaf tops and leaf necrosis, respectively. In sandy loam and loam soils, nitrogen, phosphorus and potash are applied at the rate of 80, 60 and 40 kg/ha respectively. In silt and clay loam soils, the amount of nitrogen, phosphorus and potash may be reduced to 100, 40 and 30 kg/ha respectively. A full dose of phosphorus and potash along with one-third of nitrogen should be applied at the time of sowing and rest of the nitrogen should be applied in splits after each cutting. In severe yellowing of leaves in later stages, nitrogen 40 kg/ha may be applied as side dressing. Application of nitrogenous fertilizer normally increases the leaf production of upland cress by increasing the leaf size. Moreover, potash and phosphorus improve the leaf quality by increasing leaf density.

IRRIGATION REQUIREMENT

Upland cress is a moisture-loving crop, thus, it needs to be watered regularly. The soil is kept consistently moist after sowing seeds but not fully soaked since waterlogging as well as drying of soil is detrimental to this crop. Irrigation is applied only at the base of the plant, thus, it should not be applied through a sprinkler system since it is essential to keep the foliage dry. It is not necessary to irrigate the crop in winter since irrigation requirement of upland cress is usually not too high in cool season due to low rate of evapo-transpiration during this period, but during summers, it needs a considerable amount of water for proper growth and development since the evapo-transpiration rate is high during summer months. In very hot weather, if the crop is grown in dry soil, its nips become unpleasant and bitter. The first light irrigation is given just after sowing if soil moisture in the field at the time of sowing is not sufficient, and subsequent irrigations are given as per requirement of the crop. The plants grow best if the soil is kept constantly moist. For incessant harvesting, the crop should be watered regularly.

INTERCULTURAL OPERATIONS

The intercultural operations adopted in upland cress are similar to other cruciferous vegetables. Keeping the field weed free is essential, especially for the crop grown for vegetable purposes since the weeds, particularly those which are similar in appearance to upland cress and difficult to separate, contaminate the crop, therefore, removing such weeds from the upland cress field is most crucial. Regular hoeing after each cutting or mulching with organic material can be done to suppress the weeds. Mulching will also help in keeping the soil moist, which is essential for the production of best quality upland cress.

The crop grown for fresh market hardly requires thinning since this operation is labour intensive. However, thinning is essential when the crop is sown by broadcasting. When the seedlings become large enough to handle, the extra seedlings are removed, keeping the plants at a spacing of 20 cm. Pruning may also be done to obtain quality produce. To promote new growth, the sprouts 10 days after germination are cut at ground level with the help of scissors when tender leaves of the plant just begin to have a green colour.

HARVESTING

Harvesting of upland cress depends on the purpose of its use. Usually, the crop becomes ready for harvest at 5–6 leaf stage, which is attained 7–8 weeks after sowing under favourable environmental conditions. Leaves, shoots, flower buds and seeds all are delicious and its seed oil is edible. Young tender leaves are best for use. The older leaves are cut first with scissors at ground level prior to the plants beginning to bloom, especially when the rosettes of these plants are quite low to the ground and the younger ones are left to develop. Harvesting should not be done by pulling out the plant; it is done by cutting the outer leaves and allowing the centre to continue producing new leaves. When the outer leaves begin to toughen and discolour, the larger central leaves are cut. The stems are not allowed to grow too long and untidy. The production of new leaves can be encouraged by removing the flowering stems. When the plant begins to develop flowering stalks, harvesting of leaves should be stopped since the leaves by this time become extremely bitter. For the harvesting of flower heads, it is always advisable to cut the flower heads fresh as and when needed, and they should be consumed soon after harvesting.

YIELD

The yield of upland cress varies according to variety, soil type, soil fertility, growing conditions and cultural operations adopted during cultivation of the crop. However, its average yield is about 37.5 quintals per hectare.

POST-HARVESTING MANAGEMENT

Being a leafy vegetable, upland cress cannot be stored for long periods. However, in the refrigerator, the leaves can be stored for 2–3 days in low-density polyethylene film bags. Once cut, the cress will last up to 2 days in an airtight container in the refrigerator.

INSECT-PESTS

Flea Beetles (*Phyllotreta* spp.)

Black, metallic green, or blue colour beetles are 2–3 mm long and often shiny in appearance with a broad yellow stripe running lengthwise on wing cases. The beetle jumps when it is disturbed. The beetles make small round holes in leaves, which give the foliage a distinctive *shot hole* appearance. The plants at seedling stage are more susceptible to attack by adult beetles. When the infestation is heavy, the plant growth is reduced and may be killed. The damaged areas dry up and turn pale brown.

Control
- Apply irrigation during dry spells to keep the soil moist.
- Use insect-proof net to cover the plants to exclude the adult beetles.
- Spray the crop with Rogor 0.1% or malathion 0.05%.

APHIDS (*Myzus persicae, Brrevicoryne brassicae* and *Liphaphis erysimi*)

Both nymphs and adults of this tiny insect suck cell sap from the under surface of tender leaves, reducing plant vigour. The leaves curl up and ultimately wilt. The aphids excrete honeydew on which black sooty mould develops, which hampers the photosynthetic activity. Besides, these aphids act as a vector for the transmission of virus. The severe infestation affects the yield.

Control

- Avoid excessive use of lime.
- Spray the crop with simple water with pressure.
- Remove and destroy the portion of the plant attacked by the aphids.
- Spray the crop with malathion 0.05%, monocrotophos 0.04%, methyl demeton 0.05%, dimethoate 0.03%, or phosphamidon 0.03%.

DISEASE

ANTHRACANOSE (*Colletotrichum higginsianum*)

The fungus causes small circular or irregularly shaped grey to straw-coloured dry spots on leaves, which coalesce to form large necrotic patches, causing the leaves to turn yellow and wilt. Its severe infestation may cause death of the leaves. Humid weather is favourable for the occurrence of this disease.

Control

- Rotate the crop with non-host crops.
- Treat the seed with captan, thiram, or Agrosan GN 2.5 g/kg of seed before sowing.
- Provide proper drainage facilities.
- Maintain proper sanitary conditions in the field.
- Remove all cruciferous weeds that may act as a reservoir for the fungus.
- Spray the crop with 0.2% Blitox, Difolatan, Dithane M 45, or Dithane Z 78.

ALTERNARIA LEAF SPOT (*Alternaria brassicae* and *Alternaria brassicicola*)

Both the fungi can affect the upland cress plants at all stages of growth, including seeds. Immediately after germination, dark lesions develop on collar region of the seedlings, resulting in either stunting of seedlings or damping-off disease. The infected leaves show brown lesions with raised concentric rings. The lower leaves show disease symptoms first, as they are closer to the soil and more readily infected by the fungus through rain splashes. The disease is soil- as well as seed-borne.

Control

- Follow long crop rotation with non-cruciferous crops.
- Disinfect the soil with 0.2% captan solution.
- Treat the seed with captan or thiram 2.5 g/kg of seed.
- Dip the seed in 4% sodium hypochlorite for 3 minutes.
- Avoid excessive supply of irrigation.
- Eradicate the cruciferous weed hosts from field and surrounding of the crop.
- Spray the crop with 0.2% Blitox or Dithane M 45.

BLACK ROT (*Xanthomonas campestris* pv. *campestris*)

Black rot is the most destructive disease of crucifers, infecting all cultivated species of *Brassica* worldwide. The disease may appear on plants at any stage of growth. Primarily, it occurs on above ground parts of the plant. V-shaped yellow or orange lesions spread from leaf edge to the centre.

The leaf veins become black, and in case of severe infection, the leaves start falling off. The disease is favoured by warm and humid conditions. Pathogen spreads through infected seed or splashing water and insect movement.

Control

- Rotate the crop with non-cruciferous crops every 2–3 years.
- Use certified disease-free healthy seed and seedlings.
- Treat the seed with hot water at 50°C for 30 minutes.
- Remove of cruciferous weeds that may serve as host of pathogen.
- Dip the seed with mercuric chloride solution (1:1000) for 10 minutes.
- Spray the crop with streptomycin formulation at 1000 ppm.

■■■

40 Watercress

M.K. Rana and P. Karthik Reddy

Botanical Name : *Nasturtium officinale*

Syn. *Nasturtium nasturtium-aquaticum* (L.) H. Karst.

Rorippa nasturtium-aquaticum (L.) Hayek

Sisymbrium nasturtium-aquaticum (L.)

Family : Cruciferae

Chromosome Number : 2n = 32

ORIGIN AND DISTRIBUTION

Watercress is native to Western Asia and Europe and possibly also in the highland regions of Ethiopia. It has been introduced in many African regions and is locally naturalized in mainland Africa, Madagascar and other Indian Ocean islands, mainly in mountainous regions. It has been introduced into many other tropical and temperate regions. It has been grown in many locations around the world. In the United Kingdom, commercially, it was first cultivated by the horticulturist William Bradbury in 1808.

INTRODUCTION

Watercress is commonly known as big leaf cress, water radish, water rocket and hedge mustard. It is a fast-growing aquatic or semi-aquatic perennial plant and one of the oldest known leafy vegetables consumed by human beings. In some regions, it is regarded as a weed and in other regions as an aquatic vegetable or herb. One of the best culinary aspects of watercress is its versatility. It can be used as a salad green with Romaine lettuce or fresh spinach, steamed and eaten as a vegetable, and in soups for a subtle peppery flavour. In the 1940s in the United States, Alabama was locally known as the *Watercress Capital of the World*. Today, Oviedo, Florida, in the United States, is known by that title, while Alresford in England is considered the nation's watercress capital. It has been culti-vated in Europe, Central Asia and the Americas for millennia for use as both food and a medicine.

COMPOSITION AND USES

COMPOSITION

According to the study published in Centers for Disease Control and Prevention (CDC) journal, the researchers at William Paterson University at New Jersey, watercress is labelled as the most nutrients dense food, and because of this reason, it tops the list of *powerhouse fruits and vegetables*. It contains a fair amount of manganese, iron and calcium but the active principle is phenylethyl glucosinolate (gluconasturtiin). This peppery and tangy flavoured cress is a storehouse of many natural phyto-nutrients like isothiocyanates, which have health promotional and disease prevention properties. It is one of the very low-calorie green leafy vegetables and contains negligible amount of fat. Being antioxidant rich, low calorie and low fat vegetable, it is often included in cholesterol controlling and weight reduction programmes. The nutritional composition of watercress is given in Table 40.1.

Uses

Watercress is cultivated for it leaves and shoots used for culinary and medicinal purposes. Leaves can be used raw or cooked throughout the growing season. Because of its peppery taste, it has become very popular as a salad mix. Typically, it is used in small amount like a garnish in salads but commonly used for making soup. The mature seeds, which can be dried and ground into a powder used like mustard, are also edible. If exposed to cold water and allowed to sit for 10–15 minutes, it becomes hot and spicy. The cold water activates an enzyme myrosinase, which breaks down sinigrin to produce a volatile mustard oil into known as Phenyl ethyl isothiocyanate or PEITC. PEITC has been linked to cancer prevention and is helping watercress gain prominence as a *superfood,* with strong anti-cancer properties.

Medicinal Use

Watercress helps in removing toxins from the body, boosting digestion, and has both diuretic and expectorant properties. It is also a dynamic accumulator of mineral elements like potassium, phosphorus, calcium, sulfur, iron, magnesium, sodium and also acts as a water-purifying plant. Wild watercress should not be consumed, as its plants contain pollutants that cause serious illness.

BOTANY

Watercress has creeping fleshy square stems of 30–50 cm height with many side branches. The stems being hollow allow the plant to float. The leaves are entire, simple or compound, alternate, lush-green, oblong-ovate and succulent with high moisture content. The roots are fibrous and produce runners. Its inflorescence is a raceme having very small, symmetrical, hermaphrodite, white flowers with six stamens and four petals. The fruit is about 6–25 mm in length and splits open when ripe, containing two rows of seeds.

TABLE 40.1
Nutritional Composition of Watercress (per 100 g Edible Portion)

Constituents	Contents	Constituents	Contents
Carbohydrates (g)	1.29	Zinc (mg)	0.11
Protein (g)	2.3	β-carotene (µg)	1914
Total fat (g)	0.1	Lutein-zeaxanthin (µg)	5767
Cholesterol (mg)	0.0	Vitamin A (IU)	3191
Dietary fibre (g)	0.5	Thiamine (mg)	0.090
Calcium (mg)	120	Riboflavin (mg)	0.120
Potassium (mg)	330	Niacin (mg)	0.200
Phosphorus (mg)	60	Pantothenic acid (mg)	0.310
Sodium (mg)	41	Pyridoxine (mg)	0.129
Copper (mg)	0.08	Folic acid (µg)	9.0
Iron (mg)	0.2	Vitamin C (mg)	43
Magnesium (mg)	21	Vitamin E (mg)	1.0
Manganese (mg)	0.244	Vitamin K (µg)	250
Selenium (µg)	0.9	Energy (kcal)	11

CLIMATIC REQUIREMENT

Watercress is a sun- and water-loving perennial plant that grows naturally along running waterways. Since watercress is a leafy vegetable, it is sensitive to sunlight and day length. It prefers full sunlight but can tolerate light shade also. The higher the percentage of sunshine, the faster the crop grows. Watercress grows best at a constant temperature of 10–12°C.

SOIL REQUIREMENT

The plant grows well under varying soil conditions provided the soils stay saturated with water. It thrives well in shallow aquatic or semi-aquatic conditions. Slow moving very clean water is the best condition for growing this plant. It also grows in wet muddy soils and can easily tolerate moist soils if the soil does not dry out very soon. It prefers neutral to alkaline soil with pH 6.5–8.0. However, it grows best in organic-matter-rich wet soils and tolerates a wide range of pH. If it is to be grown in containers, a soil-less potting mix containing perlite or vermiculite mixed with peat is used. When growing in pots, the potting mix is kept moist by having the pot sit in a saucer filled with water.

CULTIVATED VARIETIES

Still, no standard variety of watercress is available anywhere in the country.

SOWING TIME

Watercress seeds take approximately 7 to 14 days to germinate in soil, thus, the seeds are sown roughly 3 weeks before the occurrence of frost. Seedlings are transplanted at spacing of 15 × 15 cm. If plants or cuttings (30 cm long) are scarce they may be planted as far as apart as 30 cm each way. Close planting is preferred because it gives more uniform growth for the first harvest. As soon as the transplants have started to grow vigorously, the tips of the new growth should be clipped to induce the development of short and stocky growth.

SEED RATE

A quantity of one gram contains approximately 5,000 seeds. The seed rate of watercress has not yet been standardized.

PROPAGATION

Watercress may easily be propagated by seeds or stem cuttings. The seeds are sown just below the soil surface about 6 mm deep. The seed is sown at a spacing of 22–30 cm. The field at the time of sowing should have plenty of moisture, as its seeds require wet soil for best germination. Stem cuttings form roots quickly in rich soil along stream beds where the soil stays very wet. The seeds can germinate without stratification. As and when the seedlings become large enough to handle, the excess seedlings are thinned out by hand, keeping 10 cm distance between plants. The usual method of propagation is through cuttings and is often known to the commercial grower as the *plants* and the method of laying the runners as *planting*.

NUTRITIONAL REQUIREMENT

Watercress does not have a high nutrients requirement. However, the cultivated watercress sometimes shows symptoms of phosphorus, potassium, or iron deficiency. Phosphate-deficient plants are stunted and dark coloured. Symptoms of potassium deficiency are marginal scorching on older leaves.

Commonly in winter, the iron-deficient leaves are expressed as yellowing between the veins of newer foliage. Mixing a complete soluble fertilizer mixture with water at the recommended rates minimizes these problems. Before planting, the garden soil is amended by incorporating well-composted organic matter into the soil up to a depth of 10 to 15 cm. The nutrients requirement of watercress is met by the mineral content of irrigation water. The individual growers supplement fertilizers to watercress, but still, no standards have been established for supplying nutrients through fertilizers.

IRRIGATION

Overhead sprinkling for twelve 15-minute periods/day increased the quality of watercress grown in the warm season, and increased yield from 1.25 kg/m² to 1.96 kg/m², comparable to yields obtained in the cool season without sprinkling (2.02 kg/m²). Sprinkling in the warm season caused differences in temperature between sprinkled and unsprinkled leaves of up to 4.8°C.

INTERCULTURAL OPERATIONS

The area around watercress plants is kept weed free and light mulching is done around the plants to maintain moist soil conditions. Various algae, which compete with watercress plants for nutrients and contaminate the harvested produce by imparting a fishy odour, are commonly present in watercress beds. Adding 2–10 ppm copper sulfate efficiently controls the algae.

FLOWERING

Watercress produces small white flowers in flat-topped clusters. Flowering starts in late spring to autumn (May–October) and the leaves become hard and bitter if harvesting is not done frequently to prevent flowering. Its flowers are self-fertile and are pollinated by bees and flies.

HARVESTING

In growing areas with leeward sunlight exposure, watercress can be harvested 6 to 7 weeks after planting when the plant has grown 30 to 35 cm above the water surface. Harvesting is usually done with a sickle. Young leaves and stems taste best if harvesting is done before the start of flowering, as later, the leaves become bitter in taste. The peppery flavour is considered best when harvesting is done in spring months. The flavour deteriorates as the temperature goes above 30°C. Taking more than three cuttings should be avoided to allow the plants to grow enough foliage continuously. Periodic harvesting helps in encouraging new growth.

Seeds can be harvested from mid-summer to autumn. Under favourable growing conditions, subsequent crop can be taken by ratooning the *mother* plants. If the growing conditions are favourable, it can be grown as a ratoon crop indefinitely.

YIELD

High as far as greens are concerned. Harvesting can start at 30–45 days and may continue every 40–80 days. An average yield of 12.5 to 20 tonnes per hectare can be obtained.

POST-HARVEST MANAGEMENT

Watercress is used fresh within a few days of harvesting. It is mentioned in some reports that watercress can also be used in dehydrated form. After harvesting and packaging, the watercress is quickly chilled using vacuum cooling or by submerging in ice water.

INSECT-PESTS

Watercress has no specific insect-pest and disease problems in most production areas. However, the following insect pests are common in watercress.

BUCKTHORN APHID (*Aphis nasturtii*) and Green peach aphid (*Myzus persicae*)

Aphids are an intermittent pest of watercress. Heavy aphid infestation can cause an unmarketable appearance of watercress because the presence of a single live or dead aphid on the finished product can lead to customer rejection.

Control

- Spray the crop with simple water with pressure.
- Remove and destroy the plant portion attacked by the aphids.
- Spray the crop with imidacloprid 0.025%, malathion 0.2%, Thiodan 0.2%, monocrotophos 0.04%, methyl demeton 0.05%, dimethoate 0.03%, or phosphomidon 0.03%.

CYCLAMEN MITE (*Stenotarsonemus pallidus*)

Cyclamen mite has been a pest of watercress since the 1950s. It starts feeding on the apical shoot of the plant. The damage symptoms are shoot and leaf stunting, bronzing of leaf tissue and deformation of the entire plant with heavy mite infestations. It is widespread and found in all watercress growing areas.

Control

- Intermittent overhead irrigation is very effective in controlling cyclamen mite.
- Collect and burn severely infested plant parts to reduce further multiplication of mites.
- Spray the crop with acaricides like dicofol 0.05% and wettable sulfur 0.3%.
- Spray diazinon to plant canopy @ 1.25 kg a.i./ha.

DIAMONDBACK MOTH (*Plutella xylostella*)

Diamondback moth is the most serious insect-pest of watercress. Its caterpillars damage the plant by feeding on leaves and young growing shoots. The life cycle of moth is very rapid. Eggs are laid on undersurface of the leaves and in the growing points. Egg to adult maturation can occur in as little as 12 days but averages 21 days in most watercress growing areas. The diamondback moth is world renowned for its ability to develop resistance quickly to insecticides used to control it. The moth feeds exclusively on cruciferous host plants.

Control

- Use intermittent overhead irrigation system in combination with the parasitoid wasp (*Cotesia plutella*), which is established in all watercress growing areas.
- Spray the crop with malathion 0.1%, profenofos 0.25–0.5 kg a.i./ha, or spinosad 0.05%.
- Spray *Bacillus thuringiensis* spp. *kurstaki* directly to the plant canopy.

IMPORTED CABBAGEWORM (*Pieris rapae*)

Imported cabbageworm is an occasional pest of watercress. Its caterpillars damage the crop. Its adults are white butterflies commonly seen searching for nectar along roadside weed stands. Outbreaks often appear from the months of February through June. Leaves are damaged by larval feeding, which render the crop unmarketable.

Control

* Spray the crop with 0.1% malathion or Rogor.

SHARPSHOOTERS (*Draeculecephala californica, D. inscripta* and *D. minerva*)

Sharpshooters are minor pests of watercress. These small insects are commonly found on grass but can build to high numbers. Nymph and adult stages cause damage called *hopper burn*, which is a condition that results in localized yellowing of leaves and leaflets and can render the crop unmarketable.

Control

* Remove and destroy the attacked portion of the plant.
* Spray the crop with imidacloprid 0.025%, malathion 0.2%, Thiodon 0.2%, monocrotophos 0.04%, methyl demeton 0.05%, dimethoate 0.03%, or phosphomidon 0.03%.

SNAILS (*Cornu aspersum* and *Theba pisana*) and Slugs (*Deroceras reticulatum, Lehmannia poirieri, L. valentiana, Limacus flavus* and *Milax gagates*)

Various species of snails and slugs can be a problem on watercress when they move high up on the plant stalks at the time of harvest. They chew irregular holes with smooth edges in leaves and flowers and can clip succulent plant parts. The presence of slugs or snails on plant can lead to market rejection of the finished product. Slug and snail pressure increases during rainy periods or from excessive moisture supplied by intermittent overhead irrigation systems used by growers to control diamondback moth.

Control

* Collect and destroy snails and slugs manually.
* Use drip irrigation system in place of sprinkler system to reduce humidity and moisture on surfaces, making the habitat less favourable for these pests.
* Use metaldehyde baits containing 4% active ingredient.

DISEASES

ASTER YELLOWS (*Phytoplasmas*)

The disease can infect numerous plant hosts. It is vectored by the leafhopper, which is known to feed on a variety of host plants. Symptoms are severe chlorosis and stunting, witch's broom effect on new shoots, and in some cases, death of the plant.

Control

* Use disease-free planting material.
* Rogue out the infected plants.
* Remove all asymptomatic watercress within a radius of 90 cm.
* Spray an approved insecticide for all vegetation within 1.5 m from edge of watercress field regularly.

BLACK ROT (*Xanthomonas campestris*)

Black rot can be a serious disease during prolonged rainy periods. The disease begins as a yellow spot at the edge of leaflets. The disease is spread by splashing water droplets. The use of intermittent overhead irrigation can also spread the disease if relative humidity is high (>80%) and wind is light (<10 mph).

Control

- Remove and destroy the affected plants.
- Treat the seeds with Agrimycin 100 (100 ppm) or streptocycline (100 ppm).

CERCOSPORA LEAF SPOT (*Cercospora nasturtii*)

Cercospora leaf spot is an occasional disease caused by high humidity (>80%). Mostly older leaves are affected but the disease can sometimes be found high up on the plant if the conditions which favour the disease (heat and high humidity) are prolonged.

Control

- Remove infected leaves manually at harvesting.
- Spray the affected plants with Bavistin 0.1% or chlorothalonil 0.2%.

■■■

41 Winter Cress

M.K. Rana and Pradeep Kumar Jatav

Botanical Name	: *Barbarea vulgaris*
	Syn. B. vulgaris var. arcuata
	B. vulgaris var. brachycarpa
	B. vulgaris var. longisiliquosa
	B. vulgaris var. sylvestris
Family	: Cruciferae
Chromosome Number	: 2n = 18

ORIGIN AND DISTRIBUTION

The exact centre of origin of winter cress is not known since it occurs worldwide naturally in temperate regions. A great diversity exists in many parts of the world, including Middle East, Europe, North America and New Zealand. However, the gardens of winter cress were developed in Persia. The winter cress has been cultivated since ancient Roman times but its commercial cultivation was first time recorded in Germany in 1750 and later in Britain in 1808. However, Canada and Eurasia are considered the primary centre of its origin.

INTRODUCTION

Winter cress, also known as yellow rocket, is most commonly found growing wildly in winters but its cultivation has been noticed in kitchen gardens in the countries like Canada, Russia and North America. So far, its cultivation has not been noticed in India. Its natural habitats include cropland, fallow fields, orchards, construction sites, wastelands, moist meadows and areas along roadsides and railway lines. It is extensively grown as a trap crop around the cabbage field to trap the diamond-back moth (*Plutella xylostella*). Winter cress has a pungent, peppery taste when eaten either raw or cooked.

COMPOSITION AND USES

COMPOSITION

Winter cress leaves provide a good amount of calcium, potassium, phosphorus and vitamin A and C. It contains phytochemicals such as flavonoids, glucosinolates and saponins, which are very beneficial for human health. The nutritional composition of winter cress leaves is given in Table 41.1.

TABLE 41.1

Nutritional Composition of Winter Cress Leaves (per 100 g Edible Portion)

Constituents	Contents	Constituents	Contents
Water (%)	89	Phosphorus (mg)	20
Carbohydrates (g)	0	Potassium (mg)	112
Protein (g)	1	Vitamin A (mg)	159
Fat (g)	0	Riboflavin (mg)	0
Dietary fibres (g)	1.2	Vitamin C (mg)	15
Calcium (mg)	41	Energy (kcal)	4

USES

Winter cress is especially grown for its succulent and tender leaves, and to some extent, for its young flowering stem like broccoli. Its young tops serve as delicately flavoured vegetable in temperate regions of the world during spring. The young leaves of winter cress are used as salad just like lettuce. It is extensively used as cattle fodder but seldom as green vegetable. Both the dark and glossy green leaves and the yellow flower buds of winter cress can be eaten, however, the leaves are best when used just before flowering since after flowering, the leaves and flower buds become bitter and must be parboiled in order to reduce their bitterness before preparing any dish or freezing for future use.

MEDICINAL USES

The plant is said to possess anti-cancerous properties. The winter cress is used to prevent scurvy disease caused due to the deficiency of vitamin C and to cure peptic ulcers. The leaves poultice is used by the Europeans to treat wounds. An infusion prepared from its leaves is appetizer, diuretic and blood purifier.

BOTANY

Winter cress is a dicotyledonary annual herb for the purpose of greens production and biennial for seed production. The basal leaves are odd-pinnate with 1–4 pairs of lateral lobes and a terminal lobe, which is larger than the lateral lobs. The alternate leaves are sessile clasping the stems. Both the basal and alternate leaves are dark green, hairless and shiny on the upper surface. In the first year, it forms a rosette of basal leaves, while in the second year, it bolts upward with one or more flowering stalks, which are hairless, smooth surfaced, stout, light green to reddish purple and somewhat angular. Secondary stalks are produced in upper half of the plant, which terminate in racemes of yellow flowers, which bloom toward the apex of each raceme, consisting of four yellowish green sepals that are linear-lanceolate, four yellow petals, six stamens, of which four are long and two short, with pale yellow to light brown anthers and a single pistil with a thick style. A robust plant bears mildly fragrant flowers in great abundance from mid-spring to early summer. Each flower bears 2.5–3 cm long angular to cylindrical seedpod known as siliqua. The base of each seedpod is connected to a short slender pedicel, while at the other end it terminates in a short slender beak. The winter cress plant has a stout taproot system.

CLIMATIC REQUIREMENT

Winter cress is primarily a cool season crop resistant to frost. The best quality leaves are successfully produced in temperate climate. Most of the vegetative growth takes place during cool weather of early to mid-spring but winter cress prefers little warm conditions at flowering, pod formation and maturity of pods. The minimum temperature for its growth and development is 0.5–3°C, whereas, the optimum temperature is 15–25°C. Low soil temperature reduces the water and nutrients absorption and increases susceptibility to pest and diseases. Rainfall, high humidity and cloudy weather are not suitable for the crop, as under these conditions, the incidence of insect-pests and diseases, especially of aphids and white rust is higher. However, the crop required an annual rainfall of 40–100 cm. This crop also does not tolerate waterlogging conditions.

SOIL REQUIREMENT

Winter cress can be grown on a wide range of soils. However, the higher yield with best quality greens can be produced in well-drained sandy loam rich in organic matter soils with good moisture holding capacity. Its cultivation should be avoided in clayey and very sandy soils. The crop is moderately tolerant to acid soils having pH range from 4.8 to 7.5. However, the soils with neutral to slightly alkaline reaction are ideal for its successful cultivation.

CULTIVATED VARIETIES

Still, no popular variety has been developed in this crop. However, some identified most popular cultivars along with their characteristic features are given as follows:

WATERCRESS

This is the most common variety of cress, especially grown in flooded soil beds. The leaves are small, heart-shaped with dark green colour and slightly bitter taste.

BROAD LEAVED CRESS

It is also known as curly cress. It has oval leaves, with an attractive ornamental appearance, extra curled. The plant is compact with short stalk.

SEED RATE

Winter cress is exclusively propagated by seed. However, its seed rate is nowhere mentioned in the literature.

SOWING TIME

The sowing time considerably varies with variety and climatic conditions of that region where the crop is to be raised. The optimum sowing time ensures good conditions for growth and development of the crop and minimizes the loss caused due to the attack of insect-pests and diseases. Sowing of seeds of a particular variety at its right time is very essential to get higher yield of better quality produce. In subtropical regions, its sowing is done in autumn from September to October, and in temperate regions, in spring from March to April. The seed takes 3 to 4 days to germinate under favourable conditions. Seed sowing is done at regular intervals for continuous supply of young greens.

SOWING METHOD

The seeds are sown directly in flat seedbeds at depth of 1–1.5 cm. At the time of sowing, the soil must have ample moisture, which facilitates easy and early germination of seeds. However, in low-lying areas, raised beds are preferred over flat beds, as it facilitates drainage. Sowing can be done by broadcasting the seeds, however, line sowing gives yield higher than sowing seeds by broadcasting method. The seeds are sown sparsely to avoid overcrowding of plants. The seeds of winter cress are very small, thus, should be mixed with a small quantity of fine sand before sowing to facilitate uniform distribution of seed.

NUTRITIONAL REQUIREMENT

Among the cruciferous crops, winter cress has not yet achieved the commercial status. Therefore, the information available on its nutritional aspects is scanty. Normally, this crop needs more nitrogen as compared to phosphorus and potassium. Since the crop is similar to mustard, the fertilizer doses recommended for leafy mustard may be applied for the successful cultivation of winter cress also. Well-decomposed farmyard manure 15–20 t/ha should be incorporated into the soil at the time of land preparation. Besides, nitrogen is applied 125 kg and phosphorus and potassium each 75 kg/ha. The full dose of phosphorus and potassium along with one-third of nitrogen is applied as basal dose at the time of sowing and remaining two-third of the nitrogen is top-dressed in two splits after each cutting for the quick growth of plants.

IRRIGATION REQUIREMENT

Pre-sowing irrigation should be applied to have ample soil moisture for proper seed germination and further growth of the plants. If moisture is not enough at the time of sowing, a light irrigation should be applied soon after sowing, and subsequent irrigations should be given at an interval of 4–5 days in light soils and 8–10 days in heavy soils.

INTERCULTURAL OPERATIONS

HOEING AND WEEDING

Removing weeds at early stage of crop growth is very essential because the crop plants due to slow growth rate at early stages cannot compete with weeds, which make the harvesting operation difficult, thus, the field should be kept weed free at initial stages of crop growth. Hoeing is also beneficial for keeping the weeds down and providing good soil environment to the crop roots for their development. As the crop is shallow rooted, deep hoeing should be avoided to prevent damage to the crop roots. One hand weeding 25 days after sowing is necessary for effective weed control. Pre-plant incorporation of fluchloralin 1 kg/ha or pre-emergence application of pendimethalin 1 kg/ha is effective in controlling weeds in winter cress.

BOLTING

Bolting is an undesirable process of developing seed stalk except where the crop is raised for seed production. The tendency of plants to bolt depends on growing temperature, type of variety and many other factors. Bolting in winter cress is not a photoperiodic response though it is a low temperature response. Therefore, temperature is more responsible to manipulate the physiology of biennial plants rather than photoperiod. Winter cress is similar to many of the other members of the Crucifereae family such as wild mustard, wild turnip and wild radish. In subtropical climate, it flowers at the beginning of January.

HARVESTING

Harvesting of winter cress leaves for cooking depends on variety. Both the dark and the glossy green young leaves are picked when they are 7.5 to 12 cm long and the immature flower buds are harvested before their opening. The entire plant can be harvested 60 days after sowing before the emergence of seed stalk, whereas, the tender stem along with leaves is harvested about 180 days after sowing. Soon after harvesting, the diseased, yellow, and injured leaves are removed and the produce is tied in bundles of required size and sent to the market.

YIELD

The yield of winter cress depends on several factors like type of cultivar, growing season, soil and cultural practices followed during its cultivation.

POST-HARVEST MANAGEMENT

The leaves cannot be stored longer under ordinary conditions as they lose their moisture very soon. However, they can be stored successfully for about 2–3 weeks at 0°C and 90–95% relative humidity.

INSECT-PESTS

APHIDS (*Myzus persicae*)

The pest lives in colonies and has a higher rate of multiplication. Moist, cloudy weather and the temperature of 10–18°C favours the rapid multiplication of aphids. Both its nymphs and adults suck cell sap from under surface of the tender leaves and stem, reducing plant vigour. The leaves curl up and ultimately wilt. The honeydew excretes by aphids encourages the growth of black sooty mould, which lower down the photosynthetic activity of leaves. The severe infestation affects the yield and quality of leaves.

Control

- Remove and destroy the affected portion of the plants.
- Apply recommended doses of fertilizers.
- The crop should be sown early at optimum time of recommended for particular areas.
- Spray the crop with malathion at 0.05%, monocrotophos 0.04%, or dimethoate 0.03%.

MUSTARD SAWFLY (*Athalia proxima*)

The larvae cause maximum crop damage by making irregular holes in leaves. In case of severe infestation, the plants may completely denude, which necessitates re-sowing.

Control

- Follow crop sanitation practices.
- Handpick and destroy the larvae.
- Provide timely irrigation, which helps in killing larvae through drowning.
- Spray the crop with malathion 500 ml in 500 litres of water.

PAINTED BUG (*Bagrada cruciferarum*)

Both adults and nymphs suck the cell sap from tender parts of the plant. The infected young plants become bushy, while blotchy spots develop on leaves. This pest mostly occurs in the months of October to November.

Control

- Follow long crop rotation with non-host crops.
- Keep the field weeds free and destroy infected plant debris.

- Dust the crop with quinalphos 1.5% at 20–25 kg/ha.
- In severe infestation, spray the crop with malathion 500 ml in 500 litres of water.

Pea Leaf Miner (*Chromatomyia horticola*)

Generally, this pest occurs in February and March. The tiny maggots eat away chlorophyll part between mesoderm layers of the leaves in a zigzag fashion. The heavily infested leaves become yellow and fall down, affecting the yield adversely.

Control

- Pluck the infested leaves and burn them to kill the larvae and pupae.
- Keep the field free from weeds.
- Spray the crop with oxydemeton or dimethoate 875 ml dissolved in 875 litres of water/ha.

DISEASES

Alternaria Blight (*Alternaria brassicae*)

The disease is characterized by appearance of small light brown circular spots on cotyledonary leaves, turning black in advanced stage. Small circular brown or blackish spots increase in size and multiply rapidly, forming dark brown concentric rings on leaves. Circular to linear dark brown lesions on petioles and stem elongate at a later stage. In severe infection, blighting and defoliation of young plants occur. High humidity and warm weather with intermitted rains favour the disease development.

Control

- Collect and burn the plant debris to minimize the primary source of inoculums.
- Early sowing enables the crop to escape severe development of disease.
- Spray the crop with Mancozeb 0.2% at 10-day intervals.

Downy Mildew (*Peronospora parasitica*)

The fungus can attack the plants at any stage of growth, even at flowering. Infected plants show whitish mycelium growth on lower surface of the leaves and corresponding purplish or yellowish-brown spots appear on the upper surface of the leaves; later the whole plant is covered with whitish growth. The pathogen multiplies rapidly in wet or cloudy weather. Low temperatures of about 18–20°C and high relative humidity about 95–98% favour the disease. The disease is soil- as well as seed-borne.

Control

- Follow long crop rotation with non-cruciferous crops.
- Use disease-free seeds.
- Remove all infected leaves and destroy them.
- Treat the seed with hot water at 52°C for 30 minutes before sowing.
- Follow with prophylactic sprays of Dithane M-45 at 0.3%.
- Spray the crop with Ridomil 0.25% for effective control of disease.

White Rust (*Albugo candida*)

The occurrence of disease is more severe under humid conditions. Small, white raised pustules appear on the leaves and stem. These pustules coalesce to form large patches. Swelling and distortion of leaf petioles, stem and floral parts result in hypertrophy and hyperplasia, commonly known as stage head, which is the result of mixed infection of white rust and downy mildew under humid weather. The infected plants are dried when the infection is severe. The pathogen survives on secondary host plants. It is an air-borne disease.

Control

- Follow long crop rotation with non-cruciferous crops.
- Use certified disease-free healthy seed from reliable source.
- As far as possible, destroy the weed hosts surrounding the crop.
- Remove and destroy the affected plants and crop debris.
- Spray the crop with Dithane M-45 0.2% or Ridomil 0.25% at 10-day intervals.

Powdery Mildew (*Erysiphe* sp.)

The disease occurs mainly in dry weather condition. Greyish-white, circular, powdery patches appear on leaves and stem, which increases with increase in temperature, causing yellowing of leaves at later stage followed by defoliation of leaves. It is a seed-borne disease.

Control

- Plant the crop on normal sowing date.
- Reduce all affected plant parts to reduce the primary source of inoculums.
- Dip the seeds in hot water at 52°C for 30 minutes.
- Spray the crop with Karathane 0.5%, Calixin 0.1%, or Sulfex 0.2% at 3–4 times at an interval of 10 days.

■■■

42 Horseradish

M.K. Rana and Pooja Rani

Botanical Name	: *Armoracia rusticana*
	Syn. *Cochlearia armoracia*
	Syn. *Armoracia lapathifolia*
Family	: Cruciferae
Chromosome Number	: 2n = 32

ORIGIN AND DISTRIBUTION

Horseradish is native to Mediterranean regions but it is mainly cultivated in the United States of America such as in Pennsylvania, Oregon, Washington, Wisconsin and California. Now, it is grown throughout the temperate regions of Northern and Southeastern Europe and in Scandinavia. It is also grown in Canada and Europe to sell commercially and to a small extent in the gardens of North India and hilly tracts of South India. Ancient Greeks and Romans cultivated it for culinary and medicinal purposes. In the past, it has been used medicinally to treat backaches caused by the common cold. The horseradish sauce made by John Henry Heinz in 1869 was one of the first condiments sold in the United States. The commercial horseradish industry began to grow in the Midwest during the 19th century. Today, a large portion of horseradish is still grown in the areas surrounding Collinsville, Illinois. The town of Collinsville refers to itself as *the horseradish capital of the world*.

INTRODUCTION

In India, horseradish is commonly known as *sahijan*. The name 'horseradish' has probably been derived from the German word *meerettich,* meaning *more* or *stronger radish*. The English people have great confusion for the term *meer* with that of *mare* or *mahre*, thinking that the name is referred to a horse. The regional name of horseradish is *kren* in southern Germany and Austria. *Aremoricus*, which means *maritime* or *near the sea*, may also be a contributing factor to its botanical name since horseradish was often growing near the seaside. The generally accepted meaning for *Armoracia* that most closely fits horseradish is *wild radish. Rusticana* means *of the country, rustic*, or *rural*. It is still grown predominantly in some areas, where the growing season is short. Once introduced to a farm, horseradish roots are difficult to eradicate completely since its roots can readily develop new plants.

COMPOSITION AND USES

COMPOSITION

The cooked horseradish is high in nutritional value. It is a rich source of minerals like phosphorus, potassium, calcium and magnesium. It is richer in vitamin C than lemons or oranges. Being fat-free, the freshly grated roots are low in calorie as well as being rich in vitamin A. In addition, its roots have small amount of essential vitamins such as thiamine, riboflavin, niacin, pantothenic acid, pyridoxine and folate. The nutritional composition of horseradish is given in Table 42.1.

TABLE 42.1

Nutritional Composition of Horseradish (per 100 g Edible Portion)

Constituents	Contents	Constituents	Contents
Carbohydrates (g)	11.29	Copper (mg)	0.058
Protein (g)	1.18	β-carotene (μg)	1.00
Total fat (g)	0.69	Lutein-zeaxanthin (μg)	10.0
Dietary fibre (g)	3.30	Vitamin A (IU)	2.00
Calcium (mg)	56.0	Thiamine (mg)	0.008
Phosphorus (mg)	31.0	Riboflavin (mg)	0.024
Potassium (mg)	246.0	Niacin (mg)	0.386
Magnesium (mg)	27.0	Pantothenic acid (mg)	0.093
Sodium (mg)	314.0	Pyridoxine (mg)	0.073
Iron (mg)	0.42	Folic acid (μg)	57.0
Manganese (mg)	0.126	Vitamin C (mg)	24.9
Zinc (mg)	0.83	Energy (cal)	48.0

The pungency and aroma of horseradish is due to the presence of isothiocyanates, which are bound with sugar as either glucosinolates sinigrin or 2-phenylethylglucosinolate in cell vacuole. They are separated by hydrolysis of sugar bond with the help of membrane bound myrosinase enzyme. The intact root has no odour. However, the non-volatile form of oil upon crushing of tissues is converted into volatile form with the help of myrosinase enzyme, which develops aroma and pungency. Thus, the roots should usually be processed under refrigeration soon after dicing due to fast volatilization of oil. An ether extract of a ground root contains 76–80% allyl isothiocyanate and 16–18% β-phenylethyl isothiocyanate compounds, which are known to carry antioxidant and detoxification functions. Its essential oil is toxic and should be used with extreme caution.

Uses

The National Heart, Lung and Blood Institute advocates horseradish as part of a healthy diet. Its roots are usually grated or minced and mixed with vinegar, salt, or other flavouring substances to make sauce, which is often used with fish or other seafood and meat. The plant material is also used as an ingredient in ketchup. It is also consumed after dehydration in off season. Horseradish is normally considered as safe for human consumption as a natural seasoning and flavouring.

Medicinal Uses

The horseradish fresh roots are medicinally used as an appetizer, antiseptic, diaphoretic, diuretic, rubefacient, stimulant, stomachic and vermifuge. It is also a valuable remedy for those who are suffering from asthma, chest complaints, coughs, colic, rheumatism, scurvy, toothache, ulcers, circulation problems, venereal diseases and cancer. Peroxidase enzyme is extracted from the plant roots and used as an oxidizer in chemical tests such as blood glucose determinations. Horseradish causes strong irritation, blood vomiting and diarrhea if ingested in large quantity. Its tops and roots may be harming to livestock. The volatiles of horseradish roots have been reported to have herbicidal and germicidal properties.

BOTANY

Horseradish is a hardy herbaceous perennial but it is cultivated as an annual in many areas. The plant grows in clumps with leaves that spread out from the main taproot. It can grow up to a height of 60–90 cm or taller at flowering. The leaf shape, length and base may vary as the plant grows up to maturity. The leaf base of the individual cultivar can be heart-shaped, tapering, or somewhere in between. Leaf margins may be smooth (entire), wavy (sinuate), or lobed, while the leaf surface can have a rugose or crinkled appearance. In late spring to early summer, the off-white flowers with a sweet honey scent are borne on terminate and axillary raceme. They have four petals, standing opposite to each other in a square cross and six stamens, of which two are short and four long, and a special kind of pod called sliqua. The fruits have spherical compartments that contain four to six seeds in each compartment. The side shoots arise from the main taproot, which has a colour ranging from off-white to light tan. The roots grow underground up to a several feet, depending upon plant age.

Scientists hypothesize that *Armoracia rusticana*, which is generally reported to be sterile, may be an inter-specific hybrid of *Armoracia lacustris* and *Armoracia sisymbroides*. Japanese horseradish (*Wasabia japonica* M.), a glabrous perennial herb with creeping pungent rhizomes, is found growing wild along streams in Japan. It is cultivated and used as a condiment.

CLIMATIC REQUIREMENT

Horseradish being a hardy perennial herb capable of surviving winter temperature −20°C has wide climatic adaptation, with the exception of warmer zones of the far southern United States. It grows best at 5–24°C temperature with an annual precipitation of 50–170 cm. The crop enjoys cooler temperatures and can withstand a short spell of frost. It thrives best in full sun around 6–8 hours a day. It can tolerate semi-shaded environments of the north-temperate regions of America but it performs poorly in constant shade or wetness. It also grows well in hotter climates but it may require shade in the afternoon.

SOIL REQUIREMENT

Although horseradish can grow on any type of soil from sandy to alluvial soils, the best growth takes place in deep loam soil high in organic matter. Loose garden soils with good drainage facilities are suitable for the development of healthy roots. Heavy soils filled with clay, rocks, stones, or pebbles below the surface can restrict its root development and result in a poor harvest. The optimum soil pH for economic yield is 5.0 to 7.0.

Prior to planting, the turning of soil up to a depth of 25 to 30 cm or deeper is recommended for getting best results from horseradish crop. The soil is prepared to a fine tilth by ploughing the field with disc plough three or four times, followed by planking to make sure the field is properly levelled.

VARIETIES

Numerous cultivars of horseradish are available for use in kitchen gardens and for commercial production. Each varies in its resistance to diseases and harvesting yield. Depending on the site, one or more may develop large solid white roots with no discoloration or hollowness. The roots should be sweet, straight and smooth surfaced with no corking, cracking, or rootlet. The botanical classification of horseradish depends upon leaf shape, size and texture with many variations. A key distinct character is angle and shape of the leaf base. A smooth aboveground crown texture and low shoot number are also the desirable traits. Its two main forms are acute (tapering) and cordate (obtuse and round). Leaves can also be smooth and crinkled or have slightly wavy edges. According to shape of the leaves where they attach to the petiole, many of the common cultivars can be classified into three types. It is important to make an effort to keep planting stock sorted and separated, as roots have a similar appearance and are easily mixed. Some of the common cultivars and their classifications are listed as follows:

VARIETIES WITH HEART-SHAPED BASE

Big Top Western: Its plants have smooth-surfaced large upright leaves, which are tapered at the base. It exhibits resistance to mosaic virus and white rust. Its roots are large and of good quality; however, the outer surface of the roots is rough or corky.

Bohemian: Its plants have smooth-surfaced medium-sized leaves, which have a more rounded (60°–90°) attachment angle. It is susceptible to virus but tolerant to white rust. The roots are smooth-surfaced and of good quality but are not too large.

Sass: Its leaves are smoothly surfaced and heart shaped.

VARIETIES WITH TAPERED BASE

Common: It is a broad and crinkled leaf variety with a cordate (<60°) leaf base attached to the petiole. It is known for its high-quality and large roots; however, it is susceptible to virus and white rust.

Maliner Kren: Sometimes it is referred to as 'Common,' as it also has crinkled leaves.

INDETERMINATE VARIETIES

Swiss: This variety also has smooth leaves.

OTHER VARIETIES

Variegata: It has green leaves splashed with white colour and may be cultivated as an ornamental plant.

Wildroot: The variety has leaves with hot and strong spicy flavour.

Czechoslovakian: It is a new commercial variety with a more mild taste than other cultivars.

PLANTING TIME

Horseradish can be planted in early spring around March to April in temperate regions and September to October in subtropical regions. It is a cold-hardy plant. Planting of crowns or root cuttings can be done 4 to 6 weeks before the average last frost and after the end of heavy rains.

SEED RATE

Horseradish cuttings are called *sets*. A total of 15,000 to 22,500 sets per hectare is needed, depending on variety and plant spacing. After preparing the field properly, row and set spacing in the field are marked before the sets are planted and covered. This operation can be done with the help of tractor attached paddlewheel implement, which creates a *pillowed* or *dished* indentation in the soil, marking the set placement sites at desired spacing. The root sets are planted in shallow trenches about 10–12 cm deep, allowing 45–60 cm between plants and 30 cm between rows. The sets may also be planted horizontally on soil surface at a slight angle (30°–45°) with the head-end placed slightly higher in a dished-out pillow than the bottom end. The same set orientation or direction is maintained in several rows (usually four rows), depending on the cultivation equipment used. After planting of sets, a disc-hiller is used to cover the sets with a light soil.

An alternative system of production, the small pieces of root are dropped either by hand or with a potato planter in rows. This method of planting will result in a more carrot-like root system with several medium-sized primary and a significant number of smaller and secondary roots. The increased number of plants per hectare results in greater tonnage. The crop is grown more successfully on raised beds with black plastic and a drip irrigation system.

SOWING METHOD

Horseradish is usually propagated vegetatively by crown division and root cuttings but not from seed. Initial planting stock of few varieties can be obtained from special nurseries. The grocery stores can also be a viable source for obtaining planting material since the roots sold in grocery stores are not treated with any chemical, which may affect sprout growth. The main disadvantage of grocery-obtained stock is that the variety and characteristics may not be known, and there may also be traces of disease to the site where it was grown. The small pencil-sized roots developed from the main root cuttings, or sets, can be used for planting purposes. Since the horseradish root exhibits polarity, it is important to make identification marks showing which end of the set is the top and which is the bottom in order to plant the piece in the correct orientation. A good way of planting is to use a straight cut across the top end of the set and an oblique cut at the bottom. If planting is to be done by dividing an established crown, the plant should be dug out carefully from the soil and split into four equal pieces, each with some leaf and root.

NUTRITIONAL REQUIREMENT

About 25–30 tonnes of well-decomposed farmyard manure per hectare should be incorporated in the soil during field preparation at least 15 days before planting. While fertilizing horseradish crop, nitrogen fertilizer is generally applied in a higher amount than phosphorus and potash fertilizers since most of the soils are deficient in nitrogen, hence, its application produces a quick and most pronounced effect on horseradish crop. A balanced application of nitrogen, phosphorus and potash should be incorporated prior to planting in topsoil. Application of nitrogen 56 kg/ha along with 110–170 kg/ha each of phosphorus and potassium as a basal dose increases the crop yield. Supplemental nitrogen application of 56–85 kg/ha should be applied 8–12 weeks after planting. Excess application of nitrogen fertilizer should be avoided as it will encourage the top growth at the expense of root development, which promotes the hollow roots and also causes forking of roots, particularly when applied too late in the season. Most of the Common-type varieties are prone to this disorder. In addition, 2.5–4.0 kg/ha boron should also be applied with fertilizers. Application of sulfur 30–55 kg/ha gives beneficial effect in sulfur-deficient soil to reduce the soil pH.

IRRIGATION REQUIREMENT

Depending on moisture holding capacity of the soil, irrigation during dry periods, particularly in September and October, can improve marketable yield. Soil type does not affect the amount of total water needed but does dictate frequency of water application. Lighter soils need water application more frequently but less water is applied per irrigation. Horseradish requires approximately 2.5–5.0 cm of water, which is applied at 7-day intervals. Additional irrigation may be required during dry spells. Excessive irrigation should be avoided, as it causes root rot disease. On the other hand, a long dry spell during root development may cause the root tissues to become woody and without flavour.

INTERCULTURAL OPERATIONS

HOEING AND WEEDING

Hoeing is an essential operation at initial stages of crop growth to reduce weed competition with crop plants for moisture, nutrients, light and space and to improve soil aeration around the root system for better plant growth. At least two hoeings are essential to keep the weeds under control in the horseradish field. Shallow weeding will help in good root development. Applying organic mulch around the horseradish plant conserves soil moisture by reducing evaporation rate and also suppresses the weeds' growth. Dry leaves and compost are the effective mulching materials. If no mulch is applied around the young horseradish plants, weeds should be removed by gentle hoeing around the plants. Many growers make use of the practice of lifting and suckering their horseradish plants. Once the plant has begun to grow, it is gently lifted to expose the crown and branched roots, if any, are removed. This focuses the plant's energy on growing the main root.

HARVESTING

Horseradish makes its greatest growth in late summer and early fall, thus, harvesting is usually delayed till October or early November for higher yield. The horseradish should be harvested when most of foliage is killed by frost usually starting in November and continuing throughout the winter but the soil is not frozen and dry enough to support machinery. In commercial fields, the tops of the plants are mowed back to facilitate the digging of roots mechanically. A one- or two-row potato digger can be used to dig out the roots and remove the plant. In kitchen gardens, harvesting is performed by pulling out the roots with hands or with the help of shovels, and the foliage is removed after harvesting the roots. The small root pieces left in field during digging will sprout the following year. If sprouting of small root pieces in the field is not desired, all the root pieces should be dug up carefully to prevent the horseradish becoming a weed in the garden. In areas with mild winters, if the roots are not harvested before the ground freezes, they can be left overwinter and then dug out in early spring before they grow actively.

YIELD

Number of factors such as planting material, soil fertility, irrigation facilities, growing region, rainfall and the cultural practices adopted by the growers affect the yield of horseradish. The average yield varies from 80 to 90 q/ha.

POST-HARVEST MANAGEMENT

Well-flavoured and reasonably straight roots without side shoots and no mechanical or decay damage are sold directly in the market at a higher price. The roots to be sold in the market should be at least 20 cm long with a diameter of not less than 2.0 cm.

The smaller roots may be acceptable for processing. For small-scale processing, roots should be chopped and ground in a well-ventilated room. Its roots are very sensitive to wilting, thus, horseradish should be pre-cooled at 4–5°C immediately after harvest, using forced air at 0°C with 90–98% relative humidity. Horseradish can be stored successfully in a cold storage for about 10–12 months at 0°C and 90–98% relative humidity. The light exposed roots become green, hence, the roots should be stored in complete dark to avoid greening. Roots can also be stored at −1°C to −2°C. Freshly harvested properly washed roots can also be stored for several months in polyethylene bags or crates lined with polyethylene film.

INSECT-PESTS

Horseradish can tolerate some insect-pests without any serious effect on yield and quality of roots. Flea beetles, caterpillars, false cinch bugs and diamondback larvae may cause defoliation of leaves. Growers are often more concerned with insects that cause root damage.

FLEA BEETLES (*Phyllotreta armoraciae*)

This insect makes small holes or pits in leaves, which give a *shot hole* appearance of foliage. It comes particularly at the seedling and young plant stage before the development of roots. Its infestation can reduce plant growth. If the damage is severe, the plant may be killed.

Control

- Apply a thick layer of mulch to prevent beetles reaching to surface.
- Use diamotecoeus earth or neem oil as a barrier.
- Follow phytosanitary measures.
- Spray the crop with 0.15% malathion or 0.4% sevin at 10- to 15-day intervals.

CABBAGE LOOPER (*Trichoplusia ni*)

This insect attacks the plants at all stages of crop growth and its attack causes complete defoliation. Larvae feed on underside of developing leaves but usually do not feed at leaf margin. The outer leaves become pierced with small irregular holes. If leaves are parted, masses of greenish brown pellets of excreta are found at base of the leaves.

Control

- Remove wild mustard, pepper grass and shepherd's purse that harbour the pest.
- Clean up and dispose of the plant debris after harvesting to remove the overwintering pupae.
- Use *Bacillus thuringiensis* formulation spray to kills the younger larvae effectively.
- Spray pyrethrins 0.3%, cyfluthrin 0.05%, permethrin 0.5% and spinosad 0.1% at 7-day intervals.

DISEASES

BRITTLE ROOT (*Spiroplasma citri*)

The brittle root is most destructive virus-like disease, which is transmitted by the beet leafhopper. The symptoms of this disease include foliar chlorosis, stunting of seedlings and discolouration in the phloem ring of affected plant roots and sometimes wilting. It is also characterized by loss of root flexibility in the affected plants. The roots become brittle and discoloured and show dark brown colour with a dark ring when snapped in two parts. Carrot is a host of *S. citri*. Inoculative leafhoppers feed on carrot during seasonal migration.

Control

- Spray systemic insecticide to control the insect vectors.

CERCOSPORA LEAF SPOT (*Cercospora amoraciae*)

The disease symptoms appeared on leaves are characterized by angular to irregularly shaped chloratic lesions, which later turn greyish brown with profuse sporulation at the centre of the spot. Severely infected leaves are yellow, dried and defoliated, resulting in reduction of root yield.

Control

- Treat the seeds with hot water to eradicate the fungus prior to planting.
- Remove and destroy infected plants to prevent spread of disease.

- Avoid working with plants when they are wet.
- Spray the crop with Dithane Z-78, Fytolan or Blitox 0.25%, Bavistin 0.2%, or Benlate 0.1% at fortnightly intervals.

RAMULARIA LEAF SPOT (*Ramularia cynarae*)

The yellow-green circular patches appear between leaf veins, which become distinct lesions with irregular margins. The centers of lesions may dry out and drop from the plant, producing a shot-hole appearance. If infection is severe, the entire plant may dry out.

Control

- Rotate crops away from horseradish for a period of 3 years.
- Remove weeds and any horseradish debris from the field.
- Avoid the use of a sprinkler irrigation system.
- Remove and destroy any infected plants to prevent spread of the disease.
- Spray the crop with 0.75–1.0 liter/ha triazoles alone or mixed with carboxamides.

WHITE RUST (*Albugo candida*)

The symptoms appear as pale yellow patches on upper leaf surface followed by creamy white pustules on lower leaf surface. Often these pustules coalesce to form large irregular erupted patches. The host epidermis ruptures easily, giving a white powdery appearance to the lesion. If infection is severe, the leaves may become curled and distorted with pustules covering the entire leaf surface.

Control

- Follow long crop rotation with cereal crops.
- Keep the field weed free.
- Remove horseradish debris from the field.
- Avoid the use of a sprinkler irrigation system.
- Remove and destroy the infected plants to prevent the spread of disease.
- Spray the crop with zineb 0.2–0.3%, difoltan 0.3%, or Ridomil 0.2%.

■■■

43 Kale

M.K. Rana and N.C. Mamatha

Botanical Name : *Brassica oleracea* var. *acephala* L. (Kale)

B. o. var. *acephala* sub-var. *laciniata* L. (Curly kale)

B. o. var. *acephala* sub-var. *plana* L. (Smooth leafed kale)

B. o. var. *acephala* sub-var. *palmifolia* L. (Tree kale)

B. o. var. *acephala* sub-var. *medullosa* L. (Marrow stem kale)

B. o. var. *acephala* sub-var. *mellicapitata* L. (Thousand head kale)

Brassica oleracea var. *alboglabra* L. (Chinese kale)

Brassica oleracea var. *fimbriata* L. (Kitchen kale)

Family : Cruciferae (Brassicaceae)

Chromosome Number : 2n = 18

ORIGIN AND DISTRIBUTION

Like broccoli, cauliflower and collards, kale is a descendent of wild cabbage, a plant thought to have originated in Asia Minor and to have been brought to Europe around 600 B.C. by groups of Celtic wanderers. Soon after domestication, people in the Mediterranean region began growing this first ancient cabbage plant as a leafy vegetable. Since leaves were consumed as part of the plant, it was natural that those plants with the largest leaves would be selectively propagated for next year's crop. This resulted in slower development of larger-leafed plants, as the seed from the largest-leafed plants was favoured. By the 5th century B.C., continued preference for larger leaves had led to the development of the vegetable, which is now known as kale, which is botanically known by the name *Brassica oleracea* var. *acephala* that translates to mean *cabbage of the vegetable garden without a head*. Kale continued to be grown as a leafy vegetable for thousands of years and is still being cultivated today. However, as the time passed, some people began to express their preference for those plants with a tight cluster of tender young leaves in the centre at the top of stem. Because of this preference, these plants were selected and propagated more frequently. A continued favouritism of these plants for hundreds of successive generations resulted in the gradual formation of a more and more dense cluster of leaves at top of the plant. Eventually, the cluster of leaves became so large that it tended to dominate the whole plant, and the cabbage *head*, which is known today, was borne. Curly kale played an important role in early European food ways, having been a significant crop during ancient Roman times and a popular vegetable eaten by peasants in the Middle Ages. English settlers brought kale to the United States in the 17th century. Both ornamental and dinosaur kale are the much more recent varieties. Dinosaur kale was discovered in Italy in the late 19th century. Ornamental kale, originally a decorative garden plant, was first cultivated commercially as in the 1980s in California. Ornamental kale is now better known by the name salad savoy.

INTRODUCTION

Kale is one of the important salad vegetables grown throughout the world for a long time, far before the cultivation of cabbage. It is mainly a winter vegetable when other vegetables are scarce. It is a popular green leafy vegetable in countries like China, Taiwan, Vietnam, Korea, Netherland, Canada and Middle East countries. In India, kale is grown in the northern parts, especially Chinese border (Chinese kale). Similar to this in Kashmir, *Karam saag* is grown, which is closely related to Chinese kale. The only difference is Chinese kale being grown in tropical regions and *Karam saag* requires chilling temperature. Among different kales, curly kale is cultivated for its rosette leaves, which bear at the top of the stem, which are harvested as individual leaves or in bundle.

COMPOSITION AND USES

COMPOSITION

Kale is very high in β-carotene, vitamin K and vitamin C and rich in calcium. It is a good source of two carotenoids *viz.*, lutein and zeaxanthin. It contains sulforaphane, a chemical with potent anticancer properties. Boiling decreases the level of sulforaphane; however, steaming, microwaving, or stir-frying does not result in significant loss. The nutritional composition of kale is given in Table 43.1.

USES

Kale is widely used in salads and soups. In the Netherlands, it is very frequently used in a traditional winter *stamppot* dish called *boerenkool*. In Ireland, it is mixed with mashed potatoes to make the traditional dish colcannon. A variety of kale, *kai-lan*, is a popular vegetable in China, Taiwan and Vietnam, where it is commonly combined with beef dishes. Japanese kale is primarily used for decorative or ornamental purposes.

MEDICINAL USES

Like other mustard family vegetables, kale is also a source of indole-3-carbinol, a chemical which boosts DNA repair in cells and appears to block the growth of cancerous cells. It has been found to contain a group of resins known as bile acid sequestrants, which have been shown to lower cholesterol and decrease absorption of dietary fat. Steaming significantly increases these bile acid binding properties.

BOTANY

Kale is non-heading with curled succulent leaves. It is a biennial crop but grown as an annual for vegetable purpose. Flowering stalk arises in the centre from the base, which is thickened at base and narrows towards the tip with small yellow-coloured flowers. Inflorescence is racemose and flower is

TABLE 43.1
Nutritional Composition of Kale (per 100 g Edible Portion)

Constituents	Contents	Constituents	Contents
Water (g)	91.2	Potassium (mg)	3.47
Carbohydrates (g)	7.2	Iron (ppm)	684
Protein (g)	3.9	Sulfur (mg)	1.04
Fat (g)	0.6	Vitamin A (IU)	20000
Calcium (mg)	2.16	Thiamine (mg)	63
Phosphorus (mg)	0.62	Vitamin C (mg)	187

characterized by four petals arranged opposite to each other in a square, cross six stamens, of which two are short (having nectarines at their base). The carpels form a superior ovary with a false septum and two rows of campylotropous ovules. It bears seeds in a special kind of bicarpellary pod called siliqua. The plants may be dwarf or tall but dwarf types are most preferred. Although kale varies in colour from pale yellowish to deep green through deep steely blue to purplish red and almost black, it is usually classified on the basis of leaf form and texture, namely Scotch types, which have very curled and wrinkled leaves, and Siberian or Russian types, which are almost flat with finely divided edges, while heirloom *Lacinato* is in a class of its own. Blue-green colour is associated with greater cold tolerance.

CULTIVATED VARIETIES

There are many varieties of kale but they are usually grouped by the type of leaf, as follows:

- Curly leaved kale
- Plain leaved kale
- Rape kale
- Leaf and spear (a hybrid between curly leaved and plain leaved kale)
- Cavolo nero also known as Black cabbage, Tuscan cabbage, Tuscan (Tuscano) kale, Lacinato kale and Dinosaur kale.

In these leaf shapes, there are number of varieties with varied growth time from transplanting to harvesting *viz.*, Blue Armor (45–75 days, hybrid), Blue Curled Scotch (65 days), Blue Knight (55 days, hybrid), Dwarf Blue Curled (55 days), Dwarf Blue Scotch (55 days), Dwarf Green Curled (60 days), Dwarf Siberian (65 days), Greenpeace (65 days), Hanover Late Seedling (68 days), Konserva (60 days), Red Russian (40–60 days), Squire (60 days), Verdura (60 days) and Winterbor (60–65 days, hybrid).

Depending upon stature, kale is grouped into two types as given below:

- **Dwarf types:** Dwarf Green Curled Scotch and Dwarf Moss Curled
- **Medium tall types:** Hamburger Market and Moss Curled

However, in India, there are no released varieties by the public sector. Only local varieties and some imported varieties are grown. In Kashmir, famous variety named *Karam Saag* is cultivated on a large scale.

Scotch Kale

Dwarf Blue Curled Scotch: A squat plant, which is good for container culture, produces curly blue leaves that are good in salads when young, or cooked when mature. It is very cold hardy plant and become ready for harvest in 55 days.

Redbor: A plant with deepest red-purple leaves whose colour enhances with cold looks like gorgeous in a flowerbed or as an edging. Its flavour is sweet. It takes 28 days for baby kale and becomes ready for harvest in 55 days.

Winterbor: Its blue-green plants are 60 to 90 cm tall, extremely hardy and very productive. It takes 28 days for baby kale and 60 days for mature kale.

Siberian (Russian) Kale

Red Russian: It has blue-grey, flat, deeply cut leaves. Veins and stems are blue-green in warm weather, turning red with cold. It is one of the most tender kale, delicious raw in salads, and can be harvested at 25 days for baby kale and normally matures 50 days after sowing.

Blue-Curled Vates: Its great flavoured leaves can be used like lettuce. It is best cold weather kale, medium green in colour and matures in 60 days.

White Russian: Its taste is mild and sweet and is excellent for cool weather salads. It matures in 58 days.

HEIRLOOM KALE

Lacinato: An Italian heirloom also known as *Nero di Tosca*, *Tuscan Black*, or *Dinosaur* produces 30 to 60 cm long, 7 cm wide, slightly crinkled and deep blue-grey leaves, which are excellent for cooking. It is heat and cold tolerant and it takes 30 days for baby kale and normally matures in 65 days.

CLIMATIC REQUIREMENT

Kale is essentially a winter season crop. Being highly resistant to frost, it can withstand relatively unfavourable conditions, and thus, it is a very reliable crop. It enjoys full sunshine during cool season, and during winter, it can be planted under partial shade conditions. The optimum soil temperature for seed germination is 12–24°C. It grows quite well even in Southern Greenland but it is not commercially grown in India. A form of kale known as *Karam Saag* is grown in Jammu and Kashmir. It can be grown in northern plains and in the hills of Jammu and Kashmir, Himachal Pradesh and Uttaranchal and also in the Nilgiris in southern regions.

SOIL REQUIREMENT

The kale crop grows very well in medium-heavy and fertile soils with good moisture supply, though it can also be grown successfully in lighter soils. It can tolerate salt more than the cabbage and can also be grown at C_8 salt index. Soil pH raging from 6.0 to 6.8 is ideal for its cultivation. Planting Cole crops in the same land each year should be avoided to prevent the build-up of soil pathogens.

SEED RATE

Direct seeding as well as transplanting is practiced for growing kale. About 3.5–4.5 kg is required for direct seeding crop and 350–400 g for transplanting a hectare area.

SOWING TIME

Seeds can be sown in September–October in the plains of north India and August–September in the hills. For spring and early summer harvest, the seeds are sown during the first week of April in Virginia and warmer southern areas, and in Pennsylvania and cooler parts seeds are sown in the third week of April.

DIRECT SEED SOWING

The field is prepared after adding well-decomposed farmyard manure and fertilizers to the soil. The seeds are sown @ 3.5–4.5 kg per hectare in the rows spaced 45–60 cm apart. The seeds are usually sown after any danger of hard frost, and for a fall harvest, sowing is done in such a way that the crop would mature in cool weather. The seeds are sown at a depth of 1.2 cm, allowing 3.5 cm between plants within the row. As and when the seedlings attain three to four leaves, they are thinned to a final spacing of 15–30 cm between plants within row. The soil is kept moist during germination to prevent formation of a hard crust on soil surface, as it will hamper germination.

TRANSPLANTING

The seeds of kale are sown in raised nursery beds of 3.0 × 1.0 × 0.15 m size for raising seedlings. The soil should be loose and friable added with farmyard manure supplemented with inorganic fertilizers for better growth of seedlings in nursery. The seedlings in nursery are kept free from biotic and abiotic stresses during seedling growth. The seedlings will be ready for transplanting as and when they attain three to four leaves. The seedlings are transplanted in main field at spacing of 45–60 cm between rows and 15–30 cm between plants. The seedlings are transplanted slightly deeper than they were previously sown in the nursery. The plantings can be staggered at 2-week intervals to prolong the harvesting period.

NUTRITIONAL REQUIREMENT

The requirements of organic manure and fertilizers will depend upon the type of soil, climate and cultivar used. Kale can successfully be grown with farmyard manure 18–23 t/ha supplemented with nitrogen 120–180 kg, phosphorus 75–80 kg and potassium 125 kg/ha. Half of the nitrogen and entire quantity of phosphorus and potassium are applied as basal dose at the time of last ploughing. The remaining half of nitrogen is top dressed 4 weeks after transplanting or when the plants are half grown in case of direct sown crop. It is beneficial to spray urea at 0.5–1.0% every 3 weeks.

IRRIGATION REQUIREMENT

Crop irrigated immediately after sowing seeds in the main field in direct sown crop for better germination and soon after transplanting the seedlings. The subsequent irrigations are given at 10- to 15-day intervals. The total number of irrigations may be six or seven during the crop season and it may again vary with the variety, type of soil and season. There should be adequate soil moisture for optimal growth and development of the plant. Irrigation is especially important to help the young plants to withstand the intense sunlight and heat of summer and to supply the developing heads with sufficient water to develop quickly.

INTERCULTURAL OPERATIONS

HOEING AND WEEDING

The kale field should be kept weed free in the early stage of crop growth, as the plants in early stage may not compete with weeds. Thus, at least two shallow hoeings are beneficial and necessary for keeping the weeds down and providing good soil environment for plant development. One or two hand weedings are enough to remove weeds.

THINNING

In kale, thinning is an important operation to maintain optimum spacing between plants and to provide better conditions to the plants for their growth and development. It is also essential to reduce competition among plants for space, light, nutrients and moisture. Thinning is an essential operation done in a direct-sown crop.

MULCHING

Mulching is done with 2.5 to 5.0 cm thick layer of organic matter, keeping the mulch 2.5 cm away from the plant stem. Mulching helps in conserving soil moisture and provides nutrients to the plants and controls weeds.

HARVESTING

Kale is harvested with a knife while young and tender. For fresh market, the whole rosette tops are frequently harvested, while in tall cultivars, the lower leaves if adequately curled are harvested and packed individually. Frost favours the development of flavour and colour in kale. After harvesting, it is tied in small bundles. Since the leaves have a high respiration rate, bunches are subjected to hydro cooling, and after proper drying, they are kept in the refrigerator.

YIELD

The yield of kale varies from 10 to 25 t/ha. Yield may also vary with crop variety, environmental and soil factors and the cultural practices adopted by the growers.

POST-HARVEST MANAGEMENT

Kale is a highly perishable crop so after pre-cooling bunches are packed in polyethylene-lined crates and protected by crushed ice and kept at 0°C temperature and 95–98% relative humidity. Under such condition, it can be held up to 10–15 days in excellent condition without wilting and loss of quality. Air circulation is necessary to remove heat produced by respiration.

INSECT-PESTS

CABBAGE APHID (*Brevicoryne brassicaea*)

Large population of the insect causes the plant growth stunted or even causes death of the plant. The grey-green colour insects visible on the plant leaves are small with a soft body covered with a white waxy coating. They suck cell sap from the tender leaves and emit a honeydew-like substance, which is attacked by sooty mould.

Control

- Remove the infested leaves or shoots.
- Avoid planting of seedlings infested with aphids.
- Grow tolerant varieties if available.
- Use reflective mulches such as silver-coloured plastic.
- Spray the plants with a strong jet of water to knock the aphids from leaves.
- Spray the crop with *neem* or canola oil @ 5%.
- Spray 0.1% malathion or Thiodan at an interval of 10–15 days.

DIAMOND-BACK MOTH (*Plutella xylostella)*

Larvae with pro-legs at rear end that are arranged in a distinctive V-shape are small (1 cm) and tapered at both ends. The young larvae feed between upper and lower leaf surface and may be visible when they emerge from small holes on underside of the leaf. The older larvae make large irregularly shaped shot holes on underside of the leaves. Larvae may drop down from the plant on silk threads if the leaf is disturbed.

Control

- Apply *Bacillus thurengiensis* to kill the larvae.
- Spray 5% NSKE to control this pest.
- Insecticidal sprays of Regent / Fipronil (0.15%), Thiodan (0.2%), Polytrin-C (0.1%), cartap hydrochloride (0.05%) or fenvalerate (0.05%) at 10- to 15-day intervals.

Cabbage Looper (*Trichoplusia ni; synonym Plusia orichlacea*)

The caterpillars are pale green with white lines running down either side of their body. They are easily distinguished by the way they arch their body when moving. The adults lay eggs singly usually, on lower leaf surface closer to the leaf margin, and are white or pale green in colour. The caterpillars make small or large holes in leaves and damage often extensively.

Control

- Usually, the natural enemies keep its populations under check.
- Pick the larvae by hand and destroy them.
- Application of *Bacillus thuringiensis* effectively kills younger larvae.
- Spay neem seed kernel extract @ 5% at regular intervals.
- Spray the crop with any of the chemicals like fipronil 0.15%, Thiodan 0.2%, or fenvalerate 0.05% at 10- to 15-day intervals.

Flea Beetles (*Phylotreta* spp.)

The pest responsible for the damage is a small (1.5–3.0 mm) dark-coloured beetle that jumps when disturbed. The beetles are often shiny in appearance and make small holes or pits in leaves that give the foliage a characteristic *shot hole* appearance. Young seedlings are particularly attacked by this insect. As a result, the seedlings growth is reduced. If damage is severe, the plant is completely killed.

Control

- In problematic areas, cover the seedlings with floating row covers prior to emergence of the beetles to provide a physical barrier to protect young seedlings.
- Sow the seeds bit early to establish the plants before the beetles become a problem since mature plants are damaged less.
- Grow trap crops (cruciferous) to provide protection.
- Apply a thick layer of mulch to prevent beetles reaching the surface.
- Spray the crop with dichlorovos or Nuvan 0.1%.

Cabbage Caterpillar (*Pieris brassicae*)

The green colour caterpillar is hairy with a velvet-like appearance. It may have faint yellow to orange stripes down its back and it is slow-moving as compared to other caterpillars. The young caterpillars scrape the leaf surface, and the older larvae damage the leaves by eating from the margins inwards leaving the main veins intact.

Control

- Pick the caterpillars from plants by hand and destroy.
- Scrape eggs from the leaves prior to hatching.
- Apply appropriate insecticide like malathion or Thiodan 0.2% if infestation is very heavy.

Cutworm (*Agrotis* spp.)

The larvae of cutworm are 2.5–5.0 cm long, which may exhibit a variety of patterns and colouration but will usually curl up into a C-shape when disturbed. They attack the stems of young seedlings at the soil line. If infection occurs later, irregular holes on the surface of plant appear. The larvae causing the damage are usually active at night and hide during the day in the soil at the base of plants or in plant debris of toppled plant.

Control

- Remove all plant residues from the field after harvest or at least 2 weeks before planting.
- Fit plastic or foil collars around plant stems to cover the bottom 7 cm above the soil line and extending a couple of cm into the soil.
- Pick the larvae by hand after becoming dark and destroy them.
- Drench the soil with chlorpyrifos 0.1%.

THRIPS (*Frankliniella occidentalis* and *Thrips tabaci*)

The insect is small (1.5 mm) and slender and best viewed using a hand lens. Adult thrips are pale yellow to light brown and the nymphs are smaller and lighter in colour. If its population is high, the leaves may be distorted badly. Leaves are covered in coarse stippling and appear silvery. They are speckled with black faeces.

Control

- Avoid planting next to onions, garlic, or cereals where large thrips population can build up.
- Use reflective mulches early in growing season to deter thrips.
- Spray the crop with Confidor 0.03–0.05%, carbosulfan 0.2%, Regent 0.2%, or dimethoate 0.2%.

BEET ARMYWORM (*Spodoptera exigua*)

The young larvae are pale green to yellow in colour, while the older ones are generally darker green with a dark and light line running along the side of their body and a pink or yellow underside. The clusters of 50–150 eggs covered in a whitish scale, which gives the cluster a cottony or fuzzy appearance, may be present on the leaves. They make singular or closely grouped circular to irregularly shaped holes in leaves, and heavy feeding by young larvae leads to skeletonized leaves.

Control

- The natural enemies parasitize the larvae.
- Apply *Bacillus thuringiensis* to kill the larvae.
- Spray the crop with fipronil 0.15%, Thiodan 0.2%, Polytrin-C 0.1%, cartap hydrochloride 0.05%, or fenvalerate 0.05%.

DISEASES

DAMPING-OFF (*Rhizoctonia solani*)

The damping-off may be per-emergence and post-emergence. In pre-emergence damping-off, the seeds are attacked by the fungi in soil and cause death of the seeds before they germinate, and in post-emergence damping-off, the seedlings are attacked by the fungi at colour region. As a result, the seedlings start toppling over the ground.

Control

- Sterilize the nursery soil before sowing seeds.
- Treat the seed with captan, ceresan, Foltaf, or captafol @ 2.5 g/kg of seed.
- Avoid excessive irrigation.
- Keep the nursery bed surface dry.
- Drench of nursery beds with captan @ 2–3 g/litre of water @ 5 litre/m^2 area.

ALTERNARIA LEAF SPOT (*Alternaria brassicae*)

Small dark round to angular spots with purple-black margin appear on leaves, which later turn brown to grey. The lesions forming concentric rings become brittle and crack in the centre. Dark brown elongated lesions may also develop on stems and petioles.

Control

- Sow only pathogen-free seed.
- Follow long crop rotation.
- Spray the crop with Dithane M 45, Difolatan 80, or Daconil (Kavach) @ 2 g/litre water at an interval of 15 days.

DOWNY MILDEW (*Peronospora parasitica*)

Irregular yellow patches appear on leaves, which later turn light brown in colour. Fluffy grey growth can be seen on undersurface of the leaves.

Control

- Remove all crop debris after harvest.
- Follow long crop rotation with non-cruciferous crops.
- Spray the crop with Ridomil MZ 72 @ 0.5 g/litre of water or Dithane M 45 @ 2 g/litre of water at 10- to 15-day intervals.

ANTHRACNOSE (*Colletotrichum higginsianum*)

Small circular to irregular shaped grey to straw-colour dry spots appear on leaves. More spots may cause death of the leaves. Lesions may coalesce to form large necrotic patches causing the leaves to turn yellow and wilt. The lesions may split or crack in dry centres.

Control

- Follow long crop rotation.
- Provide good soil drainage facilities.
- Maintain phytosanitary conditions in the field.
- Remove all cruciferous weeds, which may act as a reservoir for the fungus.
- Soak the seed in hot water at 52°C for 30 minutes.
- Treat the seed with thiram or captan prior to sowing @ 2.5 to 3 g/kg of seed.
- Spray Dithane M 45, Topsin M, or Bavistin 2 g/litre of water.

BLACK ROT/LEAF SPOT (*Xanthomonas campestris*)

V-shaped brown lesions originating from edge of the leaves, black discolouration on stem and brown spots appear on leaves and ultimately the affected leaves fall down. Pathogen spreads *via* infected seed or by splashing water and insect movement. The disease is favoured by warm and humid conditions.

Control

- Grow resistant varieties.
- Use pathogen-free seed.
- Follow good sanitation practices.
- Follow long crop rotation with non-cruciferous crops.
- Keep the cruciferous weed species, which act as a reservoir for bacteria, under control.
- Give hot water treatment to seeds at 52°C for 30 minutes followed by 30 minutes' dip in streptocycline (0.01%) solution.

ROOT-KNOT NEMATODE (*Meloidogyne* spp.)

The areas of irregular and stunted plant growth can be seen in the field. The infected plants wilt during hot afternoons in periods of water-stress. Galls are formed on roots.

Control

- Rotate crops every 2 years with non-host crops.
- Avoid transferring soil from infested fields.
- Fumigate the nursery beds.
- Treat nursery bed soil with methane sodium @ 25 ml/m^2 or carbofuran @ 2 g a.i./m^2.
- Dip the seedlings bare roots in E.C. formulation of systemic pesticides like dimethoate @ 500 ppm or carbosulfan @ 1000 ppm at transplanting time.

■■■

44 Seakale

M.K. Rana and Archana Brar

Botanical Name : *Crambe maritime*

Family : Cruciferae

Chromosome Number : 2n = 18

ORIGIN AND DISTRIBUTION

Seakale is a perennial wild plant of the Northwest European seashores and cliffs of both the Mediterranean Sea and Atlantic Ocean. It was wild for thousands of years before it was first cultivated in the 16th century and became a popular garden vegetable in the 18th century in Europe and North America. It could not find a place in *modern agriculture*, as it was difficult to store or ship it well. It has also been naturalized on the West Coast of North America.

INTRODUCTION

Botanically, seakale is known as *Crambe maritime*. The word *Crambe* has been derived from the Greek word *Krambe*, meaning cabbage or crucifer and *maritima,* meaning oceanside. Since the plant can tolerate salt, it can be raised in regular gardens and on beaches. It is nearly a perfect primitive food, and it is difficult to imagine that it would not be in a primitive menu of man's food. It is very similar to leaf cabbage though its leaves are comparatively thicker. Its florets are very much like broccoli, although smaller. It is most commonly grown for its spring shoots, which are very crisp with a fresh, nutty flavour and slightly bitter taste and used like asparagus. Its roots, which are very starchy and a little sweet, can be eaten after boiling and roasting.

Seakale with its big leaves and abundant fragrant flowers has become a more popular ornamental plant in recent times. It has gained a Garden Merit Award of the British Royal Horticultural Society. Seakale is general insect nectar plant; thus, the bees and parasitoid wasps visit this plant, which is being used to grow as a groundcover.

COMPOSITION AND USES

COMPOSITION

Seakale is a very low-calorie vegetable but it provides health-benefiting vitamins such as thiamine and folic acid. Fresh seakale contains adequate amount of mineral salts *viz.*, potassium and sulfur. Like other greens, seakale contains nitrates levels (> 17 mg/100 g). The nutritional composition of seakale is given in Table 44.1.

TABLE 44.1

Nutritional Composition of Seakale (per 100 g Edible Portion)

Constituents	Contents	Constituents	Contents
Water (g)	73.1	Thiamine (mg)	0.27
Carbohydrates (g)	6.0	Folic acid (mg)	0.099
Soluble fibres (g)	0.8	Valine (mg)	0.072
Insoluble fibres (g)	0.8	Histadine (mg)	0.034
Potassium (mg)	430.0	Calorie (kcal)	16.9
Sulfur (mg)	28.0	Nitrates (mg)	17.0

Uses

Every part of seakale plant is edible. Very young leaves and thin stems are used either raw or cooked like kale, collard, or asparagus since the older leaves become tough and bitter in taste. Its roots are very nutritious and eaten raw; however, more calories are available when eaten after cooking or roasting. Like broccoli, the flower heads along with their peduncles can also be eaten in raw or cooked form. This plant is also used as fodder for cattle, sheep, swine and poultry.

Medicinal Uses

The seakale health benefits include its ability to prevent scurvy and viral infections. It helps in gaining weight, improves digestive processes and boost metabolism, thereby avoiding excess weight gain, fatigue and organ system malfunctions. It also keeps the body running smoothly and ensures that all the hormonal processes stay balanced. Like other Brassicas, its high fibre content makes it very beneficial for digestion, which helps in preventing various gastrointestinal diseases, including colon cancer. Its mineral content provides it with certain diuretic properties. Thus, its consumption promotes the flushing of toxins from the body, as it makes the kidneys more functional and helps to rid the body of excess salts, water and fat. Its high ascorbic acid content can stimulate the system to increase the production of white blood cells, and *per se*, it makes the body immune system strong and protects the body from common colds and cancer.

BOTANY

Seakale, an herbaceous plant, is about 60–90 cm tall with deciduous leaves. Its deeply lobed leaves with long stalks, purple veins and wavy edges are usually over 30 cm long, smooth and fleshy. The leaves are alternate. The lower leaves contain blades lobed with an entire toothed margin. The newly formed leaves are purple, which later become green. Morphologically, seakale looks like large foliated cabbage. The leaves are covered with a thick waxy layer, which acts as an efficient water repellent. The plant has a strong taproot system with cylindrical and vigorous roots, which show a regeneration capacity when divided. After completing its basic vegetative phase, the plant develops hundreds of 0.8–1.5 cm large white flowers, which open in spring over several weeks and release a honey-like fragrance. Inflorescence is corymbose raceme, containing a number of flowers. The flower is botanically regular, *i.e.*, actinomorphic, having four petals, four sepals and six stamens, out of which four are long and two are short. The gynoecium is fused with a single carpel. It is a highly cross-pollinated plant due to self-incompatibility. The pollination is performed by honeybees and wind; thus, the seakale has a large genetic variability, which involves morphology of vegetative and reproductive organs, earliness and anthocyanin formation in leaves. Its indehiscent fruit usually contains one seed, which germinates very slowly with a low germination power due to seed dormancy and slow water imbibition because of thick and hard integument.

CLIMATIC REQUIREMENT

Seakale usually requires little maintenance; thus, it is quite easy to grow it in gardens. The basic requirement of sunlight and water results in a healthier plant growth. These plants prefer very sunny locations, where they may get sunlight for most of the day. Its strong taproot system makes it a drought-tolerant plant. Being a hardy plant, its cultivation is possible under any environmental condition. Although it is resistant to frost, temperatures above 30°C are most preferable. High-velocity winds seem to cause it little problem.

SOIL REQUIREMENT

Seakale can be grown on a wide range of soils. Its natural environment is sandy soil. However, the best quality seakale can be produced in sandy loam soils rich in organic matter and good in moisture-holding capacity. However, for higher yield, the seakale should be cultivated in well-drained deep heavy soils rich in plant nutrients but low in organic matter. It grows well in soils with a pH around 7.0.

CULTIVATED VARIETIES

LILY WHITE

It is the most widely grown variety, producing pure white heads with good flavour. It takes 5–14 days for germination and 45–50 days to maturity after seed sowing. It gives an average yield of 200–300 q/ha.

IVORY WHITE

It is another variety of seakale, which is mostly similar to Lily White in most of the characteristics.

PLANTING TIME

Seakale is planted from mid to late spring in hilly tracks of India. The development cycle of seakale is similar to rhubarb.

PLANTING METHOD

Seakale can be propagated by root cuttings known as thongs and seeds. Propagation by seed is more affordable than thongs, which take more time to start. Its propagation by seed is much similar to cabbage. The plants through thongs look like a second-year plant. When it is propagated by seed, the plants start to deteriorate after about 7 years. Therefore, the best way is to replace few plants each year.

Root cutting is a quicker method of propagation. The cuttings can be obtained from well-established healthy plants at least 3-years-old, with no sign of rotting on the crown. Taking root cuttings from a 1-year-old plant is not desirable. Thongs should be lifted carefully in October–November after shedding of leaves. The thongs should be straight, clean, 15 cm long and of pencil thickness. To avoid planting them upside down, a straight cut across the bottom and a slanting cut across the top are made. The cuttings are tied into bundles and kept straight in a box containing sand in a cool shed until sprouting in March. These thongs before planting are soaked in water for an hour. A hole deep enough for the thong to stand vertically in the soil with the top end 2.5–5 cm below the surface is dug. The top end of the thong starts thickening in soil and then produces sprouts, which later reach the soil surface. The sprouting of thongs can be quite slow, but once they sprout, they easily outpace the plants grown from seeds. While preparing for planting, leaving the strongest central bud all the buds are rubbed off. A hole is made with a dibber and the cuttings are planted 2.5 cm below soil level, keeping 37 cm distance each way.

Seakale seeds are relatively robust; thus, they can be sown directly in most cases. They are very corky, which enables them to float at sea for several years, which is why the germination is very slow. This corky case can be removed carefully to speed up the germination process, as the seed can imbibe water more easily. The seeds are sown 2.5 cm deep in moist soil or protrays in late May. Germination usually takes place in 3–26 weeks at 15°C. The seeds take 2 weeks to germinate. The seedlings in seedbeds are thinned at the three- to four-leaf stage to make them strong. As and when the seedlings become ready for transplanting they are picked out and transplanted in well-prepared seedbeds in the last week of May. Only the healthiest seedlings are planted in the field. Plants can be cropped once they are more than 12 months old. The young seedlings are very attractive to slugs; thus, the seedlings must be protected against slugs.

NUTRITIONAL REQUIREMENT

Seakale responds well to soil amendments like farmyard manure. Thus, the crop needs regular application of fertilizers, which are to be applied from March to October. The slow-release granular fertilizers or easily water-soluble fertilizers should be applied after every 15–20 days. During autumn months, the nutritional requirement can be fulfilled with some slow-releasing fertilizers, or with some vermicompost.

IRRIGATION REQUIREMENT

For better seed germination and subsequent growth of plants, the soil at the time of sowing must have ample moisture. Depending on the season and availability of soil moisture, the crop needs irrigation at an interval of 10–15 days. It is advisable not to irrigate the field very frequently; however, care should always be taken that the field may not dry and become to compact, otherwise the root development is affected adversely. The crop is sensitive to waterlogging conditions; thus, water stagnation should be avoided. The irrigation requirement of seakale varies from region to region. In coastal areas, no irrigation is required. During the coldest months, irrigation should be applied only when the soil dries out completely, and in case of low temperature, irrigation frequency is reduced. For better growth and performance of plants, maintaining adequate humidity is also very necessary.

INTERCULTURAL OPERATIONS

HOEING AND WEEDING

Seakale is a shallow-rooted crop. Hence, deep hoeing and weeding should be avoided to keep the roots free from damage. Weeds are a problem during the first month after transplanting on account of more vacant space. Manual weeding and hoeing with hoe (*Khurpi*) are done after 2 weeks of transplanting. Top-dressing with nitrogen fertilizer followed by earthing up is carried out after 4–5 weeks of transplanting, and thereafter, spot weeding can be done if required.

In India, the chemical weedicides have not been experimented; however, pre-plant soil incorporation of weedicides recommended for crucifer crops *viz.*, fluchloralin (Basalin 48 EC) 1.0 kg/ha, or alachlor (Lasso 50 EC) 2.0 kg/ha or pendimethalin (Stomp 30 EC) 1.0 kg/ha should prove effective.

BLANCHING

Blanching is one of the most important operations, which is to be done in the month of January or early spring. The purpose of blanching is to protect the plant from sunlight exposure. This operation makes the shoots milder in flavour but decreases the nutrients content. The crowns of 2-year-old plants are covered either with special earthen pots, large plastic buckets, or with a mixture of well-decomposed garden compost and leaf mould. Dry grass, sugarcane tresses, or paddy straw can also be scattered loosely over the plants for the blanching purposes but pots are preferred over any other types of covering.

WINTER PROTECTION

Seakale plants can be forced *in situ*, in which the same plants can be grown for several years, or they can be lifted and forced indoors in warmer conditions. After dying back of crowns, the plant refuses are removed from the field and the crowns are covered with 8 cm thick layer of dry grass, traditional clay seakale pot or black polythene attached to a wooden frame during winter months, which keeps the temperature a bit high. Such covering will also exclude light. Whatever covering is used, it should be at least 37 cm high and firmly held down so that it may not blow away. This process makes the stems ready for cutting within 3 months. Using a sharp knife, the stems are cut with a little piece of root attached. Cutting is stopped in the month of May and the plants are allowed to re-sprout. They can then be blanched again in the following year.

FORCING INDOORS

The roots are dug after the first frost or lifted before the frost but left on the ground to expose them to frost. After trimming off the side roots, they are packed in boxes or large pots and the frame covered with black polythene attached to a wooden frame to exclude light. Roots can also be forced in a greenhouse or cool room.

HARVESTING

Seakale leaves and shoots at the small and tender stage are harvested in spring since the older leaves are tough enough; thus, they need a long cooking period. They are often left on the plant to allow them to remain strong to build reserve to live in the cold months through dormancy. The flowering heads are ready for cutting in summer when they are 15–18 cm long. They are harvested as per the need. If they are not consumed at once after harvest, they become poor in colour and deteriorate quickly. Its roots are lifted from the field when the plant has become dormant. Only the outer roots are harvested and the main taproot is left to continue growing. The roots take about 100 days to reach maturity. For obtaining good-size roots, 3-year-old plants are harvested. Only the blanched roots are stored for an extended period, as the stalk does not stay fresh for longer time; hence, they should be harvested and consumed relatively quickly. If the plants are allowed to flower, they will produce hundreds to thousands of seed capsules. Each flower becomes a green fruit, which dries to a corky capsule containing a single seed. After blanching, the stems having the length of 15–25 cm should be harvested by cutting them below the crown.

YIELD

The yield of seakale plant varies with variety, soil type, soil fertility, irrigation facilities, cultural practices adopted during the course of cultivation and age of the plant, since older plants have larger leaves and shoots. Improved varieties also have greater yield potential as compared to local varieties. On an average, its yield varies from 200 to 300 q/ha.

POST-HARVEST MANAGEMENT

The main crop of seakale is the spring shoots. The blanched shoots like asparagus are cut when they attain a length of 15–20 cm and have a slight hazelnut flavour. In its first and second year, seakale is slow to grow; thus, it should not be harvested until the third year. Seakale does not store very well; therefore, it should be used within a day. However, its roots can be stored in damp sand for few months before replanting.

PHYSIOLOGICAL DISORDERS

BROWN HEAD

The seakale suffers from brown head, which can be caused due to poor growth, disease and excessive temperature. It is typified by flowers that are small, turning brown and not growing. The intensity of this disorder can be lessened by ensuring enough nutrients' availability to the plant.

CRACKING

Sudden increase in soil moisture can cause cracking of heads. Cracking is particularly a problem in seakale when heavy irrigation is applied near maturity. The young leaves inside the head expand rapidly with the sudden increase in soil moisture; as a result, the heads develop cracking. There will be no head cracking with a regular and uniform supply of moisture.

TIP BURN

The symptoms are characterized by tan or light brown tissues, which may later appear as dark brown or even black. The affected tissues lose moisture and become papery in appearance. The causes of tip burning are application of high doses of nitrogen and high humidity in the atmosphere. The main cause of this disorder is calcium deficiency in soil and plant tissues particularly in margins of inner leaves. Recent evidence suggests that tip burning in cabbage is due to calcium deficiency in the tissues. Rapid plant growth, high temperature and high levels of nitrogen make the plant more susceptible to tip-burn development. Under such conditions, the plants are unable to supply sufficient calcium to actively grow inner young leaves at a critical point. Neither soil nor foliar application of calcium will be effective in reducing tip burning because calcium is fixed by the outer leaves and not further translocated to the young growing leaves. The only solution of tip burning is to grow resistance to tolerant cultivars.

HOLLOW STEM

Hollow stem in seakale is caused due to the deficiency of boron and excessive use of nitrogenous fertilizer. The hollowness due to boron deficiency appears as water-soaked areas, while the hollowness due to the excessive use of nitrogenous fertilizer is recognized as lucid white stem with no tissue disintegration sign. The incidence occurs more in crops planted at wide spacing. The plants that produce larger heads are prone to this disorder. Hollow stem is also infected with fungi, which enter through the openings on stem. To avoid this problem, grow those varieties which are resistant to hollow stem. Spraying borax 0.25–0.3% can cure the hollowness caused due to boron deficiency.

BLACK PETIOLE

This disorder has been noticed in the recent past. Being internal, this disorder cannot be seen on outer leaves. As the cabbage-heads tend to maturity, the dorsal side of inner leaves midribs or petioles turns dark or black at or near the point where petiole is attached to the core. This complex form of physiological disorder is entirely due to certain unfavourable environmental factors which play an important role in the development of heads.

INSECT-PESTS

CABBAGE WEBWORM (*Hellula undalis*)

The cabbage webworm, also known as cabbage borer, is most destructive pest of seakale. Its young caterpillars mine the leaves, while old caterpillars feed on undersurface of the rolled leaves as well as on stems and growing points. Its feeding on young seedlings causes their death, especially when the growing point is attacked. The older plants due to its attack produce new shoots and several small heads of little commercial value. Its feeding after head formation may cause stunting of head. Besides, the presence of caterpillars and their excreta reduce the produce market value. Frass accumulation at tunnel entrance along the stem is the indication of damage caused by cabbage webworm.

Control

- Follow crop rotation of at least 2–3 years.
- Plant only vigorous insect-free seedlings.
- Remove and destroy the affected plant leaves.
- Uproot and burn the infected seakale stalks.
- Introduce parasitic wasps such as braconid, ichneumonid and chalcidoid in seakale field.
- Spray 5% *neem* seed kernel extract.

APHIDS (*Myzus persicae* and *Lipaphis erysimi*)

Aphid, a soft-bodied grey-green tiny insect, lives in colonies on the undersurface of the leaves, causing them to become deformed or curled. It is a serious pest of almost all cruciferous crops and is a potentially damaging pest of seakale too. Small bleached areas on leaves are the foremost signs of its infestation. Due to its incidence, the leaves become yellowish and excessively crumpled. Aphids secrete a honeydew-like sticky substance, on which black sooty mould grows, which inhibits photosynthesis and ultimately reduces yield of the crop.

Control

- Remove and destroy the infested leaves and twigs at initial stage.
- Wash the infested plants with simple water using strong jet pressure.
- Use reflective films to deter the aphids.
- Spray *neem* seed kernel extract 5% or *neem* oil 1.0%.
- Spray the crop with non-systemic malathion 0.1%, Nuvan 0.05%, or methyl demeton 0.05–0.2%.

HARLEQUIN OR STINK BUG (*Murgantia histrionica*)

Harlequin bug is an attractive shield-shaped insect, usually with black with bright red, yellow, or orange markings. Both, its adults and nymphs damage the crop by eating all parts of the plants, including stems, fruits and seeds. It is very destructive pest of almost all cruciferous plants, feeding voraciously on juice inside the leaves, causing the plant to wilt, turn brown and often die, if its population is high. The females feed greedily before laying eggs. Its damage appears on stems and leaves, depending on the plant species. Old plants may survive with very poor growth. They especially prefer brassica family plants but when crucifers are unavailable, the bugs migrate to squashes, beans, corns, okra, tomato, etc.

Control

- Grow resistant varieties if available.
- Cover the plant rows with a lightweight floating row cover.
- Grow a few rows of strong-scented plants such as basil, celery, chamomile, garlic, mint, rosemary and sage around the crop.
- Grow a few lines of mustard, broccoli and kale as trap crop to attract the bugs.
- Handpick the bugs and egg masses and drop them into a bucket of soapy water.
- Burn the plant refuges and plant debris at the end of the crop.
- Spray the crop with insecticidal soap or spinosad at 0.1%.

DISEASES

DARK LEAF SPOT (*Alternaria brassicicola* and *Alternaria brassicae*)

The above fungi cause economic loss to seakale in different ways. The disease is characterized by small circular to slightly elongated brown to black spots of 1 cm diameter surrounded by dark to black concentric rings. The spots caused by *A. brassicae* are smaller and lighter in colour as compared to the spots caused by *A. brassicicola*. The disease is seed borne. Seed infection causes reduced germination and seedling vigour. The spots on leaves reduce the photosynthetic area and accelerate senescence. The disease-causing pathogens are responsible for major loss in seed and oil yield of seakale.

Control

- Use disease-free healthy seed.
- Disinfect the seeds before sowing by soaking in 0.2% thiram suspension at 30°C for 24 hours or in hot water at 50°C for 30 minutes.
- Apply fungicides when needed.

CLUB ROOT (*Plasmodiophora brassicae*)

The slime mould causes club root in cruciferous plants, especially in temperate zones. Although it affects the underground portion of the plant, symptoms on above-ground parts appear conspicuously only in later stages of crop growth. The disease symptoms appear as formation of small or large club-shaped swelling on roots and rootlets, which disrupts water supply that results in pale green to yellowish foliage, flagging and wilting of plants. Club root occurs predominantly in crops grown on soils with a pH below 7.0. The disease is often less serious in crops raised on heavy soils or on soils poor in organic matter.

Control

- Follow at least 4–5 years crop rotation with non-host crops.
- Plough the field repeatedly in hot summer months.
- Avoid moving infested seedlings, soil, or farm equipment to clean fields.
- Use disease-free healthy seed.
- Add hydrated lime in soil to raise pH at or above 7.2.
- Provide good drainage facilities in field.
- Use fungal isolates *viz.*, *Haplorporangium* or *Motierella*.

BLACK LEG (*Leptosphaeria maculans*)

Black leg has not been a serious problem in India, but nowadays, it has started appearing repeatedly in valley areas of Himachal Pradesh. Black pucnidia appear as small dots on hypocotyls and cotyledons of the young seedlings, resulting in their death. In later stages, the symptoms first appear as black round spots on leaves and then greyish brown oval spots with purple or black margins on stems, which may extend to entire surface. As the disease progresses, the spots enlarge and become greyish in the center with small black dots. The severely infected plants become stunted with a constricted stem at soil line.

Control

- Rotate the crop with non-crucifer crops for at least 2–3 years.
- Use disease-free seed and transplants.
- Provide hot water treatment at 50°C for 30 minutes.
- Follow field and phyto-sanitary measures.
- Destroy the residues after harvesting the crop.
- Apply benzimidazole 1 kg a.i./ha, or carboxamide or dicarboxamide 1.5 litre/ha when needed.

Downy Mildew (*Peronospora parasitica*)

The disease symptoms first appear as irregular purplish spots on leaves and stems. Under humid conditions, a purplish fluffy growth is evident on underside of the diseased spots. Small chlorotic, irregular, translucent, light green lesions usually delimited by the veins appear on leaves under cool and moist conditions. As the lesions dry out, they become necrotic, and in severe cases, the leaves abscised due to formation of abscission layer. The disease is most prevalent in cooler regions of the world.

Control

- Rotate the crop with non-crucifer crops for at least 2–3 years.
- Treat the seed with metalaxyl compound prior to sowing.
- Destroy the crop refuses after harvesting.
- Spray the crop with Ridomil MZ-72 0.25% followed by mancozeb 0.2% at 10- to 14-day intervals.

Black Rot (*Xanthomonas compestris* pv. *compestris*)

Black rot is the most destructive disease of crucifers in subtropical as well as in temperate regions and occurs worldwide. The disease first appears as chlorotic lesions along the leaf margins and the chlorotic lesions sometimes turn black. The infected leaves fall down prematurely due to the development of abscission layer in young seedlings. The infection extends through the xylem to the stalk, and consequently, the vascular bundles turn black. When the infection is severe, black spots also appear on flower stalks and siliquae. The black rot resembles yellow disease but it is differentiated by the presence of black veins, whereas, in yellow disease, the veins turn yellow or brown. The disease is seed borne.

Control

- Follow at least 3–5 years crop rotation with non-host crops to reduce soil-borne inoculums.
- Treat the seed with hot water at 50°C for 30 minutes to control the disease effectively.
- Soak the seed in 500 ppm solution of antibiotic Aureomycin, Terramycin, or streptomycin for 1–2 hours at 24°C; thereafter, the seeds are rinsed in running water for 3 minutes and re-soaked in 0.5% solution of sodium hypochlorite for 30 minutes, rinsed again and finally dried for 48 hours at 34°C since it is safer than hot water treatment.
- Spray the crop with streptomycin at 100–200 ppm.

Bacterial Soft Rot (*Erwinia carotovora*)

The first symptom on fleshy affected tissue is small water-soaked lesions, which enlarge rapidly. The affected plants show a soft, slimy, foul-smelling rot. The infected plants fail to bolt. Emission of foul smell is the foremost characteristic of the disease. It is a weak pathogen, which penetrates the plant through injured tissues.

Control

- Control other diseases for controlling this disease.
- Avoid mechanical injuries to the plants
- Spray Plantomycin or streptocycline 500 ppm in combination with copper oxychloride 2000 ppm at fortnightly intervals after appearance of first symptoms of the disease.

45 Mustard Greens

M.K. Rana, V.S. Hooda and Bunty Sharma

Botanical Name   : *Brassica juncea* L.

Family   : Cruciferae

Chromosome Number : 2n = 36

ORIGIN AND DISTRIBUTION

Himalayan region of India is the origin place of mustard green where it has been grown and consumed for more than 5,000 years. During the time of slavery, mustard greens might have become an integral part of Southern cuisine, like turnip greens, serving as a substitute for greens that were an essential part of Western African food. Besides India, Nepal, China and Japan are among the leading producers of mustard greens and significant amount of mustard greens is also grown in the United States of America as well. In India, Rajasthan, Uttar Pradesh, Madhya Pradesh, Haryana and Gujarat are the major mustard greens producing states.

INTRODUCTION

Mustard is an important oil and condiment crop. In India, it is popularly known as *Rai*, *Raya*, *Rayada*, *Laha*, *Sarson*, etc. Its tender leaves and stalk in early stages are used as green vegetable in the plains and hills of northern India, which is called mustard greens. In addition to providing excellent nutritious greens, mustard plant also produces the acrid tasting brown seeds that are used to extract vegetable oil. Mustard greens are used in a range of recipes to add a pungent and peppery flavour. In India particularly in northern parts, mustard greens are available throughout the year but the best-quality greens are readily available in the season from November to March.

Adding these brilliant green leaves of mustard to food preparations certainly enhances the beauty of any meal. Mustard greens come in a host of varieties and each has distinct characteristics. Most mustard greens are actually bright green in colour, while some other mustard greens have shades of dark red or deep purple colour. Mustard leaves may have either a crumpled or a flat texture with toothed scalloped, frilled, or lacy edges. The type of mustard greens often available in stores is *Mizuna*.

COMPOSITION AND USES

COMPOSITION

The nutritional profile of boiled mustard greens includes carbohydrates, sugars, soluble and insoluble fibres, minerals especially calcium, potassium, phosphorus, sodium, magnesium, vitamins such as vitamins A and C, fatty acids and amino acids. The nutritive value of mustard greens differs from species to species. The nutritional composition of boiled mustard greens is given in Table 45.1.

Uses

The leaves of mustard are used for preparing curries (*Sarson Saag*), which is a prominent vegetable in India and in many different cuisines ranging from Chinese to southern Americans. The young mustard greens as raw make great additions to salads. Healthy sautéed mustard greens are also served with walnuts and lemon juice, whereas, chopped mustard greens to a pasta salad gives it a little kick. Chopped tomatoes, pine nuts, goat cheese, pasta and mustard greens tossed with a little olive oil is one of the favourite combinations. Cooked mustard greens are served with beans and rice for a simple meal with a southern flavour.

Medicinal Value

Mustard greens being full of nutrients provide adequate amount of dietary fibres, protein, minerals, vitamins and health-promoting phytochemicals, *i.e.*, glucosinolates. Besides, mustard greens also provide notable antioxidants *viz.*, b-carotene (pro-vitamin A), vitamin C and vitamin E. These three nutrients work together to scavenge the super oxides, which are excessively interactive molecules that not only cause damage to the molecules with which they interact but also are linked to a host of different diseases and health conditions. Beta-carotene in conjunction with vitamin E exert a protective action against super oxides in lipid-soluble areas of the body, though vitamin C balances out the job by working in the body's water-soluble environment. In addition, mustard greens offer great benefit to its consumers suffering from asthma to cardiovascular heart disease to menopausal symptoms by providing antioxidant protection to water- and fat-soluble areas of the body. Mustard greens also provide magnesium, a mineral that helps in smooth functioning of muscle cells, like those lining the bronchial tubes and lungs, to stay relaxed rather than constricting themselves and the airways of which they are a significant part. Adding mustard greens in the diet can help in improving the health of individuals having asthmatic problem.

Mustard supplies numerous nutrients, including b-carotene, vitamin C and vitamin E, which being antioxidants contribute to a healthy cardiovascular system. The vitamin E in mustard greens

TABLE 45.1
Nutritional Composition of Boiled Mustard Greens (per 140 g Edible Portion)

Constituents	Contents	Constituents	Contents
Water (g)	132.24	Selenium (mg)	0.84
Carbohydrates (g)	2.94	Zinc (mg)	0.15
Protein (g)	3.16	Vitamin A (IU)	4243.40
Fat (g)	0.34	Thiamine (mg)	0.06
Dietary fibre (g)	2.80	Riboflavin (mg)	0.09
Calcium (mg)	103.60	Niacin (mg)	0.61
Phosphorus (mg)	57.40	Folic acid (mcg)	102.76
Potassium (mg)	282.80	Pantothenic acid (mg)	0.17
Magnesium (mg)	21.00	Vitamin C (mg)	35.42
Sodium (mg)	22.40	Vitamin E (mg)	3.00
Iron (mg)	0.98	Vitamin K (mcg)	419.30
Copper (mg)	0.12	Ash (g)	1.32
Manganese (mg)	0.38	Energy (kcal)	21.00

reduces the development of atherosclerosis by protecting low-density lipoprotein (LDL) from oxidation and decreasing platelet clumping. Dietary intake of vitamin E and C is thought to be associated with paraoxonase enzyme that inhibits the oxidation of low-density lipoprotein (bad cholesterol) and high-density lipoprotein (HDL). Beta-carotene is also able to reduce vessel spasm. In many studies, it has been shown that the patients with low intake of b-carotene may have almost twice the risk of heart attack compared to those with high intake. Low intake of vitamin C has also been found associated with higher level of low-density lipoprotein and lower level of high-density lipoprotein (good cholesterol). Mustard greens are also a good source of magnesium, vitamin B_6 and folic acid, which provide protection against cardiovascular heart disease. Its greens are good source of phenolic compounds that can prevent the cardiovascular and other chronic diseases.

Mustard greens may also be good for women going through menopause as they provide calcium, which is essentially required to support bone health during this period. Bone loss usually occurs at this stage of life. Mustard greens are helpful for women suffering from osteoporosis since they provide concentrated amount of calcium and magnesium, which have also been found to be helpful in reducing stress and promoting normal sleeping pattern. Vitamins of B group in mustard greens are helpful in reducing the build-up of homocysteine (a bad amino acid), which results in poor bone matrix and osteoporosis. Mustard greens sharpen the memory in the elderly. Vitamin E found in mustard greens decreases the occurrence of hot flushes experienced by many women around menopause.

Too much smoking or exposure to second-hand smoke may develop complications associated with emphysema and cancer due to a common carcinogen benzopyrene in cigarette smoke, which induces vitamin A deficiency in the body, however, inclusion of mustard greens in diet can help in countering such effect, thus, greatly reducing emphysema and cancer.

Mustard greens contain phytochemicals known as glucosinolates, which with the help of myrosinase enzyme are converted into indoles and isothiocyanates that appear to reduce the potential of carcinogens through their ability to modulate liver detoxification enzymes beneficially. They inhibit certain enzymes that normally activate carcinogens and induce other enzymes that help to dismantle active carcinogens. Sulforaphane, a compound formed when mustard greens are chopped, is known to trigger the liver to produce enzymes that detoxify cancer-causing chemicals, inhibit chemically induced breast cancers and induce colon cancer cells to commit. Sulforaphane also stops the proliferation of breast cancer cells, even in later stages.

Mustard oil is excellent for individuals who are suffering from heart problems as it contains the least amount of saturated fatty acids, making it safe for heart patients. Its oil also provides linolenic and linleic acid, while many other edible oils lack one of the two essential fatty acids. Among the vegetable oils, it contains very low amount of saturated fatty acids, which has the tendency to deposit in the arteries, causing heart problems. Glucosinolates, a pungent principle in mustard oil, at the same time have the antibacterial, antifungal and anticardiogenic properties, which make this oil of many medicinal utilities.

At the same time, mustard greens contain a considerable amount of oxalates, which when become too concentrated in the body fluid, they can crystallize and cause health problems so the individuals having the kidney or gallbladder stones should avoid the inclusion of mustard greens in their diet. Like oxalates, mustard greens contain goitrogens, which interfere with the functioning of thyroid gland. However, cooking may inactivate these goitrogenic compounds to some extent but not clear from the research that exactly what amount of goitrogenic compounds is inactivated by cooking, thus, the individuals already suffering from thyroid problems should avoid the consumption of mustard greens.

CULTIVATION

In India, mustard is not exclusively grown for the production of greens; it is grown as an oilseed crop in winter season. Its greens (20–30 cm top portion) are mainly removed at very young stage (40–50 days after sowing) of the crop grown mainly for seed purpose. This practice does not affect the seed yield adversely rather it enhances the seed yield due to increased number of branches. Usually, the lower 3–4 leaves are removed from bud stage (30–35 days after sowing) to pod formation stage (65–90 days after sowing) for the preparation of *Sarson Saag.*

CLIMATIC REQUIREMENT

Mustard is a cool season plant of tropical as well as temperate zones. A temperature of 25°C is needed for satisfactory germination of seeds and seedlings growth. A temperature of 20–32°C is desirable during vegetative phase. It follows C_3 pathway for carbon assimilation, therefore, it has efficient photosynthetic response at 15–20°C temperature. At this temperature, the plant achieves maximum CO_2 exchange range, which declines thereafter. A dry clear weather coupled with abundant bright sunshine is essentially needed during reproductive phase for higher production of mustard greens as well as seed yield. It is long day plant in periodic response and well adapted to areas where the annual precipitation is 350–450 mm, thus, it fits well in rain-fed cropping systems. Rain at the time of flowering is harmful. High humidity and cloudy weather are not good for this crop during winter, as it invites aphids and the crop gets spoiled completely. However, under rain-fed conditions, one to two pre-flowering rains help in boosting the grain yield. Excessive cold and frost are harmful to the crop.

SOIL REQUIREMENT

Mustard can be grown in a wide range of soils varying from sandy loam to clay loam but it performs best on well-drained loam soils rich in organic matter. This crop may not perform satisfactorily on heavy soils. It can tolerate neither water logging nor the long-term drought conditions. It can tolerate modest salinity and alkalinity but a neutral soil reaction (6.5–7.0 pH) is considered ideal for its normal growth and development. Excessive acidic or alkaline soils are not at all suitable for its farming.

Mustard can be grown in rain-fed as well as in irrigated areas. Its seeds are very small, thus, clean and fine-textured seedbeds are needed for proper germination. The field should be prepared properly by ploughing deep with soil turning plough followed by cross harrowing. Each ploughing is followed by planking to make the soil pulverized and to level the field for sowing. The weeds and previous crop stubbles should be removed from the field for proper germination of the seeds. The field should have adequate soil moisture to ensure good seed germination.

CULTIVATED VARIETIES

The improved varieties with profuse early growth and higher yield potential should be chosen for cultivation. The description of some important varieties of mustard with their suitability for different growing areas is given as follows.

RH 30

A non-shattering variety developed at CCS Haryana Agricultural University, Hisar, was released for cultivation in 1983. It is suitable for late sowing and mixed cropping. Its seeds are bold (5.5–6.0 g/1000 seeds) and brown with an average oil content of 40% and having average seed yield potential of 16–20 q/ha in crop duration of 130–135 days. It is recommended for growing under rain-fed and irrigated conditions of Haryana, Jammu, Punjab, Northern Rajasthan and Western Uttar Pradesh for normal sowing.

Laxmi (RH 8812)

A bold and black seeded variety developed at CCS Haryana Agricultural University, Hisar, was released for cultivation in 1996. It bears 40% oil content and produces bold and much branched tall plants. Its average seed yield potential is 22 q/ha in crop duration of 142–145 days. It is recommended for growing in Haryana, Punjab and Rajasthan under irrigated conditions.

Geeta (RB 9901)

A variety developed at CCS Haryana Agricultural University, Regional Research Station, Bawal, District Rewari (Haryana), was released at the national level in 2002 for cultivation under rain-fed conditions of Haryana, Delhi, Punjab and South Rajasthan. It is a tetralocular bold seeded variety with 40–42% oil content. Its average seed yield is 17.73 q/ha in a crop duration of 145–149 days.

Swaran Jyoti (RH 9801)

A variety with oil content 39–43% developed at CCS Haryana Agricultural University, Hisar, was released for cultivation in 2002. It is suitable for irrigated late-sown conditions of Haryana, Madhya Pradesh, South Rajasthan and Uttar Pradesh. It matures in 123–130 days, and its average seed yield is 13.77 q/ha.

Vasundhra (RH 9304)

A variety with oil content 38–40% developed at CCS Haryana Agricultural University, Hisar, was released for cultivation in 2002. It is suitable for normally sown irrigated conditions of Haryana, Madhya Pradesh, South Rajasthan, Uttar Pradesh and Uttaranchal. It matures in 129–137 days, and its average yield is 21.09 q/ha.

CS 52 (DIRA 343)

A variety developed at Central Soil Salinity Research Institute, Karnal (Haryana), was released at the national level for cultivation in 1997. This variety bears 40% oil content and is suitable for cultivation in sodic and saline soils of Haryana, Rajasthan and Uttar Pradesh. This variety matures in 135–145 days, and under sodic and saline field conditions, its average seed yield potential is 11.41 q/ha.

Pusa Agrani (SEJ 2)

An early maturing variety with oil content 40% developed at Indian Agricultural Research Institute, New Delhi, was released for cultivation at national level in 1997. It matures in 95 days, and its average seed yield is 17.18 q/ha. It is recommended for cultivation in Haryana, Punjab, Rajasthan, Eastern and North Eastern states.

Pusa Mahak (JD 6)

An early maturing variety developed at Indian Agricultural Research Institute, New Delhi, was released for cultivation at national level in 2004. It matures in 98–102 days and gives an average yield 5.97–10.49 q/ha. It is suitable for cultivation in Bihar, West Bengal, Orissa, Assam and Jharkhand.

Pusa Saag 1

A variety developed at Indian Agricultural Research Institute, New Delhi, gives large, broad, glabrous, smooth and attractive green leaves that contain more carotene and ascorbic acid. Its average yield is 700 q/ha in a maturity period of 35 days.

RGN 13

A longer-duration variety with oil content 41–43% developed at Agricultural Research Station, Navgaon, Rajasthan Agricultural University, Bikaner, was released for cultivation in 2002. It matures in 135–163 days and gives an average yield 22 q/ha. It is highly resistant to frost and tolerant to high temperature. It is suitable for normal sown irrigated conditions of Rajasthan.

RGN 48

A variety developed at Agricultural Research Station, Navgaon, Rajasthan Agricultural University, Bikaner, was released for cultivation in 2004. A variety bearing oil content 40–41% is suitable for normally sown rain-fed situations of Rajasthan, Punjab, Haryana, Delhi, Jammu and Kashmir and Uttar Pradesh. It matures in 138–157 days and gives an average yield 16–17 q/ha.

Varuna (T 59)

A variety developed at Chander Shekhar Azad University of Agriculture and Technology, Kanpur, Uttar Pradesh, was released at the national level for cultivation in 1975. A bold seeded (5–5.5 g/1000 seeds) fertilizer responsive tall variety with average oil content of 43% is highly stable and good for early- as well as late-sown conditions. It is suitable for commercial cultivation in entire mustard growing areas of the country. This variety takes 135–140 days from sowing to maturity and its average seed yield potential is 20–22 q/ha.

Basanti (RK 8501)

A yellow seeded variety developed at Chander Shekhar Azad University of Agriculture and Technology, Kanpur, Uttar Pradesh, was released for cultivation in 2000. It bears 40% oil content and gives an average seed yield 14.88 q/ha in crop duration of 130 days. This variety is resistant to white rust and tolerant to *Alternaria* blight diseases. It is suitable for cultivation in Uttar Pradesh.

Kanti

A variety developed at Chander Shekhar Azad University of Agriculture and Technology, Kanpur, Uttar Pradesh, was released for cultivation in 2002. This variety is suitable for multiple cropping and early sowing (mid September) in Uttar Pradesh gives an average seed yield of 20–22 q/ha in a crop duration of 100–105 days.

Maya (RK 9902)

A variety developed at Chander Shekhar Azad University of Agriculture and Technology, Kanpur, Uttar Pradesh, was released for cultivation in 2002. This variety bears oil content of 39–40% and suitable for irrigated conditions of Madhya Pradesh and Uttar Pradesh. It is resistant to white rust, and its average seed yield potential is 25–29 q/ha in a crop duration of 130–135 days.

Ashirwad (RK 01-2)

A variety developed at Chander Shekhar Azad University of Agriculture and Technology, Kanpur, Uttar Pradesh, was released for cultivation in 2005. This variety bears oil content 37–41% and gives an average seed yield of 14.37–16.84 q/ha in a crop duration of 125–130 days. It is recommended for late-sown conditions in Uttar Pradesh, Rajasthan and Madhya Pradesh.

Kranti

A variety developed at G.B. Pant University of Agriculture and Technology, Pantnagar, Uttranchal, was released for cultivation at the national level in 1982. It bears 40% oil content, matures in 125–130 days and its average seed yield is 11.07–21.23 q/ha. This variety is relatively more tolerant to *Alternaria* blight and frost than Varuna. It is recommended for cultivation in Bihar, Delhi, Gujarat, Haryana, Orissa, Punjab, Rajasthan, Uttar Pradesh and West Bengal.

Rohini (KRV 24)

A variety with oil content 43% developed at Narendra Dev University of Agriculture and Technology, Faizabad, Uttar Pradesh, was released for cultivation in 1985. It is suitable for cultivation in Madhya Pradesh and Uttar Pradesh. Its average yield potential is 22 to 28 q/ha in a maturity period of 125–130 days. Its siliquae are oppressed.

Narendra Rai

A variety developed at Narendra Dev University of Agriculture and Technology, Faizabad, Uttar Pradesh, was released for cultivation in 1990. It bears 39% oil content and is suitable for cultivation in salt-affected areas of Uttar Pradesh and Madhya Pradesh. It matures in 125 days, and it average seed yield is 13.33 q/ha.

GM 2

A bold seeded early variety developed at Gujarat Agricultural University, S.K. Nagar, was released for cultivation in 1996. It bears 38% oil content and matures in 112 days. Its average seed yield is 24.0–24.7 q/ha. This variety is resistant to powdery mildew and suitable for growing under the north Gujarat agro-climatic zone.

MR 702

Its leaves are also used as a green. Its leaves are highly nutritious. This variety gives green leaves' yield of 333.3 q/ha.

SOWING TIME

Sowing time is the most vital non-monetary input to achieve higher yield of mustard greens. The production efficiency of different genotypes greatly differs under different planting dates. In various zones, temperature and soil moisture influence the sowing time of mustard in the country. Sowing time influences phenological development of the crop *via* temperature and heat unit. Sowing at right time always gives higher greens yield due to availability of suitable environment at all the growth stages. Date of sowing also influences the occurrence of insect-pests and diseases. Timely sown crop often escapes the incidence of insect-pests and diseases. Delayed sowing affects the plant growth, flowering duration, seed formation and seed size adversely, *per se*, reducing the yield of mustard greens and seed. Too early or too late sowing has been reported to be disadvantageous. In varied agro-climatic conditions of India under different ecological situation, the sowing starts from September and continues until the first fortnight of November.

SEED RATE

Rapid seedling growth, early stem elongation, better root development, fast ground covering and early flowering are important yield determining parameters under low temperature and radiation regime. These traits can successfully be exploited if a good seed is sown at appropriate time and rate along with maintaining optimum plant population. The factors that may affect the seed rate of mustard are the variety used, soil moisture, soil type, soil reaction, soil fertility and weather conditions of the growing region. For irrigated areas, a seed rate of 3.5 kg is enough for sowing a hectare area and for rain-fed areas, a seed rate of 5.0 kg/ha is sufficient. In salt affected soils, about 25% higher seed rate should be used for higher yield.

SOWING METHOD

Sowing method depends upon land resources, soil conditions and level of management. Broadcasting, line sowing, ridge and furrow method and broad bed and furrow methods are commonly followed for the sowing of mustard. If moisture in soil is plenty, broadcasting followed by light planking gives early emergence and growth. Under conserved moisture regime, seed placement in moist horizon under line sowing becomes beneficial. Mustard is sown with seed drill in lines 30 and 45 cm apart under irrigated and rain-fed conditions, respectively, while in situations of mixed cropping, the seeds are generally sown in lines 1.8–2.4 m apart in the main crop. The seeds are placed in soil at a depth of about 4–5 cm. To avoid the incidence of root rot and wilt diseases at early seedling stage, the seeds before sowing should be treated with captan, thiram, captafol, or foltaf @ 2.5 g/kg of seed.

NUTRITIONAL REQUIREMENT

Nutrient management is one of the most important agronomic factors that affect the yield of mustard greens. Adequate nutrients supply increases the leaf size, leaf quality, number of branches, number of siliquae per plant, seed yield and oil content. The fertilizers requirement changes with climate, soil type, time of sowing and type of cropping system followed. Mustard responds well to organic and inorganic manure. Incorporation of farmyard manure improves the physico-chemical properties of the soil along with the supply of macro- and micronutrient, and ultimately increases the crop yield. Farmyard manure should be applied @ 15–20 t/ha at the time of land preparation about 3–4 weeks before sowing. No significant difference has been found in sensory quality, *i.e.*, flavour and bitterness of mustard greens, whether the crop is raised with or without manure.

Mustard responds well to chemical fertilizers especially nitrogen and sulfur. Depending on irrigation and initial soil fertility status, the crop is supplied with nitrogen 60–120 kg, phosphorus 30–60 kg, potassium 30–60 kg, sulfur 40 kg and zinc sulfate 25 kg/ha for sustainable mustard production in view of soil health and ecological balance. Under irrigated conditions, one-half dose of nitrogen along with a full dose of phosphorus, potash, sulfur and zinc sulfate should be applied at the time of sowing as basal application by placing below the seed, and the remaining half of the nitrogen should be applied at the time of first irrigation. Under rain-fed conditions, the fertilizer doses are reduced to half of the recommended dose. Applying biofertilizers (*Azotobacter* or *Azospirillum*) has also been found useful in improving growth and yield of the mustard. Inoculation of mustard seeds with efficient strains of *Azotobacter* and *Azospirillum* augments the greens and seed yield of mustard. A balanced use of fertilization at right time by appropriate method increases the nutrients use efficiency in mustard. Integration of chemical fertilizers with farmyard manure and biofertilizer may improve crop yield and soil health as well. Moreover, the supply of fertilizers especially nitrogen to the soil influences the activity of bio-fertilizers.

IRRIGATION REQUIREMENT

Mustard crop is though sensitive to shortage or excess of water, it is usually grown as a rain-fed crop in conserved soil moisture. For taking higher yield of mustard, the field should be bunded and levelled before the onset of monsoon and ploughed two to three times during monsoon followed by planking. Addition of manure in soil can improve moisture-holding capacity of the soil and *per se* helps in getting higher crop yield. How efficiently the mustard crop uses the water depends on the number of irrigations applied. In general, two irrigations, one at pre-flowering stage (30–35 days after sowing) and the second at pod formation stage (70–90 days after sowing), are sufficient to harvest a good crop; however, irrigation at vegetative, flowering and pod formation stage results in highest yield. Irrigation at vegetative stage is found to be most critical since irrigation at this stage increases the number of branches and yield of the crop. In irrigated crop, frost causes less damage. However, excessive irrigation at later stages of crop growth reduces the crop yield due to inducing secondary flowering and lodging of the crop.

Irrigation is undoubtedly important for getting optimum productivity potential of mustard but the quality of irrigation water is equally important. If the quality of irrigation water is poor, it needs certain treatments and management before being utilized for crop production since the increasing salinity levels of irrigation water applied at pre-sowing and flower initiation stage may reduce the plant height, branching pattern and pod formation. Mustard greens as winter-grown leafy vegetable have the potential in drainage water-reuse systems and even the use of moderately saline irrigation water occasionally for the mustard greens production does not adversely affect the crop quality rated through colour, texture and the nutrients content.

INTERCULTURAL OPERATIONS

WEEDING AND HOEING

Weeds are responsible for 15–30% loss in mustard greens yield. The common weeds found in mustard field are *Chenopodium album*, *Chenopodium murale*, *Asphodelus tenuifolius*, *Fumaria parviflora*, *Euphobia* spp., *Orobanche* spp., *Anagallis arvensis*, *Cynodon dactylon*, *Melilotus* spp., *Avena ludoviciana*, *Phalaris minor*, etc. Weeds compete with crop plants for nutrients, water, space and light, therefore, timely removal of weeds from the field increases the nutrients use efficiency and crop yield. The most critical period for crop-weed competition in mustard is between 25 and 30 days after sowing, thus, one hoeing with wheel hand hoe at this stage is very beneficial for better growth and development of crop plants. Hoeing also improves aeration into the soil and creates soil mulch, thus reducing moisture loss through evaporation. In mustard, no herbicide is recommended for the control of weeds. Farmers should adopt weed control with herbicides since chemicals can increase the net profit, weed control efficiency, production flexibility and reduce time and labour requirement for the management of weeds. Application of pendimethalin as pre-emergence application supplemented with hand weeding at 30 and 60 days after sowing is found to give good results.

Broomrape (*Orobanche*) is a major parasitic weed of mustard. Its infestation causes more than 25% reduction in mustard yield. Among its species, *Orobanche aegyptiaca* is the most serious parasitic weed, causing severe yield and quality loss in mustard. It is endemic in semiarid region and may reach epidemic proportion depending upon soil moisture and temperature. Application of glyphosate (Round-up) 60 ml/ha at 30 days after sowing and second spray @ 125 ml/ha at 60 days after sowing dissolved in 375 litres of water results in 50% to 80% control of broomrape weed in mustard. Doubling of spray should be avoided and irrigation is given 1–2 days before or after the spray for making it effective.

THINNING

Dense plant population reduces the yield due to the reduction in photosynthetically active leaf area caused by mutual shading, which is found more at 91–110 over 71–90 days after sowing. Specific leaf weight, crop growth rate and net assimilation rate are more adversely affected by 50% shading at 71–90 days after sowing.

Excessive high or low plant population per unit area is harmful for getting economic yield since too high plant population per unit area increases competition among the plants for nutrients, moisture, space and light, resulting in poor plant growth and low yield and too-low plant population decreases the yield due to deficit plant population per unit area, which may not be compensated, hence, maintaining optimum plant population is essential to obtain economic yield. Due to this reason, initially, more quantity of seed is sown, and later, thinning is done as soon as the seedlings become large enough to handle to maintain at a plant spacing of 10–15 cm. Gap filling can also be done with hand plough after 2 weeks of sowing.

MULCHING

Mulching is also found helpful to curb some of the weeds in mustard field by providing a shield against sunlight, reducing the soil temperature and acting as a physical barrier for the emergence of weeds. Surface covering with artificial mulch such as straw, crop residues and plastic films between mustard rows checks the weeds by inhibiting their seed germination and providing protection to plant roots against extremely low or high temperature. Black polythene mulch is also found effective in controlling weeds in mustard.

HARVESTING

Around 20–30 cm top portion of mustard plants is cut at bud initiation stage about 40–45 days after sowing without any reduction in seed yield. However, the lower 2–4 leaves per plant can be cut from vegetative (30–35 days after sowing) to maturity stage, depending on availability. Mustard greens along with stems can be plucked manually with sharp knife or serrated sickle. While harvesting the mustard greens, extreme care must be taken to prevent moisture loss.

YIELD

When the top 20–30 cm portion is removed along with stems at bud initiation stage (40–45 days after sowing), the average yield of fresh mustard greens varies from 30 to 45 quintals per hectare, while the average mustard greens yield varies from 12 to 15 q/ha when only three or four lower leaves per plant are plucked at 55–60 days crop stage.

POST-HARVEST MANAGEMENT

Mustard greens are washed in a large bowl under tap water, which dislodges the sand or dirt particles if any. It is so repeatedly washed until no sand or dirt particle remains in the water. Washing should be done in chlorinated sterile water, which is finally removed before packaging to minimize decaying during transit and marketing.

After washing, mustard greens are allowed to dry in open air before packaging, and thereafter, about 0.5 to 1 kg weight bundles are made for easy handling and sale directly in the market. Yellow, diseased, or insect damaged leaves are trimmed before making the bundles or packaging in plastic bags or crates. Each step from production to consumption may affect the microbial load on produce. Hence, it is necessary to reinforce the government recommendations for ensuring high-quality mustard greens.

Mustard greens packed in low-density polyethylene film bags may be kept fresh for about 3 to 4 days in refrigerator. It can successfully be stored at 1–5°C temperature and 90–95% relative humidity. Modified atmosphere packaging reduces the respiration rate of mustard greens, therefore, slows down the colour loss. In a study, it was found that mustard greens pickled and canned in vinegar contained lead 4.0–10.0 mµg/g, whereas, fresh or bottled products had less lead than the Canadian legal limit of 2 mµg/g.

PHYSIOLOGICAL DISORDERS

Viral-like symptoms resulting in stunted growth, cup-shaped leaves and delayed flowering some-times appear on plants. The plants near the bunds and in low-lying areas are affected more, however, the plants later automatically recover up to some extent and flowering takes place. This disorder is probably due to the effect of abiotic stress on physiology of the plants.

INSECT-PESTS

MUSTARD APHID (*Lipaphis erhsimi*)

The nymphs and adults both damage the crop by suck sap from the young tender leaves, stems, inflo-rescence and pods, resulting in stunted plant growth. They feed in clusters and prefer new succulent shoots and leaves. The affected leaves usually curl, and when the infestation is severe, the plants wilt and dry. Pod formation is severely affected. Its incidence is much more in the end of December to the first fortnight of January when the temperature is 10–20°C and relative humidity 75%.

Control
- Grow resistant variety like Aravali.
- Remove and destroy the infested portion of plant.
- Triazamate controls aphids better than dimethoate.
- In crop grown for greens, spray 0.1% malathion 50 EC, Metasystox 25 EC, or Rogar 30 EC at 7- to 10-day intervals before the aphid population reaches the economic threshold level, *i.e.*, 9–19 aphids per 10 cm shoot at 30% plant infestation.

MUSTARD SAWFLY (*Athalia proxima*)

The insect is most destructive predominantly at seedling stage. It appears in the month of October and its peak activity is found in November, however, the population disappears suddenly on the onset of winter. The insect eat the shoot epidermis greedily, resulting in drying up of seedlings and failure to bear seeds in older plants. Its grubs feed on young leaves, make holes into the leaves and later eat up the entire leaf lamina leaving only the midribs behind. When the attack is serious, complete defoliation takes place.

Control
- Grow only the resistant varieties like Aravali.
- Plough the field deep in summer to destroy its pupae.
- Sow the seeds early and follow clean cultivation.
- Apply irrigation at seedling stage.
- Spray the crop with malathion 50 EC 0.1% or quinolphos 25 EC 0.01%.

PAINTED BUG (*Bagrada cruciferarum*)

The bug is black with red and yellow lines. Its nymphs and adults suck sap from tender leaves and stems at seedling stage, resulting in poor plant growth showing pale yellow colour of the leaves, and eventually, the plants dry up. Young plants wilt and wither due to its attack. Adult bugs excrete resinous substance, which spoils the pods. When the attack is severe, it results in complete failure of the crop. It appears first in October and again in the months of March–April when the temperature is around 30°C.

Control

- Avoid the sowing of crop too early when the maximum temperature is above 30°C.
- Plough the field deep in summer to destroy the eggs of painted bugs.
- Irrigate the crop 4 weeks after sowing to reduce pest attack.
- Burn the crop refuses of mustard crop to avoid the attack on next year crop.
- Treat the seed with imidacloprid 70 WS @ 5.0 g/kg of seed.
- Spray the crop with malathion 50 EC 0.1% and repeat if required.

Leaf Miner (*Phytomyza atricornis*)

The pest attacks the adaxial surface of mustard leaves in the month of February. The young maggot feed on the leaves by making tunnels into them and cause huge damage to the crop. The attacked leaves wither and vigour of the plant is reduced. Its damage is often more prominent on the older leaves.

Control

- Remove the affected twigs and destroy them.
- Spray the crop with 5% *neem* seed kernel extract (NSKE) or 2% *neem* oil.
- In crop raised for greens, spray with 0.1% Malathion 50 EC at 7- 10-day intervals.

Bihar Hairy Caterpillar (*Spilosoma oblique)*

This polyphagous pest attacks the mustard crop in the month of October–November. It causes severe damage to the mustard crop when it appears in epidemic form. The young larvae live in clusters on lower surface of the leaves and feed gregariously mostly on under surface of the leaves. The grown-up larvae disperse and feed in isolation. They feed on entire leaf tissues leaving behind only the midribs, and in severe infestation, the whole crop is defoliated.

Control

- Plough the filed two to three times deep in summer to expose the hibernating pupae to sunlight and predatory birds.
- Grow few lines of trap crops like *Jatropha*, cowpea, castor, or *Crotalaria* near field bounds.
- Pluck the leaves having larvae in clusters and destroy them.
- Spray the crop with 0.01% quinalphos 25 EC, 0.1–0.15% phosphamidon 85 SL, or 0.1% monocrotophos 36 WSC.

Note: To avoid mortality of the pollinators, the spray of insecticides should be done in the afternoon.

DISEASES

Numerous diseases that adversely affect the quality and quantity of mustard greens and make the seed and oil unfit for human consumption have been recorded on this crop. Some of those diseases, along with their control measures, are described as follows.

Alternaria Blight/Leaf Spot (*Alternaria brassicae*)

The disease, which is characterized by the appearance of small dark brown or blackish sooty and velvety spots with concentric rings on leaves, stems and pods, is most destructive and widely spread in all mustard growing regions. The centre of spots soon dries up and drops out. When the infection is severe, the leaves, pods and upper parts of stem wither. The pathogen, which causes 10–70% damage, is seed borne and remains viable in soil on plant refuse. Warm (temperature 12–25°C) and moist (relative humidity >70%) weather with intermittent winter rains plays crucial role in the development of this disease. When the attack is severe, the pods containing shrivelled and undersized seeds turn black and may rot.

Control

- Follow long crop rotation to minimize the inoculum build-up.
- Plough the field deep in hot summer months to eliminate the fungus.
- Use balanced doses of recommended fertilizers.
- Avoid over-irrigation and drain out excess water.
- Sow the seeds timely (10–25 October).
- Use disease-free healthy seed for sowing.
- Keep the weeds under control to reduce the spread of disease.
- Burn the crop residues of previous year.
- Spray the crop with iprodione (Roval) or Diathane M-45 0.2% at fortnightly intervals.

WHITE RUST (*Albugo candida*)

The disease is very common in all mustard growing regions. The symptoms appear on all aerial parts except root as white to creamy yellow pustules of 1–2 mm diameter on lower surface of the leaves 25–35 days after sowing, which correspond to tan yellow colour on upper side of the leaves. When the attack is severe, the white pustules also appear on stems and pods. These pustules burst and liberate a white powdery mass. In later stages, deformation of floral parts takes place usually on top branches, which is known as stag-head formation. Temperature 5–12°C and relative humidity >75% with short days (2–6 h sunshine) favour the development of disease. The pathogen is seed borne and survives through oospores in soil on crop residues. The seeds may become smaller and shrivelled.

Control

- Follow long crop rotation to minimize the inoculum build-up.
- Plough the field deep in summer months to destroy the fungus.
- Use balanced dose of recommended fertilizers.
- Avoid over irrigation and drain out excess water.
- Grow resistant varieties like Basanti, JM 1 and Maya.
- Sow the seeds timely (10–25th October).
- Collect the seeds from stag-head-free plants to avoid carry-over of the oospores.
- Treat the seed with metalaxyl (Apron 35SD) 6 g/kg seed or thiram 2.5 g/kg of seed.
- Control the weeds timely to avoid disease spread.
- Destroy the crop debris, particularly stag heads of previous year's crop.
- Spray the crop with Ridomil MZ 72 WP or Diathane M-45 0.2% at fortnightly intervals.

DOWNY MILDEW (*Peronospora parasitica*)

The disease is characterized by the appearance of yellow irregular spots on upper surface of the cotyledonary leaves 10–15 days after sowing and white growth on underside of leaves opposite to spots. The crop plants due to early infection become week and feeble, and ultimately, heavy yield loss is observed. Disease is favoured by cool (temperature 10–20°C) and wet (relative humidity >90%) weather. If the infection is severe, the inflorescence becomes malformed, twisted and covered with a white power. No pod is produced on such inflorescence.

Control

- Follow long crop rotation to minimize inoculum build-up.
- Plough the field deep in hot summer to eradicate fungus.
- Use balanced dose of recommended fertilizers.
- Avoid over irrigation and drain out excess water from the field.
- Grow moderately resistant variety Narendra Ageti Rai 4 and moderately tolerant variety RLM.
- Sow the seeds timely (10–25 October).
- Collect the seeds from stag-head-free plants to avoid carry over of the oospores.
- Treat the seed with metalaxyl 6 g/kg seed or thiram 2.5 g/kg of seed.
- Keep the crop field weed free.
- Destroy the crop debris particularly stag heads of previous year crop.
- Spray the crop with Ridomil MZ 72 WP followed by Diathane M-45 0.2% at 15-day intervals.

SCLEROTINIA STEM ROT (*Sclerotinia sclerotiorum*)

The disease prevails most commonly in temperate regions. However, it has also been reported from Rajasthan, Uttar Pradesh, Uttaranchal, Madhya Pradesh, Chhattisgarh, Bihar, Haryana, Punjab, Delhi and West Bengal. The disease is soil borne and mono cropping of mustard results in the development of severe infection in a specific geographical area. It is characterized by the appearance of elongated water soaked lesions on stem, which later on are covered with cottony mycelial growth. The infected plants look whitish from a distance, especially at the base; therefore, the disease is known as white blight. Greyish white to black spherical sclerotia develop either on the surface or in the pits. In low-lying areas, the damage may be extended up to 35%. The disease results in premature ripening and may cause additional yield loss due to the shattering of siliquae. High humidity (90–95%) and temperature 18–25°C along with wind current are most favourable for the spread of disease.

Control

- Follow long crop rotation with non-host crops like wheat, barley, rice and maize.
- Plough the field deep in summer months to make the field fungus free.
- Use balanced dose of recommended fertilizers.
- Avoid excessive irrigation and drain out the excess water.
- Sow healthy seeds free from sclerotial bodies.
- Burn the crop debris of previous year along with sclerotia.
- Spray Bavistin 0.1% twice during flowering at 20-day intervals.

POWDERY MILDEW (*Erysiphe cruciferarum*)

The disease usually appears late in the season when temperature begins to rise and is commonly observed in Gujarat, Rajasthan, Haryana, Maharashtra and southern part of the country. The symptoms appear as dirty white circular floury patches on leaves, stems and siliquae. As the disease advances, the whole plant looks to be dusted with white talcum powder. Minute spherical black fruiting bodies called *Cleistothecia* can be seen on the infected parts. Infection first occurs on lowermost leaves near the soil surface through *Cleistothecia* present as primary inoculum in crop debris in the field. Yield losses depend on the stage at which the disease appears. High temperature (15–28°C), low relative humidity (60%) and low or no rainfall with wind current is most favourable for epidemic development of the disease.

Control

- Follow long crop rotation with non-host crops.
- Follow deep ploughing in summer months to destroy the fungus.
- Use balanced dose of recommended fertilizers.
- Avoid over-irrigation and provide proper drainage facilities.
- Grow resistant variety like GM-2.
- Avoid sowing of crop late.
- Burn the crop debris of previous year.
- Spray wettable sulfur 0.2%, Karathane 0.1%, or Bavistin 0.05% at the initiation of disease.

■■■

46 Maca

S.A. Ranpise and Chandanshive Aniket Vilas

Botanical Name : *Lepidium meyenii* Walp.

Family : Cruciferae (Brassicaceae)

Chromosome Number : 2n = 64

ORIGIN AND DISTRIBUTION

Maca, which is one of the Andean crops occupying a very restricted area, is native to the high Andes of Peru. It is also believed that the genus perhaps originated in the Mediterranean region, where most of the diploid species are found. It has been an important staple for the Andean Indians and indigenous people and was domesticated during the pre-Inca Arcaica period around 3800 B.C. It is found only in the Central Sierra of Peru above 4000 m mean see level where low temperature and strong wind limit the cultivation of other crops. It was found growing around Junín Lake in the Andes until late 1980. Gerhard Walkers described this species first time in 1843 and gave it a name *Lepidium meyenii*. Gloria Chacon in 1990 considered the natural maca a new domesticated species with different characteristics. It was domesticated in the Andean mountain at altitudes from 3500 to 4450 m above mean sea level in the *Puna* and *Suni* ecosystems. It is cultivated mainly in Peru, high Andes of Bolivia and to a small extent also in Brazil. Historically, it was often traded for lowland tropical food staples, such as corn, rice, tapioca, quinoa and papaya. It was also used as a form of payment of Spanish imperial taxes, as it was eaten by Inca imperial warriors before battles. Their legendary strength allegedly imparted by the preparatory consumption of copious amount of maca, fuelling formidable warriors. The area where it is found is high in the Andes, which is an inhospitable region of intense sunlight, violent winds and below-freezing temperature. The area with extreme temperature and poor rocky soil rates among the world's worst farmland, yet over the centuries, maca has evolved to flourish under such conditions. It was domesticated about 2,000 years ago by the Incas, and its primitive cultivars have been found in archaeological sites dating as far back as 1600 B.C. It has been cultivated as a vegetable crop in Peru for at least 3,000 years.

INTRODUCTION

Maca, also known as *ayak chichira*, *maca-maca*, *maino*, pepper weed and *ayak willku* with the synonyms as *Lepidium peruvianum*, is a relative of radish but resembles turnip in shape with butterscotch odour. Taxonomically, the plants most closely related to maca are rapeseed, mustard, turnip, black mustard, cabbage, garden cress and water cress. It can be divided into different groups based on root colour, which can be red, black, pink, or yellow but the purple and grey roots are highly valued for their nutritional value, especially protein and minerals. The genus *Lepidium* contains around 150–175 species. It is usually grown for its fresh fleshy nutritional taproot, which is used as a vegetable and medicinal herb. Maca, which is a dietary supplement derived from the processed tuberous roots, was grown for food by the *Pumpush*, *Yaros* and *Ayarmaca* Indians. It is the only species cultivated as a starch crop. It has been traditionally used as a folk medicine for vitality and fertility in the Andean region of Peru and has been used in both genders as well as animals and is sometimes referred to as Peruvian Ginseng despite not bearing any resemblance to the *Panax ginseng* plant of its family. For approximately 2,000 years, it has been an important traditional food and medicinal plant in its limited growing region, where it is regarded as a highly nutritious energy-imbuing food. During Spanish colonization, it was used as a form of currency.

COMPOSITION AND USES

COMPOSITION

Maca is nutritionally rich and is an excellent source of starch, sugars, protein, essential minerals like iron, iodine, potassium, sodium and magnesium and vitamin C. It is low in fibre. It has a balance of thiamine to riboflavin levels that make it an ideal food for body development. The nutritional composition of maca is given in Table 46.1.

USES

Maca is prepared and consumed in various ways, although traditionally it is used as raw salad, cooked, or roasted. The roots boiled and mashed to produce a sweet thick liquid and then dried and mixed with milk to form a sweet aromatic porridge. The cooked roots are also used with other vegetables in jams, soups, or fermented beer called maca chichi, which is consumed during dance and religious ceremonies after adding hallucinogenic products. The roots are used to produce flour for bread, cakes, or pancakes. The edible leaves are used as animal fodder.

MEDICINAL USE

Based on a long history of traditional use of maca in Peru and elsewhere, a wide array of commercial maca products have gained popularity as dietary supplements throughout the world for aphrodisiac purposes and to increase fertility and stamina. It has been used all over the world and affects both the genders. It does not work through hormones and does not increase testosterone or oestrogen. Men supplementing maca have been known to experience an increase in sperm production. It also appears to be a potent suppressor of prostate hypertrophy, with potency similar to finasteride, a synthetic drug for the treatment of enlarged prostates. It is used to cure anaemia and chronic fatigue syndrome and to enhance energy, stamina, athletic performance and fertility. In women, it is used to improve memory and cognition and to maintain hormonal balance, which creates menstrual and menopause disorders. It reduces anxiety and depression, and lowers sexual dysfunction in postmenopausal women independent of estrogenic and androgenic activity. It is also used in problems like weak bones and immune system, stomach cancer, leukaemia, HIV/AIDS, tuberculosis and erectile dysfunction. The dark colour maca roots

TABLE 46.1

Nutritional Composition of Dried Maca (per 100 g Edible Portion)

Constituents	Contents	Constituents	Contents
Water (g)	10.4	Copper (mg)	5.9
Carbohydrates (g)	59	Zinc (mg)	3.8
Protein (g)	0.70	Thiamine (mg)	280
Dietary fibre (g)	8.5	Riboflavin (mg)	650
Calcium (mg)	150	Vitamin C (mg)	8
Potassium (mg)	2050	Phenolic (mg)	5.5–7.6

contain significant amount of natural iodine, which is essential to shun the growth of goitres. Black maca being energy rich promotes stamina. It improves neuro-protection, cardiovascular health, body interaction with hormones and normalizes the blood pressure. In livestock, it is used to increase fertility, strength and endurance.

BOTANY

Maca is a perennial herbaceous plant but grown as annual. Plant with succulent edible roots and low-growing mat-like stem is 10–20 cm in height. It develops underground part, consisting of a hypocotyl or underground reserve organ (root), which can be up to 8 cm in diameter and may be white, reddish white, yellow, purple, or light grey. The roots having meaty texture may be rounded, globose and napiform. It is integrated in the root tissue and ends in a thick central root, covered with numerous thin lateral feeding roots. It grows as a rosette plant and the roots are barely visible due to dense foliage. Its leaves are small compound and pinnate, which lie close to the ground, renewing continuously from centre of the rosette with enlarged leaf sheath and composed limb with 6–9 cm length. Leaves show a dimorphism up to reproductive stage, which is more prominent in vegetative phase. The inflorescence is composed of axillary racemes with hermaphrodite, actinomorphic and off-white tiny self-fertile flowers, which are borne on a central raceme and are followed by 4-5 mm siliculate fruits, each containing two small 2.0–2.5 mm reddish-grey ovoid seeds. Being a short day plant, it blooms in second year of its growth between October and November. The species is mainly self-pollinated and autogamous. The fruit is two seeded 4–5 mm long siliqua, separated by a partition that divides the fruit into two equal parts. The siliqua is dehiscent with two carinated valves, each containing a seed. The seeds, which are ovoid, 2–2.5 mm and orange, yellow, brown, or reddish grey, are used for reproduction.

CLIMATIC REQUIREMENT

In natural habitat, maca like other root crops grows well only in cold climate with relative humidity approximately 70%. Strong winds and sunlight are also the characteristics of its native habitat. Being cold hardy plant, it grows naturally in the ecological zone located between 3800 and 4450 meters above mean sea level. For its cultivation, the optimum average temperature is 4°C–7°C with −10°C night and 22°C daytime but it can tolerate a temperature as low as −20°C. It can also be cultivated at high altitudes with intense sunlight, violent winds and below freezing temperature. However, moist and relatively sunless conditions are not suitable for its cultivation.

SOIL REQUIREMENT

Maca can be grown in relatively poor agricultural and rocky soils, however, well drained sandy loam or loam soils rich in organic matter having good moisture retentive capacity and pH 6–8 are most suitable for its cultivation. It prefers loose and well fertilized soils with plenty of organic matter. It can also be grown in highly acidic soil. Naturally, it grows in soils with poor nutrition. However, commercial crop is supplemented with farmyard manure. It can also be grown on freshly ploughed pasture land that has lain fallow or on ground under an annual rotation with another crop.

CULTIVATED VARIETIES

Over thirteen different variants of maca are produced differently, ranging from white to black but the most commercially desirable is yellow. These variants have been named due to the coloration of hypocotyls being visibly different. These variants are sometimes referred to as ecotypes of maca.

SOWING TIME

Maca seeds are sown in late winter or early spring after the last frost or in the beginning of rainy season from September to November in the southern hemisphere as a sole crop or combined with potato strips.

SEED RATE

Unlike other tuberous plants, maca is propagated by seed and approximately 3–3.5 kg seed is enough for sowing a hectare land area.

PLANTING METHOD

The seeds are sown directly by broadcasting in unprepared field. At the most, the seeds after sowing are covered with manure. The seed is mixed with sand for uniform distribution. The seedlings usually emerge in about a month after sowing. Over the next 8–9 months, its plants remain in a vegetative state until tuberous roots are fully developed, and ninth months onwards, maca starts to develop roots. At this stage, which lasts from 4 to 6 months, the plant begins to produce seeds, which will be used for sowing in following year. Maca can also be reproduced by planting its tuberous roots once it has developed its aerial parts at 70 cm spacing.

NUTRITIONAL REQUIREMENT

Maca is often grown on sites where no other crop can be cultivated or on pastures after long fallows of sheep grazing where crop gives higher yield. If it is cultivated as a succeeding crop after potato, it requires no manure or fertilizer for its cultivation. However, it is usually fertilized with manure if the crop is to be planted on fallow land often resting for a period of years to rebuild nutrients in soil. Traditionally, it is fertilized only with sheep manure, which prevents the soils from nutrients depletion.

IRRIGATION REQUIREMENT

Maca is planted in rainy season and grown as a rain-fed crop. It develops tubers in summer months when rainfall is lower, thus, applying irrigation is not essential during tuber development phase. However, supplying irrigation in first few months is indispensable for the completion of vegetative phase.

INTERCULTURAL OPERATIONS

HOEING AND WEEDING

Hoeing and weeding are necessary in early stage of crop growth for destroying the weeds, keeping the soil loose around the plant root system for better aeration and covering the tubers well with soil in field. A light hoeing 10–15 days after germination is done to break the soil crust for making the seedling emergence easy.

The weeds may also be controlled effectively by pre-planting application of Fluchloralin @ 0.7–1.0 kg/ha or post-emergence application of paraquat @ 2.0–2.5 l/ha at 3–5% emergence of crop. Depending upon the intensity of weed growth, a pre-emergence application of alachlor @ 1.0–1.5 kg/ha is also recommended.

EARTHING UP

The plants are earthed up to cover the growing and developing tubers. Generally, one earthing up is enough when the crop is 10–15 cm tall and starts tuber formation. Depending upon the variety and crop duration, a second earthing up can also be done to cover the tubers properly. This operation is done with the help of spade or ridge plough and must be completed timely as it will have direct effect on final tuber yield.

HARVESTING

Harvesting of hypocotyls, which are intended to food production as a dietary supplement, is done at the end of vegetative stage. The crop takes seven to nine months to produce tubers for harvesting, which is done in the southern hemisphere from May to mid July, before the development of aerial parts of the plant, while fruits are harvested in summer of the second year from December to February, when the plant is mature. When most foliage is yellow, it must be harvested. Harvesting should be done manually or with the help of a pickaxe. They are then stored in a cool and dark place until they are eaten. Majority of the harvested maca is dried, and in this form, the hypocotyls can be stored for several years.

YIELD

The yield of maca varies from 16 to 20 t/ha, depending on soil type, soil fertility, climatic conditions of the growing region, irrigation facility and the cultural practices adopted during cultivation. Maca is typically dried for further processing, which yields about 5 t/ha.

POST-HARVEST MANAGEMENT

The crop produce should be stored at optimum temperature to reduce post-harvest losses. It must be dried for 2–3 days until the leaves are dry, while the dry leaves must be removed to avoid contamination. Then, they are kept in plastic baskets, and a month later, they are packed in bags and stored in a cool and dry place. The seeds should also be stored in a dry place until sowing, which is done from September to November of the same year.

INSECT-PESTS

ANDEAN WEEVIL (*Premnotrypes* spp. hymeli)

This is a serious pest of maca crop since it causes more damage. Its grub enters into the developing tubers and feeds on them. Grubs are white and C shaped with prominent black head. The larval period lasts for 8–10 days and the pupa remains inside the plant debris. The adult emerges by making a hole in the tuber. The adults are shiny black weevils with prominent snout having the ability to fly. Adult also causes damage by feeding on the leaves, young shoots and flower buds. It makes small holes in flower buds and eats inside, which ultimately affects the flower opening, leading to poor yield. Early maturing crops are more susceptible to its damage than the late maturing ones.

Control
- Follow long crop rotation with non-host crops.
- Avoid planting in field where cutworm is a problem.
- Collect and destroy the infested plants along with their parts.
- Dip planting material in monocrotophos 0.5% for 30 minutes.
- Spray the crop with chloropyrifos 20 EC 0.02%, dimethoate 30 EC 0.03%, malathion 50 EC 0.05%, or quinalphos 25 EC 0.04%.

BLACK CUTWORM (*Agrotis ipsilon*)

Caterpillar cuts the stalk of seedlings as well as transplanted plants at ground level. They feed only at night and remain present in daytime either below the soil or leaves. This species occurs frequently in many crops and is one of the best known cutworms. Despite the frequency of occurrence, it tends not to appear in great abundance. Black cutworm is not considered to be a climbing cutworm, most of the feeding occurring at soil level. Its larvae feed above ground parts and consume foliage during their development but maximum occurs during the adult stage. Thus, little foliage loss occurs during early stages of development but considerable damage is done at young plant stage. A larva may cut several plants in a single night.

Control

- Handpick and destroy the larvae.
- Apply soil insecticides at pre-planting or planting time to prevent damage.
- Avoid planting following longstanding pastures, maize, meadows, Lucerne, or red clover.
- Follow ploughing in autumn and use shallow tillage.
- Use poison bait traps to monitor the larvae.
- Plant few border rows of perennial ryegrass, turf grass and Kentucky bluegrass (*Poa pratensis*) to reduce the pest attack.
- Spray the crop with chlorpyrifos @ 0.16, monocrotophos 0.24 kg a.i./ha.

DINGY CUTWORM (*Feltia jaculifera*)

The insect damages the plants and tubers, especially during night. They damage the young plants by cutting their stems, pulling all parts of the plant into the ground and devouring them particularly at 25–35 days after planting. Signs of damage on tubers are boreholes larger than those made by potato tuber moths. Larvae are up to 4.5 cm long. Paired tubercles on each body segment are approximately of the same size. They are grey in colour and may show some faint reddish-brown coloration. A dull grey stripe can be present running the length of the back bordered by a narrow dark stripe on each side. Larvae make small holes in leaves or along the leaf edges but rarely cut the stem.

Control

- Irrigate the field at subsequent intervals to bring cutworms on the surface for the predators.
- Keep green grass in heaps at suitable intervals in infested field during evening and next day early in the morning along with caterpillars to destroy them.
- Follow clean cultivation and destroy the caterpillars mechanically.
- Apply chlordane or heptachlor dust @ 50 kg/ha in soil.
- Spray the crop with Coragen 20 SC 300 ml, Tracer 48 SC 200 ml, or indoxacarb 30 WG 130 g/ha.

DISEASES

DAMPING-OFF (*Aphanomyces, Fusarium, Pythium* and *Rhizoctonia* spp.)

'Damping-off' is a term often used for sudden death of seedlings. Initial disease symptom appears as light yellowing of lower leaves tips, which gradually spreads down to the leaf blade and leaf sheath along the margin. Generally, post-emergence damping-off disease damages the seedlings more. In early stages, the middle portion of lamina remains green, while the margins become yellow. The yellowing spreads to all the leaves from bottom to upwards and is followed by drooping, withering and drying. Collar region of the pseudostem shows pale translucent brown colour, which becomes water-soaked due to the destruction of parenchyma tissues. The infected seedlings can be easily pulled out from the soil. Infection spreads from collar to the rhizome gradually.

Control

- Use healthy planting material.
- Avoid water-logging in the field.

- Uproot and destroy the diseased plants.
- Provide proper drainage and keep the field free from weeds at all times.
- Treat the tubers with hot water at 52°C for 10 minutes.
- Disinfect the soil either through solarization or by drenching with 1% formaldehyde.
- Apply *Azadirachta indica, Calophyllum inophyllum, Pongamia glabra* and *Hibiscus sabdariffa* cake in soil.
- Treat the tubers with *Trichoderma* 5 g/kg for 30 minutes or Ridomil MZ 72, Topsin M, or Apron 35 WS 2 g/litre of water for 20 minutes.
- Treat the seed with Dithane M-45, Blitox, or Captan 3 g/kg of seed.
- Spray the crop with 0.3% copper oxychloride, 0.05% copper sulfate, or 0.25% Dithane M-45 or captan 50 WP at 10- to 12-day intervals.

Fusarium Wilt (*Fusarium oxysporum* f.sp. *spinaciae*)

Initially, the affected plants show yellowing of leaves in some branches followed by drooping and drying of leaves. The entire plant dies in a few months or in a year. When affected plant is cut cross-section, dark greyish-brown coloration of wood is observed. The primary source of inoculums is soil and secondary source is water. Disease is more in heavy soil and increases with soil moisture. It is most serious disease in poorly drained soils. Warm soil temperature, poor drainage, light soils and high soil moisture favour the spread of this disease.

Control
- Follow deep summer ploughing.
- Follow long crop rotation with non-host crops.
- Maintain proper drainage facility.
- Grow resistant varieties, if available.
- Uproot and destroy the infected plants from the field.
- Apply bioagents such as *Trichoderma viride* or *Pseudomonas fluorescens* in soil.
- Treat the seed with Bavistin, Benomyl, or thiram @ 3 g/kg of seed.
- Drench the field with 25 ml/l formaldehyde, 0.3% Lanacol or Brassicol, or 2 ml propiconazole + chlorpyrifos 4 ml/litre of water.

Downy Mildew (*Peronospora* spp.)

The disease affects the roots, basal stem, leaf and seed. The symptoms are systemic infection, local lesion, damping-off and basal rot or stem gall. Chlorotic lesions appear on upper leaf surface infected with downy mildew. Purplish-brown spots appear on underside of the leaves. Abnormally thick whitish downy growth develops on lower surface of the leaves. Local foliar lesions are characterized by small angular greenish yellow spots on leaves. Systemic infection shows chlorotic leaf veins. Pathogen spreads through seed, soil and wind. Rain during seedling growth favours disease.

Control
- Follow deep summer ploughing.
- Follow clean cultivation and maintain proper field sanitation.
- Remove and destroy the affected plants.
- Avoid excessive irrigation.
- Treat the seed with metalaxyl @ 6 g/kg of seed.
- Spray the crop with 0.2% zineb, maneb, or Ridomil MZ 72 at 20, 40 and 60 days after sowing and prior to the onset of rainfall.

47 Rutabaga

Sonia Sood and Nivedita Gupta

Botanical Name   : *Brassica napus* var. napobrassica

Family   : Cruciferae (Brassicaceae)

Chromosome Number : 2n = 38

ORIGIN AND DISTRIBUTION

Rutabaga is a root vegetable originated as an inter-specific hybrid bred in Switzerland between summer turnip and winter white cabbage. Napobrassica group (rutabaga) is closely related to cole crops and this relationship was first published by Woo Jang-choon in 1935 in triangle of U in Brassica species. It was first introduced in England at the end of the 18th century and was called *the turnip-rooted cabbage*. It is primarily grown in Canada, California, Wisconsin and Minnesota. Both, white and yellow-fleshed cultivars of rutabaga exist. Its root consists of both true root and true stem. The upper portion of the stem forms a neck, which distinguishes rutabaga from turnip.

Rutabaga has a complex taxonomic history. The earliest account comes from the Swiss botanist Gaspard Bauhin, who wrote about it in his 1620 *Prodromus*. In 1753, *Brassica napobrassica* was first validly published by Carl Linnaeus as a variety of *B. Oleracea, i.e., B. oleracea* var. napobrassica in *Species Plantarum*. It has since been moved to other taxa as a variety, subspecies, or elevated to species rank. In 1768, a Scottish botanist elevated Linnaeus' variety to species rank as *Brassica napobrassica* in *The Gardeners Dictionary*, which is the currently accepted name.

INTRODUCTION

Rutabaga is an allopolyploid with chromosome number (2n) 38. It is similar in appearance to turnip but differs from it in denser roots, which are usually rounded or elongated instead of being flattened. Leaves are smooth and covered with bluish bloom, but in case of turnip, leaves are green and hairy. It has many national and regional names. The word rutabaga has been derived from the old Swedish word *Rotabagge*, meaning simply *ram root*. In the United States of America, the plant is also known as Swedish or yellow turnip. The term Swede is used instead of rutabaga in many Commonwealth Nations, including much of England, Wales, Australia and New Zealand. In Scotland, it is known as turnip and in Northeast England, turnips and Swedes are colloquially called snadgers, snaggers, or narkies.

COMPOSITION AND USES

COMPOSITION

Rutabaga is one of the most nutritive root vegetables. It contains relatively more quantities of carbohydrates, proteins, minerals such as calcium, iron and magnesium and vitamins especially vitamin B_1, B_2 and B_5. For most of the parameters studied, nutritional quality of rutabaga is better than turnip. Leaves are most nutritious than its roots. The nutritional composition of rutabaga is given in Table 47.1.

TABLE 47.1

Nutritional Composition of Rutabaga (per 100 g Edible Portion)

Constituents	Contents	Constituents	Contents
Water (g)	89.66	Iron (mg)	0.44
Carbohydrates (g)	8.62	Manganese (mg)	0.13
Sugars (g)	4.46	Thiamine (mg)	0.09
Protein (g)	1.08	Riboflavin (mg)	1.08
Fat (g)	0.16	Niacin (mg)	0.04
Dietary fibre (g)	2.30	Pantothenic acid (mg)	0.70
Calcium (mg)	43.0	Vitamin C (mg)	0.16
Magnesium (mg)	20.0	Energy (kcal)	36.0

USES

Rutabaga is a nutritive root vegetable, which is most often roasted to serve with meat, baked and boiled. It is eaten raw as salad and is used as a flavour enhancer in soups and stews. The leaves can be eaten as a leafy vegetable. The roots and tops are also used as winter-feed for livestock. In Sweden and Norway, rutabaga is served as a puree called *rotmos*. It is also a component of pickle. In United States of America, Australia and England, it is mostly eaten as part of stews or casseroles, served mashed with carrots and in soups as a flavour enhancer.

MEDICINAL USES

Rutabaga has highly significant antioxidant capacity, which varies with the agricultural techniques adopted during cultivation of the crop. Its roots due the presence of vitamin C and E act as an antioxidant, which neutralizes the free radicals produced by the body. Vitamin C works throughout the body to build up immune system and lower the risk of some kind of cancer. Manganese is a major antioxidant in mitochondria that produces energy for the body.

BOTANY

The rutabaga (*Brassica napus* var. napobrassica) is a member of the family Brassicaceae. The cultivated forms of rutabaga are dicotyledonary herbaceous annual for the purpose of root production and biennial for seed production. The storage root may be purple, white, or yellow with yellowish flesh. The *root* consists of hypocotyls, the plant part that lies between the true root and the first seedling leaves (cotyledons) and the base of the leafy stem. A rutabaga root has a swollen *neck*, bearing a number of ridges, and the leaf-base scars. Its root shape varies according to the variety, being cylindrical, conical, globular, pear-shaped, spherical, or flattened and according to colour, being white, yellow, purple, red, grey, or black. The leaves of rutabaga are bluish, thick like cabbage, smooth, waxy and grow from the part that protrudes above the ground. They emerge from the crown in a broad, low-spreading growth habit that inhibits growth of weeds. Normally, rutabaga is biennial, *i.e.*, it forms a swollen root during first year of the growth and flowering stems in second year of the growth after a cold period. In the first year of vegetative growth, the plant forms a rosette of leaves varying greatly in size and shape, which may be oblong to oval and with slightly or deeply serrated margins. The foliage arises from an extremely shortened stem, which elongates to develop into a seed stalk after the juvenile stage of the plant enters the reproductive phase. The alternate leaves formed at the internodes of seed stalk are glaucous oblong or lanceolate and entire dentate. The seed stalk may grow to a height of 90–150 cm.

Rutabaga has a typical small, yellow *Brassica* flower with four petals that form a cross and give them their designation as cruciferous vegetables. The inflorescence is a terminal raceme and flowers are hermaphrodite, bright yellow to pale-orange yellow in colour and highly cross-pollinated mainly due to their self-incompatible nature. Flowers are not formed above the unopened buds on the raceme. The colour of rutabaga flowers varies with the root flesh colour. The white-fleshed varieties produce bright yellow petals in contrast to the pale orange yellow petals of the yellow-fleshed varieties. Anthesis and dehiscence take place in the morning and the pollens remain available until early noon hours, depending upon the prevailing temperature. The siliqua (pod) dehisces longitudinally at maturity into distinct equal parts. Seeds are very small, globular and smooth in shape having uniformly brown to various shades of dark brown colour depending upon the variety. If plants are subjected to low temperature (below 5°C) at less than 10 weeks old, this will trigger the development of flowering stalks. The low temperature duration required to cause flowering depends on the cultivar being grown. However, as little as 3 to 5 nights with temperature around 3°C are believed to result in the development of flowering stalks. Field plantings or seedlings can be affected by low temperature but transplants more than 10 weeks old would require several nights of freezing temperature to induce flowering.

CLIMATIC REQUIREMENT

Rutabaga is well adapted to cool and humid growing conditions. The minimum temperature for germination is 5°C but the optimum germination takes place at 15°C to 29°C. It is frost resistant and can withstand dry period if some soil moisture is available. Cracking of roots may occur with a fast growth rate due to excessive fertilization, wide spacing and hot and humid weather. Sometimes, these growth cracks become infected with soft rot bacteria.

SOIL REQUIREMENT

Rutabaga grows best in moderately deep, highly fertile soil with pH 6.0 to 6.5. The crop thrives on moist, well-drained and moderately acidic, sandy loam and clay loam soils well supplied with organic matter. On sandy loam soils, the roots may tend to be elongated especially in dry weather with dense plant populations. The best quality roots seem to come from well-drained clay loam soils, which have not been intensively cropped. Soil *capping* can be a problem in heavy soils with poor aeration and unfavourable climatic conditions.

One deep ploughing followed by 2–3 shallow ploughings is sufficient to make the soil surface smooth, loose, friable and free from clods. The last ploughing should be followed by planking to level the field.

CULTIVATED VARIETIES/HYBRIDS

Varieties of rutabaga differ from turnip mainly in colour and shape of roots. There are white- and yellow-fleshed varieties of both the crops, although most of the turnip varieties are white-fleshed and most rutabaga varieties are yellow-fleshed. The characteristic description of few varieties is given as under:

AMERICAN PURPLE TOP

A most popular variety of rutabaga matures in 90 days. The morphological character of this variety includes deep purple crown, yellow below the crown, globe-shaped root, 12 to 15 cm in diameter with yellow flesh colour, medium sized and blue green cut leaves.

LAURENTIAN

This variety also becomes ready for marketing in 90 days. The morphological character of this variety includes purple crown, light yellow below crown, globe-shaped roots, 12 to 14 cm in diameter with yellow flesh and medium blue-green cut leaves.

AMERICAN YELLOW

A variety with small tap root and a sweet mild flavour becomes ready for marketing in 85 days. It is a yellow variety with purplish colouring in the upper portion, which is above ground.

DOON MAJOR

A variety resistant to nematode with purple top, smooth uniform shape, yellow skin, very large retains excellent texture and flavour.

HELENOR

A variety producing sweet and delicious roots with golden flesh and tinged red skin on top is good for fresh market and storage.

MAJOR PLUS

A variety maturing in 160 days is characterized by yellow-fleshed broad elliptical bulb with red skin.

MARIAN

A vigorous and very productive variety giving globe-shaped, purple-topped roots with yellow flesh and interior creamy colour becomes ready for marketing in 85–95 days.

WINTON

A variety with high leaf yield potential is characterized by its white flesh with a bronze skin colour.

JOAN

A variety with good field resistance to club root produces smooth, round, uniform and sweet roots.

SOWING TIME

Sowing date has a direct effect on the harvest date. Early sowing promotes an early harvest, hence, sowing should always be done at the right time to get higher yield. It is mainly grown in Canada where summers are cool. The roots when produced in warm weather become woody and bitter very quickly. It must be seeded two to three months before heavy frost. Rutabaga takes about 4–5 weeks longer to be mature in comparison to the turnip. In south India, it is grown during winter and early spring. The optimum time for sowing rutabaga in hills (up to 1500 m above mean sea level) with mild or rare snowfall is February to April, whereas, it is March to July in very high hills experiencing heavy snowfall. In low hills and northern plains of India, the time of sowing is October to November. Prolonged exposure of post-juvenile plants to low temperature will induce seed stalks to form later in the season.

SEED RATE

Rutabaga seed size is comparable with that of the turnip. The seed rate usually depends on variety, viability of seed, growing season, type and fertility of soil and irrigation facilities. About 225–450 g seed is required for sowing a hectare land area.

SOWING METHOD

Rutabaga has more robust growth than the turnip, and therefore, it require more space between lines and plants. The seeds are sown on ridges at a depth of 1.0–1.5 cm. The row and plant spacing of 45–75 and 10–15 cm should be adopted to get higher yield, respectively. Sowing can be done either by hands or by seed drill. Seed drill is a sowing device that precisely positions the seeds in soil and then covers them. The use of a seed drill can improve the crop yield. Although hand sowing is an efficient method, a seed drill consumes less time and is used economically as a modern agricultural implement.

NUTRITIONAL REQUIREMENT

Rutabaga responds to the application of manure and fertilizers. Well-decomposed farmyard manure is applied 30 t/ha at the time of land preparation. Besides, it is advisable to apply nitrogen 80–100 kg and phosphorus and potassium each 50–60 kg/ha. It will help to increase yield. Entire dose of phosphorus and potash fertilizer along with half dose of nitrogenous fertilizer is applied as basal application. The remaining half dose of nitrogenous fertilizer is top dressed at the time of root formation. Lime should be applied to maintain the soil pH in the range of 6.0 to 7.0.

IRRIGATION REQUIREMENT

Growth is reduced with excessive application of irrigation. Rutabaga requires an abundant supply of moisture to ensure a high quality produce. Pre-sowing irrigation is essential for better germination of seeds. The total water requirement of rutabaga crop is about 500 mm. In general, in most of the soils, the crop requires 4 cm of water after every 7 to 10 days until harvest.

INTERCULTURAL OPERATIONS

HOEING AND WEEDING

Manual weeding and hoeing is done as and when needed. Weeding should be done according to the need to keep the growth unchecked. The plants should be earthed up 25–30 days after sowing at the time of top dressing of nitrogenous fertilizer. In India, the chemical weedicides have not been experimented, however, pre-plant soil incorporation of fluchloralin (Basalin 48 EC) 1.0 kg/ha, alachlor (Lasso 50 EC) 2.0 kg/ha, or pendimethalin (Stomp 30 EC) 1.0 kg/ha controls the weeds effectively.

THINNING

Thinning is an important cultural operation in rutabaga to maintain optimum plant to plant spacing and required population. The plants are thinned to a spacing of 7.5–15 cm within the row, which will help in improving the size and quality of roots.

HARVESTING

A light irrigation applied a day before harvesting facilitates the lifting of roots. The roots are pulled with hand, and the taproot and crown are trimmed with a knife in the field. Best quality rutabagas should have a diameter of 7–9 cm. The quality and flavour of rutabaga roots are much improved when the roots are fully mature and are exposed to frost before the harvest. Immature roots have a bitter taste and if the early seeded rutabagas are left in the field until late fall, the roots tend to become fibrous and woody. Certain types of potato diggers can be readily adapted for harvesting rutabaga, but before the start of the digging operation, the tops are removed with the aid of a topping device. Rutabaga can also be harvested by using a modified sugar beet or carrot harvesting machine. Harvesting the roots under wet conditions or placing wet roots in storage should be avoided. Mechanical harvesters tend to spread bacterial and fungal organisms to crowns and wounds on roots. To minimize crown rot, rutabaga should be topped a week or 10 days before harvesting, with a hoe, knife, or mechanical topping machine. Harvesting should be completed before severe freezing.

POST-HARVEST MANAGEMENT

Harvested roots are washed and graded according to size and tied into bunches along with the tops. The rutabaga roots should ideally be uniform in shape, well formed, smooth, firm but of tender texture. It should be free from harvest damage, decay, insects and diseases. Bunched rutabaga tops should be fresh in appearance and turgid. Normally, loose produce is sent to the market, or sometimes, it is packed in gunny bags. Wrapping in low-density polyethylene (LDPE) films keeps the roots fresh and tender when stored at room temperature.

Like other root crops, rutabaga roots can be stored well at 0–1°C temperature and 90–95% relative humidity. Practically, the roots that are to be marketed are waxed. Roots are trimmed, washed, dried and dipped for one second in hot paraffin wax at a temperature of 121°C to 132°C. Waxing greatly enhances appearance of roots and markedly reduces shrivelling and loss of weight during marketing. In preparation for long-term winter storage (6 to 8 months), the tops are trimmed to within 1 cm of the body as the leaf tissue causes severe crown rot, which makes the roots unmarketable. The newly harvested rots should be protected from sunburn and freezing at all times and handled gently to avoid bruising. Controlled atmospheric (CA) storage at 1–2% oxygen and 2–3% carbon dioxide is slightly beneficial in maintaining quality of topped roots, as it helps in retarding the re-growth of shoots and rootlets in topped and tailed roots.

YIELD

Rutabaga yield depends upon varieties, growing season, soil type and fertility, irrigation facilities and the technology adopted during the course of crop cultivation. In general, the yield of rutabaga varies from 250–300 q/ha.

PHYSIOLOGICAL DISORDERS

CRACKING

Cracking in rutabaga roots occurs due to excessive fertilization, wide spacing and occurrence of hot and humid weather. Large size roots are prone to cracking more. Fluctuation in soil moisture is also found to be responsible for cracking. To prevent cracking, it is advisable to sow seeds at close spacing, supply recommended dose of nitrogenous fertilizers, maintain optimum moisture in the field and harvest the crop at right maturity stage to avoid the roots to be larger.

WHIPTAIL

Whiptail is caused due to deficiency of molybdenum. The young leaves become narrow and cupped, show chlorotic mottling, especially around the margins and develop deep patches, which ultimately affect the root growth. Applying sodium or ammonium molybdate 0.5–1.0 kg/ha at the time of field preparation, spraying crop with 0.1–0.3% ammonium molybdate with 0.1% Teepol as sticker and adding lime in soil to raise the pH to 6.5 are suggested to cure molybdenum deficiency.

WATER CORE OR BROWN HEART

Brown heart in rutabaga is due to boron deficiency. The disorder is prevalent in very acidic soils where boron is deficient and in alkaline soils, in which, boron is not available to the plants. The symptoms are characterised by the appearance of grey or brown colour in inner portion of the affected roots. The roots become unfit for consumption. Avoiding sowing in too acidic soils, maintaining adequate soil moisture to prevent drought conditions, incorporating borax @ 10–15 kg/ha in the soil before sowing and spraying the crop with boric acid (0.2%) two or three times at vegetative stage are advocated to prevent the occurrence of the disorder. Foliar sprays are essential, especially if the soil pH is over 6.0 or moisture stress occurs during 3–6 weeks period after planting. The cultivar Laurentian is more susceptible to boron deficiency than other cultivars.

INTERVEINAL CHLOROSIS

This disorder occurs due to the deficiency of magnesium. The symptoms generally start with the development of mottled chlorotic areas in interveinal tissues. On roots, the tints of orange, yellow and purple colour develop and root growth is affected adversely. Applying magnesium sulfate is essential as and when required.

FREEZING INJURY

Initiation of freezing injury takes place at −1°C temperature. Because of freezing injury, the shoots become water soaked, wilted and they turn black. Roots also become water soaked, soft and glassy. Freezing injury can be prevented by raising soil temperature through light and frequent irrigations during winter season.

INSECT-PESTS

CABBAGE MAGGOTS (*Delia radicum*)

Larvae or maggots feed by making tunnels inside the roots. Plants may be killed, weakened, or stunted and yield is reduced drastically. The severely infected plants wilt and small tunnels are formed, which render the crop unmarketable.

Control
- Practice long crop rotation with non-cruciferous crops.
- Avoid growing crops of early and late rutabagas close, together, or near early broccoli, cabbage, cauliflower, or other cruciferous crops.
- Drench the seedbeds with chlorpyrifos @ 0.04%.

APHID (*Brevicoryne brassicae*)

Aphids are tiny, soft-bodied, slow-moving insects. They live in large colonies on under surface of the leaves. A colony consists of winged and wingless adults and various sizes of nymphs. Aphids may be black, yellow, or pink but most are of various shades of green colour. Aphids feed by sucking plant sap. Saliva injected while feeding may carry plant viruses or may be toxic to the host plant. Feeding by large numbers discolours foliage, curls leaves and damages developing buds. The plants may be covered by honeydew-like sticky substance, which is excreted by the aphids.

Control
- Use of predators such as ladybird beetles and their larvae may control aphid population.
- Treat the seeds with imidachloprid or thiamethoxam @ 5 g/kg seed or spray the same @ 0.5 ml/l.
- Apply Acetamiprid 20 SP @ 0.4g/l.
- Spray the crop with *neem* oil at 1.25%.
- Application of Triazophos 40 EC@ 2ml/l is also effective.

Flea Beetles (*Phyllotreta cruciferae, Phyllotreta birmania* and *Phyllotreta chotanica*)

Flea beetles are small shiny black beetles that feed on the leaves of various cruciferous plants. Flea beetles are prevalent mostly in the spring. Adult beetles that overwinter and feed on emerging seedlings can severely damage the seedlings. On larger leaves, the damage appears as small holes. Hot sunny weather favours the adult beetles and damage is most severe during such periods. The larvae cause damage on the root surface.

Control

- Plough the field deep just after harvesting to minimize the beetles.
- Apply malathion dust (5%) @ 10–15 kg/ha.
- Spray the crop with methyl demeton @ 0.025% when attack is more.

Cutworms (*Agrotis segetum* and *Agrostis ipsilon*)

Occasionally, black and dark sided cutworms may cause considerable damage to rutabaga and turnip. They feed on the crown, and leave deep scars, or burrow into the root. Damage may occur in spring as well as in late growing season. However, the late season incidence is difficult to detect, as it is not often noticed until harvesting time.

Control

- Apply malathion @ 0.1% alone or with wheat bran and molasses as bait.
- Apply methyl parathion 2% dust @ 5 kg a.i./ha 3–4 days after planting around each plant.
- Drench the soil around the plants with chlorpyrifos 20 EC @ 2.5 litres/ha.

DISEASES

Rhizoctonia Rot (*Rhizoctonia solani*)

The fungus causes damping-off and wire stem of other crucifers as well. On rutabaga, the fungus causes a superficial dry rot. Lesions may be sunken, shiny and brown with purplish rims that may develop into large irregular black craters with faint zonation and a scabby appearance. On turnips, the lesions are circular to elliptical and often zonate and vary from grey to brown in colour. Infection may occur in field or during storage. Field infections may be more severe where insecticides are not applied to control root maggot. The amount of decay increases rapidly as the temperature is raised above 4°C.

Control

- Follow long crop rotation with non-host crops.
- Grow resistant varieties, if available.
- Plough the field repeatedly in hot summer months to kill the pathogen.
- Soak the seeds in hot water dip at 52°C for 30 minutes, and then, dip the seeds in 0.01% streptocycline for 30 minutes.
- Provide good drainage facilities in rutabaga field.
- Store the roots at 0°C–2°C for good control.
- Drench the seedbeds with Brassicol @ 2.0–3.0 kg/hectare.

Downy Mildew (*Peronospora parasitica*)

The symptoms appear as a white fuzzy fungus in patches on underside of leaves. The patches turn yellow as they enlarge. Under humid conditions, a greyish white mouldy growth appears on underside of the leaves. Later, the leaf may become papery and die. The individual lesions may rapidly enlarge and then quickly collapse with a secondary soft rot during rains. The fungus overwinters on seed, cruciferous weeds and perhaps in soil. Young leaves are less susceptible in comparison of older ones. Infection occurs more readily on lower surface than on upper ones. The disease is prevalent in cooler regions of the world and may cause internal discoloration of roots. In seed crop, downy mildew on an inflorescence can lead to stag head formation, abnormal shape and development of the inflorescence due to downy mildew.

Control

- Adopt long crop rotation with non-susceptible crops.
- Use judicious quantity of seed per unit area to avoid overcrowding of seedlings.
- Destroy the crop residue after harvesting.
- Treat the seed with Captan 2.5–3.0 g/kg of seed or with hot water at 50°C for 30 minutes.
- Destruct the cruciferous weeds and follow other sanitary measures.
- Follow prophylactic sprays of Dithane M-45 0.2%, Dithane Z-78 0.2%, or Ridomil MZ at 0.25%.

POWDERY MILDEW (*Erysiphe cichoracearum*)

The symptoms appear as a white powdery superficial patchy fungal growth on upper surface of the leaves, which later spread to underside of the leaves and petioles. It usually develops first on crown leaves, which turn light green, then yellow and finally tan colour. Yellow spots may form on upper leaf surface opposite powdery mildew colonies. Ultimately, the leaves may die and fall off. In advanced stages, leaves are distorted, twisted and retarded in growth. Older plants are affected first. Disease development is favoured by low relative humidity and water stress.

Control

- Rotate the crop with non-host cereal crops.
- Eradicate cruciferous weeds surrounding the field.
- Spray the crop with Karathane 0.1%, Dinocap 0.1%, Bavistin 0.1%, or wettable sulfur 0.2% at an interval of 8–10 days.

CLUB ROOT (*Plasmodiophora brassicae*)

The fungus affects underground portion. The symptoms on aboveground portion are apparent only in advance stages. Large spindle-shaped galls form on roots and rootlets. The hypertrophy of tissue disrupts water supply, which results in pale green to yellowish foliage, flagging and wilting of plants. The fungus causing club root in rutabaga is usually present in areas where these crops have been grown for many years. A soil pH less than 7.2 favours the occurrence of disease. The disease is often less serious on heavy soils and on soils containing little organic matter. The incidence is especially more in temperate regions.

Control

- Adopt long crop rotation with non-host species.
- Grow club root resistant varieties like *York*.
- Ploughing in summer months reduces the risk of disease by reducing the amount of primary inoculum.
- Incorporate hydrated lime @ 2.5 t/ha into the soil six weeks before planting to raise pH to 7.2.
- Sterilize the seedbeds soil with Chloropicrin, Methyl bromide, or Vapam @ 0.05% 2 weeks before planting.

BLACKLEG (*Phoma lingam*)

Early signs of blackleg appear as small spots on leaves of young plants. On stems the spots are more linear and often surrounded by purplish borders. Stem lesions at the soil line usually extend to the root system causing dark cankers. Plants may wilt abruptly and die. Black rot lesions first appear at margins of leaves. The tissue turns yellow and the lesion progresses toward the centre of the leaf, usually in a V-shaped pattern with the base of the V toward the mid rib. The veins become dark and discoloration frequently extends to the main stem and proceeds upward and downward. Black discoloration of rutabaga roots may occur.

Control

- Practice a 4-year rotation.
- Destroy cruciferous weeds as they may harbour the black rot organism.
- Do not work in the field when plants are wet.

COMMON SCAB (*Streptomyces scabies*)

Scab is a well-known disease of potato and cause severe damage to rutabaga. Infection results in circular to oval lesions scattered over the surface of the root and lesions frequently coalesce to form a band around the root just beneath the soil line. The affected tissues may consist of a tan-coloured superficial or raised layer, or tissues may become pitted and dark following secondary decay. The disease usually develops in dry soils. Scab tends to be most severe where wood ash or manure containing wood fibre has been applied to the soil.

Control

- Follow a rotation of potato, grain and one year of hay or other green manure crop.
- Maintain adequate moisture during the period of rapid roots expansion.
- Incorporate hydrated lime @ 2.5 t/ha into the soil six weeks before planting to raise pH to 7.2.
- Avoid incorporation of wood ash or manure containing wood fibre into the soil.
- Spray the crop with Mancozeb @ 0.25 for its control.

■■■

48 Gherkin

M.K. Rana, P. Jansirani, L. Pugalendhi and N. Meenakshi

Botanical Name   : *Cucumis anguria* L.

Family   : Cucurbitaceae

Chromosome Number : 2n = 14

ORIGIN AND DISTRIBUTION

Tropical America is considered as its probable origin and its true wild form is not clearly known. Gherkin, a small cucumber has been used for pickling since the 1660s. A Dutch source says that 'gherkin' is from a word for immature and that the vegetable originated in northern India and came to Eastern Europe *via* the Byzantine Empire. The Dutch suffix is perhaps the diminutive 'kin', though some regard it as a plural affix, with the Dutch word mistaken for a singular in English. The 'h' was added in the 1800s to preserve the hard 'g' pronunciation. Although naturalized in many parts of the New World, gherkin is indigenous only to Africa, in the countries Angola, Botswana, the Democratic Republic of the Congo, Malawi, Mozambique, Namibia, South Africa (KwaZulu-Natal, Limpopo, Mpumalanga), Swaziland, Tanzania, Zambia and Zimbabwe.

INTRODUCTION

Gherkin, commonly known as bur cucumber, bur gherkin, cackrey, gooseberry gourd, maroon cucumber, West Indian gherkin and West Indian gourd, is a vine that is indigenous to Africa but has become naturalized in the New World, and is cultivated in many places. It is similar and related to the common cucumber (*Cucumis sativus*) and its cultivars are known as gherkins. Botanically, it is an annual climber growing to a height of two to four meters. The stem is tender and hardy. The leaves are alternate, palmate with serrated margins. Flowers are solitary, yellow, showy, monoecious (individual flowers are either male or female but both sexes can be found on the same plant) and pollinated by insects. Fruits are similar to cucumber with wart-like (less prominent) structures. Fruits are green to dark green with agreeable cucumber flavour without any bitterness. Fruit size ranging from 5 to 15 cm can be used. The fruits are generally consumed as canned or pickled vegetable. It can be used as salads or part of a savoury dish. Mostly, the fruit is exported as processed vegetable. Apart from fruits, the young leaves and seeds are also edible. Seed has oil with a musky flavour.

COMPOSITION AND USES

COMPOSITION

One large gherkin contains 35 milligrams of potassium. One large sweet gherkin has 114 micrograms of beta-carotene or 276 IU vitamin A. One large sweet gherkin pickle contributes 16.5 micrograms vitamin K. The nutritional composition of gherkin fruit is given in Table 48.1.

TABLE 48.1

Nutritional Composition of Gherkin Fruit (per 100 g Edible Portion)

Constituents	Contents	Constituents	Contents
Water (g)	97	Sodium (mg)	5
Carbohydrates (g)	2.2–3.6	Zinc (mg)	0.2
Protein (g)	0.6	Iron (mg)	0.3
Fat (g)	0.1	Vitamin A (IU)	45
Dietary fibre (g)	0.5	Thiamine (mg)	30
Calcium (mg)	14	Riboflavin (mg)	20
Phosphorus (mg)	24	Niacin (mg)	300
Magnesium (mg)	15	Vitamin C (mg)	12
Potassium (mg)	124	Energy (kcal)	12

Uses

Gherkin is primarily grown (as a crop plant) for its edible fruit, which are used in pickling, a cooked vegetables, or eaten raw. The flavour is similar to that of the common cucumber. Its fruits are popular in the northeast and north of Brazil, where they are an ingredient in the local version of meat and vegetable stew. A gherkin pickle is prepared using a specific type of cucumber and is made with vinegar, sugar and spices. Homemade pickle recipes are available but sweet gherkin pickle can be purchased from supermarkets as well.

Medicinal Uses

Sweet gherkin pickles undoubtedly have health benefits but adding to diet can help in increasing intake of several nutrients like vitamin A, folate, calcium and iron. However, it is important to eat them in moderate quantity since they are high in sodium, with 160 mg per large pickle. Gherkin contains a moderate amount of fat and calories controlling weight, which reduces risk of health concerns like heart disease and cancer. Potassium in gherkin helps in contraction of muscles and bones and is important for digestion as well. Its deficiency may cause an irregular heartbeat, muscle problems and fatigue and stomach issues. Vitamin A in gherkin is a nutrient that supports healthy immunity by producing white blood cells, which in turn protect from viruses and bacteria that cause illness. It is also important for reproduction, vision and bone growth. Beta-carotene is a nutrient that converts to vitamin A in the body, so including sweet gherkins is a good way to increase the health benefits of vitamin A. Vitamin K in gherkin is important since it helps in blood clotting. Its intake can treat bleeding disorders such as liver disease and malabsorption. Adequate vitamin K intake is also important for bone health. Its higher intake increases bone density, which reduces the risk of osteoporosis. Gherkin has been used in folk medicine to treat ailments of the stomach. The high water content makes the gherkin a diuretic, due to which, it has a cleansing action within the body by removing accumulated pockets of old waste material and chemical toxins. Gherkin helps in eliminating uric acid, which is beneficial for those who have arthritis, and its fibre-rich skin and high levels of potassium and magnesium helps in regulating blood pressure and help promote nutrient functions. The magnesium content in gherkin also relaxes nerves and muscles and circulates the blood smoothly.

CLIMATIC REQUIREMENT

Gherkin is a warm season crop, which can be grown throughout the year in the regions where temperature fluctuations are very less. The temperature between 20°C and 35°C is ideal for gherkin cultivation. The temperature <15°C will slow down the plant metabolic activities, therefore, affecting the plant growth and development.

SOIL REQUIREMENT

Generally, gherkin can be grown on all types of soils. However, well-drained red sandy and sandy loam soils are ideal for this crop. Black and heavy soils are not recommended due to water stagnation leading to lack of root aeration and subsequent susceptibility of plants to fungal diseases. Special care has to be taken with respect to water management and intercultural operations when gherkin is grown on black soils. The pH range of the soils for gherkin cultivation is preferred to be between 6.5 and 7.0.

VARIETIES

Ajax and calypso are the popular gherkin varieties cultivated by the growers. Seeds of imported varieties are generally supplied by the multinational corporations based on contractual seed cultivation.

SEED RATE

A seed rate of 800 g is enough for the sowing of a hectare area.

FIELD PREPARATION

When a field is selected for gherkin cultivation, there should not be any cucurbitaceous and solanaceous vegetables grown as previous crop. Two to three deep ploughing followed by harrowing is given to get good tilth. Weeds and the previous crop stubbles are to be removed for better field sanitation. Big clods are broken to get uniform fine tilth of soil and better germination. Field levelling is also important for uniform supply of water and nutrients to all the plants. Long beds of 5 m length and 1 m width are prepared. Main irrigation channels of 15 cm width and 15 cm depth are made in between two beds. Sub-irrigation channels of 7 to 10 cm width and 7–10 cm depth are provided at every 5 m length distance to ensure proper supply of water to entire length of beds. Ridge and furrow system of cultivation is also followed by some growers at a spacing of 1×0.3 m. Little wider spacing of 20 cm between rows is also practiced during rainy season to ensure proper drainage.

SOWING METHOD

Gherkin is propagated by seeds, which are sown on both sides of beds, on one side of the ridges at three-fourth height of the slope of the same side where farmyard manure and basal fertilizers are applied. Sowing with only one seed per hill is sufficient at a depth of 2.5 cm. Soon after sowing, irrigation has to be given. It takes 4 to 7 days for seed germination. Gap filling has to be carried out at the places where germination has failed within 5 days after sowing to maintain required population of 24,000 to 25,000 plants per hectare and uniformity in growth. Before sowing, the seeds are treated with *Trichoderma viridi* 4 g or *Pseudomonas* 10 g or carbendazim 2 g per kg of seed as seed treatment. At the time of sowing, application of Furadan 2 g per hill is recommended to protect against sucking pests initially for a period of one month.

NUTRITIONAL REQUIREMENT

Being a short duration crop with high yield potential, gherkin demands large quantities of fertilizers. Further, it responds well to small quantities of fertilizers at frequent intervals. Timely application of recommended fertilizers with correct dosage will ensure maximum utilization of essential nutrients by plants resulting in better yield and quality of the produce.

Well rotten farmyard manure of 25 tonnes per hectare is applied at the time of last ploughing. Nitrogen 100 kg, phosphorus 75 kg and potassium 150 kg have to be applied. Full dose of phosphorus and potassium and one-third of nitrogen are applied at the time of sowing and

remaining two-thirds dose of nitrogen is applied in two equal splits 3 and 5 weeks after sowing. Micronutrient mixture of zinc, boron, manganese and magnesium is also applied in small
quantities to ensure good and quality produce. Immediately after application of fertilizers,
irrigation has to be given. Fertilizers should not be applied near the root zone. All the recommended fertilizers should be applied in a pit in between two gherkin plants. It is advised to give
top dressing of fertilizers by side placement on the small furrows opened at 10 cm distance
from the plants and cover it properly to avoid wastage. Gherkin is also grown under fertigation.
The recommended dose of fertilizers is applied every third day after sowing starting from the
fifth day.

NUTRITIONAL DEFICIENCIES

Gherkin responds well to micronutrients application so that the quality of the fruits can be maintained.

Potassium	The young leaves become pale white with spots on the leaf edges.
Manganese	Irregular interveinal yellowing of growing leaves will be seen. The stem become thin and the internodes become narrow.
Nitrogen	The lower leaves become yellow and then dry at later stages. The flower tip will bend and the fruit become small and narrow.
Molybdenum	The lower leaves become small with interveinal yellow spots and then completely turn yellow from the tip towards the stalk end.
Calcium	Interveinal yellowing of emerging leaves will be noticed and the affected older leaves curl downwards, while the young leaves curl upwards. It affects root growth and aborts flower development and stunted petiole growth.
Iron	The leaves become chlorotic except the veins. In severity, the veins will also become yellow and the leaf tips turn to pale colour. The fruits become small and pale in colour.

IRRIGATION REQUIREMENT

Gherkin is a warm season crop, which requires irrigation at regular intervals. Irrigation management is essential for the success of the gherkin cultivation. Frequent and light irrigations are
ideal for the crop growth and fruit development. Water stress during fruiting and harvesting
period would result in flower and fruit drop besides of crooked fruits. Excess irrigation affects
the root respiration resulting in stunted growth and makes the plant more susceptible to fungal
diseases.

In general, irrigation is recommended depending on the texture of the soil. For red sandy soils,
irrigation on alternate days and once in two to three days in the case of black soils depending on
the texture of the soil. Care should be taken to keep the water only in channels and not allowing to
over flow to the bunds.

INTERCULTURAL OPERATIONS

HOEING AND WEEDING

Frequent weeding has to be done as and when required to keep the field clean as weeds compete
with gherkin crop for water and nutrients. Weeds may also give shelter to some of the insect-
pests and diseases as an alternate host that affects the crop growth. When plants reach four leaf
stage, the lower two cotyledon leaves are removed and destroyed at far away places by burning

or burying them into the ground. This helps in reducing the leaf minor damage to a great extent since these cotyledon leaves serve as an initial source for the multiplication of the insect-pests and diseases. Weeding is done whenever it is necessary. First weeding is done at 10–20 days after sowing. Generally, it coincides with first top-dressing. After weeding, the fertilizers are applied and followed by irrigation. Subsequent weeding is done depending on the presence of weed population.

STAKING

Staking is highly essential for gherkin cultivation. The growing vines of gherkin are allowed to trail on a support. Two strong wooden casuarina poles of 2.0–2.5 m height and 5 cm diameter are fixed at 30 cm depth on both the edges of the lengthy beds. These two poles are inter-connected with GI wires at both 15 cm top and bottom. Further, these two poles are again supported by tying with GI wires of 1 to 2 inches thickness at angled position, which is tied to a wooden peg, plugged into soil. The main poles are fixed about 30 cm deep into the soil. In general, the spacing between two poles should not be less than 120 cm so as to hold the plants when full of fruits without any drooping nature. The individual plants and branches are tied to the GI wire using single line jute thread. Each branch has to be tied separately to the GI wire. After reaching the top of the pole, the growing stem may again be brought to the ground level so that the fruits that appear on the stem are kept in good position and not to be allowed to trail on the ground.

POLLINATION

Gherkin is a naturally cross-pollinated crop. Pollination is entomophilous and honeybees are the pollinating agents. Honeybee hives may also be placed in gherkin fields to encourage good pollination and fruit set.

HARVESTING

Good quality fruits at specific size are harvested based on the market requirements. Harvesting of tender gherkin fruits can be made from 30 days after sowing. It has to be harvested every alternate day. Fruits of 5 cm length to various size and grades are harvested. Generally, the higher grades like 160 + or 100 + worked have more quantity of export quality producer. Fruits of 5 cm length are preferred as export quality. However, fruits of 15 cm length is also harvested and presented to various kinds of products.

No. of fruits per kg	Percentage of contribution of various grades per kg	Approximate Cost per kg (Rs.)
160	30	14.00
100	25	8.50
60	25	6.00
30	15	4.00
30	5	1.00

While harvesting, care should be taken to harvest the fruits without the remnants of floral parts attached to the fruits. There should not be any physical damage to the fruits.

They are harvested using a specific technique of twisting the fruits on one side to get stalk-free fruits. Bad-quality fruits like insect damaged, rotten, virus infected, smooth skinned and crooked fruits are to be separated while harvesting. These separated fruits should not be thrown in the field, which may act as a breeding source for insects for further growth and multiplication of insects leading to great damage to the crop. They have to be dumped either into the compost pits or in small pits dug away in wastelands and covered properly with soil.

In field, the fruits are harvested in early morning hours to maintain freshness until it reaches the market or processing industry. Quality of gherkin fruits can never be enhanced after harvesting but can only be maintained, therefore, it is very important to maintain the quality at farm level until it reaches the industry. Wooden baskets and plastic crates are used for harvesting and the fruits are spread under the shade in a single thin layer on clean surface to maintain freshness of the fruits.

TEMPORARY SHED ARRANGEMENTS

Tender gherkin fruits of various sizes ranging from 5 to 10–15 cm length are harvested from the field. The harvested fruits are to be protected from the hot sun. A temporary shed of rectangular shape with shade net as covering material of all the sides including roof has to be made in a shade area of the field. One side of the shed has to be kept open. Fertilizer bag, chemical bag, or polythene material should not be used to erect the shed. Further, these shed should not be used to store or keep any chemical, fertilizer, or equipment like sprayers, etc.

INSECT-PESTS AND DISEASES

Gherkin is affected by a number of insect-pests and diseases that account to 50% decline in yield and quality, if timely control measures are not taken. The integrated pests and diseases management has to be followed to get the effective control and to reduce the input cost on chemical sprays. Many cultural, mechanical and biological control measures may be either single measure or a combination of more than one measure may be followed to control them.

- **Cultural methods:** It is advisable to keep the field clean and weed free. Planting of wind breaks like casuarina or gliricidia may be advocated. The basal two cotyledonary leaves and a basal pair of leaves are to be removed after expansion of three or four pairs of true leaves, which act as a potential source for the spread of disease or pest attack.

- **Mechanical methods:** Hand picking and physical destroying of some insect-pests like fruit borer larvae have to be done. Provision of pheromone traps is also to be encouraged. The weeds that serve as alternate host plants have to be removed periodically.

- **Biological methods:** From sowing to harvest, biocontrol agents may be sprayed or treated to control few insect-pests and diseases of gherkin. Some insects that act as natural enemies may be protected by avoiding chemical sprays against these insects.

- **Chemical methods:** Application of pesticides at appropriate stage on gherkin to be grown for 100% export helps in effective control of insect-pests and diseases with minimum risk and cost. Only approved chemicals with minimum waiting periods are to be used.

INSECT-PESTS

FRUIT BORER (*Diafia indica*)

Larval stage of this insect affects the gherkin crop throughout the year of cultivation. The first instar larvae affect the growing tip and leaves. It grinds the green leafy portion leaving the midribs to stick together so as to develop a net-like appearance. This insect bores into the developing fruits and adjacent leaves. Affected fruits will have borehole at the bottom, which declines the yield. The incidence of this pest is the highest during rainy season.

Control

- To control this serious pest, integrated control measures have to be followed.
- There should not be any cucurbitaceous vegetable grown as previous crop or neighbouring crop during the season of gherkin cultivation.
- The affected leaves, leaf tips, bud, or fruits are to be identified and removed manually.

Fruit Fly (*Bacteria cucurbitae*)

Incidence of the fruit fly attack can be noticed from March to October at various stages. It is a polyphagous insect. Both, the larvae and adult fly affect the crop. The female insect lays eggs on different parts of plant such as leaves, fruits, etc. The larvae that hatched out are feeding on developing fruits. It thrives even on dead organic matter. The affected fruits may bend, become misshaped, lead to decay and emit an obnoxious smell.

Control

- The affected fruits are to be collected and destroyed outside the field of cultivation
- Spray any systemic insecticide at regular intervals.
- Trap crop (maize) grown as windbreak also acts as barrier for the entry of adult fly into the field.
- Placing of 6 to 8 pheromone traps in gherkin field helps in attracting male flies thereby reducing the pest load. A specific fruit fly trap can be prepared using ripe mango (50 g) + grind gherkin fruits (50 g) + jaggery (10 g) + ammonium acetate (1 g) + Dichlorvos insecticide (1 ml) and placed at 6–8 places in the field to attract and destroy both male and female flies. Apart from these measures, collecting and destroying affected fruits also help in reducing the overall pest's population.

Serpentine Leaf Miner (*Liriomyza trifolii* burgess)

The adult insect is 1–2 mm long with reddish yellow colour. It lays eggs of 0.5 mm length by making tunnels between the leaf tissues thereby reducing the photosynthetic area the leaves. When the infection becomes severe, premature drop of leaves will occur.

Control

- Leaf miner can be managed by growing castor as trap crop around gherkin field.
- Removal of cotylendonary leaves at 5 leaf stage of the crop and affected lower leaves during later stages and destroying them has to be practiced.
- Spraying of any systemic insecticide is recommended as control measure for the leaf miner. While spraying any chemical, sticking agents like rice gruel, maida paste, etc. may be added for effective control of this pest. The infected leaves are to be collected and burnt outside the field.

Aphids (*Aphis* sp.)

Aphids' incidence in the field is seen on gherkin throughout the year. However, incidence will be severe during summer months. Cloudy days with high humidity and cooler months develop more population. Both, young and adult insects suck the sap from leaf, stem, flower and fruit. The affected leaves turn backward, reduce in size and turn yellow in colour at latter stage. Aphids secrete honey-like liquid, which spreads on leaves and affects the photosynthesis. Aphids spread two major diseases, *i.e.*, cucumber mosaic virus and zucchini yellow mosaic virus disease in gherkin. It is important to control this insect at early stages of attack to control the spread of the viral diseases.

Control

- Spraying of malathion, Metasystox, or Dimethoate at 0.2% concentration is recommended.

THRIPS (*Thrips tabaci* lindeman)

Thrips attack the flowers and terminal buds. It results in stunting of plant growth and transmits the serious virus disease called Tospovirus. The attack of the insect is more in early crop period.

Control

- Spraying of systemic insecticide can check the population of this pest.
- Solanaceous crops should not have been grown as previous crop in gherkin.
- Growing castor or maize as trap crop well in advance of gherkin sowing is recommended.

MITES (*Tetranychus urticae*)

Mites remain on the lower sides of the leaves and suck cell sap. The leaves become rough and yellow specks on the leaf are formed. The leaves gradually turn pale and then dry. Mites also affect the flower by sucking cell sap.

Control

- The affected leaves are to be collected and destroyed.
- Spraying of any systemic insecticide may be recommended to control the spread of some of the viral disease like Tospovirus.

NEMATODES (*Meloidogyne incognita*)

Gherkin is affected by root knot nematodes. The infection is severe during summer months affecting the roots by forming galls. The affected plant leaves become yellow, stunted in growth and later the plants dry.

Control

- The affected plants have to be removed and destroyed.
- Crop rotation has to be followed to avoid cucurbitaceous and solanceous vegetables for cultivation.
- Trap crops like tagetes, sweet potato and coriander can be grown to reduce the nematode population.
- Application of more neem cake as basal dose or nematicide may be recommended.

DISEASES

Like other cucurbitaceous vegetable, gherkin is affected by downy mildew, powdery mildew, fusarium wilt, Tospovirus, CMV and ZYMV.

DOWNY MILDEW (*Pseudoperonospora cubensis*)

This disease affects most of the cucurbitaceous vegetable. The disease prevails in areas of high humidity especially when summer rains occur regularly. It is characterized by formation of yellow angular spots on upper surface of the leaves. White purplish spores appear on upper surface of the leaves. It spreads rapidly, killing the plant quickly through rapid defoliation.

Control

- Spray the crop with mancozeb 0.2%, Difolatan 0.2%, or copper oxychloride 0.2% once in 10-day intervals to control the spread of disease.

- Avoid irrigation late in the evening and night hours.
- The infected leaves should be removed periodically and destroyed.
- Remove all the lower leaves up to 22 cm height from the ground level 30 days after sowing so as to encourage more aeration.
- Adequate drainage should be provided.
- Repeated cultivation of any cucurbitaceous vegetable has to be avoided.

Powdery Mildew (*Erisipihae ciccohrocearum* sp. *shaerothecae fuliginea*)

The disease appears during cooler months of rain-free regions and seasons. It affects all the cucurbitaceous vegetables. Symptoms of the disease appear on lower surface of the leaves. Severely affected leaves become brown, shrivelled, and finally, defoliation occurs. Affected flowers will not open at all. Fruits of the affected plants do not develop fully and remain small.

Control

- Spray mancozeb (0.2%), Difolatan (0.2%) or copper oxychloride (0.2%) once in 10 days.
- Application of more nitrogenous fertilization has to be avoided.
- Correct spacing and population has to be followed and maintained.
- Spraying of carbendazim (0.1%) or Karathane (0.2%) may be recommended as an effective control measure of this disease.
- Seed treatment and soil drenching with systemic fungicide can also give protection to young seedlings.

Fusarium Wilt (*Fusarium oxysporum* f. sp. *cucurmerinum, Fusarium oxysporum* (f) synedor)

It is a common disease affecting many cucurbits. In young seedlings, cotyledons droop and wither. In older plants, the leaves wilt suddenly and vascular bundles in the collar regions will become yellow or brown.

Control

- Chemical control of soil-borne root diseases is very difficult.
- Soil has to be drenched with captan or thiram @ 0.2% to check the spread of the disease.
- Use of disease-free seed and cultivation of resistant varieties is the best way to control the disease.
- Seed treatment with systemic fungicide like carbendazim before sowing has to be followed.
- Hot water treatment at 55°C for 15 minutes also helps in eliminating seed-borne infection.

Tospovirus

It is commonly known as tomato spotted wilt virus disease transmitted by thrips. The first instar larvae are highly effective in transmitting the disease form affected to healthy plants within 10 minutes of sap sucking. The stems of affected plants become pale yellow and the edges turn upward. The growing tips become dry and turn backwards. The infected fruits will have green circular spots with erupted centre. There will not be any side branches and the plants become stunted.

Control

- The infected leaves and plants are to be removed periodically.
- The disease spreading vector, *i.e.*, white fly, population has to be controlled by spraying systemic insecticide like imidacloprid @ 0.2% at regular intervals.
- Grow trap crops or use insect-proof net.

Cucumber Mosaic Virus (CMV)

The leaves of the virus disease affected plants become irregular in shape with yellow spots and reduced in size. The leaves become chlorotic. The leaf edges turn downwards. The internodes become narrow and stunted in appearance. In severity, there is no production of side branches, flowers and fruits. If fruits appear, they will be soft, ashy white in colour with irregular colour and size. The affected fruits become watery and unfit for consumption.

Control

- There should not be any cucurbitaceous vegetable grown in the nearby field.
- Spray systemic insecticide to control plants in which of cucurbitaceous whitefly that transmits diseases from plant to plant of malvaceous and solanceous crops.

Zucchini Yellow Vein Virus

This virus-affected plant's leaves become yellow, showing irregular blister-like growths, which leads to stunted growth. The shape of the developing fruits become half round or bent and unfit for harvest.

Control

- Growing of alternate host plants of cucurbitaceous, solanceous and leguminous vegetables are to be avoided.
- The disease-transmitting aphid population has to be checked by spraying any systemic insecticide.
- The affected plants have to be uprooted and destroyed.

■■■

49 Cassabanana

M.K. Rana and Archana Brar

Botanical Name : *Sciana odorifera* (Vell.) Naudin

Cucurbita odorifera Vell. and *Cucurbita evodicarpa* Hassk

Family : Cucurbitaceae

Sub-family : Cucurbitoideae

Chromosome Number : 2n = 24

ORIGIN AND DISTRIBUTION

Cassabanana is believed to be the native to Brazil, but now, it has spread throughout the tropical America. Historical evidences show that it was cultivated in Ecuador in pre-Hispanic times. First time in 1658, it was mentioned by European writers that it has been cultivated in Peru as a popular vegetable. In the Central America, it is grown near sea level but its fruits are carried to markets even up to the highlands. In Cuba, Puerto Rico and Mexico, it is grown mostly for the usefulness of its fruit, but in Brazil and Venezuela, it is mainly grown for its ornamental purpose.

INTRODUCTION

Cassabanana (*Sicana odorifera* Naud. and synonym *Cucurbita odorifera* Vell.), an interesting member of Cucurbitaceae family, is also known as *sikana* or musk cucumber. It is also known as *melocotonero, calabaza melón, pérsico,* or *alberchigo* in Mexico, *melocotón* in El Salvador and Guatemala, *calabaza de chila* in Costa Rica, *cojombro* in Nicaragua, *chila* in Panama, *pavi* in Bolivia, *padea, olerero, secana,* or *upe* in Peru, *calabaza de Paraguay, curuba,* or *pepino meloco-ton* in Colombia, *cura, coróa, curua, curuba, cruatina, melão caboclo,* or *melão macã* in Brazil, *cajú cajuba, cajua, cagua,* or *calabaza de Guinea* in Venezuela, *pepino, pepino angolo,* or *pepino socato* in Puerto Rico and *cohombro* in Cuba. It is not at all a banana though it is cucurbitaceous edible fruit, looking a lot like a cucumber. It is still quite tropical but nearly unheard in India to date. Cassabanana is native to the New World tropics and cultivated as an ornamental plant and for its sweet-smelling fruit.

COMPOSITION AND USES

COMPOSITION

The yellow and orange-fleshed fruits of cassabanana are rich in beta-carotene, minerals, niacin and ascorbic acid. The mature green fruit contains 92.7% moisture, whereas, the ripe fruit contains 85.1% moisture. As compared to the decrease in moisture content of fruit during ripening, there is a marked increase in the protein content, which is from 0.093 to 0.145 g per 100 g of the edible portion. As the fruit matures, the fat content decreases from 0.2 to 0.02 g. The nutritive value of cassabanana fruit is given in Table 49.1.

TABLE 49.1

Nutritional Composition of Peeled Green and Ripe Fruit (per 100 g Edible Portion)

Constituents	Peeled green fruit (including seeds)	Ripe fruit (without peel, seeds, or soft central pulp)
Moisture (g)	92.7	85.1
Protein (g)	0.093	0.145
Fat (g)	0.21	0.02
Dietary fibre (g)	0.6	1.1
Ash (g)	0.38	0.70
Calcium (mg)	8.2	21.1
Phosphorus (mg)	24.2	24.5
Iron (mg)	0.87	0.33
Carotene (mg)	0.003	0.11
Thiamine (mg)	0.038	0.058
Riboflavin (mg)	–	0.035
Niacin (mg)	0.647	0.767
Ascorbic Acid (mg)	10.0	13.9

FOOD USES

The flesh of ripe fruits is considered as cooling and refreshing when eaten raw, especially in hot summer months. The fruit, which is said to have a sweet tropical flavour, is either cooked prematurely like squash or allowed to ripen and used fresh, in drinks, or pies. However, it is mainly used for making jam or other preserves. The immature fruits are cooked as a vegetable and also used in soup and stews.

MEDICINAL USES

In Puerto Rico, the flesh of its ripe fruits is crushed and steeped in water, with added sugar, overnight at room temperature to allow it to ferment slightly. The fermented product is sipped along with its flesh to get relief from sore throat. At the same time, it is also believed that wearing a necklace of its seeds around the neck is also very beneficial for a sore throat.

The leaves of cassabanana are employed in treating uterine haemorrhages and venereal diseases. In Brazil, the infusion prepared from its seeds is also taken as a febrifuge, vermifuge, purgative and emmenagogue. In Yucatan, a decoction prepared from its leaves and flowers (2 g in 180 ml water) is prescribed as a laxative, emmenagogue and vermifuge. The seeds and flowers of cassabanana yield a certain amount of hydrocyanic acid, thus, it is advisable not to make a strong decoction from seeds and flowers.

OTHER USES

People believe that keeping its ripe fruits around the house, especially in muslin clothes repels moths especially in rainy season because of its long-lasting strong fragrance. Placing its fruits on church altars in holy weeks is considered promising by the people of Christian community. In fact, many people prefer to use its long-lasting ripe fruits as an air freshener in kitchens and lavatories due to its long-lasting agreeable and pleasant fragrance.

CASSABANANA PUREE

INGREDIENTS

- 3 cups of cooked cassabanana pulp
- 1 tablespoon unsalted butter
- 1 tablespoon of clarified butter

- Pepper one finger-seeded woman chopped into squares
- One tablespoon of finely chopped parsley
- Half cup milk
- A pinch of salt

Recipe: In a saucepan, pulp of cassabanana is taken, stirred with raw milk and beaten well with the mixer. Before adding milk, the pulp is passed through the sieve to make the pulp homogeneous, and thereafter, it is returned to heat and stirred constantly. Clarified butter is added into it after turning off the heat and stirred until it is melted completely. Then, salt, pepper and parsley are added according to taste and mixed properly. After preparation, the puree is put in a bowl and decorated with butter and finely chopped green coriander leaves.

BOTANY

Cassabanana is a fast-growing perennial herbaceous plant, requiring trellises for climbing. The vines prefer to grow in partial to full sun. Its vines climb trees of 15 m or more with the help of tendrils equipped with adhesive discs, which can adhere tightly to the smoothest surface. The young stems are covered with some hair-like structures. The leaves, which are three lobed with serrated margins deeply indented at the base, are grey in colour, rounded kidney-shaped and 30 cm wide. The leaves petioles are 4–12 cm long. Its five-lobed and urn-shaped flowers borne solitary in the axil of leaf are either white or yellow in colour. The male flowers are 2 cm long, whereas, the female flowers are about 5 cm long. The fruits of cassabanana, which are renowned and famous for its melon-like sweet and lovely aromatic strong odour, are long and cylindrical and seem just an overgrown cucumber with a very tough skin.

The striking fruit is 30–60 cm long, 7–11.25 cm thick, maroon, orange-red and dark-purple with violet tinges or entirely jet-black in colour and nearly cylindrical, ellipsoid, or sometimes slightly curved in shape. On ripening, they become glossy, smooth with firm, orange-yellow or yellow, tough, cantaloupe-like with 2 cm thick juicy flesh. The fruit pulp with fleshy core present in the central cavity is soft. The cavity is filled with seeds, which are oval, 16 mm long, 6 mm wide and light brown bordered with a dark brown stripe. In cavity, the seeds are present in tightly packed rows extending the entire length of the fruit.

CLIMATIC REQUIREMENT

Cassabanana is essentially a warm season crop mainly grown in tropical and subtropical regions. Hence, its seeds are sown after the threat of frost has passed in the spring. The optimum temperature to harvest a good yield is 27–30°C, and for proper germination, the optimum daytime temperature should be above 25°C. Since it thrives in hot summer, it does not perform much till the summer arrives. Afterwards, it needs a lengthy ripening time before the occurrence of frost. Thus, this plant is nominated to grow indoors for early starting. At low temperature, it grows slowly to start until the heat kicks it. The plant does not produce many female flowers but all of them set fruit. It grows like a melon and needs ample water and full sunshine with warm to hot temperature for the fruit to ripen. It can be grown as an annual in most of the climates during warm months.

SOIL REQUIREMENT

The soil at the planting site is amended with compost. A well-drained sandy loam or loam soil rich in organic matter is best for its cultivation. Lighter soils, which warm up quickly in spring season, are usually used for early crop. Cassabanana vines prefer to grow in soil having a pH as close to 7.0 as possible, thus, amending soil is necessary by adding lime if the pH is less than 7.0 and by adding gypsum if the pH is greater than 7.0. The required amendments are mixed well into the soil. The field should have adequate drainage facility. For direct sown crop, the field is prepared to a fine tilth by repeated ploughing and planking.

SOWING TIME

The best time for its sowing is February last week to first week of March when the danger of frost is over. In north India, early sowing may also be done by raising the seedlings in protrays and transplanting them in the field as and when the temperature becomes favourable.

SOWING METHOD

Cassabanana is very hard to find any information on its cultivation. It is generally propagated by seeds, which are not too difficult to come by since several tropical-oriented companies offer seed for sale but cuttings can also be used for the propagation of this crop. The seeds are directly sown in the field at a depth of 4–5 cm and about 30 cm apart.

Before sowing seeds in the field or protrays, they are soaked in lukewarm water for a period of 16–24 hours for fastening the germination process. For the completion of germination anywhere, the seeds take 4 to 6 weeks. For filling of the plastic trays, a mixture is prepared by mixing sand or perlite and compost in the ratio of 1:1. After filling the trays with this mixture, the seeds are gently sown and watered with a rose can, and thereafter, the seed trays are covered with plastic sheet to maintain temperature and humidity. If a plastic sheet is not available then the trays may be covered with some plastic wrap secured with a rubber band. When the seedlings attain 5 cm height, they are transplanted to a well prepared field having full sunlight.

NUTRITIONAL REQUIREMENT

Cassabanana responds well to the application of compost and fertilizer in the early spring, hence, for getting economic yield from cassabanana crop, about 20–25 tonnes of farmyard manure or compost should be mixed thoroughly into the soil at the time of land preparation and a mixture of nitrogen 50 kg, phosphorus 25 and potassium 25 kg per hectare should be applied at the time of sowing or transplanting.

IRRIGATION REQUIREMENT

At the time of sowing, the soil should have plenty of moisture since the cassabanana seeds need high humidity for better germination. If the soil is not sufficiently moist, water the soil immediately after sowing seeds. The soil is kept moist by sprinkling water regularly with a spray bottle until the germination process is completed. Over-watering the vines should be avoided since excessive moisture in soil promotes root rot.

INTERCULTURAL OPERATIONS

HOEING AND WEEDING

As and when the plants are large enough to handle, the excessive plants should be removed from the field to give proper spacing, leaving only two healthy seedlings per hill. Weeds during early stages of crop growth pose problem as they compete with crop for moisture, nutrients, space and sunlight and provide shelter to insect-pests and diseases. Hence, to keep the weeds under check, two or three shallow hoeing should be done before the vines cover the ground. However, deep hoeing should be avoided as it may damage the fibrous roots near the soil surface.

MULCHING

The soil around the vines should be mulched with bark chips or straw mulch, which improves drainage and reduces weed growth. Mulching also provides protection to the roots of cassabanana vines against summer heat and winter chill. Mulching with paddy straw, wheat straw, or sugarcane tresses may improve the number of fruits per plant.

STAKING AND TRAINING

The vine of cassabanana may be too long and cumbersome for very small gardens. It needs a lot of sprawl space or a tall trellis. Training has an additional advantage since the vine of cassabanana loves to climb, hence, providing support to the vine is necessary. A strong trellis is placed behind each seedling for climbing the vine. Training cassabanana vines over trellis is of great benefit since per unit area more number of plants can be adjusted and it allows maximization of space by controlling vine growth and prevents the fruits from rot. If vines are allowed to creep on the ground, yield is reduced 30% since the vine grows up and down on the trellises, resulting in a jumbled mess, which at least keep the fruit off the ground. Training also allows the production of fruits with more uniform shape and colour and allows the fruit to grow straight and longer. The vines training over bower system is considered the best for this crop. For the preparation of bower, bamboos or cemented, wooden, or angle iron poles are used. In early stage, bamboo sticks can be used for proper support. Training in cassabanana production is very useful since training ensures quality fruit of uniform size, rotting of fruit is very much reduced, harvesting becomes much easier, intercultural operations and plant protection measures can be easily applied, vitality of the vine is maintained for a longer period and the marketable yield per vine increases.

Brazilians train the vine to grow over arbours or they may plant it close to a tree. However, if it is allowed to climb too high up the tree there is the risk that it may smother and kill it.

HARVESTING

At maturity, it begins changing its colour from green to yellow and then from yellow to purplish or black. Flesh is orange-yellow like cantaloupe. However, rain at the time of maturity causes the fruit to split open and bugs quickly find their way inside and ruin the fruit as and when they become edible. The ripening period for this fruit is too long.

The harvesting of cassabanana fruits depends on the purpose. Generally, the fruits are harvested as and when they are ready for use. Cassabanana fruits can be kept in good condition for several months if they are stored in dry place and away from sunlight. Its fruits have very high market value, and in the market, they are sold by the weight.

INSECT-PESTS AND DISEASES

The insect-pests and diseases attacking the cassabanana crop and their control measures are similar to other cucurbitaceous crops.

INSECT PESTS

MELON APHID (*Aphis gossypii*)

The nymphs and adults of melon aphids usually attack in colony on soft growing tips but may also feed on underside of leaves. They suck the sap from the leaves and tips, causing leaf and tip curl, distortion and premature death of the leaf or growing tip. Melon aphids emit a sticky substance called honeydew on leaves, which encourages sooty mould growth, reducing the photosynthetic rate of the plant. In severe cases, the crop may be destroyed. These are the vectors of viruses.

Control

- Plant the crop in aluminium-foil-covered beds.
- Introduce beneficial insects to keep the aphid population under control.
- Spray simple water with strong jet pressure to wash the aphids.
- Fix yellow pans to trap the aphids.
- Spray the crop with bifenthrin 0.1%, malathion 0.1%, cyhalothrin, cyfluthrin, or Metasystox 0.1–0.2%, or Rogor 0.1–0.2%.

MELON THRIPS (*Thrips palmi*)

Thrips are minor and frequent pests of cassabanana in temperate regions but major and regular pests in tropical and subtropical regions. Eggs are laid into actively growing leaf tissue, developing flower buds and fruits. They can be seen on all the aboveground parts of a plant, flowers and on back of the leaves. Its damage appears as silvering and flecking on seedlings leaves. Severely infested plants are characterized by a silvering or bronzing of leaves, stunted leaves and terminal shoots and deformed fruits.

Control

- Use predatory mites to control the thrips.
- Apply irrigation through drip or furrow method.
- Use reflective mulches such as silver or grey-coloured synthetic much.
- Cover the plant rows with polytunnels.
- Spray the crop with malathion 0.1%, bifenthrin 0.1%, or cyfluthrin 0.1%.

CUCUMBER BEETLES (*Diabrotica undecimpunctata*)

The spotted, striped and banded cucumber beetles are most destructive to young cassabanana seedlings. They start feeding on plants as soon as they emerge from the soil and feed on all plant parts including flowers and fruits. Its larvae feed on roots and bore into both roots and stems of cassabanana plants. They either slow the plant growth or kill the plants and remain present throughout the growing season. They also transmit bacterial wilt. The bacterium overwinters in the intestines of beetles, and it is spread from plant to plant as the beetles feed. The infected plants eventually wilt and die.

Control

- Spray the crop with 0.4% *neem* oil.
- Spray insecticides like bifenthrin 0.2%, cyhalothrin 0.1%, or cyfluthrin 0.1%.

SQUASH BEETLES (*Epilachna borealis*)

The squash beetle is one of two species of Coccinellidae that eat plant material rather than other insects. The beetle feeds upon the leaves, while the other species, the Mexican bean beetle (*Epilachna varivestis*), a close relative of the squash beetle, is a serious bean pest. The adult of the squash beetle overwinters in crop debris. All other lady beetles are beneficial because they feed on insect pests such as aphids and scale insects.

Control

- Destroy the crop residues after harvesting.
- Spray of 0.4% *neem* oil, bifenthrin, cyhalothrin, cyfluthrin, or malathion 0.1%.

TWO-SPOTTED SPIDER MITES (*Tetranychus urticae*)

Sometines, this insect can be a serious problem in cassabanana, especially during hot and dry weather. These tiny mites suck sap from individual cell of the leaves. The damage appears as pale yellow and reddish-brown spots ranging in size from small specks to large areas on the upper sides of leaves. Their attack is very swiftly and they kill the plant or seriously stunt its growth. Spider mites because of their tiny size are difficult to notice until the vines are completely damaged.

Control

- Spray simple water with strong jet pressure.
- Introduce predatory mites and beneficial insects such as lady beetles and minute pirate bugs.
- Spray the crop with 0.4% *neem* oil, malathion 0.1%, bifenthrin 0.05%, diazinon 0.03%, or Labaycid 0.05%.

SQUASH VINE BORER (*Melittia cucurbitae*)

The squash vine borer is the most serious enemy of cucurbits. Its attack is severe when only a few plants are available in the gardens. Damage is caused by larvae tunnelling into stems, which often kills the plants, especially when the larvae feed in basal portions of the vines. Sometimes, this insect also attacks the fruits. Sawdust-like insect waste coming out of the holes in the stem and sudden wilting of a vine are the evidences of its attack.

Control

- Plough the field in hot summer months to expose it to the direct sunlight.
- Rotate the cassabanana crop to another location in the garden each season.
- Plant the crop as early as possible as the borers do not emerge until early summer.
- Destroy the infected vines to break the life cycle.
- Keep the plants well watered.
- Place soil over slit stem after removing the borer to encourage root development.
- Spray the crop with bifenthrin 0.05%, cyhalothrin 0.1%, or cyfluthrin 0.1%.

PICKLE WORM (*Diaphania nitidalis*)

The caterpillars of pickleworm make tunnels in flowers, buds, stems and fruits. However, the fruits are preferred more. Frass, *i.e.*, sawdust-like insect waste, often protrudes from the small holes in damaged fruits. Sometimes, the damaged fruits can not be recognized until they are cut open. Its attack spoils the fruits completely. Flowers, buds and at times the entire plants may be killed.

Control

- Grow early maturing varieties if available.
- Plant the crop as early as possible before the pickleworm population peaks.
- Crush the rolled sections of leaves to kill the pupae.
- Destroy the damaged fruit.
- Spray the crop with bifenthrin 0.05%, cyhalothrin 0.1%, or cyfluthrin 0.1%.

SQUASH BUGS (*Anasa tristis*)

The squash bug is one of the most common and destructive pests in kitchen gardens. This sap-feeding pest often kills the plants. Leaves of affected plants may wilt rapidly and become brittle. The squash bug is secretive in its habits. Nymphs and adults can be seen cluster near crown of the plant, beneath damaged leaves and under clods or any other protective ground cover. They run for cover when disturbed.

Control

- Grow early varieties, if available.
- Remove the eggs and nymphs.
- Remove and destroy vines and fruits after harvest.
- Spray the crop with malathion, bifenthrin, cyhalothrin, or cyfluthrin 0.1%.

Diseases

ANTHRACNOSE (*Collectotrichum* sp.)

This is very serious disease of all the cucurbits including cassabanana. Typical irregular and jagged lesions appear on leaves. The centres of older lesions may fall out, which gives the leaf a shot hole appearance. Light brown spindle-shaped lesions appear on stem, which often start on lower surface of the fruit where moisture accumulates. The disease is transmitted by seed.

Control

- Strictly follow fall tillage and crop rotation of at least 3 years without a cucurbit crop.
- Follow clean cultivation to minimize the initial inoculums.
- Destroy the crop residues after harvesting the crop to reduce inoculums.
- Spray the crop with benomyl, Bavistin, or thiophanate methyl 0.1%, Dithane M-45 or Maneb 0.2% at 10- to 12-day intervals.

FUSARIUM WILT (*Fusarium oxysporum*)

This is a most common disease of all the cucurbits including cassabanana. The infected vines wilt during day hours but recover during evening but eventually wilt permanently. Initial symptom is dull grey green appearance of leaves that precedes a loss of turgor and wilting, which is followed by a yellowing of leaves and finally necrosis. The wilting usually starts with the older leaves and progresses to the younger ones. The disease often occurs when the plant begins to vine, and wilting occurs in only one vine leaving rest of the vines unaffected. When the inoculums density is very high, the entire vine may wilt and die within a short period. The affected vines that do not die often become stunted and they give considerably less yield. Since the pathogen is soil borne, symptomatic plants often occur in clusters corresponding to the distribution of high inoculums densities in the soil. The disease is usually most severe in sandy and slightly acidic soils at 25–27°C temperature. However, higher temperature checks its infection, often in plants that are yellowed and stunted but not wilted.

Control

- Follow long crop rotation with cereal crops.
- Use disease-free healthy seed and transplants.
- Disinfect the soil by covering the soil with clear plastic sheet during hot summer months.
- Use *Gliocladium* spp., *Trichoderma* or non-pathogenic strains of *F. oxysporum* that compete with pathogenic forms for root colonization.
- Drench the soil with captan, Hexocap, or Thiride 0.2–0.3% solution.
- Apply fumigants like methyl bromide, metam-sodium chloropicrin 0.2–3% in soil.

Downy Mildew (*Pseudopernospora cubensis*)

Symptoms of downy mildew are dissimilar in different cucurbits. It affects plants of all ages. In cassabanana, angular yellow spots appear on leaves. The leaves eventually turn brown and crispy. In humid conditions, a dark feathery mould may grow on underside of the affected leaves. The disease infects only the foliage, thus, a reduction in photosynthesis early stages of growth results in stunting of plants and yield reduction. Premature defoliation also results in fruit sun scald due to exposure to direct sunlight. The disease is more prevalent in humid areas, especially when rains occur regularly in summer.

Control

- Follow long crop rotation with cereal crop.
- Repeatedly plough the field deep in hot summer months.
- Grow resistant varieties if available.
- Remove diseased plants as soon as possible.
- Grow cassabanana in humid environments or traditional polyhouses.
- Plant the crop at wider spacing to encourage airflow and reduce leaf wetness.
- Spray the crop with copper oxychloride, Dithane M-45, Daconil, or Difolatan 0.2%.

Powdery Mildew (*Sphaerothera fuliginea*)

This is one of the easier fungal diseases to identify since its symptoms are unique. Infected plants show white powdery spots on the leaves and stems. It starts with little spots but grows to encompass the whole leaves. The mildew can appear on any aboveground part of the plant but lower leaves are affected more. As the disease progresses, the spots become larger and denser. The mildew may spread up and down the length of the plant. It grows well in environments with high humidity and moderate temperature.

Control

- Follow long crop rotation with cereal crop.
- Repeatedly plough the field deep in hot summer months.
- Grow resistant varieties if available.
- Spray baking soda and liquid castile soap solution.
- Spray the crop with 0.4% *neem* oil.
- Spray the crop with Calixin 0.2%, Sulfex 0.2%, Karathane 0.2%, or carbendazim 0.1%.

Bacterial Wilt (*Erwinia tracheiphila*)

This disease causes the plants to wilt and eventually kills them. When wilted portion of the branch is cut off near the ground and touched with figure, a white milky substance will stick to the finger as it is pulled away. If the cut portion also emits the foul smell, it proves the bacterial infection. Wilting of individual leaves or vines of the plant is the characteristic symptom. The disease spreads from the leaves downward into the petioles and then the stem until the entire plant wilts and dies.

Control

- Grow resistant varieties if available.
- Cover soil below plant with reflective mulch like aluminium coated plastic mulch.
- Keep the cucumber beetles under control by using yellow sticky traps in field.
- Pull and destroy any affected plants as soon as it is observed to avoid its spreading.
- Spray the crop with kaolin clay, bifenthrin, or pyrethrin 0.1% to control striped cucumber beetle.

50 Long Melon

B.R. Choudhary, S.M. Haldhar and Suresh Kumar

Botanical Name   : *Cucumis melo* var. utilissimus (Roxb.) Duth. & Full.

Family   : Cucurbitaceae

Chromosome Number : 2n = 24

ORIGIN AND DISTRIBUTION

The centre of origin for *Cucumis* species is likely Africa for the wild species. However, the initial site of domestication for the melons is probably the Middle East Asia, and from there, the genes have contributed extensively to plant improvement. India also possesses considerable variability in melons, therefore, considered as the secondary centre of origin of melons. Long melon is cultivated throughout India and Pakistan (Punjab and Sind). It is very popular predominantly in Rajasthan, Punjab and Uttar Pradesh. Riverbeds of Yamuna, Ganga and Narmada in the North and Kaveri, Krishna and Godavari in the South India are the major growing centres of long melon. Presently, it is being cultivated throughout the world under tropical and subtropical climatic conditions.

INTRODUCTION

The cucurbits form a distinct group of species with many similarities in botany, cultural requirements and susceptibility to insect-pests and diseases. Long melon popularly known as *Kakri* or *Tar Kakri* is also one of them. It is a warm season crop chiefly grown in tropical, subtropical and milder zones of India. It is also best adapted for river bed cultivation, a kind of vegetable forcing being used in different parts of the country where the crop is grown in riverbeds during winter season to get early produce. It is valued for its tender fruits eaten raw with salt and pepper. If it is eaten without salt, it is not easily digested, and similarly, drinking water soon after eating its fruits causes indigestion. Due to its cooling effect, it is very popular in most parts of the country during summer months.

COMPOSITION AND USES

COMPOSITION

The immature fruits of long melon contain about 90% water and are not particularly high in nutrients but its flavour and texture have made it popular for use as a fresh addition to salads. It possesses both curative and nutritive value as given in Table 50.1.

USES

Long melon is a delicious fruit. It is used as salad, pickle and cooked as vegetable. It is becoming popular among the consumers due to its cooling effect during summer season. The tender and thin fruits are preferred for salad purpose and eaten with salt. The fruit is always used without peeling.

TABLE 50.1

Nutritional Composition of Fruit (per 100 g Edible Portion)

Constituents	Contents
Water (g)	90.15
Carbohydrates (g)	8.16
Protein (g)	1.84
Fat (g)	0.57
Dietary fibre (g)	0.9
Energy (kcal)	34

Full size immature fruits are desirable for the preparation of pickle. The seed cavity is removed and only the rind is used in making pickles. The fruits are also used in the preparation of various ethnic dishes. The seeds are dried, grounded into a meal and employed as an article of diet.

BOTANY

Cucumis melo L. var. utilissimus (Roxb.) Duthie & Fuller (syn. *Cucumis utilissimus* Roxb.) belonging to Cucurbitaceae family with chromosome number (2n) 24 is adopted for the plant commonly known as *Kakri* or *Tar Kakri* and closely related to musk melon. It is dicotyledonous, annual, climbing, or trailing herb. Stem is obtusely angular and scabrous having soft hairs and simple tendrils. Leaves are alternate, membranous, exstipulate, five-lobed, lobes neither deep nor acute. Both the leaf surfaces are scabrid with soft hairs. Plants are monoecious in sex form. Female flowers are axillary and solitary with 5 cm long peduncle. Male flowers with 1.2 cm long peduncle are axillary, mostly fascicled. Calyx is campanulate, five-lobed and subulate. Corolla is yellow, corolla tube turbinate, rotate or subcampanulate, five-lobed, lobes mostly oblong or ovate and acute. Petals are five in number and united. Stamens are three in number, free, inserted at the base of calyx, one- or two-celled, filaments short, connectives produced, bifid, pistillode glandular. Style is short, simple, surrounded by annular disc. Stigmas are three in number, obtuse and spherical. Ovary is inferior, ovoid or globose and ovules horizontal and numerous. The period of bud developmental stage is completed within 12–15 days in male and within 11–13 days in female bud. Fruits are green, long (45–90 cm), slender, 2.5–5.0 cm in diameter, striated or ridged, softly hairy and light green in colour when young and botanically called pepo. The fruits do not burst after ripening.

CLIMATIC REQUIREMENT

Being a warm season crop, long melon is successfully grown in tropical and subtropical regions. It cannot tolerate frost and strong winds. Freezing temperature kills the plants and cool weather below 16°C slows or stops its plant growth. The optimum temperature for germination of seed is 25–30°C and grows well at day temperature between 25°C and 35°C. Growth is virtually stopped at temperature below 10°C, and temperature above 40°C affects flowering and fruiting adversely. It tolerates cool climate better than musk melon. It shows slight photoreaction to short days for flowering. Relatively low temperature, high humidity and short day length promote femaleness but excessively high humidity increases the incidence of diseases, particularly those affecting foliage.

SOIL REQUIREMENT

A well-drained sandy or sandy loam soil rich in organic matter is ideal for its cultivation. Lighter soils, which warm up quickly in spring season, are usually used for early crop. In riverbeds, alluvial substrata and subterranean moisture of river streams support its growth since its long tap root system is adapted to grow in riverbeds. Long melon is sensitive to acidic soil reaction, hence, cannot be grown below 5.5 pH. For successful cultivation, the pH of the soil should be between 6.0 and 7.5. Adequate drainage facility should be provided for rainy season crop. Thorough preparation of land by ploughing and planking is essential. Nearly 2–3 ploughings and planking are sufficient to obtain good tilth.

CULTIVATED VARIETIES

The important varieties of long melon developed in India along with their characteristic features are given below:

ARKA SHEETAL

A high-yielding variety developed at Indian Institute of Horticultural Research, Bangalore, through pure line selection from a local collection (IIHR 3-1-1-1-5-1) collected from Lucknow is tolerant to high temperature and bears medium sized light green fruits. Its fruits are free from bitter principle. Its yield potential is 350 q/ha.

PUNJAB LONG MELON-1

An early maturing variety developed at Punjab Agricultural University, Ludhiana, bears long, thin and light green fruits. Its vines are long with pubescent, angled and light green stems.

SOWING TIME

In north plain areas for summer season crop, it is sown in the month of February–March when the danger of frost is over, while for rainy season crop, it is sown in June–July. In South and Central parts of the country where winter is mild, the crop is grown almost throughout the year. On north Indian hills, the best time for its sowing is April–May. In north India, early sowing is also done in riverbeds from November to January. To get an early crop, seeds can be sown in polyethylene bags or portrays under protected conditions. The seedlings become ready for transplanting at 2–4 true leaves stage.

SOWING METHOD

Long melon is sown on raised beds, in furrow, trenches, or pits. Trenches are prepared in well prepared field at a distance of 2 to 2.5 m. Generally, depth of trench is kept 20–30 cm maintaining 40–50 cm width. Near to these trenches, pits of 45 × 45 × 45 cm size are dug and filled with well decomposed farmyard manure, top soil and fertilizer mixture. Plant spacing considerably affects the growth and yield of long melon. The optimum plant spacing is 2–2.5 m between rows and 50–60 cm between hills. For sowing a hectare area, 2–2.5 kg seed is sufficient. In case of raised bed, 2 seeds per hill should be sown, while 3 to 4 seeds should be sown in case of pit method. In case of early sowing, the seeds are sown after sprouting. For better germination, the soil must have plenty of moisture, for that, the channels where sowing is to be done should be irrigated two days before sowing seeds.

Prior to sowing, the seeds must be treated with captan or thiram @ 2 g/kg of seed. For December–January sowing, it is also advisable to wrap the seeds in moist gunny bags and keep them for 3–4 days in warm place to initiate germination.

The important methods for sowing long melon are as follows:

i. **Shallow pit method or flat bed method:** Shallow pits of 45 × 45 × 45 cm size are dug and left open for 3 weeks before sowing for partial soil solarization. These pits are filled with a mixture of top soil, well decomposed farmyard manure or compost and recommended dose of fertilizers. After filling the pits, 2–3 seeds per hill are sown at 2–2.5 cm depth.

ii. **Deep pit or trench method:** Deep pit method is commonly practiced for raising long melon in riverbeds. In this method, the circular pits of 60–75 cm diameter and 1.5 m depth are dug at recommended distance. About 60 cm wide trenches are dug at a distance of about 2.0–2.5 m up to the depth of clay layer. Trenches are filled with a mixture of top soil, well rotten farmyard manure and recommended NPK mixture. The pre-germinated seeds are sown in trenches at a distance of 45–60 cm apart. Generally, three or four seeds are sown per hill at a depth of 2–2.5 cm.

ii. Sowing seeds on ridges: In this method, 40–50 cm wide channels are prepared manually or mechanically maintaining 2–2.5 m distance between two channels, depending upon cultivars. Seed is sown on both ridges of channel at a spacing of about 50–60 cm. Generally, two to three sprouted seeds are sown per hill in spring-summer and adequate moisture is maintained at the time of emergence. The vines are allowed to spread in between space of channels.

OFF-SEASON CULTIVATION

It is a kind of vegetable forcing where long melon is sown in December–January under plastic low tunnels to get early crop. Plastic low tunnels protect the crop from frost and advance the crop by 30–40 days over their normal season of cultivation, which ultimately fetches high price in the market. Low tunnels are flexible transparent covering that are installed over the rows of directly sown or transplanted vegetables to enhance plant growth by warming the air around the plants during winter season. It is quite cost effective technology for the growers in northern parts of the country, where the night temperature during winter season falls below 8°C for a period of 30–40 days.

Seedlings of long melon are raised in greenhouse in plastic protrays having 1.5 inch cell size in soil-less media in December–January, and 28–32 days old seedlings at 4 leaf stage are transplanted under plastic low tunnels in the open field from mid January to mid February. Nursery of these crops can also be raised even in polythene bags under very simple and low cost protected structures. It can also be transplanted even in the month of December for complete off-season production, and the crop will be ready for harvesting in the first week of February and can fetch very high price in the market. Direct sowing of seed can also be done beginning from mid December to mid January.

The seedlings are planted in rows at recommended distance on drip system of irrigation. The flexible galvanized iron hoops are fixed manually at a distance of 2.0–2.5 m. The width of two ends of hoop is kept 1 m with a height of 60–70 cm above the levels of ground for covering plastic on the rows or beds for making low tunnels. Transparent 30 micron biodegradable plastic is generally used for making low tunnels, which reflects infra-red radiation to keep the temperature of the low tunnels higher than outside. The 3–4 cm size vents are made on eastern side of the tunnels just below the top at a distance of 2.5–3.0 m to keep the temperature normal when temperature increases within the tunnels during the peak day time. These vents also facilitate pollination of crop by visit of bees. Plastic is completely removed from the plants in February–March, depending upon the prevailing night temperature in the area.

NUTRITIONAL REQUIREMENT

Fertilizer requirement depends on nutritional status of the soil. About 20–25 t/ha well rotten farmyard manure should be incorporated in the field. It should be mixed thoroughly into the soil at the time of field preparation or applied to each pit during pit preparation as the case may be. Long melon crop requires nitrogen 50–60 kg, phosphorus 60–80 kg and potash 40–50 kg/ha. Half of nitrogen and full dose of phosphorus and potash should be applied at pre-plant in the mound. The remaining nitrogen is divided in two equal parts and applied at the time of vine growth and at full bloom stage. Additional nitrogen should be placed at 20–25 cm apart from the plant row and about even with the bottom of furrow as top dressing after 25–30 days of sowing. Excessive application of nitrogen should be avoided as it results in more number of staminate flowers, which affect the fruit yield adversely. While applying fertilizers, care must be taken not to cause damage to roots. Irrigation should be supplied shortly after applying fertilizers to move them towards roots. Foliar sprays of boron 25 ppm at two or three times beginning from two to four true leaves stage to flower initiation have also been found to increase number of female flowers and yield.

IRRIGATION REQUIREMENT

Irrigation depends upon the soil, growing season and prevailing weather conditions. However, it is essential to maintain sufficient moisture in the soil. For better seed germination, the soil must have ample moisture at the time of sowing, and for this purpose, a pre-sowing irrigation is applied. During dry summer, irrigation is applied at 3- to 4-day intervals, while irrigation requirement in rainy season depends upon the occurrence of rains. It is essential to maintain optimum moisture in the field at critical stages *viz.*, vine development, pre-flowering, flowering and evenly throughout fruiting period to get high yield. Over irrigation should be avoided as it causes excessive vegetative growth and delayed flowering. Careful irrigation management is required when the crop is grown in sandy and clay-loam soil to obtain high yield.

Among the irrigation systems, drip irrigation is the most efficient method for long melon crop. The single lateral lines (12–16 mm size) spaced at 2.0–2.5 m distance within line or on line drip-pers of 4 litre/hour discharge capacity at 50–60 cm distance can successfully be used. This method saves irrigation water up to 60%, reduces fruit rot disease and increases fruit yield in comparison to channel system of irrigation.

INTERCULTURAL OPERATIONS

Thinning of plants should be done 10–15 days after sowing, retaining only two healthy seedlings per hill. Weeds pose problem during early period of crop growth and compete with crop for nutrients, moisture, sunlight and space and also act as reservoir of many insect-pests and diseases. Two or three shallow hoeing should be done to control weeds before the vines cover the ground area. First hoeing may be done about 15–20 days after sowing or planting and second 35–40 days after sowing, however, deep hoeing should be avoided as it may destroy many of the fine roots near the soil surface. Mulching is also found effective for controlling weeds.

FLOWERING

Long melon starts flowering 45 days after sowing. Anthesis takes place at around 5:30 a.m. and stigma becomes receptive 12 hours before the day of anthesis. Duration of pollen fertility is between 5:0 a.m. to 2:0 p.m., and dehiscence of anther takes place between 5"0 and 6:0 a.m. Pollination during early morning hours is more effective to get maximum fruit setting. The chief pollinating agent of long melon for natural crossing is honeybee. For effective pollination, some beehives may be placed around periphery and inside of the field.

CROP REGULATION

The excessive vine growth due to high dose of nitrogen application coupled with high temperature and high soil moisture conditions promotes staminate flowers on the vines, resulting poor fruit set and low yield. The best way to control vine growth within reasonable limits is by adjusting nitrogen fertilizer doses and frequency of irrigation. It has been well documented that short days and low temperature conditions favour the expression of more female flowers in all the cucurbits. Application of Ethrel 250 ppm thrice, first at two to four true leaves stage of the plants and subsequent two sprays at 7- to 10-day intervals helps in increasing the number of pistillate flowers and subsequently yield. In the seed production field, at least 10% plants should be kept untreated to allow usual production of comparatively high number of staminate flowers for effective cross-pollination by the pollinators.

HARVESTING

The fruits are picked when they are still tender and thin. The fruit is most flavourful when it is 12–15 inches long. The fruits do not keep well for more than one day, so they should be sold promptly. For making pickles, fruits are harvested when they attain full size but still immature.

YIELD

The average yield of long melon varies from 300 to 350 q/ha, depending on variety, season, soil, environmental factors and cultural practices followed by the growers.

PRODUCTION CONSTRAINTS

EXCESS OF STAMINATE FLOWERS AND LOW FRUIT SET

Monoecious sex form is found in long melon where staminate and pistillate flowers borne separately on the same plant. In this sex form, fruit set and yield depend on the number of pistillate flowers. Generally, staminate flowers appear on first four to six flowering nodes and later the pistillate flowers. In monoecious sex form, the proportion of staminate flowers is very high than the pistillate flowers. Therefore, a high female to male ratio is desirable to increase fruit set and finally the fruit yield.

Control

- Avoid over-irrigation and excess application of nitrogenous fertilizers, which lead to profuse vegetative growth eventually promoting the maleness.
- Application of Ethrel 250 ppm two or three times starting from two to four leaves stage to flower initiation increases the number of pistillate flowers.

POOR POLLINATION

Long melon being monoecious in sex form requires pollination to produce fruits. Several visits of pollinators on the day that a flower is open are often required to ensure appropriate pollination and fruit setting. Many fruits will appear misshapen and small when pollination is poor. Very high and low temperature can also affect pollen viability, resulting in poor pollination. If too much nitrogen is applied (resulting in excessive vegetative growth) or plants are improperly spaced, bees may have difficulty in locating the flowers.

Control

- Provide pollinators to ensure good fruit set and high yield.
- Do not spray insecticides during morning hours when flowers are open and insects are pollinating the flowers actively.

DROUGHT STRESS

Cucurbits are particularly sensitive to drought. Fruits contain 88–90% water, hence, can suffer under drought conditions. Long melon often produces long vines with many leaves and can transpire large quantities of water during hot summer days. Severe drought stress adversely affects flowering, fruit setting and fruit development, resulting in unmarketable produce. Affected fruits become soft, wrinkled, curled, distorted, or tapered at blossom end. The loss of foliage during drought will also result in sunburn of the fruits.

Control

- Avoid moisture stress during critical stages of crop growth *viz.*, vine development, pre-flowering and flowering.
- Protect the crop from hot winds during summer season.

INSECT-PESTS

HADDA BEETLE (*Epilachna vigintioctopunctata*)

The adult and grub attack the young seedlings and feed on green leaves between the veins, giving a lacy appearance to the leaves. Finally, the affected leaves dry and fall.

Control

- Collect and destroy the infested plant parts along with insect in initial stage.
- Mix 1 kg cow dung in 10 litre of water, filter the extract through gunny cloth, add 5 litre of water in filtrate and again filter it through gunny cloth. Then, spray this clear filtrate on plants.
- Spray the crop with spinosad 45 SC (0.5–0.7 ml/litre of water), dimethoate 30 EC (1.5–2 ml/litre water), or malathion 50 EC (1.5–2 ml/litre water).

APHID (*Aphis gossypii* or *Myzus persicae*)

Both nymphs and adults of this tiny insect suck cell sap from the tender leaves, reducing plant vigour. As a result, the leaves curl up and ultimately wilt. The aphids excrete honeydew on which black sooty mould develops, which hampers photosynthetic activity. Besides, these aphids act as vector for the transmission of many viruses. The severe infestation affects the yield and quality of fruits adversely. The attack is severe during March–April.

Control

- Spray pressurized water to the crop.
- Use sticky traps in the field.
- Use natural enemies of aphids, *i.e.*, lady bird beetle, lace wings, etc.
- Spray *neem* seed kernel extract (5%), *neem* oil (2%) or tobacco decoction (0.05%) for the effective control of this insect.
- Spray Imidacloprid 17.8 SL @ 0.5–0.6 ml/litre of water and repeat after 15 days, if necessary.

FRUIT FLY (*Bactrocera cucurbitae*)

The damage is caused by maggots of the fly. The adult fly punctures tender fruits and lay eggs below the fruits skin (epidermal layer). The maggots feed inside the fruits and make them unfit for consumption. Due to the attack of this pest, the fruits may also be deformed. The hot and humid weather is most suitable for its attack, and generally, it causes 50–60% loss.

Control

- Collect and destroy the infected fruits with insects.
- Stir the soil around the plants to kill the pupae.
- Follow phytosanitary measures.
- Install pheromone trap (8/ha).
- Apply bait spray containing 0.05% malathion 50 EC and 10% *Gur* or sugar in 20 litre of water.
- Spray spinosad 45 SC (0.5–0.7 ml/litre water), dimethoate 30 EC (1.5–2 ml/litre of water), or malathion 50 EC (1.5–2 ml/litre of water).

RED PUMPKIN BEETLE (*Aulacophora foveicollis*)

Both, grub and adult attack the crop at seedling stage and make holes in the cotyledonary leaves. When the attack is severe, the crop is totally destroyed.

Control

- Follow phytosanitary measures.
- Floating row covers can block the beetles from young seedlings as these row covers serve as physical barrier to the beetles. At the time of flowering, these covers must be removed to allow pollination by honeybees.
- Spray the crop with garlic, ginger, or chilli extract @ 50–100 ml/litre of water.
- Spray spinosad 45 SC (0.5–0.7 ml/litre of water), dimethoate 30 EC (1.5–2 ml/litre of water), or malathion 50 EC (1.5–2 ml/litre of water).

DISEASES

POWDERY MILDEW (*Erysiphe cichoracearum, Sphaerotheca fuliginea*)

The white powdery spots of fungal growth appear on upper surface of the leaves, stem and succulent parts of the plant. Later, theses spots become powdery and enlarge into patches. Under severe infection, the fruits may also be covered with powdery mass. The fruits remain undersized and sometimes become deformed. The humid weather is favourable for the spread of this disease.

Control

- Follow 2–3 years crop rotation.
- Remove the infected leaves and destroy.
- Spray the crop with Karathane @ 0.1% two or three times at an interval of 10–15 days just after the appearance of disease.

DOWNY MILDEW (*Pseudoperonospora cubensis*)

The first visible symptoms are the appearance of small water-soaked lesions on the leaves. These lesions appear yellow on upper surface of the leaves. Greyish downy growth of fungus develops on lower surface of the leaves.

Control

- Hot water treatment of seeds before sowing (55°C for 15 minutes).
- Rouge out the affected plants and burn or burry them.
- Follow prophylactic spray of Dithane M-45 @ 0.2% at 15-day intervals.

Anthracnose (*Colletorichum lagenarium*)

The disease-causing fungus is chiefly soil borne but it may also be seed borne if fruits are attacked and the fungal mycelium reaches the seed. Reddish brown spots are formed on the affected leaves and become angular or round when many spots coalesce. It results in shrivelling and later dying.

Control

- Follow 2–3 years crop rotation.
- Treat the seeds with Agrosan GN or Bavistin @ 2 g/kg seed.
- Spray the crop with Dithane M-45 @ 0.2%, Bavistin @ 0.1%, or copper oxychloride @ 0.2% and repeat at 7- to 10-day intervals if necessary.

Fruit Rot (*Pythium aphanidermatum* and *P. butleri*)

Water soaked lesions girdle the stem, later extending upwards and downwards. The rotting of affected tissues occur and even grown-up plants collapse. The fungus causes rotting of fruits having direct contact with soil.

Control

- Avoid flood irrigation and raise the crop on drip system.
- Use mulch to avoid direct contact of fruits with the soil.
- Treat the seeds with thiram or Bavistin @ 2 g/kg seed.
- Apply *Trichoderma* @ 5 kg/ha in soil before sowing.
- Spray the crop with Bavistin @ 0.1% or copper oxychloride @ 0.2% and repeat the spray if necessary.

Fusarium Wilt (*Fusarium oxysporum f. sp. melonis*)

The fungus is seed borne as well as a persistent soil inhabitant. Seedling injury is high at 20–30°C temperature. Wilt development is also favoured by a temperature of about 27°C. No infection occurs at temperature below 15°C and above 35°C. The plants are prone to attack at all stages of growth. Germinating seeds may also rot in the soil. The affected plants turn yellow, show wilting, and later, the whole plant dies. It causes damping-off disease of seedlings. Small leaves lose their green colour, droop and wilt.

Control

- Follow long term crop rotation.
- Treat the seed with hot water (55°C) for 15 minutes to kill the seed-borne infection.
- Treat the seeds with captan, thiram, or Bavistin @ 2 g/kg seed before sowing.
- Drench the soil around roots with captan or Bavistin @ 0.2% solution.

Mosaic virus

The young leaves develop small greenish-yellow areas and they become more translucent than those in remaining parts of the leaf. Yellow mottling is seen on leaves and fruits. Leaf distortion and stunting of infected plants occur. The virus is transmitted through sap, seed and aphids.

Control

- Use the seed collected from virus-free plants.

- Rough out the infected plants from the field as soon as they are noticed.
- Eliminate the weed hosts from the field.
- Grow barrier crops like sorghum and pearl millet around the crop.
- Spray imidacloprid 17.8 SL @ 0.5–0.6 ml/litre of water to control the vectors.

■■■

51 Kachri

M.K. Rana and G.K. Sadananda

Botanical Name   : *Cucumis melo* var. callosus L.

Syn. *Cucumis melo* var. agrestis

Cucumis trigonus

Cucumis pubescens

Family   : Cucurbitaceae

Chromosome Number : 2n = 24

ORIGIN AND DISTRIBUTION

This wild species is originated from India and adjacent countries. It has no English name, though it is widely used for medicinal purposes in Asia.

INTRODUCTION

Kachri, commonly known as wild melon, small gourd, senat seed, wild musk melon, *kachariya*, *shinde* (Marathi), *chibdin* (Konkani), or *gurmi* (Assamese), is most probably a wild variety of melon, *i.e.*, *Cucumis melo* var. agrestis. Fresh *kachri* resembles a brown-yellow small melon, which grows wildly in desert areas. It is a drought tolerant annual climber found growing copiously during rainy season in arid and semiarid regions of North-western India. It is the real earth food growing wild and becoming a protein rich vegetable for people living in the harsh arid areas of Western India, where it is hard to grow conventional vegetables. It is also found in Bengal, Punjab, parts of Maharashtra, the North Western Provinces and the Sind area (now in Pakistan). Rural people grow *kachri* with other rain-fed crops such as sorghum, maize and pearl millet for their food security. Its fruits are small and egg-shaped, and at ripening, its taste turns to acerbic sweet. It is used in its wild form and is seldom cultivated as a crop.

Kachri is available only for a short duration. During this period, very little of the total produce is being used profitably, mostly in the form of vegetable preparation, and the rest of the produce is spoiled either due to over-ripening of the fruits or due to lack of knowledge about different processing techniques to convert it to value-added durable products. The post-harvest losses in *kachri* vary from 30% to 40% due to its perishable nature and glut during harvesting time, which also reduces market value of the fruits. In times of scarcity, the dehydrated fruits are sold at a very good price in the market.

COMPOSITION AND USES

COMPOSITION

Bitter-tasting to start out, the fruit sweetens (faintly sour with a melon-y taste) as it ripens. The detailed nutritional composition per 100 g edible portion of *kachri* is given in Table 51.1.

TABLE 51.1

Nutritional Composition of Kachri (per 100 g Edible Portion)

Constituents	Contents	Constituents	Contents
Water (g)	88.3	Vitamin C (mg)	29.81
Carbohydrates (g)	7.45	Energy (kcal)	43
Protein (g)	0.28	Alkaloids	++
Fat (g)	1.28	Flavonoids	+
Dietary fibre (g)	1.21	Triterpenoids	++
Ash (g)	1.46	Anthraquinone	-
Calcium (mg)	0.09	Phytosterols	+
Phosphorus (mg)	0.003	Tannins	-
Iron (mg)	0.18		

++ = *Maximum presence of chemical constituents;*
+ = *Minimum presence of chemical constituents;*
- = *Absence of chemical constituents*

USES

Karchi is sold fresh in the markets. Unripe *kachri* tastes bitter. When it ripens it becomes sweet and a bit sour. Both raw and mature fruits are edible in any form. Fresh fruits are usually cooked with various vegetable preparations and used in salad and for making chutney, which has a piquant taste. It is one of the important components of the delicious vegetable popularly known as *panchkuta* in the desert district of North Western India. Its fruits are sliced when fresh and then the slices are dried in shade to preserve them for off season use. These dehydrated fruits are also used in vegetables, chutney, etc.

Fine brown powder of dried *kachri* is the best natural meat tenderizer. Minced lamb meat together with garlic, ginger and fresh herbs are mixed with *kachri* powder and left for 4–6 hours to make it tender. Its dry powder is extensively used as souring agent in combination with other spices *viz.*, chilli, turmeric, coriander, cumin and fenugreek to make curry powder, spice premixes, ready to use spices and mouth fresheners. In Rajasthan, the fruits are mostly used in pickled form. Dried *kachri* powder, when used in curries, adds a tangy taste. Since fresh *kachri* is rarely available outside Rajasthan, the use of *kachri* powder is popular. *Kachri* powder is used extensively in Rajasthani cuisine. *Kachri* chutney is popular in Marwari cuisine, more as a substitute for *amchur* powder.

MEDICINAL USES

The fruits are widely used for medicinal purposes in Asia. It has a triple benefit of being a coolant, tonic and stimulant. Regular use of *kachri* powder may help to cure minor skin diseases like boils, bed sores, lice, manifestation, prickly heat, itching, earache, etc. Its powder with other spices is commonly used for various therapeutic purposes to cure stomach pain, nausea, vomiting and constipation. The dehydrated fruits are coughicide, vermicide, cooling, diuretic and gastric stimulant. The fruit is apparently used to prevent insanity and remove vertigo. The seeds are astringent and useful in bilious disorder.

BOTANY

Kachri is an annual climber growing up to a length of 1.5 m. Stem is generally covered with rough hairs. Leaves are triangular, ovate, 3–5 lobed and rough with rigid hairs. Leaf stalk is 1–6 cm long.

Flowers are small, yellow, solitary, or rarely in pairs or threes. Flower clusters are carried on 5–10 mm long stalk, densely hairy. Sepal is about 1.5 mm long, subulate. Flowers are 6–8 mm long with petals ovate-oblong. Fruit, which looks like a miniature watermelon, is ellipsoid, oval-round, sometimes obscurely trigonous, smooth and hairless and 4 × 2.5 cm in size, generally with dark green stripes.

Fruits contain a mean of 180 seeds, which are small (3–8 mm length and 3–5 mm width). Hermaphroditic flowers are only present at two first nodes of ramification of order ≥2, while the male flowers are present at all nodes of all stems. Male flowers are grouped in small cymes bearing two to four flowers, whereas, hermaphroditic flowers are solitary. A higher proportion of male than hermaphroditic flowers is observed per plant (82·9±11·9%, n = 100). Male flower number varies between eight and 125 per individual, whereas, hermaphroditic flower number only reaches 2 to 15.

The plant is protandrous, with male flower anthesis preceding that of hermaphroditic flowers between 7"00 a.m. and 10"00 a.m. Male flowers are smaller (16·5±2·3 mm corolla width) than hermaphroditic flowers (20·7±3·4 mm). All flowers are pentamerous, with five yellow petals and five stamens. The hermaphroditic flowers present a trilobulate stigma on a short style. Stigma receptivity is concomitant with flower anthesis. Pollen viability is high and does not differ between male and hermaphroditic flowers (97·6±0·7 $vs.$ 97·4±0·5%, respectively). Pollen number per flower also does not differ between the two flower types (5580±669 $vs.$ 4985±821 in male and hermaphroditic flowers, respectively). Ovule number reaches 173·4±56·4.

Although floral morphology has been well documented but the breeding system remains unknown, as well as the role of male flowers for pollination success. The fruit has a bitter flavour but the plant is of potential value in breeding programmes.

CLIMATIC REQUIREMENT

Kachri requires very sunny warm days. A frost-tender annual plant, a form of melon, is closest to the wild species. It will do nothing at all until the weather warms. Cool temperature and prolonged cloudiness, even where no freezing occurs, are hard on its sensitive plants.

Similar to muskmelon, the optimum temperature for proper seed germination is above 21°C and for economic production 27–35°C. The seeds may not germinate if the soil temperature is below 18°C. However, at the time of fruit maturity and ripening, it needs high temperature from 35°C to 40°C. Short day and moderate temperature encourages femaleness. It is very sensitive to low temperature and frost. Slightly higher temperature coupled with high humidity promotes diseases like downy mildew, powdery mildew and viruses and insect like fruit fly.

SOIL REQUIREMENT

Although *kachri* can be cultivated on all types of soil from light sandy to heavy clay but prefers a well-drained humus-rich soil, which is sandy, light and friable with good water holding capacity. It will not do well in heavy clay soils. It can be cultivated in acidic, neutral and basic (alkaline) soils but cannot grow in shade. Slopes and south-facing hills are ideal sites for planting *kachri*.

CULTIVATED VARIETIES

AHK 119

An early high-yielding selection from land races of *kachri* bears small egg-shaped fruits having 50–60 g weight, 5–6 cm length and 4–4.5 cm diameter. Fruits are suitable for processing (dehydration). It flowers very early, male flowers appearing about 28 days and female flowers 40 days after sowing. Vines are medium long (1.7–2.5 m) with 5–7 primary branches. On an average, 22 fruits are borne per vine, giving a yield of 94.99 q/ha. It is drought hardy and can be grown under high temperature conditions. It can be sown in February–March as an irrigated summer crop and in June–July as a rain-fed crop.

AHK 200

A very early maturing variety possessing drought tolerance characteristics was selected from land races growing in arid region. Fruits are suitable for garnish vegetables or as salad. Big-sized fruits are oblong in shape and develop tough, light yellow skin at maturity. Fruits are 100–120 g weight, 7.5 cm in length and 5.1 cm in diameter. The vines are medium in length (1.7–2.5 m) with 5–6 primary branches. Anthesis of first male flower starts 26 days and female flowers 32 days after sowing. Fruits are ready for harvest 65 days after sowing. On an average, 20 fruits can be harvested from a single vine, giving a yield of 100–113 q/ha. The variety is suitable for sowing in summer and rainy season.

SEED RATE

The germination percentage of seed chiefly affects the seed rate. The seed rate may also vary due to growing season, method of sowing and soil properties. A seed rate of 1.5–2 kg/ha is required for maintaining optimum plant population per unit area for direct-seeding crop and 0.5 kg/ha for transplanting crop. The seeds take about a week to germinate.

SOWING TIME

The seed is capable of emerging in a wide range of environmental conditions, however, constant temperature between 25°C and 40°C is ideal for its seed germination. Similarly, its seeds are capable to germinate under considerable saline conditions. The best time for its sowing is 3 to 4 weeks after the last average frost date in spring. In order to produce the best -flavoured and robust fruits, it should be sown when the moon is in the second quarter, *i.e.*, waxing and in one of the following Zodiac signs: *Cancer, Scorpio* and *Pisces*.

SOWING METHOD

It is never cultivated, it grows between other crops as a weeds. Dried stems can be found between the sorghum, maize, pearl millet and sesame fields. Karchi grows on poor and arid grounds, which makes it an interesting crop. The cultivation practices for *kachri* are almost similar to melons. It grows well with corn and sunflowers but dislikes potatoes. The weeds fat hen and sow thistle improve the growth and cropping of *kachri*.

In warm areas, the seeds are sown directly in filed. Per hill three or four seeds are sown. As and when the plants are large enough to handle, the extra week plants are pulled out keeping one plant per hill. In colder areas, the seed are sown indoors in individual polyethylene bag or other containers which can go directly in the field. The seedlings after attaining four or five true leaves are transplanted in the field without disturbing the earth ball as the seedlings will not thrive if their roots are disturbed.

NUTRITIONAL REQUIREMENT

The crop requires moderate doses of fertilizers so an application of 15–20 tonnes per hectare well-rotten farmyard manure to the planting site is advisable. It should be thoroughly mixed into the soil at the time of land preparation or a shovelful of rich compost or manure is added per hill where seeds are to be sown. Besides, nitrogen 50 kg, phosphorus 25 kg and potassium 25 kg/ha is optimum for *kachri* cultivation. One-third amount of nitrogen and whole amount of phosphorus and potash are applied at the time of furrow or pit preparation as basal dressing near expected root zone and remaining two-thirds of the nitrogen is applied as top dressing in two equal splits after about 20 days (at start of vine growth) and 40 days (before flower initiation) of sowing followed by a light hoeing.

IRRIGATION REQUIREMENT

The field at the time of sowing should have plenty of moisture for better germination of seeds. For that, pre-sowing irrigation is good practice if there is insufficient moisture in soil. The young plants require lots of moisture so each hill is mulched well to conserve moisture for longer period. Water thoroughly at least once a week when the fruits begin to form but allow the ground to dry out a bit as they approach maturity. The crop is moderately deep-rooted, hence, requires adequate soil moisture with proper drainage facility. First irrigation is applied when the vines start spreading and the fruits start developing. The fruits ripen properly when the vines remain healthy at the time of maturity so regular irrigation at an interval of 5–7 days is required up to fruit setting stage. However, excessive soil moisture during ripening results in poor quality fruits and predisposes the plants to *Fusarium* wilt.

INTERCULTURAL OPERATIONS

HOEING AND WEEDING

At initial stage of crop growth, *kachri* is very sensitive to weeds, thus, removing weeds is an essential operation about 15–20 days after sowing when the vines start their growth. However, at later stages, retaining weeds in the field is beneficial since providing shade weeds protect the fruits from direct sun exposure, which causes sun-scald disorder.

MULCHING

Kachri responds positively to mulching with black polyethylene film, which can be spread over the prepared field few days before sowing or planting. For sowing seeds or transplanting seedlings, the holes are made at recommended spacing on the spread mulch. Black polyethylene film is mainly used to increase the soil temperature and conserve moisture. Using mulch improves the fruits quality by avoiding their direct contact with soil.

HARVESTING

Each vine usually produces three to four fruits. Size of the fruit is an indication of its maturity, and many people are able to detect the ripe fruit by its characteristic aroma, however, the stems are the surest indicator. As the fruit matures, a small crack appears around the peduncle where it joins the fruit. When the crack circles the peduncle and the peduncle itself looks waxy, the fruit will break off with a light twist. If more than light pressure is necessary to pick the fruit, still it is not ripe and should be left on the vine until it ripens. For peak flavour, the fruit is allowed to rest a day or two before serving.

YIELD

The yield of *kachri* depends on the cultivar and growing season. Normally, *kachri* gives higher yield in rainy season rather than the crop grown in spring-summer. On an average, a yield of 95–110 q/ha can be taken by following good package of practices during cultivation of crop.

POST-HARVEST MANAGEMENT

The fruits harvested at mature but unripe stage can safely be retained for 7–8 days at ambient room temperature (25–35°C). However, pre-cooling the fruits soon after harvest is necessary to slowdown the respiration rate and prolongs the shelf life. The fruits harvested at full ripe stage have a very short shelf life since they can be stored hardly for 2–3 days at ambient room temperature though the period of storage can be extended by storing at comparatively low temperature and high relative humidity.

Crushed *kachri* powder is kept in a tightly sealed glass container in a dark cool and dry place. Whole dried *kachri* may be kept safe almost indefinitely, while *kachri* powder will stay fresh for about three months.

INSECT-PESTS

The insect-pests attacking the kachri crop are similar to other melon crops, which have been given below:

RED PUMPKIN BEETLES (*Aulacophora foveicollis, Syn. raphidopalpa foveicollis*)

The most serious threat to this plant is the red pumpkin beetle. Besides feeding on leaves, stems and roots, this insect carry fungus, which causes *Fusarium* wilt. If the plant is infected by this fungus, it may suddenly wither away even in later stage.

Control
- Plough the field deep in summer months to kill its grubs.
- Follow long crop rotation excluding cucurbitaceous crops.
- Sow the seeds bit early (in November).
- Dust the crop with a mixture of 75% colloidal phosphate and 25% wood ashes two or three times a week.
- Drench the beds with chlorpyrifos 20 EC 2 ml/litre at 10- to 15-day intervals.

FRUIT FLY (*Bactrocera cucurbitae*)

The female lays eggs below the fruit surface penetrating conical ovipositor. After hatching, its maggots feed inside the fruits, causing premature dropping or rotting of fruits, which makes the fruits unsuitable for consumption.

Control
- Collected and destroyed the damaged fruits at initial stage of infection.
- Spray the crop with fenthion 0.1% at 10- to 15-day intervals starting from flowering.
- Lure containing or carbaryl 50 WP 0.15% (3 g/litre) may be used for trapping male.
- Use poison baits containing six parts ripe banana pulp and 1 part malathion 50 EC 0.1% (2 ml/litre) and 4 parts protein hydrolysate or molasses.

DISEASES

A relatively high level of resistance has been reported to two diseases that cause major crop loss (i) the melon vine decline, a fungal disease, and (ii) the melon yellows virus, (MYV). The plants attacked by diseases are discussed as follows:

FUSARIUM WILT (*Fusarium oxysporum* f. sp. *moniliforme* and *Fusarium solani*)

The disease is either soil or the seed born and found more common in sandy soil. Yellowing of lower leaves, which progresses on upper leaves, is the characteristic symptom of this disease. Soon after infection, the plant starts drooping followed by wilting, and afterward, wilting becomes permanent.

Control

- Apply *Trichoderma* at the rate of 5–8 kg/ha.
- Before sowing, treat the seed with Bavistin by dipping in 0.25% solution.

ANTHRACNOSE (*Colletotrichum orbiculare*)

Anthracnose is not a common disease of *Kachri* but appears occasionally. Symptoms appear on above ground plant parts starting from cotyledonary leaves to true leaves, stem and tendril. Light brown circular sunken spots appear on leaf, which later turns to deep brown. Elongated lesions appear on stem and circular to oval sunken lesions on fruit. Later, the infected fruits rot and infect the seeds, thus, the disease is seed borne.

Control

- Follow long crop rotation excluding cucurbitaceous crops.
- Use disease-free seed collected from healthy plants.
- Treat the seed with Bavistin (2.5 g/kg of seed) before sowing.
- Spray the crop with systemic fungicide such as Bavistin 0.1% or chlorothalonil 0.2 %.
- Burn the disease debris and keep the field clean.

POWDERY MILDEW (*Sphaerotheca fuliginea*)

White coating appears first on under surface of the leaves, which later spread to upper surface also, stem, petiole and other succulent parts. Severe infection causes stunted growth and premature defoliation. The affected plants fruits do not develop to maturity. Dry weather, moderate temperature, reduced light intensity and succulent plant growth favour the occurrence of this disease.

Control

- Spray the crop with fungicides like penconazole 0.025–0.05%, tridemorph 0.05–0.1%, or carbendazim 0.1% starting from first appearance of disease.
- Collect and burn all infected leaves.

DOWNY MILDEW (*Pseudopernospora cubensis*)

Usually, the small irregular yellowish lesions appear on leaves at initial stage. Older lesions, which become necrotic, are evidently demarcated with light yellow areas. In humid weather, white downy growth is seen on the undersurface of the leaves.

Control

- Grow in well-drained field.
- Sow seeds at wider spacing.
- Spray the crop with mancozeb or copper oxychloride @ 0.25% at weekly intervals.
- In severe cases, spray the crop with Ridomil @ 0.2% once.
- Follow field sanitation by burning crop debris.
- Collect seed from disease-free healthy fruits.

CUCUMBER GREEN MOTTLE MOSAIC VIRUS (CGMMV)

The prominent symptoms of this disease are slight clearing of veins and crumpling of young leaves followed by a light or dark green mottling together with blistering and distortion of leaves. The virus spreads naturally or through red pumpkin beetle, but the disease is never transmitted through seed. The virus is very stable and mechanically transmissible at the time of cultural operation.

Control

- Sow the crop bit early (in November).
- Avoid contamination by workers and implements.
- Spray the crop with Dimethoate or Metasystox (1 ml/l) at 10-day intervals.

■■■

52 Snap Melon

M.K. Rana and Navjot Singh Brar

Botanical Name	: *Cucumis melo* var. momordica
Family	: Cucurbitaceae
Chromosome Number	: 2n = 24

ORIGIN AND DISTRIBUTION

Melon (*Cucumis melo* L.; 2n = 2x = 24), whose origin is thought to be Africa, is a morphologically diverse outcrossing species that belongs to the Cucurbitaceae family. Based on vegetative morphology and fruit variation, melon morpho types of *Cucumis melo* as botanical variety include *Cucumis melo* var. momordica. The common name of this botanical variety is *phoot* or snap melon and its origin is mainly Asia. India is considered the main centre of diversity, where much diversity is found in germplasm. During domestication, snap melon received wide use and it was distributed from the place of origin to other parts of the world. Under National Field Repository of Central Institute of Arid Horticulture (CIAH), Bikaner (Rajasthan), 65 Accession number of snap melon germplasm are maintained after massive survey and collection. Snap melon served as a source of powdery mildew transferable to musk melon.

INTRODUCTION

Snap melon (*Cucumis melo* var. momordica) is an important minor cucurbitaceous fruit vegetable of arid region. The nutrients rich snap melon juice is now gaining popularity as squash. Snap melon is a popular cucurbit cultivated in arid and semi-arid regions of western India, particularly in Rajasthan, Uttar Pradesh, Punjab and Bihar. Its ripe fruits are used as a substitute for muskmelon due to its cooling effect. The unripe fruit locally known as *kherla*. It is also known under the name of *phoot*, *Phoot kakari* and *kakadia*, meaning splitting, which bursts naturally on ripening. Due to this, the fruit is known as snap melon and *pottu vellari* in Malayalam. The tender fruits are also used as salad and as a cooked vegetable.

COMPOSITION AND USES

COMPOSITION

The fruits have a good amount of water and dietary fibres that help in relieving constipation and other digestive disorders. It is naturally loaded with high levels of phytonutrients (alpha- and beta-carotene), folic acid, potassium, vitamins like C, which are highly required nutrients to keep the skin natural, soft, glowing, healthy and younger looking. The total soluble solids content of fruits ranges from 4° to 5°Brix and acidity content 3.5% to 6.8%, while solid to acid ratio is 1.07:1.62. Its flesh gives a banana-like aroma. The seed contains 12.5–39.1% edible oil. The nutritional composition of snap melon is given in Table 52.1.

TABLE 52.1

Nutritional Composition of Snap Melon (per 100 g on Fresh Weight Basis)

Constituents	Contents	Constituents	Contents
Water (g)	80.0	Phosphorus (mg)	0.008
Carbohydrates (g)	15.6	Iron (mg)	0.84
Protein (g)	0.37	Vitamin C (mg)	18.6
Fat (g)	0.12	Ash (g)	1.64
Dietary fibre (g)	1.34	Energy (kcal)	74.0
Calcium (mg)	0.76		

USES

The immature and mature fruits are used as vegetable and desert, respectively. Immature fruits can be cooked or pickled and used as dehydrated value added product in off-season. Mature fruits are eaten raw with salt. Though the flesh is mealy and tasteless, its ripe fruits are mostly cherished as dessert. Ripe fruits are used for making squash as a light drink or are mixed with rice powder and jaggary. Its seed kernel is used in bakery products and a traditional drink (*thandai*). It has prospects of value addition in pickles, jam, chutney, squash and as dehydrated.

MEDICINAL USES

The fruits can be used as cooling and intestine cleanser. They are also used as a first aid treatment for burns and abrasions and as a natural moisturizer for the skin. The fruits improve appetite, cure stomach pain and help in relieving constipation and vomiting. The seeds are antitussive, digestive, febrifuge and vermifuge, and its roots are diuretic and emetic. The sprouting seed produces a toxic substance in its embryo so should not be consumed.

BOTANY

Snap melon is an annual climber, which is able to climb with the help of tendrils but more commonly sprawl along the ground. Its vine length can reach 1.5 m. Fruits are small and smooth, oblong or cylindrical in shape, dark green with 15 × 10 cm size, somewhat insipid or slightly sour fleshed. The average weight of fruit is 700 g and flesh colour is whitish pink with low sugar content.

CLIMATIC REQUIREMENT

Snap melon is a warm season crop predominantly grown in tropical and subtropical regions. However, it is not grown during winter season since it cannot tolerate frost or low temperature. The optimum temperature for seed germination is 18°C to 25°C. The plants can be grown successfully at a temperature range of 15°C to 32°C with an optimum of 22°C. Short days coupled with low temperatures increase femaleness. Growth and flowering of the crop coincide with high temperatures (30–40°C) during April–May and the result is predominance of male flower and higher flower drop.

SOIL REQUIREMENT

Snap melon is grown on a wide range of soils but it grows best in well-drained sandy loam or loam soils rich in organic matter. Snap melon is sensitive to acidic soil. It grows well in neutral soils having pH from 6.5 to 7.5. The soil is prepared well to a fine tilth by repeated ploughing followed by planking.

CULTIVATED VARIETIES

PUSA SHANDAR

A first commercial variety developed at Indian Agricultural Research Institute, New Delhi, was released in 2006 for cultivation as well as for growing in home and kitchen gardens. It is an early maturing (46–48 days) variety tolerant to many diseases. Its fruits are oblong with 700 g weight and creamy white to light pink thick flesh. Its average yield potential is 385 q/ha.

AHS 10

A variety developed at Central Institute of Arid Horticulture, Bikaner (Rajasthan), was released in 1971. Its fruits are oblong with 900 g weight, 17–20 cm long and 9.7–10.5 cm diameter. Flesh is whitish pink and sweet in taste having total soluble solids 4.5–5.0%. Fruits become ready for harvesting 68 days after sowing. Each vine bears 4.0–4.5 fruits, giving an average yield of 200–220 q/ha under arid conditions.

AHS 82

A variety developed at Central Institute of Arid Horticulture, Bikaner (Rajasthan), was released in 1971. Its fruits are oblong with 900–950 g weight and 22.5–24.5 cm long. The flesh is light pink and sweet in taste having total soluble solids 4.3–4.9%. Its fruits become ready for harvesting 67–70 days after sowing. Each vine bears 4.5–5.0 fruits, giving an average yield of 225–248 q/ha under arid conditions.

SRINIKETAN SM-1

It is a landrace variety for the red and lateritic belt of West Bengal, India. Its fruits are less sweet in taste containing total soluble solids of only 4.52°Brix. The average yield potential is 106.52 q/ha.

SOWING TIME

In states like Rajasthan, West Bengal and Bihar, it is sown in the months of January to March, while in Punjab, Haryana and Uttar Pradesh, snap melon seeds are sown in the months of June and July.

SEED RATE

Snap melon seed rate depends upon the sowing distance between rows and plants. A seed rate of 2 to 3 kg is required to sow snap melon in a hectare area.

SOWING/PLANTING METHOD

Snap melon is sown either on ridges or in flat beds. In the ridge method of sowing, the ridges and furrows are made at a distance of 2.5 m with a furrow width of 40–45 cm. The depth of the furrow may be kept 15–20 cm. The seeds are sown at a distance of 75–100 cm on ridges, whereas, in the flat bed method, shallow pits of 30 cm length, 30 cm width and 30 cm depth are made at a distance of 150–200 cm between rows and 75–100 cm between plants. The pits are filled with a mixture of soil and 2–3 kg compost, and the seeds (two or three) are sown at a depth of 1.5 to 2.0 cm.

NUTRITIONAL REQUIREMENT

Nutritional requirement generally depends on soil type and fertility status of the soil. Well-decomposed farmyard manure should be incorporated @ 200–250 q/ha at the time of field preparation. In addition to this, nitrogen is applied @ 40 kg per hectare in two splits. Half a dose of nitrogenous fertilizer should be applied as basal at the time of last ploughing and the other half at the time of flowering. In fact, nutrients requirement of snap melon crop needs to be standardized.

IRRIGATION REQUIREMENT

Snap melon is a moderately deep-rooted crop and requires adequate soil moisture. Irrigation interval may be kept at 5 to 7 days, depending on climatic conditions and soil types. Vine growth, flowering and fruit development are critical stages for irrigation. Among controlled irrigation, drip system of irrigation is most suitable for cucurbitaceous crops under arid conditions. At Central Institute of Arid Horticulture, Bikaner (Rajasthan), a series of experiments was conducted to assess the feasibility of different combinations under drip and micro-sprinkler systems in cucurbitaceous crops. In conclusion, a single lateral line (12–16 ìnm size) at 1.5–2.0 m distance with on-line drippers (4 litre/h) at 50 cm distance was found to be the most suitable for supplying irrigation in snap melon crop. Increased fruit yield of about 25–30% is obtained in these crops in comparison to the channel system of irrigation under arid conditions.

INTERCULTURAL OPERATIONS

HOEING AND WEEDING

Snap melon is sensitive to weeds at initial stage of crop growth, thus, weeding should be done frequently in order to keep the weeds under control. Hoeing is done once in initial stage of crop growth for providing good soil environment to the roots for their development and keeping down the weeds. Later, the vines themselves suppress the weed growth.

Pre-plant application of fluchloralin (1.2 kg/ha) effectively controls the weeds and increases yield in snap melon.

MULCHING

Mulching reduces water evaporation, delays the drying of soil and reduces the solid thermal regime during the daytime under extremely high temperature conditions. It also checks weed population and improves microbial activity in the soil by improving the environment around the active root zone. Mulching with locally available dry hays and weeds have given significant improvement in quality fruit yield by 58% over control (5.18 kg/plant) in snap melon cultivar AHS 82.

USE OF PLANT GROWTH REGULATORS

Plant growth regulators enhance fruit setting, fruit quality and early yield under irrigated or rain fed mixed crop cultivation. Spraying GA_3 @ 20 ppm at the two to four leaf stage (18–20 days after sowing) increased yield potential by 63% over the control (5.05 kg/plant) in snap melon.

HARVESTING

Early varieties of snap melon require 46–50 days to be mature after sowing. Generally, snap melon becomes ready for harvesting 80–90 days after sowing. For vegetable and salad purposes, light green to dark green immature fruits are harvested. For table purpose, half to fully ripe fruits are harvested. An abscission layer around the peduncle is developed at fully ripe stage of fruit, which results in separation of fruit very easily from the peduncle. Most of the fruits burst when they ripen, therefore, it is popularly called *phoot*.

YIELD

The average yield of snap melon is about 150 to 200 q/ha and the seed yield varies from 1 to 3 q/ha, depending on soil fertility, irrigation facilities and crop management practices followed by the farmers. The snap melon fruits do not keep well, therefore, soon after harvesting should be sent to market or consumed.

STORAGE

Snap melon fruits are perishable in nature and can be stored for 2–4 days at room temperature. However, in cold storage at 2–4°C temperature and 85–90% relative humidity, the fruits can be stored for 2–3 weeks.

PLANT PROTECTION MEASURES

A number of insect-pests and diseases attacks snap melon crop during the growing season. Some of the insect-pests and diseases along with their control measures are given as follows:

INSECT-PESTS

RED PUMPKIN BEETLE (*Aulacophora foveicollis* lucas)

Both grubs and adults attack the crop at seedling stage. The beetle feeds on cotyledons and foliage. Grubs develop into the soil, and occasionally feed on roots, causing wilting of plants. In severe attack, the crop is completely destroyed.

Control
- Plough the field repeatedly in hot summer months.
- Spray the crop twice or thrice with Rogor 0.1% near the base of plants just after germination.
- Spray 0.01% *neem* seed kernels extract and 0.4% *neem* oil to control the beetles effectively.

SPIDER MITE (*Tetranychus cucurbitae*)

These attack the leaves of the plant and suck the plant sap at their younger stage.

Control
- Spray the crop with Dicofol 0.5% or acephate 0.1%.

FRUIT FLY (*Bactrocera cucurbitae*)

This fly is a serious problem and reduces the number of marketable fruits. In India, fruit fly is the most common, destructive and polyphagous insect of cucurbits. Adult flies puncture the tender fruits and lay eggs in/onto the fruit skin. The maggots feed inside the fruits, causing premature dropping of fruit.

Control
- Collect and destroy the damaged fruits.
- Spray the crop with malathion 0.05%.

DISEASES

Anthracnose (*Colletorichum lagenarium*)

The fungus attacks leaves, stem and fruits. Yellowish water-soaked areas, which coalesce with one another on enlarging and turn brown to black, appear on leaves. It results in shrivelling and death of leaves.

Control

- Follow 2 to 3 years' crop rotation.
- Treat the seeds with Agrosan GN or Bavistin @ 2 g/kg of seed.
- Spray the crop with Blitox 0.2%, or Bavistin 0.1%, and repeat at 7- to 10-day intervals if necessary.

Powdery Mildew (*Erysiphe cichoracearum*)

Powdery mildew first appears as white powdery spots that may form on both the surfaces of leaves, on shoots and sometimes on flowers and fruits. These spots gradually spread over a large area of the leaves and stems. Snap melon shows some resistance power among melons.

Control

- Plant the crop in sunny areas as much as possible.
- Provide good air circulation.
- Avoid applying excess fertilizer. A good alternative is to use a slow-release fertilizer.
- Overhead sprinkling may help in reducing powdery mildew because spores are washed off the plant.
- Sulfur products (0.1% Karathane) have been used at 14-day intervals.

Downy Mildew [*Pseudoperonospora cubensis* (Berk. & Cart.) rosto]

Symptoms appear as numerous irregular small yellow areas surrounded by green tissues scattered all over the leaves. Lesions grow in size and coalesce with one another. Severely infected leaves roll upward with brownish tinge that produces a blighted appearance. Greyish black downy fungal growth is observed on undersurface of leaf in humid weather.

Control

- Air circulation and sunlight exposure help in checking the disease initiation and development.
- Field sanitation by burning crop debris helps in reducing the inoculum.
- Crop should be grown at wide spacing in well-drained soil.
- Protective spray of mancozeb 0.25% at 7-day intervals gives good control.
- In severe cases, foliar spray of metalaxyl + mancozeb @ 0.2% may be given only once.

Angular Leaf Spot (*Pseudomonas lachrymans* carsner)

The disease appear on leaves, stems and fruits as small, water-soaked, angular, white to brown lesions. Later on, these spots become dry, making irregular holes in the leaves.

Control

- Follow field sanitation to avoid the occurrence of disease.
- Destroy the disease debris after harvesting the crop.
- Spray the crop with streptomycin at 400 ppm or Bordeaux mixture at 1%.

53 Oriental Pickling Melon

M.K. Rana and Manoj Kumar Nalla

Botanical Name   : *Cucumis melo* var. conomon

Family   : Cucurbitaceae

Chromosome Number : 2n = 24

ORIGIN AND DISTRIBUTION

Cucurbits are the popular name for the plants of the Cucurbitaceae family, commonly known as the gourd family. The family is represented by about 120 genera and 800 species. Oriental pickling melon is the most ancient form of melon domesticated in China. It is mentioned that it was originated from the wild melon variety agrestis in China. Oriental pickling melon is cultivated in Asian countries like India, China, Japan, Korea and Southeast Asia. Based on the molecular polymorphism studies, it was suggested that *Cucumis melo* var. makuva and conomon have been derived from small seed type melon in south Asia. Conomon is supposed to originate in India, which is the secondary centre of diversity of cultivated melon, and it was introduced to China from the West *via* the Silk Road.

INTRODUCTION

Oriental pickling melon, which is very similar to pickling cucumber in taste, flavour and processing suitability, belongs to the family Cucurbitaceae. It is a popular vegetable in Kerala and is known as *kavivellari*. Mature and ripened fruits are used as cooked vegetable and immature fruits are used for pickling purposes. Fruits are oblong to cylindrical in shape and green at immature stage but turn golden yellow on ripening. Tender fruits can be used as salads and ripened fruits had special value in rituals, especially during the Vishu festival in Kerala.

COMPOSITION AND USES

COMPOSITION

Oriental pickling melon is a valuable source of vitamins and minerals as it contains unique anti-oxidants in moderate ratio such as β-carotene, vitamin A and vitamin C. It does not contain more amount of sugar, which makes it suitable for pickling. It is also a good source of calcium, phosphorus and iron. The nutritional composition of oriental pickling melon is given in Table 53.1.

USES

In oriental pickling melon, both immature and mature fruits are used as raw or cooked, pickled or made into sweets and juices. The fully ripened fruit is rarely used in eating habits since the half-ripe fruits are used for making pickles but fully ripe fruit is no longer valuable for pickling since its flesh is too soft. The fully ripe fruits are used in non-pickling products as it has unique properties of not being sweet and possessing a strong melon-like odour. The young fruits are used to prepare a

TABLE 53.1

Nutritional Composition of Oriental Pickling Melon (per 100 g Edible Portion)

Constituents	Contents	Constituents	Contents
Water (g)	95.3	Iron (mg)	0.45
Carbohydrates (g)	3.3	Vitamin A (IU)	2.9
Protein (g)	0.9	Thiamine (mg)	0.07
Fat (g)	0.2	Riboflavin (mg)	0.07
Calcium (mg)	35	Vitamin C (mg)	8.0
Phosphorus (mg)	45	Energy (kcal)	15

special dish called *rujuk cingur* in Indonesia. The fruits can be finely diced and used as a seasoning in salads and soups. The seed is used as raw or cooked and yields a rich nutty flavoured oil.

MEDICINAL USES

Fruits contain methyl thiopropionic acid ethyl ester, which is isolated from fully ripened fruits, is an odour-producing compound, which has antimutagenic and anticarcinogenic activity. In folkloric ethno medicine, fruit is used as a moisturiser for skin and for burns, scalds and abrasions. Its seeds are used as a source of antitussive, digestive, febrifuge and vermifuge, and the roots are used as a source of diuretic and emetic.

BOTANY

The plant is trailing prostate or climbing, much branched, annual monoecious herb with hirsute stem having diameter of 5.5–8.5 mm. The tendrils are simple unbranched. Leaves are alternate, simple, sub-orbicular to ovate, shallowly five- to seven-lobed with finely dentate margin covered with villous hairs on both the surfaces. The plant form is andromonoecious, bearing yellowish flowers on 5–7 cm long peduncle. Male flowers are fasciculate with two to four flowers. Calyx and corolla are five in number and are campanulate. Stamens are yellow in colour and three in number. Perfect flowers are solitary with a short sericeous (short-pressed hairs). The flower is characterized by inferior ovary and parietal placentation. Fruit is polymorphous, ellipsoid oval, oblong and pyriform to globose elongate pepo. The fruit is not sweet, not aromatic and it is of a non-climacteric nature. The fruits are smooth and glabrescent and their colour varies from white, yellow, golden yellow to yellowish white with edible insipid to mildly sweet flesh and numerous white, elliptical, small (<8 mm) flattened seeds. Botanical groups of oriental pickling melon are listed in Table 53.2.

CLIMATIC REQUIREMENT

Oriental pickling melon is essentially a warm season crop but can be successfully grown in tropical, subtropical and temperature regions. Compared to other melons, it needs long period of warm, preferably dry weather with abundant sunshine. This is very susceptible to frost and not tolerant to shaded conditions. Excess humidity promotes disease like powdery mildew, downy mildew, anthracnose and other pests such as fruit flies. It requires a minimum temperature of 18°C during early growth period of the crop but optimal temperature required by this crop is in the range of 24–27°C.

TABLE 53.2
Botanical Groups of Melon (*Cucumis Melo*)

Groups	Subspecies agrestis	Subspecies melo
Non-sweet	*acidulus, conomon* and *momordica*	*chate, flexuosus* and *tibish*
Sweet	*makuwa* and *chinensis*	*adana, ameri, cantalupensis, chandalak, reticulates* and *inodorus*
Fragrant		*dudaim*

SOIL REQUIREMENT

A well-drained loamy soil rich in organic matter is preferred more for its cultivation, though it can be grown on a wide range of soils. The crop requires a soil pH ranging from 6 to 7. Lighter soils that warm quickly in spring usually gives early yield, but in heavier soils, the vine growth is greater and fruits mature late. Riverbeds that contain alluvial substrate and subterranean moisture of river streams support cucurbits cultivation, which is popularly known as riverbed cultivation. Those soils that are prone to cracking in summer and water logging in rainy season may not be suitable for growing oriental pickling melon.

CULTIVATED VARIETIES

SAUBHAGYA

An early variety developed at Kerala Agricultural University bears small to medium sized oblong green fruits with whitish green lines from stalk end to stylar end, which turn golden yellow at ripening. Its plants spread less, so the variety is suitable for close planting. The length of the fruit varies from 18 to 23 cm and fruit weight ranges from 1000 to 1400 g. Its average yield is 170 quintals per hectare in crop duration of about 60–70 days.

MUDICODE

A variety developed at Kerala Agricultural University bears long cylindrical fruits, which are green in colour at immature stage and turn attractive golden yellow at ripening stage. The length of the fruit varies from 27 to 35 cm and fruit weight ranges from 1.8 to 2.5 kg. The total duration of crop is about 90 days. It is very high yielding and the productivity of the variety is 300 quintals per hectare.

ARUNIMA

A variety developed at Kerala Agricultural University bears small- to medium-sized fruits. The colour of the fruit is green with creamy spots at tender stage, which turn to orange yellow at ripening stage. The average length of the fruit is 33.14 cm and the fruit weight is around 2.3 kg. The yield potential of the variety is about 270 quintals per hectare.

SOWING TIME

The ideal time for the sowing of oriental pickling melon is from January to March and from September to December. Sowing of seeds of a peculiar variety at its correct time is very essential to get higher yield and to improve the quality of fruits by eluding the unfavourable weather conditions that have contrary effect on these plant populations. For garden crops, it is sown by the middle of February, and for rainy season crop, it is seeded in June–July. In South and central India where winter is mild, it can be grown almost throughout the year.

SEED RATE

The seed rate varies according to the season, plant spacing and seed viability. Approximately 0.5 to 0.75 kg seed of oriental pickling melon is needed to sow a hectare area. Seeds are directly sown in the main field without any nursery growing.

SOWING METHOD

The land should be thoroughly prepared in order to get fine tilth by harrowing repeatedly for the preparation of seedbeds. Pits of 60 cm diameter and 30–45 cm deep are dug in seedbeds. The distance between two pits is kept 2 × 1.5 m. The top soil of the pit is mixed with well-decomposed farmyard manure and fertilizers. Seeds are soaked in water for 24 hours prior to planting for better germination. Soaking of seeds in 0.2% of Bavistin reduces the attack of soil-borne diseases. A pre-sowing irrigation of 5–6 days prior to planting is beneficial. Four to five seeds are sown in each pit at 1–2 cm depth. Sowing of seeds deeper into the soil should be avoided, as it hampers germination. The seeds usually germinate 4–5 days after sowing and water is applied with the help of a rose can. Unhealthy plants are removed 2 weeks after sowing, retaining only three vigorous plants per pit. In some cases where early crop is desired, seeds can be sown in polyethylene bags (10 × 15 cm size) of 100–200 gauges and put under cover to protect them from adverse climatic conditions.

NUTRITIONAL REQUIREMENT

Oriental pickling melon responds well to manuring and fertilization. Application of nitrogen, phosphorus and potash 100:75:50 kg/ha is found to be optimum for plant growth and development of fruits. A full dose of phosphorus and potassium along with a half dose of nitrogen is applied as basal dose at the time of sowing and remaining half of the nitrogen is applied 30–35 days after sowing. The nitrogenous fertilizer will help the plant foliage to grow as much as possible. The phosphate and potassium will help the plant to bear and develop fruits. A well-rotten organic manure 20–25 tonnes per hectare is applied at the time field preparation for thorough mixing into the soil. If the crop growth is poor, 1% urea can also be sprayed.

IRRIGATION REQUIREMENT

During initial stages, an interval of 3–4 days irrigation is given, but during flowering and fruiting, irrigation is given on alternate days since flowering and fruiting are very critical stages, thus, water stress during these periods should be avoided. Frequent irrigation is very important in spring-summer, while in rainy season, supplying irrigation is not necessary if the rainfall is well distributed between July and September. In heavy soils, it is always safe to sow seeds when the soil contains adequate moisture for seed germination. Crust formation should be avoided for better germination of seeds. When the seeds are sown in dry soil on ridges, the ridges should always be kept moist until the germination is completed. Furrow method is ideal for irrigation but if water is limited, trickle or drip irrigation system is used.

INTERCULTURAL OPERATIONS

HOEING AND WEEDING

During early stages of crop growth, the field should be kept clean by hoeing and weeding. First and second hoeing are done 20–25 and 45–45 days after sowing, respectively. Hoeing should be avoided when the plants have covered the interspaces in the field. Hoeing is necessary after the application of nitrogenous fertilizer. At later stages, large weeds are pulled out manually without disturbing the vines.

EARTHING UP AND STAKING

Oriental pickling melon is a shallow-rooted crop. Therefore, deep intercultural operations should be avoided. Earthing up is essential to cover the roots properly especially in rainy season crops. All secondary shoots should be pruned up to five nodes to improve yield and fruit quality. Staking of plant especially in rainy season is helpful in checking rotting of fruits.

MULCHING

Mulching is commonly done in crops grown in spring–summer. Mulch can be organic or black polyethylene sheet, based on availability. However, the mulching should allow the nutrients and water to penetrate to the plant root zone. It also prevents weed emergence, keeps the soil moist and increases its water-holding capacity.

POLLINATION

Oriental pickling melon is a cross pollinated crop. Insects especially bees are the main pollinators of this crop. Pollination is a problem during wet and overcast conditions. Introduction of beehives ensures good pollination thereby enhancing fruit yield.

USE OF GROWTH REGULATORS

Spraying of vines with flowering hormones at the two to four true leaf stage increases the ratio of female to male flowers and thereby enhancing the fruit yield. Their application at the two to four true leaf stage is essential for changing sex ratio, since at this stage, differentiation of sex occurs in the plant. Application of plant growth regulator such as etheral (an ethylene-releasing compound) 100–250 ppm enhances the femaleness in oriental pickling melon.

HARVESTING

Stage of harvesting depends upon the purpose for which the fruits are being harvested. Usually, the fruits should be harvested at tender stage for pickling purposes but for vegetable purposes after attaining full size and changing colour from green to golden brown. Harvesting of tender fruits for pickling is a repetitive process, thus, should be done either daily or at alternate days. Fruits are harvested by twisting with hands and snapping off the vine since pulling the fruits can damage the vine.

YIELD

The average yield of oriental pickling melon varies from 100 to 150 quintals per hectare, depending on crop variety, soil type, growing season and cultural practices adopted by the growers.

POST-HARVEST MANAGEMENT

The harvested fruits should be washed and cleaned before storing. Pre-cooling is done in order to remove field/sensible heat. Bruising and crushing during handling should be avoided as it decreases post-harvest life of the fruits. All the diseased, infected and low-quality fruits should be discarded from the lot. The fruits can be stored safely for a longer time at 10°C–12°C coupled with 95% relative humidity.

GRADING AND PACKING

The harvested fruits are cleaned and wrapped in newspaper. Thereafter, the fruits are packed in mulberry or bamboo baskets, or in gunny bags according to the availability of transportation facilities. The marketable fruits are graded based on the size, shape, colour and fruits free from handling defects and decay.

INSECT-PESTS

Aphids (*Aphis gossypii*)

Large colonies of nymphs and adults are found on tender twigs and shoots and also on the ventral surface leaves, sucking the vital sap from the tissues. The affected parts of the plant turn yellow and become wrinkled and deformed in shape and ultimately the plant dries and dies away. Fruit size and quality are also reduced. The aphids also exude a copious quantity of honeydew, on which sooty mould develops, which in turn hinders photosynthetic activity of the vines, resulting in stunted growth. The fruits covered by sooty mould appear unattractive and lose their market value.

Control

- Collect and destroy the affected parts to prevent the spread of viral diseases.
- Spray the crop with Malathion @ 2 ml, Metasystox @ 1.5 ml, or Rogor @ 1 ml/litre of water at fortnightly intervals.

Leaf Hopper (*Amrasca biguttula*)

Both the nymphs and adults suck cell sap from the tender leaves of oriental pickling melon vines, causing typical hopper burn symptoms. Initially, the margins of the leaf turn yellow and finally dry off. Fruit size and quality are also reduced. The hoppers spread the viral diseases.

Control

- Collect and destroy the affected parts to prevent the spread of viral diseases.
- Spray *neem* and garlic oil mixture at fortnightly intervals.
- Remove garden trash and other debris shortly after harvest to reduce over-wintering sites.
- Spray the crop with malathion @ 2 ml or Metasystox @ 1.5 ml/litre of water.

Red Spider Mites (*Tetranychus* spp.)

Spider mites live in colonies on the undersurface of the leaves. The larvae, nymphs and adults of mites lacerate the leaves from the undersurface. They suck cell sap, resulting in production of white patches between veins. Infected leaves turn yellow and fall off prematurely. In severe cases, intense webbing occurs, which gives a dusty appearance to undersurface of the leaves.

Control

- Proper irrigation and clean cultivation are essential to control the pest population.
- Collect and burn the severely infested plant parts to reduce further multiplication of mites.
- Spray the crop with dicofol 0.05% or wettable sulfur 0.3%.
- Spray *neem* and garlic oil mixture at fortnightly intervals on undersurface of the leaves in early morning hours.

Red Pumpkin Beetle (*Aulacophora foveicollis*)

The grubs feed on roots and underground portion of the host plants and also on the fruits that are touching the soil. The damaged roots and infested plant parts start rotting due to secondary infection by saprophytic fungi and the unripe fruits of such vines dry up. Infested fruits become unfit for human consumption. Adult beetles feed voraciously on leaf lamina, making irregular holes. They prefer tender leaves of young seedlings. Their damage may even kill the seedlings.

Control

- Repeated deep ploughing in hot summer months kills grubs and pupae as insect pupate in the soil.
- Collect and destroy of beetles in early stage of infection.
- Apply Furadan 3G Granule 3–4 cm deep into the soil near base of germinated seedlings.
- Spray malathion @ 2 ml/litre of water for effective control.
- Dust the crop with 5% malathion @ 10 kg/ha to control the pest.

Fruit Fly (*Bacterocera cucurbitae, Dacus cucurbitae*)

The maggot burrows into the fruits. As a result, the infested fruits start decaying and drop off from the plant. The flies mainly prefer tender fruits to lay eggs. The ovipositional punctures caused by adults also cause injury to the fruit, which results in distortion and malformation of fruit. The maggots feed on fruit pulp as well as on immature seeds and cause premature dropping of fruit.

Control

- Collect and destroy infested fruits on the plant.
- Grow a few lines of maize around the cucurbit field, as flies rest on such tall plants.
- Cover developing fruits with paper or polythene cover immediately after pollination.
- To prevent egg-laying, flytraps (pheromone trap) can be set up in the field with 1% methyl eugenol or citronella oil.

DISEASES

Collar Rot (*Rhizoctonia solanii* and *Phythium* spp.)

Collar rot is more serious under waterlogged conditions and during rainy season. Dark brown water-soaked lesions are formed, girdling at the base of stem occurs and finally the entire plant is decayed and dies. Cotton-like white mycelium on surface of the infected tissue occurs with advanced infections. Affected plants can easily be uprooted but the lower part of the root usually remains in the soil.

Control

- Sow the seeds on raised beds.
- Treat the seed with captan @ 3 g /kg of seed before sowing.
- Remove and deeply bury the previous crop residues.
- Drench the beds with Ridomil (0.2%) or carbendazim (0.1%).

Fusarium Wilt (*Fusarium oxysporum* and *Fusarium solani*)

In young seedlings, the cotyledons drop and wither. Older plants wilt suddenly and vascular bundles at the collar region show brown discoloration. The petiole and leaves drop and wilt. When the plant roots and basal stems are split open, dark brown or black discolouration of the vascular tissues may be seen.

Control

- Grow resistant varieties.
- Follow long crop rotation with non-host crops.

- Give hot water treatment to seeds at 52°C for 15 minutes.
- Treat the seed with carbendazim @ 3 g/kg.

Powdery Mildew (*Sphaerotheca fuliginea*)

The infected leaves show white to dirty grey spots or patches and other succulent plant parts show symptoms of greyish-white powdery patches of various size. A white powdery coat of fungal growth appears on leaf surface, generally on upper surface but sometimes also on lower surface and stem. The affected leaves become yellow and later on fall down. The vines do not set fruits properly. Powdery coating covers the entire plant parts in later stages and causes defoliation.

Control

- Irrigate the crop frequently.
- Remove and destroy the infected leaves.
- Dust the crop with sulfur @ 15 to 20 kg/ha.
- Spray Karathane 0.5%, Calixin 0.05%, or carbendazim 0.1% at fortnightly intervals.
- Spray 0.3% solution of wettable sulfur such as Sulfex.

Downy Mildew (*Pseudoperonospora cubensis*)

Water-soaked lesions appear on undersurface of the leaf lamina. Angular spots appear on upper surface, similar to water-soaked lesions. In young seedlings, discolouration occurs, and in severe cases, the whole plant dies. Lesions appear first on the older leaves and progressively appear on the younger leaves. As the lesions expand, they may remain yellow or become dry and brown. Affected vines do not set fruit properly. The incidence is very severe under wet weather conditions. The disease is soil as well as seed borne.

Control

- Grow resistant cultivars.
- Follow long crop rotation and sanitation.
- Pick and destroy the affected leaves.
- Treat the seed with hot water at 52°C for 30 minutes to kill the pathogens present on seed.
- Follow prophylactic sprays of mencozeb 0.3%, Dithane M-45 0.2%, or Dithane Z-78 0.2%.

Anthracnose (*Collectotrichum lagenarium*)

The first symptoms of anthracnose are the appearance of spots on leaves that begin as yellowish or water-soaked areas. Spots enlarge and turn brown to black. The diseased tissue dries and the centre of the spots fall out, giving a *shot-hole* appearance. Infected fruits have black, circular, sunken cankers of different sizes. Under humid conditions, pink masses of spores can be seen in the centre of these spots. Vines have brownish specks, which grow into angular to circular spots. Girdling of affected portion leads to blight symptoms.

Control

- Follow long crop rotation with non-host crops.
- Clean cultivation minimizes disease incidence.
- Treat the seed with carbendazim @ 2.5 g/kg.
- Soak the seeds in 0.1% mercuric chloride solution for 5 minutes.
- Spray Indofil M-45 0.35%, benomyl 0.2%, or carbendazim 0.1% at 10-day intervals.

54 Ivy/Little Gourd

T. Saraswathi and K. Shoba Thingalmaniyan

Botanical Name	: *Coccina grandis* (L.) Voigt.
	Syn.: *Coccinia cordifolia* (L.) Cogn.
	Cephalandra indica
	Bryonia grandis L.
Family	: Cucurbitaceae
Chromosome Number	: 2n = 24

ORIGIN AND DISTRIBUTION

The ivy or little gourd (*Coccinia grandis* L.) is indigenous to India, particularly the Eastern India where a rich gene pool is available in wild as well as in homestead gardens. The nomenclature is derived from the Latin word *coccineus*, meaning scarlet, in reference to fruit colour. In Indian scriptures, this vegetable has also been mentioned extensively as *bimba* fruit (*Coccinia*), of which ripe red fruit has been compared with woman's lips in several places like *Mahabharata* and also referred to by Kalidasa in his Sanskrit plays. It has about 35 species distributed in Tropical Africa and Asia, out of which, only single species, *i.e.*, *Coccinia grandis*, is under cultivation. The related species are *Coccinia histella* and *Coccinia sessilifolia*, which are widely cultivated in several countries, like Africa, Central America, Philippines, China, Indonesia, Malaysia, Myanmar, Thailand, Vietnam, eastern Papua New Guinea, the Northern Territories, Australia and other tropical Asian countries. Australia. In Ghana, a monoecious species is also cultivated. It has become a vigorous weed in several countries like Australia and Spain and in the US states of Hawaii, Texas and Florida. Under Indian conditions, ivy gourd has naturalized in the central and peninsular area.

INTRODUCTION

Ivy gourd, little gourd, or scarlet-fruited [*Coccinia grandis* (L.) Voigt], popularly known as *kundru*, is so far a neglected and underutilized species of the Cucurbitaceae family. It is a hardy perennial dioecious herbaceous creeper consisting distinct male and female plants and has a vast scope of growing in marginal lands where it gives fruits almost year yound. The gynodioecious sex form consisting of female and hermaphrodite flowers has also been observed in nature. In tribal areas of Orissa, Jharkhand, Chhattisgarh and Andhra Pradesh, presently, the fruits of ivy gourd are being gathered from natural forests or are being grown in homestead gardens and small pockets on commercial scale. Now, it is gaining popularity among the consumers due to its nutritive and medicinal value and year yound production potential. Owing to non-availability of standard variety, the farmers are cultivating only land races, which yield variable fruits. Ivy gourd is a great potential of export, thus, there is a need to develop insect-pest and disease resistant early varieties having bold and tender fruits with good storability and suitable for pickle preparation.

COMPOSITION AND USES

COMPOSITION

The immature fruits are rich source of carbohydrates, proteins (1.2%), several macro- and micro-nutrients, vitamins A and vitamin C. Its edible portion constitutes around 90%. Its leaves contain protein 3.3–4.9 g and vitamin A 8000–18000 IU/100 g edible portion. The immature fruits of *Coccinia adenensis* species are bitter in taste due to the presence of cucurbitacin B in the form of glycoside, which is lost rapidly during ripening. The nutritional composition of ivy gourd fruit is given in Table 54.1.

USES

The young leaves and slender tops of the stems are eaten as greens in soups, or as a side dish, often with rice. Its immature tender green fruits are mainly eaten raw in salads or as a vegetable curry by deep-frying, boiling, or steaming along with spices in eastern and southern parts of the country. It is also used in *sambar*, a vegetable- and lentil-based soup. It can also be used for various preparations of mixed vegetable and can also be candied. In Ethiopia, the species *Coccinia abysinica* is cultivated for its edible tuberous roots.

MEDICINAL USES

The medicinal properties of scarlet gourd are attributed to it under most Asian indigenous systems of medicine, in which, the fruits are used to treat leprosy, fever, asthma, bronchitis, skin diseases and bronchitis. The plant has shown antioxidant, anti-triglyceride and antibacterial activity and is useful in the treatment of jaundice. The fruit also possesses mast cell stabilizing, anti-anaphylactic and antihistaminic potential. It has also been used to treat abscesses and high blood pressure. In Bangladesh, the roots are used to treat osteoarthritis and joint pain. A paste prepared from leaves is applied to the skin to treat scabies and skin eruptions. Leaf juice is used to treat earache. The plant as a whole is laxative. Internally, it is used in the treatment of gonorrhoea. The juice of roots and leaves are used to reduce the amount of sugar in urine and improves metabolism. Its products help in regulating blood sugar levels, therefore, it is recommended for diabetic patients. Ethanol extract of the plant has shown hypoglycemic principles.

BOTANY

The long-lived climbing vine forming a very dense cover over bower is slender, much branched, semi-perennial and trained as live fence in home gardens. Its stems are thin, green, smooth, terete, slightly scabrid, longitudinally furrowed when young, dark green and 6–25 m long but as

TABLE 54.1

Nutritional Composition of Ivy Gourd Fruit (per 100 g Edible Portion)

Constituents	Contents	Constituents	Contents
Water (g)	93.5	Carotene (µg)	156
Carbohydrates (g)	3.1	Vitamin A (IU)	260
Protein (g)	1.2	Thiamine (mg)	0.07
Fat (g)	0.1	Riboflavin (mg)	0.08
Dietary fibre (g)	1.6	Niacin (mg)	0.7
Calcium (mg)	40	Ascorbic acid (mg)	1.4
Phosphorus (mg)	30	Energy (kcal)	18
Iron (mg)	1.4		

the stem becomes, it gets swollen and semi-succulent in nature. The stem develops roots where its nodes come in contact of moist soil. Its tuberous roots are succulent and are capable of storing water throughout the dry period. The alternately arranged leaves are ovate to oblong or heart to pentagon shaped with angular basal lobes. Petiole is sub-terete, narrowly furrowed on upper side, thinly pubescent, dark green and 0.74 cm long. These slightly to prominently lobed leaves are somewhat ivy-shaped in nature and usually have tiny teeth spaced along their margins. Long tendrils are present in axils of the leaves. This species produces male and female flowers on separate plants. It bears white tubular solitary flower with long peduncle in the leaf axils on 1–5 cm long stalk, which is faintly curved, shining, glabrous, yellowish green and 7–25 cm long. Calyx is campanulate, greenish white, glabrous and 1–1.25 cm long. Corolla is campanulate, limb 4–6 partite hardly down to the middle and white on the outside yellowish green veins. Lobes are broadly ovate, acute, 1.5–2.5 cm long and broad. Male flowers have three to five stamens with conduplicate, connate anthers of 0.5–0.75 cm long and connective is yellowish white in colour. Female flowers have three staminodes and are white in colour. Ovary is inferior, narrowly oblong, yellowish green, shining with several longitudinal, undulate, glabrous and 1.5–2 cm long. Stigma is three-lobed. Fruits, which develop from flowers arising in the leaf axils of new growth, are cylindrical with base suddenly narrowed over short length. The stripes on immature fruits are green, which on ripening first turn to orange colour and finally scarlet. Fruit contains many seeds that are 3–6.5 cm long and 1–1.25 cm in diameter. Seeds are transverse, oblong with a narrow base and flat.

RELATED SPECIES

- *Coccinia trilobata* (*Caudiciform coccinia*): The plant is an herbaceous dioecious of tropical Africa. Only its leaves are used as vegetables.
- *Coccinia abyssinica* (**Anchote**): Its tuberous fleshy rootstocks and young shoots are cooked as vegetable.
- *Coccinia rehmannii* : Its fruits are edible. Its starchy tuberous roots are eaten after roasting.

CLIMATIC REQUIREMENT

Among the climatic factors, relative humidity is most important for the anthesis of male and female flowers and anther dehiscence, thus, the crop requires well-distributed rainfall and high humidity for its successful cultivation. It thrives best under moderately warm and humid condition. Its cultivation is possible in areas where average temperature lies between 25°C and 35°C with annual rainfall 150–250 cm. The crop is very sensitive to frost and chilling temperature as the temperature below 15°C damages its vegetative parts. In winters, the stems and leaves are killed by low temperature, while the underground tuberous roots remain dormant in soil during December–January, which sprout again in February–March when temperature becomes normal.

SOIL REQUIREMENT

Ivy gourd is perennial in nature and keeps growing for 3–4 years in field for regular bearing, thus, it requires sandy loam or loam soil rich in organic matter for better plant growth and production of excellent quality fruits. The soil pH 5.8 to 6.8 is ideal. The highly acidic and sodic soils are detrimental to plant health. The soil should be well drained, as the crop is susceptible to waterlogging conditions. The field is prepared in May–June by ploughing two or three times to destroy the weeds, insect-pests and disease-causing pathogens. Repeated deep ploughings in summer months with harrow or cultivator followed by planking are essential before rain to get a good tilth.

VARIETIES

A large number of local varieties are popular among the farmers in their respective areas of adaptability. Ivy gourd has been classified into two distinct types, *i.e.*, bitter and sweet types, and based on fruit characters as recorded by Indian Institute of Vegetable Research, Varanasi (Uttar Pradesh), ivy gourd has been categorized into two groups:

- **Round or oval fruited type:** Basically, the fruits are round to oval in shape with light green to light-yellow colour stripes.
- **Long-fruited type:** The fruits are long with light green stripes.

Preliminary evaluations of few local collections were made at Indian Institute of Vegetable Research, Varanasi, and based on fruiting ability and fruit quality, a few desirable strains identified and recommended for commercial cultivation are VRK-05, VRK-10, VRK-20 and VRK-35.

The descriptions of few selections giving higher yield with better quality fruits are given as follows:

VRK 20

A very early strain bears light green striped fruits of 6.0 cm length and 2.7 cm diameter. The individual fruit weight is about 20 g.

VRK 31

Fruit is somewhat swollen in the middle but pointed at the upper portion. The individual fruit weight is 25–30 g.

VRK 35

A high-yielding strain bears highly striped fruits of medium size with fruit weight 15–18 g.

VRK 37

The oval-shaped fruits of this strain are round 1.90 cm in diameter with individual fruit weight 16 g.

VRK 49

Fruits are oval shaped. It is a high-yielding variety with fruit weight about 15 g.

INDIRA KUNDRU 5

A very high-yielding variety developed at Indira Gandhi Krishi Vishvavidyalaya, Raipur, has been released by State Variety Release Committee, Chhattishgarh State. The plants bear light green oval shape fruits having 4.30 cm length and 2.63 cm diameter. Its fruit yield is 22.94 kg/plant.

INDIRA KUNDRU 35

A very high yielding variety isolated at Indira Gandhi Krishi Vishvavidyalaya, Raipur, has been released by the State Variety Release Committee, Chhattishgarh State. The plants bear light green long fruits having 6.0 cm length and 2.43 cm diameter. Its fruit yield is 21.08 kg per plant.

ARKA NEELACHAL KUNKHI (CHIG-15)

A promising genotype of ivy gourd for salad purpose developed at Indian Institute of Horticultural Research, Bangalore, has a sequential fruiting habit, and fruit develops by means of vegetative parthenocarpy. Male plants are not needed for pollination, hence, more number of female plants can be accommodated per unit area. A single plant produces 20 kg fruits in one growing season.

ARKA NEELACHAL SABUJA (CHIG-33)

A high value culinary variety of ivy gourd developed at Indian Institute of Horticultural Research, Bangalore, bears dark green fruits, which store well for 2–3 days and can withstand transport shock as well. The plants are vigorous and produce high yield. The fruits contain high pulp and soft seeds.

CO 1

A perennial variety of ivy gourd developed at Tamil Nadu Agricultural University, Coimbatore, through a clonal selection from Anaikatti type is suitable for culinary purpose. It bears long green, less seeded and sweet (4.5°Brix) fruits with white stripes. Its average yield is 830.9 q/ha per year.

SULABHA

A perennial variety of ivy gourd developed at Kerala Agricultural University is high yielding (600 q/ha), having a fruit length and weight of 9.25 cm and 18.48 g, respectively.

PLANTING TIME

Ivy gourd can be planted twice, once in spring–summer and secondly in rainy season. For spring–summer crop, planting is done in February–March when temperature becomes normal, and for rainy season crop, planting is done in June–July. Planting in June–July is preferable to February–March. Under rain-fed conditions, planting is done with the onset of monsoon in pits of 30 × 30 × 30 cm at a distance of 3 m.

SEED RATE

Ivy gourd is perennial and dioecious in nature (male and female flowers appear on separate plants), thus, planting of male and female plants in the ratio of 1:9 is quite essential for better pollination and higher yield. Its seeds exhibit no dormancy but remain viable for only a short period. If the crop is raised through seeds, there is possibility of high ratio of male to female plants, which will affect the yield adversely. It is true that male and female plants in field can be identified only at flowering and fruiting stage, hence, it is propagated through semi-hard vine cuttings. Nearly 2500–3000 well-rooted and sprouted cuttings are required for planting a hectare land area.

PLANTING METHODS

The following methods can be adopted for the planting of ivy gourd.

RAISED BED METHOD

Ivy gourd is usually planted on raised beds for getting higher yield with better quality fruits. In this method, ridges and furrows are prepared either manually with spade or mechanically by using tractor drawn furrow opener. For the preparation of raised beds, 45 cm wide furrows are opened at 4.0 m distance. The plants prepared from cuttings are planted on the edges of channels.

Flat Bed Method

In this method, shallow pits of 60 × 60 × 45 cm size, which are dug at a distance of 4 m 3 weeks before sowing, are left open for partial solarization, and thereafter, each pit is filled with a mixture of soil, 4 kg well-decomposed farmyard manure or compost, urea 40 g, single super phosphate 100 g, muriate of potash 70 g, Furadan 2 g and *neem* cake 100 g. Before filling the pits, all the ingredients are mixed thoroughly.

Mound Method

In this method, 15–20 cm size raised mounds are prepared. Well-decomposed farmyard manure or compost @ 4 kg and urea 25 g, single super phosphate 75 g, muriate of potash 60 g, Furadan 2 g and *neem* cake 100 g are incorporated in the soil before filling the mounds. The plants prepared from cuttings are planted in the evening at a depth of 20 cm.

Nursery Production

Ivy gourd is generally propagated by using 30 cm long stem cuttings during July–August when vines are available and temperature is favourable for sprouting. The stem cuttings of pencil thickness (8–10 mm thick) can directly be propagated in the field. Usually, two cuttings are planted per pit. In plains, sometimes, it becomes difficult to get the planting material in June–July due to flowering and fruiting season. Therefore, when fruiting ends in the months of October and November and growers start heading back to the plant, the nursery can be prepared by collecting vines. Nursery can be raised on ground in polyhouse or in planting tubes. Stem cuttings of pencil thickness (6 mm thick) and 25–30 cm length with at least five to seven nodes are taken from 1-year-old healthy vines and planted vertically in well-prepared raised nursery beds having 20 cm height from soil surface. The planting tubes should be of 15 × 10 cm size, having four or five holes at the bottom and upper surface for better aeration and drainage. The polyethylene tubes filled with a mixture of garden soil, sand and well-decomposed farmyard manure in equal amount could be used for nursery production. A single cutting with five to seven nodes is placed in the tube at a depth of 5–6 cm. Sprouting starts 15–20 days after planting, and the plants become ready for shifting in the field in about 7–8 weeks. The sprouting can be improved by dipping the cuttings in the solution of auxin, *i.e.*, indole acetic acid or naphthalene acetic acid.

NUTRITIONAL REQUIREMENT

The fertilizer requirement of ivy gourd depends on soil type and soil fertility. However, in general, nitrogen is applied 60 kg/ha and phosphorus and potash each 40 kg/ha to get good yield. A half-dose of nitrogen along with a full dose of phosphorus and potash is applied at the time of planting and the remaining half of the nitrogen is given in four splits at monthly intervals from June to July. Well-decomposed farmyard manure should be applied @ 4 kg/plant just after pruning. In the second year, the same quantity of manure and fertilizers is applied again. The basins every year are cleaned and filled with fresh well-decomposed farmyard manure.

IRRIGATION REQUIREMENT

After planting fresh or sprouted cuttings in pits, irrigation is applied frequently for the proper establishment of plants. Thereafter, the irrigation frequency depends on the soil type, growing season and prevailing climatic conditions. Ivy gourd is very sensitive to water stagnation conditions, thus, excess irrigation should be avoided. Usually, irrigation is required at intervals of 15–20 days in spring and 5–6 days in summer, while in rainy season, if rainfall is well distributed between July and September, applying irrigation is not necessary at all. During rooting of the cuttings, adequate soil moisture is necessary. Excessive moisture in soil affects rooting, sprouting and plant growth adversely and also leads to yellowing of crop and wilting. It is essential to maintain adequate moisture in soil during flowering and fruit setting stage. Drip irrigation is considered quite useful as it not only saves 50% water but also is helpful in keeping the weeds under control.

INTERCULTURAL OPERATIONS

HOEING AND WEEDING

Ivy gourd is a perennial crop with deep-root system, thus, intercultural operations play an important role in flowering and fruiting of the crop. In early stages of crop growth, the weeds compete with crop for nutrients, moisture, light and space, hence, two or three hoeings are essential to keep the weeds down and provide better aeration into the soil near root zone for better growth and yield, but in late stages, it tends to grow in dense blankets that shade other plants from sunlight forming dense canopy that smothers vegetation underneath.

Mounting up of soil around the roots is also essential to create conducive environment for the roots. If aeration is good near root zone, the plant flourish better and gives higher yield. However, the hoeing should be done shallow to avoid damage to the feeding roots. In late stages of crop growth when the plants have trailed on bowers, weeds do not grow due to shading effects.

MULCHING

Mulching not only conserves soil moisture and regulates soil temperature but also increases yield and quality of ivy gourd fruits. Mulching with organic material not only improves the soil organic matter content but also improves soil fertility particularly in infertile sandy soils through a phenomenon of decomposition at the end of crop. Covering the soil with black polyethylene sheet suppresses the weeds. It also extends the fruiting period by regulating temperature and providing favourable conditions to the vines. Mulching with paddy straw, wheat straw, dry grass, or sugarcane tresses increases the number of fruits per plant and fresh weight per fruit and also reduces the percentage of unmarketable fruits. Mulching also promotes early fruit setting (2–3 weeks).

PREPARATION OF BOWERS

The plants of ivy gourd are either allowed to creep on the ground or trailed on bower. The vines trained on bower give yield nearly 40% higher than the vines allowed to creep on the ground. Ivy gourd being creeper bears more and quality fruits on bower, thus, it is essential to provide support to its vines either using poles or bower system. Within 2 months of planting, the vines start trailing over bower, which is prepared from cemented poles or angle iron poles of 2.0 m length by placing their base 45 cm deep into the soil at a distance of 4 m. The corner poles are supported and tightened by two another poles in a bending manner. Besides, a heavy stone, which is tied with corner poles with connecting wires of 2.0 mm thickness, is placed near base of the pole to support the wire. While fixing pole, 30–45 cm long rod is placed at upper end of the pole in such a way that 15 cm length may remain above to poles. The steel wires of 3 mm thickness are tied with stones through a nut-bolt. The wires are tightened every year after harvesting and cutting the vines. The wire of 8 mm thickness is used to connect a pole to another pole.

TRAINING

Pruning and training of vines are essential operations for obtaining higher yield of better quality fruits. The vines after attaining a length of 30–40 cm are trained with the help of coconut or thin jute twine and allowed to reach over the bower. In early stages of crop growth, bamboo sticks near the plants are inserted into the soil to provide proper support, and for training the vines on bower system, retaining two or three branches rest of the branches are pruned. The central shoots emerging from the main stem below the bower are removed in order to encourage shoots reaching on the bower. For the pruning of dead and diseased vines, the period from November to December is considered the best time since during this period, flowering and fruiting are ceased due to low temperature. After pruning the vines, Thiovit or Geigy-1250 fungicide at 0.2% concentration is sprayed at weekly intervals to protect the newly emerged shoots from fungal diseases. The new sprouts start emerging with the onset of monsoon. If more sprouts emerge out from the base, they must be pruned from the base with the help of sharp knife so that the vines reaching to bower may get sufficient strength. The shoots, which are of finger thickness, are also pruned back to two buds. All weak branched vines are also removed while pruning the shoots.

HARVESTING

Ivy gourd starts flowering and fruiting 10–12 weeks after planting and fruiting continues for 8 to 10 months. Normally, the plant starts fruiting from February onwards and continues up to November, though the highest yield can be obtained between July and November. Under mild conditions, it bears fruits almost throughout the year. However, the fruit yield is very low in regions where winter is relatively severe. The fruits are harvested at a stage when they are fully grown and start changing their colour from dark green to light green. Generally, 5 cm long oval-shaped fruits are preferred more by the consumers in the market, thus, picking is done at short intervals to get tender fruits of excellent quality. Its vines are very tender, thus, care should be taken while harvesting the fruits so that the injury to its vines can be avoided. Under north Indian conditions, it gives fruits only once a year, whereas, under south and central Indian conditions where winter is not distinct and plant growth continues, ivy gourd gives fruits year yound. The crop gives normal yield for 3 to 4 years under a condition of proper management.

YIELD

Under ideal conditions, roughly 8–10 kg fruits can be harvested per vine per year. Nevertheless, on an average, 200–300 fruits weighing 3–4 kg can be harvested from a vine in a year. Its average yield varies from 120 to 140 q/ha in the first year and 300–400 q/ha in the second and third year; however, adopting bower system its yield potential can be realized up to 1300 q/ha.

POST-HARVEST MANAGEMENT

Fruits of 5 cm length are used for making pickles, and sometimes, the mature fruits are preserved in sugar syrup. The flesh can be processed into fermented or dehydrated chips, which can easily be stored over a long period. Its fruits can be shipped to long distance markets or stored for 2 weeks at room temperature.

INSECT-PESTS

FRUIT FLY (*Bactrocera ciliatus* coq.)

The most serious insect of ivy gourd is fruit fly, the female of which lay four to ten eggs per fruit below the epidermis by introducing its stout and hard ovipositor in soft and tender fruits. For oviposition, fruits in very early stage are preferred over fruits at late stage. After hatching, the maggots feed inside the fruits, causing decaying. Infested fruits become unfit for consumption and drop quickly. In north India, it is most active from July to August.

Control

- Grow early maturing resistant variety.
- Collect and dump the infested fruits deep into the soil.
- Spray the crop with 0.1% Fenthion three to five times at 10- to 15-day intervals.
- Install poison baits containing 6 parts ripe banana pulp + 4 parts protein hydrolysate + 1 part malathion in the field to trap the male flies.
- Spray the crop with a bait mixture of 10 g malathion 50% WP with 250 g molasses or *gur* + 10 g yeast hydrolysate per litre of water.

Red Pumpkin Beetle (*Aulacophora foveciollis* lucas)

Red pumpkin beetle is the most damaging and widely distributed insect of ivy gourd. It attacks more in spring–summer when the leaves and shoots are very tender. Its incidence is much more in plain areas than in hilly tracts. Its adult female lays 50–70 orange-coloured eggs in clusters in the soil near collar region of the plant. On hatching, the grubs start feeding on tender leaves and underground portion of the plant. After emerging from the soil, the beetles feed on foliage and holes of different size. When the infestation is severe, only the veins skeleton is left.

Control

- Plough the field deep after harvesting the crop to expose the grubs to the predators.
- Follow clean cultivation.
- Collect the adults manually and dump deep into the soil.
- Spray the crop with malathion 0.5% or *neem* seed kernel extract (NSKE) at 4%.

Aphid (*Aphis gossypii*)

Aphid is the most damaging insect of ivy gourd. In northern parts of the country, it becomes very active during spring months. Its nymphs and adults live in large colonies on tender leaves and newly emerged shoots and suck cell sap from them. As a result, the infested parts turn yellow and become curled and deformed, and later, the affected part gets dried. With severe infestation, the fruit size is reduced. While attacking, the aphids exude a copious quantity of honeydew-like substance, on which the sooty mould grows very fast, which reduces the photosynthetic activity of the plant, resulting in stunted growth. It also acts as vector for the transmission of mosaic virus.

Control

- Snip and destroy the infested parts at initial stage.
- Spray the crop with Rogor or monocrotophos 0.05% at 10- to 12-day intervals.
- Spray of the crop with *neem* seed kernel extract 4% at weekly intervals.

Thrips (*Scirtothrips dorsalis*)

Thrips are wingless tiny insects, which live in soil and suck cell sap from tender portion of the plant. Both adults and nymphs cause damage. Its severe infestation is expressed as yellowish pin dot like spots on foliage. As a result, the plant becomes weak and stunted. Damage resulting from direct feeding and transmission of virus disease can occur. Thrips are often concealed inside flowers making them difficult to see. If left uncontrolled, the new shoots and flowers become deformed and stunted. Wilting and browning can also occur.

Control

- Spray the crop with imidacloprid @ 40 µg/l or monocrotophos 35 EC @ 1.5 ml/litre of water at an interval of 15–20 days.
- Spray of the crop with 4% *neem* seed kernel extract at weekly intervals.

DISEASES

FRUIT ROT (*Colletotrichum destructivum* or *Colletotrichum lagenarium* pass.)

A substantial yield loss (60–75%) has been reported in ivy gourd under field conditions due to the decaying of fruits. Pathogen attacks all aboveground plant parts especially the newly emerged shoots, fruits and tendril. The symptoms first appear as light brown circular spots on the leaves, which later turn to deep brown and become brittle. Elongated lesions are observed on stem and circular to oval sunken lesions on fruits, and subsequently, the fruits shrivel, show darkness, and finally, dry up. This disease infects fruits at both immature and mature stages, but near ripening, the fruits become more susceptible to infection.

Control

- Follow long crop rotation with cereal crops.
- Plant the crop at wider spacing.
- Maintain phyto-sanitary conditions.
- Spray the crop with penconazole 0.05%, carbendazim 0.1%, Benlate 0.2% and difolatan 0.3 % at 10-day intervals.

WILT (*Fusarium oxysporum*)

Wilt is a soil born fungal disease, which can occur at any stage of plant growth. The pink colour mycelia growth can be observed by digging the roots. Brown colour lining is developed inside the roots and translocation of nutrients is stopped. In older plants, vein clearing and leaf epinasty are often followed by stunting, yellowing of the lower leaves, formation of adventitious roots, wilting of leaves and young stems, defoliation, marginal necrosis of remaining leaves, and finally death of the entire plant. Further, the symptoms become more apparent on older plants between blossoming and fruit maturation.

Control

- Sterilize the soil through solarization.
- Destroy the disease debris.
- Follow long crop rotation with non-host crops.
- Avoid planting the crop too early.
- Grow resistant variety, if available.
- Follow field sanitation.
- Dip the seedling roots in 0.2% solution of carbendazim.
- Drench the soil with captan @ 0.2–0.3% solution.

Watermelon Mosaic Virus (WMV)

Watermelon mosaic virus (WMV) also known as marrow mosaic virus is universally prevalent all over the growing season and transmitted by sap-sucking insects like *Aphis gossypii*, *Acalymma* spp., *Diabrotica* spp., *Epilachna* spp. and *Myzus persicae* in a non-persistent manner. The symptoms appear on leaves as curling mosaic or mottling accompanied by green vein banding and reduction in leaf size. Sometimes, the petiole and internodes length become short. This virus inhibits the nutrients transmission in plant system. Since this crop is commercially propagated by stem cuttings, transmission through seed is irrelevant.

Control

- Use disease-free healthy planting material.
- Plant the seedlings at wider spacing.
- Follow phyto-sanitary conditions.
- Collect and burn the disease debris.
- Eradicate wild cucumbers, chickweeds, clovers, dandelions, fleabane, ground cherries, horse nettle, jimson weed, milkweed, motherwort, nightshades, pokeweeds, etc. from ivy gourd field.
- Spray the crop with imidacloprid @ 3 ml per 10 litres of water or Dimethoate 0.1% at regular intervals to reduce the whitefly population.
- Spray the crop with oxytetracycline hydrochloride 500 ppm to control phyllody.

■■■

55 Malabar Gourd

M.K. Rana and Naval Kishor Kamboj

Botanical Name   : *Cucurbita ficifolia*

Family   : Cucurbitaceae

Chromosome Number : 2n = 40

ORIGIN AND DISTRIBUTION

Malabar gourd originated from the highland regions of Latin America, from Mexico to Chile, where it is still widely cultivated. It also occurs in tropical Africa in the highlands of Ethiopia, Kenya and Tanzania, and is occasionally grown in Angola. In Asia, it is grown in India, Japan, Korea, China and in the highlands of the Philippines. It was introduced in India and the Mediterranean regions of Europe in the 16th century by Europeans; from there, it spread to many other parts of the world.

INTRODUCTION

Malabar gourd, also known as fig-leaf gourd and black-seeded gourd, is a type of squash grown for its edible green leaves, fruits and seeds. Its fruits are oblong, resembling a watermelon, with wide black seeds. In contrast to other cucurbits, its fruits are highly uniform in size, shape and colour. Although it is closely related to other squashes in its genus, it is considerably different from them in biochemical composition and does not cross readily with them.

COMPOSITION AND USES

COMPOSITION

Malabar gourd is a great source of minerals, particularly of potassium. Its fruits are an excellent source of vitamin A and C and a good source of carbohydrates and folic acid. The nutritional composition of Malabar gourd is given in Table 55.1.

USES

Malabar gourd is grown as a vegetable for its leaves, tender shoots, flowers and immature fruits. Young fruits are used like cucumber. The immature fruits are consumed as cooked vegetable, while its mature fruits are sweet in taste and used to make confectionery and beverages. The mature fruits are sometimes boiled and eaten. Confection is made from the flesh by boiling with crude sugar. Raw seeds are rich in oil with a nutty flavour but very difficult to use since the seeds are small and covered with a fibrous coat. The seeds are delicious when roasted and eaten like peanuts. In Mexico, its seeds are used to make *palanquetas*, a sweet similar to peanut brittle. The edible oil obtained from the seeds is rich in oleic acid. The flesh of mature fruits is impregnated with sugar for the preparation of candy and jam. Its male flowers and buds are used in soups, stews and salads. The vines and fruits are also used for fodder. The shell of mature fruit is very hard, thus, it can be used as a container.

TABLE 55.1

Nutritional Composition of Malabar Gourd (per 100 g Edible Portion)

Constituents	Contents	Constituents	Contents
Water (g)	91.60	Vitamin A (IU)	8513
Carbohydrates (g)	6.5	Thiamine (mg)	0.05
Sugars (g)	2.7	Riboflavin (mg)	0.11
Protein (g)	1.0	Niacin (mg)	0.60
Fat (g)	0.10	Pyridoxine (mg)	0.06
Dietary fibre (g)	0.5	Folic acid (µg)	16.0
Calcium (mg)	21.0	Vitamin C (mg)	9.0
Phosphorus (mg)	44.0	Vitamin E (mg)	1.06
Potassium (mg)	340.0	Vitamin K (µg)	1.1
Magnesium (mg)	12.0	Energy (kcal)	26.0
Iron (mg)	0.80	–	–

BOTANY

Malabar gourd is a monoecious perennial herbaceous climber with branched tendrils but becoming somewhat woody. Its taproot is up to 2 m long, lateral roots forming a network slightly below the soil surface. Stem with numerous long runners is smoothly five-angled to rounded, prickly or spiny, often rooting at nodes. Leaves are alternate, simple, without stipules, blade circular-ovate to nearly reniform in outline, sinuate to lobed with obtuse sinuses and 18–25 cm in diameter. Margins are toothed to entirety. Flowers are solitary, unisexual, regular, pentamerous, up to 7.5 cm in diameter and yellow to pale orange. Calyx and corolla are campanulate with short-tube. Male flowers are short, thick and columnar androecium and filaments with trichomes more than 1 mm long. Female flowers are short and ridged pedicel with inferior ovary. Fruit is large, globose to cylindrical berry, 15–50 cm long, green with white stripes and blotches. Rind is hard and smooth. Flesh of mature fruits is white, coarse, tough and fibrous with many seeds, which are oblong-ellipsoid, flattened, 1.5–2.5 cm long, without a spongy epidermis, usually black, but sometimes, they pale buff-coloured. The test weight is 180–250 g. Seed germination is epigeal. Flowering starts under short-day conditions 30–60 days after emergence of the seedlings and is continuous. The ratio of male to female flowers is about 20:1 but it is influenced by growing conditions, *i.e.*, long days and high temperature favour the male flower production, whereas cool temperature and short day conditions generally favour the production of female flowers. Production of sticky pollen is abundant. Anthesis and pollination take place early in the morning. Insects, mainly bees, are the major pollinating agent.

CLIMATIC REQUIREMENT

Malabar gourd is a warm season crop mainly grown in tropical and subtropical regions. It has resistance against cold more than any other *Cucurbita* species but frost during night kills the plant. In tropics, it can be grown up to 1000 m above mean sea level. The optimum soil temperature for seed germination is 15°C to 35°C. The plants can be grown successfully at a temperature range of 15°C to 32°C with an optimum of 22°C. Short days and low temperatures increase femaleness.

SOIL REQUIREMENT

Malabar gourd crop can be grown on a wide range of soils; however, it grows best on well-drained sandy loam to loam soil rich in organic matter. It is sensitive to alkaline soil and the pH of the soil should be 6.0 to 7.5 for its successful cultivation. The soil is prepared well to a fine tilth by repeated ploughing followed by planking.

CULTIVATED VARIETIES

In India, no registered variety of Malabar gourds is available but farmers grow the local cultivars and germ plasm material.

SOWING TIME

The sowing time is mainly depends on soil temperature and rainfall. In northern plains of India for spring-summer crop, it is usually sown in the mid of February to first week of March, and for rainy season, it is sown in the last week of June to first week of July.

SEED RATE

The seed rate of Malabar gourd depends upon soil fertility, irrigation facilities and sowing method. A seed rate of 2 to 3 kg is required to sow a hectare land area. It may also be vegetatively propagated since the vine nodes have the root regeneration capacity.

PLANTING METHOD

Malabar gourd is sown on either ridges or flat seedbeds at a distance of 2×2 m, giving a population of 2500 plants per hectare. Seeds are sown at a depth of 2–5 cm, depending on soil texture. Seeds take 5–7 days to germinate after sowing.

NUTRITIONAL REQUIREMENT

Nutritional requirement generally depends on soil type, fertility status of the soil, previous crop grown and irrigation facilities. A well-decomposed farmyard manure @ 20–25 t/ha should be incorporated during field preparation to enrich the soil with organic matter. In addition to this, Malabar gourd requires nitrogen 80–100 kg, phosphorus 40–50 kg and potash 80–90 kg/ha. The entire dose of phosphorus, potash and half of the nitrogen fertilizers should be applied as basal dose at the time of last ploughing. Rest of the nitrogen is top-dressed about 4 weeks after sowing.

IRRIGATION REQUIREMENT

Soil moisture is an important factor for good germination of cucurbits. In spring–summer crop, the frequency of irrigation is very important. Malabar gourd requires light but frequent irrigation from germination to the development of fruits. Intervals of irrigation may be kept 5 to 7 days, depending on climatic conditions and soil types of the growing region.

INTERCULTURAL OPERATIONS

HOEING AND WEEDING

Weeds at initial stage of crop growth may pose problems, as the weeds in Malabar gourd field compete with crop plants for nutrients, moisture, light and space; hence, weeding should be done frequently in order to keep the weeds under control. At least one shallow hoeing is beneficial and necessary for keeping the weeds down and providing good soil environment to the roots for their development. The weeds near maturity should not be removed from Malabar gourd field, as they produce shade on fruits, which protects the fruits from sunscald.

To control the weeds chemically, fluchloralin, trifluralin (0.75–1.5 kg/ha), or bensulide (5–8 kg/ha) may be applied as pre-plant soil incorporation 2 weeks before sowing.

MULCHING

Mulching in spring–summer crop is beneficial to reduce water evaporation from the soil surface, suppress weed growth and to moderate diurnal and seasonal temperature fluctuations. The use of plastic mulch, row covers, leaves, or grass clippings helps in controlling weeds and retaining soil moisture, which increases yield as well as quality of the crop produce.

HARVESTING

The harvesting of young Malabar gourd leaves is done at weekly intervals, starting from 6 weeks after emergence of seedlings to the end of the vegetative growth period. The fruit matures in 30–40 days after pollination. The immature fruits are harvested by cutting the fruit stalk from 2 months after sowing to end of the crop.

YIELD

The yield of Malabar gourd largely depends upon variety, soil fertility, growing season and cultural practices adopted during cultivation of crop. Up to 10 mature fruits may be harvested per plant. The weight of immature fruit is about 1 kg and weight of mature fruit is 3–8 kg. However, the average yield of leaves ranges from 200 to 220 q/ha and young fruit yield is 400 q/ha. Regular picking of leaves decreases the fruit yield. Its average seed yield is about 500 kg/ha.

POST-HARVEST MANAGEMENT

The leaves should be consumed within 2 days after harvesting, while the young fruits can be kept for a few weeks at ambient room temperature. However, the mature fruit can be stored for 2 years or more in a dry place at room temperature and becomes sweeter with storage.

PLANT PROTECTION MEASURES

A number of insect-pests and diseases attack the Malabar gourd crop during the growing season. Some of the insect-pests and diseases along with their control measures are given as follows:

INSECT-PESTS

RED PUMPKIN BEETLE (*Aulacophora foveicollis* lucas)

Both grubs and adults attack the crop at seedling stage. The beetle feeds on cotyledons and foliage. Grubs develop into the soil, and occasionally, feed on roots, causing wilting of plants. In severe attack, the crop is destroyed.

Control
- Plough the field repeatedly in hot summer months.
- Spray the crop twice or thrice with Rogor 0.1% near base of the plants just after germination.
- Spray the crop with 0.4% *neem* seed kernels extract and 0.01% *neem* oil.

Leaf Miner (*liriomyza trifolii* l.; *sativae* blanch)

The leaf miner usually attacks during the months of March and April. It causes damage by making mines in leaves, especially in mature leaves. The eggs hatch within 2–3 days, and the larvae feed on leaves making zigzag tunnels between lower and upper epidermis and scrap the chlorophyll and leaf tissues.

Control

- Pluck old leaves severely infested by leaf miner and destroy them.
- Spray the crop with 4% *neem* seed kernel extract at weekly intervals.
- Spray the crop with dimethoate 30 EC @ 1 mL/litre of water at 15-day intervals.

Fruit Fly (*Bactrocera cucurbitae*)

In India, fruit fly is the most common, destructive and polyphagous insect. It is a serious problem and reduces the number of marketable fruits. The adult flies puncture tender fruits and lay eggs below the fruit skin. The maggots feed inside the fruits causing premature dropping of fruit. The mature attacked fruits develop a water-soaked appearance. The larval tunnels provide venue for the entry of bacteria and fungi that cause the fruit to rot. These maggots also attack young seedlings, succulent taproots of stems and buds of host plants.

Control

- Collect and destroy the damaged fruits.
- Spray the crop with Malathion 0.05% or dimethoate 30 EC @ 1 mL/litre of water.

DISEASES

Fusarium Root Rot (*Fusarium solani*)

The symptoms usually noticed in the field as wilting of the leaves. The rot appears as a light brown water-soaked area, which becomes progressively darker. It begins in the cortex of the root, causes cortex tissue to slough off and eventually destroys all the tissues except fibrous vascular strands. Infected plants break off easily about 2–4 cm below the soil surface. Plants showing symptoms develop numerous sporodochia and macroconidia (spores), giving the mycelia a white to pink colour on the stem near the ground surface. The pathogen is both seed and soil borne.

Control

- Use disease-free healthy seed.
- Follow long crop rotation and clean cultivation.
- Avoid root injury during intercultural operations.
- Apply *Trichoderma* (5 kg/ha) during soil preparation.

Anthracnose (*Colletotrichum lagenarium*)

The fungus attacks the leaves, stem and fruits. On leaves, symptoms appear as rough, circular, light brown to reddish brown spots, which later turn black in colour and become brittle. It results in shrivelling and death of leaves. The spots on fruits are roughly circular, sunken and water-soaked with dark borders. The favourable temperature for the spread of disease is 22–27°C with relative humidity 100%.

Control

- Follow 2 to 3 years crop rotation with non-host crops.
- Use disease-free healthy seed.
- Treat the seeds with Bavistin @ 2 g/kg of seed.
- Spray the crop with Blitox 0.2%, or Bavistin 0.1% at 7- to 10-day intervals.

POWDERY MILDEW (*Erysiphe cichoracearum*)

The symptoms of the disease appear as white powdery spots that may form on both the surfaces of leaves, shoots and sometimes on fruit. These spots gradually spread over a large area of leaves and stems. The affected leaves turn completely yellow, curl, die and fall down. The disease is favoured by moderate to warm temperature (25°C) and 95% relative humidity.

Control

- Collect and burn the infected plant debris.
- Spray the crop with carbendazim 0.05% or hexaconazole 0.05% at 7- to 10-day intervals.

DOWNY MILDEW (*Pseudoperonospora cubensis*)

Symptoms appear as numerous angular lesions that are limited by the leaf veins. Lesions grow in size and coalesce with each other. Initially, the lesions are light green in appearance and become chlorotic and finally necrotic as host plant cells die. The severely infected leaves roll upward with brownish tinge that produces a blighted appearance. Greyish black downy fungal growth is observed on under surface of the leaf in humid weather.

Control

- Follow field sanitation by burning crop debris to reduce the inoculums.
- Grow crop at wider spacing in well-drained soil.
- Spray the crop with 0.25% mancozeb or Ridomil MZ 72 at 7-day intervals.

ANGULAR LEAF SPOT (*Pseudomonas lachrymans* carsner)

The symptoms appear as small irregular water-soaked spots on the surface of cotyledons, leaves, stems and fruits. On older leaves, the spots spread until they are confined by the veins, giving them a characteristic angular appearance. After few days, the spots dry and turn yellow-brown, and their centres fall out, leaving angular shot holes and tattered leaves. Stems and petioles are also affected by water-soaked areas that later dry to form a whitish crust. On fruits, small water-soaked, angular, white to brown lesions appear, and later on, these spots dry and crack, revealing a chalky tissue below. Infected plants grow poorly because of the reduced photosynthetic area. Affected fruits become unmarketable. The optimum temperature for the development this disease ranges from 24°C to 27°C.

Control

- Follow field sanitation to avoid the occurrence of disease.
- Destroy the disease debris after harvesting the crop.
- Spray the crop with streptomycin at 400 ppm or Bordeaux mixture at 1%.

56 Stuffing Cucumber

M.K. Rana and Chandanshive Aniket Vilas

Botanical Name : *Cyclanthera pedata* (L.) Schrad.

Family : Cucurbitaceae

Chromosome Number : 2n = 32

ORIGIN AND DISTRIBUTION

Stuffing cucumber, also known as slipper gourd, *Meetha karela*, or botanically known as *Cyclanthera pedata*, is a native to Andean South America. The genus *Cyclanthera* containing 30 species is considered a native to warm-temperate and tropical America. It is also believed that the first site of stuffing cucumber domestication was somewhere in the mountainous regions of Peru, but later, it was geographically distributed among the Andean countries and afterwards it spread towards Southeast Asia and Africa. In Africa, its cultivation is restricted to highlands of East Africa. It is usually cultivated from Mexico to Peru, Ecuador and also in the Old World tropics. The stuffing cucumber is grown in many parts of Central America, South America and the Andes. It is also widely grown in the highlands of Colombia, Caribbean Islands and Bolivia. This annual climber has also been grown for some time in mountainous habitats in Bhutan, Nepal and Northeastern India for subsistence and commercial uses.

INTRODUCTION

Stuffing cucumber, a member of the Curcubitaceae family, is a tropical herbaceous climbing plant, which is often grown as a vegetable in Indian hills, commonly in Manali and Sikkim. It is used as a substitute of cucumber and is a popular vegetable in the Andes region. In India, it is known as *meetha karela*, and locally, it is known as *achojcha*. In Bhutan, it is known as *kichipoktho*. The plant is also known as *caiba, caygua, concombre, grimpant, korila, kaikua, lady's slipper*, slipper gourd, or wild cucumber. The fruit has a big cavity that can easily be stuffed. Hence, it is called *pepino de rellenar* in South America. In Peru, the plant is known by its Spanish name *caigua* or *caihua*. Its indigenous Quechua name is *achocha* or *achoccha*. Its large size as compared to its closely related wild species suggests that it is a fully domesticated crop going back for many centuries, evidence coming from ancient Peruvian ceramics depicting the fruits. Its fruits are pretty accepted for edible and medicinal purposes. It is quite popular in many parts of the country like Northeastern states, Orissa, Bihar and West Bengal. The crop is also considered ideal as a fence as it is tolerant to shade.

COMPOSITION AND USES

Composition

Stuffing cucumber fruits are a good source of carbohydrates, calcium, phosphorus, vitamin A, vitamin C and a fair source of protein, iron, thiamine, riboflavin and niacin. Its fruits also provide a good amount of energy. The nutritional composition of slipper gourd is given in Table 56.1.

TABLE 56.1

Nutritional Composition of Slipper Gourd (per 100 g Edible Portion)

Constituents	Contents	Constituents	Contents
Water (g)	94	Vitamin A (mg)	15
Carbohydrates (g)	44	Thiamine (mg)	0.05
Protein (g)	0.7	Riboflavin (mg)	0.04
Fat (g)	0.1	Niacin (mg)	0.29
Dietary fibre (g)	0.7	Vitamin C (mg)	14
Calcium (mg)	13	Ash (g)	0.8
Phosphorus (mg)	20	Energy (kcal)	19
Iron (mg)	0.8		

USES

Stuffing cucumber is usually grown for its immature fruits eaten raw or pickled. The immature fruits and tender vine tips are used in a variety of culinary preparations. The young shoots and fresh leaves may also be eaten as greens. The mature fruits are prepared as stuffed peppers, stuffed with meat, fish, or cheese and then baked. Older fruits are eaten after removal of seeds and boiling or cooked. They can be stuffed in much the same way as the marrows. Its taste is similar to that of a cucumber. The fruit is about 6–15 cm long and 6 cm wide. In north eastern states of India, it is used to prepare pickles.

MEDICINAL USE

Stuffing cucumber has the status of being anti-inflammatory and hypocholesterolaemic. A tea prepared from its seeds is used for controlling high blood pressure. It has several herbal properties *viz*., fruit juice lowers cholesterol, reduces blood pressure, cleans arteries, balances blood sugar, increases urination, relieves pain and hypertension, reduces inflammation and aids in digestion, arteriosclerosis and circulatory problems. The fruits and the seeds are also recommended for gastrointestinal disorders. The leaves of stuffing cucumber are considered hypoglycemic and used as a decoction to control diabetes. The fruits are boiled in milk and gargled to cure tonsillitis. The fruit and leaves both are boiled in olive oil and used externally as a topical anti-inflammatory and analgesic ointment. The roots are analgesic, antidiabetic, antiparasitic, hypotensive and hypoglycemic and used in circulatory problems and to clean teeth. Its roots decoction is used to reduce the gastrointestinal disorders, high blood pressure, high cholesterol and intestinal parasites.

BOTANY

Stuffing cucumber is an annual plant, which grows up to a height of 3.5–4.5 m and flowers from August to September. The wildly grown stuffing cucumber vine bears small fruits used as a vegetable. Leaves are 10–12 cm wide, alternate, palmately, 3–5 foliolate, or simple but very deeply lobed, stipules absent, petiole 1–8 cm long, leaflets or lobes elliptical, sinuate-serrate and divided into 5–6 lobes or leaflets. The plant sex form is monoecious and the flowers are unisexual, regular and pentamerous, thus, they are pollinated by insects. The male flowers borne axillary on 10–20 cm long panicles and female flowers borne solitary and sessile having no stalks with inferior one celled ovary. It is a climbing plant, which uses its long tendrils for climbing. The plant develops several branches. It produces a small pale green, semi-flattened indehiscent oblong shaped berry. The fruits are 10–15 cm long and 5–8 cm wide, sometimes with soft spines and many seeds. The fruits have a curved bottom and hollow from inside. The endocarp is spongy white and the mesocarp is succulent. The central cavity has a parietal placenta, where the seeds remain attached. Seeds are 1.5 cm in diameter, black, square shaped and wrinkled.

CLIMATIC REQUIREMENT

Stuffing cucumber grows well under the same conditions as preferred by the other cucurbits. It prefers a temperature range of 14°C to 27°C for sustained growth and development. Seed germinates well at a temperature range of 15–20°C. During early growth, it requires a minimum temperature of 18°C. The plant grows well in very warm moist climate with sunny days and sheltered position but has a good tolerance for colder conditions as well. It can grow at an elevation up to 2000 m above mean sea level. It can also be found growing prolifically in mountainous valleys in Northeastern India up to 1500 m elevation.

SOIL REQUIREMENT

Like other cucurbits, stuffing cucumber thrives well on a variety of soils but prefers well drained sandy loam soil rich in organic matter. It can also be cultivated on acidic and basic alkaline soils. The optimum soil pH is 6–7 but the plants can tolerate a soil pH up to 8. It grows best in well prepared loose fertile soil having a depth of 20–40 cm.

CULTIVATED VARIETIES

For Indian conditions, no specific improved variety is available, thus, only the high yielding indigenous collections are preferred by the growers to earn higher prices in the market depending on the requirement and suitability of the area but there are some introduced varieties of stuffing cucumber, which are recommended for cultivation. In the United Kingdom, several species are available to grow by seeds. *Fat Baby* (*Cyclanthera brachystachya*) bears solitary fruits, which are covered with soft fleshy spines. *Ladies Slippers* (*Cyclanthera pedata*) usually bears fruits in cluster of two sets of smooth fruits in a pair. It is a large-fruited variety bearing 10–20 cm long fruits used for stuffing. This variety is likely to be more productive when grown in a greenhouse or poly-tunnel rather than outside.

PLANTING TIME

The planting of stuffing cucumber depends on climatic conditions of the growing region. It is very easy to grow crop from seeds. For spring–summer crops, the best time for sowing its seeds in plains is January–February after the last expected frost and the plants are protected from frost by using cloches until they are growing away well, whereas, for main season crop (rainy season), the seeds are sown in early April to May.

SEED RATE

For planting a hectare land area, 3–4 kg of seed is required. The seeds before planting should be treated with captan or thiram at the rate of 3–4 g/kg of seed.

SOWING METHOD

Generally, stuffing cucumber is propagated by seeds. Direct seeding is the most common method used for its propagation. The field is prepared to a fine tilth by repeated ploughing with harrow followed by planking. In spring–summer and rainy season, the crop is sown on ridges of 15–20 and 30 cm width made at 1.5–2.5 m distance, respectively. For maintaining optimum plant density per unit area, 2–3 seeds are sown per hill. After completion of germination process, retaining only one strongest seedling per hill, the other seedlings are pulled out if all the seeds germinate.

NUTRITIONAL REQUIREMENT

Stuffing cucumber requires a balance of nutrients through organic and inorganic sources. Fertilizer application rates depend on soil type, fertility level, crop grown in previous season and soil organic matter content. Well-decomposed farmyard manure 15–18 t/ha is incorporated into the soil at the time of land preparation. Besides, nitrogen, phosphorus and potash are applied at the rate of 180, 110 and 120 kg/ha, respectively. Fertilizer application consists of a basal application followed by side dressings. In heavy soils, the entire quantity of phosphorus and potash along with one-third dose of nitrogen is applied at the time of sowing either by broadcasting or by banding a few centimetres deep and to the side of plant row in the beds. The rest of the nitrogen is applied in two or more splits as side dressings. The first side dressing is done when the plants have formed four to six true leaves and subsequent side dressing is done at fortnight intervals.

IRRIGATION REQUIREMENT

The soil at the time of planting should have plenty of moisture. The plant is adapted to a wide range of rainfall but regular irrigation is needed to ensure economic yield. Proper water management is essential for obtaining maximum yield. The stuffing cucumber crop requires plenty of water throughout its growing period. However, the irrigation should be applied in such a way that the water may not come in contact of its stem; otherwise it may decay. Subsequent irrigations may be applied at an interval of 5–7 days in spring–summer and 10–15 days in wet season, depending on moisture-retention capacity of the soil.

INTERCULTURAL OPERATIONS

HOEING AND WEEDING

Hoeing is an essential operation, which should be done 1 and 2 months after planting to make the soil loose around the plant roots necessary for proper aeration, keep the field weed free and to reduce competition between crop plants and weeds. Weeding during early growth phase is beneficial since the crop may not compete with weeds at early stages due to its slow growth rate. If labour is not available for hoeing and weeding, diuron herbicide may be applied at the rate of 1.6 kg/ha as pre-emergence to control the weeds.

STAKING

Staking is an important operation for better growth of the plants and harvesting good quality produce. Stuffing cucumber grows very fast and vines elongate rapidly within two weeks of planting. Its plant is a weak climber, thus, it needs support for its growth, as staking increases fruit yield and size, reduces fruit rot and makes the sprayings and harvesting easier. The vines trailed on bower continue to give yield for 6–7 months as against 3–4 months when the vines crawl on the ground without support. The vines on bower are less susceptible to insect-pests and diseases as they do not come in direct contact of soil. When the vines are to be trailed on bower system, the seeds should be sown on the edges of furrows opened up at 2.5 m. Cemented poles of 3 m height are pitched on both the ends of alternate furrows at a distance of 5 m. These poles are connected to each other with steel wires, which along the furrows are further connected with cross wires fastened at 45 cm distance so as to form a network like structure. Seeds are sown at a distance of 1 m along the furrow and slightly covered with fine soil. The vines reach a bower height in about 1.5–2 months, thus, the vines at initial stages are trailed on twines until they reach the bower. Once the vines reach a bower height, they trail on bower with the help of tendrils.

MULCHING

Mulching with black polyethylene film soon after sowing is beneficial to the crop as it conserves soil moisture, regulates soil temperature around the plant roots and also suppresses weed emergence.

HARVESTING

Stuffing cucumber starts giving fruits about 3 months after sowing. Its fruits grow rapidly, thus, harvesting must be done at very short interval. The fruits are preferably harvested when they have attained two-third of their full size and the seeds are still tender. The fruits at harvesting stage are partially green with some spines. The amount of fruits to be harvested depends on the size of fruits at final harvest. If the fruits are smaller in size, the more will be the total number of fruits per plant since the vines will have more carbohydrates to bear more fruits per plant. Usually, a vine gives between 8 and 20 fruits per plant. Avoiding damage to the vines, harvesting is done manually through scissors so that the vines may give fruits constantly for a longer period.

YIELD

The yield of stuffing cucumber depends on variety, soil type, soil fertility, irrigation water availability, growing season, climatic conditions of the growing region and package of practices followed by the growers. Its average marketable yield is 80–100 q/ha; however, the yield of improved varieties is around 200–300 q/ha.

POST-HARVEST MANAGEMENT

After picking, the fruits should not be exposed to direct sunlight. While harvesting, care must be taken to avoid bruising to the fruits as it will depreciate their market value. Its fruits have shorter shelf life, thus, they are no longer retained, although soon after harvest they are sent to the market for sale. The damaged and deformed fruits are separated from the lot to avoid rotting of the fruits, and the marketable fruits are carefully packed in boxes or in plastic bags containing holes for ventilation. The shelf life of its fruits can be extended to 2–3 weeks at 10–13°C temperature and 85–90% relative humidity. Stuffing cucumber is sensitive to low temperature. Therefore, its storage below 10°C should be avoided as it may cause chilling injury. Storing its fruits at a temperature above 13°C should also be avoided, as it will result in yellowing and splitting of the fruits.

INSECT-PESTS

APHIDS (*Aphis gossypii*)

Being a sucking pest, it sucks cell sap from tender leaves, on which, it secretes honeydew, a sticky sugary substance, which encourages the growth of black sooty mould, affecting the photosynthesis of plants. It transmits virus, which causes yellowing and curling of the leaves. It attacks the plants mostly at early stage of crop growth. High humidity and high temperature favour the occurrence of this pest.

Control

- Grow resistant variety if available.
- Set up yellow sticky traps in the field.
- Spray the crop with Malathion 0.05% or Confidor 0.3% at intervals of 10–12 days.

RED PUMPKIN BEETLE (*Aulacophora foveicollis*)

The adult beetles emerging from the soil feed on cotyledonary leaves and leaf lamina, causing perforation and irregular holes on foliage and seedling mortality. Adult beetles also cause damage to its roots and stems as well. They may also damage the fruits by puncturing pinholes into them.

Control

- Plant the crop bit early.
- Follow clean cultivation by removing the plant refuses.
- Plough the field deep after harvesting the crop to kill the grubs in the soil.
- Spray the crop after germination with 0.1% dichlorovos or dimethoate 0.03% after 10 days followed by two sprays at 10- to 15-day intervals.

FRUIT FLY (*Bactrocera cucurbitae*)

Fruit fly is the most destructive and polyphagous pest of stuffing cucumber. Its maggots feed inside the fruits and adult fly punctures the tender fruit and lays eggs below the epidermis. The eggs hatch into maggots, which tunnel deep into the fruit pulp and feed on it, *per se*, the whole fruit is completely spoiled. The attacked fruits get distorted, rotten and dropped down.

Control

- Grow early maturing variety.
- Collect and destroy all infected fruits to kill the maggots.
- Plough the field deep after harvesting to kill the pupae in soil.
- Spray fenthion 0.05% or fenitrothion 0.1% at initiation of female flower.
- Spray the crop with malathion 50 EC 2 ml with jaggery 20 g/litre of water at weekly interval.

EPILACHNA BEETLE (*Epilachna duodecastigma*)

The adults are green with green streaks on the body. Both its grubs and adults feed on the leaves causing skeletonized patches on them. Leaf tissue is eaten between the veins. The leaves may completely be stripped to the midrib and small areas eaten out, as a result, shallow holes may develop on the fruit surface. In severe infestation, the whole plant may die.

Control

- Handpick and destroy the grubs and adults at initial stage of infestation.
- Irrigate the crop at regular interval to minimize the pest population.
- Use of *Bacillus thuringiensis* is effective to control the beetle.
- Add *neem*, *mahua*, or ground nut cake to suppress the pest population.
- Spray the crop with fenvalerate 0.05% or quinolphos 0.2% at 10- to 15-day intervals.

DISEASES

ANTHRACNOSE (*Colletotrichum lindemuthianum*)

Anthracnose disease infests all plant parts at any growth stage but the fungus usually affects the developing shoots and expanding leaves. Anthracnose lesions typically develop along major leaf veins. The characteristic small tan, brown, black sunken lesions or tarlike spots appear on leaves and ripe fruits developed under favourable cool and wet conditions. Later, these spots expand and merge to cover the whole affected area. If leaves are infected at very young stage, they become curled and distorted. The disease also produces cankers on petioles and stems, which causes severe defoliation and rotting of roots and fruits. The fungus spreads from one to another season through seeds and plant debris.

Control

- Rotate the crop with non-host cucurbitaceous crops.
- Follow clean cultivation by destructing the infected plant debris.
- Treat the seed with Dithane Z-78 2 g/kg of seed.
- Spray crop with Dithane Z-78 0.15% or Bavistin 0.1%.

FUSARIUM WILT (*Fusarium oxysporum*)

The disease spreads through the infected soil and it may also be carried to the next season through the infected seeds. The infected plants show yellowing and wilting of leaves, and finally, the entire plants wilt and die. Vascular tissue inside the root shows browning but the pith remains unaffected. The fungus is seed borne and also survives in soil.

Control

- Grow resistant variety.
- Follow long crop rotation with non-cucurbitaceous crops.
- Sterilize the field soil through solarisation.
- Plant the crop bit late.
- Drench the soil with Bavistin 0.1% or thiram 0.2%.

POWDERY MILDEW (*Spharerotheca fuliginea*)

Initially, the disease manifests on plants as white powdery fungal growth on leaf surface. Usually, both sides of the leaf show fungal growth. White dirty green patches appear on the leaves, petioles and stems, which later cover the whole surface. Leaves become yellow, chlorotic and shrivelled, and later, defoliation occurs. In severe cases, plants become stunted. The infected plant bears no fruit or the set fruits become very small, and the flavour is affected adversely. The disease spreads during high temperature and dry weather.

Control

- Grow resistant varieties.
- Follow field sanitation.
- Demolish the infected plant debris after harvesting the crop.
- Dust the crop with Sulfex 2.5 kg/ha.
- Spray the crop with Karathane 0.1%, dinocap 0.1%, carbendazim 0.1%, or wettable sulfur 0.2%.

DOWNY MILDEW (*Peronospora destructor*)

The early symptoms of downy mildew are the appearance of yellow spots on upper surface of the leaves, while bluish white fluffy growth develops on lower surface of the leaves. Under humid conditions, violet growth of the fungus appears on lower surface of the leaves, which later become pale green and finally collapse. As the leaf spot dies, the fluffy growth darkens to grey in colour. Infected leaves and branches may be distorted and die. The affected fruits in storage are soft and immature. The disease infestation is severe during rainy season.

Control

- Use disease-free healthy planting material.
- Strictly follow a crop rotation of at least 3 years.
- Plant the crop at wider spacing.
- Remove infected plant debris and keep the field clean.
- Spray the crop with Dithane M-45 0.2%, chlorothalonil 0.2%, or nitrophenol 0.05% twice at 10- to 15-day intervals.

MOSAIC DISEASE (*Potyvirus*)

The leaves of virus-infested plant become small, misshapen and blistered. The mosaic infected plants become stunted and bear no or very few small and deformed fruits. The leaves are mottled with downward curling. The disease is generally transmitted by white fly and aphids.

Control

- Follow long crop rotation with non-cucurbitaceous crops.
- Use highly reflective mulches to deter aphids.
- Keep the field weed free and eradication of weed hosts.
- Spray mineral *Krishi* oil 1.5% at 10 days interval.
- Spray the crop with phosphomidon 0.02%, or methyl dematon 0.05%
- Incorporate Furadon or phorate 10 G granules 1.5 kg a.i./ha in soil at the time of seed sowing followed by 2–3 foliar sprays of dimethoate 0.05%, or Metasystox 0.02%, or Nuvacron 0.05 at 10-day intervals.

■■■

57 Pointed Gourd

Dipika Sahoo and J. Debata

Botanical Name   : *Trichosanthes dioica* Roxb

Family   : Cucurbitaceae

Chromosome Number : 2n = 2X = 22

ORIGIN AND DISTRIBUTION

Trichosanthes, being of old-world origin among the cultivated plants, is a large genus, which has about 44 species, of which 22 are found in India. Among these, *Trichosanthes dioca* and *Trichosanthes anguina* are two important cultivated species. So far, the centre of origin of *Trichosanthes* is not precisely known even though most authors agree on India or the Indo-Malayan region as its original home. Assam-Bengal, including Bangladesh, is considered the primary centre of origin of pointed gourd due to rich diversity of this crop in this region. Wild forms of pointed gourd are found throughout north India. Semi-wild populations of *Trichosanthes dioica* occur in Assam plains and Brahmaputra valley in India.

INTRODUCTION

Pointed gourd is commonly known as *parwal* (Hindi), *kovakkai* (Tamil), *thondekayi* (Kannada), *potol* (Bengal, Odisha and Assam) and *paror* (Maithili). Colloquially, it is often called green potato. It is intensively cultivated cucurbitaceous vegetable in the eastern parts of India, particularly in Odisha, West Bengal, Assam, Bihar and some parts of Madhya Pradesh, Gujarat, as well as Eastern Uttar Pradesh. It is also grown on a limited scale in hilly tracts of south Indian states like Andhra Pradesh, Tamil Nadu and Karnataka. In Bihar and eastern Uttar Pradesh, the crop is grown in *diara* lands near the rivers. Apart from India, it is commercially grown in Myanmar, Sri Lanka, Bangladesh and China. Its immature and tender fruits are available for 8–10 months from February–March to October–November except few winter months, but under protected condition, it is cultivated round the year.

COMPOSITION AND USES

COMPOSITION

Pointed gourd is known as the *king of gourds* because of having higher nutrients content than other cucurbits. Its protein content is 10 times higher than that of bottle gourd and four times higher than that of snake gourd, ridge gourd and wax gourd. Similarly, pro-vitamin A and vitamin C content is higher than many other cucurbits. Among the cucurbits, the protein, mineral, fibre and calcium content of its leaves are highest (5.4 g, 3.0 g, 4.2 g and 531 mg per 100 g, respectively), and it also provides energy of about 55 kilocalories. The bitter principle cucurbitacin content is very less in pointed gourd fruits but is commonly found in roots, shoots and leaves. The nutritional composition of pointed gourd fruit is given in Table 57.1.

TABLE 57.1

Nutritional Composition of Pointed Gourd Fruit (per 100 g Edible Portion)

Constituents	Contents	Constituents	Contents
Water (g)	92.0	Potassium (mg)	83.0
Carbohydrates (g)	2.2	Magnesium (mg)	9.0
Protein (g)	2.0	Iron (mg)	1.7
Fat (g)	0.3	Vitamin A (IU)	255.0
Dietary fibre (g)	3.0	Thiamine (mg)	0.05
Minerals (g)	0.5	Riboflavin (mg)	0.06
Calcium (mg)	30.0	Niacin (mg)	0.5
Phosphorus (mg)	40.0	Vitamin C (mg)	29.0

USES

Pointed gourd is one of the most nutritive and wholesome vegetables. It immature tender fruits are generally consumed in several culinary preparations and pickles. Like ash gourd, its fruits are also used for the preparation of sweets. The fruits filled with milk cake and dipped in sugar syrup are a famous sweet in north India. The newly emerged tender shoots with its leaves are used as potherb in soups and other preparations, particularly in West Bengal.

MEDICINAL USES

Fruits are easily digestible, diuretic and laxative. It invigorates the heart and brain and is significantly effective in several circulatory disorders. The fruit is particularly recommended during convalescence. It is also recommended for bronchitis, biliousness, high fever and nervousness. The fruit helps in proper renal functions and prevents constipation. The leaves of pointed gourd are a preventive antidote of bile disorders and worms.

BOTANY

The genus *Trichosanthes* has chromosome number 24. However, the cytological studies have shown that *T. dioica* and *T. anguina* have the chromosome number 22, while polyploidy series of 22, 44 and 66 chromosome number have also been recorded in *Trichosanthes palmate*. Jaffery in 1980 divided the family Cucurbitaceae into two sub-families, *i.e.*, (i) Zananioideae and (ii) Cucurbitoidae and later subdivided into eight tribes, of which, the genus *Trichosanthes* comes under the tribe Trichosantheae. However, Chakraborty in 1982 classified the family into three tribes and the tribe Cucumerineae comprises the genus *Trichosanthes*.

Pointed gourd is a dioecious semi-perennial cucurbitaceous vegetable crop, remaining dormant during winter season. Roots are tuberous in nature with long taproot system. Soon after the onset of spring-summer season, the sprouts come from the underground tuberous roots. The leaf lamina is cordate, ovate and oblong with narrow basal lobes at base. Normally, a node in the male plant bears a leaf, one or two solitary male flowers on distinct pedicels, a simple or bifid tendril, a vegetative shoot and a glandular bract, while the female plants also bear the same parts except one or two (sometimes even four) female flowers on very short and indistinct peduncles. Pistillate flowers have one to five carpels usually three thick and short styled terminated by bi-lobed or divided pistillate stigma. Flowers are tubular and white with 16–19 days initiation to anthesis time for pistillate flowers and 10–14 days for staminate flowers. Stigma remains viable for approximately 14 hours and 40–70% of the flowers set fruit. Anthesis takes place during night. Generally, night temperature favours anthesis and fruit set in the early hours of morning when the insects visit them. The maximum pollen germination takes place during anthesis between 6:00 p.m. and 11:00 p.m. with the peak at 9:00 p.m. to 9:30 p.m., and then, it retards gradually. Therefore, hand pollination must be completed as early as possible in the morning hours preferably within a time of 5"00–5:30 a.m. There is no definite test to identify male and female plants at the young stage on morphological traits except the flowers. Fruit, the main economic part, is botanically the *pepo*, the edible portion being mostly pericarp with a little mesocarp. Based on shape, size and striation, fruits can be grouped into four categories, *i.e.*, (i) 10–13 cm long and dark green with white stripes, (ii) 10–16 cm long and thick dark green with very pale green stripes, (iii) 5–8 cm long, roundish and dark green with white stripes and (iv) less than 5–8 cm long and small tapering with green stripes. This perennial vine crop survives for long time through giving rise to sprouts from the tuberous roots even left uncared.

CLIMATIC REQUIREMENT

Pointed gourd is a warm season crop, thus, thrives well under hot or moderately warm and humid conditions, and the optimum temperature for proper growth ranges from 25°C to 35°C. Abundance of sunshine and fairly high rainfall favour good crop yield. Regeneration of new sprouts is generally inhibited below 20°C temperature and severe cold below 5°C is detrimental for the crop growth. Vine growth becomes highly restricted during winter, which starts again along with sprouting from fleshy root at the onset of spring. However, pointed gourd can be forced during winter months (November to February) in different riverbeds or river basins, popularly called as *Diara* lands in Uttar Pradesh and Bihar that wipe low temperature effect of the winter. Pointed gourd is susceptible to cold and frost. It can withstand water stress but not waterlogging conditions. In plains, pointed gourd cultivation is very intensive and remunerative too.

SOIL REQUIREMENT

The pointed gourd crop can be grown in a wide range of light textured soils having good drainage facility. However, a well-drained sandy-loam to loam soil with slightly acidic reaction (pH 6–7.0) is ideally suited for this crop. Alluvial soils of riverbanks are preferred for successful cultivation. Heavy soils are not suitable for its cultivation but can also thrive well with addition to organic matter or by raising and incorporating green manure crop well in advance. The soil should be prepared up to a good tilth with a deep ploughing followed by cross harrowing and planking.

CULTIVATED VARIETIES

Extensive clonal variation in this crop exists in West Bengal, Assam, Tripura, Bihar and the eastern part of Uttar Pradesh since heterozygous nature and asexual propagation lead to large number of diverse cultivars on farmers' field. Fruits of cultivars are marketed under different local names without any standardization in nomenclature. Some of such popular local cultivars in three important pointed gourd-growing states are given as follows:

- **West Bengal:** Damodar, Kajli, Kajli Bombai, Kajli Damodar Chandra, Sandhamani, Hilli, Guli, Shampuria, Dhanpa, etc.
- **Uttar Pradesh:** Dandali, Kalyani, Guli, Bihar Sharif, etc.
- **Bihar:** Dandali, Nimia, Hilli, Santokhwa, etc.
- **Madhya Pradesh:** Sel.-1 and Sel.-2

Asexual propagation provides unique advantages and opportunity in breeding this crop because single outstanding clone selected from a population may form the basis of a new variety. Some improved clonal selections developed by the State Agricultural Universities and Research Institutes are presented as follows:

SWARNA REKHA

A variety developed at the Horticultural and Agro-Forestry Research Programme (HARP), Ranchi, through clonal selection from germplasm of Champaran district of Bihar produces elongated, light green-stripped soft seeded fruits with blunt ends. The fruits are 5–8 cm long, solid, thin-skinned and good for vegetable as well as sweets preparations. The yield potential of this variety is 230–280 quintals per hectare on vertical staking.

SWARNA ALAUKIK

A variety developed at HARP, Ranchi, through clonal selection from germplasm collected from Bhagalpur district of Bihar produces 8–10 cm elongated fruits having greenish-white stripes and tapering on both ends with long shelf life and is suitable for sweet preparation. Its yield potential is 200–250 quintals per hectare on vertical bower system.

FAIZABAD PARWAL 1

A variety developed at Narendra Dev University of Agriculture and Technology, Faizabad, through clonal selection produces very attractive, round, green fruits and the average yield being 150–170 q/ha. It is recommended for commercial cultivation in Uttar Pradesh and adjoining parts of Bihar.

FAIZABAD PARWAL 3

A variety developed at Narendra Dev University of Agriculture and Technology, Faizabad, through clonal selection produces spindle-shaped and green-striped fruits, which are excellent for culinary purpose. Its average yield potential is 125–150 quintals per hectare.

FAIZABAD PARWAL 4

A variety developed at Narendra Dev University of Agriculture and Technology, Faizabad, through clonal selection produces spindle-shaped and light green-fruits with tapering ends. It is recommended for reclaimed sodic soils and the bower system for cultivation.

RAJENDRA PARWAL 1

A variety developed at Rajendra Agricultural University, Bihar, through improved clonal selection is highly suitable for *diara* land cultivation. Its fruits are large with green stripes. The average yield potential is 140–150 quintals per hectare.

RAJENDRA PARWAL 2

A variety developed at Rajendra Agricultural University, Bihar, through improved clonal selection is highly suitable for *diara* land cultivation. Fruits are big and dark green with whitish stripes. It is suitable for cultivation under eastern Indian conditions. Its average yield potential is 150–170 quintals per hectare.

KASHI ALANKAR

A high yielding variety developed at Indian Institute of Vegetable Research, Varanasi, is recommended for cultivation in Uttar Pradesh, Bihar and Jharkhand. Fruits are green and spindle-shaped. The average yield potential is 180–200 quintals per hectare.

Kashi Suphal

A high yielding variety developed at Indian Institute of Vegetable Research, Varanasi, is recommended for cultivation in Uttar Pradesh, Bihar and Jharkhand.

Arka Neelachal Kirti

A variety developed at Indian Institute of Horticultural Research, Bangalore, produces dark green fruits with prominent white stripes, tapering at both ends and fruit weight 50 g with 280–290 fruits per plant. It is moderately tolerant to *Fusarium* wilt. With better management practices, it produces yield of 10–12 kg/plant and 300–400 quintals per hectare.

Konkan Harita

A variety developed by Dr Balasaheb Sawant Konkan Krishi Vidyapeeth, Dapoli, produces dark green 30–45 cm long fruits tapering at both the ends. Its average yield is 10–12 fruits per plant.

Chhota Hilli

Fruits are medium long, oval to spindle-shaped, swollen in the middle, greenish with prominent white stripes, blunt end and bulged at the stock.

Dandli

Fruits are medium-sized, egg-shaped, light green, stock end dispersed, striped slightly and grooved towards distal end.

Hilli

Fruits are oblong, greenish and white striped and tapering towards distal end with dispersed neck.

Shankolia

Fruits are medium long, resembling those of shankh or shell. They are tapered towards ends, greenish with white stripes, slightly beaked towards distal end and bulged towards the stalk.

CHES Elite Line

Fruits are very attractive, dark green, large, striped with tapering ends, early maturing, weighing 60–80 g each and can be successfully stored for 4–6 days under ambient room conditions. It is well adapted to uplands of Bihar, Odisha and Uttar Pradesh. It performs well in Sambalpur region of Odisha, Midnapore area of West Bengal and in parts of Assam. It is resistant to fruit fly infestation. Its average yield potential is 250–300 quintals per hectare.

CHES Hybrid 1

A first pointed gourd hybrid developed in the country produces solid green striped fruits weighing 30–35 g each. It is resistant to fruit fly. Its average yield potential is 280–320 quintals per hectare.

CHES Hybrid 2

A high yielding hybrid produces dark green striped fruits with an average fruit weight of 25–30 g. Its average yield potential is 300–400 quintals per hectare.

PLANTING TIME

Pointed gourd is mainly cultivated as an irrigated crop in summer and a rain-fed crop in rainy season. The ideal time of planting is February–March in uplands and mid-November in *diara* lands. In upland conditions with irrigation, it is grown as a semi-perennial crop. Planting time of vine or root cuttings should be adjusted in such a manner that regeneration and sufficient vine growth may be completed at a temperature range of 25–35°C before the onset of low temperature in winters, *i.e.*, the dormant season. However, earlier planting in north India and *diara* lands just after rainy season is not suitable due to high soil moisture conditions, causing poor establishment due to rotting of planting materials. The planting of roots or vine cuttings after the onset of winters delays sprouting and establishment of vines. The critical physiological vine maturity of around 100–110 days after planting that coincides with rise in atmospheric temperature is ideal for the initiation of fruiting, which favours both early and total yield. Under South Indian conditions, the vine cuttings are planted in either July–August or September–October in pits at a spacing of 2 × 2 m.

Consecutively, three successful crops can be taken from a single planting. In ratoon cropping generally practiced in plain areas, the vines are pruned 15 cm from the ground level during October after the fruiting is over and recommended dose of farmyard manure and fertilizers are applied by the end of winter as top-dressing after loosening soil around the hills.

PROPAGATION

Pointed gourd is propagated vegetatively both by root and vine cuttings. Very small or very big cuttings from both root and vines are never ideal. Root cuttings of 15 cm and vine cuttings of 60 cm length with 10–15 nodes are considered the best propagating materials.

NURSERY RAISING

Vine of 1 m long is collected and a coil is made. The nodal portion of the coil is treated with rooting hormone. The treated portion of the coil is inserted in polythene bags filled with a potting mixture in such a way that two or three nodes may remain outside the bag. Water is applied with rose can and they are put in shade or plastic house. Sprouting can be seen 10 days onwards. These are then transplanted in the main field after 3 months. This method can ensure 100% rooting and producing a large number of plants of a variety at one place. They can easily be transported to the main field or distant places. Vine cuttings of about 50–60 cm long made from basal portion of the vine are planted at a spacing of 2 m between rows and plants with a success rate of over 90%.

The vines of pointed gourd being semi-perennial in nature grow long, thus, propagation by layering is possible only when crop is retained in the field. Layering may be done after fruiting is completed and successful layers can be separated and used for new planting.

For propagation from root cuttings, mature roots of 3- to 4-year-old plants are cut into pieces with three nodes each. The root cuttings are dipped in the solution prepared with copper sulfate and planted in the main field. Regular watering is required until it produces new sprouts. Root suckers can ensure 100% sprouting as propagation by this method is easier and faster but the availability of roots is a major problem.

Propagating pointed gourd from vines requires a definite length of cuttings, requiring huge quantity of planting materials to cover the field. However, such inconvenience may be mitigated by micro-propagation. Rapid *in vitro* multiplication in pointed gourd could be possible by culturing shoot tip and nodal explants on a Murashige and Skoog medium containing indole acetic acid and indole butyric acid, and plantlets should be acclimatized and established *ex vitro* with varying efficiency (75–95%) in different genotypes. Good establishment of regenerated plantlets (over 70%) in soil medium could also be achieved using physiologically mature seeds as explants cultured in different hormone (benzyl adenine, naphthalene acetic acid, etc.).

Pointed gourd can also be propagated by seed. The seed propagation is most important for evolving new cultivars through breeding. Its seed rate is 20 kg/ha, however, for commercial production, propagation by seed is not advisable because of poor seed germination and viability as they are recalcitrant in nature. The seed is not recommended as commercial propagating material because of dioecious nature of pointed gourd, producing male and female plants in equal proportion with long juvenile phase or vegetative growth phase, which results in flowering after 2 years.

PLANTING METHOD

At the time of planting, a minimum of 10–15% male plants are essential to ensure the supply of sufficient pollens required for fertilization and fruit set. For every 10–15 female plants, one male plant is a must for effective pollination, in turn of higher fruit yield. The female plants in the periphery of male plants set more fruits. Both the vine and root cuttings are generally planted directly in the main field. Sometimes, the vine cuttings are buried in sand medium to initiate rooting, particularly for planting them in riverbeds. Pointed gourd can be planted on the ground and the vines are trained over the bowers or trellises, however, the ground culture on beds is more common. After thorough ploughings, raised beds of 15 cm height, 3 m width and convenient length are prepared, maintaining a spacing of 60–75 cm between two beds, which serves the irrigation cum drainage channel. Digging pits and subsequent filling with soil and farmyard manure in equal proportion prepare the mounds spaced by 60 cm on both sides of the bed at close proximity of the channel.

The fruits are less prone to rot disease and more consumers acceptable when grown on trellis made at 60 cm height with bamboo, ropes and wires. Mounds are prepared in the same manner. Harvest span and net return from the crop can also be increased through practicing trellis system of pointed gourd cultivation. The vine cuttings may be planted on mounds by adopting following methods:

- **Lunda or lachhi method:** In this system, mature vines of 1.0–1.5 m length having 8–10 nodes per cutting are taken and folded into a figure of eight and placed flat in the pit or mound and pressed 3–5 cm deep in the middle into the soil. To enhance sprouting, fresh cow dung may be applied over central part of the pit.

- **Moist lump method:** In this system, a lump of moist soil encircles the mature vine of 60–90 cm long leaving both ends of 15 cm length free and is buried in soil 10 cm deep into well-prepared pits. The nodes below the soil produce roots and those that are exposed gives rise to sprouts.

- **Straight vine method:** The mature vine cuttings are planted end to end horizontally 15 cm deep into the furrows about 2 m apart. Before planting, the furrows are filled up with a mixture of farmyard manure.

- **Ring method:** Mature vine cuttings are coiled into a spiral or ring-shaped structure, which is directly planted on the mounds covering one-half to two-thirds of the ring under the soil. The exposed portion produces new sprouts.

- **Rooted cutting method:** In this system, the cuttings from mature vines are first defoliated and then planted in the nursery mainly in sand medium to initiate easy rotting. The cuttings become ready for transplanting within 2 to 3 months. Very small cuttings less than 60 cm should be avoided.
- **Trellis method:** In this method, pointed gourd is planted in the pits at a spacing of 2.0 m × 0.6 m, and the vines are trained over the trellis made of iron or bamboo of 2 m height facilitating effective pollination, which results in increased harvesting span and yield. Fruits are less prone to rot disease, easily harvested and have more consumer acceptance. However, in riverbeds, the crop is planted every year.

PLANT POPULATION DENSITY

In trellis system, 6500 cuttings are required to plant a hectare land area at a spacing of 1.0 m × 1.5 m, whereas, in flat bed planting system, 4500–5000 cuttings are required to plant at a spacing of 1.0 m × 2.0 m. Similarly, under river bed cultivation, 3500–4000 cuttings are required to plant at a spacing of 1 m × 2.5 m. Pointed gourd is a long duration crop of about 8 months, thus, a very high-density planting affects its growth and yield adversely. Mounds in bed at a spacing of 2 m × 2 m are generally practiced accommodating about 2500 plants per hectare.

In case of seed propagation, about 20–25 kg seed per hectare is sown one meter apart in hills. Although red brown beetle (*Carpophilus dimidiatus*) and small ants (*Componotus compressus*) contribute significantly to cross pollination, hand pollination in female flowers is widely practiced nowadays due to paucity of pollinator population and environmental pollution, and even results in a twofold increase in fruit set over natural pollination.

NUTRITIONAL REQUIREMENTS

Pointed gourd, being a long duration crop, responds very well to nutrient management. Application of 20–25 t/ha of well-decomposed farmyard manure at the time of pit or mound preparation ensures sustained vegetative growth and yield. Based on soil test, a nitrogen dose of 150 kg, phosphorus 60 kg and potassium 80 kg/ha is advocated for better crop of pointed gourd. One-third dose of nitrogen and a full dose of phosphorus and potash fertilizers are applied as basal application at the time of pit or mound preparation and the rest amount of the nitrogen is applied in two splits, one at the onset of flowering and the other after 1 month. No fertilizer is applied to pointed gourd grown in riverbeds. Nutritional requirement for major pointed gourd growing states is given in Table 57.2.

In case of prolonged fruiting, nitrogen top-dressing should be continued at monthly intervals. Excessive nitrogen under frequent irrigation promotes excessive vegetative growth, especially in heavy soils, which reduce the propensity of flowering in both male and female plants.

TABLE 57.2
Nutritional Requirement for Major Pointed Gourd

State	Quantity of fertilizers (kg/ha)		
	Nitrogen	Phosphorus	Potash
Bihar	60	40	40
Uttar Pradesh	60	60	40
Odisha	150	60	80
West Bengal	32	20	20

IRRIGATION REQUIREMENT

Among cucurbits, pointed gourd is treated as a drought hardy crop, and thus, it can withstand soil moisture stress. However, the crop is highly susceptible to waterlogging conditions, hence, taking utmost care for drainage of excess water from the field, particularly during *Kharif* season, is essential. *Kharif* crops usually require no irrigation, except during the period of long dry spell, but during winter and summer months, irrigation is necessary at sprout elongation stage for proper growth and development. Generally, the pointed gourd field is irrigated once in 5 or 6 days in summer, depending upon the soil, location, temperature, crop growth stage, etc. Frequent irrigation also promotes excessive vegetative growth, especially in heavy soils. The early crop requires frequent irrigations and in dry weather at least once in 7–10 days. Irrigation water of 111 cm is optimum for good growth and yield. It is observed that irrigation at IW/CPE of 1.8 in combination with straw mulch also saves 42.5 cm water. Irrigation water should be restricted to the base of the plant or root zone without wetting the vines, especially when flowering, fruit set and fruit development are in progress. Frequent wetting of stems, leaves and developing fruits promote rotting disease of the vines and fruit.

INTERCULTURAL OPERATIONS

In pointed gourd, shallow intercultural operations should be done carefully during initial stage of crop growth, as it increases aeration in the root zone and conserves soil moisture. At least three to four hoeings are required, which should be done just before the application of fertilizers. Once the vines start trailing on the ground, intercultural operations become very difficult unless the plants are trained on bowers. Therefore, a minimum of two hoeings are essential in each month for good fruiting. *Cynodon dactylon*, *Chenopodium album*, *Echinochloa colona*, etc. are the major weed species associated with pointed gourd. Application of herbicides, like Gramoxone @ 1 litre a.i. or Fernoxone @ 0.8 litre a.i. per hectare as post-emergence spray along with mulching can effectively control the weed growth and ensures maximum total yield. The top portion of the pointed gourd plants dies during each winter and new shoots come up in spring. Towards the end of winter, the soil nearer to the plant root zone should be loosened and well-decomposed manure should be applied followed by irrigation.

Intercropping can be done with different vegetable crops, like beet leaf, radish, coriander, fenugreek, cauliflower, pea, etc. during initial stages of crop growth during winter season for better land use and greater economic return as the main crop goes to dormancy.

RATOONING

Pruning of pointed gourd vines helps in increasing fruit yield. Therefore, its vines should be pruned before winter, as the crop passes through the dormant phase in winters. Pruning can be carried by leaving one feet long vine in the basin. Pruning should be done when the vines are well established in October and November leaving 30–35 cm long stem. The crop being semi-perennial continues to live for 3–4 years and as the shoot portion dies in each winter, the new shoots come in spring. Towards the close of winter, soil is loosened, manure is applied and the fields are irrigated. The yield goes down after the third year and replanting is necessary.

PHYSIOLOGICAL DISORDERS

UNFRUITFULNESS IN POINTED GOURD

Being a dioecious perennial cucurbitaceous vegetable, both male and female pointed gourd plants are produced separately. Hence, the required number of male plants should be in the population of female plants to ensure an adequate population for better pollination and fertilization and ultimately higher fruit setting.

A common problem frequently observed in pointed gourd is that the pistillate flowers on female plants are shed due to lack of pollination and fertilization. In severe cases, ovary of the unfertilized flower may grow a bit due to parthenocarpic stimulation, which also abscises after few days. The unfruitfulness problem is more severe during high temperature period, *i.e.*, summer months, primarily because of low population of pollinating agents, *i.e.*, honeybees and beetles.

Remedial Measures

Male plants must be grown in the field along with female plants in the ratio of 10–15 male plants per 85–90 female plants to ensure adequate pollination and fertilization. Application of plant protection chemicals should be avoided during morning hours, as it hinders the movement of pollinating agents. Hand pollination may be done successfully to achieve adequate fruit set. Hand pollination to the female flowers should be done in the early morning hours because stigma receptivity decreases with the advancement of the day. Usually 7–8 female flowers can satisfactorily be hand pollinated with the pollens as one male flower. Growing of bitter gourd or bottle gourd in the adjoining of pointed gourd field as pollen of bitter gourd or bottle gourd stimulate successful parthenocarpic development of pointed gourd fruits. Spaying of water or plant growth regulators around the pointed gourd field during high temperature time reduces the temperature and increases the pollinating agents for better pollination, fertilization and fruit set.

HARVESTING

Fruits are available from late January onwards to October. Vines generally start fruiting 90–100 days after planting. October planted crops start fruiting from February and continues up to July–August and even up to September–October if the new flushes come with the monsoon rains. For tender, immature, soft-seeded fruits, harvesting is done at 8- to 12-day intervals, depending on cultivars, or twice a week. Delay in harvesting leads to the development of hard seeds and reduces further fruiting capacity of the vine by affecting successive female flower production and fruit setting.

YIELD

Fruit yield varies from place to place depending on soil and climatic conditions. Ratoon crop, if managed properly, generally gives higher yield compared to the first year crop. The fruit yield increases subsequently and declines the fourth year onwards. Fruit yield varies from 60 to 80 q/ha during first year and 100–140 q/ha during second year.

POST-HARVEST MANAGEMENT

In spite of its popularity as a vegetable, it is difficult to store pointed gourd well for longer period under ambient room temperature conditions. Dipping the freshly harvested tender fruits in kinetin 50 ppm, gibberellic acid 20 ppm, naphthalene acetic acid 20 ppm, or chemicals like sodium benzoate 200 ppm, or potassium metabisulfite 1900 ppm for 10 minutes followed by air drying and storing at 27–31°C in a zero energy cool chamber with 94–96% relative humidity increases the shelf life by 4 days with almost non-shrinkage and very low yellowing. Post-harvest dipping of fruits in 1% salicyclic acid, 4% citric acid and 10% waxol increases the keeping quality of fruits.

INSECT-PESTS

FRUIT FLY (*Dacus dorsalis*)

Fruit fly is the most serious and destructive insect of pointed gourd. Its adult fly lays eggs singly or in clusters of 4–12 in flower or young developing or ripening fruits with the help of sharp ovipositor of female, which consequently start rotting and ripening even at young stage. Eggs hatch in 2–9 days and maggots feed on internal contents of fruit causing rotting. Pupation is in ground at a depth of 1.5–15.0 cm. Infestation is more during rainy season. Direct control of maggots is unfeasible since they remain inside the developing fruits but the adult flies can be controlled.

Control

- Destroy the affected fruits.
- Install pheromone traps @ 10 traps per hectare.
- Use poison baits prepared with 20 g Malathion powder and 200 g molasses in 2 litres of water during fruit setting period.
- Spray the crop with fenthion 0.05% + 5% jaggery at fruit formation or ripening stage at an interval of 10–15 days.

ROOT KNOT NEMATODE (*Meloidogyne incognita*)

Root knot nematodes are polyphagous pests that are most serious in light sandy soils. Infested plants become stunted and show premature wilting. The roots have profuse brownish galls or root knots. While the infestation is severe, the entire plant may wilt and die.

Control

- Plough the field repeatedly in hot summer months to reduce the population of root-knot nematode.
- Apply organic amendment in soil with oil cakes like *neem*, *mahua*, *karanj*, groundnut, cotton, mustard, linseed, sesame, etc. @ 250 kg/ha 3 weeks prior to planting of cuttings.
- Follow marigold inter-copping with pointed gourd to lower the gall formation significantly due to release of α-terthienyl.
- Incorporate chopped leaves of *Ricinus communis*, *Calotropis procera*, *Leucaena leuco-cephala*, or *Melia azedarach* in nursery beds to reduce infection and increase sprouting.
- Burn the dried crop residues on field surface after harvesting
- Apply potassium fertilizer to reduce the root-knot intensity.
- Treat the root and vine cuttings with carbofuran 0.05% for 6 hours followed by shade dry-ing before planting to reduce nematode incidence significantly.

RED PUMPKIN BEETLE (*Raphidopalpa foveicollis*)

Beetles eat the leaf lamina, causing defoliation particularly at cotyledon stage of crop. Grubs feed on underground stem and root portion of plants, causing holes or galleries and result in drying of plants. Red pumpkin beetle is also a serious pest puncturing the young fruits, leading to their decaying.

Control

- Deep ploughing soon after the previous crop exposes and kills the grubs and pupae.
- Apply Furadon 3G granules @ 2 kg/ha 3–4 cm deep in soil near base of germinated seedlings.

WHITE MEALY BUG (*Ferrisia virgata*)

White mealy patches are seen on the leaflets, petioles, tender shoot portions and fruit pedicels. The nymphs and adult females of white mealy bugs suck cell sap from lower surface of the leaves, tender shoots and fruits. The severely affected portion dry up and die. Ants and sooty moulds are associated with this pest infestation.

Control

- Soak 100 g fish oil resin soap in 10 litres of hot water (80–90°C) for 4 h to dissolve it and then dilute it to 50 litres and spray.
- Apply Profenofos 50% EC 2 mL/litre of water along with detergent 1 g/ litre followed by spray of *neem* pesticides 2–3 mL/litre of water.

EPILACHNA BEETLE (*Epilachna septima*)

Epilachna beetle is also seen in pointed gourd. Adult flies feed on foliage causing holes and defoliation. A large number of yellow-coloured thorny grubs are seen on undersurface of leaves and feed on chlorophyll resulting in skeletonisation of leaves. Cigar-shaped bright yellow eggs are laid in clusters by the adults.

Control

- Collect and destroy the egg masses, grubs and adults in early stages when infestation is less.
- Spray contact insecticides like quinalphos 0.05% if damage is serious.

DISEASES

ROT DISEASE (*Pythium* spp. and *Phytophthora* spp.)

Under water stagnation conditions, the incidence of shoot and fruit rot caused by soil borne fungi is very severe, and as a result, a white cottony growth develops on the tender shoots, leaves and fruits. The spread of fungi is aggravated through rain splash and irrigation water. The other soil borne fungus (*Macrophomina phaseolina*) attacks the collar region, and as a result, the affected portion starts rotting gradually and is more rampant in the crop preceded by jute.

Control

- Grow the crop on raised beds with good drainage to avoid water stagnation.
- Spray the crop with Blitox @ 4 g/litre of water at 7- to 10-day intervals.

DOWNEY MILDEW (*Pseudoperonospora cubensis*)

The disease is prevalent in areas of high humidity, especially during rainy season. Symptoms appear as water-soaked lesions on undersurface of leaf lamina and angular yellow-coloured spots on upper surface corresponds to the water-soaked lesions on undersurface. Such leaves may die quickly. Disease usually spreads very fast in moist weather.

Control

- Pluck and destroy the affected leaves followed by chemical spraying.
- Spray the crop with Dithane M-45 0.2% on under surface of leaves at 7- to 10-day intervals.

CUCUMBER GREEN MOTTLE MOSAIC VIRUS (CGMMV, *Tobamovirus*)

The leaves of affected plants show slight vein clearing, crumpling of young leaves, light or dark green mottling, blistering, distortion of leaves and stunting of plants. It is transmissible by contact and mechanical inoculation. The virus is carried over through season by infected crops, weeds and contaminated seed materials.

Control

- Use disease-free planting material taken from healthy plants.
- Destroy the infected crop plants and weeds to check the carryover of virus.
- Disinfect the tools and workers' hands.
- Treat the seed with dry heat at 70°C for 3 days.
- Spray the crop with imidacloprid (5 mL/15 litre of water) to control aphids, which are the virus vectors.

■■■

58 Spine Gourd

S.K. Bairwa and M.K. Rana

Botanical Name : *Momordica dioica* (Roxb. Ex Willd.)

Family : Cucurbitaceae

Chromosome Number : 2n = 28

ORIGIN AND DISTRIBUTION

Spine gourd, also known as teasel gourd or small bitter gourd, *kakrol, kankro, kartoli, kantoli, kantroli, ban karola*, and in Sanskrit as *karkotak*, is essentially a native of tropical regions in Asia with extensive distribution in China, Japan, South East Asia and Polynesia, as well as in tropical Africa and South America. As many of its species have been found to grow wildly in India, Bangladesh, Srilanka, Myanmar, Malay, etc., it indicated that this region might be the origin of spine gourd. It has been considered as under-utilized minor cucurbitaceous crop, which is cultivated in Orissa, Maharashtra, West Bengal, Bihar, Karnataka, Chhattisgarh and few regions of Rajasthan in India. The plant sometimes is found growing wild and commonly in hedges. It is often cultivated for its fruits, which are used as vegetable.

INTRODUCTION

Spine gourd, a semi-wild species of flowering plant in gourd family, is an important minor cucurbitaceous vegetable in Bangladesh and the Indian subcontinent, which is closely related to bitter gourd. However, it is not bitter in taste as the bitter gourd is. It is found available in cultivated or semi-wild form in dry and moist forests. For annual cultivation, it requires lowland. It is dioecious perennial in nature and has tuberous roots. It is consumed by tribal groups, living around the natural forests, especially at higher altitudes, where the native folk consume it as a daily vegetable. It has commercial importance and is exported and used locally. It has many advantages, like having a high market price and good nutritional value, and keeping quality longer. However, this vegetable could not gain much popularity until it was discovered to have a high nutritional and medicinal value. It is a tuberous climber found in Western Ghats of Maharashtra and in other regions of India to Ceylon, ascending to 1525 m in the Himalaya, Malaya, Bangladesh and China. West Bengal and Karnataka are two major states where spine gourd is grown commercially. In West Bengal, it is largely cultivated in the Malda and Nadia districts, and in Karnataka, it is distributed frequently in the wet forests of Western Ghats and Deccan (Belgaum, Chikmagalur, Dharwad, Dakshina and Uttar Kannada, Hassan, Mysore and Shimoga).

COMPOSITION AND USES

COMPOSITION

Spine gourd is an important vegetable with high food value. Its fruits provide a good amount of protein and iodine and contain moisture 84.1%, carbohydrates 7.7 g, protein 3.1 g, fat 1.0 g, fibre 3.0 g, minerals 1.1 g and small quantity of essential vitamins like ascorbic acid, carotene, thiamine, riboflavin and niacin. The fruit also contains alkaloid, flavonoids, glycosides and amino acids. Its ash contains a trace of manganese. The nutritional value per 100 g of edible portion is given in Table 58.1.

USES

Spine gourd is a cucurbitaceous summer vegetable. The unripe green fruits, young twigs and leaves of this crop are extensively used as raw vegetable or cooked vegetable. The fruits are fried and sometimes eaten with meat or fish. The tuberous roots of female plants are also eaten as a vegetable. Seed oil is used as an illuminant.

MEDICINAL USES

Studies show that this plant is used in the prevention and cure of numerous ailments and in human food. Both male and female spine gourd has sex specific medicinal properties. Traditionally, it is used as sedative, astringent, febrifuge, antiseptic, anthelmintic, spermicidal and also used in urinary infection, bowel affections, and for snake bites and scorpion stings. The decoction prepared from the leaves of female plant is used as an aphrodisiac to eradicate intestinal parasites and to cure fever, asthma, and hiccups. The root of female plant is used in the form of paste to heal bleeding piles, and also used with benefit in head troubles, kidney stones and jaundice. Its roots' aqueous and ethanol extract is found effective in causing significant post-coital anti-fertility activity. This plant also gives gastro-protective and ulcer healing activities. Its aqueous extract is found to paralyze earthworms within hour up to 91.6%. The steroidal glycosides isolated from its roots reveal a moderate antibacterial and a poor antifungal activity. The fruit is considered pungent, bitter, hot, stomachic and laxative and plays a role in curing biliousness, asthma, leprosy, bronchitis, fever, tumours, urinary discharge, excessive salivation and heart disease. Juice of the fruit is a domestic remedy for inflammation. Powder obtained from dried fruit is used to induce sneezing, leading to nasal clearing. Ethanol extract of fruit has shown nephroprotective activity. Studies indicate that it possesses antioxidant, hepatoprotective, hypoglycaemic, antibacterial, anti-inflammatory, anti-lipid peroxidative, hypoglycaemic, analgesic and antifeedant properties. Its fruits are quite useful in controlling diabetes and help in building natural immunity of a human body. Its seeds ethanol extract possesses nephroprotective and curative activities without any toxicity due to its antioxidant

TABLE 58.1
Nutritional Composition of Spine Gourd Fruit (per 100 g Edible Portion)

Constituents	Contents	Constituents	Contents
Water (g)	84.1	Phosphorus (mg)	42.0
Carbohydrates (g)	7.7	Iron (mg)	4.6
Protein (g)	3.1	Carotene (mg)	162
Fat (g)	1.0	Thiamine (mg)	0.05
Dietary fibre (g)	3.0	Riboflavin (mg)	0.18
Minerals (g)	1.1	Niacin (mg)	0.6
Calcium (mg)	33.0	Energy (kcal)	52.0

activity and offers a promising role in the treatment of acute renal injury caused by nephrotoxin-like gentamicin. Its seed oil solvent extract is found to cause 100% mortality of mustard aphid in 24 h at 4% concentration due to the presence of alkaloid momordicin in its oil.

BOTANY

The cultivated forms of spine gourd are perennial dioecious climbers with tuberous roots. Its stem is slender, branched, furrowed, glabrous and shining. It shows a wide genetic diversity in shape, size and colour of leaves. Tendrils are simple, elongate, striate and glabrous. Its small leaves are membranous, broadly ovate in outline, variable, cordate at base, more or less deeply three- to five-lobed and distinctly denticulate. Male flowers' peduncle is solitary, slender, usually pubescent near the top, bract cucullate, strongly nerved, calyx lobes distinct, linear lanceolate and petals wholly yellow. Female flower peduncle is nearly as long as males, usually with a small bract near the base and ovary clothed with long soft papillate. Anthesis takes place during evening hours from 6.00 p.m. to 7.00 p.m. and the flower takes 7–22 minutes to open. The fruits are round to oval with soft spines, green when unripe and bright yellow to orange when ripe. The ripe fruits open in three irregular pieces. Fruiting and flowering continue for a long period. Fruit contains 10–20 seeds, which are slightly compressed enclosed in a red pulp and smooth-surfaced with bright red aril.

CLIMATIC REQUIREMENT

Spine gourd is essentially a warm season crop grown successfully in the plains and hills of tropical and subtropical regions. It requires warm and humid climate for its best growth. Plenty of sunshine and low humidity are the ideal conditions for its normal growth. The optimum temperature for its cultivation ranges from 25°C to 40°C. It is susceptible to frost and strong winds. During winter season, the plants go to dormant condition.

SOIL REQUIREMENT

Spine gourd can be grown on a wide range of soils except light soils. However, a well-drained sandy loam soil rich in organic matter is most suitable for it cultivation. The pH of the soil should be between 6.0 and 7.5. It is susceptible to waterlogging conditions, thus may not be cultivated on heavy soils. The field is prepared to a fine tilth by ploughing two or three times followed by planking.

CULTIVATED VARIETIES

Spine gourd has two different types of variety male and female as well as fruiting and fruitless variety. Since this is underexploited crop and consumed by the ethnic people living near natural forest, no improvement work has been done so far on this crop to develop varieties. Hence, improved varieties are not available in this crop, but some local cultivars are naturally grown by the farmers.

INDIRA KANKODA I (RMF 37)

A variety developed at Indira Gandhi Agricultural University RMD College of Agriculture and Research Station, Ambikapur (Chhattisgarh), is resistant to all major pests. It bears dark green very attractive fruits, which become ready for harvest in the first year in 75–80 days when the crop is propagated by seeds and in the second year in 35–40 days when the crop is raised through tubers. It gives three to four pickings of green fruits and up to 5–6 years. Its single fruit weight is about 14 g and average green fruit yield in first year is 8–10 q/ha, second year 10–15 q/ha and third year 15–20 q/ha.

SOWING/PLANTING TIME

Spine gourd is sown twice, once in spring–summer and second in rainy season. In tropical plains, the planting is normally done in February–March for spring-summer crop and in June–July for rainy season crop. In tropical hills, its sowing is done in April.

SEED RATE

If propagated by seed, about 2.5–5.0 kg seed is required for sowing a hectare area. The seed should be disease-free so it should be extracted from good quality disease-free ripe fruits.

PROPAGATION

Spine gourd is propagated vegetatively through tuberous roots or their cuttings, which are under cultivation in very few regions of India and Bangladesh. The mature, healthy, disease-free and vigorous mature tuberous roots are selected for propagation. Female plant tubers are larger than the tubers of male plants. The root cuttings are planted in pits at a line to line spacing of 2 m and plant to plant of 60–90 cm. If staking is practiced, this spacing may be reduced according to cultivation requirement. For every nine females, one male plant is retained in plant population for getting higher fruit set and yield from this crop.

If the crop is to be propagated by seeds, 2–3 seeds are placed at a depth of 1–2 cm in the pits prepared on raised beds. If more seeds per pit are to be used, thinning is to be done as and when the seedlings are large enough to handle.

NUTRITIONAL REQUIREMENT

The general practice is to incorporate 15–20 tonnes well-rotten farmyard manure or compost per hectare into the soil at the time of field preparation. Every year, when the fruiting is over and the plants become dormant, the field is deeply hoed and 15–20 tonnes organic manure per hectare is applied to the field. Generally, chemical fertilizers are not applied in this crop but it will be useful to apply nitrogen @ 50–60 kg/ha as topdressing in two equal splits, first at 30–35 days and second at 45–60 days after transplanting.

IRRIGATION REQUIREMENT

Irrigation is supplied by running water through the channels or pits from irrigation channel or drip channel. First irrigation is applied soon after sowing by watering the beds lightly to facilitate seed germination. The subsequent irrigation frequency depends on soil factors and prevailing weather conditions. In dry weather, irrigation is applied at an interval of 3 to 5 days. Flowering and fruiting are very critical stages and any stress during this period affects the yield and quality of fruits adversely. In rainy season, the crop does not require irrigation. Heavy rainfall causes rooting of tubers and death of the plants. Hence, providing proper drainage facilities to drain out the excess water is essential.

INTERCULTURAL OPERATIONS

HOEING AND WEEDING

For growing spine gourd successfully, no special intercultural operation is required. During early stages of crop growth, the pits and channels need to be kept weed free. For this purpose, two or three hand weedings or hoeings are needed. It is a shallow-rooted crop, and hence, shallow hoeing should be practiced. A light earthing-up can also be done after application of chemical fertilizers. No chemical weedicide is recommended to control weeds in this crop.

Staking

For getting higher yield from spine gourd, providing staking to the vines after application of basal dose of fertilizer is essential so that the vines when start spreading may get support for their normal growth. As each node bears flowers after 8–10 nodes from the base, excess lateral branches, if any, may be pinched off for allowing the plant to reach bower height at the earliest. When plants start emerging vines, they may be trailed to bower by erecting small twigs in pits. Erection of a bower or *pandal* is a costly operation, and nearly 20% of its production cost is for making bower alone. The height of bower is adjusted as 1.5 to 2.0 m and is usually made of bamboo poles, galvanized iron wire and thin coir or plastic wire. Bower, once erected, can be utilized for raising at least three crops.

HARVESTING

The fruits become ready for harvesting from July to October. As in most cucurbitaceous crops used as vegetable, the fruits are always picked when they are still young and tender. The immature and fully develop fruits are harvested 30–35 days after flowering. Frequent harvesting at every alternate day or at intervals of 2–3 days is recommended to avoid losses due to oversized and over-mature fruits. Fruits are allowed to ripen only when seeds are to be extracted for propagation purposes. If that is the case, the fruits are harvested only when the fruit has changed colour from green to yellow to orange and the fruit pulp has entirely turned red with mature seeds.

YIELD

The average yield of spine gourd is 80–100 fruits (2–3 kg) per plant, depending upon soil and natural environmental factors and cultural practices adopted by the growers. The average yield of spine gourd is approximately 125 q/ha.

POST-HARVEST MANAGEMENT

Fruits can be stored 48–96 hours at ambient room temperature. Fruits covered with moist jute cloths are usually stored at cool and aerated place, and from time to time, water is sprinkled on these jute cloths.

INSECT-PESTS

Aphids (*Aphis* sp.)

Both nymphs and adults of this tiny insect suck cell sap from under surface of the tender leaves, reducing plant vigour. The leaves curl up and ultimately wilt. The aphids excrete honeydew, on which, black sooty mould develops, which hampers photosynthetic activity of the plants. Besides, these aphids act as a vector for the transmission of virus. The severe infestation affects the fruit yield.

Control

- Spray the crop with simple water with pressure.
- Spray the crop with Malathion 0.05%, methyl demeton 0.05%, dimethoate 0.03%, or phosphomidon 0.03%.

FRUIT FLY (*Dacus cucurbitae* coquillett)

This is a very serious pest of cucurbitaceous crops. The adults lay eggs into the tender fruits and ovary of female flower. The maggots after hatching start feeding inside the fruits. The infected fruits become deformed and start rotting.

Control
- Pick and destroy the affected fruits by digging into the soil.
- Dust the crop with malathion 0.05%.
- Use poison traps and bait containing malathion 0.05% and 10% *gur* in 20 litres of water.

RED PUMPKIN BEETLES (*Aulacophora foveicollis* lucas)

Grub and adult both attack the crop at seedling stage and make holes in the cotyledonary leaves, and ultimately, the fruit yield is affected.

DISEASES

POWDERY MILDEW (*Erysiphe cichoracearum* DC. *Sphaerotheca fuliginea* pollacci)

The disease first appears as round white floury spots on under-surface of the older leaves. The spots enlarge, increase in number, coalesce on upper surface of the leaves and eventually cover both sides of the leaves. The fungus also attacks the stems and floral parts of the plant.

Control
- Follow long crop rotation with non-host crops.
- Irrigate the crop frequently.
- Treat the seed with hot water at 50°C for 30 minutes.
- Spray the crop with Karathane 0.5%, Sulfex 0.3%, or carbandizam 0.15% three or four times at an interval of 10 days starting just after the appearance of disease.

DOWNY MILDEW (*Pseudoperonospora cubensis* berkely & Curtis)

In young seedlings, discoloration occurs, and in severe cases, the whole plant dies. In mature plants, purplish or yellowish-brown spots appear on upper surface of the leaf, while fluffy downy growth is observed on lower surface of the leaf. The infected leaves turn yellow and later die. During flowering, blackish patches appear on the seed stalks; as a result, the seed production is affected adversely. Moisture and temperature both are important factors for the reproduction of pathogen and further spread of this disease. The incidence is very severe under very wet weather conditions.

Control
- Follow long crop rotation.
- Treat the seed with hot water at 50°C for 30 minutes to kill the pathogens present on seed.
- Spray mencozeb 0.3%, Dithane M-45 0.2%, or Dithane Z-78 0.2%.

ANTHRACNOSE (*Colletotrichum orbiculare* Berk. & Mont.)

Small yellowish or water-soaked areas, which enlarge rapidly and turn brown, appear on the leaves. It results in shrivelling and death of leaves. Later, the symptom appears on fruits also.

Control

- Follow long crop rotation.
- Use high-quality seed from a reliable source.
- Treat the seed with 0.15% carbendazim solution.
- Remove affected plants and crop debris, and destroy them.
- Spray Dithane M-45 0.2%, Dithane Z-78 0.2%, or Blitox 0.2% and repeat it at 10- to 15-day intervals.

Mosaic Virus

The infected plants show a mottled mosaic leaf pattern. The fruits on diseased plants are mottled and often distorted, malformed and whitish. It is rapidly transmitted by sucking insect-pests especially aphids that suck cell sap from the infected plants and transmit virus to healthy plants.

Control

- Use healthy and virus-free tuberous roots for planting.
- Uproot affected plants from the field and destroy them to check further spread of the disease.
- Spray malathion 0.05%, methyl demeton 0.05%, dimethoate 0.03%, or phosphomidon 0.03% and repeat it at 10- to 12-day intervals.

■■■

59 Sweet Gourd

M.K. Rana, Baseerat Afroza and Sugani Devi

Botanical Name : *Momordica cochinchinensis*

Vernacular name : Baby jackfruit

Family : Cucurbitaceae

Chromosome Number : 2n = 28

ORIGIN AND DISTRIBUTION

Sweet gourd is a Southeast Asian fruit found throughout the region from Southern China to Northeastern Australia, mostly Vietnam. Its centre of origin is probably in Indo-Burma region. In India, it is a wild or less cultivated vegetable occurring in Andaman Islands, northeastern states and Eastern India. The crop is cultivated in Bihar, Jharkhand, Odisha and West Bengal by tribal people in their home gardens. However, it is most popular in Assam. It is available to some extent in the Garo hills of Meghalaya, Madhya Pradesh and Gujarat. It is also grown as a vegetable in China and Japan.

INTRODUCTION

Sweet gourd is an underutilized perennial dioecious cucurbit vegetable. It is variously known as red melon, baby jackfruit, spiny bitter gourd, sweet gourd, or cochinchin gourd in English but is commonly called as *gac*. It is also known as *mùbiēguǒ* in Chinese. Sweet gourd is highly valued for its nutritional and medicinal qualities, wide range of adaptability and economic value in the regions where it grows. It is particularly important due to the absence of bitterness in fruits.

COMPOSITION AND USES

COMPOSITION

Sweet gourd is a nutritious and mildly delicious vegetable that has many medicinal uses. It contains substantial amount of lycopene, beta-carotene and protein. Relative to its mass, it contains up to 70 times the amount of lycopene found in tomatoes. It has also been found to contain up to 10 times the amount of beta-carotene in carrots or sweet potatoes. The nutritional composition of sweet gourd fruit is given in Table 59.1.

USES

In sweet gourd, the flesh and oily sacs are quite palatable and have very little and mild taste. It is slightly sweet or not sweet and it may not be described as delicious. When the seeds or oil sacs are cooked with rice, they impart a lustrous appearance and oil-rich, mild nutty flavour to the rice.

TABLE 59.1

Nutritional Composition of Sweet Gourd Fruit (per 100 g of Edible Portion)

Constituents	Contents	Constituents	Contents
Edible portion (%)	100	Phosphorus (mg)	38
Water (%)	90.0	Iron (mg)	3.5
Carbohydrates (%)	6.4	Beta-carotene (µg)	5.1
Protein (mg)	0.6	Thiamine (mg)	145
Fat (mg)	0.1	Riboflavin (mg)	0.20
Dietary fibre (g)	1.6	Vitamin C (mg)	48
Calcium (mg)	27	Energy (kcal)	29

Sweet gourd is typically served at ceremonial or festive occasions in Vietnam such as Vietnamese New Year and weddings. It is most commonly prepared as a dish called *xoi gâc*, in which, the aril and the seeds are cooked in glutinous rice, imparting both their colour and flavour.

MEDICINAL USES

Other than the use of its fruits and leaves for culinary dishes, sweet gourd is also used for its medicinal and nutritional properties. Sweet gourd is considered an esteemed fruit from heaven and is prized for its ability to promote longevity, vitality and health. In Vietnam, the seed membranes are said to aid in the relief of dry eyes, as well as to promote healthy vision. Similarly, in traditional Chinese medicine, the seeds of sweet gourd, known as Mandarin Chinese, are employed for a variety of internal and external purposes. In west, the chemical analysis of fruit has suggested high concentration of several important phytohormones. Most recently, the fruit has begun to be marketed outside Asia in the form of juice and dietary supplements because of its high phytonutrient content.

The unripe fruit acts as an appetizer and astringent. It is traditionally used for wound healing, to improve eye health and to promote normal growth in children. Roots of sweet gourd are used in bleeding piles, bowel infection and urinary complaints. Two isolated cyclotides, *i.e.*, MCoT-I and MCoT-II, may have properties to inhibit trypsin.

The sweet gourd fruit contains highest content of beta-carotene of any known fruit or vegetable. It has also high lycopene content. Due to its high content of beta-carotene and lycopene, sweet gourd is often sold as a food supplement in soft capsules. Moreover, the carotenoids present in it are bound to long chain fatty acids, resulting in what is claimed to be more bioavailable form. Recent research suggests that sweet gourd contains a protein that may inhibit the proliferation of cancer cells.

BOTANY

Sweet gourd is a perennial vine crop with a thick stem and herbaceous roots. The tendril is unbranched. The petiole shows two to four glands at the base and on the middle part. The blade is oval-cordate or oval-round, 10–20 cm in length and width and divided into three to five to the middle or undivided with palmate veins. It is dioecious in nature, bearing male and female flowers on separate plants. A male flower occurs solitarily, showing at the top of the peduncle a large bag-form round kidney-shaped bract covered with pubescences on both sides. The calyx tube is funnel-shaped and displays yellow tooth-like glands at the base. The corolla bears black speckles at the base, the stamens are five in number, united in two pairs, and each stamen is two-lobed and twisted.

A female flower occurs solitarily from leaf axil, with one style and three stigmas, which are united. Anthesis takes place in the morning (7:00–8:00 a.m.). Anthesis time is 05.50 hr in female and 06.20 hr in male flowers under Coorg conditions. Dehiscence begins at 22.45 hr, reaches peak at 23.00 hr and completes by midnight. The stigma remains receptive for 18 hr before and after anthesis. Ovary is inferior, unilocular and has parietal placentation. Fruit stalk is short (9–10 cm). Fruit is uniformly covered by small, green, thick projections turning scarlet red on ripening and splitting open exposing the bright red arillate seeds. Mature fruits are large, globular, dark orange, weighing 350–600 g, 15–18 cm long and 8–9 cm across. Fruiting occurs only once a year. The number of seeds may vary from 20 to 35. The seed is egg-shaped or somewhat triangular, blackish brown when dry, toothed marginally and furrowed on both ventral and dorsal surfaces.

CLIMATIC REQUIREMENT

Sweet gourd is a tropical vine plant. However, it can be grown as an annual plant in cooler climates as well. Under Indian conditions, it is cultivated as a Kharif crop. Sweet gourd requires a temperature range of 25°C to 35°C and average annual rainfall of 1500 to 2500 mm for normal development of the plant. The atmospheric day temperature above 25°C is ideal for seeds germination. Soil temperature ranging from 10°C to 25°C is required for its seed germination but the ideal range lies between 18°C and 22°C. Long warm and dry weather with abundant sunshine is favourable for proper plant growth. The plant does not resist frost and needs to be protected if grown during cooler months. Excessive humid weather conditions may promote diseases like powdery mildew, anthracnose and pests like fruit flies.

SOIL REQUIREMENT

Sweet gourd can be grown on many kinds of soil, out of which, sandy loam or loam soils with high organic matter are the best. It grows well on soils of shallow to medium depth (50–150 cm) and prefers acidic to alkaline soil with pH of 6–7.5. It cannot tolerate high salt concentration. The soil should be well drained. Lighter soils are preferred for early yield since in heavy soils, vine growth is more and fruit maturity is delayed. Such soils crack in summer and become waterlogged during rains and hence become unfit for the cultivation of this crop. This crop is very much sensitive to waterlogging and may not tolerate freezing temperature. However, like other cucurbits, it requires good moisture during vegetative growth and flowering. Deep soils can support the vines for a longer period.

Senescence of aerial parts and dormancy of tubers for 3–4 months are reported during winter, whereas, in Andaman Islands, the plant thrives well throughout the year, provided sufficient moisture is available for growth.

CULTIVATED VARIETIES/HYBRIDS

Not much research work has been done on the improvement of sweet gourd. Hence, there is unavailability of improved varieties. In all parts of country, only local races are being grown. Wild and semi-wild domesticates are maintained on farm by tribal settlers in Andaman Islands. At Central Horticultural Experiment Station (CHES), Bhubaneswar, three genotypes of Andaman and Nicobar Islands are being maintained in field gene bank.

SOWING TIME

Reports indicate that rainy season is ideal season for cultivation. The seeds could be sown inside protected places in early winter in a warm spot with bright light and transplanted outside after last frost in spring. Under greenhouse conditions, the seedlings are raised in polybags in November and transplanted in field in December. The best time for planting tubers is February–March. The stem cuttings are planted in August–September.

SEED RATE

A seed rate of 3 to 5 kg is required for raising sweet gourd seedlings enough for planting a hectare area. In case underground roots are used, it would require about 20 to 50 kg of such tuberous roots for planting a hectare area.

PROPAGATION

Sweet gourd is a dioecious perennial climber and can be propagated by seeds, tuberous roots, stem cuttings and apical shoots. The crop once planted is allowed to remain in the same field for several years and is considered as a ratoon crop. The following methods may be adopted for its planting:

By seeds: When seeds are sown under optimum conditions of soil, water and light, they will germinate in 7–14 days. Seeds have prolonged dormancy besides low germination rate (50%) and long juvenile period extending to more than a year. Moreover, in case of seed propagation, it is impossible to predetermine sex of plants. The seedlings are also raised in polybags, and after attaining appropriate size, they are transplanted in the field at a spacing of 2 × 3 m.

By underground roots: Sweet gourd can be propagated by underground tuberous roots for getting an early crop. Plants that are 2–3 years old develop more number of tuberous roots. Due to dioecious nature of crop, tuberous roots of male and female plant should be collected separately, which can be recognized easily by seeing flowers and fruits. The propagating tuberous roots should be 40–50 g. The large tubers may be cut into pieces keeping at least two eyes in each piece. The cut tubers must be treated with 0.2% Dithane M-45 and kept in shade for 30 minutes.

By stem cuttings: A better method to raise such crop is to use rooted vine cuttings as propagules. Two nodes stem cuttings of sweet gourd are ideal for multiplication but stem should be free from insect-pests and diseases. These cuttings should be planted in sand beds at the time when temperature and humidity are conducive.

PLANTING METHOD

Pits of 60 × 60 × 60 cm are dug in well-ploughed soils during monsoon showers at a spacing of about 2 m between pits and 5 m between adjacent rows (for single trellis system of training) or 5 × 5 m (for pandal system). Pits are filled with farmyard manure and topsoil. Planting is generally done in June–July.

NUTRITIONAL REQUIREMENT

Sweet gourd responds well to the application of manure and fertilizers. The dose of manure and fertilizers can not be generalized for all situations although depends on soil fertility level, soil type, climatic conditions of the growing region, etc. However, well-rotten farmyard manure 25 kg may be filled per pit at the time of planting. Application of 150 g single super-phosphate and 50 g muriate of potash to basin and top dressing with urea @ 15 g twice at the four- to five-leaf stage and 30 days later are generally recommended for good yield. Excessive application of nitrogen should be avoided to prevent enormous vine growth and keep the female to male flower ratio high. The fertilizer application must be completed well before the fruit set.

IRRIGATION REQUIREMENT

First irrigation should be given immediately after planting and subsequent irrigations should be given at weekly intervals or as required. Care should be taken to irrigate the crop at critical stages of irrigation, *e.g.*, at flowering and fruit development stages. The frequency of irrigation is very important to maintain good soil moisture at the root zone, which promotes rapid development of taproots and uptake of nutrients. However, too frequent irrigation promotes excessive vegetative growth and harms the vines. Hence, the crop needs to be irrigated once in 5 to 6 days, depending upon temperature, soil type and location. Application of irrigation water should be restricted to the root zone only, preventing the vines, leaves, or flowers from getting wet, which would otherwise lead to the occurrence of diseases.

INTERCULTURAL OPERATIONS

WEEDING

Weeds should be kept low by regular weeding. Mulching may be done at early stage of crop for controlling weeds and conserving moisture. Mulching may also be practiced for summer or winter protection. Plastic mulching like black plastic mulch may be beneficial in controlling weeds. Besides, mulching conserves soil moisture and increases soil temperature, which in turn improve vigour, yield, earliness and quality of fruits.

TRAINING

Usually, the plants are trailed to thatched sheds, fences, or treetops and pollination is left to nature. For optimum yield, a strong bower system is required. Bowers should be prepared at a height of 1.5 m by placing cemented or bamboo poles followed by criss-cross wire netting. Training on bower or staking must be started just after 30 cm of vine length from ground level. Once the crop anchors well, the branches spread easily because of tendrils. Fruit quality improves due to exposure to sun and yield increases because of vigorous growth, more number of branches and proper distribution of mineral nutrients. Kniffin system is also advisable for better yield and quality.

POLLINATION MANAGEMENT

Sweet gourd is dioecious in nature. The seedling progenies segregate into male and female in a ratio of 1: 1. However, the male to female plants must be maintained in ratio of 1: 9 by uprooting surplus male plants. One to two healthy male plants are retained as pollen source for an effective pollination and fruit production. First flower develops about 3–4 months after sowing and vines bloom for about 6 months. Generally, the male plants flower earlier than female plants.

PLANT GROWTH REGULATORS

Self-fertilization in dioecious plants has been made possible through modifying flower sex by applying growth regulators. Application of silver nitrate ($AgNO_3$) at different concentrations (200–800 mg per litre of water) induced male organs in female flowers when sprayed at the two- to four-leaf stage. Spraying of $AgNO_3$ on female plant, 10–15 days before bud initiation, at a concentration of 500–1000 ppm induced hermaphrodite flowers.

HARVESTING

The plant starts flowering after about 2 months of planting of root tubers. After 8–10 nodes from base, each node bears flowers. This flowering continues up to July–August and sometimes up to September. Fruits reach harvestable maturity by 15–20 days, beyond which the seeds become hard and skin leathery. For consumption purpose, the fruits are harvested at tender stage when seeds are in immature stage. Raw fruits have an agreeable taste and are not at all bitter. It cannot be picked green and shipped like bananas. Fruits can be stored safely for 1 week at room temperature.

YIELD

Generally, a plant produces about 30 to 50 fruits in one season weighing about 3 to 5 kg, which comes out to be 50 to 80 quintal per hectare.

POST-HARVEST MANAGEMENT

Usually, the cultivation of sweet gourd is restricted to the areas where these are actually consumed. Mostly, these are consumed fresh, and hence, not much post-harvest management is needed. However, the sweet gourd fruits, being smaller, may be packed in baskets and transported to the nearby markets.

INSECT-PESTS

EPILACHNA BEETLE (*Epilachna* sp.)

Adult and larvae both feed voraciously by scrapping chlorophyll from epidermal layer of leaves causing characteristic skeletonization of the lamina. Sometimes, larvae also attack the roots.

Control

- Dust the crop with 0.2% carbaryl or 0.05% malathion.

FRUIT FLY (*Dacus* sp.)

Maggots of this fly damage the young developing fruits severely. The adult female flies puncture the epidermis of ovaries or soft and tender fruits with their stout and hard ovipositor and lay eggs below epidermis. The eggs hatch into maggots, which feed inside the fruits and cause rotting.

Control

- Grow resistant varieties.
- Pinch off the affected flowers.
- Spray Thiodon @ 2 ml/litre of water.
- In case of severe infestation, bait spray is recommended.

APHIDS (*Aphis* sp.)

These are the small green insects, causing damage to the plants by sucking sap from their tender leaves. In young seedlings, the cotyledonary leaves crinkle and the plant withers. In older plants, the leaves turn yellow and ultimately the plant loses its vigour and yield.

Control

- Grow cultivars resistant to aphids.
- Spray the crop with malathion @ 0.1% or Metasystox @ 0.1 to 0.2%.

NEMATODES (*Meloidogyne* sp.)

Nematode infestation results in poor growth and stunted plants.

Control

- Grow nematode resistant varieties.
- Follow long crop rotation with non-host crops.
- Treat the seed with carbofuran 6% W/W.

60 Snake Gourd

Pradyumna Tripathy and Monalisha Pradhan

Botanical Name : *Trichosanthes cucumerina* L.
Synonym: *Trichosanthes anguina* L.

Family : Cucurbitaceae

Chromosome Number : 2n = 22

ORIGIN AND DISTRIBUTION

Snake gourd had probably originated from India or the Indo-Malayan region. The crop was first domesticated in India and still wild forms of the crop are available in India. *Trichosanthes* is a large genus having about 44 species, including about 21 species available in India. About 40 species of *Trichosanthes* occur in East, and Southeast Asia, tropical Australia and Fiji.

INTRODUCTION

Snake gourd, which is popularly known as serpent gourd, serpent cucumber, viper gourd, padwal and *chichinda*, is an annual herbaceous viny cucurbit. It is monoecious in sex form. The flowers, having peculiar features, are delicate, white and very aromatic, and fringed with fine hairs (*trichosanthes* means hairy flowers). It has extremely long fruits of 30–190 cm, cylindrical, slender and pointed at both the ends of the fruits. In general, two types of fruit shapes are observed in snake gourd such as long narrow fruit, which is tapered from top to bottom, while the other type has the same width from top to the bottom. Usually at widest portion, the fruit diameter ranges from minimum of 4.0 to 10.0 cm having fruit surface colour of greenish-white and stripped. The fruits at maturity become inedible, orange-red in colour and rich in fibre content, having seeds packed with red spongy pulp. The fruits at maturity become extremely bitter in taste, hence, unsuitable for consumption.

Snake gourd is cultivated in Southeast Asia, China, Japan, West Africa, Latin America, the Caribbean, tropical Australia, etc. In India, the crop is grown commercially in Kerala, Tamil Nadu, Karnataka and undivided Andhra Pradesh as well as sporadically in eastern Uttar Pradesh, Bihar, West Bengal, Odisha, Bihar, Gujarat and Maharashtra.

COMPOSITION AND USES

COMPOSITION

In general, the fruits of snake gourd are an important source of carbohydrates, proteins, minerals and vitamins A, B and C. Besides, the tender fruits, the young and tender leaves of snake gourd are also a good source of nutrients particularly protein, carbohydrates as well as minerals. The average nutritional composition of snake gourd fruits is presented in Table 60.1.

TABLE 60.1

Nutritional Composition of Snake Gourd Fruit (per 100 g Edible Portion)

Constituents	Contents	Constituents	Contents
Water (g)	92.0–94.6	Iron (mg)	0.3–1.7
Carbohydrates (g)	3.3	Vitamin A (µg)	96
Protein (g)	0.5–2.0	Thiamine (mg)	0.04–0.05
Fat (g)	0.3	Riboflavin (mg)	0.05–0.06
Dietary fibre (g)	0.8	Niacin (mg)	0.3–0.5
Minerals (g)	0.5	Vitamin C (mg)	29
Calcium (mg)	26–30	Energy (kcal)	18–20
Phosphorus (mg)	20–40	–	–

Uses

The edible part of snake gourd has long immature fruits, which are used both as salad and cooked as vegetable in various ways as well. Generally, the long fruits produce an unpleasant odour, which usually disappears after cooking. The young shoots, tendrils and leaves are also consumed as greens.

Medicinal Uses

Snake gourd fruits have a cooling effect on the body on consumption. The fruit induces appetite and acts as a tonic. Being low in calorie, the fruits are highly suitable for sugar patients. The juice extracted from young tender leaves is used as a home remedy against the treatment of jaundice as well as heart disorders such as palpitation and heart pain due to physical exertion. Both the roots and seeds are used for the treatment of diarrhoea, bronchitis and fever. Overall, the important health benefits derived by the consumption of snake gourd include its ability to improve the strength of immune system, detoxify the body, improve digestive processes of the body, reduces the heart diseases, diabetes, improves the quality of hair, etc. However, excess consumption of snake gourd seeds should be avoided as it may cause nausea, upset stomach, indigestion, gastric problems, etc.

BOTANY

Snake gourd (*Trichosanthes cucumerina* L.) is an annually grown cucurbitaceous crop. Stems are normally slender, branched and angular with 2 cm diameter. Leaf blades are often wider than long with three or more large lobes. Leaf blades are about 5–19 × 6–15 cm, petioles about 2–5 cm long and the leaf margins are finely denticulate. The plant is of monoecious nature. Tendrils are usually branched (3-branched), tips equipped with haustoria-like structures. One or two glands usually visible on underside of the leaf blade close to the basal margin. Lateral veins are about 4 to 6 on each side of the midrib. Generally, the male raceme produces 8–15 flowers on a peduncle of 8–10 cm and pedicel of about 5–12 cm and erect having slightly spreading bracteate. Calyx tube (hypanthium) is about 20 mm long, lobes recurved, about 1.5–2.5 mm long. Petals are about 15 mm long including the fringed tips, divided into long narrow lobes particularly at the apex. Stamens are three, two stamens with bilocular anthers and one stamen with a unilocular anther. Anthers are about 2 mm long and filaments about 0.5–1.5 mm long, free from one another. Anther locules are sinuous (shaped like a horizontal S). Pollen grains are white or translucent. On the other hand, female flowers are solitary with a peduncle up to 1.0 cm long and a narrow fusiform ovary. The corollas of female flowers are perhaps slightly shorter, about 10 cm long. Staminodes are small and inconspicuous. Style is as long as the hypanthium. Fruits are extremely long, varying from 30.0 to 190.0 cm, cylindrical, slender and pointed at both the ends. Seeds are normally oblong shaped and finely rugulose, having undulated margins.

CLIMATIC REQUIREMENT

Snake gourd being tropical in origin thrives well in hot or moderately warm and humid climate. Abundance of sunshine and fairly high rainfall favour good crop yield. High rainfall coupled with prolonged cloudiness promotes high incidence of fruit rot disease and in turn drastically reduces the yield. The crop is susceptible to frost and cold especially temperature below 10°C. The optimum temperature for better growth and development of snake gourd is around 25–35°C. Low temperature <20°C and high temperature >35°C are harmful for its growth, flowering as well as fruiting, hence, snake gourd cultivation <20°C and >35°C should be avoided.

SOIL REQUIREMENT

Snake gourd can be grown almost on any type of soil except very heavy soil. However, the ideal soil for snake gourd cultivation is well-drained sandy loam rich in organic matter with pH range of 6.5–7.5. The land with good drainage and plenty of sunshine are preferred for this crop. The field should be prepared to a fine tilth before sowing by ploughing 4–5 times followed by planking. All the weeds and previous crop stubbles should be removed from the prepared field for making clean seedbeds.

CULTIVATED VARIETIES

CO 1

An early maturing cultivar developed at Tamil Nadu Agricultural University, Coimbatore through selection from a local type of Alangulam, District Tirunelveli of Tamil Nadu, bears 10–12 fruits per vine. The fruits become ready for harvest within 70 days after sowing. The average fruits length is 160–180 cm having dark-green colour with white stripes. The fruits possess good cooking quality. The average fruit weight is about 500–750 g. The average fruit yield is about 180 quintals per hectare in crop duration of 135 days.

CO 2

A variety developed at Tamil Nadu Agricultural University, Coimbatore through selection from a local cultivar of Coimbatore district is suitable for high density planting. Fruits are usually very short of about only 30 cm and light greenish-white in colour. The average yield is about 350 quintals per hectare in a crop duration of 105–120 days.

CO 4

An early maturing variety developed at Tamil Nadu Agricultural University, Coimbatore bears 160–190 cm long dark-green fruits with white stripes and light green flesh. It gives marketable fruits in about 70 days after sowing. The variety has the potential to produce about 10–12 fruits per vine, weighing 4–5 kg.

MDU 1

A F_1 hybrid developed at Tamil Nadu Agricultural University, Coimbatore, by crossing between Pannipudal (a short-fruited type) and Selection 1 from Thaniyamangalam, is suitable for high density planting due to less spreading plant habit. The fruits are medium long (about 70 cm) and green with white stripes. The average fruit yield is about 320 quintals per hectare in crop duration of about 140 days.

PKM 1

A vigorously growing mutant induced from H.375 at Tamil Nadu Agricultural University KVK, Periyakulam, is suitable for growing round the year. Fruit colour is dark green with white stripes on the outside and light green inside. The average fruit size is about 155 to 170 cm and fruit weight is 500 to 800 g. The average fruit yield is about 240 quintals per hectare.

PLR (SG) 1

A variety developed at Tamil Nadu Agricultural University Vegetable Research Station, Palur, through a pure line selection from white long type is suitable for cultivation under irrigated conditions only. Its fruit has excellent cooking quality due to less fibre and high flesh content and does not twist due to maturity. It has a yield potential of 350–400 q/ha with 30.50% increase over CO 1. This variety can be cultivated during June–September, November–March and April–May. It is highly adoptable to Tiruvannamalai, Cuddalore, Villupuram, Vellore, Kancheepuram, Tiruvallur and Perambalur districts of Tamil Nadu. It thrives best in well-drained organic matter and rich sandy loam soil.

PLR (SG) 2

A variety developed at Tamil Nadu Agricultural University Vegetable Research Station, Palur, has excellent cooking quality due to less fibre content. Fruits are fleshy with attractive white colour and preferred in the local and super markets. An average single fruit weighs 600 g. Short fruit enables easy handling and long-distance transport.

APAU Swetha

A variety developed at Acharya N.G. Ranga Agricultural University, Hyderabad, through selection from a local type of snake gourd. The fruits are long, having a whitish background with green stripes. The average fruit yield is about 240 quintals per hectare.

Baby

A variety developed through a natural selection from the locally collected material of Peruva, Ernakulam district, bears 30–40 cm long elongated attractive uniformly white fruits. The average fruit weight is about 470 g with yield potential of 400 quintals per hectare in 8 to 10 pickings. First picking starts in about 55 days after sowing and total crop duration ranges from 120 to 140 days. It can be grown all through Kerala but mainly suited for cultivation in the central zone of the State.

TA 19

A variety developed at Kerala Agricultural University, Vellanikkara, through selection from a local cultivar. Fruits are about 60 cm long having light green with white stripes. The average fruit weight is about 600 g. Fruits are harvested in about 65–70 days after sowing.

Kaumudi

A cultivar developed at Sugarcane Research Station, Tiruvalla, Kerala Agricultural University, Vellanikkara, through selection from local types bears 100 cm long fruits with average fruit weight of 1.34 kg. Fruits are uniformly white with acute tip. Average fruit length is about 100 cm and breadth about 9.0 cm with 1.3 kg average fruit weight. The yield potential is about 500 quintals per hectare.

MANUSREE

An early high-yielding variety developed at Agricultural Research Station of the Kerala Agricultural University, Mannuthi, through a selection from local collection from Kurupanthara is suitable for growing in warm humid tropics of Kerala, especially in Thrissoor, Ernakulam and Palakkad. It bears 65–70 cm long attractive white fruits with green stripes at the pedicel end. It has a yield potential of 600 quintals per hectare with average fruit weight of 750–800 g.

KONKAN SWETA

A variety developed at Konkan Krishi Vidyapeeth, Dapoli (Maharashtra), bears about 90–100 cm long fruits having a whitish surface with greenish stripes. Its fruit has good flesh if harvested timely; otherwise the fruit becomes hollow. Average fruit yield is 100 to 150 quintals per hectare and a crop duration of 120–130 days.

SOWING TIME

Being a cucurbitaceous crop, snake gourd is also sown twice in a year, preferably February to March as a summer crop, while during June to July as a rainy season crop. However, snake gourd can also be grown successfully in the winter season during October to November in the regions of mild winter as in the case of southern states of India.

SEED RATE

In general, about 4.0–6.0 kg seed is enough for sowing a hectare land area. However, seed sowing in pits at a spacing of 2.0 m × 2.5 m requires about 1.5 kg seeds per hectare. In general, seeds should be collected from larger fruits as the healthier seeds produce vigorous seedlings, in turn higher fruit yield. In order to protect the seeds or seedlings, the seed should be treated with *Trichoderma viride* @ 4 g, *Pseudomonas fluorescens* 10 g, or carbendazim 2 g/kg of seed.

SOWING/PLANTING METHODS

SHALLOW PIT OR FLAT BED METHOD

In case of shallow pit method of planting, pits of 60 × 60 × 45 cm size are dug at least 3 weeks before sowing seeds. The pits are then filled with a mixture of soil, compost (4–5 kg/pit) and recommended dose of nitrogen, phosphorus and potassium. Carbofuran @ 1.5 g/pit should also be thoroughly mixed before sowing seeds. After filling the pits, water is applied for better germination of seed. Per pit, at least three or four seeds are sown at a depth of 2–3 cm. The recommended spacing of 1.5–2.5 m × 60–120 cm should be adapted for better growth and development of snake gourd.

SOWING OF SEEDS ON RAISED BEDS OR RIDGES

In this method, usually channels of 40–50 cm width should be prepared either manually or mechanically with a spacing of about 2.0–2.5 m between two channels. Seeds are then sown on both the edges of channel at least with a spacing of about 1 meter. In order to maintain the optimum plant population for higher fruit yield, at least 2 to 3 seeds per hill should be sown followed by light irrigation for obtaining uniform germination. The vines are usually allowed to spread over the raised portion in between the channels for obtaining better growth and development. Using this method, 3500–4500 plants per hectare land area can be accommodated.

PLANT POPULATION DENSITY

In snake gourd, population density generally varies from 3500 to 10,000 plants per hectare, depending upon varieties, season of cultivation as well as the cultural methods adopted. However, under high-density system, adopting suitable training practices, about 8,000–10,000 plants can be accommodated in a hectare land area.

NUTRITIONAL REQUIREMENT

Snake gourd responds well to both organic manure and fertilizers application. Farmyard manure should be applied @ 10–15 t/ha at the time of land preparation or applied in pits with soil at least two weeks before planting. It is always essential to grow the crop with optimum nutrients management based on soil test report. In the absence of soil test report, the recommended dose of nitrogen 50–70 kg, phosphorus 30–40 kg and potash 30–40 kg/ha should be applied. A half dose of nitrogen with full dose of phosphorus and potassium should be applied before sowing seeds. The remaining half of the nitrogen should be top-dressed when the plants start bearing fruits. Although snake gourd responds well to nutrients application, especially nitrogen, but applying excessive nitrogen under frequent irrigation conditions should be avoided, as it promotes excessive vegetative growth, especially in heavy soils, which increases the male flower production proportionately than female flowers, thereby reduction in fruit yield.

IRRIGATION REQUIREMENT

Snake gourd is treated as one of the moisture-sensitive crop, which suffers badly with both excess and moisture stress conditions. Hence, in order to obtain a good crop of snake gourd, the crop should be irrigated judiciously throughout the growing period. During the early period, light irrigation should be given more frequently for uniform seed germination and subsequent vegetative growth. However, snake gourd requires controlled irrigation for better vegetative as well as fruit yield. Generally, irrigation is applied once in 3 to 5 days in summer, depending upon soil, location, temperature, crop growth stage, etc. During rainy season, irrigation is generally not required; rather, drainage is important. Frequent irrigation also promotes excessive vegetative growth, especially in heavy soils. Irrigation should be restricted to the base of the plant or root zone without wetting the vines, especially when flowering, fruit set and fruit development are in progress. Frequent wetting of stems, leaves and developing fruits promotes decaying of vines and fruits. The critical stages of irrigation for snake gourd include immediately after seed sowing, active vegetative growth stage, flowering and fruit setting stage as well as fruit development stage. Moisture stress during these stages of crop growth significantly reduces the fruit yield of snake gourd.

INTERCULTURAL OPERATIONS

HOEING AND WEEDING

In general, in most of the cucurbits including snake gourd, it is essential to keep the field weed free in order to obtain better growth and development of the crop. The snake gourd field should be free from weed flora during the initial growth stage by application of herbicides such as pre-emergence application of either pendimethalin or Alachlor @ 1 kg/ha, which will effectively control the weed flora under field conditions. Subsequently, if weeds pose problem in snake gourd crop, one manual weeding may be done in order to keep the weed population low.

TRAINING OF VINES

In general, the snake gourd prefers proper training for better growth and yield. The crop has the potential to produce very long and large-sized fruits, hence, more prone to infections when coming in contact with wet soils. In order to improve the fruit yield and quality, it is generally advocated to trail the vines on bowers or pendals. This crop is usually grown by training the vines over low trellises or bowers of 60 cm height, made up with bamboos and ropes or wires. The major advantages of training vines of snake gourd are lower incidence of diseases, particularly vine and fruit rot, fruits become attractive in appearance, shape and size, longer availability of fruits, higher yield per unit area due to accommodation of more plants, higher return from the quality produce, etc. The most important aspect being the nature of fruits of snake gourd, *i.e.*, due to production of long, slender and soft fruits, training should be done so that the fruits may grow downward, increase in size and remain in a straight shape. Fruits with coiling or twisting types are not desirable due to non-preference of the consumer's point of view with the reduction of fruit yield. Overall, the snake gourd performs better when the plants are trained over bower system than other systems.

USE OF PLANT GROWTH SUBSTANCES

Being a monoecious cucurbit, the proportion of female flowers can be increased by spraying plant growth regulator, preferably after pruning the vines, either Ethrel @ 150 ppm or potassium naphthenate @ 0.1–0.2% at the two- and four-true leaf stage.

HARVESTING

Depending upon the variety, season and cultural practices adopted, snake gourd starts fruiting 60 to 80 days after sowing seeds. The fruits should be harvested when they are still tender and about half to two-thirds of their full size, depending on the variety under cultivation. The fruits will be at a tender stage when harvested 15–20 days after anthesis. In snake gourd, flowering and harvesting usually remain continued for 1–2 months, depending upon the growing season and cultivars. For obtaining higher fruit yield with high quality, the harvesting should be done at 5- to 6-day intervals, and generally, six to eight pickings are possible.

YIELD

In general, the local cultivar yields about 6–10 fruits per vine while that of the improved cultivar is up to 50 fruits per vine in a cropping season. The average fruit weight varies from 300 to 1000 grams. Depending upon the variety, season and cultural practices adopted, the average fruit yield varies from 100 to 250 quintals per hectare.

POST-HARVEST MANAGEMENT

Due to the presence of high moisture, snake gourd fruits have poor shelf life and can be stored no more than 1–2 days at ambient room temperature without much reduction in quality. Therefore, it is always essential to send the harvested fruits to markets as quick as possible. The fruits are usually packed either in baskets or in other containers for marketing. Care should be taken to protect the harvested fruits from excessive moisture loss and injuries during transportation. Snake gourd fruits can be stored effectively for about 2 to 3 weeks at a temperature of 15–18°C with relative humidity of about 80–85%.

INSECT-PESTS

Fruit Fly (*Bactrocera cucurbitae*)

Among the important cucurbits cultivated, snake gourd is treated as the most preferred crop by the fruit fly; when it infests at early stages of crop growth, the fruit yield is drastically reduced. The fly completes incubation, larval and pupal stages within 3 weeks, hence, it is very difficult to control effectively. Adult females usually lay 20 to 30 eggs, which are cigarette-shaped with white colour on tender fruits. Soon after hatching of the eggs, the white maggots feed inside the fruits, causing damage to the fruits. The typical symptoms of fruit fly infestation include deformity, rotting, and in severe cases, premature dropping of fruits. The infested fruits are identified by the presence of resinous liquid, which oozes out of the punctures made by the flies for oviposition on the developing fruits. The infested fruits start rotting due to the flies' feeding action, which invoke a secondary infection of different microorganisms. The incidence is greater in a wet climate. Their activity becomes severe during rainy season, as more than 80% fruits are damaged.

Control

- Grow resistant or tolerant variety, if available.
- Use maize as a trap crop.
- Collect and destroy the infested plant debris.
- Plough the field deep during land preparation to bury the infected plant material.
- Follow deep ploughing in summer to expose the pupae to reduce pest population.
- Apply fishmeal traps poison baits (5 g fishmeal + 1 ml of dichlorvos) @ 50 numbers/ha.
- Apply poison baits containing 20 g malathion 50% WP or 50 ml diazinon 20% EC with 500 g molasses + 20 g yeast hydrolysate.
- Spray the crop with cypermethrin 0.025% or fenthion 0.05% along with 5% jiggery.

RED PUMPKIN BEETLE (*Aulacophora foveicollis* lucas)

Shiny yellowish-red, orange, or black-coloured adults of this polyphagous pest lay eggs in moist soils around the plant. The beetles attack most of the cucurbits at seedling stage, especially at cotyledonary leaf stage. However, snake gourd is less affected by the beetle. The yellow-brown flying adults infest plants immediately after germination and defoliate their leaves. On hatching, the grubs feed on the roots and the underground portion of the host plant and also on fruits touching the soil. Adult beetles feed voraciously on leaf lamina, making irregular holes. Young seedlings and tender leaves are mainly preferred by the adults, and damage at this stage may cause death of the seedlings.

Control

- Follow clean cultivation.
- Collect and destroy the adult beetles.
- Follow deep ploughing after harvesting to kill the grubs in the soil.
- Use entomopathogenic fungi *Beauvaria bassiana* as biocontrol agent (2×10^6 cfu/g).
- Spray the crop with 0.4% *neem* seed kernels extract with 0.01% *neem* oil at seedling stage.
- Spray the crop with chloropyrifos or quinalphos 0.05%.
- Dust ash mixed with kerosene oil to repel the pest, but heavy dusting should be avoided as it hampers plant growth.

LEAF MINER (*Liriomyza trifolii*)

This is an introduced polyphagous pest and destructive to almost all cucurbitaceous crops including snake gourd. The tiny adult flies lay their eggs within the leaf tissues. After hatching, the maggots feed inside the leaf lamina, making irregular mines. The incidence starts after 15 days of seed germination. The mines at the beginning remain very thin, which gradually widen with the enlargement of maggots, and after complete development within the leaf tissues, they come out and pupate in soil. The loss in severely infested crop is vital. The crop seems to be burnt. Usually, the infestation of leaf miner increases during March–April.

Control

- Collect and destroy the infected plant parts after 7 days of germination.
- Spray the crop with 0.4% *neem* seed kernels extract at weekly intervals.
- Apply *neem* cake @ 250 kg/ha immediately after germination.
- Spray the crop with imidacloprid 0.3 ml or thiomethoxam 0.4 ml per litre of water.

Root Knot Nematodes (*Meloidogyne incognita, M. hapla* and *M. javanica*)

In general, all the cucurbits including snake gourd are highly susceptible to nematode infestation *viz.*, root-knot nematode. These nematodes are normally long, slender and microscopic organisms, which prefer to attack the root system of the crops. These plant parasitic worms incite root galls, causing poor and stunted growth and drastic reduction of yield.

Control

- Grow resistant variety, if available.
- Plough the field deep in hot summer months.
- Follow long crop rotation with non-host crops.
- Apply *neem* cake @ 12–15 t/ha.
- Apply granular insecticides like aldicarb 10G @ 25 kg/ha, Thimet 10G @ 20 kg/ha, or Furadon 3G @ 40 kg/ha.

DISEASES

Anthracnose [*Colletotrichum lagenarium* (Pass.) ellis & Halsted]

This is the most harming disease of cucurbits like muskmelon, pumpkin, squash and gourds. The fungus is highly prevalent in high humidity, especially when summer rain occurs regularly, affecting mostly the above-ground plant parts. Foliar symptoms vary considerably depending upon the severity of infection. The disease is characterized by the formation of yellow, more or less angular spots on the upper surface of the leaves. White-purplish spores appear on upper surface of the leaves. The disease spreads rapidly, killing the plant quickly through rapid defoliation. The fungus attacks the young fruits and tips of veins and stops plant growth.

Control

- Grow resistant variety, if available.
- Use disease-free healthy planting material.
- Collect and destroy infected leaves.
- Apply well-decomposed farmyard manure with *Trichhoderma harzianum* 1 t/ha.
- Treat the seed with carbendazim or captan 2.5 g/kg of seed.
- Apply 250 kg/ha *neem* cake immediately after germination and at the time of flowering.
- Spray the crop with chlorothalonil 0.2%, mancozeb 0.25%, or Ridomil 72 MZ 0.25% at 10-day intervals.

Downy Mildew [*Pseudoperonospora cubensis* (Berk. & Curt.) rosto.]

This is one of the most destructive foliar fungal diseases attacking the snake gourd and other cucurbits. The typical symptoms of downy mildew appear as numerous small irregular yellow lesions surrounded by green tissues scattered all over the affected leaves. Subsequently, if not controlled, gradually the lesions grow in size and coalesce with each other. As a result, the older lesions become necrotic, which are clearly demarcated with slightly yellowish areas. The severely infected leaves of downy mildew roll upward with brownish tinge that produces a blighted appearance. Greyish black downy fungal growth is usually noticed on the undersurface of the affected leaves.

Control

- Use disease-free healthy planting material.
- Strictly follow proper field sanitation.
- Plant the crop at wider spacing.
- Remove and burn infected plant debris.
- Spray the crop with mancozeb 0.25%, or Ridomil 0.2% at 10- to 15-day intervals.

POWDERY MILDEW [*Sphaerotheca Fuliginea* (Schlecht.) pollacci.]

The disease causes maximum damage in warm and dry areas where moisture is present as dew. In India, this disease is found in almost all states and causes losses in terms of quality of produce as well as fruit yield of different cucurbits, including snake gourd. The infection of powdery mildew disease first appears on upper side of the leaves, as well as the stem, as white to dull white powdery growth, which subsequently further increases, covering most of the leaf surface, resulting in serious reduction in photosynthetic rate. In due course of time, the powdery mildew disease causes premature withering, drying and finally defoliation of leaves. In severe cases, the infected fruits may be covered with a white powdery mass, and they may drop off subsequently. In addition, the infected fruits also deteriorate in quality.

Control

- Grow resistant varieties, if available.
- Use disease-free healthy seed.
- Maintain phyto-sanitary conditions in field.
- Follow long crop rotation with non-cucurbitaceous crops.
- Destruct the infected plant or plant parts and weeds that act as alternate host for the pathogen.
- Spray the crop with Bavistin 0.1%, Benlate 0.2%, Karathane 0.1%, wettable sulfur 2%, or Topsin M 1 g/litre of water at 15-day intervals.

FUSARIUM ROOT ROT (*Fusarium solani* f.sp. *cucurbitae* Snyder & Hansen)

The disease is more destructive at places where soil temperature is high enough for the pathogen to grow but unsuitable for the plants. In severe cases, the fruit yield loss may be as high as 80%, if not controlled effectively. The seedlings may die at any growth stage after germination onwards due to shrinkage of stem below the collar region. The symptoms of fungal infection are usually observed in root and on basal leaves. However, typical symptoms of wilt are usually observed at both flowering and fruiting stage. Due to infection, the plants wilt gradually and the fruits become shrivelled before attaining full size. Sometimes, the plants may not show wilting but remain severely stunted and do not bear any fruits. The pathogen is both soil and seed borne in nature, hence, difficult to control.

Control

- Grow resistant varieties, if available.
- Collect and destroy infected plants.
- Use disease-free healthy planting material.
- Avoid flood irrigation and maintain proper irrigation schedule.
- Follow crop rotation with non-cucurbitaceous crops like onion, garlic, radish, beetroot, etc.

- Apply mustard cake, mushroom compost, or poultry manure to reduce fungus.
- Treat the seed with fungicides such as captan or thiram @ 2–3 g/kg of seed.
- Drench the soil with Bavistin 2 g/l at flowering and fruiting stage at 10-day intervals.

Mosaic

The virus, namely, cucumber mosaic virus (CMV), infects the crops belonging to the Cucurbitaceae family, including snake gourd. The infected leaves initially show vein-clearing symptoms. Later, the subsequent leaves show a conspicuous vein banding consisting of a narrow border of dark green parallel with each of the main veins of the leaf, the remnants of the leaf remaining usually lighter green. The affected leaves become smaller than the normal size. The fruits from infected plant exhibit various deformations and dark green lines. Fruits are blistered, small and misshaped. The virus is transmitted by sap, seed and by several species of aphids. The aphids can acquire and transmit the virus within 5–10 seconds. Even a single aphid can transmit the virus.

Control
- Grow resistant varieties, if available.
- Follow crop rotation with non-cucurbitaceous crops.
- Use healthy disease-free seed for sowing.
- Use barrier crops like sorghum or maize around the field.
- Plant the crop bit early to avoid aphid population.
- Destroy the infected plant debris before planting.
- Spray the crop with dimethoate 0.05%, imidacloprid 0.035%, or phosphamidon 0.05% at weekly intervals.

61 Winter Squash

M.K. Rana and Naval Kishor Kamboj

Botanical Name : *Cucurbita maxima* Duch.

Family : Cucurbitaceae

Chromosome Number : 2n = 40

ORIGIN AND DISTRIBUTION

Winter squash (*Cucurbita maxima*) is a native of South America. The historically record tends to confirm the evidence from archaeology. *Cucurbita maxima* was domesticated in South America and its distribution was limited to that continent until the arrival of Europeans in the New World. The wild form, parent of the cultivated winter squash, *Cucurbita andereana,* is a small- to medium-sized gourd that grows wild in distributed areas such as roadways, fence rows, watercourses, etc., in northern Argentina, Bolivia and possibly Paraguay. *Cucurbita andereana* hybridizes readily with *Cucurbita maxima* in reciprocal crosses with no loss of fertility in F_1, F_2 and succeeding generations. It has been domesticated in North America for at least 10,000 years.

INTRODUCTION

There is a lot of confusion in correct identification of the pumpkin and squashes. The common *sitaphal* grown throughout India is *Cucurbita moschata.* The summer squash or *Chappan kaddu* (*Cucurbita pepo*) is cooked when green and is grown in summer only. Winter squash (*Cucurbita maxima*) is grown mostly on the hilly tracts of Karnataka, Maharashtra and Tamil Nadu. Winter squash is also known in English as true squash, *Hubbards* (all types), Giant pumpkins, Banana, Boston Marrow, Buttercup, Delicious (all types), Turban, Marblehead and in Hindi *Vilayati kaddu* or *Halwa kaddu.* Winter squash is a monoecious annual climber. The shape of fruit is round, medium-sized and the colour is yellow to bright orange with light specks. The leaves are rounded and lobed. The seeds are large, bright solid, white or brown with a slanting seed scar.

COMPOSITION AND USES

COMPOSITION

Winter squash is a great source of minerals, particularly of potassium and manganese. It provides low calorie but good amount of dietary fibre. Fruits of winter squash an excellent source of vitamin A and vitamin C and good source of folic acid, omega 3 fatty acids, carbohydrates, vitamin B_1 (thiamin), tryptophan, vitamin B_3 (niacin), vitamin B_5 (pantothenic acid) and vitamin B_6 (pyridoxine). The nutritional composition of winter squash is given in Table 61.1.

TABLE 61.1

Nutritional Composition of Winter Squash (per 100 g Edible Portion)

Constituents	Contents	Constituents	Contents
Carbohydrates (g)	8.59	Vitamin A equivalent (µg)	68
Protein (g)	0.95	Lutein zeaxanthin (µg)	38
Fat (g)	0.13	Thiamine (mg)	0.03
Calcium (mg)	28	Riboflavin (mg)	0.062
Iron (mg)	0.58	Niacin (mg)	0.5
Magnesium (mg)	14	Pantothenic acid (mg)	0.188
Manganese (mg)	0.163	Folic acid (µg)	24
Potassium (mg)	350	Pyridoxine (mg)	0.156
Phosphorus (mg)	23	Vitamin C (mg)	12.3
β-carotene (µg)	820	Energy (kcal)	34

Uses

Winter squash is harvested and eaten at its mature stage, when the seeds become mature fully and the skin turns into a tough rind. At this stage, most varieties of this fruit can be stored for use during winter. It is generally cooked before eating, and Buttercup squash is one of the most common varieties of this winter squash: it can be roasted, baked and mashed into soups, and for a variety of filler uses, much like pumpkin. It is extremely popular, especially as a soup in Brazil and Africa.

Medicinal Uses

Winter Squash has some excellent antioxidant phytonutrients to serve up. Among them, α-carotene, β-carotene, lutein, zeaxanthin and β-cryptoxanthin are the main components. The α- and β-carotenoids found in squash are known for their support to the immune system, anti-inflammatory effects, blood sugar regulation and prevention of type II diabetes and for their cancer-preventative properties. Many of the phytochemicals and nutrients in winter squash are well studied in the laboratory. Winter squash also contains dietary fibre, which can act in several ways to lower the risk of cancer, including helping in controlling weight. Excess body fat increases the risk of several different cancers but dietary fibre can increase the feeling of fullness.

CLIMATIC REQUIREMENT

Winter squash is a warm season crop and sensitive to cold and frost. It cannot tolerate frost, thus, it should be sown when danger of frost is over. The minimum temperature for seed germination is 16°C, with a maximum of 38°C. The plants can be grown successfully at a temperature range of 20°C to 32°C. The best crop growth occurs at 18°C to 25°C, with a minimum of 10°C and a maximum of 32°C.

The environmental factors, *i.e.*, temperature, light intensity and photoperiod, play a key role in the determination of flower gender/sex expression in cucurbits. Cool conditions generally favour the production of female flowers. The overall mean temperature is most important in sex expression. However, night temperature plays a significant role since warm nights lead to increased male flower production at a given temperature compared to warm days. The temperature effect may occur during the differentiation of flower primordia or the development of flower to anthesis. Low temperature may inhibit development of male flowers after differentiation, resulting in precocious female flowering. In brief, those conditions that encourage the buildup of carbohydrates and that reduce the amount of vegetative growth tend to favour female flower expression. However, conditions such as high temperature, low light intensities, high fertility (especially nitrogen levels) and close spacing, which promote stem extension and reduce carbohydrate buildup, which also increases maleness.

High light conditions, generally, favour the production of female flowers, while shading or low light intensity of course delays the onset of female flowers. Prevalence of low light intensity in association with high temperature predominantly favours the production of male flowers. The effect of the photoperiod on sex expression appears to be less striking than the temperature and light intensity. A short photoperiod tends to favour the production of female flowers. Cooler temperature and shorter photoperiod increase the proportion of female flowers than when the same cultivar is planted in contrasting conditions, and where the conditions conflict, temperature tends to be more important. Monoecious plants produce female flowers earlier under cool and long-day conditions more than in hot-short day conditions.

SOIL REQUIREMENT

Winter squash crop can be raised on a wide range of soils, however, a deep, well-drained sandy loam or loam soil rich in organic matter is considered the best. Winter squash is sensitive to acidic and alkaline soils with high salt concentration. The pH of the soil should be 6.0 to 6.5 for its successful cultivation. The soil is prepared well to a fine tilth by repeated ploughing followed by planking.

CULTIVATED VARIETIES

ARKA SURYAMUKHI

An improved variety released by Indian Institute of Horticultural Research, Bangalore, bears small round fruits (1–1.5 kg), with flat ends and deep orange and creamy white streaks on the rind. Flesh is firm and orange yellow in colour. It has good keeping quality and transportability. It carries resistance against fruit flies.

ALICE

A winter squash cultivar with bush-type growth habit (*Bu* gene) is recommended for cultivation in Brazil. It has been selected from an F_6 population of Coroa x Zapallo de Tronco. It flowers 45–50 days after germination and fruits can be harvested 45–50 days later. Its fruit length to diameter ratio is 0.75 to 0.80, its weight about 2–2.5 kg/fruit and it is slightly ribbed. Peel and pulp colour is intense orange with little fibre content. In three regional trials, this new cultivar yielded commercial fruits 250–300 q/ha.

KHERSONSKAYA

This drought-resistant winter squash variety released in 1982 was obtained from a cross between Ispanskaya 73 and Volzhskaya Seraya 92. The plant is of medium vigour, producing dark grey fruits with light grey stripes. Fruits are round and flattened, weighing 7.5 kg on average, with yellow, sweet, juicy flesh. Seeds are oval, glossy and cream-coloured, with a test weight of 390–400 g. Khersonskaya is resistant to *Colletotrichum lagenarium* and has moderate resistance to powdery mildew.

In the Korea Republic, the summer season production of *Cucurbita maxima* cultivars (Evis, Hamaguri, Miyako, Yako and Bamhobak) was investigated, and it was found that the marketable yield was highest in Evis (28.9 t/ha) and Evis fruits contained the highest content of β-carotene and carbohydrates.

In winter squash, Boston Marrow, Butter Cup, Green Hubbard and Winter Banana are some of the well-known cultivars in the United States of America.

The other varieties are Golden Hubbard, Blue Hubbard, Golden Delicious, Winebago, Boston Marrow, Buttercup, Turban, Pink Banana, Warren and Mammoth Chill.

SOWING TIME

In northern plains of India, usually, it is sown in the months of January to April. It can also be sown from October to November to take an early crop. In this case, plants are overwintered under some artificial protection. In hills, the optimum sowing time is April–May.

RAISING OF NURSERY UNDER PROTECTED CONDITIONS

Low temperature is limiting factor for early production of crop during spring–summer. Winter squash is very susceptible to low temperature (<10°C). However, its seedlings can be raised during last week of December or first week of January under protected conditions. Seedlings are raised in polyethylene bags of 15 × 10 cm size and thickness of 200 gauges. The bags are provided with a hole in the bottom and sides to drain out excess water and to provide aeration to the roots for their better growth. Bags are filled with a mixture of one part well-rotten farmyard manure and one part garden soil. In each bag, two or three healthy seeds are sown during last week of December or first week of January. As soon as the seedlings attain a three or four true leaf stage and the temperature outside becomes favourable, they are transplanted in the main field.

SOWING/PLANTING METHOD

Winter squash seeds are sown at a distance of 3.0–3.5 m row to row and 1.0 m plant to plant. This way around 3000 plants can be accommodated in a hectare area with a seed rate of 4–6 kg. In spring–summer season, the pits of 45 × 45 × 45 cm are dug 2–3 weeks before planting at a spacing of 3.0–3.5 × 1 m and filled with a mixture of farmyard manure (3–4 kg/pit) and topsoil; however, in rainy season, either the raised beds or the ridges are made to facilitate drainage. Usually, about 40 cm wide furrows are made at a distance of about 2.0–3.0 m. The seeds are sown at a plant to plant distance 75–100 cm. At least two or three sprouted seeds are sown per hill to avoid gaps created due to failure of germination.

NUTRITIONAL REQUIREMENT

Nutritional requirement generally depends on soil type, fertility status of the soil, previous crop grown and crop spacing. A well-decomposed farmyard manure @ 20–25 t/ha should be incorporated during field preparation to enrich the soil with organic matter. In general, winter squash grown in average fertile soil needs nitrogen 100–120 kg, phosphorus 80–100 kg and potash 40–60 kg/ha. The entire dose of phosphorus and potash and half dose of nitrogenous fertilizers should be applied as basal at the time of last ploughing. The rest of the nitrogen is top-dressed about 4 weeks after planting.

IRRIGATION REQUIREMENT

Soil moisture is an important factor for good germination of cucurbits. In spring–summer season crop, the frequency of irrigation is very important. Winter squash requires light but frequent irrigation from germination to the development of fruits. Intervals of irrigation may be 5 to 7 days or so depending on climatic conditions and soil types. Vine growth, flowering and fruit development are critical stages for irrigation. In rainy season, proper drainage is equally important to drain out the excess rainwater.

INTERCULTURAL OPERATIONS

HOEING AND WEEDING

Weeds may pose a problem in initial stages of crop growth so weeding should be done frequently in order to keep the weeds under control. At least one shallow hoeing is beneficial and necessary for keeping the weeds down and providing a good soil environment to the roots for their development. To control the weeds chemically, Fluchloralin, Trifluralin (0.75–1.5 kg/ha), or Bensulide (5–8 kg/ha) may be applied as pre-plant soil incorporation 2 weeks before sowing. Butachlor (1.0 kg/ha) or Chloramban (2–3 kg/ha) as pre-emergence and Naptalam (2–4 kg/ha) as post-emergence after weeding gives an excellent control of weeds in winter squash crop.

THINNING

In winter squash, thinning is an important operation to maintain optimum spacing among plants and to provide better conditions to the plants for their growth and development. It is also essential to reduce competition among plants for space, light, nutrients and moisture. Thinning must be done to take out all the weak and feeble seedlings to maintain proper spacing. Only a single healthy plant per hill should be allowed to grow.

MULCHING

Mulching can reduce water evaporation from the soil surface, suppress weed growth, restrict the leaching of fertilizers and moderate the diurnal and seasonal temperature fluctuations. The use of plastic mulch or row covers can overcome the chilly and frosty weather conditions and result in an early crop production, and use of reflective films as mulch also increases yield.

HARVESTING

Winter squash requires from 3 to 4 months to be mature. The fruits should be well matured before they are harvested. The fruits are often allowed to remain in the field until frost kills the vines or the vines die due to natural causes.

The winter squash should be carefully handled during harvest and storage. Cuts or bruises may provide a point of entry for decay organisms, which can cause considerable loss from rot in a short time. Wound areas are capable of healing over by producing suberized tissue, which offers protection against the entrance of rot organisms. Suberized tissue develops best at a relatively high temperature and in a moist atmosphere. Winter squash fruits should be harvested 3–4 weeks after anthesis to ensure best quality and longest shelf life. When fruits are harvested at the proper stage, their quality is improved in the first 2–3 months under ventilated and shaded storage conditions.

POST-HARVEST MANAGEMENT

Winter squash is cured for a period of 10 days with artificial heat at 27–30°C and a relative humidity of 80%. After 10 days, the fruits should be placed in storage with low humidity and at temperature between 10°C and 15°C. It is essential that the surface of the fruits should be kept dry during the storage period. Temperatures above 15°C tend to maintain too high a rate of respiration. Thus, high temperatures cause a considerable loss through shrinkage. Winter squash should be stored in single layers on shelves for best results and not heaped on one another. They may be kept for long period if handled properly.

YIELD

The fruit yield of winter squash depends upon variety, soil fertility, growing season and cultural practices adopted during cultivation of crop; however, the average yield ranges from 200 to 250 q/ha.

INSECT-PESTS AND DISEASES

The winter squash crop is attacked by a number of insect-pests and diseases during the growing season. Some of the insect-pests and diseases along with their control measures are given as follows:

INSECT-PESTS

FRUIT FLY (*Bactrocera cucurbitae, B. tau* and *B. diversus*)

This pest causes damage by puncturing tender fruits. The adult fly is reddish brown with lemon yellow marking on its thorax. An adult female lays 20–30 cigarette-shaped white eggs on both tender and ripe fruits. After hatching, the white maggots start feeding inside the fruits, which causes deformity, rotting and premature dropping of fruits. More than 50% of fruits can be damaged partly or fully by fruit flies.

Control

- Collect and destroy the infested plant parts and fruits.
- Behavioural control of flies through methyl eugenol (0.5%) baited trap along with insecticides reduce adult population to some extent.
- Foliar spray of carbaryl (0.1%) at the tender fruit stage (up to 5 days after fertilization) gives good control of fruit flies.
- Bait spray with malathion (50 EC) 20 ml + yeast hydroxylase (10 g) + 20 litre water + 500 g molasses gives effective control of fruit flies.

RED PUMPKIN BEETLE (*Aulacophora foveicollis* lucas)

The beetle feeds on cotyledons and foliage. Grubs develop into the soil, and occasionally feed on roots, causing wilting of plants. The adult beetles emerge from the soil and makes typical shot holes on foliage. The grubs also damage the underground portion of plants by boring into the roots, stems and sometimes into the fruits touching the soil.

Control

- Plough the field repeatedly in hot summer months.
- Spray the crop twice or thrice with Rogor 0.1% near the base of plants just after germination.
- Spray 0.01% *neem* seed kernels extract and 0.4% *neem* oil to control the beetles effectively.

LEAF MINER (*Liriomyza trifolii* l.; *sativae* blanch)

The leaf miner usually attacks during the months of March–April. It causes damage by making mines in the leaves, especially in mature leaves. The eggs hatch within 2–3 days, and the larvae feed on leaves, making zigzag tunnels between the lower and upper epidermis and scraping the chlorophyll and leaf tissues.

Control

- Pluck the old leaves severely infested by leaf miners and destroy them.
- Spray the crop with 4% *neem* seed kernel extract at weekly intervals.
- A foliar spray of dimethoate (30 EC) @ 1 ml/litre is advantageous for the control of leaf miners.
- A foliar spray of chlorpyrifos @ 1.5 kg *a.i.*/ha weekly or 15-day intervals resulted in the reduction of pupae on squash plants.

ROOT KNOT NEMATODES (*Meloidogyne incognita, M. hapla* and *M. javanica*)

Squash is also infested with nematodes that are long, slender and microscopic and attack the plants' root system. Infective larvae enter the roots and form galls, which adversely affect the root system, and consequently, reduce the plant growth and yield.

Control

- Soil application of carbofuran 3G (2–4 kg/ha), Phorate (10 kg a.i/ha), or Aldicarb (2–4 kg a.i./ha) during field preparation is effective to control the nematode population.
- Soil fumigation with D-D mixture (200–400 litre/ha) is very effective to control nematodes.
- Soil application of *neem* cake @ 80–100 kg/ha is also advantageous.

DISEASES

ANTHRACNOSE (*Colletorichum lagenarium*)

The fungus attacks the leaves, stem and fruits. Reddish brown spots are formed on the affected leaves and they become angular or round when many spots coalesce. It results in shrivelling and death of leaves.

Control

- Follow 2- to 3-year crop rotation.
- Treat the seeds with Agrosan GN or Bavistin @ 2 g/kg of seed.
- Spray the crop with Dithane M-45 0.2%, Difolatan 0.2%, Blitox 0.2%, or Bavistin 0.1% and repeat at 7- to 10-day intervals if necessary.

DOWNY MILDEW [*Pseudoperonospora cubensis* (Berk. & Cart.) *rosto*]

It is a serious and most destructive disease of winter squash. Symptoms appear as numerous irregular small yellow areas surrounded by green tissues scattered all over the leaves. In time, lesions grow in size and coalesce with one another. Old lesions become necrotic, and these are clearly demarcated with slightly yellow areas. Severely infected leaves roll upward and have a brownish tinge that produces a blighted appearance. Greyish black downy fungal growth is observed on undersurface of leaf in humid weather.

Control

- Air circulation and sunlight exposure help in checking the disease initiation and development.
- Field sanitation by burning crop debris helps in reducing the inoculums.
- Crop should be grown at wider spacing in well-drained soil.
- A protective spray of mancozeb 0.25% at 7-day intervals gives good control.
- In severe cases, a foliar spray of Metalaxyl + mancozeb @ 02% may be given only once.

POWDERY MILDEW [*Sphaerotheca fuliginea (Schlecht.) pollacci.*]

The infection appears first on upper side of the leaves and stem as a white to dull white powdery growth, which quickly covers most of the leaf surface and leads to a seriously disadvantaged photosynthetic rate. In time, all the above-ground parts are infected. Finally, the lesions turn brown and necrotic. The affected leaves become yellowish, dry and get defoliated. The fruits do not develop property, and sometimes are covered with white powdery masses.

Control

* The fungicidal spray of penconazole (0.05%) can very well control the disease.
* Spray the crop with Karathane 1.5 ml/litre, carbendazim 0.1%, Calixin 0.1%, or Sulfex 2.5 g/litre.

FUSARIUM ROOT ROT (*Fusarium solani f. sp. cucurbitae* Snyder & Hansen)

The symptoms include vascular browning, gummosis and tyloses in xylem vessels of mature plants, and subsequently, the whole plant is wilted. The pathogen is both seed and soil borne.

Control

* Use only disease-free healthy seed.
* Follow long crop rotation and clean cultivation.
* Avoid root injury during intercultural operations.
* Grow resistant varieties in disease-infested areas.
* Soil application of *Trichoderma* (5 kg/ha) during soil preparation is advantageous.

ANGULAR LEAF SPOT (*Pseudomonas lachrymans carsner*)

The disease appears on leaves, stems and fruits as small water-soaked spots. On leaves, they enlarge up to a diameter of about 3 mm, becoming tan on the upper surface and gummy or shiny on the lower surface. Lesions attain an angular shape as they are delimited by veins. The necrotic centres of leaf spots may drop out. On stems, petioles and fruits, the water soaked areas are covered with white crusty bacterial exudates. As the fruits begin to mature, the brown lesions in fleshy tissues beneath the rind develop and the discolouration continues along the vascular system, which extends to the seeds.

Control

* Follow long crop rotation as the preventive measures to reduce the inoculums.
* Grow resistant/tolerant cultivars to reduce the risk.
* Follow field sanitation to avoid the occurrence of disease.
* Destroy the disease debris after harvesting the crop.
* Treat the seed with hot water at 52°C for 30 minutes or mercuric chloride (1: 1000) for 5–10 minutes.
* Spray the crop with streptomycin at 400 ppm or Bordeaux mixture at 1%.

■■■

62 Chayote (Chow Chow)

Ruchi Sood and M.K. Rana

Botanical Name   : *Sechium edule* (Jacq.) *Swartz.*

Family   : Cucurbitaceae

Chromosome Number : 2n = 24

ORIGIN AND DISTRIBUTION

Chayote [*Sechium edule* (Jacq.) Swartz.] is indigenous to Southern Mexico and Central America. All *Sechium* species are distributed in Central America, where the cultivated species was domesticated. *Sechium edule* is a species which was undoubtedly domesticated within the cultural areas of Mesoamerica and specifically in the regions lying between Southern Mexico and Guatemala. It was introduced into the Antilles and South America between the 18th and 19th centuries. The first botanical description mentioning the name *Sechium* was in fact done in 1756 by P. Brown, who referred to plants grown in Jamaica. During this period, the chayote was introduced into Europe when it was taken to Africa, Asia and Australia, while its introduction into the United States dates from the late 19th century.

The close relatives to *Sechium edule* are as follows:

- The so-called wild forms of *Sechium edule*, the taxonomic positions of which are unresolved since they are distributed in an apparently natural way in the Mexican states of Veracruz, Puebla, Hidalgo, Oaxaca and Chiapas.
- *Sechium compositum* is a species restricted to Southern Mexico (Chiapas) and Guatemala.
- *Sechium hintonii* is a species endemic to Mexico, until recently considered to be extinct, and which grows in the states of Mexico and Guerrero and possibly in Jalisco.
- A new species of *Sechium* selection grows in the north of Oaxaca State.

INTRODUCTION

Chayote is little-known underutilized fruit consumed locally as a vegetable. However, it is gaining popularity as a food crop throughout the world due to its inherent qualities. It is very important food crop in tropical America, the West Indies and in other tropical regions.

Chayote is an important perennial and climbing vegetable crop grown mostly at high altitudes of the Eastern and Western Himalayas. In India, it is grown in northern and southern areas like Darjeeling, lower areas of Himachal Pradesh and in Karnataka. It is a popular vegetable in Northeastern hilly region, is commonly called squash and grows abundantly without much care and attention in the high hills of Meghalaya, Manipur, Mizoram, Nagaland and Sikkim. If the plant is cared for properly, it produces fruits enough for a family of four to five persons.

COMPOSITION AND USES

COMPOSITION

The fruits of chow-chow are a rich source of carbohydrates and vitamins, especially vitamin A and vitamin C. However, the edible parts of *Sechium edule* have a lower fibre and protein content. Its calorific value and carbohydrate content is high, chiefly in the case of the young stems, root and seed, while the micro- and macronutrients supplied by the fruit are adequate. The fruit and particularly the seeds are rich in amino acids such as aspartic acid, glutamic acid, alanine, arginine, cysteine, phenylalanine, glycine, histidine, isoleucine, leucine, methionine (only in the fruit), proline, serine, tyrosine, threonine and valine. The composition of chayote is similar to that of zucchini, containing 90–92% water, 1% protein, 3–4% carbohydrates and traces of lipids. The nutritional composition of chayote fruit is given in Table 62.1.

USES

The fruit, stems and young leaves as well as the tuberized portions of the roots are eaten as a vegetable, both alone and plain boiled and as an ingredient of numerous stews. Unripe fruits are used as a vegetable. They can be sliced and eaten raw as salad or prepared in various ways by stewing, stuffing, frying, boiling, or cooked in many other ways like asparagus. Its dried fruits can be used as a vegetable after a certain time. The cooked fruits can be creamed, baked or made into fritters, sauces and tarts, puddings, salads or used like potatoes with other meats and vegetables. The fruits can also be used for the preparation of pickles, candy and flour. It is also consumed in the form of jams, juices, sauces, pastes and other sweets. The tubers of 2- or 3-year-old plants are dug up and used in the same way as the sweet potatoe. Because of its softness, the fruit has been used for children's food. The tender leaves and tendrils are also reported to be consumed. Fruits, vines and tubers are also excellent fodder for livestock. In India, the fruits and roots are not only used as human food but also as fodder. If the harvested chayote is abundant, it is cheaper to use it as food for pigs than the usual commercial feed. The woody stems furnish a fine fibre. Because of their flexibility and strength, the stems have been used in the craft manufacture of baskets and hats.

MEDICINAL USE

Infusions of its leaves are used to dissolve kidney stones and to assist in the treatment of arteriosclerosis and hypertension, and the infusions of fruits are used to alleviate urine retention, while the pulp is sometimes used to soothe rashes.

BOTANY

The chayote is an herbaceous perennial monoecious climber with thickened roots and slender branching stems up to 10 m long. The stems are angular, grooved and glabrous. Several stems grow

TABLE 62.1
Nutritional Composition of Chayote Fruit (per 100 g Edible Portion)

Constituents	Contents	Constituents	Contents
Water (g)	93.00	Vitamin A (IU)	0.09
Carbohydrates (g)	5.40	Thiamine (mg)	0.03
Protein (g)	0.90	Riboflavin (mg)	0.04
Fat (g)	0.30	Nicotinic acid (mg)	0.50
Dietary fibre (g)	0.70	Vitamin C (mg)	11.00
Minerals (mg)	0.40	Energy (kJ)	101.0

simultaneously from a single root in the cultivated plants. The roots thicken towards the base and appear woody, while towards the apex, there are many thin firm herbaceous branches. Its leaves are on sulcate petioles of 8 to 15 cm length. They are ovate-cordate to sub-orbicular measuring 8–18 × 9–22 cm, slightly lobate with three to five angular lobes and have minutely denticulate margins and three to five divided tendrils. Both blade surfaces are pubescent when young and later become glabrous. Flowers are unisexual, normally pentamerous coaxillary and with ten nectarines in the form of a pore at the base of the calyx. The staminate flowers grow in axillary racemose inflorescence, which is 10 to 30 cm long, and the groups of flowers are distributed at intervals along the rachis. The calyx is patelliform and 5 mm wide, the sepals triangular and 3 to 6 mm long and the petals triangular, greenish to greenish white and measuring 4–8 × 2–3 mm. There are five stamens, and the filaments are fused almost along their total length, forming a thickened column, which separates at the apex into three or five shoot branches. The pistillate flowers are normally on the same axilla as the staminate flowers. They are usually solitary but are occasionally in pairs, and the ovary is globose, ovoid or piriform, glabrous and unilocular. The perianth is in the staminate flowers but has slightly different dimensions. The styles are fused in a slender column, and the nectarines are generally less evident than in the staminate flowers.

Pollination is entomogamous. Among the most efficient pollinators are species of native bees of the genus *Trigona*, chiefly in areas of medium and low altitudes, which are free from pesticides and *Aphis mellifera*. Secondary pollinators include wasps of the genera *Polybia*, *Synoeca* and *Parachartegus*.

The fruit is solitary or rarely occurs in pairs. It is fleshy and sometimes longitudinally sulcate or cristate. It is of varying shapes and sizes as well as number and type of spines. It has white and yellowish or pale green to whitish flesh, which is bitter in the wild plants and non-bitter in the cultivated ones. The light green pear-shaped fruit with deep ridges lengthwise may weigh as much as one kilogram but most often is from 170 to 340 g. The chayote fruit differs from its multi-seeded relatives, in that it contains only a single flat edible seed. Fruits may be slightly grooved and its skin may be prickly or smooth.

The seed is ovoid and compressed with a soft and smooth testa. The fruit is viviparous, *i.e.*, the seed germinates inside the fruit even when it is still on the plant.

CLIMATIC REQUIREMENT

The crop is successfully grown in tropical and sub-tropical areas. It thrives well at a temperature range from 18°C to 22°C. Chayote vines need full sun. It grows well in the plains where the temperature is moderate during the cool season. It can also be grown hills at altitudes of 1200–1500 m elevation in high rainfall areas. The annual rainfall from 1500 to 2500 mm is best suited for its cultivation. In hills, the plants remain dormant during winter months. It is susceptible to frost, freezing temperature and heat too. Chayote is grown as an annual in areas where freezing weather is common. At least 5 or 6 months of growing season is required to get fruits. In frost-free regions, the vines are cut back to 2 m or so before the new growth begins in spring.

SOIL REQUIREMENT

Well-drained sandy loam soil rich in organic matter is best suited for its successful cultivation. It grows well on slightly acidic soil having a pH between 5.5 and 6.5.

VARIETIES

Considerable variations in size, shape, spininess and fibre content of the fruit are found to exist in the types under cultivation. Two distinct types are generally available, namely the green fruited and the white fruited. Improved varieties are lacking and the local cultivars like Round White, Long White, Pointed Green and Creamy Green varieties are grown. In the region of Northeastern hill, four different types are cultivated, *i.e.*, (i) big green spiny fruited, (ii) medium green smooth fruited, (iii) medium yellow smooth fruited and (iv) small green smooth fruited. Commercial varieties available in the international markets are mainly Florida Green and Monticello White.

SEED RATE

Unlike other cucurbits, which are propagated by seeds, chayote is propagated by using the entire fruit. It is also viviparous, with a sprout developing from embryo of the seed while the fruit is still attached to and growing on the vines.

SOWING TIME

In the plains, the crop is planted in July–August and during April–May in the hills. The vines start flowering after 3–4 months of planting.

SOWING METHOD

The crop is propagated both sexually and vegetatively. Generally, the whole fruit is planted as a seed. It can also be propagated by tuberous roots. The crop is raised by planting fruits or tuberous roots directly in the field or first it is planted in a nursery and then transplanted in the main field. The shoots are removed from the crown and rooted in sand and the rooted cuttings are then established in pots before being moved to the field. November, with mild temperature and high humidity, is the best month for the production of cut-plants in chayote through vine cuttings under Chitwan conditions.

NURSERY MANAGEMENT

The seed is not separated from the fruit, and hence, the whole fruit is planted in a polyethylene bag containing a good planting medium of topsoil and manure in a ratio of 3:1. The fruit is buried horizontally in the soil exposing the sprout ends. Adequate moisture is provided for early sprouting. The sprouts come out within a week and bear three to four leaves. When plants grow up to the height of 30 cm they are transplanted at the main site. The fruits and tuberous roots are also planted on raised nursery beds and they are lifted at the transplantable stage.

Prior to transplanting, the pits of 45 cm are dug and filled with a mixture of topsoil and well-rotten farmyard manure or compost in a ratio of 3:1. As far as possible, 2 g of Furadan granules should also be mixed in the planting mixture as it checks any insect cutworm present during the planting. On next day, the plants are taken out from the polythene bags without disturbing the earth ball and transplanted individually in a pit by burying the whole soil ball and pressing well around it. The basin around the plant is prepared and water is applied.

The raising of nursery plants in polythene bags is necessary in the areas where there is risk of frost burning and chilling. Transplanting can be performed very conveniently during the frost-free season.

PLANTING *IN SITU*

Under mild temperatures, the whole fruit is planted directly in the field. The pits are dug as mentioned above and fruits are placed horizontally and covered with soil thinly and thereafter light irrigation is applied. Since it is a perennial vine crop, the distance between two rows of 2 to 3 m and between two vines of 1 to 2 m is maintained. However, the planting distance can be increased when the soil is fertile.

NUTRITIONAL REQUIREMENT

Chayote grows best with little nitrogen but requires potassium. Excessive use of nitrogen stimulates vine growth and produces few fruits. At the initial planting time, application of 10 kg farmyard manure, urea 250 g and 500 g each of single super-phosphate and muriate of potash mixed with soil in each pit is recommended. However, later, it requires 10 kg farmyard manure or 1.5 kg *neem* cake, 100 g urea, 100 g single super-phosphate and 50 g muriate of potash per plant every year. Sometimes, additional application of 100 g of urea per plant may be required during the rainy season when plant shows poor growth. In Maharashtra, application of 950 g of ammonium sulfate and 450 g of super phosphate is recommended per vine to be given in two doses before flowering. In Nilgiri hills, each pit receives an application of farmyard manure (two or three baskets) at planting and again a similar application 6 months later. These applications are repeated in the second and third years.

TRAINING OF VINES

The vines are mostly trained on trellis or pandal of 2 m height. The vines cover the pandal in about 5–6 months. Sometimes, the vines are allowed to trail on the ground, but in such cases, it is better to have mulching with dry grasses on the ground to protect the fruit from damage when these come in contact with moist soil. However, trailing of vines on the ground should be avoided.

PRUNING

The plants are pruned 1–1.5 m above the ground level in winter (December–January) when they are dormant. The plants will sprout after winter and produce fruits during July to December. In Maharashtra, a second pruning is done in May (summer season) to have another harvest. The plants in the mild climate of Bangalore do not require any pruning, and fruiting continues over a long period. After pruning, it is advisable to apply manure and fertilizers to each plant as mentioned earlier.

IRRIGATION REQUIREMENT

Chow-chow requires a large quantity of water and should be copiously irrigated in regions of low rainfall and during periods of drought. A constant and copious supply of water is essential for good growth, and at no stage must the plants be allowed to lack moisture. Frequent irrigation is required in the summer season. During rainy season, the crop may be irrigated according to the rainfall and particularly during dry periods when the rains are delayed. In freeze-prone areas, a mulch of straw or hay of about 10 cm thick should be used to protect the roots from the frost.

INTERCULTURAL OPERATIONS

Chayote does not suffer a lot from weed competition, as it covers the soil surface and checks the weeds' growth, but at initial stages, one or two weedings should be done with the help of a *khurpi*. Light hoeing and weeding are practiced during the early stages of vine growth and later when the plants have been pruned.

HARVESTING

The fruits are available throughout the year except between February and March. The peak season for harvest is from October to December and May to June. Generally, the vines bear fruit in abundance during October–November. The fruits become ready for harvesting within 120 days or so, and harvesting continues for 2 to 3 months. They reach the marketable size in about 30 days from flowering and are picked before they are fully mature. It was observed that out of four female flowers in an inflorescence, usually a single fruit is set. Sometimes, there is a setting of two fruits but these fruits are rare and the retention of all four fruits is rarest. Though fruit setting is high, fruit drop occurs immediately after fruit setting. Fruits are picked when they are tender. The fruits are easily harvested by cutting them from the peduncle. For planting, mature fruits are harvested. Generally, the fruits become ready for harvesting 30–35 days after anthesis. The plants continue to give good yield for 3–4 years, after which replanting will be required.

YIELD

Since it is perennial, the best production is obtained 2–3 years after the plant is established. One plant can produce more than 300 fruits per year. The average yield of fruit is about 25–35 tonnes per hectare per year. A well-managed crop may yield about 400–500 q/ha in a year.

POST-HARVEST MANAGEMENT

Once the fruits have been collected and selected, they are put in wooden boxes and sent to the packers. The fruits are packed in cardboard boxes and plastic bags with antitranspirants and fungicides, which are then sent to the market in refrigerated containers. Fruits stored below 8°C remain in good condition for planting for as much as 6–8 weeks, although shrivelling and decaying are common. The fruits can easily be shipped and stored for 2 to 4 months at moderate temperature.

PHYSIOLOGICAL DISORDERS

CHILLING INJURY

The fruits are sensitive to chilling, hence, should be kept at 10–15°C for marketing. Chilling injury is manifested in a variety of symptoms, including surface and internal discolouration, pitting, water-soaking, failure to ripen, uneven ripening, development of off-flavour and heightened susceptibility to pathogen attack.

Control

- In order to avoid chilling injury, the fruits should be shipped at 10–12°C, which confers a storage life of several weeks.

INSECT-PESTS

PICKLE WORM (*Diaphania nitidalis*)

This is a major pest. The larvae burrow into the tissues of blossoms or buds. Some burrow deep into the stem and complete their growth, thus causing injury to the vines, but the greatest injury results from their burrowing into the fruit.

Control

- Destroy the waste fruits and vines by burning or composting.
- Dust the crop with 0.75% Rotenone, 50% Cryolite, or 5% Methoxychlor as soon as the plant comes up.

SPOTTED CUCUMBER BEETLE (*Diabrotica undeciumpunctata howardi*)

The adults of this insect feed on tender foliage of the crop. The larvae feed on the roots and cause great damage.

- Apply Nemagon @ 30 litres per hectare.
- Use soil fumigant D-D. @ 200 litres per hectare.

DISEASES

ANTHRACNOSE (*Colletotrichum orbiculare*)

Leaves, flowers and fruits are infected with the attack of anthracnose. Light brown spots appear on the leaves, which later turn to deep brown. This disease is promoted by high humidity and moist weather and is severe in rainy seasons.

Control

- Grow resistant varieties.
- Use only certified seed.
- Rotate crops every year.
- Spray 0.1% carbendazim at 10-day intervals.

DOWNY MILDEW (*Pseudoperonospora cubensis*)

Yellow, more or less angular spots appear on upper leaf surface with purplish spores appearing on lower leaf surface. The disease spreads rapidly killing the plant quickly through rapid defoliation. Hence, fungicidal sprays need to be started at initial stage of the disease development.

Control

- Avoid overcrowding of plants.
- Avoid overhead irrigation.
- Water the plants from base.
- Spray 0.3% Ridomil MZ or copper oxychloride at 10-day intervals.

POWDERY MILDEW (*Erysiphe cichoracearum*)

Plants infected with mildew initially develop white powdery spots or blotches on lower and upper leaf surface. The severely infected crop fail to set fruits. Yield loss increases with the increase of disease severity of the crop.

Control

- Grow resistant varieties.
- Spray the crop with 0.1% carbendazim or Thiovit 0.2%.

FUSARIUM WILT (*Fusarium* sp.)

The disease is characterized by sudden wilting of leaves in older plants and the vascular bundles at collar region become yellow or brown. In young seedlings, cotyledons droop and wither.

Control

- Grow wilt resistant varieties.
- Drench the field with captan 0.2 to 0.3% solution.

■■■

Control

- Same as for the pickle worm.

STRIPED CUCUMBER BEETLE

It is a serious pest and attacks the plant as soon as it emerges, destroying the leaves and eating into the stems. The main injury is done by overwintering adults that attack the young tender plants. These beetles also carry bacteria of cucumber wilt, while its larvae burrow into the roots, causing the plants to wilt.

Control

- Dust as soon as the plants come up.

FRUIT FLY [*Bactrocera (Dacus) cucurbitae coquillett*]

The fruit fly attacks the young fruits of chayote, and as a result, the infested fruits drop off. The low susceptibility of chayote fruits to fruit fly is associated with the dense and long spines on the fruits. The spineless smooth fruits are most susceptible to the attack of fruit flies.

Control

- Collect the damaged fruits and destroy.
- The fly population is low in hot day conditions and it is peak in rainy season. Hence, adjust the sowing time accordingly.
- Plough the field in hot summer months to expose the pupae.
- Spray 0.01% *neem* oil, preferably using a rocker sprayer or foot sprayer.
- Use polythene bags fishmeal trap with 5 g of wet fishmeal + 1 ml dichlorvos in cotton. Per hectare, about 50 traps are required. Fishmeal + dichlorvos soaked cotton are to be renewed once in 20 and 7 days, respectively.

COTTON APHID (*Aphis gossypii glov.*)

This insect becomes serious on chayote during spring. Both the adults and nymphs suck sap from tender shoots and young leaves of the plants. The heavily infested plants become stunted.

Control

- Application of disulfoton granules @ 0.1 g a.i./plant is found effective throughout the season.
- Monocrotophos, omethoate, formothion, methomyl, pirimicarb and phosphamidon applied as foliar sprays are the most effective insecticides.

ROOT KNOT NEMATODE

It is one of the serious pests of chayote. The nematode invades the plant roots and form galls on them. These irregular swellings are sometimes confused with club root. The smaller roots are mostly affected and knots are formed nearer the tips.

Control

- Follow long crop rotation.
- Fumigate the affected soil with methyl bromide, chloropicrin, or chlorobromopropene.

DISEASES

POWDERY MILDEW (*Podosphaera xanthii*)

Symptoms include greyish white powdery growth consisting of epiphytic mycelia and conidia mostly on the upper surface of the leaves. Symptoms on the lower surface are less conspicuous, and on the corresponding upper surface, the yellow spots are present. Older leaves are found to be more susceptible. Defoliation is observed in severely infected plants.

Control
- Use tolerant lines.
- Foliar spray of Penconazole @ 0.05%, Tridemorph @ 0.1%, or carbendazim @ 0.1% gives good control of the disease.

∎∎∎

63 Zucchini

Ruchi Sood, Surbhi Gupta and M.K. Rana

Botanical Name : *Cucubita pepo L.* var. *cylindrica* Paris

Family : Cucurbitaceae

Chromosome Number : 2n = 40

ORIGIN AND DISTRIBUTION

Zucchini is the most popular type of summer squash that can be found in dark, medium and light green colours as well as yellow-orange. The world 'zucchini' is the Italian name for the plant, which is why it is more common name in Italy and in North America, Australia and Germany, while courgette is the French name for this vegetable and is more commonly used in France, Ireland, the United Kingdom, Greece, New Zealand, The Netherlands, Belgium, Portugal and South Africa. Zucchini generally has a similar shape to a ridged cucumber although some round varieties are also available.

There is a lot of confusion in common epithets like squash, pumpkin, marrow, etc. and more than one common name is ascribed to any one botanical species; however, in horticulture, squash cultivars are divided into two groups, *i.e.*, *bush* and *vining* type. The bush type, commonly called summer squash under the species *Cucubita pepo*, does not store well. The vining type includes winter squash (*Cucubita maxima*) and pumpkin (*Cucubita moschata*).

According to archaeological recordings, *Cucubita pepo* appears to be one of the first domesticated species. The oldest remains have been found in Mexico, in the Oaxaca valley (8750 BC to AD 700) and in the caves of Ocampo, Tamaulipas (7000 to 500 BC). Its presence in the United States also dates back a long time, as the recordings in Missouri (4000 BC) and Mississippi (1400 BC) indicate. This species may have been domesticated at least on two occasions and in two different regions, *i.e.*, in Mexico and in the eastern United States, in each case having *Cucubita fraterna* and *Cucubita texana,* respectively, as possible progenitors.

Eight groups of edible botanical varieties of *Cucubita pepo* are known:

- Pumpkin (*Cucubita pepo* L. var. *pepo* L. Bailey) includes cultivars of creeping plants, which produce spherical, oval, or oblate fruit, which is rounded or flat at the ends. The fruit of this group is grown for eating purposes after ripening and sometimes it is used as fodder.

- Scallop (*Cucubita pepo* L. var. *clypeata* Alefield) has a semi-shrubby habit; the fruit ranges from flat to almost discoidal with undulations or equatorial margins and it is eaten before maturity.

- Acorn (*Cucubita pepo* L. var. *turbinata* Paris) is a shrubby and creeping plant with fruit, which is obovoid or conical, pointed at the apex and longitudinally costate-grooved. The rind is soft, thus, the fruit can be eaten in the ripe state.

- Crookneck (*Cucubita pepo* L. var. *torticollia* Alefield) is a shrubby type with yellow, golden, or white fruit, which is claviform and curved at the distal or apical end. Generally, it has a verrucose rind. It is eaten unripe since the rind and the flesh become hard after ripening.

- Straightneck (*Cucubita pepo* L. var. *recticollis* Pans) is a shrubby plant with yellow or golden fruit and a verrucose rind similar to that of var. *torticollia*.

- Vegetable marrow (*Cucubita pepo* L. var. *fastigata* Paris) has creeper characteristics as a semi-shrub and has short cylindrical fruit, which is slightly broader at the apex, with a smooth rind, which hardens and thickens on ripening and which varies in colour from cream to dark green.

- Cocozzelle (*Cucubita pepo* L. var. *longa* Paris) has cylindrical long fruit, which is slender and slightly bulbous at the apex. It is eaten in the unripe state and one of its most common names is cocozzelle.

- Zucchini (*Cucubita pepo* L. var. *cylindrica* Paris) is the most common group of cultivars at present. Like the previous group, the zucchini group has a strong affinity with vegetable marrow and its origin is recent, in the 19th century. Its plants are generally semi-shrubby and its cylindrical fruit does not broaden or broadens only slightly. It is eaten as a cooked vegetable in unripe state.

With regard to traditional cultivars, it is common to see a fair representation of cultivars with characteristics similar to those of each of the commercial groups in one single field cultivated by Mesoamerican peasants. The question of the origin of cultivars native to the Mayan area in the middle and low areas of Chiapas and the Yucatán peninsula still has to be resolved. These cultivars, whose fruits are without or with unpronounced ribs and have rather rounded and oval seeds, are grown from a little above sea level to nearly 1800 m.

The distribution of *Cucubita pepo* outside America is possibly the best documented of this genus. It is known that some cultivars reached Europe approximately half a century after 1492 and it is even said that others originated on that continent. In contrast to *Cucubita pepo's* long-established presence in the Old World, it seems that its arrival in South America was very recent. At present, the fruit of some cultivars, *e.g.*, zucchini and cocozzelle, has a nutritional and commercial role in several regions of the world.

INTRODUCTION

Zucchini is a kind of summer squash, meaning the fruits have soft, edible skin. Summer squash is fast-growing, as it can be harvested in 35 to 55 days. Zucchini squash is highly rated for economic value, based on the period between planting and harvest, as well as the number of fruits produced per square metre. On an average, a plant yields 1.5 to 4.5 kg of young fruits. In recent years, zucchini has surpassed other types of summer squash in popularity as a fresh and cooked vegetable. The crop is cultivated for its tender fruits, which are available in different shapes, colours and sizes for use as a cooked vegetable. Fruits when mature become unfit for consumption. It is one of the earliest cucurbits reaching market and its cultivation is confined to areas near markets. It is grown in limited scales in Punjab, Delhi and Uttar Pradesh.

COMPOSITION AND USES

COMPOSITION

Zucchini is a rich source of nutrients, especially the natural antioxidants beta-carotene, folic acid, and vitamins C and E. It also contains healthful minerals, including potassium, iron, calcium, magnesium, phosphate, copper and zinc. One-half cup of boiled zucchini has only 18 calories, 0.3 g fat and 1.0 mg sodium. Each half-cup serving provides 0.8 g protein, 3.9 g carbohydrate and 1.3 g dietary fibre. The nutritional composition of zucchini is given in Table 63.1.

TABLE 63.1

Nutritional Composition of Zucchini (per 100 g Edible Portion)

Constituents	Contents	Constituents	Contents
Water (g)	95.3	Iron (mg)	0.4
Carbohydrates (g)	2.9	Zinc (mg)	0.2
Protein (g)	1.2	Vitamin A (IU)	340
Fat (g)	0.14	Thiamin (mg)	0.07
Dietary fibre (g)	1.2	Riboflavin (mg)	0.03
Calcium (mg)	15.0	Folic acid (µg)	22.0
Potassium (mg)	32.0	Vitamin C (mg)	9.00
Magnesium (mg)	22.0	Energy (kJ)	59.0

Uses

Zucchini when used as food is usually picked when its length is <20 cm and the seeds are still soft and immature. Mature zucchini fruits can be as much as 1 m long but the larger ones are often fibrous and with the flowers attached are a sign of truly fresh and immature fruits, which are especially sought by many people. Zucchini has a stronger and zestier flavour than the summer crooknecks and straightnecks. A favorite way to enjoy zucchini is to eat it raw either in a salad or as a party dip. Its main nutritional contributions are vitamins A and C.

Unlike cucumber, zucchini is usually served as cooked. It can be prepared using a variety of cooking techniques, including steamed, boiled, grilled, stuffed and baked, barbecued, fried, or incorporated in other recipes such as soufflés. It can also be baked into bread, zucchini bread, or incorporated into a cake mix. It has a delicate flavour and requires little more than quick cooking with butter or olive oil, with or without fresh herbs. The skin is left in place. Quick cooking of barely wet zucchini in oil or butter allows the fruit to boil and steam partially, with the juices concentrated in the final moments of frying when the water has gone, prior to serving. It can also be eaten raw, sliced, or shredded in cold salad and baked into bread similar to banana bread, as well as lightly cooked in hot salads, as in Thai or Vietnamese recipe.

Both male and female flowers are edible. Firm and fresh blossom that are only slightly open and with intact pedicels or pedicels removed are harvested for cooking. Before cooking, the pistils are removed from the female flowers and stamens are removed from the male flowers. There are varieties of recipes, in which the flowers are deep-fried as fritters or used for tempura, stuffed, sauted, baked, or used in soups. The use of seeds as dried fruit is common in some areas of Mesoamerica and almost unknown in others. The seeds are a good source of protein and oil, and their industrial preparation and marketing should be investigated.

Medicinal uses

Research has shown that its seeds also contain traces of cancer-preventing substances, known as protease trypsin inhibitors, which inhibit the activation of viruses and carcinogens in the human digestive tract.

BOTANY

The plant has the bush habit rather than the vining habit of the winter squashes. However, within the bush habit, there is a wide range of variations in general plant characters, primarily in density and arrangement of leaves. The leaves of zucchini are quite large, with more notches per leaf than crookneck and straightneck squash. Its leaves are also characterized by having light greenish grey splotches and streaks on the leaf surface. These light markings are sometimes mistaken for a mildew problem. Leaves are alternate, simple, prickly with speculate bristles, large, broadly trian-gular without or usually with deep acute lobes 20–30 × 20–35 cm with or without white blotches and with denticulate to serrate denticulate margins. Tendrils have two to six branchlets or are simple and little-developed tendrils in the semi-shrubby types.

Flowers are bright yellow pentamerous, solitary and borne in leaf axiles. Male flowers have 7–20 cm peduncle, a campanulate calyx of 9–12 mm linear sepals, a tubular/campanulate 5–10 cm long corolla, which is divided into five lobes for up to one-third or more of its length and three stamens. The female flowers are sturdy, sulcated pedicels of 2–5 cm, ovary is glo-bose, oblate, ovoid, cylindrical, rarely piriform, smooth, ribbed, or verrucose and multilocular, and the calyx is very small. The fruit is very variable in size and shape, *viz.* smooth to heav-ily ebbed, often verrucose and rarely smooth, with a rigid skin varying in colour from light to dark green, plain to minutely speckled with cream or green contrasting with yellow, orange, or two-coloured. The flesh is cream to yellowish or pale orange; it ranges from soft and not bitter to fibrous and bitter and has numerous seeds, which are narrowly or broadly elliptical or rarely orbicular, slightly flattened and 3–20 × 4–12 mm.

CLIMATIC REQUIREMENT

Zucchini is a warm season crop that grows well in nearly all climates. Since it is a short season crop, it thrives in cooler climates better than other cucurbits such as cantaloupes and watermelon. Its seeds should not be sown until the danger of frost is past. The optimum soil temperature for the germination of seeds is about 30°C and little or no germination occurs below 15°C. The growers often plant before the optimum temperature is attained in order to get an early harvest. The use of plastic mulch, row covers, row orientation and/or transplants can be utilized to overcome the cold temperature conditions and produce an early crop.

SOIL REQUIREMENT

A well-drained sandy loam soil with high organic matter content and a pH of 6.0 to 6.5 is most suitable for its successful cultivation. For an early crop, a lighter soil that warms rapidly should be chosen. Poorly drained soils on which waterlogged conditions are likely to occur should be avoided for its cultivation. Sandy soils require more frequent watering and fertilization than heavy textured clay soils so are not suitable for its cultivation.

VARIETIES

Zucchini is represented by several cultivars. It is usually dark green but may also be yellow or light green and its fruits have a shape similar to a cucumber. There are a few cultivars with round or bottle-shaped fruits. The most commercially grown zucchini cultivars are hybrids because they are usually heavier producers than the open pollinated varieties.

Fruits of this member of the Italian marrow squashes grow not only in cylindrical shapes but also in round and intermediate shapes. Fruit colour varies from green so dark as to be near black to lighter shades of green both with and without stripes, all the way to tones of yellow. Many fruits are highlighted with various degrees of speckling. Cylindrical fruits range in average size from 12–15 cm of Caserta to the longer varieties such as cocozelle, which reaches 35–40 cm in length. The average diameter of most of the varieties fruits is 7.5–10 cm.

The varieties can be grouped based on colour of the fruit. The following categories are gen-erally recognized, *i.e.*, (i) very dark (green-black) such as Blacknini, (ii) dark (dark green) such as Ambassador and Greenbay, (v) grey-green such as Caserta and (vi) yellow such as Goldzini, Sunburst, Gold Coast and Goldsmith.

Varieties may also be classified based on bush habit, open habit, where the leaves are more sprawling and less cluttered and dense habit where the leaves are crowded upright (closed). Varieties with open habit are Burpee Hybrid, Ambassador, Blackjack, El-Dorado, Grey, Ball's Zucchini and Caserta and Blackini (semi-open). Good examples of the closed-bush type are Hyzelle, Hyzini, Black Zucchini, Seneca Gourmet, Black Eagle, Blackee, Burpee Fordhook, Long White Vegetable Marrow and Mexican Globe.

Newer varieties include the golden zucchini and the globe or round zucchini. The golden variety is somewhat milder in taste than the dark green. The globe variety is about the size of a softball, about 7.5 cm in diameter and perfect for stuffing.

SEED RATE

The seed rate of zucchini is about 6–8 kg/ha. The seeds of this crop are large, creamish white in colour and have 75–85% germination under ideal conditions.

SOWING TIME

The seeds are sown in the last week of February or the first week of March when night temperatures become around 18–20°C. Nursery can also be raised in protected structures in plug trays of 3.5 cm cell size in the months of December to February.

SOWING METHOD

Zucchini is sown on raised beds of 1.2 to 1.5 m width and commercially propagated through seeds, which are sown directly in the field or by raising a nursery in the polyhouse. Bush varieties are sown at a spacing of 90–120 cm between rows and 45–75 cm between plants. Seedlings are transplanted in a single row on each bed at a planting distance of 50 cm on a drip system of irrigation. Distance between the rows in usually kept at 1.5 to 1.6 m.

RAISING OF NURSERY

The seedlings of zucchini can also be raised in a greenhouse in plastic pro-trays having 3.5 cm cell size in soil-less media in the months of December or January the seedlings that are 28–32 days old and at the four-leaf stage are transplanted under row covers or plastic low tunnels in the open field from mid January to mid February, when the night temperature is very low in northern parts of the country.

A nursery of zucchini can be also be raised even in polythene bags in very simple and low-cost protected structures like walk-in tunnels or in locally available plastic trays in soil-less media as per need of the area. It can be transplanted even in the month of December for complete off-season production, and this crop will be ready for harvesting in the first week of February and can fetch a very high price in the market. The transplanting can be performed very conveniently during a frost-free season.

NUTRITIONAL REQUIREMENT

About 15 to 20 tonnes of farmyard manure and nitrogen 40 to 60 kg, phosphorus 50 to 60 kg and potash 60 to 80 kg/ha will be sufficient to meet the nutritional requirement of this crop. Farmyard manure should be incorporated into the soil at the time of land preparation, and a half-dose of nitrogen and full dose of phosphorus and potash should be applied at the time of last ploughing or planting, and the remaining half of nitrogen should be applied at the time of vining.

IRRIGATION REQUIREMENT

Adequate soil moisture is very important, especially during flowering and fruit development. The soil should remain moist. Drying slightly between watering is not a problem but the soil should never be allowed to dry completely nor should it remain too soggy. Waterlogging should be avoided at all times. Zucchini roots extend deep into the soil profile, therefore, it is rarely necessary to irrigate daily. Overwatering can contribute to disease problems or lower the fruit quality. During dry spells, irrigation is applied once a week. Very sandy or soils with low-moisture holding capacity may require two irrigations per week. Drip irrigation is recommended for more efficient water use and also for disease and weed management. Watering through a drip irrigation system keeps the crop not only weed free but also helps in preventing foliar fungal diseases by delivering water without wetting the leaves.

During hot sunny afternoons, it is normal for the large succulent leaves to wilt slightly, even in an adequately watered crop. Wilting that persists into the evening or overnight indicates either yield-limiting soil moisture deficit, or pest or disease problems.

INTERCULTURAL OPERATIONS

Zucchini requires frequent weeding. The first weeding may be performed 15–20 days after sowing and a total of three weeding operations are required to keep the weeds down and to get an economic yield from this crop.

HARVESTING

For tender produce, the fruits are harvested at their immature stage when the rind is soft and seeds are underdeveloped. Fruits should be 15 to 20 cm long and 3.5 to 7.5 cm in diameter. Some varieties may be edible even after attaining full size. Delayed harvesting results in further development of fruit, as a result, the rind becomes tough and seeds become hard, which make the fruits of poor quality. It is important to keep zucchini picked; if not, the fruit production becomes slow. On the other hand, if there are so many zucchini fruits that even the neighbours have had enough, leave one or two fruits on the plant to slow it down.

Be sure that the fruit is cut from the plant at the stem between the fruit and the main stem. Attempting to pull zucchini off will usually damage the entire plant. For edible flowers, harvest early in the morning before they close, place them with their bases in water and store in the refrigerator until they are used.

YIELD

The average yield is 15–25 tonnes per hectare under good crop management.

POST-HARVEST MANAGEMENT

Early shipment is made in baskets, hampers and crates. Zucchini fruits are sometimes wrapped in tissue paper to safeguard against damage and blemishes. Some careful growers ask their workers to wear cotton gloves to avoid injury to the fruits caused by fingernails, and the transport vehicle may be lined with quilts and rugs.

Zucchini fruits are very tender and normally should not be stored for more than a week. They can be held 3 to 4 days at 7°C temperature and 95% relative humidity in an ethylene-free atmosphere. Since the fruits are sensitive to ethylene, they should not be stored or shipped with ethylene-producing produce. Some cultivars are more sensitive to chilling.

Decay caused by fungal and bacterial pathogens can cause significant post-harvest losses in zucchini. The incidence of decay increases in fruits that have physical injury or chilling stress. The common post-harvest diseases include *Alternaria* rot, bacterial soft rot, cottony leak, *Fusarium* rot, *Phythopthora* rot and *Rhizopus* rot. However, *Alternaria* rot can be especially pronounced following chilling injury.

PHYSIOLOGICAL DISORDERS

Zucchini fruits are susceptible to several physiological and environmental disorders that limit production or affect fruit quality. Most of the disorders are poorly understood and can be induced by many conditions related to nutrition, environments, or cultural practices.

PRECOCIOUS FEMALE FLOWERING

This occurs when zucchini is sown too early. Plants exposed to cool conditions commonly produce many more female flowers, which can affect fruit set due to a lack of pollen-bearing male flowers. If the cultivar grown is parthenocarpic, then it is not a serious problem.

BLOSSOM END ROT

This disorder is associated with irregular watering. A sunken brown to black spot develops on the blossom end of the fruit, and it is caused by a localized calcium deficiency.

BITTER FRUIT

Bitter fruit can occur but its cause is not clearly understood. Some growers think that it is a mutation or out-crossing with wild types during seed production. Often only one or two plants may be affected by this disorder and if these are identified, rogue them out of the planting.

CHILLING INJURY

Zucchini is chilling-sensitive, thus, it should not be exposed to temperatures below 5°C. Its fruits show chilling injury damage as pitting on 93% of their surface after 12 days at 2°C temperature, and a 2- to 2.5-fold increase in putrescine levels in both skin and pulp was detected during storage (from 162 and 38 to 320 and 98 nmol g^{-1} of fresh weight, respectively). Abscisic acid concentration increased mainly in the skin. Treatment with CO_2 prior to storage at 2°C was found to be effective in reducing chilling injury. With 5% CO_2, the increase in putrescine and abscisic acid levels was lower than in control fruits, while exposure to 40% CO_2 resulted in a decrease of these values. Spermidine and spermine levels decreased during storage, regardless of treatment. Thus, the 40% CO_2 treatment was more effective than the 5% CO_2 in reducing both chilling injury and the putrescine and abscisic acid changes found in control chilling-injured fruits.

WATER LOSS

Zucchini fruits are very susceptible to water loss. Shrivelling may become evident with as little as 3% weight loss. Pre-cooling and storage at high relative humidity will minimize the weight loss. Its fruits can be waxed but only a thin coating should be applied. Waxing provides some surface lubrication that reduces chafing during transportation. Its fruits' skin is very tender, thus, skin breaks and bruises can be a serious source of water loss and microbial infection.

INSECT-PESTS

Fruit Fly (*Bactrocera cucurbitae coquillet*)

The adult fly punctures the tender fruit and lays eggs below the epidermis. The eggs hatch into maggots, which make tunnels deep into the fruit pulp and feed on it, and thus, the fruits are completely damaged. The attacked fruits are distorted, rot and drop down.

Control
- Plough the field deep in hot months for killing the pupae.
- Remove and destruct the infested fruits.
- Apply *neem* cake @ 250 kg/ha immediately after germination and repeat at flowering followed by sprays of *neem* soap 1% or PNSPE 4% at 10 days after flowering.
- Erect cue-lure (parapheromone trap), three per acre, to attract and trap male fruit flies.
- Spray the crop with indoxacarb 0.5 ml/litre of water.

Red Spider Mite (*Tetranychus neocaldonicus andre*)

Mites are very small insects covered with fine webs and found on the lower surface of the leaves. Both nymphs and adults suck cell sap from tender parts, resulting in yellowish specks on the upper surface of the leaves; as a result, the leaves become pale and dry up.

Control
- Spray *neem* or *pongamia* soap at 1% on lower surface thoroughly.
- Alternately, spray Dimethoate 30 EC @ 2 ml, Ethion 50 EC @ 1 ml, or wettable sulfur 80 WP @ 3 g/litre of water.

Thrips (*Thrips palmi karny*)

Thrips can be seen on all the above-ground parts of a plant, most commonly on flowers and on the backs of the leaves. Thrips feeding damage appears as silvering and flecking on leaves of the seedlings. Heavily infested plants are characterized by a silvered or bronzed appearance of leaves, stunted leaves and terminal shoots and scarred deformed fruits.

Control
- Apply *neem* cake (once immediately after germination and again at flowering).
- Alternately, spray Dimethoate 30 EC @ 2 ml, Ethion 50 EC @ 1 ml, or wettable sulfur 80 WP @ 3 g/litre of water.

Red Pumpkin Beetle (*Auacophora foveicollis lucas & A. lewsii baly*)

The grubs of red pumpkin beetle found in soil feed on the underground roots and stems of the host plants, forming holes or tunnels, resulting in wilting and death of the plants. The adult beetles emerge from the soil and feed on cotyledonary leaves and leaf lamina, causing perforation and irregular holes and seedling mortality. They may also damage the fruit by puncturing fine holes in it.

Control
- Collect and destroy the pest mechanically if population is low.
- If the pest incidence is very severe, spray indoxacarb 14.5 SC @ 0.5 ml or chlorpyrifos 20 EC @ 2.5 ml/litre of water.

Root-knot Nematodes (*Meloidogyne incognita*)

These minute eelworms cause root galls or swellings in initial stages of the crop growth. The first above-ground symptoms of pest are the presence of poor growth, yellowing of foliage, stunting of plants and dropping of flowers and fruits. During later stages, the entire root system is galled, plants become weak and eventually they die.

Control

- Treat the seed with bio-pesticide *Pseudomomnas flurescens* @ 10 g/kg of seed.
- Apply carbofuran 3 G @ 1 kg a.i./ha at sowing and repeat after 45 days.
- Apply 2 tonnes of farmyard manure enriched with *Pochonia chlamydosporia* and *Paecilomyces lilacinus* per acre before sowing, along with 100–200 kg of *neem* or *pongamia* cake.

DISEASES

Downy Mildew [*Pseudoperonospora cubensis (Berkley and Curtis) rost.*]

Appearance of angular yellow spots on upper surface of the leaves with a greyish mouldy growth on surface of the leaves is the major symptom. The spots enlarge with yellowing of leaves, which die later. During moist weather, downy pale grey to purple mildew covers the lower surface of the leaves. Diseased leaves soon wither, cup upward and turn brown.

Control

- Destroy the plant debris after harvest.
- Keep the field free from all wild cucurbit weeds.
- Avoid flood irrigation while applying irrigation.
- Spray Ridomil MZ 72 (0.2 or 0.3%), Aliette (0.1%), Syllit (0.1%), Blitox (0.2%), mancozeb (0.2%), Dithane M-45 (0.3%), or Dithane Z-78 (0.3%).

Powdery Mildew [*Erysiphe cichoracearum dc and Sphareothica fuliginea (Schl.) poll*]

White or grey patches or spots appear on leaves, petioles and stems. Leaves become shrivelled, and later, the defoliation occurs. Plants are chlorotic and stunted in severe cases. The fruits become small. The disease spreads at high temperatures during dry weather.

Control

- Destroy the plant debris after harvest.
- Control the weeds, particularly those which harbour the pathogen.
- Avoid flood irrigation while applying irrigation.
- Spray Benalate (0.1%), Thiophanate M (0.05%), Calixin (0.05%), Karathane (0.05%), Bavistin (0.1%), or Sulfex (0.2%). At least three sprays at 5- to 6-day intervals are needed.

Anthracnose [*Colletotrichum lagenarianum (Pass.) ellis and Halsted*]

Round reddish brown spots appear on leaves, starting from veins. These spots coalesce and leaves may be blighted. The centres of these spots fall off, giving a shot-hole appearance. Brown to black elongated streaks appear on stems and petioles. The affected fruits have brown to black water-soaked depressed round spots, which turn ash-coloured in the centre.

Control

- Follow long crop rotation with non-host crops.
- Plough the field deep immediately after harvest.

- Treat the seed with thiram, Bavistin, or Dithane M-45 (2 g/kg of seed).
- Keep the field free from all wild cucurbit weeds.
- Spray the crop with difolatan (0.2 %), Bavistin (0.2%), Benomyl (0.1%), thiophanate phenyl (Topsin) (0.1%), or Kavach (0.1%).

FUSARIUM ROOT ROT OF SQUASH (*Fusarium solani f. sp. cucurbitae* Snyder & Hansen)

Sudden wilting of plants with brown cortical decay is the main symptom of this disease. The taproot is replaced by lateral roots and clogging of the xylem vessel of the infected taproot. If the infection is severe, the plants become stunted and they bear no fruit.

Control

- Follow crop rotation with crops like garlic, radish, onion, beetroot and lucerne.
- Sterilize the soil through solarization and add soil amendments like *neem* and mustard cake.
- Avoid flood irrigation.
- Grow the crop on raised beds.
- Drench the soil with 0.2% Brassicol.
- Treat the seed with thiram, Vitavax, or Brassicol (3 g/kg of seed).
- Spray the crop with Ridomil MZ 72 (0.25%).

ANGULAR LEAF SPOT [*Pseudomonas syringae pv. lachryans (Smith and Bryan)*]

Water-soaked irregular to angular spots of white to brown colour appear on leaves, stems, blossoms and fruits. Later, these spots become dry and fall off with shot holes in the leaves. In zucchini, particular black or brown spots appear with a yellow halo. During favourable weather conditions, the spots progress on the leaf lamina, which become angular near the veins.

Control

- Use seed produced in dry areas.
- Follow long crop rotations, avoiding cucurbitaceous crops in rotation.
- Collect and destroy the infected plant debris.
- Treat the seed with mercuric chloride solution (1:1000) for 5–10 minutes.
- Spray the crop with streptomycin (400 ppm) or Bordeaux mixture.

VEGETABLE MARROW MOSAIC VIRUS, CUCUMBER MOSAIC VIRUS (CMV), PAPAYA RING-SPOT VIRUS-W (PRS-W) OR WATERMELON MOSAIC VIRUS (WMV), (WMV-2) IN SINGLE OR MIXED INFECTIONS

Vein clearing in leaves, yellowing, blistering in patches, puckering, reduction in lamina and deformation, mosaic in young leaves and shortening of internodes with crowding of young shoots are conspicuous symptoms of the disease. Severely infected plants do not produce fruits.

Control

- Spray the crop with oils like castor, groundnut, or paraffin (1–2%).
- Treat the seed with hot air (70°C for 2 days) or with hot water (50°C) for 60 minutes.

64 Chinese Water Chestnuts

M.K. Rana and Naval Kishor Kamboj

Botanical Name : *Eleocharis dulcis* (Burm. f.)

Family : Cyperaceae

Chromosome Number : 2n = 10, 16 and 18

ORIGIN AND DISTRIBUTION

Chinese water chestnut is native to China and widely cultivated in flooded paddy fields in southern China and parts of the Philippines. The plant grows wild, especially in the Yangtze valley of China and in Indonesia in fresh water and brackish swamps. The plant is universally grown in tropical, subtropical and warm-temperate wetlands of Madagascar, Central Africa, India, Malaysia South East Asia, Japan, Hong Kong, Polynesia, southern United States of America, Hawaii and other Pacific islands. In many parts of India, wild stands occur in shallow lakes and swamps up to 1000 m above mean sea level. About 14 species are are available in India but only two species are economically important. Primarily, it was cultivated in humid areas of China, but later, it was imported to the United States, where attempts were made to cultivate it, with limited success. Its plant thrives in water-inundated areas like ponds, flooded fields, swamps, marshes and in mud of shallow lakes. In China, it is extensively grown for its corms. It became known to the West only in the 17th century. In Fiji, the plant since the 1990s has become endangered and it is deeply missed by the Fijian women, as they have been using this plant for hundreds of years for weaving mats, indigenously known as *kuta*. Water chestnuts are sold in markets all over Southeast Asia.

INTRODUCTION

Chinese water chestnut, a perennial rhizomatous aquatic herb, is a traditional vegetable of Asia, often grown as an annual crop. Its stems grow to a height of 1.5 m and die off after the first frost or 6–7 months after planting, depending on climatic conditions. It is a grass-like grown in flooded water for its edible corms. It produces two types of corm, *i.e.*, (i) used only for propagation and (ii) used for storage, which is actual the edible water chestnut. The turnip-shaped corms are 2.5–5 cm in diameter, four- to six-ringed, oblate, crisp, juicy, sweet (sugar greater than 8°Brix) and starchy with burgundy to black skin, enclosing bleach-white flesh within it. The corms develop at a depth of 15–25 cm beneath the soil at the tips of the stolons. Its skin is difficult to peel. The plant produces numerous flowers, which are very small and occur on tips of the culms. Flowers are usually produced before the plant reaches peak vegetative growth.

COMPOSITION AND USES

COMPOSITION

The corms of Chinese water chestnut are rich in starch with very large grains. The corms are also a good source of carbohydrates, dietary fibre, riboflavin, pyridoxine, potassium, copper and manganese. The nutritional composition of Chinese water chestnut is given in Table 64.1.

TABLE 64.1

Nutritional Composition of Chinese Water Chestnut (per 100 g Edible Portion)

Constituents	Contents	Constituents	Contents
Water (g)	77.3	Sodium (g)	3.0
Carbohydrates (g)	23.94	Vitamin A (IU)	202
Sugars (g)	4.8	Thiamine (mg)	0.14
Protein (g)	1.4	Riboflavin (mg)	0.2
Fat (g)	0.2	Niacin (mg)	1.0
Dietary fibre (g)	0.9	Pantothenic acid (mg)	0.479
Calcium (mg)	18.0	Pyridoxine (mg)	0.328
Phosphorus (mg)	63.0	Folic acid (μg)	16.0
Potassium (mg)	584.0	Vitamin C (mg)	4.0
Magnesium (mg)	22.0	Vitamin E (mg)	1.2
Manganese (mg)	0.331	Energy (kcal)	97.0

Uses

The Chinese water chestnut corms are consumed raw or cooked and used in many local dishes such as omelettes, soups, salads, meat, fish dishes and even in sweet dishes in China, Indo-China, Thailand and other parts of Southeast Asia. The larger corms are widely eaten raw and considered a delicacy in China, where they are eaten raw as a substitute of fresh fruit due to their crisp apple-like flesh. After drying, they are ground into a flour, which is used to make water chestnut cake. In Indonesia and the Philippines, the corms are usually made into chips. Its stems are used for making sleeping mats and skirts. Its foliage is used as cattle feed or as mulch.

Medicinal Uses

The juice extracted from its fresh corms contains antibiotic called *puchin*, which is effective against *Staphylococcus aureus*, *Escherichia coli* and *Aerobacter aerogenes*. However, its medicinal properties due to the presence of the antibiotic are destroyed during canning but fresh corms remain crisp and retain puchin even after cooking. Due to this reason, the Chinese consider fresh water chestnut superior to its canned products. Its flavour is a blend of apple, chestnut and coconut.

BOTANY

Chinese water chestnut is upright aquatic sedge, as it grows under flooded conditions. The above-ground parts of the plant consist of green cylindrical stems with very small-sized, almost invisible leaves. The tube-shaped leafless stems are erect, terete, tufted, 40–200 cm tall, 3–10 mm in diameter, longitudinally striate and distinctly crosswise septate. The intersepta are 5–12 mm long, hollow, smooth and greyish to glossy dark green. Leaves reduced to some bladeless basal sheaths are 3–20 cm long, membranous, oblique or truncate at the apex and reddish brown to purple. Due to smaller leaves, photosynthesis in the plants is carried out by the culms or stems. In addition to roots, the underground parts comprise two types of modified stem, *i.e.*, (i) rhizomes and (ii) corms. Rhizomes, horizontally growing stems, grow laterally below the soil from the parent corm and produce either daughter plants or corms. These corms are swollen tips of the underground elongated stolons with dormant buds and are the marketable part of plant. Inflorescence with numerous flower spikelets is a single, terminal, cylindrical, 1.5–6.0 × 3–6 mm thick or somewhat thicker than the stem, and the apex is obtuse to acute. Glumes are numerous, oblong, 4.0–6.5 × 1.7–3.2 mm in size and densely imbricate. Flowers are bisexual with a perianth of six to eight filiform, unequal and white to brown bristles. Stamens are three, the anthers linear and 2–3 mm long, the style 7–8 mm long and bifid or trifid, and the enlarged base persistent in fruit, which is an obovoid nut (achene), 1.5–2.2 × 1.2–1.8 mm in size and shiny yellow to brown.

CLIMATIC REQUIREMENT

Chinese water chestnut for its cultivation requires a long warm growing season with at least 220 frost-free days. It grows quickly during warm summer months by forming new plants from underground rhizomes. With the onset of cool weather, the rhizomes stop producing new plants. The crop can successfully be grown in area above mean sea level up to 1350 m. A soil temperature of 14–15.5°C is necessary for the sprouting of corms. Plants perform best at temperature between 30 and 35°C during growth and about 25°C at the time of corms formation.

SOIL REQUIREMENT

The Chinese water chestnut thrives best in 5–20 cm deep, nutrients-rich clay soil with plenty of well-composed manure or other organic matter added a few weeks before planting. It prefers slightly acidic to neutral soils with pH 5.0 to 7.0 for its successful cultivation. Since water chestnuts are harvested manually to avoid damage to their fragile skin, it is important to use soil free from stones, woody plant material, etc. The soil is prepared well to a fine tilth by repeated ploughing followed by planking.

CULTIVATED VARIETIES

HON MATAI

An improved variety recommended for cultivation in Australia gives a marketable yield of 88 q/ha with average corm weight of 12.1 g and total soluble solids 8.6°Brix.

MATAI SUPREME

A cultivar with a yield potential of 120 q/ha bears corms of 12.9 g weight and total soluble solids 8.6°Brix.

SINGAPORE

A cultivar with a yield potential of 103 q/ha produces corm of 12.4 g weight with total soluble solids 8.5°Brix.

SHU-LIN

A variety with a yield potential of 46 q/ha bears corm of 11.4 g weight with total soluble solids 9.4°Brix, which are much higher than that for the other cultivars.

PLANTING TIME

The planting time of Chinese water chestnut varies with the climatic conditions of the growing region, but in general, its corms are planted in March–April and the seedlings are transplanted in June–July.

SEED RATE

Seed rate usually varies with corm size, planting time and climatic conditions of the growing region, but as usual, approximately 500 kg of corms are sufficient to plant a hectare land area.

PLANTING METHOD

Chinese water chestnut crop can be established by either planting corms directly in the field or raising seedlings in a nursery by planting a small quantity of corms.

SELECTION OF CORM FOR PLANTING

The corms, which are high in sugars, are preferred for planting, as they have high energy, which is quickly available to the buds for sprouting. Sugar content can be maximized by storing the seed corms at 4°C temperature for 4 months. Seed corms stored below 4°C temperature show chilling injury and their viability is also affected adversely. As far as possible, larger corms are selected for planting since larger corms have the greater reserve energy and nutrients for the seedlings.

TRANSPLANTING OF SEEDLINGS

Seedlings are raised in plant pots, netted trays, or wooden boxes. In trays, the corms are planted 2 cm deep and 2 cm apart with the shoot end facing upwards. The placement of corms and orientation of apical shoot are not critical, especially if the corms have shoot lengths less than 10 mm. Corms sown in trays or boxes will tend to form a dense intermingled root mass, which must be either separated or cut at transplanting time. While root stirring or cutting will induce more transplanting shock compared to seedlings raised individually in pots, this intricacy of roots can be minimized by having fewer corms in the box or by transplanting the seedlings at an earlier age. The potting media should be friable with good water retention capacity and should possibly be sterilized to make it free from pathogens and weed-seed contamination. Water is supplied through sprinklers. Seedlings become ready for transplanting within 3–4 weeks, as the growth rate in semi-tropical regions is rapid, whereas, in temperate regions, the growth of seedlings even under plastic is slower and they become ready for transplanting in 2–3 months. The transplanting is done at 75 cm distance between rows and 75 cm between plants. Propagation by transplanting generally results in maturity 6 weeks earlier than planting by corms. However, commercial propagation from seed corms is avoided since seed corms are slow in emergence, show poor emergence and require 2 years from planting to the development of normal size corms. Propagation through transplanting is simple, quick and cheap to establish the crop.

DIRECT CORM PLANTING

Direct planting of seed corms in flooded fields has advantages over seedling transplanting since they do not experience transplanting shock. However, it has been observed that the crop raised through direct-planting of corms in the warm-temperate region grows slower during first 2 months compared to the crop raised through seedlings transplanting. Weed management can become a critical issue with a slow-growing crop. The other complications include dealing with non-sprouting corms, which need to be replaced, and damage by pests, e.g., waterfowls. In the direct planting method, corms are planted by hand in a prepared flooded field at a spacing of 75 × 75 cm. The seed corms are usually planted at a depth of 2–3 cm. Higher planting density tends to increase total yield through greater number of corms but it reduces the corm size, while lower planting density has the potential to produce bigger corms.

NUTRITIONAL REQUIREMENT

High productivity in all commercial crops, including water chestnut, can only be achieved if plants are supplied with a balanced dose of fertilizers. Soil analysis is essential in order to calculate fertilizer requirement precisely since over-application can reduce quality of the corms. In general, for the production of 400 q/ha fresh biomass, a Chinese water chestnut crop needs nitrogen 180–200 kg, phosphorus 80–100 kg and potash 250–300 kg/ha. The full dose of phosphorus and potash along with a half-dose of nitrogen should be applied just before planting. The rest of the nitrogen is applied in the standing crop at the onset of the corm-enlargement phase, which may be indicated in above-ground stems at the onset of flowering. If a higher dose of nitrogen is applied before planting, the water chestnut crop produces excessive stem biomass at the expense of yield. Therefore, not more than half of the total nitrogen should be applied before planting.

IRRIGATION REQUIREMENT

Like a paddy crop, Chinese water chestnut needs to be completely submerged with a controlled level of water for most of its life, but at the time of harvesting, it is preferable to drain out the water. The soil should be submerged several centimetres to facilitate the planting operation. It is important to maintain a uniform level of water across the field. After planting, the field is flooded with at least 12.5 cm of water for 24 hours and then allowed to drain out naturally. As the plant reaches a height of 20–30 cm, the field is flooded again up to 10–20 cm, which should be maintained throughout the growing season. Lack of water for long time may result in crop failure. The water demand of Chinese water chestnut varies during the growing season, and it is bit lower after transplanting and much greater during the peak growing season. Therefore, water height can safely be lower at the beginning of the growing season. The pH of the water should be around 7.0. Lime may be added if the water is too acidic and gypsum may be added if the pH of the water is too high, although the crop has a good tolerance for slightly saline water.

INTERCULTURAL OPERATIONS

HOEING AND WEEDING

Weeds are opportunistic by nature and they germinate anywhere suitable for their growth. Therefore, crop establishment is rare against these weeds. Controlling weeds in Chinese water chestnut is not so simple. Thus, efforts should be made to manipulate growing conditions that favour the crop but depress the germination of weed seeds. Since no selective herbicide is registered for this crop, a greater dependence is on mechanical and manual weeding. Manually, the weeds should be kept under control until the establishment of daughter plants. While preparing seedbeds, it is desirable to kill all existing weeds to reduce the risk of weed establishment within the field. Weed control during the first 3–4 months after direct planting or 2 months after seedling transplanting is the most critical period. The number of weedings depends on the weed seed reservoir in soil as well as in water. In general, Chinese water chestnut requires three or four weedings. If mulch is applied, only a single weeding is normally required.

MULCHING

Applying a 2 cm thick layer of straw mulch to moist soil before planting of corms or transplanting of seedlings effectively suppresses the germination of weed seeds and conserves soil moisture in Chinese water chestnut. *Azolla* that fixes environmental nitrogen into the soil is also used as a living and floating mulch under low nitrogen conditions.

HARVESTING

The crop becomes ready for harvesting when the corms have turned brown or plants have been killed off by frost. The irrigation is stopped 3–4 weeks before harvesting for the drying of the field, and the corms are dug out carefully without any injury or breakage. Harvesting is done manually using garden forks and spades.

YIELD

The yield depends on several factors such as variety, climatic conditions of the growing region and package of practices followed during the cultivation of the crop. The yield of Chinese water chestnut in general varies from 200 to 400 q/ha with a maximum yield of 280 q/ha.

POST-HARVEST MANAGEMENT

Sorting and grading after harvesting assure the supply of quality corms to the market. The damaged corms are discarded during sorting and the sound corms are graded according to size, skin colour and internal quality. The best quality standard corms are dark-brown to black in colour though dark red-burgundy colour corms are also acceptable by the consumers. Based on corm diameter of Chinese water chestnut, the Australian Aquatic Vegetable Growers Network has established three official grades, *i.e.*, 25s (25–32 mm), 32s (>32–40 mm) and 40s (>40 mm), in the Australian fresh market.

Sprouting, *i.e.*, the development of new stems from corms, is the most reasonable cause of quality loss in corms. When the sprouts are longer than 1.5 cm, they become unacceptable to most of the consumers, hence, it is desirable to have negligible shoot growth for optimum corm quality, which can be achieved by ensuring that the crop is not over-mature. As soon as possible, the harvested corms are shifted to cold storage. Chinese water chestnut can successfully be stored for 6 months at 4°C temperature.

INSECT-PESTS AND DISEASES

Chinese water chestnut crop is attacked by a number of insect-pests and diseases during the growing season. Their description along with control measures are given as follows:

INSECT-PESTS

WATERFOWL (*Lissorhoptrus ozyzophilus*)

Waterfowl damages the crop at every stage, especially when the plants are younger, *i.e.*, about 4 months after planting. However, the corms are more susceptible to damage soon after sowing or transplanting.

Control
- Prevent the crop from waterfowl by covering with nets or nylon lines (triangular) across the field.
- Use flashing lights or shock guns to repel waterfowl.

PLANT HOPPER (*Nilaparvata lugens*)

Nymphs and adults live together at base of the plant above the water level and they suck cell sap from the stem and leaf sheath. At the early stage of infestation, circular yellow patches appear which soon turn brownish due to the drying up of plants. Hoppers are recognized by the presence of a honeydew-like substance and dirty moulds on the infected area.

Control

- Drain out water from the field for 3–4 days in early stage of infestation.
- Apply nitrogen in split doses to reduce plant hopper buildup.
- Spray the crop with phosphamidon 40 SL 1000 mL/ha, monocrotophos 36 SL 1250 mL/ha, Phosalone 35 EC 1500 mL/ha, methyl demeton 25 EC 1000 mL/ha, Acephate 75 SP 625 g/ha, or chlorpyrifos 20 EC 1250 mL/ha at 7- to 10-day intervals.

DISEASES

RUST (*Uromyces species*)

Rust attacks the Chinese water chestnut in warm climates about 5 months after planting. The symptoms initially appear as circular to elongated yellow or white slightly raised spots on upper and lower surface of the leaves. These spots enlarge and form reddish-brown or rust-coloured pustules, break through the leaf surface and release rust-coloured powdery spores. Severe infection causes browning, drying and dropping of leaves prematurely.

Control

- Follow crop rotation of at least 2 to 3 years to break the disease cycle.
- Remove and burn the infected plant debris.
- Spray the crop with sulfur-containing fungicides as band when the disease appears.

■■■

65 East Indian Arrowroot

M.K. Rana and Sanjay Kumar

Botanical Name : *Tacca leontopetaloides* (L.) Kuntze

Family : Dioscoreaceae

Chromosome Number : 2n = 30

ORIGIN AND DISTRIBUTION

East Indian arrowroot has a paleotropic origin from Western Africa to Southeast Asia, mainly Malaysia, where it gives flowers and fruits year round. It is naturally distributed from Western Africa to Northern Australia through Southeast Asia. It was also widely distributed as a wild plant to most of the tropical Pacific islands of the world, including tropical Asia, tropical Africa, Northern Australia, New Guinea, Polynesia, Fiji and the Indian subcontinent regions and intentionally brought to tropical Pacific islands with early human migrations. Once established, it propagates itself and becomes a wild plant.

INTRODUCTION

East Indian arrowroot, also known as Polynesian arrowroot and *tacca* plant, is a large perennial deciduous herb with tuberous tubers, which are the primary and economic produce of this herb. In appearance, its tubers are similar to potato tubers and used as a nutritious edible starch. The tubers have eyes, yellowish skin and dull whitish flesh with bitter taste due to the presence of the bitter principle *taccalin*, which is a little bit poisonous. In their raw state, the fresh tubers are almost inedible, but after a long processing, they can be made edible due to denaturation of this alkaloid. It is quite popular for its starchy tubers. Its starch grains are semi-hemispherical and very similar in many respects to that of cassava and cornstarch. In past, its tubers were regarded as a staple food in famine and the chief source of edible starch of many Pacific islands like Polynesia and those in Northern Australia and Western Africa. However, the demand for its starch has never been so high, and nowadays, its starch has completely been replaced by cassava and cornstarch because of the bitter taste of its tubers, requiring a lengthy and laborious processing for making the starch edible. Therefore, it has become a widespread volunteer wild herb found growing near natural streams. However, it is still cultivated on a small scale for local medicinal uses and as an ornamental plant by many inhabitants of low islands. Further attention is needed to grow it as an important food crop on a large scale.

COMPOSITION AND USES

COMPOSITION

East Indian arrowroot tubers are well recognized as a good source of carbohydrates and minerals, especially calcium, but known to have only a small amount of protein and no vitamins. The starch content in its tubers varies from 20% to 30%, depending on location, soil conditions and stage of harvesting. The principle amino acids present in its protein are arginine, leucine, lysine, valine, and glutamic and aspartic acids. Its flour has about 86% carbohydrates and less than 0.2% protein. The nutritional composition of East Indian arrowroot flour is given in Table 65.1.

TABLE 65.1

Nutritional Composition of East Indian Arrowroot Flour (per 100 g Edible Portion)

Constituents	Contents	Constituents	Contents
Moisture (%)	12.1	Phosphorus (mg)	7.2
Carbohydrates (g)	85.74	Iron (mg)	0.55
Protein (g)	0.18	Riboflavin (mg)	0.0
Fat (g)	0.05	Niacin (mg)	0.0
Dietary fibre (g)	0.0	Ash (g)	1.89
Calcium (mg)	58.0	Energy (kcal)	346

Uses

The East Indian arrowroot tubers were formerly used as a staple and famine food in parts of West Africa and as a source of starch in the Pacific area. The tubers are eaten after roasting and used for the extraction of starch, which is extracted after peeling, grating, soaking in fresh water, pounding and washing repeatedly to remove bitterness, sieving and drying. Its starch is used for making breads and a variety of puddings and as a laundry starch for fabrics. Its flour is mixed with mashed taro and coconut cream to prepare different puddings. Its fine crystal starch makes it easily digestible, thus, it is a favourable food for weak and sick people as well as small children. The leaf stalks and flower scapes are used for making an excellent straw used in Europe for the manufacturing of lightweight hats and the stem fibres for woven mats. Fresh starch is also used for pasting the thin layers of beaten bark of paper mulberry clothes.

Medicinal Uses

East Indian arrowroot is often used in traditional medicines in the Pacific Islands. A mixture of its starch and red clay is taken with fresh water to cure or dysentery, to stop internal haemorrhaging in the stomach and applied as a poultice to wounds to stop bleeding. Its roots are also used as a thickener in remedial preparations, and starch from its roots is rubbed onto sores and burns. The crushed plant stalks are rubbed onto bee and wasp stings to avoid swelling of muscles. The squeezed sap from the roasted stem is used as an eardrop to relieve earache.

BOTANY

East Indian Arrowroot is a dioecious perennial plant having male and female plants separately. The female plant bears more and larger tubers as compared to the male plant. The plant has a very good resistance against drought due to its tuber formation ability. Under drought conditions, the above-ground portion of the plant is withered entirely and the tubers remain dormant until the new growing season. Under favourable conditions, new palmately incised young leaves arise from the underground tubers in early spring months and the tubers mature from October to January. The entire cycle takes about 8–10 months including 2–4 months of a dormancy period. Such plants do not flower until 2–3 years. The true stem consists of an underground tuber, and leaves and inflorescence are the only parts that appear above the ground. The tubers developed at the end roots are like potato tubers, having 1.25–10 cm diameter, a thin brown skin and whitish starchy interior. A new tuber is formed about 10–15 cm below the soil surface close to the tuber of the previous year. The number of tubers per plant ranges from 1 to 30, with numbers between 10 and 20 being the most common. The leaves are generally bipinnate or tripinnate and basal, arising directly from an aerial stem. Leaves are large, deeply divided, 30–70 cm long and 120 cm wide. Petiole is variable in size up to 1 m long and longitudinally ribbed. The leaf upper surface has depressed veins and the undersurface is shiny with bold yellow veins. The leaf colour ranges from light green to dark green. Inflorescence is 1.5 m tall and 30–40 cm above the top of the leaves, arising from an underground tuber. A single plant develops one or two flower stalks. A cluster of flowers, called an umbel, is 4–10 cm in diameter, containing about 20 flowers per umbel. The individual flower is small (1–3 cm in diameter) with six or more leafy bracts of faint greenish colour and about 20–40 purple-coloured thread-like trailing bractlets. Tepals persist at the apex. The stigma is three-lobed with each lobe divided into two segments. Fruit is a berry, which is 23–30 mm in length, 17–21 mm in diameter and ovoid in shape with a smooth yellowish skin and a six-ribbed cross section. The fruit emits a strong fruity odour on cutting. Fruit remains on the plant for quite a long time and contains a large number of flattened, ovoid, or ellipsoidal seeds, which are 3–8 mm long, 1.5–3.0 mm thick and slightly kidney-shaped with a polygonal cross section. The seeds are extracted by splitting the fruit after it falls over the ground and dries. The seeds, with soft whitish cores, are dark brown in colour.

CLIMATIC REQUIREMENT

East Indian Arrowroot grows well in a warm and humid tropical climate. It can be cultivated in regions up to about 300 m elevation with an annual precipitation of 1250–2500 mm. It requires medium to low light and partial shade, thus, it is usually found growing in coastal vegetation, between coconut groves and lagoon shores, but it does not grow in too-dense plantations or too-shaded areas. The optimum temperature for its successful cultivation is 26°C –31°C. Temperature below 20°C and above 30°C affects the plants growth adversely. Short days favour early tuberization in East Indian Arrowroot.

SOIL REQUIREMENT

East Indian Arrowroot can be grown on a wide range of soils. However, medium-textured sandy loam soils rich in humus near seashores are considered best for its cultivation. For obtaining higher yield, the soil should be deep, stone free and well drained but good in water-holding capacity. The ideal soil pH for its normal growth and development is 6.0–7.8.

VARIETIES

The East Indian arrowroot varieties have been categorized into two groups, i.e., (i) green stem type and (ii) purple stem type. The green stem type varieties produce a mixture of tubers with a white and purple interior, while the purple stem type varieties produce tubers with brown skin and a yellow to white interior.

PLANTING TIME

The tubers or their divisions are planted on the onset of monsoon showers in the last week of June or first week of July.

SEED RATE

Traditionally, East Indian arrowroot can be intercropped with coconut plantation, papaya, banana and breadfruit but it is the most suitable intercrop in a coconut plantation. No information on seed rate is available anywhere in the literature. About 15–20 quintals of rhizomes are enough for planting a hectare land area.

PLANTING METHOD

East Indian arrowroot is vegetatively propagated by using small tubers, which form at the base of the plant or their divisions. Before planting, the tubers are dipped in 4–8% solution of ethylene chlorohydrin or cow dung slurry and dried in shade to break the bud dormancy of tubers. During harvesting of the previous season's crop, only the large tubers from female plants are picked up and the small tubers from male plants are left in the field. The leftover tubers sprout in the next season. These sprouted seedlings are taken out and used for the planting of next year's crop in a different field. The seedlings are usually transplanted in rainy season.

In another method of planting, one to two small tubers of about 2.5–3.0 cm diameter are placed in a 15 cm deep pit often dug at a distance of 90 cm between rows and 45 cm between plants. Thereafter, no special care is needed since the plant is hardy and spreads just like a weed. However, the field before planting is prepared well to a fine tilth by repeated ploughing followed by planking.

NUTRITIONAL REQUIREMENT

East Indian arrowroot usually requires a high amount of organic matter in soil. Well-decomposed farmyard manure of 10–15 t/ha is incorporated in the soil at the time of field preparation. Besides, nitrogen is supplied @ 100 kg, phosphorus 70 kg and potash 90 kg/ha. A full dose of phosphorus and potash along with one-third dose of nitrogen is applied as basal application at the time of planting. The remaining two-thirds of nitrogen is applied 30 and 60 days after planting at the time of first and second intercultural operation.

IRRIGATION

Generally, East Indian arrowroot does not require much irrigation, as the crop is drought tolerant. Pre-planting irrigation in pits where the tubers are to be planted is essential for better sprouting of tubers. The first irrigation is given after completion of sprouting. If the tubers are planted in dry pits, the soil of the pits should be kept moist until the emergence of sprouts. Thereafter, irrigation is supplied once in 15–20 days, depending on soil and weather conditions. However, care should be taken to avoid water stagnation in the pits, as it may cause decaying of tubers. Irrigation is stopped a few days before harvesting the tubers; however, a light irrigation 2–3 days prior to harvesting makes the digging operation easy.

INTERCULTURAL OPERATIONS

HOEING AND WEEDING

The weed species viz., *Ipomea* spp., *Wedelia biflora*, *Vigna marina* and *Polypodium scolopendria* compete with crop plants for nutrients, moisture, space and light. Light hoeing and earthing-up twice are the important practices in Polynesian arrowroot to keep the field weed free, especially during initial stages of plant growth. First hoeing is done a week after sprouting of tubers and the second 30 days after first hoeing. The most critical period of competition is August to October since during this period, the crop plants are completely covered by *Vigna* and *Wedelia* species. For keeping the field weed free in the next season, the weeds are completely burnt including the above-ground portion of the crop plant at the end of crop maturity or the weeds are pulled out by hands.

MULCHING

Mulching with dry grass, compost, or black polyethylene sheet immediately after planting is beneficial for increasing the tubers yield, as mulching ensures quick and uniform sprouting and provides protection against excessive moisture loss from the soil. It provides the most favourable conditions to the plants for the development of tubers.

STAKING

The leaves of staked plants receive more sunlight due to full exposure and therefore synthesize more assimilates through photosynthesis. As a result, the staked plants give a yield higher than the non-staked plants. Usually, bamboo sticks are used to stake the East Indian arrowroot plants.

HARVESTING

The crop usually takes approximately 8–10 months after planting. Yellowing of leaves and the dying back of plants are the indices used in East Indian arrowroot to judge the right stage of harvesting. The crop normally matures from the end of November to the start of February. Only the large tubers of female plants are dug up, and the smaller tubers are left behind in the field to produce new seedling plants. Before digging of tubers, the above-ground vegetation of the plant is burnt for clean harvesting, however, this practice makes the identification of male and female plants impossible since the tubers are to be harvested only from the male plants.

POST-HARVEST MANAGEMENT

The tubers of East Indian arrowroot are converted into flour to be used for preparing a variety of puddings. For this purpose, the tubers are peeled and grated, and the resultant pulp is washed in water several times to remove bitterness and then sieved. The aqueous starch solution is collected and the starch grains are allowed to settle down. Thereafter, the starch is separated and dried in the sun for further use.

INSECT-PESTS

The East Indian arrowroot crop is attacked by the following insect-pests:

APHIDS (*Aphis* sp.)

The tiny nymphs and adults of this polyphagus insect damage the plants by sucking cell sap from the surfaces of both tender leaves and shoots, which causes curling and ultimately drying of leaves. Aphids excrete a honeydew-like sugary substance, on which the black sooty mould grows, which eventually obstructs the photosynthetic activity of the plant and thereby reduces the tuber yield.

Control
- Remove and destroy the infested twigs and leaves in early stages.
- Wash the plant with water using a strong jet pressure pump.
- Fix yellow sticky traps in the field @ 300–400 stickers per hectare.
- Spray the crop with 0.05% Malathion or dimethoate two or three times at 10-day intervals.

MITES (*Tetranychus* sp.)

Occasionally, mites attack the plant leaves, however, they do not cause great damage. These tiny acarids are found in large numbers on the lower surface of the leaves under a protective web. Both nymphs and adults suck cell sap from the tender leaves and make the plant weak. The severely affected leaves show a brown or reddish brown colouration.

Control
- Dust the crop with fine sulfur @ 20 kg/ha.
- Spray the crop with 0.3% Phosalone or Dicofol.

SCALES (*Aspidiella* sp.)

Nymphs and adult females of this insect infest the leaf petiole and tubers. In severe cases of infestation, the attacked plants show drying due to continuous sucking of cell sap. The infested tubers start shrivelling, affecting the quality, viability and marketability of tubers.

Control
- Use pest-free tubers for planting purpose.
- Dip the seed tubers before planting in 0.1% monocrotophos.

ROOT-KNOT NEMATODE (*Meloidogyne* sp.)

The soil-borne eelworms are very destructive tuber parasites. The nematodes attack the plant roots, where they produce swelling or galls. As a result, malformation and reduction in tuber size occur. The infested plants become stunted and yellow in colour.

Control
- Plough the field deep in hot summer months to expose the soil to high temperature for killing the nematodes.
- Use nematode-free tubers for propagation to reduce field infestation effectively.
- Incorporate *neem* or castor cake or leaves powder @ 400 g/m^2 to reduce the infestation.

DISEASES

Many fungal diseases attack the East Indian arrowroot at various stages of plant growth. Among the fungal diseases, anthracnose and cercospora leaf spot are the most destructive ones.

ANTHRACNOSE (*Colletotrichum* sp.)

Light brown pinhead spots appear on leaves, which later turn to deep brown. Under humid conditions, the spots are dotted with pink conidia. The spots coalesce and form large lesions, causing shrivelling and scorching of foliage. In advanced stages, the leaves and stems completely dry up and the plant fails to form a tuber. Infected plant debris serves as a primary source of inoculums and secondary spread takes place through air-borne dispersal of conidia. The disease incidence is rapid under high humidity and warm conditions, particularly during a period of heavy summer rains.

Control
- Remove of plant debris from the field and follow with clean cultivation.
- Spray the crop with 0.2% Dithane M-45 or 0.1% Bavistin three or four times at 10-day intervals.

Cercospora Leaf Spot (*Cercospora taccae*)

Brown spots surrounded by yellow rings appear on leaves. The severely infected leaves as a result drop off prematurely. The fungus is caused by air-borne spores.

Control
- Spray the crop with 0.1% Bavistin or Benlate two or three times at 12- to 15-day intervals, starting from 45 days after planting.

■■■

66 Chekkurmanis

M.K. Rana and A.K. Pandav

Botanical Name	: *Sauropus androgynus* (L.) Merr.
	Syn. *S. albicans* Blume.
	S. gardnerianes Wight.
	S. zeylanicus Weight.
	S. indicus Weight.
	S. sumatrans Mig.
Family	: Euphorbiaceae

ORIGIN AND DISTRIBUTION

Chekkurmanis originated from hot and humid lowland forest of Borneo. The Indo-Burma region is considered its primary centre of origin. It was introduced to India in the 1950s and then found in cultivated and wild state in Sri Lanka, Southern China and throughout Southeast Asia. In India, it is found in Sikkim, Himalayas, Khasi, Abor and Arka hills at 1200 m elevation and in western Ghats of Kerala from Wayanad northwards at altitudes of 300 to 1200 m. It grows wild in the evergreen forests of western Ghat and southern parts of Kerala. It was first located in Malabar in 1953, and during 1955–56, it was introduced to the Agricultural Research Station of Tamil Nadu Agricultural University, where it became popular as *Thavarai Murangai*.

INTRODUCTION

Chekkurmanis, also known as *katuk*, star gooseberry, or sweet leaf, is a shrub grown in some tropical regions as a leafy vegetable. In Malayalam, it is called *Madura Keera*. It is most popular in South and Southeast Asia. It is one of the important perennial leafy vegetables grown in southern India. The major chekkurmanis growing states in India are Tamil Nadu, Kerala and Sikkim. Apart from India, the crop is also grown in Indonesia, Malaysia and Singapore. In India, chekkurmanis is a minor vegetable crop. Hence, the statistics regarding area, production and productivity are not available in records. It is popularly known as a *multivitamin- and multimineral-packed* leafy green vegetable due to its high nutritive value and high vitamin content.

COMPOSITION AND USES

COMPOSITION

Chekkurmanis leaves provide a good amount of carbohydrates, protein, fat, calcium, phosphorus and iron. It also contains an appreciable quantity of carotene, thiamine, riboflavin and vitamin C. The nutrient content of leaves is usually higher in more mature leaves than in newly formed leaves. The nutritional composition of chekkurmanis leaves is given in Table 66.1.

TABLE 66.1

Nutritional Composition of Chekkurmanis Leaves (per 100 g Edible Portion)

Constituents	Contents	Constituents	Contents
Water (g)	73.6	Phosphorus (mg)	200
Carbohydrates (g)	11.6	Vitamin A (IU)	9510
Protein (g)	6.8	Thiamine (mg)	0.48
Fat (g)	3.2	Riboflavin (mg)	0.32
Dietary fibre (g)	1.4	Niacin (mg)	2.6
Minerals (g)	3.4	Vitamin C (mg)	247
Calcium (mg)	570	Energy (kcal)	103
Iron (mg)	28	–	–

Uses

The young shoots, fresh and tender leaves, often together with flowers and fruits of chekkurmanis, are consumed raw in salads as well as cooked to add to rice, egg dishes, soups and casseroles. They are also used for the preparation of several dishes like *idlies* and *dosa*. The leaves have a pleasant taste, similar to fresh garden peas and slightly nutty. The leaves retain their dark green colour and firm texture on cooking and are served in restaurants as *Sayor Manis*. They are also used for giving a light green colour to pastry and fermented rice in the Dutch East Indies and Java for the preparation of soup. The green colour is obtained by rubbing the leaves and pressing them out. In Malaysia, the top 15 cm stem tips are harvested and sold to restaurants and even exported to other Far Eastern markets. The small white or purplish fruits are sometimes comfitted into sweetmeat. The leaves are also used as cattle and poultry feed. It can easily be cultivated in kitchen gardens, thus, it is worth including this plant in home gardens and institutional gardens.

Medicinal Uses

The chekkurmanis possesses several medicinal properties and it is useful for women after childbirth to stimulate milk production and recovery of the womb. In traditional medicine, a decoction of root is used against fever and other urinary ailments. The juice of leaves is used against eye troubles. It has huge potential for areas of the tropics where vitamin A deficiency is a problem, especially among malnourished children. It can overcome many nutritional disorders. It is an excellent source of vitamin A, B, C, carotenoids and phytochemicals, which act as an antioxidant. It used as an effective medicinal herb in the treatment of diabetes, cancer, inflammation, microbial infection, cholesterol and allergy due to its antioxidant effect.

BOTANY AND FLORAL BIOLOGY

Chekkurmanis is a slow-growing glabrous perennial shrub with 2–3.5 m height, terete main branches and thin lateral branchlets. Leaves are variable in size (2.3–7.5 × 1.2–3.0 cm), biseriate, sessile, alternate, petiole 0.2–0.4 cm long, membranous and very short-stalked, lamina ovate or oblong, obtuse at both ends and entirely glaucous and glabrous. Usually, the plants bear monoecious, minute, pedicelled and more or less intensely red-coloured flowers in clusters. The inflorescence is an axillary fascicle, dense unbranched and first producing one or two female flowers, afterwards several male ones. The flowers are without corolla but with more or less 40 intensely red-coloured persistent calyxes. The calyxes of male flowers are disc form, fleshy, more or less sinute or six-lobed, green with a red centre or entirely red, 0.6–1.0 cm in diameter, stamen three-connate into a trifid column, and the calyxes of female flowers are six clefts more than halfway down, fleshy, dark red or yellowish with dark red dots, 0.65–0.8 cm broad and the segments bise-riate, the outer ones patent. The ovary is shortly turbinate, trigonous, green and three-celled and the stigmas are three, thickened at the base. The fruits are small, resting on the dark red calyxes that are spherical in shape/depressed globose, faintly six angular capsule, dull white or slightly tinged with rose/white or purplish in colour and about 1.2–1.7 cm in diameter.

The fruit, dehiscing with valves, when ripe dehisces longitudinally at maturity into distinct equal parts and contains viable seeds. The crop is highly cross pollinated and *entomophilous*, the reason for cross pollination being monoecious, along with the protogynous nature of the flower. The genus *Sauropus* consists of several other species, of them, some species are related to *Sauropus and rogynous*. The four related species, *i.e.*, *Sauropus retroversus* Wight, *S. assimilis* Thw, *S. rigidus* Thw and *S. quadraangularis* Muell, have also been identified.

CLIMATIC REQUIREMENT

Being a perennial crop, it thrives best under high humid conditions of tropical regions with equable temperature, *i.e.*, neither very high (40°C) nor too low (freezing or near freezing). It can be grown luxuriantly at lower elevations of 500 m above mean sea level though it is found growing at higher elevations, up to 1,200 m above mean sea level, too.

Although chekkurmanis is moderately drought-tolerant, it is not an appropriate choice for an area with an extended dry season of 7–8 months, unless irrigation facilities are available or heavy mulch-ing is practiced. It grows profusely under shade conditions and produces broad leaves, whereas, under full sunlight, it produces narrow leaves.

SOIL REQUIREMENT

The chekkurmanis can be grown on a wide range of soils. However, the growth and yield are greater when it is cultivated in organic matter–rich, well-drained sandy loam or semi-laterite soils with fairly good drainage facilities. It is often grown as a live fence or hedge around a kitchen garden. It will not grow as vigorously in areas with poor or depleted soils, unless a considerable quantity of organic matter or other nutrients are incorporated in soil.

CULTIVATED VARIETIES

Earlier no attention had been given to this crop, and no germplasm has been collected on this crop, but now, the breeding work is in progress, thus, no distinct variety is available to date. Generally, broad-leaved and narrow-leaved chekkurmanis are sold in the market. However, this difference in leaf shape is not genetically governed; it is due to shading effect. When the plants are grown under shade, they produce broad leaves, and under full sunlight, they produce small and narrow leaves.

SEED RATE

The seed rate undoubtedly varies with soil fertility, germination percentage, irrigation facilities and climatic conditions of the growing region. Under normal conditions, about one lakh cuttings of chekkurmanis are required for planting a hectare land area.

SOWING/PLANTING TIME

The sowing/planting season is usually determined by the amount of precipitation occurred at a particular place and the availability of irrigation facilities. The cuttings are usually planted in shallow furrows at least 15 days earlier than the onset of monsoon during, *i.e.*, April–May or May–June in different zones.

PROPAGATION METHODS

Chekkurmanis is propagated by direct seeding or by using 6- to 12-months-old herbaceous stem cuttings. Both the seeds and the cuttings are short-lived, thus, both should be planted as soon as possible after collection.

Direct seeding: This method is also preferred when plenty of seed is available but labour is limited. Seeds are viable for only 3–4 months when kept in dry and cool places. Even when the seeds are viable, the germination will probably be 50% at best. Seed should be collected just as the seedpods begin to crack. Germination will be higher if the seed coat is pulled off just prior to planting. Although the seeds germinate in 2–4 weeks, they continue to germinate up to 5 months. Plants propagated by this method take more time to flower than planting by stem cuttings.

Planting by cuttings: Propagating the crop through soft wood and semi-hard wood cuttings is much easier and more common since the crop raised through seeds requires more time to flourish. The cuttings of 20–30 cm length are taken from healthy mature branches with five or six nodes and planted under shade in polybags of 15 × 10 cm size containing a mixture of manure and garden soil. Before planting the cuttings, all the side branches and leaves from the cuttings are removed, and the bottom end of the cuttings should be within 3 cm of a node. The cuttings are usually planted at a line-to-line distance of 60 cm and plant-to-plant distance of 10–15 cm, keeping a depth 12–18 cm, with a single node exposed to the atmosphere for sprouting. It takes 20–25 days for rooting. After rooting, the plants are gradually exposed to full sunlight to make them hard.

MICRO-PROPAGATION

For chekkurmanis, micro-propagation has also been developed using uninodal explants in Murashige and Skoog (MS) medium supplemented with various concentrations of cytokinins, *i.e.*, benzyl adenine and kinetin. Shoot induction is observed in 0.1 mg/litre benzyl adenine and 0.1 mg/litre kinetin after 25 days, and a greater number of shoots is achieved in a shoot induction medium. Rooting is induced from shoots in an MS medium supplemented with various concentrations of Indol-3-Butyric Acid (IBA) and Naphthalene Acetic Acid (NAA). All the shoots induce roots in 0.5 mg/litre indol-3-butyric acid and 0.2 mg/litre naphthalene acetic acid medium after 25 days. Rooted plants are transferred to soil and successfully acclimatized.

NUTRITIONAL REQUIREMENT

If the crop is raised in pits of 30 cm^3 size, 5 kg of well-decomposed farmyard manure is added in addition to 25 g each of urea, superphosphate and muriate of potash. Liquid manure, which may be obtained locally by placing cow dung in a gunny bag in an earthen or cemented vessel of water and soaked for 48 hours, then drained and diluted three or four times, is applied at fortnightly intervals for its growth. Application of 30 g of 7:10:5 NPK mixture per plant supplemented with 1% urea spray after each clipping enhances the leaf yield significantly. However, it is better to avoid the use of chemical fertilizers and to instead apply well-rotten farmyard manure or compost at the rate of 5 t/ha. When the plants attain a height of about one metre, they are nipped to develop lateral branches. Afterward, clipping of tender leaves and shoots is done at monthly intervals. The manure and fertilizers are applied regularly to develop lateral branches.

IRRIGATION REQUIREMENT

Although chekkurmanis is a drought-resistant crop, irrigation during summer increases the number of cuttings. After every clipping, it is essential to apply irrigation regularly during rainless periods so that the new growth may emerge.

INTERCULTURAL OPERATIONS

Chekkurmanis possesses medicinal properties, hence, weeds do not pose much problem in this crop, even though removal of weeds periodically in the early stages of the crop is essential to reduce competition with the crop for nutrients, light, moisture and space, resulting in better growth of the crop.

PRUNING

Pruning of chekkurmanis plant is usually undesirable since it increases the cost of cultivation, but this operation is essential for effective cultivation because if the plant is left unpruned, it grows so tall that it falls over to the ground. Chekkurmanis should be pruned to maintain a height of the shrub of about one metre. Pruning stimulates growth of new shoots and encourages constant production of new foliage. The plant if left untrimmed in its early stage, it reaches the size of a small tree. It can be trained as a hedge or even on bowers or on trellises. In Malaysia, production of new growth of stem tips is forced through appropriate pruning, fertilization, irrigation and use of shading cloth.

USE OF PLANT GROWTH REGULATORS

In chekkurmanis, the growth substances are mainly used to induce roots in cuttings. It is observed that dipping the bottom end of cuttings in 50 ppm solution of IAA or IBA hastens rooting in cuttings.

HARVESTING

Chekkurmanis crop becomes ready for clipping succulent leaves 3–4 months after planting and when the plants attain a height of 60–90 cm. The subsequent clippings are done at fortnight intervals. Another harvest is possible, if the plants are manured and irrigated timely. Plants are usually trimmed to a height of 1–1.5 metre to facilitate easy harvest and to give a pleasant look to the garden.

YIELD

The average yield of chekkurmanis varies from 300 to 500 q/ha, depending upon the soil fertility, irrigation, manuring and climatic conditions. A single plant gives a yield of fresh green biomass about 1–3 kg per plant per year.

INSECT-PESTS AND DISEASES

Chekkurmanis is not so seriously damaged by insect-pests and diseases but some insects like aphids, scales, Chinese rose beetle and slugs sometimes do considerable damage to the crop. The most destructive insect-pests are given as follows:

APHIDS (*Aphis malvae* koch)

These are tiny sucking-type green-colour insects that attack tender growing leaves. They suck cell sap from the tender foliage due to which the plant growth is restricted. Plants become weak and the quality of leaves is reduced; as a result they become unfit for consumption after infestation.

Control
- Cut and destroy all the infested leaves during the early stage of an attack.
- Spray Malathion 50 EC at 0.1% or Metasystox 25 EC at 0.025%.

CHINESE ROSE BEETLE (*Adoretus sinicus*)

The Chinese rose beetle feeds at night and attacks the leaves. The adult rose beetle is reddish brown and about 1.25 cm long. Beetles come out at dusk and feed for a couple of hours before retreating to the leafy humus at the base of nearby plants.

Control

- Chinese rose beetle is attracted to dim light and repelled by bright light so a flashlight is provided in the field.
- Pick the beetles by hand for few weeks and drop them into soapy water.

SNAILS (*Bensonia monticola*) and Slugs (*Anabenus schlagintweiti*)

These pests feed on the leaves and tender parts of the crops and damage the growing tip of plants. Slugs that girdle the stems and kill the plant can be a problem in rainy season.

Control

- Use poison bait of metaldehyde and bran (1:25) in 12 litres of water to trap the insects.
- Alum solution @ 2% also acts as a repellant to snails and slugs.

DISEASES

Chekkurmanis is not attacked by any pathogen to cause seviour disease.

■■■

67 Tapioca (Cassava)

T. Saraswathi and L. Pugalendhi

Botanical Name : *Manihot esculenta* Crantz

Family : Euphorbiaceae

Chromosome Number : 2n = 36

ORIGIN AND DISTRIBUTION

Cassava or tapioca has many Spanish names like manioc or yucca. It is believed to be a native of Brazil. Tapioca was introduced into India by the Portuguese in the 17th century. Cassava is mainly cultivated in Brazil, Nigeria and Ghana. *Esculenta* is the only species of *Manihot* under cultivation and not found in a wild state.

INTRODUCTION

Globally cassava is grown in an area of 18.51 million hectares producing 202.65 million tonnes of tubers with a productivity of 10.95 t/ha. It is grown in 102 countries. The African continent is the foremost producer, having 66.21% of the total cassava area and producing 53.37% of the world cassava, as it is a staple food in many of the African countries. Even though the area of production is greater in Africa, its production is low due to lower productivity (8.82 t/ha) than the world average.

Among all the cassava-growing countries in the world, Nigeria has the largest area under cassava (22.25%), with an annual output of 38.18 million tonnes. Indonesia, Thailand, Vietnam and India are the major countries growing cassava in Asia. India acquires significance in the global cassava scenario due to its highest productivity in the world (27.92 t/ha), cultivated in an area of 240,000 hectares producing 6.7 million tones.

In India, 60–70% of the total cassava production is used commercially to produce sago, starch, dried chips, flour, etc. Human consumption of cassava is common in Kerala, Tamil Nadu and in northeastern states like Assam, Meghalaya, Gujarat, Maharashtra and West Bengal as sago.

Cassava is a crop of the tropics. Although it is cultivated in India in 13 states, it is concentrated in the southern peninsular region and to a certain extent in the northeast region of the country. The crop is cultivated in the southern states of Kerala, Tamil Nadu and Andhra Pradesh owing to the favourable climate and efficient utilization. Among the southern states, Kerala has a major share in area (1.04 lakh hectare) under cassava (43.33% of the cassava area in India) and the maximum production goes for human consumption. Tamil Nadu has an area of 95,000 ha (40% of the total area under cassava in India) and 60% of cassava production is utilized industrially to produce starch, sago and other value-added products. Andhra Pradesh has 8.1% of the cassava area, and it is utilized exclusively for industrial purposes. The remaining area of cassava is concentrated in the northeast region of the country. The nutritional composition of tapioca is given in Table 67.1.

TABLE 67.1

Nutritional Composition of Tapioca (per 100 g Edible Portion)

Constituents	Contents	Constituents	Contents
Water (g)	60	Beta-carotene (µg)	8
Carbohydrates (g)	38	Vitamin A (IU)	13
Sugar (g)	1.7	Thiamine (mg)	0.09
Protein (g)	1.4	Riboflavin (mg)	0.05
Fat (g)	0.28	Niacin (mg)	0.85
Dietary fibre (g)	1.8	Pantothenic acid (mg)	0.11
Calcium (mg)	16	Pyridoxine (mg)	0.09
Phosphorus (mg)	27	Folic acid (µg)	27
Potassium (mg)	271	Vitamin C (mg)	20.6
Magnesium (mg)	21	Vitamin E (mg)	0.19
Sodium (mg)	14	Vitamin K (µg)	1.9
Iron (mg)	0.27	Saturated fatty acids (g)	0.07
Zinc (mg)	0.34	Monounsaturated fatty acids (g)	0.08
Copper (mg)	0.10	Polyunsaturated fatty acids (g)	0.05
Manganese (mg)	0.38	Energy (kJ)	670
Selenium (µg)	0.7	–	–

USES

The cassava tuber is boiled and eaten with or without sugar. The tuber is a good source of gapleck and sago. From gapleck, starch is manufactured, which is then made into flour and used in the preparation of biscuits, cookies and bread. Sago is used in the preparation of *guinata-an*, a native delicacy. The cassava starch is also made into paste, which serves as glue.

MEDICINAL USES

A head compress of pounded cassava leaves is applied for relief of headache or fever. Solution formed by boiling the bark of cassava stems helps in expelling intestinal worms and serves as medicine for rheumatism. Pounded cassava tuber is applied on ulcerated wounds as medicine. The juice from the pounded tuber is used as an antiseptic. Cassava starch is placed on affected parts to relieve ill effects of rashes and prickly heat. In folk medicine, the cassava plant is promoted for treating snakebites, boils, diarrhoea, flu, hernia, inflammation, conjunctivitis, sores and several other problems, including cancer.

Cassava plant can produce the poisonous substance cyanide as a way to fend off animals trying to eat them. Chewing the plant causes it to release an enzyme called linamarase, which converts a compound in the plant called linamarin into cyanide. However, this ability may be useful as a form of gene therapy. First, the gene for linamarase could be selectively put into cancer cells. If linamarin were then introduced into the body, cancer cells would break it down and release cyanide only in the area around the cancer cells, killing them. Since normal cells would not have the linamarase gene and would not be able to convert linamarin into cyanide, they would not be affected.

MORPHOLOGICAL CHARACTERS

PLANT TYPE

Cassava, generally cultivated as an annual crop, is a perennial with indeterminate growth habit because of the overlapping morphological characters and other biochemical characteristics. Plant type is the most important in the selection and grouping of clones in the germplasm. Based on the type of branching, the clones could be categorized into erect and spreading, and the erect types could be further divided into branches and unbranched types.

FLOWERING

In cassava, flowering occurs only in branched plants and profuse flowering in spreading and branching plants. Since its tuber is economic produce that is a vegetative part, earliness cannot be correlated with flowering. However, flowering and fertility are the most important parameters when adopted in hybridization or in gene pool development. Cassava is generally female-fertile, however, the sterile male is widely prevalent.

FLORAL MORPHOLOGY AND BIOLOGY

Flowering in cassava is frequent and regular in some cultivars, but in other cultivars, it is rare or non-existent. Flowers are borne on terminal panicles, with the axis of the branch being continuous with that of the panicle inflorescence. It is monoecious and each inflorescence has male flowers at the terminal, while the female flowers occur closer to the base. Female flowers are usually larger than the male ones. Each flower whether male or female has five united sepals with a red or yellow tinge. There are no petals. The male flower has 10 stamens arranged in two whorls of five each. The anthers are small with free filaments. Each flower has an orange-coloured nectar-bearing gland inside. The female flower has an ovary mounted on a ten-lobed glandular disc. The ovary has three locules and six ridges with a length of 3–4 mm. Each locule contains a single ovule. The stigma has three lobes, which unite to form the single style. In each inflorescence, the female flowers open first, while the male flowers do not open until about 7–8 days after females open. Cross-pollination is therefore the rule and insects are the main pollinating agents. After pollination and subsequent fertilization, the ovary develops into a young fruit. It takes 3 to 5 months after pollination for the fruit to mature.

STIGMA RECEPTIVITY AND POLLINATION

In cassava, generally anthesis takes place at about noon, even though a few clones have been noticed to show anthesis by 11:00 a.m. and weather conditions are also known to play some role in anthesis. The presence of nectar in early morning, the ability to attract bees even at 10:00 a.m. if the forced-open flowers are left as such and the liberation of pollen well before anthesis in a number of clones indicate the possible presence of stigma receptivity much earlier to anthesis. The capsule set could be achieved at 6:00 a.m. on the day of anthesis and seed set was also high up to 2:30 p.m. There was a significant reduction in fruit set after 3:30 p.m. The seed set tapered to 1.5% at 5:00 p.m. The observations on fruit set and seed set indicated that the stigma is highly receptive from early morning and the intensity remains almost at the same level up to 2:30 p.m., and thereafter, it starts declining and only trace receptivity is observed after 5:00 p.m.

In cassava, when pollination is affected before anthesis, it can easily be done by exerting pressure with a needle on the perianth of female flowers. In male flowers, a slight pressure while parting the perianth is sufficient to release the pollen and a mere touching the stigma with male flowers is adequate to achieve successful pollination. By adopting this method, 200 flowers can easily be pollinated in 1.30 hours by two people.

The mature fruit is a capsule and globular in shape with a diameter of 1 to 1.5 cm. It has six narrow longitudinal wings. The endocarp is woody and has three seeds. When the fruit is mature and dry, the woody endocarp splits explosively to release and disperse the seeds.

The cassava seed is exactly similar to castor seed. It is ellipsoidal and 1–1.5 cm long. It has brittle testa, which is grey, mottled with dark blotches. The caruncle is large at the micropylar end of the seed. The germination of the seed can be improved by soaking in 0.5% KNO_3 solution for 24 hours.

TUBER

Cassava tuber is a root tuber. Tuber formation in cassava is under photoperiodic control. Under short day conditions, tuberization takes place readily but when the day-length is greater than 10–12 hours, tuberization is delayed and subsequent yield is reduced. For this reason, cassava is most productive between latitudes 15°N and 15°S. At higher latitudes, the best growing season (summer) corresponds to the time of long photoperiods, which do not stimulate tuberization, and hence, yield is poor. When cassava is grown at altitudes above 1000 m, growth is slow and yield is very poor.

TUBER FORMATION

Tuber formation commences from the eighth week after planting. Tuberization process involves essentially the onset of secondary thickening in some of the adventitious roots, which were previously fibrous in nature. Usually, the secondary thickening commences at the proximal part of the root and progresses towards the distal portion, thus, the tuber is fattest at the proximal end, in which, the thickening had occurred for a longer time, and it tapers slowly towards the distal end.

STARCH

Cassava produces the highest dry matter among the root and tuber crops, and in germplasm, it ranges from 24% to 52% and dry matter is directly related to starch content. The starch content varied from 14% to 36% (on a fresh weight basis).

HYDROCYANIC ACID

The toxic principle in cassava is hydrocyanic acid (HCN) present in all plant parts. The HCN content varies in clones, and hence, they are often classified as *bitter* and *sweet* depending on the amount of cyanoglucosides present in the flesh of tubers. The bitter clones are more prevalent among the exotics and record about 7.5% (above 150 ppm), while it was only 0.9% among the indigenous. Though there is a correlation between high HCN and bitterness, a number of clones are found to be non-bitter even though they contained 50–100 ppm, suggesting that the amount of sugar might well have some effect in determining the bitterness of tuber.

DROUGHT-TOLERANT MECHANISM

In cassava, the mechanism of drought tolerance varies. As the moisture stress increases and the atmospheric temperature raises (especially during the afternoons when sunlight is strong), the leaves will be specifically drooping in such a way that leaf lamina of both upper and lower surfaces are not perpendicular to sunlight. As the drought is more pronounced, unlike other plants, cassava does not increase the synthesis of abscisic acid. The stomata are also not completely closed. It adapts in such a way that it partially closes the stomata so that the exit of moisture or water vapour is almost completely arrested, at the same time the entry of carbon dioxide from atmosphere is not interfered with. The net result is increased photosynthetic efficiency, *viz.* high carbon dioxide intake and low transpiration. Non-accumulation of abscisic acid in the petiole results in retention of the lamina for longer duration of the plant's life cycle.

CLIMATIC REQUIREMENT

Cassava is a plant of lowland tropics. It does the best in a warm and moist climate where mean temperature ranges from 25°C to 29°C. It exhibits poor growth under cold climates, especially at temperatures below 10°C when the growth of the plant is totally arrested. It cannot withstand frost at any time during its active growth period. This crop does the best when the rainfall is 1000–1500 mm per year and well distributed.

CULTIVATED VARIETIES

Cassava breeding research is in progress at Central Tuber Crops Research Institute (CTCRI), Thiruvananthapuram, and the State Agricultural Universities of Tamil Nadu and Kerala. Four varieties *viz.*, CO 1, CO 2, CO 3 and CO(Tp)4 were released for general cultivation from Tamil Nadu Agricultural University. The Horticulture Department of Tamil Nadu released one variety, MVD-1, in 1993. The Kerala Agricultural University has released two short-duration varieties, *viz.* Nidhi in 1993 and KMC-1 (Vellayani Hraswa) in 1998. The following are the varieties released by CTCRI, Thiruvananthapuram.

H 97

It is a hybrid between a local variety Manjavella and a Brazilian seedling selection. The tubers are conical and stout, yielding 250–350 q/ha. The tuber flesh is white with 27–29% starch content and matures in 10 months.

H 165

This is a comparatively early maturing non-branching hybrid with mature leaves showing drooping nature between two indigenous cultivars, *viz.* Chadayamangalam Vella and a clone similar to Kalikalan. The tubers are relatively short and conical with cream-coloured rind, yielding 330–380 q/ha in a crop duration of 8–9 months. It is predominantly adapted to Tamil Nadu and Andhra Pradesh for its industrial potential.

H 226

A hybrid between a local cultivar Ethakkakaruppan and the Malayan introduction M_4, it is susceptible to cassava mosaic disease (CMD). Plants are tall, occasionally branched type, and their leaves have a characteristic green colour. The tuber rind is light purple and the skin is cream with purple patches. The tuber yield is 300–350 q/ha in a crop duration of 10 months. It is the predominant variety cultivated in Tamil Nadu and Andhra Pradesh for its industrial potential.

Sree Visakham (H 1687)

It is a hybrid between an indigenous accession and a Madagascar variety S-2312. Visakham is predominantly a non-branching and tall type, having compact tubers with yellow flesh due to high carotene content (466 IU/100 g). The crop duration is 10 months and the tuber yield is 350–380 q/ha with 25–27% starch.

Sree Sahya (H 2304)

It is a hybrid involving five parents of which two are exotic and three indigenous. The tubers are long-necked with light brown skin, cream-coloured rind and white flesh. Tuber yield is 350–400 q/ha.

Sree Prakash

It is a relatively short-stature plant, generally non-branching and with high leaf retention. The tubers are medium-sized and -necked and the tuber yield is 350-400 q/ha. The duration is 7–8 months. Tubers possess good culinary quality and gives a starch content of 29–31%.

Sree Rekha (TCH 1)

An erect branching top cross hybrid of cassava, *viz.* TCH 1, with excellent cooking quality, was released for general cultivation in Kerala under the name Sree Rekha. The average yield and starch content are 480 q/ha and 28%, respectively.

Sree Harsha

This is a triploid clone developed by crossing a diploid with an induced tetraploid clone of *Sree Sahya*. Plants are tall, stout and erect non-branching with tubers of good cooking quality and high starch content (38–41%). Its average yield is 350-400 q/ha in a crop duration of 7–8 months.

Sree Prabha (TCH 2)

A semi-spreading top cross hybrid of cassava, *viz.* TCH 2, with good taste and quality, was released for general cultivation in Kerala under the name Sree Prabha. The average yield and starch content are 350–400 q/ha and 26%, respectively, in a crop duration of 10 months.

Sree Jaya

This medium-tall plant with a short-duration clonal selection is suitable for low land cultivation as a rotation crop in paddy-based inter-cropping systems. It produces conical tubers with brown skin and purple rinds and good cooking quality. Its average tuber yield is 260–300 q/ha in a crop duration of 6 months. It is susceptible to CMD (cassava mosaic disease).

Sree Vijaya

A short-duration early maturing variety is suitable for low land cultivation as a rotation crop in a paddy-based inter-cropping system. Plants are erect branching type. Tubers are with cream-coloured rind and light yellow flesh due to high carotene content. Its average yield is 250–280 q/ha with starch content of 27–30% and a crop duration of 6–7 months. It is susceptible to mites and scale insects.

Sree Athulya

It is a triploid variety with a tall, erect, branching stout stem with a tuber yield of 380 q/ha and starch content of 30%.

Sree Apoorva

It is a triploid variety. The matured stem will be yellowish brown in colour. The average tuber yield is 380 q/ha with a starch content of 29%.

SREE PADMANABHA

It has been introduced from IITA, Nigeria, through CIAT, Cali, Colombia. It is a late-branching variety suitable for irrigated conditions. It is resistant to CMD. Its yield potential is 380 q/ha with a starch content of 25%.

NIDHI

A short-duration variety developed at Kerala Agricultural University, Kerala, it is tolerant to CMD. Its plant stem is greyish white, its petiole white with a red shade and its skin light pink. Its average yield is 251 q/ha in a crop duration of 5–6 months.

VARIETIES RELEASED BY TAMIL NADU AGRICULTURAL UNIVERSITY, COIMBATORE

CO 1

It is a selection from the Tiruchirapalli (Tamil Nadu) collection. Tubers are branching type, whitish brown-skinned and white flesh with a starch content of 35%. Tuber yield is 299.7 q/ha in a crop duration of 8½ to 9 months.

CO 2

It is a selection from the Thanjavur (Tamil Nadu) collection. Tubers are branching type, brown with creamy white flesh and suitable for consumption and the starch industry. Starch content is 35%. The plants flowers profusely. Yield potential is 353.7 q/ha in a crop duration of 8–8½ months. It has a field tolerance to CMD.

CO 3

It is a selection from seedling progenies of a collection from IITA, Ibadan, Nigeria. Tubers are branching type and dark brown with white flesh and starch content is 35.6%. Its tuber yield is 365.0 q/ha in a crop duration of 8–8½ months.

CO (TP) 4

It is a clonal selection from the seedling progenies of the hybrid SM 1679 (CM 2766-5 x CM 4843) obtained from CIAT, Colombia, South America. Plants are erect with top-branching habit. Tubers are dark brown-skinned with a cream rind, white flesh and a starch content of 40%. Tuber yield is 506 q/ha in a crop duration of 10 months.

PLANTING TIME

Generally, cassava planting coincides with the beginning of rainy season. Since it is not strictly a wet season crop, under irrigated conditions planting may be done at any time. The results of a month-wise planting experiment at CTCRI showed that the best time of cassava planting is April–May, i.e., with the onset of south-west monsoon. Planting in August–September also produces higher yield compared to other timings. An experiment on planting season trial conducted in Tamil Nadu revealed that planting in September results in higher tuber yield where the onset of monsoon is in October–November. June planting is best in Andhra Pradesh if grown as rain-fed and if irrigation could be provided. March planting can also be taken up under Hyderabad conditions. In northeastern states of India, where pronounced winter climate is experienced, cassava planting may be done from February to April after cold weather.

PLANTING MATERIAL AND PLANTING

Traditionally, cassava is propagated vegetatively through stem cuttings called setts taken from the previous season's crop and so the stems are often to be stored for 2 to 3 months. Studies at CTCRI showed that sprouting is observed better when stakes are derived from stems stored with leaves intact. Stems kept in a vertical position give better sprouting as compared to those in a horizontal position. Further, storing stems beyond 60 days results in loss of viability. The setts obtained from the middle portion of stem are ideal for planting. The hard basal portions of the stem are not suitable for planting as they have more lignified tissues, which may not support the emerging buds. The top tender portions of stems desiccates quickly under dry conditions so they are equally undesirable for planting. The effect of sett size ranging from 10 to 50 cm was also evaluated at CTCRI and it was found that planting 20 cm long setts at a depth of 5 cm results in higher yield. However, for gap filling, 40 cm long setts are ideal. The setts should be treated for better establishment as well as drought and disease resistance. Dip the setts in carbendazim 0.02% solution for 15 minutes before planting. Plant the setts vertically with buds pointing upward on the sides of ridges and furrows. For rain-fed conditions, treat the setts with potassium chloride 0.05% solution. To avoid nutrients deficiency, treat the setts with micronutrients, *viz.* zinc sulfate and ferrous sulfate each @ 0.5% for 20 minutes. Dip the setts for 20 minutes in *Azospirillum* and phosphobacteria each at 30 g/ litre for better nutrient absorbance.

RAPID MULTIPLICATION THROUGH FIELD TECHNIQUES

These techniques are developed for minimizing the bulkiness of planting material for economizing the use of the same to realize maximum returns and to increase the multiplication rate. Use of minimum nodes has been found to increase multiplication rate in cassava. By inducing sprouting in single-node or even half-node cuttings, the multiplication rate is found to have risen to 300 times or more. Cassava is usually multiplied using 15–20 cm long stakes. Instead of such long setts, the sett size can be reduced to 8 to 10 cm thereby doubling the number of available setts from a given number of source plants. These mini setts carry three to four nodes, which are optimum, needed to lead to a healthy plant. The setts are planted closely (500 setts in one square metre) in nursery beds and sufficient soil moisture retained to induce sprouting. The sprouted setts (settlings) can be transplanted in the field on the 25th day to attain a normal crop yield.

The transplanted settlings are to be provided with sufficient soil moisture for the initial 5 days for establishment. An added advantage of the nursery technique is that it can be used for screening setts for the presence of CMD or any other diseases. The settlings expressing disease symptoms are culled out in the nursery and only healthy settlings transplanted to the field. Screening and roguing out of infected plants is continued in the field to isolate healthy plants, which form the planting material for the ensuing season.

PLANTING METHOD

Cassava can be planted by mound method, ridge and furrow method, flat method and pit followed by mound method. Comparison of these systems of planting revealed the superiority of the ridge and furrows system.

The mound method may be adopted in soils having clay content and restricted drainage, whereas, the ridge method may be followed on slopes to prevent soil erosion. In the mound method, stakes are planted on 25–30 cm high mounds taken on already prepared land. In ridge method, ridges of 25–30 cm height are prepared across the slope on which cassava setts are planted. The ridge and furrow method of planting is suitable for irrigated cassava cultivation under Coimbatore conditions. The flat method of cultivation may be followed in levelled lands having good drainage. In the pit method followed by mound method, the pits (30 × 30 × 25 cm) are opened first and soil is mixed with cattle manure and reformed into mounds. Cuttings need not be planted until the monsoon becomes active in the region. Under the high rainfall zone of the Kanyakumari District of Tamil Nadu, the mound method of planting of cassava gives better yield.

SPACING AND PLANT POPULATION

Cassava varieties are of three types, branching, semi-branching and non-branching. Branching and semi-branching types require more spacing and hence should be planted 90 × 90 cm apart, whereas non-branching types require 75 × 75 cm spacing. The recommendation for different states is as shown in Tables 67.2 and 67.3.

IRRIGATION REQUIREMENT

Cassava is usually raised as a rain-fed crop. However, in parts of Tamil Nadu and in some other non-traditional areas, cassava is cultivated under irrigation conditions. On the day of planting, irrigation is given followed by two irrigations at an interval of 3–5 days until the plants establish. Further irrigations are scheduled depending upon the distribution of rainfall. In Tamil Nadu, irrigation at 8-day intervals gives the higher net profit. At CTCRI, the response to supplementary irrigation by high-yielding hybrids of cassava was studied and found that cassava responded positively to supplementary irrigation at 20 mm per week, which was approximately equivalent to 70% of the weekly pan evaporation in the locality.

RESPONSE OF CASSAVA TO DRIP IRRIGATION

With the objective of economizing irrigation water through the drip system, field studies were conducted for 3 years at the Agricultural Research Station, Bhavanisagar, during 1996–2000 with the cassava variety MVD-1. The study revealed that the fresh root yield was at par when irrigated at 100% and 75% of open-pan evaporation through drip system.

TABLE 67.2
Recommendation for Different States

States/Region	Spacing	
	Branching	Semi-Branching/Non-Branching
Kerala, Andhra Pradesh	90 × 90 cm	75 × 75 cm
Maharashtra, Gujarat and Chhattisgarh	90 × 75 cm	–
	90 × 75 cm	75 × 75 cm

TABLE 67.3
Spacing Differs According to the Season and Soil Fertility

Method of cultivation	Soil fertility	Spacing (cm)	Population (ha)
Rain-fed	All types of soils	60 × 60 cm	27777
Irrigated	Low fertile	75 × 75 cm	17777
	Medium fertile	75 × 90 cm	14814
	Highly fertile	90 × 90 cm	12346

NUTRITIONAL REQUIREMENT

Cassava is adapted to soils of low fertility though it responds well to manure and fertilizers and higher yield can be realized through timely and proper nutrient management practices. Generally, most of the nutrients' deficiency symptoms appear when the concentration goes below the critical level but some nutrients like nitrogen, magnesium and potassium manifest the symptoms at an early stage.

NUTRIENTS REMOVAL BY CASSAVA

It is generally considered that cassava exhausts the soil and removes large amounts of nitrogen and potassium. A comparison of the nutrients removal by cassava in relation to other major crops showed that cassava extracts large quantities of elements, particularly potassium. Experiments conducted at CTCRI on nutrient removal by cassava indicated that considerable variations existed between cultivars in their nutrient removal pattern. Most of the varieties under favourable conditions, capable of producing 30 tonnes fresh tubers per hectare, removed 180–200 kg nitrogen, nitrogen 15–22 kg nitrogen and 140–160 kg potassium per hectare.

INDEX LEAF AND CRITICAL NUTRIENT CONCENTRATION

The Youngest Fully Expanded Leaf (YFEL) petiole or leaf blade at the 3–4 month stage is the best indicator of the nutritional status of the plant shown in Table 67.4.

NUTRIENTS DEFICIENCY DISORDERS

NITROGEN

Nitrogen-deficient cassava plants become shorter and less vigorous than normal plants. Under severe conditions, plants lack vigour, become chlorotic, mostly the lower leaves become yellow initially, and then, the entire foliage becomes chlorotic. The upper leaves are generally much paler than normal green leaves.

TABLE 67.4

Critical Nutrient Concentration (CNC) in the YFEL of Cassava at 3–4 Month Stage

Nutrient	Deficiency	Sufficiency range	Toxicity
N(%)	<4.7	5.1–5.8	–
P (%)	<0.30	0.36–0.50	–
K (%)	<1.0	1.3–2.0	–
Ca (%)	<0.65	0.75–0.85	–
Mg (%)	<0.27	0.29–0.31	–
S (%)	<0.24	0.26–0.30	–
B (%)	<20	30–60	<100
Cu (μg g^{-1})	<5	6–10	<15
Fe (μg g^{-1})	<100	120–140	<200
Mn (μg g^{-1})	<45	50–120	<250
Zn (μg g^{-1})	<25	30–60	<120

Source: Howeler (1983). Centro Asteracional de Agricultura Tropical (CIAT), Cali, Colombia, 28P.

Phosphorus

The notable symptoms of acute phosphorus deficiency under field conditions are the chlorosis that develops on lower leaves, and such leaves cannot maintain their turgor. They hang down from the petioles, and in severe cases, the stem becomes thinner and the petioles shorter, with narrow foliar lobes and fewer lobes than the normal leaves. Under severe phosphorus deficiency, the plants develop dark yellow or orange-colour lower leaves, which later become necrotic and shed, resulting in poor yield.

Potassium

Potassium deficiency is characterized by a marked reduction in plant height, petioles become shorter and leaves smaller. Under severe conditions, small purple or brown spots appear on older leaves, which curl upwards from the margins, and in some cases, downward from the tip. As the deficiency intensifies, chlorotic areas develop on tip and edges of lower leaves and lead to marginal necrosis. Chlorosis followed by necrosis and premature shedding of middle leaves are observed due to avoidance of potassium in nutrient combination.

Magnesium

The characteristic symptom is the interveinal chlorosis of the lower leaves beginning from the tip and edges of the foliar lobes and extending inward between the central and secondary veins. The symptoms usually appear from the third month of planting and continue until harvest, and some cultivars like Sree Visakham are more susceptible to magnesium deficiency. Moreover, when cassava is cultivated continuously in the same field, the occurrence of magnesium deficiency is very common.

Calcium

Burning and curling of tips of younger leaves is the characteristic symptom of calcium deficiency. Generally, cassava is tolerant to acid soils having low calcium content because of the physiological adaptation of cassava to tolerate low calcium levels. Under field conditions, excess lime may lead to the induction of iron, potassium, magnesium, manganese, or copper deficiency.

Sulfur

Sulfur-deficient leaves are pale green to yellow in colour and very similar to nitrogen-deficient leaves except that the symptoms normally appear on upper leaves but later extend over the whole plant, and the leaves remain small.

Boron

Deficiency of boron causes dwarfing of young plants and slight chlorosis of the younger leaves. Initial symptoms are seen on the roots, which show suppressed lateral root development, and sometimes, death of the root tip. This is followed by the emergence of short lateral roots with swollen ends. Boron-deficient plants tend to be short due to reduction in internode length and mildly affected leaves develop chlorosis, consisting of numerous minute light spots, which appear mainly at tips and margins of the leaf lobes. A distinctive symptom of boron deficiency is the development of lesions from which a brown gummy substance exudes on upper petioles and stems. The stem lesions may after coalesce and develop into cankers.

IRON

Iron deficiency is characterized by chlorosis of the younger leaves. Initially, the interveinal tissues are chlorotic, the veins remaining green, but in severe cases, the veins lose their green colour and the whole leaf including the petiole become yellow and later almost white. Plant height is severely reduced and leaves are smaller but not deformed. High levels of manganese, copper, or zinc in soil solution can induce iron deficiency. Usually, iron deficiency is rare under field conditions but in saline alkaline soils having pH >8, calcium-induced iron deficiency is very common. In alkaline soils, it is accompanied by deficiency of manganese and zinc. Excessive liming or high phosphorus fertilization can also induce iron deficiency.

MANGANESE

Manganese deficiency produces an interveinal chlorosis of young, recently unfolded leaves. The boundary between the green and chlorotic area tends to be rather diffuse, the veins being surrounded by a definite zone of non-chlorotic interveinal tissue. The affected leaves may become completely chlorotic.

ZINC

Zinc deficiency causes a characteristic interveinal chlorosis of the younger leaves. Small white or light yellow chlorotic patches develop between the veins, and then, successive leaves produced become smaller and more chlorotic. The leaf lobes become narrow, light green to white in colour with the margins cupped upwards. Leaf tips turn necrotic under severe conditions. Since the growing point is most affected, zinc deficiency drastically reduces plant growth and yield. Cassava is found exceptionally susceptible to zinc deficiency and its continuous monoculture definitely leads to zinc deficiency. Zinc deficiency is prevalent in both acid and alkaline soils.

COPPER

The main symptoms of copper deficiency are chlorosis and deformity of young leaves. Often, the leaf tips become necrotic and the leaves are either cupped upwards or curled downwards. The number of lobes per leaf may be reduced in three or even one. Under severe deficiency, plants die back from stem apices and grow again from the base, producing a bush plant with one or more dead stems projecting from the top. Leaves in the middle of the plant have abnormally long drooping petioles. Severe copper deficiency can markedly reduce root development. Such plants become susceptible to water stress.

MANAGEMENT OF NUTRIENT DEFICIENCIES

Magnesium, zinc, boron and molybdenum deficiencies can be ameliorated by the application of fertilizers or amendments such as magnesium sulfate, zinc sulfate, borax and ammonium molybdate @ 20.0, 12.5, 10.0 and 1.0 kg/ha, respectively. Magnesium sulfate may be applied at the time of planting. Apply borax or sodium borate 1–2 kg/ha or by soaking stakes for 15 minutes in 0.5–1% borax solution. Cassava can tolerate calcium deficiency, however, application of calcium oxide 2 t/ha can increase the soil pH by 2 units. Dipping the stakes in 2–4% solution of zinc sulfate for 15 minutes prior to planting or foliar application of 1–2% solution of zinc sulfate is effective. To combat the lime-induced chlorosis, spraying a mixture of 1% ferrous sulfate + 0.5% zinc sulfate and 2% urea is suggested. Application of sulfur 10–20 kg/ha as elemental sulfur or gypsum or use of sulfur containing fertilizers like ammonium sulfate, single super-phosphate, or potassium sulfate can rectify sulfur deficiency. Though manganese and copper deficiencies are rarely observed, in case of occurrence, soil application of manganese oxide or manganese sulfate or foliar spray of manganese sulfate and soil application of copper as copper sulfate at 2.5–3.5 kg/ha can be resorted to.

MANURE AND FERTILIZERS/BIO-FERTILIZERS

Cattle manure or compost 10–15 t/ha is incorporated into the soil during the preparation of land or while filling up the pits; 1 kg organic manure per pit is recommended. Studies conducted at CTCRI revealed that *in situ* green manuring with cowpea, incorporation of crop residue and use of recently available organic manure like press mud, mushroom spent compost, sawdust compost and coir pith compost can substitute farmyard manure for cassava production. Integrated application of cattle manure or compost 12.5 t/ha as basal dose and NPK 100:50:100 kg/ha is usually recommended for high-yielding varieties under Kerala conditions. For M4 and local varieties, half the above dose is sufficient. The fertilizer recommendation for different states as shown in Table 67.5.

Low input management practices comprising of green manuring with cowpea, indigenous rock phosphate (20–21% P_2O_5) and use of bio-fertilizers, namely *Azospirillum* and *Phosphobacterium* at 2 kg/ha each, can reduce the requirement of nitrogen and phosphorus to 50% of the recommended dose and maintain the soil nutrient status.

Application of AM fungi and phosphobacteria (10 kg ha^{-1}) along with half the recommended dose of phosphorus and full dose of nitrogen and potassium and farmyard manure (12.5 t/ha) recorded a tuber yield of 28.03 t/ha, and it was comparable to that of the recommended dose for NPK and farmyard manure. For acid laterite soils, lime application 2000 kg calcium oxide per hectare at the time of land preparation is beneficial to increase the yield and improve the quality of tubers.

INTERCULTURAL OPERATIONS

Inter-cultivation to suppress weeds in initial stages of crop growth and to encourage tuber development is an essential requirement in cassava production. Light raking of the soil, followed by earthing-up twice at 2 months after planting, had significant beneficial effect on cassava yield. Inter-culturing in cassava is usually done with hand tools or with the help of animal-drawn hoes. Experiments on herbicidal weed control under Indian conditions are lacking but in Malaysia, Thailand and in some of the South American countries, herbicides are used for weed control in cassava on large farms. Most recent research on weed control in cassava has shown that in addition to an application of selective pre-emergence herbicides, at least one inter-culturing is necessary for higher yield. In Tamil Nadu, pre-emergence herbicide application of pendimethalin 1 kg a.i./ha followed by two hand weedings are suggested.

TABLE 67.5
Fertilizer Recommendation for Different States

State		Fertilizer dose NPK (kg/ha)
Chattisgarh	Rain-fed	60:60:60
	Irrigated	150:60:150
Tamil Nadu	Rain-fed	50:65:125
	Irrigated	90:90:240
Kerala	–	100:50:100
Gujarat and Maharashtra	–	100:50:100

MANAGEMENT OF CASSAVA IN CROPPING SYSTEM

In most of the small farms cassava is found involved in intercrop and sequential cropping systems. In small farms, cassava is generally intercropped with legumes like groundnut, cowpea, black gram and onion in south India. In Kerala and Tamil Nadu, the cassava is also raised as intercrop in coconut gardens. Research on cassava intercropping has shown that land equivalent ratio above one is frequently obtained, when cassava is intercropped with short-duration crops such as beans, groundnut, cowpea and some vegetables. This is possible mainly due to the slow initial growth of cassava, which is a long-duration crop that takes 9–10 months for harvest. It takes 3 to 3½ months to develop enough canopies, and the available sunlight, water and nutrients between rows can be effectively utilized for short-duration intercrops. Select only bushy types of intercrop which mature within 100 days. Legumes are the best suited intercrops in cassava. Intercrops like groundnut (bunch type TMV-2 and TMV-7) and French bean (cv. Contender) were found to be promising. Onion, coriander, short-duration pulses and short-duration vegetables can also be grown as intercrops. These crops can give an additional income within 3–4 months. Intercrops also help to control weeds and add organic matter and nitrogen to the soil. Slight modifications in the agronomic practices become necessary, depending on the intercrop. Cassava should be planted in the months of May–June at the recommended spacing. Seeds of intercrop are dibbled immediately after planting cassava. Basal dressing of cassava should be given as per schedule for pure crop. Apply recommended dose of NPK to the intercrops about 30 days after sowing followed by light inter-culturing. Recommended dose of fertilizers is top-dressed immediately after harvest of intercrops and earthing-up is done. When cassava is grown under tree crops in general, the growth and yield of tree crops are not negatively affected but the productivity of cassava is reduced by shading from the perennials.

INSECT-PESTS

Soft-scale (*Aonidomytilus albus*)

Among the pests that infest cassava stems, the most serious one is the mussel-shaped soft scale. At present, the pest is prevalent throughout southern India. Though the incidence is sporadic and localized under field condition, it is very severe in storage throughout Kerala where the stems are stored after harvest as planting material. The scales encrust the stem in large numbers and multiply rapidly under defective storage. They desap the stem and destroy the latex content leading to quick drying-up of planting material.

The infestation in the field spreads through the contaminated planting material and the infested setts do not sprout properly. When sprouted, the main twig breaks and side shoots appear in the form of a bunch. In some places, 20–40% mortality of the seedlings occurs due to infestation. The infected plants become weak, the stems are dried up and tubers become unpalatable. The infestation occurs in dry situations and aggravates during prolonged moisture stress situations.

Control

- For effective control, only scale-free stems should be selected for storing and planting.
- Dipping the setts in dimethoate or methyl demeton (0.05%) solution for 10 minutes is effective for better establishment of the seedlings.
- Stacking the stems in a vertical position and keeping them in the shade of trees to facilitate sufficient aeration is effective to reduce multiplication of scales.

MEALY BUG (*Pseudococcus filamentosus* ckll.)

Another important pest that infests the planting material is the mealy bug which is sporadic in occurrence and at times becomes serious. The nymphs and adults colonize in large numbers on the leaves, petioles and stems. The honeydew secretion of the mealy bugs causes sooty mould. The infested stems dry up soon. As the infested stems can be conspicuously seen, it is easy to detect them and destroy them.

Control
- Spraying the stems with dimethoate or methyl demeton (0.05%) is effective.

TERMITE (*Odontotermes* sp.)

Termites are polyphagous pests attacking a number of plants and attack the planted setts and young seedlings. Bark of the setts is eaten away and the pith is tunnelled, resulting in quick drying and mortality of setts. In severe cases, the setts are covered with termitoria. The damage is more serious during dry periods in the mound method of planting than in flat planting. Although termite incidence is sporadic and localized, it assumes serious proportion in a prolonged dry season, especially in laterite and red soil areas, and hence, warrants control measures.

Control
- Drenching around the planted setts with monocrotophos (0.05%) has been found effective for control of termites on cassava.

WHITEFLY (*Bemisia tabaci* genn.)

The whitefly is economically important as a vector of CMD. In fact, it is the only known vector of cassava mosaic diseases.

TETRANYCHID SPIDER MITES (*Tetranychus telarius l., T. cinnabarinus boisd., T. neocaledonichus, Eutetranychus orientalis klein. and Oligonychus biharensis hirst.*)

Tetranychid spider mites are the most serious pests of cassava. Altogether 22 different species of spider mites infest cassava all over the world. In India, four species of mites comprising two distinct groups do extensive damage to the crop.

The red mites (*T. cinnabarinus* and *T. neocaledonichus*) usually feed on the lower surface of the leaves. The infestation first starts from the mature lower leaf upwards showing yellowish specks along the main leaf vein, and subsequently, it spreads apart producing characteristic blotching and elongated streaks along the leaflets. During severe infestation, they cover the surface of both the mature and young leaves and form a web. Severely infested leaves wither, dry and fall off quickly from bottom upwards before senescence.

Significantly, a higher population of mite complex in the rain-fed situation indicates that moisture stress is the predisposing factor for greater multiplication of mites of cassava. Under irrigated conditions, the leaf turgidity is increased and the microclimate is also changed by the increase of relative humidity around the leaf canopy.

Control

- Spray of Dicofol 0.05% is found to be highly effective in controlling mite populations.
- Alternatively, when the labour is cheap and pesticides are not available, spraying water at 10-day intervals can be resorted.

DISEASES

CASSAVA TUBER ROT (*Phytophthora drechsleri, P. tropicalis, P. erithroseptica, P. cryptogea and P. palmivora*)

The disease is characterized by the appearance of dark-coloured round to irregular-shaped water-soaked lesions (15–30 mm diameter) on mature tubers in the field. White mycellial mats of the fungus develop around these lesions. On advancement of infection, the lesions enlarge, causing internal browning, oozing of internal fluids and shrivelling of the tubers. The infected tubers emit a characteristic foul smell and rot with in 5–7 days, depending on the soil conditions. However, no external symptoms are visible on leaves and stem of infected plants. Hence, early detection of the disease is very difficult.

This disease has been reported infecting cassava plantations in Africa and tropical South America, where it causes yield loss up to 80%. High rainfall, especially during the months of September–November, when actual bulking of tubers occurs, favours the disease incidence. Heavy black soil, presence of a hard soil pan below the surface soil, poor drainage conditions, water stagnation during rainy periods, lack of field sanitation, presence of infected tubers leftover in the field after harvest, flat bed method of planting and indiscriminate use of water through flood irrigation are some of the predisposing factors favouring the disease occurrence. It is a soil-borne fungus activated with high soil moisture.

Control

- Crop rotation with other crops, especially millets, once in 2 years helps to reduce the population of pathogen in the soil.
- Remove infected tubers from the field immediately after harvest, collect them in a heap in a corner of the plot and burn them. This practice also reduces the future inoculum for the next crop.
- Deep ploughing of the plot using a subsoiler (single plough) up to a depth of 20–24 inches helps to break the hard soil pan developed just below the surface. This will improve the drainage and prevent water stagnation, thereby reducing the occurrence of disease.
- Plant the setts on top of the ridges instead of planting along the sides of ridges to avoid high moisture levels at the root zone during the early growth period.
- Application of organic amendments like good quality neem cake (minimum 5% N) at the rate of 250 kg/ha at the time of planting helps to widen the C:N ratio of the soil and thereby increases the population of saprophytic micro flora including antagonistic organisms.
- Incorporate efficient biocontrol agents like *Trichoderma harzianum* into the soil. Mix 1 kg of *Trichoderma* culture in sorghum grains with 100 kg of dry cow dung/acre, incubate it for 2 weeks in room conditions and apply on top of the ridges by mixing with topsoil before planting. In gardens with poor drainage, setts treatment or soil application of *Pseudomonas fluorescence* will be effective.
- Avoid water stagnation during rainy periods by providing sufficient drainage facilities. Irrigation should be restricted to the minimum, depending on the soil conditions, especially from the 6-months stage until harvest, just to maintain the crop as well as optimum tuber bulking.

Cassava Mosaic Disease

Cassava mosaic disease (CMD) is first reported in India in 1956. It is widely distributed in Africa and the Indian subcontinent and it causes severe yield loss. It occurs in all the cassava growing areas of India and particularly more prevalent in South India. Cassava is mainly propagated by stem cuttings, and use of CMD-infected plant stem cuttings spreads the disease. It is also transmitted by the whitefly vector *Bemisia tabaci* (Gennadius) in India.

Cassava mosaic disease induces the chlorotic mosaic on leaf, often with distortion of lamina, visible as the leaf unfurls, and causes severe stunting, especially of those plants propagated from infected plant material. The CMD-affected plant exhibits splitting of tubers at the time of maturity under irrigated conditions. The disease significantly reduces stem girth, plant height, middle leaflet length and width in different cassava varieties and also reduces the rate of photosynthesis. Yield reduction ranges between 16% and 42%. Tagging the healthy plant as mother plants for source of cuttings at the 10-month stage reduces CMD incidence. Also use of a balanced dose of NPK such as 90:15:75 kg/ha enhances the growth of cassava without increasing CMD severity.

Control

- Spray the crop with Rogor (0.03%), Metasystox (0.05%), or parathion (0.02%).
- Roguing out the infected plant from the main field is practiced by the farmers and reduces the primary spread of CMD but more rouging-outs reduce the main field population and affect the yield.
- Planting the disease-free cassava seedlings (15 days old) in the main field by raising cassava cuttings in a nursery bed and discarding the cuttings, which are showing CMD symptoms at the initial stage, is also recommended by CTCRI.
- Providing wide spacing in a cassava crop is recommended to reduce the secondary spread of CMD by the whitefly vector.

Brown Leaf Spot (*Cercospora henningsii allescher.*)

The symptoms appear mostly on mature leaves as chlorotic spots on both sides of the leaves. The spots soon enlarge and turn into dark green areas, which finally attain a uniform brown colour with distinct dark margins. Mature spots are either circular or irregular in shape and measure 3–10 mm in diameter. A definite yellow halo is always seen surrounding the spots in certain varieties. The infected tissues become necrotic, break and fall off, leaving shot holes in certain varieties. In a single leaf, 1–100 spots occur and under favourable weather conditions several spots coalesce and the leaves turn yellow, dry up and drop down. The most conspicuous effect of the disease is the excessive premature defoliation of leaves in susceptible varieties.

The three main climatological factors, namely, rainfall, relative humidity and temperature, play an important role in the disease buildup. The rainfall and relative humidity are more directly related to the disease intensity. Disease intensity is usually observed minimally during summer (March–May) which is preceded by minimum amount of rainfall and relative humidity from December–February. Maximum disease incidence is observed during August–September and again during November–December. These periods are preceded by periods of heavy rainfall and relative humidity during southwest and northeast monsoon in the main cassava-growing tracts. Excessive application of nitrogen and phosphorus increased the disease severity while potash reduced it.

Control

- Application of nitrogen and phosphorus alone or in combination increases the disease severity, while potassium decreases it, therefore, a balanced fertilizer dose with all the three major nutrients (NPK) would help to reduce the severity of the disease.
- Destroy the inoculum sources.
- Significant differences in varietal resistance have been reported in India and elsewhere. Though absolute resistance has not been found, the varieties such as H 97, Sree Visakham and Sree Sahya are moderately resistant, and Sree Prakash is highly resistant to this disease.
- Spraying Benlate (0.1%) or Cercobin (0.2%) at monthly intervals, starting from the third month onward, is effective in reducing the spot intensity and premature leaf fall and in increasing the tuber yield.

ANTHRACNOSE [*Colletotrichum manihots chevaugeon (Syn. gloeosporium manihotis P. henn), colletotrichum gloeosporioides penz. and Colletotrichum capsici (Syd.) bulter and Bisby.*]

Anthracnose, also known as dieback or wither tip, is generally considered as a minor disease but during favourable conditions, this may become serious, killing the terminal shoots and showing typical dieback symptoms. The disease appears during the prolonged rains and normally disappears after the cessation of the rains. Young shoots are normally most susceptible to this disease.

The first symptom is the appearance of brown, circular, elongated or irregular lesions on the leaves which gradually enlarge and coalesce. Severely infected leaves crinkle and finally dry, wither and fall off. Light to dark brown spots appear on young and tender shoots and petioles also. The infected areas turn black and the parts above them wither. As infection spreads, the bases of the petioles shrivel; as a result, the leaves droop, turn yellow and dry up. The infection spreads along the nodes to the stem and the branches also and dries ultimately.

Control

- The best way to control the disease is to use healthy planting materials and to avoid planting during high rainfall seasons. Since potash deficiency in the soil tends to make the plant more susceptible to anthracnose, an additional dose of potassium fertilization may help to reduce the disease severity. Pruning and burning of infected branches may reduce the inoculums available and thereby the secondary spread of the disease.
- Spread of the disease could be checked by spraying copper fungicides like Bordeaux mixture and Milop (copper oxychloride).

SETT ROT (*Diplodia natalensis pole-evan*)

The characteristic symptoms are the black discolouration and necrosis of vascular strands spreading from cut ends of the stem. Blisters are produced on the epidermis, which affect the tissues below. The tissues turn dark brown or black. The blisters rupture and reveal masses of black confluent of pycnidia.

Control

- Use of healthy planting material is the best means of control. Fungicides, such as Benlate (0.1%), Bavistin (0.1%), Vitavax (0.1%) and thiram (0.2%) have been found to be very effective in checking the disease in the field.
- The cuttings are to be dipped in one of the fungicidal solutions for 15 minutes, dried in shade and then planted.

Dry Rot of Tapioca Tuber [*Fusarium solani (Mart.) sacc.*]

The infected tubers are usually hard. The colour of the pulp, instead of being white, varies from brown to reddish brown, and dark brown streaks are seen inside the tubers; one or more tubers are affected in each plant and the diseased tubers are unfit for marketing and consumption.

Control

- Grow the crop in well drained sandy loam soil.
- Avoid growing tapioca very close to rivers or streams.
- Use disease resistant or tolerant varieties.
- Follow long crop rotation with non-host crops.
- Use disease free healthy planting.
- Dip the planting material in the solution of Borex, Captan, or thiobendazole-based chemicals.
- Practice good farm sanitation.
- Avoid flooding or water logging in field.
- Avoid planting in soils that are poor in nutrients.
- Collect plant debris with fungi and destroy particularly after harvest.
- Harvest the crop early since early harvesting prevents the incidence of rots.

68 Baby Corn

Sonia Sood and Nivedita Gupta

Botanical Name : *Zeam ays* var. saccharata

Family : Gramineae (Poaceae)

Chromosome Number : 2n = 20

ORIGIN AND DISTRIBUTION

The word 'corn' is a generic term for a grain or cereal, commonly used in United States of America, Canada and Australia. For diversification and value addition of maize as well as growth of food processing industries, the recent development is growing maize for vegetable purpose, which is commonly known as *baby corn*. *Zea mays* originated in the western hemisphere; however, commercial cultivation began nearly 200 years ago in United States of America, which is said to be the native place of maize. It was domesticated from its progenitor, *i.e.*, the wild teosinte (*Zea mays* var. mexicana). The early Americans spread this crop from southern Mexico to Argentina and Canada. Later, Columbus took it back from the new world to Spain, and within an era, it spread all over the world.

Baby corn development in Thailand began in 1976 with the open-pollinated varieties, and subsequently, the hybrids were developed. In the world, Thailand is the largest producer of baby corn, estimated to account for 80% of the world's trade. Hence, based on history, Thailand is believed to be the centre of origin.

INTRODUCTION

Baby corn belongs to the tribe Maydeae of the family Poaceae (Graminae). It has a speciality of sugary form of the common corn, as it has one or more genes that retard the conversion of sugar to starch in the endosperm. The simply inherited endosperm gene(s) increase sugar content and decrease starch content. Eight single recessive endosperm genes that have been utilized to develop commercial cultivars or hybrids includes sugary (*su*), sugar enhancer (*se*), shrunken-2 (*sh2*), brittle (*bt*), brittle-2 (*bt-2*), amylose-extender (*ae*), dull (*du*) and waxy (*wx*).

Baby corn is primarily used in Asian cuisine, and now, its consumption is highest in Asia. Countries widely producing baby corn other than Thailand are Sri Lanka, Taiwan, China, Zimbabwe, Indonesia, South Africa and United States of America. In India, it is cultivated in Meghalaya, Uttar Pradesh, Haryana, Maharashtra, Karnataka and Andhra Pradesh. In Tamil Nadu also, baby corn is becoming popular among the farmers of near towns and cities. Both public and private sectors are developing composites and giving seeds to the growers. Baby corn is becoming popular among urban people, and its demand is increasing day by day.

Baby corn has increased sugar and decreased starch in its endosperm, with a thinner pericarp, which imparts flavour to the corn and soft texture of endosperm. Different varieties or hybrids have the different kernel colour, *i.e.*, white, yellow, or bicolour. For fresh consumption, it should have a medium-sized ear with straight rows of narrow and deep kernels filled to the tip and a glossy green husk. It is typically eaten both raw and cooked, including whole-cob, in contrast to mature corn, whose cob is too tough for human consumption. Maize (*Zea mays* L.) is the third most important cereal crop next to rice and wheat and has the highest production potential among the cereal crops. It has immense yield potential and is therefore called a *miracle crop* and also the *queen of cereal crops*. Baby corn is a small young corn ear harvested at the stage of silk emergence. This vegetable has a great potential for processing as a canned product. Canned baby corn as an export to Thailand, Japan and Europe has a good future. Young cob corn has a short growth, wide range of adaptation and does not need intensive cultivation, and four or more crop cycles can be taken.

COMPOSITION AND USES

COMPOSITION

Baby corn has edible tender cobs, which are good source of proteins, lipids and pro-vitamin A due to light yellow in colour. Light-yellow coloured tender cobs are more nutritious than yellowish-brown and white cultivars. It has nutritive value similar to that of non-legume vegetable such as cauliflower, tomato, cucumber and cabbage. Its ears are completely free from pesticides residues since the ears are tightly wrapped with several layers of husk. The nutritive value of baby corn and sweet corn is given in Table 68.1.

USES

In India, only a fresh whole ear is eaten after boiling in water and generally it is not roasted as in field corn for fresh consumption. Light yellow-coloured, sweet and tender ears of baby corn are used as a salad vegetable by adding lime juice and black pepper powder either alone or along with raw onion and tomato. It is a potential ingredient for mixed vegetables, vegetable *biryani* and Chinese food preparations. Other value-added products like soup and pickle are also served in big hotels. In China, Manchurian is a kind of food prepared from baby corn. In addition to significant nutrition, it provides diversity to meals because of its crispy texture and flavour. The whole ear or cut kernels are used for canning or freezing.

Bably corn green plants, which contain protein 13.2%, fat 4.4%, fibre 34.8% and ash 6.5%, form a balanced fodder for cattle, thus, baby corn production provides an opportunity for farmers to develop dairy industry for the diversification of agriculture.

TABLE 68.1
Nutritional Composition of Baby Corn and Sweet Corn (per 100 g Edible Portion)

Constituents	Baby Corn	Sweet Corn	Constituents	Baby Corn	Sweet Corn
Water (g)	89.1	75.96	Phosphorus (mg)	86.0	46.0
Carbohydrates (g)	8.2	9.02	Iron (mg)	0.11	0.2
Sugar (g)	1.3	3.22	Vitamin A (IU)	64.0	150.0
Starch (g)	6–7	–	Thiamine (mg)	0.50	0.30
Proteins (g)	1.9	3.27	Riboflavin (mg)	0.08	0.05
Fats (g)	0.2	1.18	Vitamin C (mg)	11.0	4.4
Calcium (mg)	28.0	3.5	Energy (kcal)	18.0	86.0

BOTANY

Baby corn is a monocot, monoecious, annual and highly cross-pollinated crop. Mostly pollination occurs through the agency of wind. Like field corn, it has a terminal staminate inflorescence on its main stem, known as tassels, and its pistillate inflorescence is the ear (cob) with thin thread-like stigmas (silks) protruding from the tip of the ear. A single plant produces one to several ears. Baby corn is a young tender ear of specific group of maize plants harvested before fertilization when silks have just emerged.

The plant of baby corn is generally protandrous. The spikelet at some distance from the tip of the tassel opens first and the anthesis proceeds downwards. In each spikelet, the upper floret opens before the lower floret. It takes around 14 days for all the staminate florets to open. However, pollination and anthesis are not preferable in baby corn since the economic produce is young, tender and finger-like unfertilized cobs, except for seed production.

CLIMATIC REQUIREMENT

Baby corn is a warm season crop but can be grown under different climatic conditions from tropical and subtropical to temperate regions. It can be grown at 10–40°C temperature but the best temperature for its plant growth is 21–30°C. Chilling injury occurs below 5°C temperature. Being C_4 plant, it is adapted to high light intensities since low light due to high planting density impairs ear development. Although the rate of plant development is primarily determined by temperature but since it is a quantitative short-day plant, photoperiod can potentially influence the leaf number and time of flowering. Short days tend to reduce leaf number and accelerate emergence of tassels and silks. Some of the cultivars do not form ears in temperate region until the day length decrease to 12–14 h. The tropical varieties keep their vegetative phase continued. The day length less than 8 hr and temperature below 22°C can also delay flowering. Cultivars bred for long days and cool temperature yield less when grown under warm temperature and short days. Warm days and cool nights are the optimum growing environment for maximum productivity. In India, baby corn is cultivated in almost all parts of the country wherever the field corn is cultivated.

SOIL REQUIREMENT

Baby corn grows well in almost all types of soil but well-drained deep clay to light sandy loam soils rich in organic matter with adequate water-holding capacity is ideal. A pH range of 7.5–8.5 is preferred for optimum growth. The crop is very sensitive to waterlogging. Care should be taken to assure that water does not stagnate on the soil surface for more than 4–5 hr. The agronomic requirement of baby corn is similar to grain maize. The field should be ploughed thoroughly with a disc plough once followed by cultivator ploughing twice, after spreading farmyard manure or compost until a fine tilth is obtained. It facilitates the sowing and provides aeration as well as better seed-soil contact to improve germination and further growth.

VARIETIES/HYBRIDS

The cultivars recommended under Indian conditions are MEH 114 (hybrid), VL 42 (open-pollinated), Parkash (hybrid), Kesari (composite), COBC 1 (synthetic) and Golden Baby (hybrid). The details of these hybrids or varieties are given as follows:

MEH 114 AND VL 42

These varieties have been developed by the Directorate of Maize Research, IARI, New Delhi. Both are early varieties and they produce multiple and uniform ears. The yield of MEH 114 is 1720 kg/ha and VL 14 is 1913 kg/ha.

PARKASH

A hybrid developed at Punjab Agricultural University, Ludhiana, produces ears with ear length 9.3 cm and girth 1.45 cm. Its average yield is 15.5 q/ha.

KESARI

A hybrid developed at Punjab Agricultural University, Ludhiana, produces two or three ears per plant and has a yield potential of 14.2 quintals per hectare with an average ear length and girth of 7.5 and 1.3 cm, respectively.

SOLAN SUGAR BABY

A variety developed at Dr. YS Parmar University of Horticulture and Forestry, Nauni, Solan is a dual-purpose variety and used as both baby corn and sweet corn. It becomes ready for harvesting 45 days after sowing and yields about 12 quintals per hectare.

COBC 1

A synthetic variety developed by the Department of Millets, Centre for Plant Breeding and Genetics, TNAU, Coimbatore, produces tender cob yield of 66 q/ha under well-irrigated conditions. The cobs have standard size of 9–10 cm suitable for the export market.

GOLDEN BABY

This is a hybrid developed by Pro-Agro Seeds India Limited, New Delhi. Its plants are medium tall (2.25–2.5 m) and fast-growing. Each plant produces two or three uniform-sized (8–10 cm long) ears. Tender ears become ready for harvesting 47 days after sowing and harvesting continues for 8–10 days.

SEED RATE

The seed rate varies with sowing time, soil fertility, irrigation facilities and the variety or hybrid used for cultivation. High density of plants per unit area is recommended. About 25–40 kg seed is required for sowing a hectare area. Seeds should be well treated with fungicide, *i.e.*, Bavistin or thiram @ 3–4 g/kg of seed, to prevent disease attack at the seedling stage.

SOWING TIME

Baby corn is cultivated in *Kharif* and *Rabi* season in different areas. It can be sown year round in southern India, whereas, in northern India, it can be sown from February to November. The time of seed sowing is May–June in Peninsular India, end of June to mid July in the Indo-Gangetic plains and up to late August in Punjab. Baby corn in *Rabi* season is sown in October to mid November in Bihar, Andhra Pradesh, Karnataka and Tamil Nadu. It is sown in spring in late January to the end of February in Bihar and *tarai* area of Uttranchal. It can be grown through transplanting in furrows in December–January in northern India. For this purpose, nursery should be raised in November. Generally, August to November planting yields the best quality baby corn. However, baby corn sowing during the second week of May followed by the third week of April produces the highest yield, particularly in hilly regions of India.

SOWING METHOD

Seed sowing is done by the ridge and furrow method; generally sowing is done on the southern side of the ridge by dibbling at a depth of 4 cm along the furrow, in which fertilizers are placed. Seeds should be covered properly with soil. Per hill, one or two seeds should be sown. Plants require more nutrients and light energy to synthesize and supply more metabolites to cobs for the development of kernels. The seeds are sown 2.5–4.0 cm deep in flat beds at 45 × 25 cm spacing for rainy and late-sown crops. In the rainy season, it is mandatory to raise the crop on ridges to avoid waterlogging near the plant root zone for better root respiration and ions uptake. For producing baby corn, two methods, *i.e.*, (i) as a primary crop and (ii) as a secondary crop for planting of baby corn, sweet corn or field corn are used:

- **Baby corn as primary crop:** The variety is planted to produce only baby corn.
- **Baby corn as secondary crop:** A sweet corn or field corn variety is planted primarily to produce either sweet corn or field corn. The choice of the variety, planting density and fertilizer rates are different in these two methods. Economic yield may be separated into three components, *i.e.*, (i) husk yield, (ii) young ear yield (baby corn) and (iii) standard ear yield.

The crop can be sown by the following methods:

Drilling method: It is the most common method of sowing baby corn. This method is adopted by big farmers; seed drills are used for sowing seeds. Sometimes, under high soil moisture conditions, it becomes difficult to run the seed drill because the sprouts are chocked with wet soil and blank rows are observed. This method is well suited for sandy soils and is very quick.

Hill sowing: In this method, the seeds are dropped with the help of corn planters, which are placed and pressed at an appropriate place for dropping four seeds at one place. It is technically known as a hill sowing. It needs relatively less quantity of seed.

Sowing seeds behind the plough: Shallow furrows are opened with the help of a country plough and seeds are dropped in them followed by planking to cover the seeds. After the emergence of seedlings, thinning is done to maintain the required plant population.

NUTRITIONAL REQUIREMENT

Baby corn is a nutrients-exhaustive crop and prefers light soil for its cultivation, which further aggravates the nutrients deficiency, as they are lost quickly after fertilizer application in the field. Therefore, fertilizer management is of paramount importance in crop cultivation. The quantity of manure and fertilizers to be applied depends upon the soil fertility and the variety or hybrid used for cultivation. Besides the application of farmyard manure 12.5 t/ha, it is advisable to apply nitrogen 75 kg, phosphorus 60 kg and potassium 20 kg/ha. The entire dose of phosphorus and potash fertilizers and a half-dose of nitrogenous fertilizers are applied as basal application just before sowing. The remaining half-dose of nitrogenous fertilizer is top-dressed 25–30 days after sowing when plants become 60–75 cm tall. Baby corn suffers from deficiency of zinc and shows symptoms like broad bands of white or light yellow patches with reddish veins around the midrib. It can be overcome by basal application of zinc sulfate @ 15–20 kg/ha at intervals of 10–12 days, starting from the appearance of the first deficiency symptom.

IRRIGATION REQUIREMENT

The water requirement depends on rainfall and weather conditions. About 50% of the water requirement is within 30–35 days after tassel formation. There should not be any water shortage at cob development stage.

Being a C_4 plant, it has high water use efficiency, and also as a shallow-rooted crop, it is not tolerant to drought conditions. First irrigation should be given immediately after sowing and subsequent irrigations should be given at 7- to 10-day intervals. Irrigation should be ensured at the most sensitive stages for water stress, *i.e.*, young seedling stage, knee-high stage, silking stage and picking stage. During winter (mid December to mid February), the soil should be kept wet to avoid frost injury. Waterlogging conditions should be avoided, as waterlogging reduces plant height, ear development and ultimately yield. When long dry spells occur at the tasseling to silking stage of the crop, then crop may be irrigated by stored-water harvesting practices for better baby corn production as shown in Table 68.2.

INTERCULTURAL OPERATIONS

THINNING AND GAP FILLING

Leaving only one healthy and vigorous seedling per hill, rest of the seedlings are pulled out after 12–15 days of sowing to give proper spacing among the plants. Where seedlings have not germinated, the pre-soaked seeds are dibbled @ 2 seeds per hole and immediately followed by irrigation.

HOEING AND WEEDING

Baby corn is a shallow-rooted crop. Hence, deep hoeing and weeding should be avoided to keep the roots free from damage. Weeds are a problem during the first month after sowing because of more vacant space. Manual weeding and hoeing with a hoe (*khurpi*) are done after 2 weeks of sowing. Top-dressing with a nitrogenous fertilizer followed by earthing-up is carried out 25 days after sowing, and thereafter, spot weeding can be done, if required.

DETASSELING

Detasseling is considered to be the most important and mandatory operation to prevent pollination and fertilization. It is a process of removing male flowers (tassels) before shedding of pollen grains. The tassels are removed by holding firmly with one hand and giving an upward jerk. Though it is laborious, it is essential to avoid pollination and to get quality baby corn acceptable in the international market. This operation increases the baby corn yield by 6.2–8.9%.

INTERCROPPING

Baby corn is remunerative, with 20 intercrops, namely, potato, green pea, French bean for green pods, beet leaf, cabbage, cauliflower, beetroot, green onion, garlic, fenugreek, coriander, kohlrabi, broccoli, lettuce, turnip, radish, carrot, celery, gladiolus, etc. Since the season is long, farmers can

TABLE 68.2
Irrigation Schedule for Baby Corn (DAS*)

Irrigation schedule (in DAS*)	Light soils	Heavy soils
First	Immediately after sowing	Immediately after sowing
Second	3–4	3–4
Third	10	12
Fourth	18	21
Fifth	25	30
Sixth	32	40
Seventh	40	50
Eighth	47	58
Ninth	55	65
Tenth	62	–

*days after sowing

utilize this lean period and get additional income through intercropping in baby corn. There is no adverse effect of intercrops on baby corn yield and *vice-versa*; rather, some of the intercrops help in improving soil fertility and protect the baby corn crop from cold injury. In general, short-duration varieties of intercrops are preferred for intercropping. The recommended dose of fertilizers of inter-crops should be applied in addition to the recommended dose of fertilizers for the baby corn crop.

WEED MANAGEMENT

The crop is heavily fertilized and sparsely grown, which attract a severe weed infestation, result-ing in a drastic reduction in yield of baby corn. The important grassy weeds are *Echinochlos colona*, *Elesine indica*, *Cyperus rotundus* and *Cynodon dactylon*, whereas, *Amaranthus viridis*, *Celois argentea*, *Cleome viscose*, etc. are most common broad-leaved weeds associated with baby corn crop in the field. Two manual weedings would be sufficient for keeping the baby corn field weed free, while mechanical weeding is good for soil aeration and root respiration; but continuous rain during *kharif* season poses a serious problem, in which the soil becomes wet and agricultural operation cannot be done. Under such situations, Atrazine or Simazine herbicide may be applied immediately after sowing of the crop. The rate of application varies from 1.0 to 1.25 kg a.i. ha^{-1} in light soils and 1.25 to 1.50 kg a.i. ha^{-1} in heavy soils dissolved in 500–600 litres of water as pre-emergence application. Herbicide should be applied only when the soil has sufficient moisture. If the broad-leaved weeds are posing a problem, a post-emergence application of 2,4-D should be applied @ 1.5 to 2.0 kg a.i. ha^{-1}.

HARVESTING

Immature cobs or ears of sweet corn at milk and before early dough stage should be harvested because in over-mature ears, the flavour and sugar in kernels will deteriorate. The ears are harvested about 18–24 days after pollination when kernels are still immature. The early varieties will mature in 65 days and late ones in 110 days after sowing.

The baby corn should be harvested at a very early stage when ears are very immature, very small without pollination and kernels have not developed. The ratio of husk-ear to dehusked-ear is 7:1. In Thailand, very small ears of commercially grown baby corn are harvested.

Baby corn ears are hand-picked as soon as the corn silks emerge from the ear tips, or a few days after. Corn generally matures very quickly, thus, the baby corn must be harvested timely and carefully to avoid ending up with more mature corn ears. The ears become ready for harvesting 47–50 days after sowing. Picking of cobs should be started from the day of silk emergence before fertilization or when the silk attains a length of about 1.3 cm for processing purposes. The harvesting stage is very critical, because a single day delay in the harvesting of cobs deteriorates the quality of cobs and degrades mar-ket acceptability. Therefore, it is advisable to assess the right size of young cobs by an organoleptic test at first picking and later by visual observation. The proper harvesting stage is when the cobs attain a size of 5–10 cm length, 1.0–1.5 cm diameter and 7–8 g weight. The baby ears should be picked early in the morning to avoid the change in the texture of the ears. The cobs should be harvested using secateurs to avoid mechanical injury. A total of six or seven pickings can be done in a crop duration of 65 days.

YIELD

The yield of baby corn depends on cultivar, soil fertility, growing season, irrigation facilities and cultural practices adopted during cultivation. Usually, early-maturing hybrids or varieties give a higher yield than late-maturing ones. With adequate irrigation facilities, the hybrids on an aver-age give a dehusked yield of about 12.5 q/ha and 250–300 q/ha green fodder, which can be fed to domestic animals or ploughed in the field as green manure.

POST-HARVEST MANAGEMENT

Post-harvest handling operations for baby corn crop are relatively simple but require pre-cooling and cold-storage facilities to prevent quality loss after harvest. Baby corn can become diseased as a result of secondary infection caused by mechanical damage and broken kernels caused by poor trimming and bruising. Handling systems that minimize mechanical damage prevent water loss and allow rapid cooling to reduce such problems.

Pre-cooling is the method of removing field heat to preserve the crop's best quality. The temperature of the ears should be lowered to 10°C by immersing them into ice water soon after harvesting, since low temperature reduces the rate of respiration, moisture loss and degradation of taste and quality of the ears. To prevent spreading of post-harvest diseases from mechanically damaged cobs, it is recommended to use chlorinated water in the form of sodium hypochlorite or calcium chloride @ 100–150 ppm as a disinfectant in chilled water.

After pre-cooling, the cobs are dehusked manually and subjected to grading. Based on ear length, the cobs are graded either manually or mechanically. The grades of baby corn are (i) large size, 11–13 cm long with 1.4–1.5 cm diameter; (ii) medium size, 7–11 cm long with 1.2–1.4 cm diameter and (iii) small size, 4–7 cm long with 1.0–1.2 cm diameter.

Moisture-retention film bags are generally used for the packaging of baby corn before shipment. The film should be perforated to prevent development of off-flavour inside the package during transportation. These are further packed into well-ventilated plastic crates.

Refrigeration soon after harvesting helps in retaining sweetness. Dehusked baby corn is better for refrigeration, and can be stored for 1 week without losing its quality. If baby corn has to be stored frozen, the husked ears are placed in boiling water or steam for 30–45 seconds, cooled properly by keeping in freezer bags or ziplock bags and then placed in a refrigerator. Ears can optimally be stored at 0°C to slow down the respiration rate and maintain quality. Low temperature reduces the rate of sugar conversion to starch. It should be stored at 5–7°C with a relative humidity of 90%.

Baby corn can be marketed as fresh husked or dehusked young cobs, canned products and pickles. It can be processed to improve its shelf life. The main processing methods which can be used to improve the shelf life are canning, dehydration and freezing. Baby corn can be canned in 3% brine, 2% sugar and 0.4% citric acid, stored for months together and transported to far-off places. Baby corn candy and pickle are in great demand in the United States of America and some European countries. India could be an important country for export of baby corn in future.

As baby corn is highly perishable, exports are made by air. Packages should be transported from the packing facility to the airport in cool trucks, especially if the corn has been cooled. In all cases, trucks should be covered to prevent contact with wind, rain and sun.

INSECT-PESTS

Armyworm (*Spodoptera frugiperda* and *Pseudaletia unipuncta* haworth)

The dark green 5 cm long armyworm with white stripes on its sides and down the middle of its back is a serious pest of corn in the early growth stages; armyworms attack and feed on tender leaves. At the crop's later stage, the insect-pest destroys the centre of the plant shoot during the daytime.

Control
• Spray ethyl parathion or methyl parathion 0.3% at weekly intervals.

Corn Rootworm Larvae (*Viabrotica undecimpunctata*)

This pest feeds by making tunnels inside the roots of the plant. As a result, the plant grows poorly or not at all, lodges after heavy rains and dies frequently. Corn rootworm control is extremely difficult as the insect-pest lays eggs frequently in the field after the corn plants are well grown.

Control

- Follow late sowing to minimize vegetative growth.
- Apply chlorpyrifos or methyl parathion 0.3%, or Carbofuran granules 1.0–1.5 kg a.i./ha a few days after germination.

Stalk Borer (*Papaipema nebris*)

The stalk borer is a major pest of baby corn found throughout India and attacks at every stage of crop growth. It mainly occurs during monsoon season. Its caterpillars damage the crop by making bores into the stem and ears. A borer that has left the whorl and entered the stalk cannot be controlled. It lays eggs 10–25 days after germination on the lower side of the leaves. The larva enters in the whorl and causes damage to the leaves.

Control

- Destroy the stubbles, weeds and other alternate hosts of stem borers by ploughing the field after harvest.
- Grow a few lines of cowpea around the baby cornfield to reduce the incidence of the stalk borer.
- Release *Trichogramma chilonis* @ 100000/ha 10–15 days after sowing.
- Spray the crop 2–3 weeks after sowing with synthetic pyrethroids like fenvalerate 20EC @100 ml/ha, cypermethrin 10EC @10 ml/ha, or deltamethrin 2.8EC @ 200 ml/ha.

Shoot Fly (*Atherigona* sp.)

In southern India, it is a serious pest but it also appears on the spring–summer maize crop in northern India. It attacks mainly at the seedling stage. The tiny maggots creep down under the leaf sheaths until they reach the base of the seedlings, and then, they cut the growing point or central shoot, which results in dead heart formation.

Control

- Sow the crop before the first week of February to protect the crop from shoot fly infestation.
- Treat the seed with imidacloprid @ 6 ml/kg of seed.

Corn Leaf Aphids (*Rhopalosiphum maidis* fitch)

Tiny insects suck cell sap from tender foliage and exude honeydew, on which sooty mould grows, and reduce photosynthetic activity of the plant, ultimately affecting the yield adversely. Affected plants show stunted growth and silky black appearance.

Control

- Apply phorate or carbofuran granules @ 1.0–1.5 kg a.i./ha to the soil at the time of sowing.
- Spray the crop with monocrotophos 0.03% or Dimecron 0.05% at weekly interval.

Jassids (*Amrasca biguttula*)

Nymphs and adults suck cell sap from tender leaves and inject some toxic substances, which leads to upward curling of the leaves. Jassids cause serious damage from May to October. Severe infestation causes burning of leaves, which later fall down.

Control

- Apply phorate 10G granule @ 1.5 kg a.i./ha to the soil at the time of sowing.
- Spray the crop with quinalphos 0.5% at weekly intervals.

Leaf Roller (*Marasmia trapezalis*)

Leaf roller damage in baby corn is caused by caterpillars, which are glossy green in colour and become pink when fully grown up. The larvae fold the leaves by silken threads and hide themselves inside the rolled leaves. The surface of the attacked leaf becomes papery in appearance.

Control

- Avoid excess use of nitrogenous fertilizers.
- Use light traps in baby corn field.
- Release *Trichogramma chilonis* @ 5 ml/ha/week 37, 44 and 51 days after sowing followed by sprays of any one of the following insecticides: *i.e.*, fenthion 80 EC @ 200 ml, dichlorvos 76 WSP @ 100 ml, *neem* seed kernel extract 5%, or monocrotophos 36 WSP @ 400 ml/ha.

Sting Nematodes (*Belonolaimus nortoni*)

Nematodes damage the plant at its young stage. Severely infected seedlings die and surviving seedlings become chlorotic and stunted. The below-ground roots show deep necrotic lesions and root tips become thick and stubby in structure.

Control

- Maintain adequate inter-row spacing.
- Adjust garlic in crop rotation.
- Use marigold as intercrop to reduce the population of nematodes.
- Apply carbofuran 3G or Aldicarb at 25 kg/ha during field preparation.
- Fumigate soil with Nemagon or D-D mixture at 20–25 kg/ha.

Other Emerging Pests

Recently, some other non-traditional pests are also causing damage to baby corn crop *viz.*, the larvae of American bollworm (*Helicoverpa armigera*), which cause damage to the cob in the southern part of India, while the chaffer beetle (*Chiloloba acuta*) feeds on maize pollens, which adversely affects pollination in the northern parts of India.

DISEASES

Downy Mildew (*Sclerophthora rayssiae*)

The affected plant leaves show long and narrow yellowish parallel stripes, and purplish downy growth appears mainly on lower surface of the leaves. The favourable conditions are low temperature (21–23°C), high relative humidity (98%), waterlogging and light drizzling. The affected plants become pale, stunted and bunchy, and the leaves may die. As the infection proceeds, purple irregular spots also develop on stem. The disease is soil as well as seed borne and the pathogen survives over-wintering either on perennial hosts or on infected plant debris.

Control

- Follow long crop rotation with pulses.
- Plough the field deep in hot summer months.
- Use healthy and disease-free seeds.
- Provide proper drainage facilities.
- Remove and destroy the infected leaves and disease debris from the field.

- Treat the seeds with hot water at 52°C for 30 minutes to kill the pathogen present on seed.
- Treat the seeds with metalaxyl compound (Apron 35 SD) @ 6 g/kg of seed.
- Spray 0.2% Dithane Z-78 after 20 days of sowing and repeat the spray at 10- to 12-day intervals.

CORN SMUT (*Ustilage zeae*)

The disease is apparent as black puffy masses on ears, tassels and other parts of the plant. Early open-pollinated yellow varieties are more susceptible to corn smut.

Control

- Remove and destroy the infected plant parts as early as possible to check further spreading of pathogen spores.
- Treat the seeds with Vitavax @ 2.5–4.0 g/kg of seed.

RUST (*Puccinia sorghii*)

The pathogen causes circular to oval or elongated cinnamon-brown powdery pustules, which are scattered over both surfaces of the leaves. As the plant matures, the pustules become brown to black owing to the replacement of red uredospores by black teliospores. Cool temperature and high relative humidity favour the development of this disease.

Control

- Remove alternate hosts (*Oxalis corniculata* and *Euchlaena mexicana*) from baby corn field.
- Spray the crop with mancozeb 0.25% at 10- to 12-day intervals.

TURCICUM LEAF BLIGHT (*Helminthosporium turcicum*)

The disease affects the plant at its young stage. Plants have a burnt appearance due to the formation of long elliptic greenish brown lesions, especially on the lower surface of the leaves. The surface is covered with olive green velvet masses of conidia and conidiophores. The disease is prevalent in cooler conditions coupled with the high humidity of Jammu and Kashmir, Himachal Pradesh, Sikkim, West Bengal, Meghalaya, Tripura, Assam, Rajasthan, Uttar Pradesh, Uttarakhand, Bihar, Madhya Pradesh, Gujarat, Maharashtra, Andhra Pradesh, Karnataka and Tamil Nadu. Long elliptical greyish green or tan lesions (2.5–15 cm) appear on lower leaves, progressing upward.

Control

- Treat the seeds with captan, thiram, or Agrosan GN @ 3 g/kg of seed.
- Spray the crop with 0.25% maneb, zineb, or Dithane M-45 at 8- to 10-day intervals.

BACTERIAL LEAF STRIPE (*Pseudomonas andropogonis*)

The bacteria usually affect the leaves below the ear, which shows narrow, linear lesions extending up to the leaf blade.

Control

- Grow only resistant hybrids or varieties.

CRAZY TOP (*Sclerophthora macrospora*)

Crazy-top infection results in excessive tillering, rolling and twisting of upper leaves, and sometimes, it is coupled with proliferation of tassels into a mass of leafy structures. The infected plant produces no pollen grain since the flower part in the tassel gets completely deformed.

Control

- Provide adequate soil drainage.
- Keep the grassy weeds under control.

CHARCOAL ROT (*Macrophomina phaseolina*)

The fungus attacks the roots of seedlings and young plants. The affected plants first exhibit wilting symptoms. The stalk of infested plants has greyish streaks and becomes weak. The pith becomes shredded, and greyish black minute sclerotia develop abundantly on vascular bundles. Shredding of interior of stalk often causes the stalks to break at the crown region. High temperature and low soil moisture are favourable for the occurrence of this pathogen.

Control

- Follow long crop rotation with crops that are not natural host of the fungus.
- Keep the crop field irrigated from the emergence of ear head to the harvesting of crop.
- Treat the seed with Bavistin or captan @ 2 g/kg of seed.

BACTERIAL STALK ROT/SOFT ROT (*Erwinia dissolvens*)

Infection starts from the collar region and spreads up to a height of 75 cm from the ground level at any stage of the crop growth. The diseased plants show typical wilt symptoms, and they collapse to the ground because the organism dissolves the middle lamella of the cell wall that results in the disintegration of tissues. The plant shows white oozing at the infected parts, which may be seen by the naked eye. Poorly drained heavy soils along with hot and humid weather favour the attack of this disease.

Control

- Provide efficient drainage by selecting a light textured soil for growing the crop.
- Avoid injury to the plants at the time of weeding and top dressing of urea.
- Apply bleaching powder along the rows @ 10–25 kg/ha.

MAIZE CHLOROTIC MOTTLE VIRUS (MCMV) and MAIZE DWARF MOSAIC VIRUS (MDMV)

The synergistic infections of these viruses cause lethal necrosis. Infected upper leaves show bright yellow mottling or mosaic symptoms, which later turn dull yellow. Death of leaves occurs from the top downwards. Ears are reduced in size, with small or aborted kernels.

Control

- Grow only resistant varieties.
- Follow long crop rotation with non-host crops such as soybean, sorghum or small grain crops.

69 False Yam

Raj Kumar and Flemine Xavier

Botanical Name : *Icacina senegalensis* A. Juss

Synonym *Icacina oliviformis* (Poiret) J. Raynal

Family : Icacinaceae

Chromosome Number : 2n = 48

ORIGIN AND DISTRIBUTION

False yam has changed in terms of classification to be known as *Icacina oliviformis* and *Icacina* in English. *Icacina oliviformis* first appeared in literature in 1813 when J. Poiret validly published its description as *Hiretlla olivaeformis* Poiret belonging to the *Rosaceae* family. In 1823, Adrien de Jussieu described the same species as *Icacina senegalensis* Adr. Juss, and J. Raynal corrected this misnomer in 1975. *Icacina senegalensis* is indigenous to West and Central Africa in the savannah areas of Senegal, the Gambia, Northern Ghana, Guinea and parts of the Sudan.

INTRODUCTION

The wild plant simultaneously produces three types of food *viz.*, (i) a fruit, which is enjoyed as a snack; (ii) a seed, which is utilized as a staple and (iii) a tuberous root, which is eaten as emergency food when other crops have failed and communities are threatened with famine. Though the species is wild, millions of people rely upon its different products like fruits, seeds and tubers during various seasons. The plants grow so densely and yield so luxuriantly that tonnes of fruits can be collected in a day during the season, even from wild stands. It is popular as a medicinal plant in its area of availability and also sometimes cultivated locally for the edible tubers eaten. The fruit ripens at the end of the dry season when other food-producing wild plants have generally run out of produce. This makes it an important food store for the hungry. False yams store very well, making them an important backup staple.

False yam plant is best known for its edible starchy tubers, which look more like large turnips or beetroot. Sometimes, they are ground into a flour to extract starch. In areas where crop failures are common, false yam could be cultivated as source of commercial starch. In fact, it is regularly cultivated by some rural farmers in Senegal, though it is generally harvested from wild shrubs. False yam, the common English name for the crop, is unfortunate because it gives farmers, agricultural policymakers and investors a sense that it is not a legitimate crop and instead is generally considered a weed. False yam in local language is referred to as *barkanas* or *kouraban* in Senegal, *manankaso* in Gambia, *takwara* in Ghana and *basouna* in West Africa. However, some researchers have begun calling the plant by its genus name, *Icacina*.

COMPOSITION AND USES

COMPOSITION

False yam is preferred by millions of people because of its highly nutritious fruits, seeds and tubers. Flour manufactured from the tubers is an excellent source of starch, energy, protein, calcium and vitamin A. In addition, a bitter toxic principle reported to be a gum resin is present in good range. The seed flour is considered a rich source of minerals like phosphorus, magnesium, calcium and iron. The root contains significant amounts of protein, calcium and iron. The nutritional composition of false yam tuber and seed flour is given in Tables 69.1 and 69.2.

USES

False yam tubers are used as a source of starch or flour with a nutty flavour, which is used as a substitute of cassava flour. Its plum-like ripe fruits, having thin sweet gelatinous pulp with a pleasant flavour, are often eaten fresh by children and their seeds are dried and pounded to yield flour, especially eaten at the time of food scarcity; however, the seeds before use are subjected to some degree of boiling or fermentation to reduce the anti-nutritional factors present in them. Since the phytochemical components in seeds are below the recommended toxic levels, the overall nutritional value of seeds will not be toxic to human beings. Its roots containing a significant amount of protein are eaten when yams are in short supply. Dry roots are used to strain out the powdered mass and boiled with millet or beans to form a paste, which is a highly palatable food and said to be good enough for guests.

TABLE 69.1

Nutritional Composition of False Yam Tuber Flour (per 100 g Edible Portion)

Constituents	Contents	Constituents	Contents
Moisture (g)	11.7	Vitamin A (IU)	3622
Carbohydrates (g)	74.5	Thiamine (mg)	0.04
Protein (g)	10.3	Riboflavin (mg)	0.18
Fat (g)	0.7	Niacin (mg)	1.4

TABLE 69.2

Nutritional Composition of False Yam Processed Flour (per 100 g Edible Portion)

Constituents	Contents	Constituents	Contents
Moisture (g)	12–13	Sodium (mg)	18.37
Carbohydrates (g)	72–73	Zinc (mg)	10.10
Protein (g)	8	Copper (mg)	8.22
Fat (g)	0.1	Vitamin A (IU)	3622
Crude fibre (g)	1.61	Thiamine (mg)	0.04
Calcium (mg)	26	Riboflavin (mg)	0.18
Phosphorus (mg)	126.45	Niacin (mg)	1.4
Potassium (mg)	7.54	Ash (g)	0.5
Magnesium (mg)	10.7	Energy (kcal)	354
Iron (mg)	2.7	–	–

The leftover substrate of false yam tubers during processing is a good option as a soil amendment, which when added to soil increases the crop performance. Its leaves are used as a container for carrying cashew nuts. The burnt leaves are used to blacken pottery.

MEDICINAL USE

False yam tubers are reported to have medicinal properties. It is traditionally used for the treatment of malaria. A decoction of newly cut leaves is used in the treatment of internal haemorrhages, coughs and all chest affections, as an antidote to snake bites and feverish states, but it is normally taken while the patient is going to sleep at night. A decoction of its twigs and roots is given to adults for general debility of an undiagnosed origin. The leaves are used externally in baths and washes. The leaves after heating are applied topically to points of pain, particularly in elephantiasis, as an analgesic.

The bitter toxic compounds, *i.e.*, *icacinon* and *icacinols*, contained in tubers prevent pests and other disease-causing organisms from damaging the shrubs, and thus, its decoction is found to be effective in controlling sweet potato weevil, when applied as soil drenching on ridges.

BOTANY

False yam is a shrubby perennial variable in form, which produces glabrous or pubescent erect leafy shoots from a large underground long-rooted fleshy tuber. The aerial stems are light green and may reach about 1 metre height. The leaves are simple, ovate or obovate, pointed or rounded at the apex, 5–10 cm long and 4–7 cm broad, light green when young but become leathery and dark green on the upper surface and dull green on the lower surface. Flowers are inconspicuous, usually white or cream and pedunculate, with erect corymbose cymes, collected into a terminal leafless panicle, or the lower peduncles arising from the axis of reduced leaves. The calyx is in five divisions, the pointed lobes are bright green and the corolla is composed of five narrow white or creamy-white petals covered with silky hairs. The fruit, covered with very short hairs, is a bright red ovoid berry, approximately 2.5–3 cm long and 2–2.5 cm wide. It contains a thin layer of white pulp approximately 0.2 cm thick, surrounding a single spherical or ovoid seed. Inside each fruit is a small round edible single seed at the centre of the fruit. Dried seeds are incredibly hard, thus, not damaged by rodents. Its tubers, resembling large turnips or beetroot, show considerable variation in size, ranging from 30 to 45 cm in length with a diameter of about 30 cm and weighing from 3 to 25 kg. The tubers are greyish in colour with a thin skin enclosing white flesh, which is usually speckled with yellow spots that correspond to bundles of xylem.

CLIMATIC REQUIREMENT

False yam grows best in tropical areas having moderate rainfall of 800–1000 mm. It mainly grows in full sunlight. However, it thrives best in an average temperature range of 25°C to 30°C. These tropical plants cannot tolerate frost. It requires annual rainfall of 1200–1300 mm. It grows successfully up to mean sea level of 800 m. It tolerates adverse conditions such as drought and temperature. High light intensity encourages the crop growth development. A short day length is favourable for root growth, while a long day length is ideal for vine growth.

SOIL REQUIREMENT

False yam is an upland crop, thus, it prefers a well-drained sandy loam or silt loam fertile soil for its cultivation. Optimum yield can be obtained from sandy loam but the acceptable yield can also be obtained from clay loam soils, particularly those high in organic matter. Stony and highly compacted soil should be avoided for growing false yam. It requires a soil pH range of 6.5–7.5 for its good growth and development.

CULTIVATED VARIETIES

False yam is a crop, which is grown on small scale, thus, no improved variety is available. Farmers generally use the locally available traditional cultivars.

PLANTING TIME

The plant usually occurs wild and is seldom cultivated. However, it is occasionally planted in Africa on the onset of rainy season. The crops can be grown any time of the year, but for best results, planting should be done at the beginning of or just after the rainy season, when it can obtain sufficient moisture for its growth. Planting is done from March to June. However, planting time depends upon the release of tuber dormancy and the start of rain in the area.

SEED RATE

For its successful cultivation and higher yield, about 18,500–19,500 plants are required to plant a hectare land area.

PLANTING METHOD

False yam is propagated by division of seed tubers, which are usually planted in a well-prepared field at the beginning of rainy season. Tubers are planted manually on ridges. Generally, one seed tuber is planted per hill, keeping a distance of 120 cm between rows and 45 cm between plants.

NUTRITIONAL REQUIREMENT

False yam crop in a hectare area removes nitrogen about 128 kg, phosphorus 17 kg and potassium 162 kg from the soil, which represents more or less its fertilizer requirement. However, like other yams, its fertilizer requirement is probably similar to that of taro. The recommended amount of fertilizer is applied in two splits: 1 month after emergence and 2 months after the first application. Fertilizers are applied about 10 cm away from the plants in bands of 2–3 cm width. False yam responds well to organic manure in the form of compost, which is mixed well with the soil at the time of field preparation or placed just below the spot where tuber pieces are planted.

IRRIGATION REQUIREMENT

False yam is a drought-resistant plant grown mostly in West and Central Africa. It requires ample moisture throughout its growing period, particularly from 14 to 20 weeks after planting when tuber bulking occurs rapidly. Irrigation should be provided in areas where the dry season is longer than 3–4 months.

INTERCULTURAL OPERATIONS

HOEING AND WEEDING

The number of hoeings and weedings depends upon several factors like the use of pre-sprouted tuber pieces, application of mulching and rate of weed growth. If tuber pieces without pre-sprouting are used and the field is not mulched, three to five hoeings and weedings are needed before the crop canopy covers the space between rows to partially suppress weed growth. If pre-sprouted tuber pieces are used and the field is mulched, at the most two hoeings and weedings are performed about 2 months apart with the help of hand tools. In unmulched fields, animal-drawn implements are used for weeding when the plants are small.

The plant often becomes a weed as it is very vigorous and rapidly invading in newly cleared and cultivated land. Its eradication is fraught with difficulty because of its large or very large tubers, which may reach 30–50 cm diameter or 25 kg in weight and set 30 cm deep in the soil, with long creeping roots.

EARTHING-UP

As the tubers grow and enlarge rapidly towards the end of the growing period of the plants, some tubers tend to push up soil and get exposed to the sun. Heavy rains also expose the tubers to the atmosphere. The exposed tubers should be covered with soil to prevent them from greening.

MULCHING

Mulching of the field is preferred to reduce soil temperature, conserve soil moisture and suppress weed growth. Dry coconut fronds, corn stalks, paddy straw and other similar materials may be used as mulch. If paddy straw or similar material that rots readily is used, the mulch should be made about 10 cm thick so that it may not rot completely within 4 or 5 months.

HARVESTING

False yam becomes ready for harvest when its foliage starts yellowing or drying up. Only the basal leaves turn yellow and the rest remains green even when new sprouts have already emerged from the base of the plant. The yellowing or drying-up period of the foliage usually starts late. Tubers, especially those intended for using as planting material, are harvested later.

The tubers are harvested by hand as and when required. Owing to their larger size and the fact that they can penetrate to about 30 cm or more below the surface, they are difficult to dig out, hence, at one time, the plant was nicknamed *abub ntope*, meaning breaking hoe. Sometimes, the tubers are harvested earlier even before the yellowing of foliage either for family consumption or catching an early market price. A hand tool is used to dig around the tuber to loosen it from the soil, and then, the tuber is lifted and clinging soil particles are removed. The vines before digging the tubers are cut at the base. In sandy soil, a sturdy stick sharpened at one end is sometimes used to dig out the tuber. For varieties with deeply buried tubers, especially in clay soils, a tractor-driven digger is used for the harvesting of tubers.

The digger is used like a shovel. Whatever implement is used to harvest the tubers, it is important that care should be taken to avoid injury to the tubers. After cleaning and discarding the diseased tubers, they are placed in rattan or bamboo baskets or wooden crates lined with soft cushioning material like banana leaves, paper, or paddy straw. The tubers are arranged in the container in two to four layers, depending upon tuber size, and a soft cushioning material is placed between layers and in the spaces between tubers in a layer. The container is then covered with paper or banana leaves, and a string net is woven over the mouth of the container if the tubers are to be transported immediately to the market. No cover is provided for the container if the tubers are to be transported to a nearby storage place.

YIELD

On an average, false yam gives a tuber yield of 20–30 q/ha, although in some parts of West Africa, it can give a tuber yield of 200 q/ha.

POST-HARVEST MANAGEMENT

Since false yam tubers contain a bitter toxic principle, *i.e.*, cyanogenic glycosides, a thorough processing including soaking in water for several days and drying is needed. Its seeds are immersed in boiled water for 4 hours. The water is changed at hourly intervals to ensure adequate removal of the bitter principle and then drained out using a perforated plastic basket. Finally, the seeds are dried in the sun for 3 days before milling to form boiled false yam meal. In the second method, false yam seeds are totally submerged in water for 7 days and the water is changed regularly every 2 days. On the seventh day, the water is totally drained off and the seeds are dried in the sun for 3 days before milling to form soaked false yam seed meal. In the third method, the false yam seeds can be roasted for 1 hour in a heated metal pot using coal as a source of heat, and then, the seeds are cooled and milled to form roasted false yam seed meal.

INSECT-PESTS AND DISEASES

GREEN SPIDER MITE (*Mononychellus* spp.)

The infestation of mites is mostly observed during warm and dry periods of the season. Nymphs and adults suck cell sap, and whitish grey patches appear on leaves. Symptoms appear on leaves as small yellow or brown spots. When the infestation is severe, the leaves become yellow and growth is affected adversely. Affected leaves become mottled, turn brown and fall. Under severe infestation, the top canopy of the plant is covered by white cottony webbing, which can be seen on the underside of the leaves. The mite-infested plants can be identified from a distance by a characteristic mottling symptom produced on the upper surface of the leaf.

Control
- Spray the crop with water during dry periods to maintain a humid atmosphere.
- Use insecticidal oil like *neem* oil or cottonseed oil 0.4% at 8- to 10-day intervals.
- Spray the crop with sulfur 80 WP 2 g or Dicofol 18.5 EC 2.5 ml, or Spiromesifen 22.9 SC 400 ml/500 litres of water.

YAM BEETLES (*Heteroligus* spp.)

Adults feed on tubers after entering through newly emerged shoots, reducing tuber yield. They usually infest the crop during high rainy season. The adult beetle feeds on young leaves and tubers. The infested plant gets distorted and severe scarring leads to underdeveloped tubers to reduce the yield drastically.

Control
- Follow deep summer ploughing to expose pupae and adults.
- Use light traps to trap the adult beetles.
- Pick off beetles by hand and destroy.
- Keep the field weed free.
- Spray the crop with Thiodicarb 0.09%.

WEEVIL (*Palaeopus costicollis* marshall)

The weevil grubs feed on tubers by making tunnels, leading to tuber decaying. Its attack causes the leaves to turn yellow, wither and die prematurely. In heavily infected plantations, production is reduced. Adults feed on dead plants and live under newly cut or rotten tissue. The borers spread from season to season through planting material.

Control
- Use a weevil-free planting material.
- Use fallow fields and intercropping with non-host crops.

- Follow clean cultivation and proper field sanitation.
- Harvest the crop a bit early.
- Collect and destroy the adult beetles.
- Treat the tubers with hot water at 52°C for half an hour.
- Spray the crop with chloropyrifos 20 EC @ 0.02%, dimethoate 30 EC @ 0.03%, or malathion 50 EC @ 0.05%.

DISEASES

POWDERY MILDEW (*Sphaerotheca* spp.)

Initially, its attack is characterized by the formation of white patches on both sides of the leaf. These patches originate as minute discoloured specks from which a powdery mass radiates on all sides, and later, the symptoms resemble a heavy dusting of greyish white chalk on the plant. The entire foliage gets covered by this powdery mass, leading to yellowing and defoliation. It can attack the leaves, stems, or roots of the false yam. If heavily infected, it will lead to reduction in crop yield. This disease becomes severe under a condition of high humidity and moderate temperature.

Control

- Cut and remove infected plant parts and destroy them.
- Maintain field and crop sanitation.
- Avoid over fertilization as new growth is more susceptible.
- Apply a slow-release fertilizer that provides more controlled growth.
- Spray the crop with 0.25% Dithane M-45, or 0.1% carbendazim.
- Spray crop with copper oxychloride 50WP 0.25%.

ANTHRACNOSE (*Colletotrichum gleosporoides*)

This is a fungal infection, which can affect the plants in any geographic region. Small dark brown spots or black lesions appear on leaves, which may be surrounded by a chlorotic halo, leading to necrosis and withering of leaves. Leaves and shoots of the false yam become black, resulting in small and deformed tubers.

Control

- Plant the crop at wider spacing.
- Collect and destroy all the infected plant parts by burning.
- Spray the crop with fungicides Chlorothalonil 2 g/litre followed by mancozeb 2 g/litre.
- Spray the crop with benomyl 0.1%, Bordeaux mixture 1%, carbendazim, difenconazole, thiophanate methyl 0.1%, Prochloraz 45 EC 0.75 ml/litre, or hexaconazole 1 ml/litre.

BACTERIAL LEAF SPOT (*Pseudomonas syringae*)

Initial symptoms of bacterial leaf spot are small water-soaked spots on both sides of the leaves. The lesions, often limited by leaf veins, are angular, square, or rectangular in appearance. These water-soaked lesions rapidly turn brown and with ageing may dry out and become papery and tan. Lesions tend to be relatively small and restricted to leaves. However, in the field, the disease is usually found only on older leaves that are protected by the plant canopy, except where sprinkler irrigation is used. Under favourable conditions, bacterial blight lesions may coalesce and cause considerable foliage blighting.

Control

- Use disease-free healthy planting material.
- Give hot water seed treatment at 52°C for 25 minutes.
- Disinfect the dirty transplant trays.
- Avoid excessive application of nitrogen fertilizers and sprinkler irrigation.
- Spray the crop with 0.5 g streptomycin sulfate or Bacterinashak + 2.5 g copper oxychloride per litre of water, mancozeb 0.25%, or chlorothalonil (Kavach) 0.1% at intervals of 10–15 days.

Mosaic Virus (Vector *aphis gossypii*)

The disease is characterised by typical mosaic symptoms on leaves, which later become mottled and distorted. The earliest symptoms appear on young leaves as light green or yellowish streaks and bands, giving a mottled appearance.

Control

- Use only disease-free healthy planting material.
- Keep the field and nearby area weed free.
- Remove and destroy the affected plants.
- Spray the crop with Triazophos 40 EC 2.0 ml, malathion 50 EC 2.0 ml, or oxydemeton methyl 25 EC 2.0 ml/litre of water at intervals of 10–15 days.

■■■

70 Chinese Artichoke

M.K. Rana and Neha Yadav

Botanical Name   : *Stachys affinis*

Family   : Lamiaceae or Labiatae

Chromosome Number : 2n = 34

ORIGIN AND DISTRIBUTION

Chinese artichoke, a member of the mint family, is native to northern provinces of China and has been eaten by Chinese and Japanese for millennia. It is well known across Asia where it is used as a pickle, especially in Japan where it is served during Osechi or Japanese New Year. In starting, this crop was not popular outside its native place, but by the mid 1800s, it achieved some importance in Central Europe. In 1882, it was imported to Paris by a French horticulturalist and eventually to the village of Crosne, where it was cultivated on large areas, and hence, its French name is *crosne*. In Japan, it is known as *chorogi*. The genus *Stachys* means *an ear of corn/grain* in reference to the floral spike, as it resembles an ear of corn or a head of wheat. The species name comes from a Latin word *affinis*, meaning *related* in reference to its growth habit, which is very similar to the related species lamb's ear (*Stachys officinalis*). This crop was not popular in Europe until the 1920s, after which it became virtually unknown outside France or Asia. However, in recent years, it has begun to appear in the markets.

INTRODUCTION

Chinese artichoke, also known as crosne, knot root, artichoke betony, chorogi, or *Stachys sieboldii*, is a delicious and easily digestible herbaceous perennial tuber crop like potato and Jerusalem artichoke but not related to either of these crops. The plants form a mat of attractive spearmint-like leaves during summer. It is less cultivated as a root vegetable due to its small, convoluted and indented tubers with whitish-brown or ivory-white skin. It is very difficult to peel out the tubers since they are spiral shaped. The skin of the tuber is thin so it can easily be removed by scrubbing under cold running water or more easily by placing the tubers with salt and shaking. The flesh is white and tender, flavour is delicate and delicious and taste is nutty due to the presence of essential oils. It can be used as an edible ornamental plant. This plant is also used to attract bees for better pollination in cross-pollinated crops.

COMPOSITION AND USES

COMPOSITION

Chinese artichoke is a nutritious vegetable containing carbohydrates, starch, lipid and cellulose with some protein. It is a good source of calcium, phosphorus, potassium and vitamins such as vitamins C and K. Its roots also contain a fair amount of selenium, iron, zinc, copper, manganese, magnesium, thiamine, riboflavin, niacin and pantothenic acid. Its tubers also contain alkaloids, tannins and glycosides. The nutritional composition of Chinese artichoke is given in Table 70.1.

TABLE 70.1

The Nutritional Composition of Chinese Artichoke Root (per 100 g Edible Portion)

Constituents	Contents	Constituents	Content
Moisture (g)	84.94	Zinc (mg)	0.49
Carbohydrates (g)	11	Manganese (mg)	0.256
Sugar (g)	1.6	Selenium (μg)	0.2
Protein (g)	3	Thiamine (mg)	0.072
Fat (g)	0.15	Riboflavin (mg)	0.066
Dietary fibres (g)	5	Niacin (mg)	1.046
Calcium (mg)	44	Pantothenic acid (mg)	0.338
Phosphorus (mg)	90	Pyridoxine (mg)	0.116
Potassium (mg)	370	Folate (μg)	68
Sodium (mg)	94	Vitamin C (mg)	1.7
Magnesium (mg)	60	Vitamin K (μg)	14.8
Iron (mg)	1.28	Energy (kcal)	47

USES

The Chinese artichoke is a traditional ingredient of Far Eastern cuisine. The tubers are eaten either raw as salad or after boiling, frying, stir-frying, steaming, roasting, braising, or sautéing like potato and Jerusalem artichoke. It is also popular as chopped garnish. One of the popular ways to serve its tubers is to simply blanch them in boiling water for a few minutes, and thereafter, they are sautéed and eaten after adding butter or extra virgin olive oil. In China, they are often pickled in vinegar or brine because of their perishable nature, as their flavour is lost in a few days. The leaves can also be used after cooking.

MEDICINAL USES

Chinese artichoke leaves and roots are used in Chinese traditional medicines for the treatment of a number of ailments like urinary and respiratory tract inflammation. Its tubers are dried and powdered to use as pain reliever or pickled in rice or sorghum wine to treat cold and influenza. Being a low-calorie food, it is good for overweight individuals. Its constituents also have laxative, digestive, antispasmodic, hypotensive and mineralizing properties. It stimulates bile activity, hence, it is recommended against bile and liver ailments. Its water-soluble fibre lowers the cholesterol level, thus, it reduces the risk of heart attack and cancer. Being an excellent source of folic acid, it reduces the possibility of premature delivery and low birth-weight babies and the risk of cardiovascular disease. It helps in strengthening the immune system, and being rich in antioxidants, it protects the cells from free radicals damage responsible for cancer, heart attack and Alzheimer disease.

BOTANY

Chinese artichoke, a flowering edible herb with dark green foliage and well-developed roots, can grow up to a height of 45–50 cm. The plant that tends to spread over the ground is compact and bushy in growth habit. Its leaves are similar to lemon balm (a member of the same family), but unlike lemon balm, they do not have any aroma. The stem is thin, hairy, aerial and rectangular with opposite oval green leaves and long rough stalks with serrated edges. Its tubers, the edible part of the plant, developed on underground creeping stems, are small, 4–6 cm long and 2 cm wide. The creamy white tubers shine like a pearl after cleaning. The roots are brittle and can break easily. The flowers borne on terminal clusters are attractive and dark magenta, purple, or pink in colour. The flowers in tight inflorescence are hermaphrodite and are pollinated by insects. The corolla is pinkish crimson in colour.

CLIMATIC REQUIREMENT

Chinese artichoke is a frost-sensitive crop as it kills the tops but not the tubers. The crop can be cultivated in regions with tropical and subtropical climates. The crop needs warm and humid conditions with full sun exposure for its successful cultivation. Although the plants can tolerate high temperature, the optimum temperature range for its good vegetative growth varies from 20°C to 25°C. It flowers rarely in temperate regions and hardly bears seeds.

SOIL REQUIREMENT

Chinese artichoke can be grown on a variety of soils but moisture retentive, medium to deep fertile light sandy loam soils rich in organic matter are ideal for its cultivation, as it is a rhizomatous plant. It prefers well-drained neutral to slightly acidic soils with a of pH 6.6 to 7.5. The plants can not tolerate waterlogged conditions.

CULTIVATED VARIETIES

Chinese artichoke is one of the crops for which no improved variety is available, thus, only the local strains are used for its production.

SOWING TIME

The sowing time depends upon the climatic conditions of the growing region. It is quite easy to grow Chinese artichoke throughout the year if basic care is provided. In subtropical climates, its sowing is done in the end of fall or starting of spring season, while in tropical climates, it can be grown throughout the year.

SEED RATE

The quantity of seed varies with growing season and region. About 400 kg tubers are enough for planting a hectare land area.

PLANTING METHOD

Chinese artichoke is mainly propagated by tubers, which are planted in groups of 3 to 4 at a depth of 7.5–10 cm, keeping row spacing 45 cm and plant spacing 22.5–30 cm. It can be planted on flat beds or ridges. While planting on ridges, the tubers are planted halfway down on inner sides of the ridges. Planting Chinese artichoke near other crop plants should be avoided, as its plants are vigorous in growth, thus, may smother their neighbouring plants, and planting near perennials can make the uprooting of tubers difficult.

The plant can also be transplanted when its tubers are dormant during fall or winter. The tubers should not be planted too deep, as they can decay due to unavailability of oxygen. The plants can be transplanted during their growth stage but the transplanted plants fail to grow after transplanting for the rest of the growing season, unless transplanted during early spring. Planting should be done when the temperature is mild and chances of frost occurrence have passed, since Chinese artichoke is a tender plant.

NUTRITIONAL REQUIREMENT

Chinese artichoke responds well to manure and fertilizers application. At the time of field preparation, farmyard manure is applied @ 25–30 t/ha, nitrogen 30 kg, phosphorus 60 kg and potash 50 kg. Later, nitrogen 30 kg and potash 50 kg/ha are top-dressed 40–45 days after planting, at the time of earthing-up. Top-dressing with fertilizers and mulching encourage its plant growth.

IRRIGATION REQUIREMENT

Chinese artichoke needs enough and a steady supply of moisture from planting until harvesting, as low soil moisture reduces size of the tubers and makes them bitter. At the time of planting, the field must have enough soil moisture for better sprouting of tubers. If planting is done in dry field, first irrigation is applied soon after planting, and thereafter, the irrigation should be applied regularly, as drying of soil makes the tubers dormant. Soil can be kept moist through mulching. A total of six to nine irrigations are required before the first monsoon and three irrigations after monsoon during September, October and November.

INTERCULTURAL OPERATIONS

Chinese artichoke, being a perennial plant, requires a bit of thinning every year. Once established, the crop needs little maintenance. However, weeding and earthing-up are helpful for its better growth. A total of two or three weedings are required to keep the field weed free.

HARVESTING

In warmer climate, Chinese artichoke becomes ready for harvesting from early to mid autumn, usually by October. The crop usually takes 5–7 months to develop tubers, hence, the first harvesting can be done 8 months after planting when the foliage has turned brown and dried out. Ground cracking is also noticed when the tubers are fully mature, as the tubers form at a very shallow depth. The harvesting operation is similar to that for potato crop. The tubers that are out of the ground should be covered with soil to avoid their withering. The tubers are harvested using spades when they are actually required, since they have very short shelf life due to their tendency to oxidize. The tubers are very delicate and tend to dry out within a week after harvesting. The tubers should not be left in the ground in hope that they will sprout and regrow in the following year. The tubers should be lifted from the field before they are used for the planting of next crop.

YIELD

The number of factors *viz.*, the material used for planting, growing season, soil type and fertility, irrigation facilities and cultural practices adopted during the cultivation of crop affect the yield of Chinese artichoke. On an average, the Chinese artichoke gives a yield of 100–150 q/ha.

POST-HARVEST MANAGEMENT

Chinese artichoke may not be stored since it becomes soft quickly. Tubers can be left in the soil until needed or stored for about 1 month in the refrigerator in water containing lemon juice. The storage potential of artichoke is usually less than 21 days, as the visual and sensory qualities deteriorate swiftly. Unwashed tubers can also be stored well in plastic food containers in the refrigerator. Packaging should be so designed that it may prevent physical damage to the produce and be easy to handle. Hydrocooling, forced air-cooling and package icing are the common methods of artichokes pre-cooling. Relative humidity of >95% coupled with a controlled or modified atmosphere offers moderate to little benefit to maintain artichoke quality. Oxygen 2–3% and carbon dioxide 3–5% in the atmosphere delay discoloration of bracts and the onset of decay by a few days at a temperature around 5°C. Oxygen below 2% may result in internal blackening of artichokes.

INSECT-PESTS

Aphids (*Capitophorous elaeagni*)

The soft bodily tiny insects, which are usually green, yellow, pink, brown, red, or black, depending on the species and host plant, live in a colony on the underside of the leaves and stems of plant. They generally not move very quickly when disturbed. They suck sap from tender parts of the plant. Under heavy infestation, the leaves become yellow and distorted with necrotic spots on them and plant growth becomes stunted. Aphids secrete a sticky sugary substance called honeydew, on which sooty mould grows, which reduces the photosynthetic activity of the plant and yield. Aphids also transmit virus from unhealthy to healthy plants.

Control
- Remove the infested plant parts if aphid population is limited.
- Install reflective mulches such as silver-coated plastic to deter aphids from feeding on plants.
- Spray water with a strong jet pressure to wash the aphids from leaves.
- Spray the crop with insecticidal soap or oils such as neem cake or canola oil.
- Spray the crop with malathion 0.1%, monocrotophos, dimethoate, or methyl demeton 0.05% at 20 and 35 days after sowing.

Cutworms (*Agrotis* spp.)

Cutworms are dull brown caterpillars that are found on or just below the soil surface or on lower parts of the plants. They are smooth skinned and are most active at night. Its larvae cut the young plants at ground level. They make holes in tubers and damage them.

Control
- Remove all plant residues of previous crop.
- Keep the field weed free.
- Handpick the larvae and destroy them.
- Drench the soil with chloropyrifos 0.1%.
- Apply phorate 4–6 kg/ha before sowing seeds.

Slugs (*Limax maximus* and *Limax flavus*)

Slugs feed on leaves and soft tissues of the plant and make irregular holes with smooth edges in leaves and flowers. They can trim succulent plant parts and can feed on fruits and young plant bark. They are mainly the pests of seedlings and herbaceous plants but they are also the serious pests of ripening fruits that are close to the ground. They also feed on foliage and fruits of some trees. Silvery mucous trails appear on the plant due to damage caused by slugs.

Control
- Keep the preceding crops and seedbeds weed free, reducing sources of food and shelter.
- Expose the slugs and eggs to freezing temperature and hot dry conditions.
- Handpick the slugs thoroughly at regular interval.

- Install slug traps in the field.
- Keep the seedbeds firmly hard to reduce slug activity.
- Follow seedbed cultivations repeatedly to cause death of the slugs.
- Apply predatory nematodes *Phasmarhabditishermaphrodites* in advance of damage.
- Install baits for effective control of slugs.
- Use ferric phosphate–based slug pellets.
- Use natural repellents like citrus rinds, used coffee grounds, eggshells and sharp sand.
- Use molluscicide pellets, *e.g.*, metaldehyde and carbamate types such as methiocarb.
- Spray the crop with thiacloprid 200 g/litre or the mixture of water and ammonia in a 1:1 ratio.

ROOT-KNOT NEMATODE (*Meloidogyne hapla*)

Root-knot nematodes are plant-parasitic nematodes. The larvae infect plant roots, causing the development of root-knot galls that drain the plants' assimilates and nutrients. The infected plants become stunted, wilted, or chlorotic. It persists in soil for years. The degree of root galling generally depends on three factors, *i.e.*, cultivar, nematode population density and host plant species. Severe infection results in reduced yield and affects consumer acceptance. Infection of young plants may be lethal, while infection of mature plants causes decreased yield.

Control
- Follow long crop rotation with non-susceptible plants such as corn.
- Grow the crop in a nematode-free area.
- Avoid carrying soils through implements from infected to healthy field.
- Apply carbofuran 1 kg a.i/ha along with *neem* cake 200 kg/ha and urea 23 kg/ha.
- Treat the seed with carbofuran 3% w/w.

DISEASES

The Chinese artichoke is a rustic plant, which is hardly affected by diseases, but under certain weather conditions, its plants can be attacked by some of the diseases.

POWDERY MILDEW (*Erisyphe heraclei*)

The disease affects all the plant parts. A greyish powdery growth first appears on upper and lower surfaces of leaves and thereafter on petioles, flowers stalks and bracts, which later turn yellow, curl, dry up and fall off. Leaves become chlorotic, and severe infections can cause the flowers to become distorted. Early infection causes more yield reduction than late infection. Dry weather conditions are favourable for the occurrence of the disease. The infection is most severe in shaded areas.

Control
- Use seeds collected from pathogen-free plants.
- Treat the seed with Bavistin/Sulfex or hot water prior to sowing.
- Avoid application of higher doses of fertilizers.
- Apply protective spray with Bavistin 0.1%, Benlate 0.1%, wettable sulfur 0.3–0.5%, Sulfex 0.2%, or Karathane 0.1%.

BACTERIAL CROWN ROT (*Erwiniachry santhemi*)

The symptoms appear as wilting and a slow or rapid collapsing of plants. Leaves become rosette and wilt and die. Seedlings may also die. The roots can appear water-soaked and brown instead of white.

A water-soaked lesion can often appear at the base of the stems, which become hollow, and death of plants takes place. A black discoloration appears in stem. The fungus is soil borne and survives in soil for several years.

Control

- Follow crop rotation of at least 2 years with non-susceptible hosts such as corn.
- Destroy the infected plant debris.
- Maintain good drainage facilities in artichoke field.
- Treat the seed with Bavistin 3 g/kg seed or Vitavax 6 g/kg seed.
- Spray with Bordeaux mixture 1% or copper oxychloride 0.4%.

ARTICHOKE CURLY DWARF VIRUS (ACDV)

The disease symptoms vary among the crops but often appear as straw-coloured and diamond-shaped lesions on leaves. Leaves and buds become distorted with dark necrotic spots or patches. Some lesions have distinct green centres with yellow or tan borders and others appear as concentric rings of alternating green and yellow or tan tissue. Plant growth and vigour are reduced. The diseased plants may be widespread across the field but the highest disease incidence is often observed on field edges.

Control

- Use certified disease-free healthy seed.
- Remove and destroy all the infected plants.
- Eliminate weeds in and around the field.
- Spray the crop with *neem* seed kernel extract 0.4%, methyl dematon 0.075%, or spinosad 0.1% at 10- to 15-day intervals.

■■■

71 Agathi

A. Vijayakumar and R. Sridevi

Botanical Name	: *Sesbania grandiflora* (L.) Pers
	Syn. *Agati grandiflora* (L.) Desv.
	Sesbania aegyptiaca (Pierre) Pers
	Sesbania formosa F. Muell.
	Sesban coccinea (L. f.) Poiret
Family	: Leguminosae (Fabaceae)
Chromosome Number	: 2n = 14, 24

ORIGIN AND DISTRIBUTION

The exact origin of *Sesbania grandiflora* is not known (India or Indonesia has been suggested) but it is considered native to many Southeast Asian countries. It is very closely related to the endemic Australian species *Sesbania formosa*. This information supports the supposition that *Sesbania grandiflora* originated in Indonesia. It is widely distributed throughout the tropics, from southern Mexico to South America, and has been planted in southern Florida and Hawaii. It has been cultivated for at least 140 years in West Africa and more recently in East Africa.

INTRODUCTION

Agathi, more commonly known as hummingbird tree or august tree, is a fast-growing tree that grows primarily in hot and humid tropical areas of the world. It is usually grown as a green manure and forage crop, valued for its leaves and flowers. This quick-growing soft-wooded tree has ornamental, food and fodder values. Its bark yields good fibre. It is grown as a standard for pepper and betel vines as a shading plant for coconut seedlings and as windbreak for bananas. It is a multipurpose tree, which has food, fodder and medicinal values. It can be used for ornamental plantings, living fences, windbreaks, gum and tannin production, and pulp and paper production. It is the main (and sometimes the only) component of ruminant diets. This species, apparently frost-sensitive, seems limited to the tropics. In India, it is grown in parts of Punjab, Assam, Tamil Nadu and Kerala.

COMPOSITION AND USES

Composition

Agathi leaves provide a good amount of protein, fat, fibre, ash, calcium, phosphorus, iron, vitamin A, thiamine, riboflavin, niacin and ascorbic acid. Its flowers are also a good source of protein, fat, carbohydrates, fibre, ash, calcium, phosphorus, sodium, potassium, βeta-carotene, thiamine, riboflavin, niacin and ascorbic acid. Seeds contain crude protein, fat, total carbohydrates and ash. The seed oil contains palmitic, stearic, oleic and linoleic acids. The nutritional composition of *agathi* leaves is given in Table 71.1.

TABLE 71.1

Nutritional Composition of *Agathi* Leaves (per 100 g Edible Portion)

Constituents	Contents	Constituents	Contents
Water (g)	73.1	Vitamin A (IU)	9000
Carbohydrates (g)	12	Thiamine (mg)	0.21
Protein (g)	8.4	Riboflavin (mg)	0.09
Fat (g)	1.4	Niacin (mg)	1.2
Calcium (mg)	1130	Vitamin C (mg)	169
Phosphorus (mg)	80	Ash (g)	3.1
Iron (mg)	3.9	Energy (kcal)	93

USES

The tender leaves, green fruits and flowers are eaten alone as a vegetable or mixed into curries or salad. The leaves can be consumed in the form of juice or cooked spinach. Flowers may be dipped in batter and fried in butter. Its tender portions serve as cattle fodder, however, its overeating is said to cause diarrhoea. Ripe pods apparently are not eaten. The inner bark can serve as fibre, and the white soft wood, not too durable, can be used for making cork. The wood is used like bamboo in Asian construction. The tree is grown as an ornamental shade plant and for reforestation.

MEDICINAL USES

All parts of the august tree are utilized for medicinal purposes in Southeastern Asia and India, including preparations derived from the roots, bark, gum, leaves, flowers and fruits. Leaves are used as tonic, diuretic, laxative, antipyretic and chewed to disinfect mouth and throat. Flowers are used for cooling purposes and improving appetite and in headache, dimness of vision, catarrh and headache. Its astringent and bitter bark is used for cooling, as tonic, anthelmintic, febrifuge and in diarrhoea and smallpox. Its bitter and acrid fruits are a laxative and used in fever, pain, bronchitis, anaemia, tumours, colic, jaundice and poisoning. Thin stem is used as an ingredient for good health medicines in *Siddha* and *Ayurveda*. Roots are used as an expectorant and for rheumatism, painful swelling and catarrh.

BOTANY

Agathi is a fast-growing perennial, deciduous, or evergreen legume tree up to 15 m tall and 30 cm in diameter. Its lifespan is about 20 years. Roots are normally heavily nodulated with large nodules. The tree can develop floating roots and aerenchyma tissue. Stems are tomentose and unarmed. Bark is light grey, corky and deeply furrowed and the wood is soft and white. Leaf is pinnately compound, up to 30 cm long including a petiole 7–15 mm long. The rachis is slightly pubescent or glabrous. Leaflets are 20 to 50, in pairs opposite to alternate on the same leaf, oblong to elliptical, 12–44 × 5–15 mm size, rounded to obtuse to slightly emarginate at the apex, glabrous or sparsely pubescent on both the surfaces. Stipels are filiform, 0.75–1 mm long, pubescent, persistent, and the stipules broadly lanceolate, 8 mm long and early deciduous.

The flowers are large (7–9 cm long) and borne on an unbranched pendulous inflorescence. Two varieties of *Sesbania grandiflora* are recognized, including variety *grandiflora*, which has white flowers, and variety *coccinea*, which has rose-pink or red flowers. In shape and arrangement, the flowers are similar to pea flowers with five petals that are differentiated into a standard, wing and keel petals. The standard petal is usually upright, the wing petals spread out on either side of the flower, and the keel is boat-shaped, and in this species, it is curved down and away from the flower. It bears raceme axillary 2–4 flowered with rachis up to 65 mm long, peduncle 15–35 mm long, tomentose, pedicels 15–18 mm long and pubescent, bracts lanceolate, 3–6 mm long and early deciduous, flowers white, yellowish, rose-pink, or red, calyx 15–22 mm long, enclosed in young buds, splitting or breaking at anthesis, the basal part persistent in the fruit, standard up to 10.5 × 8 cm, no appendages at the claw, wings up to 10.5 × 3 cm without a basal tooth, staminal tube 10–12 cm long, curved for most of its length and ovary and style glabrous.

The fruit is a thin pod, which can be up to 60 cm long. Pods are linear to slightly falcate, 20–60 × 6–9 mm with broad sutures, 15–50 seeded, septa 7.5–10 mm apart, glabrous, hanging vertically and indehiscent. Seed is subreniform, 5–6.5 × 2.5–3 mm in size and dark brown in colour with a weight of 17,000–30,000 seeds per kilogram.

CLIMATIC REQUIREMENT

Agathi is well adapted to hot and humid environments but it does not grow well in the subtropics, particularly in areas with cool season minimum temperature below 10°C. It is outstanding in its ability to tolerate waterlogging and is ideally suited to waterlogged or flooded environments seasonally. When flooded, the plants initiate floating adventitious roots and protect their stems, roots and nodules with spongy aerenchyma tissue. It is best adapted to regions with annual rainfall of 2000–4000 mm but been grown successfully in semi-arid areas with 800 mm annual rainfall and up to 9 months' dry season. It is tolerant to flooding over short periods. It is adapted only to the lowland tropics up to 800 m, occasionally to 1000 metres. Ranging from tropical dry to tropical moist forest life zones, *agathi* is reported to tolerate annual precipitation of 4.8 to 22.5 dm and annual temperature of 24.3°C to 26.7°C.

Agathi is frost sensitive and intolerant to extended periods of cool temperature but poor in shade tolerance. Its rapid early growth and erect growth habit usually enable it to access sunlight by overtopping neighbouring plants.

SOIL REQUIREMENT

Agathi is tolerant to a wide range of soils including soils that are alkaline, poorly drained, saline, or low in fertility. It is reported to tolerate drought, heavy soils, poor soil and water-logging conditions. It tolerates saline and alkaline soils and has some tolerance to acidic soils down to water pH 4.5. It is well adapted to heavy clay soils. However, its tolerance to highly acid and aluminium saturated soils is not known. It requires a soil pH of 6.6 to 8.5 for good growth.

CULTIVATED VARIETIES

No cultivar of *agathi* has been formally released. There are two varieties of *agathi* leaves. One variety bears white flowers and the other one bears red flowers called red august tree leaves. The common variety is white-flower august tree leaves.

SEED RATE

The germination percentage of seed chiefly affects the seed rate. The seed rate may also vary due to growing season, method of sowing and soil properties. Seed rate is 7.5 kg/ha if sown at a row-to-row spacing of 100 cm and plant-to-plant 100 cm. It can also be propagated vegetatively by using stem and branch cuttings but this is not a widespread practice. For wood production, *agathi* can be planted very densely, *i.e.*, over 3000 plants per hectare are planted in Australia and India.

SOWING TIME

Sowing of seeds of a particular variety at its right time is very essential to get a higher yield from *agathi* since unfavourable conditions have an adverse effect on its yield-contributing components. In subtropical areas, sowing is done in November–December and in tropical areas from October until January. In general, *agathi* is sown in India in June and July. Cuttings should be set out in the beginning of rainy season.

SOWING METHOD

Agathi is easily propagated by direct seeding. It is not hard seeded so the seeds usually germinate well without scarification. Pretreatment could involve either scratching or nicking the round end of each seed, avoiding the cotyledon, or soaking in cold or tepid water for 24 hours. Its seed germination is 85–90%. In the nursery, plants can be raised in polythene bags. These plants can be established in the field within a month by planting them in 20–25 cm holes at a spacing of 1.2 × 1.2 m. Seeds can be sown directly into well-worked soil at the beginning of rainy season. The field is burned to obtain ash and sowing may be done in patches or lines afterwards. *In vitro* tissue culture has also been reported and methods to raise plants from hypocotyl and cotyledon explants have been developed.

NUTRITIONAL REQUIREMENT

Being tolerant of low fertility, no fertilizer is generally required. *Agathi* responds to lime in strongly acidic soils and phosphorus (about 20 kg/ha) in low-fertility soils. If *agathi* seeds are inoculated with the correct strain of *Azorhizobium caulinodans* (ORS571) it can fix a large amount of nitrogen (50–200 kg/ha) often in a period of only 6–8 weeks. The farmyard manure should be incorporated into the soil at the rate of 12.5 t/ha during field preparation. Nitrogen, phosphorus and potash fertilizers (30, 15 and 33 kg/ha, respectively) are applied in the field. Application of superphosphate as a basal dose followed by two applications of nitrogen and potash through urea and muriate of potash at monthly intervals is advantageous to the crop.

IRRIGATION REQUIREMENT

Agathi is irrigated every 4–5 weeks with one to two buckets of water per plant, which means when the soil has been dry for days. If the season is very hot and arid, the quantity of irrigations can be intensified although leaving the substratum soaked with water must be avoided. Particular attention should be given to those plants which have recently been placed in shelter. Over the year, trees develop a conspicuous radical apparatus and need watering only if the climate is particularly dry.

INTERCULTURAL OPERATIONS

HOEING AND WEEDING

The removal of weeds is useful because these unwanted plants compete with the crop for space, light, water and nutrients. Weeding should usually be done in an early crop stage, as this is more practical and more effective when the weed plants are smaller. Often, weeding has to be repeated two or three times, whenever weeds start to grow again. Besides helping to get a healthier crop, weeding has some other advantages. It helps to alter the microclimate below the plants. Sun and wind can penetrate deeper in a weeded crop and reduce the humidity. This can have a positive impact on pest populations and some diseases. Weeding also helps in loosening the soil. Water can infiltrate more rapidly and roots of the cultivated plants can develop better. Weeds can be controlled mechanically or by using chemicals (herbicides).

THINNING

In *agathi*, thinning is an important operation to maintain optimum spacing between plants and to provide better conditions to the plants for their growth. It is also essential to reduce competition among plants for space, light, nutrients and moisture. Thinning also improves air circulation around seedlings and so helps in reducing the likelihood of some plants from developing diseases. Crowding by weeds limits air movement around foliage and slows drying after rain, dew or overhead irrigation. Crowded plants are harder to treat for insect or disease problems because sprays or dusts have a hard time penetrating dense foliage. Poor coverage translates directly into poor control.

EARTHING-UP

Either this operation could be done manually or, in the case of mechanical cultivation, a ridger is used for earthing-up, depending upon the spacing adopted. It provides sufficient soil volume for root proliferation, promotes better soil aeration, helps in controlling weeds and provides a sound anchorage or support to the crop, thus preventing lodging.

HARVESTING

Agathi may not tolerate frequent and complete defoliation as it causes high mortality rates. Initially, the side branches of a tree may be cut, leaving the main growing point untouched. After the tree has reached a height of 3 m or more, it can be cut back to a height above 1.5 m. First cut is taken after 8 months and subsequent harvests at an interval of 60–80 days. When cultivated for fodder, *agathi* is usually cut when the plants are 1 m tall. Indonesian foresters harvest the plants of species growing for fuel wood on 5 years' rotation. A hectare area can yield 3 m³ of stacked fuel wood in a 2-year rotation. After the plant is harvested, the shoots are sprout with such vigour that they seem irrepressible. The tree's outstanding quality is its rapid growth rate, particularly during its first 3–4 years.

YIELD

Forage yield of *agathi* depends very much on soil fertility and the management imposed. When side branches are lopped periodically, a tree can yield up to 27 kg of fresh green leaves per year. Wood yield of 20–25 m³ per hectare per year is commonly obtained in Indonesia. For green manure, yield of 55 t/ha green material has been obtained in Java in 6–7 months. Green fodder yield of 100 tonnes per year is obtained from a hectare area.

POST-HARVEST MANAGEMENT

The harvested *agathi* leaves are used as food for human consumption or feed purposes for cattle. The leaves are the source of green leaf manure. The leaves are used as such freshly for food, feed and green leaf manure purposes so no post-harvest management practice is done.

INSECT-PESTS

WEEVIL (*Alcidodes buko*)

The weevil causes serious damage to young crops both in its adult and larval stage. It makes holes through leaves and bores the stem, causing gall-like swellings.

Control

- DDT (0.05%) and BHC dust (5%) are cheap and effective.
- Pyrethroids (0.2%) may be repellent against the weevil, particularly at low application rates.
- Chlorpyrifos (0.05%) and Parathion (0.2%) are also effective against the grub.

DROSOPHILID FLY (*Protostegana lateralis*)

Drosophilid fly is a serious pest in Tamil Nadu. The maggots bore into the tender shoots of mature plants, causing a gradual wilting of affected parts. The maggots (larvae) tunnel and feed within the pod. Infested pod will usually drop to the ground.

Control

- Practice sanitation techniques.
- Use *Drosophilid* fly lure.
- Use traps for monitoring the levels of male fly.
- Follow long crop rotation.

TREE HOPPER (*Otinotus oneratus*)

Tree hopper infests *agathi* from July to February. The primary injury is caused by the females cutting into the bark to lay their eggs. Egg laying results in slits in the bark. The feeding results in honeydew like substance, which allows the fungus sooty mould (black mycelium) to grow and turn the plant blackish.

Control

- Dormant oil sprays can be used to kill overwintering eggs.
- Use of biopesticides like azadirachtin or pyrethrin.
- Use of pesticides like chlorpyrifos or imidacloprid.

STEM BORER (*Azygophleps scalaris*)

The larvae of the stem borer make tunnels through the main stem of the host plant and also feed on roots and eat the pith region without damaging the epidermis. The plant becomes weak and breaks off at the slightest jerk.

Control

- Uprooting the stumps immediately after harvesting and burning them may prove effective means of control.

SEED CHALCID (*Bruchophagus mellipes*)

Larvae of this insect infest and damage the seeds. Chalcid-damaged seeds can be detected by looking for small round holes in mature pods through which the adults have emerged. Green seeds containing a larva are distinguished by an off-colour, and a portion of the seed will have an irregular surface. Cutting the seed open will verify the presence of a larva. The seed is no longer viable.

Control

- Fumigation by treating the larvae with carbon disulfide indicated that lowering of pressure by 200 mm Hg from the atmospheric pressure resulted in substantial increase in mortality of the pest.
- Insecticides such as Fenvalerate are currently available for the control of seed chalcid.

Root Knot Nematode (*Melodogyne incognita*)

Root-knot nematode symptoms on plant roots are dramatic. As a result of nematode feeding, large *galls* or *knots* can form throughout the root system of infected plants. Severe infections result in reduced yield. Severely affected plants will often wilt readily. Since galled roots have only limited ability to absorb and transport water and nutrients to rest of the plant, severely infected plants may wilt even in the presence of sufficient soil moisture, especially during the afternoon. Plants may also exhibit nutrient deficiency symptoms because of their reduced ability to absorb and transport nutrients from the soil.

Control

- Cover crops such as sudangrass and marigolds actually produce chemicals that are toxic to nematodes.
- Fumigants (such as 1,3-dichloropropene, methyl bromide, or dazomet) are commonly applied as pre-plant treatments to reduce nematode numbers but they must thoroughly penetrate large soil volumes to be effective.
- In addition to broad-spectrum fumigants, nervous system toxins, including oxamyl and fenamiphos, have been shown to be extremely effective for controlling root-knot nematodes.

DISEASES

Seedling Blight (*Colletotrichum capsici*)

This disease is characterized by formation of elongated or oblong cankers on collar region of the affected seedlings. Seedling blight occurs when seedlings are attacked before or shortly after emergence. There may be brown spots on the coleoptile, roots and culms. Seedlings may be killed or stunted by early infections. Damping-off often occurs when the disease is seed-borne.

Control

- The cankers, controlled with a Bordeaux spray, have black bristle-like tufts of setae.
- Avoid deep seeding as this increases disease severity.
- Fungicide seed treatment will reduce seedling blight.
- Maintaining adequate and balanced soil fertility helps in decreasing the severity of prematurity blight and supports vigorous root and shoot growth.

Grey Leaf Spot (*Pseudocercospora sesbaniae*)

The most diagnostic disease symptom is an oblong leaf spot, the centre of which may have a grey felt-like growth of sporulation after extended periods of warm moist conditions. A leaf spot may appear olive green to brown with a dark brown border when sporulation is not present. When the disease is severe, stands may appear thin and generally unthrifty. Closer inspection will reveal several leaf spots, many of which may have coalesced to turn entire blades or shoots brown.

Control

- Management practices aimed at reducing stress and extended periods of leaf wetness may help in reducing disease severity.
- Irrigation frequency should be reduced to allow foliage to dry between irrigation events.

Rusts (*Uromyces poonensis*)

The initial symptoms are yellow spots on leaf surface. The spots develop into oval to elongate reddish-brown powdery and elevated lesions that contain a powdery mass of orange to reddish-brown spores (pustules) on upper and lower leaf surfaces. The severely affected leaves often turn yellow and fall prematurely.

Control

- Use disease-free seed.
- Crop rotation can break the disease cycle because many rusts are host specific.
- All diseased and dead material should be removed at the end of the growing season.
- Difenoconazole and copper-based fungicides may be used to control rust.

Angular Black Leaf Spot (*Protomycopsis thirumalacharii*)

On leaves, the bacterium causes small, angular, water-soaked areas, which later turn brown or straw-coloured. Leaf lesions are delimited by the veins, thus, the lesions appear angular. The affected leaf tissue often dries and drops out, leaving irregularly shaped holes in the leaves. Heavily infected leaves may turn yellow.

Control

- Use pathogen-free seed.
- Follow long crop rotation.
- At the first sign of disease, apply fixed copper + maneb at label rates and repeat the sprays every 7 days.

Sesbania Mosaic Virus

Sesbania mosaic virus is reported in India and is spread from infected growing trees. This disease is transmitted by the vector aphid. The virus causes decreased nitrate reductase activity and interferes with nodulation in India. Plants infected with virus may show mottling, yellowing and malformation of leaves and pods. Infected plants may be stunted and bunchy, and seeds may be aborted, smaller, or malformed.

Control

- Use virus-free seeds.
- Spray insecticides to control aphids.
- Uproot affected plants from the field and destroy them to check further spread of the disease.

72 Chinese Potato

M.K. Rana and Mekala Srikanth

Botanical Name	: *Solenostemon rotundifolius* Poir.
	Syn. *Plectranthus rotundifolius* Poir.
	Coleus parviflorus Benth
Family	: Labiatae
Chromosome Number	: 2n = 64

ORIGIN AND DISTRIBUTION

Chinese potato is supposed to have originated in Kenya or Ethiopia in East Africa from where it spread throughout tropical Africa, Guinea and then into Southeast Asia including India, Sri Lanka, Malaysia and Indonesia. It is cultivated on a small scale in many West African countries like Burkina Faso, Ghana, Mali, Niger, Nigeria and Togo. It is still popular in the middle belt and north-eastern regions of Nigeria, especially around the states of Bornu, Taraba, Nasarawa, Jos Plateau and Kaduna. In India, its commercial cultivation is confined to southern parts of the country, *i.e.* Karnataka, Kerala and Tamil Nadu. In most of the areas where this crop is grown, it is cultivated as a garden or subsistence crop for family consumption, and in local markets, it is available seasonally.

INTRODUCTION

Chinese potato, also known as Hausa potato, native potato, country potato, Madagascar potato, frafra potato, Sudan potato, Zulu round potato, coleus potato, *innala, kembili, ketang, koorka, ratala, saluga, tumuku* and *vatke*, is a member of the family Labiatae, which consists of heterogeneous assemblage of over 200 species, comprising ornamental, medicinal plants and edible tuber crops. Its local names also vary with geographic locations. In Ghana, it is known as Salanga potato and in Nigeria Hausa potato. It is a small herbaceous bushy annual with a succulent stem and aromatic leaves. It is grown for its sweet tubers, which are usually round or slightly elongated in shape with thin skin. Formerly, it had considerable importance as a staple food in tropical Africa, but later, it was replaced largely by other starchy food crops, such as cassava and potato, thus, its production has declined to such an extent that it has almost disappeared in many areas. The taste of its tubers is similar to Irish potato and trifoliate yam and can be eaten as a main starchy staple or in combination with legumes, rice and vegetables.

COMPOSITION AND USES

COMPOSITION

Chinese potato is a rich source of carbohydrates, which are present in the form of sugar, starch and dietary fibres. It contains calcium and iron in good amounts, protein, vitamin C and niacin in moderate amounts and riboflavin and thiamine in trace amounts. It is also good in amino acids *viz.*, arginine, aspartic acid and glutamic acid. The tubers are low in calories. The nutritive composition of Chinese potato has been given in Table 72.1.

TABLE 72.1

Nutritional Composition of Chinese Potato (per 100 g Edible Portion)

Constituents	Contents	Constituents	Contents
Water (g)	75.6	Iron (mg)	6.0
Carbohydrates (g)	21.9	Thiamine (mg)	0.05
Protein (g)	1.3	Riboflavin (mg)	0.02
Fat (g)	0.2	Niacin (mg)	1.0
Dietary fibre (g)	1.1	Vitamin C (mg)	1.0
Calcium (mg)	17.0	Energy (kcal)	394

USES

Chinese potato tubers, which are somewhat sweet in taste and release aroma during cooking, are very much similar to potato in ways of cooking and eating. The tubers can be eaten raw and in cooked, baked, or fried form. The tubers can be used as a substitute of potato and are usually cooked in a curry and eaten with rice. In West Africa and South India, tubers are commonly peeled, diced, boiled and prepared as a spicy vegetable dish to accompany rice, *naan*, yam, etc. The tubers can be used as a base for a variety of South Indian dishes, such as *thoran*, *mezhukkupuratti*, *ularthi-yathu*, *erisseri*, *sukke*, *upkari* and *humman*. The tubers are roasted and eaten as a snack. In Ghana, an alcoholic drink has also been brewed from its tubers. Sometimes, its leaves are eaten as potherb.

MEDICINAL USES

Chinese potato tubers possess an elite flavour and taste and have medicinal properties due to the presence of flavonoids, which help to lower the cholesterol level in blood. In Africa and Asia, it is commonly used in the treatment of dysentery, eye disorders, sore throat and haematuria.

BOTANY

Chinese potato is an erect low-growing semi-succulent bushy plant with foliage reaching a height of about 30 cm. The leaves are thick, serrated or oval, growing in opposite pairs on succulent square-sectioned stems, which branch out from the base of the plant. Alternating pairs of leaves grow at right angles to each other and the leaves have an aromatic smell resembling that of mint. The leaves of some varieties have a variegated pattern of pigmentation. The plant bears blue, purple, pale violet, or pinkish white small hermaphrodite flowers on terminal racemes or false spikes. The stems and racemes of most varieties are green but those of some varieties are brownish green or purple. The dark-brown tubers produced in clusters at the base of the stem are small, oblong or oval-shaped and about 5 cm length. The tubers generally have dark grey or greyish brown skin and white or cream-coloured flesh. However, the varieties with considerable variation in size, shape and skin and flesh colour of the tubers are also cultivated in Africa. The skin colour of tubers may be red or creamy and the flesh deeper reddish yellow, dark brown, light grey, or reddish purple. The cultivated varieties produce few or no seeds. The tubers of South Asian varieties are generally somewhat larger as compared to the tubers of African varieties. Chinese potato tubers are very much similar to potato in appearance but comparatively small in size.

CLIMATIC REQUIREMENT

Chinese potato is a crop of tropical and subtropical regions and grows well in regions receiving an annual rainfall between 700 and 1000 mm. It requires very good rainfall for its proper growth and cannot withstand drought conditions. A low temperature at night favours the development of tubers. The optimum temperature required for its successful cultivation is 20–25°C with relative humidity 75–90%. In India, it is grown as a monsoon crop.

SOIL REQUIREMENT

Chinese potato prefers a well-drained friable sandy loam soil with good organic matter content. Heavy clay soils are not suitable for its cultivation. It grows well in moist habitats but not in water-logged soils since waterlogging conditions cause tuber deformities, which reduce the total yield considerably. The optimum soil pH for its successful cultivation is 5.2–7.0.

For the preparation of the field, two or three light ploughings followed by planking are required to make the soil loose, friable and porous. All the stubbles of previous crop are removed from the field to prepare clean ridges. Care should be taken to provide drains or a gentle slope for good drainage.

CULTIVATED VARIETIES

Sree Dhara

A compact bushy variety developed at Central Tuber Crops Research Institute, Trivandrum, Kerala, bears a cluster of dark brownish heteromorphous tubers, which possess an aromatic flavour. Starch content of the tubers is 19.5%. It has a yield potential of 250–280 q/ha in a crop duration of 150–160 days.

Suphala

A photo-insensitive mutant variety developed at Kerala Agriculture University, Thrissur, Kerala, through tissue culture matures in 120–140 days after planting and is adapted for year-round cultivation. The average yield of tubers is 160–180 q/ha.

Nidhi

A variety developed at Regional Agricultural Research Station, Pattambi, Kerala, through clonal selection from NBPGR accession CP 79 matures in 150–160 days after planting.

PLANTING TIME

The optimum time for the planting of tubers in a nursery is March to May, and in the main field, planting is done in the months of July to October.

SEED RATE

The seed rate varies according to the growing season, growing region and cultivation system. The plant density in different regions varies from 20,000 to 60,000 plants per hectare. Chinese potato is propagated by vine cuttings. Depending on varieties, planting time, irrigation facilities and field conditions, about 170 to 200 kg of tubers are required to raise enough seedlings for planting a hectare land area.

RAISING OF NURSERY

For raising seedlings, the nursery soil is thoroughly prepared to make it friable. An area of 500–600 m² is sufficient to produce cuttings required for planting in a hectare area. Well-decomposed farmyard manure of 125–150 kg is thoroughly incorporated in nursery soil. The nursery beds of

10 m length, 90 cm width and 15 cm height are prepared. Healthy seed tubers having a weight of about 15–20 g are planted about 4 cm deep in rows at 15 × 5 cm spacing. The tubers are covered with soil or compost after planting. Water is sprinkled with a water-can over the beds after planting and thereafter on alternate days. To encourage a good vine growth, urea is applied @ 5 kg/500 m² area as top dressing 3 weeks after planting. The tubers sprout within 10–15 days and give rise to a cluster of sprouts, which become ready after 2–3 months for transplanting to the main field. The cuttings of 10–15 cm length are taken from the top portion of the vine for planting in the main field.

PLANTING IN MAIN FIELD

Chinese potato is usually grown on ridges, except in very well-drained soils. The herbaceous vine cuttings of 10–15 cm length taken from the nursery are planted from July to October on 60–90 cm wide ridges, which are prepared 60 cm apart. The cuttings are planted on ridges at a spacing of 30 cm either in vertical or horizontal position at a depth of 4–5 cm by exposing the terminal bud to the atmosphere. Horizontal planting of the vines ensures quick establishment and promotes tuber formation.

In Sri Lanka, three methods of planting are commonly followed *viz.*, (i) ordinary planting, in which about 15 cm long cuttings with three or four leaves at the top end are planted 7 cm deep in rows down the ridges 22 cm apart; (ii) coiled planting, in which about 22 cm long cuttings are used and about 12 cm of the more mature portion is coiled and planted in holes about 7 cm wide and 5 cm deep and (iii) horizontal planting, in which about 30 cm long cuttings are placed horizontally across the ridge, two at a time, in opposite directions and almost touching each other. About 22 cm of the cuttings should remain inside the ridge and 7 cm outside the ridge, and about 7 cm distance is maintained between each pair of cuttings. Of these three methods, coiled planting has been reported to give the best results.

NUTRITIONAL REQUIREMENT

Nutritional requirement usually depends on soil type and fertility status of the soil. Well-decomposed farmyard manure of 20–25 t/ha should be applied at the time of field preparation. In addition, nitrogen is applied @ 60 kg, phosphorus 60 kg and potash 100 kg/ha to obtain a good yield. The entire dose of phosphorus and potash along with a half-dose of nitrogenous fertilizer should be applied as a basal dose at the time of the last ploughing or ridge preparation. The rest of the nitrogen is top-dressed at the time of earthing-up, about 45 days after planting.

IRRIGATION REQUIREMENT

The frequency of irrigation depends on soil type and climatic conditions of the growing area. Chinese potato responds well to regular irrigation, especially in dry season. In the nursery, irrigation is applied immediately after planting, and thereafter, even moisture should be maintained for better growth of the planting material. In the field, the first irrigation should be given before planting or soon after planting for better establishment of the crop. When rain is infrequent, the crop is irrigated at frequent intervals to maintain optimum moisture in the field. The plant is susceptible to waterlogging, thus, heavy irrigation should be avoided, and in wet conditions, the tubers are prone to deformities like branching, making them more difficult to peel and susceptible to tuber rots. In general, Chinese potato crop is irrigated at an interval of 7–10 days.

INTERCULTURAL OPERATIONS

HOEING AND WEEDING

A Chinese potato field should be kept weed free in the early stages of crop growth since the plants in early stages may not compete with weeds due to poor crop growth rate. Thus, at least two shallow hoeings and hand weedings are beneficial and necessary for keeping the weeds down and providing a good soil environment for tuber development. However, in the plant's late stages, no weeding is required, as the large crop canopy suppresses the weed growth. A light mulch of dried ground leaves, compost, or grass also helps in suppressing weeds and conserving soil moisture.

EARTHING-UP

Earthing-up is the other important cultural operation, in which the soil is mounted around the root zone to ensure better growth of tubers, as earthing-up ensures enough space to the growing tubers and avoids damage by rodents and insect-pests. Earthing-up is done 45 days after planting at the time of top-dressing of nitrogenous fertilizers to cover a portion of the vine with soil to promote tuber formation and development. This operation also indirectly helps in controlling weeds.

HARVESTING

Chinese potato requires 140–150 days to be mature. The tubers become ready for harvesting when the leaves begin to wither and vines dry up. Harvesting cannot be delayed since the mature tubers deteriorate rapidly if left in the soil. Harvesting is normally done by pulling out the plants and digging out the leftover tubers in the field. After harvesting, the tubers are separated from the plant and the crop residues are destroyed by burning.

YIELD

The yield of Chinese potato depends upon variety, soil fertility, growing season and management practices adopted during cultivation of the crop. The average yield of Chinese potato ranges from 7 to 15 t/ha, but under ideal conditions, it may give yield up to 18–20 t/ha.

POST-HARVEST MANAGEMENT

After harvesting and trimming, the soil particles adhered with tubers are removed by washing and thereafter, the other non-tuberous parts are removed. The tubers are cured for few days under shade for reducing moisture and healing the injuries, if any. The tubers of Chinese potato may not be stored well for longer periods due to decaying caused by pathogenic organisms or enzymes-mediated deterioration. In traditional farmsteads, many methods have been devised to store its tubers. The tubers can be stored in sand in a well-ventilated shed or on the ground under a tree where it is cooler than in the open. In this way, the tubers can be stored well for 2 months without any quality deterioration. The tubers can also be stored in sealed earthen vessels or sacks using straw as a cushioning material but the tubers by this method may not be stored longer in hotter regions.

INSECT-PESTS AND DISEASES

Chinese potato crop is attacked by a number of insect-pests and diseases during the growing season. Some of the insect-pests and diseases along with their control measures are given as follows:

INSECT-PESTS

APHIDS (*Myzus persicae*)

The soft-bodied polyphagous aphids damage the plants by sucking cell sap from tender parts and spreading viral diseases. When infestation is severe, the leaves turn yellow and distorted, necrotic spots appear on leaves and plants become stunted. Aphids secrete a honeydew-like sticky sugary substance, which encourages black sooty mould and other saprophytic fungal diseases.

Control

- Use resistant varieties, if available.
- Use silver coloured plastic mulch to deter the aphids feeding on plants.
- In initial stages, remove infected parts of the plant or rogue out the infected plants.
- Spray the crop with 4% *neem* extract, Malathion 0.1%, monocrotophos 0.04%, or methyl demeton 0.025%.

Cutworm (*Agrotis* spp.)

The larvae of cutworm are 2.5–5.0 cm long, which may exhibit a variety of patterns and colouration and usually curl up into a C-shape when disturbed. They attack the stems of young seedlings at the soil line, especially in the young nursery stage. If infection occurs later, irregular holes appear on plant surfaces. The larvae causing damage are usually active at night and hide during the day in the soil at the base of the plants or in plant debris of the toppled plant.

Control

- Remove all crop residues from the field after harvest or at least 2 weeks before planting.
- Cover the plant stem 7 cm above the soil line with plastic or foil collars extending a couple of cm into the soil.
- Pick and destroy the larvae by hand after dark.
- Drench the soil with chloropyrifos 0.1%.

Thrips (*Frankliniella occidentalis*)

The insect is small (1.5 mm) and slender and best viewed using a hand lens. Adult thrips are pale yellow to light brown, and the nymphs are smaller and lighter in colour. If the population is high, the leaves become distorted badly and are covered with coarse stippling. As a result, they appear silvery and are speckled with black faeces.

Control

- Avoid planting next to onion, garlic, or cereal crops.
- Use reflective mulches in early growing season to deter thrips.
- Spray the crop with Confidor 0.03–0.05%, Regent 0.2%, or dimethoate 0.2%.

Two-spotted Spider Mite (*Tetranychus urticae)*

The mites may appear as tiny moving dots on the webs or underside of the leaves. They are best viewed using a hand lens. Upon attack, the leaves become mottled or speckled and then bronzed. The leaves as a result of web covering turn yellow and drop down from the plant. Spider mites can thrive in dusty conditions. Water-stressed plants are more prone to the attack of spider mites.

Control

- Rogue out the infected plants from the field.
- Wash away the spider mites by spraying water with strong jet sprayer.
- Dust the crop with the sulfur @ 20–25 kg/ha.
- Spray the crop with 0.25% detergent suspension, 4% *neem* extract, cypermethrin 0.1%, dimethoate 0.2%, quinolphos 0.03%, or monocrotophos 0.05%.

Leaf Folding Caterpillars (*Hymenia recurvalis*)

The caterpillars damage the leaves by feeding on the green portion of the plant, leaving behind a thin parchment tissue. The damage results in reduced leaf area due to feeding and folding of leaves, which in turn affects photosynthesis of the plants. The pest is very common in Tamil Nadu and west coastal areas.

Control

- Use clean and healthy planting material.
- Follow long-duration crop rotation.
- Collect and destroy the folded leaves along with egg clusters, larvae and pupae.
- Spray the crop with 5% *neem* seed kernel extract, 0.03% dimethoate, or 0.04% monocrotophos twice at 10-day intervals.

Vine Borer (*Nupserha vexator* p.)

The grubs feed on the stem by boring and tunnelling, which result in arresting nutrient transmission and in the destruction of tissues. Infected plants show yellowing and drying of leaves. The grub also attacks the tubers found underneath soil. The attack of the pest is severe from October to February.

Control

- Use clean and healthy planting material.
- Follow long-duration crop rotation.
- Remove infested vines along with tubers from the field.
- Spray the crop with 5% *neem* seed kernel extract, 0.03% dimethoate, or 0.04% monocrotophos twice at 10-day intervals.

Root-knot Nematode (*Meloidogyne Incognita*)

The nematode infection causes suppression of root growth and formation of galls, resulting in stunted growth and wilting of plant. The infected tubers become malformed and hypertrophied in size due to heavy galling, rendering them unsuitable for consumption and marketing. In the case of heavy infection, the tubers rot even before harvest. The infested tubers swell in size with irregular surfaces, which result in cracking of their skin. Starch content of the tubers is also considerably reduced. The root-knot nematode damage often leads to crop failure.

Control

- Grow resistant varieties, if available.
- Use clean and healthy planting material.
- Follow long-duration crop rotation.
- Use Sree Bhadra variety of sweet potato as trap crop.
- Apply neem cake in soil @ 80–100 kg/ha or *Trichoderma harzianum* and *Pseudomonas fluorescens* @ 2.5 kg/ha.
- Apply carbofuran 3% granules @ 20 kg/ha near the root zone under wet conditions.

DISEASES

Downy Mildew (*Peronospora belbahrii*)

The disease symptoms appear as numerous yellow spots on the top surface of the leaves. As the disease progresses, a fuzzy growth develops on the underside of the leaves. The infected areas enlarge and their centres turn tan to light brown with a papery texture. When the disease is severe, all the leaves turn yellow and die.

Control

- Grow resistant varieties, if available.
- Follow long crop rotation with cereal crops.
- Use disease-free healthy planting material.
- Grow crop at a wider spacing in well-drained soil.
- Spray the crop with 0.25% Dithane M-45 or Dithane Z-78.

Leaf Spot (*Colletotrichum* spp.)

The disease symptoms appear as brown to black spots or patches on leaves. As the infection increases, small clusters of back-fruiting bodies develop on the leaves. Infected leaves become yellowish brown in colour. Wet conditions encourage the spread of disease.

Control

- Follow field sanitation to avoid the occurrence of disease.
- Collect and burn the infected plant debris.
- Spray the crop with carbendazim 0.05% or hexaconazole 0.05% at intervals of 7–10 days.

Rust (*Puccinia* sp.)

Rust is the most serious foliar disease spread by wind. The symptoms appear mainly on leaves and other above-ground plant parts. Small, dusty, bright orange, yellow, or brown pustules appear on the undersides of leaves, which later shrivel and fall off. The newly emerged shoots become pale and distorted, large areas of leaf tissue die and leaves may fall down.

Control

- Follow at least 2–3 years' crop rotation with non-host crops.
- Grow resistant cultivars, if available.
- Use disease-free planting material.
- Treat the planting material with hot water at 44°C for 10 minutes.
- Maintain good field sanitation by destroying diseased debris and weeds.
- Spray the crop with Dithane M-45 0.2% or Bayleton 0.05% at intervals of 10–12 days.

■■■

73 Adzuki Bean

M.K. Rana and Naval Kishor Kamboj

Botanical Name : *Phaseolus angularis* Wight.

Family   : Leguminosae

Chromosome Number : 2n = 22

ORIGIN AND DISTRIBUTION

Adzuki bean is native to Indo-Burma and China, where it is still a major crop. About 1000 years ago, it was introduced from China to Japan, and now, it is the second largest legume crop after soybean. It is also cultivated in Korea, New Zealand, India, Taiwan, Thailand, United States of America and the Philippines. In the United States, it is grown in Florida, Minnesota and California. The presumed wild ancestor of the cultivated adzuki bean is *Vigna angularis* var. nipponensis. This wild species is distributed across a wide area in Japan, Korea, China, Nepal and Bhutan.

INTRODUCTION

Adzuki bean is a small red-brown bean, frequently used in Japanese desserts. It is round in shape with a sharp point at one end and has a sweet nutty flavour compared to other beans. It is an annual vine widely grown throughout East Asia and the Himalayas for its small (approximately 5 mm) beans. The cultivars most familiar in Northeast Asia have a uniform red colour. However, white, black, grey and mottled varieties are also known.

COMPOSITION AND USES

COMPOSITION

Adzuki bean like other legumes is a powerhouse of nutrients, as it is a rich source of carbohydrates, protein, fibre, calcium, phosphorus, potassium and magnesium but a poor source of fat, iron, zinc, vitamin A, thiamine, riboflavin, niacin and pantothenic acid. The nutritional composition of adzuki bean is given in Table 73.1.

USES

The young tender pods of adzuki bean are used as a cooked vegetable, a stuffing in mushroom caps, candied bean, a source of flour, a beverage base, a substitute of coffee and as an addition to rice and sweet soups. Its flour mixed with wheat flour is used to make noodles. The beans are also popped like popcorn. In the United States of America, the sprouted seeds are often eaten. It is mixed in sour cream or yogurt. Its flour is used to make shampoo, facial cream and as an ingredient in culture media.

TABLE 73.1

Nutritional Composition of Green Pods of Adzuki Bean (per 100 g Edible Portion)

Constituents	Contents	Constituents	Contents
Water (g)	66.29	Manganese (mg)	0.573
Carbohydrates (g)	24.77	Iron (mg)	2.0
Protein (g)	7.52	Zinc (mg)	1.77
Fat (g)	0.10	Vitamin A (IU)	6.0
Dietary fibre (g)	7.3	Thiamine (mg)	0.11
Calcium (mg)	28.0	Riboflavin (mg)	0.64
Phosphorus (mg)	168.0	Niacin (mg)	0.71
Potassium (mg)	532.0	Pantothenic acid (mg)	0.43
Magnesium (mg)	52.0	Energy (kcal)	295

MEDICINAL USES

The seeds of adzuki bean are used to treat kidney problems, breast cancer, constipation, abscesses, tumours, threatened miscarriage, retained placenta and for the improvement of blood circulation and urination. It also helps in improving bladder function and is thus effectively used in treating problems of bladder infection. The leaves are used to lower fever and the sprouts are used to prevent threatened abortion caused by injury. Being a rich source of fibres, it is also useful in eliminating wastes and preventing the body from absorbing many harmful substances.

BOTANY

Adzuki bean is an annual, usually bushy and erect herb up to 90 cm tall, sometimes climbing or prostrate and rooting at the nodes. Its taproot is 40–50 cm long. Leaves are alternate, tri-foliolate; stipules small, peltate, often bifid with basal appendages; stipels lanceolate; leaflets lanceolate to ovate, 5–10 cm × 5–8 cm, acuminate and entire to three-lobed. Inflorescence is an axillary false raceme with 2–20 flowers. Peduncle is long at lower nodes to very short at upper nodes. Flowers are papilionaceous, bisexual and the pedicel short, bearing an extrafloral nectary at its base. Bracteoles are longer than the calyx, and the calyx are campanulate with short teeth. Corollas are 15–18 mm long, bright yellow, standard orbicular, wings oblong and keel turned towards right with a horn-shaped spur on left side. Stamens are 10, 9 fused and 1 free. Ovary is superior, shortly hairy, style abruptly bent in upper part, hairy on one side near top and stigma lateral and discoid. Pods are cylindrical, 5–13 cm × 0.5 cm size, pendulous, slightly constricted between the seeds, nearly glabrous, pale yellow, blackish or brown and with 2–14 seeds. Seeds are cylindrical with rounded ends, flattened, 5–7.5 mm × 4–5.5 mm, smooth, wine red, occasionally buff, creamish and black or mottled. Seed germination is epigeal, leaving the cotyledons below the soil surface. Plants generally mature in 110 to 120 days after sowing.

CLIMATIC REQUIREMENT

Adzuki bean grows well in warm and moist climates. The optimum temperature for its proper growth and development is 26.5°C to 30°C. Soil temperature of 15.5°C or above favours rapid germination and vigorous seedling growth. The crop does not grow well if the temperature falls below 10°C or goes above 40°C. A lower temperature tends to delay flowering. It grows well in regions having an annual precipitation ranging from 530 to 1730 mm. Day length is the key factor in adzuki bean production, as it is sensitive to day length, requiring short days for flowering.

SOIL REQUIREMENT

Adzuki bean can be grown in a wide range of soils but medium-textured fertile sandy loam soils high in organic matter are considered best for its cultivation. Sodic and saline soils inhibit germination of seeds. The ideal soil pH for the growth of adzuki bean is 6.0–7.5. Waterlogging is injurious to the crop.

The land should be prepared thoroughly to a fine tilth by repeated ploughings with a harrow. Care should be taken to provide drains or a gentle slope for good drainage. All the stubbles of previous crop are removed from the field to prepare clean seedbeds.

CULTIVATED VARIETIES

Wase

An early-maturing small-seeded variety suitable for growing in late spring is recommended for cultivation in Japan. It matures in 70 days after sowing.

Late Tamba

A most popular red-seeded bush-type variety suitable for growing in late spring to early summer is recommended for cultivation in Japan. It matures in 90 days after sowing.

Doyou

A large-seeded variety suitable for growing in early summer is recommended for cultivation in Japan. It matures in 90 days after sowing. It is mainly grown for its dried red seeds.

Dainagon

A large-seeded variety suitable for growing in early summer is recommended for cultivation in the Kyoto Prefecture, Japan. Its seeds stay firm and a good bean shape after cooking. It matures in 90 days after sowing.

The other important cultivars recommended for cultivation in Japan and China are Japanese Red, Chinese Red, Adzuki Express, Takara, Minoka, Hikari, Erimo and Blood Wood.

SOWING TIME

The sowing time of adzuki bean varies with available irrigation facilities and weather conditions of the growing area, but in general, the seed sowing is done from May to July.

SEED RATE

Seed rate of adzuki bean usually varies with numerous factors such as seed size, seed weight, plant spacing, cropping season, germination and purity percentage of the seed. Generally, the seed rate of 12 to 18 kg is required to sow a hectare land area. The seed before sowing should be treated with Thiram or Captan 3 g/kg of seed to protect the seed and seedlings from different diseases.

SOWING METHOD

The sowing of adzuki bean is done on flat beds in lines at a row spacing of 30–45 cm and plant spacing of 5–10 cm with the help of a seed drill or behind the plough. The depth of sowing should not be more than 3–4 cm under optimum moisture conditions.

NUTRITIONAL REQUIREMENT

Adzuki bean responds well to the application of farmyard manure and fertilizers. A well-decomposed farmyard manure should be incorporated into the soil @ 25–30 t/ha at the time of field preparation. Being a legume crop, it requires less nitrogen. Under normal soil conditions, it requires nitrogen 40 kg, phosphorus 30–40 kg and potash 20–30 kg/ha. A half-dose of nitrogen along with a full dose of phosphorus and potash is applied as basal dose at the time of sowing, and the remaining half of the nitrogen is top-dressed at the time of flowering.

IRRIGATION REQUIREMENT

Good irrigation management is essential for obtaining higher yield from adzuki bean since it is very susceptible to waterlogging conditions, which result in poor plant growth and reduced nodule formation. Regular irrigation is needed to ensure good establishment of plants. Moisture stress conditions lead to indeterminate growth and flower abortion, resulting in poor yield and seed quality. About three to five irrigations are needed during the growing season at intervals of 10–12 days, depending on climatic conditions and type of the soil of the growing region. The flowering and pod-filling stages are very critical for irrigation.

INTERCULTURAL OPERATIONS

HOEING AND WEEDING

In adzuki bean, weeds grow profusely because of wider plant spacing and slow plant growth, especially at the early stages, competing with crop plants for nutrients, moisture, light and space, which adversely affect the crop yield. The weed infestation during rainy season is too much, thus, timely control of weeds in an adzuki bean field is necessary. One or two hand weedings are generally needed to keep the weeds under control. Besides, pre- or post-emergence herbicides can be applied to control weeds in adzuki bean. Pre-sowing application of fluchloralin 2 litres per hectare or pre-emergence application of alachlor 1.5 kg/ha, or Pendimethalin 2 kg/ha effectively controls the weeds in an adzuki bean field.

MULCHING

In summer months, under a condition of low moisture stress, mulching with straw is very effective for good plant growth, as it conserves soil moisture and suppresses weed growth.

HARVESTING

The adzuki bean pods for fresh market are picked as soon as the seeds appear faintly outlined in the pod. Picking is usually done at 5- to 6-day intervals to avoid toughening of pods due to high fibre and wax content. For grain purposes, the crop becomes ready for harvesting 115–120 days after sowing when more than 70% of the pods turn yellowish brown and moisture content in grains becomes 12–15%. Delay in harvesting results in shattering of seeds in the field. The entire plant with dry pods is pulled and stacked in a well-ventilated place to dry the seeds completely, which occurs 1–2 weeks after harvesting. Thereafter, the seeds are separated from the pods by threshing and winnowing and stored in refrigerated airtight containers for the next planting season.

YIELD

The yield depends on several factors such as variety, soil type, climatic conditions of the growing region and package of practices followed during the cultivation of crop. The average dry seed yield varies from 5 to 6 q/ha.

INSECT-PESTS AND DISEASES

Adzuki bean crop is attacked by a number of insect-pests and diseases during the growing season. Their description along with control measures is given as follows:

INSECT-PESTS

Aphid (*Aphis craccivora*)

The nymphs and adults both attack the young plants and suck cell sap from growing tips and later from the whole plant. While feeding, an aphid produces a considerable amount of honeydew upon which sooty mould grows. The black sooty mould reduces photosynthesis and ultimately crop yield. Its attack causes curling of plant leaves and reduces plant vigour, yield and quality of the produce.

Control

- Spray the crop with 0.03% dimethoate or 0.05% phosphamidon.
- Apply carbofuran granules 1.0 kg a.i./ha.

Bean Pod Borer (*Maruca vitrata*)

The caterpillars are pale yellow with a pale brown head. The larvae cause damage to the corolla and caterpillars distort the pods by making round holes. The larvae eat seeds within damaged pods either partially or completely. Infected plants are converted to a mass of brownish frass, and defoliation occurs in early stages.

Control

- Remove and destroy all the affected parts.
- Spray the crop with 0.5% *neem* seed kernel extract.
- Spray 0.03% dimethoate or 0.04% monocrotophos twice at 10-day intervals.

Red Spider Mite (*Tetranychus urticae*)

Adults and nymphs of red spider mites cause damage to the plants by feeding on the undersides of the leaves, which show a blotchy characteristic due to the extraction of chlorophyll. The infested leaves also have a yellowish speckled appearance, and later turn brown and then dry up.

Control

- Spray the crop with 0.5% *neem* seed kernel extract, wettable sulfur 0.3%, malathion 0.05%, dicofol 0.035%, or Ethion 0.005%.

DISEASES

White Mould (*Sclerotinia sclerotum*)

Symptoms appear as small, circular, dark green, water-soaked lesions on pods, leaves and branches, which enlarge and become slimy. Flowers are covered with white cottony fungal growth, which is visible during periods of high humidity. The affected leaves and entire plant die and fall down.

Control

- Follow a crop rotation of at least 2–3 years.
- Use disease-free healthy seed.

- Plant the crop at a wider spacing.
- Soak the seeds in Bavistin or benomyl 0.2% solution for 30 minutes.
- Spray the crop with Bavistin 0.1% or mancozeb 0.25% at 10- to 15-day intervals.

ALTENARIA LEAF SPOT (*Altenaria aiternata*)

The disease symptoms appear as small irregular-shaped brown lesions on leaves, which later turn large oval grey-brown with concentric rings. When several lesions coalesce, a large portion of the leaf area becomes necrotic. Premature defoliation of the lowest leaves occurs. On pods, small water-soaked, reddish-brown lesions develop that merge into long streaks. Cool and wet weather favours the germination and spread of spores.

Control

- Follow at least 3–4 years' crop rotation with non-host crops.
- Use disease-free healthy seed.
- Treat the seed with carbendazim 0.2%.
- Spray the crop with mancozeb 0.25% or copper oxychloride 0.3% and repeat at intervals of 10–14 days.

POWDERY MILDEW (*Erysiphe polygoni*)

The fungus of powdery mildew attacks all parts of the plant. Initially, the symptoms appear as yellow spots on the upper surface of the older leaves. These spots develop the characteristic powdery growth and spread to the underside of the leaves and other parts of the plant. On pods, small, moist, circular spots develop into white powdery masses of pathogen mycelium and spores. The affected leaves turn completely yellow, curl, die and fall down.

Control

- Grow resistant varieties, if available.
- Spray the crop with systemic fungicide like Karathane, Bavistin, Benlate 0.2%, or Sulfex 0.5% at 10- to 15-day intervals.

COMMON BLIGHT (*Xanthomonas campestris* pv. *phaseoli*)

Common bacterial blight first appears as small translucent water-soaked spots on leaves. As the spots enlarge, the tissue dies, leaving brown spots with yellow margins. These lesions are often large and irregular in shape. On pods, the symptoms appear as water-soaked sunken lesions, which later turn brownish red. Bacteria may also infect the vascular system and kill the main stem and branches. Common blight is seed borne and is favoured by warm temperatures (28°C) and high humidity.

Control

- Follow long-duration crop rotation.
- Collect the infected plant debris and destroy them.
- Use disease-free healthy seed.
- Treat the seed with streptocycline 0.01% and captan 0.25%.
- Spray the crop with Blitox 0.35% as and when the disease appears.

■■■

74 African Yam Bean

M.K. Rana and N.C. Mamatha

Botanical Name   : *Sphenostylis stenocarpa* (Hochst Ex. A. Rich.) Harms.

Family   : Leguminosae (Fabaceae)

Chromosome Number : 2n = 18, 20, 22 and 24

ORIGIN AND DISTRIBUTION

African yam bean is native to Africa. The centre of diversity spreads from the west to the eastern and southern parts of Africa. Diversity of crop and its wild relatives exist in Africa. The crop is found growing wild throughout tropical Africa, and is cultivated in Central and Western Africa, especially in Nigeria, Cote d'Ivoire, Ghana, Togo, Gabon, Zaire and central Africa. The crop is relatively unknown outside the African continent.

INTRODUCTION

African yam bean, also known as yam pea, tuber bean, haricot *igname, pomme deterre dumossi, afrikanische yam bohne,* or *knollenbohne* and some indigenous lingual synonyms in Africa namely, *akidi, ekulu, sunmunu, otili, kutreku, kulege, akitereku, apetreku, girigiri, kutonoso, roya, efik, nsama, ibibio, bitei, sesonge, gundosollo, sumpelegu, tschangilu, sese, sheshe, gilia-bande, pempo,* or *mpempo.* The six mostly occurring botanical synonyms of African yam bean are *Sphenostylis ornata, Sphenostylis congensis, Sphenostylis katangensis, Vigna katangensis, Vigna ornate* and *Dolichos stenocarpus.* This is one of the important tuberous legumes of tropical Africa, grown for both seeds and tubers, which are an important source of protein and starch.

The genus *Sphenostylis,* comprising seven species, occurs in dry forests and in open or forested savannas in tropical and southern Africa. Among all, *Sphenostylis stenocarpa* is the most widely distributed and morphologically variable species in the genus as well as an economically important vegetable crop.

COMPOSITION AND USES

COMPOSITION

African yam bean is regarded as a non-fatty food, which contains carbohydrates 50–75%, protein 20–25%, oil around 1% and fibre 5–6%, providing nearly 400 calories per 100 g dry weight. Although protein is produced in copious quantities, it is not necessarily an ultimate measure of a proteinaceous food. More important is the nutritional quality of protein since a protein lacking certain minor components is nutritionally next to useless. Protein quality is judged by the presence of a few essential amino acids, as many plant proteins, being deficient in lysine and methionine, are poor in nutritional quality. In contrast, African yam bean seed protein contains those two amino acids in abundance, *i.e.,* lysine 7–8% and methionine 1–2%. Adding to the tubers' nutritional contribution is their 63–73% content of carbohydrates and their 3–6% fibre. The starch alone has been put at 65–70% and about 370 overall calories. The nutritional composition of African yam bean seeds is given in Table 74.1.

TABLE 74.1

Nutritional Composition of African Yam Bean Seeds (per 100 g Edible Portion)

Constituents	Contents	Constituents	Contents
Water (g)	81	Potassium (mg)	43.6
Carbohydrates (g)	85.3	Magnesium (g)	43.2
Protein (g)	0.6	Iron (g)	12.6
Fat (g)	8.25	Fibre (g)	1.1
Phosphorus (mg)	28	Total ash (g)	2.2

USES

African yam bean is grown primarily for its dry seeds and tubers. The seeds are milled into flour, which is processed into a paste with water and some condiments, then boiled after wrapping in plantain leaves and eaten as *turbani*. The flour mixed with cassava flour may also be cooked into a paste and eaten with soup or sauce. Seeds after boiling alone or in soups are commonly served with yam, maize or rice. The shelled beans from mature green pods are eaten after boiling. The tubers are eaten both raw and cooked.

BOTANY

African yam bean, a perennial climbing bush, is usually grown as an annual. The herbaceous vine grows to a height of 1–3 m. Its red stems are often strongly branched, bearing trifoliate leaves up to 14 cm length and 0.2–5.5 cm broad. The individual leaflets are oval with pointed tips and smooth edges. The root system produces branches vigorously, and some roots thicken into the storage organs, known as tubers. These spindle-shaped tubers outwardly resemble sweet potato but their taste is more like potato. In general, they are 5–25 cm in length, each weighing 50–300 g. The smallest are normally kept aside and used for planting the next crop. The butterfly-shaped flowers born on 30 cm long racemes having 2.5 cm long twisted petals are probably pollinated by insects. The flowers may be of pink, purple, greenish white, yellowish white, red, magenta, lilac and blue colour. Most blooms develop into 20–30 cm long and about 1 cm wide pods, containing 20–30 large (up to a cm long) seeds per pod. They are pointed and subdivided inside by fine transverse walls. The pods become brownish at maturity. The seed colour varies from white to brown or black. Some are speckled or marbled in brown and white, and their veined seeds are known as well. The hilum has a brown border.

CLIMATIC REQUIREMENT

African yam bean, being a crop of tropical and subtropical regions, needs a warm and dry climate for its growth and development. It grows better in regions where annual rainfall ranges from 800 to 1400 mm and where the temperature comprises between 19°C and 27°C. It is little affected by altitude but flourishes well at elevations from sea level to 1800 m. It is sensitive to frost as well as high temperature.

SOIL REQUIREMENT

The crop can be cultivated successfully on a wide range of soils, even in poor soils, but it thrives best on deep loose sandy loam or loam soils with good organic matter content and proper drainage facilities. Fertile sandy soils are said to be highly suitable, and the optimum soil pH for its successful cultivation is 5 to 6.

CULTIVATED VARIETIES

African yam bean is not cultivated in India on a commercial scale. Hence, no breeding work has been done yet for the improvement of this crop. The farmers grow only the local landraces, though some accessions have been given to those landraces.

PLANTING TIME

African yam bean is generally sown during the start of the rainy season between May and July. In tropical areas, it can be sown throughout the year, and in subtropical areas, it is sown in late spring. In warm temperate areas, the crop can be sown at the end of spring frost, but in this region, bottom heat is required for germination of seeds. The crop is unsuitable for areas with a short growing season unless grown in a greenhouse. In cooler regions, the crop can be grown where it has a frost-free period of at least 5 months. In India, the seeds are sown in June–July and for a late crop in September.

SEED RATE

The seed rate varies according to the growing season, growing region and cultivation system. The seed rate in different regions varies from 20 to 60 kg/ha, however, the optimum seed rate is 35–40 kg/ha. There are some reports that the crop can also be propagated by tubers.

SOWING METHOD

African yam bean can be propagated by seeds, small tubers, or root pieces, but commercially, seed is used for planting. Soaking of seeds before sowing in lukewarm water overnight helps in softening of seed coat and speeding up the germination. The field is prepared thoroughly by ploughing repeatedly with a harrow to a fine tilth for making smooth seedbeds and ensuring an even germination. The seeds are usually sown at 5 cm depth on ridges of 15 cm height made 60 cm apart maintaining 30 cm spacing between plants.

NUTRITIONAL REQUIREMENT

African yam bean plants prefer soil of moderate fertility since they fix atmospheric nitrogen into the soil, but an insufficient quantity of potassium in soil affects the tuber growth and quality adversely. Well-decomposed farmyard manure 10–20 t/ha should be incorporated into the soil at the time of land preparation. Besides, nitrogen, phosphorus and potassium are applied, each at the rate of 60 kg/ha. The entire dose of phosphorus and potash along with a half-dose of nitrogen fertilizer should be incorporated into the soil before sowing seeds and the remaining half of the nitrogen should be supplied in the form of top dressing at about 40–45 days after sowing.

IRRIGATION REQUIREMENT

Water is one of the essential components required for growth and development of the crop. The total water requirement of the crop varies with soil type and climatic conditions of the growing region. If the seeds are sown in dry soil, the first irrigation is applied immediately after sowing. Once the crop is well established, appropriate soil moisture must be maintained and care must be taken not to over-irrigate the crop, as excessive soil moisture can damage the crop by increasing the chance of fungal infections. The crop grown in rainy season may not require irrigation if the distribution of rain is normal, but in a September-sown crop, irrigation is given at regular intervals to maintain adequate soil moisture for proper development of the tubers.

INTERCULTURAL OPERATIONS

HOEING AND WEEDING

Hoeing and weeding are the most important operations in crop husbandry. The field should be kept weed free during the early stages of crop growth since weeds compete with crop plants for nutrients, moisture, space and sunlight. Hoeing is done manually with a spade at least twice during the first couple of months since making the soil loose around the root system improves aeration into the soil, which is essential for the *rhizobium* bacteria to absorb nitrogen from the atmosphere and to provide nitrogen to the plant.

EARTHING-UP

Mounting up of soil on both sides of the ridges is also essential to provide enough space to the growing tubers and also to cover them, as the exposed tubers are damaged by rodents and insect-pests. This operation also indirectly helps in controlling weeds in a yam bean field.

PRUNING AND STAKING

An important cultural operation in African yam bean is reproductive pruning, which helps in obtaining maximum yield, as excessive flowering reduces the tuber yield. It is considered important to achieve bigger tubers. During pruning, leaving a few flowers for seed production, the main growing flower shoots are rigorously pruned as many as four times with the first cut being made about 2 months after planting so that the plant may pump its energy into underground parts for their development. However, the effect of this operation depends on cultivar, season and climate. In some areas, not only the reproductive shoots are removed but also the top half of the vegetative parts is pruned. In most of the cultivars, two reproductive prunings are sufficient to ensure a good yield. The plants are normally supported by trellises or stakes, but as far as yield is considered, no significant difference between supported and unsupported crops has been reported.

CROPPING SYSTEM

In some parts of tropical America, African yam bean plays an important role in crop rotation. When it is grown consecutively in the same field for two seasons, it produces a higher yield in the second than the first season. In the third season, maize, beans, onions, etc. should be included in the crop rotation instead of African yam bean since this rotation in Africa has been proven more effective for sustaining land productivity. A long crop rotation of at least 3–4 years should be followed to restore the lost fertility and to wipe out the pathogen inoculums.

HARVESTING

The duration required by the crop to be mature depends on the type of cultivar, soil type, climatic conditions of the growing region and consumer preference. The pods become ready for harvesting 5 months after planting. The tubers develop more slowly, normally taking 5–8 months to reach a harvestable size. Most are dug up toward the end of the rainy season. They can be harvested a bit earlier if local preferences encompass small- or medium-sized tubers, or they may be left in the field for a while after rainy season. Tubers are picked up manually after loosening the soil around the plant with a spade or crossbar mounted on a tractor, and then, the vegetative tops are removed with a pair of shears after harvesting the tubers and left in the field for later use as forage or organic fertilizer. Finally, the tubers are collected, cured, sacked and stacked for shipment.

YIELD

The yield also varies with the cultivar, soil type, climatic conditions of the growing region and the cultural management. The average yield of African yam bean is 13.69 q/ha.

POST-HARVEST MANAGEMENT

The post-harvest operations followed in the African yam bean involve the cleaning of tubers by washing, trimming (removing roots and stem parts) and sterilization and bleaching by dipping in concentrated chlorinated water. The tubers, after cleaning adhered soil particles and curing in shade, are packed in sacks made of plastic wire or jute. The tubers can be stored for 1 to 2 months at 12.5–17.5°C temperature. The tubers are sensitive to colder temperature, thus, they should not be refrigerated. Prolonged storage may alter the starch to sugar ratio where starch content decreases and sugar content increases with storage period, requiring research on this aspect. After removing the vegetative tops, the small holders tend to leave the tubers underground until eaten or sold. Starch, protein and other components under such conditions still need investigation, as the tubers of American species become much sweeter and more valuable when aged.

INSECT-PESTS

Rarely, the pests may be a serious problem in African yam bean. The pests that affect other leguminous crops in that area affect the African yam bean also. A number of insect-pests which have been reported to cause damage to different parts of the African yam bean are as follows:

Aphid (*Aphis* sp.)

Aphids are the polyphagous pest of all leguminous crops. They may be winged or wingless and are the vectors of viral diseases. They damage the plant by sucking cell sap from almost all tender parts. The affected plants become weak and leaves become yellow and curl downwards. Aphids secrete a honeydew-like sticky substance, which invites black sooty mould and other fungal diseases. A short spell of rainfall coupled with high humidity increases its population.

Control

- Install yellow sticky traps @ 12 per hectare.
- Spray soap water suspension @ 25 ml liquid detergent per litre of water.
- Spray the crop with 4% *neem* extract, Malathion 0.1%, monocrotophos 0.04%, or methyl demeton 0.025%.

Bean Pod Borer (*Maruca testulalis*)

This is a serious pest of African yam bean in the tropics. Its larvae attack the flowers, buds and young pods by webbing them. They bore into the pods and feed on young immature seeds. The symptoms are characterised by small holes on flower buds, flowers and pods. The reported yield loss ranges from 20% to 80%.

Control

- Collect and destroy the damaged pods and larvae.
- Spray the crop with dimethoate 0.05%, tobacco decoction 3%, or *neem* oil 3% to minimize the borer population.

Hairy Caterpillar (*Ascotis imparata* and *Spilosoma obliqua*)

The adult lays eggs in clusters over the surface of the leaves. The caterpillars feed gregariously on green portion of the leaves during early stage, leaving them skeletonised. In severe cases, it results in defoliation, which ultimately affects the plant growth.

Control

- Collect and destroy the larvae in early stage of the crop growth.
- Spray the crop with quinalphos 0.05%, monocrotophos 0.15%, or fenvalerate 0.05% at intervals of 10–12 days.

Root-knot Nematode (*Meladogyne arenaria*)

Root-knot nematodes are polyphagous and widely distributed in the country. Infected tubers show warts and give a bitter taste. This pest causes reduction in tuber yield and quality and invariably reduces marketable yield. The above-ground symptoms consist of foliage discolouration (more pale than normal), dwarfing the plants and wilting.

Control

- Strictly follow long crop rotation with non-host crop species.
- Apply *neem* cake along with recommended fertilizers.
- Apply nematicides like Furadan 20 kg/ha or Cadusafos 1% in soil.

DISEASES

Bacterial Blight (*Xanthomonas* sp.)

Bacterial blight, a most serious seed-borne disease, usually affects the crop in areas with high humidity and high temperature. Spattering rains favour the spread of this disease. The disease appears as small circular water-soaked spots on the upper surface of the leaves, which coalesce to form large spots surrounded by a chlorotic area on older leaves. A black longitudinal streak also develops on the main stem. The affected portion becomes flaccid, the necrotic spots turn brown, and finally, the affected leaves fall down.

Control

- Remove all the plant debris from the field before sowing.
- Grow resistant varieties, if available.
- Dip the seed in 0.2% aqueous solution of streptomycin for 3 hours.
- Give hot water treatment to seeds for 10 minutes at 56°C temperature.
- Spray the standing crop twice with 250 ppm solution of streptocycline.

Alternaria Leaf Spot (*Alternaria* sp.)

The disease appears as dark brown circular to angular spots with grey centres and a reddish border on the leaves, which spread rapidly to form circular lesions under humid conditions, and gradually they cover the entire surface of the leaves. In severe cases concentric rings of dark brown conidiophores are developed and the leaves usually drop off.

Control

- Follow long crop rotation with non-host crops.
- Grow resistant varieties, if available.
- Incorporate micronutrients in soil at the time of sowing.

- Destroy the infected plant debris.
- Spray the crop with 0.2% zineb or Dithane Z-78 twice at 15-day intervals.

Rust (*Phakopsora pachyrhizi*)

Rust is the most serious disease spread by wind. The symptoms appear mainly on leaves and other above-ground plant parts as small, round, reddish brown rust pustules, especially on the lower surface of the leaves, which as a result shrivel and fall off.

Control

- Follow at least 3–4 years' crop rotation with non-host crops.
- Grow early-maturing cultivars to avoid rust until harvesting.
- Maintain good field sanitation by destroying diseased debris and weeds.
- Spray the crop with Dithane M-45 0.2% or Bayleton 0.05% at intervals of 10–12 days.

Sincama Mosaic Virus (SMV)

This is the most serious viral disease of Africa yam bean. The leaves of infested plants show irregular chlorosis, young shoots become brittle and seed setting is reduced due to reduced pollen fertility. Tuber yield as well as quality is affected markedly. The disease is transmitted through mechanical wounding and insect attack. The aphids and possibly spider mites are capable of transmitting virus from infected to healthy plants.

Control

- Remove all infected plants and weeds from the field.
- Apply Disulfotan or Phorate 10 G granules 1.5 kg/ha at the time of seed sowing.
- Spray the crop with monocrotophos 0.05% or dimethoate 0.05% at intervals of 10–12 days to control vectors.

Witches' Broom Disease

This disease is probably caused by mycoplasma-like organisms. The symptoms are excessive branching, dwarfing of leaves and deformation of flowers. It is not transmittable through sap inoculations but some reports indicate that the disease may possibly be transmitted by sucking insects such as whiteflies (*Orosius argentatus*), aphids and mealy bugs.

Control

- Apply carbofuran, fensulfothion, or disulfoton 1.5 kg a.i/ha in soil at the time of sowing.
- Remove the infected plants as soon as noticed.
- Spray the crop with monocrotophos 0.05%, dimethoate 0.05%, or imidacloprid 0.1% at intervals of 10–12 days.

■■■

75 Broad Bean

*M.K. Rana and
Chandanshive Aniket Vilas*

Botanical Name : *Vicia faba* Linn.

Family : Leguminosae (Papilionaceae)

Chromosome Number : 2n = 2x = 12, 14

ORIGIN AND DISTRIBUTION

Europe and Asia are said to be the origin of broad bean. Faba bean (minor) is an ancient small-seeded relative of the Chinese broad bean (*Vicia faba major*). The oldest seeds of *Vicia faba* were found in Jericho dated at 6250 BC. The crop is grown in the Mediterranean region where it is consumed commonly. In Europe, it is cultivated primarily as a livestock feed. Britain, where both winter and spring types are grown, is the largest producer of this bean.

INTRODUCTION

Broad bean, also known as faba bean, horse bean, or bakla bean, is a minor legume crop culti-vated on a small scale in India but widely cultivated in Latin America. The plant with a square, hollow, erect single stem, strong taproot and compound leaves grows up to a height ranging from 30 to 100 cm, depending on the variety (dwarf or giant type). Broad bean, which is pollinated by insects, bears clusters of four white, black blotched flowers in the axils of leaves. The plant bears upright large (up to 10 cm long and 1–2 cm wide) fleshy green pods in clusters of five or more in the axils of leaves along the stem. Each pod contains three to four oblong or oval seeds. Flowering starts in 45–60 days after sowing, and the crop requires 110–130 days to be mature. It is cultivated on a small scale mainly in Uttar Pradesh, Punjab, Haryana, Kashmir, Rajasthan, Karnataka, Madhya Pradesh and Bihar.

Broad bean, being a legume crop, fixes nitrogen into the soil. It increases yield of cereal crops since it reduces the incidence of cereal cyst nematode (CCN), which affects the root system of cereal crops. In some people, its pollens and green pods cause allergy or illness, which is known as *favism*.

POLLINATION

Faba bean has a limited ability for self-pollination and yields benefit from insect pollinators to maximize seed set. If the population of native bees is not high, the use of bee colonies can increase the yield by 40%. At least one hive per 2.5 hectares is needed to introduce, if bees in mid afternoon are less than one per square metre since faba bean flowers open in mid afternoon and close again at dusk. The bees become active as the days lengthen and the air temperature increases.

CHINESE BROAD BEAN

The Chinese broad bean (*Vicia faba major*) is cultivated on large area in China, Europe, the Middle East, North Africa and South America. However, it can also be seen growing in limited area in Canada. Chinese broad bean matures significantly earlier than most of the current faba bean varieties and gives yield equal to faba bean.

COMPOSITION AND USES

COMPOSITION

Stachyose, raffinose and verbascose are the main oligosaccharides found prevalent in faba bean. These molecules restrain glucose and galactose residues, which persist in sugar metabolism pathways. They ferment in digestive tracts and produce methane and other gases, causing discomfort and abdominal pains. The nutritional composition of broad bean leaves is given in Table 75.1.

USES

Young tips of plant or pods or shelled seeds at the green tender stage are cooked as a vegetable. Immature pods can be steamed and served as whole. Broad bean is the most common fast food eaten with onion and tomato sauce in Egypt and surrounding areas. The boiled beans are a delicious food after blending into a spicy dip with lemon, garlic, chilli and yogurt. Dried beans are powdered and spiced as the basis for falafel mix. Its dried seeds can also be used as a feed for livestock. Its crop residue offers a valuable source of feed for cattle since its stubbles contain grains remaining even after harvest.

CLIMATIC REQUIREMENT

Faba bean is well adapted to moist areas and performs well under relatively cool growing conditions. Being a cold-hardy plant, it can withstand a temperature up to 4°C, hence, it can be grown

TABLE 75.1
Nutritional Composition of Broad Bean Seed (per 100 g Edible Portion)

Constituents	Contents	Constituents	Contents
Carbohydrates (%)	59.87	Thiamine (mg)	0.45
Starch (%)	35–39	Riboflavin (mg)	0.19
Sugars (%)	5.0	Niacin (mg)	2.4
Crude protein (%)	30.57	Cystine acid (%)	0.25–0.31
Protein (%)	24–30	Lysine (%)	1.48–1.61
Crude fat (%)	3.36	Methionine (%)	0.17–0.19
Lignin (%)	2.65	Ash (%)	3.61
Crude fibre (g)	2.37	Energy (kcal)	328
Iron (mg)	4.2	–	–

successfully in autumn–winter in North plain areas and at high altitudes in summer months. The prolonged cool weather in spring is ideal for the development of pods. It can tolerate frost better than other legume crops. However, the optimum temperature for its growth is 15–20°C, especially at flowering and pod formation stage. The flowers abort when temperature exceeds 27°C and are also particularly sensitive to hot dry conditions at pod formation stage since hot dry spells result in wilting of plants and affect the seed setting adversely. A thin plant stand often occurs under dry conditions but the crop is able to compensate the poor plant stand if moisture conditions are improved throughout the growing season. The crop can be grown in dry areas where the average annual rainfall is 400 mm though the crop yield is reduced.

SOIL REQUIREMENT

Broad bean can be cultivated on all types of soil but well-drained organic matter rich loam soil with pH range 6.5 to 7.5 is preferred the most. Red-brown, black, grey clays and alluvial loam soils are also best suited for its cultivation. However, growing broad bean in highly fertile soils (fields with high levels of available nitrogen) should be avoided as the vegetative growth may be excessive at the expense of seed production. This crop can tolerate waterlogging conditions better than other legume crops even though growing faba bean on poorly drained soils should be avoided as it favours the development of seedling and root rot disease. It may not be cultivated successfully in light-textured soils unless assured irrigation facilities are available since faba bean tends to have a shallow root system. It can tolerate salinity to some extent but may not tolerate acidic soils. If soil pH is below 5.0, lime should be incorporated to raise the soil pH.

Faba bean should not be grown repeatedly on the same land. A rotation of 3 to 4 years should be followed with cereal crops but the crop should not follow oilseeds or other legume crops in the rotation because of the potential increase in soil-borne diseases. A major benefit of rotating pulse crops with cereal crops is the interruption of pest cycles since cereal diseases are different from pulse crops. Land should be prepared thoroughly by ploughing the field repeatedly followed by planking. The stubble of previous crop should be collected or burnt in order to obtain a smooth seedbed.

CULTIVATED VARIETIES

The variety should be selected accordingly to achieve a balance between adaptability to environment, disease reaction, agronomic performance and marketability. The description of some important varieties is given below:

Pusa Sumeet

The plants of this variety are 75 cm tall having on an average 10 branches per plant. Plant bears attractive dark green 100 pods in clusters. Its pods are 6 cm long and 1.3 cm thick and its average yield potential is 180 q/ha.

Pusa Udit

A dual-purpose broad bean variety bears extra long, flattish and light green pods. The variety is suitable for packaging and transport. It gives yield 88.52% higher than the earlier released variety Pusa Sumeet.

Sutton-White Seeded

A variety suitable for autumn or spring sowing produces plants about 30 cm high and pods about 13–15 cm long containing five seeds per pod.

SOWING TIME

Broad bean is sown twice in a year, once in the months of February–March and another in September–October. Sowing in rest of the months reduces the yield potential considerably. Flower dropping may be a serious problem in broad bean if flowering takes place during a period of high temperature.

SEED RATE

The plant density requirements vary with variety and sowing time. For small-seeded varieties, 20–30 plants per square metre are recommended, and for medium-seeded varieties, 18–23 plants per square metre are more appropriate. Lower seed rate is used for early sowings (February–March) and the higher seed rate for late sowings (September–October). In dry areas, the optimum plant population is 44 plants per square metre. The germination of broad bean is less than 75%. Hence, it is advisable to sow some extra seeds at a place which may later be used to fill in gaps. For sowing a hectare area, about 70–100 kg seed having germination of 80% is needed. The seed rate can be calculated by using the following formula, which takes into account the seed size and germination level of the seed used:

$$\text{Seeding rate}\left(\text{kg / ha}\right) = \frac{\text{Plant density (Plants/m}^2) \times 100 \text{ seeds weight}\left(\text{g}\right)}{\text{Per cent germination}} \times 10$$

SEED TREATMENT

Being a legume crop, broad bean can fix a significant amount of nitrogen from atmosphere to the soil using a symbiotic relationship with *Rhizobium* bacteria, and for fixation of nitrogen, the seed or soil must be inoculated with an appropriate strain of *Rhizobium*. If sowing is to be done on virgin land, inoculation of seed before sowing is considered cheap insurance to maximize nodulation and *per se* increase the nitrogen fixation. It is recommended to use seed from a healthy crop. If seed is inoculated with *Rhizobium*, it must not be treated with a fungicide.

INOCULATION

The seed or the soil is inoculated with *Rhizobium* bacteria, which later enter the root hairs and induce nodules onto the roots. The plant provides shelter to the bacteria living inside the nodules, and in lieu of that, the bacteria provide nitrogen to the plant by converting the atmospheric nitrogen into plant-useable forms. However, the utmost benefit may be obtained if the nitrogen level in soil is low, and soil moisture and temperature are adequate for normal development of plant roots since high nitrogen in soil delays the onset of nodulation and inhibits nitrogen fixation since the plant will use the soil nitrogen rather than the nitrogen fixed by the bacteria. The *Rhizobium* bacteria can survive in the soil for a number of years, however, the most efficient nitrogen-fixing bacteria may not be among those that survive in soil.

SOWING METHOD

Broad bean seeds are bigger in size, thus, they can be sown individually. The seeds are sown in 15 cm wide shallow furrows (5–7 cm) into a firm seedbed at a spacing of 25 × 75 cm. In each furrow, seeds are sown in two rows in a zigzag manner along the furrow. The seed is placed in one row opposite the gaps in other row so that the plants may not shade one another. It can also be sown in a single row at 45 × 15 cm spacing. Its seeds take 10–15 days to germinate after sowing, thus, it is advisable to soak the seeds overnight prior to sowing. Its seeds are highly susceptible to mechanical damage during handling. Dry seeds are brittle and can easily split, leading to reduced germination, hence, it is advisable to handle the seed carefully. Germination will be better if seeds are sown under cover.

NUTRITIONAL REQUIREMENT

Broad bean has an excellent nitrogen-fixing capacity. Hence, it does not require extensive use of nitrogenous fertilizer. Soil testing is advised prior to sowing for specific nutrient recommendations. In general, higher rates of fertilizer in broad bean should be avoided. Farmyard manure is incorporated into the soil at the rate of 10 t/ha at the time of field preparation. Besides, nitrogen is applied 15–20 kg, phosphorus 40–50 kg and potassium 30–40 kg/ha at the time of sowing. Being a legume crop, broad bean is a relatively high user of phosphorus.

IRRIGATION REQUIREMENT

Broad bean is well adapted to production under irrigation conditions. Pre-sowing irrigation is essential to apply to have ample moisture in the field at the time of sowing for better germination of seeds. If soil moisture is scanty in the field at the time of sowing, irrigation has to be applied soon after sowing, which is followed by light irrigation on the third day. Since this crop responds well to irrigation, irrigations should be applied lightly at intervals of 7 to 12 days, as waterlogging will cause temporary growth reductions and affect the yield adversely. Though broad bean is a hardy crop, it cannot withstand long periods of drought, hence, irrigation intervals should be as short as possible.

INTERCULTURAL OPERATIONS

HOEING AND WEEDING

Broad bean is a poor competitor with weeds that compete with the crop for nutrients, moisture, space and light, hence, hoeing and weeding at early stages of the crop when the plants are small are essential to remove the weeds and to provide a good environment to plant roots for better growth. Perennial weeds should essentially be controlled in the years prior to broad bean production.

STAKING AND PINCHING

The plants of giant varieties need support, which may be provided with bamboo or wooden sticks against wind. The bamboo sticks are placed down into the soil on both sides of the double rows closer to the plants at 1 m interval, and then, the plants 30 to 60 cm above the ground are tied around the stakes with plastic string. The side shoots arising from the base of the stem are removed to keep the plants erect. Dwarf varieties need less space and less staking. On attaining five or six flowers, the growing points are pinched out at the top of the stems. This operation encourages the formation of pods and discourages aphids, which feed on growing points and young leaves.

HARVESTING

The spring-sown crop becomes ready for harvesting in 3–4 months and the autumn-sown crop in 6–7 months. Very young tender pods are preferred the most by the consumers in the market, hence, the pods are harvested for table purposes or fresh market when they are green and tender, and the remaining pods on the plant are used after maturity as dry-shell beans. As and when the pods become finger-thick, they are culled frequently by twisting them downward.

As the pods mature, the lower leaves become dark and drop down and the bottom pods turn black and dry from the bottom to the top of the plant. Broad bean shatters if left standing until full maturity. For dry-shell beans, the crop is usually harvested when about 25% of the plants in the field have the lowest one to three pods turning dark. By this time, the topmost pods should be developed absolutely and the middle pods turning to a lighter green colour. At this stage, the moisture content of the most mature seeds may be > 40% and the seeds in the upper part of the plant may be > 60%. A light swathe should be used as the crop may take about 3 weeks to dry in the swathe. Straight-cutting broad bean will most likely lead to high shattering loss. Shattering may be reduced by harvesting the crop early in the morning under a condition of high humidity. Leaving harvest until the beans are too dry should be avoided as it will result in pod shattering. After harvesting, the crop residues are harrowed to make use of captured nitrogen.

YIELD

Depending on growing season, about 70 to l00 q/ha green pods and 18–30 q/ha dry shell beans may be harvested if good cultural practices are adopted.

POST-HARVEST MANAGEMENT

The seed coat of broad bean is prone to cracking, so to reduce seed coat cracking, the beans should be dropped down from the least possible distance when transporting broad bean seed. When stored in a bin, broad bean seeds often respire, hence, extra care should be taken to avoid spoilage due to build-up of moisture inside the bin. The seeds before storing in a bin should be dried up to a safe moisture level, *i.e.*, 12–16%. A maximum of 32°C temperature is advised for drying the beans to avoid cracking the seed coat.

PHYSIOLOGICAL DISORDERS

FROST INJURY

Broad bean seedlings are tolerant to frosts. As the maturity of beans often starts from the bottom to the top of the plant, early frost injury may be much greater on the top pods of the plant. The pods developing closer to the ground may have little frost injury so the top green pods should be harvested earlier for fresh market. The occurrence of frost during early pod-filling stage can cause discoloration and deformation of pods and seeds, which later become water-soaked, and they may not stay firm for a longer time. The intensely damaged pods will have a rubbery wilted appearance.

INSECT-PESTS

Aphids, caterpillars and nematodes are the most common pests attacking the broad bean crop. The major insect-pests of broad bean and their control measures are discussed as follows:

APHID (*Aphis* sp.)

The minute insects that suck cell sap from tender parts of the plant are grey or black in colour. As their population becomes abundant, they start attacking the developing pods, causing great reduction in growth and yield. Minimal damage occurs from direct feeding of aphids. However, heavy or prolonged periods of infestation can cause stress on plants, resulting in yield loss. Aphids may also transmit viruses. A moisture-stressed crop is more susceptible to aphid infestation, especially when the atmosphere is dry and when warm weather occurs in spring and autumn.

Control

- Apply systemic insecticide such as phorate or aldicarb 10G @ 10–15 kg/ha or carbofuran 3G @ 30–33 kg/ha at the time of sowing.
- At early stage of crop when there is no pod, spray dimethoate 30 EC or methyl demeton 25 EC @ 0.1–0.2%.
- At pod bearing stage, spray the crop with malathion 50 EC @ 0.2% or DDVP 0.1% at fortnightly interval after harvesting the edible pods.

NATIVE BUDWORM (*Heliothis punctiger*)

The caterpillar causes damage to maturing seeds in pods. The newly hatched caterpillars are so small (1–2 mm) that they are easily overlooked when the crop is inspected, but when they mature (40–50 mm long), they have a yellow-white stripe down each side of the body and a dark stripe down the centre of the back. They become a serious pest of beans and first feed on the surface of pods, bore into them and then feed on seeds.

Control

- Spray the crop with Endosulfan @ 0.2% at fortnightly intervals.
- Spray the crop with Hexavin 50 WP @ 2.25 kg/ha at fortnightly intervals.

RED-LEGGED EARTH MITE (*Halotydeus destructor*)

It is a black-bodied mite with red legs, which damages the seedlings as they emerge; as a result, leaves turn silvery and then brown and then shrivel. Mites feed gregariously by rasping surface cells with their mouth parts and causing significant sap to leak out. Damage to leaves in winter and spring reduces plant growth and affects seed production.

Control

- Follow long crop rotation.
- Use natural enemies.
- Tillage fallowing is effective.
- Treating seed with a systemic insecticide before sowing.
- Spray the cop with dimethoate, omethoate or synthetic pyrethroids after the emergence of 60% of the plants.

LUCERNE FLEA (*Sminthurus viridis*)

It is a small (2.5 mm), wingless, light green hopping insect. It chews through leaves in layers, resulting in "window pane"–like holes. The crop shrivels and becomes stunted.

Control

- Spray the crop with carbaryl, chlorpyrifos, omethoate and isofenphos for effective control.

DISEASES

Broad bean is subject to a number of diseases that can reduce yield and quality. The infection can come from a variety of sources, *i.e.*, seed, soil and crop residues. The disease caused by these sources can be minimized through preventative management discussed as follows:

SEED AND SEEDLING ROOT ROT (*Fusarium* spp., *Rhizoctonia solani, Pythium* spp. and *Aphanomyces euteiches*)

Under northern Indian conditions, the seeds may rot inside the soil, particularly in rainy season due to excessive moisture, and the seedlings are killed before they emerge out of the soil due to the attack of above fungi, which cause poor stand of the crop.

Control

- Follow a long crop rotation excluding legume crops.
- Use well-drained soils for its cultivation to avoid excessive moisture.
- Treat the seed with Emison, captan, foltaf, captafol, or bavistin @ 2.5–3.0 g/kg of seed.

ASCOCHYTA LEAF AND POD SPOT (*Ascochyta fabae*)

The disease is seed- and residue-borne, which is characterized by grey-brown spots on leaves, stem and pod. It may spread to the seed also. When the infection is severe, seeds developing in the pods become discoloured and shrivelled. Small fruiting bodies may also appear on leaves after rain and stems may collapse if symptoms are severe.

Control

- Use varieties that are resistant to this disease.
- Avoid areas which have been under legume crops in the previous 2–3 years.
- Take care not to introduce the disease by sowing infected seed.
- Spray the crop with carbendazim, benomyl, or mancozeb 0.2% at 10-day intervals.

ALTERNARIA BLIGHT (*Alternaria alternata*)

The disease is characterized by dark brown circular to angular spots with grey centres and dark margins on the leaves, which gradually cover the entire leaf surface. The symptoms can be confused with chocolate spot. *Alternaria* spots can be distinguished from *Ascochyta* blight as the spots have a brown margin containing obvious concentric rings but do not produce black fruiting bodies (pycnidia) on a grey centre.

Control

- Follow a long crop rotation with non-host crops.
- Destruct the infected plant debris to check the spread of disease infection.
- Spray the crop with 0.1% bavistin or 0.2% Indofil M-45 at intervals of 10–12 days.

ANTHRACNOSE (*Colletotrichum truncatum*)

The disease symptoms are dark brown to black sunken lesions appearing along the leaf veins on the underside of the leaves and pods. The infection may extend to seeds also, which carry the fungus from season to season. The fungus survives on debris of infected plants. The stems of fatally affected plants may break and topple over to the ground. It is a common problem during rainy season.

Control

- Follow a crop rotation of at least 4 years.
- Collect seed from healthy fruits and a disease-free area.
- Treat the seed with carbendazim or benomyl 0.25%.

- Keep the field clean by burning disease debris.
- Spray chlorothalonil 0.2% just after noticing infection.
- Spray systemic fungicides such as benomyl, bavistin, or thiophanate M (0.1%).

CERCOSPORA (*Cercospora zonata*)

The disease usually affects the leaves but may also affect stems and pods. Rough circular cherry-red to dark-red spots variable in size first appear on the lower leaves early in the growing season and can swiftly expand to 15 mm and coalesce with adjacent lesions under favourable conditions. Severe infection can result in extensive defoliation of plants. The infected pods at maturity show black sporulation of fungus.

Control

- Spray the crop with 0.2% Indofil M-45 at 10-day intervals starting soon after the appearance of disease.

CHOCOLATE SPOT (*Botrytis fabae*)

The disease is characterized by reddish or chocolate brown spots on leaves and reddening of stems. The spots may enlarge and merge, forming a black mass on the leaves, which is followed by defoliation and lodging. The disease is residue- or seed-borne. Dense crop sowing, waterlogging and wet weather favour the occurrence of this disease.

Control

- Follow a crop rotation of 2 or more years with non-host crops.
- Avoid buckwheat in the rotation.
- Use seed collected from disease-free fields.
- Follow wider spacing to maximise air flow around the plants.
- Avoid damp and humid sites for its cultivation.
- Destroy infected plant debris at the end of the crop season.
- Eliminate common vetch from the vicinity of the crop.
- Spray the crop with carbendazim, benomyl, or mancozeb 0.2% at 10-day intervals.

RUST (*Uromyces viciae-fabae*)

The disease, which is more common on the undersurface of the leaves, may prove destructive by defoliation of the plant since the disease mostly attacks the leaves but rarely the stem and petioles. The leaves have a rusty appearance, and as a result, the leaves turn yellow, dry and fall off. Occasionally, it may also infect the pods.

Control

- Follow a crop rotation of at least 2 years with non-host crops.
- Follow wider spacing between lines and plants within line.
- Remove weeds from the crop field to improve aeration and destroy the disease debris.
- Spray the crop with Indofil M-45 or Daconil (0.2%) at the beginning of flowering or the onset of disease symptoms at intervals of 10–12 days.

76 Jack Bean

M.K. Rana and Manoj Kumar Nalla

Botanical Name	: *Canavalia ensiformis* (L.) DC
	Canavalia ensiformis var. truncate Ricker
	Dolichos ensiformis L.
Family	: Leguminosae
Chromosome Number	: 2n = 22

ORIGIN AND DISTRIBUTION

Jack bean is almost certainly a native of the West Indies and the adjacent mainland. In Jamaica, where it first became well known, it is called the horse bean or the overlook bean. *Canavalia* is a pan-tropical genus encompassing 48 species, which is believed to have originated in the New World based on the large genetic diversity of species in the fossil record. It closely resembles sword bean (*Canavalia gladiate*), an Old World cultivated bean species and the predominant African wild species (*Canavalia virosa*). In South Africa, it has been called the sunshine bean. From time to time, it is exploited under various names, including Pearson bean, coffee bean, wonder bean, giant stock-bean Wataka bean and Gotani bean. Some authors consider both jack bean and sword bean to be derivatives of the African wild species because of the close resemblance among all three species. Now, it is widely distributed in the tropics and subtropics although it is regarded as a minor vegetable. It is cultivated on a limited scale in the northeastern region of India.

INTRODUCTION

Jack bean is a legume crop, which is used for animal fodder and human nutrition. It is also the source of concanavalin A and is a low-priced source of protein-rich foodstuffs that have been important in alleviating protein malnutrition. The Indian subcontinent may be the area of greatest dependence on pulses. However, pulse production in India could not keep pace with population growth. The use of less-known grain legumes provides one possibility, which is not used to an important extent by human population. Among these, one that offers an excellent possibility is its seed. The mature seeds are consumed by the Indian tribal sects Kurumba, Malayali, Erula and other Dravidian groups after cooking. In Western countries, this legume is used as a cover crop and the roasted seeds are ground to prepare a coffee-like drink. It is considered one of the few pulses that grow well on the highly leached and nutrient-depleted lowland tropical soils. It can be grown relatively easily and produce high yield in the regions of low altitude, high temperature and high relative humidity.

COMPOSITION AND USES

COMPOSITION

The proximate composition of jack bean has been investigated earlier and the seeds contain 43.1–60.30% carbohydrates, 24.25–32.23% crude protein, 1.8–9.60% crude lipid, 4.65–10.00% crude fibre and 2.00–4.64% ash. The nutritional composition of jack bean is given below in Table 76.1.

USES

The plant is cultivated on a small scale. Its young green pods are eaten as a cooked vegetable. The young leaves may be cooked and eaten as a potherb. Its seeds sometimes are roasted and used as a coffee substitute. In Indonesia, steamed flowers and leaves are used as flavourings. The dried beans are good source of protein and starch. The whole plant is used for fodder. It has been investigated as a potential source of urease enzyme and fair source of lectin, which is used in biotechnology in lectin affinity chromatography. It can be grown in soils with high lead concentration and has potential to be used for restoration of lead-contaminated soils. The growth-inhibiting properties of jack bean forage is attributed to the presence of toxic amino acid, canavanine, which interferes with the human body's use of arginine and the presence of proteins concanavalin A and B that interfere with nutrients absorption in the animal and human digestive systems. However, heat treatment eliminates the toxic effects of these growth-inhibiting substances.

MEDICINAL USES

Jack bean is a source of anti-cancer agents such as trigonelline and canavanine. In developing nations, its seeds have been promoted as a potential source of affordable and abundant protein. It is used to cure hiccups, vomiting, abdominal swelling, lumbago (lower back pain) due to kidney deficiency, asthma with sputum, etc.

BOTANY

Jack bean is a twining plant that grows up to a eight of 1 m. It is also a semi-erect bushy annual plant. Leaves are trifoliate and shortly hairy. It is of two types, *i.e.*, one that climbs and thoroughly covers the soil and another that has a bushy growth habit and does not climb at all. The plant begins flowering after 4–5 months and then produces seedpods continuously for at least the next year. The flowers are pink to purple in colour. The pods are up to 36 cm long, each containing 10–14 seeds that are white and smooth surfaced with a brown seed scar, which is about one-third of seed length. Its roots have nodules, which fix nitrogen into the soil. It has deep roots, which make it drought resistant. Under favourable conditions in the tropics, it is sometimes said to be more or less twining. The stems are rather coarse and become woody toward the base. Its thick leaves have a bitter taste.

TABLE 76.1
Nutritional Composition of Jack Bean (per 100 g Edible Portion)

Constituents	Contents	Constituents	Contents
Water (g)	10–11	Calcium (mg)	380.5
Carbohydrates (g)	45.8 to 65.4	Phosphorus (mg)	107.4
Protein (g)	23.0 to 35.3	Iron (mg)	631
Fat (g)	2–3	Energy (kcal)	1470 to 1910
Dietary fibre (g)	5.0 to 11.4	–	–

The flowers are purple and the earliest flower is borne near base of the stem. Because of this reason, most of the pods touch the ground. The roots of the plant are well tubercled.

CLIMATIC REQUIREMENT

Jack bean grows at moderate temperature of 20°C to 30°C. A distinguishing characteristic of this plant is its ability to grow continuously under severe environmental conditions, even in nutrient-depleted, highly leached and acidic soils. Being drought-resistant, it can grow in poor droughty soils but does not grow well in excessively wet soil. It drops its leaves under extremely high temperature and may tolerate light frost. Optimum growth occurs in full sunlight but can tolerate shade to some extent. Preferred environmental conditions for jack bean are found in the humid lowland tropics. It can grow through dry season at 600 meters above mean sea level and can serve to shade the soil during hot summer to prevent loss of organic matter. Below 500 meters mean sea level, it often stops growing without rain and may even drop its leaves if soil is dry and temperature is exceptionally high. It can be used as a green manure crop up to 1600–1800 meters above mean sea level. It is tolerant to a wide range of annual rainfall from 650 to 2000 mm. It can be grown relatively easily and produce high yield in the regions of low altitude, high temperature and high relative humidity.

SOIL REQUIREMENT

Jack bean is well adapted to poor quality soils. The long taproot system allows the plants to draw moisture from deeper layer of the soil, surviving drought in low rainfall environment. Unlike most pulses, it is tolerant of leached and nutrient-depleted lowland tropical soils. It also tolerates acid soil reaction (pH 4.3–6.8) relatively well. It is more tolerant to waterlogged or high salinity soil conditions than most pulses. Its adaptations to marginal soil conditions, including its nitrogen-fixing capabilities, make it a species with promise for reclamation of weakened tropical soils.

CULTIVATED VARIETIES

Unlike other pulses, no rich and diverse gene pool of jack bean is available. Therefore, only limited inbred strains of jack bean have been developed by Agricultural Experiment Stations that focused on selecting cultivars with low toxicity. Both viny and bushy varieties exist in this crop. Based on seed colour, two types of cultivar are available. The white-seeded varieties are bushy in nature, whereas, the red seeded varieties are trailed over pandals.

Based on flowering time, the varieties can be classified into three groups, *i.e.*, (i) cultivars with early season flowering, taking 58 days after sowing, (ii) cultivars with mid-season flowering, taking 84 days after sowing and (iii) cultivars with late season flowering, taking 90–110 days after sowing. The cultivars with early and mid-season flowering have a shorter flowering period than the cultivars with late season flowering. Most of the bush types are early flowering cultivars, whereas, most of the climbing types are late flowering cultivars. When grown as a cover crop or green manure, the crop is terminated mechanically with a roller-crimper or chemically when it first begins to flower.

SEED RATE

The seed rate is mainly affected by the germination percentage. It may also vary due to growing season, method of sowing and soil properties. Usually, a seed rate of 40 to 60 kg is enough for sowing a hectare area.

SOWING TIME

Jack bean seeds are usually sown in the months of May–June and September–October. If mild climatic conditions prevail round the year, it can be grown throughout the year.

SOWING METHOD

Ridge and furrow method is ideal for its sowing. After thorough mixing of farmyard manure in the field, ridges and furrows are prepared at a recommended distance. If it is to be grown as a green manure crop, the seeds are usually sown by broadcasting. If it is to be cultivated as a pulse crop, the seeds are properly spaced. The seeds of pole type varieties are sown at a spacing of 40 × 30 cm and bush type varieties at 60 × 60 cm. One or two seeds are sown per hill at a depth of 2 cm.

NUTRITIONAL REQUIREMENT

Farmyard manure is applied at the rate of 5 tonnes per hectare. The nitrogen, phosphorus and potash mixture (7:10:5) may be applied as basal dose and top dressing at several splits. Applying fertilizers by band placement at 7–10 cm depth and 5–7 cm away from the seed is a good practice. Half of the nitrogen along with a full dose of phosphorus and potash are applied as basal dose at the time of making ridges and furrows, and rest half of the nitrogen is applied one month after completion of germination. The application of higher doses of nitrogen is said to reduce the yield. Jack bean also responds to the application of micronutrients. The foliar application of boron, copper, molybdenum, zinc, manganese and magnesium is effective in enhancing quality and pod yield. Seeds are inoculated with rhizobium culture before sowing as it helps in quick root nodulation, which in turn fixes atmospheric nitrogen.

IRRIGATION REQUIREMENT

Jack bean is sensitive to an oversupply of water. Therefore, excessive watering and water logging conditions should be avoided. Pre-sowing irrigation is essential for proper seed germination. The critical stages for irrigation are flowering and pod setting. Further irrigation is given as per need of the crop. The frequency of irrigation depends upon growing season, type of soil and organic matter content in the growing medium.

INTERCULTURAL OPERATIONS

HOEING AND WEEDING

The weed-crop competition starts 2–3 weeks after sowing, thus, for the harvesting of higher yield, first hoeing should be done 2–3 weeks after sowing and second at flower initiation, however, root injury should be avoided during hoeing, hence, the hoeing should be done shallow and carefully. The hoeing operation should be followed by light earthing-up to strengthen the plants and to encourage the root growth.

A pre-emergence application of Alachlor @ 2–2.5 kg/ha is recommended for controlling weeds. Another pre-emergence application of weedicides like Stomp @ 3 litre/ha or Goal @ 750 ml/ha can control the weeds most effectively for about 45 days.

STALKING

Wooden or iron poles are used for the support of twining varieties for obtaining higher yield with best quality pods. Stalking also helps in making the harvesting operation convenient.

HARVESTING

The seeds should be harvested by hand to prevent shelling. Foliage and stems, which are green in colour, turn reddish on the onset of maturity. Green immature pods are harvested 90–120 days after sowing and the pods for the extraction of mature seeds are harvested 180–300 days after sowing.

YIELD

A single peduncle bears one to three pods and a plant about four to seven pods. The bush type cultivars with early flowering give 9–10 seeds per pod, whereas, bush type cultivars with mid and late season flowering give 11–12 seeds per pod. The average yield of green tender pods is 3–4 tonnes per hectare. The average yield of mature seeds is 10–15 kg per plant and about 7–9 q/ha. The yield of green herbage is 40 to 50 t/ha. Most jack bean seeds are white in colour and weigh 1.5–2 g each.

INSECT-PESTS

In jack bean, the toxicity of seeds makes them unpalatable to insects and protects them from insect attack. The thick seed coat resists the pests to attack the mature seeds during storage. Stem borers and leaf eating beetles poise problems in early stage of crop growth.

Aphid (*Aphis craccivora*)

Nymphs and adults both suck cell sap from ventral surface of tender leaves, growing shoots flower stalks and pods. The infested leaves turn pale yellow and the flower buds fall down, whereas, the pods become shrivelled and malformed. The vine growth is retarded and ultimately the yield is adversely affected.

Control

- Spray the crop with Rogor, Metasystox, Thiodan, or Malathion 0.1%, or dichlorvos 0.05%.

Whitefly (*Bemisia tabaci*)

Nymphs are oval shaped, scale-like and greenish-white in colour. Its adults are 1 mm long, covered completely with white waxy bloom. They suck cell sap from tender parts of the plant and transmit viral diseases. The affected leaves become shrivelled and curled gradually.

Control

- Use 100-mesh agro-net to avoid entry of whitefly into the nursery.
- Spray the crop with Metasystox or Rogor 0.25% in early stages of crop.

Pod Borer (*Helicoverpa armigera*)

The caterpillars roll the leaves and web these with the top shoots. Later, the caterpillars bore into the pods and feed on seeds. If flowers and pods are not available, the larvae feed on foliage.

Control

- Follow deep ploughing during hot summer months.
- Follow long crop rotation to reduce pest population.
- Adopt clean cultivation to expose the resting pupae.

- Apply high quality Helicoverpa Nuclear Polyhedrons Virus (HNPV).
- Spray the crop with Thiodan or malathion @ 2 ml/litre of water, Cypermethrin (25 g a.i. per hectare), or Permethrin (200 g a.i./hectare)

BEAN WEEVIL (*Acanthoscelide obtectus*)

The larvae bore into the seeds and feed until maturity. The emerging adults chew their way out of the larval chambers, exposing holes in the seeds. Its damage reduces the food value as well as the germination potential of seeds.

Control
- Fumigate the seeds with phosphine gas available in the form of tablets named Celphos or Phosfume; apply @ 1–2 tablets per tonne of seed or per cubic meter of space.

LEAF MINER (*Liriomyza trifolii*)

The black and yellowish female lays about 300 to 400 eggs thrusting these into leaf tissues with the help of a sharp and pointed ovipositor. Its infestation can easily be recognized by the presence of shiny whitish streaks on the leaves against the green background. Its adults puncture the leaves and feed on exuding sap. The flowering and fruiting capacity of infested plants are adversely affected.

Control
- Spray the crop with dichlorvos or phosphamidon 0.25 to 0.5% or *neem* cake extract or *neem* oil emulsion 0.25%.
- Spray Rogor or Metasystox 0.1% or Thiodan 0.2%.

DISEASES

Like other beans, this crop is also attacked by several diseases but anthracnose, bacterial blight and powdery mildew are the most destructive diseases of jack bean.

ANTHRACNOSE (*Collectotrichum lindemuthianum*)

Leaves, stems and pods are susceptible to infection. Small slightly sunken reddish-brown spots form on pods and rapidly develop into dark-sunken large lesions. In moist weather, masses of pink spores develop on these lesions. Black sunken spots, similar to those on the pods, are developed on the stem and peduncles. Infection of the leaves causes blacking along the veins particularly on the undersurface of the leaves.

Control
- Avoid growing jack bean for at least 2 years on the same land.
- Treat the seed with Bavistin or Thiram 3 g/kg of seed.
- Spray the crop with mancozeb or chlorothalonil 0.2% at weekly intervals from the first leaf stage until the pods mature.

BACTERIAL BLIGHT (*Xanthomonas Campestris pv. Cyanospsidis*)

The bacterial blight organism attacks the shoots while they develop in early spring, causing brown spots on the leaves, petioles and stems. In the beginning, the spots on the leaves are olive green and water soaked, but later, they turn brown with yellow margins. Immature leaves turn black and die quickly. Infection of young twigs causes girdling and blackening of infected areas. The blossoms become limp dark brown and blighted. Formation of white floury patches on both sides of leaf, stem and pods take place.

Control

- Give hot water treatment to seed for 10 minutes at 56°C.
- Spray the crop with streptocycline 100–200 ppm and Agrimycin 100–500 ppm.

POWDERY MILDEW (*Erisiphe polygoni*)

In advanced stages, all aerial parts are covered with a powdery mass. In severe cases defoliation occurs. Dry weather at 24–32°C favours the powdery mildew disease. The patches appear as minute discoloured specks, from which a powdery mass radiates on all sides. The white mass consists of mycelium and spores of fungus.

Control

- Grow early growing varieties.
- Remove and destroy the diseased debris from the field.
- Spray 0.5% wettable sulfur like Thiovit, Sulfex, etc. as soon as the disease appears.
- Spray the crop with Karathane 40 EC @ 200 ml/ha at 10-day intervals.

■■■

77 Lima Bean

Sonia Sood and Nivedita Gupta

Botanical Name   : *Phaseolus lunatus* L.

Family   : Leguminosae (Fabaceae)

Chromosome Number : 2n = 22

ORIGIN AND DISTRIBUTION

Lima bean is believed to be a native of the tropical fertile valleys of Central and South America and originated in or near Guatemala. It has two main centres of diversity, *i.e.*, (i) Central America (Mexico and Guatemala) for the small-seeded forms and (ii) the Andean region (Peru and Ecuador) for the large-seeded forms. It has been widely cultivated in Peru for more than 7,000 years. Lima bean spread to the United States of America, Central America, Brazil and Peruvian coastal region. Later in the post-Columbian period, it was taken to other countries by the early European explorers. From Spain, it went to various countries of Asia. It was brought to Africa from Brazil through the slave trade. Since lima bean could withstand humid tropical weather conditions better than most of the beans, it became an important crop in the areas of Africa and Asia. In India, it was introduced from Europe, where it is cultivated in Karnataka, Andhra Pradesh, Maharashtra, Tamil Nadu, Gujarat and West Bengal.

INTRODUCTION

Lima bean is one of the important members of leguminous group of vegetables, which is not closely related to the French bean (*Phaseolus vulgaris*) or runner bean (*Phaseolus coccineus*). It is incompatible with the French bean but both have similar cultural requirements. It is also known as butter bean or double bean, sieva bean, Burma bean, guffin bean, haba bean and Rangoon bean. The largest area and production of lima bean is in the United States of America followed by the Malagasy Republic and Peru. In India, it is not grown on a large commercial scale and it is not as common as the French bean. It is grown in the hilly areas of the country.

Wild lima beans probably support themselves by climbing on other plants. Climbing lima bean varieties are known as pole beans because using a pole to support their growth is very popular. However, the varieties developed for commercial production have more determinate growth habits. Smaller bush types are necessary for mechanical harvest.

COMPOSITION AND USES

COMPOSITION

Lima bean is the most nutritive vegetable among the leguminous crops because it contains fat-free high quality protein and is a good source of cholesterol-lowering dietary fibre. Its immature pods are an excellent source of carbohydrates, protein, calcium, phosphorus, iron, thiamine, riboflavin and niacin and a very good source of manganese, potassium, copper, magnesium, antioxidants and plant sterols. The pods are very low in fat and hence serve as a good source of energy. Although it has high nutritive value, it is not commonly grown in India. The nutritive value of lima bean is given in Table 77.1.

TABLE 77.1

Composition of Lima Bean (per 100 g Edible Portion)

Constituents	Green	White	Constituents	Green	White
Moisture (%)	93	92	Zinc (mg)	0.8	0.9
Carbohydrates (g)	23.6	20.9	Copper (mg)	0.3	0.2
Protein (g)	6.8	7.8	Manganese (mg)	1.3	0.5
Fat (g)	0.3	0.4	Thiamine (mg)	0.1	0.2
Dietary fibre (g)	5.3	7.0	Riboflavin (mg)	0.1	0.1
Calcium (mg)	32.0	17.0	Niacin (mg)	1.0	0.4
Phosphorus (mg)	130.0	111.0	Selenium (mg)	2.0	4.5
Potassium (mg)	510.0	508.0	Vitamin A (mg)	0.1	0.2
Magnesium (mg)	74.0	43.0	Vitamin E (mg)	0.1	0.2
Sodium (mg)	17.0	2.0	Vitamin K (mg)	6.0	2.0
Iron (mg)	2.5	2.4	Energy (kcal)	345	338

Uses

Lima bean is also one of the ancient and most valuable vegetable crops cultivated in home, market and truck gardens for canning purposes and shipment. It is grown commercially for its tender green pods used as a green vegetable and dry beans as pulse. It is widely used in a variety of preparations because of its delicate flavour. Like other beans, its pods are usually used in a variety of dishes including soups, salads, stews and casseroles. Its delicate green pods are also fried with other vegetables like carrot, corn and potato. Succotash is a popular dish in North America prepared from lima bean, corn, pepper and ground beef. Its immature pods are marketed as fresh, frozen, or canned. Most of the cultivars grown for the fresh market are usually flat podded, while the cultivars grown for processing are round podded. In India, it is grown for vegetable purposes, while in the United States of America, it is cultivated in large quantities, especially for processing for off-season use when green pods are not available.

Medicinal Use

Lima bean is carminative, diuretic and beneficial in treating kidney problems. It is considered good for the bladder, burns, cardiac problems, diarrhoea, dropsy, dysentery, rheumatism and itching. Isoflavones present in its pods provide protection against breast cancer. It also helps in lowering the cholesterol levels in blood and regulating diabetes. The presence of fibres also helps in stabilizing glucose metabolism and in preventing constipation. Lima beans are rich in iron, which helps in proper blood circulation as well as in replenishing lost energy. It also helps in improving the immune system in terrible diseases like HIV or AIDS. However, eating raw or sprouted lima bean seeds may cause stomach cramping, diarrhoea and vomiting, and eating a large quantity of undercooked beans releases cyanide, which can impair tissue oxygenation and cause severe illness. Lima beans are a good source of trace minerals like copper. Copper is necessary for reducing the risk of inflammatory conditions, like rheumatoid arthritis. Additionally, it promotes certain enzymatic actions, which are required to maintain elasticity of blood vessels, ligaments and joints.

BOTANY

The plant is an herbaceous species, which includes both annual (determinate bush type) and perennial (indeterminate climbing) types. It has a taproot system with poor nodule formation capacity. It is generally grown as an annual crop. The lima bean varieties are of two types, *i.e.*, (i) bush types and (ii) pole or vine types. The bush type of variety grows to a height of about 60 cm and tends to have smaller seeds. Pole types have large seeds and pale green pods that vary from 7.5 to 20 cm in length, depending upon the variety. Leaves are commonly composed of three leaflets (trifoliate) and the flowers are white.

In bush type varieties, inflorescence develops simultaneously, while in pole type varieties, the flowers appear one after another for a long time and form loose racemes or clusters. Flowers that are either white, yellowish, or pinkish-purple are papilionaceous. The androecium is in a diadelphous condition (9+1). Vexillary stamens are free from the base upwards. The ovary is monocarpellary, unilocular and with many ovules. The pods of the lima bean are flat, oblong and slightly curved. Kidney-shaped seeds are present inside the pod. The mature seeds are generally cream or green in colour, although in certain varieties white, red, purple, brown, or black seeds are present.

CLIMATIC REQUIREMENT

Lima bean requires a dry cool climate and grows well at temperature range of 15–25°C. When the temperature goes above 26°C, pollination and fertilization are impaired with the result that the pod formation is severally reduced. The optimum temperature for seeds to germinate is 15–30°C. The mean temperature <10°C shows poor seed germination, slow growth and delayed maturity, while temperature >30°C during flowering may cause a high percentage of flower drop and abortion of ovule, resulting in poor yield. During early growth, it requires a minimum temperature of 21°C. It can grow at an elevation up to 2000 m above mean sea level. In the plains, it can be grown as a winter vegetable while in the hills it is grown during the spring and rainy season. It does not tolerate frost, high temperature and high rainfall. It is a long duration crop and is retained in the field for 9 months. The large-seeded varieties are generally better adapted to low temperatures and higher humidity than are the small-seeded varieties.

SOIL REQUIREMENT

Lima bean can be grown on a wide range of soils from light sandy loams to heavy clays. Soil should have good moisture-holding capacity and be friable enough to enhance uniform germination and emergence. The ideal soil pH for its normal growth is from 6.0 to 6.5. The crop is very sensitive to higher concentrations of aluminium and manganese in soil. Light soils are good for early yield, while heavy soils are good for higher yield of lima bean. Slightly acidic soils are also preferable, while extremely acidic and alkaline soils are not good for its cultivation.

CULTIVATED VARIETIES

Almost all varieties of lima bean grown in India are introductions from abroad. The varieties are generally grouped as pole, semi-pole, bush, or dwarf types.

- **Pole types:** Florida Butter, Challenger, King of the Garden and Carolina Butter
- **Semi-pole types:** Hopi and Wilbur
- **Bush types:** Baby Potato, Purple Bush, Ford Hook 242, Henderson Bush, Baby Ford Hook and Burpee Bush

KODAIKANAL 1

An improved variety developed at TNAU Horticultural Research Station, Kodaikanal (Tamil Nadu), through selection from a local type has bolder seeds with a matured seed yield of 12 q/ha. It is pole type growing to a height of 2.4 m and producing pods in clusters. The pods are 11.6 cm long, each containing 5–6 beans. The crop will be ready for harvest after 100 days and harvest will continue up to 140 days. Its green pods' yield is 3.5 t/ha in three or four pickings.

SOWING TIME

Direct sowing of pole lima bean should be done between the second fortnight of May to the second fortnight of June. In plains of north India, lima bean is sown twice in a year, *i.e.*, (i) August to September and (ii) January to February. In hilly tracts, sowing is done in March for bush types and in June for pole type varieties. In south India, sowing is done from September to November. In the northeast and central zones, it is sown in the end of October.

SEED RATE

Lima bean is propagated by seeds, which are sown about 2–4 cm deep. A seed rate of 20–35 kg for small-seeded varieties and 45–50 kg for large-seeded varieties will be sufficient for sowing a hectare land area. Before planting, the seeds should be treated with Captan or Thiram at the rate of 3–4 g/kg of seed.

SOWING METHOD

Direct seeding is the most common method used for its propagation. The field is prepared to a fine tilth by repeated ploughing with a harrow followed by planking. The field is divided into beds of convenient size. The seeds are sown, eyes down, at a spacing of 60 × 40 cm in the case of pole and semi-pole types and at 35 × 25 cm in the case of bush types.

In Maharashtra, generally, this crop is taken after a heavily irrigated and manured crop like sugarcane. After harvesting of sugarcane, the field is ploughed to a fine tilth and seeds are sown in hills @ two or three seeds per hill at a distance of 180 × 120 cm during July. For a successful crop, use of seed-borne disease-free material is highly recommended. Bush types are usually planted at a row distance of 85–90 cm and plant distance of 7.5 to 10 cm.

NUTRITIONAL REQUIREMENT

Lima bean, being a legume crop, should fix atmospheric nitrogen into the soil but it lacks in biological nitrogen fixation because of poor nodulation and being very sensitive to fertilizer injury. However, it responds well to initial application of farmyard manure and fertilizers. Well-rotten farmyard manure should be incorporated into the soil @ 25 t/ha at the time of field preparation. It requires a lesser amount of nitrogen as compared to other legume vegetables since higher doses of nitrogen promote foliage growth but delay fruiting. Soil pH, crop nutrients requirement and residual nutrients influence the fertilizer recommendation. In general, it requires nitrogen 40 kg, phosphorus 50 kg and potash 50 kg/ha. A half-dose of nitrogen along with a full dose of phosphorus and potash is applied as a basal dose at the time of sowing, and the remaining half of the nitrogen is top-dressed at the time of flowering. Excessive nitrogen fertilization promotes luxuriant growth and inhibits pod formation. Phosphorus rapidly becomes fixed into the soil and does not leach, thus, the full dose of phosphorus should be applied at the time of sowing.

IRRIGATION REQUIREMENT

Lima bean is a shallow-rooted crop and requires frequent light irrigations. The plant is adapted to a wide range of rainfall but regular irrigation is needed to ensure economic yield. An effective irrigation schedule promotes the maximum yield, uniform germination, plant development, harvest maturity, pod quality and higher yield of lima bean. To ensure adequate moisture requirement for proper vegetative growth and pod formation, irrigation may be given at intervals of 1–2 weeks, depending upon the soil and prevailing weather conditions. On an average, the crop needs about 150–450 mm irrigation. An adequate moisture in the soil is essential at the time of sowing to ensure better seed germination. About six or seven irrigations are required during the growing season at intervals of 5–10 days. The critical stages for irrigation are blooming and pod set stages. Water stress at flowering and pod setting shall result in dropping of flowers, young pods, or both and reduced yield.

INTERCULTURAL OPERATIONS

HOEING AND WEEDING

Lima bean is a short-duration crop, needing 50–60 days from sowing to maturity. Weeding during the early phase of crop growth is very crucial. After germination, the first 30–40 days are the critical period for crop weed competition. Thus, two light hoeing and weedings are found beneficial during this period to reduce competition with weed plants. However, during hoeing, care should be taken not to injure the plant roots.

In addition to hand-weeding, pre-plant application of herbicides is effective for the control of annual grasses and small-seeded broadleaf weeds. In India, the chemical weedicides have not been experimented, however, pre-sowing soil incorporation of weedicides is recommended for French bean *viz.*, fluchloralin 2 litres/ha or pre-emergence application of oxadiazon 0.5 kg/ha, Alachlor 1.5 kg/ha, or pendimethalin 2 kg/ha should prove effective.

Hoeing is an essential operation, which should be done one and two months after planting to make the soil loose around the plant roots necessary for proper aeration, keep the field weed free and to reduce competition between crop plants and weeds.

STAKING

Lima bean, being a weak climber, needs support for its growth. Staking also helps in increasing pod yield and pod size, reduces pod rot and makes cultural operations and harvesting easier. The vines trailing on a bower continue to give yield for 6–7 months as compared to the vines crawling on the ground without support gives yield 3–4 months early. The vines on the bower are less susceptible to insect-pests and diseases as they do not come in direct contact with soil. When the vines are to be trailed on bower system, the seeds should be sown on the edges of the furrows opened up at 2.5 m. Cemented poles of 3 m height are pitched on both ends of alternate furrows at a distance of 5 m. These poles are connected to each other with steel wires, which along the furrows are further connected with cross wires fastened at 45 cm distance so as to form a network-like structure. Seeds are sown at a distance of 1 m along the furrow and covered with fine soil. The vines reach a bower height in about 1.5–2 months. A single bamboo stick of 1.5–2 m length is inserted into the soil near the plant to give support when the plants start developing vines.

MULCHING

In summer months, under low-moisture stress, mulching with straw is very effective for good plant growth. It conserves soil moisture, suppresses weed growth, increases water-holding capacity of the soil and regulates soil temperature around the plant roots. In winters, the use of transparent and black polythene mulch increases the soil temperature, promotes earliness and increases fresh green pods yield.

USE OF GROWTH REGULATORS

Use of plant growth regulators in lima bean improves the plant growth, induce early flowering, pod set and pod yield. Spray of gibberellic acid 50–200 ppm increases the plant height, number of leaves and setting of pods. Spray of para-chlorophenoxy acetic acid (PCPA) at 2 ppm or beta-naphthalene acetic acid at 5–25 ppm improves pod set and total economic yield.

HARVESTING

Lima bean harvesting depends on the purpose of use. Bush type varieties become ready for harvesting in 60 to 80 days after sowing, while pole bean varieties are harvested 85 to 90 days after sowing. It requires 60–90 warm and frost-free days to reach a harvest maturity depending upon type and variety. After flowering, the crop becomes ready for harvesting in about 7–10 days. Flowering starts 80–85 days after sowing and pod formation takes place 1 month later. First harvest is done 4 months after sowing, *i.e.*, by November, and it extends up to March. A total of 12–14 pickings are possible at intervals of 15–16 days. The pods for culinary purposes are picked at the immature green stage when the pods are tender and non-fibrous. The time of harvesting is a compromise between yield and quality as delayed harvesting increases the yield but decreases the pod quality. With the advancement of maturity, the pods lose their green colour, tenderness and firmness rapidly and become fibrous. The pods at the immature stage are harvested manually by pinching the pedicle slightly above the pod and pulling gently. After the beans are picked, they are immediately transferred to a shaded area, and the over-matured, unfilled, diseased pods and the pods that have lost their green colour as well as trash and dirt are removed from the beans lot. For the extraction of dry seeds, the crop is harvested when most of the pods have turned yellow. Harvesting at the pre-dry stage does not affect the seed yield but reduces the seed quality.

YIELD

The yield of lima bean varies considerably due to variety, temperature, soil moisture and the packaging practices adopted by the growers. In very hot summers, the yield can be reduced significantly due to flower drop and pod abortion. On an average, the yield of green tender pods is 50–100 q/ha. The shelled green beans' yield is about 8–11 q/ha.

POST-HARVEST MANAGEMENT

Lima bean is highly perishable as the respiration of freshly harvested pods is very high. The freshly harvested produce is hydro-cooled as soon as possible by spraying or immersing the harvested produce in cold or chilled water in order to avoid deterioration of pods and to preserve their quality. Lima beans are usually packed in wire-bound boxes but locally mesh bags are generally used to send them to markets. The fresh unshelled pods can be stored only for a very short period under ambient room conditions but can be kept fresh for few days in a refrigerator. Desiccation of pods is a common post-harvest problem under low humidity conditions. Therefore, its hydro-cooled unshelled green pods can be stored for several weeks at temperature 4–5°C and 95% relative humidity.

Green pods can be processed as frozen or canned. However, the pods for processing must be long, straight, tender, medium-sized with thick walls and small seeds, stringless and disease-free. The pods should not split or slough off and discolour during cooking. The harvested pods before processing are cleaned using a shaker and blower and then they are subjected to size grading and removing stems and blossom ends. Thereafter, the pods, with the help of mechanical cutters, are cross cut into pieces of 1.0–1.5 cm length. The shreds after blanching in steam or water at 99°C for 2–3 minutes are quickly cooled and then sorted, packed and frozen or canned. The cut beans are filled in cans containing 2% brine, thereafter, the cans are heated in a steam exhaust box for 5 minutes and hermetically sealed. Lima beans come in several forms, *i.e.*, fresh, dried, large and in variety of distinctive colour patterns, thus, they should be graded accordingly.

INSECT-PESTS

Pod Borer (*Etiella zinckenella* Treitscke)

The old larvae feed from outside and characteristically leave part of the body exposed. The feeding hole is usually clean and circular with faecal frass usually deposited away from the hole. The young larvae feed within flower buds, flowers and pods. Damaged flowers abort and do not set pods. Larval damage also results in several partially eaten seeds that reduce the quality and yield. Mature larvae are about 4 cm long and vary in colour between yellowish-green, green, brown and black but, they all have a characteristic marking of pale and dark bands on each side of the body. One larva may damage several pods and cause them to wilt. The pods may have shrivelled or half-eaten seeds or bear no seeds at all.

Control

- Avoid relay cropping of beans into established alternate hosts such as tomato, cotton, okra, etc.
- Collect and destroy all damaged pods and larvae.
- Use natural enemies such as parasites and predators like ants to reduce the pest population as they feed on small larvae and pupae.
- Spray the crop with neem seed-kernel powder 5% followed by triazophos 0.05%, Monocrotophos 36 SL 0.05%, Quinalphos 0.2%, methyl demeton 25 EC 0.1%, or dimethoate 30 EC 1%.

Bean Stem Maggots or Bean Fly (*Ophiomyia phaseoli*)

The pest is widely distributed throughout beans and other leguminous crop growing areas. Severe damage is indicated by wilting and dying of seedlings. The larvae mine into leaves, petioles and stems. They also feed on outer layers of the stem at the collar region of the young plants causing death of the plants. The presence of this pest is indicated by small shiny black flies about 2 mm in length with clear wings that reflect a metallic blue colour in sunlight. They lay eggs in the leaf tissue or directly in the stem. Early signs of attack are egg-laying punctures on the primary leaves. Bean flies are generally active during rainy season. Alternate wild leguminous plants and crops such as cow pea and soy bean as well as volunteer bean plants are other sources of infestation.

Control

- Grow resistant and early-maturing varieties.
- Grow beans in more healthy and fertile soils.
- Collect and destroy all infected plants to kill the maggots.
- Use rice straw mulch to enhance adventitious root formation.
- Plough the field deeply after harvesting to kill the pupae in soil.
- Apply carbofuran or phorate 0.25 kg a.i./ha in soil.
- Spray systemic insecticides monocrotophos or acephate 0.5% before planting to protect the young seedlings.
- Spray the crop twice with dimethoate 0.1% after completion of germination at weekly intervals.

Epilachna Beetle (*Epilachna duodecastigma*)

The adults are green with green streaks on the body. Both grubs and adults feed on the leaves causing skeletonized patches on them. Leaf tissue between the veins is eaten. The leaves may completely be stripped off to the midrib and small areas eaten out; as a result, shallow holes may develop on the fruit surface. In severe infestation, the whole plant may die.

Control

- Handpick and destroy the grubs and adults at the initial stage of infestation.
- Irrigate the crop at regular intervals to minimize the pest population.
- Use *Bacillus thuringenesis* formulations to control the beetle effectively.
- Add *Neem*, *Mahua*, or ground nut cake to suppress the pest population.
- Spray the crop with fenvalerate 0.05% or quinolphos 0.2% at intervals of 10–15 days.

DISEASES

FUSARIUM WILT (*Fusarium oxysporum f.sp. phaseoli*)

The infected plants show yellowing and wilting of leaves, causing the entire plants to wilt and die. Vascular tissue inside the root shows browning but the pith remains unaffected. However, at seedling stage, the infection causes stunting, wilting and death of the plants. The fungus can cause water-soaked lesions on pods. The pathogen penetrates vascular tissue of the root and hypocotyls, producing a reddish discoloration throughout the root, stem, petioles and peduncles. Later, infection causes yellowing of lower leaves and may progress to the upper leaves, causing premature defoliation. The disease is favoured by low humidity and high temperature, especially when the plants are wounded. Infected soil is the most common source of infection. The disease spreads through the infected soil and it may also be carried to the next season through the infected seeds, plant debris, irrigation water and contaminated farm equipment.

Control

- Grow resistant variety if available.
- Follow long crop rotation with non-host crops.
- Sterilize the field soil through solarisation.
- Use organic amendments like farmyard and green manure to improve the soil fertility.
- Follow deep ploughing to destroy the infected bean residues and plant debris.
- Drench the soil before sowing seeds with Bavistin 0.1% or Thiram 0.2%.
- Treat the seeds with captan, Semisan, Thiram, or Benomyl 2–3 g/kg of seed.

FUSARIUM ROOT ROT (*Fusarium solani f.sp. phaseoli*)

Generally, its symptoms are observed on the hypocotyls and primary roots 1–2 weeks after emergence of the seedlings. As the infection increases, the lesions coalesce, and in advanced stages, they become brown and extend up to the soil surface. Lesions may be accompanied by longitudinal fissures or cracks. There is frequently killing and detaching of primary and lateral roots, which result in hollowness of the lower stem.

Control

- Plough the field deeply to destroy the infected plant material.
- Allow the previous crop to be well decomposed prior to planting.
- Grow resistant variety if available.
- Follow long crop rotation with cereal crops.
- Fumigate the soil with formaldehyde.
- Treat the seed and soil with Apron or Ridomil.
- Plant lima bean on ridges or raised beds to facilitate rapid plant growth.

RHIZOCTONIA ROOT ROT (*Rhizoctonia solani*)

Initially, disease shows circular to oblong dark sunken cankers delimited by brown margins on hypocotyls and taproots. Later, these cankers enlarge and become red, rough, dry and pithy. The cankers girdle the stem, resulting in retarded plant growth. Minute brown sclerotia may develop on or in the cankers. Lesions also can develop on pods in contact of moist soil, causing pod rotting and seed discoloration.

Control

- Follow long crop rotation with cereal crops.
- Grow resistant variety if available.
- Destroy the infected crop debris.
- Treat the seed with Bavistin 3 g/kg seed or Vitavax 6 g/kg of seed.

ANTHRACNOSE (*Colletotrichum lindemuthianum*)

Anthracnose is favoured by cool and wet weather. The fungus becomes inactive during hot and dry weather. Symptoms include brick red to purple lesions running along leaf veins. Infected pods may have tan to rust-coloured sunken cankers. This fungus overwinters on diseased bean plants left in the field after harvest and can live in the seed as long as they remain viable. Spores will survive for more than 2 years in old bean debris under field conditions. Once the disease is brought into the field through seed, it can spread through splashing rain drops, insects, equipments, or by the people moving in the field when the bean plants are wet.

Control

- Rotate the crop with non-host cucurbitaceous crops.
- Grow resistant varieties such as G 2333, NAT 002, NAT003 and AB 136.
- Use disease-free healthy seed.
- Treat the seed with Thiram, Ziram, Arsan, or Cerasan @ 2–3 g/kg of seed.
- Follow clean cultivation by destructing the infected plant debris.
- Remove and bury the plant debris from field soon after harvest.
- Spray the crop with Dithane Z-78 0.2% or Bavistin 0.1%.

ANGULAR LEAF SPOT (*Phaseoriopsis griseola*)

Angular leaf spot is a serious disease of beans. Its infection and development are favoured by moderate temperature of 20–25°C and humid conditions. It may affect all above-ground parts of the plant. Symptoms are more prominent during late flowering and early pod formation stage. The disease develops round lesions, which usually appear as brown spots with a tan or silvery centre. The spots initially occur on major veins, giving it an angular appearance. The lesions eventually cover the entire leaf, which becomes yellow before aborting prematurely. Lesions can be observed on the underside of the leaf and appear slightly paler than those on the upper leaf surface. Lesions on stems and petioles appear elongated and dark brown. Lesions on pods are irregular to circular black sunken spots with reddish brown centres and may be similar to those caused by anthracnose. Infected pods may bear poorly developed, shrivelled, or discoloured seeds. The main sources of infection are plant debris or residue and infected volunteer plants. The disease spreads through contaminated equipment, water and rain.

Control

- Plough the field deeply to bury and destroy the pathogen-infected debris.
- Strictly follow long crop rotation with non-host crops.
- Grow resistant cultivars if available.
- Treat the seed with fungicides like captan, Captafol, Caftap, Foltaf, or Bavistin 2.5 g/kg of seed, etc.

Ascochyta Leaf Spot (*Ascochyta phaseolorum*)

Symptoms appear first on leaves as large dark grey to black spots. Later, the infected area converts to zones followed by concentric rings around the spot containing black pycnidia. Blackening of nodes is the characteristic symptom of disease infection. It can girdle the stem and kill the plant. Under severe infection, extensive blight and premature leaf drop may occur. Flower infection can lead to end rot of pods and cause extensive cankers. Pod infection often results in seed infection, which can be transmitted to the next crop.

Control

- Plough the field deeply to bury and destroy the pathogen-infected debris or residue.
- Rotate the crop with cereal crops.
- Grow resistant variety if available.
- Sow pathogen or disease-free healthy seeds.
- Follow wider spacing to improve aeration and reduce disease infection.
- Mulch the crop to reduce water splash and spread of inoculums during rainy season.
- Treat the seed with zineb, Benomyl, or Dithane M-45 @ 3–4 g/kg of seed.

Leaf Rust (*Uromyces fabae*)

Initially, leaf spots enlarge to form reddish brown or rust coloured pustules and give a rusty appearance to anything they contact. Severe infection may cause the leaves to turn brown, curl upwards, dry and abort prematurely. Green pods and occasionally stems and branches, may also become infected and develop typical rust pustules. Pod setting, pod filling and seed size can be reduced if early infection is severe. A severely damaged bean field looks as if it has been scorched. Severely infected leaves turn yellow, wilt and then drop off from the plant. Stems and pods may also be infected. It is favoured by cool to moderate temperature with moist conditions that result in prolonged period of free water on leaf surface for more than 10 hours. Rust can be distinguished from other leaf spots. The spores from the leaf spots rub off onto your fingers, while blights do not. The major sources of infection are infected bean debris and infected volunteer plants. The disease is mainly spread by wind.

Control

- Strictly follow a crop rotation of at least 3 years.
- Grow rust-resistant varieties like JP-4.
- Use disease-free healthy planting material.
- Remove potential sources of rust spores, *e.g.*, alternate hosts and volunteer plants.
- Spray the crop with Bavistin 0.1%, Benomyl 0.1%, Calixin 0.1%, or Diathane M-45 0.2%.

Powdery Mildew (*Erysiphe polygonii*)

The disease is mainly favoured by warm temperatures (20–24°C), low humidity and shade. Initially, the disease manifests on plants as white powdery fungal growth on leaf surfaces. Usually, both sides of the leaf show fungal growth. White dirty green patches appear on leaves, petioles and stems, which later cover the whole surface. Leaves become yellow, chlorotic and shrivelled, and later, defoliation occurs. In severe cases, plants become stunted. The infected plant bears no fruit or the set fruits become very small, and the flavour is affected adversely. Seeds get infected through pod infection and may carry the infection to the next crop if planted. The major source of infection is contaminated seed and infected plant debris. The disease spreads through wind, rain and insects.

Control

- Follow long crop rotation with non-hosts such as maize and other cereal crops.
- Grow resistant varieties if available.
- Follow field sanitation.
- Use disease-free healthy seed.
- Grow the crop at wider spacing to increase aeration.
- Plough the field deeply to destroy the diseased plants after harvest.
- Dust the crop with Sulfex 2.5 kg/ha.
- Spray the crop with wettable sulfur, lime-sulfur, Dinocap, Benomyl, Triforine, Karathane 0.2%, or carbendazim 0.1%.

CHARCOAL ROT/ASHY STEM BLIGHT (*MACROPHOMINA PHASEOLINA*)

Generally, infection occurs at cotyledons and hypocotyls at the soil level. It produces symptoms like black sunken cankers, which have concentric rings with sharp margins. In severe cases, the growing tips of the plant may die and the stems break at the canker affected areas. The infected younger plants show black wounds and stems show ash-like wounds on diseased stems, thus, the name has been given ashy stem blight. Severe infection on older plants may cause stunting, leaf chlorosis, wilting, premature defoliation, rotting of the stem and roots, and death of plants occur. Warm and humid conditions are favourable for the occurrence of disease.

Control

- Follow 2–3 years' crop rotation with non-hosts crops.
- Grow resistant variety if available.
- Use disease-free healthy seed.
- Flood the soil to suffocate the pathogen.
- Fumigate the soil with Chloropicrin or methyl iodide.
- Treat the seed with Thiram, Ziram, Captan, Captaf, Captafol, or Foltaf 3–4 g/kg of seed.
- Rogue out and burn the infected plants.

COMMON BACTERIAL BLIGHT (*XANTHOMONAS PHASEOLI*)

Initially, the symptoms appear as water soaked spots on the underside of leaves. The leaf spots enlarge and merge to form large brown irregular lesions surrounded by narrow yellow patches. Spots may begin to coalesce and the leaves become yellow. The stem may rot at first node where cotyledons are attached, which causes the plant to break from that point. Pods have sunken circular spots, which are water-soaked initially but later dry with a reddish brown narrow border. Under wet conditions, yellow slimy bacterial exudates ooze out of the lesions and form a crust. Pods may shrivel and infected seed may rot. Bacteria also invade seed and may remain dormant until germination. Even a trace of infected seed when planted can initiate severe infection in whole field. The major source of infection is seed. Other sources of infection include infected dry bean debris or residue on the soil surface, and the infected volunteer plants may be another source. The disease is favoured by warm to high temperature and high humidity and it spreads by water splash, windblown rain, overhead irrigation, insects and contaminated equipment and clothing.

Control

- Plough the field deeply in summer months to ensure destruction of pathogen.
- Grow the resistant varieties if available.
- Follow long crop rotation with non-host crops.
- Use disease-free healthy seed.

- Treat the seed with copper sulfate, copper hydroxide, or streptomycin @ 3–4 g/kg of seed.
- Avoid the field operations when leaves are wet to reduce the spread of infection.
- Spray the crop with Basicop 53WP 0.03% and streptomycin 400 ppm before the infestation spreads.

Halo Blight (*Pseudomonas syringae pv. phaseolicola*)

The disease is favoured by cool temperatures of 16–20°C and followed by moist cloudy conditions. It multiplies rapidly in the presence of dew. It is not common in regions or seasons with high temperature. The most characteristic symptoms appear first as small water-soaked pinprick spots on the underside of leaves. The tissues surrounding the spots gradually become yellow green, resembling a halo, thus, the name is halo blight. Infected branches and petioles bear greasy spots, which develop a reddish discolouration. The stem may rot at the first node where cotyledons were attached and cause the plant to break. The spots soon turn reddish brown and infected pods show signs of water soaked lesions, which turn to reddish discolouration. Under humid conditions, whitish to yellow bacterial exudates appear on the lesions. Seed may become shrivelled, rotten, or discoloured, but sometimes, no visible symptoms may be seen at all. Major sources of infection are infected seed and plant debris. The disease spreads through rain water splash and farm equipments.

Control

- Plough the field deeply to bury and destroy the pathogen-infected debris.
- Strictly follow at least 2–3 years' crop rotation with a non-host crop.
- Grow resistant varieties if available.
- Use disease-free healthy seed.
- Avoid movement of workers in irrigated field.
- Follow recommended rogueing practices.
- Spray the crop with copper hydroxide 0.1%, mancozeb 0.25% with urea 2%, streptomycin 400 ppm, Ferbam 0.25%, or copper oxychloride 0.2–0.3%.

Lima Bean Mosaic (Lima bean mosaic virus – Cucumovirus group)

Symptoms appear as vein-clearing of leaves and puckering of leaf lamina. Leaf size is greatly reduced and the margins slightly curled. Later, the leaflets have a distortion of margin and etching and become misshapen. Patchy mosaic mottle appears on leaves, resulting in poor pod setting with only a few small seeds. The disease is sap transmissible and transmitted by aphids (*Aphis craccivora, A. gossypii, Myzus persicae* and *Acyrthosiphum pisum*) and seed.

Control

- Keep the aphids under control to reduce the spread of disease.
- Apply carbofuran 1.5 kg a.i./ha in soil at sowing time.
- Spray the crop with monocrotophos 0.05% or phosphamidon 0.02% at 10-day intervals.

Yellow Mosaic of Lima Bean (Yellow mosaic virus)

Of the various viral diseases affecting the legume crops, yellow mosaic virus is one of the most destructive and widely distributed viruses in India. Initially, small alternating green and bright yellow patches or spots appear on the young leaflets. Slowly, the area of yellow discoloration increases and the entire leaf may turn yellow. Sometimes, the green areas are slightly blistered and the leaves show a slight puckering and reduction in size. The leaves become papery white and thin.

Diseased plants have stunted growth. Pods on infected plants are small size, turn yellow and distorted. Early infection causes death of the plant before seed set. Infection causes reduction in plant yield and quality. The disease leads to heavy crop loss from 50% to 70%, especially if the disease occurs in early stages of the crop growth. The disease is transmitted by whitefly (*Bemisia tabaci* Genn.).

Control

- Grow a barrier crop such as sorghum, maize, or pearl millet.
- Remove the host weeds from the surrounding of the lima bean field.
- Use seeds collected from disease-free plants.
- Treat the seed with insecticides like carbosulfan 30 g/kg of seed or monocrotophos 5 ml/kg of seed before sowing.
- Remove and destroy all the infected plants from the field to avoid secondary spread.
- Apply carbofuran 1.5 kg a.i./ha in soil at sowing time.
- Spray the crop with methyl parathion 0.02% or Confidor 0.05% at 10-day intervals.

■■■

78 Pigeon Pea

M.K. Rana and Prashant Kaushik

Botanical Name   : *Cajanus cajan* (L.) Millspaugh

Family   : Leguminosae (Fabaceae)

Chromosome Number : 2n = 22

ORIGIN AND DISTRIBUTION

The origin of pigeon pea has remained unclear for a long time. Early literature indicates that the crop originated in either Africa (Zanzibar or Guinea) or India. The presence of its seeds in historical tombs indicated that it was cultivated in Egypt around 2000 BC. The availability of high diversity among germplasm led Vavilov (1939) to conclude that India is the primary centre of origin of the cultivated pigeon pea. De (1974) and van der Maesen (1980) also reported that the cultivated pigeon pea originated in India. In time, the pigeon pea made its way to Africa, where it obtained the name Congo pea, and from Africa *via* the slave trade to the Americas.

INTRODUCTION

Pigeon pea is credited to be one the most suitable crops for subsistence agriculture that needs minimum external inputs. It is cultivated in a wide range of cropping systems and so is its usage. In the northern India, its dehulled split cotyledons are cooked to make *dal*, while in the southern parts of the country, its usage as *sambar* is very popular. Also in some parts of India, including Karnataka and Gujarat, the use of immature shelled seeds is very common as a fresh vegetable. Besides, in the tribal areas of various states, its use as a green vegetable is very common. The recipes prepared with green pigeon pea seeds are nutritive and tasty and are consumed with rice as well as *chapati*. During off-season in southern and eastern Africa, southern America, and the Caribbean islands, the whole dry seeds are used in making porridge, while in the crop season, its immature pods are used as fresh vegetable. It is known to produce a reasonable quantity of food even under unfavourable production conditions mainly due to its qualities such as drought tolerance, nitrogen fixation, and a deep root system.

COMPOSITION AND USES

COMPOSITION

The green pigeon pea seeds have high crude fibre, fat and digestible protein. The cotyledons contribute about 90% proteins, 95% fat, 86% carbohydrates, 83% minerals and most of the phosphorus of the whole seed. The nutritional composition of pigeon pea is given in Table 78.1.

TABLE 78.1

Nutritional Composition of Tender Pods of Pigeon Pea (per 100 g Edible Portion)

Constituents	Contents	Constituents	Contents
Water (g)	65.10	Phosphorus (mg)	164.0
Carbohydrates (g)	16.90	Iron (mg)	1.10
Protein (g)	9.80	Niacin (mg)	3.00
Fat (g)	1.00	Thiamine (mg)	0.32
Dietary fibre (g)	6.20	Riboflavin (mg)	0.33
Minerals (g)	1.00	Vitamin C (mg)	25.0
Calcium (mg)	57.0	Energy (kcal)	116.0

Uses

Pigeon pea is one of the major pulses grown and used in the forms of split pulse. It is consumed in various ways, commonly being cooked with vegetables and spices. Green peas in the form of frozen or canned products are also available for use as vegetable in the markets of the United States of America and Europe. Its nutritious broken seeds, husks and pod shells are fed to the cattle, and the dry stems make it a popular household fuel, particularly in rural areas.

Medicinal Uses

Pigeon pea is also being used as an integral part of traditional folk medicine in India, China and some other developing countries. Its dried roots are sometimes used as an alexeritic, anthelminthic, expectorant, sedative and vulnerary. In Brazil, pigeon pea leaves are infused for coughs, fevers and ulcers.

BOTANY

The vegetable pigeon pea plant is erect and branching. There are two major growth patterns in pigeon pea, *i.e.*, (i) the determinate types and (ii) the indeterminate types. The determinate types bear pods in clusters at the top of the canopy. They cease their plant growth after the induction of flowering and their pods mature more or less uniformly. On the contrary, in indeterminate types, the terminal bud remains vegetative throughout their life cycle and they bear flowers in axillary clusters. In general, the indeterminate types tolerate biotic and abiotic stresses better than determinate types.

The stem is woody and the leaves are trifoliate and compound. The plants grow into woody shrubs and 1–2 m tall when harvested annually. It may also attain a height of 3–4 m when grown as a perennial plant. Vegetable pigeon pea has a taproot system, which is nodulated by the cowpea group of *Rhizobia*, mainly on the upper 30 cm of the root system. It has a strong woody stem. During vegetative phase, the xylem parenchyma and the medullary rays contain starch. This disappears during reproductive phase, showing that the reserves are mobilized for pod development. The branching pattern depends on genotype, whether determinate or indeterminate type, and spacing between rows and plants. At a wide spacing, it may form a bush, and at narrow spacing, it may remain compact and upright. The leaves are pinnately trifoliate with lanceolate to elliptical leaflets that are acute at both ends and are spirally arranged. The leaflets are borne on a rachis, which is swollen at the base (pulvinus). In most of the cultivars, flowers are borne on terminal or auxiliary racemes (4–12 cm) and carried on a long peduncle.

Flowering proceeds acropetally (in the direction of apex) both within the raceme and on the branch. The flower colour varies from yellow to yellow with purple veins and from yellow to diffused red. Stamens are 10, diadelphous (in two bundles 9+1) with nine fused in a column and one free. The ovary is superior, subsessile and flattened dorsoventrally with a long style. In a fully developed bud, anthers surround the stigma and dehisce a day before the flower opens. The duration of flower opening depends on the climatic conditions. Usually, it varies from 6 to 36 hours. Fertilization occurs on the day of pollination. It is an often cross-pollinated crop. The extent of cross-pollination ranges from 3% to 40%, with an average of 20%. A plant produces many flowers, of which only 10% set pods. Pods are of various colour, *i.e.*, green, purple, dark purple, or mixed green and purple. Pods with deep constrictions in shape are beaded, while others are somewhat flat. The seeds per pod range from two to seven and sometimes up to nine. The seeds are in separate locules and the cross walls develop during the first week after fertilization. The pod wall develops more rapidly than the young seeds. Seed development is visible 7 days after pollination. A pod is formed 15–20 days after fertilization.

CLIMATIC REQUIREMENT

Vegetable pigeon pea is predominantly a crop of tropical areas mainly cultivated in semi-arid regions of world. The isoclimes similar to India exist in western Africa and southern Sudan, having a suitable environment for growing vegetable pigeon pea. It is also cultivated in sub-humid regions of Uganda, the West Indies, Myanmar and the Caribbean regions. It can be grown between 14°N and 28°N latitude, with a temperature ranging from 26°C to 30°C in the rainy season (June to October) and 17°C to 22°C in the post-rainy (November to March) season. Mean annual rainfall ranges from 600 to 1400 mm, of which 80% to 90% is received in rainy season. Under dry conditions, it matures fast, while under humid conditions, it remains in vegetative phase for a longer period and therefore matures late. It is highly sensitive to frost and low radiation at pod development, therefore, flowering during monsoon and cloudy weather leads to poor pod formation.

SOIL REQUIREMENT

The vegetable pigeon pea crop can successfully be grown on almost all kinds of soil such as black cotton soil, red loamy, sandy and clay soils. However, for prolific growth, loamy soils are ideal for its cultivation. Moreover, it is adaptable to soil pH ranging from 5 to 8. In acidic soils, 2–4 t/ha of lime is incorporated 3–4 weeks before sowing to neutralize the acidity. Maintaining proper drainage facility is necessary for good root and nodule development. A ridge and furrow system or contour beds are useful in preventing waterlogging by draining excess surface water and in preventing soil erosion.

Land preparation for pigeon pea requires at least one ploughing during dry season followed by two or three ploughings with a harrow. The *summer* ploughing helps in minimizing harming microbes, the weed flora and conserving moisture. Organic manure may be applied 2–4 weeks before sowing and mixed thoroughly.

CULTIVATED VARIETIES

Vegetable pigeon pea cultivars should have long pods and large seeds. Cultivars are generally grown as a normal field crop but immature pods are harvested at an appropriate stage for use as vegetable. This practice is more prevalent around cities where green pods can readily be marketed at attractive prices. After harvesting green pods, the crop is left for producing dry seeds. Such dual-purpose varieties are very profitable for peri-urban farmers. Cultivars with white seed coats are preferred because the cooking water remains clear when such seeds are cooked. Sweetness of fully grown immature seed is also a preferred trait. In hilly tribal areas of India, a large number of large-seeded landraces are traditionally grown for vegetable purposes. Some of the important varieties are given as follows:

BP 2060

An indeterminate variety gives 4–5 pickings when grown as rain fed and 5–7 pickings under irrigated conditions. The pods are harvested 130–135 days after sowing.

ICP 7035

A variety developed from the germplasm collected from *Bedaghat Township* located near Jabalpur city in Madhya Pradesh, India, has a sugar content as high as 8.8 %. Its flowers are dark red that produce purple colour pods.

ICP 7035

The seeds are large and purple with a mottle pattern. It produces an excellent quality of vegetable. Its pods are 7-8 cm long, and on an average, each pod contains six seeds.

T 15-15

A dual-purpose variety widely is grown in Gujarat state for both green and dry seed. It possesses high protein content.

ICPL 87

A dual-purpose variety produces pods for a relatively longer time and allows two to four pickings within a year.

In southern India, the large seeded lines such as 'HY-3C' and 'TTB-6' are also popular as vegetable.

SOWING TIME

The sowing time varies considerably in relation to the type of variety and climatic conditions of the growing region where the crop is to be raised. Vegetable pigeon pea is mostly grown as a *Kharif* crop. For that, the sowing generally starts with the commencement of monsoon in June–July. In areas where winter is not severe, it can be grown as a *Rabi* crop, usually sown in September–October. This will ensure good plant growth and canopy development.

SEED RATE

The germination percentage of seed chiefly affects the seed rate. The seed rate may also vary due to growing season, method of sowing and soil properties. Seed rate varies from 1.5 to 5.0 kg/ha when the crop is grown in mixture and 15 to 20 kg/ha when grown as a pure crop.

SOWING METHOD

Pigeon pea is known to be highly sensitive to environmental factors, especially photoperiod and temperature, which results in different phenologies. Early maturing vegetable type varieties are short in stature and thus require high density (200,000 to 300,000 plants per hectare) in contrast to long duration indeterminate types, which require 40,000–50,000 plants per hectare for economic yield of the crop.

For different agro-ecological zones, specific agronomic practices are required, and it may be improper to recommend a single production package of practices for different locations. The seeds are placed about 3–5 cm deep and covered firmly with soil. For short-duration types, the row to row spacing of 50 cm at low latitudes and 60 to 75 cm at high latitudes with plant to plant spacing of 25 to 30 cm can be adopted. For medium- and long-duration types, the seed should be sown to maintain row to row and plant to plant spacing of 100 and 50 cm, respectively.

NUTRITIONAL REQUIREMENT

Vegetable pigeon pea, being a leguminous crop, needs inoculation with effective *Rhizobium* culture, which increases its pod yield ranging from 19% to 68%. It responds well to phosphorus application as it increases root nodulation. Under Indian conditions, vegetable pigeon pea crop can successfully be grown with farmyard manure 15–20 t/ha supplemented with nitrogen 20–25 kg, phosphorus 40–60 kg and potassium 20 kg/ha. The farmyard manure should be incorporated into the soil during field preparation. The entire dose of phosphorus, potash and a half dose of nitrogen fertilizers should be placed 5–7 cm below the seed depth before sowing the seeds, and the remaining half of the nitrogen should be supplied in the form of top-dressing at the time of the second or third irrigation, when the pods start developing. Nitrogen and phosphorus are useful fertilizers to reduce the percentage of under-developed seeds.

IRRIGATION REQUIREMENT

For better seed germination and subsequent growth of the plants, the soil at the time of sowing must have ample moisture, and for this purpose, pre-sowing irrigation is applied. Sometimes, the seeds do not germinate due to the lack of moisture. Under such a situation, a light irrigation may be given to obtain early and higher percentage of germination. It is, although, advantageous only in light soils but not in heavy soils as it hinders germination in heavy soils, where the soil surface after drying forms a hard crust before the sprouts come out of the soil. Therefore, in heavy soils, it is always safe to sow the seeds when the soil contains enough moisture for seed germination. When the seeds are sown in dry soil on ridges, the ridges are always kept moist until germination is completed. In case the field is irrigated before sowing, the first irrigation is given just after thinning. Depending on the season and availability of soil moisture, the crop needs irrigation at an interval of 10–15 days.

INTERCULTURAL OPERATIONS

WEEDING

Slow seedling growth of pigeon pea makes the crop prone to weed competition for nutrients, light, moisture and space during the first 6 weeks of growth. In general, three weedings manually at the right intervals are sufficient to control most of the weeds. Alternatively, spraying of a pre-emergence weedicide such as Basalin or Prometryn each @1.5 litres per hectare followed by two manual weedings have been found effective in managing weeds.

HARVESTING OF PODS

Near townships and cities where marketing is easy, the immature pods are harvested for sale as a fresh vegetable. Since for vegetable purposes, fully grown bright green seeds are preferred, the pods are harvested just before they start turning their dark green colour to light green colour. Pigeon pea is a perennial plant that allows continuous flowering and pod formation. Under sufficient moisture conditions, flowering and harvesting of green pods takes place in more than one flush. For vegetable purposes, the green tender pods are harvested.

The pods are harvested for different purposes. The developing pods are harvested and their green seeds shelled and used as a vegetable in India and in some South Asian and African countries. Near cities where they can be readily marketed, they are harvested for sale as a vegetable. Fully developed, bright green seed is preferred, thus, the pods should be harvested just before they start losing their green colour. It is important to remember that the appearance of pods at this stage varies between cultivars. Green pods used as a vegetable are commonly picked by hand, but they may be mechanically harvested for large-scale processing *i.e.*, for canning and freezing.

YIELD

The average yield of green pods varies from 300 to 600 q/ha, depending on crop variety, soil and environmental factors and cultural practices adopted by the growers.

POST-HARVEST MANAGEMENT

Harvesting, processing of green seeds and storage are important aspects of post-harvest management in vegetable pigeon pea. Each involves a set of traditional practices dependant on growing region and consumption. After harvesting, the green pods, shelling, canning and freezing are important operations. Canning and freezing involve several operations including cleaning, blanching and filling cans and polyethylene bags. Although, processing of green-seeds is a common practice in several African countries as well as in the West Indies and the Dominican Republic. In the Dominican Republic, over 80% of the seeds produced is canned in brine and exported to the United States and other Latin American Countries. Frozen seeds are used for vegetable purposes and in several other dishes. Commercial vegetable pigeon pea is commonly processed into canned or frozen peas.

INSECT-PESTS

BORER (*Helicoverpa armigera*)

Young larvae feed voraciously on the developing buds and flowers, while the mature larvae bore into the pods and feed on the grains. The adult moth has a V-shaped speck on the light brownish forewings and a dark border on the hind wings. It lays spherical yellowish eggs singly on tender parts of the plants. Larva is greenish with dark grey lines laterally on the body and pupates in the soil. Its life cycle completes in a period of 28–40 days.

Control
- Spray the crop with emamectin benzoate 5 SG 220 g/ha in 375–500 litres of water after harvesting of consumable pods.
- Spray Nuclear Polyhedrosis Virus (NPV) @ 250 Larval Equivalent (LE)/ha.
- Release egg parasite (*Trichogramma* spp.) and egg larval parasite (*Chelonus blackburnii*).

POD-FLY (*Melanagramyza obtuse* Malloch)

Pod-fly is most destructive pest causing considerable yield loss in North and Central India. The eggs are laid by the female fly in the developing grains, which make them unfit as seed. Pupation takes place within the pod itself without leaving external symptoms of damage. The adult fly emerges through the pod by making a pin-sized outlet. Thus, the concealed feeding habit causes more loss to the crop without the farmers noticing and taking up timely control measures.

Control
- Grow tolerant varieties.
- Spray neem oil 20 ml/10 lit or neem seed kernel extract 5%.

Besides these, several pod bugs, caterpillar, leaf and pod webbers and pod borers are reported to cause damage to flowers, pods and seeds. The root knot nematode *Meloidgyne incognita* has also been reported.

DISEASES

WILT (*Fusarium udum* Butler)

This disease causes about 15–25% mortality in crop, and under severe conditions, it may also increase to 50%. Symptom of disease appears as black streak in the collar region just below the soil surface. In advanced stages, more streaks are seen that are several centimetres high. The affected plants start wilting and withering, and the affected tissues turn black in a later stage.

Control
- Grow wilt-resistant varieties.
- Use disease-free fields with no previous record of wilt.
- Follow long crop rotation with non-host crops.

POWDERY MILDEW (*Oidioposis taurica* and *O. ciltorie*)

Symptoms usually appear as white patches on the plant surface, mostly on the leaf. They generally enlarge in area and thereby cover the whole leaf lamina.

Control
- Spray wettable sulfur @ 2 g/litre of water.

LEAF SPOT [*Alternaria alternate* (Fr.) Keissilar]

The disease causes severe damage to the seed crop. The symptom is characterized by brownish black spots with prominent black dots. *Alternaria tenuissima* (Nees ex Fr.) Wiltshire has also been recorded. It causes spot on young leaves resulting in their premature fall of pods and seeds are infected.

Control
- Grow resistant varieties.
- No concrete measure is available.

BLIGHT (*Phytophthora dreschsleri* var. cajani)

Symptoms appear as irregular to circular grey spots on young leaves. These increase in size and coalesce and the infection spreads to the stem where dark brown lesions are formed. These enlarge and girdle the stem, and the plant above the girdled regions dries up.

Control
- Plough the field deep in summer months.
- Provide proper drainage facilities.
- Seed treatment with metalaxyl (Apron) 4 g/kg of seed.

BACTERIAL LEAF SPOT AND STEM CANKER (*Xanthomonas campastris* var. cajani)

Symptoms generally appear as small round water-soaked areas on the leaves. The spots later become quadrilateral lesions surrounded by a halo where bacterial exudates dry up. Stem infection occurs in young plants. This pathogen also produces a dark brown canker on the main stem and side branches.

Control

- Select well-drained field.
- Take seed from healthy plants.
- Spray the crop with antibiotics like streptocycline 100 ppm at 10 days interval.

LEAF MOSAIC OR STERILITY MOSAIC DISEASE

It is a viral disease of pigeon pea transmitted by grafting and eriophyd mite (*Accceria cajani*). The mosaic pattern appears mostly on young leaves in the form of wild, scattered joint green and yellow spots. Leaves become small and cluster near leaf tips.

Control

- Grow resistant varieties.
- Spray miticide to check the spread of disease.
- Spray Metasystox @ 0.1% to control the mite vectors in the early stages of plant growth.

YELLOW MOSAIC

The disease first appears in the form of yellow diffused spots scattered on the leaf lamina, not limited by veins and veinlets. Such spots slowly expand, and in later stages of disease development, the affected leaflets show broad yellow patches alternating with green colour. Sometimes, the entire lamina turns yellow. Leaf size is conspicuously reduced in early infections. In peninsular India, the disease incidence is relatively higher in late-sown pigeon pea. The disease is transmitted by white flies (*Bemisia tabaci*).

Control

- Rogue out the infected plants up to 40 days.
- Remove weed hosts periodically.
- Spray systemic insecticides to control the vector.

■■■

79 Rice Bean

M.K. Rana, P. Karthik Reddy and Chandanshive Aniket Vilas

Botanical Name : *Vigna umbellata* (Thunb.) Ohwi & Ohashi

Family : Leguminosae (Fabaceae)

Chromosome Number : 2n = 22

ORIGIN AND DISTRIBUTION

The Indo-China region is probably considered the centre of rice bean origin and diversity, as it is found naturally growing in India and Central China. From Southern China through the North of Vietnam, Laos and Thailand, it was distributed to Burma, India and Nepal and was also introduced to Egypt by the Arabs. In tropical Africa, it is grown in West Africa, East Africa and the Indian Ocean islands, and less frequently in Central and southern Africa. Furthermore, there are reports that its cultivation is common in Queensland, Honduras, Brazil and Mexico too. It is now cultivated in tropical Asia, Fiji, Australia as well as in the Americas.

INTRODUCTION

Vigna umbellata is closely related to *Vigna angularis*. *Vigna minima*, a wild species from tropical Asia, is even more closely related to *Vigna umbellata*. Within *Vigna umbellate*, two types have been distinguished, commonly designated as varieties, *i.e.*, (i) var. *gracilis*, the wild type with slender branchlets, narrow leaflets and long peduncles, found from India to Malaysia, the Philippines and Central China and (ii) var. *umbellata*, the cultivated type cultivars identified based on maturity period and seed colour. In Madagascar, two types, *i.e.*, (i) yellow rice bean and (ii) red rice bean, are distinguished. It is believed that rice bean evolved from its wild form, *i.e.*, *Vigna umbellata* var. gracilis, which is typically small leaved, fine-stemmed, freely branching, photoperiod sensitive with indeterminate growth habit, sporadic and asynchronous flowering and strongly dehiscent pods containing small and hard seeds. It is known for its diverse distribution and range of adaptation from humid subtropical to warm and cool temperate climate. It is a multipurpose crop and mainly used for human diet with a smaller proportion for fodder and green manuring. It forms an important part of cereal-based diet, as its dried grains are rich in protein, minerals and vitamins. Additionally, it carries social and cultural values in some communities in the country. Promotion of this crop could play an important role in improving diet and food security for people. It is a neglected summer legume crop grown under diverse conditions with no additional inputs. It thrives well on rain fed *tars*, drought-prone areas, exhausted soils and marginal lands. It is generally grown as a mixed crop or intercrop with maize. It is also cultivated along rice bunds and terrace-margins in the mid hills. It is one of the minor legumes grown by farmers and its area under cultivation is too low as compared to other legumes such as lentil, black gram and chickpea.

COMPOSITION AND USES

COMPOSITION

Rice bean is highly nutritious. Its dry seeds are a good source of carbohydrates, protein, minerals and vitamins. Its protein is rich in limiting amino acids methionine and tryptophan. The seeds are also rich in other amino acids including valine, tyrosine and lysine. The seeds contain vitamins such as thiamine, riboflavin, niacin and ascorbic acid. A laboratory study has also proven that it is a good source of various nutrients. It also contains various anti-nutritional factors, *i.e.*, phytic acid or phytate, polyphenols and fibres that reduce the uptake of micronutrients, in particular iron and zinc. The nutritional composition of rice bean in given in Table 79.1.

USES

Rice bean is a crop with multiple uses as it is locally used for the preparation of several food items. The seeds, which are the primary produce, are usually taken as a soup or as a pulse (*Dal*) with rice. The grains from young pods are used as a vegetable. *Batuk, Bara, Biraunla, Masyaura, Kwanti, Rot, Furaula* and *Khichadi* are other major local recipes prepared from it. They are ground to make nutritive flour for including in children's food. Young pods, leaves and sprouted seeds are boiled and eaten as a vegetable. Sometimes, its young pods are eaten raw. It has cultural and sacred significance in Nepalese society. *Batuk* and *Bara* are used during marital ceremonies and other social functions in Magar and Newar communities, respectively. The Newar community has a tradition of preparing soup (*Kwanti*) from a mixture of nine legume grains during the festival of *Janai-Purnima*, and rice bean constitutes one of these grain legumes. *Khichadi* is a traditional dish prepared from a mixture of rice and black gram or rice bean on the eve of *Maghe-Sankranti*, a festival celebrated during mid January. During *Gaura Parba*, a festival celebrated widely in far western regions of Nepal, rice bean is one of the five grains used in preparing *Biruda*, an offering which is made to the festival deity.

Rice bean foliage, which is a high-class nutritious fodder, is known to increase livestock milk production, and its seeds are used as cattle feed. Some farmers include this crop in their cropping system to increase soil fertility. It is also used as a cover crop in hills, as it is a deterrent against soil erosion. Rice bean is used as a living fence and biological barrier.

MEDICINAL USES

Rice bean is primarily used in different herbal medicines and the red type is used in Chinese traditional medicine, sometimes in combination with female ginseng (*Angelica sinensis*). Nepalese

TABLE 79.1
Nutritional Composition of Rice Bean Pods and Seeds (per 100 g Edible Portion)

Constituents	Pod	Seed	Constituents	Pod	Seed
Water (g)	78.6	13.3	Sodium(g)	0.07	–
Carbohydrates (g)	24	60.7	Zinc (mg)	0.88	–
Protein (g)	4.6	20.9	Manganese (mg)	0.11	–
Fat (g)	3.8	0.9	Copper (mg)	0.30	–
Dietary fibre (g)	3.4	4.8	Thiamine (mg)	–	0.49
Calcium (g)	1.52	200	Riboflavin (mg)	–	0.21
Phosphorus (g)	0.35	390	Niacin (mg)	–	2.4
Iron (mg)	152.3	10.9	Energy (kcal)	151	327
Magnesium (g)	0.36	–	–	–	–

categorized it as a cold food as it keeps the people cool in summer months. Simultaneously, it keeps the people warm during winters, however, in Eastern Nepal, it is considered warm food, and old and sick people are advised not to eat it in the hot season, as the people with weak digestive systems may not digest it easily and develop stomach problems from eating it.

A peptide isolated from its seeds has shown strong antifungal activity against *Botrytis cinerea*, *Fusarium oxysporum*, *Rhizoctonia solani* and *Mycosphaerella arachidicola* fungi. Moreover, it has shown mitogenic and anti-HIV-1 reverse transcriptase activity.

BOTANY

Rice bean is an annual climbing herb, forming a thick mat with about 3 m long grooved and branched stems, usually clothed with fine deciduous deflexed hairs. It has an extensive root system with a taproot that can go as deep as 1.0–1.5 m. Leaves are trifoliate, stipules lanceolate and 0.5–1.5 cm long, petiole 5–10 cm long, leaflets broadly ovate to ovate-lanceolate, 5–10 × 1.5 cm, entire or 2–3 lobed and the lateral leaflets with unequal sides, membranous and almost glabrous. Inflorescence with 20 cm long peduncle is an erect 3–10 cm long axillary false raceme containing 5–20 flowers usually in pairs. Flowers are self-compatible but cross pollination also occurs. Bright yellow flowers on 5 mm long pedicel are bisexual papilionaceous with 4 mm long campanulate calyx, five-toothed bright yellow corolla, standard 1.5–2 cm in diameter, large and broad wings, enclosing keel with a curved beak and a conical pocket on one side, stamens 10, 9 connate into a tube, upper one free and ovary single celled superior with broadened and curved style. Its slender pods containing 6–8 seeds are 6–12 cm long, deflexed, glabrous and green when young and black brown at maturity. Generally, seeds are elongate, smooth surfaced, brown, speckled and mottled.

CLIMATIC REQUIREMENT

Rice bean can be grown in diverse environments due to its wider adaptation. Normally, it is sown in late May to the end of June in Nepal. The middle hills 700–1400 m with warm south facing facets are suitable for its cultivation, although there is evidence that it can thrive best at altitudes as high as 2000 m. A temperature range of 25–35°C and average rainfall of 1000–1500 mm per annum is optimal for healthy vegetative growth and proper pod development. Heavy rain during flowering period affects the pod development and hampers its production. Most varieties are photoperiod-sensitive, tend to be late in flowering and produce vigorous vegetative growth when grown under conditions of ample water and warm temperature in subtropical regions. Rice bean as a post-rainy season crop an in upland rice-and-maize-based system has shown promise in the low hills of Nepal. It is typically suited to humid tropical lowlands but some cultivars are well adapted to subtropical and temperate conditions. It is fairly tolerant to drought but susceptible to frost. A day length of less than 12 hours is required to initiate flowering and seed production; when grown under long day conditions, the crop produces masses of vegetation but little or no seed.

SOIL REQUIREMENT

Rice bean can be grown in a wide range of soils, including shallow, infertile, or degraded soils but red soils, which are moderate in fertility status, are considered best for its cultivation. Indeed, high soil fertility may hinder pod formation and reduce seed yield. The optimum pH is 6.8–7.5. It is best adapted to drought-prone sloping areas and flat rain-fed *tars*. It is grown preferably in marginal lands. It is a drought-tolerant crop but susceptible to waterlogging or excess moisture conditions, which ultimately kill the plants. Hence, it requires proper drainage for its healthy growth and development. Rice bean cultivation on highly fertile soil promotes vegetative growth rather than an increment in yield. It does not perform well under shade or in north-facing parts.

CULTIVATED VARIETIES

Since it is an underutilized rare vegetable, no standard variety has been developed so far. The farmers are growing only the local races that are adapted to their region.

SOWING TIME

In India, rice bean is normally grown as a *kharif* crop, which is sown in June–July and harvested in October–November and since the crop sown during *kharif* gives maximum seed yield, which decreases gradually with delay in sowing. In Asia, it is mostly grown as an intercrop with maize, as its seedlings grow vigorously and establish themselves quickly.

SEED RATE

The seed rate depends on several factors such as variety, foil fertility, growing season, irrigation resources, climatic conditions of the growing region and sowing method. A seed rate of 30–37 kg/ha is required if sowing is done by the broadcasting method and 15–22 kg/ha if sowing is done by the dibbling method. In India, a normal seed rate is 40–50 kg/ha if grown for seed production or 60–70 kg/ha if grown as a catch crop for fodder. The seed test weight varies from 30 to 120 g.

SOWING METHOD

Rice bean is propagated by seed sown by either broadcasting or dibbling. Usually, it is sown by broadcasting as an intercrop with maize. However, if it is sown on rice bunds or in kitchen gardens, dibbling is the most appropriate method. Attractive well-mature grains free from pest infestation should be selected for sowing. Immature, shrivelled and old grains should not be used as seed. Pods for the extraction of seed are usually collected from middle portion of the plant since the upper portion generally has smaller pods with poorly formed seeds. As rice bean has a problem of hard seeds, it is common practice among farmers to soak the seeds overnight (10–12 hours) before sowing. This is beneficial not only to sort out the hard seeds but also to promote better seed germination.

NUTRITIONAL REQUIREMENT

Farmers usually do not apply any chemical fertilizer to this crop since rice bean, being a leguminous crop, maintains soil fertility by fixing atmospheric nitrogen through symbiosis with *Rhizobium* strains. In the case of intercropping with maize, fertilizers applied to maize are sufficient for rice bean. For the post-rainy season crop or sole crop, 5 tonnes of well-decomposed farmyard manure and inorganic fertilizers, *i.e.*, nitrogen, phosphorus and potash at the rate of 20, 40 and 20 kg/ha, respectively, could be beneficial for better production. Application of nitrogen and phosphorus at the rate of 40 and 60 kg/ha, respectively, resulted in the highest leaf area index, crop growth rate, number of pods per plant, number of seeds per pod, test weight and grain yield.

IRRIGATION REQUIREMENT

Rice bean is well adapted to a wide precipitation range of 650–2000 mm. It can also be grown in annual rainfall regimes as low as 400 mm per annum. However, annual rainfall regimes of 750–1100 mm are preferable for fodder production. Extended waterlogging or poor drainage should be avoided, as it cannot tolerate excessive soil moisture, which affects the plant growth adversely and favours fungal infection. Since the crop is grown in the rainy season, normal rainwater is sufficient for its proper growth and development.

INTERCULTURAL OPERATIONS

Weeds are not a major problem in this crop. However, 2–3 weedings could be beneficial at about 6 weeks after sowing seeds and before flowering. In the case of mixed and inter cropping, weeding provided for the main crop is adequate for its proper growth and development. During rapid vegetative growth, farmers commonly practice cutting off the tip of the plant, which promotes more lateral branching and higher grain yield.

HARVESTING

Rice bean normally becomes ready for harvesting in 120–150 days after sowing. The young leaves, pods and seeds may be harvested from 40 to 130 days after sowing. For forage, it may be harvested 70–80 days after sowing but yield is higher if harvesting is done 120–130 days after sowing. The higher the altitude, the longer the crop duration. The early varieties are ready for harvesting along with maize. However, medium and late varieties require an additional 3 to 4 weeks. It can be harvested when approximately 75% of the pods turn brown.

For obtaining a good yield, it is advisable to harvest the pods in two to three or more pickings as needed. The viny habit and shattering of pods make the rice bean harvesting difficult. Leaves, green pods and mature seeds are usually harvested manually. To prevent seed loss from pod shattering, it is worth collecting the pods during early morning or late afternoon. The common practice by farmers is to harvest the whole plant when the majority of the pods are dry and then sun-dry the plants on the threshing floor for 3–4 days. Thereafter, the pods along with whole plant are beaten with sticks to collect the seeds.

Rice bean for fodder should be harvested when the pods are immature, as the leaves drop easily when the plant reaches maturity. Rice bean as a green manure crop can be ploughed with disc plough at about 30 days after sowing.

YIELD

Optimum sowing time and seed rate are key factors affecting the rice bean yield. Its average seed yield is only 2–3 q/ha. Its low yield is often due to a very short crop duration. Fresh fodder yield is 35 t/ha. Average yield of green forage is about 2.2–3.5 t/ha, while the yield of dried beans may average 200–800 kg/ha and yields up to 2.2 t/ha.

POST-HARVEST MANAGEMENT

Rice bean if dried properly can be stored for longer periods compared to any other grain legume under ambient room conditions. Thus, before storage, rice bean seeds are dried in the sun for 3–4 days so that fungal and insect attack may be reduced. Farmers usually store seeds in a mud bin, as it protects seeds from external moisture. Mixing some *neem* leaf powder with rice bean seeds during storage is a local method of protecting seeds from storage grain pests.

INSECT-PESTS

Relatively, rice bean is a pest-free crop and even immune to most storage pests, including bruchids, which generally damage other pulses during storage. Unlike other grain legumes, rice bean plants are not attacked by insects. On the other hand, since rice bean is not a priority crop, farmers do not spend much time in managing insects attacking a rice bean crop. Some of the insects commonly noticed in rice bean crops are discussed as follows:

Pod Borer (*Helicoverpa armigera* Hubner *and Lampides boeticus* L.)

Larvae of *Helicoverpa armigera* are more destructive as compared to its adults. A full-grown larva has a pale body lined with rows of conspicuous black spots on its dorsal surface. Eggs are laid in small clusters of 10–15 on leaves, flower buds and flowers. Larvae feed inside the webbed mass of leaves, flower buds and pods. Caterpillars of *Lampides boeticus* are green or light green in colour and damage rice bean after pod formation and seed development.

Control

- Follow deep summer ploughing.
- Follow proper recommended agronomic practices.
- Collect and destroy the infected plant parts as the larvae live in well-protected webs.
- Spray *B. bassiana* 109 spores/ml within an interval of 3–4 days.
- Spray the crop with 5% *neem* seed kernel extract, Malathion 0.1%, Quinalphos 25% EC, or Margosom 0.05% or Nemarin 0.15%.

Aphid (*Aphis craccivora koch*)

Aphids are dark-purple to black insects, which are usually wingless, although some may have clear membranous wings. They live in colonies, which contain nymphs and adults of different sizes. As the newly born nymphs increase in size, moulting occurs and the white exoskeleton is left on the leaf surface. Development of nymphs is rapid and many generations occur each season. Aphids feed by inserting their stylet into the plant tissue, sucking plant sap in large amounts. Mould may develop on the excreta of aphids. Heavily infested plants develop yellow curled foliage, distorted pods, wilt and they may be killed.

Control

- Treat the seeds with imidacloprid 75WS or thiamethoxam 30FS @ 5 g/kg of seed.
- Apply carbofuran granules 1 kg/ha in soil at the time of sowing.
- Spray the crop with dimethoate 0.03%, phosphamidon 0.05%, oxydemeton methyl 25EC 0.25%, or imidacloprid 0.2%.

Pod Weevils (*Apion clavipes*)

Weevil adults are small and black in colour. The younger ones (caterpillars) feed on immature seeds inside the pod. Adults feed on leaves and flowers as well. The adult emerges by making the holes in the pods. The males and females are equal in number and adult longevity ranges from 30 to 40 days. The adults are shiny black weevils with prominent snout having a good ability to fly. The adult also causes damage by feeding on the leaves, young shoots and flower buds. On young leaves, it makes small holes and nibblings in the flower buds, which affect flower opening, ultimately leading to poor pod set. The adult weevils are also found feeding on young pods.

Control

- Use disease-free healthy planting material, if available.
- Follow proper agronomic practices.
- Maintain proper field sanitation.

- Use entomopathogenous fungus *Beaveria bassiana* 109 spores/ml within 3–4 days after planting.
- Spray the crop with 5% neem seed kernel extract, dimethoate 0.03%, or monocrotophos 1.7 ml/litre of water.

Leaf Folder (*Hedylepta indicata* Fab.)

Leaf folder insects can attack rice bean leaves starting from its early vegetative stage. They fold the leaves and form a web of many leaves. Longitudinal and transparent whitish streaks appear on the folds of the damaged, tubular leaves. Sometimes, leaf tips are folded earlier than the basal part of the leaf. Heavily infested fields appear scorched with many folded leaves.

Control

- Use resistant varieties, if available.
- Follow proper plant population in the field.
- Use balanced dose of fertilizers.
- Remove grassy weeds from field and borders.
- Use recommended biological agents.
- Spray the crop with monocrotophos @ 1.7 ml/l and dichlorvos @ 2.5 ml/l of water.

Soybean Hairy Caterpillar (*Spilarctia casignata* Kollar)

Young larvae feed gregariously on chlorophyll mostly on the lower surface of the leaves, due to which, the leaves look papery and brownish yellow in colour. The insects begin to attack rice bean plants during the last week of August.

Control

- Collect and burn the infested plant parts, eggs and young larvae.
- Apply pheromone traps @ 2 traps per hectare.
- Apply nuclear polyhedrosis virus at 250 LE (larval equivalents) per hectare with sugar 2.5 kg/ha in evening hours.
- Spray the crop with Deltamethrin 0.5%, Cypermethrin or fenvalerate 0.25%, or Margosom or Nemarin 0.15% at 10 days interval.

Banded Blister Beetle (*Mylabris pustulata* Thunberg)

Blister beetles vary with species in shape, size (2–2.5 cm long) and colour, which is grey to black or with paler wing margins, metallic, yellowish striped, or spotted. Most of them are long cylindrical narrow-bodied beetles, which have heads wider than their first thoracic segment. Mouthparts are specialized for chewing. Blister beetle species feed on flowers and foliage of a wide variety of crops including rice bean. Adults usually occur in loose groups or swarms and are more detrimental.

Control

- Plough the field deep during the hot summer.
- Follow manual hand picking and destroy adult beetles using insect nets.
- Spray the crop with botanical insecticide *neem* oil 0.4%.
- Spray the crop with synthetic pyrethroids 0.1%, or deltamethrin, cypermethrin, or fenvalerate 0.25% at 10-day intervals.

Green Stinkbugs (*Nezara* spp.)

Both nymphs and adults of green stinkbugs suck cell sap from leaves, stems and pods with their needle-like mouthparts. The degree of damage depends, to some extent, on developmental stage of the plant when it is pierced by stinkbugs. Thus, the whole plant becomes weak as well as withered. Immature fruit and pods punctured by bugs become deformed as they develop. Seeds are often flattened and shrivelled, and seed germination is reduced.

Control

- Use cowpeas and beans as a trap crop.
- Collect and destroy the nymphs and adults.
- Remove and destroy the infected plant debris.
- Spray the crop with deltamethrin, cypermethrin, or fenvalerate 0.25% at intervals of 10–15 days.

DISEASES

Rice bean, a grain legume, is least susceptible to diseases. However, rust, *Cercospora* leaf spot, powdery mildew, *Rhizoctonia* blight and bacterial blight are the common diseases of rice bean as described below:

Rust [*Uromyces appendiculatus* (Pers.) Unger]

Rust pustules are evident on leaves of rice bean but they rarely appear on the stem and pods. Initially, the rust pustules are whitish, minute and slightly raised. In the later stage, the pustules become reddish-brown, and they are seen distinctly on the underside of the leaves. If rust attacks the leaves during the crop's vegetative phase, premature leaf drop may occur and ultimately affects the yield directly. If rust appears at the crop's maturing stage, it has less impact on yield.

Control

- Use resistant varieties, if available.
- Follow long crop rotation of at least 3–4 years.
- Collect and destroy the infected plant debris to reduce inoculums.
- Spray the crop with 0.2% Mancozeb 75 WP or 0.1% hexaconazole followed by two more sprays at 10-day intervals.

Cercospora Leaf Spot (*Cercospora* spp.)

Small round circular to angular brown shapeless dry spots initially occurring on the older, lower leaves become larger and ultimately the plant defoliation takes place. The leaves begin to turn bright yellow, orange or red. As the summer starts severely, many infected plants drop down.

Control

- Follow long duration crop rotation of at least 3–4 years.
- Use disease-free healthy planting material.
- Follow plucking and subsequent burning of diseased leaves.
- Spray the crop with Dithane M-45 or copper oxychloride 0.3% before or immediately at the appearance of dry spots.

Powdery Mildew [*Oidiopsis taurica* (Lev.) Salmon]

When the fungus attacks, a powdery coat appears on leaves, stems and pods. Severely affected plant parts get shrivelled and distorted. The affected leaves turn yellow, then brown, curl and thereafter dry before the defoliation occurs. Usually, pods are not formed. If formed, they are malformed, small and reduced in number. It is prominently observed in areas with temperature from 20°C to 35°C coupled with high humidity.

Control

- Grow resistant varieties, if available.
- Follow slightly late planting.
- Maintain proper field and crop sanitation.
- Spray the crop with Karathane, Sulfex, Cosan, Bavistine, or Benlate 0.25% at intervals of 10–15 days.

RHIZOCTONIA BLIGHT (*Rhizoctonia solani*)

The symptoms start with the development of small, irregular, water-soaked, pale greenish spots with damp appearance on lower leaves, where from, it extends towards the apex. During high humidity, the spread of disease is very fast and the spots cover greater parts of the leaf blade and stem. This is a very conspicuous and destructive phase of the disease. The leaflets and pods shrivel, turn brown and dry up. The affected parts of the plant rot rapidly in wet weather. If the pods are infected at the initial stage, the seeds are not formed, and in case of late infection, the seeds are poorly formed. In some cases, the plants are killed before flowering.

Control

- Follow long crop rotation of at least 3–4 years.
- Collect and burn the infected plants.
- Follow late season planting and wider spacing.
- Treat the seed with Bavistin at the rate of 1 g/kg of seed.

BACTERIAL BLIGHT (*Pseudomonas* spp.)

The pathogen is seed-borne, which is the main source of primary inoculums in new areas. In the beginning, small, round or irregular, flat, water-soaked spots surrounded by a greenish-yellow zone appear on the leaf. The infection spreads along the veins and veinlets, turning them brown and necrotic and causes distortion of the leaflets.

Control

- Follow long crop rotation of at least 3–4 years.
- Use disease-free healthy planting material.
- Collect and destroy infected plant debris.
- Soak the seeds in 1000 ppm solution of streptocycline, Plantomycine, or Agrimicine before sowing.

■■■

80 Scarlet Runner Bean

M.K. Rana and Chandanshive Aniket Vilas

Botanical Name : *Phaseolus coccineus* L.

Family : Leguminosae (Fabaceae)

Chromosome Number : 2n = 22

ORIGIN AND DISTRIBUTION

Scarlet runner bean is believed to be originated in the tropical highlands of fertile valleys of Mexico and Central America. Warm areas of Central and Southern America are considered the primary centre of origin and Peru, Bolivia and Ecuador regions the secondary centre of origin. This species has been cultivated in the high parts of Mesoamerica for many centuries. In pre-Columbian Mexico, the people of Anahuac cultivated it extensively and ensured its distribution. Its introduction into southern Colombia and Europe, where it is known as scarlet runner bean, could have occurred in the 17th century before reaching other parts of the world such as the Ethiopian highlands, and thereafter, it was brought from Europe to India, where it was cultivated in Karnataka, Maharashtra, Tamil Nadu, Gujarat, Andhra Pradesh and West Bengal. It has been found in archaeological remnants only in Mexico in Durango and Puebla and wild only in Tamaulipas. Although the archaeological information available is very limited, it could be assumed that its Mexican domestication took place in humid high zones.

INTRODUCTION

The genus *Phaseolus* has five species, namely, *coccineus*, *Darwinians*, *formosus*, *glabellus* and *griseus* but only *Phaseolus coccineus* (scarlet runner bean, also known as Oregon runner bean) is grown as a food as well as an ornamental plant. Like other major beans, scarlet runner bean is also one of the ancient and most valuable vegetable crops cultivated in home or market gardens and also in truck gardens for canning purpose. It is a warm season crop grown as a vegetable in India. It is widely used in a variety of preparations because of its delicate flavour, thus, grown commercially for its tender green pods used as a green vegetable and dry beans as pulse. Its immature pods are marketed as fresh, frozen, or canned. In India, it is grown for vegetable purpose, while in other countries, it is cultivated in large quantities especially for processing and off-season use when green pods are not available. The varieties having red flowers and multi-coloured seeds are often grown as ornamental plants.

COMPOSITION AND USES

COMPOSITION

Scarlet runner bean is a nutritious vegetable as it contains many important nutrients that help in proper development of our body. Its immature pods are an excellent source of carbohydrates, protein, calcium, phosphorus, magnesium, carotene, vitamin C and dietary fibre. It is a very good source of iron, niacin, manganese and a good source of thiamine and riboflavin. In addition, scarlet

TABLE 80.1

Nutritional Composition of Scarlet Runner Bean (per 100 g Edible Portion)

Constituents	Contents	Constituents	Contents
Water (g)	93.5	Iron (mg)	1.2
Carbohydrates (g)	3.2	Zinc (mg)	0.2
Protein (g)	1.6	Carotene (μg)	145
Fat (g)	0.4	Thiamine (mg)	0.06
Dietary fibre (g)	2.6	Riboflavin (mg)	0.03
Calcium (mg)	33	Vitamin C (mg)	18
Phosphorus (mg)	34	Energy (kcal)	22
Magnesium (mg)	19		

runner bean is rich in antioxidants. The pods are very low in fat and energy. It also contains traces of the poisonous lectin and phytohaemagglutinin, which are also found in common beans. The nutritional composition of scarlet runner bean is given in Table 80.1.

Uses

Scarlet runner bean seeds can be used fresh or as dried beans and pods are used as a vegetable and in many other cuisines. Its tuberous roots are boiled or chewed as candy by the Indians in Central America. Like other beans, the immature shelled seeds are consumed after boiling, steaming, baking and saucing. They can be added to salads, soups, stews, etc. Its flowers having a bean like flavour are also used in salads. Young leaves are used as a potherb. Its delicate green pods are also fried with other vegetables like carrot, potato and corn. The appealing colour of pods and seeds make them chef's favourite for garnishing many dishes and adding flavour and colour at the same time. It is used as a fodder to feed cattle. Scarlet runner bean is widely grown as an ornamental crop for its attractive flowers.

Medicinal Use

Scarlet runner bean is diuretic and beneficial in kidney problems. It is considered good for cardiac problem, diarrhoea, itching and rheumatism. Its pods are good for individuals with diabetes, provide protection against breast cancer and help in lowering cholesterol levels in blood. The presence of fibres also helps in stabilizing glucose metabolism and preventing constipation.

BOTANY

Scarlet runner bean is a quick-growing perennial plant but grown as an annual. The stem, which emerges from a fleshy edible tuberous roots, is vigorous and of several metres (2.5–4.5 m) in height. It has a taproot system. Its leaves are typical alternate dark green with purple tinged veins, tri-foliolate, triangular stipules, petiole 8–10 cm long, rachis 2.5–4 cm long, stipels 5 mm long, ovate-rhombic leaflets, base truncate, apex acute and thinly pubescent to glabrescent three leaflets. Many-flowered inflorescence is axillary or terminal raceme with 11–16 cm long peduncle and 10–16 cm long rachis. Flowers on 0.5–1.5 cm long pedicel are bisexual, papilionaceous and campanulate. Corolla are scarlet, pink, or white in standard hood shaped circular or broadly obovate and coiled keel. Stamens are 10, 9 fused and 1 free, and ovary is superior, long, finely pubescent and style coiled with collar of hairs below the stigma. Its large inflorescence is 20 cm long with scarlet white, red, or more rarely bicolour flowers borne in clusters like sweet peas. Pods are slender and a foot long. Runner bean is a self-compatible and cross-pollinated species assisted by its extrorse stigma and nectaries. Pollination takes place through insects and birds. The pods are knife-shaped, linear, lanceolate, straight or slightly curved, laterally compressed, beaked, rough with small oblique ridges and normally green in colour. The pale green pods are 10 to 20 cm long depending upon variety. The crop can easily be distinguished because of its large seeds with small, narrow and elliptical hilum. Seed shape is similar to lima bean seeds. Their colour varies from shining black to violet-black mottled with deep red. It carries out hypogeal germination. It has a fleshy root, which is divided and generally fusiform, which allows cotyledonary young shoots to resprout over several seasons.

CLIMATIC REQUIREMENT

Scarlet runner bean is a crop of tropical climates. It can be cultivated most profitably at an altitude of 1500–2800 m. As compared to other *Phaseolus* species, it is more tolerant to cool conditions but temperature below 5°C causes damage to its plants. The ideal temperature for its successful cultivation is 12–22°C. However, temperature above 25°C inhibits its growth and pod development. Soil temperature is equally important for uniform seed germination and seedling development. Seed germinates well at a temperature range of 15–20°C. The mean temperature <10°C shows poor seed germination, slow growth and delayed maturity, while the temperature >30°C during flowering may cause a high percentage of flower drop and abortion of ovules, resulting in poor yield. Rain during flowering also causes flower drop and spread of leaf spot disease. Scarlet runner bean is extremely susceptible to drought and requires a well-distributed rainfall throughout the growing period. It is successfully grown in areas with an average annual rainfall of 400–2600 mm. It needs high relative humidity for setting seeds. Scarlet runner bean comprises short-day and day-neutral types.

SOIL REQUIREMENT

Scarlet runner bean can be grown on a wide range of soils but medium-textured fertile sandy loam soils good in organic matter are considered best for its cultivation. However, the vegetative growth will be luxurious in soils having a high amount of organic matter. Soil should have good drainage facilities but must be fair in moisture-holding capacity and friable enough for uniform germination and seedling emergence. Soil containing heavy clay should be avoided, as it hampers root penetration in soil. Light soils are preferred for early yield and heavy soils for higher yield of scarlet runner bean. Slightly acidic soils are most preferable but the extremely acidic and alkaline soils are not good for its cultivation. Waterlogging at any stage affects the crop yield adversely.

CULTIVATED VARIETIES

No specific improved variety of scarlet runner bean is available in the country. Only local selections with high yield and good pod size are available. The introductions that are suitable for cultivation under Indian conditions are Butler (stringless), Painted Lady (red with white flowers), Kelvedon Wonder (early with long pods), Sunset (pink flowers) and Scarlet Emperor (scarlet flowers with good pods).

SOWING TIME

The sowing time of scarlet runner bean depends upon weather conditions of the growing area. It is grown in *Kharif* and *Rabi* season in different parts of the country. In areas where the climate is mild, it is grown as *Kharif*, *Rabi* and spring–summer crop. The pole type varieties should be sown from the second fortnight of May to the second fortnight of June. In the plains of north India, it is sown twice in a year, *i.e.*, (i) August to September and (ii) January to February. In hilly tracts, sowing is done in March for bush types and in June for pole types. In south India, sowing is done from September to November, and in the northeast and central zone, it is sown in the end of October.

SEED RATE

The seed rate of scarlet runner bean depends on varieties, soil, environmental and cultural factors. For planting a hectare land area, a seed rate of 20–35 kg is enough. The seeds before planting should be treated with Captan or Thiram at the rate of 3–4 g/kg of seed.

SOWING METHOD

Generally, scarlet runner bean is propagated by seeds. Direct seeding is the commonest method used for its propagation. The field is prepared to a fine tilth by repeated ploughing with a harrow followed by planking. The spacing is kept 15–30 cm × 90 cm for bush types and 15–30 cm × 150 cm for pole-type varieties. The seeds are sown 1.5–2.5 cm deep in heavy soil and 2.5–3.5 cm in light soils.

NUTRITIONAL REQUIREMENT

Scarlet runner bean, being a legume crop, should fix the atmospheric nitrogen into the soil but it lacks biological nitrogen fixation because of poor nodulation. Hence, it responds well to initial application of farmyard manure and fertilizers. In any given area, actual fertilizer requirements vary with growing season, physical and chemical properties of soil, residual nutrients of previous crop, irrigation facilities, rainfall and method and time of application. Well-decomposed farmyard manure @ 25–30 t/ha should be incorporated into the soil at the time of field preparation to improve the physical conditions of the soil. Besides, it requires nitrogen 30–40 kg, phosphorus 40–60 kg and potash 40–50 kg/ha. A half-dose of nitrogen along with a full dose of phosphorus and potash is applied as a basal dose at the time of sowing, and the remaining half of the nitrogen is top-dressed at the time of flowering. It requires less nitrogen as compared to other legume vegetables since higher doses of nitrogen cause luxuriant vine growth, inhibition of pod formation and delayed pod maturity. Phosphorus and potash become quickly fixed in soil and do not leach down, thus, both the fertilizers are applied as a basal dose.

IRRIGATION REQUIREMENT

Adequate soil moisture is essential at the time of sowing to ensure better and uniform seed germination. Scarlet runner bean, being a shallow-rooted crop, requires irrigation at frequent intervals. Although the plant is well adapted to a wide range of soil-moisture, regular irrigation ensures economic yield. On an average, the crop needs about 100–450 mm irrigation. About six or seven irrigations at intervals of 5–10 days are required during the growing season. Blossoming and pod set are the critical stages for irrigation. Hot and dry soil conditions at these stages result in severe dropping of flowers without setting pods. Application of excessive irrigation is harmful to this crop. If the crop is sown in a dry field, irrigation may be applied immediately after sowing for a uniform crop stand.

INTERCULTURAL OPERATIONS

HOEING AND WEEDING

Weeds not only reduce the yield and quality of the produce but also reduce efficiency of the harvester. After germination, the first 30–40 days are critical for crop–weed competition since it is short duration crop, requiring only 50–60 days from sowing to maturity. Applying pre-sowing irrigation allows the weed seeds to germinate, which later keeps the seedbeds weed free. Absence of weeds in the early stages hastens the plant growth due to no competition with weeds for nutrients, moisture, space and light, thus, two light hoeings and weedings during this period should be done to make the soil loose around the plant roots, which is necessary for proper aeration, keeping the field weed free and reducing competition with weed plants. However, during hoeing, care should be taken not to injure the plant roots.

Besides manual weeding, pre-plant soil application of herbicides is effective primarily for controlling annual grasses and small-seeded broadleaf weeds. Pre-emergence herbicides are applied to soil surface after planting but before the emergence of crop and weeds. Post-emergence herbicides are applied as the crop emerges and until it reaches a two to three trifoliate leaf stage. Pre-sowing application of Fluchloralin 2 litres/ha or pre-emergence application of Oxadiazon 0.5 kg/ha, Alachlor 1.5 kg/ha, Pendimethalin 2 kg/ha, or Diuron 1.5 kg/ha effectively controls the weeds.

MULCHING

In summer under a condition of low-moisture stress, mulching with straw is very effective for good plant growth since it conserves soil moisture and suppresses weeds growth, increases the water-holding capacity of the soil and regulates soil temperature around the plant roots. In winter, the use of black polythene mulch controls weeds, increases soil temperature, promotes earliness and increases fresh green pods' yield with a drip irrigation system. Pole-type scarlet runner bean responds well to black polyethylene mulch used soon after sowing.

STAKING

Staking is an important operation for adjusting more number of plants per unit area, better growth of the plants and harvesting good-quality produce. Its plant is a weak climber, thus, it needs support for its growth. The plants of pole type varieties are provided support with bamboo poles or trellis. Staking increases yield and size of pods, reduces fruit rot and makes the pesticides spraying and harvesting easier.

HARVESTING

Harvesting of scarlet runner bean depends on the purpose of its use. It requires 60 to more than 90 frost-free days to reach harvest, depending upon type and variety. The crop becomes ready for harvesting when the pods and seeds are bright green in colour, which usually occurs approximately 7–10 days after flowering. Picking of maturing beans prolongs the vines' life. The pods for fresh market should be harvested at an immature stage, when they are tender and well filled but before they become fibrous and waxy. The pods at the immature stage are harvested manually by pinching the pedicle slightly above the pod and pulling gently. A total of 12–14 pickings are possible at an interval of 15–16 days. All trash, leaves, stems, peduncles, dirt and diseased pods should also be separated from the lot before packaging and sending them to the market for sale. Scarlet runner beans are usually packed in wire-bound boxes but locally mesh bags are generally used by the farmers for their packaging.

Soon after picking, the pods are sifted under shade. All unfilled, over-mature pods and the pods that have begun to lose their green colour should be removed from the lot. For dry seeds, the crop is harvested when most of the pods have turned yellow. Harvesting at the pre-dry stage does not affect the seed yield but reduces the seed quality.

YIELD

On an average, the green pods yield is 2–9 q/ha with wide variations depending on sowing density and rainfall. If the crop is grown with recommended dose of fertilizers, the yield varies from 10 to 20 quintals per hectare.

POST-HARVEST MANAGEMENT

Pre-cooling of fresh pods soon after harvesting is essential operation to remove field heat and to avoid the post-harvest losses that cause deterioration of pods. Its pods are highly perishable due to their high moisture content and respiration rate, thus, they can be stored only for a short period under ordinary room conditions. Generally, the pods of scarlet runner bean can be stored for 10–15 days at 0°C temperature coupled with 90% relative humidity. The harvested green pods are cleaned using a shaker and blower and then subjected to size grading and removing stems and blossom ends. Thereafter, the pods with the help of mechanical cutters are crossed cut into pieces of 1.0–1.5 cm length, and then, the pods are sorted, packed and frozen or canned. The dry beans are stored in a cool and dry place away from high temperature and high humidity.

INSECT-PESTS

APHIDS (*Aphis gossypii*)

Both larvae and adults suck cell sap from tender leaves, stems, flowers and pods, which makes the plant growth stunted. They secrete honeydew, a sticky sugary substance, which invites black sooty mould, affecting the photosynthetic activity of the plants. It attacks the plants mostly at early stage of crop growth. It transmits a virus which causes yellowing and curling of the leaves. Its infestation is severe under warm and humid conditions.

Control

- Grow resistant variety, if available.
- Set up yellow sticky traps in the field.
- Spray the crop with malathion 0.05% or Confidor 0.3% at intervals of 10–12 days.

POD BORER (*Helicoverpa armigera*)

Mature larvae are about 4 cm long and vary in colour, *i.e.*, yellowish-green, green, brown, or black, but they all have a characteristic marking of pale and dark bands on each side of the body. The older larvae feed outside the pod, characteristically leaving part of the body exposed. The feeding hole is usually clean and circular with frass usually deposited away from the hole. The young larvae feed within flower buds, flowers and pods. The damage caused by larvae also results in several partially eaten seeds that reduce the quality and yield. A single larva may damage several pods.

Control

- Plant the crop bit early.
- Avoid relay cropping of beans in established alternate host plants such as tomato, okra, etc.
- Use natural enemies such as parasites and predators.
- Spray the crop with Dimethoate 0.05% at intervals of 10–15 days.

BEAN STEM MAGGOTS OR BEAN FLY (*Ophiomyia phaseoli*)

The pest is widely distributed all over beans growing area. Its damage is indicated by wilting and dying of seedlings. Its attack disrupts nutrients transportation, causing death of the taproot. Young seedlings under stress wilt and die within shortest time. Older and vigorous plants may tolerate its damage but they become stunted. The flies are shiny black and about 2 mm long with clear wings that reflect a metallic blue colour in sunlight. They lay eggs in leaf tissue or stem. An early sign of its attack is egg-laying punctures on primary leaves. The flies are usually active during rainy season. Their damage is aggravated by environmental stresses such as infertile soils, drought or moisture stress and the presence of soil-borne diseases.

Control

- Grow resistant and early maturing varieties, if available.
- Collect and destroy all infected plants to kill the maggots.
- Grow beans in more healthy and fertile soils.
- Plough the field deep after harvesting to kill the pupae in soil.
- Mulching with rice straw enhances formation of adventitious roots, which tolerate its damage.
- Spray the crop with systemic insecticides *viz.*, monocrotophos or acephate 0.5%, diazinon 0.03%, or quinalphos 0.05% at 10-day intervals.

DISEASES

FUSARIUM WILT (*Fusarium oxysporum f.* sp. *phaseoli*)

The disease is characterised by yellowing and wilting of leaves, resulting in wilting and dying of entire plant. Later, infection causes yellowing of lower leaves, which progresses to upper leaves, causing premature defoliation. The infection at seedling stage causes stunting, wilting and death of seedlings. Vascular tissue inside the root shows browning but the pith remains unaffected. The pathogen penetrates the vascular tissue of the root and hypocotyls, producing a reddish discoloration throughout the root, stem, petioles and peduncles. The fungus can cause water-soaked lesions on pods. Infected soil is the most common source of infection. It may also be carried to next season through the infected seeds, plant debris, irrigation water and contaminated farm equipments.

Control

- Grow resistant variety, if available.
- Follow long crop rotation with non-host crops.
- Sterilize the field soil through solarisation.
- Plant the crop bit late.
- Follow deep ploughing to eradicate the infected bean residues and plant debris.
- Treat the seeds with fungicides like Cerasan, Thiram, or Benomyl 2 g/kg of seeds.
- Drench the soil with Bavistin 0.1% or Thiram 0.2%.

FUSARIUM ROOT ROT (*Fusarium solani f.* sp. *phaseoli*)

Generally, its symptoms are observed on the hypocotyls and primary roots 1–2 weeks after the emergence of seedlings. As the infection increases, the lesions coalesce, and in advanced stages, they become brown and extend up to soil surface. Lesions may be accompanied by longitudinal fissures or cracks. The primary and lateral roots are frequently killed and detached. Death of the primary root results in hollowness of the lower stem.

Control

- Grow a resistant variety.
- Follow long crop rotation with cereal crops.
- Fumigate the soil with formaldehyde (1 part dissolved in 100 parts of water).
- Plough the field deep during land preparation to bury the infected plant material.
- Remove previous crop residues prior to sowing.
- Use raised seedbeds or ridges for the planting of crop.
- Treat the seed with Captan or Thiram 2 g/kg of seed.
- Drench the seedbeds with carbendenzin 0.1%.

Bean Anthracnose (*Colletotrichum lindemuthianum*)

Anthracnose is a common disease of beans, especially in regions with frequent rainfall. Initially, the symptoms appear as dark brown to black lesions along the veins on the underside of the leaves, petioles, stems and pods. Pod lesions are typically sunken and are encircled by a slightly raised black ring with reddish border. Under severe infection, young pods may shrivel and dry prematurely. The fungus infects the seeds in pods causing discoloration and distortion of seeds. Such infected seeds, when planted, serve as the source of infection to the succeeding crop. The disease spreads through rain splash and movement of insects, animals and man, especially when plant foliage is moist. The disease infests all plant parts at any growth stage but the fungus usually affects the expanding leaves.

Control

- Grow resistant varieties.
- Use certified disease-free seed for planting.
- Follow at least 2 years' crop rotation with non-host crops.
- Follow clean cultivation by destructing the infected plant debris.
- Avoid overhead watering or splashing soil onto the plants when watering.
- Burn the residues of previous crop.
- Treat the seed with Captan, Thiram, Ziram, Arsan, or Cerasan 4–5 g/kg of seed.
- Spray the crop with Dithane Z-78 0.2% or Bavistin 0.1%.

Angular Leaf Spot (*Phaseoriopsis griseola*)

The disease forms round lesions, which usually appear as brown spots with a tan or silvery centre. The spots initially occur on major veins, giving them an angular appearance. These lesions eventually cover the entire leaf, which becomes yellow before aborting prematurely. Symptoms are more prominent during late flowering and early pod formation stage. Lesions can also be observed on the underside of the leaf and appear slightly paler than those on the upper leaf surface. Lesions on stems and petioles appear elongated and dark brown. Lesions on pods are sunken irregular to circular black spots with reddish brown centres and may be similar to anthracnose disease. Infected pods may develop shrivelled or discoloured seeds. The main sources of infection are plant debris or residue and infected volunteer plants. The disease spreads through contaminated equipments, water and rain. Humid conditions and moderate temperature of 20–25°C favour the disease.

Control

- Grow resistant cultivars, if available.
- Follow crop rotation with non-host crops.
- Destroy the infected plant debris and crop residues.
- Plough the field deep to bury the pathogen-infected residues.
- Treat the seed with Captan or Thiram 2 g/kg seed.
- Spray the crop with streptomycin sulfate 100 ppm + copper oxychloride 0.25% at an interval of 15 days.

Ascochyta Leaf Spot (*Ascochyta phaseolorum*)

Symptoms appear first on leaves as large dark grey to black spots. Later, the infected area converts into zones followed by concentric rings around the spot containing black pycnidia. Blackening of nodes is a characteristic feature of infection on stems. It can girdle the stem and kill the plant. Under severe infection, extensive blight and premature leaf drop may occur. Flower infection can lead to end rot of pods and cause extensive cankers. Pod infection often results in seed infection, which can be transmitted to the next crop.

Control

- Grow resistant varieties, if available.
- Follow long crop rotation with cereal crops.
- Bury the infected plant debris or residues.
- Sow disease-free healthy seed.
- Follow wider spacing to improve aeration and reduce disease infection.
- Mulch the crop to reduce water splash and spread of inoculums during rainy season.
- Treat the seed with fungicides such as Captan or Thiram @ 3–4g/kg of seed.

POWDERY MILDEW (*Erysiphe polygoni*)

Initially, the disease manifests on plants as greyish-white powdery fungal growth on upper leaf surface. Both sides of the leaves usually show fungal growth. The infected area becomes dark brown with white dirty green patches and subsequently the white mould may cover all parts of the plant. Leaves become yellow, chlorotic and shrivelled, and later, premature defoliation may occur if the disease is not controlled. In severe cases, plants become stunted. The infected plant bears no fruit or the set fruits become very small. The major source of infection is contaminated seed and infected plant debris. The disease spreads by wind, rain and insects. Powdery mildew is mainly favoured by warm temperature, low humidity and shade.

Control

- Grow resistant varieties, if available.
- Follow long crop rotation with non-host crops such as maize and other cereal crops.
- Use disease-free healthy seed.
- Follow field sanitation.
- Demolish the infected plant debris after harvesting the crop.
- Dust the crop with Sulfex 2.5 kg/ha.
- Spray the crop with Karathane 0.1%, Dinocap 0.1%, Benomyl 0.1%, Bavistin 0.1%, or wettable sulfur 0.2%.

DOWNY MILDEW (*Peronospora parasitica*)

The early symptoms of downy mildew are the appearance of yellow spots on upper surface of the leaves, while bluish white fluffy growth develops on the lower surface of the leaves under humid conditions, which later become pale green and finally collapse. The infected leaves and branches may be distorted and die. The affected fruits in storage are soft and immature. The disease infestation is severe during rainy season.

Control

- Use disease-free healthy planting material.
- Strictly follow a crop rotation of at least 3 years.
- Plant the crop at wider spacing.
- Remove infected plant debris and keep the field clean.
- Spray the crop with Dithane M-45 0.2%, Chlorothalonil 0.2%, or Nitrophenol 0.05% twice at 10–15 days interval.

WHITE MOULD (*Sclerotinia sclerotiorum*)

The disease attacks old plant parts such as leaves or injured plant tissue, cotyledons, flowers, pods and seeds. Initially, the symptoms appear under cool and moist conditions as greyish water soaked lesions, followed by a white mould growth, accompanied by a watery soft rot. Black sclerotia form in and on infected tissue. The infected tissue then becomes dry and light coloured, and wilting of the plant may be observed. Seeds are damaged through pod infection, which spreads through infected soil, seeds and plant debris. The disease also spread through furrow irrigation. The host plants are sugar beet, tomato, cauliflower, lettuce as well as some grasses.

Control

- Grow resistant variety, if available.
- Sterilize the field soil through solarisation.
- Burn the infected plant debris or residues.
- Follow wider spacing of plants to increase aeration.
- Apply mulch to avoid contact of pods with soil.
- Follow post-harvest flooding to suffocate the pathogen.
- Expose diseased plant debris to the sun to desiccate and kill sclerotia.
- Spray the crop with Mancozeb 0.2%, Thiram, or Captan 0.2% at intervals of 10–15 days.

Leaf Rust (*Uromyces appendiculatus*)

Leaf spots enlarge to form rusty brown pustules on lower surface of the leaves and give a rusty appearance to anything that comes in its contact. It is mainly a disease of bean leaves under humid conditions. Severe infection causes the leaves to turn brown, curl upwards, dry and drop prematurely. Green pods, occasionally the stems and branches, also become infected and develop typical rust pustules. If infection is early and severe, the pod setting, pod filling, yield and seed size are reduced. A severely infested bean field gives a scorched appearance. Wind is mainly responsible for spreading the disease. Cool to moderate temperature and moist conditions favour the disease.

Control

- Grow resistant varieties, if available.
- Follow long crop rotation with cereal crops.
- Plant the crop at wider spacing.
- Remove alternate hosts and volunteer plants.
- Adjust sowing time to avoid disease.
- Spray the crop at bloom stage with chlorothalonil, sulfur, or maneb 0.5% at weekly interval.

Charcoal Rot (*Macrophomina phaseolina*)

Generally, the pathogen infects the cotyledons and hypocotyls at soil level. It produces symptoms like black sunken cankers with sharp margins and concentric rings. In severe cases, the plant's growing tips may die and the stems break at canker-affected regions. The infected younger plants show black wounds and ash-like wounds on diseased stems; that is why the name ashy stem blight has been given. In later stages, flakes of charcoal colour develop on old grey wounds. Severe infection in older plants may cause stunting, leaf chlorosis, wilting, premature defoliation, rotting of stem and roots and death of plants. The infection is more pronounced on one side of the plant. Infected seeds shrivel and lose their colour. The disease is found throughout the tropics and subtropics and has a wide host range. It is most devastating in areas of unpredictable rain and high temperature.

Control

- Grow resistant variety.
- Follow at least 2–3 years' crop rotation with non-host crops.

- Use disease-free healthy planting material.
- Use recommended phyto-sanitary measures to control the disease.
- Rough out and burn the infected plants.
- Treat the seed with Thiram, Captan, Captaf, Captafol, or Foltaf 3–4 g/kg of seed.
- Fumigate the soil with Chloropicrin or methyl iodide.
- Flood the soil to suffocate the pathogen.

Common Bacterial Blight (*Xanthomonas phaseoli*)

The symptoms appear as water soaked spots on underside of the leaves, which later enlarge and merge to form large brown irregular lesions surrounded by a narrow yellow patch. The stem may rot at first node and cause the plant to break. Sunken circular spots, which are water-soaked initially but later dry with a reddish brown narrow border, develop on pods. Under wet conditions, yellow slimy bacterial exudates ooze out of the lesions and form a crust. Pods may shrivel, and the infected seed may rot. The primary source of infection is seed. The disease spreads through water splash, windblown rain, overhead irrigation, insects and contaminated equipments and clothing. Warm and humid conditions favour the disease.

Control

- Grow the resistant varieties.
- Use disease-free healthy seed.
- Follow long crop rotation with non-host crops.
- Plough the field deeply in summer months to ensure destruction of pathogen.
- Treat the seed with copper sulfate, copper hydroxide, or streptomycin 3–4 g/kg of seed.
- Avoid the field operations when leaves are wet to reduce the spread of infection.
- Spray the crop with streptomycin 200 ppm.

Halo Blight (*Pseudomonas syringae pv. phaseolicola*)

The characteristic symptoms appear first as small water-soaked pinprick spots on undersides of the leaves. The tissues surrounding the spots gradually become yellow-green, resembling a halo, thus, it is known as halo blight. Infected branches and petioles bear greasy spots, which soon turn reddish brown, and infected pods show signs of water-soaked lesions, which turn into reddish discolouration. Under humid conditions, whitish to yellow bacterial exudates appear on the lesions. Seed may become shrivelled, rotten, or discoloured. The prime sources of infection are infected seed and plant debris. The disease spreads through water splash, rainwater and farm equipments. Cool temperatures ranging from 16°C to 20°C followed by moist cloudy conditions favour the disease. Its multiplication is rapid in the presence of dew.

Control

- Use disease-free healthy seed.
- Grow resistant varieties, if available.
- Follow at least 2–3 years' crop rotation with non-host crops.
- Plough the field deep to bury and destroy the pathogen infected debris.
- Rogue out the infected plants as early as possible.
- Avoid movement of workers in irrigated field.
- Spray the crop with copper-based fungicides Blitox 50 WP 0.2% at 10-day intervals.

BLIGHT (*Rhizoctonia solani*)

The characteristic symptoms first appear on primary leaves as small grey-greenish to dark brown necrotic spots. The spots expand rapidly and irregularly under favourable conditions. Finally, the spots cover the whole leaf and give the plant a burnt appearance. The necrotic leaf tissues are covered with brown sclerotia and light-brown hyphae, which form a web of mould over the plant and lead to defoliation. The infection on young pods appears as light brown, irregular-shaped lesions, which may coalesce and kill the pods. Lesions on older pods appear as dark brown, circular, lightly zonate and sunken with a dark border. The disease spreads through wind, rain, man, or agricultural implements. The disease occurs mainly in regions with high humidity and high to moderate temperature.

Control

- Use seed collected from disease-free plants.
- Follow at least 2 years' crop rotation with non-host crops like maize.
- Plough the field deep to bury the infected crop debris and residues.
- Grow crop at wider spacing to maximise air circulation.
- Use mulch to avoid the spread of inoculums.
- Spray the crop with Maneb, Bavistin, Benomyl, or Captafol 0.2%.

BEAN COMMON MOSAIC VIRUS (BCMV) OR BEAN COMMON MOSAIC

Both the diseases are characterised by typical symptoms, which are light to green or yellow to dark green mosaic patterns on leaves accompanied by puckering, rolling, distortion of the affected leaves, mottling, downward curling and malformation of leaves. The virus-infested plants leaves become small, misshapen and blistered. In severe infection in early growing season, it may cause leaf and flower distortions, stunting of plants and delayed maturity and deformation of pods. The disease is generally transmitted by white flies and aphids.

Control

- Use only the resistant varieties.
- Use disease-free seed for sowing.
- Keep the field weed free and eradicate the host weeds.
- Plant the crop a bit early to escape the aphid population.
- Destroy the infected plant debris before planting.
- Spray mineral *Krishi* oil 1.5% at 10 days interval.
- Apply Furadon or Phorate 10G granules 1.5 kg/ha at the time of seed sowing followed by two or three foliar sprays of dimethoate 0.05% to control vectors.

81 Snap Bean

M.K. Rana and Naval Kishor Kamboj

Botanical Name : *Phaseolus vulgaris* L.

Family : Leguminosae

Chromosome Number : 2n = 22

ORIGIN AND DISTRIBUTION

Warm regions of Central (Mexico and Guatemala) and South America are considered the primary centre of origin, while Peru, Bolivia and Ecuador regions are considered the secondary centre of snap bean origin. According to archaeological records, it has been cultivated for at least 7000 years in Mexico and was first domesticated in South America, from where it was taken to Central America, which has the most crop diversity. The wild bean *Phaseolus aborigineus*, which has a very wide geographical distribution ranging from northern Mexico to northern Argentina, is the probable progenitor of cultivated *Phaseolus vulgaris*. It was introduced into England at the end of the 16th century (1594) and in Europe in the 16th and 17th centuries. It was brought from Europe to India in the 17th century, where it is cultivated in states like Karnataka, Andhra Pradesh, Maharashtra, Tamil Nadu, Gujarat and West Bengal.

INTRODUCTION

Snap bean (*Rajmah*) also known as common bean, French bean, kidney bean, garden bean, fresh bean, navy bean, haricot bean, China bean, marrow bean, string bean and wax bean, is one of the most important leguminous vegetable crops. It is grown for its tender green pods used as a green vegetable and dry beans as pulse, depending to the stage of harvest. Its immature pods are marketed as fresh, frozen or canned. It is one of the most valuable vegetable crops grown in home or market gardens and also in truck gardens for canning purposes and shipment. Most of the cultivars grown for processing are round-podded, while the cultivars grown for fresh market are usually flat-podded. In India, it is grown for vegetable purposes, while in the United States of America, it is cultivated in large quantities for use in the off-season when green pods are not available and especially for processing.

COMPOSITION AND USES

COMPOSITION

Snap bean is a nutritious vegetable, as it contains a high amount of protein, vitamins and minerals. The immature pods are a good source of protein, carbohydrates, fibre, calcium, phosphorus, iron, and potassium. The pods are very low in fat. The dried beans are rich in protein and closely compared to meat. The nutritional composition of snap bean is given in Table 81.1.

TABLE 81.1

Nutritional Composition of Green Pods of Snap Bean (per 100 g Edible Portion)

Constituents	Contents	Constituents	Contents
Water (g)	90	Zinc (mg)	0.24
Carbohydrates (g)	7.1	Vitamin A (IU)	668
Protein (g)	1.8	Thiamine (mg)	0.08
Fat (g)	0.1	Riboflavin (mg)	0.11
Dietary fibre (g)	2.7	Niacin (mg)	0.75
Calcium (mg)	37.0	Pantothenic acid (mg)	0.225
Phosphorus (mg)	38.0	Pyridoxine (mg)	0.141
Potassium (mg)	211.0	Folic acid (µg)	33.0
Magnesium (mg)	25.0	Vitamin C (mg)	16.3
Manganese (mg)	0.216	Vitamin K (µg)	14.4
Iron (mg)	1.03	Energy (kcal)	31.0

USES

Snap bean pods at their immature, delicate and tender green stage are consumed as a vegetable and at their mature stage as dry pulse. Additionally, the newly emerged tender shoots and leaves are also used as a potherb in many parts of the world. Green bean casserole, a dish prepared with green beans, is very popular in the southern United States. Its green pods are also served battered and fried with other vegetables like carrot, corn and potato.

MEDICINAL USES

The snap bean is anti-diabetic, carminative, diuretic, an emollient and a kidney resolvent. It is considered good for the bladder, burns, cardiac problems, depurative diarrhoea, dropsy, dysentery, hiccups, rheumatism, sciatica, tenesmus, eczema and itching.

BOTANY

The snap bean plant has a taproot system with poor nodule formation. Its leaves are trifoliate, stems dwarf, bush type, or pole type, depending upon growth habit, *i.e.*, determinate and indeterminate. In bush type varieties, inflorescence develops simultaneously, while in pole type varieties, the flowers appear one after another for a long time. Flowers that are white, yellowish, or pinkish-purple are papilionaceous. Androecium is in diadelphous condition (9+1). Vexillary stamens are free from the base upwards. Ovary is monocarpellary, meaning unilocular with many ovules. The pods are 13 cm long or more and flat or round in shape. The mature seeds having white, black, red, or purple colour are kidney shaped or elongated.

CLIMATIC REQUIREMENT

Snap bean is a cool-season crop but it can also grow well under mild warm climatic conditions with annual rainfall 50–150 mm. In the plains of India, it is cultivated as a winter crop, whereas, in hilly tracts, it is grown in spring–summer. Most of its varieties are photo-insensitive except some semi-pole types, which are short-day types. Its plants are sensitive to frost, high temperature and high rainfall. The optimum atmospheric temperature for its successful cultivation ranges from 20°C to 25°C. The mean temperature <10°C shows poor seed germination, slow growth and delayed maturity, while temperatures >30°C during flowering may cause a high percentage of flower drop and abortion of ovule, resulting in poor yield. Pouring rains during flowering also causes flower drop and spread of leaf spot diseases, however, pole type varieties can generally be grown in areas of heavy rainfall. Soil temperature is also important for uniform seed germination and seedling development. The favourable soil temperature range for seed germination is 18°C to 24°C. Waterlogging at any stage affects its yield adversely.

SOIL REQUIREMENT

Snap bean can be grown in a wide range of soils but medium-textured fertile sandy loam soils good in organic matter are considered best for its cultivation. However, soils having high amounts of organic matter promote more vegetative growth. Soil should have good drainage facilities to prevent soil-borne diseases and for proper root respiration but must be fair in moisture-holding capacity and friable enough to enhance uniform germination and emergence. Light soils are good for early yield, while heavy soils are good for a higher yield of snap bean. The ideal soil pH for the growth of snap bean is 5.5–6.0. Slightly acidic soils are preferable, while extremely acidic and alkaline soils are not suitable for its cultivation. The crop is very sensitive to high concentrations of aluminium and manganese in soil. An electrical conductivity of 1 dS m^{-1} is threshold limit for snap bean cultivation.

Field preparation is the beginning of the planting operation and is important in obtaining a good stand of crop as it supports uniform germination and plant growth. Soil should be prepared well before sowing. To obtain the required tilth, field is prepared by ploughing two or three times with harrow followed by planking. All the stubbles of the previous crop are removed from the field to prepare clean seedbeds.

CULTIVATED VARIETIES

Based on the growth habit, snap bean varieties can be classified into two groups, *i.e.*, (i) pole type varieties, the plants of which are tall with longer internodes, requiring support and (ii) bush type varieties, plants of which are dwarf with short internodes, requiring no support. In the areas of very high rainfall, generally, the pole type varieties are preferred for cultivation. The varieties based on presence or absence of strings on the pod suture, snap bean varieties can be classified into two groups, *i.e.*, (i) string type, pods of which contain string and (ii) stringless type, pods of which contain no string. Depending upon the use, the snap bean varieties are grouped into four classes, *i.e.*, vegetable bean, dry shelled bean, processed bean and seed bean. Some commercially grown important varieties are given as follows:

Bush Type

Contender

A most popular early maturing bush-type variety resistant to mosaic and powdery mildew disease introduced from the United States of America that bears pink flowers and round, green, slightly curved and stringless pods. Dry seeds are light brown in colour. It is ready for first picking of green pods in 50–55 days after sowing. The average yield of fresh green pods is 200 q/ha.

Bountiful

A bush-type variety introduced from the United States of America is suitable for growing in September to February in South India. It bears green, tender pods in cluster, which remain edible for 4–5 days after harvesting. The average yield is 100–120 q/ha.

Premier

A black-seeded bush-type variety susceptible to mosaic virus and wilt introduced from Sweden is well adapted for late sowing and suitable for growing in hilly tracts. The pods are 11–13 cm long and flat, which mature in 55–60 days after sowing. Its green pods yield is about 75–90 q/ha.

Pant Anupama

An early-maturing bush type variety moderately resistant to mosaic virus and angular leaf spot disease developed at G.B. Pant University of Agricultural and Technology, Pantnagar, is suitable for growing in hills and plains. The pods of this variety are round, soft, slimy, green and stringless with straight and upright growth habit. It is ready for first picking in 60 days after sowing and gives fresh green pods yield of 100–120 quintals per hectare.

Pant Bean 2

An early bush-type variety moderately resistance to common mosaic virus and suitable for long-distance transport developed at G.B. Pant University of Agricultural and Technology, Pantnagar, through a cross between Turkish Brown and Contender, is suitable for growing in plains and hills. Pods are straight, round and stringless. The average yield of fresh green pods is 80–100 q/ha.

Giant Stringless

An early-maturing highly productive bush-type variety introduced from Sweden bears green, medium-sized, slightly curved, tender, meaty and stringless pods. The seeds of this variety are glossy and yellowish-brown.

Arka Komal (Selection 9)

A bush type variety with good keeping and transport quality developed at the Indian Institute of Horticultural Research, Bangalore, through a pure line selection from IIHR 60, bears straight, flat, green, tender and stringy pods. It takes about 65–70 days from sowing to harvesting. The yield of fresh green pods is about 200 q/ha. Seeds are light brown, oblong and large.

Arka Sharath

A bushy and photo-insensitive variety developed at the Indian Institute of Horticultural Research, Bangalore, is suitable for cultivation in *Kharif* and *Rabi* season. It bears round, stringless, smooth pods suitable for steamed beans. Pods are round on cross section, crisp and fleshy with no parchment. It has a pod yield potential of 180–190 q/ha in crop duration of 70 days.

Arka Suman

A bushy photo-insensitive early variety with good cooking quality and resistance against rust developed at the Indian Institute of Horticultural Research, Bangalore, through pedigree selection from a cross between Pusa Komal and Arka Garima. Plants of this variety are erect with pods above the canopy. Pods are medium-long, tender, fleshy and crispy without parchment. It gives a pod yield of 180 q/ha in crop duration of 70–75 days.

Arka Suvidha

A bush-type photo-insensitive variety resistant to rust developed at the Indian Institute of Horticultural Research, Bangalore through pedigree selection from a cross between Blue Crop and Contender bears straight, oval, light green, fleshy, stringless and crisp pods. On an average, it gives a pod yield of 190 q/ha in a crop duration of about 70 days.

Pusa Parvati

A mutant bush-type variety resistant to mosaic virus and powdery mildew disease developed at Indian Agricultural Research Institute, New Delhi, through X-ray irradiation of a wax-podded American variety EC 1706 is suitable for cultivation in hilly tracts. It bears dark green, tender, meaty, attractive, straight, flat, 12–14 cm long pods. It is ready for first picking in 50–55 days after sowing. The fresh green pod yield is 200 q/ha.

VL Boni 1

A bush-type variety developed at Vivekananda Parvatiya Krishi Anusandhan Sansthan, Almora (Uttarakhand), is suitable for growing in hilly tracts and bears medium-long, round, straight, stringless pale green pods. It matures in 45–60 days after sowing and gives a pod yield of 100–110 q/ha.

Jampa

A bush-type variety introduced from Mexico bears flat, soft, stringless pods in clusters. It is ready for first picking in 60 days after sowing.

Yercaud 1

A dual-purpose bush-type variety tolerant to root rot, rust, yellow mosaic and anthracnose diseases developed at Tamil Nadu Agricultural University, Horticultural Research Station, Yercaud, through pure line selection from a local indigenous, is grown for its green pods and seeds. It bears slightly flat, green, 15.1 cm long pods. The yield of green pods is about 90–100 q/ha. Its dry seeds are bold and dark purple.

Ooty 2

A bush-type variety with good cooking quality developed at Tamil Nadu Agricultural University, Horticultural Research Station (Ooty), through a single line selection from PV 15-1, bears fleshy, stringless, pale green pods with high protein content (22.38%). It has a green pod yield potential of 140 q/ha in a crop duration of 90 days.

Kashi Param (IVFB 1)

An early maturing bush type variety developed at Indian Institute of Vegetable Research, Varanasi, bears dark green, fleshy, round, 14–15 cm long pods. Plants of this variety are 70 cm long with dark green leaves. The average yield of green pods is 120–140 q/ha.

POLE TYPE

Kentucky Wonder

A pole-type stringless variety introduced from the United States of America bears curved, green, soft, flat, meaty, 20 cm long pods. The pods are borne in cluster and they mature in 65–75 days after sowing. The fresh green pod yield is 100–125 q/ha. Seeds are light brown in colour.

Pusa Himalata

An early-maturing stringless variety developed at IARI Regional station, Katrain, Kullu Valley, bears round, fleshy, straight, light green, 14 cm long pods. It becomes ready for first picking in 60 days after sowing.

Phule Surekha

A variety resistant to anthracnose, leaf crinkle, yellow mosaic and wilt disease developed at Mahatma Phule Krishi Vidyapeeth, Rahuri, through selection from Jama Improved, is suitable for cultivation in *Kharif*, *Rabi* and spring–summer seasons. It bears flat and light green, 9–10 cm long pods.

SVM 1

A dual-purpose variety resistant to angular leaf spot developed at Dr. Y.S. Parmar University of Horticulture and Forestry, Solan, through inter-specific hybridization between Contender and PBL 257, is grown for its green pods and seeds. It bears green, round, 13–14 cm long, stringless pods, which become ready for first picking in 65–75 days after sowing. The fresh green pod yield is 105–125 q/ha.

Lakshmi

A white-seeded variety tolerant to angular leaf spot disease developed at Dr. Y.S. Parmar University of Horticulture and Forestry, Solan, through hybridization between Contender and Local pole type varieties, is suitable for growing in the hilly tracts during the rainy season. It bears round, green, attractive, 13–14 cm long, stringless pods in a cluster of three. The first picking of green pods is in 65–70 days after sowing with an average yield of 120–150 q/ha.

TKD 1

A pole type variety developed at Tamil Nadu Agricultural University, Coimbatore (Tamil Nadu), through a selection from a cross between two pole type varieties in F_4 generation, is suitable for cultivation in hilly tracts of Tamil Nadu. It bears flat and pale-green pods. The fresh green pod yield is 60–80 q/ha.

VL Lata Bean 12

A pole type variety developed at Vivekananda Parvatiya Krishi Anusandhan Sansthan, Almora (Uttarakhand), bears pods in a cluster. Its plants are 110 cm tall, and it is ready for first picking in 65–70 days after sowing.

VL Lata Bean 17

A pole type variety developed at Vivekananda Parvatiya Krishi Anusandhan Sansthan, Almora (Uttarakhand), bears green, long, stringless pods and requires 60–70 days for first picking.

KKL 1

A pole type variety suitable for cultivation in hilly tracts bears 28 cm long pods with low fibre content. Seeds are white and flat. The potential yield of fresh green pods is 70 q/ha.

Ooty 1

A variety moderately resistant to leaf spot, anthracnose disease and pod borer developed at Tamil Nadu Agricultural University, Horticultural Research Station, Ooty, bears fleshy, stringless pods with high protein content (17.15%). It has a green pods yield potential of 140 q/ha in a crop duration of 80 days.

SOWING TIME

The sowing time of snap bean depends upon the weather conditions of the growing area. It is grown in *Kharif* and *Rabi* seasons in different parts of the country. In areas of mild climate, the snap bean is grown as *Kharif*, *Rabi* and spring–summer crop. In plains of north India, snap bean is sown twice in a year, *i.e.*, (i) August to September and (ii) January to February. In hilly tracts, sowing is done in March for bush types and in June for pole type varieties. In south India, sowing is done from September to November. In the northeast and central zone, it is sown in the end of October.

SEED RATE

Snap bean is a direct seeded crop and seed rate usually varies with seed size, seed germination percentage, variety, soil type, season and climatic conditions of the growing region. Generally, the seed rate for bush type cultivars is 60–75 kg/ha, whereas, for pole type cultivars, it is 30 kg/ha. Before sowing, the seed should be inoculated with *Rhizobium phaseoli* culture, the nitrogen-fixing bacteria, for better nodulation. About 30 g of culture is mixed with jaggery for coating of one kg of seed.

SOWING METHOD

The seeds of bush type snap bean are sown by dibbling on one side of the ridge in a ridge-furrow system of sowing. Seeds may be dibbled by hand or drilled mechanically at a depth of 2 to 3 cm. Sowing can also be done by broadcasting method but it is not an appropriate method as the vines interfere with cultural operations. Pole type varieties are sown in hills. At least three or four seeds are sown per hill, and later, thinning is accomplished. The bush type varieties are sown at a row spacing of 30–40 cm and plant spacing of 5–10 cm. However, the pole type varieties are sown at a row spacing of 90–100 cm and plant spacing of 22–30 cm. For a continuous harvesting of beans, sowing should be done at regular intervals.

NUTRITIONAL REQUIREMENT

Snap bean, being a legume crop, should fix atmospheric nitrogen into the soil but it lacks biological nitrogen fixation because of poor nodulation. However, it responds well to an initial application of farmyard manure and fertilizers. Well-decomposed farmyard manure should be incorporated into the soil @ 25–30 t/ha at the time of field preparation, as the application of farmyard manure improves the physical properties of the soil. It requires less nitrogen as compared to other legume vegetables since higher doses of nitrogen promote vine growth and delay fruit maturity. In general, it requires nitrogen 40 kg, phosphorus 60 kg and potash 40–50 kg/ha. A half-dose of nitrogen along with a full dose of phosphorus and potash is applied as a basal dose at the time of sowing, and the remaining half of the nitrogen is top-dressed at the time of flowering.

IRRIGATION REQUIREMENT

Snap bean is a shallow-rooted crop, thus, requires frequent irrigation. An effective irrigation schedule promotes uniform germination, plant development, harvest maturity, pod quality and higher yield of snap bean. An adequate moisture in the soil is essential at the time of sowing to ensure better seed germination. About six or seven irrigations are required during the growing season at intervals of 5–10 days. Application of excessive irrigation is harmful to this crop. However, sufficient moisture is needed at flowering and pod setting since moisture stress during this period may cause dropping of flowers. On an average, the crop needs about 150–400 mm irrigation.

INTERCULTURAL OPERATIONS

HOEING AND WEEDING

Controlling weeds in snap bean crops is essential, as weeds not only reduce the yield and quality of the produce but also reduce the efficiency of the mechanical harvester. Since it is a short-duration crop, requiring only 50–60 days from sowing to maturity, weeding during early phase of crop growth is very crucial. Application of pre-sowing irrigation allows the weed seeds to germinate, *per se*, the seedbeds will be later weed free. When the weeds in the early stages of crop remain absent, snap bean can have rapid growth without any competition with weeds during growing season. After germination, the first 30–40 days are the critical period for crop–weed competition. Thus, two light hoeings and weedings are found beneficial during this period to reduce competition with weed plants. However, during hoeing, care should be taken not to injure the plant roots. In addition to hand-weeding, pre- or post-emergence herbicides application can control weeds in a snap bean field. Pre-sowing application of Fluchloralin 2 litres/ha or pre-emergence application of oxadiazon 0.5 kg/ha, Alachlor 1.5 kg/ha, or pendimethalin 2 kg/ha effectively controls the weeds.

MULCHING

In summer months, under a condition of low-moisture stress, mulching with straw is very effective for good plant growth, as it conserves soil moisture and suppresses weeds growth, while in winters, the use of transparent and black polythene mulch increases the soil temperature, promotes earliness and increases fresh green pods yield.

STAKING

The pole-type varieties require staking since growing plants vertically will allow a greater number of plants per unit area, avoiding contact of pods with moist soil, which causes deterioration, and makes the picking operation easy. Usually, cemented or bamboo poles, wooden sticks, or any locally available material can be used for providing mechanical support to snap bean plants. A single bamboo stick of 1.8–2 m length is inserted into the soil near the plant to give support when plants start developing vines.

USE OF GROWTH REGULATORS

Use of plant growth regulators in snap bean improves plant growth, promotes early flowering, fruit set and pod yield. Spray of commercial gibberellic acid 50–200 ppm increases the plant height, number of leaves and setting of pods. Spraying snap bean with para-chlorophenoxy acetic acid (PCPA) 2 ppm or beta-naphthalene acetic acid at 5–25 ppm improves pod set and total yield.

HARVESTING

Snap bean harvesting depends on the purpose of use. The crop will be ready for harvesting when the pods and seeds are bright green in colour, which usually occurs approximately 7–10 days after flowering. The pods for culinary purpose are picked at the immature stage when the pods are tender and non-fibrous. The time of harvesting is a compromise between yield and quality, as delayed harvesting increases the yield but decreases the pod quality. With the advancement of maturity, the pods lose their green colour, tenderness and firmness rapidly and become fibrous. The bush-type varieties are harvested 50–55 days after sowing and they provide about three pickings, while the pole type varieties are harvested 60–75 days after sowing and provide about five pickings at intervals of 3–5 days. For dry seeds, the crop is harvested when most of the pods have turned yellow. Harvesting at the pre-dry stage does not affect the seed yield but reduces the seed quality. The pods at the immature stage are harvested manually by pinching the pedicle slightly above the pod and pulling gently.

YIELD

The green pod and seed yield depend on several factors such as variety, soil type, climatic conditions of the growing region and packaging practices followed during cultivation of the crop. The average green pod yield of bush type varieties is 60 q/ha and pole type varieties 120 q/ha. The dry seed yield varies from 12 to 18 q/ha.

POST-HARVEST MANAGEMENT

The freshly harvested pods should be pre-cooled as soon as possible in order to avoid deterioration of the pods and to preserve their quality. Its pods are highly perishable due to their very high rate of respiration and moisture content, thus, they can be stored only for a very short period under ambient room conditions. Desiccation of pods is a common post-harvest problem under conditions of low relative humidity. Therefore, its green pods should be hydro-cooled rapidly and stored at 4–5°C temperature and 95% relative humidity. Cottony leak (*Pythium butleri*), antraconose (*Colletotrichum lindemuthianum*), soil rot (*Rhizoctonia solani*) and watery soft rot (*Sclerotinia* spp.) are the most common post-harvest diseases that cause spoilage of pods.

Green pods can be processed as frozen or canned. However, for processing, the pods must be long, straight, tender, medium-sized with thick walls and small seeds; free of string, fibre and disease; and bright green and firm-skinned after blanching. The pods should not split or slough off and discolour during cooking. The leucoanthocyanins of pods are converted to anthocyanins during cooking, resulting in undesirable browning of pods.

The harvested pods are cleaned using a shaker and blower and then subjected to size grading and removing the stems and blossom ends. Thereafter, the pods, with the help of mechanical cutters, are cross cut into pieces of 1.0–1.5 cm length. The shreds after blanching in steam or water at 99°C for 2–3 minutes are quickly cooled and then sorted, packed and frozen or canned. The cut beans are filled in cans containing 2% brine, thereafter, the cans are heated in a steam exhaust box for 5 minutes and hermetically sealed.

INSECT-PESTS AND DISEASES

Snap bean crop is attacked by a number of insect-pests and diseases during the growing season. Their description along with control measures are given as follows:

INSECT-PESTS

APHID (*Aphis craccivora*)

The nymphs and adults both attack the young plants and suck cell sap from growing tips and later from the whole plant. It causes curling of leaves, reduces plant vigour, yield and quality of the produce. Aphids secrete a honeydew-like sugary substance, which encourages the growth of sooty mould, which inhibits photosynthesis and ultimately reduces the yield of the crop. It also transmits mosaic virus.

Control

- Spray the crop with 0.03% dimethoate or 0.05% phosphamidon.
- Apply carbofuran granules 1.0 kg a.i./ha.

RED SPIDER MITE (*Tetranychus urticae*)

Adults and nymphs of red spider mites cause damage to the plants by feeding on the underside of the leaves, which show a blotching characteristic due to the extraction of chlorophyll. The infested leaves also have a yellowish speckled appearance, and later turn brown and then dry up. Under heavy infestation, mites spread all over the leaves. The damage starts during May–June when the weather is dry.

Control

- Several natural enemies like coccinellids, thrips, Chrysopids and Phytoseiids control the mite.
- Spray the crop with 5% *neem* seed kernel extract.
- Spray of wettable sulfur 0.3% or Ethion 0.005% reduces the population of mites.
- Spray the crop with malathion 0.05%, Dicofol 0.035%, or Ethion (0.005%) as soon as the attack is noticed.

Stem Fly (*Ophiomyia phaseoli*)

The larvae cause damage by making mines inside the leaf lamina, veins, petioles and stem, which cause a girdling effect and death of the plant. It pupates at the base of the stem thus blocking food uptake. As a result, the plants show sudden wilting at the time of flowering.

Control
- Remove and destruct all affected plants.
- Spray the crop with 5% *neem* seed kernel extract.
- Spray 0.03% dimethoate or 0.04% monocrotophos twice at 10-day intervals.

Black Bean Bug (*Chauliops nigrescens*)

Black bean bug is the most damaging pest of snap bean. Both adults and nymphs suck sap from the lower surface of the leaves, as a result, the chlorophyll content gets reduced, which ultimately affects the photosynthesis, quality and yield of the crop. The damaged leaves gradually dry up and fall off. Its adults appear in June and continue until final harvest.

Control
- Spray the crop with monocrotophos 0.05%, dimethoate 0.03%, oxydemeton methyl 0.025%, or Lindane 0.05% for the control of this pest.

DISEASES

Fusarium Root Rot (*Fusarium solani* sp. *phaseoli, Rhizoctonia solani* and *Pythium* sp.)

Fusarium root rot is characterized by reddish lesions on the taproot and hypocotyl, which later turn brown. Brick red lesions of variable size develop on below-ground stems and taproot. In severe infection, the entire root system is destroyed. The disease is favoured by soil compaction, short crop rotation, warm temperature (22°C to 32°C), high soil moisture and low soil pH.

Control
- Follow long duration crop rotation with cereal crops.
- Use disease-free healthy seed.
- Disinfect the soil through solarization.
- Soak the seed in Bavistin or Benomyl 0.2% solution for 30 minutes.
- Spray the crop with Bavistin 0.1% or mancozeb 0.25% at intervals of 10 to 15 days.

Anthracnose (*Colletotrichum lindemuthianum*)

The disease affects all above ground parts of the plant. Symptoms appear as small, dark brown to black lesions on cotyledons of the young seedlings and oval or eye-shaped lesions on stems, which turn into brown sunken spots with purple to red margins. The most characteristic symptoms appear on the pods as dark brown spots, which become circular and sunken with raised, reddish, or yellowish rust-coloured margins. It is a seed-borne disease. The fungus can survive in the soil on crop debris and infect the crop in the following season. The disease is most severe in areas of high rainfall.

Control
- Follow a crop rotation of at least 2–3 years.
- Use disease-free healthy seed.
- Soak the seeds in Bavistin or Benomyl 0.2% solution for 30 minutes.
- Spray the crop with Bavistin 0.1% or mancozeb 0.25% at intervals of 10 to 15 days.

ANGULAR LEAF SPOT (*Phaeoisariopsis griscola*)

The disease infects all aerial plant parts including leaves, petioles, stems and pods but the symptoms are most recognizable on leaves. Lesions on leaves usually appear as brown spots with a tan or silvery centre that are initially confined to tissue between major veins, which give it an angular appearance. In some of the varieties, a yellow halo occasionally surrounds lesions and eventually the underside of the leaf and appears slightly more pale than those that appear on the upper surface of the leaflet. Severely spotted leaves show premature senescence and drop out. Lesions on the stem and petioles appear dark brown and elongated. Lesions on pods are circular, reddish brown usually surrounded by a dark border and sunken. Primary inoculum comes from seed or infested residue.

Control

- Follow crop rotation of at least 2–3 years with non-leguminous crops.
- Grow resistant varieties if available.
- Use disease-free healthy seed.
- Soak the seed in Bavistin or Benomyl 0.2% solution.
- Spray the crop with hexaconazol 0.05%, mancozeb 0.25%, or carbendazim 0.05%.

ALTENARIA LEAF SPOT (*Altenaria aiternata*)

The disease symptoms appear as small irregular-shaped brown lesions on the leaves, which later turn large oval grey-brown with concentric rings. When several lesions coalesce, a large portion of the leaf area becomes necrotic. Premature defoliation of the lowest leaves occurs. On pods, small water-soaked, reddish-brown lesions develop that merge into long streaks. Older leaves and pods are more susceptible than young leaves and pods. Cool and wet weather favours the germination and spread of spores.

Control

- Follow at least 3–4 years' crop rotation with non-host crops.
- Use disease-free healthy seed.
- Treat the seed with carbendazim 0.2%.
- Sow the seeds at wider spacing.
- Spray the crop with mancozeb 0.25% or copper oxychloride 0.3% and repeat at intervals of 10–14 days.

BEAN RUST (*Uromyces phaseoli*)

Rust affects the leaves, stem and sometime pods also. The symptoms appear as dry reddish, yellowish, or orange spore masses or pustule primarily on the lower surface of the leaves. The pods may also show disease symptoms. The favourable temperature for the spread of disease is 17–27°C with relative humidity of 95%.

Control

- Follow at least 3–4 years' crop rotation.
- Destroy the disease debris from infected crop field to reduce inoculums for next season's crop.
- Spray the crop with 0.25% mancozeb or 0.1% hexaconazole as the disease appears.

POWDERY MILDEW (*Erysiphe polygoni*)

The fungus of powdery mildew attacks all parts of the plant. Initially, the symptoms appear as yellow spots on the upper surface of the older leaves, and these spots develop the characteristic powdery growth, which spreads to the underside of the leaves and stem. The affected leaves turn completely yellow, curl, die and fall down. On pods, small moist-looking circular spots develop into white powdery masses of pathogen mycelium and spores. Powdery mildew is favoured by warm dry days and cool moist nights.

Control

- Grow resistant varieties if available.
- Spray the crop with 0.5% Sulfex or other formulations of sulfur.
- Spray the crop with systemic fungicide like Karathane, Bavistin, or Benlate 0.2% at intervals of 10 to 15 days.

ASHY STEM BLIGHT (*Macrophomina phaseolina*)

In the early stage of infection, it is characterized by damping-off and killing of young seedlings and appearance of black cankerous lesions on cotyledons. In the latter stage, the collar portion of the plant and vascular bundles of the roots become brown, resulting in dry root rot.

Control

- Use disease-free healthy seed.
- Apply adequate quantity of organic matter in the field.
- Treat the seed with captan or thiram 2–3 g per kg of seed.

FLOURY LEAF SPOT (*Mycovellosiella phaseoli*)

Symptoms of the disease appear as floury patches on the underside of the leaves followed by necrosis. The upper surface of leaves shows brown spots with yellow margins.

Control

- Use disease-free healthy seed.
- Collect the infected plant debris and destroy them.
- Spray the crop with Bavistin 0.1% at intervals of 10 to 15 days.

COMMON BLIGHT (*Xanthomonas campestris* pv. *phaseoli*)

Common bacterial blight first appears as small translucent water-soaked spots on the leaves. As the spots enlarge, the tissue dies leaving brown spots with a margin. These lesions are often large and irregular in shape. On pods, the symptoms appear as water-soaked sunken lesions, which later turn into brownish-red. Bacteria may also infect the vascular system and kill the main stem and branches. Common blight is seed borne and is favoured by warm temperatures (28°C) and high humidity.

Control

- Follow long-duration crop rotation.
- Collect the infected plant debris and destroy them.
- Use disease-free healthy seed.
- Treat the seed with streptocycline 0.01% and captan 0.25%.
- Spray the crop with Blitox 0.35% as and when the disease appears.

BEAN COMMON MOSAIC

It is the most damaging viral disease of snap bean in India. The tri-foliolate leaves usually show irregular-shaped light-yellow and dark-green patches. Infected leaves are narrower and longer than normal ones, and show malformation and downward curling. Early-infected bean plants are usually yellowish and dwarfed. Infected pods become chlorotic, shortened and possess a glossy sheen. The virus is transmitted by vectors like *Aphis gossypii*, *Aphis craccivora*, *Myzus persicae* and *Brevicoryne brassicae*.

BEAN GOLDEN MOSAIC

The symptoms appear as scattered, slightly discoloured patches on leaves, which gradually turn bright yellow, and later, the whole leaf turns golden yellow. Infected pods become malformed, stunted and show mosaic spots, while the seeds become discoloured, malformed and small. The virus is transmitted by white flies (*Bemisia tabaci*).

Control

- Grow resistant varieties if available.
- Use virus-free healthy seed.
- Rogue out the infected plants and destroy them.
- Adjust the sowing time to grow the crop in periods free of aphids and white flies.
- Apply Phorate 10 G 20 kg/ha in the soil at the time of sowing.
- Spray the crop with phosphamidon 0.05% or methyl demeton 0.02% at 10-day intervals to control the virus vectors.

■■■

82 Sword Bean

M.K. Rana and Mahesh Badiger

Botanical Name   : *Canavalia gladiata* (Jacq.) D.C.

Family   : Leguminosae (Fabaceae)

Chromosome Number : 2n = 22, 44

ORIGIN AND DISTRIBUTION

Sword bean is native to the Indo-Malayan region. Tropical Africa is its secondary centre of origin. The cultivated species is believed to have originated from *Canavalia virosa*, a wild species occurring principally in Africa. It is cultivated in South and Southeast Asia, especially in India, Sri Lanka and Burma. It is used for local consumption in humid tropics.

INTRODUCTION

In developing countries, malnutrition due to lack of protein-rich food and other neutraceuticals is a major problem to be addressed. Supply of protein can be met by exploration and exploitation of alternative legume sources. Sword bean is one such inexpensive underexploited legume species known by several local names in India *viz.*, *Makhan Shim* in West Bengal, *Tlvardi* or *Tarvardi* in Gujarat, *Lal Khadsumbal* in Hindi, *Shombi Avare* in Karnataka and *Yerratamma* in Andhra Pradesh. In Tamil Nadu, it is called *valavaraikkay* or *valavarangai*, meaning the vegetable that looks like a sword. It can sustain production even in the changing climatic scenario because of its desirable characteristics such as being tolerant to drought, low pH, salt, sand, shade, virus and waterlogging, thus, making it able to grow on less fertile soils too. Being a legume crop, it also fixes atmospheric nitrogen into the soil. It is best fit in different cropping systems like green manuring, cover crop and intercrop in plantation crops. It is a minor leguminous vegetable grown in tribal areas of the country where subsistence agriculture is practiced and in some parts of South India.

COMPOSITION AND USES

COMPOSITION

Sword bean pods provide a good amount of carbohydrates, protein, calcium, phosphorus, vitamin C and vitamin B also. The protein content in the mature seed is comparatively higher than in most legumes and the amino acid profile is comparable with that of casein protein. The nutritive value of young green pods is given in Table 82.1.

USES

Sword bean is cultivated for its tender pods, which are used as a cooked vegetable. The fully grown immature seeds are also consumed as a cooked vegetable like other beans. Sometimes, the young leaves and flowers are also eaten. In Tanzania, there is a phrase that *eating sword bean* means *being happy*. Dry seeds can also be eaten after cooking. However, the mature or dried seeds must

TABLE 82.1
Nutritional Composition of Sword Bean Pod (per 100 g Edible Portion)

Constituents	Contents	Constituents	Contents
Water (g)	87.2	Iron (mg)	2
Carbohydrates (g)	7.8	Carotene (µg)	24
Protein (g)	2.7	Thiamine (mg)	0.08
Fat (g)	0.2	Riboflavin (mg)	0.08
Dietary fibre (g)	1.5	Niacin (mg)	0.5
Calcium (mg)	60	Vitamin C (mg)	12
Phosphorus (mg)	40	Energy (kcal)	44

be cooked thoroughly as they contain slightly higher toxic protein substances. Roasted seeds are used as a substitute for coffee in Central America. The seed is used as feed for cattle and chicken. It is further grown as a forage crop, cover crop and as an ornamental climber on fences and houses.

MEDICINAL USES

In Korea, it is used for the treatment of vomiting, abdominal dropsy, kidney related lumbago, asthma, obesity, stomach-ache, dysentery, coughs, headache, epilepsy, schizophrenia, swelling and inflammatory diseases. However, its limited widespread utilization could be due to the presence of anti-nutritional factors like canavalin, concanavalin A, α-amylase inhibitor, trypsin inhibitor, urease and HCN, which can be removed upon a long cooking. Because of this, the boiling water is drained off to remove poisonous substances or peeling and using the seeds as broad bean. On the other side, it is commercially exploited as a source of urease enzyme, which is used in clinical laboratories for the *in vitro* determination of urea in human blood.

BOTANY

Sword bean (*Canavalia gladiata*) belongs to the genus *Canavalia* and family Fabaceae with chromosome number (2n) 22 and 44, indicating the presence of polyploids. The probable progenitor species *Canavalia virosa* is found to occur in some parts of the northeastern region (Manipur, Mizoram and other areas) of India. *Canavalia polystacha* grown in China is closely related to *C. gladiata*. The other species *C. ensiformis*, commonly known as jack bean, and having significant importance as a minor leguminous vegetable like that of sword bean, is also grown in some parts of the world in kitchen gardens. The jack bean and sword bean are very similar but differ mainly in the length of the seed hilum (scar). The hilum of sword bean is more than one-half the length of the seed, whereas, that of the jack bean is only about one-third as long as the seed. Sword bean is an annual or perennial climber with a terate glabrous stem. Leaves are pinnately trifoliate, acute and glabrous. Flowers are papilionaceous, white (bluish-purple in *C. virosa*) and borne on an axillary raceme. Usually, the flowers open in an acropetal manner. Stamens are 10 (6 long and 4 short), slightly curved and projected straight after the complete opening of flower. Anthers are dorsifixed, tetra-locular and generally burst longitudinally 30-40 minutes prior to anthesis. Gynoecium is monocarpellary. The existence of morphological protandry, *i.e.*, the anthers dehisce at a young bud stage prior to stigma receptivity, and temporal separation of pollen release and stigma receptivity act as a barrier to selfing, thus promoting out crossing to some extent. Tripping of stigma by pollinators is prerequisite for the sticking and germination of pollen grains.

CLIMATIC REQUIREMENT

Sword bean is well adapted to tropical climates. Being a hardy crop, it is tolerant to drought, low pH, salt, sand, shade, virus and waterlogging, thus making it able to grow under varied agro-climatic conditions. It performs well under moderately high temperature of 20–30°C. It can be grown up to 1600 m above mean sea level. It is tolerant to a wide range of rainfall conditions, *i.e.*, 700 to 4200 mm. A long taproot system allows the plants to use deep soil moisture reserves, thus, it is a source of vegetable in drought-prone areas. Although it can tolerate some shade but it needs full sunlight for its optimum growth and development.

SOIL REQUIREMENT

Being a hardy crop, it can be grown in all types of soil and is well adapted to poor-quality soils. It comes up well on loose friable soil rich in organic matter. Unlike most beans, it is tolerant of leached and nutrients-depleted lowland tropical soils. Like other legume crops, sword bean also has nitrogen-fixing capacity and better tolerance to salinity and soil acidity (pH 4.3-6.8) than most of the legumes. Thus, it is a species with promise for the reclamation of poor tropical soils.

VARIETIES

Based on seed colour, two types of sword bean variety are available, *i.e.*, (i) white-seeded varieties that are bushy in nature and grown at closer spacing and (ii) red-seeded varieties that are pole type, trailed over pandals and grown at wider spacing. Still, its improved varieties are not available in India but an introduced variety called SBS-1 is available.

SBS-1

A short-duration (110–120 days) variety with erect, dwarf and bushy growth habit is a vegetable-cum-grain type. It is highly nutritious with 25.9% protein, free from beany odour and delicious in taste. It is resistant to major insect-pests and diseases.

SOWING TIME

The usual time for its sowing is May–June and September–October. In North Indian plains, it is sown during June–July, and in South India, the variety SBS-1 can be sown at any time, as it is photo-insensitive.

SEED RATE

Sword bean is propagated by seed. About 25–40 kg seed is required to sow a hectare area. In the lands where beans are grown for the first time, inoculation of seed with *Rhizobium* species helps in quick nodulation on roots, which fix atmospheric nitrogen into the soil.

SOWING METHOD

Actually, sword bean is grown on a small scale only by the small holding farmers near houses, and the vines are allowed to climb on walls, fences and trees. The soil before sowing is made free of clods and weeds and brought to a fine tilth by ploughing the field three or four times followed by planking. Pole-type varieties are sown at 4 × 3 m and bush type at 60 × 30 cm spacing. Generally, one or two seeds are sown per pit.

MANURIAL REQUIREMENT

Although sword bean is a legume crop, it responds well to the application of fertilizers. Fully decomposed farmyard manure is applied @ 25 t/ha before sowing at the time of land preparation and mixed thoroughly into the soil by repeated ploughings. Application of nitrogen 63 kg, phosphorus 100 kg and potash 75 kg/ha is recommended. Half of the nitrogen along with the entire dose of phosphorus and potash fertilizer should be applied at the time of sowing, and the remaining half of nitrogen should be applied 3 weeks after sowing. Application of VAM and *Rhizobium* species mixture along with farmyard manure has been reported to increase the nodule number and size.

IRRIGATION REQUIREMENT

Pre-sowing irrigation is advantageous for uniform emergence of sword bean seeds. One irrigation 2 to 3 days prior to sowing and a light irrigation 2 to 3 days after sowing help in better germination and crop stand. It grows well under rain-fed conditions but still responds well to irrigation. A long taproot system allows the plants to use deep soil moisture reserves. The crop should be irrigated at intervals of 1 week. Flowering and pod development are the critical stages for irrigation supply. As usual, irrigation is scheduled depending on the atmospheric conditions.

INTERCULTURAL OPERATIONS

HOEING AND WEEDING

Sword bean is a legume crop, thus, at least one or two shallow hoeings are necessary and beneficial for providing a good soil environment to the roots for their development and absorbing atmospheric nitrogen. To avoid the problem of weeds at the early stage of crop growth, a hand weeding at 20 to 25 days after sowing should be done.

STAKING AND TRAINING

In pole-type varieties, staking of vines promotes growth, flowering and pod development. Trailing of the vines over pandals is also in practice. In homestead gardens, the vines are allowed to trail over the roof.

HARVESTING

Sword bean is a fast-growing crop. Its young pods become ready for harvesting 3–4 months after sowing. The pods are picked when they are tender, 12.5–15 cm long and before they swell and become hard. Sometime, the young leaves and flowers are also eaten. Mature seeds are consumed thorough cooking and draining off the water.

YIELD

The yield depends on variety, soil fertility, growing season and cultural practices adopted by the farmers during cultivation of the crop. The green pod yield varies from 40 to 50 q/ha and seed yield from 20 to 30 q/ha.

INSECT-PESTS

STEM FLY (*Ophiomyia phaseoli*)

This insect also attacks cowpea, French bean, etc. Its maggots mine the leaf, bore inside the petiole and make tunnels inside the tender stem, which ultimately cause a girdling effect and death of the plant. The affected plants become yellow and dry up.

Control

- Remove and destroy the affected plants.
- Spray the crop with Rogar 1 ml/litre of water at intervals of 10–12 days.
- Spray the crop with *neem* seed-kernel extract 5% to reduce the incidence.

Pod Borers (*Muruca testulalis* and *Lampides boeticus*)

A number of borers damage the pods. Round holes are produced in corolla of the flowers by the larvae. Flowers are converted to a mass of brownish frass within a day. The larva thrusts its head alone inside the developing pods and leaves the rest of its body hanging out. The pods are distorted by larger larvae and affected by frass.

Control

- Remove and destroy the infested pods at initial stage.
- Spray the crop with Monocrotophos 0.04% at intervals of 10–12 days.
- Apply nuclear polyhedrosis virus (NPV) to suppress the pest.
- Spray the crop with *neem* seed-kernel extract 5% to reduce the incidence.

Aphid (*Aphis craccivora*)

This pest is a major constraint in the cultivation of French bean, cowpea, Indian bean, etc. Its nymphs and adults attack young plants, suck sap from the growing tips and later from the whole plant. It causes curling of leaves and reduces plant vigour, yield and quality of pods.

Control

- Apply carbofuran granules 1 kg a.i./ha at the time of sowing.
- Spray the crop with 0.03% dimethoate (Rogar 1.5 ml/litre of water) at intervals of 10–12 days.

Leaf Miner (*Spilasoma obliqua Euproctis*)

Larvae feed by making tunnels in the leaves. Small blister-like mines are seen on the upper leaf surface. As the feeding advances, the mines increase in size and the entire leaflet becomes brown, rolls, shrivels and dries up. In severe cases, the affected crop gives a burnt appearance. In later stages, the larvae web the leaflets together and feed on them, remaining within the folds.

Control

- Follow long crop rotation with non-leguminous crops.
- Use resistant varieties for cultivation.
- Spray the crop with 0.03% dimethoate (Rogar 1.5 ml/litre of water) at intervals of 10–12 days.
- Use larva parasites, i.e., *Bracon gelechidae* Ashm. and *Elasmus brevicornis* Gah.

Root-knot Nematodes

The root-knot nematode enters the roots, causing characteristic root galls. As a result, plant growth becomes stunted and yield is decreased severely. The aerial symptoms consist mainly of stunted plant growth and yellowing of leaves. Nematodes attack in the seedling stage leads to pre- and post-emergence damage, resulting in reduced crop stand.

Control

- Follow long crop rotation with non-host crops.
- Plough the field deep two or three times in hot summer months.

- Apply *neem* seed cakes in the field during land preparation.
- Apply granular insecticides like furadon 3 kg/ha at the time of sowing.

DISEASES

ANTHRACNOSE (*Colletotrichum lindemuthianum*)

It is a seed-borne disease, and its severity is more in high rainfall areas with a subtropical climate. The fungus attacks stem, leaves and pods. Necrotic spots appear on leaves, which become thin and papery and produce a pinkish slimy mass. The characteristic symptoms, such as dark brown sunken lesions with raised, reddish, or yellowish margins, appear on pods.

Control

- Use disease-free healthy seed.
- Follow clean cultivation.
- Dip the seeds in Cerasan 0.15% solution for 30 minutes.
- Spray the crop with 0.1% Bavistin at 10-day intervals.

ASHY STEM BLIGHT (*Macrophomina phaseolina*)

In early stages, it causes damping-off and killing of young seedlings. Black cankerous lesions also appear on the cotyledons. In later stages, the collar region of the plant becomes brown and vascular bundles of the roots turn brown, resulting in dry root rot.

Control

- Use disease-free healthy seed.
- Treat the seed with Captan @ 2–3 g/kg of seed.
- Apply adequate organic matter in the field.

BACTERIAL BLIGHT (*Xanthomonas campestris* pv. *fuscans*)

It is the most serious disease of French bean, cowpea and cluster bean too. The infected leaves show irregular sunken yellow to brown necrotic spots with a narrow yellow halo. In severe cases, defoliation occurs. Dark green water-soaked spots may appear on the pods, which later become yellow or brown and dry.

Control

- Grow resistant variety.
- Use disease-free healthy seed.
- Dip the seed in 0.01% streptocycline solution for 30 minutes.

POWDERY MILDEW (*Plasmopara halstedii*)

French bean, cluster bean and Indian bean are also affected by this disease. The fungus attacks almost all parts of the plant. Slightly discoloured specks are formed. From these specks, a greyish white powdery growth of mycelium and spores spreads over the leaves, stem and even fruits. In severe cases, defoliation occurs.

Control

- Grow resistant varieties.
- Spray the crop with 0.5% Sulfex or wettable sulfur.
- Spray systemic fungicides like Kerathane or Calyxin 0.2% at weekly intervals.

Mosaic virus

It is the most serious disease of French bean and cowpea also. It is seed-borne as well as transmitted by virus vectors such as aphids and white flies.

Control

- Use disease-free healthy seed.
- Remove diseased plants as soon as they are noticed.
- Spray the crop with systemic insecticides, like Rogar or Metasystox 2 ml/litre of water at intervals of 10–12 days.

∎∎∎

83 Tepary Bean

M.K. Rana and Pradeep Kumar Jatav

Botanical Name	: *Phaseolus acutifolius* var. tenuifolius A. Gray
	Phaseolus tenuifolius A. Gray
Family	: Leguminosae (Papilionaceae)
Chromosome Number	: 2n = 22

ORIGIN AND DISTRIBUTION

Tepary bean is native to the southwestern United States and northern Mexico as it has grown there since ancient times. It was distributed from Arizona and New Mexico to Guanacaste, Costa Rica, dry subtropical slope of the Pacific. The studies reveal that tepary bean was domesticated in Tehuacán Valley, Mexico, 2300 years ago, and its isozyme analysis suggests that its domestication took place in a single geographic province, with the Mexican states of Jalisco and Sinaloa being the potential candidates. Its wild types were distributed from the southwestern United States to Guatemala with the hub of their distribution in northwestern Mexico. Its cultivation dwindled strongly after the Second World War, but nowadays, the crop is regaining interest of people for its cultivation. It was introduced into Africa, Asia, Australia and Madagascar between the First and Second World War, and now, it is being cultivated there. Its cultivation is also recorded in Morocco, Algeria, South Africa, Swaziland and Lesotho.

INTRODUCTION

Since early 1900s, tepary bean has been mainly grown in Mexico and Arizona (United States). Its large-scale commercial production was tried but efforts were deserted due to its unfavourable morphological characteristics as compared to the common bean, changes in eating habits and lack of information on its performance. Presently, it has gained importance in semi-arid parts of tropical Africa, *i.e.*, Sudan, northeastern Kenya, Uganda and Botswana, where most other legume crops are not successfully grown due to drought and where short-duration crops are often needed to fit in rotation, but its production is mainly done for domestic consumption.

COMPOSITION AND USES

COMPOSITION

Tepary bean is a very nutritious vegetable, which provides an ample amount of protein, carbohydrates, phosphorus, calcium and vitamins to the human diet. The seeds of other pulses are low in sulfur-containing amino acids like methionine and cystine. Similar to cowpea and French bean, tepary bean contains anti-nutritional factors such as trypsin inhibitor and phytic acid but only the lectin activity is exceptionally high, which can easily be denatured by cooking. However, cyanogenic glucosides are not found present. The consumption of its uncooked flour has been found to cause death of living beings within 3–4 days but soaking and cooking of seeds for a long period gets rid of toxicity. Tepary bean has a strong typical flavour and odour and thus is less tasty than other common beans. Its dry seeds during storage become very hard and take a long time to cook. The seed coat of white cultivar seeds is more permeable than the seed coat of black-seeded cultivars, resulting in a shorter cooking time. Its pods and stems contain 8% water, 4.1% protein, 0.5% fat, 43.6% N-free extract, 37.0% fibre and 6.8% ash. The nutritive value of dry tepary bean seeds is given in Table 83.1.

USES

Tepary bean is chiefly cultivated for its mature dry seeds, which are eaten after boiling, steaming, frying or baking. Its seeds are used in stews and soups, and mixed with whole-grain maize. The dry seeds in Uganda are usually boiled and then coarsely ground before being added to soup. Intermittently, it is eaten as a green bean or as bean sprouts. Its leaves are also considered edible but they are comparatively tougher than the leaves of the common bean, and they take a longer time to cook. After removing seeds, the leftover pods and stems are used as cattle feed and the seeds are used as supplementary feed for chickens. It is also grown as a fodder or green manure crop in the United States. It may be grown as a cover crop and an intercrop in low-density orchards and in the agro-forestry system.

BOTANY

Tepary bean can easily be distinguished from other bean species due to its epigeal germination. It is a bushy annual herb, plants of which are climbing, trailing or with more or less erect stem up to 4 m length. The plant root system is adventitious. The leaves are alternate and 2–3 mm long. It has three foliolate and lanceolate stipules appressed to the stem. The petiole length is 2–10 cm and the leaflets are ovate to oval in shape. Inflorescence is an axillary raceme. Flowers are bisexual, consisting of 3–7 mm long pedicle, 3–4 mm long campanulate calyx, the upper two lobes united into one and the lower three triangular, white, pink, or pale purple corolla, half-reflexed standard up to 1 cm long, wings up to 1.5 cm long, keel narrow, coiled stamens 10, 9 fused and 1 free, 0.5 cm long superior ovary, style with a thickened terminal coil and collar of hairs below the stigma. Fruits or pods are straight or slightly curved, 5–9 × 0.5–1 cm size, rimmed on margins with short but distinct beak, hairy when young and two- to nine-seeded. The seeds are of 4–7 × 2–5 mm size, globose to oblong, with white, yellow, brown, purple, black or variously speckled and dull in colour.

CLIMATIC REQUIREMENT

Tepary bean is tolerant to drought, heat and a dry atmosphere, thus, it is suitable for cultivation in arid regions. The factors that are responsible for drought tolerance are sensitive stomata, closing at a relatively high water potentials and an extensive root system. It requires a mean day temperature of

TABLE 83.1
Nutritional Composition of Dry Tepary Bean Seed (per 100 g Edible Portion)

Constituents	Contents	Constituents	Contents
Water (g)	8.6	Calcium (mg)	112
Carbohydrates (g)	67.8	Thiamine (mg)	0.33
Protein (g)	19.3	Riboflavin (mg)	0.12
Fat (g)	1.2	Vitamin C (mg)	0.0
Dietary fibre (g)	4.8	Niacin (mg)	2.8
Phosphorus (mg)	310	Energy (kcal)	353

17–26°C for its proper growth and development and the night temperature should not fall below 8°C. It can successfully be grown in regions with an annual rainfall of 500–1700 mm. However, when the annual rainfall exceeds 1000 mm, the vegetative growth is usually luxurious, which causes a big reduction in its seed yield. Rain at flowering time is very harmful, as it affects pollination adversely. In subtropical areas, the crop-growing season is very short, while in humid regions, it may be cultivated round the year. Some of its cultivars are photosensitive as they require either short or long days for flowering, whereas, some of the cultivars seem day neutral. The crop can be cultivated from 50 to 1920 m above sea level. However, the best crop can be taken at medium altitudes.

SOIL REQUIREMENT

Although it can be grown on all types of soil, for obtaining a higher yield, well-drained medium-textured light sandy loam soils high organic matter with pH 6.7 to 7.1 are preferred. However, reasonable yield can be obtained from sandy soils with pH 5–7. It cannot tolerate waterlogging conditions, thus, heavy clays are unsuitable for its cultivation. It is moderately tolerant of saline and alkaline soils, which may not be physiological but result from its ability to escape salinity due to its deeper root system than the common bean.

CULTIVATED VARIETIES

Generally, two forms occur, one with small round, white or black seeded and another with large-sized angular seeded, which may be white, greenish white, grey, dark yellow, black, purple-mottled, or coffee colour. Two cultivars, *i.e.*, white- (Redfield) and dark-yellow-seeded, developed through mass selection. Although the cultivated and wild varieties do not have a definite habitat but a desert environment is necessary for their cultivation. The wild varieties are generally climbers (2–4 m in length) with few shoots. The cultivated varieties are of two types, *i.e.*, (i) the indeterminate shrubby varieties with short leaves and (ii) the indeterminate creepers with long leaves, which climb if they find support.

Based on shape of leaflet and seeds, three botanical varieties *viz.*, *Phaseolus acutifolius* var. acutifolius, *Phaseolus acutifolius* var. tenuifolius and *Phaseolus acutifolius* var. latifolius, have been distinguished. *Phaseolus acutifolius* var. acutifolius and *Phaseolus acutifolius* var. tenuifolius comprise wild types from the southwestern United States and northwestern Mexico, whereas, *Phaseolus acutifolius* var. latifolius comprises both the wild and cultivated types. However, the isozyme and AFPL analyses show no clear-cut differentiation between the var. acutifolius and var. tenuifolius.

SOWING TIME

The seeds are sown during September to November in Western and South India, during February–March in North Indian plains, while it is grown from April to October in the hills.

SEED RATE

The seed rate of tepary bean depends on climatic factors, soil types, genotype, method of sowing and the cultural practices followed during cultivation. Generally, it is propagated by seed. It has less seed variability as compared to kidney bean. The test weight (1000 seeds weight) of cultivated genotypes is 100–220 g and wild genotypes 15–50 g. In general, the seed rate varies from 15 to 20 kg/ha. However, when grown for hay, it is about 70 kg/ha.

SOWING METHOD

The seeds are sown directly in flat seedbeds at a depth of 2.5–10 cm. The soil at the time of sowing must have ample moisture, which facilitates easy and early germination of seeds. However, in low-lying areas, sowing on raised beds is preferred over sowing in flat beds, as it facilitates drainage. Seeds are sown either by broadcasting or in rows 60–90 cm apart with 10–45 cm spacing between plants within the row. However, in Kenya, it is sown at 60 × 30 cm spacing. Sometimes, it is sown on mounds, where two to four seeds are sown per mound. The seeds are sown sparsely to avoid overcrowding of plants. The seed before sowing should be treated with Thiram or Captan to disinfect the seed for better germination. The seed may also be treated with *Rhizhobium* culture for better nodulation.

Tepary bean can be grown either a sole crop or intercropped with cereal (sorghum, pearl millet, or maize), vegetable (*Allium*, *Brassica*, *Capsicum*, or *Cucurbita* spp.), or other legume crops. However, in the United States and Mexico, it is sometimes sown as a mix crop with the common bean, thus providing greater yield stability than the common bean alone.

NUTRITIONAL REQUIREMENT

Since the crop has not yet attained commercial status, very little information is available on its nutritional aspects. However, being a legume crop, its response to the application of nitrogenous fertilizer is poor due to the inherent capacity to fix atmospheric nitrogen in its root nodules. Thus, only a small quantity of nitrogen fertilizer is applied as a starter dose. It has been observed that the nitrogen fixation in legumes is done by symbiotic association of *Rhizhobium* bacteria in their root nodules, as in symbiosis it produces a considerable quantity of cytokinins. The application of phosphorus fertilizer in legumes enhances the formation of nodules in their roots. About 25–30 tonnes of well-decomposed farmyard manure should be applied at the time of land preparation to improve aeration in heavy soils and water-holding capacity of light soils. Generally, 40–50 kg of nitrogen, 60 kg of phosphorus and 50 kg of potash are required for successful crop production. A full dose of phosphorus and potash along with a half dose of nitrogen should be applied at the time of sowing and the remaining half of the nitrogen is applied at the time of flowering.

IRRIGATION REQUIREMENT

Applying irrigation at regular interval has a significant effect on colour, firmness and yield of pods and seed yield of tepary bean. Generally, the crop requires about six or seven irrigations at regular interval for successful cultivation of the crop. There should not be any water stress during flowering and pod formation. Pre-sowing irrigation should be applied to have ample soil moisture for proper seed germination and further growth of the plants. If moisture is not enough at the time of sowing, a light irrigation should be applied soon after sowing; however, excessive watering is harmful to the crop.

INTERCULTURAL OPERATIONS

HOEING AND WEEDING

Hoeing and weeding are very essential at early stages of crop growth since weeds compete with crop plants for nutrients, moisture, sunlight and space and act as secondary host for insect-pests and diseases. Hoeing is also beneficial for providing good aeration into the soil, which is very essential for the survival and efficiency of *Rhyzobium* bacteria to fix atmospheric nitrogen into the soil. Weed infestation is more if the crop is sown at a wider spacing but it facilitates hoeing. On the other hand, sowing at a narrow spacing reduces weed population in the field but it does not allow hoeing; thus, weeds are removed manually. Two hoeings should be done until the crop plants are well established and to prevent weed competition with the crop. Pre-emergence application of Benatazone at 2 kg/ha along with Thiobencarb at 5 kg/ha is effective for controlling weeds in a tepary bean field.

HARVESTING

The pods of teary bean should be harvested at the right stage of maturity, as early or delayed harvesting causes substantial yield loss. Usually, the pods do not mature all together, thus, the pods are harvested for fresh market as and when they attain marketable size but before they become waxy and fibrous.

For grain purpose, the crop is harvested as soon as the pods start changing their colour from green to brown. Delayed harvesting may cause shattering of pods if they are left to dry up in the field. Normally, the crop becomes ready for harvesting 2.5–3 months after sowing. The plants are harvested by hand when they reach maturity and are left to dry in the field for a few days before threshing. Beating the dried plants with wooden sticks is a common practice to thresh tepary bean. After cleaning, the seeds are packed in plastic or tetra packs and stored in a cool and dry place to maintain their viability up to 3 years.

YIELD

The yield depends upon cultivars, soil types, sowing density, rainfall and cultural practices adopted. In dry-land farming, the yield is about 5–8 q/ha, whereas under an irrigated condition it is about 9–17 q/ha, but on an average, the tepary bean gives a dry seed yield of 4.5–6.7 q/ha. When the crop is grown for fodder purposes, the yield is about 55–110 t/ha.

POST-HARVEST MANAGEMENT

More so than other pulses, tepary bean seed can also be stored well and it hardly needs storage pest control.

INSECT-PESTS AND DISEASES

Generally, the tepary bean is not affected by pest and diseases in semi-arid regions of the country. However, the following pests and diseases attack when it grown during periods of high humidity.

INSECT-PESTS

BEAN APHID (*Aphid crassivora*)

Adult and nymphs of bean aphids suck the sap from tender part of the plants, which results in curling of leaves, twisting of twinges and shedding of flowers and developing fruits. Aphids secrete a honeydew-like sugary substance, on which sooty mould grows, which lowers down the photosynthetic rate of the plant and reduces yield. Aphids transmit the common bean mosaic virus.

Control
- Set up yellow sticky traps in the field.
- Spray the crop with monocrotophos 0.1%, dichlorvos or Nuvan 0.1%, dimethoate 0.2%, or Confidor 0.3%.

BEAN WEEVIL (*Bruchus* sp.)

This pest infests the crop in the field as well as during storage. In field conditions, it is more serious in hilly areas. The elongate yellow eggs are laid on small green pods. On hatching, the larvae bore into the seeds through pods. They develop inside the pods in 30–35 days, constructing an exit tunnel to come out and pupate. In storage, it damages the quality for both consumption and sowing purposes.

Control

- In their early stage, collect and destroy the larvae.
- Spray the crop with choropyriphose 0.2%.
- In storage, fumigate the dry seeds under airtight condition with methyl iodide, Phosfume, or Celphose available in the form of tablets.

STEM FLY (*Ophiomyia phaeseoli*)

The adult lays eggs on epidermal tissues of the plants. The maggots on hatching mine into the leaves, petioles and stems, which cause wilting of plants, as they bore inside the stem. They also feed on the outer layers of the stem at the collar region of the young plants, causing death of the plants.

Control

- Apply carbofuran in soil 2 kg a.i./ha.
- Spray the crop with Diazinon 0.03%, quinalphos 0.5% or dimethoate 0.1% at least two times just after seed germination at weekly intervals.

DISEASES

ANTHRACNOSE (*Colletotrichum lindemuthianum*)

It is the most common disease of tepary bean. The occurrence of disease is more severe in subtropical to temperate regions than in the tropical regions. The disease is characterized by the development of black sunken canker-like lesions surrounded by yellow-orange margins on the pods. If the infection is severe, dull salmon-coloured ooze is formed in the centre. Similar spots are also found on the stems of young seedlings and cotyledons, resulting in mortality of seedlings when disease is severe. The disease is spread through seed.

Control

- Follow long crop rotation with non host crop.
- Use disease-free healthy seed.
- Avoid overhead irrigation to reduce spread of the disease.
- As far as possible, destroy the weed hosts surrounding the crop.
- Remove and destroy the affected plants and crop debris.
- Seed treatment with thiram and captan 1:1 at 0.3% concentration for one kg of seeds.
- Spray the crop with Dithane Z-78 0.2%, Difolatan 0.2%, Captan 0.15% at 10-day intervals.

RUST (*Uromyces appendiculatus*)

This disease is observed as severe, especially in high humid areas. It infects all types of bean. Generally, symptoms are characterized by the development of orange rust pustules on above ground plant parts but more frequent on under surface of the leaves. The rust pustules are small, round and light brown to orange in colour; however, a few are dark brown, which comprises *teliospores* formed at a later stage of disease development.

Control

- Grow resistant varieties, if available.
- Follow long crop rotation with non-host crops.
- Provide good sanitation to the field.
- Adjust sowing time to avoid the disease.
- Remove and destroy all the infected leaves and residues.

- Dust the crop with sulfur 25–30 kg/ha.
- Spray the crop with Dithane M-45 0.2% or Bayleton 0.05%.

Fusarium Rot or Dry Root Rot (*Fusarium* sp.)

The disease is confined to certain areas. The infection is restricted to the base of the plants including all the roots. The vascular bundles of the infected plants turn to dark brown and the matured leaves fall off, resulting in wilting, and finally, death of plants occur in severe cases. The infection is more severe at the flowering stage of plants. The pathogen is soil borne and has wide legume host plants.

Control
- Follow long crop rotation with non-host crops.
- Use disease-free healthy seed.
- Remove and destroy all the affected plants and crop debris.
- Treat the seed with Captan, Captaf, Captafol, Foltaf, or Bavistin 2 g/kg of seed.
- Drench the seedbeds with Captan, Captaf, Captafol, Foltaf, or Bavistin 2 g/litre of water.

Charcoal Rot and Grey Stem Blight (*Macrophomina phaseolina*)

This disease is characterized by occurrence of damping-off and death of young seedlings. The small black cankerous lesions appeared on cotyledons and stem. In severe cases, decaying of roots takes place and the whole plant turns black. The poor drainage system, crop management and nutrient deficiency increase the disease severity.

Control
- Follow long crop rotation with non-host crops.
- Use disease-free healthy seed.
- Remove and destroy the affected plants and crop debris.
- Soil drenching should be done with suitable fungicides.
- Spray Carbendazin at 1.25% an intervals of 10 days.
- Treat the seed with Captan, Captaf, Captafol, Foltaf, or Bavistin 2 g/kg of seed.
- Drench the seedbeds with Captan, Captaf, Captafol, Foltaf, or Bavistin 2 g/litre of water.

Powdery Mildew (*Erysiphe polygoni*)

This is the common disease of legume vegetables. The disease is characterized by greyish white powdery growth on leaves, which later covers the whole plant in advanced stages. In severe cases, the whole plant is dried up. The occurrence of disease is more severe under cool and dry conditions. The pathogen survives on secondary host plants. It is an air-borne disease.

Control
- Follow long crop rotation with non-host crop.
- Use disease-free healthy seed.
- Destroy the weed hosts surrounding the crop.
- Remove and destroy the affected plants and crop debris.
- Spray the crop with Karathane, Sulfex, Cosan, or Calixin 0.2% at 10-day intervals.

Common Bacterial Blight (*Xanthomonas campestris* pv. phaseoli)

This is the most destructive bacterial disease of tepary bean. The initial symptoms are characterized by the appearance of irregular sunken red to brown spots surrounded by a yellowish halo. At a later stage, these spots coalesce and form patches, which result in falling of leaves at the premature stage. In severe cases, cankerous lesions may develop on pods, and the whole plant gets blighted.

Control

- Follow long crop rotation with non-host crops.
- Dip the seed in hot water at 52°C for 30 minutes.
- Spray the crop with streptocycline at 0.3%.

VIRAL DISEASES

The most common viral diseases of tepary bean are bean common mosaic virus (BCMV) and bean yellow mosaic virus (BYMV).

BCMV

The symptoms appear as chlorotic patches, crinkling of young leaves, and their petioles' length is reduced, resulting in falling of leaves. In advanced stages of the disease, chlorosis and mottling with definite dark and light green areas on leaf blades of trifoliate leaves develop. Plants become bushy and pods deformed. It is transmitted by *Myzus persicae* and also through seeds.

BYMV

Small yellow patches develop on the leaves which may spread to the entire surface at a later stage of infection. Mild mosaic symptoms appear on the trifoliate leaves. Therefore, leaves become rough, leathery, light-coloured and crinkled with downward curling. The plants become extremely dwarf and stunted. The premature flower dropping takes place. The pods are distorted with deformed seeds. Early infection causes severe loss of pod yield. The virus is transmitted by *Myzus persicae*.

Control

- Use of disease-free healthy seed.
- Removal of all infected plants from the field.
- Complete eradication of weed hosts.
- Apply Carbofuran, Disulfotan or Phorate @1.5kg/ha at the time of seed sowing.
- Spray the crop with dimethoate or monocrotophos at 0.002%.

■■■

84 Soybean

A.T. Sadashiva and M.V. Bharathkumar

Botanical Name : *Glycine max* (L.) Merill.

Family : Leguminosae (Papilionaceae)

Chromosome Number : 2n = 40

ORIGIN AND DISTRIBUTION

Soybean plant was first recorded at around 200 B.C. as a medicinal plant in China. It was domesticated in the eastern half of Northern China around the 11th century, and then, it was introduced from China to the United States of America, Japan, Korea, and South and Southeast Asian countries at different times. Numerous landraces are still cultivated, particularly around Shanghai and Jiangsu. Though soybean was introduced from China to Japan at an early date; its first recorded use is mentioned in the *Engishiki*, a Japanese guide to trade agricultural commodities during 927 A.D. The other commercial vegetable soybean-producing countries include Korea, North America, Argentina, Australia, Israel, Mongolia, New Zealand and Thailand. In Bhutan, Brazil, Britain, Chile, France, Germany, Indonesia, Malaysia, Nepal, Philippines, Singapore and Sri Lanka, it is cultivated at a very small scale in home gardens.

INTRODUCTION

Green vegetable soybean, also known as Edamame, translated as *beans on branches* in Japanese, *mao dou* in Chinese, meaning hairy bean, and as sweet bean in the United States of America, is a nutritious and tasty vegetable or snack food with a sweet nutty flavour. It is the immature green form of edible soybean. Edible or food-grade soybean differs from field soybean by being larger-seeded, milder-tasting, more tender, with a smoother texture, more pleasant flavour, sweeter, more digestible due to lower trypsin-inhibitor levels and with easy cooking characteristics due to higher phytic acid levels. It also contains a lower percentage of gas-producing starch. Otherwise, the nutritional value of edible and field soybean is comparable. In India, a large population has protein deficiency, as the majority of Indians are vegetarian, excluding meat as an option to obtain protein. Thus, vegetable soybean, producing the highest protein per unit area among the legume crops, can be a good substitute of meat since soybean protein is equivalent to animal proteins in quality and higher than other plant proteins food articles.

COMPOSITION AND USES

COMPOSITION

Vegetable soybean is an excellent source of protein, fat, vitamin A, calcium, phosphorus, dietary fibre and phytoestrogens (plant-produced oestrogens) and a good source of total lipids, iron and energy. Normally, the legume vegetables are said to be lacking in sulfur containing amino acids and poor in digestibility but soybean is the only vegetable food that contains all eight essential amino acids required by the human body. The nutritional composition of vegetable soybean is given in Table 84.1.

TABLE 84.1

Nutritional Composition of Vegetable Soybean (per 100 g Edible Portion)

Constituent	Content	Constituent	Content
Water (g)	68	Dietary fibre (g)	4.80
Carbohydrates (g)	11	Calcium (mg)	60
Sugar (g)	2.48	Phosphorus (mg)	194
Protein (g)	13	Iron (mg)	2.1
Fat (g)	6.8	Zinc (mg)	1.32
Total lipids (g)	4.73	Energy (kcal)	100

Uses

Vegetable soybean can be a good substitute for green peas or lima beans used in the preparation of various dishes in India. In China, the shelled beans are stir-fried with other ingredients. In Japan, pods are boiled in salted water and the beans are squeezed from the pod directly and the shell is discarded.

Medicinal Uses

Though vegetable soybean is a rich source of fat, most of the fat present in it is unsaturated and beneficial since the polyunsaturated fat of soybean includes linolenic acid (omega-3 fatty acid). The presence of omega-3 fatty acid makes it special, as soybean is one of the very few plant sources containing this essential fatty acid. Omega-3 fatty acid forms it an essential dietary nutrient that helps in reducing the risk of both heart disease and cancer.

BOTANY

Vegetable soybean is an erect bushy annual having a taproot system with many adventitious roots, which consist of nodules on their surface, and they become extensive at maturity. Stems form as a result of hypocotyl elongation of the seed axis. Seed germination is epigeal, in which the cotyledons are pushed above the soil surface by the rapid elongation and growth of hypocotyls. It has four types of leaves, *i.e.*, (i) cotyledon or seed leaves that emerge with seedlings, (ii) two simple primary leaves, (iii) trifoliate leaves and (iv) prophyllus leaves. The terminal bud on indeterminate stems continues to grow and produces an axillary raceme type of inflorescence, which will bear evenly distributed pods on branches with a diminishing frequency towards the top of the stem. The growth of terminal bud in determinate stems is ceased when it bears inflorescence. This type of stem has both axillary and terminal racemes and pods are found in dense clusters along the stem. Soybean flower is a typical papilionaceous. Being cleistogamous, pollination may occur a day before anthesis within the bud itself, and hence, it ensures self-pollination and fertilization. Pods are straight or slightly curved and vary in length from 2 to 7 cm or more in some varieties. The number of pods varies from two to more than 20 on a single inflorescence and up to 400 on a single plant.

VARIETIES

Varieties differ in their seed size, colour of pod and seed and presence or absence of pubescence on the pods. In Asia, people prefer a variety producing bright green large pods with grey or light brown pubescence and light-coloured three beans per pod. The varieties grown widely in the United States of America and other western countries include Besweet 2020, Butterbean, Envy and a USDA variety Moon Cake. Japanese classify the soybeans as fall or summer types. Fall-type varieties are sensitive to day length and summer types are sensitive to temperature. Fall-type varieties, *viz.*, Kinshu, Tsurunoko and Yuzuru, are sown in early spring and they take 105 or more days to complete their growing period, while the summer-type varieties, *i.e.*, Okuhara, Sapporo-midori, Osodefuri and Shiroge, Fukura, Mikawashima and Yukimusume, are sown in spring and they are harvested immature 75–100 days after sowing.

CLIMATIC REQUIREMENT

Vegetable soybean needs temperature range of 15°C–32°C for seed germination but the plant grows rapidly at higher temperature. Temperature below 10°C leads to poor plant growth and above 40°C is harmful for flowering, seed formation, or even to the seed quality. The crop for its successful cultivation requires 60–65 cm annual rainfall. Drought at flowering or just before flowering results in dropping of flowers and pods. Rains during maturity impair the seed quality.

SOIL REQUIREMENT

Vegetable soybean can be grown on most of the soil types ranging from light to black cotton soil. However, sandy loam soil with good amount of organic matter is considered the best for its cultivation. The pH of the soil should be between 6.0 and 7.0.

SOWING TIME

Under forced conditions, in carbon dioxide-enriched heated greenhouses, its seeds are sown in November, and in north India in open fields, it is sown twice once in last week of February to first week of March and second in the month of June–July on the onset rain. The seeds can be sown in protrays in early spring months for raising seedlings. These seedlings 25–30 days later are transplanted in small plastic tunnels in field for early production. In south India, it can be sown any time in the year.

SEED RATE

The seed rate of vegetable soybean depends on variety, seed viability, soil type and fertility, growing season and irrigation facility. On an average, a seed rate of 20–30 kg is enough to sow a hectare area.

SOWING METHOD

The seeds of vegetable soybean are usually sown directly in the field, but sometimes, the crop may be raised through transplanting the seedlings grown in protrays a bit early. In temperate countries, transplants are used in forced and early production systems. The spring crop is sown at a spacing of 30–45 × 2.5 cm and *Kharif* crop at a spacing of 45–60 × 5–10 cm. The sowing depth should be kept 2–3 cm since sowing at any greater depth will reduce emergence. The seeds take 1–2 weeks to germinate. The plant tops in seeded fields are nipped after the primary leaf stage to increase the number of branches and pods.

Vegetable soybean being a legume crop fixes atmospheric nitrogen into the soil. If the field has never been used for raising soybean crop, inoculating the seed with *Bradyrhizobium japonicum*, a beneficial bacterium, will ensure the formation of nitrogen-fixing nodules on the plant roots.

Vegetable soybean is capable of fixing atmospheric nitrogen into the soil. When it is sown first time in the new field where no legume crop has been grown in the previous seasons, it is advisable to inoculate the seed with nitrogen fixing bacteria before sowing for quick nodulation and promoting nitrogen fixing capacity of the plants. For inoculation, the seeds are mixed with 10% jaggery solution and then the packet of *Bradyrhizobium japonicum* culture is spread over the seeds and mixed thoroughly. Thereafter, the seeds covered with wet cloth are retained in shade for 6–7 hours, which is essential for the completion of inoculation. The seeds should not be exposed to sunlight since direct sunrays are lethal to the inoculated bacteria. Later, the inoculated seeds are sown usually in the field.

NUTRITIONAL REQUIREMENT

Being a member of Leguminaceae family, it derives nitrogen from the atmosphere through symbiotic nitrogen fixation but the nitrogen fixed by this process cannot meet the plant need for its growth and development, hence, it is essential to supply nitrogen through external sources. Phosphate nutrition helps in profuse rooting, better nodulation and promoting nitrogen fixing capacity of the plants. Potassium deficiency has an adverse effect on physiological process in soybean plants, as it increases the nitrogen use efficiency of the plants. It has been found that nitrogen, phosphorus and potassium of 60–80, 60–80 and 40–60 kg/ha, respectively will be appropriate doses for soybean crop. Applying fertilizers at the time of sowing in a band about 5 cm to the side of the seed and 5 cm below the seed increases the fertilizers use efficiency of the plants.

IRRIGATION REQUIREMENT

In a *Kharif* crop, irrigation is not needed if the rains are well distributed but applying irrigation during dry spells is essential. The crop essentially requires irrigation at critical stages, *i.e.*, seed germination, flowering, pod initiation and pod filling stages. However, the crop during summer can be grown only under assured irrigation conditions. If the soil at the time of sowing is too dry, pre-sowing irrigation is essential since irrigating the field after sowing can cause a hard crust to form on soil surface, which prevents the emergence of seedlings.

INTERCULTURAL OPERATIONS

HOEING AND WEEDING

Controlling weeds in early stages of crop growth is necessary to prevent competition between young vegetable soybean plants and weeds for nutrients, moisture, space and light. At least two hoeings are usually compulsory to keep the weeds under control. Mechanical cultivation between rows and hand cultivation between plants is an efficient weed control method. The need for weed control is reduced with the progress of season and the closing of plant canopy, which smothers the weeds thereafter. An herbicide mixture of alachlor 1.5–2 kg a.i./ha along with paraquat 0.75 kg a.i./ha is applied to control the weeds.

STAKING

The plants of pole type vegetable soybean are supported either by individual bamboo stick for each plant or even by using the trellis. Straight rough-surfaced bamboo sticks of 2 to 2.5 m height are used to stake the plants since it is one of the easiest ways, and hence, the name is pole types. A trellis is another popular way to stake the plants of pole type beans, in which, moveable fences of 1.5 to 1.8 m height are used for staking bean plants. The seeds of pole-type vegetable soybeans are sown at the base of trellis at about 7.5 cm spacing.

HARVESTING

The vegetable soybean pods are usually harvested 100 to 120 days after sowing at an interval of 3–4 days since the pods approach maturity very soon. The right stage of harvesting is when the pods are still bright green, tender and tight with fully developed immature green seeds, usually at 85% pod fill. If the pods turn yellow, the right stage for the harvesting of pods has passed away, since at this stage the beans become starchy, losing their sweet and nutty flavour. The appearance of pod is extremely important in the Asian market since the bright green pods of about 6.5 cm length covered with white or very light brown fine hairs, containing two or more bright green seeds with light-colour hilum fetch the highest price in the market. Vegetable soybean is marketed in three main ways, *i.e.*, (i) whole plants, (ii) pods only and (iii) beans only, but sometimes, it is also marketed as bunches of pods, called *hands*.

- **Whole plants:** The whole plant is harvested by cutting at about 5 cm above the ground and bunching stalks together in groups of 4–6 plants. After removing top leaves and small damaged pods, the whole plants along with pods are packed in 10 kg wooden boxes or cartons. Japanese consumers believe that this method preserves the quality of pods well, and *per se*, the pods fetch a better price in the market.

- **Pods only:** The marketable pods are picked from the stalks, packed in plastic net bags and sold in the market. However, in this method, speedy harvesting and packaging is crucial to maintain freshness of the pods.

- **Beans only:** The seeds (beans) are shelled from the green pods and marketed as fresh or frozen beans.

YIELD

The late maturing varieties generally are tall statured with more number of nodes, pods and fresh green pod yield at harvest stage than the early maturing varieties. On an average, vegetable soybean gives a fresh green bean yield of about 29.2 quintals per hectare.

POST-HARVEST MANAGEMENT

Air cooling, vacuum cooling and hydro cooling are the effective methods to remove field heat from vegetable soybean pods. After precooling, the vegetable soybeans can usually be stored at $0°$ temperature and 95% relative humidity to maintain green pod colour, flavour and firmness. Under such conditions, the vegetable soybean pods can retain their flavour and appearance for up to 2 weeks.

INSECT-PESTS

Gram Pod Borer (*Helicoverpa armigera*)

The young larvae feed on young leaves so voraciously that they skeletonize the leaves, and later, they feed on flowers and pods, which results in defoliation. Larva feeds on the pod with the head alone thrust inside and the rest of the body hanging out. Presence of holes on pods, absence of seeds in pods and defoliation in early stages are the symptoms of its attack.

Control
- Plough the field deep in hot summer months.
- Install pheromone traps at a distance of 50 m @ 5 traps/ha.
- Erect bird perches @ 50 per hectare.
- Clip terminal shoots on 100 days of crop growth.
- Set light traps (1 light trap/2 ha) to kill moth population.
- Dust the crop with chlorpyrifos 1.5% DP or quinalphos 1.5% @ 25–30 kg/ha.
- spray or
- Spray the crop with Fenvalerate 0.4%, chlorpyrifos 1.5% DP 1200 ml/ha or quinalphos 25 EC 1.0 litre/ha.

Stem Fly (*Melanagromyza sojae*)

The adult fly lays eggs on leaves. After hatching, its yellowish maggots bore the nearest leaf vein and then reach the stem through petiole and bore down the stem. If the infected stem is opened by splitting, distinct zigzag reddish tunnel can be seen with maggot or pupae inside it. The maggots feed on cortical layers of the stem and may extend to taproot, killing of the plant.

Control

- Follow long crop rotation with non-host crops.
- Plough the field deep in hot summer months.
- Avoid pre-monsoon sowing.
- Use optimum seed rate and plant spacing.
- Remove and destroy the damaged plant parts.
- Apply phorate 10G @ 10 kg/ha or carbofuran 3G @ 30 kg/ha in soil at the time of sowing.
- Spray the crop once or twice with 0.03% dimethoate 30 EC or 0.05% quinalphos 25 EC.

Soybean Aphid (*Aphis* spp.)

The nymphs and adults of aphids suck cell sap from the tender leaves, twigs and pods, and they emit honeydew like sugary substance, which encourages the growth of sooty mould. As a result, the photosynthesis is stopped, which causes reduction in yield. Its infestation transmits virus, which causes puckering and yellowing of leaves. The infested plant becomes stunted with reduced pod and seed counts.

Control

- Remove and destroy the infested portion of the plant.
- Wash the aphids with strong jet pressure pump.
- Dust the plants with cow dung ash.
- Natural enemies like the lady beetles, pirate bug and parasitic wasps can be employed in controlling aphid damage on vegetable soybean.
- The antagonistic fungal agent *Beauvaria bassiana* (Botanigard) can be used to parasitize on aphids.
- Spray the crop with 0.05% quinalphos 25 EC, oxydemeton methyl 25 EC, or dimethoate 30 EC 0.2% at the age of 35–40 days and repeat after 15 days if needed.

Whitefly (*Bemisia tabaci*)

Both nymphs and adults damage the plants by sucking juice from new tender growth, causing the plant growth stunted, leaf curling, yellowing and reduced yield. Plants become weak and susceptible to diseases. They also secrete a honeydew-like sugary substance, which encourages the growth of black sooty mould, *i.e.*, *Capnodium* spp., which further reduces the photosynthetic ability of the plant and affects the pod yield adversely.

Control

- Yellow sticky traps are effective in monitoring the adult white fly population.
- Spray the crop with clay suspension as asphyxiants where the crop is in small area and incidence of sucking insects is less.
- Spray imidachloprid 0.03% and thiamethoxam 0.05–0.07%, alternatively.

DISEASES

Vegetable soybean and grain soybeans share the same range of disease. However, since vegetable soybean is harvested at green stage, growers can avoid many of the late season problems that occur with grain soybeans. Some of the important diseases of vegetable soybean are as follows:

Alternaria Leaf Spot (*Alternaria tenuissima*)

Brown necrotic spots with concentric rings appear on foliage, which coalesce and form large necrotic areas. These infected leaves later in the season dry out and drop prematurely. Seeds of affected plants become small, shrivelled and dark-coloured. Irregularly spreading sunken areas occur on the seed.

Control
- Use disease-free healthy seed.
- Destroy the crop residues after harvesting the crop.
- Treat the seed with thiram and carbendazium (2:1) at 3 g/kg of seed.
- Spray the crop with mancozeb or copper fungicide at 0.25%, or carbendazim 0.1%.

Anthracnose or Pod Blight (*Colletotrichum truncatum*)

The disease symptoms on cotyledons appear as dark brown sunken cankers, and thereafter, irregular brown lesions appear on leaves, stems and pods. In advanced stages, the infected tissues are covered with black fruiting bodies of fungus. Under high humidity, the symptoms appear on leaves are veinal necrosis, leaf rolling, cankers on petioles and premature defoliation. Infected seeds become shrivelled, mouldy and brown.

Control
- Rotate the crop with cereal crops.
- Use disease free healthy seed.
- Completely remove plant residue from the field soon after harvest.
- Destroy the last year's infected stubbles.
- Treat the seed with thiram, captan, or Bavistin 3 g/kg.
- Spray the crop with 0.25% Dithane M 45 or 0.1% carbendazim.

Ashy or Stem Blight or Dry Root Rot (*Macrophomina phaseolina*)

The disease occurs when the plants are grown under moisture stress, nutrient deficiencies, nematode attack, or soil compaction. It is a most common basal stem and root disease of the vegetable soybean plant. The lower leaves become chlorotic followed by wilting and drying. The diseased tissues generally develop greyish discolouration. The sclerotia look like black powdery mass, hence, the disease is known as charcoal rot. Blacking and cracking of roots are the most common symptoms. Temperature ranging from 25°C–35°C favours the occurrence of the disease.

Control
- Rotate the vegetable soybean with cereal crops.
- Plough the deep in hot summer months.
- Destroy the last year's infected stubbles.
- Treat the seed with *T. viride* @ 4 g, *P. fluorescens* @ 10 g, carbendazim, or thiram 2 g/kg of seed.
- Drenching the infested spots with carbendazim 0.1%, or *P. fluorescens/T. viride* 2.5 kg/ha.

COLLAR ROT / SCLEROTIAL BLIGHT (*Sclerotium rolfsii*)

Infection usually occurs at or just below the soil surface. It is characterized by sudden yellowing or wilting of plants. Light brown lesions, which quickly darken, enlarge until the hypocotyl or stem is girdled. Leaves turn brown, dry and often cling to the dead stem. Numerous tan to brown spherical sclerotia form on infected plant material.

Control

- Plough the field deep in hot summer months.
- Strictly follow long crop rotation with maize or sorghum.
- Destroy the infected stubbles after harvesting the crop.
- Treat the seed with *T. viride* 4 g or *P. fluorescens* 10 g/kg of seed.
- Drench the infested spots with Captan, Quitozene, Thiram, or carbendazin at 2.5 kg/ha.

MUNG BEAN YELLOW MOSAIC VIRUS (MYMV)

The disease is characterized by conspicuous systemic bright yellow mottling of leaves. The yellow area are scattered or occur in indefinite bands along the major veins and rusty necrotic spots appear on yellow areas as the leaves mature.

Control

- Rogue out and burn the infected plants as early as possible.
- Spray the crop with monocrotophos or methyl demeton 25 EC 500 ml/ha twice at 15 and 30 days after sowing.

■■■

85 Velvet Bean

M.K. Rana and Neha Yadav

Botanical Name	: *Mucuna pruriens* (L.) DC.
Family	: Leguminosae (Fabaceae)
Chromosome Number	: 2n = 22

ORIGIN AND DISTRIBUTION

Velvet bean, a native to southern Asia, has been introduced into the Western Hemisphere *via* Mauritius. It is an underutilised wild legume, which is now widespread in tropical and subtropical regions. It is cultivated in Asia, America, Africa and the Pacific Islands, where its pods are used as a vegetable for human consumption and leaves as animal fodder. During the eighteenth and nineteenth centuries, it was grown widely as a green vegetable in foot and lower hills of the eastern Himalayas and in Mauritius. It was introduced to southern states of America in the late nineteenth century as an ornamental species, and from there, it was reintroduced to the tropics in early part of the twentieth century.

INTRODUCTION

Velvet bean, commonly known as Bengal velvet bean, Florida velvet bean, Mauritius velvet bean, Yokohama velvet bean, cowage, cowitch, lacuna bean, Lyon bean, donkey eye, monkey tamarind and buffalo bean, is a widely naturalized and cultivated species, belonging to the family Fabaceae, which includes approximately 150 species of annual and perennial legumes. It is one of the most popular fast-growing green crops known in tropics because of its great potential as both food and feed. It is a climbing annual or sometimes short-lived perennial legume, which can establish very quickly without requiring complete soil preparation. It has the potential to fix atmospheric nitrogen into the soil *via* symbiotic relationship with soil *Rhizobium* bacteria. Its heavy roots and vine growth make it an excellent green manure and cover crop that greatly improves soil fertility by adding 10 tonnes organic matter and nitrogen 331 kg/ha. It is a good source of dietary protein due to its high protein content in addition to its digestibility, which is comparable to other pulses such as soybean, rice bean and lima bean. It can be used to reclaim weed infested and degraded soils. Its plant is notorious, as its young foliage, pods and seeds when come in contact cause itching, which is known as mucunain. Scratching of the exposed area should be avoided, as it will transfer the chemical responsible for itching to all other touched areas. When the subject starts scratching tends to scratch vigorously and uncontrollably, which is why, the local people in northern Mozambique refer it as a *mad bean*. The seedpods are also known as *Devil Beans* in Nigeria.

COMPOSITION AND USES

COMPOSITION

Velvet bean is an exceptional plant, as it is a good source of crude protein, essential fatty acids, starch and certain essential amino acids. Its silage contains 11–23% crude protein and 35–40%

crude fibre, and the dried beans contain 20–35% crude protein. Raw velvet bean seeds contain approximately 27% protein and are rich in minerals. The nutritional composition of velvet bean is given in Table 85.1.

It also contains various anti-nutritional factors such as protease inhibitors, total phenolics and oligosaccharides such as raffinose, stachyose and verbascose. The main phenolic compound is *L*-dopa (5%), which is toxic, and it contains hallucinogenic tryptamines too.

Uses

Velvet bean is used as food, feed (forage and seeds) and environmental services, thus, it has a great value in agricultural and horticultural use. Certain ethnic groups in many countries used it traditionally as a food source. The young leaves, pods and seeds are edible and used in several food specialties including *tempeh*, a fermented paste made of boiled seeds in Indonesia. Cooked fresh shoots can also be eaten as vegetable. However, its leaves and seeds contain toxic substances that must be removed before consumption through series of soaking and boiling. They must be dipped for at least 30 minutes to 48 hours before cooking and the water should be changed several times during dipping so that *L*-Dopa (the toxic substance) may leach out, making the leaves and seeds suitable for use. Eating unprocessed velvet bean in large quantity can be toxic to non-ruminant mammals including humans. In Central America, Guatemala and Mexico, it has been used after roasting and grinding as a coffee substitute for at least several decades and its seeds are widely known in these regions as *Nescafe*. In south India, the ethnic group, the Kanikkars, conventionally consumes its dry seeds after repeated boiling to remove anti-nutritional factors.

Sometimes, its dry leaves are used for smoking. Its foliage can be used as pasture, hay or silage only for ruminants, whereas pods and seeds after grinding into a meal are fed to ruminants and monogastrics.

Growing velvet bean reduces soil erosion and controls nematodes. The plant exhibits low susceptibility to insect-pests due to the presence of toxic compounds. It can also be used as a biological control for problematic weeds, as the plant has allelopathic effect, which may suppress competing plants. The plant also has a positive effect on soil moisture.

Medicinal Uses

Velvet bean is a popular medicinal plant, which has long been used in traditional *Ayurvedic* medicines for the treatment of many diseases including Parkinson's disease. All parts of this plant possess therapeutic properties. Its constituents have analgesic, aphrodisiac, anti-inflammatory, neuro-protective, anti-epileptic, anti-neoplastic and anti-helminthic activity. It contains some cyclitols with anti-diabetic effects. Its seeds also contain a kind of substance, which counteracts the

TABLE 85.1
Nutritional Composition of Velvet Bean (per 100 g Edible Portion)

Constituents	Contents	Constituents	Contents
Water (g)	75	Iron (mg)	15.9
Carbohydrates (g)	5.2	Copper (mg)	20
Protein (g)	15	Magnesium (mg)	2.8
Fat (g)	1	Manganese	70
Dietary fibre (g)	27	Zinc (mg)	71
Calcium (mg)	10	Sodium (mg)	0.1
Potassium (mg)	1.9	Energy (kcal)	1100

harming effects of snake venom. The methanolic extracts of its leaves have anti-microbial and anti-oxidant properties due to the presence of some bioactive compounds such as phenols, polyphenols and tannins, and also have keratinocytes, which help in the treatment of skin diseases. It is used to treat nervous disorders and arthritis. Its paste if applied on scorpion bite absorbs the poison. Its seeds have been used in *Unani* medicines for the treatment of many dysfunctions.

BOTANY

Velvet bean, a widely naturalized and cultivated summer legume, is a climbing shrub, with a long vine that grows to a length of 6–18 m. In young stage, the plant is almost covered with fuzzy hairs. However, it is almost completely free from hairs in older stage. It has a taproot with numerous 7- to 10-m-long lateral roots. The roots are fleshy, usually well nodulated and produced near soil surface. The stems are slender and slightly pubescent. The leaves are slightly pubescent, alternate and trifoliolate with rhomboid ovate 5- to 15-cm-long and 3- to 12-cm-broad leaflets. The sides of the leaves are often heavily grooved and the tips are pointed. The inflorescence is a drooping axillary raceme that bears many white to dark purple flowers. Sepals are longer or of the same length as the shuttles. Corollas are butterfly shaped. It bears pods in clusters of 10–14, which are 10–13 cm long and 1–2 cm wide. They are stout and curved, containing 2–6 seeds, covered with greyish-white or orange hairs, which may cause irritation to the skin. Seeds are oblong ellipsoid, 1.2–1.5 cm long, 1 cm broad and 0.5 cm thick with 100 seeds dry weight 55–85 g. The seed coat is hard and thick with colour from glossy black to white or brownish and black mottling.

CLIMATIC REQUIREMENT

Velvet bean being a warm season crop requires warm and humid climate coupled with high light intensity and annual rainfall of 650–2500 mm. It can be grown up to an altitude of 2100 m above mean sea level. A temperature of 20–30°C is optimum throughout its growing period and night temperature of about 21°C has been reported to stimulate flowering. Low night temperature of about 10°C ceases growth of the plants, as they are dependent on *Rhizobium* for nitrogen fixation, which increases with increasing night temperature and is poor below 18°C. The plants are susceptible to frost and require a frost-free period of 180–240 days. A short exposure of 24–36 hours to temperature below 5°C kills the plants. It is a minor leguminous crop, of which, many cultivars are suitable for wet regions of the tropics, while others, *e.g.*, the Mauritius velvet bean, is suitable for dry land farming.

SOIL REQUIREMENT

Velvet bean can be grown on wide range of soils from sandy to clay but it thrives best on well-drained light-textured sandy loam and slightly clay soils with a pH of 5–6.5. It can tolerate drought but cannot survive in wet or waterlogged soils. Most of its species exhibit reasonable tolerance to a number of abiotic stresses, including drought, low soil fertility and high soil acidity.

CULTIVATED VARIETIES

Certain cultivars, which are tolerant to temperate conditions, have been developed through breeding. The main difference in cultivated and wild species is in the characteristics of pubescence on pod, seed colour and days taken to pod harvesting. Only two species are cultivated type, *i.e.*, (a) the true velvet bean—*Mucuna pruriens* var. utilis, which has medium-sized seeds, and (b) the horse bean—*Mucuna sloatiei*, which has larger seeds with an extremely hard seed coat. However, there are some botanical varieties of *M. pruriens*, which were formerly treated as separate species. The important ones are given as under:

- *Mucuna pruriens* var. hirsuta
- *Mucuna pruriens* var. pruriens
- *Mucuna pruriens* var. sericophylla
- *Mucuna pruriens* var. utilis: the true velvet bean is non-stinging variety
- *Stizolobium deeringianum* or *Muncuna deeringiana*: Deering, Florida or Georgia velvet bean is of some importance as a cattle fodder, particularly in the United States and parts of South America.
- *Mucuna utilis* or *Stizolobium utile*: Bengal velvet bean is grown in India.
- *Mucuna aterrima* or *Stizolobium aterrimum*: Mauritius velvet bean is grown in Mauritius, Australia, Brazil and west India often as a rotating crop with sugarcane or as a drought resistant cover crop.
- *Stizolobium hassjoo*: Yokohama velvet bean is an early maturing type and a less vigorously growing than the other velvet bean. It is thought to have originated in Japan, where it is known under the names *Osharuka-mame* and *Hasshomame*.
- *Mucuna nivea*, *Stizolobium niveum*, *Mucuna cochinchinensis* or *Carpogon niveum*: Lyon bean is sometimes cultivated as a vegetable for its immature pods in the Philippines and Southeast Asia.

SOWING TIME

The sowing time depends on the growing region. In north plain areas, sowing is done in February to March for summer crop and June to July for rainy season crop. In hills, sowing is done in April to July.

SEED RATE

The seed rate varies with variety, soil fertility, growing region and season, irrigation facilities and viability of the seed. For sowing a hectare area, about 22 kg seed is enough in summer season and 12–15 kg seed in rainy season. The test weight of velvet bean seed is 1 kg.

SOWING METHOD

Velvet bean having taproot system is a direct-seeded crop, thus, requires well-prepared seedbeds free from weed seeds. It can also be sown in roughly prepared field provided the seed is properly covered with soil or sowing is followed by rain. However, better results can be obtained if the seeds are sown in well-prepared moist seedbeds. Seeds are sown either by broadcasting, dibbling or in furrows at a spacing of 1 × 1 m with two seeds per hill. Drills can also be used for the sowing of seeds. One or two inter-row cultivations improve early development of crop.

NUTRITIONAL REQUIREMENT

Velvet bean being a leguminous crop does not require much nitrogen but responds well to phosphorus. Farmyard manure about 15–20 t/ha should be incorporated into soil at the time of land preparation. Nitrogen 20–25 kg along with phosphorus and potassium each 50–70 kg/ha is applied as basal dose at the time of sowing.

IRRIGATION REQUIREMENT

The rainy season crop does not require any irrigation except dry period. The summer crop is irrigated once in 7–10 days. The crop is sensitive to waterlogging conditions, thus, water stagnation in velvet bean should be avoided.

INTERCULTURAL OPERATIONS

Light hoeing and weeding are necessary to check the growth of weeds during early stages of crop growth because weeds compete with crop plants for nutrients, moisture, space and light. Initially, velvet bean is slow to start growth, but after establishment, the crop smothers weeds effectively. Being legume crop, it needs proper soil aeration for the normal activities of *Rhizobium* bacteria present in root nodules, which can be improved by loosening the soil around the root system, thus, hoeing in legume crops is an essential operation. Total —two to three weeding are enough to keep the weeds down in entire growing season.

HARVESTING

Velvet bean becomes ready for harvesting as soon as the pods start turning their colour from green to dark brown. In tropical regions, many of its cultivars grown for seed purpose become ready for harvesting in 180–270 days after sowing, and for forage, it is harvested 90–120 days after sowing when the pods are still young. The Yokohama variety generally matures within 110–120 days; however, harvesting 120 days after planting results in high biomass yield and nutritive value. The crop is liable to shatter, hence, only the mature pods are often handpicked and dried for several days before threshing, which can be done by hand or machine.

YIELD

The crop gives good yield even under dry land and low soil fertility conditions, which do not allow the profitable cultivation of most other food legumes. On an average, velvet bean yields green material 20–35 t/ha and seed 2.5–33 q/ha, depending on variety, climatic conditions, soil fertility and cultivation practices adopted by the farmers.

INSECT-PESTS

APHID (*Aphis craccivora* and *Myzus persicae*)

This is a serious pest for beans. Its nymphs and adults suck cell sap from the growing tips and later from the whole plant. It causes curling of leaves and reduces plant vigour, yield and quality. The plant growth becomes stunted due to its attack. Under heavy infestation, the leaves become yellow and distorted with necrotic spots on them. They secrete a sticky sugary substance called honeydew, which encourages the growth of sooty mould, reducing photosynthetic activity of the plant. It also transmits mosaic virus.

Control

- Remove and destroy the infested plant parts.
- Spray water with strong jet pressure to wash the aphids from the plant.
- Apply carbofuran at 10 kg a.i./ha in soil at the time of seed sowing.
- Spray the crop with Malathion 0.1%, Monocrotophos, Dimethoate or methyl demeton 0.05% at 20 and 35 days after sowing.

LEAF MINER (*Phytomyza atricornis*)

The larvae of leaf miner cause serious damage to plants in late season by making tunnels in the leaves and feeding on them. The infested leaves wither and dry up. Flowering and fruiting are reduced much.

Control

- Collect and destroy the infected plant parts.
- Apply carbofuran 3G at 0.5% kg a.i./ha in soil at the time of seed sowing.
- Spray the crop with dimethoate 0.2%, phosphamidan 0.1%, or monocrotohos 0.15% at fortnightly interval.

STEM FLY (*Ophiomyia phaseoli*)

This is a polyphagous pest attacking all the beans. The larvae mine into the leaves, petioles and stems, which cause girdling effect and death of the plant. They also feed on outer layer of stem at collar region of the young plants, causing withering and ultimate drying of the affected shoots, thus reducing the bearing capacity of the host plants. The adults puncture the leaf, causing injury to it and the injured parts turn yellow. The affected leaves become yellow and dry up. The damage is more severe on seedlings than on the grown up plants.

Control

- Collect and destroy the affected plant parts.
- Apply carbofuran 25 kg/ha in soil at the time of sowing.
- Spray the crop with 5% *neem* seed kernel extract to reduce the incidence of stem fly.
- Spray the crop three times with 0.1% oxydemeton methyl 25 EC or dimethoate.

CATERPILLAR (*Anticarsia gemmatalis*)

The caterpillar is the most damaging foliage-feeding pest that occurs in the late summer months and can cause great damage to velvet bean and other legume crops if not managed. Newly hatched larvae strip the leaf, beginning with the lower epidermis and mesophyll. They continue feeding until the end of the second instar, eating all the soft leaf material and leaving only the veins intact. Eventually it consumes the entire leaf. Once the upper leaves and lower leaves have been consumed, foliage in the middle and lower canopy is consumed and complete defoliation may result. It may also attack tender stems, buds and small bean pods.

Control

- Grow early maturing varieties and sow the seeds bit early.
- Grow few lines of trap crop around the crops.
- Collect and destroy all the infected plants.
- Apply helicoverpa nuclear polyhedrosis virus (H-NPV) and *Bacillus thuriengenesis* formulations as a biological control agent.

POD BORER (*Helicoverpa armigera*)

This is the most destructive polyphagous insect. Adult moth has a V-shaped speck on light brownish forewings and a dark border on hind wings. It lays spherical yellowish eggs singly on tender part of plants. Larva is greenish with dark grey lines laterally on the body and pupates in the soil. Its life cycle completes in a period of 28–40 days. Its young larvae feed voraciously on the developing buds and flowers, while mature larvae bore into the pods and feed the beans. When tender leaf buds are eaten, symmetrical holes or cuttings can be seen upon unfolding of leaflets. Sometimes, its damage causes 80% yield loss.

Control

- Plough the field deep in hot summer months.
- Follow crop rotation with sorghum, maize, pearl millet and sugarcane to minimize the infestation.

- Intercrop one row of red gram for every five or six rows.
- Collect and destroy all the infested pods.
- Install pheromone trap at 5 per hectare.
- Apply H-NPV 250 LE/ha or *Bacillus thuringiensis* formulations 1 kg/ha.
- Use *Trichogramma chilonis* 1 lakh/ha or *Chrysoperla carnea* 50,000/ha at 40 and 50 days after sowing.
- Spray the crop with 5% *neem* seed kernel extract, 0.4% Monocrotophos, 0.01% Fenvalerate, 0.05% Meracaptothion or 0.3% Thiodan at fortnightly interval.

DISEASES

POWDERY MILDEW (*Erysiphe polygonii*)

This is a common fungal disease of beans, which attacks almost every part of the plant. Initially, the infection appears on leaves as small-darkened areas, later becoming white powdery growth, which spreads top stems and other parts of the plant. When the infestation is severe, the leaves become chlorotic and fall off. Pods are reduced in size and distorted in shape. Plant growth is reduced due to defoliation.

Control

- Grow resistant varieties, if available.
- Spray the crop thrice with systemic fungicides such as Karathane, Bavistin, Benlate, Cosan, Calixin or wettable sulfur 0.2% at weekly interval as soon as disease appears.

ANTHRACNOSE (*Colletotrichum lindemuthianum*)

This is also a common fungal disease of beans, which attacks stems, leaves and pods. Black sunken lesions with pinkish spores develop on pods under favourable cool and wet conditions. Such lesions may also appear on cotyledons, stems and petioles. Diseased seeds and infected plant debris carry the fungus from one season to another.

Control

- Rotate the crop with non-host crops.
- Collect and destroy the infected plant parts.
- Use disease-free healthy seeds.
- Soak the seed in 0.15% solution of Captan or Thiram for 30 minutes.
- Spray the crop with 0.1% Bavistin or Benlate at 10 days interval.

ASHY STEM BLIGHT (*Macrophomina phaseoli*)

In early stages, it causes damping-off disease and death of the seedlings. Black sunken lesions appear on cotyledons and stem of young seedlings. In later stages, collar portion of the plant become brown and vascular bundles of the roots turn brown, resulting in dry root rot. Yellowing and drooping of foliage takes place. It is a soil-borne disease. This disease is severe under warm and humid conditions.

Control

- Destroy all infested plant debris left from previous year.
- Use disease-free healthy seeds.
- Apply organic matter in the field.
- Avoid excessive application of irrigation.
- Follow phyto-sanitary measures.
- Treat the seed with Thiram or Captan at 2–3 g/kg.

■■■

86 Winged Bean

G.L. Sharma and K.L. Patel

Botanical Name	: *Psophocarpus tetragonolobus* (L.) DC
Family	: Leguminosae (Fabaceae)
Chromosome Number	: 2n = 2x = 26

ORIGIN AND DISTRIBUTION

Winged bean has been grown since ancient time in Papua New Guinea and South East Asia where it was generally considered as *poor man's crop*. Papua New Guinea is considered as the native place of winged bean. However, Thailand, Malaysia, Philippines, Sri Lanka, Vietnam, Southern Myanmar and Southern India (Tamil Nadu and Kerala) are assumed to be the secondary centre of origin due to the availability of natural variability. Winged bean is widely cultivated in the tropics, especially in Myanmar, India, Malaysia, Indonesia, Thailand, Bangladesh, Ghana, Nigeria, West Africa, West Indies and South Florida. In India, its cultivation is spread to some extent in Kerala, Tamil Nadu, Karnataka, Maharashtra, Goa, Orissa, West Bengal, Mizoram, Manipur and Tripura.

INTRODUCTION

Winged bean, also known by various vernacular names like asparagus pea, four-angled bean, four-cornered bean, Goa bean, Manila bean, Mauritius bean and winged pea, is a tropical legume plant. This vegetable plant attracts curiosity because of two main reasons, *i.e.*, firstly; its every part is used for one or other types of food preparation and second peculiarity is very high content of fat and protein, which have created much enthusiasm among the researchers in mid-eighties for its potential utilization. Therefore, it was suitably designated by many fancy names such as *Supermarket on a stalk*, *God-sent vegetable*, *Soya's rival*, etc. by various workers.

COMPOSITION AND USES

COMPOSITION

Based on chemical composition of dry seeds, pods, stalks and leaves of winged beans, it has been concluded that its seeds have a high range of protein (27.8–36.6%) and fat (14.8–17.9%), which are found similar to soybean. The seeds also contain high phosphorus, calcium and magnesium. Among all the parts studied except seeds, its leaf is said to be highest in protein content (33.7%). The protein and fat content of pods decreased as the pods ripen. The calcium content in leaf is much higher than in the other parts. The contents of lysine, aspartic acid, glutamic acid and leucine are large, while the sulfur-amino acid content is small. The nutritive value of winged bean pods and tuberous roots is given in Table 86.1 and Table 86.2.

TABLE 86.1

Nutritional Composition of Winged Bean Pods (per 100 g Edible Portion)

Constituents	Contents	Constituents	Contents
Water (g)	76.93	Sodium (mg)	3.0
Carbohydrates (g)	7.9	Iron (mg)	0.2–12.0
Crude protein (g)	1.9–3.0	Vitamin A (IU)	340–595
Fat (g)	0.1–3.4	Vitamin B (mg)	0.06–0.24
Dietary fibre(g)	0.9–2.6	Niacin (mg)	5–12
Calcium (mg)	53–236	Vitamin C (mg)	21–37
Phosphorus (mg)	26.60	Ash (g)	0.4–1.9
Potassium (mg)	205		

TABLE 86.2

Nutritional Composition of Winged Bean Roots (per 100 g Edible Portion)

Constituents	Contents	Constituents	Contents
Water (g)	56.5	Phosphorus (mg)	30
Carbohydrates (g)	30.5	Iron (mg)	0.5
Crude protein (g)	10.9	Manganese (mg)	10
Fat (g)	0.4	Zinc (mg)	1.3
Dietary fibre (g)	1.6	Ash (g)	1.7
Calcium (mg)	25	Energy (kcal)	150

Uses

The peculiarity of winged bean is that apart from stalk virtually every part of the plant, *i.e.,* underground roots, leaves, flowers, pods, and seeds, is usable. The beans are cooked as vegetable but the other parts are also edible.

The tender apical leaves can be consumed raw like spinach or cooked as leafy vegetables. The flowers can be utilised to prepare a mushroom-like dish. The light blue flowers are also used as food colouring agents (rice and pastry colorant). The tender and immature pods are the most widely eaten parts of the plant. The immature pods can be eaten like green beans. They can be eaten either raw in salads or in soups, stews and curries. The immature seeds can be utilised like green peas and the mature dry seeds are used just like soybean, as sprouts or can be made into curd known as tofu. The roasted seeds have also been reported as coffee substitute. The roots are modified in the form of tubers that can be eaten like potatoes. The root tubers can be used after baking, frying or boiling and peeling. Besides, they can also be steamed, roasted or made into chips. A winged bean *milk* and *flour* are used as dietary treatments for protein-deprived children.

Medicinal Uses

Many medicinal properties are attributed to various plants parts and in different forms of applications under indigenous system of medicine such as leaf infusion for the treatment of eye and ear infections, dyspepsia, leaf paste for boils, seeds for venereal diseases, tubers as poultice for vertigo, etc.

BOTANY

Winged bean is a twining vine plant with climbing stems and attains a height of 3–4 m under a condition of proper support. The leaves are pinnate or palmate to trifoliate. The leaflets are more or less triangular, tapering to an acute point having length ranging from 7.6 to 15 cm. Flower colour varies according to varieties, ranging from white and pale sky blue to reddish brown. The floral inflorescence is axillary raceme, which is generally up to 15 cm long and bears 3 to12 flowers. The individual flower is about 2.5 cm long. The bracteoles are ovate. Calyx is about 1.5 cm long and persistent, whereas, corollas are white to pale blue in colour, which are large and exerted in shape, standard petals are reflexed and keel petals are incurved. Stamens are monodelphous and anthers are uniform. Ovules are numerous and style is long, incurved and bearded along its length, whereas, stigma is globose.

The pods are very distinctive. The flowers have four leafy wings with frilly edges running lengthwise on the pods. The pod is square with four corners tapering out and forming thin wings to decipher it the name of *winged bean*. The pods are generally pale green in colour having length of 15–22 cm and width of about 2.5 cm at the time of maturity. The fully matured pods turn brown in colour and split open generally with a loud popping noise. This sound signifies the name of genus *Psophocarpus* as *Psophocarpus* is Greek word for *noisy fruit*. The seeds of winged bean are round in shape and resemble the seeds of soybeans. Some cultivars produce large tuberous roots from which they can sprout under suitable conditions even though the stem is dried. This resprouting helps to perpetuate the winged bean plants. Winged bean roots contain numerous and exceptionally large nodules and one plant may produce up to 440 nodules, which may weigh up to 600 mg. These nodules are pale brown in colour, smooth surfaced, semi-globose and oblate in shape and can fix nitrogen up to 120–250 kg/ha. The nodulation starts two weeks after emergence of the seedlings, which is affected by environmental factors. The superior type of nodulation in winged bean is attributed to the high amount of protein content in plants.

CLIMATIC REQUIREMENT

Being tropical plant, winged bean grows best in warm and humid climate and requires short day light duration for flowering. This can be grown up to an altitude of 2400 m above mean sea level depending on its adaptations. However, its usual cultivation is confined to 10°S to 20°N latitudes. Winged bean plant does not produce flowers under long day conditions and starts flowering when a shorter day length arrives. However, recently, the day-neutral genotypes have also been developed. Therefore, it is very significant to take into consideration the photoperiod requirements for successful cultivation of winged bean. The optimum temperature for proper growth ranges from 25°C to 30°C. Continuous temperature more than 32°C or below 18°C inhibits flowering even under otherwise suitable short-day conditions. Hence, temperature also plays important role in addition to day length for the induction of flowering. The optimum temperature for tuber formation is 13°C to 24°C, whereas, high temperature promotes vegetative growth at the cost of tuber formation.

SOIL REQUIREMENT

Winged bean can be grown on a wide range of soil types, ranging from light (sandy), medium (loamy) and heavy (clay) soils. However, it prefers well-drained sandy-loam soils with fair moisture holding capacity. The winged bean has a symbiotic relationship with soil bacteria, which form nodules on the roots and fix atmospheric nitrogen. It does not grow well in very alkaline or very acidic soils though it prefers a slightly acidic soil with a pH limit as low as 4.3. However, the pH range of 5.2–5.5 is considered optimum for successful cultivation of winged bean. It requires a well-distributed annual rainfall of 1500 mm. It withstands neither prolonged drought nor the water logging conditions.

CULTIVATED VARIETIES

Numerous locally grown cultivars are available in different parts of winged bean growing regions of the world as elite genotypes and land races. These cultivars have wide genetic and morphological variations and grown commercially as per local demands. In India, Indian Institute of Horticultural Research, Bengaluru has identified three genotypes namely, IIHR-21, IIHR-60 and IIHR-71. The other Indian cultivars mentioned in the literature include WBC-2 (Meghalaya) and Revathy.

The Indian germplasm repository of winged bean is being maintained at National Bureau of Plant Genetic Resources (NBPGR), New Delhi and Kerala Agricultural University, Vellanikkara, Thrissur (Kerala). However, the Institute of Plant Breeding, University of Philippines, Los Banos is designated as International Gene Bank and International Institute of Tropical Agriculture (IITA), Nigeria as sub-centre for winged bean germplasm.

SEED RATE

Winged bean is generally propagated through seeds, when grown as annual crop on commercial scale. However, perennially, it is grown by allowing the storage tuberous roots to develop shoots naturally when optimum conditions come after the winter die off of the vines. The seed rate varies according to spacing, genotypes and purpose of the crop, whether it being cultivated for green beans, seeds or tubers. The winged bean has a test weight (1000 seeds weight) of 250 g. The crop grown for tuber purpose is generally planted at closer spacing. Hence, taking into accounts the abovementioned factors, the seed rate of winged bean varies from 10 to 35 kg/ha.

SOWING TIME

The time of sowing of winged bean generally depends on the rainfall pattern, incidence of proper photoperiod requirements and purpose of the crop cultivation. The sowing of winged bean in south India is done from the end of July to October. However, in Northeastern states of India, it is generally sown during the month of February to March.

SOWING METHOD

Winged bean seed has hard seed coat, which may decrease and delay the germination due to impermeability of seed coat to water. Soaking of seeds in water for 1–2 days will result in swelling of seeds by the absorption of water and some seeds may not swell at all due to the hard seed coat. Planting of only swelled seeds in moist soil is recommended for proper germination. The germination of hard seeds can be improved by developing scratches on their surfaces either with sand paper, hard coke or by breaking of hard coat with sharp knife. This scarification treatment will allow moisture to get into the seed so as to enable the germination process. This treatment may result in germination of up to 90%.

The spacing for the sowing of winged bean is generally recommended as about 1.0 m between rows and 25–70 cm between plants. The seeds should be sown up to a depth of 2–3 cm. The seeds can be sown in pits, flat beds, ridges or channels, depending of soil type and local cultivation practices.

NUTRITIONAL REQUIREMENT

The studies on nutrient management practices in winged bean are very few. This crop should be provided with nitrogen 40 kg, phosphorus 100 kg and potash 40 kg/ha. The half dose of nitrogen and full dose of phosphorus and potash should be given as basal application, and the remaining half dose of nitrogen is applied at the time of flowering as top-dressing. Moreover, in Kerala, the recommended dose of nitrogen, phosphorus and potash for winged bean cultivation is 50, 100 and 50 kg/ha, respectively, however, the rate of nutrients application should be based on soil test values, and primary, secondary and micronutrients must also be taken into account for a balanced nutrient management.

IRRIGATION REQUIREMENT

Winged bean can tolerate a short period of drought but it requires regular irrigations for better production. The winged bean thrives best in well-drained soil and water logging is harmful for its successful cultivation. The quantum and frequency of irrigation depend on soil type and stage of the crop growth. However, the winged bean crop should be irrigated at weekly interval. In case of proper water management with modern techniques, the drip system of irrigation can also be used for winged bean cultivation

INTERCULTURAL OPERATIONS

HOEING AND WEEDING

Since the growth of winged bean in early stages is slow, the field should be kept free from weeds to avoid crop-weed competition for nutrients, space, moisture and light. Hence, 2–3 hand weddings and hoeing are generally required to achieve proper crop stand. Mulching either organic or plastic can also be utilised to suppress weed growth and conserve soil moisture as well.

STAKING

The winged bean vines are of twining growth habit, hence, just after few day of germination, they essentially require proper support from the beginning itself. If the vines are allowed to creep on the ground, the yield is reduced to even less than half when compared to properly staked plants. The staking can be provided by thick bamboo poles or a Y-shaped trellis connected by wires to get higher yield.

HARVESTING

The mature pods may grow as long as 22 cm, although they taste better when picked at tender, small and immature stage, generally not more than 15 cm long. The pods of commercially grown winged beans are generally harvested when they are 10–15 cm long. If harvesting is delayed, the pods become over ripe, they become stringy and attain hardness, but on the other side, the well-developed immature seeds inside pods can be eaten like garden pea.

YIELD

The green (immature) pods yield on an average ranges from 10 to 15 t/ha. However, green pod yield as high as 35 t/ha can be harvested with optimum management practices. Its dried seed yield may be from 800 to 1500 kg/ha. When the crop is grown for tuberous roots, its yield ranges from 5.5 to 12 t/ha.

POST-HARVEST MANAGEMENT

The freshly harvested green pods are of perishable nature because of their high moisture content. Hence, they should be marketed within 24 hours after picking. The pods can be stored at 10°C temperature and 90% relative humidity. The leaves can be preserved for longer duration after proper drying. The storability of its dried seeds is better as compared to the seeds of other pulses due to less damage caused by storage grain pests because of their hard seed coat. Its tuberous roots are normally consumed soon after digging but they can be stored up to 2 months.

INSECT-PESTS AND DISEASES

The incidence of serious insect-pests and diseases on winged bean is still very low, which is not in the range of high economic loss. However, the incidence of some insect-pests and diseases observed by various workers has been described below:

INSECTS-PESTS

The winged bean is attacked by many types of insect-pest. However, the extent of damage varies according to many factors. The major insect-pests are bean pod borer (*Maruca testulalis*), bean fly (*Ophiomyia phaseoli*), black bean aphid (*Aphis craccivora*), pod borer (*Heliothis armigera*), long-tailed blue (*Lampides boeticus*), southern green stinkbug (*Nezara virudala*), etc. Besides, the incidence of two species of mite, *i.e.*, broad mite (*Polyphagotarsonemus latus*) and cyclamen mite (*Steneotarsonemus pallidus*), has also been reported.

DISEASES

The major diseases affecting winged bean are false rust, leaf spot, collar rot, viral diseases and root-knot nematode. Their descriptions are given below:

FALSE RUST (*Synchytrium psophocarpi*)

In rainy season, the incidence of this disease is comparatively high. Winged bean related species *Psophocarpus scandens* is immune against false rust disease.

LEAF SPOT (*Pseudocercospora phosophocarpi and P. scandens*)

The incidence of this disease has been observed in Indonesia, Papua New Guinea, Philippines and Malaysia. In India, winged bean is free from leaf spot disease.

LEAF ANTHRACNOSE (*Colletotrichum gloeosporioides*)

This disease is not found to cause appreciable yield loss in winged bean crop.

COLLAR ROT (*Macrophomina phaseolina, Fusarium semitectum, Fusarium equiseti, Fusarium moniliforme and Rhizoctonia solani*)

The incidence quantum of collar rot disease depends on vegetative growth of the crop, depth of seed sowing and type of the soil.

VIRAL DISEASES

Three viral diseases affecting winged bean plants, *i.e.*, Psophocarpus necrotic mosaic virus, Psophocarpus ring spot mosaic virus and Psophocarpus leaf-curl disease, are usually observed, which sometimes produce very heavy damage. The aphid (*A. craccivora*) is the major vector for the transmission of Psophocarpus ring spot mosaic virus.

ROOT-KNOT NEMATODE (*Meloidogyne incognita and Meloidogyne javanica*)

Both these nematode species severely damage the roots and tubers of winged bean. The root-knot nematode incidence adversely affects both pod and root tuber yield.

Control

- The various insect-pests and diseases of winged bean can be controlled by the methods adopted to manage similar insect-pests and diseases of other legume vegetable crops.

■■■

87 Yam Bean

M.K. Rana and N.C. Mamatha

Botanical Name : *Pachyrhizus erosus* (L.) Urban

Family : Leguminosae (Fabaceae)

Chromosome Number : 2n = 22

ORIGIN AND DISTRIBUTION

Yam bean has originated in tropical America. Archaeological evidence reveals that this crop was grown by the early civilizations of Mexico and Central America, such as the Aztecs and the Mayas. The Spaniards probably introduced the yam bean to Southeast Asia *via* the Philippines in the sixteenth century. Since then, the cultivation spread to Indonesia and the rest of the Far East as well as into parts of the Pacific.

INTRODUCTION

Yam bean, also known as Mexican potato, Mexican turnip, Mexican water chestnut, Mexican yam bean, Chinese potato, *ahipa, saa got, mishrikand, sakalu* or *Jicama*, is one of the important crops among the neotropical legume genera with edible starchy tuberous roots, which is the only legume extensively cultivated both on small scale in kitchen gardens for its sweet and crisp tubers and on large scale for export purpose. The genus includes five species, of which, two are wild and three are cultivated types. The cultivated species are Mexican yam bean (*Pachyrhizus erosus*), Andean yam bean (*Pachyrhizus ahipa*) and Amazonian yam bean (*Pachyrhizus tuberosus*). Another species *Pachyrhizus palmatilobus*, which is locally known as *Jicama de leche*, has less agreeable taste. It is one of the popular vegetables grown in many parts of Central American, South Asian, Caribbean and some Andean South American regions. It is widely cultivated in India, Mexico, China, Singapore, the Philippines, Hawaii and Indonesia. In India, it is grown in parts of Bihar, Orissa, West Bengal, Tripura, Assam and Eastern Uttar Pradesh.

COMPOSITION AND USES

COMPOSITION

Yam bean is a rich source of carbohydrates, which are present in the form of sugar, starch and dietary fibre. It contains good amount of potassium and vitamin C, moderate amount of calcium, iron, niacin, riboflavin and thiamine, trace amount of protein and lipids and very low amount of saturated fat, cholesterol and sodium. The flavour is sweet and starchy, which comes from the oligofructose inulin, also called fructo-oligosaccharides, which are prebiotics – *a food for probiotics*. The nutritional composition of yam bean is given in Table 87.1.

TABLE 87.1

Nutritional Composition of Yam Bean (per 100 g Edible Portion)

Constituents	Contents	Constituents	Contents
Water (g)	84–92	Vitamin A (IU)	21
Carbohydrates (g)	8.82	Thiamine (mg)	0.020
Protein (g)	0.72	Riboflavin (mg)	0.029
Fat (g)	0.19	Niacin (mg)	0.200
Dietary fibre(g)	4.9	Pantothenic acid (mg)	0.135
Calcium (mg)	12	Pyridoxine (mg)	0.042
Potassium (mg)	150	Folic acid (μg)	12
Iron (mg)	0.60	Vitamin C (mg)	20.2
Copper (mg)	0.048	Vitamin K (μg)	0.3
β-carotene (μg)	13	Energy (kcal)	38

USES

Yam bean tubers are very much similar to potato in taste as well as in ways of eating. Its young tubers are eaten raw with salt and lemon, boiled, cooked or pickled and useful addition to stir fries, salads, soups and stews. It is cooked the same way as the potato. It can be used as cubed, chopped and sliced into fine sticks and in a variety of cuisines and seasonings along with other vegetables and fruits. The young immature pods can also be eaten, provided they are cooked properly since the mature pods and seeds are poisonous.

MEDICINAL USES

The health benefits of yam bean are attributed to its strong supply of dietary fibre, vitamin C and other minerals. Its tubers help in reducing weight, increasing immunity and improving digestive, circulatory and nervous systems.

BOTANY

Yam bean is a tropical and subtropical vigorous climbing legume with hairy stem, which grows to a length of 6 m. The stem is woody at the base. The alternate leaves are made up of three leaflets with large teeth, and the outline of leaflets varies from dentate to palmate. The flowers are bluish violet or white. The species is defined by smooth petals, the numerous flowers (4–11) per lateral inflorescence axis, *i.e.,* complex racemes and 8–45 cm inflorescence length. Its tubers are broad spindle-shaped or turbinate with easily peelable skin, which is yellow and papery, whereas the inside flesh is creamy white with a crisp texture, which resembles raw potato. Its roots can attain a length of up to 2 m and weight up to 20 kg. The pod, which is 8–15 cm long, curved and hairy, contains 8–11 flattened seeds. The seed colour varies from olive-green to brown or reddish brown but they are almost black and the shape may be flat and square to rounded but never reniform. Clear taxonomic distinctions between wild and cultivated genotypes are difficult to make. However, wild material may be identified unequivocally by smaller leaf size, the increased hairiness of leaves, the smaller and often elongated and irregular shape of the tuberous root and the dark brown colour of the tuber surface.

CLIMATIC REQUIREMENT

Yam bean being a crop of tropical and subtropical regions needs warm and dry climate for its growth and development. Therefore, warm places like coastal areas in Papua New Guinea are most suitable for its successful cultivation. It grows well up to about 70 m altitude in the tropics. It is sensitive to frost, thus, it requires 9 months frost free period for a good harvest of large tubers or growing it commercially. The crop performs better under the day and night temperature of 30°C and 20°C, respectively. Its plants require 11–13 hours daylight for tuber formation. Therefore, it is suitable for commercial cultivation in zone 10 and 11.

SOIL REQUIREMENT

Yam bean can be grown successfully on a wide range of soils provided they are well drained. Sandy loam, loam and alluvial soils are ideal for its cultivation. Heavy soils are not desirable at all, as they are prone to water logging and cause deformation of tubers. The soil should be fertile, loose friable and rich in organic matter. The optimum soil pH for its successful cultivation is 5.2–7.0.

CULTIVATED VARIETIES

Generally, the local cultivars with small sized tubers (200–300 g) are grown by the farmers. The Mexican type cultivars with big tubers (500–700 g) are available but these are not usually preferred as their tubers often crack and less sweeter than the local types. Only one improved variety released for commercial cultivation in India is available. The characteristic description of cultivar is as follows:

RAJENDRA MISHRI KAND-I

A variety developed at Rajendra Agricultural University, Dholi (Bihar) in 1999 through a seedling selection from germplasm is recommended for cultivation in North Bihar and West Bengal region. Its tubers are round and conical shaped with brownish white skin. Tuberisation is shallow and tubers are of good quality. The variety is resistant to major insect-pests and diseases and suitable for intercropping with *Kharif* maize and *Arhar*. Its average yield per plant is 2–3 tubers and per hectare is 400–550 quintals in crop duration of 110–140 days.

PLANTING TIME

In tropical areas, planting can be done throughout the year. In subtropical areas, it is sown in late spring when the soil gets sufficiently warmed up. In warm temperate areas, which have at least 5 months of frost-free period, the crop can be sown at the end of spring frost but in this region, bottom heat will be required, as the seeds of yam bean need a warm soil to germinate. It is unsuitable for areas with a short growing season unless grown in a greenhouse. In cooler regions, the crop can be grown where it has a frost-free period of at least 5 months but the tubers grown under such conditions will be smaller. In India, the seeds are sown in June to July and for a late crop in September. The September sown crop can easily be intercropped with maize or other crops. However, September sown crop produces smaller tubers with lower yield.

SEED RATE

The seed rate varies according to the growing season, growing region and cultivation system. The seed rate in different regions varies from 20 to 60 kg/ha. However, the optimum seed rate is 35–40 kg/ha. There are some reports that the crop can also be propagated by tubers.

SOWING METHOD

Yam bean is commercially propagated by seeds. Soaking of seeds before sowing in lukewarm water for 12 hours helps in softening of seed coat and speeding up of germination. The field is prepared thoroughly by ploughing with harrow to a fine tilth for making smooth seed beds and ensuring an even germination. The seeds are usually sown at 5 cm depth on ridges of 15 cm height made 45–60 cm apart maintaining 15–20 cm spacing between plants.

NUTRITIONAL REQUIREMENT

Yam bean plants prefer soil of moderate fertility since higher nitrogen levels in soil will cause excessive vegetative growth, while insufficiency of potassium in soil will affect the tuber growth and quality adversely. Well-decomposed farmyard manures 10–20 t/ha should be incorporated into the soil at the time of land preparation. Besides, nitrogen is applied at the rate of 40 kg, phosphorus 40 kg/ha and potash 80 kg/ha. The entire dose of phosphorus, potash and half dose of nitrogen fertilizer should be incorporated into the soil before sowing seeds and rest half of the nitrogen should be supplied in the form of top dressing at about 40–45 days after sowing.

IRRIGATION REQUIREMENT

Water is one of the essential components required for growth and development of the crop. The total water requirement of the crop varies with soil type and climatic conditions of the growing region. If the seeds are sown in dry soil, the first irrigation is given immediately after sowing. Once the crop is well established, suitable soil moisture must be maintained and care must be taken not to over irrigate the crop, as excessive soil moisture can damage the crop by increasing the chance of fungal infections. The crop sown in rainy season may not require irrigation if the rain distribution is normal, but in September sown crop, irrigation is given at regular intervals to maintain adequate soil moisture for proper development of tubers.

INTERCULTURAL OPERATIONS

The yam bean field should be kept weed-free during early stage of crop growth since weeds may compete with crop for nutrients, moisture, space and sunlight. Weeding is done manually at least twice during first couple of months. Mounting up of soil on both sides of the ridges is also essential to provide enough space to the growing tubers and also to cover them, as the exposed tubers are damaged by rodents and insect-pests. This operation also indirectly helps in controlling weeds in yam bean field.

Removing reproductive shoots called reproductive pruning is another most important intercultural operation to obtain maximum yam bean yield, as excessive flowering will reduce the tuber yield. However, the effect of this operation depends on cultivar, season and climate. In some areas, not only the reproductive shoots are removed but the top half of the vegetative part is also pruned. In most of the cultivars, two reproductive pruning will suffice to ensure a good yield.

HARVESTING

The duration to reach maturity differs with the type of cultivar, soil type, climatic conditions of the growing region and consumer preference. Harvesting is done when the tuberous roots have attained the marketable size, depending on the consumers' preference. Mexican cultivars are harvested 5–7 months after planting. In Thailand, the tubers are harvested 4.5–6 months after planting, as the consumers like small tubers, and in India, the tubers are harvested 3.5–4.5 months after planting since the early harvested tubers fetch better price in the market. Harvesting should not be delayed, as it adversely affects tuber quality, *i.e.,* the tubers become fibrous and cracked. The most preferred tuber size is around 700 g. Harvesting can be done by loosening the soil of ridges by spade or tractor equipped with a crossbar. Manual harvesting is a more common practice with small farmers. After making the ridge soil loose, the tubers are picked by hand. Before harvesting, the tops, which can be used as forage hay, are removed with a pair of shears and left to dry in the field.

YIELD

The yield also varies with the cultivar, soil type, climatic conditions of the growing region and the cultural management. The average yield of yam bean varies from 400 to 600 q/ha.

POST-HARVEST MANAGEMENT

The post-harvest operations followed in yam bean are washing, trimming (removal of non-tuberous part of the root and basal part of the stem) and dipping in high-concentration chlorine solution, which gives sterilizing and bleaching effects. The tubers after cleaning adhered soil particles and curing in shade are packed in jute sacks. The tubers can be stored for 1–2 months at 12 –16°C temperature. Colder temperature will damage the tubers, thus, they should not be refrigerated. Prolonged storage will alter the starch to sugar ratio where starch content decreases and sugar content increases with storage period.

INSECT-PESTS

A number of insect-pests are reported to cause damage to different parts of yam bean.

Spotted Pod Borer (*Maruca vitrata*)

This is a serious pest of vegetable legumes in tropics. Its larvae attack the flowers, buds and young pods by webbing them. They bore into the pods and feed on young immature seeds. The symptoms are characterised by small holes on flower buds, flowers and pods. The yield loss due to the attack of this insect has been reported from 20% to 80%.

Control

- Collect and destroy the damaged pods and larvae.
- Spray the crop with dimethoate 0.05%, tobacco decoction 3% or *neem* oil 3% to minimize the borer population.

Hairy Caterpillar (*Ascotis imparata* and *Spilosoma obliqua*)

The adult female lays eggs in clusters over the surface of leaves. The caterpillars feed on green portion of the leaves during early gregarious stage leaving them skeletonised. In severe cases, it results in defoliation, which ultimately affects the plant growth.

Control

- Collect and destroy the larvae in early stage of the crop growth.
- Spray the crop with Quinolphos 0.05%, Endosulfan 0.2%, Monocrotophos 0.15% or Fenvalerate 0.05% at 10–12 days interval.

Root Knot Nematode (*Meladogyne arenaria*)

Root knot nematodes are polyphagous and widely distributed in the country. Infected tubers show wart formation and give bitter taste. It causes reduction in tuber yield and quality and invariably reduces marketable yield. The above ground symptoms consist of foliage discolouration to pale than normal, dwarfing of plants and wilting.

Control

- Strictly follow long crop rotation with non-host crop species.
- Apply *neem* cake along with recommended fertilizers.
- Apply nematicides like Furadon 20 kg/ha or Cadusafos 1% in soil.

DISEASES

RUST (*Phakopsora pachyrhizi*)

It is the most serious disease spread by wind. The disease symptoms appear mainly on leaves and other above ground plant parts. Small, round, reddish brown rust pustules appear on the lower surface of the leaves, which later shrivel and fall off.

Control

- Follow at least 3–4 years crop rotation with non-host crops.
- Grow early maturing cultivars, which may avoid the rust until harvesting.
- Maintain good field sanitation by destroying diseased debris and weeds.
- Spray the crop with Dithane M-45 0.2% and Bayleton 0.05% at 10–12 days interval.

SINCAMA MOSAIC VIRUS

This is the most serious viral disease of yam bean. The disease symptoms are irregular chlorosis on leaves, young shoots becoming brittle and the seed set is reduced due to reduced pollen fertility. Tuber yield as well as quality is affected considerably. The transmission of the disease appears to be restricted to contamination through mechanical wounding and insect attack. The aphids and possibly spider mites are capable of transmitting the disease.

Control

- Remove all infected plants and weeds from the field.
- Apply Disulfotan or Phorate 10G granules 1.5 kg/ha at the time of seed sowing.
- Spray the crop with Monocrotophos 0.05% or dimethoate 0.05% at 10–12 days interval to control vectors.

WITCHES' BROOM DISEASE

This disease is probably caused by mycoplasma like organisms. The symptoms are excessive branching, dwarfed leaves and deformation of the flowers. It is not to be transmittable through sap inoculations but there are some indications that the disease may possibly be transmitted by sucking insects such as whitefly (*Orosius argentatus*), aphids and mealy bugs.

Control

- Apply Carbofuran, Fensulfothion or Disulfotan 1.5 kg a.i/ha in soil at the time of sowing.
- Remove the infected plants as soon as noticed.
- Spray the crop with Monocrotophos 0.05% or dimethoate 0.05% at 10–12 days interval.

■■■

88 Yard Long Bean

M.K. Rana and Neha Yadav

Botanical name : *Vigna unguiculata* var. sesquipedalis

Family : Leguminosae (Fabaceae)

Chromosome Number : 2n = 22

ORIGIN AND DISTRIBUTION

Yard long bean, a legume crop relative to cowpea or black eye pea, is native to Southeast Asia. The botanical epithet *unguiculata* for yard long bean means *furnished with a claw*. It is cultivated widely in tropical and subtropical regions, in warmer parts of Southeastern Asia, Thailand and Southern China. It is a conventional crop of Africa, which has a potential to improve nutrition, boost food security, foster rural development and support sustainable land care. It is also extensively cultivated in India for its pods, containing number of pea like seeds known as *chowlee* by *Hindus* and forms a considerable article of food.

INTRODUCTION

Yard long bean, commonly known as asparagus bean, snake bean, long bean, cow bean, pea bean, long-podded cowpea, Chinese long bean, *Daugok, Thuafakyao, Kacangpanjang, Vali* and *Eeril*, is a variety of cowpea grown primarily for its long immature green pods. Most of its common names refer to length of the pod, which grows as long as 35–75 cm. It can tolerate drought, hot weather and poor soil conditions. Unlike cowpea, it bears delicious stringless pods, which are eaten like green beans. It is a self-pollinated crop but cross-pollination may also take place by insects. The plant attracts many pollinators, specifically various types of yellow jacket and ant.

COMPOSITION AND USES

COMPOSITION

Yard long bean is a nutritious vegetable as it is a good source of carbohydrates, calcium, phosphorus, potassium, magnesium, vitamin C and vitamin K and a fair source of protein, dietary fibres, iron, manganese and most of the B group vitamins. The nutritional composition of yard long bean is given in Table 88.1.

USES

Yard long bean is used similar to other green beans. Its crisp tender and delicious pods are eaten both fresh as well as cooked. They are best when young and slender, as the pods at this stage have an excellent gourmet flavour. In many Southeast Asian countries, its young pods are consumed as a staple vegetable. Its tender pods are sometimes cut into small pieces and used in omelettes. They can also be cooked with rice, added to soup or used fresh in salads. It is also used in Chinese and

TABLE 88.1

The Nutritional Composition of Yard Long Bean (per 100 g Edible Portion)

Constituents	Contents	Constituents	Contents
Carbohydrates (g)	8.35	Manganese (mg)	0.205
Protein (g)	2.8	Thiamine (mg)	0.107
Fat (g)	0.4	Riboflavin (mg)	0.11
Dietary fibre (g)	5	Niacin (mg)	0.41
Calcium (mg)	50	Pantothenic acid (mg)	0.055
Phosphorus (mg)	59	Pyridoxine (mg)	0.024
Potassium (mg)	240	Folate (µg)	62
Sodium (mg)	4	Vitamin A (µg)	43
Magnesium (mg)	44	Vitamin C (mg)	18.8
Iron (mg)	0.47	Vitamin K (µg)	14.8
Zinc (mg)	0.37	Energy (kcal)	47

Kerala cuisine. Its pods can be pickled in vinegar and added to *dals*, curries and cooked salads in Malaysian and Indonesian cuisine. In West Indian dishes, the pods are often stir-fried with potatoes and shrimp. In the Philippines, they are widely eaten stir-fried with soy, garlic, or hot pepper sauce. In Orrisa (India), the pods are used to make a variety of dishes, especially a sour dish cooked with mustard sauce and lime. The immature seeds separated from the pods are cooked as vegetable curry. Relatively, the smaller seeds are better in taste.

Medicinal Uses

Yard long bean is one of the ancient cultivated vegetables. It is regarded as a low calorie food, and for this reason, it is consumed by the obese individuals to lose weight. Its constituents have anti-inflammatory property and relive inflammation during arthritis and asthma. Since it is green vegetable, it contains phytochemicals and antioxidants, thus, it plays a significant role in combating cancer and holding up ageing. Its consumption provides cardiovascular defense and lowers the blood pressure.

BOTANY

Yard long bean is a tall climbing herbaceous annual with long stringless pods. Its vine as long as 2.7–3.6 m easily crawls over shrubs, arbours and buildings. Its stems are square in shape, usually smooth, violet in colour and often twine about the nodes. Its leaves are trifoliate, green, oval and smooth edged. The flowers with five petals are large and pale pink to violet-blue in colour. Its plant continues to grow after flowering and fruiting. The plant starts flowering 5 weeks after sowing and bearing pods in pairs about 60 days after sowing and keeps bearing for a long time. Anthesis takes place early in the morning between 6:30 and 9:00 a.m. The process of opening corolla takes 45–60 minutes. Dehiscence of anthers is much earlier and it varies from 10:00 p.m. to 12:45 a.m. For hybridization, emasculation is done 20 hours before anthesis, however, the safest time for emasculation is morning hours preceding day of anthesis. Stigma becomes receptive from 12 hours before blooming to 6 hours after anthesis. The pods will grow from open flower to suitable length in about 10–12 days. Its pods are pendulous, long, thin, fleshy and 30–70 cm long, depending on the variety, with each pod containing several edible seeds. The pods at maturity become constricted and they bend. Seeds are elongated and kidney shaped.

CLIMATIC REQUIREMENT

Yard long bean, mainly a warm-season crop, can be grown in a wide range of climatic conditions but it thrives best in warm and humid climate with consistent rainfall. Its plant survives in extreme humidity and heat but it is very sensitive to cold and frost. It needs full sunlight for its better growth and development but it can also be grown in partial shade. A soil temperature of about 15°C or higher is ideal to ensure good germination. A temperature of 25–30°C is optimum throughout its growing period. It can tolerate heat, low rainfall and arid conditions.

SOIL REQUIREMENT

Yard long bean can be grown on a wide range of soils from sandy to clay soils but it thrives best in well-drained sandy loam soils with a pH range of 5.5 to 6.8, enriched with organic matter. It can also be grown in weak acid or alkaline soils. It is sensitive to waterlogging as it causes root rot. If heavy rainfall or waterlogging is not a problem, it can also be cultivated on heavy clay soils.

CULTIVATED VARIETIES

So many varieties of yard long bean usually differentiated based on mature seed colour and pod characters, *i.e.* pale green pod, more slender dark green and a deep brownish red type, are available for planting. Some varieties have mainly been developed for their suitability to slightly warmer, wetter, slightly cooler or drier climate.

ARKA MANGALA

A pole type, vigorous and photo-insensitive variety developed at Indian Institute of Horticultural Research, Hessaraghatta (Bangalore) is a new selection for high yield and has maximum number of pods per plant. This variety has been mainly developed for fresh market. Pods are stringless, can be snapped very easily and are without parchment layer. Its pods, which mature in 60 days after sowing, are green with smooth surface and its green pod yield is 25 t/ha in crop duration of 90–100 days.

STICKLESS WONDER

An unusual dwarf variety matures in 54 days after sowing. Its vines grow to a length of about 72 cm but do not require trellising. Plants start flowering early (40 days), but like other bushy varieties, it does not grow as long as the taller varieties.

LIANA

A day neutral variety starts producing pods early in the season and matures in 70 days. It is recommended as a fall crop in warm climate.

CHINESE RED NOODLE

A variety similar to purple podded cultivar in characteristics matures in 95 days after sowing but its pods are more flavourful with a crunchier texture.

PURPLE PODDED

A variety suitable for cultivation in hot season matures in 90 days after sowing. Its pods retain most of their colour when stir-fried.

Yard Long (White Seeded, Black Seeded, Red Seeded or Extra Long) are the other varieties that mature in 90 days after sowing. The other varieties are Dwarf, Snake, Gita, Orient Wonder, Taiwan Black, Thai 3, etc.

SOWING TIME

In tropical regions, yard long bean can be grown round the year. In subtropical regions, it is usually sown twice, once in February to March as spring-summer crop and another in June to July as rainy season crop.

SEED RATE

The seed rate of yard long bean varies with variety, soil fertility, growing conditions and irrigation facilities. For sowing a hectare area, about 20–25 kg seed is required.

SOWING METHOD

Yard long bean is a direct seeding crop, thus, its sowing requires well-prepared fine seedbeds, for which, the field should be ploughed 2–3 times and levelled properly by planking. The seed takes 6–10 days to germinate but soaking seeds in water before sowing enhances germination. Per hill, 1–2 seeds are sown in rows at a depth of 2–5 cm, depending on soil moisture, soil type and growing season. At the time of sowing, the field must have plenty of soil moisture for better seed germination. The row spacing of 60 and 90 cm is kept for dwarf and pole type cultivars, respectively and plant spacing of 40 cm for both types of cultivar. Flat and raised seedbeds are used for the sowing of snake bean in spring-summer and rainy season, respectively.

Yard long bean can also be grown by transplanting but should be done in such a way that disturbance to the root system can be avoided. The spacing between rows and plants is kept the same.

NUTRITIONAL REQUIREMENT

Yard long bean, a true legume, enriches the soil by fixing atmospheric nitrogen in nodules on its roots. With the help of nitrogen fixing bacteria, the plant makes its own food, but even then, it responds well to the application of fertilizers. Well-decomposed farmyard manure is applied at 15–20 t/ha in soil at the time of land preparation. As it is a legume crop, its nitrogen requirement is low but phosphorus requirement is high for nodulation and potash for the utilization of nitrogen. Nitrogen is applied at 25 kg, phosphorus 75 kg and potassium 60 kg/ha. Half dose of the nitrogen fertilizer along with entire dose of phosphorus and potash should be applied at the time of final land preparation and another half dose of nitrogen should be applied 15–20 days after sowing along with weeding and earthing up.

IRRIGATION REQUIREMENT

Yard long bean is sensitive to low as well as excessive moisture, thus, it requires light irrigation at regular interval as compared to other legume vegetables. Heavy irrigation and waterlogging should be avoided. The yield potential is reduced considerably and pods become short and fibrous under a condition of low soil moisture, whereas, overwatering leads to decaying of roots, thus, it is essential to maintain optimum soil moisture. The frequency of irrigation usually depends on soil type and climatic conditions of the growing region. During rainy season, the crop does not require any irrigation. If the rain frequency is less, one to two irrigations may be given at 10–12 days interval. However, the summer crop needs irrigations at 4–6 days interval or once in a week depending upon the weather conditions. Irrigation until the germination is completed should be avoided since immediate watering can result in decaying of seeds. Hardening of plants by restricting irrigation during pre-flowering stage is beneficial to avoid excessive vegetative growth and to induce early flowering. Once the plant starts flowering, very light irrigation should be given since frequent rains or excessive irrigations during reproductive phase induce vegetative growth at the expense of fruiting.

INTERCULTURAL OPERATIONS

HOEING AND WEEDING

Shallow hoeing during early stages of crop growth is necessary to check the weed growth and to improve aeration into the soil. In addition to earthing up, two hoeing are required to keep the weeds under control. At later stages, no weeding is required as the crop canopy is thick, which smoothers the weeds. Light earthing up along with fertilizer application is also highly advantageous for better growth of plants, as it facilitates better root growth and prevents lodging of young seedlings. Nipping of apical region is also a common practice to induce branching, flowering and fruiting especially when there is rain during flowering and fruiting phase. The vines should be encouraged to trail on trellises or bower system.

STAKING

Yard long bean bears long tender green pods. Hence, its plants need a traditional support system to avoid the contact of pods with soil. For this purpose, the long bamboo strips are fixed into the ground next to bean plants to provide them support. The bean plants will grow around the stake.

Another method of staking for pole beans is to create a tee pee, which is a support system made from three or more strips hammered into the ground at an angle so that all of the poles may tilt inwards and meet near the top, forming a tee pee shape. The tee pee is developed first, and then, several bean seeds are sown at 30 cm of each stake used to create tee pee. Most of the tee pees made from bamboo stakes are light in weight, economical and very sturdy. A trellis can also be made using two stakes and a length of wire. Both the stakes are hammered length wise into the ground along the bean plant rows. The bean seeds are sown along the bottom of wire mesh. The bean vines will grow up and into the trellis. In addition, a simple trellis is formed using staking strings. About four stakes are hammered into the ground and wind heavy twine around the stakes to form a cat's cradle, moving up at least 10–15 cm on the stakes. The seeds are sown under the lines made by the twine.

TOMATO CAGE AS BEAN SUPPORT

Tomato cages are large-holes wire cones or cylinders of heavy gauge. These cages are fixed firmly into the ground deeply with the spike feet, and then, the bean seeds are sown around base of the tomato cage. Since pole beans grow to a length of 15–20 cm, they will outgrow the cages and the tops of the beans will hang off the side.

RECYCLED MATERIAL AS BEAN SUPPORT

Using recycled materials in the garden offers not only an economical approach but a clever and fanciful too. An old ladder, for instance braced against the garage wall, with the ladder feet is firmly embedded in the ground. Pole bean seeds are sown at the foot and the vines are allowed to find their way right up the ladder. Old discarded broom and mop handles can be recycled into pole bean support. Trellises are made using old pieces of lattice and bits of wire into a variety of bean supports by keeping the points in mind that pole beans are vine types, thus, they grow only where they can climb up and over they can be invasive and heavy when they produce bean pods. The trellis should be tall enough to provide support to long vines. The base of the support is always anchored, either by hammering sticks into the ground or using some other means. The bean seeds are sown after placing the support so as not to disrupt the root system of plants.

BAMBOO CANES FOR BEAN SUPPORT

Bamboo canes: These have the advantage of being perfectly straight up to relatively long length and are very resistant to weathering, lasting for years.

Wall or Fencing: Various kinds of wire are employed for forming fence to be used successfully for the climbing of bean vines. The climbing plants need a kind of support for the coiling of tendrils around it or for trailing them up a wall. Solid fence usually requires netting for sprawling the vines on it.

ARCHES FOR GROWING BEANS ON RAISED BEDS

Arching is one of the most beautiful designs for growing the bean vines up over an arch between beds or over a walkway, especially when the bean plants are at flowering stage. However, making arches is a costly affair and unless they are movable, they limit crop rotation to the legume.

HARVESTING

For vegetable purpose, the pods of yard long bean becomes ready for harvesting when the pods are tender and before they attain full length and become fibrous and when the outline of seeds is just visible on outside of the pods. However, the best harvest stage depends on variety and market requirement. The first picking can be done 8–10 weeks after sowing. If harvesting is done regularly, the plants continue to produce pods prolifically for many weeks. Scissors, secateurs or knife should be used for picking to minimize damage to the developing pods and flowers and to prolong the shelf life of flower stalk too. Twisting of pods carefully near the stem end is also an effective method of picking. The pods may grow up to 54 cm or more but it is better to pick them when they are 25–45 cm long. Harvesting is usually done at least twice a week, otherwise, the pods become too large and tough, though in hotter climate, picking should be done on alternate days, which may extend the harvesting period up to 45 days under good agricultural practices. While picking, all the marketable sized pods should be picked since the pods that are allowed to full maturity, *i.e.* pods are hardened and seed swollen, will exhaust the plant. In dwarf type varieties, 2–5 harvests are possible. If the pods are to be harvested for seed extraction, they are allowed to reach full maturity or dried on plant. The pods on drying split and release seeds.

YIELD

The number of factors such as growing season, soil type and fertility, irrigation facilities and cultural practices adopted during the cultivation of crop affect the yield of yard long bean. On an average, yard long bean gives a pod yield of 150–180 q/ha and seed yield of 4.5–6 q/ha.

POST-HARVEST MANAGEMENT

The pods of yard long bean can stay fresh for a maximum of 3 days at ambient room temperature but they start shrivelling due to loss of moisture, and for this reason, they are stored for 5–6 days in refrigerator at 4–8°C temperature and 80–90% relative humidity and for 7–10 days if packed in low density polyethylene film bags, which prevent shrivelling of pods. They can be stored up to 4 weeks at 2–4°C temperature and high humidity. However, pitting and russet symptoms develop at temperature below 2°C. The pods can be stored longer, if blanched and frozen.

INSECT-PESTS

APHIDS (*Myzus persicae* and *Aphis gossypii*)

The soft bodily minute insects live in colony on underside of leaves and stems and suck cell sap from them, which become yellow and distorted with necrotic spots on them. The plant growth becomes stunted. It also attacks the pods. Aphids secrete a sticky sugary substance called honey-dew, on which, sooty mould grows, which reduces the photosynthetic activity of the plant and yield.

Control

- Grow aphid tolerant varieties if available.
- Pick and destroy the affected plant parts.
- Spray the crop with insecticidal soap or oils such as 0.4% *neem* cake or canola oil.
- Spray the crop with Malathion 0.1%, Metasystox 0.025%, Monocrotophos, Phosphamidon, Rogor or Methyl demeton 0.05%.

RED SPIDER MITES (*Tetranychus urticae*)

The mites remain in colonies under thick silky webs and suck cell sap from lower surface of the leaves, resulting in chlorotic specks and reduced yield. They produce a speckled silvery appearance on leaves.

Control

- Spray emulsified *neem* oil like Econeem or Neemazal 0.5% on lower surface of leaves.
- Spray the crop with Dicofol 0.25% or Ethion 0.05%.

POD BORER (*Etiella zinckenella*)

The young caterpillars are greenish white, whereas the full-grown larvae are of rose colour with purplish tinge. The caterpillar enters the pod and feed on young immature seeds inside the pod.

Control

- Collect and destroy all damaged pods along with larvae.
- Spray the crop with Quinalphos or Chloripyriphos 0.05%, or Cypermethrin 0.0125%.

DISEASES

RUST (*Uromyces appendiculatus* or *fabae*)

The infection first appears as small pale lesions, which become yellow with a small dark centre on lower surface of leaves. These spots enlarge and produce brick-red rust spores to spread the disease further. Lesions develop black spores later in the season. Stem, tendrils, stipules and pods also show disease symptoms. Symptoms begin to appear within 10–15 days of initial infection. Mild temperature and humid conditions are favourable for the development of disease. The earlier the infection time, the greater is the potential for yield reduction.

Control

- Rotate the crop with non-host crops.
- Keep the field and the surroundings weed free.
- Destroy the diseased plant debris.
- Spray the crop with Bavistin or Benomyl 0.1%, Dithane M-45 or Dikar 0.2%, or Bayleton 0.25%.

ANTHRACNOSE (*Colletotrichum lindemuthianum*)

The fungus affects all above ground parts of the plant. First water soaked lesions appear on petioles and lower surface of leaves, and leaf veins are elongate, angular and brick red to purple, becoming dark brown to black. They may also appear on upper leaf surface. Pod lesions are tan to rusty brown and develop into sunken cankers surrounded by a slightly raised black ring with a reddish brown border. The disease is favoured by 12–21°C temperature with an optimum of 17°C and high humidity. The disease is most severe on susceptible plants under frequent rainfalls accompanied by wind and splashing rain. The fungus, which survives on crop debris, can be carried through seed, air and water from one to another season.

Control

- Rotate the crop with non-host crops to reduce the inoculums.
- Destroy all the infected plant debris.
- Treat the seed with Dithane Z-78 2 g/kg of seed or dip in 0.125% Captan solution for half an hour.
- Spray the crop with Thiram 0.2%, Dithane Z-78 0.15% or Bavistin 0.1%.

ROOT ROT (*Fusarium solani* f.sp. *phaseoli, Rhizoctonia solani* and *Pythium* sp.)

Root rot is identified by reddish lesions on taproot and hypocotyls, which later turn brown. The lesions enlarge with age, become darker and rough-textured. The severely infected plants are stunted with yellow leaves. Branch roots are killed. Longitudinal cracks may develop in older hypocotyl lesions. The fungus can cause a brick red discolouration of central part of the lower stem. Soil compaction, short crop rotation and moisture stress encourage the spread of disease.

Control

- Rotate the crop with non-host crops.
- Improve soil drainage and avoid waterlogging.
- Destroy the infected crop debris.
- Treat the seed with Bavistin or Captan 3–4 g/kg of seed.
- Spray the crop with Bordeaux mixture 1% or copper oxychloride 0.25%.

POWDERY MILDEW (*Erysiphe polygoni*)

The infection on leaves appears initially as darkened areas, later becoming white powdery growth, which spreads to stems and other parts of the plant. In severe infection, the leaves become chlorotic and fall off. Pods are reduced in size and distorted in shape. Plant growth is reduced due to defoliation.

Control

- Grow resistant cultivar, if available.
- Maintain good field sanitation.
- Destroy all the infected plant debris.
- Spray the crop with Karathane 0.2%, Bavistin 0.1%, Benlate 0.1%, Sulfex 0.2%, Elosal 0.5% or Calixin 0.05%.

BACTERIAL BLIGHT (*Xanthomonas campestris* pv. *phaseoli*)

The symptoms first appear as small translucent water-soaked spots on leaves. As these spots enlarge, the tissue dies leaving brown spots with a narrow yellow margin. Lesions are often enlarged and irregular in shape. In some cases, a yellowish discharge may be observed. Water-soaked sunken lesions can also be present on petioles and pods. The pod lesions later turn brownish-red. Bacteria may infect the vascular system and kill the main stem and branches. In severe cases, defoliation takes place. Seeds become discoloured and shrivelled.

Control

- Grow resistant varieties, if available.
- Remove and destroy all the infected plants.
- Dip the seed in streptocycline solution of 1000 ppm for 2 hours or treat the seed with streptocycline sulfate 2.5 g/kg of seed.

COWPEA MOSAIC VIRUS

The characteristic symptoms of virus include mosaic mottling with slight crinkling, leaf distortion and blistering and reduced leaf size. The diseased plant will remain green for a longer time and set fewer pods. Plant growth becomes stunted.

Control

- Grow resistant varieties if any.
- Adjust sowing time and use trap crop.
- Spray the crop with Malathion 0.1%, Metasystox 0.025% or Rogor, Phosphamidon, Methyl demeton or Monocrotophos 0.05% at 10 days interval for controlling the virus vectors.

■■■

89 Roselle

M.K. Rana and R. Sridevi

Botanical Name : *Hibiscus sabdariffa* var. sabdar-iffa L. (Jamaican sorrel or Roselle)

Syn. *Hibiscus digitatus* Cay.

Hibiscus sabdariffa var. altissima (Jamaican sorrel or Roselle)

Hibiscus sabdariffa var. *rubra* (Indian or red sorrel or Roselle)

Family : Malvaceae

Chromosome Number : 2n = 72

ORIGIN AND DISTRIBUTION

Roselle also known as Indian or Jamaican sorrel is probably native to Africa, where it may have been domesticated in Sudan about 6000 years ago, first for its seed and later for leaf and calyx production. In the seventeenth century, the vegetable types were introduced to India and the United States of America. Selection for fibre production took place in Asia, where cultivation has been reported from the beginning of the twentieth century, *e.g.,* in India, Sri Lanka, Thailand, Malaysia and Java. It is now found throughout the tropics. In tropical Africa, it is especially common in the Savanna region of West and Central Africa. It is often found as an escape from cultivation. However, apparently the truly wild plants of *Hibiscus sabdariffa* have been collected from Ghana, Niger, Nigeria and Angola.

INTRODUCTION

Roselle is locally known as *Mesta* or *Meshta* in the Indian Subcontinent. In many tropical regions, it is a popular vegetable, which is cultivated for its leaves, stems, seeds and calyces. The dried caly-ces are used to prepare tea, syrup, jams, jellies and beverages. It is one of the important fibre crops, which stands next to jute in production. In India, Mesta is grown in larger parts covering areas from Karnataka to Tripura including Maharashtra, Andhra Pradesh, West Bengal, Bihar, Orissa and Meghalaya. In Tripura and Meghalaya, the Mesta is grown in highlands either as a pure crop or as mix vegetable with rice. In West Bengal and Bihar, it is grown in marginal lands. In Orissa, it is grown in the hilly districts of Koraput and Kalahandi. However, Andhra Pradesh has maximum area under Mesta in the country. Roselle is a flexible plant with a number of uses. It is intercropped with crop staples such as sorghum and sesame, or planted along field margins, as it requires little care.

COMPOSITION AND USES

COMPOSITION

Roselle plant parts like leaves, fresh raw calyces and seeds are a good source of carbohydrates, cal-cium, phosphorus and contain an appreciable amount of iron, thiamine, riboflavin and niacin too. The nutritional composition of Roselle parts is given in Table 89.1.

TABLE 89.1

Nutritional Composition of Roselle Parts (per 100 g Edible Portion)

Constituents	Leaves	Fresh raw calyces	Seed
Moisture (g)	85.6	82.6	8.2
Carbohydrates (g)	9.2	11.1	51.3
Protein (g)	3.3	1.6	19.6
Fat (g)	0.3	0.1	16
Dietary fibre (g)	1.6	2.5	11
Calcium (mg)	213	160	356
Phosphorus (mg)	93	60	462
Iron (mg)	4.8	3.8	4.2
β-carotene (µg)	4135	285	–
Thiamine (mg)	0.2	0.04	0.1
Riboflavin (mg)	0.45	0.06	0.15
Niacin (mg)	1.2	0.5	1.4
Vitamin C (mg)	54	14	Trace
Energy (kcal)	43	44	411

Uses

Roselle is a multi-use plant as it is used in food, animal feed, nutraceuticals, cosmeceuticals and pharmaceuticals. Tender young leaves are eaten as a vegetable, especially with soup or salads and as a seasoning in curries. The red calyces surrounding the fruit can be used to brew non-alcoholic beverages and as colouring reagent for jelly, jam, beverages and other foodstuffs. The stem is utilised in the extraction of fibre. Mesta fibre blended with jute is used in the manufacturing of jute goods namely, cordage, sackings, hessian, canvas, rough sacks, ropes, twines, fishing nets, etc. The stalks are used in making paper pulp, structural boards, as a blend for wood pulp and thatching huts. The seeds are used as feed meal for fish and domestic animals. The seeds contain 18–20% oil, which is used in soaps and other industries. Recently, the Mesta fibre has also been used to develop some furniture like chairs, stools and other products like lamp shades.

Medicinal Uses

Roselle leaves, stems, fruits and seeds are used in traditional medicines as it has certain therapeutic properties. It is used as an astringent, general tonic and diuretic. Taking internally in the form of herbal tea is useful in soothing colds, clearing a blocked nose, clearing mucous, promoting kidney function, aiding digestion, reducing fever and suppressing high blood pressure. The leaves are a source of mucilage used in pharmaceuticals and cosmetics. Extracts are often used medicinally to treat colds, toothache, urinary tract infections and hangovers. Leaves are applied as a poultice to treat sores and ulcers. A root decoction can be used as a laxative.

BOTANY

Roselle is an annual erect shrub that takes 5 months from planting to harvesting. It can also be regarded as a perennial plant. Species grown for their fibre are tall with fewer branches, sometimes growing to more than 3–5 m high. Culinary varieties are many-branched, bushy and generally 1–2 m tall. Stem is glabrous to sparsely pubescent, sometimes sparsely prickly, green or reddish. The plant has a strong taproot system. Leaves are alternate and simple, stipules narrowly lanceolate to linear, up to 1.5 cm long, petiole 0.5–12 cm long, blade shallowly to deeply palmate,

3–5 or 7 lobed, sometimes undivided, up to 15 cm × 15 cm, margin toothed, glabrous or sparsely pubescent, sometimes with a few prickles on midrib and palmately veined with a distinct nectary at the base of midrib. The stems, branches, leaf veins and petioles are reddish purple. The flowers are borne on very short peduncles in the axils of upper leaves, the epicalyxes are made up of 10 linear fleshy bracteoles and the calyx is eight lobed, becoming large and fleshy after flowering.

Flowers are solitary in leaf axils, bisexual, regular, five merous, pedicel up to 2 cm long, articulate, epicalyx segments 8–12, united at base, subulate to triangular, free part 0.5–2 cm long, calyx campanulate, up to 5.5 cm long, becoming fleshy in fruit, lobes nearly glabrous to hispid hairy with a nectary outside, petals free, obovate, up to 5 cm × 3.5 cm, pale yellow or pale pink, often with dark red-purple centre, stamens numerous, united into a column up to 2 cm long, pink, ovary superior, five-celled and style with five branches. The large flowers have pale yellow petals (sometimes suffused with pink or red) and a dark red eye. The flowers are usually borne singly in the leaf axils. The sepals are present at base of the large flowers, and fruits vary from dark purple to bright red (sometimes white) at maturity and are quite fleshy. The calyx increases from 1 to 2 cm in length before the flower is fertilized, then to about 5.5 cm (occasionally longer) at maturity. Some forms of Roselle contain a pigment that gives a brilliant red colour to culinary products made from the plant and the other forms are completely green. Edible types of Roselle are usually succulent and have well-developed lateral branches, lacking a hairy covering. Flowering is induced as the days become shorter and the light intensity decreases, beginning in September or later depending on the photoperiod. Flowers are red to yellow, with a dark centre containing short peduncles and have both male and female organs.

The flowers are usually self-pollinating. The fruits 2 or 3 months after pollination may begin to ripen. The seedpods begin ripening near the bottom and proceed to the top. The fruit is an ovoid capsule up to 2.5 cm length, almost glabrous to appressed-pubescent, enclosed by the calyces and many-seeded. Seeds are reniform, up to 7 mm length and dark brown. Germination is epigeal and cotyledons are round shaped with 2.5 cm × 3 cm size.

CLIMATIC REQUIREMENT

The warm and damp tropical climate is suitable for Roselle plants and exceptionally susceptible to frost and mist. The temperature range within which Roselle thrives is between 18°C and 35°C, with an optimum of 25°C. Growth of the plant is ceased at a temperature of 14°C. In tropical and subtropical regions, an altitude of 900 m above sea level is suitable for its cultivation. Annual rainfall between 400 and 500 mm is necessary throughout the growing season. It is a short-day plant, which is very sensitive to photoperiod.

In first 4–5 months of its growth, it requires a daily light period of 13 hours. Flowers do not appear if the day length is more than 13 hours, while flowering is excellent when day length is shorter than 12 hours. It can tolerate flood and heavy winds. Dry weather is also well tolerated and is desirable in the latter parts of growth. Rain or high humidity at the time of harvesting and drying can downgrade the quality of calyces and reduce the yield.

SOIL REQUIREMENT

Being a crop with a deep root system, it requires adequate soil depth and is rather drought resistant. The crop can grow in a wide range of soil types, the best being retentive to moisture and friable loams. It can also tolerate poor soils but grows best in well-drained soils. Although it may grow in a variety of soils including new and old alluvium and lateritic loam, it requires loamy soils rich in organic matter. The acidic soils are not suitable without proper amendment. It requires a permeable soil, a friable sandy loam with humus being preferable. It is well adapted to a soil reaction from 4.5 to 8.0. However, *sabdariffa* develops chlorosis in high pH soils. Soil preparation should be deep, about 20 cm and thorough.

CULTIVATED VARIETIES

HS-4288

A variety developed at CRIJAF, Barrackpore through selection from a cross between RT-1 (non-bristled) and RT-2 (bristled) takes 180 days to flower. The plants are green with red pigmentation at the nodes. The leaves are deeply lobed with at least five lobes in each leaf. The upper surface of the petiole is red and the lower surface of the petiole is green. The seeds are of coffee colour and reniform. It can be sown from mid-April to May. Its dry fibre yield is 18–20 q/ha. The variety is suitable for cultivation in the states of West Bengal, Assam, Tripura and Bihar.

HS-7910

A variety developed at CRIJAF, Barrackpore through a selection from a cross between HS-4288 and RT-1 takes around 180 days to flower. The plants are green with irregular flush of red pigment, which further extends with maturity of the crop. The variety is devoid of bristles on the stem. The leaves are deeply lobed and slightly more reddish than HS-4288. The seeds are reniform and of coffee colour. It can be sown during the month of May and yields up to 18–20 q/ha of dry fibre. It is most suitable for cultivation in the states of West Bengal, Assam, Orissa and Bihar.

AMV-1

A variety developed at Agricultural Research Station, Amadalavalasa (Andhra Pradesh) is the result of selection made from an indigenous material. The plants are of crimson purple colour with base completely green up to the height of 35–40 cm. The stem is bristled and leaves are deeply-lobed. The petiole is purple in colour. It takes 180–200 days to flower depending much upon the time of sowing. The seeds of brown in colour. The variety is sown normally between mid-April and early May. The dry fibres yield of 15–17 q/ha is obtained. The variety is suitable for cultivation in states of Andhra Pradesh and Orissa.

AMV-2

A variety developed at Agricultural Research Station, Amadalavalasa (Andhra Pradesh) is the result of selection made from an indigenously collected material. The stem is pink with green base and is fully bristled. The leaves are deeply lobed. The plants take 180–200 days to flower. The seeds are brown and of smaller size. The variety can be sown in early May and is suitable for cultivation in states of Andhra Pradesh and Orissa. Its fibre yield is 17 q/ha.

AMV-3

A selection from a cross between AMV-1 and ER-79 has pink stems with bristles. The leaves are deeply lobed. The plants take 180–200 days to flower. The seeds are brown and smaller in size. It is suitable for early May sowings in the states of Andhra Pradesh and Orissa. Its fibre yield is 18 q/ha.

AMV-4

A variety developed at the Agricultural Research Station, Amadalavalasa (Andhra Pradesh) is also the result of selection from a cross between AMV-1 and AS-163. The stem is pink with bristles. The leaves are deeply lobed. The variety flowers in 180–200 days after sowing. The seeds are slightly bigger than the seeds of AMV-1 and AMV-2. The variety is suitable for May sowing in the state of Andhra Pradesh. Its dry fibre yield is 20 q/ha.

RICO

This variety was named in 1912. The plant is low growing and spreading with simple leaves borne over a long period and the lobed leaves are mostly three parted. The flower has dark-red eye and golden-yellow pollen. Mature calyx is 5 cm long and 3.2 cm wide and the bracts are plump and stiffly horizontal. It is the highest producer of calyces per plant. It is used for juice and preserves of calyces, and its herbage is rich-red.

VICTOR

A superior selection from seedlings grown at the Subtropical Garden in Miami in 1906 that produces 2.13-m tall plants with more erect and robust growth habit. The flower has dark-red eye and golden-brown pollen. It blooms somewhat earlier than cv. Rico. Calyces are as long as the calyces of Rico but slender and more pointed at the apex. Its bracts are longer, slender and curved upward. It is used for juice and preserves of calyces, and its herbage is rich-red.

ARCHER

Sometimes, it is called *white sorrel* resulting from the seeds sent to Wester by A.S. Archer of Antigua island. It is believed to be of the race *albus*. Edward Long referred to *white* as well as red Roselle as being grown in most gardens of Jamaica since 1774. The plant is as tall and robust as the plant of *Victor* but has green stems. The flower is yellow with a deeper yellow eye and pale-brown pollen. The calyx is green or greenish-white and smaller than in the two preceding but the yield per plant is much greater. Juice and other products are nearly colourless to amber. The green-fruited Roselle is grown throughout Senegal, especially in the Cape Vert region, mainly for use as a vegetable.

UKMR-1

In April 2009, Universiti Kebangsaan Malaysia (UKM) launched UKMR-1. It was developed using an Arab variety as a parent in a mutation-breeding programme, which started in 2006. The plant height is 111 cm and leaf colour is green. The calyx and flower colour is red.

UKMR-2

In April 2009, UKM launched UKMR-2. It was developed using an Arab variety as a parent in a mutation-breeding programme, which started in 2006. It possesses purplish-red pigmentation on its stems and leaves. Its plant height is 108 cm. Its flower is light red in colour and calyx colour deep red.

UKMR-3

In April 2009, UKM launched UKMR-3. It was developed using an Arab variety as a parent in a mutation-breeding programme, which started in 2006. Its plant height is 132 cm. The leaf colour is green and calyx colour light green.

UMKL-1

A variety as popularly known as Terengganu was developed by Malaysian Agricultural Research and Development Institute (MARDI) and University of Malaya (UM) in 1993 and promoted as a cash crop for commercial planting in Terengganu by the Department of Agriculture (DOA) since 1993 to replace tobacco. Its plant height is 124 cm and calyx colour is red.

SEED RATE

Roselle is commonly propagated by seeds but it is also readily propagated by cuttings. Seed rate of 13–15 kg under broadcast and 11–13 kg under line sowing is enough for planting a hectare area.

SOWING TIME

In India, sowing is done at the beginning of rainy season. However, for seed purpose, it is sown late in the month of July or even in August. In tropical areas, it is sown in early spring. It needs at least 5 months frost-free to bear flower. If grown mainly for herbage, the seed can be sown as early as in March. The planting of *Hibiscus sabdariffa* L. may be done in the month of October under lowland conditions, aiming at increased yield, and by this time, better quality calyces can be obtained. The adequate harvest time avoids quality loss due to the senescence of calyces or attack of pathogens.

SOWING METHOD

For its sowing, two approaches, *i.e.,* (i) sowing directly in the field and (ii) sowing in nursery beds, are adopted. Seedlings are first raised in nursery beds and transplanted in the field when they attain a height of 7.5–10 cm. In India, Mesta is normally sown by the broadcasting method. However, under the extension and transfer of technology programme, line sowing has been advocated. While sowing, it is essential to adjust distance from seed to seed in the field. This can be managed by adjusting seed drills. For broadcasting sowing, the seeds are scattered in the field in crosswise manner and then the laddering is done to cover the seeds. Generally, a sowing depth of 2.5 to 3 cm is maintained for good germination. Proper soil moisture at the time of sowing is also very essential. Soil moisture of 20–32% is good for the germination of seeds. A spacing of 25–30 cm between rows and 7–10 cm between plants is recommended for good yield. In case of broadcast sowing, plant to plant spacing is maintained at 12–15 cm by thinning.

NUTRITIONAL REQUIREMENT

Balanced application of fertilizers at the appropriate time is very important for the crop since a balanced fertilizer dose also helps in improvement of the soil's physical properties as well. A proper balance between organic and inorganic doses is also important for proper plant growth.

Organic manure in the form of farmyard manure at 10 to 12 t/ha is recommended, but now a days, it is difficult to get so much compost. Therefore, it is advised to apply compost at least 4 to 5 t/ha for better growth of the crop. It is advisable to incorporate organic manure at the time of land preparation.

Application of inorganic fertilizer to the crop of Mesta is also important. Nitrogen is essential for cell enlargement and for division of cells at a faster rate, phosphorus helps in proper root development and potassium is most important in inducing drought tolerance. This element also induces resistance against insect-pests and diseases. Additionally, it helps in the development of proper fibre cells. A fertilizer dose of nitrogen, phosphorus and potash at the rate of 40:20:20 kg/ha is recommended for higher fibre yield. The full dose of phosphorus and potash along with one-third dose of nitrogen is applied at the time of sowing or transplanting and the remaining dose of nitrogen is applied as a top dressing into two splits, one at the time of the first weeding and the other half after 6 weeks of sowing.

IRRIGATION REQUIREMENT

Since it is a rain fed crop, it is mainly dependent on rains. Usually, no irrigation is applied to this crop. However, it is advisable to irrigate the Mesta field for better yield. Excess rainfall can destroy the crop, thus, waterlogging should always be avoided. The water requirement of Mesta is about 50 cm, thus, it is desirable to give one or two irrigations at an interval of 15 to 20 days. Mature plants are highly drought resistant but may require water during dry periods when soil moisture is depleted to the point of wilting.

INTERCULTURAL OPERATIONS

HOEING AND WEEDING

If weeding is not done at the proper time, whole crop is adversely affected. The growth is slow in the initial stages and picks up at later stages. Therefore, it is essential to keep the field clean up to nearly 40 to 45 days. After that, crop grows and covers the weeds. The first weeding is done at the age of 3 weeks and the second weeding is done after 5 weeks of crop age. It is always advisable to attempt thinning at the time of first weeding. After each weeding, hoeing is attempted with the help of a wheel hoe to loosen the soil. Now a days, the weeds can be controlled by using herbicide. Although there is no herbicide recommended for Mesta crop, but application of Basalin has given slightly better results. The Basalin (flucholoralin) at 2 L/ha as pre-sowing (3 days before sowing) is recommended for controlling weeds in Mesta field.

THINNING

Removing excess seedlings produces healthier plants and higher yield by allowing them plenty of growing room so that they can receive all the proper growth requirements, *i.e.*, moisture, nutrients, light and space, without having any competition with other seedlings. When thinning is done, air circulation around the plants is improved. Crowded plants limit air circulation, which can lead to fungal diseases, especially if the foliage remains wet for extended periods. As and when two or three leaves have developed, the seedlings are thinned out by 50%. The thinning operation is done by hand. Two weeks later, the plants are thinned to two plants per hole.

HARVESTING

For the calyces of fruits, about 3 weeks after tile onset of flowering, the first fruits are ready for picking. The fruit consists of large reddish calyces surrounding the small seed pods. Capsules are easily separated but need not be removed before cooking. Harvesting is timed according to the ripeness of seed. The wet red fleshy calyces are harvested after the flower has dropped but before the seed pod has dried and opened. The more time the capsule remains on the plant after the seeds begin to ripen, the more susceptible the calyx is to sores, sun cracking and general deterioration of quality. Harvesting is usually done by hand. Special care must be taken during the harvesting operation to avoid contamination by extraneous materials. At no cost should the calyces come in contact with the ground or other dirt surfaces. Clean bags or containers should be used to transport from the field to the drying location. In addition to avoiding contamination, the time between harvest and drying should always be kept at a minimum. Different harvesting methods are in use today. The crop is harvested approximately every 10 days until the end of the growing season. The calyces are separated from the seedpod by hand, or by pushing a sharp edged metal tool through fleshy tissue of calyces separating them from the seedpod.

For fibre purposes, the crop takes about 3–4 months from planting to harvest. Fibre quality is best if harvesting is done just at flowering time. Harvesting is done normally by cutting the plants close to the ground with the help of a sickle. In some areas, the plants of Mesta are also uprooted. Such plants take more time to ret and the quality of fibre is adversely affected. The hardy root portion takes more time for retting and the extraction also becomes difficult in such cases. Harvesting time is very important in bast fibre crops like Mesta. Harvesting at the proper stage gives higher yields as well as better quality of fibre. If the plants are harvested prematurely in the early stage, the quality of fibre is good but the fibre yield is poor. In case the plants are harvested late, the fibre yield is better but the quality of fibre goes down. Therefore, it is necessary to properly understand the course of development of the plant from the seedling stage to the growth stage. The best time to harvest Mesta for fibre is when the plants are left with 10 or 12 flowers. The exact time of harvesting depends on the growth and the accumulation of cellulose and hemicelluloses. The taller plant will have some of the lower fruits mature without dependence on the reserves in the fibres, while the shorter plant may not have sufficient reserves to mature even a single fruit, and therefore, it should be harvested before the first capsule is mature.

YIELD

For leafy branches, a yield of up to 200 q/ha from three cuttings has been reported. Fresh calyx yield ranges from 40 to 65 q/ha, or about 8 to 12 q/ha when dried to 12% moisture content. A single Roselle plant may yield as many as 250 calyces, or 1–1.5 kg fresh weight. The average fibre yield from Roselle is 15–25 q/ha, depending on cultivar and management practices followed by the grower during cultivation of the crop. Seed yield ranges from 2 to 15 q/ha.

POST-HARVEST MANAGEMENT

Roselle calyces are air dried prior to marketing. Drying in the sun can lead to reduced quality. Adequate ventilation is important for drying. Plastic sheets are placed on the ground to avoid contamination with the soil, which also strongly reduces its value. Drying by artificial heating is capital-intensive. While drying, the temperature must remain below 43°C. After harvesting, the plants are sorted out according to the thickness of their stems. This is followed by bundling of plants in a convenient size of 25 to 30 cm in diameter. These bundles are kept standing in the field for 2 or 3 days for shedding of leaves, which simultaneously also helps in shrinking and rupturing of the bark, which helps in the entrance of retting microorganisms. In some areas, the Mesta plants are even kept for 2 to 3 months after harvesting for want of sufficient retting water. These plants are retted when water is available or can be decorticated, and the ribbons may be retted in less volume of water. Retting is the process, which helps in removal of fibre from the stem with the help of the action of chemicals or microbes present in the retting water. Two types of retting are followed in jute and Mesta, *i.e.*, (i) chemical retting and (ii) biological retting. The biological retting is of three types, *i.e.*, (i) stack method, (ii) ribbon method and (iii) steep method. In mechanical extraction of fibre, decorticators are also used for decortications of fibre from the stem.

INSECT-PESTS

MEALY BUG (*Maconellicoccus hirsutus*)

In this case, the tip of the plant is crowded with leaves and further growth of the plant is completely checked. Infested growing points become stunted and swollen. Heavy infestation can cause defoliation and even death of the plant.

Control

- *Cryptolaemus montrouzieri* is used successfully to reduce large populations of *M. hirsutus*.
- Spray the crop with Parathion (0.04%) or Fenitrothion (0.04%).
- Spray the crop with insecticidal soap (2%) containing potassium salts of fatty acids, which penetrates and damages the outer shells of soft-bodied insect-pests, causing dehydration and death within hours.

Flea Beetle (*Nisotra orbiculate* and *Podagrica* spp.)

This is a beetle which feeds on the leaf lamina. Although it is not a serious insect, it needs proper control. Flea beetle larvae also attack the plant root system, which may make the plant more susceptible to other pests and diseases that will kill it. The adults damage the leaves and growing points. Flea beetles feeding causes small holes on plant leaves.

Control
- Follow long crop rotation strictly.
- Weeding prevents additional food sources for the larvae, which feed on plant roots.
- Add a thick layer of mulch, which inhibits the ability of larvae to come up from the ground when they become adult.
- Apply carbofuran at 1.0 kg/ha for the effective control of flea beetles.

Cotton Bollworm (*Earias biplaga, Earias insulana*)

Cotton bollworm larvae are very damaging as they bore into the unripe fruits. Terminal shoots of young plants are also bored causing death of the tip and subsequent development of side shoots and branches. Flower buds are shed after being bored by the caterpillars.

Control
- Introduce predators like trichogrammatoidea lutea in Mesta field.
- Several pyrethroids are currently being used.

Cotton Stainer (*Dysdercus superstitiosus*)

Cotton stainer sucks sap from calyces, causing brown spots. Adults and nymphs feed on seeds. The feeding activities of cotton stainers produce a stain, which primarily is a result of the bug puncturing the seeds causing a juice to exude that leaves an indelible stain. Feeding by puncturing calyces usually causes reduction in size, or the fruiting body may abort and drop to the ground.

Control
- Use natural enemies such as Tachinids (parasitic flies) and predatory bugs, *e.g.*, Assassin bugs.
- No calyces or other host plant debris that could serve as breeding material should be left in the field.
- Small heaps of seeds, fruits or bits of sugarcane can be used as baits to attract cotton strainers, and then, the insects can be killed with a spray of soapy or scalding hot water.

Spiral Borers (*Agrilus acutus*)

Spiral borer makes a spiral shape gall at the basal region of the stem approximately 5 cm in length leading to some reduction in nutrients uptake, causing the gall region to become weak, and as a result, the stem cannot withstand any sudden jerk resulting from windy, rainy weather and cause considerable loss of yield.

Control
- Grow resistant varieties.
- Encourage biological control agents like braconid, eulophid, mymarid, scelionid, chalcid, pteromalid and trichogrammatid wasps, ants, lady beetles, staphylinid beetles, gryllid or green meadow grasshopper.
- Apply carbofuran 3G at 30 kg/ha 20 days after sowing and earth up.

Root-knot Nematodes (*Heterodera radicicola*)

Root-knot nematodes can be a significant problem when planted year after year. Infested plants in these patches may be small and limited in number. Depending on the nematode population, plants in these patches may be damaged and show severe stunting. Leaves on infected plants wilt and die.

Control

- Follow long rotations with non-host crops.
- A sequence of a legume green-manure crop followed by Roselle and then corn is suggested.
- Broad-spectrum fumigants, nervous system toxins including oxamyl and fenamiphos are extremely effective against root-knot nematodes.

DISEASES

Foot and Stem Rot (*Phytophthora parasitica* var. *sabdariffae*)

This fungus is both soil and water borne and the infection starts when there is water stagnation in the field. It is a serious disease in India. The lower portion of the plant (stem) turns black and the plant dies. It has been observed that the varieties with pigmented stems are less affected by this disease.

Control

- Avoid excessive nitrogen fertilization.
- Control weeds in summer fallow land.
- Spray the crop with carbendazim (0.1%), propiconazole (0.15%) or tridemorph (0.15%) as soon as the first sign of disease is observed.

Leaf Spot (*Phoma sabdariffae*)

This disease is predominantly found in India. The leaf lobes that are attacked by the fungus start rotting. First, the leaf tip starts blackening, and thereafter, the whole leaf becomes black. The affected leaves start falling, and ultimately, the plant growth is adversely affected.

Control

- Grow resistant varieties.
- Use a thick layer of mulch to cover the soil.
- Spray the crop three to four times with carbendazim 0.3%, copper oxychloride 0.3%, mancozeb 0.25% or Captan 0.2% at 15 days interval.

Foliar Blight (*Phyllosticta hibisci*)

This fungus produces round black pycnidia in culture. Field symptoms of the disease include water soaked necrotic spots on the young foliage. Pycnidia are often formed on the upper segment of the leaf spot in a ring around the centre of the lesion. Spots on infected leaves may spread along the major leaf veins. As disease progresses, leaf petioles and stems may become infected resulting in premature defoliation.

Control

- Grow resistant varieties.
- Avoid hoeing and weeding when the foliage is wet.
- Spray the crop twice or thrice with Streptomycin sulfate + Tetracycline mixture 100 g + copper oxychloride 1.25 kg/ha at 10 days interval.

Fusarium Wilt (*Fusarium oxysporum*)

Areas of reduced or patchy plant stand can be seen in affected fields, usually spreading in the direction of irrigation flow. Yellowing, wilting, defoliation and death of plants are typical symptoms of the disease. The vascular tissue of affected plants exhibits a brown or chocolate colour discolouration through the entire main stem.

Control

- Grow resistant varieties.
- Follow crop rotation with maize, cotton or sorghum.
- Avoid over-irrigation and excessive use of nitrogen.
- Control weeds to reduce spread of disease.

Powdery Mildew (*Oidium abelmoschii*)

Powdery mildew first appears as white powdery spots that may form on both sides of leaves, on shoots and sometimes on flowers and calyces. These spots gradually spread over a large area of the leaves and stems. Leaves infected with powdery mildew may gradually turn completely yellow, die and fall off, which may expose the calyces to sunburn.

Control

- Grow resistant varieties.
- Grow crop in full sun.
- Follow good cultural practices.
- Spray the crop with 0.2% wettable sulfur.

■■■

90 Arrowroot

M.K. Rana and Chandanshive Aniket Vilas

Botanical Name   : *Maranta arundinacea* L.

Family   : Marantaceae

Chromosome Number : 2n = 48

ORIGIN AND DISTRIBUTION

Arrowroot is indigenous to tropical America and is widely distributed among the tropical countries, where it is distributed in India, Indonesia, Sri Lanka, Philippines and Australia. It is found growing wild in some parts of India and cultivated as rain-fed crop in Kerala, Assam, Orissa, Bihar and West Bengal.

INTRODUCTION

Arrowroot is an erect herbaceous tropical perennial plant belonging to Marantaceae family. The genus *Maranta* includes around 20 species. Among them, the only species *Maranta arundinacea*, commonly known as *West Indian Arrowroot*, is cultivated for its edible rhizomes. The plant grows to a height of about one and a half meter and produces long, fleshy and profoundly thick cylindrical rhizomes, which give corn-like taste after boiling. The arrowroot is useful for its high-quality digestible starch, which is smooth and highly viscous in nature.

COMPOSITION AND USES

COMPOSITION

Arrowroot is a good source of carbohydrates and supplies a variety of minerals such as calcium, phosphorus, magnesium and a fair source of zinc, iron, thiamine and riboflavin. Its mature rhizomes contain good amount of vitamin A and energy. The nutritional composition of tannia is given in Table 90.1.

USES

Arrowroot rhizomes are consumed fresh after boiling, roasting or making flour, which is used as stabiliser in ice creams and thickener in soups, desserts, cakes, jellies and acidic food, such as Asian sweet and sour sauces. Raw rhizomes are used in cosmetics and alcoholic drinks. Its starch is used as face powder and for the preparation of special glues. Its fully expended mature leaves are used for the packaging of meat and fish. The starch produced from its rhizomes is used as a substitute of Agar in potato dextrose medium. It is also used in making carbonless paper used for computer print outs, cardboards, cushions and wallboard. The fibrous material, which remains after the extraction of starch, is used as feed for cattle or as manure. The dry powder of rhizomes containing an easily digested starch is a source of purest form of natural

TABLE 90.1

Nutritional Composition of Arrowroot (per 100 g Edible Portion)

Constituents	Contents	Constituents	Contents
Water (g)	16.5	Magnesium (mg)	25
Carbohydrates (g)	83.1	Iron (mg)	1.0
Protein (g)	6.2	Zinc (mg)	0.63
Fat (g)	0.1	Copper (mg)	0.12
Dietary fibre (g)	0.1	Vitamin A (IU)	19
Calcium (mg)	10	Thiamine (mg)	0.14
Phosphorus (mg)	20	Riboflavin (mg)	0.05
Potassium (mg)	454	Vitamin C (mg)	1.9
Sodium (mg)	26	Energy (kcal)	334

carbohydrates. Its rhizomes are a good alternative source of rice and also an effective substitute of corn in broiler ration. Arrowroot provides an excellent source of mulch, which can be used as a weed barrier and as a windbreak.

MEDICINAL USE

Arrowroot rhizomes are used to treat wounds and ulcers. Its starch is used in the treatment of intestinal disorders and also employed in the preparation of barium meals and pharmaceutical tablets because of its rapid disintegration property. Its starch is easy to digest. Hence, it is a good replacement of breast milk for infants. Due to its easy digestible nature, it is a good diet for convalescents especially for those suffering from bowel complaints as it has demulcent property.

BOTANY

The arrowroot inflorescence is terminal bearing with stalked white flowers that are entomophilous in nature. These flowers born in pair and develop within 3 months but do not set seeds. Plants are erect, slender and monocotyledonous in nature. Seldom, it produces seeds. Flowers are zygomorphic and about 2 cm long having three-lobed calyx and corolla. Ovary is inferior with one ovule. Leaves are large, fleshy and bright green, lanceolate, about 20 cm long and 10 cm wide. The white, yellow or blue rhizomes, which are 30–40 cm long, occur in clusters of two or three below the soil surface. The vegetative shoots produced by mother rhizomes store starch in their basal portion. Distal end of mother rhizome gives rise more number of shoots. The mature rhizomes are shorter in length, bulging at middle and longer towards end.

CLIMATIC REQUIREMENT

Arrowroot is a short-day plant and grown up to an altitude of 900–1000 m above mean sea level. Hot and humid tropical and subtropical areas with a temperature range of 20–30°C and minimum annual rainfall of 100–150 mm are ideal for its cultivation. It gives optimum yield where the rainfall is evenly distributed. High temperature, high humidity and partial shade favour good plant growth. Growth is much slower in cold areas, and often, the leaves become brown and ultimately die. Mild frost kills its leaves but not its rhizomes. However, a short dry period in winter will not affect its plant growth.

SOIL REQUIREMENT

Well-drained, slightly acidic and loam soils with pH range from 5.5 to 6.5 are most suitable for the cultivation of arrowroot. It may also grow well in sandy soils provided they have good water-holding capacity as soil moisture without waterlogging conditions is an important factor for its growth and development. Clayey soils are generally avoided for its cultivation as rhizome development is poor in such soils, and they become deformed and tend to break during harvesting.

CULTIVATED VARIETIES

The Creole and Banana type cultivars are commonly grown in India. Creole type cultivars although have good keeping quality but they are comparatively longer and thicker, which makes their harvesting difficult, while the Banana-type cultivars that are shorter and thicker are easy to harvest but poor in keeping quality, thus, they should be consumed within a short period. Generally, local cultivars with yellow rhizomes are grown in India. However, the cultivars with blue rhizomes give higher starch yield as compared to cultivars with yellow rhizomes.

PLANTING TIME

Planting of arrowroot generally starts at the beginning of rainy season. In southern parts of the country, it can be planted from second fortnight of May to first fortnight of June depending on the onset of rain.

SEED RATE

The seed rate usually depends on variety, corm size, soil fertility, irrigation facilities and growing season. Generally, 30–35 quintal mother rhizomes are required for planting a hectare area.

PLANTING METHOD

Arrowroot is propagated either by using suckers emerging from the sprouted rhizomes or 4 to 7-cm-long rhizome pieces containing two to four buds. Disease-free healthy rhizomes with at least one sprouted bud are used as a planting material. The rhizome pieces (sets) are planted in manure-incorporated pits and the rhizomes separated from the clump are planted in off season at a distance of 30–45 cm in nursery. The field is well prepared to a fine tilth by ploughing repeatedly before the onset of monsoon and raised beds of 50 cm length, 50 cm width and 15–20 cm height are made. The healthy suckers emerged from the planted rhizomes are uprooted and cut into 10-cm-long pieces (called as bits) after removing foliage but keeping their roots intact and then they are planted in the main field at 40 × 80 cm distance and 5.0–7.5 cm depth. Planting is done just before the onset of monsoon, and after planting, the rhizomes are covered with farmyard manure and straw mulch.

NUTRITIONAL REQUIREMENT

The crop responds well to farmyard manure and fertilizers. It is essential to incorporate organic matter in sandy soil to improve water-holding capacity and in heavy soils to improve aeration. Incorporation of farmyard manure or compost 10 t/ha is advocated for a successful cultivation of arrowroot. Besides, nitrogen is applied at the rate of 150 kg, phosphorus 75 kg and potash 150 kg/ha. Full dose of phosphorus and potash along with one-third dose of nitrogen is applied at the time of planting and remaining two-third dose of nitrogen is applied in two equal splits 1 month after planting and at the time of corm formation. Nitrogen is important for overall development of the crop and potassium increases the protein and starch content in rhizomes.

IRRIGATION REQUIREMENT

Sufficient soil moisture throughout the growing period is important for the optimum yield of arrow-root. The crop is normally grown as a rain-fed crop. However, if dry spell occurs during initial 3–4 months, supplementary irrigation at weekly interval is necessary. No irrigation is needed when there is enough moisture in soil especially at early stages of crop growth.

INTERCULTURAL OPERATIONS

Field should be kept clean and free from weeds during first 3–4 months of planting for better plant growth. While weeding, earthing up should also be done. Mulching with dried leaves reduces the weeds, conserves soil moisture and ultimately improves the crop yield.

HARVESTING

Maturity is indicated by yellowing, wilting and drying of foliage. The crop attains economical maturity 10–11 months after planting; however, the starch content is found maximum when harvesting is done 11–12 months after planting. Harvesting should not be delayed as delayed harvesting reduces the starch content of rhizomes by increasing sugar content. Hence, harvesting should be done at appropriate stage to avoid the conversion of starch into sugar.

The plants are dug out with the help of spade or by passing a plough close to the ridges, exposing the tuberous roots and the rhizomes are separated from the leafy stem. Harvesting may also be done with the use of fork or by pulling out the whole plant in sandy loam soils. Recently, mechanical harvesters have also been introduced, allowing faster harvesting. Harvesting involves the breaking off of rhizomes from the shoot. The rhizomes should be dug out without any injury. The cleaned rhizomes are either marketed or dried and stored. The bigger sized rhizomes are mainly used for the extraction of starch through further processing and small rhizomes are used as a planting material for the planting of next crop.

YIELD

The rhizomes yield varies from 40 to 100 q/ha under rain-fed conditions and 120 q/ha under irrigated conditions. On an average, 1–1.5 kg of rhizomes can be obtained from a single clump of plants. Planting can be reproductive consecutively for 5–7 years through the re-growth of rhizomes portions left during harvesting.

POST-HARVEST MANAGEMENT

If the soil is sandy, the rhizomes can stay longer *in situ* in the soil. As it is resistant to rough weather, rotting of rhizomes is not observed under ordinary storage conditions. The rhizomes are therefore stored by traditional method. At some places, the rhizomes are stored in dark places by embedding in dry sand layers. However, physical damage to rhizomes during harvest enhances the chances of deterioration under normal storage conditions, therefore, the injured rhizomes should be separated from the lot before storage.

INSECT-PESTS

Aphids (*Aphis gossypii* and *Pentalonia nigronervosa*)

It is a serious sucking pest of arrowroot. It sucks cell sap from leaves and causes yellowing and curling of leaves. Its nymphs and adults secrete honeydew-like sugary sticky substance, which encourages black sooty mould, hampering the photosynthetic activity of leaves. Mostly it attacks the plants at early stage of crop growth. High humidity and high temperature favour the pest infestation.

Control

- Grow resistant varieties.
- Spray the crop with Malathion 0.5%, or Rogar 0.25% at 12–15 days interval.

THRIPS (*Heliothrips indicus*)

The pest is commonly observed in the month of February to May. It sucks cell sap from tender parts of the plant and causes white specks with black dots. The affected leaves become brown and distorted. Dry weather is favourable for its rapid multiplication, thus, it becomes very destructive during dry weather.

Control

- Follow long crop rotation with non-host crops.
- Keep the field and surroundings weed free.
- Collect and destroy the affected parts and plant debris.
- Apply Phorate in soil at 2 kg a.i./ha.
- Spray the crop with dimethoate 0.2% or Malathion 0.2% at 15 days interval.

DISEASES

BACTERIAL LEAF SPOTS (*Pseudomonas cepasia*)

Small dark brown to black spots appear on both upper and lower surface of the leaves. Later, the spots become angular to irregular. Veins become yellow. In severe cases, whole field is seen as burnt and blighted. Warm moist weather causes considerable damage to the foliage.

Control

- Follow long crop rotation.
- Maintain good sanitation in the field.
- Use disease-free planting material.
- Dip the corms in 0.2% solution of Thiram or Captan before planting.
- Spray the crop with 0.3% copper oxychloride, 0.25% Diethane M-45, or 0.25% Captan 50 WP at 10–12 days interval.

BANDED LEAF BLIGHT (*Colletotrichum gloeosporiodes* and *Glomerella cingulata*)

Brown spots encircled with yellow halo appear on the apex and edges of leaf. The spots enlarge rapidly by the formation of successive discontinuous unusual zones. White superficial growth of mycelium quickly covers both the surfaces of leaf. The affected leaves become completely necrotic and rotten by secondary organism. The disease becomes troublesome during the period of high rainfall.

Control

- Spray the crop with 0.2% Zineb, Maneb or copper fungicide prior to the onset of rainfall.

Bacterial Wilt (*Xanthomonas marantae*)

The disease appears at all stages of plant growth. Yellowing and stunting of plants and finally their wilting and death are the major symptoms of disease. Vascular tissues show yellow to brown discoloration and pith becomes dark brown. The disease is soil borne.

Control

- Use healthy propagating material.
- Areas where the disease occurs should be avoided for planting the crop.
- Dip the corms in 0.2% Captan solution before planting.

Mosaic virus

The major symptoms of mosaic virus observed in the field are leaf mottling, chlorosis, cupping, puckering, vein banding and necrosis. It is transmitted either mechanically or through aphids but not by contact. Leaves are affected badly resulting in yield reduction.

Control

- Avoid intercropping with susceptible hosts.
- Use disease-free planting material.
- Keep the field weed free.
- Rogue out the diseased plants from the field.
- Spray the crop with Malathion 0.5% to control the vectors.

■■■

91 Japanese Arrowroot

A.T. Sadashiva and M.V. Bharathkumar

Botanical Name : *Pueraria lobata*

Syn. *Pueraria hirsuta*

Syn. *Pueraria thunbergiana*

Syn. *Pachyrrhizus thunbergianus*

Syn. *Dolichos hirsutus*

Syn. *Dolichos lobatus*

Pueraria tuberosa

Pueraria triloba

Family : Leguminosae (Fabaceae)

Chromosome Number : 2n = 24

ORIGIN AND DISTRIBUTION

Pueraria lobata is supposed to be native to China or Japan region of eastern Asia, where it is still cultivated for the edible roots. At one time, it had been a staple food in southeast Asia and Oceania. In the United States, it was introduced as a food plant, but later, it became popular as a crop for controlling erosion and quickly spread out of control. Now, it has become a weed in some of the areas. In India, it was introduced in the 1920s but it was not as successful as *Pueraria tuberosa*, which has a long history of cultivation. Both these species are now widely cultivated in the tropical and subtropical regions, predominantly as cover a crop in regions prone to soil erosion. Japanese arrowroot is more widely cultivated in China, Japan and southeastern States of America.

INTRODUCTION

Japanese arrowroot, also known as kudzu, kudzu vine, arrowroot vine, *Fen-ko* in China and *Gmadhi hulu* and *Kudazinila teigi* in India, is regarded first and foremost as a multipurpose perennial crop, being especially valuable in soil conservation because of its deep and strong root system and heavy ground cover; however, it is also grown as a protein-rich forage or green manure crop. Its root tubers are an additional advantage. Because of its exceptional ability to cover ground, vegetation and trees quickly that it is difficult to eradicate, it has become an extremely troublesome weed in some southern states of America. However, the rapid growth of its vine makes it important for the production of biomass in an energy programme and its highly prolific biomass production quality has enabled it to perform well in the process being developed for producing methane-involving systems to treat waste water with higher plants.

COMPOSITION AND USES

COMPOSITION

The dry roots of Japanese arrowroot contain crude fibre, as much as 28% and starch about 40%. Its roots also contain carbohydrates in the form of sucrose, glucose and fructose. The root starch contains moisture 16.5%, total carbohydrates 83.1 g, protein 0.2 g, fat 0.1 g, calcium 35 mg, phosphorus 18 mg, iron 2.0 mg, sodium 2 mg, ash 0.1 g and energy 340 kcal/100 g. The cooked leaves contain moisture 89.0%, carbohydrates 9.7 g, protein 0.4 g, fat 0.8 g, dietary fibre 7.7 g, calcium 34 mg, phosphorus 20 mg, iron 4.9 mg, thiamine 0.03 mg, riboflavin 0.91 mg, niacin 0.8 mg and energy 36 kcal/100 g. The nutritional composition of fresh roots of *P. lobata* is presented in Table 91.1.

USES

The root tubers of Japanese arrowroot can be used either in fresh or dried form after cooking in a manner similar to other root crops. The fresh roots are used for the extraction of starch (*kudzu powder*). Instead of arrowroot, corn, or potato starch, its starch is highly regarded and used as a crispy coating in deep fried foodstuffs. It is used as a base in soups, sauces, jellied salads, desserts and noodles. The starch can be used as a substrate for the lysine-enriched baker's yeast and ethanol-fermentation process. Its flowers are consumed after cooking or used for making pickles, whereas its stems and fresh young shoots have a taste like a cross between bean and pea and are eaten either raw or after cooking.

Japanese arrowroot plant foliage is also a palatable fodder for farm cattle in the form of hay, pasture or silage, especially in the off season. Due to its unique aroma and delicate flavour, it is used to make excellent quality gels with exceptional clarity. The stalks are the source of crude fibre, which has been used for centuries in Japan, Korea and China for the manufacturing of *grass cloth*, which is valued for its texture and durability. Its fibre is still occasionally used for making fishing nets. Its leaves can be used as a substitute for tobacco and the essential oil extracted from its plant can be used as a tobacco additive, as the Japanese arrowroot roots contain the compounds such as megastigma trienones, 3-hydroxy-13-ionone and 3-hydroxy-13-damascone, which were previously reported from tobacco leaves. Because of extremely high biomass production, it is used in energy production.

The Japanese arrowroot plant because of its extensive root system is grown for controlling soil erosion and rebuilding depleted soils, and being a member of the Leguminosae family, it fixes atmospheric nitrogen into the soil through the actions of root bacteria. Its vine extract is stated to contain traces of a gibberellic acid-like substance, and the extract of its tubers has been reported to be active against plant pathogenic fungus, *i.e., Helminthosporium sativum*.

TABLE 91.1

Nutritional Composition of Japanese Arrowroot (per 100 g Edible Roots)

Constituents	Contents	Constituents	Contents
Water (g)	68.6	Dietary fibre (g)	0.7
Carbohydrates (g)	27.8	Calcium (mg)	15
Starch in fresh roots (g)	10	Phosphorus (mg)	18
Starch in dry roots (g)	40	Iron (mg)	0.6
Protein (g)	2.1	Ash (g)	1.4
Fat (g)	0.1	Energy (kcal)	113

MEDICINAL USES

The Japanese arrowroot vine, known as *Ge Gen* in China, is considered one of the 50 elementary herbs used in medicine. The compounds *daidzin* and *daidzein* isolated from its roots and flowers are safe and effective in treating alcohol abuse, as these compounds suppress the desire for alcohol, whereas, the existing medicines given for the treatment of alcohol addiction cause toxins build-up. This plant is often given in combination of *Chrysanthemum* × *morifolium* in treating alcohol abuse. Due to these compounds, the flowers and roots work as an antidote, antiemetic, antipyretic, antispasmodic, demulcent, diaphoretic, digestive, febrifuge, hypoglycemic and hypotensive. In China and Japan, its roots and flowers decoction is given to cure alcoholism, fever, cough and cold, diarrhoea, dysentery and acute intestinal obstruction or its root powder is taken orally for the same purpose in Japan. It is often useful in combination with *Cimicifuga foetida* for the treatment of migraine and measles. The *puerarin* present in its roots increases blood flow to the coronary artery and protects against acute myocardial ischemia caused by pituitrin injection. The stems having galactogogue property are also applied as a poultice to incipient boils, swelling, sore mouths, etc. and its seeds are used in the treatment of hangover and dysentery, whereas its leaves are styptic.

BOTANY

Pueraria is a small genus of perennial twining deciduous woody climbing shrub reaching a height of 10–30 m with tuberous roots. *Pueraria lobata* plant, a vigorous vine with stolons rooting at the nodes, crawls on the ground or climbs over trees. The leaves are trifoliate with entire or slightly trilobed leaflets, pubescent, 5–12 cm long and 4–10 cm broad. The flowers borne densely on 20- to 50-cm-long pubescent racemes from June to September are mauve to violet in colour and fragrant. The 5- to 10-cm-long pods containing 8–20 seeds are hairy, flat and oblong-linear, and they mature from September to January.

Pueraria tuberosa is just similar to *P. lobata* but its leaf margins are entire, racemes slightly longer, flowers blue to purplish-blue and pods are constricted between three and five seeds but somewhat shorter. In both the species, roots are elongated and often tuberous, and may be as long as 1 m with diameter 40 cm, weighing up to 40 kg; however, the roots of this large size are relatively uncommon. They may be tapering, cylindrical or a variety of irregular shapes.

CLIMATIC REQUIREMENT

Pueraria lobata performs best in a moderately humid subtropical and warm temperate climate. Although it is tolerant to an annual rainfall of 97–214 cm and an annual mean temperature of 12.2–26.7°C, its growth is favoured by an annual rainfall of at least 1000 mm and a summer temperature around 27°C, whereas, *P. tuberosa* thrives best in tropics on a wider range of conditions up to 1200 m above mean sea level, except in very humid or arid conditions. Being a deep-rooted species, Japanese arrowroot is drought tolerant. The crop requires full sunlight for its growth and development but fails to grow in shade. Slight frost kills the above ground parts but severe frost kills the entire plant. In the cotyledon stage, the plant can withstand temperatures as low as −7°C but it loses its tolerance to cold as the plant develops three to four leaves.

SOIL REQUIREMENT

Japanese arrowroot can grow under less favourable soil conditions but not on very poor sandy or poorly drained heavy clay soils. It prefers well-drained deep loamy soils for better growth and development. The plant is able to survive in frosted and shallow soils but its roots cannot develop fully. It can tolerate both acidic and alkaline soils with pH between 5 and 8. It can be grown on poor acidic soils deficient in calcium and phosphorus but responds well to manure and fertilizers. Plants cannot withstand water logging on any soil.

CULTIVATED VARIETIES

Although many cultivars have been developed, Kudze 23 is the only cultivar-which gives more crowns. Its finer leaves and stems make it valuable feed for all kinds of livestock.

PLANTING TIME

Japanese arrowroot is a warm weather plant, growing from early spring until late fall. Its seeds are sown in nursery in early or late spring as and when the soil becomes warm, depending on the locality. The runners, cuttings or crowns removed from old plants are planted in well-prepared nursery or in pots during spring-summer and young shoots are shifted to field in the next spring-summer. In Britain, where the seeds are sown in a greenhouse, Japanese arrowroot can be grown as annual.

SEED RATE

Pueraria lobata is poor in seed setting but *P. tuberosa* produces seeds abundantly. Japanese arrowroot can be propagated by seed as well as stem cuttings or rooting runners. However, vegetative propagation is more commonly used, especially the 1–2 years old rooted cuttings (crowns) for the propagation of *P. lobata*. Cuttings can be taken from soft and hard woody vines but it requires special care and conditions. About 1250 crowns of Japanese arrowroot are sufficient for planting a hectare area. Where plentiful seed is available, the crop can be raised by sowing seeds at 1 kg/ha directly in the field.

PLANTING MATERIAL

One to two years after planting the runners, cuttings or crowns, the young shoots are removed along with some underground parts of the stem, preferably with some roots already formed and are planted in well-prepared field in summer, if growth is sufficient. Otherwise, they are grown in pots for a year and planted out late in the following spring. The shoots that are potted up will usually develop new roots from their nodes, which later can be used for planting purposes.

The hard seed coat is pricked using a scarification machine or the seeds before sowing are soaked before sowing in lukewarm water for 12 hours to ensure better germination. After shade drying, the seeds are sown very thickly in early spring. The seed takes 2 weeks to germinate. Germination of 70% is considered excellent.

PLANTING METHOD

Seeds are sown in a well-drained nursery in rows 1 m apart, planting 15–25 seeds per 20 cm of row at 0.6–1.3 cm depth. Seedlings require 4 months for developing four to six true leaves and one or more roots of 1.3 cm diameter and 15 cm length. Seedlings of preferred size are transplanted into individual pots, and after the last expected frosts, they are planted in an open field. For proper establishment of seedlings, the field is prepared to a fine tilth by repeated ploughing and planking. Planting is often done in small pits. Rooted stem cuttings, crowns, or seedlings can be planted in holes large enough to hold the roots comfortably and to protect the plants from drying during early days of planting. The crown buds are kept at or above ground level and covered very lightly with soil. In a vegetatively propagating crop, spacing of 3–10 m between rows and 1.3–3.3 m between plants is kept, accommodating about 1250 plants per hectare, and in a direct seeding crop, spacing of 100 cm is maintained between both rows and plants. The plants grow up to 6 m the in the first year and make good temporary screens. The young plants are covered with a frame until they start growing well.

NUTRITIONAL REQUIREMENT

Like most leguminous crops, Japanese arrowroot has a symbiotic relationship with nitrogen-fixing bacteria (*Rhizobium* spp.) in its root nodules and can almost double the concentration of nitrogen in the top 1–6 cm soil. Hence, the seeds or the transplants should be inoculated with the right strain of bacteria to insure maximum root nodulation and nitrogen fixation. On poorly fertile soils, applying a moderate quantity of manure and superphosphate at the time of planting is beneficial for promoting root nodulation. In boron deficient soils, borax should be applied at the rate of 30 kg/ha at planting time otherwise the lack of boron will be a limiting factor in its economic yield. In an established crop, manure should be applied at the rate of 10 tonne and superphosphate 400–600 kg/ha every second or third year.

IRRIGATION REQUIREMENT

The soil should be kept moist until the crowns are established by applying light irrigations at a frequent interval. Thereafter, the crop does not require much irrigation because of its very deep root system. Only the protective irrigation can be given as and when required by the crop. However, water logging in any case should be avoided because plants grown under such conditions fail to grow in the vegetative phase, flowering and to develop the root tubers.

INTERCULTURAL OPERATIONS

HOEING AND WEEDING

Regular hoeing is necessary in first 1 or 2 years for the plant to produce numerous stolons, which form roots and produce tubers at their nodes. Japanese arrowroot is itself a weed but when it is cultivated as a crop, it is necessary to keep the field weed-free, thus, other unnecessary plants or weeds in the first year should be removed either by two to three hand weedings, hoeing or by the use of suitable herbicide. In successive years, the luxurious plant growth will itself smother the weeds by interfering with sunlight and reducing the availability of nutrients to the weeds.

STAKING OF VINES

Japanese arrowroot plants, being a climber, need support for its growth, which can be achieved using wooden, iron or cemented poles of suitable length or even by growing some hard woody plants adjacent to Japanese arrowroot plants, to which the vine can twine.

HARVESTING

Japanese arrowroot can be harvested in several ways, depending on its usage. In 2–3 years of planting, it covers the ground, if not pruned too often as the plant is long lived. Its foliage is palatable coarse hay, which is fed to all kinds of livestock. The foliage should be harvested when the vines and ground are dry, and thereafter, the foliage is left in a swath for several hours before windrowing. Following morning when the dew is off, the cut plants are put in small stacks, and in the afternoon, it is put in the barn. The Japanese arrowroot pasture is important as a reserve feed in a drought period. However, the plants, until the third year, should not be used for grazing. If the growth is vigorous, it may be grazed lightly in the second year. For good production, the pasture should be divided into two sections for grazing in rotation. The plant mixing with grass can also be used for making silage. This mixture should contain about 60% moisture and it must be handled quickly to avoid its drying.

The starchy root tubers of both the *Pueraria* species are harvested from autumn to spring when they attain a size of 30–60 cm × 25–30 cm and weight 35 kg or even more in older plants. Sometimes, its tuberous roots may be connected with the main roots or each other by thin stringy roots, thus, they should be dug out carefully from the soil manually. The depth of roots and their irregular pattern do not allow harvesting by mechanical means. The flowers are harvested just before they are fully open and thereafter dried for later use.

YIELD

Forage yield of 5 t/ha is realized from good plant stands on fertile soils. About 25,000 plants or crowns per hectare can be harvested from an ordinary well-established plantation. However, the plant number may be doubled under ideal conditions. The planting stock can be left in the field until needed. However, if the large number of plants is to be handled for commercial sale, it is necessary to dig them before the planting season. The commercially valuable tuberous root yield is 5–7 t/ha.

POST-HARVEST MANAGEMENT

Much of the material used in processing comes from harvesting the roots or vines from the wild plants. There is a small export trade from eastern Asia to the United States of Japanese arrowroot powder (starch), mainly available at *health food* outlets. The fibrous nature of Japanese arrowroot tubers makes them unattractive for consumption as human food, and difficulty in harvesting prevents its production for commercial purpose. However, the high starch quality makes it popular for use in gourmet food in eastern Asia, and efforts are being made to develop value added products in the United States and its roots provide a vitamin-rich starch for the production of ethanol and yeast. There are cottage industries in Japan to prepare Japanese arrowroot powder, popularly known as kudzu powder, and few small commercial plants are there to produce starch for export, which is extremely labour intensive and a time-consuming process.

The vines are also used as a source of high tensile strength fibre for textiles, clothing and wallpaper. One year old 2- to 2.7-m-long straight plants are harvested in mid-summer. After removing the leaves, their stems are steamed until the fibre is stripped off, then, the soft fibre is cooked with lye for 2 hours and tough fibre for 4 hours. Thereafter, the fibre is put in a ball mill for 3 hours. The resulting fibre is greenish or cream in colour and can be used as a ground cover plant in a sunny position. A strong fibre from the stems is used to make ropes, cables, coarse cordage and textile. The fibre, which is 2–3 mm long, is used for making paper.

INSECT-PESTS AND DISEASES

The cultivated forms of Japanese arrowroot are not available, thus, only the wild plants are used for various purposes. The wild plantation appears to be relatively free from serious insect-pests and diseases. However, yellow leaf mould (*Mycovellosiella puerariae*) has been reported to attack Japanese arrowroot vines.

■■■

92 Miner's Lettuce

M.K. Rana and Neha Yadav

Botanical Name	: *Claytonia perfoliata*
Family	: Portulaceae (Montiaceae)
Chromosome Number	: 2n = 12

ORIGIN AND DISTRIBUTION

Miner's lettuce is native to the western coastal and hilly regions of North America, where it is currently found growing naturally in California except lower desert areas. It was introduced into the United Kingdom by the eighteenth century where it spread extensively. It was widely naturalized in Western Europe where it was introduced in the eighteenth century by the naturalist Archibald Menzies, who brought it to Kew Gardens in 1794. It is also found growing from British Columbia south along both sides of the Cascade Mountains to Baja California and east to Arizona, Utah, Wyoming and the Dakotas. It is mostly found prevalent in shaded forests among fir, pine and oak trees, including public lawns, forests, roadsides, wetlands, bluffs and ravines and colonizes in disturbed areas, particularly those that experience fires in previous season. It also can be seen growing in virgin fields of wheatgrass and bluegrass. It was formerly known as *Claytonia perfoliata*, and in England and Europe, it is still known by this botanical designation. Elsewhere, it is known as *Montia perfoliata*. The species got its name due to its use as a fresh salad by the miners at Gold Rush in California in 1849. Occasionally, it is confused with purslane.

INTRODUCTION

Miner's lettuce, commonly known as Indian lettuce, is a species in the genus *Claytonia*, which contains about 43–88 species of flowering plants with slightly varying characteristics. In England, it is occasionally known as spring beauty and in Ireland as Plúiríne arraigh. The miner's lettuce is in fact a misleading name. It is neither a lettuce nor a purslane although both belong to the family *Portulaceae*. The generic name *Claytonia* was assigned to it on the name John Clayton—a botanist in the sixteenth century, whereas its specific name *perfoliata* was designated due to its perfoliate leaves, which encircle the stem completely and they are attached at the base of the plant. It is mainly cultivated as a salad crop but it may also be a better source of animal food, providing a grazing source for flocking birds, quail, doves and cattle, whereas birds eat its fruits. It sprouts most commonly in the spring in sunlight areas after first heavy rains of the season.

COMPOSITION AND USES

COMPOSITION

Miner's lettuce is a good source of carbohydrates and protein and a variety of minerals such as calcium, magnesium and a fair source of selenium, zinc, iron, thiamine and riboflavin. It contains high amount of vitamin A and vitamin C. The nutritional composition of miner's lettuce is given in Table 92.1.

TABLE 92.1

The Nutritional Composition of Claytonia (per 100 g Edible Portion)

Constituents	Contents	Constituents	Contents
Water (g)	84	Zinc (mg)	0.6
Carbohydrates (g)	3.6	Selenium (mcg)	0.9
Protein (g)	2.0	Vitamin A (IU)	7092
Fat (g)	0.4	Thiamine (mg)	0.1
Calcium (mg)	38.0	Riboflavin (mg)	0.1
Potassium (mg)	459	Niacin (mg)	0.4
Sodium (mg)	4.0	Pantothenic acid (mg)	0.1
Magnesium (mg)	13.0	Pyridoxine (mg)	0.3
Iron (mg)	2.2	Folic acid (mg)	12.0
Copper (mg)	0.1	Vitamin C (mg)	38.2

Uses

Miner's lettuce is consumed raw as salad and cooked or boiled as a leafy vegetable like spinach but it is not as delicate as the other kinds of lettuce. The leaves contain a mucilaginous texture, thus, it is quite pleasant in salad. However, it is better to use young leaves since the older leaves become bitter in taste particularly in hot dry summer. Its node-like roots having a nutty flavour can also be eaten. Its flowers are used as a garnish in salad. Its raw stalks and bulbs are also consumed as salad. The larger leaves of miner's lettuce may be used as a wrap for grilling the fish or tossed into stir-fries, sandwiches and pasta dishes. The boiled and peeled roots with chestnut flavour are used in flavouring food. Its raw greens because of their easiness to melt in mouth have gained popularity in the foodie greens, attracting the chefs.

Medicinal Uses

Miner's lettuce being rich in vitamin C prevents scurvy disease and cold in humans. The decoction of its powdered roots is given to children suffering from convulsion. The plant being laxative prevents constipation and purifies the blood and lymph system. Its constituents have anti-rheumatic property, thus, the poultice of its mashed plant is applied to rheumatic joints. The plant juice can be taken as an energizing spring tonic and an effective diuretic. Being high in chlorophyll and antioxidants, it acts as detoxifier and helps in eliminating environmental toxins and heavy metals stored in the liver. It contains omega fatty acids, which act as an anti-inflammatory and help in counteracting the pro-inflammatory effects of fats and oils.

BOTANY

Miner's lettuce is a fast-growing broad leaf with short-lived perennial but usually grown as a winter annual. The plant having a semi-erect habit grows up to a height of 40 cm and normally produces a single stem, which has one main leaf branching off it along with several small flowers and seedpods. The two stem leaves are found at distal end of the stem, where they adjoin together to form a 0.6- to 5-cm-wide disc, through which, the stem passes, hence, the species was given a name *perfoliata*, meaning the stem passing through the leaf. This is a key diagnostic characteristic for this species. The leaves are simple oblanceolate, 3–14 cm long and 0.5–1.3 cm broad with a 6- to 20-cm-long petiole but first true leaves form a rosette at base of the plant and are 0.5–4 cm long with 20 cm long petiole. The basal leaves are linear-spatulate to long petiolate with oblanceolate to rhombic-obovate blades up to 3 cm wide. The mature leaves are of different shapes, *i.e.*, from football to triangular kidney shaped with round or pointed tips.

The foliage colour is dark green. The flower stalk appears to grow through a circular cuplike structure (bract) that looks like a leaf and surrounds the entire stem. Flowers are hermaphrodite, protoandrous, 0.7–1.4 cm in diameter, 5–40 in number and are white or pink in colour, having five short petals, two long green petals and five stamens. The petals are 2–6 mm long and reflect ultraviolet light. The plant flowers from February to May, which remain opened for 3 days, although the five stamens on each flower remain active only for a day. Its inflorescence is raceme type. The cotyledons are usually bright green, rarely purplish or brownish-green, succulent, long and narrow. Its seeds are shiny black in colour, 0.2–0.3 cm in diameter and oval to circular shaped with a white appendage at the point of attachment. Though the crop is self-pollinated, a few degree of pollination is also performed by insects.

CLIMATIC REQUIREMENT

Miner's lettuce prefers cool and moist climate for its better growth and development. The optimum soil temperature for seed germination is 10–12°C and for growth 12–29°C. Being very hardy plant, it can tolerate temperature up to −15°C. The crop is moderately tolerant to frost but its green leaves are susceptible to hard frost in winter, thus, it needs a frost-free period of at least 6 weeks. Its leaves turn red and dry out under a condition of extreme heat. The crop for its better growth and development needs full sunlight, but in hotter months, it can also be grown in partial shade. In early summer, the crop stand and quality are found best in shaded areas, especially in the uplands.

SOIL REQUIREMENT

Miner's lettuce can be grown on a wide range of soils from light to heavy, including sand, gravel road tar, loam, rock crevices, scree and river silt but it prefers well-drained sandy loam or peat soil rich in organic matter though it can be grown in nutritionally poor or low-fertility soils. It can also be grown in acid (6.1–6.5), neutral (6.6–7.5) or basic soils (7.6–7.8) but it prefers soils with a pH range of 6.5–7. The soil should be at least 15 cm deep with good water-holding capacity for its better growth and development.

CULTIVATED VARIETIES

Several open pollinated local selections without proper nomenclature are available. These selections are distinct in terms of morphological characters, yield and quality. However, the farmers prefer to grow high-yielding local collections. *Claytonia perfoliata* belongs to a polyploidy group of considerable complexity with several geographical subspecies given as under:

- *Claytonia perfoliata* subsp. *perfoliata*: This subspecies is grown in Pacific coastal United States and southwest Canada.

- *Claytonia perfoliata* subsp. *intermontana*: This subspecies is grown in interior western United States.

- *Claytonia perfoliata* subsp. *mexicana*: This subspecies is grown in coastal southern California and Arizona south through Mexico to Guatemala. It flowers from February to May.

- *Claytonia perfoliata* subsp. *utahensis*: This is a local subspecies in Utah, which flowers from June to July.

SOWING TIME

In north plain areas, miner's lettuce is sown in last week of August to first week of September as autumn–winter crop, but in hilly tracts, it is sown in March to April as spring–summer crop, but under protected conditions, it can also be sown from August to December. Succession planting is done from early to mid-autumn for winter harvest.

SEED RATE

The seed rate in general depends on variety, seed viability, growing region, soil conditions and irrigation facilities. Under normal sowing conditions, a seed rate of 2.8–3.5 kg is sufficient for sowing a hectare area.

SOWING METHOD

Miner's lettuce is propagated through seed, which is sown directly in well-prepared raised seedbeds, which are made 1 m wide and 15–30 cm high from the ground level, maintaining 22–30 cm distance between these beds. Sowing is done at 0.5–1 cm depth, keeping 20–30 cm between rows and 2–5 cm distance between plants. Though the seeds are tiny, the sowing should be done sparsely to avoid thinning. If large quantity of seed is used per unit area, thinning is compulsory to do. The seed usually germinates within 7–10 days. When the seedlings are large enough to handle, they are thinned to a spacing of 5–8 cm.

The crop can also be raised through transplanting, if the seedlings are grown in protrays. The seedlings become ready for transplanting in 4–5 weeks. Miner's lettuce having taproot system cannot tolerate minor injury to its roots, thus, the seedlings should be transplanted carefully without disturbing the root ball.

NUTRITIONAL REQUIREMENT

Miner's lettuce is normally grown on poor soil and it does not require fertilizers in large quantity for its successful cultivation. The crop is mainly grown on small scale. Hence, the growers use fertilizers as per their convenience and crop requirement, depending on the requirement and condition of the crop. Since it grows under wide soil and climatic conditions, well-decomposed farmyard manure is quite beneficial for the crop. Information on fertilizers requirement of the crop are not available anywhere in the literature, thus, it is not possible to give fertilizers doses.

IRRIGATION REQUIREMENT

Miner's lettuce requires enough moisture from sowing to harvesting for its successful cultivation and for producing abundant and large leaves, therefore, irrigation should be applied at regular interval and the soil should not be allowed to dry up between two irrigation schedules. Lack of soil moisture will either cease the plant growth or reduce the size of leaves as well as their quality. If the crop is sown in dry soil, irrigation is applied just after sowing for better seed germination and establishment of seedlings.

INTERCULTURAL OPERATIONS

HOEING AND WEEDING

The field should be kept weed free, especially in initial stages of plant growth, to avoid crop–weed competition for nutrients, moisture, space and light. Shallow hoeing after each cutting is advisable to harvest quality leaves. Usually, two to three hand hoeing are enough to keep the weeds down and to facilitate soil aeration for proper root development. Pre-emergence application of Fluchloralin 1.5 kg a.i./ha coupled with one-hand weeding 30 days after sowing is also effective to control the weeds.

When the seedlings are large enough to handle, the plants are thinned to 5–7 cm distance if repeated cuttings are to be taken or 10–15 cm distance if merely the individual leaf is picked. This practice is normally followed to increase the size of leaves as well as their quality.

HARVESTING

Miner's lettuce becomes ready for cutting about 40 days after sowing. The greens with best flavour can be obtained when the leaves are cut frequently at young stage and not allowed to be over mature since the plant at young stage forms a basal rosette with slightly succulent, hairless and bright green leaves. The crop gives cuttings from October to December as soon as the leaves are big enough to cut, or sometimes until as late as March until the plant starts flowering. Usually, it gives four to five cuttings during this period. After the plant has branched, the heart-shaped leaves are clipped in pairs. The plants after cutting grow rapidly for constant cuttings. The last couple of cuttings in the spring give the best greens because they produce tiny white edible flowers through the centre of the leaf. The greens are harvested early in the morning when the dew is still present on the leaves. For juice purpose, the leaves are harvested along with long petioles so that more juice may be extracted.

YIELD

Yield of the miner's lettuce varies from region to region and species to species. So far its yield is concerned, no information is available anywhere in the literature.

POST-HARVEST MANAGEMENT

Miner's lettuce can be stored only for few days at ambient room temperature as well as in refrigerator since it tends to deteriorate quickly after harvest. Immediate after harvesting, pre-cooling is done in cold water, as it will help in preserving its flavour and texture.

INSECT-PESTS

APHIDS (*Aphis gossypii*)

Nymphs and adults damage the crop by sucking cell sap from tender leaves, causing yellowing and curling of leaves and death of plants if attacked at young stage. They secrete honeydew-like sugary substance, which encourages the growth of black sooty mould, reducing the photosynthetic activities of leaves and crop yield. High humidity and high temperature favour the pest infestation. Mostly, it attacks the plants at early stages of crop growth. Aphids act as virus vector often assisting in the introduction of diseases like lettuce mosaic.

Control

- Grow resistant variety, if available.
- Use bio-agents such as *Coccinella septumpunctata*, *Menochilus sexmaculatus* and *Chrysoperla carnea* to keep the pest under check.
- Use natural predators such as lady beetles, lacewings, damsel bugs, fly maggots, parasitic wasps and birds.
- Spray the crop with 0.4% *neem* oil, Malathion 0.5%, Rogar 0.25% or phosphamidon 0.03% at 10–12 days interval.

CATERPILLARS (*Spodoptera litura*)

The freshly hatched larvae feed gregariously on leaves and scrape the chloroplasts. At later stages, they feed voraciously on foliage at night and hide usually in the soil around base of the plants during the daytime. Sometimes, the feeding is so profound that only petioles and branches are left behind.

Control

- Grow resistant variety, if available.
- Grow castor or sunflower along the border as a trap crop.
- Collect and destroy the caterpillars in clusters.
- Set up pheromone traps at 2 traps/ha.
- Apply nuclear polyhedrosis virus (NPV) at 250 LE (larval equivalents)/ha with sugar 2.5 kg/ha.
- Spray the crop with Quinalphos 20 EC 0.05%, Dichlorvos 0.5%, Indoxacarb 0.1% or Spinosad 0.05% at 10 days interval.

Leaf Miner (*Liriomyza langei*)

Leaf miner makes winding tunnels within leaf beneath the epidermis. When the attack is severe, it produces small black specks within leaf mine, and its larvae feeding within leaf tunnels produce numerous black feces. Leaf epidermis is pulled back from a leaf mine to reveal two larvae feeding within the leaf. Its adult causes scarring symptoms on leaves. Its adult female lays eggs in upper leaf surface, which in turn become maggots. Although its damage restricts plant growth, resulting in reduced yield and loss of vigour but healthy plants can tolerate considerable damage.

Control

- Follow long crop rotation with non-host crops.
- Grow resistant varieties, if available.
- Incorporate compost and other soil amendments before sowing.
- Maintain plant health using fertilizers and proper watering.
- Use floating row covers to prevent the adult from laying eggs on leaves.
- Fix yellow sticky traps in field at different locations to trap the egg-laying adults.
- Cover the soil surface with polyethylene sheet to prevent the larvae from reaching the ground and pupating.
- Use parasiticii wasp *Diglyphus isaea* or spray the crop with *neem* oil 0.4%.

Thrips (*Frankliniella occidentalis*)

Nymphs and adults lacerate the surface of leaflets and suck the oozing sap, resulting in appearance of white patches on lower surface of the leaves and distortion of young leaflets. Severe infestation causes stunting of plants. It may affect the entire plant at all growth stages. It is the most damaging pest that attacks the crop widely.

Control

- Grow resistant variety, if any.
- Uproot and destroy the severely infected plants.
- Spray *neem* oil 1 L+ 1 kg soap powder mixed in 200 L of water twice at 10 days interval.
- Spray the crop with Malathion 0.5%, Rogar 0.25% or methyldemeton 0.25% at 10–12 days interval.

Beetle (*Lamprima aurata*)

Beetle is a soil-inhabiting pest. Its larvae hatch in the soil and often feed on the plant roots. These are very active in sandy and sandy loam soil. Beetles cause damage during germination of seeds in the field. They burrow in soil by digging out soil particles, which cover the tender seedlings and disturb the root system. The damage becomes more serious at irrigation time or during rain.

Control

- Grow resistant variety, if any.
- Drench the field a week after sowing with 0.04% Chlorpyriphos 2 L/m^2.
- Apply *neem* leaves extract 10% to reduce the beetle population.
- Spray the crop with Chlorpyriphos, Fenitrothion or Quinalphos 0.05% at 10–12 days interval.

Root Weevil (*Otiorhynchus* spp.)

Root weevil is a destructive pest that invades the root system of a healthy plant and then proceeds to eat the plant roots. Larval root weevils will look like white grubs or worms and live in soil. Adult weevils are beetle-like insects that can be black, brown or grey. The plant leaves become irregular, as someone has been taking bites out of the edges. The damage appears in the night, as the root weevils come out to feed at night.

Control

- Handpick and destroy the adults off the plant at night.
- Use predatory beetles to control the pest.

Spider Mites (*Tetranychus urticae*)

Adults are reddish brown or pale in colour, oval shaped and very small. Mites live in colonies, mostly on underside of the leaves and feed by piercing leaf tissue and sucking up the plant cell sap. Feeding marks show up as whitish grey dots on leaves. Affected leaves become mottled, turn brown, dry up and drop off. Under severe infestation, the top plant canopy is covered by webbing of mites. The infested plants can be identified from the distance by the characteristic mottling symptoms produced on upper surface of the leaf, often found on under surface of leaves. The infestation of mites is mostly observed during warm and dry period of the season.

Control

- Follow long crop rotation with non-host crops.
- Follow pruning of leaves, stems and other infested plant parts.
- Use insecticidal soap, *neem* oil or botanical insecticides to infested areas.
- Spray the crop with wettable sulfur 80WP 0.2%, Dicofol 0.5% or Spiromesifen 0.05% at 10 days interval.

Slugs and Snails (*Lehmannia valentiana*)

Slugs and snails attack the young tender greens and can voraciously feed on seedlings almost as soon as they emerge in the field. During daytime, they hide in weeds, plant debris, stones, boards, ground cover and anything close to the ground.

Control

- Follow long crop rotation with non-host crops.
- Grow resistant varieties, if available.
- Use trap crops such as nasturtiums, begonias, fuchsias or geraniums.
- Collect and destroy the snails and slugs from inhabitable areas.
- Use drip irrigation to reduce the humidity and soil moisture.

- Keep the surrounding area clean.
- Spray the crop with *neem* oil 0.4%, dimethoate 0.03%, Fenitrothion or Quinalphos 0.05% at 12 days interval.

Root Knot Nematode (*Meloidogyni incognita*)

Root knot nematode is a microscopic parasite that invades plant roots in soil. This pest has several species but their effect on plants is same. Nematode attacking the plants at seedling stage leads to pre- and post-emergence damage, resulting in reduced plant stand in the field. Root knot nematode enters the roots, causing characteristic root knots or galls. As a result, the root system becomes deformed or hairy, which prevents the absorption of water and uptake of nutrients from the soil. This results in yellowing of leaves and stunting of plant growth.

Control

- Follow long crop rotation with non-host crops.
- Plough the field two to three times deep in hot summer months.
- Grow resistant varieties, if available.
- Drench the field soil with 0.1% formaldehyde up to a depth of 15 cm.
- Apply Nemagon 30 L/ha with irrigation before sowing.

DISEASES

Rust (*Puccinia mariae-wilsoniae*)

The disease is commonly known as the spring beauty rust, which grows on all sides of the plant, producing scattered clusters of reddish-brown sori on surface. The aecia (specialized reproductive structures) are cup shaped and have a conspicuous peridium. The spores produced by the aecia (aeciospores) are translucent hyaline covered with minute warts. The telio spores are about 0.1–1.0 mm in diameter and dark brown.

Control

- Follow long crop rotation with non-host crops.
- Grow resistant varieties, if available.
- Remove and destroy the affected plant parts.
- Maintain good sanitation in the field.
- Spray the crop with Mancozeb 75 WP 0.2%, or extract of garlic bulb or *Eucalyptus* leaf 2.0%, at 50 and 70 days after sowing.

Beet Yellows Virus

The characteristic symptoms of beet yellows virus in Miner's lettuce are chlorosis of older leaves and red-rimmed necrotic spots on systemically infected younger leaves, which show vein clearing, then inter-veinal yellowing develops and the plants growth becomes stunted. In greehouse, the young plant leaves often show vein clearing and *vein-etch*, whereas the older leaves become yellow, thickened and brittle usually with numerous small red or brown necrotic spots. The virus is transmitted by multiple aphid species and their all instars transmit virus but adults are the more efficient. It is readily transmitted by aphid inoculation and with difficulty by manual inoculation of sap.

Control

- Use bio-agents such as *Coccinella septumpunctata*, *Menochilus sexmaculatus* and *Chrysoperla carnea* to keep the aphids under check.
- Introduce natural predators such as lady beetles, lacewings, damsel bugs, flower fly maggots, parasitic wasps and birds in field to control the virus vectors.
- Spray the crop with 0.4% *neem* oil, Malathion 0.5%, Rogar 0.25%, or phosphamidon 0.03% at 10–12 days interval to kill the aphids.

■■■

93 Lotus Root

M.K. Rana and Shiwani Kamboj

Botanical Name : *Nelumbo nucifera* (Sacred or Indian lotus)

Nelumbo lutea (American or yellow lotus)

Family : Nelumbonaceae

Chromosome Number : 2n = 16

ORIGIN AND DISTRIBUTION

Lotus, an ancient water plant, was found growing about 135 million years ago in many low and watery areas of the Northern Hemisphere. Asia and North America have been suggested as the centre of origin of sacred lotus (*Nelumbo nucifera*) and American lotus (*Nelumbo lutea*), respectively. Sometimes, the Caribbean is also believed as the centre of origin of American lotus. It is found distributed in the extreme north of North America and the extreme northeast of Asia near the Shahalin Islands. According to recent archeological studies, lotus also has a history of at least 7000 years in China, and as per the literature, lotus had been found growing along the Yantze River and Yellow River area before that time. About 70 years ago, many lotus seeds were found buried for nearly 1000 years. Now, it is widely distributed in China with more than 100 different cultivars.

INTRODUCTION

Lotus, also known as Indian lotus, sacred lotus, East Indian lotus, oriental lotus, lily of Nile and bean of Kashmir, is an important economic herbaceous perennial aquatic plant, each part of which has an economical value. The lotus with beautiful leaves and flowers standing high above the water is found to be most dramatic pond ornament to date. It is as popular as renkon in Japan. It has held high esteem in far eastern regions, especially in Chinese and Japanese cultures for centuries. The lotus grows in Asia, Oceania, North and South America and Australia. However, China, Japan, India, the Philippines and Indonesia are the major growing countries in the world. The lotus is symbolic in both Hindu and Buddhist religions. In Hindu religion, its flowers are found associated with creation mythology and the unfolded petals are considered a symbol of fertility, beauty and prosperity. In Buddhist religion, it is believed that their lord was born on a lotus leaf, and in Buddhist symbolism, the wheel-like *mandala* based on the form of lotus blossom represents the cycle of life, death and rebirth. Both Hindu and Buddhist deities are symbolised as seated in the lotus flower and lotus motifs adorn sculptures, buildings, temples and paintings. It is the national flower of India and Vietnam. In addition to the flowers, almost all parts such as roots, stems, leaves, seeds, etc. are also used widely in Asian cooking. The use of lotus root as food has a history of three to five thousand years. The immature seedpods and leaves are used in culinary preparations. All parts of the lotus plant in one or another form are used in Chinese medicines also.

COMPOSITION AND USES

COMPOSITION

The starchy lotus roots are very nutritious, as they provide a good amount of carbohydrates, proteins, lipids, calcium, potassium and phosphorus. It is also a good source of vitamins and dietary fibres. Its roots also provide an appreciable amount of energy. The nutritional composition of lotus roots is given in Table 93.1.

USES

The lotus plant is used in many ways in our daily life. Almost all lotus parts such as roots, young leaves, flowers and seeds are edible. In China, its easily digestible starchy roots that are good for people of all ages are eaten as a staple food either raw, cooked or pickled. The lotus roots can be processed into flour and used in cooking, making bread or for the thickening of soups. The seeds are also starchy and can be consumed raw, dried, roasted, pickled, boiled, candied or used as a paste in cooking though the bitter germ must be removed before eating. The lotus seeds or nuts can be consumed raw or dried and popped like popcorn. Its immature roots and seeds that are crispy and sweet in taste are used for flavouring summer delicacies. Its edible leaves are used as a flavouring agent and for wrapping food and spicy mixtures (rice, meat, fruit, etc.) for steaming. In south India, the lotus stem is sliced and marinated with salt to dry, and the dried slices are fried and used as a side dish. In Sri Lanka, the sliced lotus stem curry is a popular dish called as *Nelum Ala*.

In Myanmar, a unique fabric is produced from the lotus plant fibres. The distinctive dried seed heads, which resemble the spout of watering cans, are widely sold throughout the world for decorative purposes and for dried flower arrangement. The flowers are beautiful and fragrant during summer months, which are used in landscaping and gardens.

MEDICINAL VALUES

All parts of the lotus plant, in one or another form, are used in Chinese medicines. In India, its honey is considered as a tonic known as *Padmammadjhu* or *Makaranda*, which is used to treat eye disorders. Its roots extract has anti-diabetes and anti-obesity properties. The leaves are known for their refrigerant, astringent and diuretic actions. Therefore, they are used in diverse applications such as diarrhoea, high fever, haemorrhoids and leprosy. In Korea, the traditional liquor prepared from its leaves and flowers is useful in reducing oxidative stress and the risk of chronic diseases because of its antioxidant properties. Ripe lotus seeds are also used as a spleen tonic and in the treatment of chronic diarrhoea due to their astringent action. In many Asian countries, the seeds are used as an antidepressant. The sun-dried lotus plumules are used for nervous disorders, insomnia, hypertension and high fever with restlessness. The flower receptacles contain a small amount of alkaloid

---

TABLE 93.1
Nutritional Composition of Lotus Roots (per 100 g Edible Portion)

Constituents	Contents	Constituents	Contents
Water (g)	81.42	Potassium (mg)	363
Carbohydrates (g)	16.02	Iron (mg)	0.90
Protein (g)	1.58	Thiamine (mg)	0.13
Dietary fibre (g)	3.10	Niacin (mg)	0.3
Calcium (mg)	26	Vitamin C (mg)	27.4
Phosphorus (mg)	78	Energy (kcal)	66

Nelumbine, which is used to stop bleeding and to eliminate stagnated blood. The stamens assist kidney functions and are useful in the treatment of female leucorrhoea and male sexual disorders.

BOTANY

Lotus is an aquatic herb with a submerged horizontal stem. It is an earliest dicot plant but it has certain characteristics of monocot plant. The leaf vein system and the two cotyledon leaves are some characteristics of dicot plant. The lotus plant cotyledons are covered with a shell-like structure, the characteristics of which are similar to a monocolyledon plant. Its nodes and hairy roots are also characteristics of monocots. Based on current studies, it may be suggested that the ancient lotus plant is a dicot but has a close relationship with monocots too. The stem contains large spaces that allow air storage for the submerged plant structures.

The plant has its roots firmly in the mud and sends out long stems, with which, their leaves are attached. The leaves, sometimes, and flowers are always raised above the water surface. It has a hairy and indefinite root system, which grows underground at the internode of the stem. Around each internode, there may be 5–8 bunches of roots and each bunch may have 7–21 strings. The roots during growing season may be white or light purplish but turn dark brown when they are mature. The leaves have long stalks that join them in the centre of the blade and they either float on the surface or emerge above the water.

The beautiful flagrant flowers emerge above the waterline and open in the morning and remain closed overnight. They keep this opening and closing process continuous for 2 days. Their petals fall down in the afternoon on the day of opening. During opening, a large number of yellow stamens can be seen. The base of the flower is known as the receptacle, which expands and becomes fleshy during maturation and surrounds the flower ovary. The fruit of lotus is known as *torus*. The fruit is a conical pod, containing seeds in holes. Its mature seeds are 8 mm long, hard shelled and brown in colour.

CLIMATIC REQUIREMENT

Lotus is found growing at an altitude as high as 2000 m above sea level. For better growth, a temperature greater than 15°C is required for at least 6 months but the maximum root growth takes place at a temperature above 20°C. In addition, the plant, at high temperature, produces more root branches and flower development is stimulated, but simultaneously, susceptibility to fungal diseases is also increased. Lotus never flowers in winter since the plant enjoys warm weather coupled with sunlight and is highly susceptible to cold and drought. Long day conditions accelerate root elongation and upright leaf production, while short-day conditions promote root enlargement and inhibit upright leaf production.

SOIL REQUIREMENT

Lotus can adapt to most soils if there is no hardpan but lakes or ponds containing large amounts of organic matter in their bottom are most suitable for its cultivation since soil with appreciable amounts of organic matter provides nutrients, good binding sites for nutrients, texture and promotes root penetration. For commercial production, heavy clay loam soils are most suitable for better growth of aquatic plants. Sometimes, heavy soil may be used so that the plants may not float but it limits the use of most potting soil mixes found at garden centres. In heavy clay soils, roots cannot penetrate easily and harvesting also becomes difficult. On the other hand, sandy soil lacks a binding site for nutrients and induces unpleasant flavour in the roots, therefore, the optimal soil for its cultivation is a soft silt loam free from particulate matter. It can grow at pH 5.6–7.5 but the optimum pH is 6.5.

CULTIVATED VARIETIES

ASIATIC

The leaves of this variety are large and pea green in colour. The flowers with delicate fragrance are fine, pure white and of large size. It is one of the best white-flowered lotuses. It can tolerate very cold climates with ice and snow. The plant may reach to a height of 90–120 cm.

CHAWAN BASU

The name Chawan Basu refers to rice bowl. This is a beautiful lotus for any small to medium pond or container. Flowers are ivory white with deep pink margins and veins on the petals. The planting depth is 8–30 cm. The plant may reach to a height of 60–70 cm.

CHINESE DOUBLE ROSE

The flowers of Chinese Double Rose are of deep rose colour, double. It flowers freely, resembling that of a peony. The planting depth should be 8–30 cm. The plant reaches to a height of 45–60 cm.

MOMO BOTAN

This is one of the best blooming lotuses. The flowers are double rosy pink and stay open later in the day. It has a long blooming season. This cultivar is good for growing in containers and small to medium ponds. The plant may reach to a height of 45–60 cm.

PERRY'S GIANT SUNBURST

Large, cupped, very fragrant, double, yellow flowers appear in summer on stiff stems above the foliage. Each flower blooms for about 3 days, opening in the morning and closing at night each day. Flowers are followed by nut-like fruits that are imbedded in flat surface of a turbinate receptacle, which resembles the shape of a watering rose can. The planting depth of this variety should be 8–45 cm.

SACRED PINK

This variety possesses great significance in eastern religions. The planting depth may be kept 8–45 cm. The plant may reach to a height of 90–120 cm.

SEED RATE

Lotus can be propagated either sexually through seeds or asexually through stem division; however, the original characteristics of mother plant can be preserved only by asexual propagation, flowers can be enjoyed and its roots can be harvested for the same season. On the other hand, it is propagated by seeds mainly for breeding new cultivars but the seeds should be fully mature for better germination. The amount of seed or stem pieces used may vary with varieties and the purpose for which the crop is being growing. About 3800 seeds are required for sowing a hectare area, and if propagated vegetatively then approximately 7000 stems weighing about 30–35 quintals are required to plant a hectare area.

SOWING TIME

The sowing time varies considerably according to the purpose of growing, type of variety and climatic conditions of growing region. Sowing of seeds of a particular variety at its right time is very essential to obtain higher yield of better quality roots since unfavourable environmental conditions have adverse effect on yield. In addition, if crop is planted too early, the temperature may be too low and the seed stem section may suffer cold injury. If the lotus is planted too late, the apical bud may have already sprout and have small coin leaves, which may be injured during planting. The planting should be done when the average temperature is between 15°C and 25°C. Mostly, the seeds can be sown in March to April and planting is done in the start of June. Flowering can be expected in the same year if the lotus is planted before June.

SOWING/PLANTING METHOD

The preparation of a good seedbed is essential for better seed germination. The length of seedbed may vary according to space available and purpose of growing. To hasten germination, the seed coat should be clipped manually at the tip without any injury to the cotyledons. The seeds are then soaked for 3–5 days in water at a temperture of 15–25°C. The soaking water should be changed every day, and thereafter, the soaked seeds are placed in a row 1.0–1.5 m apart, and the distance between the seeds should be 1.0–1.5 cm. After sowing seeds, the bamboo frames should be constructed above the seedbeds, and the entire planting should be covered with plastic for maintaining the high temperature inside. Water should also be added to cover the seedbeds up to a depth of 3–5 cm. When the seedlings in the seedbeds have four floating leaves, they can be transplanted in the field.

During transplanting, the proper amount of mud should be carried around the roots and care should be taken to avoid the breaking of the fragile leaf stalk. In several regions of South China, two crops can be taken in the same year. Fleshy stems from the first crop can be planted in June for second crop. The main or branched stems and secondary stems may be used for planting in field and tubs, respectively.

Early-maturing varieties are planted closer than the late ones. The planting in the field is done closer than the planting in a lake or pond. For planting in deep water, two to four stem pieces are tied together with straw, and thereafter, a hole is made with the help of foot and the bundle of pieces is pushed into the mud with a planting fork. After inserting the bundle, more mud is stacked on it to cover stem pieces. Additionally, another method may be to place two to three stem pieces in a straw bag. Some rock pieces may also be placed in the bag and the entire bag is sunk into the lake.

NUTRITIONAL REQUIREMENT

Lotus crop for flower production does not usually require side dressing unless the leaves are turning yellow during the growing period or the size of the leaves is smaller than normal. Some side dressing of fertilizers may be necessary if the crop is raised for root or seed production. The amount of fertilizers applied depends upon stage of crop, *e.g.*, the crop at root formation and maturity requires more potassium and less nitrogen. For getting higher root yield (45–50 t/ha) well-decomposed farmyard manure should be incorporated into the soil at the time of field preparation. Application of *neem* cake 1 tonne, diammonium phosphate and muriate of potash each 250 kg/ha is required for getting higher yield. Liquid fertilizers through a large pipe containing a mixture of urea 1%, dipotassium hydrogen phosphate 0.01% and boric acid 0.05% may also be used. The time of application may be once at the time of one or two standing leaves appearance and second when the standing leaves have ended.

IRRIGATION REQUIREMENT

The water requirement of lotus varies with different stages of growth and purpose of growing. The crop for root production requires more water than the crop for flower or seed production. Generally, water level at the time of planting should be about 5–10 cm, and as the standing leaves grow, it should be raised up to 30–60 cm. Shallow water raises the temperature and is beneficial to younger plants. The water should also be lowered to about 5 cm at the time of root maturity. In summer, the temperature and the evaporation rate increase, thus, higher water levels are needed.

INTERCULTURAL OPERATIONS

For obtaining economic yield, the lotus field should be kept weed-free since weeds compete with crop plants for nutrients, light, moisture and space. Hence, during the crop period, two to three weedings should be done manually. For keeping the field weed free, first weeding should be done at the appearance of seedlings, second when young leaves start floating and third after the standing of leaves.

The lotus plants should be protected from prevailing winds by planting windbreaks on the sides of the lake or pond. In large lakes, the farmers may plant other aquatic plants several metres away from the lotus planting as a buffer against waves created by strong winds. Stakes near the flowers, which are tied to them, can prevent their breaking.

HARVESTING

The time and method of harvesting depends on the environmental conditions, variety and purpose of growing. Cream or light brown lotus roots attaining a size of 60–120 cm length and 6–9 cm diameter are ready for harvesting. The crop takes 120–150 days after planting for attaining the appropriate size. The roots are harvested manually by pulling out by hand from the deep mud. While pulling out the roots, care should be taken that some of the roots may be left in the soil for the next crop. If the crop is planted in June, both seeds and edible roots can be obtained in the same season.

YIELD

The average yield of lotus root varies from 20–30 t/ha, depending on crop variety, soil, environmental factors and cultural practices adopted by the growers.

POST-HARVEST MANAGEMENT

The roots after harvesting should be washed properly but the presence of mud is not harmful. The packinghouse operations such as sorting, grading, trimming and packaging should be done before sending them to the market. The fleshy roots cannot be stored for long periods at room temperature after removal from the soil but refrigerated storage can prolong their shelf life. They can be stored for 1 month in winter and only for 10–15 days in spring or summer months.

The lotus seeds after harvesting should be dried in the sun and stored in cool, dry place and they should be protected from rain and high humidity during transportation.

INSECT-PESTS

WINGED APHID (*Aphis nerii*)

A sucking-type insect that causes severe damage to the above waterline foliage. The attack of winged aphid causes wilting, curling and yellowing of the foliage which decreases the yield. The pest spends autumn, winter and spring on fruit trees like peach or plum trees but as soon as the leaves of the lotus come out, the pest returns to the lotus plants.

Control

- Wash off the aphids with a pressure stream of water.
- Remove the infected parts of the plant.
- Use organic slow-release biodegradable fertilizers.
- Spray the crop with Phosphamidon 0.05%, dimethoate 0.03% or imidacloprid 0.2%.

WHITEFLIES (*Trialeurodes vaporariorum*)

The larvae of whitefly are oval shaped and adults resemble a moth. The whiteflies feed on the underside of the leaves. These also suck cell sap from the leaves which causes stripping and yellowing of leaves take place. The whiteflies are also responsible for the transmission of viruses from unhealthy to healthy plants.

Control

- Grow the crop during the period when the whitefly population is less.
- Apply Furadon at 1.5 kg a.i./ha in the nursery 15 days before uprooting the seedlings.
- Use nylon nets to protect the crop from whitefly.
- Spray the crop with dimethoate or Metasystox 0.15%, Fenitrothion 0.05% or Confidor 0.3% at 10–15 days interval.

LEAF ROLLERS (*Archips argyrospila*)

The leaf roller is also a destructive pest of lotus. The pest causes severe damage to the lotus plant by cutting leaf edges and rolling them inside. The larvae also feed and cut down the leaves, causing a great loss of yield.

CADDIS FLY (*Pycnopsyche sp.*)

The larvae resemble tiny worms, therefore, known as caddis worms and adults look like brown moths. The larvae of caddis fly feed on the flower buds, roots and stems.

Control

- Remove infested parts as soon as the pest appears.
- Clean up all floating debris to eliminate the places for hiding and breeding of pests.
- Spray simple water with strong jet pressure to remove caddis flies.
- Apply *Bacillus thuringienesis*, naturally occurring bacteria to control caterpillars effectively.

Note: The lotus leaves have fine hairs, which retain the sprayed chemical longer, thus, application of insecticides may harm or even kill the leaves.

DISEASE

FUSARIUM WILT (*Fusarium bulbigenum*)

The fungus attacks the roots and collar region of lotus plant. The affected parts show brown discolouration and become wilted. The vascular tissues inside the root show browning but the pith remains unaffected. The growth of the plant becomes stunted. The plant ultimately wilts and dies. The disease is soil-borne but can be seed-borne also.

Control

- Follow long crop rotation with non-host crops.
- Use disease-free healthy planting material.
- Remove and destroy all the affected plants along with their roots.
- Treat the planting material with Captan, Captaf, Captafol, Foltaf or Bavistin 2 g/kg of seed.
- Drench the seedbeds with Captan, Captaf, Captafol, Foltaf, or Bavistin 2 g/litre of water.

Leaf Spot (*Cercospora nymphaeacea*)

The pathogen attacks the leaves and stem of the lotus plant. The affected leaves and stem have large circular, oval, or oblong spots with grey centres. In severe infection, the whole leaves turn yellow which decreases yield. The pathogen survives in plant debris and on and in the seed.

Control

- Avoid planting on infected soil.
- Treat the propagating material with 0.2% solution of Captan.
- Pick and destroy the infected plant to avoid the spread of disease.
- Remove and destroy infected plants.
- Spray the crop with Dithane Z-78, Fytolan or Blitox 0.25%, Bavistin 0.2%, or Benlate 0.1% at 10–12 days interval.

■■■

94 Oca

M.K. Rana and Chandanshive Aniket Vilas

Botanical Name : *Oxalis tuberosa* Molina.

Family : Oxalidaceae

Chromosome Number : 2n = 88

ORIGIN AND DISTRIBUTION

Oca is a tuberous plant that originated in the Andes Mountains of South America but its cultivation is not known for how long it has been cultivated there. As per the literature, it has been grown for several thousand years before the arrival of Pizarro who brought it to the Inca Empire at the end for its tubers. It is planted in the Andean region at 2800–4000 m above mean sea level. The greatest diversity of cultivated and wild form was found in the central region of Peru and northern Bolivia. Its cultivation extended due to the migration of pre-Columbian population to Venezuela, Argentina and Chile. In Mexico, it was introduced about 200 or 300 years ago and also distributed towards New Zealand and Europe, where it was introduced as a competitor of potato long before 1830, but there, it could not be established as a permanent crop. It has been known to exist in New Zealand since 1860, and in last two decades, oca cultivation has gained popularity therein.

INTRODUCTION

Oca is a perennial plant cultivated commercially in central and southern Andes. After potato, oca is the second most widely grown tuber crop in Peru and Bolivia where its large edible tubers are considered important because of their high nutritive value and health benefits. In New Zealand, it is grown on large scale for its tubers commonly known as New Zealand yam though it has no botanical relationship with yam. In United Kingdom and North America, many growers cultivate it successfully in their kitchen gardens. In South America, it is also known by many names, *i.e. yam, macachin or miquichi, apilla, hibia* and *cuiba*, but in parts of the world, it is most commonly known as New Zealand Yam.

COMPOSITION AND USES

COMPOSITION

Oca is a goldmine of nutrients having a wide range of macro- and micronutrients. It is an excellent source of carbohydrates, protein, calcium, phosphorus, iron, zinc, thiamine, riboflavin, niacin and essential amino acids. It is a good source of dietary fibre and vitamin C, and a prominent source of anthocyanin and flavonoids. The nutritional composition of oca is given in Table 94.1.

USES

All parts of oca plant are used as vegetable, however, the most consumable part is its tubers, which are eaten raw or cooked, sliced into sandwiches, boiled, mashed, fried, baked and steam grilled. It

TABLE 94.1

Nutritional Composition of Oca (per 100 g Edible Portion)

Constituents	Contents	Constituents	Contents
Water (g)	87	Iron (mg)	12.5
Carbohydrates (g)	10.5	Retinol (µg)	1.0
Protein (g)	0.8	Thiamine(mg)	0.5
Dietary fibre (g)	8.0	Riboflavin (mg)	0.95
Calcium (mg)	17	Niacin (mg)	1.0
Phosphorus (mg)	28	Vitamin C (mg)	40
Zinc (mg)	1.8	Energy (kcal)	30

is also used in sour creams, candies, soups and stews. Leaves are eaten raw due to their sour flavour and tangy lemon taste similar to sorrel. The yellow trumpet shaped flowers are consumed as a green salad in summer. Its tuber flour is used for the preparation of porridge and dessert. The tubers are dried in sun to make them sweeter and used as dried figs and stewed fruits, after parboiling or roasting and to prepare roasted meat product *pachamanca*. Oca tubers are also used to prepare a product *khaya* and pickle in vinegar. Its stems because of their more acidic nature than its tubers are used in pies as a substitute of rhubarb or gooseberry. It can also be used readily as a substitute of potato and other root vegetables and as an alternate healthy snack for children. Oca flavour is like potato, chestnut, apple and celery, thus, it is used as a flavouring agent. Because of its attractive appearance, it is used as a good decorative vegetable. Its highly nutritious leaves are efficiently used as feed for pigs.

MEDICINAL USE

The sour flavour of oca is due to the presence of oxalic acid, which is an anti-nutritional factor and causes kidney stone and gout, hence, its excessive use should be avoided. The tubers of most oca varieties contain lesser oxalic acid, thus, they do not pose any usual risk of kidney stone and gout to most of its consumers, even then, its consumption should probably be avoided by those who are prone to kidney stone. Its leaves contain a greater amount of oxalic acid, thus, should possibly be not consumed in large quantity. It is very low in calorie but high in nutrients, thus, it is an excellent food for obese people. The *Oxalidaceae* family plants contain oxalic acid, which binds calcium to form oxalates. Although these oxalates usually pass through the urine, but the susceptible individuals who do not digest oxalic acid easily may crystallize oxalates and form kidney stones. Being a good source of vitamin C, it helps in maintaining immune system and preventing early ageing.

Oca roots secret a compound *Harmine*, which is insecticidal and is used by the plant as an insect control due to its self-defence mechanism.

BOTANY

Oca, a member of *Oxalidaceae* family, is similar to many other *oxalis* species with thick succulent stems and small flowers. A compact attractive bushy plant with clover like leaves 20–30 cm high is an erect herb at initial stage, which becomes prostrate towards maturity. In cool and moist weather, the plants grow fast up to a height of 45 cm with 90 cm spread. The leaves are trifoliate with petiole length 2 to 9 cm. The inflorescence consists of four or five flowers, having five pointed green sepals, five purple striped yellow petals and 10 stamens in two groups of five along with a pistil shorter or longer than the stamens. Its flowers exhibit tristylous heterostyly, which is responsible for self-incompatibility, thus, the plant rarely bears fruit. Their loculicidal capsules dehisce spontaneously, making the seed harvesting difficult. The flower structure facilitates cross-pollination, which takes place by insects. Plant bears waxy ribbed bright colour tubers of 25–150 mm length and 25 mm width in cluster on stolons close to the base of plant in a radius of 130 cm that are used as a propagating material. Its variegated tubers are claviform ellipsoid and cylindrical having buds on whole surface. Skin and flesh colour may be white, creamy, yellow, orange, pink, red or purple.

CLIMATIC REQUIREMENT

Being tropical crop, oca for its successful cultivation requires warm and humid climate with temperature range 10°C to 12°C and average rainfall 700 to 800 mm per annum. For sustained growth and development, it requires temperature <28°C since higher temperature results in wilting of plants. It grows better in summer as compared to winter. Cool climate and frost are not suitable for its cultivation as frost injury during winter kills the plants. Cold weather is not good in early stages of crop growth as it forces the crop towards early maturity. In warm summer climate, plant completes most of its growth. It is suitable for growing well in some of the temperate regions, tolerating high altitudes. It needs a long growing season to ensure a decent harvest. Tuberisation in oca is dependent on day length, forming tubers when the day length shortens in autumn.

SOIL REQUIREMENT

Oca can be cultivated even on relatively poor soils but it prefers loose and well-drained slightly acidic light soils with pH range from 5.5 to 7.5 for its higher tuber yield. Heavy clay soils after drying restrict the development of tubers and create problem in harvesting, hence, undesirable for oca cultivation. In highly sandy soils, the crop produces long and thin tubers, which are not desirable. Soil compaction reduces the tuber yield due to poor aeration. Highly fertile soil favours luxuriant vine growth, which affects the development of tubers adversely.

CULTIVATED VARIETIES

Several local selections without proper nomenclature are available. These selections are distinct in terms of acidity, flavour and skin colour, *i.e.* creamy white or purple. However, high yielding local collections with less acidity are preferred by the farmers to grow.

Apricot and Golden are the new varieties suitable for cultivation. Both the varieties have a taste similar to its namesake as apricot. These varieties are slightly smaller, yellow coloured and sweeter in taste. Mainly the pink colour varieties are available but red, golden, orange and pink colour varieties are also available and each of the varieties has a slightly distinct taste.

PLANTING TIME

The planting time depends on climatic conditions of the growing region. Planting is usually done when the soil temperature lies between 17°C and 35°C. In cool areas, it is planted in spring, and in warmer areas, it is planted at the beginning of wet season. Generally, its tubers are planted in spring after frost for its normal growth. If planting is done too early in spring, the tubers will not begin to sprout until the soil has warmed sufficiently. It can also be planted in first fortnight of August with mild summers. In spring, the stored mini and large tubers are used to plant directly.

SEED RATE

The quantity of planting material required varies with the size of tubers and conditions of the growing region. In general, about four to five tubers are enough to plant a square meter area and 40,000 to 50,000 tubers for planting a hectare area.

PLANTING METHOD

Oca is primarily propagated vegetatively by planting tubers. Generally, large tubers are used for planting to get higher tuber yield but cut pieces containing at least two buds are also used as a planting material. Where tubers are scares, sprouts from tubers, which are pulled out from the soil are planted individually because of their easy rooting. Tubers are planted horizontally on ridges at a depth of 5–10 cm keeping 40–60 cm distance between rows and 30 cm between plants. Like potato, the tubers prior to planting are kept in a well-ventilated place where they may easily be exposed to light to avoid lanky and weak growth of sprouts.

Oca may also be propagated by seed, but commercially, it is rarely practiced due to low seed germination rate (60%) and difficulty in production of healthy seedlings as the seed takes long period to germinate. However, its germination rate can be improved by soaking the seed in water for 24 hours. Oca seed has a short period of dormancy with small lifespan. Its germination power is reduced after a year. The longest and highest seed quality can be maintained at relatively low temperature. For raising oca seedlings, the seed is sown in nursery about three and a half months before the last frost. The seedlings after attaining a height of 10–13 cm are transplanted in the growing area or in pots.

Oca can be grown in pots where very limited space is available in the garden. However, the plant growth in pots is comparatively poor. The planting in pots can be carried out during August month to harvest a higher tuber yield.

NUTRITIONAL REQUIREMENT

Oca grows well with very low production inputs. Traditionally, it is planted in a field after the harvesting of potato crop, therefore, it benefits from residual nutrients of potato crop. If the soil is poor in organic matter, 10–15 t/ha well-composted farmyard manure, which ensures good crop growth, may be incorporated into the soil at the time of land preparation. Heavy use of manure should be avoided since only a small amount of farmyard manure may supply nutrients sufficient for good growth and development of oca crop.

Higher doses of nitrogenous fertilizers result in heavy foliage growth and delayed flowering. In moderately fertile soils, nitrogen may be applied at the rate of 80–100 kg, phosphorus 60–80 kg and potash 50–60 kg/ha. Half dose of nitrogen along with full dose of phosphorus and potash is applied at the time of ridge or bed formation and the remaining half dose of nitrogen is applied a month after planting at the time of earthing up.

IRRIGATION REQUIREMENT

Oca needs enough moisture for optimum vegetative growth and leaf production. At the time of planting, the soil must have ample moisture for homogeneous sprouting of tubers. If the planting is done in dry soil, supplementary irrigation is applied immediately after planting. Subsequent irrigations may be given at frequent interval depending on moisture retention capacity of the soil. Irrigation at tuber initiation stage is most indispensable to apply, as it will promote the production of large size tubers. Applying irrigation during dry period protects the young shoots from slugs and snails. Proper water management is essential for obtaining higher yield of high quality tubers.

INTERCULTURAL OPERATIONS

The cultural practices adopted for oca cultivation are similar to potato. Weeding in oca field is essential to reduce competition between weeds and crop plants, therefore, the field should be kept weed free. Hoeing will make the soil loose around the roots, which will improve aeration in soil and help in increasing the harvestable yield. However, deep hoeing should be avoided as it may damage the shallow root system.

Earthing up is an essential operation for proper development of tubers. Like potato, mounting up of soil around base of the plant provides better soil environment for normal development of the tubers. Some plants may form small aerial tubers on the stem. These tubers are picked off before the occurrence of heavy frost.

Mulching with compost and dry grass around the plant during summer will keep the soil moisture conserved, which will support the plant for its growth. In winter, the plants should be protected against frost by covering them with paddy straw, sugarcane tresses, or transparent polyethylene sheet.

HARVESTING

Oca is harvested when the leaves start turning their colour from green to yellow. However, some of the farmers prefer to retain their crop in field to avoid frost injury to the tubers. The tuber formation starts four months after planting under short day and low temperature conditions and gives highest yield when harvesting is done 6 month after planting. Oca plants are often killed by frost but the tubers are over wintered as long as they are not frozen. Tuber maturity is affected by day length. As a rule, longer the growing season, better is the yield and size of tubers. The tubers of 8–12 cm length and 2–3 cm diameter are considered best. Harvesting is done at proper soil moisture. The foliage is removed before digging the tubers. After digging, the tubers are lifted carefully and cured properly.

YIELD

Oca yield depends on selection of variety, growing season and management practices adopted by the growers. On an average, its yield is 50 tubers of varying size per plant and 70–120 q/ha. However, a tuber yield of 350–550 q/ha can be obtained if adequate inputs with virus free propagation material are used.

POST-HARVEST MANAGEMENT

Consuming tubers soon after harvest should be avoided, as their taste is extremely unpleasant due to high acid content. Therefore, the tubers before eating are exposed to light for 2 to 3 weeks to reduce acidity quickly and to improve tuber quality. The sweet varieties tubers contain relatively low acid, thus, they require no processing.

The harvested tubers are graded and separated as per the market demand. After proper washing and drying, they are stored in trays or gunny bags in a well-ventilated cool shady place for several months without any sprouting. The large healthy tubers are stored in dry sand or sawdust in a cool dark place for planting next crop, whereas the medium and small tubers are used as food as they are easier to prepare. Although oca tubers can be stored well at 20°C but their appearance and food quality degrade significantly as the tubers become dry and soft, however, the tubers can safely be stored for more than a year in cold stores at 4°C temperature coupled with 90% relative humidity. Unlike potato, its storage is easy as its tubers do not turn green on exposure to light but light may subsequently degrade some of the nutrients.

INSECT-PESTS

Oca Root Weevil (*Premnotrypes* and *Cylydrorhinus* spp.)

It is the most destructive pest causing huge loss in field as well as in storage. Its grubs cause damage by making tunnels into tubers and adults feed on soft tissues of plants. Larval feeding causes physical damage to the tubers. These weevils often destroy the entire crop. Oca itself has a self-defence mechanism of resistance against different weevils but it becomes ineffective in severe infestation.

Control

- Keep the field fallow.
- Follow long crop rotation with non-host crops.
- Grow resistant varieties.
- Use *Beauveri abrogniartii* fungus as a bio-control agent.
- Use weevil free tubers for planting.
- Mount up soil on both sides of the ridges 30 and 60 days after planting.
- Install sex pheromone trap in the field.
- Spray Fenthion or Fentrothion 0.05% at monthly interval.
- Apply phorate 2 kg a.i. at 2, 6 and 10 weeks of planting.

GRASS GRUBS (*Costelytra zealandica*)

It is a major pest of lawn and pasture in New Zealand. Initially, it feeds on grass roots in lawns, but later, it moves to oca tubers that grow fairly close to soil surface. The grubs are pale grey or cream colour and adults are velvety brown. The grubs when disturbed have the tendency to curl in C shape. The adults emerge in spring and lay 40 eggs.

Control

- Keep the field fallow.
- In lawns, use toxic chemicals to treat grass grubs.
- Use *Neem* seed cake.
- Plant tubers bit deeper to control grass grubs, which live in top 15 cm soil.

AUSTRALIAN CLICK BEETLE (*Conoderus exsul*)

The larvae of click beetles are generally known as wireworms because of their hard, shiny, cylindrical body. When the beetle lands on its back, it uses a click mechanism between two parts of the shell to jump up and right itself with a click sound, that is why, the name click beetle has been given. Its larvae cause damage by making tunnels into tubers.

Control

- Prepare the field to a fine tilth.
- Remove the tunnels around the field.
- Mulch the crop with loose material such as spent oats.
- Grow mustard as a green manure crop to reduce the attack.

APHIDS (*Aphis gossypii*)

Its nymphs and adults suck cell sap from the tender leaves and secrete honeydew like sugary substance, which encourages the growth of black sooty mould, hampering the photosynthesis, which results in reduction of yield. It transmits virus, causing the leaves to curl up. The plants are attacked by this insect at any stage of crop growth. High humidity and high temperature favour its incidence.

Control

- Grow resistant variety.
- Set up yellow pan traps in the field.
- Spray the crop with Malathion 0.05% at interval of 10–12 days.
- Spray mineral oil, Phosphomidon 0.05%, Dimethoate 0.05% or Methydemeton 0.05% at 10 days interval.

THRIPS (*Heliothrips indicus*)

The tiny insect sucks cell sap, resulting in white specks with black dots. The attacked leaves become brown and distorted. Dry weather is favourable for its rapid multiplication.

Control
- Grow resistant variety.
- Follow long crop rotation with non-host crops.
- Apply Phorate 2 kg a.i./ha in soil.

SLUGS AND SNAILS (*Arion hortensis* agg.)

Both, slugs and snails make irregular holes with smooth edges in leaves and flowers. Typically, they attack the stems and tubers at ground level and penetrate the soil. Oxalic acid in leaves is itself a natural defence mechanism against the slugs and snails.

Control
- Use snail baits.
- Collect and destroy the slugs and snails.
- Use copper or Metaldehyde tape adhesive as a barrier.
- Use suitable plastic barrier such as crushed eggshells, lime, wood ash, wood shavings or sawdust.

ROOT KNOT NEMATODES (*Meloidoginy* spp.)

The root knot nematode is widely distributed in oca growing areas. It is a serious problem in light sandy soils. The infected plants may not absorb soil nutrients and moisture efficiently. Stunted plant growth, leaf chlorosis, wilting and death of plants are the major symptoms if the infection is very severe. Its infection causes rotting of tubers in storage and reduces marketability and storability.

Control
- Follow long crop rotation with trap crops.
- Grow nematode resistant varieties.
- Avoid planting of nematode infested tubers.
- Add organic amendments like groundnut, *neem* or *karanj* cake, in soil.
- Apply higher doses of potash fertilizer.
- Turn the soil two to three times at 10 days interval in summer.
- Sterilize the nursery beds through soil solarisation in hot summer months.
- Fumigate the field with Nemagon before planting the tubers.

RATS (*Rattus rattus* W.)

Due to its good taste, the rats damage the tender tubers closer to soil surface during very harsh weather, forming characteristic holes.

Control
- Use rat traps.
- Install baits containing bromodialone 0.5% at 10 g twice at an interval of 12 days.
- Fumigate the hiding places using Aluminium phosphide tablets.

DISEASES

STEM ROT (*Pythium ultimum*)

Stem rot occurs at ground level, usually just browning the stem but occasionally withering it all the way through causing the foliage to die. The lower stem portion becomes soft and black as a result the stem often falls off. It may affect only few stems or the entire plant. The disease is most severe in warm climate, where it may actually infect the tubers and degrade them in storage. In cool climate, the pathogen survives in infected tubers in storage. The tubers should not be taken from infected plants for planting in next season.

Control

- Follow a crop rotation of at least 3 years.
- Use disease-free healthy planting material.
- Avoid bruising during harvesting, handling and transportation.
- Store only healthy tubers in the sand or saw dust for planting.

LEAF BLIGHT (*Phytophthora infestance*)

The disease appears as pale green to brown round spots on leaves at late mature stage. Later, the spots spread and coalesce rapidly over the whole leaf and form large lesions. Light brown lesions appear on stems and petioles. High humidity and high temperature favour the occurrence of this disease. The fungus survives on plant debris in field for several years.

Control

- Grow resistant variety.
- Follow field sanitation.
- Destroy infected plant debris after harvesting the crop.
- Spray the crop with 0.25% Ridomil, Maneb or Zineb at 10 days interval.
- Apply *Tricoderma viridae* as a bio-control agent.

MOSAIC VIRUS DISEASE

The disease appears as mottling and blistering of leaves. The leaves become small and misshapen and plants stunted. The infected plants yield less of small and deformed tubers. The disease is generally transmitted by white fly.

Control

- Use disease-free healthy planting material.
- Keep the field weed free.
- Follow proper field sanitation.
- Apply Furadon 1.5 kg a.i./ha in soil.
- Spray the crop with Nuvacron 0.05% at 10 days interval.

95 Endive

M.K. Rana and D.S. Duhan

Botanical Name	: *Cichorium endivia* L. var. endivia
Family	: Compositae (Asteraceae)
Chromosome Number	: 2n = 18

ORIGIN AND DISTRIBUTION

Endive was probably grown first in South Asia in Mediterranean region, where the wild relatives of endive, *i.e.*, *Cichorium pumilum* Jacq. and *Cichorium calvum* Sch.Bip. ex Asch., are found growing. Hence, both the species can be considered as the progenitor of endive. *Cichorium endivia* was probably known to the old Egyptians but no archaeological evidence has been found. It spread to Central Europe in the sixteenth century and is now grown throughout Europe and North America. It is also grown throughout the tropics including tropical Africa though it is mostly of minor importance there.

INTRODUCTION

Endive also known as escarole is a cool-weather green vegetable with a distinct sharp taste. It may be seen in the produce bins of most stores, but it is expensive. The curly endive, also known as *Frisee*, cannot tolerate hot weather too well but can tolerate some hard frost. Therefore, in South, it is grown in winter where the temperature is mild and in North as a spring or fall crop only. For kitchen gardens, the seedlings of endive are raised in flats indoors like head lettuce and later transplanted in the garden. Like other greens, endive tastes best when it grows quickly and steadily with the supply of adequate water and fertilizers.

COMPOSITION AND USES

COMPOSITION

Endive is a powerhouse of nutrients as it is a rich source of vitamin K and a good source of vitamin A, vitamin C, folate, thiamine, pantothenic acid, riboflavin, calcium, iron, potassium, manganese, copper, zinc and dietary fibre. It also contains a bit of protein. *Cichorium endivia* contains inulin and intybin, which cause the typically bitter taste and are said to stimulate appetite. More than 100 volatile compounds have been identified in endive, among them hexenal (E)-2-hexenal and β-ionone contribute intense odour. The nutritional composition of endive root is given in Table 95.1.

USES

Endive is most commonly consumed as a fresh green in salads, for which, curl-leaved types are often preferred. It is used as a substitute for lettuce in the tropics as it is more resistant to diseases. Broad-leaved types (escarole) are used as a cooked vegetable. Sometimes, endive is used for making pickles but not used in a processed form. The plant roots are often ground and baked, and used as

TABLE 95.1

Nutritional Composition of Endive Roots (per 100 g Edible Portion)

Constituents	Contents	Constituents	Contents
Water (g)	94.0	Zinc (mg)	0.8
Carbohydrates (g)	3.4	Manganese (mg)	0.8
Protein (g)	1.3	Vitamin (IU)	2167
Fat (g)	0.2	Thiamine (mg)	0.08
Dietary fibre (g)	3.1	Riboflavin (mg)	0.08
Calcium (mg)	52	Niacin (mg)	0.4
Magnesium (mg)	15	Folic acid (µg)	142
Phosphorus (mg)	28	Vitamin C (mg)	13
Iron (mg)	0.8	Energy (kcal)	17

a substitute of coffee or as an additive. However, after consuming its leaves, smoking or drinking should be avoided for at least 6 hours since the vitamin A present in the plant may react with tobacco and alcohol.

MEDICINAL USES

The vitamin K present in Radicchio's endive is essential for blood and bone health. The plants high in insulin content lower down the glucose levels and reduce bad cholesterol low-density lipoprotein (LDL) levels, as it is high in dietary fibre. It can aid weight loss, improves skin health and keeps the mucus membranes healthy because of the antioxidant value of vitamin A and beta-carotenes present in the plant. The vitamin A in endives also helps in improving eyesight and protects the body from certain types of cancer. The B group vitamins present in endives such as folic acid, thiamine, niacin and pantothenic acid encourage the metabolism of carbohydrates and fats. It helps in increasing and stimulating the appetite, as its leaf juice improves the secretion of bile juice. Being high in water content (95%), it acts as a natural diuretic, and being high in dietary fibre, it could also be used as a mild laxative. People suffering from anorexia and other eating disorders can benefit from endives as it can treat long-term digestive problems caused by starvation and fasting. Being rich source of vitamin E, it smoothens and moisturizes the skin. Regular use of endives also clears up the skin problems such as acne. Its leaf juice mixed with other fresh vegetable and fruit (parsley, celery, carrot and apple) juices is consumed daily to tone up the skin and to treat anaemia, constipation and respiratory problems such as asthma. The blend endive juice with carrot and celery or parsley and spinach juice is consumed for an effective treatment of eye problems such as glaucoma or cataract.

BOTANY

Endive is an annual or biennial erect herb up to 170 cm tall producing taproot and rosette of large leaves when young, forming a loose head. Leaves are alternate, simple or pinnatifid, sessile, broad measuring 45 × 18 cm, slightly crumpled, margin entire or dentate (escarole type) or very narrow, deeply pinnatifid and strongly curled (curly leaved type), progressively smaller upwards on stem, slightly pubescent or glabrous, pale to dark green or yellowish and sometimes reddish along midrib. Inflorescence is a head, sessile or on up to 20–cm-long apically thickened peduncle, involucres with outer row of five bracts (7–10 × 2–5 mm) and inner row of eight bracts (8–12 × 1–3 mm). Flowers are 15–20 per head, all ligulate, corolla up to 2 cm long, blue, sometimes white, five-lobed at apex, stamens five with anthers fused into a tube, ovary inferior, single-celled, style slender and hairy with two slender stigmatic lobes. Fruit is an obovoid to cylindrical achene (2–3 × 1–1.5 mm) and brown with pappus of one to three rows of small persistent membranous scales.

Endive forms a dense rosette of leaves in the first year. Upon bolting, the stem elongates and becomes branched. The inflorescence bears many heads. Flowers usually open only in the morning and wither after about 6 hours. Most of the cultivars are self-pollinating but some of the cultivars are cross-pollinated by insects.

CLIMATIC REQUIREMENT

Both the species of endive are easy to cultivate. They are more tolerant of high temperature than lettuce and can be grown from cool temperate areas to tropical lowlands, though in the tropics, better results are obtained above 500 m altitude. It can tolerate mild frost but the optimum temperature for its growth is 15–18°C. At high temperatures, the leaves become fibrous. Endive requires long days for flowering and rarely flowers in the tropics. Vernalisation in some cultivars at temperature below 10°C gives an additional stimulus to flowering and can occur already during ripening of seed but also during storage of seed and from sowing onwards. The endive crop is more tolerant of heat than lettuce.

SOIL REQUIREMENT

Endive prefers loose and deep soils that are sufficiently fertile, especially the top 20-cm layer, so that the roots may be able to penetrate the soil to a depth of at least 30 cm. The most preferable pH of the soil is 6.5–7.8. Unlike most other root crops, mildly stony soil is also acceptable. The forking affects the roots quality but does not necessarily affect the chicon quality. The soil should be free from previously grown crop residues, as they hamper the penetration of roots into soil, which causes forking. Incorporation of fresh manure should also be avoided due to the amount of available nitrogen that causes excessive leaf growth.

CULTIVATED VARIETIES

There are three types of endive, *i.e.*, (i) the upright Batavian or escarole with large broad leaves, (ii) the curly or fringed *Frisee* with a rosette of delicately serrated leaves and (iii) cutting type. Broad-leaved types are robust and useful for winter cropping. Curly varieties are used for summer cropping. Its outer leaves can be cooked as greens. The varieties of third group are hardly grown anymore. These three groups of cultivars are distinguished within *Cichorium endivia* as mentioned in following list:

- **Batavian Group:** This group is also called Escarole Group, Scarole Group or *Cichorium endivia* var. latifolium Lam. The plants of this group have broad, almost entire, rather flat leaves, forming a loose head. Some well-known cultivars of this group are Batavian Broad-Leaved, Escarole, Deep Heart Fringed and Growers Giant.

- **Curled-leaved Group:** This group is also called Curled Endive Group, *Frisée* Group or *Cichorium endivia* var. crispum Lam. The plants of this group have narrow, deeply pinnatified, strongly curled leaves, forming a loose head. Some of the well-known cultivars of this group are Salad King, Green Curled, Frisée de Ruffec and Grosse Pancalière.

- **Cutting Group:** This group is also called Small Endive Group, *Endivia* Group or *Cichorium endivia* var. endivia. The plant of this group has very small leaves that do not form a head. The cultivars belonging to this group are hardly grown anymore.

The other popularly grown varieties are given below:

Wallone AGM: A high-yielding variety with large, vigorous, green, finely cut leaves is good for cuttings and it gives more cuttings.

Despa AGM: An attractive *Frisee* type variety is resistant to bolting.

Frenzy AGM: A variety with compact and uniform finely cut leaves is used in top restaurants.

Lassie AGM: A well-flavoured high-yielding variety has very deeply cut leaf.

Natacha AGM: A tasty, hardy Batavian or Escarole type variety is suitable for winter sowing and resistant to bolting.

Golda AGM: A short-leaf Batavian type variety, which is hardy for winter cropping, does not need blanching.

Cornet de Bordeaux: An old variety, which is very tasty and extremely hardy.

The other popular varieties are Frisan, Galia, Green Curled Ruffec, President, Tosca, Tres Fine Endive, Broad-leaved Batavian, Coral, Florida Deep Heart, Full Heart Batavian, Sinco and Sinco Escarole.

SOWING TIME

The endive seeds are sown as early as 4–6 weeks before the average date of last frost in spring. The seeds for raising seedlings indoors can be sown 8–10 weeks before the average last frost. The sowing time is so adjusted that the crop may become ready for harvest before the temperature reaches higher than 30°C. The crop is sown successively after every 2 weeks beginning in midsummer. In mild-winter regions, endive is grown in autumn, winter and spring.

SEED RATE

The seed rate depends on viability, soil fertility, irrigation facilities available and growing season. The seed of endive is very small, having test weight of 1.3–1.6 g. About 500-g seed is required for planting a hectare area.

SOWING METHOD

Endive is one of the most difficult vegetables to grow, requiring a two-step growing process before it is ready for harvesting. As usual, it is propagated by seed. If the seeds are to be sown in a nursery, six seeds are sown per 30 cm row made 45 cm apart, or three rows may be adjusted on a 1.5-m wide bed. The seeds are sown at a depth of 6 mm and 2.5–5 cm apart in a row at a distance of 45–60 cm. Thinning is an important operation because overcrowded seedlings may develop seed stalk. Thus, after emergence, the seedlings are thinned to 5 cm apart. In heavy soils, it is always safe to sow seeds when the soil contains enough moisture for seed germination.

Seedlings become ready for transplanting about 1 month after sowing, when they have four to six true leaves. At this stage, they are transplanted in the field where the first growth takes about 150 days, where the chicory grows from seed into a leafy green plant with a deep taproot. The planting distance is kept 25–40 × 25–40 cm with the widest spacing for broad-leaved cultivars. Dense planting favours self-blanching but increases the risk of rot. At harvest, the tops of the leafy chicory plant are cut off and then the roots are dug up and placed in cold storage where they enter a dormancy period. As the demand necessitates, the roots are taken out from the cold store for their second growth, which takes 28 days in dark, cool and humid forcing rooms, similar to a mushroom growing facility. Control over the initiation of this second growing process allows for the year-round production of endive.

NUTRITIONAL REQUIREMENT

For adequate nutrients supply, the soil from the field is analysed for chemical composition. Nitrogen rates may need to be adjusted depending on crop, planting date, soil type and weather conditions. Based on the analysis, nitrogen may be applied at 50–100 kg, phosphorus 100–150 kg and potassium 50–150 kg/ha. Full dose of phosphorus and potash along with half dose of nitrogen is applied at the time of sowing or transplanting preferably by placing 5 cm to the sides and 5 cm below the seed or plant roots and rest of the nitrogen is applied in two split doses, *i.e.*, 1 month and 2 months after transplanting.

IRRIGATION REQUIREMENT

Because of shallow root system, the crop prefers a uniform and frequent supply of water for its tender growth. Total 20–30 cm water is necessary, depending on seasonal variation, planting season and variety. Soil type does not affect the amount of total water needed but it affects the frequency of water application. Light soils require irrigation more frequently as compared to heavy soils. Depending on season and availability of soil moisture, irrigation is applied at an interval of 10–15 days. It is advisable not to irrigate the field too frequently; however, the care should always be taken that the field may not dry and become compact too much, otherwise the root development is affected adversely. An irregular supply of moisture may cause discolouration of young leaves margins. The plant is sensitive to waterlogging conditions.

INTERCULTURAL OPERATIONS

HOEING AND WEEDING

Weeding and hoeing are done regularly throughout the summer to retain moisture and porosity. The endive field in early stage should be kept weed-free since in early stages, the plants may not compete with weeds due to poor growth. Thus, at least two shallow hoeing are beneficial and necessary to keep the weeds down and provide good soil environment to the roots for their development. One weeding and one earthing up during early stage of crop growth are necessary to promote root development. Another hoeing and earthing up are done before bolting to provide support to the seed plants.

BLANCHING

About a week before harvesting, the heads may be tied up in order to blanch them and to moderate their bitterness. Blanching can also be achieved by covering each plant with a pot or container to exclude light for about 10 days. There are self-blanching cultivars, especially when planted densely.

HARVESTING

Endive becomes ready for harvest in 60–100 days after planting. Harvesting can start when the heads have reached a marketable size (250–400 g). It is done by cutting the heads from the roots, removing the outer, discoloured or damaged leaves and placing the heads upside down in containers.

YIELD

Per crop per hectare, 200 quintals of marketable produce can be obtained.

POST-HARVEST MANAGEMENT

Endive can be retained fresh only for 1 day at ambient room temperature. However, the healthy heads can be stored for up to 2 weeks at 0–1°C temperature and 90–95% relative humidity. Endive when grown in the tropics, it often becomes very bitter in taste; however, this bitterness can be reduced by packing the individual head tightly in plastic foils and storing them for 4–5 days at 7°C temperature.

INSECT-PESTS

Aphids (*Myzus persicae, Nasonovia ribisnigri and Brachycaudus helichrysi*)

Small soft-bodied insects on underside of the leaves and stems of plant are usually green or yellow in colour but may also be pink, brown, red or black, depending on species and host plant. If aphid infestation is heavy, it may cause the leaves yellow and distorted with necrotic spots on leaves. As a result, the growth becomes stunted. They secrete sticky sugary substance called honeydew, which encourages the growth of black sooty mould on the plants.

Control
- Grow aphid-tolerant varieties.
- Use reflective mulches such as silver coloured plastic to deter aphids from feeding on plants.
- Insecticidal soaps or oils such as *neem* or canola oil are usually best to control aphids.

Thrips (*Frankliniella occidentalis and Thrips tabaci*)

Insect is small (1.5 mm) and slender and best viewed using a hand lens. Adult thrips are pale yellow to light brown and the nymphs are smaller and lighter in colour. If population is high, the leaves become distorted. Leaves covered in coarse stippling may appear silvery and leaves speckled with black faeces. The insect transmits viruses such as tomato spotted wilt virus. Once acquired, the insect retains the ability to transmit virus for the remainder of its life.

Control
- Avoid planting next to onion or garlic where very large population of thrips can build up.
- Use reflective mulches early in growing season to deter thrips.
- Hot-pink sticky cards have been found to be the most attractive colour for trapping thrips.
- Spray Spinosad at 1–1.5 mL/L of water if thrips become problematic.

Leaf Miner (*Liriomyza* spp.)

Heavy mining by the insect can result in white blotches on leaves. As a result, they drop from the plant prematurely. Early infestation reduces yield. Adult leaf miner is a small black and yellow fly, which lays its eggs in the leaf. Larvae after hatching feed on leaf interior. Mature larvae drop from leaves into soil to pupate. The entire life cycle can take as little as 2 weeks in warm weather. The insect may go through 7–10 generations in a year.

Control
- Plant seedlings free from damage.
- Remove plants from soil immediately after harvest.
- Spray the crop with selective insecticides, such as Spintor and Proclaim at 2 mL/L of water.

Flea Beetles (*Epitrix* spp. and *Phyllotreta* spp.)

The pest responsible for damage is a small (1.5–3.0 mm) dark coloured beetle that jumps when disturbed. The beetles often shiny in look make small holes or pits in leaves that give the foliage a characteristic shot hole appearance. Young plants and seedlings are particularly susceptible. Plant becomes stunted. Severe infestation may kill the plant. Younger plants are more susceptible to flea beetle damage than older ones. During winter season, flea beetles survive on nearby weed species, in plant debris, or in the soil. The insects may go through a second or third generation in 1 year.

Control

- Plant the crop early to allow establishment before the beetles become a problem.
- Grow cruciferous plants as a trap crop.
- Apply a thick layer of mulch, which may help in preventing the beetles reaching the surface.
- Spray Spinosad, Bifenthrin or Permethrin 2 mL/L of water.
- Spray the crop with Actara at 100–200 g/ha.

DISEASE CONTROL

Damping-off (*Pythium* spp. and *Rhizoctonia solani*)

The disease is most common in poorly drained soils having excessive moisture especially at bit high temperature. Densely sown nurseries are more prone to damping-off disease, which may be pre-emergence (seed decaying) and post-emergence (rotting of collar region due to the attack of fungi, leading to collapsing and death of seedlings).

Control

- Sterilize the soil by drenching with 0.2% Captan solution.
- Collect seeds from disease-free plants.
- Sow the seeds shallow.
- Treat the seed with Captan at 2–3 g/kg of seed.
- Stop irrigating the nursery.
- Increase air circulation.
- Remove all the infected seedlings from the nursery.

Downy Mildew (*Bremia lactucae*)

The disease appears as light green to yellow angular spots on upper surface of the leaves, which are delineated by leaf veins. White fluffy growth of pathogen develops on underside of these spots. As the disease spreads, the upper side of the leaves may also show downy white fungus growth. With time, these lesions turn brown and dry up. Older leaves close to the ground usually are attacked first. Severely infected leaves may die. Rarely, the pathogen can be systemic, causing dark discolouration of stem tissue. It may also cause losses in transit. The disease is favoured by cool rainy weather. Night temperature of 5–10°C and day temperature of 12–20°C with relative humidity of 98% are ideal for its development. As the temperature increases, the disease starts disappearing.

Control

- Grow resistant varieties.
- Apply irrigation in such a way that leaves surface may remain dry.
- Spray the crop with Dithane M-45 at 0.2%.

ANTHRACNOSE (*Microdochium panttonianum*)

Small grey to straw colour circular or irregular-shaped dry spots appear on leaves. Intense spots may cause the leaf to die. Lesions when come together form a large necrotic patch, causing the leaves to turn yellow and wilt. Lesions may split or crack in dry centres. Fungus survives on plant debris and related weeds. Disease emergence is favoured by warm moist conditions.

Control

- Follow long crop rotation.
- Treat the seeds with hot water (52°C) for 30 minutes.
- Soak the seeds in 0.2% Thiram solution prior to sowing.
- Follow phytosanitary conditions in the field.
- Provide good drainage facilities.
- Spray Maneb, Zineb or mancozeb 0.2% at flowering and fruiting time.

BACTERAL SOFT ROT (*Erwinia* spp.)

Water-soaked lesions, which expand to form a large rotten mass of cream colour, appear on under-surface of the leaves. Lesions on surface usually crack and exude slimy liquid, which turns tan, dark brown or black on exposure to air. The bacteria spread easily through agricultural implements and irrigation water. Bacteria enter the plant system through wounds. Disease occurrence is favoured by warm moist weather.

Control

- Follow long crop rotation strictly with non-host crops.
- Harvest the heads when they are completely dry.
- Avoid any kind of injury to heads during harvesting.
- Spray the crop with copper bactericides (Cuprofix at 800–1000 mL/ha) 2 weeks prior to the occurrence of disease.

BOTTOM ROT (*Rhizoctonia solani*)

Small red to brown spots appear on lower leaves, usually on underside of midrib, which may expand rapidly, causing the leaves to rot. Amber colour liquid may ooze from the leaf lesions. As the stems rot, the leaves become slimy and brown and collapse. A tan or brown mycelial growth may be visible in infected tissue. Fungus survives on crop debris in soil and disease is favoured by warm and moist weather.

Control

- Grow varieties with erect growth habit to reduce leaf contact with soil.
- Plough the field deep in hotter months.
- Follow crop rotation strictly.
- Avoid irrigation close to harvest.
- Spray the crop with Endura 70 WG at 500 g/ha.

SCLEROTINIA BLIGHT (*Sclerotinia sclerotiorum*)

The outer leaves wilt, which spreads inwards until the whole plant is affected. Soft watery lesions appear on leaves. As a result, the leaves collapse and lie on soil surface. Black fungal structures may develop on infected leaf tissue and soil surface. The fungus can survive in soil for 8–10 years.

Control

- Plough the field deeply.
- Adjust cereal crops in rotation.
- Keep the weeds under control.
- Avoid dense growth by planting in adequately spaced rows.
- Spray Fluazinam (Omega) or Boscalid (Endura) at 2–3 g/L of water in advance before the appearance of disease.

Septoria Blight (*Septoria lactucae*)

Small irregular chlorotic spots, which enlarge and turn brown and dry out, appear on oldest leaves. The dead lesions may fall out of leaves, creating holes and chlorotic halos. If the plant is severely infected, the lesions may coalesce, forming large necrotic patches, which later cause wilting of leaves and ultimately death of the plant. The fungus survives in infected seed and in crop debris. Disease spreads in humid weather through splashing water. Wild lettuce is an important overwintering site for the survival of fungus.

Control

- Use disease-free seed.
- Grow endive in areas where Septoria is uncommon.
- Plant the crop in low rainfall areas.
- Give hot water treatment at 52°C for 30 minutes before sowing.

96 Rhubarb

M.K. Rana and D.S. Duhan

Botanical Name : *Rheum officinale* Baillon

Family : Polygonaceae
(Buckwheat family)

Chromosome Number : 2n = 44

ORIGIN AND DISTRIBUTION

Rhubarb is a strange plant with an interesting history but no record of common culinary rhubarb is found prior to the 1800s. In the early nineteenth century, its widespread consumption began in Britain with its popular adoption as an ingredient in desserts and winemaking. Since then, its popularity grew to a peak just before the Second World War in what Foust called rhubarb mania. It was always more popular in Britain and in the United States than elsewhere but it also achieved notable popularity in Australia and New Zealand. After the Second World War, its production was also resumed in England but not at *mania* level as in the United States. However, forcing is still popular in Yorkshire England—the English are leaders in rhubarb production.

INTRODUCTION

The word *rhubarb* is of Latin origin. The ancient Romans imported rhubarb roots from unknown barbarian lands, which were beyond the Vogue River also known as the *Rha River*. *Rha* was first adopted to mean rhubarb. Imported from barbarians across the *Rha*, the plant became *Rha barbarum* and eventually *rhabarbarum*, the Latin word for rhubarb plant. The modern English word *rhubarb* derives from *rhabarbarum*, which is sometimes called a *pie plant*. The medicinal rhubarb consists of dried rhizome and roots of *Rheum officinale* Baillon, *Rheum palmatum*, of related species grown in China (Chinese rhubarb), of *Rheum emodi* or *Rheum webbianum* which is native to India, Pakistan or Nepal (Indian rhubarb).

Scientifically, it is a herbaceous perennial with leaves growing off the top of a thick rhizome. Its species can be grouped into two broad categories, *i.e.*, (i) the first category collectively called Victoria, the stalk exteriors of which vary in colour from pure green to light red, while the interior is green and the varieties in this category tend to be larger, more robust and relatively disease resistant, and (ii) the second category consists of pure red rhubarb. The stalks of this category plants are deep red from inside to outside. They tend to be smaller, less robust and more prone to diseases. In general, the yield per plant per year of red rhubarb is almost half that of Victoria plants. The Victoria group is by far the more common type of rhubarb in the United States. The red colour of its leaf stalk is due to the presence of anthocyanin pigment.

COMPOSITION AND USES

COMPOSITION

One of the important reasons why people grow and eat rhubarb is actually because of its nutritional value, as it is packed with minerals, vitamins, organic compounds and other nutrients that make it ideal for keeping the human body healthy. Some of its precious components are dietary fibre, protein, vitamin C, vitamin K, B group vitamins, calcium, potassium, magnesium and manganese. In terms of organic compounds, it is a rich source of polyphenolic flavonoids such as β-carotene, lutein, and zeaxanthin. The nutritional composition of raw rhubarb stalk per 100 g edible portion is given in Table 96.1.

USES

Many consider rhubarb as a fruit as it is used as a dessert or an ingredient in sweet dishes due to its unique sweet taste, but botanically, it is considered a vegetable. It is generally used as a fruit in desserts and jams. The leaf stalks are actually its most commonly used parts. The leaves of this plant are poisonous so it is never ingested. It is also used for colouring hair due to being a strong dye that can produce a golden colour. It can also be used to clean burnt pots and pans as well as a repellent for any of the leaf-eating insects.

MEDICINAL USES

In history, rhubarb has been used medicinally more rather than culinary. Indeed, its widespread culinary use began only two centuries ago and the medicinal use began at least 5000 years ago when the Chinese used the dried roots as a laxative. In Western civilization, its use was first documented 2100 years ago when rhubarb roots were an ingredient in numerous Greek and Roman medicines. Dried rhubarb roots are astringent. The astringent effects closely follow the cathartic impact, which made the rhubarb roots a popular laxative in old days. Though uncommon, the dried rhubarb roots are still sold as a laxative. Although the medical efficacy of rhubarb roots significantly varies with variety, as some of the varieties have no laxative value. In traditional Chinese medicine, it is extensively used to treat constipation, diarrhoea, jaundice, gastrointestinal haemorrhage, menstrual disorders, conjunctivitis, traumatic injuries, superficial supportive sores and ulcers.

TOXIC SUBSTANCE

Health hazards are unlikely if consumed in modest portions, not more than a few times per week. A common anti-nutritional component in *Rheum* species is the mutagenous anthraquinone but no

TABLE 96.1

Nutritional Composition of Raw Rhubarb Stalk (per 100 g Edible Portion)

Constituents	Contents	Constituents	Contents
Water (g)	93.6	Zinc (mg)	0.1
Carbohydrates (g)	4.5	Manganese (mg)	0.1
Protein (g)	0.9	Vitamin A (IU)	102
Fat (g)	0.2	Thiamine (mg)	0.02
Dietary fibre (g)	1.8	Riboflavin (mg)	0.03
Calcium (mg)	86	Niacin (mg)	0.30
Magnesium (mg)	12	Folic acid (mg)	7
Phosphorus (mg)	14	Vitamin C (mg)	8
Iron (mg)	0.2	Energy (kcal)	21

evidence of mutagenicity has been detected in rhubarb leaf stalks. Anthraquinone concentration is found higher in the leaf blades and roots.

Rhubarb leaves are toxic and have no safe culinary use for human beings due to a high concentration of oxalic acid, an organic poison and corrosive found in many plants but present in relatively large amounts in rhubarb leaves. The other toxins may also exist in its leaves. However, based on oxalic acid content, 5 kg of leaves provide a lethal dose to a human although it varies with variety and growth stage. Hence, the people are advised not to consume the leaves in any quantity.

BOTANY

Rhubarb has been derived from several *Rheum* species. The perennial tufted herb is up to 1.5 m tall with a woody rhizome and fleshy roots. Leaves in a rosette are simple, large, alternate and gradually smaller on flowering stem, stipules united to form a large, whitish, membranous sheath, petiole up to 1 m long and often more than 2 cm in diameter, fleshy, flat on upper side, obscurely grooved or rounded with sharp margins on underside, green often tinged with red or pink, blade broadly ovate or cordate, base cordate, apex obtusely rounded, margins undulate or crispy and irregularly ciliate, palmately three to seven veined and pubescent on the veins beneath. Inflorescence is a large panicle with many flowers, which are bisexual, small, greenish white, sepals in two whorls of three and free. Stamens are nine, ovary superior, single-celled and styles are three. Fruit is an ovoid nut, which is 1 cm long with three membranous wings.

CLIMATIC REQUIREMENT

Rhubarb is a cool season perennial crop, which is very winter hardy and resistant to drought. The crop is produced from crowns consisting of fleshy rhizomes and buds. A sunny and open area is most suitable for its cultivation. Following a season of growth, its crowns become dormant and a temperature below 5°C is required to stimulate bud break and subsequent growth. The first shoots appearing in spring are edible petioles and leaves. These emerge sequentially as long as the temperature remains cool (<32°C). As the temperature increases, the top growth is suppressed, even appearing dormant in periods of extreme heat. With declining temperature in later summer, the foliage growth resumes. Once planted, the plant remains productive for 8–15 years.

The climate of the Northern United States and Canada is well suited for its cultivation. However, it cannot be grown successfully in southern regions of the United States, though there are some exceptions. Although it can tolerate heat, it will not grow well at high temperatures, rather it will produce only thin spindle-shaped leaf stalks with poor colour. In hot days (>32°C), the plant wilts very quickly.

Rhubarb can be grown in partial shade but cannot be grown at high temperatures and in areas prone to frost since a severe frost will have a damaging effect on rhubarb and may cause the oxalic acid from leaves to enter the stalks. For this reason, the stalks affected by a hard frost should be removed and placed on a compost heap. The decomposition of compost in the heap will smash the oxalic acid successfully.

SOIL REQUIREMENT

Rhubarb can be grown in all kinds of soil but well-drained, highly fertile soils rich in organic matter are best suitable for its cultivation. It grows well in moist soil but it will not be happy when its feet are too wet. It can thrive well even in light soils if rich in humus. Though the crop can tolerate a soil pH as low as 5.0, the maximum yield can be obtained from a soil with pH range from 6.0 to 6.8. If the pH is very low, lime is applied to maintain soil pH in the range 6.0–6.8, as with all crops, the calcium level should be 3 tonnes per hectare. The soil before planting is loosened by ploughing the field repeatedly followed by planking. Thereafter, the beds are prepared in early spring after adding at least a 7 cm layer of well-rotten farmyard manure. Poultry manure and good compost are an excellent mix.

CULTIVARS

Early Champagne: It is a strong reliable cultivar with even cardinal red colour throughout the length of stem, green flesh and pink at edges.

Early Cherry: It is a green-fleshed variety with thick stem having a bright reddish-pink base and red flecking on green towards top.

Hawke's Champagne: It is an early reliable variety with even currant red colour throughout the length of the stem. Leaves are heart shaped, attractive and vigorous. Flesh is pale green to pink.

Timperley Early: A frost susceptible variety with deep red base, passing to light green with red flecking is good for forcing. It does not hold colour well. Its flesh is green with reddish tinges. It has been disappointing in Harlow Carr taste tests.

Victoria: A late type old variety with cardinal red stems with flecking at the top is very popular. Its stems are very thick and the flesh is red with green tinge.

Some other well-known cultivars are Early Red, Prince Albert, Linneus, Oregon Red Giant and Crimson.

SOWING TIME

Rhubarb that requires low temperature (5°C) to break winter dormancy and renew growth should be planted in early spring. It can also be grown in winters through forcing, which accounts for its availability early in the year when other crops are short in supply. Early forced rhubarb has a distinctive bright pink colour and delicate flavour.

SEED RATE

Rhubarb is normally propagated by crowns consisting of woody and fleshy rhizomes containing a fibrous root system and buds. The healthy vigorous 3–4 year old crowns are divided into pieces obtaining two or more buds per propagation piece. The 5–6 year old crowns yield 8–10 good quality propagation pieces. Usually, its seed are not used to establish commercial production field since seeded varieties lack the uniformity of colour and petiole size, which are necessary for commercial production.

SOWING METHOD

Rhubarb can be propagated by seed or by planting the seedlings purchased from local gardens. The crop raised through seed will take a year or more to produce stalks, and even then, the plants are not guaranteed to be true to type. It is advisable to buy year old plants, which are known as *crowns*, that have been divided from strong disease-free plants. The seeds are sown outdoors in shallow drills or indoors in trays or boxes during spring. The transplants are thinned to 15 cm apart, and when they begin to touch, they are transplanted about 90 cm apart.

Growing rhubarb from seed is one of the easiest ways of cultivating rhubarb but growing rhubarb from crowns results in better reliability. Except *Glaskin's Perpetual*, no rhubarb variety can be harvested in the first year.

PROPAGATION

Rhubarb is usually not propagated by seed since the seedlings developed through seeds may not be true to type. Thus, for starting new plantings, quality nursery stock is recommended because it is free from virus, root rot, crown rot and weeds. Plants developed through tissue culture are vigorous and true to type. If it is not possible for plants to develop through tissue culture, vigorous and disease-free plants are selected before the divisions are to be made. Plants are selected in either the fall or early spring while they are dormant and stored in a cold place to protect them from successive freezing, thawing, drying and damage to the buds. It is also good to divide established crowns about once in every 5 years if they have become weak or overcrowded. For doing so, the crowns are dug up carefully and then divided into sections, ensuring that each section is having at least one bud along with a 5 cm root. About four to eight new root sections can normally be obtained from a crown.

Near planting in the spring, each plant is cut into as many divisions as possible, containing one or two well-developed buds and as large a piece of the adjacent root material as possible. The divisions are kept protected from drying out before and during planting, which is done from October to March while the plants are dormant. Crown pieces should be planted 8–15 cm below the soil surface. The crop intended to mechanical harvesting is planted at a spacing of 46 cm in rows and 1 cm plant to plant, maintaining a plant population of 17,784 plants per hectare, and the crop intended to hand-harvesting is planted at a spacing of 60 cm × 180 cm keeping a plant density of 8892 plants per hectare.

While deciding where to place rhubarb, its height is taken into account because its plants grow more than 90 cm tall with a spread of 120–150 cm. The holes of 45–60 cm depth and 60 cm diameter are dug and filled with a mixture of soil and equal amount of compost, leaving a depression for the crown. The crown is set in the hole and the buds are covered with 2.5 cm of soil. Thereafter, water is applied and covered with a mixture of compost and straw, hay or shredded leaves, leaving a space in the centre so that the crown may remain uncovered. The soil is kept moist but not the soggy.

FORCING

For an early harvest of tender and pink rhubarb, the crowns are covered in December or January with a layer of straw to exclude light. Stems will be ready to pull 2–3 weeks earlier than uncovered crowns. For an earlier harvest, some roots are lifted in November. Ideally, the lifted roots are left outside for up to 2 weeks prior to potting to expose them to more cold as it is needed to overcome dormancy, and thereafter, they are potted up with compost and brought into a cool room or greenhouse at 7–16°C. The crates are covered with black polythene to exclude light. In crates, the roots are kept damp but not wet.

NUTRITIONAL REQUIREMENT

Rhubarb responds well to the application of fertilizers. The quality of a crop harvested depends largely on the care and fertilization received. A green manure crop is desirable a year before planting. Well-decomposed farmyard manure is applied at 20–45 tonnes per hectare before planting or as early as possible in the spring for commercial cultivation. Manure is an extremely valuable source of organic matter as it helps in conserving moisture, preserves the soil structure and makes the nutrients readily available.

In subsequent years, nitrogen is applied at 170 kg/ha along with a required quantity of phosphorus and potash. Phosphorus is most important in the establishment year, thus, the field selected for rhubarb should be high in phosphorus. Potash is added only when potash in soil is not adequate but is usually required if high yield is to be taken and farmyard manure has not been used. Nitrogenous fertilizer should be applied as side-dressings in three splits, *i.e.*, (i) before the growth starts in the spring, (ii) after the start of growth and (iii) after harvesting for regrowth of the plants. If farmyard manure is applied, the nitrogen rates are reduced in the first 2 years. Magnesium and sulfur are needed in sandy soils poor in organic matter content. Epsom salts may be used as foliar spray or gypsum and dolomitic limestone may be applied as basal at the time of planting. Boron necessary for good health of buds and roots may be applied at 1–2 kg/ha. However, care should be taken against burning of foliage or roots in the establishment year. Manure is generally applied in the fall season and nitrogen, phosphorus and potash in the spring either by broadcasting, side-dressing or through fertigation.

IRRIGATION REQUIREMENT

Application of irrigation is essential to maintain vigorous growth. Because of a strong root system, it is tolerant to moderate dry soils. However, irrigation should be performed during periods of summer drought. Irrigation requirement for the crop depends on soil type, mulch and environmental conditions of the growing region. In general, the crop needs light irrigation at an interval of 7–10 days, particularly during July when temperature is high in the summer months. The crop in well-drained light soils requires irrigation more frequently than the crop in heavy soils. Mulching reduces the irrigation requirement as it reduces the surface moisture loss. Irrigation can effectively keep the rhubarb cool.

INTERCULTURAL OPERATIONS

A clean planting site should be used for the cultivation of rhubarb since no herbicide is registered for controlling weeds in this crop. Perennial weeds must be eliminated before establishing rhubarb planting. Annual weeds are best controlled through a good shallow hoeing, and planting distance between rows should be designed such that the tillage equipment can be accommodated. Deep tillage over the crown should be avoided, as most buds are concentrated near the soil surface. The herbicide can be applied directly over the crowns. The absence of leaves attached to crowns minimizes herbicide uptake, yet perennial weeds are fully exposed to maximum contact of herbicide.

HARVESTING

Stems are usually harvested in 5 weeks. Established plantings can be harvested 8–10 weeks or longer after planting if the plant growth is vigorous. A rule of thumb for harvesting rhubarb is not to remove more than two-thirds of the existing stalks at any given time. Choose stalks that are at least 30 cm long. Harvesting is done by pulling and twisting the leaf stalks upwards so that they may be separated cleanly from the rhizome. The stalks should not be cut since the remnant stalk base may decay and create oasis:entry points for disease causing pathogens. Adult plants can be harvested weekly or when sufficient new leaves have developed, leaving at least four leaves for photosynthesis and new production. Usually, the tough old leaves and small young ones are left. After a harvest period of some months, a rest period is recommended in order not to exhaust the plants.

YIELD

In temperate regions, a mature plant gives on an average 1.5–3 kg of leaf stalks per season, but in the tropics, it gives 1–1.5 kg of stalks per season as the plants usually have less vigorous growth. A good commercial yield is 150 q/ha and an exceptional yield is about 180 q/ha. The red varieties usually yield about 50% of the green types.

POST-HARVEST MANAGEMENT

Stalk quality is highest due to lowest fibre content in the early spring. The leaf blades should be trimmed off and the stalks are bunched in ½–1 kg units for sale. The harvested stalks may be stored for 2–4 weeks at 0–2°C and 95% relative humidity. However, the quality for culinary use or freezing is considered best when stalks are first harvested versus after storage.

PHYSIOLOGICAL DISORDERS

Flowering: Some of the cultivars can be prone to flowering. The flower stalks should be removed as soon as they appear to avoid weakening of crowns. Flowering may usually be a problem after wet summers or where high doses of nitrogen are applied.

Thin (weak) stems: The weakening of stems indicates losing of crown vigour, meaning that crown needs to be divided. Increased feeding may help in producing thick and stout stems.

Stems splitting or exuding sticky sap: This problem is sometimes caused by late frost, indicating erratic growth due to seasonal conditions. Cool or dry periods followed by moist or mild weather causes the stem to split especially at the time of new rapid growth. The problem can be controlled by mulching.

Green or poor quality stems: Warm and dry summer conditions can give rise to stalks of poor colour with bad taste. While the days are comparatively cool and moist, harvesting should be done earlier.

Slow or no growth: Rhubarb stops growing if the temperature goes above 32°C. The disorder develops in hot summer months. Growth can also slow or stop under a condition of drought stress so watering may help.

INSECT-PESTS AND DISEASES

Rhubarb is relatively free of insect pests and diseases. The problem of pests arises with newly planted crops. The main threat is leather jackets, especially if the preceding crop was turf, a weedy or fallow crop. These large grubs come up at night and eat away any growth causing the set to die off. Slugs also cause similar damage.

INSECT-PESTS

POTATO STEM BORER (*Hydraecia micacea* esper)

The caterpillars, which when fully grown are about 3.5 cm in length and pinkish-white in colour. The first stages of insect attack only weeds and couch grass in particular. Later, they move into rhubarb plants with thicker stems. They may move from stem to stem by boring into the centre of stalk. The adult moths lay eggs on the stems of grasses in August. The eggs hatch in the following spring. Damage can be expected in June and early July. Serious infestation can lead to the crop being unmarketable.

Control
- Destroy couch grass and other weeds around the rhubarb plantation.
- Burn the debris of affected fields in early spring.
- Use poison baits containing 50 mL malathion + 200 g gur + 2 L water.
- Grow few borderlines of corn as trap crop.

Tarnished Plant Bug (*Lygus lineolaris*)

Adult tarnished plant bugs are 5–6 mm in length and light brown to reddish brown in colour. They occur throughout the season and are very active and quick moving and damage the plant by feeding on its young leaves. They pierce the stalk with their mouthparts and cause wilting and distortion of the leaves. Tarnished plant bugs are mainly a pest of new plantings.

Control

- Keep the field and adjacent areas weed free.
- Avoid planting adjacent to legume crops.
- Spray imidacloprid at 0.1% or cypermethrin at 0.15%.

Slugs (*Arion distinctus*)

Slugs may be a problem in a crop planted on poorly drained heavy soils and in weedy situations. Slugs feed at night by rasping the surface of stems, leaving unsightly scars, which reduce the salability of the stem.

Control

- Provide good soil drainage.
- Keep the weeds under control.
- Remove leaves and trash from the field at harvesting.
- Avoid the use of manure and/or mulches in areas prone to slug damage.

DISEASES

Leaf Spots (*Cercospora cichoria*)

Circular or angular spots that are variable in size, having beige centres surrounded by a red zone appear on foliage. The affected tissue dies, leaving large ragged holes on the foliage. Fungi overwinter in infected plant debris and in infected propagation stock.

Control

- Remove and destroy the leaves following the first heavy frost.
- Remove stems along with spotted leaves at first harvesting.
- Avoid mechanical injuries during harvesting.
- Avoid overhead watering.
- Keep the leaves as dry as possible.
- Apply Dithane M-45 80 WP (mancozeb) at 2 g/L of water.

Botrytis Rot (*Botrytis cinerea*)

The fungus may cause a leaf, stem and crown rot of forced rhubarb. Disease intensifies where there is poor air circulation and high humidity. The fungus becomes established on dead leaves or stems at the base of the plant. A heavy grey growth of the fungus occurs and numerous spores soon appear on the material. Botrytis rot spores are easily spread by air movement. Acidic soils have a negative influence on calcium uptake in the plant. Calcium is an integral part of the cell wall. Insufficient calcium uptake is directly responsible for a structurally inferior plant, which predisposes the plant to infection by this fungus.

Control

- Follow phytosanitary conditions strictly in the field.
- Apply lime in acidic soils to keep the calcium level high in plants.
- Spray the crop with captan at 0.2% at 10–12 days intervals.

Root or Crown Rot (*Phytophthora* spp.)

The affected plant leaves turn yellow to red and collapse. The crowns when sectioned exhibit a brown-black decay. Large roots lack small feeder roots and may have large brown-black lesions.

Control
- Use healthy propagation stock.
- Grow the crop in well-drained field.
- Remove and destroy the diseased plants.

Virus

Several viruses are known to occur in rhubarb. Reports from British Columbia and the United Kingdom indicate that turnip mosaic, arabis mosaic and cherry leaf roll viruses are the most common viruses, which have wide host ranges and cause mottling and ring spotting of leaves. They may be introduced in infected planting stock.

Control
- Obtain the plants from healthy nursery stock.
- Avoid planting rhubarb near virus-contaminated crops.

■■■

97 Sorrel

Akhilesh Sharma

Botanical Name	: *Rumex acetosa* (Garden sorrel or common sorrel)
	Rumex scutatus (French sorrel)
Family	: Polygonaceae (Buckwheat family)
Chromosome Number	: 2n = 14 (female), 2n = 15 (male)

ORIGIN AND DISTRIBUTION

Garden sorrel is a native of Europe and southwestern Asia, whereas French sorrel has its origin in the mountains of southern France and the southern and central regions of Europe and southwest Asia. It is a perennial herb, which is found in grassland habitats throughout Europe from the northern Mediterranean coast to the north of Scandinavia and in parts of Central Asia. It occurs as an introduced species in parts of North America. It was collected from its wild form until the French gardeners brought it under cultivation by improving flavour and texture of the leaves in the late 1600s. Sorrel is a genus containing closely related plants, which are also called sorrel of one or another kind such as wild sorrel, sheep sorrel, grass leaf sorrel, Indian sorrel, maiden sorrel, green sorrel, red sorrel etc. In the ancient world, sorrel was an extremely well-liked culinary herb and was used in the form of an antiseptic in conventional folk medicine. Since the fourteenth century, it has been extensively used in the form of a vegetable and salad plant in the West.

INTRODUCTION

Sorrel is cultivated as herbaceous vegetable plant primarily as a culinary herb for its leaves, which are used in salads and cooking but is now merely a wild food plant. The leaf of this plant resembles that of spinach and the plant is well known for strong acidity in Europe and the United States. There are different types of sorrel, of which, garden sorrel (*Rumex acetosa*) and French sorrel (*Rumex scutatus*) are most common. These two types are commonly confused but the garden sorrel bears lots of large arrow-shaped leaves, which emerge from a dense basal clump and have a reddish tinge at young stage, whereas French or Buckler-leaf sorrel has much smaller leaves, which are green and appear like a shield having mild flavour, less acidity and grows about a height of 60 cm. The leaves of both types have a lemony lettuce flavour. It is a common plant in grassland habitats and is cultivated as a garden herb or leafy vegetable.

COMPOSITION AND USES

COMPOSITION

Sorrel is a rich source of potassium, iron and vitamins A, B_1 and C. The flavour of sorrel leaves is mainly due to oxalic acid, which is harmless when consumed in small amount but its consumption should be avoided by people with a history of kidney stones. It contains many flavonoides including vitexin, rutin and kaempferol in its aerial parts. The nutritional composition of sorrel is given in Table 97.1.

TABLE 97.1

Nutritional Composition of Sorrel (per 100 g Edible Portion)

Constituents	Contents	Constituents	Contents
Water (%)	90	Manganese (mg)	0.59
Carbohydrates (%)	2.6	Zinc (mg)	0.22
Protein (%)	2.6	Copper (mg)	0.30
Fat (%)	0.5	Iodine (mg)	0.007
Dietary fibre (%)	0.6	Beta-carotene (mg)	11
Calcium (mg)	43	Vitamin A (mg)	1.38
Phosphorus (mg)	44	Thiamine (mg)	0.08
Potassium (mg)	400	Riboflavin (mg)	0.08
Magnesium (mg)	46	Nicotinic acid (mg)	0.40
Iron (mg)	9.2	Vitamin C (mg)	124

USES

Sorrel is valued for its medicinal and culinary uses. Its leaves, which are usually eaten raw as salads, in sandwiches, cooked into creamy sorrel soup or sauces, have a tangy, acidic, sour-lemony flavour. The citrusy sour flavour of sorrel leaves makes them a great accent herb for Thai-inspired dishes. The youngest leaves have the best flavour. The Caribbean people use sorrel for jams, chutneys and make a popular sorrel drink, which is served during Christmas time. It is also used as a colourant for some food, beverage and pharmaceutical industries.

MEDICINAL USES

Owing to high level of vitamin C content, it was rightly believed to protect against scurvy disease. It is also used in herbal medicine for its diuretic properties. Decoction of this plant has been used as a folk medicine in Korea to treat arthritis, gastritis and gastric ulcers and as a substitute for rhubarb to cure gastrointestinal problems. It is also used to cure cancer. The extract of sorrel has been reported to have heat-cleaning, diuretic, insecticidal, antimicrobial and anticancer activities. It has astringent properties, which are supposed to cleanse the blood.

BOTANY

Sorrel is a slender herbaceous perennial plant with deep-rooted system. It grows to a height of 30–80 cm and produces juicy stems. The basal leaves are 7–15 cm in length with long petiole having a membranous sheath formed of fused stipules. These leaves are oblong, slightly arrow head or spike-like shaped, while the terminal ones are sessile, which frequently becomes crimson coloured. It bears whorled spikes of greenish flowers in the initial stages, which turn reddish and then purplish with the increase in size. Being dioecious plant, it bears unisexual flowers, as the male and female flowers appear on different plants. Male flowers consist of six stamens on short filaments, while female flowers have three styles with branched stellate stigma. Flowers, which consist of three inner and outer tepals each of red to yellowish in colour, are borne on raceme near top of the stem. Flower stalks are jointed close to the flower. The seeds, when ripe, are round, brown, three-edged achene and shining.

CLIMATIC REQUIREMENT

Sorrel is a cold-hardy plant, which prefers sunny or partially shaded conditions for its growth. However, it grows luxuriously in abundant sunshine. Good seed germination takes place at soil temperature of 18–24°C, and the vegetative growth can be best achieved at 16–20°C. It can withstand frost but the growth is declined below 5ºC. The crop is very sensitive to warm temperature, which in combination with long days causes the plants to bolt.

SOIL REQUIREMENT

Sorrel can be grown on a wide range of soils with reasonable fertility and moisture retention ability. However, it prefers sandy loam to humus soils with good fertility and adequate drainage. For getting higher leaves yield, sorrel should be grown in friable loam soils rich in organic matter. This crop likes slightly acidic soil having pH range of 5.5–6.8. It flourishes better in iron rich soils.

CULTIVATED VARIETIES

In India, it is not a common leafy vegetable and is grown on a limited scale, mostly in kitchen gardens. Even in Europe and other countries, different local strains are grown for both home gardening and commercial cultivation. *Profusion* is a cultivar, which mutated naturally in France and produces only leaves without forming seed, and hence, the plant always has fresh and ready to pick leaves.

SEED RATE

Sorrel is not grown on commercial scale, though, 20–25 kg of seed is sufficient to raise the crop in a hectare area.

SOWING TIME

The sowing time differs in relation to the climatic conditions of particular region. For attaining good leaf growth, the seeds of local cultivars can be sown in September to October in northern plain areas and during April to May in temperate zone at high altitude.

SOWING METHOD

It is easy to propagate sorrel either from the rooted crowns or by sowing seeds directly in the field. However, it is important to sow seeds first in light and rich warm soil to raise the seedlings. The seedlings are allowed to establish for a full season and then the root cuttings are planted in next season at 30 cm apart. In every 4–5 years, the young vigorous crowns are lifted and replanted to a new spot in early spring or autumn season.

The seeds should be sown directly in flat seedbeds in early spring or autumn by dibbling at 1–1.5 cm depth and 15 cm apart. The seeds may be mixed with fine sand or ash before sowing to facilitate uniform distribution of seed. After sowing, the soil is raked in such a way that it may cover the seeds properly. When the plants attain a good growth, thinning is done by maintaining a spacing of 30 cm between rows and plants. Sowing can also be done by broadcasting the seeds but later proper spacing is provided by thinning out the extra plants.

NUTRITIONAL REQUIREMENT

In general, sorrel grows naturally and does not require much application of synthetic fertilizers. However, for higher productivity, it needs high nitrogen application along with optimum doses of phosphorus and potassium. About 20–25 tonnes of well-decomposed farmyard manure should be applied at the time of field preparation in a hectare of area, and at the time planting, the crop can be supplemented with 20–25 kg nitrogen and 30–50 kg phosphorus and potassium each. Later, after each cutting, 20–25 kg nitrogen per hectare can be top dressed at least twice during whole growth period of the crop.

IRRIGATION REQUIREMENT

It is essential to apply irrigation prior to seed sowing to have ample moisture to get better seed germination and subsequent growth of the plants. If moisture in the field is deficit, a light irrigation may be applied to obtain early and good germination. However, there may be crust formation in heavy soils, which may cause hindrance in seed germination, and therefore, it is important to sow the seeds under sufficient moisture conditions in such soils. Proper moisture is always maintained for better growth of the plants but wet conditions should be avoided. The crop may be irrigated at 10–12 days interval depending on the season and prevailing soil moisture conditions.

INTERCULTURAL OPERATIONS

HOEING AND WEEDING

In first year, the plant establishes itself, and hence, it is essential to keep weeds under control by hoeing or hand-weeding. Sorrel is closely related to the yellow dock weed (*Rumex crispus*), which should be controlled otherwise it may deteriorate quality of the produce. Frequent and shallow cultivation kills the weeds before they become a problem. To keep the fields weed free and to loosen the soil around the plants root system for better aeration, two to three hoeing and weeding are required. Hoeing operation should be done carefully to avoid any injury to the root system and developing plants. Plants are often neglected once the harvesting season is over but it is good to take care of plants by removing weeds and continuous watering, which may prolong its life for 8–10 years.

THINNING

In sorrel, thinning is done to maintain proper spacing between plants to have better growing conditions by reducing the competition among plants for nutrients, light, moisture and space. Thinning is normally done when the plants are 5 cm tall to maintain a spacing of 15 cm between plants in initial stages and finally 30 cm by doing another thinning operation. In the later years, garden sorrel grows into a small patch of several plants, and thinning is done by removing the weak plants and retaining the healthiest plants to maintain high production of early leaves.

FLOWERING

Pre-mature flowering, *i.e.*, bolting, is a problem in sorrel, which is undesirable and reduces the quality and total production. High temperature in combination with long days causes bolting in plants. Dioecious plants in sorrel produce two different kinds of male plants, *i.e.*, (i) extreme males and (ii) vegetative males and females. The extreme male plants are small with very little vegetative growth and tend to bolt quickly. Inversely, vegetative males and females form flowering stalk late and produce considerably more foliage, which makes them the preferred plants type for commercial cultivation. The extreme males are eliminated from commercial strains by selection.

HARVESTING

Leaves can be harvested any time after a few months of its growth but the younger leaves are tend to be almost tasteless and gradually gain their characteristic and desired acidity and flavour. The tender young basal leaves are the best for harvesting for culinary purposes as the coarser and older foliage develop bitterness. Therefore, it is important to harvest the leaves more frequently and use them on the same day for best flavour. The newly seeded crop takes 35–40 days to reach small size leaves and 2 months for good size leaves for harvesting. Sorrel becomes ready for harvesting when the leaves attain a length about 10 cm. Tender leaves are best for eating. Hence, it would be imperative to cut the leaves at regular intervals to have steady supply of young and tender leaves. On commercial scale, it produces tangy and edible leaves approximately 4 months after thinning. Sorrel plant grows into sizable clumps by attaining height of 30–45 cm and width up to 60 cm.

YIELD

The average yield of sorrel greens varies from 150 to 200 q/ha.

INSECT PESTS

There is no major insect pest affecting this crop though aphids, thrips, armyworm and root knot nematode may sometimes damage the sorrel plant to some extent.

APHIDS (*Myzus persicae, Brevicoryne brassicae* and *Lipaphis erysimi*)

These small creatures feed by sucking the sap from tender parts of the plant. In small numbers, they are not a serious problem. However, under favourable conditions, they multiply rapidly by developing large colonies and suck sap to the extent that the growing shoots become stunted and misshapen. It excretes honeydew, on which, black sooty mould develops that hinders with photosynthetic activity of plant. They may also transmit virus diseases.

Control
- Remove and destroy the portion of the plant attacked by the aphids.
- Spray the crop with simple water with pressure.
- Apply malathion (0.05%) with the appearance of pest at 15 days interval. Stop spraying at least 7 days before harvesting.

THRIPS (*Thrips tabaci*)

Plants infested with thrips develop spotted appearance on the leaves, which turn into pale white blotches due to drainage of sap. The adults hibernate in soil, on grass and other plants in the field.

Control
- Foliar application of malathion 0.05% or cypermethrin 0.01% is effective.

Armyworm (*Spodoptera exigua*)

The larvae damage the young seedlings completely. The larvae may skeletonize the leaves of older plants or may completely consume them, while the stems are seldom eaten. However, the older plants may recover. The mature larvae are bright green to purplish green to blackish in colour, while the most common phase is light olive green with a darker strip down the back and a paler stripe along each side.

Control
- Use well-decomposed manure for reducing the incidence.
- Control broad leaf weeds.
- Dispose of crop residues rapidly after harvest.
- *Bacillus thuringiensis* is more effective when used in early larval stage.
- Collect and destroy the larvae after flooding of fields/beds
- Drench the beds with chlorpyriphos 0.04%.
- Spray cypermethrin (0.01%) on crop foliage and soil surface.

DISEASES

The crop is relatively disease-free. However, powdery mildew, leaf spot and *Phytophthora* foot rot may appear to some extent and cause damage.

POWDERY MILDEW (*Erysiphe* sp.)

Greyish-white powdery patches appear on the leaves and stem, which cause yellowing of leaves followed by rapid defoliation. Dry conditions favour the development of powdery growth.

Control

- Treat the seed with hot water at 50°C for 30 minutes.
- Spray the crop with Karathane 0.05% or Sulfex 0.2% at least twice at an interval of 10 days immediately after the appearance of disease.

LEAF SPOT (*Stemphylium* spp.)

The symptoms appear first as numerous small and circular white to yellow spots on older leaves, which further progress to the younger leaves. The spots often coalesce to become irregular shaped and change to an olive black. The older spots develop a narrow darker coloured margin with a tan centre and the older leaves die, due to severe infestation eventually all leaves may die.

Control

- Treat the seed with hot water at 50°C for 30 minutes.
- Spray the crop with Dithane M-45 0.25% twice at an interval of 10–15 days after the appearance of disease symptoms.

CROWN ROT (*Phytophthora* spp.)

The affected plants fail to develop leaf or plant often wilt or die abruptly due to rotting of roots and crown. The initial symptom appears when plant looks draught stressed. The leaves may turn dull green or yellow. Plants growing in poor drained soils are more prone to collar rot infection. The central core of conducting tissue in the taproot turns to be yellow or brown above the rot lesion, which further extends into the lower stem. It is a soilborne disease, which is favoured by wet conditions.

Control

- Follow good water management.
- Provide good soil drainage.
- Treat the seed with carbendazim at 3 g/kg of seed before sowing.
- Treat the seed with hot water at 50°C for 30 minutes.
- Spray or drench the crop with carbendazim at 0.1%.

98 Ceylon Spinach

M.K. Rana and Navjot Singh Brar

Botanical Name   : *Talinum triangulare* (Jacq.) Willd

Family   : Portulacaceae

Chromosome Number : 2n = 24, 48, 72

ORIGIN AND DISTRIBUTION

Tropical America, where Ceylon spinach is found growing naturally on roadsides, waste places and forests edges, is considered as its probable origin. It has now become a weed with pantropical distribution. It has been recorded for several countries in West and Central Africa but it is claimed to have a South American origin. Africa may possibly be the origin of Ceylon spinach, as several *Talinum* species including the closely related species *Talinum portulacifolium* (Forssk.) Schweinf occur in Africa. It is grown in West Africa, South Asia, Southeast Asia and the warmer parts of North and South America. In Brazil, it is cultivated along the banks of Amazon River and is consumed mainly in Para and Amazonas states. Ceylon spinach or waterleaf is very popular in Nigeria where it is used in many dishes. From Sri Lanka, the herb was first introduced into South India and its cultivation was started in Tamil Nadu for its edible leaves.

INTRODUCTION

Ceylon spinach, also known as waterleaf, Florida spinach, potherb fame flower, Surinam purslane and sweet heart, is a perennial herb with long caulescent stem. It is a cosmopolitan weed commonly found throughout the humid tropics but considered as a good alternative for purslane (*Portulaca oleracea* L.). Its leaves are high in moisture content (almost 90.8 g per 100 g of edible leaf). Its fleshy green leaves, succulent stem and young shoots are consumed as vegetable. Ceylon spinach is relatively tolerant to drought conditions, as it tends to adopt a crassulacean acid metabolism (CAM) pathway, resulting in efficient utilization of available moisture, carbon dioxide assimilation during night and increased growth.

Ceylon spinach is a mucilaginous vegetable with high oxalate content, the presence of which is a drawback since more than 90% of it is present in soluble form and can induce kidney stone if consumed in large quantity. However, cooking denatures nearly half of the soluble oxalate. Its presence in diet inhibits absorption of calcium and iodine. Its stickiness is presumably due to its high pectin content, which is associated with dietary fibre.

COMPOSITION AND USES

COMPOSITION

Ceylon spinach provides carbohydrates, protein, fat, fibre, minerals and vitamins in good amount. It contains an appreciable amount of flavonoids, alkaloids, saponins, low level of toxicants like tannins and substantial amount of bioactive compounds. The nutritional composition of Ceylon spinach is given in Table 98.1.

TABLE 98.1
Nutrition Composition of Ceylon Spinach (per 100 g Edible Portion)

Constituents	Contents	Constituents	Contents
Moisture (g)	90.8	Vitamin A (µg)	3
Carbohydrates (g)	3.7–4.0	Thiamine (mg)	0.08
Protein (g)	1.9–2.4	Niacin (mg)	0.30
Fat (g)	0.4–0.5	Pyridoxine (mg)	0.18
Dietary fibre (g)	0.6–1.1	Vitamin C (mg)	31
Calcium (mg)	90–135	Ash (g)	2.4
Phosphorus (mg)	0.18	Energy (kcal)	105
Iron (mg)	4.8–5.0		

Uses

The soft leaves and shoots of Ceylon spinach are usually consumed after cooking. They are although watery but should not be cooked for a long time. In West Java, it is also added raw to salads in the Sudanese cuisine. In Southeast Asia, it is sometimes grown as an ornamental pot plant or an edging plant in home gardens.

Medicinal Uses

Ceylon spinach significantly contributes to human health; thus, it should be included in daily diet. The bioactive substances present in Ceylon spinach are non-nutrient plant components, which are beneficial for health; flavonoids found naturally in Ceylon spinach are polyphenol antioxidants. Saponins, which present in Ceylon spinach, have favourable effects on blood cholesterol levels, cancer, bone health and stimulation of body immune system. In South America, its crushed plant is applied as a poultice on contusion, inflammation and tumours, and its decoction is used for sore eyes and to aid recovery from shocks and depression.

BOTANY

Ceylon spinach, an erect, glabrous perennial herb, grows up to a height of 80–100 cm. The plant is strongly branched with fleshy and swollen roots. Its stem is succulent and obtuse angular to terete. Leaves are alternate, simple, almost sessile, succulent, 3–15 cm long, 1–6 cm broad, stipules absent, blade obovate to spatulate, base long and tapering, apex rounded to notched, mucronate and entire venation pinnate and indistinct. Inflorescence is a terminal cyme on a 12-cm-long triangular stalk. Flowers are bisexual, regular, 1-cm-long pedicel recurving in fruit, two sepals, free with three prominent veins, five petals, free, obovate, up to 1 cm long, pink, stamens numerous, ovary superior, single celled, style slender with three-branched stigma exceeding stamens and open in the morning. It bears flower round the year and is mainly self-pollinating. Fruit is 4–7 mm long, three valved, globose to ellipsoid elastically dehiscent capsule with many seeds. Seed is 1 mm long, compressed, globose-reniform, tuberculate and shining black. Ceylon spinach is a fast growing herb and once it is established, easily reseeds itself.

CLIMATIC REQUIREMENT

Ceylon spinach thrives best under humid conditions at a temperature of about 30°C; however, the growth is very fast in rainy months. Its growth is considerably slow in dry season. It grows well in cloudy weather and under the shade. It can also grow in full sunshine but the plants remain smaller.

Both high temperature (>35°C) and drought affect the number of leaves, leaf area, stem size and number of branches negatively. It is well adapted throughout the country but a wet tropical environment with partial shade is best suited for its cultivation. It can successfully be cultivated up to 1000 m above mean sea level.

SOIL REQUIREMENT

Ceylon spinach can be cultivated on a wide range of soils but it grows best on well-drained fertile, moderately deep, friable soils rich in organic matter. Like other leafy vegetables, it prefers dark colour soils with pH from 6.5 to 7.5.

CULTIVATED VARIETIES

Ceylon spinach is an underutilized crop at world level, which is why, no research work has yet been done for its genetic improvement, but now, the crop is drawing the attention of researchers for systemic research work. Presently, some germplasm lines have been preserved for future reference, one of which is given below:

LCH 42

This line was identified and authenticated by Dr. D. Narasimhan at the Centre for Floristic Research, Department of Plant Biology and Plant Biotechnology, Madras Christian College, Chennai. A herbarium voucher specimen (LCH 42) was preserved for future reference in the Loyola College Herbarium.

SOWING TIME

Ceylon spinach being a shade loving vegetable is grown in May to June and September to October in South India. Most dark leafy vegetables grow when the weather is cool, thus, it is best to plant waterleaf during the fall before the first frost. Late August to October is probably a good time to plant it in Northeast region. If seeds are to be used for its propagation, the seeds are sown inside in July and then transferred outside in September.

PLANTING METHOD

Ceylon spinach can be propagated by seeds or cuttings taken from the wild plants, however, propagation by 10–15-cm-long cuttings is best for harvesting the side shoots. It is advisable to remove the lowest pair of leaves before planting. Its tiny seeds (about 0.25 g test weight) can be collected only from the fruits that have turned yellow since collection of seeds from fully ripe fruits is difficult as they shatter when touched. Seeds taken from green fruits will not be viable but the seeds taken from nearly mature fruits before their dehiscence will have an acceptable germination rate after proper drying. The seed before sowing is thoroughly mixed with fine sand for uniform distribution in a well-prepared nursery or seedbed. The seed takes about 5 days to germinate after sowing and subsequent growth is very fast if adequate moisture is maintained in soil. Seedlings are usually transplanted when they are 3 weeks old. A plant density of 10–25 plants square metre is maintained, depending on harvesting method and crop duration.

Ceylon spinach is usually intercropped with other vegetables but it can also be grown as a sole crop at a line and plant spacing of 15 cm. Planting at closer spacing reduces competition with weeds. In fertile soils or with adequate fertilizer application, the line and plant spacing may be increased to 25 cm. The delicate seedlings must be protected from blazing sun by providing shade with agro-net.

NUTRITIONAL REQUIREMENT

The nutritional requirement of Ceylon spinach generally depends on soil type, fertility status of the soil, crop grown in previous season and irrigation facilities. Well-decomposed farmyard manure should be incorporated into the soil at 20–50 t/ha at the time of field preparation. The crop also responds well to nitrogenous fertilizer application, thus, sufficient quantity of nitrogen fertilizer must be applied to the soil at the time of last ploughing to raise a healthy crop. Application of nitrogen 50–70 kg/ha is advisable to get maximum yield of Ceylon spinach. Nitrogen is applied as top dressing in equal split doses first 20–25 days after planting and second, third and fourth after every cutting of the leaves or shoots. In fact, the nutrients requirement of Ceylon spinach crop still needs to be standardized.

IRRIGATION REQUIREMENT

The irrigation requirement depends on soil type, growing season, wind current direction and wind velocity. Ceylon spinach usually prefers moist environment, thus, the soil after planting is kept moist but not waterlogged. Growth is most profuse when the water content in soil is closer to field capacity. After planting, a light irrigation is required daily during the first week and thrice a week when the plants have covered the soil surface completely.

INTERCULTURAL OPERATIONS

HOEING AND WEEDING

Ceylon spinach is itself a weed in cultivated field and homestead gardens but when it is cultivated as a vegetable crop, the farmers worry about it since the other weeds compete with it for nutrients, moisture, space and light, hence the removal of a weeds in early stages of crop growth is essential. Its roots are shallow and the plant is easy to uproot, therefore, hoeing should be done very shallow to avoid damage to its roots. Timely hoeing and keeping the weeds under control not only reduce the crop-weed competition but also improve the aeration in crop root zone.

MULCHING

Mulching in Ceylon spinach field reduces water evaporation and promotes numerous beneficial organisms like earthworms that keep the soil healthy, eliminate stress in shallow-rooted plants, provide nutrients to the plants and help in the establishment of newly transplanted seedlings. It also checks weeds and improves microbial activity in the soil by improving environment around the roots.

HARVESTING

The first harvesting can be done 3 weeks after planting, and thereafter, the shoots can be harvested at an interval of 1–2 weeks for a period of 2 months. However, the leaf size is decreased with the increase in plant age and number of harvests. The best quality leaves can be obtained only up to third harvesting. At the most, four shoot cuttings can be taken since thereafter the plant growth starts declining. The crop is harvested by cutting the stem just above the ground level, which allows the plant to regenerate faster than harvesting only the top portion and side shoots. When the first cutting is taken late and lower part of the stems becomes brown and drop their leaves, it is advisable to cut the shoots just above the ground in order to obtain a better quality shoots in next cutting. A few weeks after the commencement of rainy season, the yield from the wild plants in the market is sold at highly competitive price since cultivation at this stage is no longer feasible due to high labour cost though a rainfed crop can give cuttings bit longer. Under favourable conditions, harvesting may be done 15–20 times a year at an interval of 2 weeks. However, it is advisable to plant the crop after every 6 months.

YIELD

The greens yield of Ceylon spinach depends on several factors such as planting material, method of planting, soil type and fertility, sowing time and cultural practices adopted during the course of crop cultivation. On an average, the Ceylon spinach greens yield varies from 100 to 600 q/ha and seed yield from 1 to 3 q/ha.

POST-HARVEST MANAGEMENT

Ceylon spinach is highly perishable and the shoots start withering only a few hours after harvesting. The greens in fresh form may not be stored longer under ambient room conditions but it is possible to store the shoots in a plastic bag in refrigerator for several days. Its greens can be dehydrated in shade at low temperature and the dehydrated product can safely be stored for longer period for offseason use.

INSECT PESTS AND DISEASES

A number of insect pests and diseases attack the Ceylon spinach crop during the growing season. Some of the insect pests and diseases along with their control measures are given in the following section.

INSECT PESTS

Aphid (*Myzus persicae*)

The nymph and adult both suck cell sap from tender parts of the plant. The affected plants remain stunted, leaves curled and plant vigour is reduced. They excrete honeydew like sugary substance, on which the black sooty mould grows, which checks photosynthesis, and ultimately, it reduces the crop yield.

Control
- Follow clean cultivation.
- Set up yellow pan traps in the field.
- Introduce natural parasites and predators such as lady beetles and their larvae, lacewing larvae, and syrphid fly larvae to control aphids.
- Spray the crop with dimethoate or methyl demeton 0.03% at 10 days interval.

DISEASES

White Leaf Spot (*Pleospora* spp.)

The disease appears as light yellow specks, which later develop into spots and coalesce, resulting in shrivelling and drying of leaves. Severe leaf spotting can result in premature leaf loss. Disease symptoms may appear on crops earlier in season under stress due to lack of moisture, insufficient nitrogen or severe competition with weeds.

Control
- Follow long crop rotation with non-host crops.
- Plant the seedlings at proper spacing.
- Adopt good cultural practices to reduce stress caused due to weed competition and nutrients deficiency.
- Maintain phyto-sanitation conditions by removing infected plant materials.
- Spray the crop with Blitox 0.2% or Bordeaux mixture 1% at 15 days interval.

Leaf Mosaic

The leaves are mottled with yellow, white and light and dark green spots, which appear to be elevated, which gives the leaves a blister-like appearance. The plant growth is slow. Consequently, the plants often become stunted. Mosaic virus is mostly spread by insects, especially aphids and leafhoppers.

Control

- Grow plants in floating row cover.
- Use aluminium foil to deter the leafhoppers.
- Keep the field weed free, as some of the weeds may be the host for leafhoppers.
- Rogue out the affected plants and burn them.
- Spray malathion 0.1%, methyl demeton 0.2% or phosphamidon (0.05% 10 days interval to control virus vectors.

Common Blight (*Xanthomonas axonopodis*)

Symptoms include appearance of white and irregular spots on leaf. Dark green spots appear on underside of the leaves. The spots later turn brown or reddish on upper side of the leaves and eventually become black, rendering the shoots unsaleable.

Control

- Rogue out the affected plants at an early stage.
- Soak the seeds in solution of streptocycline 0.01% + captan 0.25% before sowing.
- Spray the crop with copper oxychloride 0.3% at 7–10 days interval.

■■■

99 Purslane

M.K. Rana and Mahesh Badiger

Botanical Name : *Portulaca oleracea* L. (Common or green or garden purslane)
Portulaca sativa L. (Golden purslane)

Family : Portulacaceae

Chromosome Number : 2n = 52

ORIGIN AND DISTRIBUTION

The actual site of origin is not known for certain and several temperate areas of the Northern Hemisphere are expected including Eurasia, Europe, Western Asia, China, India and the sub-desert areas of Northern Africa. However, it seems to have originated in the Western Himalayan area, and then spread towards the south of Russia and Greece, perhaps as far back as 4000 years ago. Mexico and Australia are considered as the centres of diversity of purslane.

INTRODUCTION

Purslane (synonyms Verdolagas [Mexico] and *Sanhti* or *Punarva* [India]) is an annual potherb, having a long history of use for human food, animal feed and medicine. In many parts of the world, it is still regarded as a *weed with nutritional potential*. Its young stems and leaves are succulent and edible with a slightly acidic and salty taste similar to spinach.

COMPOSITION AND USES

COMPOSITION

Purslane has been described as a *power food of the future* because of its high nutritive and antioxidant properties. Purslane contains more omega-3 fatty acids (300 to 400 mg/100 g of fresh leaves) than any other leafy vegetable plant. Succulent leaves and stems also contain dietary minerals, such as calcium, phosphorus, iron and potassium, as well as vitamins, mainly vitamins A, C and E and some amount of vitamin B and carotenoids. However, its nutritional status varies with growth stage. Generally, roots and leaves contain more iron and manganese than stems. A general estimate of its nutritive value is presented in Table 99.1.

USES

Purslane is eaten extensively as a potherb, cooked and added in soups and salads around the Mediterranean and tropical Asian countries. Its pleasant nips add spice to a meal. In salads, its leaves are delicious, and yellow leaved ones are used for garnishing dishes. It can be enjoyed as a condiment after pickling. The dried herb can be boiled and is made into tea/soups. The people in China use it in stir-frying dishes and call it *carti choy*. Purslane has potential as an animal feed, in aquaculture and in the food processing industry too as a soup thickener. However, due to oxalic acid and nitrate content, excessive consumption of purslane is not recommended.

TABLE 99.1

Nutritional Composition of Purslane (per 100 g Edible Portion)

Constituents	Contents	Constituents	Contents
Water (g)	90.5	Iron (mg)	14.8
Carbohydrates (g)	2.9	Vitamin (IU)	1320
Protein (g)	2.4	Thiamine (mg)	0.10
Fat (g)	0.6	Riboflavin (mg)	0.22
Dietary fibre (g)	1.3	Vitamin C (mg)	29
Calcium (mg)	111	Vitamin E (mg)	12.2
Phosphorus (mg)	45	Energy (kcal)	27

MEDICINAL USES

There is a newer interest in cultivation of purslane as a crop since its identification as one of the best sources in the plant world for omega-3 fatty acids (α-linolenic acid in particular), which offers protection against cardiovascular disease, cancers and a number of chronic diseases. It is listed in the World Health Organization as one of the most used medicinal plants and it has been given the term *Global Panacea*. It has diuretic, anti-ascorbic, anti-pyretic, anti-asthma and anti-inflammatory properties.

BOTANY

Purslane belongs to the family Portulacaceae and genus *Portulaca*. *Portulaca* includes more than 40 species, which are scattered in tropical and subtropical regions. Purslane is a very variable species due to its diploid, tetraploid and hexaploid nature. The cultivated form of purslane, *i.e.*, *Portulaca oleracea* (garden or green purslane) and *Portulaca sativa* (yellow leaves, less hardy than the green purslane but possessing the same qualities) is hexaploid with 2n = 54 and has more erect taller and vigorous plants with broad leaves and larger seeds. However, cultivated species can be distinguished from the cylindrical or terate leaved group that includes *Portulaca grandiflora*, *Portulaca okinawensis*, *Portulaca pilosa* and *Portulaca grandifolia* on the basis of leaf margin, kranz anatomy, axillary hairs, seed morphology and floral characters. *P. pilosa* is commonly grown as an ornamental flower in pots. Cultivated forms are more upright and vigorous than those growing in the wild. Purslane is a succulent and well-branched herb, leaves are spirally arranged, flowers are yellow, stamens 7–10, style with arms, fruit ovoid, capsule, seeds numerous, tiny and black in colour.

CLIMATIC REQUIREMENT

Purslane is a halophytic C_3 plant, susceptible to frost and comes up well under tropical and sub-tropical conditions. For proper germination of seed and growth, high temperature is essential. It is adaptable to both dry and saline conditions, making it a prime candidate to form edible landscape in areas with dry conditions and salty soils. It grows well in loam and heavy loam soils containing good amounts of organic matter. The soil should be loose, friable and of fine tilth.

VARIETIES

Seeds of two horticultural varieties, *i.e.*, *Purslane* and *Green Purslane*, are sold by several seed companies.

SEED RATE

Purslane reproduces from seeds and stem fragments. It is generally cultivated by seeding. Seeds of 1–1.5 kg are required to sow a hectare area.

SOWING TIME

Purslane is susceptible to frost, thus, its sowing in North India is done in the last week of February to first week of March and in Southern India in January to February. In Southern India, early sowing can also be done to have an early crop.

SOWING METHOD

The field is prepared to fine tilth by ploughing two to three times followed by planking. Seeds being very small are mixed with fine sand for uniform distribution during sowing. The seed is generally sown by broadcasting. After sowing, the seeds are covered with a thin layer of soil. Dense sown crop in lines at 15–20 cm distance promotes erect growth of shoots and produces a higher yield. Being highly salt tolerant, it can also be grown as an intercrop in saline environments of fruit orchards.

NUTRITIONAL REQUIREMENT

The purslane crop is well adapted to different climatic conditions and no efforts have been made to standardize its agronomical practices. Growing in soil supplied with farmyard manure qt 10–12 tonnes per hectare produces soft and succulent leaves. After each cutting, a top dressing of nitrogenous fertilizer enhances the growth of new flush but excess application of the nitrogen stimulates the nitrates content in the leaves.

IRRIGATION REQUIREMENT

A light irrigation is given just after sowing, and later on, it is irrigated three to four times during the season, preferably after every cutting.

INTERCULTURAL OPERATIONS

Due to dense planting, it comes to harvest in 4–5 weeks after sowing and the crop itself smothers the weeds.

HARVESTING

First harvesting is done when the plants attain a height of 12–15 cm. After this, two to three cuttings are taken at fortnightly intervals. The crop remains in a productive condition from February to May.

YIELD

Yield of purslane depends on the number of harvests. Generally, 15–20 tonnes yield of fresh leaves can be obtained from a hectare area.

POST-HARVEST MANAGEMENT

Purslane is not so sensitive to chilling injury. Therefore, it can be stored successfully for 3–5 days at 0–1°C temperature and 95% relative humidity. It is considered a very good raw material for the food industry because during processing, different methods of drying do not affect its nutritional composition.

INSECT PESTS

PURSLANE SAWFLY (*Schizocerella pilicornis*)

It is a natural enemy of common purslane. Overwintering prepupae or pupae in the soil develop into adults in April or May, after which, the insect completes several generations (number is unknown) on purslane until the larvae develop into overwintering forms in September and October. Larvae are miners only in the thick leaves of purslane. Its adults do not harm the crop.

Control

- By the time it develops sufficient numbers to have an impact on the crop, seed development takes place as it is a short duration crop.
- Spray the crop with malathion 0.05%.

LEAF MINING WEEVIL (*Hypurus bertrandi*)

It is also a natural enemy of common purslane. Larvae of the weevil, like those of the sawfly, mine the leaves. Adult weevils damage the leaves by actively feeding on the margins and epidermal surfaces. The adults also feed on stems and developing seed capsules. Common purslane is the only reported host plant of this weevil.

Control

- Spray the crop with malathion 0.05%.

DISEASES

BLACK STEM OF PURSLANE (*Dichotomophthora portulacae*)

This disease causes stem blackening and constriction at the soil line. Black lesions occasionally develop on the upper stem parts. Infected roots become dark and constricted. Spread of pathogen in the stems and roots may result in death of purslane plants.

Control

- Spray the crop with Bavistin, mancozeb, Blitox or Blitane at 0.2% to control this disease.

■■■

100 Pepino

M.K. Rana and Archana Brar

Botanical Name : *Solanum muricatum* Ait.

Family : Solanaceae

Chromosome Number : 2n = 2x = 24

ORIGIN AND DISTRIBUTION

Pepino, a cultigen with no known wild ancestor, is presumed to be native to the temperate Andean regions of Columbia, Peru and Chile though it is not known in the wild form and its domestication is unknown. It is a domesticated native to the Andean region known as the *Sierra*, which is home to several solanaceous fruit crops. The Spanish Conquistadors reported that it was being cultivated on the coast in the Moche Valley of Peru. It is also well known in South American countries, New Zealand and Australia. The cultivation of pepino is extensively described by chroniclers in the kingdom of Peru, which in addition to Peru also included the present Ecuador, the North of Chile and the high plateau of Bolivia. Its introduction into France is said to be carried out by Andre Thouin, head of gardening at the Royal Gardens at Versailles in 1785. In 1887, Immer introduced it in Russia as a greenhouse plant in Saint Petersburg. From there, it spread to many other places in Russia, as far west and south as the Caucasus. In Italy, the experimental samples coming from the Canary Islands were grown at the beginning of this century. It was introduced into Holland in 1916 by a farmer but its cultivation was abandoned. It was reintroduced in 1948 through cuttings from Egypt. Recently, it has been introduced to the Nilgiri Hills in Tamil Nadu from the Northern Andes. Attempts have been made to produce commercial cultivars and to export fruit to New Zealand, Turkey and Chile. Morphologically, this plant is a member of the *Solanum* section *Basarthrum*, and as such, it is allied to a number of Andean wild species. Previously, pepino might have existed in the form of wild plant, and now, it may be represented only by the cultigen, however, if its ancestors are existing, the three wild species, *i.e.,Solanum basendopogon* from Peru, *Solanum caripense* from Coasta Rica through Peru and *Solanum tabanoense* from Colombia and Ecuador emerge as the most likely progenitors. *Solanum caripense* also might have been a direct ancestor of pepino or might have hybridized subsequent to its origin with the pepino to yield some of the haplotype variation. Similarly, *Solanum cochoae* might have hybridized with pepino, however, there is no DNA character supporting the involvement of *S. basendopogon* in its origin.

INTRODUCTION

Pepino is a nutritious vegetable-cum-fruit species of evergreen shrub, which is grown for its sweet edible fruits. Although its fruits look like a melon but it belongs to the same family as that of well-known crops including tomato, potato, peppers and eggplant, *i.e.*, Solanaceae. It is known around the world by many names like melon pear, melon shrub, mellow fruit, pepino dulce and sweet pepino. Its fruit on ripening is highly juicy and moderately sweet. It is more closely related to the tomato but looks and tastes more a cross between a cucumber and a sweet melon. It is enjoying increasing exposure in the international market. In Europe, the demand for pepino has significantly increased due to its high nutritional value, flavour and attractive appearance.

Pepino is a good option for kitchen gardens and for intercropping in newly planted orchards. It has also been identified as a potential crop for greenhouse cultivation. It is a little known crop from the tropical and subtropical Andes esteemed for its edible fruits, which are juicy, scented and sweet. When ripe, it has a flavour similar to muskmelon, however, when it is not fully ripe, it has a cucumber-like scent and flavour. In recent years, it is grown as a commercial crop in some Andean countries, New Zealand and Australia. In Spain, Italy, France, the Netherlands, the United States, Israel, Turkey, Korea, China and India, the experiments related to pepino cultivation have also been undertaken.

COMPOSITION AND USES

COMPOSITION

Pepino juice recovery is around 94–98%, depending upon cultivars. A medium serving (100 g) of its fruit provides energy 80 kcal and dietary fibre 5 g (Table 100.1). The fruits are rich in minerals and vitamin C but low in starch and soluble sugars, which consisted of sucrose, fructose and glucose with sucrose accounting for 50% of the total sugars. The level of glucose and fructose decreases during ripening, whereas sucrose concentration increases as the ripening progresses. A discernible reduction has also been noticed in content of protein, ash and fat while fruit turns from raw to mature.

Citric acid made up more than 90% of the non-volatile organic acids. Malic acid and traces of quinic acid are also present. Aspartic acid accounts for 70% of the total free amino acids. Cellulose, pectin and hemicellulose represent 71%, 17% and 11% of the cell wall polysaccharides, respectively. During maturation, there is a continuous increase in the levels of sucrose, citric acid and the dicarboxylic amino acids. The vitamin C level present in pepino fruit is much higher than found in citrus fruits. However, an increase of drying temperature results in a considerable reduction of both vitamin C and total phenolic content. A good and important feature of its fruit is the absence of oxalates in it. The nutritional composition of pepino fruit is given in Table 100.1.

The flavour of its fruits is described as a delicious blend of cucumber and honeydew melon. The major flavouring substances present in pepino seed oil are 3-methyl-2-buten-1-ol, 3-methyl-3-buten-1-ol, 3-methyl 2-buten-1-yl acetate, 3-methyl-3-buten-1-yl acetate, butyl acetate and (Z)-6-non-1-ol. In addition to the above flavouring volatiles, 3-methyl-3-buten-1-yl 3-methylbutanoate, 3-methyl-2-buten-1-yl 3-methylbutanoate, 3-methyl-2-buten-1-yl 3-methyl-2-butenoate and 3-methyl-3-buten-1-yl 3-methyl-2-butenoate have been identified as minor components.

TABLE 100.1
Nutritional Composition of Pepino Fruit (Values per 100 g of Edible Portion)

Constituents	Contents	Constituents	Contents
Moisture (%)	91.45	Sucrose (%)	4.00–5.50
Juice content (%)	94.75–98.50	Glucose (%)	2.20–3.00
Dry matter (%)	7.05–8.92	Fructose (%)	1.40–1.90
Starch (mg)	22.80–97.50	Total soluble sugars (%)	6.00–10.80
Protein (g)	0.12–0.47	Ascorbic acid (mg)	38.30–68.80
Fat (%)	0.09	Titratable acidity (mg/100 g)	78.00–106.00
Dietary fibre (g)	5.0	pH	5.30–5.60
Ash (%)	0.05	EC (dS/m)	3.32–3.38

FOOD USES

Pepino, a neglected solanaceous crop, that has elicited an increasing interest from exotic fruit markets can be consumed in various ways like green and cooked vegetable, in fruit and spinach salads, fresh fruit as dessert and delicious fruit juice and squash. Pepino is served after peeling and slicing. Contrary to other fruits, the original flavour in pepino is retained during canning process, and hence, it has a better preference among consumers.

MEDICINAL USES

Pepino is said to have some medicinal properties. It is a good antiscorbutic since it contains vitamin C at higher levels than normally found in most fruits. Its juice mixed with pink ointment is profitable against kidney heat. Aqueous extract of its fruits could attenuate the progression of diabetes due to its anti-inflammatory, antiglycative and antioxidative effects. However, its antioxidant activity is decreased at low temperature.

BOTANY

Pepino belongs to the genus *Solanum*, the species *muricatum* and the family Solanaceae. In 1789, Aiton gave the name *Solanum muricatum*, and it seems that the specific name *muricatum* was given in reference to the rough or warty character of the adventitious roots that spring from base of the stems. Pepino, a perennial erect shrub growing to a height of 150 cm, is a diploid species with chromosome number 24 (2n = 2x = 24), but as the plant is frost sensitive, it is grown as an annual in many areas. Its plant is profusely branched, and its lanceolate leaves with slightly undulating margins and tapering apex are entire in early stages but tri- or penta-foliate thereafter. It has inflorescence with 8–20 purplish to white flowers, which usually set 0–2 fruits, although up to seven fruits per inflorescence may be observed. The long stalked fruits borne in clusters are medium to large in size (15–600 g) and either round, ovoid or elongated. The young fruit is pale green to white in colour, however, as it ripens the attractive purple stripes appear along the fruit's length and the background colour changes to a light golden yellow. The ripening fruits are tender and juicy but vary greatly in shape, colour and size. The fruit on ripening gives off an exotic fragrance from the stem end. A good-quality pepino fruit will have smooth skin, sweet smell and will be as firm as a partially ripened plum. The number of seeds per fruit varies from 10 to 200 but many cultivars develop parthenocarpic fruit.

CLIMATIC REQUIREMENT

Being a relatively hardy species, it grows in its native place at altitudes ranging from close to sea level to 3000 m. However, it performs best in a warm and relatively frost-free climate. The plant can survive at as low as –2.5°C, if the freezing temperature does not prolong too much, though the plant drops many of its leaves. However, the optimum temperature range for its growth and development is from 15°C to 25°C. At flowering, high temperature (>25°C) may cause flower dropping and results in poor fruit set. Although the species is a perennial but its sensitivity to chilling temperature, insect pests and diseases force the growers to replant the crop every year.

Low yield as a result of occasional lack of fruit set may hamper its introduction as a new crop in areas with mild-winter climate. Sensitivity to high temperature, lack of pollen release, competition with vegetative growth, low light intensity in certain types of greenhouse and the peculiar characteristics of each cultivar have been identified as the main causes of unfruitfulness. However, parthenocarpy meaning no need of pollination for setting fruits can overcome poor fruit set caused due to absence of pollination. Parthenocarpic clones set seedless fruits and give a consistently higher yield than the non-parthenocarpic ones, especially in environment unfavourable for pollination and fruit set, ensuring crop production and yield stability. Therefore, the development of parthenocarpic clones could contribute to an increase in yield and also to an improvement in stability without modifying the environment, or a need for the application of additional cultural practices.

SOIL REQUIREMENT

Pepino grows well and yields high in well-drained fertile loamy soils having pH about 5–6 but satisfactory yield can be obtained in soils considered poor for other crops and in conditions of moderate salinity. Salinity stress inhibits growth and development causing osmatic stress and yield losses. It requires constant moisture for good fruit production but intolerant to moist soils, hence, providing good drainage facility is very essential. Established bushes show some tolerance to drought stress but this typically affects yield. Its introduction to arid and semiarid areas where increase of water and soil salinity is one of the major problems in vegetable production can represent an alternative for the diversification of vegetable crop production and contribute to a more sustainable horticulture and higher stability of yield and farmers' income. Hence, pepino has an excellent prospect for its cultivation in areas affected by salinity.

CULTIVATED VARIETIES

In New Zealand, the cultivars like Suma, Comeraya, Miski, Lincoln Gold and El Camino are commercially grown. In the Andean region, cultivars *per se* are not distinguished instead characteristic types are distinguished by fruit characteristics and/or the regions where they are grown. Some varieties are perennials and require 18–24 months to produce fruit, while others are annuals requiring 90–100 days. At Horticultural Research Station, Tamil Nadu Agricultural University, Ooty, nine accessions (SMu 1 to SMu 9) have been identified based on the plant and fruit characteristics, which showed significant variation in yield.

PLANTING METHOD

The seeds of pepino are fertile and are able to produce vigorous offspring but this crop is primarily propagated by stem cuttings since they establish easily without dipping in any rooting hormone. It is propagated in a manner similar to its relatives such as tomato and peppers. Planting should be done in well-prepared field at a spacing of 90 cm between rows and 75 cm between plants.

GREENHOUSE CULTIVATION

The crop is also well adapted to greenhouse cultivation (minimum temperature 15°C and maximum temperature 25°C), training the plants up to 2 m tall and obtaining yield two to three times higher as compared to those obtained from outdoor fields.

The shoot tips from healthy plants are rooted in mixture containing vermiculite:peat (1:1). Ten days old-rooted cuttings are transplanted in pots in a 1:1 (v/v) mixture of sandy soil (96% sand, 4% clay and pH 7.65) and perlite. The individual pots are drip fertigated two to four times a day at 2 L/hr.

The bushy plants are thinned to two stems vertically supported by individual strings and all lateral buds are removed. All the trusses above the fourth or fifth inflorescence are pinched for the first and second season, respectively. In first year, all the sets are allowed to reach maturity. In second year, only three fruits are allowed per cluster. The ripe fruits with yellow skin and calyx folded upwards are harvested daily.

NUTRITIONAL REQUIREMENT

For getting higher yield, a well-rotten farmyard manure or compost at 25 t/ha should be incorporated into the soil at the time of land preparation. Nitrogen, phosphorus and potash fertilizers may be applied each at 135 kg/ha. Full dose of phosphorus and potash along with one-third dose of nitrogen is applied at the time of transplanting and remaining two-third dose of nitrogen is supplied in two equal splits 1 month after transplanting and at the time of flowering. For bumper harvest, the crop may be fertigated with 17.5 kg/ha each of nitrogen and potash once in every 2 months.

IRRIGATION REQUIREMENT

First irrigation should be applied immediately after planting, and subsequent irrigations are given based on the intensity of rainfall. The crop is sensitive to water logging conditions, thus too much irrigation may be damaging to the crop. In dry period, giving irrigation at frequent interval is necessary but irrigation must be light. Salinity or irrigation with saline water and ethephon sprayings improve earliness and fruit quality as they increase soluble solids content, dry matter and titratable acidity.

INTERCULTURAL OPERATIONS

HOEING AND WEEDING

At initial stage, the seedlings are sensitive to weed infestation, but at later stage, the crop can easily compete with low-growing weeds. Hence, the field should be kept free from weeds by weeding frequently. The plants are staked at fruit set to prevent lodging and subsequent soiling and rotting of fruits. Mulching the basins with dry grass will also help in preventing soiling of fruits. In the planted crop, after the harvests are over, the plants are pruned at 15–20 cm from the ground and supplied with farmyard manure 12.5 t and nitrogen, phosphorus and potassium each 67.5 kg/ha for sprouting of the plants.

PRUNING AND TRAINING

In pepino, pruning and training are very important part of vine management. Pepino plants when pruned and trained at four branches give higher yield and fruit number, without affecting fruit weight, total soluble solids content or fruit firmness. It grows naturally upright by habit, thus can be cultivated as a freestanding bush. Sometimes, its plants are pruned and trained on trellises, and occasionally, additional support is provided to withstand the weight of fruits from pulling the plant down. For ratooning in next season, the shoots should be pruned to 15–20 cm from the ground level followed by fertilizer application for new growth of the plant.

HARVESTING

Being a fast-growing crop, it bears fruits within 4–6 months after planting. The fruits should be harvested with immense care since any physical injury to the fruit may lead to decaying. The stage of harvesting should be chosen as per the market demand. Generally, the fruits are harvested at full ripe stage for table use, partially ripe, *i.e.*, yellow green fruits, for the local market and the light green colour fruits, which are physiologically mature but still unripe, could be harvested for distant markets since such fruits are suitable for cold storage purpose. In ripe pepino fruits, there is an increase in weight loss, ethylene production, respiration rate and 1-aminocyclopropane-1-carboxylic acid (ACC) levels as well as decrease in fruit firmness and Hue angle may occur, if these fruits are exposed to chilling temperature during storage.

POST-HARVEST MANAGEMENT

Pepino fruits show a non-climacteric type of respiratory activity during post-harvest period and ethylene production is not detectable. Treatment with propylene accelerates colour change and fruit softening. Fruit produces little ethylene at harvest but does not produce ethylene autocatalytically in response to exogenous propylene. Fruit respiration is increased transiently in response to propylene, but in control fruits, the respiration rate is increased only slowly during post-harvest ripening. During prolonged storage at 20°C, the fruits to produce ethylene at a rate over 0.2 µL kg^{-1} h^{-1} are only those, which subsequently develop rots. Nevertheless, fruits show a colour evolution from a green stage to a normal ripe, light yellow green colour.

The quality of mature pepino fruits can be maintained for up to 21 days under controlled atmosphere storage conditions coupled with 5–10°C. However, the ripe pepino fruits may be sold in local fresh markets as a ready to eat fruit. This crop is exported mainly by using sea freight and often reveals a poor external and sensory quality with a short shelf life when arriving at designated market. Sucrose ester as a post-harvest surface coating can prolong shelf life of pepino fruits. However, it is less effective than the use of polyethylene foodtainer, which reduces fresh weight loss, transpiration rates and fruit softening. In pepino, this coating also delays ripening, as indicated by a higher retention of fruit colour changes and inhibition of carotenoids synthesis.

INSECT PESTS AND DISEASES

Similar to other relatives, *i.e.*, tomato, eggplant, tomatillo and peppers, pepino is extremely prone to the attack of insect pests and diseases as described below:

INSECT PESTS

Aphids (*Aphis gossypii* Glover and *Myzus persicae* Sulzer)

Aphids are called the mice of the insect world since they multiply so quickly and provide food for so many creatures. In small number, aphids are not a problem, but under favourable conditions, they can multiply rapidly and create large colonies. The nymphs and adults of this tiny insect attack the tender parts of the plant in dense clusters on the stems, leaves or new plant growth and feed by sucking sap from plants. In such cases, they remove so much sap from the growing shoots that they become stunted and misshapen. They also transmit virus diseases. The various aphid species attack a huge variety of crops but they are particularly common on the Brassicas.

Control
- Grow few rows of daisy and carrot family plants around the pepino field, as they are good sources of food for aphids.
- Use yellow sticky traps to monitor aphid movement into the field.
- Pinch off the foliage where aphids are densely concentrated and dump into the soil.
- Wash the aphids from the plants with a strong jet pressure of water.
- Introduce the beneficial insects such as ladybugs or lacewings in pepino field.
- Use insecticidal soap containing natural fats and plant oils 0.1%.
- Spray the crop with 0.4% neem oil, 0.1% Malathion, Metasystox 25 EC or Rogar 30 EC.

Whitefly (*Bemisia tabaci* Gennadius)

Whitefly is a serious pest of pepino. These tiny flying insects feed underside of the leaves and suck cell sap, leaving behind a sticky substance called honeydew, which later becomes a medium for sooty mould growth, reducing the photosynthetic rate of the plant. The affected leaves become yellow and curl down. Its severe infestation may cause defoliation. The nymphs and adults transmit viruses and mycoplasma. The pest is more serious during dry season and its activity decreases with the onset of rain and start of winters since its multiplication depends on atmospheric temperature and humidity.

Control

- Cover the nursery beds with agro-net when raised in open field.
- Adopt proper cultural operations and follow clean cultivation.
- Limit the irrigation frequency to reduce the population of whitefly.
- Fix yellow sticky traps to monitor and suppress infestations.
- Introduce natural predators such as ladybugs, lacewings or whitefly parasites.
- Spray the crop with Dimethoate, Phosphamidon, Monocrotophos 1.5 mL/L or Fenitrothion 0.05%.

FLEA BEETLE (*Epitrix hirtipennis*)

This is tiny black chewing type insect. Its adults eat foliage and create numerous small round holes in leaves and fruits of many plants including pepino, while its larvae feed on roots. A potentially devastating flea beetle attacks on both sides of the leaves. It attacks first the early plant growth in spring. Most serious early in the growing season, this injury eventually kills infested leaves. Young plants with little leaf area sometimes suffer badly and seedlings occasionally die. In addition, the flea beetles may transmit early blight. This beetle overwinters as adult among debris in or near field of host plants. They resume activity in spring and feed on weedy hosts until crop hosts are available. Eggs, deposited in soil near the bases of host plants, may require a week or more to hatch. Grubs feed on or in roots and lower stems for 3–4 weeks before pupating.

Control

- Destroy the weeds and debris from the field.
- Fix yellow sticky traps to monitor larvae and capture the adults.
- Use row covers in early season since young plants are more vulnerable to damage.
- Dust the plants with diatomaceous earth to control adults feeding on foliage.
- Spray the crop with 0.1% Bifenthrin, Cyfluthrin, Cyhalothrin, Esfenvalerate, Permethrin, imidacloprid or Spinosad.

LEAF MINER (*Liriomyza* spp.)

Being polyphagous pest, leaf miner attacks various kinds of plant. The adult female lays very tiny eggs inside the leaf tissue just below the leaf surface. Small yellow maggots hatch and feed inside the leaf tissue, leaving long, slender, winding, white tunnels (mines) through the leaf. The damage can be limited in initial stages of infestations but increase as leaf miner increases in numbers. Damage may cover so much of the leaf that the plant is unable to function and yields are noticeably decreased.

Control

- Add soil amendments like compost or farmyard manure.
- Cover the soil under infested plants with polyethylene film to prevent larvae from reaching the ground and pupating.
- Use floating row covers to prevent fly stage from laying eggs on leaves.
- Use yellow sticky traps to catch egg-laying adults.
- Spray *Bacillus thuringiensis* formulations on crop.
- Spray the crop with *neem* oil 0.4% to reduce the leaf infestation by leaf minor.
- Spray of Cryomazine followed by Dimethoate 0.1% to control leaf miner effectively.

Spider Mites (*Tetranychus urticae*)

Spider mites are very minute insects. They tend to congregate on underside of the leaves, usually beginning on bottom leaves and working their way to upper leaves. They damage the plants by inserting their stylet mouthparts into individual plant cells and withdrawing cellular juice. The affected portion of plant shows pin dot like pale yellow spots and crop gives sickly appearance. The coalescence of dead cells results in appearance of bronze and brown spots on leaves and plants. Most mite species overwinter as eggs on the leaves and bark of host plants.

Control

- Remove and burn the already infected plants.
- Use natural predators such as those that feed on spider mites.
- Pick off the mites or eggs infected parts and destroy them.
- Spray plants using a stream of water to knock mites off the underside of leaves.
- Spray acaricides formulations like garlic extract, clove oil, mint oil, rosemary oil or cinnamon oil.
- Dust the crop with 20–25 kg/ha sulfur or spray the crop with 0.05% wettable sulfur or Rogor at 0.15% as and when the mites appear.
- Spray the crop with Malathion 0.1%, Bifenthrin 0.05%, Diazinon 0.03% or Labaycid 0.05%.

Thrip (*Thrips tabaci*)

Thrip that has a very extensive host range is very small slender insect. It possesses piercing and sucking mouthparts and causes damage by extracting the contents of individual epidermal cells leading to necrosis of tissue. Feeding of thrips causes distortion of plant growth, transmission of tomato spotted wilt virus curling of leaves, deformation of flowers and white to silvery patches on emerging leaves that often have tiny black faecal specks in them. The immature stage of thrips acquires virus but the adults transmit this virus from plant to plant. A long dry weather is favourable for its rapid multiplication. The losses caused by thrips may be from 25% to 50%.

Control

- Strictly follow long crop rotation with non-host crops.
- Avoid planting pepino next to onion, garlic or cereals since thrips often build up to large numbers on these crops.
- Place pink or blue sticky traps around susceptible plants as thrips have an affinity for bright pink and blue colour.
- Remove and destroy infested leaves, fruit and weeds and plough down the old crops.
- Incorporate phorate 10 G 1 kg a.i./ha in nursery and field soil.
- Apply Imidacloprid 0.1%, Nuvacron 0.1%, Dimethoate 0.03% or Monocrotophos 0.04%.

Fruit Borer (*Helicoverpa armigera*)

Fruit borer feed on pepino leaves and fruit. Distorted leaves often result when they feed upon the tips of leaves in the developing bud. Larvae may also bore into stalks or midribs. The larvae enter the fruit at the stem end soon after hatching, causing extensive direct damage and promoting decay. The larvae are cannibalistic, thus there is rarely more than one larva per fruit. Larvae usually complete development in a single fruit but when the fruits are small, they may feed on several fruits.

Control

- Grow African marigold with pepino seedlings a rows ratio of 1:16.
- Setup pheromone traps with Helilure at 12 per hectare.

- Collect and destroy the infected fruits along with larvae.
- Release *Trichogrammapretiosum* at one lakh per hectare per release at an interval of 7 days starting from flower initiation stage.
- Spray the crop with *neem* seed kernel extract (*Azadirachta indica*) 2–5%, pongamia (*Pongamia glabra*) or mahua (*Madhuca indica*) 5%.
- Spray of chemical pesticides such as Phosalone, Monocrotophos, Dichlorvos or Oxydemeton methyl 0.1%.

ROOT-KNOT NEMATODE (*Meloidogyne incognita* and *M. javanica*)

Nematodes attack the crop roots. Stunting, reduction in leaf size and yellowing due to formation of galls in roots are the common symptoms of its attack, however, the typical and diagnostic symptoms are the production of galls in roots, which affect the activity of roots and reduce the crop yield, depending on variety, moisture, temperature and initial population of nematodes in the soil. The infection may start at seedling stage. The rate of photosynthesis is considerably reduced. The nematode population increases due to monoculture.

Control

- Plough the field repeatedly in hot summer months.
- Flood the field for longer time to cause death of the nematodes.
- Follow marigold, mustard, garlic, paddy, maize or sorghum-based crop rotation.
- Treat the seed with Carbofuran 1% against nematode.
- Fumigate the infested beds with methyl iodide at 0.49 kg/10 m^2.
- Incorporate organic amendments like *neem* cake, groundnut cake, *mahua* cake, castor cake, *Calotropis* leaf or *Pongamia* leaf at 500 g/m^2 in soil to reduce root-knot nematode population.
- Apply Carbofuran 3G at 2–4 kg a.i./ha, Phorate 10G at 10–15 kg a.i./ha or Aldicarb 25 kg/ha at the time of field preparation.

DISEASES

DAMPING-OFF (*Rhizoctonia, Pythium, Fusarium* and *Phytophthora* species)

A number of fungi attack the seeds before or after germination in the nursery. The seedlings in groups due to the attack of fungi at collar region topple over the ground roughly in circular patches. It results in death of some seedlings in any given population. Roots sometimes rot completely. High soil moisture and moderate temperature favour the development of this disease.

Control

- Follow long rotation for nursery beds.
- Drench the nursery soil with 0.2% copper oxychloride about 2 weeks before seed sowing.
- Treat the seeds with Captan, Bavistin, Foltaf or Captafol 2–3 g/kg of seed before sowing in nursery.
- Sow the seeds sparsely in nursery and avoid excessive watering.
- Drench the nursery beds with 0.2% Captan, Foltaf or Captafol solution.

EARLY BLIGHT (*Alternaria solani*)

This is also a fungal disease, which produces leaf spots and also causes stem lesions and rotting of pepino fruits. Despite the name early, the symptoms usually occur on older leaves. Its infestation causes significant yield reduction. The appearance of circular or irregular dark spots on lower leaves is one of the first symptoms of this disease. Eventually, the spots enlarge into a series of concentric rings surrounded by yellow area. The entire leaf, which is completely killed, will drop off the plant. The infection typically progresses upward from base of the plant. Early blight can result in extensive defoliation, exposing the fruit to sunscald and reducing yield.

Control

- Follow long crop rotation with cereal crops.
- Use black plastic mulch to protect the plant from inoculums splashing from soil to the lower leaves.
- Avoid irrigation through overhead sprinkling system to avoid splashing of spores from one leaf to another.
- Remove lower leaves up to a height of 20–30 cm of the plant to avoid infections through splashing of spores from the soil.
- Remove the infected leaves as soon as they are noticed in the field.
- Spray the crop with Difolatan 0.2%, Dithane M-45 0.2%, Benlate 0.1% or Bavistin 0.1%.

LATE BLIGHT (*Phytophthora infestans*)

This is a fungal disease, which attacks the older leaves first and then spreads to the fruit. Irregular greasy greyish lesions develop on upper surface of the older leaves. These lesions expand quickly, and in moist weather, a downy growth may develop on underside of the leaves opposite to the lesions. The leaves dry and shrivel. On fruits, greasy spots develop, which gradually cover the entire fruit, and as a result, the fruit disintegrates. High rainfall and cool temperature (10–20°C) favour the disease development.

Control

- Strictly follow a long crop rotation with non-host crops.
- Use disease-free pepino seedlings for transplanting.
- Practice good crop sanitation and destroy diseased plants promptly.
- Provide support to the plants to avoid contact of leaves with soil.
- Remove the lower leaves up to a height of 15–20 cm and avoid irrigation through sprinkler system.
- Apply potassium silicate fertilizer can help to make the seedlings resistant to fungal diseases and some insect pests.
- Spray the crop with 0.25% Dithane M-45, Blitox or Ridomil MZ at 12–15 days interval, starting from the appearance of symptoms.

ANTHRACNOSE (*Colletotrichum capsici*)

The fungus is widely distributed in most regions where cucurbits are grown. The pathogen is usually associated with leaf tip dieback. Leaves and maturing fruits show dark brown circular pinpoint dots. Downward drying of tender shoots and fruit tip is the conspicuous symptom of this disease. However, the typical symptoms are circular or angular sunken lesions on leaves and fruits with concentric rings of acervuli. Lesions may coalesce under severe disease pressure. Anthracnose can cause extensive pre- and post-harvest damage as well as pre-harvest symptoms on leaves and stems. Often, the symptoms of the post-harvest disease do not develop until the fruit is ripe. The infected fruits are shrivelled prematurely. The disease becomes very severe in wet weather at maturity of the crop.

Control

- Follow a long crop rotation with non-host crops.
- Always collect seeds from healthy fruits.
- Uproot and destroy the infested plants.

- Destroy the disease debris from the field.
- Treat the seeds with Thiram, Bavistin or Benomil 2–3 g/kg of seed or dip the seeds in 0.25% solution of Captan for half an hour.
- Dip the seedlings in 0.1% solution of Bavistin before transplanting.
- Spray the crop with Dithane M-45, Dithane Z-78, Benomil or Blitox 0.2%, or Bavistin 0.1–0.2% at 15–20 days interval.

CERCOSPORA LEAF SPOT (*Cercospora capsici*)

Large circular or oval spots with light grey centres and reddish-brown margins appear on leaves, stems, petioles and fruit peduncles, and later, the entire leaves turn yellow and drop down. The spots later become tan with a dark ring and a yellowish halo around the ring, resulting in a *frogeye* appearance. The affected centres of lesions dry and often drop out as they age. When numerous spots occur on the foliage, the leaves turn yellow and may drop or wilt. Defoliation is often serious, exposing the fruits to sunscald disorder. Heavy dew favours the germination of fungus spores. The disease is most severe during period of warm temperature (20–25°C) and excessive rain or overhead irrigation. The disease spreads through splashing water, wind, rain, implements, tools, workers and contact with infected leaf. The fungus survives in or on seed and old affected leaves in soil. Spores survive in infected debris for at least one season.

Control

- Follow at least 2 years rotation with non-host crops.
- Keep the solanaceous weeds under control during the rotation period.
- Collect seeds from disease-free crops.
- Disinfectant the seed by dipping in hot water at 52°C for 30 minutes if collected from infected plants.
- Plant the seedling at wider spacing to allow good air circulation and to avoid extended period of leaf wetness.
- Uproot and destroy the infected plants along with neighbouring plants.
- Promptly destroy the infected plant tissues by burning or deep ploughing after harvest.
- Before transplanting, dip the seedlings in 0.1% solution of Bavistin.
- Spray the crop with 0.1–0.2% Bavistin, 0.125% Difenoconazole or 0.2% Brasicol as and when the symptoms start appearing.

CERCOSPORA LEAF SPOT (*Cercospora solani*)

The disease is characterized by angular to irregular chlorotic lesions, which later turn greyish-brown with profuse sporulation at the centre of the spot. The older leaves near the ground show infection spots first, which are circular to polygonal, yellowish brown and often coalesce to form large patches. Severely infected leaves drop off prematurely, resulting in reduced fruit yield. On some susceptible varieties, large spots with concentric zonations of growth are produced and these may develop confusion with infection spots of *Alternaria solani*. The disease appears in latter part of December and it spreads through airborne conidia.

Control

- Grow resistant varieties, if available.
- Avoid growing beans in the same area for 2–3 years.
- Use disease-free healthy seed for planting.
- Remove all debris from the field after harvest.
- Spray the crop with 1% Bordeaux mixture, 0.125% Difenoconazole, 0.2% copper oxychloride, 0.1% Bavistin or 0.25% Zineb.

Fruit Rot (*Phomopsis vexans*)

The pathogen affects all above ground plant parts. Spots usually appear first on stems or leaves of seedlings. It girdles the seedling stems and kill them. The brown to grey circular 2.5 cm diameter spots with narrow dark brown margins appear on leaves. The disease spots on fruits are much larger. The affected fruits become first soft and watery but later they become black and mummified. Centre of the spots becomes grey and black pycnidia develop. The fungus survives in the infected plant debris in soil. It is seed borne. The spores spread through rain splashes, agricultural implements and insects.

Control

- Plough the field deep in hot summer months.
- Follow a long crop rotation with cereal crops.
- Give hot water treatment to seeds at 50°C for 30 minutes.
- Avoid excessive watering and provide proper drainage.
- Collect and destroy all the diseased plant debris.
- Spray the crop with Difolation, Captan, Zineb, Blitox or Copper Sandoz 0.2–0.3% at 10–12 days interval starting from the nursery.

Bacterial Wilt (*Pseudomonas solanacearum*)

Wilting of leaf, yellowing of the foliage, stunting of growth and finally collapse of the entire plant are characteristic symptoms of the disease. Lower leaves may droop first before the wilting occurs. The vascular system becomes brown. Bacterial ooze comes out from the affected parts. Plant show wilting symptoms at noontime will recover at night but die soon.

Control

- Strictly follow crop rotation with cruciferous vegetables to reduce the disease incidence.
- Avoid areas with poor drainage (Bacterial wilt).
- Use disease-free healthy transplants for planting.
- Treat the seed with hot water at 50°C for 25 minutes.
- Remove plant debris and weeds serve as reservoir.
- Keep the field clean by uprooting and destroying the infected plants.
- Spray the crop with Agrimycin 100 (Streptomycin sulfate + Oxytetracycline) 200 ppm, Streptocycline 25 ppm or Streptomycin Sulfate 100 ppm as and when the symptoms appear and repeat the spray five times at 15 days interval.

Tomato Spotted Wilt Virus

Plants infected with tomato spotted wilt virus exhibit bronzing of upper side of the young leaves, which later develop distinct necrotic spots. Leaves may be cupped downward. Sometimes, tip die-back may occur. On ripe fruits, chlorotic spots and blotches appear, often with concentric rings. The green fruits show slightly raised areas with faint concentric zones.

Control

- Grow resistant varieties of pepino with *Sw-5* gene.
- Use thrips and virus free seedlings from greenhouse.
- Manage thrips on seedlings before planting.
- Avoid planting near crops infected with spotted wilt virus.
- Remove infected plants at seedling stage.
- Keep weeds in and around the fields under control.
- Promptly remove and destroy old plants and other host crops after harvest.

Tomato Mosaic Virus

The foliage of affected pepino plants shows mottling, with alternating yellowish and darker green areas, the latter often appearing thicker and raised giving a blister-like appearance. The leaves tend to be fern-like in appearance with pointed tips and younger leaves may be twisted. The infected fruits become distorted. Yellow blotches and necrotic spots may develop on both ripe and green fruits and there may be internal browning of the fruit wall. In young plants, the infection reduces the fruit set and may cause distortions and blemishes.

Control

- No chemical products are available to cure or protect plants.
- The best factor in controlling and reducing infection is to practice sanitation. Remove any infected plants, including the roots.
- The coat of the mosaic virus reacts with proteins in milk, so milk can be used to inactivate the virus.
- Be aware of cross contamination. Tools and containers must be sterilized.
- Wash hands thoroughly after removing diseased plants, and especially before handling healthy ones.
- Never handle plants without washing hands after the use tobacco products. Remember that the soil can harbour this virus for many years to come.

■■■

101 Tree Tomato

M.K. Rana and Navjot Singh Brar

Botanical Name : Cyphomandra betacea Sendt.

Family : Solanaceae

Chromosome Number : 2n = 2x = 24

ORIGIN AND DISTRIBUTION

The general belief is that tree tomato is native to the Andes of Peru and surrounding countries of South America where it is found growing wild. It is cultivated and naturalized in Venezuela and grown in the highlands of Costa Rica, Guatemala, Jamaica, Puerto Rico and Haiti. It has also been cultivated in kitchen gardens for many years in Queensland and Australia and a practical crop in the highlands of the Australian part of New Guinea. It was introduced into New Zealand in 1891 and its commercial growing on a small scale began about 1920. It has been accepted at an early date in East Africa, Asia and the East Indies. Its commercial production occurred in the 1930s but its popularity grew during the Second World War. A promotional campaign for its cultivation was launched in 1961. This small industry grew until 1967 when annual production reached a peak of 2000 t. In 1967, its commercial name was switched to *tamarillo* to distinguish it from the common tomato. Today, its commercial production is restricted to a handful of countries *viz.*, Argentina, Brazil, Colombia and Ecuador. Australia, California and few countries in Africa and Asia are other *tamarillo* growing regions.

In the late 1800s, the explorers brought its seeds to India where they were grown throughout the hilly tracts of Himachal Pradesh, Assam and West Bengal, Maharashtra, Uttaranchal, Nagaland and even throughout the country on a limited area. In South India, it is produced in the Nilgiri Hills, as this is one of the only areas, which is cool enough to sustain the tree tomato crop. *Tamarillo* is a subtropical fruit growing between 300 and 2250 m elevations in India.

INTRODUCTION

Tree tomato is a small tree or shrub, bearing egg-shaped edible fruits. Among 30 species, *Cyphomandra betacea* is the important one. Although its fruits look like a medium-sized tomato, it is not a true tomato. Its taste is somewhat like a tomato but it is usually eaten with sugar or boiled to make a popular refreshing drink. Among its various regional names, *i.e.*, caxlan pix (Guatemala), tomate de palo (Honduras), arvore do tomate, tomate de arvore (Brazil), lima tomate, tomate de monte, sima (Bolivia), pepino de arbol (Colombia), tomate dulce (Ecuador), tomate cimarron (Costa Rica) and tomate francés (Venezuela, Brazil), the name tamarillo was adopted, and in New Zealand in 1970, it became the standard commercial designation for this fruit. Tree tomato is not difficult to grow but its fruits ripen unevenly on the tree. The constant pruning makes it one of the most laborious fruits for the farmers to grow. In areas where the common tomato is not grown commercially, tree tomato is used to prepare stews, thus replacing the common tomato.

COMPOSITION AND USES

COMPOSITION

The tree tomato fruit contains minerals, especially calcium, phosphorus, potassium, magnesium and iron. It is a good source of vitamin A, B, C, E and K and high levels of fibre but it is low in calories. It is also an important source of protein, carotene and pectin. The nutritional composition of tree tomato is given in Table 101.1.

USES

The tree tomato is chiefly cultivated for its edible fruits, which are a source of food and a potential raw material for the processing industry to prepare syrup and chutney in Auckland, New Zealand. Its ripe fruits are more often eaten raw as a dessert after removing skin, as it is bitter in taste. The fruits are put in boiled water for a short period so that the skin can be removed easily. Honey is then prepared with cinnamon and cloves, the peeled fruits are added and left to boil until it reaches a suitable consistency. When the fruits are orange in colour, they are used to prepare a sauce together with *Capsicum* in Peru. For making sauce, the fruits are lightly grilled to remove skin (epicarp) and then ground with a large green pepper and salt. This spicy sauce is eaten as an appetizer. After peeling and slicing, the fruits are added to stews or soups, which are served after sprinkling sugar and perhaps with a scoop of vanilla ice cream. The fruit slices can be mixed in salads or served as sandwich filling. The chopped fruits are blended with cream cheese and used as sandwich spread. Being high in pectin, its fruits are used for making jelly but the fruit is oxidized and discoloured during processing if special treatment is not given. The uncut peeled fruits are cooked with sugar to use on ice cream. The peeled fruits may be pickled or used as a substitute of tomato in hot pepper sauce.

MEDICINAL USES

The tree tomato contains vitamin A, which is imperative for improving eyesight, maintaining skin health and supporting red blood cells. The fruits high in vitamin C content assist the body with immunity, bone health and wound healing. The fruits are a natural remedy for respiratory diseases and anaemia. In Ecuador, after warming, the leaves are wrapped around the neck to treat a sore throat. Its fruit pulp is used by the Colombians for making a poultice to treat inflamed tonsils. Jamaicans, based on their belief, refer to it as *vegetable mercury*, which is curative to liver problems. Its protein has an antimutagenic effect since it reduces the oxidative damage and inhibits uric acid formation. Its fruit has a high level of anthocyanin, which protects the body from diabetes, symptoms of ageing, certain cancers and neurological diseases. Its fruits also contain lycopene, which is a compound that wards off degenerative diseases, boosts heart health and aids the skin to endure ultraviolet radiation.

TABLE 101.1
Nutrition Composition of Tree Tomato Fruits (per 100 g Edible Portion)

Constituents	Contents	Constituents	Contents
Water (g)	82.7–87.8	Vitamin A (IU)	540
Carbohydrates (g)	10.3	Thiamine (mg)	0.038–0.137
Protein (g)	1.5	Riboflavin (mg)	0.035–0.048
Fat (g)	0.06–1.28	Niacin (mg)	1.10–1.38
Dietary fibre (g)	1.4–4.2	Vitamin C (mg)	23.3–33.9
Calcium (mg)	3.9–11.3	Ash (g)	0.61–0.84
Phosphorus (mg)	52.5–65.5	Energy (kcal)	35
Iron (mg)	0.66–0.94		

BOTANY

The tree tomato is a small, partially woody, attractive and fast growing brittle tree having a shallow root system and grows to a height of up to 5.5 m but rarely as much as 7.5 m. It has a single stem, which is monopodial and branched into two or three branches at a height of 1–1.5 m. Its leaves are evergreen, alternate, more or less heart shaped at the base, ovate, pointed at the apex, 17–30 cm long, 12–19 cm wide and delicately hairy with conspicuous coarse veins. Inflorescence is caulinar opposite, the leaf and flowers are 1.4 cm long. The calyx persists with the fruit and the corollas are pinkish white and rotate campanulate with reflexed apices. Stamens, which are shorter than the corolla, are connivent. Anthers are yellow and dehisce through two apical pores. The style emerges between the anthers. The long-stalked fruits borne singly or in cluster of 3–12. Fruit is 5–10 cm long and 4–5 cm wide, smooth, egg-shaped but pointed at both the ends and capped with persistent conical calyx. Skin colour is deep purple, blood red or orange to yellow with faint dark longitudinal stripes. Flesh colour may be orange red, orange to yellow or cream yellow. Its seeds are thin, flat, circular, larger and harder as compared to those of the true tomato and they are distinctly bitter in taste.

The plant survives nearly 10–11 years. Flowering begins with the branching of the main stem about 8–10 months after sowing in its permanent location. The first inflorescence is produced adjacent to the point of branching on the main stem and the following ones at the end of branches around their respective branching. Flowering continues and the number of inflorescences is directly proportional to the number of branches. The plant is evergreen and forms new leaves constantly but the lower leaves later fall down, leaving the main stem and lower part of the branches leafless.

CLIMATIC REQUIREMENT

The tree tomato prefers subtropical climate with rainfall between 600 and 4000 mm and annual temperature between 15°C and 20°C. The optimum soil temperature for germination is 24–29°C. It can be cultivated in areas about 300–2300 m above mean sea level. In cool areas, it grows well even at lower elevations. Light frost hampers the crop by destroying the small branches and newly formed leaves. However, continual heavy frost can kill even a well-established tree, thus the establishing and even the well-established plantation should be protected against strong cold waves.

SOIL REQUIREMENT

The tree tomato cannot tolerate highly compact soil with low oxygen content. It prefers deep lateritic soil with good fertility and water retention capacity. It is sensitive to water logging conditions in the field, as water stagnation even for few days may kill its plants. Therefore, a well-drained fertile soil rich in organic matter is ideal for its successful cultivation. It grows well in soils having pH from 5.5 to 8.5.

CULTIVATED VARIETIES

No improved cultivar of the tree tomato is available but some local strains *viz.*, dark red, yellow and purple strains are grown by the growers based on their colour preference. Generally, the red-fruited strains are grown for fresh market and yellow fruited for preservation purposes because of their superior flavour and aroma. In Sikkim, local red- and yellow-fleshed varieties are grown. Some of the popular varieties grown in different parts of the world as well as in India are given below:

Ecuadorian Orange
A cultivar with medium-sized, egg-shaped and orange colour fruit is suitable for culinary purpose. The fruit pulp is light orange, creamy textured and less acidic.

Goldmine

A recently introduced superior cultivar originating in New Zealand that bears very large golden-yellow fruits with highly flavoured flesh, less acid content and superb eating quality.

Inca Gold

The plant of this cultivar bears yellow fruits with less acid content. The fruit flavour when cooking is just like apricot.

Oratia Red

The plant of this cultivar bears oval to round shape large red fruits with sharp acid flavour. The fruits are excellent for making jam and other processed products.

Rothamer

An exotic cultivar originated in San Rafael, California bears unusual large fruits (over 3 oz.) with bright red skin colour, golden-yellow flesh, sweet flavour and dark red colour seeds. It is a vigorous and heavy bearing variety, its fruits mature from December to April.

Ruby Red

The plant of this cultivar bears large red fruits with dark red, tart and flavourful pulp. The fruits are good for culinary use. After picking, if the fruits are allowed to ripen for 1–3 weeks, they become less acidic. This cultivar is generally grown in New Zealand for export purposes.

Solid Gold

The plant of this variety bears large and oval shaped fruits with golden-orange skin, less acidic soft pulp and acceptable flavour. The fruits are good for culinary use.

Yellow

The plant of this variety bears a large plum sized and shaped fruit with yellowish-orange skin. The flesh is yellow with a milder flavour than the red-fruited variety. The yellow form is the oldest in cultivation in New Zealand.

SOWING TIME

The planting time of the tree tomato varies with the climatic conditions of the growing region, but in general, in subtropical areas, the crop is planted in the spring season and in temperate areas, planting is done in the early spring season. The best possible planting time for the tree tomato is when the night temperatures stay consistently above 10°C.

PROPAGATION

The tree tomato crop can be propagated either by seed or by planting cuttings directly in the field. Direct seeding in the field produces a profusely branched erect tree, which is suitable for sheltered locations, while planted cuttings develop into a shorter bushy tree with low-lying branches, which are suitable for exposed windy sites.

RAISING OF SEEDLINGS AND TRANSPLANTING

The seed for raising this crop should be extracted either from red fruits with black seed pulp or from yellow fruits with yellow seed pulp. Before sowing, the seeds are extracted from fully ripened fruits, washed properly, dried in shade and then placed in a freezer for 24 hours to accelerate germination. Thereafter, the seeds are sown in protrays containing growing media. The potting media should be friable with good water retention capacity, *i.e.*, coco peat, vermiculite and perlite. After sowing, take 4–6 days to germinate.

The seedlings, after attaining a height of 8–10 cm, are transplanted in a well-prepared field or pits at a row spacing of 3.0 m and plant spacing of 1.0–1.5 m. In windy unprotected locations, closer spacing, *i.e.*, 60 cm between rows and 30 cm between plants is adopted. Irrigation is applied immediately after planting through sprinklers or a drip irrigation system.

DIRECT CUTTING PLANTING

The cuttings of 75–100 cm length and 10–30 mm thickness are selected for planting in the field. The cuttings are taken from 1 to 2 year old twigs for planting purposes. Before planting, all the leaves are removed from the cuttings. For planting of the cutting, the spacing varies with soil fertility, environmental conditions and management practices to be adopted. In the direct planting method, cuttings are planted in a well-prepared field at a spacing of 3.0 m between rows and 2.5 m between plants.

NUTRITIONAL REQUIREMENT

The tree tomato is a nutrients exhaustive crop, therefore, judicious application of fertilizers is necessary to obtain a good yield. Well-decomposed farmyard manure is applied at 20 kg/pit. Additionally, nitrogen, phosphorus and potash are applied each at 0.25–1.0 kg/plant. These fertilizers along with farmyard manure are thoroughly mixed into the top soil at the time of transplanting. Subsequent doses of nitrogen, phosphorus and potash fertilizers are essential to apply year after year for good crop growth and yield. An additional dose of nitrogen, phosphorus and potash, each at 1–1.5 kg/plant, is applied in the fifth or sixth year of crop growth.

IRRIGATION REQUIREMENT

Being a shallow-rooted crop, tree tomato cannot tolerate prolonged drought conditions, and thus, the crop requires frequent irrigation with good drainage facility. The soil should have sufficient moisture during the hot summer months to sustain plant growth and to improve fruit size and yield. After planting, a very light irrigation is applied daily during first week of crop growth. In general, light irrigation is applied at 7–8 days interval in summer months and 10–15 days interval in winter months.

INTERCULTURAL OPERATIONS

HOEING AND WEEDING

Being a shallow-rooted crop, a light hoeing is beneficial to eliminate weeds from the tree tomato field. Deep hoeing around the plants should be avoided to prevent damage to its roots.

MULCHING

Mulching in the tree tomato field conserves soil moisture by reducing water evaporation. It also checks weed growth and improves microbial activity in the soil by improving the environment surrounding the plant roots. Mulching with compost, dry grasses, paddy straw or sugarcane tresses is beneficial, as it provides nutrients to the plants after decomposition. It also helps in the establishment of newly transplanted seedlings.

WINDBREAK

Tree tomato is a delicate and shallow-rooted crop. Therefore, it requires protection against strong winds, as strong winds can easily break the branches. Windbreaks are positioned perpendicular to the air current before setting out the plantation in order to protect young plants against strong winds and cold waves. Staking can also be accomplished to prevent swaying and to minimize root disturbance.

PRUNING

The tree tomato plants should be pruned to a height of 0.9–1.2 m a year after planting to encourage branching. Annual pruning is done to remove branches that have given fruits in previous season, as the fruits are normally produced on new growth. Pruning helps in controlling plant and fruit size and facilitates easy harvesting. Light pruning leads to the production of medium-sized fruits, while heavy pruning leads to the production of large-sized fruits. Judicious pruning can help in extending the fruiting season.

HARVESTING

The tree tomato plant usually begins to bear fruits 1.5–2 years after planting. The peak production can be obtained 4 years after planting, and with good management practices, it continues for 5–6 years. The fruits become ready for harvesting when they have developed yellow or red colour with longitudinal stripes as per the characteristic of a variety. The fruits from the tree are harvested simply by pulling the fruits with a snapping motion and keeping the stem attached. The fruits do not ripen at one time, thus several pickings are necessary to complete harvesting.

YIELD

The yield of tree tomato depends on several factors such as planting material, method of planting, soil type, fertility status of the soil, sowing time and cultural practices adopted by the growers during the course of crop production. On an average, a tree produces a fruit yield of 15–20 kg annually.

POST-HARVEST MANAGEMENT

After harvesting, the fruits, depending on size, are graded into small, medium and large sizes and packed in paper-lined wooden boxes for marketing. The fruits, because of their hard skin and firm flesh, can be transported long distances without bruising. The fruits can be stored for 9 weeks at temperatures of 3.0–4.4°C and 90–95% relative humidity. However, temperatures below 3°C causes chilling injury to the fruits.

INSECT PESTS AND DISEASES

A number of insect pests and diseases attacks tree tomato crop during its life cycle. Some of the insect pests and diseases along with their control measures are given below:

INSECT PESTS

APHID (*Myzus persicae*)

The nymphs and adults both suck cell sap from the growing tips and the underside of the tender leaves. The affected plants remain stunted, leaves curled and plant vigour is reduced. They excrete a honeydew like sugary substance, which promotes the growth of sooty mould, inhibiting photosynthesis and reducing yield of the crop.

Control

- Set up yellow pan sticking traps in the field.
- Remove infested culls and weed species around the field that harbour the aphids.
- Introduce natural parasites and predators such as lady beetles and their larvae, lacewing larvae and syrphid fly larvae to control aphids.
- Spray the crop with Dimethnoate or Methyl demeton 0.03% at 10 days interval.

TREE TOMATO WORM (*Neoleucinodes* sp.)

The larvae of tree tomato worm bore into the fruit, causing it to spoil prematurely and bringing fruit losses around 40–60%. Young shoots and leaf petioles show signs of wilting and drooping due to insect attack.

Control

- Remove and burn the badly infected shoots and leaves.
- Follow sanitation by removing freshly damaged fruits during each harvest.
- Spray the crop with 4% *neem* seed kernal extract or 0.01% Cypermethion at 10–15 days interval.

WHITEFLY (*Bemisia tabaci*)

These white tiny insects are more active during the dry season and their activity decreases with the onset of rains. They suck sap from tender parts of the plant. Numerous chlorotic spots develop on the leaves of affected plants. Severe damage results in yellowing and curling of leaves. When the plant is shaken, numerous small white adult whiteflies flutter out and quickly resettle. It also acts as a vector to transmit tamarillo mosaic virus (TaMV) or other viral disease.

Control

- Follow clean cultivation and uproot the alternate hosts.
- Maintain good sanitation in areas of winter and spring host crops.
- Remove weeds from the surroundings of the tree tomato field.
- Spray the crop with Imidacloprid or Methyl demeton 0.1% at 10–12 days interval.

ROOT-KNOT NEMATODES (*Meloidogyne incognita, M. javanica* and *M. hapla*)

Root-knot nematodes usually cause distinctive swellings, which are called galls on the roots of affected plants. They feed and develop within the galls, which can grow as large as 2.54 cm in diameter on some plants but usually are much smaller. The formation of these galls interferes with the flow of water and nutrients to the plant. These galls can crack or split open and allow the entry of soilborne disease-causing microorganisms. The aboveground symptoms of its infestation include wilting during hottest part of the day even with adequate soil moisture, loss of vigour, yellowing of leaves and other symptoms similar to a lack of water or nutrients. Infected plants produce fewer and smaller leaves and fruits. Damage is most serious in warm days, irrigated and in sandy soils.

Control

- Grow resistant varieties if available.
- Soil solarization provides effective control of nematodes.
- Apply *neem* cake in soil at 80–100 kg/ha.
- Apply Carbofuran 3G at 2–4 kg, Phorate 10 kg a.i. or Aldicarb 2–4 kg a.i./ha in soil.
- Fumigate the soil with DD mixture 200–400 L/ha to control of nematodes.

DISEASES

Powdery Mildew (*Erysiphe* sp. and *Oidium* sp.)

The most damaging disease of the tree tomato is powdery mildew, which may cause serious defoliation. The disease first appears as white powdery spots on both sides of the leaves, shoots and sometimes on flowers and fruits. These spots gradually spread over a large segment of the leaves and stems. The affected leaves turn completely yellow, die and fall down.

Control
- Plant the crop in sunny areas as much as possible.
- Avoid the use of excessive doses of fertilizers.
- Spray the crop with systemic fungicide like Karathane, Bavistin or Benlate 0.2%, Sulfex 0.5% at 10–15 days interval.

Tamarillo Mosaic Virus

The symptoms appear as yellow-green mottling on leaves and sometimes on fruits. The infected leaves are usually distorted and smaller than normal size. The number of branches and flowers is reduced. Mosaic virus is mostly spread by insects, especially aphids and whitefly.

Control
- Grow resistant varieties if available.
- Use virus-free healthy seed.
- Control the weeds, as they serve as hosts for the disease.
- Rogue out the affected plants and burn them.
- Spray the crop with Confidor or Methyl demeton 0.1% at 10 days interval to control vectors.

■■■

102 Arracacha

M.K. Rana and Sanjay Kumar

Botanical Name   : *Arracacia xanthorrhiza* Bancroft

Family   : Umbelliferae (Apiaceae)

Chromosome Number : 2n = 44

ORIGIN AND DISTRIBUTION

Arracacha is native to the Andean region of South America, which includes the countries of Peru, Ecuador, Colombia, Bolivia and Venezuela. The most closely resembling wild species of arracacha are known from Peru and especially from Ecuador. The linguistics of its vernacular names, *i.e.*, *racacha*, *virraca* and *arrecate* also provide clues to its Andean origin. It is perhaps one of the oldest cultivated Andean plants, its domestication having preceded that of potato. Outside of the Andes, it is cultivated on a small scale in the Antilles, Central America, Africa, Sri Lanka and on a commercial scale in Southern Brazil, where it is industrialized.

INTRODUCTION

Arracacha, a root vegetable, is an intermediate between carrot and celery, while its root aroma is like a delicate blend of celery, parsley, celeriac and roasted chestnut, which is slightly sweet and nutty. Its starchy taproots are creamy white and of carrot root size. It is a major food crop of South America. Most of the umbelliferous crops from the old world are biennials and seed propagated but arracacha is a vegetatively propagated perennial crop and cultivated as a secondary global food crop for millions of people because of its versatility in cuisine, productivity, peculiar aroma and fine starch. Short post-harvest life of storage roots (6–7 days) and longer crop duration (12–14 months) are the limiting factors in its production, however, low input requirements in its cultivation and role in human nutrition and processing industry make it a regular foodstuff in urban markets, as it is consumed by a majority of people. If the crop duration is reduced and post-harvest life of its roots are improved through genetic improvement or by developing better storage technologies, its area and production can further be increased.

COMPOSITION AND USES

COMPOSITION

The roots, rootstock, cormels and leaves contain 38%, 17%, 41% and 4% of the total plant dry matter content. It is quite popular for its starchy roots, as these are a good source for ascorbic acid, vitamin A and calcium. The major part of its root dry matter is carbohydrates, about 95% of which is starch and 5% is sugar (mainly sucrose) but the root is a poor source of protein with an average of about 1% on fresh or 4% on dry weight basis. The starch content in roots, which is free from undesirable substances like oxalates, mucilage, isothiocyanates and astringent principles, varies from 10% to 25%. The main carotenoid present in its roots is β-carotene, which is responsible for a wide range of root colours from white over cream to yellow and orange, depending on its concentration. The yellow flesh roots contain large amount of carotenoids. The chemical composition of rootstock and cormels is similar to those of the storage root but the rootstock is somewhat more fibrous and superior in nutritional quality, as it has protein 1.3 and calcium 2.1 times higher than the roots. The nutritional composition of arracacha roots is given in Table 102.1.

Uses

Every part of the arracacha plant is used as human and animal nutrition, however, the starchy roots are principally used for human consumption and the rootstock and aerial plant parts (cormels, petioles and leaf blades) for domestic animals. Its use as food for direct consumption or in the form of processed products can be explained in terms of starch content and quality, colour and flavour, as it adds unique colour to dishes and processed products. The young stems are used as a salad or cooked vegetable and leaves as a flavouring agent. The stump or crown of the roots, containing about 9% protein, is used to feed dairy livestock. Traditionally, its roots are processed as a culinary analogue for potato, *i.e.*, roasted, boiled, soups, pastas, gnocchi, chips, flakes, fries etc. In the Andes region, its fine-grain starch is used for making biscuits, coarse flour and as a thickener for baby food and instant soups. Its roots require a light cooking to soften its tissue and gelatinize its starch, *per se* rendering it more digestible. Its fries resemble potato French fries but frying reduces its typical aroma. Its chips are superior to that of potato chips in quality and acceptance due to their better crispness, appearance, light sweetness, lower fat absorption and attractive bright orange colour.

Arracacha is also used for the preparation of the fermented product *guarapo*. For this purpose, its roots are ground and allowed to cool for a night. Thereafter, the ground mass is passed through a cloth or screen, water and raw sugar are added and the mixture is left to ferment for 3 days. For starch production, the roots are washed, peeled and grated over a perforated metal sheet with nails to provide an abrasive surface. The resulting pulp is suspended in water and subjected to several cycles of washing and settling until clean starch sediment is obtained.

Medicinal Uses

Arracacha roots are known to possess immense medicinal properties such as carminative, diuretic, anti-stomachache, tonic and digestive, thus being used in traditional medicines. Its easily digestible starch makes it a perfect base in soups for every person including young, elderly and especially those having weak digestive systems. Its leaves are used as a flavouring agent to conceal and modify

---

TABLE 102.1

Nutritional Composition of Arracacha Roots (per 100 g Edible Portion)

Constituents	Contents	Constituents	Contents
Moisture (g)	83	Iron (mg)	9.5
Total solids (g)	26.0	Vitamin A (IU)	1760
Carbohydrates (g)	24.9	Thiamine (mg)	0.08
Protein (g)	0.96	Riboflavin (mg)	0.04
Fat (g)	0.26	Niacin (mg)	3.45
Dietary fibre (g)	0.85	Pyridoxine (mg)	0.03
Calcium (mg)	65.0	Ascorbic acid (mg)	23.0
Magnesium (mg)	64.0	Ash (g)	1.30
Phosphorus (mg)	55.0	Energy (kcal)	416
Potassium (mg)	2.40		

the disagreeable offensive odour and nauseating qualities of other medicines in pharmaceutical preparations. Its starch is used as a thickener in remedial preparations too.

BOTANY

Arracacha is a monoecious perennial plant growing to a height of about 1 m and with four distinctive parts, *i.e.*, (1) storage roots, (2) central rootstock, (3) aerial stems or cormels and (4) leaves. The roots emerging from the stem are conical to cylindrical-shaped storage roots (5–25 cm long and up to 8 cm in diameter) and long fine filamentous roots, which can easily be differentiated 2–3 months after planting. The individual storage root weight is about 100–300 g. The storage root is well connected with rootstock through *necks*, which usually break easily at harvest. The storage root does not regenerate shoots, thus can be used as a propagule. The storage roots resemble a short carrot with lustrous off-white skin and white, yellow or purple interior. The root cortex contains numerous longitudinal oil ducts lined with secretory cells. In open-cut roots, the ducts exude yellow resinous oil with a typical umbelliferous odour. These oil ducts in wild species are more numerous and give an astringent taste. Owing to root enlargement, the rhizodermis experiences enormous dilation and a special mechanism of cork formation has evolved to allow for the continuous renewal of root skin. Rhizodermic cork is formed in tangential layers separated by several layers of periclinally dividing cells. As the new layers are formed from the underlying phloem tissue, the outer layers scale off. Rootstock is a highly swollen and compressed central enlarged stem structure developing from the propagule tissue, carries shoots and has storage roots inserted beneath soil surface. The aerial stems or cormels are very peculiar structures. Cormels are derived from stem tissue and are composed of internodes, nodes and scars left by shed leaves, therefore, the cormels have a segmented structure. Each cormel carries on its apical nodes 3–5 petioled leaves, which persist for short duration. The leaf consists of a long petiole with a weakly developed basal sheath and the typical bipinnate blade. The leaves having 30–60 cm length are similar to parsley and vary from dark green to purple. In most of the growing regions, the plant rarely flowers but flowering can be induced by dehydrating the rootstock until 35% weight loss or by giving appropriate flowering stimuli. Its inflorescence is a terminal compound umbel with 8–14 rays, which carry 16–30 umbellets. On outer umbellets, most of the flowers are hermaphrodite with 3–5 perfect flowers per umbellet and 15–25% of the total flowers in the umbel. The inner umbellets consist of staminate flowers only. Its flowers are actinomorphic (radially symmetric) with 2–5 mm diameter. The epigynous perfect flower has no sepals (asepalous flower), five petals, five stamens and two carpels, each with only one ovule, hence, only one seed develops per carpel and two seeds per flower and fruit. The petals of immature flowers are green, but at anthesis, they always turn maroon. The styles emerge from an epigynous disk that functions as a nectary. The styles are enlarged to form more or less conical stylopodium. Pollen dehiscence coincides with the peak of nectar production from nectaries on epigynous disk. Anthers and petals are now caducous. During the entire flower development, the petals remain curled. Flowering progresses gradually from periphery of the umbellet to the central staminate flowers and from outer to inner umbellets. Total flowering time of an umbel is about 15 days and the flowering period of any given shoot is 1–2 months. Pollination is performed by insects. The fruit is schizocarp consisting of two achene or mericarp (two seeds) with bicarpellar inferior ovary. At maturity (8–10 weeks after pollination), the fruits do not shed though they remain connected to the carpophore.

CLIMATIC REQUIREMENT

Arracacha grows well in cool but frost-free humid tropical climates. It can be cultivated in regions up to about 900–3000 m elevation with an annual precipitation of 1000–1200 mm. It can also

tolerate some shade but obviously high starch production would depend largely on high sun exposure. The optimum temperature for its successful cultivation is 14–21°C. Lower temperature delays
ripening of the roots and affects foliage growth, while higher temperature seems to reduce root size.
Short days favour early rooting in arracacha.

SOIL REQUIREMENT

Arracacha can be grown on a wide range of soils. However, medium-textured sandy loam soils rich
in humus are considered best for its cultivation. For obtaining higher yield, the soil should be deep,
friable, stone free and well drained but good in water holding capacity. The ideal soil pH for its
normal growth and development is 6.3–6.8.

VARIETIES

The arracacha varieties have been categorized into three groups, *i.e.*, (1) white rooted, (2) yellow
rooted and (3) purple rooted. The yellow rooted varieties produce roots with highest β-carotene
content.

In Colombia, many varieties have been identified, of which, *Paliverde* is the most common, producing roots 10–15 t and stumps 4–7 t/ha. *Paliamarilla* is grown to a lesser extent, and *Palirrusia*
and *Palirroja* are the rarely grown varieties.

PLANTING TIME

In temperate regions, the crop is planted in spring, whereas in tropical and subtropical regions, the
crop is planted at the onset of the rainy season.

SEED RATE

Approximately 25,000–30,000 offshoot cuttings and about four quintals of crown divisions are
enough for planting a hectare land area.

PLANTING METHOD

Arracacha can be propagated by using offshoots (entity of cormel and attached leaves), crown divisions and seed. However, traditionally, offshoots are preferred as the propagule since the propagules can be produced on-farm year after year without degeneration of storage roots. The offshoots,
depending on age and development, have a few to several dozen of buds, each of which has the
capacity to sprout and form a new shoot. Apical parts of the cormels with basal part of the petioles
serve as the propagule, which consisting of only apical part of cormel results in a plant with fewer
shoots, less foliage, smaller aerial parts and storage roots with higher dry matter content. To prepare
the propagule, the cormel-offshoot is detached from the rootstock, and its leaves are trimmed back
leaving only a few centimetres length of petiole. Then, about two-third to three-fourth of the cormel
are cut off in a slanting manner to stimulate root development. Before planting, the propagules are
left for 2–3 days to permit the cut surface to dry.

Usually, the rooted propagules are planted since planting rooted propagules improves early plant
development and brings about a number of benefits, the most significant being rapid crop establishment, more homogeneous plant canopy, reduced crop duration and higher yield. Pre-rooting of propagules, which can conveniently be achieved on small plots with improved soil substrate, requires about
45–50 days. The crop planted by this method takes 6–7 months as compared to crop duration of

8–10 months for unrooted planting material. Planting density varies from 15,000 to 30,000 plants per hectare when planting is done at 80 cm spacing between rows and 40 cm within rows.

Crown is the aerial portion and rootstock is the portion that remains after removing the storage roots. After attaining the optimum cropping stage in autumn, the roots and herbaceous material are removed. The crown is then separated bilaterally with effort, ensuring that there is at least one node for new foliage to sprout. The resulting unrooted buds are left in a cool and dry place for about 3–4 days for *healing* purpose. After a light callusing, they are potted separately in a well-drained open compost mix and left to overwinter somewhere relatively protected like a greenhouse and kept moist until spring conditions become suitable to plant them in the field. Often a shallow lattice-like pattern is sliced into the underside of divisions to ensure numerous and even root production. A dozen of propagules can be obtained from a medium-sized crown.

Arracacha plant rarely set seed. The self-seed is collected between November and January and sown in greenhouse in spring months. The seed is covered lightly with a thin layer of compost, which is kept moist until the completion of germination, which is often poor, less than 50%. When the seedlings are large enough to handle, they are pricked out into individual pots and allowed to grow for the first year in greenhouse, and then they are shifted in the field in late spring after the last expected frost.

NUTRITIONAL REQUIREMENT

Arracacha is very slow-growing crop, requiring 10–12 months to be mature, even then, it requires much less fertilizer inputs. The crop can successfully be cultivated with moderate organic matter content in the soil. Well-decomposed farmyard manure of 10–15 t/ha is incorporated in the soil at the time of field preparation. In addition, nitrogen is applied at 130 kg, phosphorus 200 kg and potash 100 kg/ha. The full dose of phosphorus and potash along with one-third dose of nitrogen is applied as basal application at the time of planting. The remaining two-thirds of nitrogen is applied 30 and 60 days after planting at the time of first and second intercultural operations. Application of borax at 20 kg/ha every second year gives the highest root yield. Applying nitrogen higher than the recommended amount produces copious foliage growth, harvest indices drop and crop duration is prolonged.

IRRIGATION

Arracacha crop requires frequent irrigation. The soil should be kept moist until the emergence of sprouts. Thereafter, irrigation is applied at an interval of 12–14 days, depending on soil and weather conditions. Applying irrigation is stopped a few days before harvesting the roots for making the tubers mature, however, a light irrigation 2–3 days prior to harvesting makes the digging operation easy.

INTERCULTURAL OPERATIONS

HOEING AND WEEDING

Hoeing and weeding during the early stage of crop growth is essential as weeds compete with crop plants for nutrients, moisture, space and light and affect the crop growth adversely. In late stages, the crop develops a large canopy and suppresses weed growth. Hence, no weeding is required when the crop has developed a large canopy area. Total —two to three hoeings are required at the start to keep the field weed free. Mounting up of soil around the plant stem is essential to provide better space and soil environment for the development of storage roots, which is done simultaneously at the time of hoeing and weeding, however, some farmers are of the opinion that repeated hilling suppresses the development of storage roots and promotes only foliage growth.

Arracacha can successfully be intercropped with maize, beans and coffee and included in a rotation. It generally follows potatoes and other vegetables or it can be combined with maize (five rows of maize or coffee with one row of arracacha in Colombia). The sowing time coincides with the beginning of maize ripening and it is then left in the field for up to 2 years.

HARVESTING

The crop being slow growing usually becomes ready for harvesting 10–12 months after planting but some young immature roots can also be harvested 5–8 months after planting. Some improved varieties require merely 7 months of growth. Harvesting is done by pulling up the plants along with roots. The roots are easily broken away from the plant, thus must be picked promptly as they are prone to turning woody. The storage roots must reach the consumers within a week of harvest, as they have a short shelf life. The remaining crown is divided into rootstock mostly for animal feed and cormels. The entire crowns are often left in a heap for few days or weeks until needed for the preparation of propagules.

YIELD

The marketable yield usually includes the storage roots, but because of the highly variable dry matter partitioning of the plant, the total of roots and rootstocks would be a more appropriate measure for the capacity of the plant to build up starchy dry matter. At harvest time, there can be as many as 10 roots per plant. The edible roots yield is 2–3 kg/plant and 80–200 q/ha. Preventing the plant from flowering or pinching off the inflorescence further encourages the yield of storage roots.

POST-HARVEST MANAGEMENT

The storage roots of arracacha are highly perishable, which restrains its commercial exploitation. A few days after harvest, the roots lose their luster, become unattractive and develop brown spots and *Rhizopus*, *Penicillium*, *Aspergillum*, *Nigrospora*, *Mucor* and *Syncephulastrum* infections. The shelf life can be extended barely a week at 25°C, the main cause of deterioration being rapid weight loss and subsequent decaying, however, the deterioration can be delayed over several weeks, either by reducing storage temperature (3–12°C) or adopting measures that prevent root desiccation, *i.e.*, wrapping individual root in low-density polyethylene or shrink films. Burying the harvested roots keep them fresh for 3 weeks. Leaving arracacha roots unwashed has also proven to enhance shelf life but marketing usually requires a washing operation. Shelf life can also be doubled by gamma irradiation.

INSECT PESTS AND DISEASES

Arracacha is generally regarded as a robust crop with only a few insect pest and disease problems if it is appropriately rotated. If not rotated suitably, insects and diseases can cause significant damage to the crop.

INSECT PESTS

TWO-SPOTTED SPIDER MITE OR ACCARID (*Tetranychus urticae*)

Sporadically, mites attack the plant leaves but do not cause great damage. These tiny accarids are found present in large number on the underside of the leaves under a protective web. Nymphs and adults suck cell sap from tender parts of the plant, causing yellowing of the leaves. The severely attacked leaves show necrotic spots. If the infestation is severe and not controlled suitably, complete defoliation may occur.

Control

- Introduce predators like lady beetles, ghost ants, larva of the minute pirate bug (*Orius insidiosus*) and predatory mites (*Phytoseiulus persimilis, Mesoseiulus longipes, Neoseiulus californicus, Galendromus occidentalis* and *Amblyseius fallicus*) in the field.
- Spray the crop with 0.3% Phosalone or Dicofol at 5 days interval during summer or 7 days interval during winter.

CHISA BEETLE (*Conotrachelus cristatus*)

A beetle pest locally called *chisa* is increasingly limiting arracacha culture. Its larvae mine the root and cause up to 40% yield loss.

Control

- Handpick and destroy the bright colour egg masses, larvae and adults at initial stage of attack.
- Spray the crop with 0.05% Dimethoate at 10 days interval.

MOTH (*Agrotis ipsilon*)

The moth lays eggs on the underside of the leaves, commonly feeds on seedlings at ground level, cuts off the stem and sometimes drags the plants into their burrows. Most of the plant is not consumed but merely eaten enough to cause it to topple. Since larvae burrow near the host roots, they sometimes feed on roots and the below-ground stem. Because of the nature of their feeding on young plants, they can cause great damage to the crop in newly planted fields.

Control

- Destroy the weeds from the outlying areas to reduce the cutworm population as weed hosts are often the preferred sites of oviposition and food for the younger larvae.
- Plough the field deep between crops to turn up the larvae and pupae to the soil surface making them exposed to the predators and sun.
- Flood the infested field to control in some cases, depending on the crop.
- Handpick and destroy the larvae in small gardens.
- Use pheromone traps for adults and yellow traps to capture moth.
- Apply granular insecticides directly to the soil or apply high-volume sprays (at least 1000 L/ha) of insecticides.

NEMATODES (*Meloidogyne* sp.)

The soilborne eelworms are very destructive root parasites. The nematodes attack the plant roots, where they produce swellings or galls. As a result, malformation and reduction in root size occur. The infested plants become stunted and yellow in colour.

Control

- Plough the field deep in hot summer months repeatedly to expose the soil to high temperature for killing the nematodes.
- Adjust antagonistic or non-host crops like marigold in rotation.
- Incorporate *neem* or castor cake or leaves powder at 400 g/m^2 to reduce the population.
- Fumigate the soil with DD mixture at 75 L/ha to reduce nematode infestation.
- Apply granular insecticides like Furadan at 40 kg/ha in two halves at planting and earthing up.

DISEASES

BLACK ROT (*Ceratocystis fimbriata*)

Black rot is the most destructive disease of arracacha both in the field and in storage. Its attack is observed in roots as well as in cormels. In the field, it causes yellowing of older leaves, followed by wilting and death of the plant. On roots, the symptoms appear as small slightly black spots, which enlarge and cover the entire root. The diseased root will have an unpleasant taste and colour when cooked.

Control

- Follow long crop rotation with non-host crops.
- Use disease-free planting material.
- Dip the planting material in 0.2% solution of fungicides such as Thiabendazole, Benomyl or Ferban to kill fungus spores contaminating the root surface.

COLLAR ROT (*Sclerotinia sclerotiorum*)

This is also the most damaging disease, especially in conditions of high soil moisture. It causes the plant collar region and roots to rot, leading to total crop loss. At first, small brown colour lesions appear at collar region, which gradually enlarge and encircle the stem base and tissue becomes necrotic, resulting in rotting of the new sprouts.

Control

- Follow long crop rotation with non-host crops.
- Maintain field sanitation by removing crop debris to avoid the multiplication of pathogen.
- Treat the planting material with 50 ppm solution of Vitavax and Plant Vax.
- Drench the field with Vitavax and Plant Vax 50 ppm for its effective control.

LEAF SPOT (*Cercospora* sp. AND *Septoria* sp.)

This disease is prevalent in all the warm and humid tropical regions. The disease is characterized by the appearance of yellowish-brown spots, which gradually turn deep brown with irregular margins. The spots coalesce and form larger patches, covering major portion of the leaf. Hot and humid conditions are very congenial for the occurrence of disease, and during dry weather, disease incidence is at its minimum.

Control

- Follow field sanitation to reduce the incidence of the disease.
- Spray the crop with Dithane M-45 or Zineb 0.25% at 15 days interval, at least three sprays commencing from the first appearance of the disease.

SOFT ROT (*Rhizopus* sp., *Fusarium* sp. and *Phoma* sp.)

Soft rot is the most widely distributed and commonly occurring storage disease. The infected roots quickly turn soft and moist at one or at both the ends, become stringy and emit a characteristic fermented smell.

Control

- Avoid injury to the roots during harvest and transit to reduce the disease intensity.
- Cure the roots properly just after harvest to reduce the disease incidence.
- Spray the foliage of seed tuber with DCNA to reduce the occurrence of the disease.

BACTERIAL SOFT ROT (*Erwinia* sp.)

The plants usually remain dwarf and stunted with curling of leaves. Brown-black slimy areas appear on the leaf petiole. The surface of the affected roots tissue becomes discoloured and depressed, and the internal tissue turns soft, resulting in disintegration of the whole root. Control of the disease is not always effective but sanitary practices in production, storing and processing can be adopted in order to slow the spread of disease.

Control

- Rotate the crop with non-susceptible cereal crops to limit soft rot infection.
- Use disease-free healthy planting material.
- Provide appropriate drainage facilities for proper aeration into the soil.
- Control the soilborne insect vectors to limit the spread of disease in the field.
- Avoid injury to the plant tissues as much as possible.
- Remove all the plant debris from the storage warehouse and disinfect the walls and floors with formaldehyde or copper sulfate.
- Store the produce under conditions of low humidity and temperature and maintain adequate ventilation.

BACTERIAL SPOT (*Xanthomonas compestris* pv. *arracaciae*)

The disease is characterized by the appearance of small water-soaked lesions with distinct borders on leaves as the first symptom. Afterwards, the spots become angular and brown to black in colour. The centre of the spots becomes dry and then falls out.

Control

- Use disease-free healthy planting material.
- Spray the crop with a mixture of Streptocycline 1 g and copper oxychloride 20 g dissolved in 10 L of water to check secondary spread of the disease.
- Spray the crop with 0.02% Agrimycin-100 along with 0.3% Blitox or Dithane M-45.

VIRUSES

The viruses infecting the arracacha crop are AVA (arracacha virus A and nepovirus), AVB (arracacha virus B, nepovirus), potyvirus AP-1, carlavirus AV-3 and PBRV/A. All these viruses are transmitted by the contact of affected plant sap with that of the healthy one. Plants show yellow-green mottling on leaves, reduction in leaf size and stunted growth.

Control

- Use virus-free healthy planting material.
- Rogue out the infected plants from the field.
- Wash all the tools and implements with 3% solution of trisodium phosphate before using these again.

■■■

103 Celery

M.K. Rana and Naval Kishor Kamboj

Botanical Name : *Apium graveolens* L.

Family : Umbelliferae (Apiaceae)

Chromosome Number : 2n = 22

ORIGIN AND DISTRIBUTION

The centres of origin and domestication of celery are still uncertain. The places of origin and diversity of celery extend from Sweden to Algeria, Egypt, Abyssinia, Asia, Caucasus, Baluchistan and the mountainous regions of India. It was first cultivated for its medicinal uses in Egypt in 400 BC and also by the Greeks and the Romans. It was first mentioned in France in 1623 for its cultivation as a food plant and introduced in the United States in 1806. The *Apium* species are widely distributed in the Mediterranean basin, South Africa, New Zealand, South America and Australia.

INTRODUCTION

Celery is an important salad crop and grown for its fleshy leaf stalk. It ranks second among important salad crops after lettuce. In India, celery is a minor vegetable crop and not commercially grown. However, quite a large area in Punjab and Uttar Pradesh is under this crop for the production of its seeds, which are used for condiments, culinary and medicinal uses. In Himachal Pradesh, Karnataka, Tamil Nadu and Kerala, celery is grown on a limited scale as salad crop. It is commercially grown in Europe, the United States, South America, Australia, New Zealand and South Africa and is available in the market throughout the year.

COMPOSITION AND USES

COMPOSITION

Celery is an important food source of conventional antioxidant nutrients, including vitamin C, β-carotene and manganese. Leaves of celery are more nutritive than stalks, especially in vitamin A, protein and calcium. Its leaves are a rich source of phytochemicals such as flavonoid antioxidants (apigenin, luteolin, zeaxanthin and beta-carotene), phenolic compounds (apiole) and carotenoids (limonene). Traces of copper and arsenic were reported in the tuberous root. Its seeds contain 2–3%

essential oil and 17–18% fatty oil. The essential oil has *d*-selenene, sedlanolide and sedanoic acid anhydride contributing to its flavour and 60% of *d*-limonene. The nutritional composition of celery stalks is given in Table 103.1.

USES

The leaves are used in salad or cooked as vegetable and leaf stalks. The petioles are eaten as salad or used for the preparation of sauces, soups, puree, stews, vegetable juices etc. Sometimes, it is taken as fried and spice. Its seed is used as condiment in European countries and spice in India and seed oil is used for flavouring tinned food and sauces. Its seed oil is also valued as a fixative and used for making perfumes. It is also used in pickles.

MEDICINAL USES

Celery is a great choice for balancing body weight as it provides low calories and dietary fibre. It reduces inflammation and pain of joints, lung infection, asthma or acne and relieves stress. The minerals in celery, especially magnesium and the essential oil in it soothe the nervous system. It regulates the body's alkaline balance, thus, protecting from problems such as acidity. It is good for the digestive system. The high water content of celery combined with insoluble fibre in it makes it a great tool for easy passage of stool. It contains *good quality* salts as they are organic, natural and essential for health. Its leaves stalks can reduce low-density lipoprotein (bad cholesterol) by the presence of its butylphthalide component. Raw celery reduces high blood pressure. Active compounds called phthalides in celery have been proved to boost circulatory health. It helps in boosting up sex life due to the presence of two pheromones in its stalk, *i.e.*, *androstenone* and *androstenol*, by boosting arousal levels. Celery can combat cancer by the flavonoid present called luteolin that inhibits the growth of cancer cells, especially in the pancreas. Celery also helps in protecting the eyes and prevents age-related degeneration of vision due to the presence of vitamin A. The seed has carminative and nerve stimulant properties. It is used as a neuro-tonic in domestic medicine.

BOTANY

Celery is a biennial herbaceous plant, usually 60–120 cm high with flowers borne in compound umbels. Annual cultures are grown in India mainly for seed purpose. The root is succulent and well developed with numerous lateral roots. Stems are branched, angular, jointed and light green in colour. Leaves are oblong, 7–18 cm long, pinnate or trifoliate and leaf stalk 15–38 cm long. It throws up a flowering head in later part of autumn producing a mass of fruits. The flowers are white or greenish white, very small on compound umbels. They produce flattened small fruits, which are commonly called seeds, although the true seed is contained within. Among the smallest vegetable

TABLE 103.1

Nutritional Composition of Celery Stalks (Per 100 g Edible Portion)

Constituents	Contents	Constituents	Contents
Water (g)	93.5	Iron (mg)	4.8
Carbohydrates (g)	3.5	Vitamin A (IU)	867
Protein (g)	0.8	Thiamine (mg)	0.12
Fat (g)	0.1	Riboflavin (mg)	0.05
Dietary fibre (g)	1.2	Niacin (mg)	0.3
Minerals (g)	0.9	Vitamin C (mg)	6.0
Calcium (mg)	30.0	Calories (kcal)	18.0
Phosphorus (mg)	38.0		

seeds, its seeds are about 1–2 mm in length and grey brown with five lighter, longitudinal ridges. When the umbels dry, the crop is harvested and threshed. The 10 g of celery seeds have a count of about 26,450 and a pleasingly crisp texture but a slightly bitter taste.

CLIMATIC REQUIREMENT

Being a cool season crop, it requires a cool and humid climate. It is a perennial crop but is grown as an annual crop. It can successfully be cultivated in areas with low humidity and plenty of sunshine. It can also be grown in dry regions having irrigation facilities. A dry weather at maturity is favourable for seed crops. The ideal temperature for growing celery is 15–21°C. When the soil temperature goes above 25°C, the seeds become dormant. Seed germination is best at 15–20°C. Higher temperature causes bitterness in leaves.

SOIL REQUIREMENT

Celery crops require a deep, fertile, well-drained silt loam or loam soil rich in organic matter. It is slightly sensitive to acidic soils and does not perform well in soils having pH below 5.5 and above 6.7. The optimum soil pH for its successful cultivation is 5.5 to 6.7.

Being a shallow-rooted crop, it grows well in pulverized soil; thus, the soil is prepared well to a fine tilth by ploughing the field repeatedly followed by planking.

CULTIVATED VARIETIES

There are usually two types of varieties, *i.e.*, (1) self-blanched/yellow cultivars and (2) green-leaved cultivars. The demand for self-blanched or yellow cultivars is very low in the market due to their lesser vitamin A content, whereas, the green-leaved cultivars are gaining more popularity due to their high vitamin A content.

SELF-BLANCHED OR YELLOW CULTIVARS

The self-blanched cultivars are *Cornell 19* (with thick petioles but susceptible to brown spot), *Cornell 619* (with thick petioles and resistance against fusarium yellows), *Michigan Improved Golden, Supreme Golden, Golden Plume* or *Wonderful, Golden Self-blanching, Golden No. 15, Florida Golden* and *Emperor of Jeen.*

GREEN CULTIVARS

The green cultivars are *Utah* (dark green type), *Utah 5–70* (resistant to brown spot disease) and *Utah 10-B* (susceptible to boron and magnesium deficiencies).

The Summer Pascal type green cultivars are *Giant Pascal* and *Emerson Pascal* (resistant to early and late blights and tolerant to bolt)

The slow bolting type green cultivars are Slow Bolting Green No. 12 and Slow Bolting Green No. 96.

The other important green type cultivars recommended by Indian Agricultural Research Institute Regional station, Katrain (Himachal Pradesh) are as follows:

Wright Grove Giant: A medium late cultivar with high-yielding potential produces tall and large white stalks of fine quality.

Ford-Hook Emperor: A late cultivar with dwarf and stocky plants produces solid, white, thick, broad and tender stalks.

Standard Bearer: An early maturing cultivar with medium tall plants has stem medium, pink coloured with white longitudinal streaks. The stalk is solid with good size and flavour.

SOWING TIME

In plains and lower hills, usually, the seed sowing is done in the month of August and September. In temperate regions, the sowing of crop is done in March to April and August to September.

SEED RATE

To sow a hectare area, a seed rate of about 1.0–2.0 kg is enough, depending on varieties, growing regions, sowing time, irrigation facilities available and field conditions.

RAISING OF NURSERY

For raising seedlings, the nursery soil is thoroughly prepared to make it friable. Well-decomposed farmyard manure is applied at 10 kg/m². The nursery beds of 10 m length, 90 cm width and 15 cm height are prepared. The nursery beds are sterilized with 1% formaldehyde solution 10–15 days before sowing. After drenching with formaldehyde solution, the beds are covered with polyethylene sheet for 3–5 days and thereafter left open for 5–6 days to avoid toxic effect of formalin.

Seeds are soaked in water overnight before sowing in nursery for better germination and sown by broadcasting or in rows. Sowing of seeds in rows is preferable because it facilitates watering, thinning, weeding operations and due to better growth of seedlings. Seeds are sown at a spacing of 5–10 cm and 1–1.5 cm depth. The seeds are covered with fine soil or compost after sowing. Water is sprinkled over the beds immediately after sowing with water-can and thereafter on alternate days. Dry spell and absence of watering will affect germination adversely. The seeds after sowing take 20–25 days to germinate. As far as possible, nursery beds should be provided with partial shade during germination. Before transplanting, the seedlings should be hardened by withholding irrigation to obtain better stand of crop.

TRANSPLANTING

When the seedlings become 40–60 days old and attain a height of 8–10 cm, they are uprooted carefully. Only the healthy (disease-free) seedlings are transplanted. The distance between rows and plants is kept as 60 cm and 10–20 cm, respectively.

NUTRITIONAL REQUIREMENT

Well-decomposed farmyard manure should be incorporated into the soil at 20–25 t/ha at the time of field preparation. Besides, nitrogen 150–200 kg, phosphorus 80–100 kg and potash 150 kg/ha are applied to obtain a good yield. The full dose of phosphorus and potash and half dose of nitrogenous fertilizers should be applied as basal before transplanting. The rest of the nitrogen is top-dressed in two splits at 40 and 80 days after transplanting. At the time of top-dressing with nitrogen, the plants are earthed up to cover the leaf stalk for blanching followed by irrigation.

The deficiency of minor elements such as calcium, magnesium and boron causes various disorders in celery. Calcium deficiency causes *black heart*, which shows a tip-burn on young leaves. This can be prevented by foliar application of 0.5% solution of calcium nitrate or calcium chloride. The deficiency of magnesium results in leaf chlorosis. Soil application of magnesium sulfate at 12 kg/ha prevents leaf chlorosis. Boron deficiency causes cracking of stem, which can be controlled by applying borax at 12 kg/ha at the time of last ploughing.

IRRIGATION REQUIREMENT

Celery is a shallow-rooted crop, thus, it requires frequent irrigation with good drainage facility. The soil should have sufficient moisture for its better growth. Light irrigation should be applied at 7–8 days interval in summer and 10–15 days interval in winter.

INTERCULTURAL OPERATIONS

HOEING AND WEEDING

Good quality celery crop depends on cultural operations followed from time to time. Since it is a long duration crop, weeds pose a major problem. Frequent light hoeing and weeding are required to control the weeds. Two to three light hoeings are sufficient. Besides hand weeding, pre-planting application of Fluchloralin at 1–1.25 kg/ha or Linuron at 5 kg/ha effectively controls the weeds.

BLANCHING

Excluding sunlight on the leaf stalks while the plants are still growing makes them lack chlorophyll, and the process is known as blanching, which is done either by wrapping paper around the leaf stalks or by covering the petioles with soil, which keeps the petioles pale in colour. The plants are earthed up when they are about 40 cm tall, after removing suckers. Two earthings are allowed at fortnightly intervals and only tips of leaves should be visible after the final earthling up. Blanching improves the nutritive value, flavour and tenderness of edible portion.

HARVESTING

The crop is ready for harvesting 120–140 days after sowing or 75–90 days after transplanting. The plants are cuts just below the soil surface with the help of a sharp knife so that the complete set of leaves may come out. Thereafter, the plants are trimmed and prepared for sending to the market.

YIELD

The yield varies with variety, growing region, sowing time, irrigation facilities and field conditions. However, on an average, its yield ranges from 35–60 t/ha.

POST-HARVEST MANAGEMENT

Celery can be stored at 0°C temperature and 95–98% relative humidity for about 2–3 months.

INSECT-PESTS AND DISEASES

Celery crop is attacked by a number of insect pests and diseases during the growing season. Some of the insect pests and diseases along with their control measures are given below.

INSECT PESTS

CELERY LEAF MINERS (*Liriomyza trifolii and Liriomyza langei*)

Small (5 mm long) white maggots mine tunnels through the leaves, leaving behind the brown blotches and blisters. Celery leafminers attack on plants from late spring onwards. These insects cause stunted growth and bitterness.

Control

- Follow long crop rotation with non-host crops.
- Remove and burn badly infected leaves.
- Remove crop debris from the field.
- Crushing the maggots within their tunnels is also effective.
- Spray phosphamidon (0.03%) or dimethoate (0.003%) at 15 days interval.

CELERY FLY (*Euleia heraclei* linn.)

The larvae eat inner portion of the leaves very fast. The affected plants turn brown and often die.

Control

- Spray the crop with 0.025% Quinalphos.

APHIDS (*Cavariella aegopodii, Myzus persicae, Dysaphis apiifolia* and *Aphis gossypii*)

Small soft-bodied insects attack the underside of the leaves or stems, which as a result turn yellow or pink, brown, red or black depending on species and host plant. If aphid infestation is heavy, it may cause yellowing or distortion of leaves, necrotic spots on leaves or stunted shoots. Aphids secrete a sticky, sugary substance called honeydew, which encourages the growth of sooty mould on the plants.

Control

- Spray the crop with Dimecron (0.05%), Endosulfan (0.05%) or Malathion (0.1%).
- Natural enemies, *i.e.*, lady beetles, lacewings, damsel bugs, flower fly maggots, certain parasitic wasps, birds and fungal diseases, play a very important role in controlling aphid populations.

DISEASES

DAMPING-OFF (*Pythium* sp.)

Young seedlings are affected just below the surface level and the affected plants fall down on the ground.

Control

- Provide good drainage.
- Treat the seeds with Captan or Thiram at 2.5 g/kg of seed.
- Drench the nursery with Dithane M-45 0.2% or Bavistin 0.1%.

ROOT ROT (*Pythium polymastum*)

The affected plants remain stunted and leaves rosette. Finally, the whole plant wilts and dies.

Control

- Follow long-term crop rotation.
- Uproot the affected plants and destroy them.

■■■

104 Celeriac

M.K. Rana and Neha Yadav

Botanical Name	: *Apium graveolens* var. rapaceum
Family	: Umbelliferae (Apiaceae)
Chromosome Number	: 2n = 22

ORIGIN AND DISTRIBUTION

Celeriac is native to the temperate Mediterranean Basin. It is found growing wild in all temperate countries and is widely cultivated in Germany and other European countries, North America, North Africa and Southwest Asia where it is considered a religious plant and also having medicinal properties, which are described in some of the greatest civilizations including Italy, Greece and Egypt. It was known as *Selinon* to Greeks. This vegetable has also been described in Homer's *Odyssey* of 850 BC. It was first recorded as a food plant in France in 1623. French and Swiss botanists around 1600 BC gave description about celeriac, and thereafter it achieved importance as a crop. It became popular in Europe by the seventeenth century. Now, it is cultivated all around the world as a root vegetable. Celeriac and celery were bred from the same wild plant.

INTRODUCTION

Celeriac, also known as turnip-rooted celery, knob celery or celery root, is although a biennial to perennial less-grown herb but cultivated as an annual crop. It is a very hardy plant with bulbous hypocotyl and grown for its large swollen roots. Its shape is similar to turnip but taste is identical to celery. The foliage also looks like celery but the difference is that celeriac develops root at ground level resembling turnip and has a similar growth habit. The root grows to a diameter of 10 cm or more. The plant has dark green foliage, producing brown-skinned roots with white flesh. It requires a very long growing season of 200 days, thus only very few farmers grow it. Initially, its taste and odour were unpleasant but these disagreeable traits have now been eliminated in cultivated varieties. In ancient times, it was grown as a medicinal crop, but now, it is predominantly cultivated as a food crop. In India, it is cultivated in the foothills of Northwestern Himalayas and the outlying hills of Punjab, Himachal Pradesh and Uttar Pradesh. It is mostly cultivated in Punjab (Amritsar, Gurdaspur and Jallundhar) followed by Uttar Pradesh (Ladhwa, Saharanpur districts) over an area of about 5000 ha. However, about 90% of the total produce comes from Punjab.

COMPOSITION AND USES

COMPOSITION

Celeriac contains many health-benefiting nutrients. It is a good source of carbohydrates, proteins, calcium, phosphorus, potassium and vitamins such as vitamin C and K. Its roots also contain a fair amount of calcium, iron, zinc, copper, manganese, magnesium, thiamine, riboflavin, niacin and pantothenic acid. The nutritional composition of celeriac is given in Table 104.1.

TABLE 104.1

The Nutritional Composition of Celeriac Root (per 100 g Edible Portion)

Constituents	Contents	Constituents	Contents
Water (g)	88	Iron (mg)	0.7
Carbohydrates (g)	9.2	Zinc (mg)	0.33
Sugar (g)	1.6	Manganese (mg)	0.158
Protein (g)	1.5	Vitamin B (mg)	0.165
Fat (g)	0.3	Thiamine (mg)	0.05
Dietary fibre (g)	1.8	Riboflavin (mg)	0.06
Calcium (mg)	43	Niacin (mg)	0.7
Phosphorus (mg)	115	Pantothenic acid (mg)	0.352
Potassium (mg)	300	Vitamin C (mg)	8
Sodium (mg)	100	Vitamin K (µg)	41
Magnesium (mg)	20	Energy (kcal)	42

USES

Celeriac roots and leaves are both eaten as a vegetable all over the world, however, it is mainly grown for its roots, which are eaten either raw as a salad or after cooking. It is also used as garnish and stuffing, in casseroles, gratins and baked dishes. Its outer surface is very rough and tough, thus the peel is removed by either peeler or sharp knife before use. Its flavour is just like celery. It is therefore added as a flavouring agent in soups, stews and tomato juice. It is used in many recipes after blanching in boiled water. Its seeds and essential oil are also used for flavouring.

MEDICINAL USES

Celeriac roots and seeds have excellent medicinal properties. It acts as diuretic, carminative, stimulant, anti-inflammatory, anti-rheumatic, tonic and its constituents also have antispasmodic, sedative and anticonvulsant actions. It is good for breaking down and flushing out kidney stones. It helps in menstrual periods and promoting milk in breast-feeding mothers. It strengthens the nervous system and relieves indigestion. Being a low-calorie food, it is good for overweight individuals. Its potassium lowers blood pressure and the risk of strokes, and potassium in combination with magnesium helps men with erectile dysfunctions as it promotes blood flow and dilates the arteries. Its seeds are also used as tonic and an aphrodisiac. Its roots containing phosphorus are beneficial to nervous, lymphatic and urinary systems. Its water-soluble fibre lowers cholesterol, thus it reduces the risk of heart attack and cancer. Being an insect repellent, it repels cabbage white butterfly, thus it can be a good companion crop for Brassicaceae family plants.

BOTANY

The celeriac plant with dark green foliage and well-developed succulent roots can grow to a height of 1 m, while the wild plants with aromatic pinnately divided leaves may grow up to 60–90 cm height and 30–45 cm wide. The plant being biennial in growth habit forms a basal rosette of leaves in the first year and bears off-white flowers in dense umbels (an umbrella of short flower stalks) in the second year with the start of spring. Leaves are oblong, 7–18 cm long and pinnate or trifoliate. The plant may naturalize in landscape through self-seeding. The rhombic leaves that grow in a rosette are 3–6 cm long and 2–4 cm broad on a branched central light green stem, having angular ribs. The fruit is made up of two united umbels, each containing a 1.5–2 mm long and wide oval seed having a pleasingly crisp texture, subtle flavour and a slightly bitter taste.

CLIMATIC REQUIREMENT

Celeriac being a temperate crop requires cool weather with moderately well-distributed rainfall for its successful cultivation. The plants grow best in cool moist locations and can tolerate some shade but cannot tolerate heat. Its growth habit and requirements are similar to the celery crop. It requires constant moisture for its growth and can tolerate temperature between 5°C and 27°C with optimum temperature ranging from 15°C to 27°C. The plant is susceptible to frost and tends to bolt if exposed to chilling temperature (<10°C) continuously for 1–2 weeks. The seeds germinate best when the soil temperature is around or over 15°C. It requires a long growing season with bright sunny days for good harvest. Northern and Central India, including the hills having a cold and dry climate, are most suitable for its cultivation.

SOIL REQUIREMENT

Celeriac is a moisture-loving plant, thus needs soils rich in organic matter and retentive to moisture. It prefers well drained and neutral to slightly acidic sandy loam soil with pH 6.0–6.8. If the fertility level is lower than the ideal, it results in tough, stringy and stunted plant. Therefore, it is essential to grow celeriac in nutrient-rich soil. Being a root vegetable, it grows well in pulverized soil. Therefore, the soil should be prepared well to a fine tilth by repeated ploughing followed by planking.

CULTIVATED VARIETIES

At present, no Indian cultivar of celeriac is available since no breeding work has still been done on crop improvement. Its cultivation is restricted to temperate zones. Hence, in India, only the imported cultivars are cultivated on a very small scale. The descriptions of some important cultivars are given as follows:

Monarch AGM: The roots of this variety are round, irregular or flattened ball shaped with smooth skin and white succulent flesh. Its plant is 60 cm tall with dark green foliage.

Alabaster: This is a high yielding variety resistant to bolting. It takes 120 days to be mature.

Prinz: This variety is resistant to leaf diseases and bolting. Its small-to-medium roots are flattened round in shape with smooth surface and aromatic flesh. Its plant foliage is shorter than the foliage of other varieties.

Diamant: Its medium-sized roots are flattened, round or irregular in shape with smooth surface and white flesh.

Ibis: Its round and regular shape roots are uniform and small with smooth surface.

Kojak: The roots of this variety are round, irregular or flattened ball shaped with smooth surface and white clean flesh.

Giant Prague: This variety takes 120 days from sowing to maturity.

Some other varieties of celeriac are Bergers White Ball, Snow White, Goliath Diamant, Anita, Dolvi, Ofir, etc.

SOWING TIME

The sowing time depends upon climatic conditions of the growing region. Its sowing is usually done when soil temperature is around 15–20°C. In temperate regions, it is mainly sown in spring, and in warm winter regions, sowing is done in autumn and early winters at least 10 weeks before the average last frost date in spring so that it may mature in cool weather.

SEED RATE

The quantity of seed varies depending upon growing season and region. A seed rate of about 200–250 g is enough for sowing a hectare land area. Celeriac seed can stay viable for several years, if stored in a cool and dry place.

SOWING/PLANTING METHOD

Celeriac is propagated by seeds. The seeds, before sowing, are soaked overnight in lukewarm water for better germination. After proper drying, they are sown in well-prepared seed or nursery beds at a depth of 2 cm. The soil should be kept moist until the emergence of seedlings, as its seed germination is slow. It is always better to sow seeds in a nursery since its seeds are very tiny and hard to handle in outdoor fields. Seedlings emerge out 2 weeks after sowing with 80% germination. Seedlings become ready for transplanting as and when they attain a height of 15 cm with 7–8 cm root or 60 days old to ensure a good plant stand in field. The seedlings are transplanted in the field at a row spacing of 45–60 cm and plant spacing of 10–15 cm. Before transplanting, the seedlings at more than five leaf stage can be hardened off. If the seedlings after completion of basic vegetative phase are exposed to a temperature below 10°C, the plants are likely to bolt.

NUTRITIONAL REQUIREMENT

Celeriac is a nutrients exhaustive crop, thus it responds well to manure and fertilizers for giving good yield. Farmyard manure is applied at 20–30 t/ha in the field at the time of land preparation. Additionally, nitrogen is applied at 100–150 kg, phosphorus and potash each at 40–60 kg/ha. Half dose of the nitrogen along with the full dose of phosphorus and potash is applied at the time of transplanting and the remaining half of the nitrogen at 1 and 2 months after transplanting in equal halves. It also responds well to other nutrients like calcium, magnesium and boron, as calcium deficiency in soil results in black heart disorder Magnesium deficiency causes chlorosis and boron deficiency results in wacked stem.

IRRIGATION REQUIREMENT

Celeriac needs enough moisture from sowing to harvesting. It is a shallow-rooted crop, thus uniform and regular irrigation is applied to keep the top few centimetres of soil moist at all times since lack of soil moisture will stop plant growth. At the time of sowing, the soil should have ample moisture for good germination. In transplanted crop, the first irrigation is applied immediately after transplanting, and thereafter irrigation is applied at fortnightly interval in early parts and weekly interval near seed maturity. Irrigation requirements in general depends upon soil type, growing season and climatic conditions of the growing region.

INTERCULTURAL OPERATIONS

HOEING AND WEEDING

Celeriac is a slow-growing crop, thus controlling weeds in the early stages of crop growth is essential to avoid competition with crop plants for nutrients, moisture, light and space. At least two weeding cum hoeings are given for harvesting a bumper crop. However, the hoeing should be done carefully to avoid damage to its roots since it is a shallow-rooted crop. Use of Basalin 2–2.5 L/ha as pre-planting application controls the weeds and reduces the cost on weeding.

BLANCHING

Blanching is an important and essential operation. As the roots develop, the side roots are snipped off and the soil is mounted up around the swollen bulbous roots and stalks just below the leaves to avoid exposure to sunlight. This process will keep the exposed roots blanched white and flesh brown. However, some of the varieties are self-blanched. Blanching is also done to prevent bitterness and hardening of celery stalks. This operation can be done regularly as the plants grow, or about 2 weeks prior to harvest. Alternatively, newspaper can be wrapped around the stalks to block out sunlight.

HARVESTING

Celeriac roots are supposed to be ready for harvesting as and when they attain a diameter of about 7–10 cm or slightly larger (12 cm) depending on the growing region. Though light frost increases the flavour of roots, it is always better to harvest the roots, especially of self-blanched varieties, before the occurrence of first hard freeze, provided they are mounded and mulched to prevent them from freezing, which damages the roots. The other varieties can be harvested throughout fall and winters. The roots can be left in the ground for about a month following the first fall frost, provided they are thickly mulched to provide protection against freezing. The roots are harvested with the help of garden fork and the foliage is cut with a sharp knife or sickle close to the knobby roots. The roots at the time of harvesting must be tender, which can be kept by a process known as blanching. Usually, smaller bulbous roots are more tender than the larger ones. Unblanched roots are comparatively tougher after maturity but good in flavour.

For seed extraction, the crop is harvested when 80% of the umbels turn from light to dark brown. Usually, the fruits mature by the end of March or early April. The seed stalks should be harvested early in the morning to avoid seed shattering. The seed stalks after harvesting are allowed to dry in sun over threshing floor for a considerable period since prolonged exposure to sun during the drying period may result in loss of oil. Thereafter, it is threshed, winnowed and graded with the help of a sieve. The seed after proper drying can be stored in a cool and dry place for 1–3 years without much loss in oil content and oil quality.

YIELD

The celeriac root and seed yield varies with varieties, soil type and fertility, irrigation facilities, the growing region and cultural practices adopted during the cultivation of crop. On an average, celeriac gives a root yield of 150–200 q/ha and a seed yield of 10–13 q/ha.

POST-HARVEST MANAGEMENT

Celeriac roots, after clipping off all leaves, can be stored up to 1 week in the refrigerator or for 2–3 months in a cool and moist place. The roots can also be retained underground in the field where the soil does not freeze. The roots can be stored in a freezer but they become soft and unusable upon thawing. They can also be stored well in a root cellar, shallow container full of slightly moist clean sand or any storage location that stays around over the winter.

PHYSIOLOGICAL DISORDERS

BLACK HEART

Black heart is a disorder resulting from calcium deficiency in the plant, which is associated with excessive or scarce moisture in soil. Celeriac grown in hot weather can frequently develop a calcium deficiency in the centre of the crown, causing growth retardation. This disorder can also develop when environmental and soil conditions stimulate rapid plant growth, such as heavy fertilization, rain or irrigation. High soil nitrogen, low calcium-to-nitrogen ratio and high levels of potassium, which reduce the absorption of calcium, may also be associated with this disorder. Initially, the water soaked spots appear on younger leaves, which later wilt and turn black. When the disorder occurs, the tips of the affected leaves become black at the end of the petioles. If the disorder persists for a longer period, the growing point dies and the plant becomes unmarketable. The disorder affects the market value of celeriac.

Control

- Use resistant/tolerant varieties, if available.
- Apply lime and calcium fertilizers before planting.
- Cover the surface with mulch to conserve soil moisture.
- Supply irrigation at regular interval.
- Avoid excessive use of nitrogen, especially the ammonium-based fertilizer.
- Maintain good drainage conditions in the field.
- Spray of calcium nitrate or calcium chloride 0.5% before the symptoms appear.

STEM CRACKING

Cracking of stem is caused due to the deficiency of boron in the soil, which is more intensified in soils containing excessive lime, preventing uptake of boron by the plant roots. High doses of potassium and ammonium-based nitrogenous fertilizers are also associated with this disorder, which is promoted more by drought. The disorder is easily identifiable, as crosswise brown cracks with raised margins appear on ribs of the petioles. When the deficiency is pronounced, the plant becomes stunted and bushy with stiff and fragile petioles. The leaves of affected plants may show patches of brownish discoloration. On the inner side of stalks, the brown translucent spots are elongated in shape. The growing point is often shattered and a dry rot gradually destroys the petioles from the heart to the base. Later, the affected plants may be attacked by saprophytes.

Control

- Grow tolerant varieties.
- Maintain a slightly acidic soil reaction.
- Apply a balanced dose of fertilizers along with micronutrients.
- Apply borax or sodium borate 10–15 kg/ha at last ploughing.
- Spray borax at 0.25–0.50% on plants if cracking is severe.

CHLOROSIS

Chlorosis may develop due to the deficiency of either manganese or magnesium. Manganese deficiency is found more common in alkaline soils, where it is unavailable to the plants due to its presence in unavailable form. Yellowing symptoms usually appear first on older leaves. Its deficiency causes chlorosis between the leaf veins and eventually affects the entire leaf. Its severe deficiency can affect overall growth of the plant. Magnesium deficiency develops in the form of a very obvious chlorosis on older leaves, which later extends to youngest leaves. Chlorosis begins at leaf margins and progresses steadily until the affected leaves turn yellowish white. When the deficiency is severe, the leaves turn brown, dry out, die and abscise.

Control

- Grow celeriac in a slightly acid soil.
- Avoid excessive liming and add gypsum to lower down the soil pH.
- Apply adequate doses of fertilizers including manganese or magnesium sulfate.
- Spray the crop with manganese or magnesium sulfate at 0.2%.

INSECT PESTS

Aphids (*Cavariella aegopodii, Myzus persicae, Dysaphis apiifolia* and *Aphis gossypii*)

The soft-bodied tiny insects, which are usually green, yellow, pink, brown, red or black, depending on species and host plant, live in a colony on the undersurface of leaves and stems of the plant. Its nymphs and adults generally do not move very quickly when disturbed. Under heavy infestation, the leaves become yellow and distorted with necrotic spots on them, and plant growth becomes stunted. It also attacks the fruits. Aphids secrete a sticky sugary substance called honeydew, which supports sooty mould growth, reducing photosynthetic activity of the plant and ultimately yield.

Control

- Grow aphid tolerant varieties if available.
- Remove the infested plant parts if aphid population is limited.
- Spray simple water with a strong jet pressure to wash the aphids from leaves.
- Install reflective mulch such as silver-coated plastic to deter aphids from feeding on plants.
- Spray the crop with insecticidal soap or oils such as *neem* cake or canola oil to cure the plants.
- Spray the crop with Malathion 0.1%, Monocrotophos, Phosphamidon, Dimethoate or Methyl demeton 0.05% at 20 and 35 days after sowing.

Celery Leaf Miner (*Liriomyza trifolii*)

The insect has a very broad host range, including many vegetables, ornamental crops and weeds. Its small larvae make tunnels in leaves leaving behind brown blisters. The adult leaf miners feed on flowers. Additionally, the adult females puncture the leaves with their stylets and feed on the plant sap that accumulates at feeding puncture, while its males, which live for only a few days and are unable to puncture the leaf, feed on the sap left by the females. Its attack leads to early senescence of outer petioles, late maturity of the crop and reduced flowering, fruiting and yield. Its severe infestation may prevent plant growth, *per se*, the plants become stunted and leaves turn upward, wither and finally die.

Control

- Grow the crop in mesh house or cover the crop with nylon mesh net.
- Transplant only the insect free seedlings.
- Remove and destroy all the affected leaves.
- Spray the crop with Malathion at 0.1%, Phosphamidon or Dimecron 0.1%, Dimethoate 0.2% or Monocrotophos 0.15% at fortnightly interval.

Armyworm (*Pseudaletia unipuncta*)

The attack is symptomized by single or closely grouped circular to irregularly shaped holes on foliage, and heavy feeding by young larvae leads to skeletonized the leaves. Shallow dry wounds may be present on fruit and clusters of 50–150 eggs on leaves. These egg clusters are covered in a whitish scale, which gives the cluster a cottony appearance. Young larvae are pale green to yellow in colour, while older larvae are generally darker green with a dark and light line running along the side of their body and a pink or yellow underside. Insect can go through three to five generations a year.

Control

- Install light traps with a sex pheromone in field to capture the adults.
- Apply natural enemies, which parasitize the larvae.

- Apply *Bacillus thuringiensis* on the crop.
- Spray the crop with Chloropyriphos at 0.1% or Malathion at 0.1%.

Slugs (*Limax maximus*)

Slugs feed on a variety of living plants as well as on decaying plant matter. These insects feed on soft tissues and leaves of the plant. They make smooth edged irregular holes in leaves and flowers and snip tender plant parts. They also feed on young plant bark and fruits. Since they prefer succulent foliage or flowers, they are principally the pests of seedlings and herbaceous plants but may also cause serious damage to ripening fruits that are in contact of moist soil. In the absence of herbaceous plants, they survive on foliage and fruits of some woody trees. Silvery mucous traces appear on the plant due to the damage caused by slugs.

Control

- Plough the field repeatedly in hot summer to expose the slugs and their eggs to blazing sun and natural predators.
- Remove weeds from celeriac field and surroundings, thereby reducing sources of food and shelter.
- Handpick and destroy the slugs regularly in the evening.
- Install slug traps in celeriac field.
- Use environment friendly molluscicide slug pellets containing ferric phosphate.
- Install slug baits containing metaldehyde and carbamate types such as methiocarb in field.
- Spray the crop with Splendour (thiacloprid) 100–500 g dissolved in 500 L of water.

Root-Knot Nematode (*Meloidogyne hapla*)

The common symptoms of nematode infection are stunting, reduced leaf size, yellowing and sickly appearance of the plant and the diagnostic symptoms are the formation of knots in roots. Its infection causes severe yield loss depending on variety, moisture, temperature and initial nematodes population in the soil. Its host range is very wide and infection starts at the seedling stage. The nematodes persist in the soil for years.

Control

- Rotate the crop with non-susceptible crops such as corn and African marigold.
- Grow resistant varieties, if available.
- Grow celeriac in a new field to avoid its infection.
- Avoid carrying soil from infested fields to the new site.
- Fumigate the infested beds with methyl iodide 500 g/10 m^2 area.
- Apply carbofuran 1 kg a.i./ha along with *neem* cake 200 kg/ha and urea 23 kg/ha in soil.
- Treat the seed with carbofuran 3% w/w.
- Apply Aldicarb 10 G 4 g, Furadan 10 G 7 g or Thimet 10 G 10 kg a.i./ha at the time of field preparation.

Pin Nematodes (*Pratylenchus hamatus*)

The pin nematodes are so microscopic that they cannot be seen with naked eye. They live in soil and feed on celeriac roots. They insert their stylets into the plant roots and suck food from them. Although the nematodes individually are very small, they are present in such numbers that they cause stunting and yellowing of the celeriac plant.

Control

- Strictly follow rotation with non-susceptible crops such as corn and African marigold.
- Plough the field repeatedly in hot summer months.
- Flood the field for longer period.
- Avoid carrying any soil from infected field to a healthy field.
- Fumigate the infested beds with methyl iodide 500 g/10 m^2 area.
- Add neem cake, castor cake or Calotropis leaf 500 g/m^2 area.
- Apply Aldicarb 10 G 4 g, Furadan 10 G 7 g or Thimet 10 G 4 g/m^2 area.

DISEASES

DAMPING-OFF (*Pythium* spp. and *Rhizoctonia solani*)

The disease may be pre- and post-emergence damping-off. In pre-emergence damping-off, the seeds fail to germinate due to the rotting of seeds, and in post-emergence damping-off, there is rapid death of the seedlings after emergence, which is caused by girdling of stem at soil line. The disease is favoured by conditions which slow down seed germination. The fungi spread through contaminated soil, seed, irrigation water or equipments in use. High moisture in the soil favours the multiplication of fungi causing damping-off disease.

Control

- Follow long crop rotation with non-host crops.
- Sterilize the soil by drenching with formaldehyde 1%.
- Grow the crop on well-drained raised beds.
- Increase air circulation and avoid overwatering of plants.
- Remove the affected plants along with soil.
- Treat the seed with Captan, Thiram, Ceresan or Bavistin 3 g/kg of seeds before sowing.

FUSARIUM WILT (*Fusarium oxysporum* f. sp. apii)

The disease in the field is recognized as stunting, yellowing and wilting of plants, which die prematurely. When the plant is transected, the vascular tissues inside the stem show dark brown to black discoloration. The disease can lead to death of the susceptible cultivars plants. The fungus is soil borne and can persist in soil for a longer period. The fungus can survive on weeds that act as collateral or alternate host for the fungus and other crop plants without exhibiting symptoms.

Control

- Grow the moderately resistant varieties.
- Use disease-free healthy seeds.
- Follow 2–3 years rotation with cucurbits and onions to reduce the soil inoculums.
- Remove collateral or alternate hosts from the surrounding of celeriac field.
- Drench the soil with Bavistin 0.1%, Benomyl 0.1% or Thiram 0.2%.

ROOT AND CROWN ROT (*Rhizoctonia solani* and *Pythium* spp.)

The disease symptoms appear as wilting and slow or rapid collapsing of plant. Leaves become rosette and wilt and die. Seedlings may also die. The roots become water soaked and brown instead of white. A water-soaked lesion can often appear at base of the stem. The stems become hollow and death of the plants takes place. The fungus is soil borne and survives in the soil for several years.

Control

- Follow 2–3 years rotation with cereal crops to prevent the buildup of inoculums.
- Destroy the infected plant debris.
- Provide good drainage facilities in the field.
- Treat the seed with Captan, Thiram, Captafol, Foltaf or Bavistin 2 g/kg of seed.
- Spray the crop with Bordeaux mixture 1% or copper oxychloride 0.2%.

EARLY BLIGHT (*Cercospora apii*)

The disease appears as large irregular, dark-brown to black concentric rings or angular spots surrounded by chlorotic halos on the lower older leaves and stems. The leaf spots are usually bounded by the veins. In severe cases, the whole leaf turns yellow, dries up and drops off. Dark black lesions also appear on the leaf petioles and stems, and dark sunken leathery lesions on the stem end of fruits. It is caused by a fungus, which may survive in soil for several years, but on seed, it may not survive for more than 2 years.

Control

- Grow varieties tolerant to early blight.
- Follow long crop rotation with cereal crops.
- Maintain proper field sanitation.
- Dip the seeds for 30 minutes in hot water (52°C) to disinfect them.
- Treat the seed with Captan, Thiram or Bavistin 3 g/kg of seed before sowing.
- Spray the crop with Ridomil 0.1%, Chlorothalonil, Thiophanate-methyl or Dithane M-45 0.2%.

LATE BLIGHT (*Septoria apii*)

The disease occurs on foliage at any stage of crop growth. It is a common fungal disease of celeriac, which appears as small, circular, tan leaf and stem spots. Usually, there are small black dots scattered across the tan spots. Gelatinous threads of spores are exuded from these dots during wet weather. At the later stage, the plants die. Cool moist weather favours the occurrence of the disease but the primary source of inoculums is the infected soil.

Control

- Follow at least 2–3 years crop rotation with non-host crops.
- Use disease-free healthy seed.
- Soak the seeds with hot water at 48°C for 30 minutes.
- Sow seeds at wider spacing to avoid overcrowding of plants.
- Remove the affected plants from celeriac field.
- Spray the crop with Ridomil 0.25% followed by Dithane M-45 or Blitox 0.25% at 10–12 days interval.

CELERIAC LEAF SPOT (*Septoria apiicola*)

The causal organism usually first invades the leaves, producing small brown spots on older leaves, and thereafter, it spreads to younger leaves. As the disease worsens, the spots on the leaf blades expand and produce a dark purplish-red oval border around a tan centre. The spots enlarge until the entire width of the leaf blade is blighted. The leaf-spotting or leaf-blighting phase is less damaging than the melting-out (crown and root-rot) phase of the disease. In the melting-out phase, the crowns and roots are damaged, causing severe thinning of the plants.

Control

- Follow a long crop rotation with non-host crop.
- Treat the seed with 0.05% Bavistin solution.
- Avoid excessive use of nitrogenous fertilizer.
- Destroy the residues after harvesting the crop.
- Spray the crop with Dithane M-45, Blitox or Difenoconazole 0.2% at 15 days interval.

PINK ROT (*Sclerotinia scerlotiotum*)

The disease is recognized as soft brown lesions with pink margins at the base of the stalks, which cause the entire stalk to rot, turn brown and collapse. Large black fungal fruiting bodies are visible on the infected tissues, when plants are infected at a young stage. The fungus causes damping-off with a watery rot visible on the seedling stems at the collar region. Fungus can survive in the soil for up to 10 years and its appearance is favoured by high soil moisture at least for 2 weeks.

Control

- Plough the field deep in hot summer months repeatedly.
- Trim back the foliage to promote air circulation to control the pathogen.
- Avoid excessive application of irrigation.
- Use drip irrigation up to 5–8 cm depth in soil.
- Spray Thiram 0.2% or Benomyl at 50 g/100 L of water.

DOWNY MILDEW (*Peronospora umbelliferum*)

Pale-green yellow spots appear on the upper surface of the leaves, which first appear on the lower leaves and then spread to the upper leaves and flower stalks. The under surface just beneath the discoloured area on the upper surface is covered with a white to greyish violet downy growth of the fungus under humid conditions. The white fluffy growth on the underside of leaves and lesions become darker as the crop matures and the infected plants finally collapse. The young leaves may drop off if the incidence takes place in the young stage. The disease spread is favoured by prolonged leaf wetness. The pathogen can survive in soil and seed either as dormant mycelium inside the seed or as contaminated oospores.

Control

- Rotate the crop with non-umbelliferous crops.
- Use disease-free healthy seeds.
- Avoid overcrowding of plants.
- Remove all the affected plant parts.
- Avoid excessive application of irrigation.
- Spray the crop with Ridomil 0.15% followed by Dithane Z-78 0.2% at 15 days interval.

POWDERY MILDEW (*Erysiphe heraclei*)

The fungus affects all the plant parts. The greyish powdery patches appear first on the upper and lower surface of the leaves and thereafter on petioles, flowers stalks and bracts. The affected leaves turn yellow, curl, dry up and fall off. Leaves become chlorotic and severe infections can cause the flowers to become distorted. The fungus spreads to long distances by air. Early infection leads to heavier yield loss than late infection. Dry weather conditions favour the incidence of the disease. Additionally, the infection is most severe in areas with shade.

Control

- Grow tolerant varieties if available.
- Use pathogen-free healthy seed.
- Avoid excessive use of fertilizers.
- Treat the seed with fungicides or hot water at 50°C for 15 minutes prior to sowing.
- Spray the crop with Karathane 0.1%, Bavistin 0.1%, Benlate 0.1%, Wettable sulfur 0.3–0.5% or Sulfex 0.2%.

BACTERIAL BLIGHT AND BROWN STEM (*Pseudomonas cichorii*)

Circular to angular small water-soaked spots appear on leaf blades, which become necrotic and may develop chlorotic halos. Later, the lesions become dry and rust coloured with brown stem symptoms, which include the development of a brown discoloration of petioles and brown streaks along the petioles. Closer plant spacing between plants favours the spread of bacteria, due to which, the bacterium causes severe disease symptoms at temperature above 30°C.

Control

- Grow resistant varieties, if available.
- Use disease-free healthy seeds.
- Keep the field fallow to reduce the inoculums.
- Uproot and destroy the infected plants along with soil.
- Trim the diseased leaves to reduce mechanical transmission.
- Treat the seed with streptomycin sulfate 2.5 g/kg of seed.
- Soak the seed in Streptocycline 1000 ppm solution for 2 hours.
- Treat the seed with hot water at 50°C for 30 minutes prior to sowing.
- Spray the crop with copper oxychloride (Blitox) 0.2%.

ASTER YELLOWS (Phytoplasma)

Aster yellow of celeriac is caused by a bacterium-like organism called Phytoplasma, which also causes the disease on aster, carrot and lettuce. The celeriac plant infected with aster yellows has an abnormal number of leaves, which are yellow, twisted and stunted. The roots become slender with an abnormal number of fine hairy roots. The aster yellow causing organism is transmitted by leafhoppers.

Control

- Avoid growing aster, carrot and lettuce near celeriac field.
- Destroy the weeds in celeriac field and its surroundings.
- Spray the crop with insecticides like 5% *neem* seed kernel extract or oil to control the leafhoppers.

CELERY MOSAIC VIRUS

Vein clearing and dark green mottling of leaves between veins and on petioles, twisting, curling, or cupping of leaves and stunting of young plants are the major symptoms of celery mosaic virus (CeMv), which is transmitted within 30 seconds of feeding by several species of aphids. The virus symptoms typically develop within 10 days of primary infection.

Control

- Grow resistant cultivars, if available.
- Use virus-free healthy seeds.
- Keep the field weed-free.
- Rogue out and burn the infected plants as soon as noticed.
- Remove umbelliferous weeds surrounding the field giving shelter to virus vectors.
- Control virus vector by spraying insecticide like Dimecron or Nuvacron at 0.05–0.1%.

■■■

105 Garden Chervil

M.K. Rana and Pooja Rani

Botanical Name	: *Anthriscus cerefolium* L. (Garden chervil)
	Anthriscus sylvestris L. (Wild chervil)
	Anthriscus caucalis L. (Bur chervil)
	Chaerophyllum bulbosum L. (Turnip-rooted chervil)
Family	: Umbelliferae (Apiaceae)
Chromosome Number	: 2n = 22

ORIGIN AND DISTRIBUTION

Garden chervil originally came from the Southern Europe, as it is mainly cultivated in temperate parts of the world. Later, the Romans extended its cultivation to a wider area. It was introduced in north Mediterranean regions by the edict Charlemagne giving the name *Capitulare de villis*. In the first century, Pliney used this herb as a seasoning. Recently, it is cultivated and used in the United States. It is now popular in Central and Western Europe and is widely grown in California and New Mexico.

INTRODUCTION

Since long back, *Anthriscus cerefolium* L. Hoffm., commonly known as garden chervil, is a warmth-giving herb belonging to the Apiaceae family. The name chervil has been derived from the Latin word *Chaeophyllum*, meaning festival herb. Chervil was once called *myrrhis* for its volatile oil extracted from its leaves, which has an aroma similar to the bibilical resinous substance *myrrh*. In ancient time, the plant was believed to symbolize sincerity, sharpen the wit and a new life. Its flavour and fragrance resemble the *myrrh* brought by the wise men to the baby Jesus. Since chervil symbolizes new life, it is linked with Easter ceremonies in some parts of Europe. Bowls of soup prepared from chervil leaves are traditionally served on Holy Thursday.

COMPOSITION AND USES

COMPOSITION

Garden chervil is a great source of carbohydrates, proteins, fats and fibres. It is rich in vitamins A, B, C and D. In terms of minerals, it is rich in calcium, phosphorus, potassium, magnesium, manganese and iron. The plant contains flavonoids as major constituents and essential oil (0.3% in fresh herb and 0.9% in seeds) in trace amount. It also contains methyl chavicol (estragole) and hendecane (undecane). The nutritional composition of this herb is given in Table 105.1.

TABLE 105.1

Nutritional Composition of Chervil (per 100 g Edible Portion)

Constituents	Contents	Constituents	Contents
Water (g)	7.2	Magnesium (mg)	130
Carbohydrates (g)	49.1	Iron (mg)	31.95
Protein (g)	23.2	Thiamine (mg)	0.93
Fat (g)	3.9	Beta carotene (mg)	0.38
Dietary fibre (g)	11.3	Vitamin A (mg)	50
Calcium (mg)	1436	Vitamin C (mg)	5.4
Phosphorus (mg)	450	Vitamin D (mg)	0.68
Potassium (mg)	4740	Energy (kcal)	237

USES

Chervil leaves are used in salads or as a flavouring agent in cooked food such as soups, butter sauces, omelettes etc. Both leaves and stems are used for cooking, and whole sprigs make it a delicate and decorative garnish. Its delicate flavour is easily lost by heating and drying, thus, it should be added at the end of cooking process to keep its flavour intact. Its flavour can easily be retained by preserving it in vinegar. The turnip-rooted chervil possesses a sweet carrot like root used to flavour soups and stews. This herb serves as a good source of syrup for gaining energy.

MEDICINAL USES

Chervil herb has been used for several medicinal purposes by herbalists. An infusion of its fresh leaves is used as a skin cleanser and lotion to clean the skin. A tea often made from its leaves improves blood circulation and reduces cellulites. Chervil drink has been used as an expectorant, diuretic, stimulant and dissolver of congealed blood. Being rich in iron and zinc, it prevents and cures anaemia. It is used in the form of spring tonic to purify the kidney and liver. The juice extracted from its leaves is used to treat arthritis, dropsy and skin problems. It is also used to cure stomach disorders. It is known to be valuable for the treatment of poor memory as well as mental depression.

BOTANY

Garden chervil is a hardy annual herb that grows to a height of 25–70 cm and width of 30 cm. The lacy, fern like, light green leaves are opposite, compound and bipinnate; the leaflets are subdivided again into opposite and deeply leaflets. The lower leaves are pointed and upper leaves are sessile with stem sheaths. The stems are finally grooved, round, branched, light green and hairy. The plant bears small white delicate flowers arranged in tiny umbels that grow in compound umbels. The flowers of this herb bloom during May to August and seeds ripen from June to July. The flowers are hermaphrodite and pollinated by insects. The plant is self-fertile. The herb has a thin white taproot system. The oblong fruit is 0.50–0.75 cm long, segmented and beaked. The seeds are long and pointed with a conspicuous furrow from end to end. It forms a rosette of sweetly scented delicate ferny foliage and is an ideal potherb.

CLIMATIC REQUIREMENT

Chervil, being a hardy crop, may thrive in much cooler climates provided with warm locations, but in spite of a cool weather crop, it is susceptible to frost, thus, it should be grown in a sheltered area. Its seeds germinate best at soil temperature of 20°C. At a temperature above 21–24°C, the plants grow slowly and bolt to flower at an early age. Cool temperature is critical to ensure good root growth. Unless shade and special cooling are provided, it is difficult to grow chervil in summer months. It is an absolute low temperature requiring crop and grows successfully under low light or partial shade conditions, thus, the plants should not be exposed to direct sunlight. It cannot grow well during hot summer, but in temperate regions, it can be grown as a summer crop due to prevailing low temperature, however, under such conditions, it requires partial shade.

SOIL REQUIREMENT

Chervil can be grown well on a variety of soils but moist and humus-rich loamy soils with good drainage facilities are most suitable for its cultivation. For better growth, especially of turnip-rooted chervil, the soil pH should be between 5.5 and 7.0. It has a wider adaptability, thus, grows in all parts of the world where the soil is rich in organic matter. It can also be grown in soils that contain less organic matter but should be well drained. However, nitrogen rich soils are more preferable for its cultivation. The best soil physical conditions for the crop can be provided by medium-deep ploughing for harvesting good yield.

VARIETIES

There are two types of chervil, *i.e.*, (1) plain and (2) curly type. The popular varieties of chervil grown in home gardens are French, Brussels Winter, Crispum, Great Green and Simple.

SOWING TIME

In temperate regions, the seeds are sown in March to April, whereas in tropical and subtropical regions, it is sown in October. In Southern plains, the seeds are generally sown in autumn but may not germinate until spring. In Northern plains, the seeds may be sown in autumn, which germinate in spring. Autumn sowing can be made for the production of fresh leaves throughout the winter if cultivated in a cool greenhouse. Tuberous-rooted chervil seeds do not germinate under ordinary conditions because of seed dormancy, which can be broken by subjecting the moist seeds to chilling temperature, *i.e.*, 4°C for 10 weeks (stratification), thus, the seeds are sown at the beginning of winter. However, this process induces flowering under long day conditions. Since this is a biennial crop for seed production, bolting does not take place even under long day conditions without the completion of vernalization process, which, in general, decreases the yield but when the vernalized seeds are sown at 20°C they give maximum yield.

SEED RATE

The seed rate usually varies with seed germination percentage, soil and environmental factors and method of sowing. In general, about 3–4 kg seed is required for sowing a hectare land area if sowing is done in lines.

SOWING METHOD

The crop is propagated by seeds so the seeds are directly sown in the field containing ample moisture. Raised beds are also recommended for culinary herbs. The seeds, before sowing, are embedded in damp sand for a week to improve and fasten the germination process since its germination is slow if the seeds are sown directly without soaking. The seeds are sown 30 cm apart at a depth of

0.5–1.0 cm in well-prepared seedbeds by using a shallow drill. The plant to plant spacing within the row is kept at 20 cm. As and when the seedlings attain a height of 7–8 cm, the plants are thinned to 8–10 cm apart. Sowing may also be done by broadcasting the seeds evenly if the ground is properly levelled. The seed, before sowing, is mixed with a small quantity of sand for uniform and even distribution of seeds in the beds. The seeds take about 7–14 days to germinate if the soil moisture and temperature are adequate. Chervil can be grown well in a container of 20–25 cm diameter and 30 cm depth. Most of the herbs perform better in very sunny windows when grown indoors but chervil grows better in indirect light. Outdoor pots can be kept on shaded decks.

NUTRITIONAL REQUIREMENT

Chervil plants grow well organically with the application of well-decomposed farmyard manure or leaf mould at 8–10 t/ha. However, inorganic fertilizers may also be applied to obtain higher yield as the application of fertilizers influences the size and quality of leaves and roots. It is essential to apply a balanced dose of fertilizers for its fast growth since heavy application of fertilizers, especially those containing large amounts of nitrogen, decreases the content of essential oil present in the green foliage. Nitrogen is applied 80–100 kg, phosphorus 70–80 kg and potash 60–80 kg/ha. Half dose of the nitrogen along with the full dose of phosphorus and potash is applied at the time of planting by band placement 5 cm away from and 5 cm below the seeds and the remaining half of the nitrogen is applied 3–4 weeks after thinning or transplanting.

IRRIGATION REQUIREMENT

The irrigation of the crop depends on the soil types and growing season. Being an herbaceous crop, it requires frequent irrigation for its growth and development. It may not grow well under hot and dry conditions, thus, regular watering is essential for obtaining a higher yield. In the case of insufficient soil moisture, a light irrigation should be given before sowing to ensure good germination. Thereafter, the irrigation is scheduled as per the needs of the crop. The soil is kept moist either by covering the furrows with cheesecloth or sprinkling water with a soaker hose.

INTERCULTURAL OPERATIONS

HOEING AND WEEDING

Hoeing is an essential operation at initial stages of crop growth to reduce weed competition with the crop plant for moisture, nutrients, light and space and to improve soil aeration around the root system for better plant growth. At least two hoeings are essential to keep the weeds down in the chervil field. If there is a problem of labour and heavy weed infestation, the chemicals such as influtalin/ethafluralin (1.1 kg/ha), linuron (1 kg/ha) or thiobencarb (6–8 kg/ha) can also be used to control the weeds. The flower stalks should be removed as soon as they appear since this practice causes rapid shoot growth to produce dense foliage.

INTERCROPPING

In tropical areas, chervil can be intercropped with radish, lettuce and broccoli for improving growth and flavour, as these crops provide partial shade to the chervil plants. Intercropping also prevents the crop from insect pest and disease infestation. It can also be intercropped with snakeroot or *sarpagandha* (*Rauvolfia serpentina*), mint (*Mentha arvensis*) or clary sage (*Salvia sclarea*).

HARVESTING

Depending on types and purpose, whether it is being grown for salad or vegetable or for obtaining seed, its harvesting should be done at an appropriate time and stage. At the time of maturity, the leaves tend to turn a purple bronze colour. The chervil leaves also lose their pungent taste during this stage so that only young green leaves may be used.

For salad or vegetable purposes, the leaves are harvested at a young stage before the development of the flower stalk. If they appear, the flower stalks should be removed. After the cutting of the required leaves, the whole plant should be cut down to the ground to give dense foliage before harvesting to get maximum shoot growth. The best time for picking its leaves is early morning after the evaporation of dew. The flowers may also be used in cooking. The leaves become ready for cutting in about 6–8 weeks after sowing. If the crop is to be raised for seed production, the plants are left in the field until the seeds become mature.

YIELD

Depending upon soil fertility, growing season and irrigation facilities, the average greens yield varies from 25 to 30 q/ha and seed yield from 5 to 6 q/ha. However, the turnip-rooted chervil gives an average roots yield of about 100 q/ha.

POST-HARVEST MANAGEMENT

After harvesting, the leaves and stems are dried on wire racks in a cool, dry and ventilated place at a safe moisture level. When the leaves are brittle and dried, they are stored in air tight containers. Under such conditions, it can be kept safe for 1 year. Fresh leaves may also be chopped and frozen in ice cube trays. Storage of turnip-rooted chervil at 6–8°C temperature increases starch conversion in its roots and prevents fungus infection. Proper storage is essential to preserve the delicate flavour and natural green colour of leaves. The best colour of its leaves can be retained in plastic bags containing 0.5% carbon dioxide and $0.5 \text{ mol m}^{-2} \text{s}^{-1}$ photosynthetic flux densities in controlled atmosphere storage at 5°C.

INSECT PESTS

Aphids (*Myzus persicae, Brevicoryne brassicae* and *Lipaphis erysimi*)

These tiny insects are found in large number on tender parts of the plant. They suck cell sap from the tender tissues, causing the plant to become weak. They excrete honeydew, on which, black sooty mould develops, which hampers photosynthetic activity of the leaves. Additionally, this acts as a vector for the transmission of virus. The severe infestation reduces the growth, yield and quality of crop plants.

Control
- Avoid excessive use of nitrogenous fertilizer.
- Spray simple water with pressure to remove the insect.
- Remove and destroy the affected portion of plant.
- Spray malathion (0.5%) two or three times during infestation.

DISEASES

ROOT ROT (*Fusarium* spp.)

The disease is characterized by severe stunting and rosetting of leaves, ultimately followed by wilting and death of the plant. Due to the attack of this fungus, the lateral roots and cortical tissues of the taproot start rotting.

Control
- Follow long crop rotation with cereal crops.
- Remove and burn the affected plants.
- Treat the seed with Agrosan GN at 3 g/kg of seed.
- Drench the soil around the root system of infected plants with 0.1% Bavistin.

POWDERY MILDEW (*Erysiphe polygoni, Erysiphe tryphi* and *Leveillula taurica*)

The disease appears at early seedling and flowering stage. The symptom includes the appearance of dull white powdery patches on leaves and their petioles. As a result, the leaves turn yellow and finally drop down. The disease is transmitted by wind and insects.

Control
- Collect and burn the disease plant refuse.
- Spray the crop with wettable sulfur, Karathane, Thiovite, Calexin or Sulfex 0.02%, Bavistin 0.1% or Tridemorph 0.05% two to three times at weekly interval.

CHERVIL YELLOW DISEASE (*Phytoplasma* spp.)

The disease appears on curled type chervil. The symptoms include reddening, chlorosis, mottling, dwarfing and malformation of auxiliary shoots of curled chervil due to the infection of virus. The plants infected at an early stage become rosette like in appearance.

Control
- Control of aphids and whitefly.
- Spray the crop with Nuvan or malathion 0.05%.

■■■

106 Florence Fennel

M.K. Rana and Chandanshive Aniket Vilas

Botanical Name : *Foeniculum vulgare* var. azoricum L.

Family : Umbelliferae (Apiaceae)

Chromosome Number : 2n = 22

ORIGIN AND DISTRIBUTION

Florence fennel has followed the Italian civilization, hence it is considered native to Italy. It grows wildly in temperate Europe, but it is usually considered indigenous to the Mediterranean region where it has been cultivated since the seventeenth century and from where it spread eastwards to India. It is also found growing wild in many parts of the world on dry soils near the seacoast and riverbanks. The ancient Greeks and Romans cultivated it for its aromatic fruits and succulent bulbous stems for food, medicine and insect repellent. Its cultivation began with the Greeks and Romans who spread it to Europe, Middle East and Asia. Its tea was believed to give courage to the warriors prior to battle and thus became famous in the era of Greeks and Romans. Greek mythology provides evidence that Prometheus used its stalk to carry bonfire from Olympus to earth. The king Charlemagne encouraged its cultivation on its imperial farm. In England, it is known since the eighteenth century when the nurseryman, Stephen Switzer of London, brought it from Italy and first included it in his seed list. Here it has been available for more than 200 years but not grown extensively as a vegetable.

INTRODUCTION

Florence fennel is known by many names *viz.*, bulb fennel, bulb anise, carosella, fetticus, fino, finocchio, finochio, Florence fennel, Italian fennel, Roman fennel, sweet anise and sweet fennel. It is different from fennel herb and cultivated for its bulb eaten as a vegetable. Its every part is aromatic and can be used for culinary purpose. Mostly, it is grown in kitchen gardens due to its adaptability, aromatic nature and also as home decorating crop. It is a cool season annual herb grown for its tender stalks and round bulbous base. The crispy white bulb consists of swollen stem base of the leaves.

COMPOSITION AND USES

COMPOSITION

Florence fennel bulbs provide a good amount of minerals such as calcium, phosphorus, potassium, magnesium and vitamins such as riboflavin, pantothenic acid and vitamin C. It is also a good source of fibre but a poor source of calories. The nutritional composition of Florence fennel is given in Table 106.1.

TABLE 106.1

Nutritional Composition of Florence Fennel (per 100 g Edible Portion)

Constituents	Contents	Constituents	Contents
Water (g)	93	Iron (mg)	0.75
Carbohydrates (g)	7.2	Zinc (mg)	0.2
Protein (g)	1.2	Manganese (mg)	0.2
Fat (g)	0.2	Vitamin A (IU)	134
Dietary fibre (g)	3.1	Thiamine (mg)	0.01
Calcium (mg)	49	Riboflavin (mg)	0.65
Phosphorus (mg)	50	Pantothenic acid (mg)	0.23
Potassium (mg)	414	Pyridoxine (mg)	0.04
Magnesium (mg)	20	Folic acid (µg)	27
Sodium (mg)	52	Vitamin C (mg)	12
Selenium (µg)	0.7	Energy (kcal)	130

USES

Florence fennel produces attractive green delicate foliage and golden flowers that add beauty to the kitchen garden. Its bulbs are used in salads, vegetables and chicken soups, grilled fish dishes and served with puddings. Its lacy leaves are used fresh as a garnish and in salads. Its complex and delicate flavour mixes well with cream sauces, dips and soups. Its bulb is used in the preparation of pickles and cakes and fresh or dried leaves to prepare egg and fish dishes. It contains a compound called anethole, which is widely used as rice flavouring.

MEDICINAL USES

Florence fennel is used in pharmaceutical preparations due to its useful properties, phytonutrients, antioxidants and lots of vitamin C content. Its leaves are diuretic, carminative and its roots are purgative. It is also used in the preparation of alcoholic combinations, which originated as a medicinal elixir and popular alcoholic drink.

BOTANY

Florence fennel, a member of the parsley family, is a perennial herb but grown as an annual crop. It is a group of cultivars with inflated edible bulbous structure forming due to overlapping of a thick leaf base, which is aromatic and sweet in taste. Its plant is stocky, which grows to a height of 120–150 cm and looks like celery with fleshy stalks and feathery leaves. It produces a flat-topped cluster of 20–50 tiny bright golden flowers on short pedicels, known as a compound umbel with 13–20 rays. It has a perennial thick rootstock with many cylindrical branched stems bearing leaves with very fine segments.

CLIMATIC REQUIREMENT

Florence fennel is a crop of temperate regions. It requires cool temperatures and short days for its growth, development and production of best quality bulbs. Seeds germinate quickly in sunny locations. It can tolerate some heat and light frost for short periods. It does best when the crop reaches maturity in cool weather. Best flavour is achieved when it is grown in full sun at soil temperature 10–21°C and daytime air temperature 15–21°C. Higher temperature coupled with longer days accelerates bolting. The crop matures during the hot summer. Rainfall during maturity deteriorates the quality of bulbs and stems. Prolonged cloudy weather is conducive to diseases like blight, powdery mildew and the pest-like aphid.

SOIL REQUIREMENT

Florence fennel requires well-drained loamy to clay soils, which is rich in organic matter with a pH range 5.5–8 but a pH range 5.5–7.0 is the most ideal for its cultivation. It thrives best in black cotton soil having proper drainage facilities. Highly porous sandy soils are not suitable for its cultivation. Heavy soils are more desirable than light soils for higher yield.

CULTIVATED VARIETIES

Only few bulb producing varities are available. Their description is given as under:

TRIESTE: It requires 90 days from planting to maturity.

ZEFA FINO: A slow bolting short duration variety of Europe requires 65 days to produce the highest bulb weight. Its white bulbs are firm, large and sweet with aromatic flesh.

ORION: A hybrid with compact thick leafstalks and soft small bulbs is resistant to bolting.

PERFECTION, CANTINO AND AMIGO: All three are the high yielding varieties with resistance against bolting.

FENNEL OF PARMA: A late maturing variety sown in late July to August that matures in the early winter season. Its bulbs fetch high prices in the market.

SOWING TIME

Florence fennel is a photosensitive plant. For bulb production, it can be sown twice, once in March to April for spring-summer crop and second in July to August for autumn-winter crop. Its sowing, in early spring, should be avoided since the plants are more likely to bolt. Hence, the best sowing time for bulb production is the autumn season. However, in kitchen gardens, it can be sown in spring as early as 2–3 weeks before the last frost date. Seeds need a minimum of 15°C for optimum germination. Usually, its seed take 7–10 days to germinate. Sowings after midsummer have a better chance than those in the spring, producing fat tender juicy bulbs due to wetter weather at bulb maturity and short day length, as long days induce bolting.

SEED RATE

Florence fennel requires 10–12 kg of seeds for sowing 1 ha area. Generally, 3–4 kg seeds are enough to raise seedlings sufficient for planting 1 ha area.

SOWING METHOD

Florence fennel is propagated by seeds. Its sowing depends on the purpose of cultivation whether it is being grown for leaves, bulbous stalks, stems or seeds. For successful germination, the seeds are sown in dark at a soil temperature of 20–30°C.

For bulb production, the field is prepared to a fine tilth after incorporating recommended quantity of farmyard manure and seeds are sown directly in flat seedbeds about 1 cm deep at a row spacing of 60–90 cm and plant spacing of 25–30 cm. Its seedlings are sensitive to frost but larger plants can tolerate temperature as low as 4°C, thus, seedlings should be protected from late severe frost.

TRANSPLANTING

The seedlings are raised in protrays containing 4–6-cm-deep cells in early spring to avoid bolting. At least three seeds are sown per cell at a depth of 1–2 cm, and later, the extra seedlings are removed by keeping a single seedling per cell. The seedlings are transplanted in the field as and when they attain a height of 5–10 cm or an age of 4–6 weeks. The seedlings from the protrays should be removed without disturbing the roots. Spacing between rows is kept 45 cm and between plants 15–30 cm for its successful cultivation and higher bulb yield, as overcrowding of plants encourages bolting. Transplanting is useful in areas where the growing season is usually too short.

NUTRITIONAL REQUIREMENT

Florence fennel responds well to farmyard manure than fertilizers. The nutrient rich well-decomposed organic material is incorporated at the time of field preparation for improving aeration into the soil and better bulb formation, however, side dressing with compost in the middle of crop increases the bulb yield. Farmyard manure is applied at the rate of 10–15 t/ha. Additionally, nitrogen is applied at the rate of 90 kg, phosphorus 40 kg and potash 90 kg/ha. One-third dose of nitrogen along with entire dose of phosphorus and potash is applied at the time of sowing or transplanting and the remaining dose of nitrogen is applied in two equal splits at 30 and 60 days after sowing. Florence fennel requires a moderate amount of nitrogen but low doses of phosphorus and potassium.

IRRIGATION REQUIREMENT

Florence fennel requires regular irrigation to keep the soil sufficiently moist throughout the growing season. Irrigation requirement depends on soil type and climatic conditions of the growing area. If moisture in soil at the time of sowing is enough, irrigation in early stages of crop growth can be avoided. If sowing is done in dry soil, first light irrigation is applied just after sowing and second light irrigation is applied at 10–12 days after the first irrigation to facilitate germination. Subsequently, the irrigations are applied at an interval of 12–15 days, depending upon soil type, stage of crop and weather conditions. The crop should not face any moisture stress particularly at bulb development stage, as it may affect the size and yield of bulb adversely. If dry spell occurs during initial stages, it may cause stem splitting or bolting, thus, to ensure succulent bulbs and prevent bolting, supplementary irrigation at a weekly interval is necessary. Drip system is an ideal method to keep optimum moisture in the soil.

INTERCULTURAL OPERATIONS

In direct sown crops, slow crop growth at initial stages, wider row spacing and frequent irrigations favour the growth of weeds, which cause serious reduction in bulb yield due to competition between weeds and crop plants for nutrients, moisture, light and space. First hoeing and weeding should be done 30 days after sowing when the seedlings are about 5–10 cm tall and second 50 days after sowing. Earthing up at the time of bulb development is beneficial to keep the plants stable and blanch the overlapped leaves bases, which make bulbs.

Mulching with dried leaves, straw or lawn cuttings is an important operation to reduce the soil evaporation and to prevent weed growth. It also helps in maintaining the soil conditions favourable, which ultimately improves the bulb quality and yield. Leaving the strongest seedling at each position, thinning in direct seeding crops is much better than transplanting as Florence fennel rarely recovers from transplanting shock and gives another good reason to bolt.

HARVESTING

Florence fennel requires 90–120 days to harvest bulbous stalk, which becomes ready for harvesting when the plant is near to flower and when the bulb diameter is at the size of 7–10 cm or small tennis ball and firm to touch. The bulbs are usually harvested before the occurrence of frost as it will damage the bulbs. The leaves may be picked at any stage for seasoning or garnish purpose.

Florence fennel develops a thick white base known as an apple. Blanching is done by mounting up soil around swollen base of the plant 8–10 days before harvesting so that the bulbous stalk may become white and succulent. A fork is used to loosen the soil around the bulbous stalk, and then, it is cut with a sharp knife just above the taproot at ground level to avoid bruises and damage to the bulb. The stems are trimmed off 2-3 cm from the bulb. Timing is important for harvesting the bulbs. Only those bulbs, which have attained the appropriate size, are harvested, thus, it can be harvested continuously for several weeks but not year round, as it is seasonal.

YIELD

The yield of Florence fennel varies with variety, soil type, growing conditions and the cultural practice adopted during crop cultivation. On an average, the bulb yield varies from 7 to 8 q/ha.

POST-HARVEST MANAGEMENT

The properly matured bulbs after grading and sorting can be kept fresh for several weeks in an ordinary cool and dry place and for 2–3 months in a moist place adjunct with refrigeration. The best storage conditions for the storage of Florence fennel are 0°C and 95% relative humidity. Its stalks and leaves can be frozen or dehydrated. The dried leaves are stored in hermetically sealed containers.

INSECT PESTS

Aphid (*Macrosiphum euphorbiae*)

It is a sucking type pest but not serious in nature. Its nymphs and adult suck sap from the tender leaves, causing curling of leaves and secrete a honeydew-like sugary substance, which encourages black sooty mould, reducing photosynthetic activity of the plant. Florence fennel is usually not attacked by aphids, but sometimes, it attacks the crop at seedling stage. Higher temperature and humidity support the pest infestation but leaf damage may result in smaller bulbs.

Control
- Grow resistant varieties.
- Spray the crop with malathion 0.5% or Rogar 0.1% at 12–15 days interval.

Thrips (*Heliothrips indicus*)

It sucks cell sap from tender parts of the plant, causing white specks with black dots. The affected leaves become brown and distorted. When infestation is severe, the entire plant may die or wilt. Dry weather is favourable for its rapid multiplication, thus, it is very destructive during dry weather.

Control
- Follow long crop rotation.
- Keep the field weed free.

- Collect and destroy the affected plants.
- Apply phorate in soil at 10 kg a.i./ha.
- Spray the crop with dimethoate 0.2% or malathion 0.2% at 15 days interval.

STRIPED FLEA BEETLE (*Phyllotreta striolata*)

The larvae and adult beetles feed on leaves or the bulbs, making small holes or pits into the base of leaves. These insects are the most damaging at seedling stage. However, the mature plants are more tolerant to its attack and rarely suffer severe damage. Even its small population can destroy the seedlings. Its attack makes the leaves and bulbs unfit for consumption and results in low bulb yield.

Control

- Follow long crop rotation.
- Grow resistant varieties.
- Remove host weeds and volunteer plants.
- Spray methomyl 0.5% and permethrin 0.5% at 10 days interval.

BLACK CUTWORM (*Agrotis ipsilon*)

The attack of cutworm is sporadic but increases at warm temperature. The female worms lay eggs on leaves and stem near the soil surface. The newly hatched larvae temporarily feed on leaves, but later, they drop on soil surface and burrow underground. The larvae emerge at night and cut the stem at or just below the soil surface. A single larva is capable of damaging several plants in limited time and its large population destroys the entire crop. Recently thinned crop is especially sensitive to its attack, which occurs in fields that have previously been under alfalfa crop.

Control

- Follow long crop rotation with non-host crops.
- Plough the field repeatedly 2 weeks prior to planting.
- Use *Bacillus thuringiensis* for its control.
- Avoid growing in field that has previously been under alfalfa crop.

DISEASES

STEM ROT (*Sclerotinia* spp.)

Stem rot can cause a sizeable damage to this crop. The fungus causes damping-off at seedling stage, and in older plants, it attacks the lower portion of the stem, causing watery soft rot of swollen portion. Eventually, a white cottony mould is developed over the infested tissue. Its severe infestation causes wilting of the whole plant. The fungus bears black sclerotia on infested tissue and soil surface. It is a soilborne disease since the fungus survives in the soil for a long time, especially in wet weather. Its sclerotia spread through agricultural implements, soil and plant tissue.

Control

- Follow long crop rotation with trap crops like corn or beets.
- Avoid excessive irrigation, as wet conditions favour *Sclerotinia*.
- Keep the field weed free.
- Drench the field with 0.2% captan solution.
- Destroy the infested plant debris.
- Use *Bacillus subtilis* bioagent to control *Sclerotinia*.

LATE BLIGHT (*Septoria apiicola*)

The disease appears as small chlorotic spots on stalks and leaves. Eventually, the black fruiting bodies develop in the centre of spots, which continue to increase in size and coalesce together to form a large lesion. Later, these lesions become dry and papery. The fungus survives for several years on plant refuse. The disease spread through irrigation water, rain and infected seed. Favourable field conditions for late blight can result in huge crop loss. Moderate temperature with high humidity favours the disease development.

Control

- Follow long crop rotation.
- Use disease-free seed.
- Maintain good sanitation in the field.
- Treat the seed with copper fungicide.
- Treat the seed with hot water at 52°C for half an hour.
- Spray the crop with Bavistin 0.1% or Indofil M-45 0.2–0.3%.

POWDERY MILDEW (*Erysiphe polygoni*)

The fungus attacks all parts of plant except roots. It first attacks the leaves and later spreads to stem and flowers. It is characterised by the development of dull white powdery mass on leaves and stems. Severe infestation may cause defoliation. The crop is usually attacked by powdery mildew near flowering particularly in the cloudy weather.

Control

- Collect and burn the diseased plant refuse.
- Dust the crop with fine sulfur powder at the rate of 20–25 kg/ha in evening hours.
- Spray the crop with Karathane 0.2%, Calixin 0.1%, Benlate 0.1%, Bavistin 0.1%, Sulfex 0.2%, Dinocap 0.05% or Thiovit 0.3% at 10–15 days interval.

BACTERIAL SOFT ROT (*Erwinia carotovora*)

Initially, water soaked spots appear on plant, which later turn brown. The bacterium first attacks the leaves and then bulbous stalks, destroying the entire field. The disease spreads rapidly under humid conditions. The bacterium dissolves the middle lamella and causes the inner cell contents to shrink. Bacterial soft rot can occur in the field as well as in storage especially at warm temperature. Wounds provide venue for the entry of bacterium. The portion of plant damaged by insects or frost is particularly susceptible to bacterial soft rot. The disease spreads through agricultural implements, insects, rain, irrigation, humans and animals.

Control

- Follow long crop rotation.
- Maintain good sanitation in the field.
- Use disease-free planting material.
- Dip the seeds in 0.2% solution of thiram or captan before planting.
- Spray the crop with Diethane M-45 0.25%, copper oxychloride 0.3% or Captan 50 WP 0.25% at 10–12 days interval.

Mosaic Virus

The disease is characterized by leaf chlorosis, mottling, puckering, necrosis, vein banding and brown discoloration of the leaves. Leaves, stems and bulb are affected badly, resulting in reduction of yield. The disease spreads through agricultural implements, seed or aphids.

Control

- Avoid intercropping with susceptible hosts.
- Use disease-free planting material.
- Keep the field weed free.
- Rogue out the infected plants.
- Spray the crop with malathion 0.2% to control the vectors.

■■■

107 Parsley

M.K. Rana and Sachin S. Chikkeri

Botanical Name : *Petroselinum crispum* (Mill.)
Nym. var. crispum (Curled parsley)

Syn. *Apium petroselinum*

Petroselinum sativum Hoffm.
(Hamburg parsley)

Petroselinum hortense Hoffm.

Petroselinum neapolitanum
(Italian parsley)

Family : Umbelliferae (Apiaceae)

Chromosome Number : 2n = 22

ORIGIN AND DISTRIBUTION

Parsley is believed to be originated from the Mediterranean region (especially in Sardinia Island of the Mediterranean Sea) and later it acclimatized in all parts of Europe. It was evidently quite popular among Greeks and Romans, as it is mentioned by several authors and was described in different forms. Theophrastus in 322 BC described two types of parsley, *i.e.*, (1) one with densely crowded leaves and (2) the other with open broad foliage. Columella in 22 AD and Pliny in 79 AD both described curled- and broad-leaved types. Parsley was said to be sacred to Pluto, and in the ancient days of Greece and Romans, a wreath of parsley was presented to the winners of the many games.

INTRODUCTION

Parsley is a popular biennial herb but cultivated commercially as annual in many parts of the world for its attractive and aromatic leaves. Its plant has altered significantly through cultivation. It is cultivated in Germany, France, Holland, United Kingdom and other European countries. It is also cultivated in the United States and Canada. In India, it is grown as a minor crop in Punjab, Himachal Pradesh and Uttar Pradesh in the northern region and in Nilgiri hills, Kodaikanal and Ooty in the southern region. It is widely used in Middle Eastern, European and American cooking.

COMPOSITION AND USES

COMPOSITION

Parsley is a good source of vitamin C, vitamin A and minerals like iron. It is not important food for contributing to human nutrition since only a small amount of parsley is eaten. One tablespoon of chopped parsley has zero calories with no fat. This serving of parsley provides 6% of RDA of vitamin C, 4% of RDA of vitamin A and 2% of RDA of iron. All parts of the plant contain essential oil responsible for the characteristic aroma and flavour of parsley. The main components are myristicin, limonene and 1,3,8-*p*-menthatriene and the minor components are sesquiterpines. The leaves, stems and fruits contain glucoside apiin, which on hydrolysis forms apigenin, glucose and sugar apiose. The fruit (seed) oil contains apiol (parsley camphor) and alpha pinene with small amount of myrusticine, aldehydes, ketones and phenols. The fruit yields up to 20% of greenish fatty oil with a peculiar odour and disagreeable sharp flavour. The oil has a high content of petroselinic acid (up to 76%). The nutritional composition of leaves is given in Table 107.1.

USES

Parsley is one of the popular garnishes for salads, sandwiches and cooked dishes, and flavouring for soup and pasta. Fleshy leaf stalks and petioles are generally used for consumption. Leaves are also dried and powdered for culinary flavouring when fresh leaves are not available. In addition to leaves, the stems are also dried and powdered and used as culinary colouring and dye. The swollen root of turnip-rooted parsley is eaten as a cooked vegetable. It is also used extensively in various cosmetics, decorations and medicines. Parsley is added to bath water to sooth and cleanse the skin. It is also used in shampoo, perfume, soap and skin lotions.

MEDICINAL USES

Parsley as an herb possesses medicinal value for curing a number of ailments. It is a diuretic and has ecbolic, carminative, emmenagogue and antipyretic properties. It is supposed to be an aphrodisiac. Parsley cleans the blood, maintains elasticity of blood vessels and speeds up oxygen metabolism. It treats stomach ailments, menstrual problems, arthritis and colic. It aids in digestion and in suppressing the odours of onion and wine because of its high clorofics content.

BOTANY

Parsley belongs to the dicotyledonous family. It is an herbaceous medium growing biennial plant. It grows to a height of 23–46 cm and spreading of 15–23 cm. It produces a rosette of divided leaves on a short stem in the first year and branched flowering stalk with inconspicuous yellow flowers and small dry fruits in the second year. The lower leaves are bi- or tri-ternately divided. The dark green upper leaves are divided pinnately into featherlike sections and can be flat or curled, depending on variety. Leaves are green and flowers are greenish yellow with five petals borne on a seed stalk in a head referred to as an umbel. Umbels are small and white, similar to carrot and celery. It has thin spindle-shaped roots, which produce erect, grooved, glabrous, angular stems. Seeds are smooth, ribbed and ovate.

TABLE 107.1

Nutritional Composition of Parsley (per 100 g Edible Portion)

Constituents	Contents	Constituents	Contents
Water (g)	79–89	Iron (mg)	6.2
Carbohydrates (g)	6.33	Vitamin A (IU)	1800
Protein (g)	3.7–5.2	Thiamine (mg)	0.09–0.2
Fat (g)	0.79	Riboflavin (mg)	0.18–0.6
Calcium (mg)	138	Vitamin C (mg)	133
Phosphorus (mg)	58	Energy (kcal)	36

CLIMATIC REQUIREMENT

Parsley is well adapted to a wide range of climatic regions from tropical to cold temperate areas. However, very low temperature causes severe damage to parsley under outdoor conditions. Temperature is considered the most important climatic factor for its production. It performs best in a warm temperate climate. It is reasonably tolerant to frost but its growth becomes significantly slow at low temperature. It can grow well in a temperature range of 7.2–15°C. However, for best growth and quality, it requires a temperature range of 15–18°C. Its seeds may germinate at a minimum of 10°C and maximum of 32°C, however, the normal germination takes place at a temperature range of 23–28°C.

SOIL REQUIREMENT

Parsley grows best on well-drained sandy loam to medium loam soils. A rich moist soil with good drainage facilities and pH 5.3–7.3 is amendable for its cultivation. Heavy, clayey soils are not suitable for its cultivation. The highest overall yield can be harvested under best soil physical conditions obtained by medium to deep ploughing. Soil compaction affects its yield adversely.

CULTIVATED VARIETIES

A number of parsley cultivars is available for commercial cultivation. Based on morphological traits, parsley varieties are classified into five groups, *i.e.*, (1) plain-leaved, (2) curl-leaved, (3) fern-leaved, (4) celery-leaved (Neapolitan) and (5) turnip-rooted. Among all of these, curl-leaved are most commonly used followed by plain-leaved and turnip-rooted. The plain-leaved parsley has good flavour as the curl-leaved parsley but it is not so attractive, hence, little grown.

CURL-LEAVED VARIETIES

Moss Curled (Double Curled): A variety with 30.5 cm tall plants produces vigorous, compact, very dark green, finely cut and deeply curled leaves. It is frost resistant and becomes ready for harvesting in about 70 days.

Evergreen: A variety selected from Moss Curled variety produces coarsely cut leaves.

Triple Curled: A strain selected from Moss Curled variety produces slightly shorter and closely curled leaves.

Extra Triple Curled: A strain selected from Moss Curled variety produces very finely cut and very closely curled leaves. The growing period is about 75 days.

Paramount: A variety with 25.5 cm tall plants produces uniform, triple curled and very dark green leaves. The growing period is about 85 days.

PLAIN-LEAVED VARIETIES

Plain (Single): A variety with spreading plant habit produces deeply cut flat leaves. The growing period is about 72 days.

Dark Green Italian: A variety produces heavy and glossy green leaves.

TURNIP-ROOTED VARIETIES

Hamburg: A turnip-rooted cultivar is grown only for its enlarged fleshy edible taproot. The foliage is plain-leaved. The roots are parsnip like as well as turnip-shaped.

Neapolitan or Celery-Leaved Parsley: It is cultivated for its leaf stalks, which are blanched and eaten like those of celery.

SEED RATE

Parsley seeds are very slow in germination, which takes 25–30 days under field conditions. For the transplanted crop, about 250 g of seeds of good germination power are required for raising enough seedlings for planting a hectare area, and for direct sown crop, 3.250 kg of seed is required.

SOWING TIME

The time of sowing alters with the variety type, soil and weather conditions of the region to be grown, purpose of growing and other parameters. For continuous supply of leaves, it is sown thrice, *i.e.*, (1) in early February, (2) in April or early May and (3) in July or early August. The crop is sown in July or early August under protected conditions for the winter supply. It can be sown in February for summer supply and for drying purposes but seeds take several weeks to germinate, often up to a month. The main sowing is generally done in April as the seed germinates quickly and it is available for cutting throughout the summer. A mid-August sowing will furnish good plants for placing in cold frames for winter use. For seed production, parsley is sown in late April or early May. However, in plains, parsley is grown for leaves during September to November. The sowing time for lower hills is October, mid hills August to September and high hills March to April.

SOWING METHOD

The cultivation of parsley includes both transplanting and direct sowing in the field. As the seeds are slow in germination, it is often sown in greenhouses, hotbeds or in specially prepared beds in an open field. If the seed is soaked in water for a day, it hastens germination. The main field is prepared well after two to three ploughings and bringing the soil to a fine tilth. Seed is sown 6 mm deep in rows at a row spacing of 40–45 cm and plant spacing of 10 cm. An even broadcast sowing is preferable, if the beds are properly levelled. For transplanting, when seedlings attain a height of 5–8 cm, they are planted in the field in late evening with a row spacing of 40–45 cm and plant spacing of 10–15 cm.

MANURIAL REQUIREMENT

The rate of fertilizer application depends on soil type and crops grown in previous seasons. On well-drained light-textured soils, higher fertilizer rates are frequently used. The manure requirement for average soils is 10–12 t/ha and NPK at 100:80:40 kg/ha. The half dose of nitrogen and full dose of phosphorus and potash should be applied at the time of land preparation. The remaining half of nitrogen should be top dressed in two splits at an interval of 30 and 60 days after germination.

IRRIGATION REQUIREMENT

Irrigation requirements of parsley is similar to those of other leafy vegetables. Overhead sprinklers and drip irrigation are successfully used to apply water to the growing crop. The best growth of parsley crop is obtained when the plants are grown under constant water flow from 3 weeks after sowing. Plant being more prone to hot and dry conditions, the field should be watered generously. Maintenance of uniform moisture in the field is also a concern in cultivation of this crop since water stress causes reduction of leaf growth and development, which restricts yield. Mulching around the plant is also advocated to conserve soil moisture.

INTERCULTURAL OPERATIONS

The field should be kept free from weeds by weeding and hoeing during initial stage of crop growth since parsley is a slow growing plant in the beginning. During the second weeding, thinning should be practiced in the direct sown crop and extra plants should be removed after keeping a desirable distance between plants. Thermal weed control is also followed in this crop.

HARVESTING

Parsley leaves are ready for harvesting in about 90–100 days, depending upon varieties. A few leaves at a time may be removed from each plant or the entire bunch of leaves may be cut for use. Harvesting can be done either manually or by machines, depending upon the crop quality. However, manual harvesting is preferred more but it is labour intensive. This is done to obtain the lowest amount of crop damage, which is acceptable for fresh marketing. In parsley, only a few leaves are picked from the plant and this method is followed to take marketable produce continuously for several weeks. Parsley must be cut at least 2.5–4 cm above the crown if multiple cuttings are to be taken. Although the parsley leaves are used most commonly in fresh condition, they can also be used after dehydration because their characteristic flavour and green colour can be retained even after dehydration.

Seed umbels are harvested four to five times, depending upon umbel maturity. The ripe fruits of parsley shatter easily, thus it is important that the crop should be cut at a stage to minimize loss, usually just before the primary umbels shatter. Harvesting in rainy season is avoided. The umbels are harvested when seed in the umbels is completely mature.

YIELD

The greens and seed yield of parsley crop depends upon cultivars, season, soil type, agro-climatic conditions of the growing region and cultural practices followed during the crop period. In commercial plantations, 2.5 thousand dozen bunches of leaf yield can be obtained at each cutting from a hectare area. Generally, two to five cuttings can be obtained from each planting. The average yield of parsley per plant is 425–625 g from four cuttings. On an average, it gives a leaf yield of 125 and 140 q/ha when spacing is maintained 30 cm between rows. The average seed yield of parsley is 5.6 q/ha though the genetic potential of this crop is 18 q/ha.

POST-HARVEST MANAGEMENT

Washing should be done thoroughly after harvesting and all the yellow and damaged leaves are trimmed off. The stalks are trimmed and tied into a small bunches by using rubber bands, and thereafter, they are packed in baskets. Crushed ice is placed in the packs to maintain crispness and fresh appearance and also to prevent heating and decaying during transit. A controlled atmosphere of 10% oxygen and 11% carbon dioxide helps in retaining green colour and salability. Parsley can be stored for 2 months at 0–3°C temperature and 95% relative humidity. Like carrot or parsnip, the roots of Hamburg parsley can be stored for several months with the top removed at around 0°C temperature.

INSECT PESTS

Green Peach Aphid (*Myzus persicae*) and Potato Aphid (*Macrosiphum euphorbiae*)

Aphid is the major insect causing damage to the parsley crop. Green peach aphids feed on the ventral side of mature leaves but form larger colonies on young leaves. Potato aphids are also similar to green peach aphids but they are larger and form small colonies on the lower surface of newly formed leaves. Their population reaches a maximum during the month of November and again during February and March. The major damage occurs during the maturing stage. Its severe infestation exhausts the plant and reduces the plant vigour or even kills the plant.

Control
- Use predators like ladybird beetle larvae, lacewing larvae, syriphid fly larva or parasites.
- Spray the crop with 4% *neem seed kernel extract* , 0.1% imidacloprid or Malathion 50 EC.

Carrot Weevil (*Listronotus oregonensis*)

Adults of carrot weevil are the most damaging pest of parsley. The pest overwinters in crop debris, which is left after harvesting. The pest becomes active in new parsley field from mid-May to the end of June but its activity is observed more from late May to mid-June. The adult female lays eggs in the stems. After hatching, its larvae make tunnels inside the petiole towards the root or drop to the soil then burrow into the root.

Control
- Spray the crop with 0.1% Malathion 50 EC, Thiodicarb, Methomyl or Indoxacarb.

Aster Leafhopper (*Macrosteles quadrilineatus*)

Feeding itself causes little observable damage but it transmits different viral diseases, the most notably Aster yellows. It is less significant as a vector in arid areas, where population levels may be depressed by adverse timing of seasonal rainfalls.

Control
- Remove weeds because they may harbour the pest.
- Place oat straw mulch or aluminium mulch around the plants.
- Promptly pick and destroy the diseased plants to prevent further spread.

Western Black Flea Beetle (*Phyllotreta pusilla*)

The adult flea beetles are 1 mm long, shiny, hard beetles with enlarged hind legs, allowing them to jump like fleas. Most of the damage is caused by its adults. Its larvae feed on the undersides of leaves, making irregular shaped holes or feed on roots but this activity is not of economic concern. Its heavy infestation can kill or stunt the seedlings. Older plants rarely suffer economic damage although their lower leaves may be damaged. The pest is most common in spring but can occur any time, especially in field, which is surrounded by weeds.

Control
- Plant the crop as late as possible.
- Remove weeds along field margins and deeply disk plant residue in infested fields after harvest.
- Remove old crop debris and other surface trash to deprive overwintering beetles.
- Spray insecticidal soaps at cotyledon stage for partial control.
- Spray the crop with Spinosad 0.05%, Bifenthrin 0.2% or Malathion 0.1%.

Potato Leafhopper (Empoasca fabae)

The adult potato leafhopper is a tiny, yellowish-green, wedge-shaped insect, about 3 mm long. Both adults and nymphs are extremely active and suck sap from underside of the leaf, petioles and sometimes young stems but the most serious damage is caused by the nymphs. During sucking, the hopper also injects a toxin that causes stunting of growth, leaf curl and whitening of veins. Leaf curling is the most obvious symptom and is usually accompanied by reduced internodal growth.

Severe hopperburn can stunt growth significantly, reducing yield substantially and weakening the plant for damage by other organisms.

Control
- Heavy rains may dislodge nymphs from plants temporarily reducing their impact.
- Use predators like ladybird beetles, green lacewings, parasitic wasps, damsel bugs, big-eyed bugs or assassin bugs.
- Grow resistant varieties in areas heavily affected by potato leafhopper.
- Soil application of neonicotinoids, acetamiprid, clothianidin, imidacloprid, thiacloprid or thiamethoxam provides superior long-term vine protection.

DISEASES

CERCOSPORA BLIGHT (*Cercospora apii*)

The disease first appears on older leaves, causing circular brownish spots with concentric rings, which may further appear on stem also. It is a seedborne disease.

Control
- Treat the seed with Thiram or Captan at 2.5 g/kg of seed.
- Provide hot water treatment to seed at 50°C for 30 minutes.
- Spray the crop with copper oxychloride 0.2% at an interval of 10–15 days.

SARTORIAL BLIGHT (*Sartorial Apia*)

The disease is caused by fungus and the symptoms appear as small circular water soaked areas, later turning black on tips and margin of leaves and may spread to other parts of the plant.

Control
- Follow long crop rotation with non-host crops.
- Use disease-free healthy seed.
- Avoid growing flat-leaf parsley varieties that are generally more susceptible to the disease.
- Grow curly-leaf type varieties that are resistant.
- Use drip or trickle irrigation rather than overhead sprinklers.
- Spray the crop with Dithane Z-78 0.2% at an interval of 10–15 days.

BACTERIAL LEAF SPOT (*Pseudomonas apii*)

The disease is caused by bacteria and the symptoms appear as circular reddish spots with pale yellow borders on leaves. It is a soilborne disease.

Control
- Follow long crop rotation with non-host crops.
- Spray the crop with Plantamycin 100 ppm regularly.

PARSLEY MOSAIC

The virus is transmitted by green peach aphids, which causes the parsley's leaves to develop a yellow, light or dark green mottled appearance. Necrotic areas can also develop. When the infestation is severe and occurs in early plant growth stage, it can decrease the plant vigour. If young plants are infected, outer leaves become yellow and the plants may be severely stunted.

Control

- Use only disease-free seed.
- Grow resistant cultivars.
- Keep the field weed free as weeds can serve as hosts for viral diseases.
- Spray the crop with 0.1% Malathion 50 EC to control the vectors.

■■■

108 Parsnip

M.K. Rana and V.P.S. Panghal

Botanical Name   : *Pastinaca sativa*

Family   : Umbelliferae (Apiaceae)

Chromosome Number : 2n = 22

ORIGIN AND DISTRIBUTION

Parsnip is native to Europe and Asia. It was used by the ancient Greeks and Romans for medicinal and food purposes. Parsnip was introduced into North America in the early 1600s and grown by the early colonist and Indians. Although parsnip tastes better after a bit of frost, it apparently comes from the Mediterranean area, from where the Romans brought it to rest of the world.

INTRODUCTION

Parsnip is a root vegetable closely related to carrot and parsley. It is left unpeeled and steamed or boiled to bring out its maximum flavour potential. It ranks as an important source of food and offers a wide complementary with other tasty food items. Compared to carrots, parsnip is a minor crop in most grown countries. It is a biennial crop but grown commercially as an annual crop. Its long tuberous root has cream-coloured skin and flesh and can be left in the ground when it matures as it becomes sweeter in flavour after winter frost. It is probably grown in gardens much more than being commercially present in supermarkets. It was a starchy vegetable of choice before the potato replaced it. In sunlight, handling stems and foliage can cause a skin irritation. It is cultivated mainly in temperate regions worldwide and occasionally in cooler parts of the tropics, including Eastern and Southern Africa. It is an underutilized crop, which is not recognized by the Food and Agriculture Organization (FAO). It is grown mostly in the United Kingdom, Ireland and the United States because of its British and Irish immigrants. It is still grown a bit in Germany, Italy, Portugal, Spain and Russia.

COMPOSITION AND USES

COMPOSITION

This creamy-white root vegetable is low in calories, fat and sodium, is naturally cholesterol-free and has minerals, particularly potassium and vitamins. Most of the B group vitamins are present in it but vitamin C, to a great extent, is lost during cooking. Since most of the minerals and vitamins are found close to the skin, many of them are lost unless the roots are finely peeled or cooked whole. It also contains antioxidants and both soluble and insoluble dietary fibre comprising cellulose, hemi-celluloses and lignin. Parsnip roots and leaves contain compounds, *i.e.*, furano and coumarin, which may cause inflammation of the skin. The nutritional composition of parsnip is given in Table 108.1.

TABLE 108.1
Nutritional Composition of Parsnip (per 100 g Edible Portion)

Constituents	Contents	Constituents	Contents
Water (g)	80	Potassium	375
Starch (g)	6.2	Iron (mg)	0.59
Carbohydrates (g)	18	Thiamine (mg)	0.09
Protein (g)	1.2	Riboflavin (mg)	0.05
Fat (g)	0.2	Vitamin C (mg)	17
Calcium (mg)	36	Vitamin K (µg)	22.5
Phosphorus (mg)	71	Energy (kcal)	75

USES

The fleshy, aromatic and slightly mucilaginous root is eaten as a cooked or fried vegetable. Its roots can also be fried or thinly sliced and made into crisps. It is also used in making soups and adding flavour to stews. The seeds, which taste similar to dill, are occasionally used as a condiment. It is used as the carrot roots are used but it has a sweeter taste, especially when cooked. It can also be used for preparing wine. A roast dinner is incomplete without a roast parsnip. It adds a whole new dimension to stews and casseroles too.

Parsnip is not only a valuable item of human food but equally preferred by fattening pigs. Horses also eat it readily when given with bran. It is also fed to milch cows in winter to give as much as good milk and yield butter as well flavoured as when feeding on grass. It is usually cooked but can also be eaten raw. The roots can be baked, boiled, pureed, roasted, fried or steamed.

MEDICINAL USES

The consumption of parsnip has potential health benefits as it contains antioxidants such as falcarinol, falcarindiol, panaxydiol and methyl-falcarindiol, which have anticancer, anti-inflammatory and antifungal properties. Its high fibre content may help in preventing constipation and reducing cholesterol levels in blood. A wrapping of roots is applied to sores and inflammations for treating skin diseases. The leaves have diuretic properties.

BOTANY

Parsnip produces a tough creamy white root, tapering somewhat from the crown and an erect stem arises from it. The stem is about 20 cm long and the leaves divide into several pairs of leaflets, each 2–5 cm long and up to 2 cm wide. Leaves are finely toothed at their margins and softly hairy, especially on underside. The flowers are yellow and borne in umbels at the ends of the stems, similar to carrot. The flowers of the cultivated parsnip are a deeper yellow colour than those of the wild types.

CLIMATIC REQUIREMENT

Parsnip requires cool (16–20°C) conditions for optimum root maturity and quality. The crop is scorched above 30°C temperature. Its roots are able to tolerate frost. However, cold weather may result in root loss due to development of seed stalk (bolting). Rising temperature causes the root quality to decline rapidly and temperature above 30°C is likely to result in root damage. The optimum soil temperature for parsnip seeds to germinate is between 10°C and 21°C. In general, germination is slow, even under ideal field conditions, and where suboptimal conditions prevail, emergence may require more than 4 weeks. After germination, the soil surface should not be allowed to dry out otherwise the seedlings may die off.

SOIL REQUIREMENT

The effective rooting depth of parsnip is about 35–50 cm, thus grows well in deep sandy or sandy loam soils. Acid soils should be limed before planting. Like other root crops, heavy and stony soils are not suitable for parsnip. Deep, loose and fertile soils with good water-holding capacity and pH 6 or above are suitable for the development of long and straight roots. On mineral soils, a major problem in parsnip production is stand establishment so careful attention is given while preparing the seedbeds. Soil obstructions make the seedling emergence difficult thereby leading to a more variable plant stand and may also result in misshapen roots. For making the soil fine and free from clods and debris, two to three deep ploughings followed by planking are required.

CULTIVATED VARIETIES

ALBION

It develops evenly tapered wedge-shaped roots of about 33 cm length, which are about 6 cm wide at the top. Foliage height goes up to 75 cm. The skin is smooth, white and slow to discolour. It is ideal for organic farming. It is resistant to parsnip canker and other diseases.

ARCHER AGM

It is very high yielding with smooth-skinned roots, which are wedge-shaped with average length of 33 cm and width of 7 cm, good flavoured and resistant to canker. Foliage height is about 80 cm.

GLADIATOR AGM

It is a high-yielding variety with wedge-shaped good flavour roots and suitable for growing in heavy soils. Foliage height is 70 cm and root length is 28 cm and width is 7 cm.

PALACE AGM

It is a high yielding canker resistant variety with wedge-shaped quality roots having shallow crown. Foliage height is 70 cm and root length is 26 cm and width is 6 cm.

The other important varieties of parsnip are Tender and True AGM, Hollow Crown, Guernsey, Offenham, Andover, Harris Model, Gladiator and Javelin.

SOWING TIME

At low altitude, it is sown in the early spring after the threat of frost is over, and at high altitude, it is sown in the field from April until mid-May. In plain areas, its seeds are sown October onward.

SEED RATE

Parsnip seeds usually retain viability only for 2 years. Therefore, it is advisable to purchase new disease-free seed each year from a reliable seed distributor. A total of 3.3–4.6 kg of seed is enough to sow a hectare area.

SOWING METHOD

Cuttings from strong roots are planted in the late autumn. These will overwinter well and produce new parsnip in the next growing season. At optimum temperature, germination takes about 14 days, but in cold conditions, it can take up to 4 weeks. Establishment is better in cooler months. The seed is more difficult to germinate than that of carrot seed and requires regular moisture for establishment. Parsnip may be seeded with a nurse crop such as cereal rye, which germinates quicker than parsnip and protects the seedlings from cold winds. The seeds of cereal rye are sown by broadcasting in the field when the parsnip has attained three true leaves. The spacing between the rows is adjusted according to the machine used for harvesting the crop. The seeds are sown at a depth of 5–20 mm in rows 30–40 cm apart. If the seeds are sown too deep, emergence is reduced and if they are sown too shallow, the soil moisture necessary for germination may be lacking. In areas where high velocity winds are a problem, attempts should be made to protect the sown field from soil dispersal. Blowing soil may result in unevenly covered seeds, which may lead to patchy uneven stands.

NUTRITIONAL REQUIREMENT

The nutritional requirement of parsnip is similar to those of beets and carrots. Adequate supply of nitrogen, phosphorus and potash is important for optimal parsnip development. On mineral soils, nitrogen is recommended 110 kg/ha, out of which, 70 kg/ha is supplied as pre-plant application and 40 kg/ha as a side dressing. However, the quantity of nitrogen applied to muck soils is reduced to 60 kg/ha. Nitrogen fertilization should be monitored carefully since its excessive use may result in luxurious foliage growth, which increases susceptibility to disease causing pathogens. If manure is applied or a soil-improving crop such as a legume precedes the planting of parsnip, the nitrogen dose should be reduced. If manure is applied, it should be well decomposed before the parsnip is sown, as the incorporation of fresh or undecomposed manure may cause root deformity. The farmyard manure may be applied at 10 t/ha, which will supply nutrients and help in conserving soil moisture. The doses of phosphorus and potash depend on soil analysis test.

The crop could be susceptible to boron deficiency in new muck soils. Severe boron deficiency may cause a breakdown in leaf and root tissue so borax may be applied at 15–20 kg/ha.

IRRIGATION REQUIREMENT

Irrigation in general depends on soil type, organic matter content in soil and growing season. Parsnip, because it is a shallow rooted crop, requires higher moisture than other vegetables. In summer, sufficient moisture may not be supplied with overhead rotary sprinklers. In warmer months, water is applied twice in morning and evening, and in cooler months, it is applied only once in either the morning or evening. The soil is kept moist in order to avoid splitting of roots. Irrigation of 25–40 mm is provided at an interval of 7–10 days or as and when necessary. Soluble fertilizers may be supplied through fertigation. The soil should not be allowed to dry completely but watering should be reduced after the development of first leaf to encourage the roots to grow deeper.

INTERCULTURAL OPERATIONS

HOEING AND WEEDING

At early stages of seedling growth, the crop is quite vulnerable to being crowded out by invading weeds, which makes the weeding an essential operation during the first month after emergence. If mechanical methods are employed for controlling weeds, the hoeing should be done shallow, as deep hoeing may damage the feeding roots developing near the soil surface.

THINNING

The density of plant per unit is an important factor since this will determine size of roots at harvest. Some of the growers use large quantity of seed to ensure optimum plant population per unit area, which may sometimes cause overcrowding of plants. To reduce competition among the overcrowded plants, thinning operation is essentially practiced. As and when the plants become large enough to handle easefully, they are spaced to a distance of 15 cm by removing the extra seedlings.

CROP ROTATION

The best possible rotation is to include parsnip every 4 years to avoid diseases such as *Rhizoctonia* and canker. If the field is disease-free, one crop of parsnip may be taken on same piece of land each year. In general, it should be avoided if possible as it increases the possibility of diseases. Too many crops of carrots, celery and parsley should not be included in the rotation as these belong to the parsnip family. In rotation, leafy vegetable crops should be included to avoid the problem of diseases.

MULCHING

Soil mulches are often used to moderate soil temperature, control weeds and erosion, conserve moisture and to protect the plants from insect pests and diseases. Sawdust, paddy straw, dry grasses, sugarcane tresses, aluminium foil or black polyethylene films can be used as soil mulch. In addition to affecting soil temperature, opaque mulches control weed growth. In summer months, black polyethylene films become hot, causing burning of plants touching them during the daytime. Aluminium foils are also used as mulch that disrupts aphid flights and decreases the chances of aphid-transmitted diseases.

HARVESTING

Parsnip although is a long-season crop typically needing 110–130 days to reach maturity, but in certain areas, roots are lifted and sold after 95 days of sowing. It is harvested in a similar way as the carrot roots. Removing tops prior to harvesting in fields where excess foliage growth has occurred facilitates harvesting. The roots are lifted by hand manually or by carrot harvester. Its roots and foliage possess compounds that induce skin inflammation. Therefore, long sleeves hand gloves are used while handling parsnip. Like other root crops, some of the varieties roots are lifted in late August and sold fresh. However, for winter storage, the roots are harvested in October and November. Sometimes, the roots of varieties very resistant to freezing damage are left in the field over the winter and then harvested in March to April. However, overwintered parsnip roots will not tolerate soil thawing and freezing, which may occur in April. The roots subjected to such conditions quickly lose quality and become cracked and pithy. Overwintered roots should be dug up as early in the spring as possible before sprouting, as it leads to a loss in root quality. Parsnip is a delicious root crop that tastes best when harvested in early winter, after the soil has turned cold.

YIELD

The yield varies with soil types, varieties used, growing season and growing purpose, stage of harvesting, weather conditions, irrigation facilities and incidence of insect pests and diseases. In general, on an average, it yields about 200–300 q/ha.

POST-HARVEST MANAGEMENT

The roots are topped and washed before being graded and then packed into bags or cartons 10 kg capacity, which have a plastic sheet on the bottom and sides. Roots should not be allowed to dry out fully, as this may cause root discolouration. The roots are graded according to length and quality. Leaves are removed and only straight, sound, white roots are sent to the market.

Healthy roots may be stored in cold store for 4–6 months at 0°C temperature and 90–95% relative humidity but not possible to store the roots with ethylene producing crops since ethylene may impart a bitter flavour to the root tissue. The sweet flavour develops when the roots are exposed to near freezing conditions since the starch in the root is converted to sugars at this temperature. Mature roots, which have not been exposed to low field temperature, may be stored only for 2 weeks at 0–1°C temperature to develop flavour. Extremely low temperature in storage may damage the root tissues.

PHYSIOLOGICAL DISORDERS

Parsnip freezes at -l.7°C. The centre of the core of frozen parsnip may become water soaked with reddish-brown areas developing in the cambium layer. The roots may become dark brown after freezing and subsequent thawing and exposing to room temperature for a short period. Surface browning is a significant problem, which is largely associated with bruising and abrasion injury during harvesting. The roots grown in sandy soils are more susceptible to browning, however, the cultivars are available that are resistant to browning. Surface browning also increases with length of storage. Waxing done to reduce moisture loss increases browning of roots.

INSECT PESTS

APHIDS (*Cavariella aegopodii*)

Small soft-bodied insects on underside of leaves are green or yellow in colour. If the infestation is severe, necrotic spots will develop on leaves as a result the plant becomes stunted and distorted. Its nymphs and adults secrete a sticky sugary substance called honeydew, which encourages the growth of sooty mold on the plants, which hampers the photosynthetic activity of plant.

Control
- Remove the infested leaves or shoots manually.
- Use tolerant varieties if available.
- Use reflective mulch such as silver coloured plastic to deter the aphids from feeding on plants.
- The plants can be sprayed with a strong jet of water to knock aphids from leaves.
- Spray the crop with insecticidal soaps or oils such as *neem* or canola oil.

CUTWORMS (*Agrotis* spp., *Peridroma saucia* and *Nephelodes minians*)

This is a polyphagous pest and has a wide range of hosts. Its larvae are 2.5–5.0 cm long, exhibiting in a variety of patterns and colouration but it usually curls up into a C-shape when disturbed. The seedlings are cut off at soil surface. If the infection occurs at later stage, the larvae eat the fruits surface and make irregular holes. Larvae usually become active at night and hide during the day in the soil at the base of the plants or in plant debris of toppled plant.

Control
- Remove all the residues of previously grown crop from soil after harvest.
- Pick the larvae manually from the field.
- Apply Monocrotophose at 0.2%.

BEET ARMYWORM (*Spodoptera exigua*)

The young larvae are pale green to yellow in colour, while the older larvae are generally dark green in colour having a dark and light lines running along the side of their body and a pink or yellow on underside. Insect can go through three to five generations in a year. Young larvae feed on leaves and make irregular shaped holes in foliage. Heavy feeding leads to skeletonize the leaves and shallow dry wounds on fruits. The adult lays eggs in cluster of 50–150 on the leaves covered in a whitish scale, which gives the cluster a cottony appearance.

Control

- Introduce natural enemies like *Bacillus thuringiensis*, which parasitize the larvae.
- Spray the crop with Cypermatherin 25 EC at 0.5%.

CARROT RUST FLY (*Psila rosae*)

The adult insect is a small dark coloured fly and its larvae are white maggots about 1 cm long. Scarring of taproots caused by tunnels filled with a rust colour mush is the common symptom on roots.

Control

- Cover the crop plants with agro-net.
- Harvest parsnips in blocks.
- Do not leave parsnips in field over winter.
- Apply Diazinon 20G at 11 kg/ha.
- Spray the crop with UP-Cyde 2.5 EC at 280 mL/ha dissolved in 550 L of water.

CARROT WEEVIL

Adult carrot weevils are dark brown with three light stripes on the thorax. They are about 6 mm long. The adults overwinter in grasses on borders of the field. Its female lays eggs when the plants are at second true leaf stage. The larvae are white legless grubs. They feed on the taproot. Usually, the damage is restricted to upper third portion of the root. Only a single generation is known to occur per year.

Control

- Follow a long crop rotation with cereal crops.
- Avoid planting in fields adjacent to last year's carrot crop.
- Apply Imidan 70 WP at 1.6 kg/ha.

DISEASES

PARSNIP CANKER (*Itersonilia perplexans*)

Small brown necrotic lesions with pale green halos appear on leaves. In later stages, these lesions may coalesce to form large necrotic lesions and becomes grey to black. Red-brown cankers may develop on root shoulders with rough texture. Disease appears late in growing season and is favoured by cool and wet weather.

Control

- Follow long crop rotation.
- Sow resistant cultivars such as *Avon Resister* and *Archer*.
- Avoid sowing seeds too early in the year.
- Provide proper drainage.
- Avoid damage to the roots.
- Remove weeds and crop residues after harvest.
- Protect the roots from carrot fly.

POWDERY MILDEW (*Erysiphe heraclei*)

Powdery growth appears on leaves, petioles, flowers stalks and bracts. Leaves become chlorotic and severe infection causes the flowers to become distorted. The spores of fungus can spread to long distances through air. Disease becomes more prominent under high humidity, moderate temperature and in shady areas.

Control

- Grow tolerant varieties.
- Avoid excess fertilization, especially nitrogenous fertilizers.
- Spray the crop with Karathane, Calxin, Cosan or Sulfex at 0.2%.

DOWNY MILDEW (*Peronospora umbellifarum*)

The disease appears as white fluffy growth on underside of the leaves and yellow spots on upper surface of leaves. Later, these spots become darker in colour. Young tender leaves are affected first and disease spread is favoured by prolonged leaf wetness.

Control

- Follow crop rotation with non-umbelliferous crops.
- Sow disease-free seeds.
- Avoid overcrowding of plants.
- Spray the crop with Chlorothalonil, Cymoxanil, Mancozeb, Maneb or Metalaxyl at 0.2%.

CAVITY SPOT (*Pythium* spp.)

Sunken, oval, grey lesions appear across the root. Outer layer of root ruptures and develops dark, elongated lesions. Small vertical cracks may form the cavities on roots. Fungi can persist in soil for several years and disease outbreaks are associated with wet soils.

Control

- Grow crop in disease-free field.
- Follow long crop rotation.
- Avoid excessive manure and fertilizers in the field.
- Treat the seed with Metalaxyl or Thiram at 2.5 g/kg of seed.

DAMPING-OFF (*Pythium* spp. and *Rhizoctonia solani*)

Rapid death of seedling prior to emergence, slow seed germination and collapse of seedlings after emergence from the soil are the prominent features of damping-off disease. Water-soaked reddish lesions girdling the stem at the soil line appear commonly. Fungi can spread through water and contaminated soil.

Control

- Avoid planting parsnips in poorly drained wet soils.
- Plant crop on raised beds.
- Sow disease-free high-quality seed that germinates quickly.
- Treat the seeds with Captan or Bavistin at 2–3 g/kg of seed prior to sowing.

■■■

109 Zedoary

M.K. Rana and Neha Yadav

Botanical Name   : *Curcuma zedoaria*

Family   : Zingiberaceae

Chromosome Number : 2n = 64

ORIGIN AND DISTRIBUTION

Zedoary, a perennial herb, is native to Southeast Asia, although the exact distribution of this species is not known as it has been dispersed along with human migrations throughout its history. It has been in cultivation since prehistoric times and spread and become naturalized throughout India, Southeast Asia, Southern China, Sri Lanka, Indonesia and the Philippines. The plant is found naturally grown in wet forests of tropical and subtropical regions. Arabs introduced the plant into Europe around the sixth century but its use as a spice in the West today is extremely rare, having been replaced by ginger. It is also found growing in subtropical regions of Eastern Nepal. Zedoary is though cultivated throughout India, it is mainly found in regions of Eastern Himalayan and Karnataka. *Curcuma zedoaria* was already defined, however, Gottlieb Friedrich Christmann published this name authentically and William Roscoe reclassified it into today's legal botanical systematics in 1807.

INTRODUCTION

Curcuma genus contains approximately 92–96 species. All plants of Zingiberaceae family share a common trait that their flowers produce just one true stamen and have food storing rhizomes with a *gingery* or *lemony* scent. Its root has a fragrance reminiscent of mango, which is why, it is called *amb halad* (amb means mango) but its flavour is more akin to ginger, except for its bitter taste. It is also known as white turmeric. The other common names are zedoary in English, *kachur* in Hindi, *karchur* in Sanskrit, *shatkachuro* in Gujarati and *Meitei Yaingang* in Manipuri. Forests are its natural habitat, where it flowers spontaneously but seldom flowers when it is cultivated in gardens. It has been cultivated for its rhizomes as a good source of food, spice and medicine for more than 4000 years. The plant bears rhizomes having a thin brown skin and a bright orange hard interior. Its roots are pungent in taste, warming and slightly aromatic.

COMPOSITION AND USES

COMPOSITION

Zedoary is a good source of essential oil, starch and fibre. The chemical composition of zedoary is given in Table 109.1.

TABLE 109.1

The Nutritional Composition of Zedoary (per 100 g Edible Portion)

Constituents	Contents	Constituents	Contents
Moisture (g)	6	Cineol (%)	9.6
Starch (%)	12–13	A-camphor (%)	4.2
Fibre (%)	18–19	D-borneol (%)	1.5
Ash (%)	1–1.01	Sesquiterpenes (%)	10
D-alpha-pinene (%)	1.5	Sesquiterpene alcohols (%)	48
D-camphene (%)	3.5		

Uses

Zedoary rhizomes are used both in raw and cooked forms. Sometimes, its leaves are used for culinary purposes, especially for cooking fish and the tender young buds as a salad ingredient. Like turmeric, the swollen rhizomes are edible but somewhat more bitter in taste than turmeric, thus less commonly used as a spice. Its rhizomes are extensively used for the extraction of starch, which is similar to that of arrowroot and is utilized in cottage industries in India for the preparation of baby food and milk puddings, which are easily digested by children and cooling and soothing for those recovering from illness. In Thai cuisine, it is used in several dishes with other herbs and vegetables and raw in salads. In Indonesia, its roots are ground to a fine powder, which is added to curry pastes, whereas in India, it is used fresh or for making pickles. The rhizomes of wild plants are eaten after washing during scarcity of food. In India, its roots are grated to add flavour to pickles. In Java, the leaves are used for flavouring several food items. In the ninth to thirteenth centuries, the sliced and dried rhizomes were shipped from India to European countries for the extraction of starch and oil.

The rhizomes are also used in preparation of liquors. Its light yellow essential oil isolated from its dried rhizomes through steam distillation is used in perfumery and soap making.

Medicinal Uses

The zedoary rhizomes are widely used in medicines mainly because of its bitter principles, *i.e.*, *curcuminoid* [1,7-bis(4-hydroxyphenyl)-1,4,6-heptatrien-3-1] and sesquiterpenes (*procurcumenol* and *epipro curcumenol*) to cure various diseases. The most efficacious one is *Rimpangnya*. It is used in Chinese traditional medicines to treat various cancers and as a constituent of several *Ayurvedic* preparations like *Dasamula Rishtam*, *Valiya Rasnadi Kashayam* etc. largely used to cure stomach pain and ulcers. It helps in increasing appetite, toning up of uterus, inhibiting the growth of cancerous cells (anti-neoplastic) and curing many ailments like arthritis, asthma, cough, dysentery, dyspepsia, dysuria, erectile dysfunction, fever, gonorrhoea, hiccups, piles, skin-related disorders, tastelessness, vomiting, weakening of heart muscles and worm infestation. Its roots paste is used in application on any inflammation, wounds and skin diseases. Its root powder relieves pain and swelling associated with sprains. It regularizes menses and normalizes the digestive tract and respiratory disorders. Its juice is used in anti-periodic pills, urinary tract infection and urine-related disorders. It improves blood circulation and acts as anti-contusion and anti-diarrhoea. It helps in abdominal cramps, amenorrhoea-abdominal pain and rheumatic pain. It acts as an aphrodisiac and restorative agent, stimulant and is useful for flatulence and colic. It is also used as anti-venom against the Indian Cobra venom.

Zedoary has no toxic effect when consumed in normal dosage in normal human beings but one of its negative impacts is that it reduces the ability of white blood cells and it is not good for pregnant and lactating mothers.

BOTANY

Zedoary, a deciduous robust herbaceous perennial plant with thick fleshy branched rhizomes, which are light yellow from outside and bright yellow from inside, 30 cm long and 4 cm in diameter, grows up to a height of 120 cm. The plant has both vertical aerial Pseudostems composed of long succulent interlocked leaf petioles giving rise clasping leaves and horizontal underground stem known as ovoid tuber, which bears several short thick horizontal rhizomes and several tuberous roots. Its leafy or flowering shoots arise from the ends of rhizome branches. Each shoot has about five leaves in two rows on opposite sides of the shoot. The leaves are alternate and can be solid green or variegated with a red central blotch. They are lanceolate and petiolate with entire margins and parallel venation. The ovate leaves with purple spots are 30–60 cm long and 18 cm wide, narrowing at the base. Flowering may occur early in the growing season, just before the leaves unfurl or along with them late in the growing season, depending on the species. The pale-yellow flowers are borne on 15 cm tall spikes in clusters of four to five in the axils of bracts, which are green at lower end of the spike, tipped with purple in the middle region and entirely purple at the uppermost end. The bracts near top of the spike are colourful and showy but do not have florets, which are held lower down on the spike amongst less-showy bracts. Its flower has six stamens although five of these are sterile. These five sterile stamens are fused to form a liplike coloured structure, which resembles a petal. Its ovary is a tri-locular and triangular to ovate shaped capsule, which splits at maturity to release seeds. Each capsule contains many oval- or spear-shaped seeds, which are surrounded by a fleshy covering.

CLIMATIC REQUIREMENT

Zedoary can be grown in diverse tropical conditions from sea level to 1500 m elevation at a temperature range of 20–35°C, but most of the species being winter hardy can tolerate temperature as low as 1°C. Because it is a warm requiring plant, it goes through a dormant period and dies back to the rhizome in winters, which emerges slowly in spring. It requires hot and humid climate for its normal growth and development with an annual rainfall of about 90–125 cm. It mainly prefers shady conditions for its successful cultivation and partially shady conditions on moist soils. The plant needs heat to trigger flowering, thus it performs well in warm climates. In mild climates, it may grow well but never bloom.

SOIL REQUIREMENT

Zedoary can be grown on different types of soil from light black, loam and red soils to clay loams but it gives optimum yield in a well-drained sandy soil rich in humus with loamy subsurface at a depth of 25 cm. It can also be grown successfully on poorly drained soils where other tuber crops fail to develop tubers.

CULTIVATED VARIETIES

Because of minor crop, no breeding work has been done so far to develop improved varieties. Therefore, only the wild forms are used for cultivation on a small scale.

SOWING TIME

The sowing time depends on the growing season, growing region and availability of irrigation facilities. Sowing is usually done in April to May on the onset of pre-monsoon showers and June to August for rainy season crop.

SEED RATE

The seed rate in general depends on variety, seed viability, growing season, soil conditions and irrigation facilities. On an average, about 11 quintals rhizomes are required to plant a hectare land area.

PLANTING METHOD

Zedoary, being a vegetatively propagated crop, is planted by rhizomes. Well-developed disease-free healthy mother rhizomes either whole or split weighing 35–44 g are used for planting purpose. The rhizomes are first planted in nursery in the month of February to March, and in the month of June to July, they are planted in field keeping a spacing of 45 cm between rows and 22 cm between plants within a row. Planting can be done in many ways but planting on flat beds, which are subsequently earthed up into ridges after the second application of fertilizer, gives better results than planting in flat beds or on ridges. After planting, the field is usually covered with thick mulch and kept weed free. Since the plant is shade loving, *Sesbania grandifolia* or castor should be raised along the field borders.

NUTRITIONAL REQUIREMENT

Zedoary responds well to applied manure and fertilizers for giving good yield. Farmyard manure is applied 25 t/ha in the field at the time of land preparation. Additionally, nitrogen is applied 100 kg and phosphorus and potash each 50 kg/ha. Full dose of phosphorus and potash along with one third of nitrogen is applied at the time of planting and remaining two thirds of the nitrogen is applied in two equal halves at the time of first and second earthing up.

IRRIGATION REQUIREMENT

Zedoary needs enough and a steady supply of moisture from planting to harvesting. Germination, rhizome initiation (90 DAP) and rhizome development (135 DAP) are the critical stages for irrigation. The first irrigation should be applied immediately after planting and subsequent irrigations are given at an interval of 7–10 days. Sprinkler and drip irrigation system can also be employed for better water use efficiency and obtaining higher yield. Moisture stress at rhizome bulking stage affects the growth and development of zedoary adversely.

INTERCULTURAL OPERATIONS

HOEING AND WEEDING

Light hoeing and weeding are necessary at early stages of crop growth to keep the weeds under check since weeds compete with crop plants for nutrients, moisture, space and light. Weeding may be done thrice at 60, 120 and 150 days after planting, depending upon weed intensity. Earthing up is done twice at 45 and 90 days after planting to prevent exposure of rhizomes to light.

HARVESTING

Zedoary normally takes about 10 months to be ready for harvesting. Yellowing and withering of leaves are the indication of maturity. The crop is usually dug out manually. The finger rhizomes, which are used for replanting, should be separated carefully from the mother rhizomes.

YIELD

The zedoary yield varies with varieties, soil type and fertility, irrigation facilities, growing region and the management practices adopted by the farmers. On an average, it gives a root yield of 75–120 q/ha.

POST-HARVEST MANAGEMENT

Zedoary rhizomes are usually used for the extraction of starch. The rhizomes are shredded into a pulp and steeped for 24 hours in 10 times their volume of water with frequent stirring. The starch slurry is filtered off, repeatedly washed with pure water, centrifuged and then dried at 50°C. The starch recovery is about 83% and a starch of approximately 94% purity may be obtained by treating with dilute sulfuric acid or alkali during washing process. Generally, the production of zedoary for industrial use as source of starch is not considered economically viable because of its low-yield production.

INSECT PESTS

Spider Mites (*Tetranychus urticae*)

Mites mostly live in colonies on under surface of the leaves and cause damage by sucking cell contents. Presence of light dots on leaves is the characteristic sign of mites attack. If feeding continues, the leaves turn yellowish or reddish and drop off. When mite infestation is severe, the top plant canopy is covered by webbing of mites. The infestation can also be identified by characteristic mottling on upper leaf surface. Mite infestation is mostly observed during warm and dry period of the season.

Control

- Follow long crop rotation with non-host crops.
- Apply water with high speed at regular interval to remove mites.
- Follow pruning of leaves, stems and other infested plant parts.
- Spray the crop with insecticidal soap or *neem* oil 0.5%.
- Use natural enemies like ladybird beetle, minute pirate bugs, big-eyed bugs, lacewing larvae and *Frankliniella occidentalis*.
- Spray the crop with wettable sulfur 80 WP 0.2% or Dicofol 0.5%.

Slugs and Snails (*Lehmannia valentiana*)

Slugs and snails are similar in structure and biology except that the slugs lack the snail's external spiral shell. They are active at night and on cloudy or foggy days. During daytime, they hide in weeds or plant debris. They feed on a variety of living plants as well as on decaying plant material. They attack the young tender greens and feed voraciously on seedlings as soon as they emerge in the field. Presence of silvery mucous trails on foliage confirms their damage.

Control

- Follow long crop rotation with non-host crops.
- Grow resistant varieties, if available.
- Use beer-baited trap to attract slugs and snails.
- Use barriers like copper flashing and screen to keep them away from the crop.
- Collect and destroy the snails and slugs from inhabitable areas.
- Use drip or sprinkler irrigation to keep humidity and soil moisture low.
- Keep the surrounding area clean by removing weeds etc.
- Use trap crops such as nasturtiums, begonias, *California poppy*, impatiens, lantana, fuchsias or geraniums.
- Spray the crop with *neem* oil 0.4%, Dimethoate 0.03%, Fenitrothion or Quinalphos 0.05% at 12 days interval.

MEALY BUG (*Pseudococcus* spp.)

Mealy bugs are sucking pests found in warmer areas. These soft-bodied wingless insects appear as white cottony masses on leaves, stems and fruits. They suck cell sap from tender parts of the plant by inserting stylet. Damage is not significant at low pest level but they cause yellowing and curling of leaves and make the plant weak when they are more in number. They secrete honeydew like sticky substance, which encourages the growth of sooty mould, which reduces the photosynthetic activity of the plant.

Control

- Avoid over watering and fertilization.
- Spray water with strong jet pressure to wash the mealy bugs from the plant.
- Use beneficial insects such as ladybugs, lacewing and *Cryptolaemus montrouzieri*.
- Prune out light infestations with a Q-tip dipped in rubbing alcohol.
- Spray insecticidal soap containing potassium salts of fatty acid, which penetrates and damages the outer shells of insect within an hour.
- Spray the crop with *neem* oil 0.4% or an aqueous solution containing 50% v/v isopropyl alcohol and 1% w/v sodium dodecyl sulfate.

ROOT-KNOT NEMATODE (*Meloidogyne* spp.)

Root-knot nematode, a microscopic parasite, invades the plant roots in soil. Its attack is more severe in sandy soils. It enters the roots at seedling stage and forms galls on them as swiftly as a month prior to transplanting. Its infestation affects the absorption of water and uptake of nutrients by the roots and reduces the plant vigour. Finally, the plants become yellow and wilt in hot weather.

Control

- Follow long crop rotation with non-host crops.
- Plough the field two to three times deep in hot summer months.
- Grow resistant varieties, if available.
- Drench the field soil with 0.1% formaldehyde up to a depth of 15 cm.
- Apply Nemagon 30 L/ha with irrigation before sowing.

DISEASES

SOFT ROT (*Pythium aphanidermatum*)

The disease is seed as well as soilborne but the symptoms can be seen in July. Yellowing first appears on lower leaves and then proceeds to upper leaves. Roots arisen from affected rhizome start rotting. The rotten parts attract other fungi, bacteria and insects particularly rhizome fly. During rainy season, the disease spreads rapidly from infected to healthy rhizomes.

Control

- Rotate the crop with non-host crops.
- Improve soil drainage and avoid water logging.
- Destroy the infected crop debris.
- Treat the rhizomes with Mancozeb 0.3% for 30 minutes prior to storage and at the time of sowing too.
- Drench the beds with 0.3% Mancozeb, 1% Bordeaux mixture or 0.25% Copper oxychloride.

Leaf Spot (*Phyllosticta zingiberi*)

Small spindle to oval-shaped spots appear on younger leaves. The spots have white papery centres and dark brown margins surrounded by yellow halo. The spots later increase in size and coalesce to form larger lesions, which eventually decrease the photosynthetic area. When the infestation is severe, the entire leaves dry up.

Control

- Grow the crop in partial shade.
- Spray the crop with 0.3% Zineb, 1% Bordeaux mixture or 0.25% Copper oxychloride – three to four times at an interval of 15 days with the initiation of disease.

Bacterial Wilt (*Ralstonia solanacearum*)

Bacterial wilt, which appears from July to August, is the most serious disease of zedoary. The leaf margins of affected plant turn bronze and curl backward. The whole plant wilts and dies. The base of the infected Pseudostems and rhizome emits foul smell. When the Pseudostems are cut and immersed into clean water, milky exudates will ooze out from the cut end.

Control

- Use disease-free healthy rhizomes.
- Dip the planting material in 1000 ppm solution of *Streptomycin* before planting.
- Remove the affected plants along with rhizomes and drench the soil with 1000 ppm solution of *Streptomycin*.

■■■

110 Drumstick

B. Varalakshmi and Deven Verma

Botanical Name : *Moringa oleifera* Lam.

Family : Moringaceae

Chromosome Number : 2n = 28

ORIGIN AND DISTRIBUTION

Drumstick is native to the southern foothills of the Himalayas in northwestern India, as it is found growing wild in the sub-Himalayan tract from river *Chinab* eastwards to *Sarda* and in *Tarai* tract of Uttar Pradesh in India. Its wild forms can be seen in the *Tarai* belts of Uttar Pradesh and Bihar. It has long been cultivated in rural areas throughout the Indian subcontinent as well as much of Southeast Asia.

The maximum diversity of drumstick is found in the *Horn of Africa*. With nine species, it is the centre of diversity, and eight of the drumstick species found here are endemic (not found elsewhere in the world). It is found in northern Somalia, Arabia and Red Sea coasts north to Dead Sea. The highest concentration of *Moringa* species can be seen in Mandela district in the extreme of northeast Kenya. *Moringa stenopetala*, a fast growing round shrub like tree with lush green foliage and sparse flowering, is a native to Africa. Although widespread in Africa, drumstick is seldom cultivated intensively in Africa. There are about 13 species of *Moringa* within the monogeneric family Moringaceace. *Moringa oleifera* (the drumstick tree) is the only species, which has been studied extensively and is widely distributed in Egypt, Sri Lanka, Thailand, Malaysia, Burma, Pakistan, Singapore, Cuba, Jamaica, Nigeria, Kenya and the Philippines. In addition, it has been introduced to the West Indies and tropical America from Mexico to Peru, Paraguay and Brazil.

INTRODUCTION

Drumstick, also known as miracle tree, Ben oil tree, West Indian Ben, *sahjan* etc., is an important perennial multipurpose vegetable grown widely in India. Among the ethnic population, it is popularly known as *Ganigana, Mullakkai, Murrugi, Sahjan* and *Muringa*. It is grown for its tender pods and leaves and even flower buds, which are used for culinary purposes. It is a delicacy in the south Indian households and most popular for its distinct, appealing flavouring fruits. Pods are rich in vitamins and minerals like iron, calcium, magnesium, potassium, sulfur, chlorine etc. The crop is grown in homesteads for family use or cultivated commercially for market. Tender leaves and flowers are comparable to that of *Colocasia* in vitamins and minerals and have great role for combating malnutrition of urban and rural masses. In India, it is very popular in South Indian states, especially Karnataka, Tamil Nadu and Kerala. In Malaysia, its seeds are used as nut-rich proteins, vitamins and minerals. Basically, it is a perennial crop, but recently, few short duration varieties, which are referred to as annual types, have been developed.

Drumstick is known by several names such as horseradish tree arising from the taste of a condiment prepared from true horseradish roots (*Armoracia rusticana*), drumstick tree arising from the shape of the pods, Benz olive tree in Haiti and Mothers best friend and Malunggay in Philippines. It is one of the most nutritious crops of the world and rated as a high nutrients supplement vegetable. Immature green pods are probably the most valued containing all the essential amino acid and widely used part of the tree.

COMPOSITION AND USES

COMPOSITION

Drumstick is a nutritious tree since its leaves contain high amount of vitamin A (four times more than carrot), vitamin C (seven times more than orange), protein (5–10%), calcium (four times more than milk) and potassium (three times more than banana). Mature seeds of *M. oleifera* contain about 40% oil of an excellent quality (73% oleic acid) but *Moringa. peregrina* is known to produce higher quantity of oil (50%), which can be extracted from seeds at home. Mature seeds are roasted, mashed and placed in boiling water for 5 minutes. After straining and sitting overnight, the oil floats to the surface. The oil contains high levels of unsaturated fatty acids, especially oleic acid up to 67.80%. The dominant saturated acid is behenic acid up to 6.81% and stearic acid up to 5.86%. The nutritional composition of pods, leaves and dried leaf powder are given in Table 110.1.

USES

Fresh green leaves and tender pods of drumstick are used extensively in cooking in Asia, Africa and Caribbean cuisine. Only tender growing tips and young leaves are generally used as greens in

TABLE 110.1
Nutritional Composition of Drumstick Leaves and Pods (Per 100 g Edible Portion)

Constituents	Pods	Leaves	Leaf powder
Water (%)	86.5	75.9	7.5
Carbohydrates (g)	3.7	12.5	38.2
Protein (g)	2.5	6.7	27.1
Fat (g)	0.1	1.7	2.3
Dietary fibre (g)	4.8	0.9	19.2
Minerals (g)	2.0	2.3	–
Calcium (mg)	30	440	2003
Phosphorus (mg)	110	70	204
Potassium (mg)	259	259	1324
Magnesium (mg)	24	24	368
Sulfur (mg)	137	137	870
Copper (mg)	3.1	1.1	0.57
Iron (mg)	5.3	7.0	28.2
Oxalic acid (mg)	10	101	1.6%
Vitamin A (mg)	0.11	6.8	16.3
Thiamine (mg)	0.05	0.06	2.64
Riboflavin (mg)	0.07	0.05	20.5
Pyridoxine (mg)	0.20	0.8	8.2
Vitamin C (mg)	120	220	17.3
Vitamin E (mg)	–	–	113
Calories (kcal)	26	92	205

cooking. However, mature leaves can be dried, powdered and stored for many days to be used like that of dried fenugreek leaves in the recipes. Its leaf powder is being used in sauces. Livestock relish the foliage so much that in some regions drumstick is an important fodder, *e.g.* water buffaloes in India are fed the chopped up leaves and branches, which are said to boost milk production. Ethanol extracts of drumstick leaves at 0.04–0.08% is known to increase nodulation and nitrogenase activity in leguminous crops.

The flowers can also be consumed just like the leaves or in various soups. The flowers, which occur round the year in some places but are more often seasonal, are cooked as a vegetable, and sometimes, they are steeped in boiling water to yield a fragrant tea. In Oaxaca, Mexico, poor people have adopted the tree solely as a source of white flowers for decorating churches and houses on religious festival days.

The immature seed pods, *called drumsticks*, are commonly consumed in South Asia in the form of curries. The seeds sometimes removed from more mature pods are eaten like peas or roasted like nuts. Young seedlings are pulled up, boiled and eaten whole.

The multipurpose drumstick seed oil known as *Ben oil* is used for cooking, watch lubrication, lighting and in perfumes and soap industries. Its seed oil is often used for cooking and in rare cases used for even lubrication purposes. The refined seed oil is clear and odourless and does not develop rancidity. The seed cake remaining after oil extraction may be used as a fertilizer. Its seed oil also has potential for use as a biofuel.

The seed powder of drumstick that has efficiency similar to alum is used to purify the dirty or cloudy drinking water by coagulation. It is toxic to guppies, protozoa and bacteria. The toxic effect is believed due to the presence of 4-L-α-1-rhamnosyloxybenzyl isothiocyanate, a glycosidic mustard oil. The powder from ground up seed and the press cake left over from oil extraction can be used for the treatment of turbid water since suspension of ground seed is used as primary coagulants. Their action at high turbidities is almost as fast as alum but slower at low and medium turbidities. However, the doses should not exceed 250 mg/L. The concentration of seeds used for nutrition, medicine or water purification is not harmful for human health. The solid matter can be removed as impurities using drumstick kernel powder. The coagulant property of seeds is due to a series of low molecular weight cationic proteins. These water-soluble proteins attach to and bind between the suspended particles in water forming larger agglomerated solids. These flocculated solids then allow settling down prior to boiling and subsequent consumption of water. However, the drumstick seeds should be used as a coagulant in water and wastewater treatment only after an adequate purification of the active proteins.

The trunk is gaining importance as a raw material for papermaking. The wood has suitable characteristics for pulp, paper, cellophane and textile production. Its wood is rated good for firewood and charcoal and has the property to resist termites. The stem gum exudates are used in calico printing and medicinally. The roots contain pterygospermin, which inhibits growth of gram-positive and gram-negative bacteria. The starchy roots of *M. peregrina* are eaten as food.

Medicinal Uses

As per *Ayurveda*, the leaves of drumstick tree prevent 300 diseases. All parts of the tree are considered medicinal and are used in the treatment of ascites, rheumatism, snakebites and as a cardiac stimulant. It has many medicinal properties such as abortifacient, antidote, bactericide, diuretic, ecbolic, estrogenic, expectorant, purgative, rubefacient, stimulant, tonic, vermifuge and vesicant. The alkaloids, moringine and moringinine are present in drumstick, the latter being responsible for many of the medicinal properties of the plant. Aqueous extract of roots and leaves are known to have anti-malarial and anti-implantation properties. Its leaf extract is also known to have anti-carcinogenic property as the extract has the ability to inhibit mutagenicity caused by nitrosodi-ethanolamine (NDEA, a plant environment carcinogen). In a long-term strategy, drumstick can successfully be utilized to combat vitamin A- and vitamin C-associated diseases, particularly in rural India. Such a type of approach has successfully been carried out on pre-school children in 20 villages (3585 households) of Andhra Pradesh, India. The drumstick seed oil is also known to enhance the absorption of nutrients and beta-carotene.

Drumstick seed powder can be toxic to animals and particularly to fish. This toxicity may be used in pond management to control predators of cultured fish. Its leaves have been found free from trypsin inhibitors. Saponin content was relatively high (up to 8%). Its leaves and twigs have been reported to contain limited amounts of cyanogenic glucosides in a few studies.

BOTANY

Drumstick is a very fast-growing deciduous perennial tree with slender and drooping branches, which are brittle with corky bark. It can attain a height of 5 m or more within a year of planting. The perennial drumstick grows up to a height of 10–12 m with an open umbrella-shaped crown, whereas, annual types attain merely a height of 5–6 m. Its trunk bark is smooth and dark grey. Its leaves are feathery, pale green, compound, tripinnate, 30–60 cm long with many small leaflets, 1.3–2 cm long, 0.6–0.9 cm wide, lateral ones somewhat elliptic and terminal ones obviate and slightly larger than the lateral ones. Its flowers are zygomorphic, hermaphrodite and gullet shaped. They are sweetly fragrant and produced in loose axillary panicles, which are about 15 cm long. The individual flower stalks are slender and up to 12 mm long. The flower consists of five pale green sepals and five white petals, which are little longer than the sepals. The five stamens bear anthers, while another five stamens are without anthers and have a slender style. Under North Indian conditions, annual drumstick flowers twice in a year, once in August to September and again in December to January, the latter resulting in fruit set. However, rare flowering may be seen throughout the year. Pods of 30–120 cm length and 1.8 cm width containing about 20 seeds embedded in pith, tapering at both the ends and nine-ribbed are initially light green, slim and tender, eventually becoming green, firm and pendulous and split lengthwise into three parts when dry. Pods usually attain edible maturity 15–20 days after fruit set. At edible maturity, the annual drumstick attains a pod length of 60–90 cm, perennial 45–55 cm and *Barahmassi* 30–40 cm. Seeds are dark brown with three papery wings.

In northern Indian plains, anthesis takes place between 7.00 and 13.00 hours. Each flower produces an average of 23,525 pollen grains with an average of 523 pollen grains per anther sac. Flowers favour cross-pollination due to delayed stigma receptivity (protandry condition). Biochemical studies of stigma reveal that some extra proteins and esterase contribute towards its receptivity. Stigmas remain receptive for about 4 hours and 12–36 hours after flower opening in annual and year-round types, respectively. Crossing between the annual and the year-round types is successful in both the directions. Pollination in the evening or morning is equally effective but it is consistently lower and often ineffective when carried out at noon. Natural population yields an average of 10.28% fruit, and successful pollination of the flowers requires visitations of a large number of insects of Thysanoptera, Hymenoptera, Lepidoptera and Coleoptera order. Among these, *Xylocopa* (carpenter bees) is the most common pollinating agent. In annual drumstick, huge flowering and sufficient pollen production can be seen during rainy season but pollen grains did not dehisce due to their very sticky nature, resulting in very less pod setting.

BIODIVERSITY IN DRUMSTICK

The genus *Moringa* has 13 species mostly spread over South East Asia and Africa. They are given under:

Moringa drouhardii found in Madagascar
Moringa concanensis found in India

Moringa arborea found in northeastern Kenya
Moringa hildebrandtii found in Madagascar
M. oleifera found in India
Moringa borziana found in Kenya and Somalia
Moringa ovalifolia found in Namibia and extreme southwestern Angola
M. peregrina found in Horn of Africa, Red Sea and Arabia
Moringa longituba found in Kenya, Ethiopia and Somalia
M. stenopetala found in Kenya and Ethiopia
Moringa pygmaea found in northern Somalia
Moringa rivae found in Kenya and Ethiopia
Moringa ruspoliana found in Kenya

The above 13 species of *Moringa* can be fit into three broad categories that reflect their life form and geography given as under:

1. **Bottle trees:** These are massive trees with bloated water-storing trunks and small radially symmetrical flowers.
   M. drouhardii found in Madagascar
   M. hildebrandtii found in Madagascar
   M. ovalifolia found in Namibia and extreme SW Angola
   M. stenopetala found in Kenya and Ethiopia

2. **Slender trees:** These are trees with a tuberous juvenile stage and cream to pink slightly bilaterally symmetrical flowers.
   M. concanensis found in India
   M. oleifera found in India
   M. peregrine found in Red Sea, Arabia and Horn of Africa

3. **Trees, shrubs and herbs of Northeast Africa:** The eight *Moringa* species found in northeast Africa span the whole range of life form variation found in *Moringa*. All the eight species except *M. peregrina* are endemic to northeast Africa. These species are tuberous adults or tuberous juveniles maturing to fleshy-rooted adults, having colourful bilaterally symmetrical flowers.

GEOGRAPHY OF THE *MORINGA* SPECIES

ARABIA AND INDIA: 3 SLENDER TREES

The three species that occur in the Red Sea area, Arabia and Indian subcontinent are all slender trees. This group includes the family's best-known and most economically valuable species *M. oleifera*. Now cultivated in all the countries of the tropics, *M. oleifera* seems to be native to sub-Himalayan India. In India, over 85 types of drumstick have been developed to reflect local tastes in pod size and shape. Many of these types are grown for 1–3 years and then replaced with new trees on a regular rotation, almost as if they were biennials.

M. concanensis is mainly from India and Pakistan, barely reaching into Bangladesh. *M. peregrina* is the widely distributed species, growing from the Dead Sea area sporadically along the Red Sea coasts to northern Somalia and around the Arabian Peninsula to the mouth of the Persian Gulf.

SOUTHERN HEMISPHERE: 3 BOTTLE TREES

The three species restricted to the southern hemisphere are all bottled trees. Namibia and Angola in southwestern Africa are home of *M. ovalifolia*, while *M. drouhardii* and *M. hildebrandtii* are endemic to Madagascar.

Highest *Moringa* Diversity in the Horn of Africa

With nine species, the Horn of Africa is the centre of *Moringa* diversity and includes a variety of life forms. *M. peregrina* is found in northern Somalia, Arabia and the Red Sea coasts north to the Dead Sea. *M. stenopetala* is widely cultivated but only known from five localities in the wild. The densest concentration of *Moringa* species is in Mandera District, in the extreme northeast of Kenya, where *M. arborea*, *M. longituba*, *M. rivae* and *M. ruspoliana* are found, though the species do not grow intermingled with one another. The other two species, *i.e.*, *M. borziana* and *M. pygmaea*, are distributed in Kenya and north Somalia.

The detailed description of the 13 *Moringa* species is given below:

1. *Moringa oleifera* Lam.

This species is native to India, Arabia and possibly Africa and the East Indies. It is widely cultivated and naturalized in tropical Africa, tropical America, Sri Lanka, India, Mexico, Malabar, Malaysia and the Philippine Islands. It thrives in subtropical and tropical climates, flowering and fruiting freely and continuously. It grows best on a dry sandy soil and is drought resistant. It is commonly and incorrectly known under the names *M. aptera* and *M. pterygosperma*.

The drumstick crop has not been systematically improved but several cultivars with individual qualities are known in India. The first cultivar *Bombay* has curly fruits and is considered among the best, the second cultivar *Jaffna* is noted for having 60–90 cm long fruits, and the third cultivar *Chavakacheri Murunga* is known for its exceptionally elongated pods (90–120 cm long). Finally, there are the so called *baramassi* varieties, which tend to flower continuously because they have been selected through the ages to provide buds, flowers and fruits for food throughout the year. There is also a dwarf variety known as *PKM 1* bred for the short stem that makes the harvesting of pods easy, which are exceptionally long. Some Indian varieties are showing exceptional promise at the International Institute of Tropical Agriculture in Nigeria.

2. *Moringa concanensis* Nimmo

This species has a very strong central trunk, which is covered with an extremely distinctive layer of very furrowed bark that can be more than 15 cm thick. The flowers also have distinctive green patches at tips of the petals and sepals. It occurs in tropical dry forest from southeastern Pakistan almost to the southern tip of India. It has recently been found in western Bangladesh. Despite this wide range, *M. concanensis* habitat is under heavy pressure from agriculture and urbanization.

3. *Moringa peregrina* Forssk. ex Fiori

This species, which is a native of the Red Sea region, has a peculiar growth habit. When its seedlings start coming out, they have broad leaflets and a large tuber. Through many dry seasons, the shoot dies back below ground to the tuber. As the plant gets older, the leaves get longer and longer but the leaflets get smaller and smaller and more widely spaced. Adult trees produce leaves with a full complement of tiny leaflets, only to drop them as the leaf matures. However, the naked leaf axis remains, giving the tree a wispy look similar to *Tamarix* or *Cercidium microphyllum*. The pink zygomorphic flowers are sweetly scented and contrast with the blue leaves.

4. *Moringa drouhardii* Jumelle

Moringa drouhardii with its bloated bright white trunks is a conspicuous tree species in the southern Malagasy dry forest. It grows in scattered stands that can number hundreds of individuals, usually on limestone. It grows extremely fast, surpassing 3 m in its first year.

5. *Moringa hildebrandtii* Engler

M. hildebrandtii is a beautiful tree with a massive water-storing trunk that can grow to a height of 20 m. This bloated trunk makes it strongly resemble the more well-known but unrelated baobab trees (*Adansonia*) that also occur in Madagascar. The pinnately compound leaves can be up to a metre long, and the leaf rachis and stem tip of young plants is often a distinctive deep red.

6. *Moringa ovalifolia* Dinter ex Berger

After *M. oleifera*, this is perhaps the most familiar species of *Moringa*, growing in such well-visited Namibian parks as Etosha and Namib-Naukluft. Its bloated white trunks standing out on otherwise bare hillsides have earned it the name *ghost tree*. The species is found from central Namibia to southwestern Angola, usually on very rocky ground.

7. *Moringa stenopetala* (Baker f.) Cufodontis

Most of the research on applied uses of *Moringa* other than *M. oleifera* focus on this species. It is an important food plant in southwestern Ethiopia, where it is cultivated as a crop plant. There are readily accessible populations of this species in the island of Lake Baringo, in the Rift Valley in Kenya. However, the distribution of the species through its range is very poorly known. After Lake Baringo, it is known from only four other localities, all around Lake Turkana, a huge lake that reaching hundreds of miles into Kenya from the Ethiopian border.

8. *Moringa arborea* Verdcourt

This species was discovered by Allan Radcliffe-Smith and Peter Bally in 1972. They found a single tree growing in a rocky canyon in northeastern Kenya near the Ethiopian border. Its tree is very beautiful, especially when covered with its large sprays of pale pink and wine red flowers. The young fruits resemble a yard-long string bean. Like other species of *Moringa*, locals use the tree for medicine, especially the roots, which are thick and fleshy and pungent smelling.

9. *Moringa borziana* Mattei

This small *Moringa* occurs from southern Kenya to the region of Kisimayu in southern Somalia. Usually, it bears only one or two stems, which typically reach about waist level. These shoots seem to die back to the tuber every few years, but occasionally, the plant may grow into a small tree. Its plant grows to a height of 3 m in the course of 2 months. The tuber is often very large and can reach more than one and a ½ m below the surface. Its flowers are greenish cream to yellow with brown smudges on the petal tips, and it has an almost sickeningly sweet odour.

10. *Moringa longituba* Engler

This species does not have any confusion with any other *Moringa* species since no other species has such bright red flowers or the petal and sepal bases fused to form a long tubular hypanthium. Usually, it has a large tuber deep underground with one small shoot reaching knee-high above the soil. If an individual is well-sheltered under a tree, it can scramble through the branches for over 3 m. At the beginning of this century, the botanist Engler placed this species in its own subgenus *Dysmoringa*.

M. longituba grows in northeastern Kenya, southeastern Ethiopia and much of Somalia. Like other *Moringa* species in the Horn of Africa, *M. longituba* is used medicinally, particularly for treating intestinal disorders of camels and goats, for which, the root is given internally.

11. *Moringa pygmaea* Verdcourt

M. pygmaea is from northern Somalia. It is a delicate tuberous shrub or herb with very tiny leaflets on its tripinnate leaves. Its flowers are yellow in colour.

12. *Moringa rivae* Chiovenda

This is the most taxonomically frustrating taxon in the family. It is native from southern Lake Turkana to Mandera District in Kenya and throughout southeastern Ethiopia. The subspecies *rivae* is recognized as having creamy petals with brownish tips and short fruits. The subspecies *longisiliqua* has yellow flowers and very long fruits. These are large shrubs to a height of 3 m with greatly swollen roots. The subspecies *longisiliqua* has an arborescent habit to over 6 m tall. *M. rivae* is very similar to *M. arborea* and *M. borziana*.

13. *Moringa ruspoliana* Engler

One of the most morphologically divergent species in the family, *M. ruspoliana* occurs from northern Somalia to southeastern Ethiopia. It is easily distinguished from all other species in the family by having simply pinnate leaves. The leaflets are the largest in the family, reaching 15 cm in diameter and are the thickest and toughest leaflets in the family. This species also stands out for having the largest flowers in the family, reaching 3 cm long. The flowers are pink with green bases. When young, it forms a thick taproot. As the plant ages, the root swells and becomes more globose. It begins to send out thick side roots, usually four. The adult plant is a small tree to 6 m tall with an octopus-like system of long fleshy roots.

CLIMATIC REQUIREMENT

Drumstick is a tropical plant and performs well up to an altitude of 600 m above mean sea level. Growth of the plant becomes restricted if it is grown above 1600 m altitude, however, the adaptability of the tree has been demonstrated at altitudes of 1200 m in Mexico and above 2000 m in Zimbabwe. *M. stenopetala* in Ethiopia may be seen regularly up to an altitude of 1800 m. It is predominantly a crop of dry and arid regions. The most optimum temperature range is 25–35°C. The temperature above 40°C causes flower drop. It is susceptible to frost. *Moringa olerifera* grows in ecologies ranging from subtropical dry to moist through tropical very dry to moist forest zones. Drumstick is reported to tolerate the annual precipitation of 760–2250 mm and annual temperature of 18.7–28.5°C. The optimum temperature range for seed germination is 20–30°C. However, the maximum seed germination takes place at a temperature of 25°C. In subtropics, it can tolerate light frost, as it is capable of recovering from freeze.

Drumstick is well adapted to hot and humid conditions. In certain areas, it thrives under annual rainfall exceeding 3,000 mm. It is especially adapted to dryness and resists several months of drought, probably because its swollen roots store water, *e.g.* it has endured blistering heat, desiccating drought and depleted dunes in places such as Sudan and the Sahel.

SOIL REQUIREMENT

Drumstick can be grown successfully on all types of soil but sandy loam soils are specifically found to be more suitable. It can tolerate clay soils too but not the water logging conditions. Water stagnation in field even for a day may kill the plant. Thus, the field should have good drainage as the tree cannot survive under prolonged flooding conditions. It can successfully be grown at a soil pH of 5.0 to 8.0. *M. oleifera* plants can survive up to exchangeable sodium percentage (ESP) of 41%. Higher ESP of 60% may cause injury symptoms at 120 days after planting and plant is died after 240 days.

CULTIVARS

The varieties of drumstick are classified as annual and perennial ones. The annuals are short-duration varieties, which are propagated by seeds, while perennials are propagated by stem cuttings. The characteristic traits of some of the important varieties are given as below.

BHAGYA

A variety developed at University of Horticultural Sciences, Bagalkot (Karnataka) has fast-growing dwarf plants with average height of 2.5–3.0 m. Branching starts from the ground level. It starts flowering in 130–140 days from seeding and pods attain edible harvest stage in 40–50 days from flowering. Pod length is 65–70 cm with average pod weight of 154.75 g. This is a high-yielding variety (300–350 pods in the first year and 800–1000 pods per year in the subsequent years). It is suitable for high density planting.

CHAVAKACHERI MURINGAI

An ecotype of Jaffna muringa bears pods as long as 90–120 cm.

CHEMMURUNGAI

Another ecotype of Jaffna muringa is high-yielding variety, which flowers and fruits throughout the year. The tips of the pods are red. The tree is medium-sized, bearing long pods.

JAFFNA

This is a Yazphanam type muringa introduced from Sri Lanka. Its fruits are 60–90 cm long with soft flesh and good taste. This type can yield 400 pods from second year of planting, which increases to 600 pods per tree per year third year onwards.

KATTUMURUNGAI

A variety developed at Anna Farm of the State Agriculture Department located at Kudumiyamalai in Tamil Nadu through pure line selection is an annual drumstick. When it is propagated by seeds, it yields around 400–500 fruits per plant annually. The fruits are only 25–30 cm long. As the plants are like shrubs, harvesting is very easy. After first harvest, the plants are headed back leaving 1 m above ground and used as ratoons. Ratooning is done for 2–3 years. It starts bearing fruits from the sixth month onwards.

KODIKALMURUNGAI

It is cultivated predominantly in Bekel Vine Gardens of Tiruchirapalli districts of Tamil Nadu. The pods are shorter (20–25 cm in length) and thick fleshed. The pods and leaves are very tasty. Trees are short statured with smaller leaves.

PALMURUNGAI

It is preferred for its thick pulp and tasty pods.

PKM 1

An annual bush type variety developed at Tamil Nadu Agricultural University, Coimbatore flowers 90–100 days after planting and yields about 275 fruits annually, the total fruit weight being 33–35 kg. The fruits are 65–70 cm long and 6.3 cm in diameter, weighing 160 g. A ratoon crop can be taken consecutively for 3 years.

PKM 2

A good yielding variety developed at Tamil Nadu Agricultural University, Coimbatore requires more water. The immature pods of this variety are greenish in colour and good in taste. The length of each pod is 45–75 cm. Each plant can yield 300–400 pods.

PUNAMURUNGAI

This variety is grown in kitchen gardens of Tirunelveli-Kattabomman and Kanyakumari districts.

YAZPHANAM MURINGA

A variety grown commercially in Salem and Madurai districts of Tamil Nadu as a single irrigated or rain fed crop is propagated by stem cuttings and comes to bearing within 8–9 months, the yield being 250–300 fruits per plant annually. It is grown along fences in households with no proper irrigation and fertilization. It is grown as perennial in homestead gardens and it can last for more than 5 years.

ROHIT

A variety recommended for Maharashtra state starts fruiting 4–6 months after plantation and gives commercial yield up to 10 years (per year two seasons). Pods are dark green in colour, 45–60 cm in length, pulp is soft and tasty and keeping quality is very good. From a single plant, one can get 40–135 pods of about 3–10 kg. A 5-year-old plant gives 585–650 pods of about 45–50 kg per year. Its average pod yield is 17–30 t/ha in well-drained soil under irrigated conditions.

DHANRAJ

A dwarf type variety developed at University of Agricultural Sciences, Dharwad (Karnataka) is suitable for intercropping with coconut plantation. Harvesting starts 9–10 months after seed sowing. The pod length is 35–40 cm and a 2-year plant can yield 250–300 pods per year.

COIMBATORE 2

A variety developed at Tamil Nadu Agricultural University, Coimbatore is more popular in Gujarat state. The pods are dark green in colour and tasty and length of the pods is 25–35 cm. Each plant yields 250–375 pods. Each pod is bulky. Each plant yields the pods for 3–4 years. If the pods are not harvested from the plant earlier, the market value of the pods is lost.

The other local cultivars are Chavakkacheri Muringai, Kattu Muringai, Pal Muringai, Poona Muringai and Kodigal Muringai.

PLANTING TIME

Drumstick can be propagated through seeds or limb cuttings. Perennial type drumsticks are propagated by stem cuttings after the harvesting of pods from March to June, though the annual type varieties perform best under the conditions of north Indian dry plains when propagated by seeds from June to August. In southern parts of the country, it is sown in November to December with assured irrigation.

SEED RATE

About 1800–2000 seedlings of drumstick are required for planting a hectare of land area, and for the production of seedlings sufficient for a hectare area, a seed rate of 600–700 g is enough. In general, the seed rate for dwarf varieties is 250 g/ha, tall varieties 100 g/ha and annual type 500–625 g/ha.

PLANTING METHOD

Drumstick is propagated by direct seeding, transplanting and using year-old stem cuttings. Annual drumstick is propagated by either direct seeding or transplanting, whereas, perennial ones are mostly propagated by hard stem cuttings. Direct seeding is also preferred when plenty of seed is available and labour is limited. The methods of drumstick propagation are given below:

- **Direct seeding:** This method is preferred for leaf, pod and seed production in annual drumstick. However, the plants raised from seeds produce fruits of inferior quality. Always fresh seed is used for sowing purposes since storing seeds for more than 4 months results in poor germination, which starts 4 days after sowing and lasts up to 9 days. Black seeds stored in polyethylene bags comparatively have higher germinability and seed vigour than brown or white seeds. Seed treatment with potassium nitrate at 0.1% improves seedling vigour and branching. Good germination can also be obtained by immersing seeds in water for 24 hours at room temperature. The seeds are sown at a depth of 2.5 cm in 30 cm wide and 30 cm high bunds during the last week of June to August. Generally, two seeds are sown per hill, and 2 weeks after germination retaining one vigorous seedling per hill, the poor ones are gently removed. For pod production, a spacing of 2.5 m × 2.5 m or 2.5 m × 2.0 m between raised beds and plants is kept, and for leaves production, a spacing of 1 m between rows and 50 cm only between plants is sufficient to promote leaf growth. Plants propagated by this method take comparatively longer time to flower than transplanted ones.

- **Transplanting:** This is most commonly used method for the propagation of annual type drumstick. Transplanting consists of two steps, *i.e.*, (1) seedling production and (2) field planting.

- **Seedling production:** Seedlings can be grown in divided trays, pots, poly bags or in seed-beds. Plastic trays with 25–50 cells of 3–4 cm width and depth can be used for raising seedlings. Cells should be filled with either peat moss, commercial potting mixture or a pot mixture prepared from soil, compost or rice hulls and vermiculite or sand. Poly bags or poly tubes of 15 × 10 cm size containing 0.5–1.0 kg soil by volume may also be used for raising seedlings. For filling pots, a mixture of soil, sand and compost in a ratio of 2:1:1 should be used. Usually, two seeds are sown in each poly bag and 1 week after germination, the weak seedling is removed. Seedlings become ready for transplanting 30–35 days after sowing when they reach about 30 cm in height with 6–7 compound leaves. If the seedlings are to be raised in raised nursery beds, the soil should be solarized by moistening and covering with polyethylene sheet or by burning 3–5 cm layer of organic residues on the bed. Per hill, 2 to 3 seeds are sown at a spacing of 15 cm in rows spaced 25 cm apart. In nursery, the seedlings become ready 40–45 days after sowing. While taking out the seedlings from the nursery, damage to seedling roots should be avoided.

- **Field planting:** Seedlings age for transplanting varies from 30 to 45 days, depending upon method of raising seedlings. The seedlings raised in poly bags become ready about 7–10 days earlier than the seedlings grown in nursery beds. Perennial drumstick requires 7–10 days more than the annual ones. Drumstick can be transplanted any time from June to August, though the best time for planting is the first week of July to the mid of August. Planting is done on 25–30 cm high bunds or beds at a spacing of 2.0–2.5 m. If the crop is to be retained for 2–3 years, spacing may be increased. Before planting, the pits of 50 cm × 50 cm × 50 cm should be filled with well-decomposed farmyard manure or compost. Immediately after planting, a light irrigation is applied but care should be taken to avoid water logging.

- **Stem cutting:** This method is usually adopted for the propagation of perennial drum-stick. Hard stems of 8–12 months old with 5–15 cm diameter can be used for propagation. Cuttings of 50–150 cm length are considered most suitable for planting. Before planting, the cuttings are dried in shade for 3 days to avoid rotting of soil-inserted cut end and to ensure better sprouting. It may be planted directly in the field, poly bags or in the nursery. When planted in the field, about one third of the stem (30–50 cm) should be buried into the soil. Before planting, 5 kg of compost is added to each pit to encourage root development. Applying irrigation regularly, the soil is kept moist but not wet. Cuttings planted in nursery beds become ready for transplanting after 2–3 months.

- **Tissue culture:** *In-vitro* propagation of drumstick can be achieved successfully using nodal segment of plant in medium containing 2% sucrose + 0.8% agar and supplemented with 1 mg basal medium with naphthalene acetic acid 0.5 mg/L. Micropropagation from hypocotyls and cotyledonary explants is achieved successfully. Initially, multiple shoot regeneration is induced on MS medium containing 3% sucrose and 100 mg L-asparagine, 100 mg L-glutamine and 150 mg M-inositol per litre supplemented with naphthalene acetic acid 0.2 mg + kinetin (M_1) 2 mg or naphthalene acetic acid 0.1 mg + benzyl adenine (M_2) 5 mg. The regenerated shoots when transferred to M_2 supplemented with gibberellic acid at 0.2 mg/L grow faster and develop roots.

One week before planting, pits of 45 cm × 45 cm × 45 cm size are dug at 2 m × 2 m spacing for dwarf and annual varieties and 4 m × 4 m for tall varieties. The farmyard manure is applied at 15 kg/pit. A 60-cm circular irrigation basin is made around the pit. The pits are closed. Two seeds are sown 3 cm deep in each pit in the centre. Seeds germinate 7–9 days after sowing. Seeds can also be sown in polyethylene bags. Seedlings, 1-month-old, are planted in the field.

NUTRITIONAL REQUIREMENTS

Drumstick grows well in most soils without addition of fertilizer. Once the plant is established, the extensive and deep root system of drumstick is efficient in mining nutrients from the soil. However, for optimum growth and yield, 3–5 kg farmyard manure or compost should be added in trenches of 50 m² at the time of planting. The nutrients requirement of a year old drumstick is 250:125:125 g NPK per plant. One-fourth amount of nitrogen, full phosphorus and half dose of potassium should be applied at planting and the rest of the dose of nitrogen should be applied in two equal splits at 45–60 days and during flowering, whereas, remaining half dose of potassium should be applied before flowering in the month of February to March. In some parts of India, during rainy season, 15 cm deep ring trenches are dug about 10 cm away from the tree base and filled with green leaves, manure and ash, and then, covered with soil. This practise promotes high dry matter yield.

IRRIGATION REQUIREMENT

Drumstick is a drought tolerant crop. It does not require too much irrigation but the pit should be watered at the time of sowing or transplanting and on the third day of sowing. During germination, the seeds should get optimum moisture. In established crops, irrigation at 10–15 days interval is sufficient.

In northern plains of India, irrigation is not necessary during monsoon season (June to September), however, irrigation during very dry conditions is supplied regularly for the first 2 months. A well-rooted drumstick tree can tolerate drought and needs irrigation only when persistent wilting is evident, however, for better growth and yield, drumstick should be irrigated once in a month during winter season and twice in a month from February to June. Flowering and pod development are the critical phases so sufficient soil moisture should be maintained during flowering and pod development stages. In arid conditions, flowering can be induced through irrigation.

INTERCULTURAL OPERATIONS

HOEING AND WEEDING

Drumstick is a fast-growing tree, thus, weeds pose no serious problem with the crop. One hand weeding near the plant at 30–45 days after planting is advised to ward off weeds in early growth stages. Before flowering, two to three hoeings manually or with cultivator are beneficial for better plant growth. If planting is done at 2.5 m row spacing, it can be done effectively with a tractor or power tiller. Intercropping with sunflower is found beneficial for the control of weeds. The drumstick tree is reported to be highly competitive with eggplant and sweet corn, which can reduce pod yield up to 50%. In heavy pod load condition, it is necessary to prop up the branch to prevent it breaking off.

PINCHING OF TERMINAL TIPS

In annual drumsticks, first pinching should be performed at 60 days after sowing in terms of improving the biometric parameters of the crop. When seedlings reach a height of 60 cm in the main field, the top 10-cm growing tips should be trimmed with fingers to encourage shoot growth. About a week later, the secondary branches begin to appear on main stem below the cut. When they reach a length of 20 cm, they are again cut back to 10 cm making a slanting cut with downward facing. Tertiary branches may also be pinched in the same manner. Pinching is done four times in about a period of 3 months before flowering as it encourages the tree to become bushy and develop a strong frame for maximizing the yield. In the absence of pinching, the tree has the tendency to grow vertically like a mast with sparse flowers and fewer pods only at the top.

TRAINING AND PRUNING

As the plant attains a height of 70–75 cm, the shoot tips are nipped to encourage side branching. If plant stem is weak and very tall, it is susceptible to heavy wind and branches may break easily. At the same time, nipping reduces the plant height and facilitates easier harvesting. Mounting of soil is done up to a height of 30–40 cm above ground level to provide better support. Annual pruning encourages auxiliary and lateral branches. In young planting of drumstick, intercropping with okra can be done to get better profits.

In annual drumsticks, usually severe pruning is done only once at the end of the harvest. In this, the main trunk of the tree is cut back to about 90 cm from ground level. About 2 weeks later, 15–20 sprouts appear below the cut, however, only—four to five robust branches are allowed to grow and the remaining sprouts are pinched off while they are young. The pinching process can be repeated as done for new seedlings to make the tree bushy. After taking the second crop, the tree can be replaced by new seedlings for obtaining higher yield.

In perennial drumsticks, the old, dead and infected scaffold branches are cut back to 50 cm every year at harvest. Once in 4–5 years, the main trunk of tree should be cut back to 1 m from ground to allow re-growth.

USE OF PLANT GROWTH SUBSTANCES

In the development of drumstick pods, endogenous level of auxins and cytokinins plays a vital role. In the initial stage up to a length of 30–35 cm, the pods grow very rapidly because of the presence of auxins, $i.e.$, indole acetic acid and indole aceto nitrile. The auxin activity in pods rises to a peak when the pods are 25–30 cm long and falls down to a very low level at maturity (45–50 cm length). Similarly, the cytokinin levels increases up to 1.0 µg/g pod weight until the pods are 15–20 cm long, and then, reach a plateau. Cytokinin concentrations again reach a peak (up to 6.6 µg/g pod weight) when the pod approaches a length of 35–40 cm.

HARVESTING

Drumstick is usually harvested when the pods are still young (about 1 cm in diameter) and snapped easily. Older pods develop a tough exterior but white flesh and seeds remain edible until the ripening process begins. In north India, pods are generally harvested at a tender stage, while in south India, the pods are harvested at fully mature pulpy stage since the pods at this stage are preferred for the preparation of *Sambhar* and *Rasam*. Harvesting in local perennial cultivars is done during March to April, while the harvesting in annual cultivars, is done from March to May. In *Baramassi* cultivars, pods may be harvested 2–3 times in a year. In annual cultivars too, some pods may be picked from June to September under north Indian conditions.

The annual drumstick types come to harvest after 6 months of sowing. Perennial types propagated through limb cuttings take 8–9 months for first bearing and harvesting. The pods become ready for harvesting about 2 months after flowering and the period of harvest continues for 2–3 months. In annual cultivars, each tree bears 200–350 pods or sometimes even more, depending upon varietal capacity. In perennial types, the yield will be low during first 2 years, *i.e.*, 100 pods per plant per year, but thereafter, the yield increases to 500–600 pods per plant, depending upon the cultivar. Setting up four to six beehives per hectare during flowering season to ensure good pollination helps in increasing yield. Under south Indian conditions, drumstick bears flowers twice in a year. The main harvesting season is March to June, while lesser yield can be obtained during September to October.

If the flower clusters on the tree are more in number, the yield is low. An individual inflorescence carries 19–126 flowers, average being 53 but it results in only one fruit. Normally, three to five fruits per inflorescence are obtained. The fruit set ranges from 1.0–2.8%, which varies from tree to tree. The fruit length varies from 25–100 cm. Each fruit weighs around 230 g containing 10–20 seeds each on an average. Annual drumsticks start bearing fruits from sixth month onwards, providing 200 fruits per plant per annum. Its average yield is 52 t/ha.

The ratoon crop can be obtained by pruning its trees to a height of 1 m after harvesting. The plants start bearing fruits 4–5 months after heading back, thus, three ratoon crops can be taken. After ratooning, once every 6 months, 30:30:30 g NPK per plant along with 25 kg farmyard manure is applied.

For seed production and oil extraction, pods are allowed to dry and turn brown on the tree. Pods should be harvested before they split open and fall to ground. Annual drumsticks should be harvested at physiological maturity stage about 100 days after anthesis to get seeds of high quality. Delay in harvesting date beyond 100 days results in deterioration of seed quality.

Seeds can be stored in well-ventilated sacks in dry and shady place for about 6 months. Drumstick seeds have no dormancy period, thus, can be planted as soon as they attain maturity. For storage, drumstick seeds should be dried to a moisture content of 8%. Seed viability decreases progressively as the storage duration increases after 2 months; however, the seeds treated with Captan at 2 g/kg of seed and stored in polyethylene bags show 40% germination even after 12 months compared to 4% in untreated control.

YIELD

The yield and quality of drumstick depend on climatic conditions of the growing region, soil type, soil fertility, irrigation facility and spacing, and the market rates depend on quality and demand in the market. On an average, an annual drumstick yields about 4–5 kg edible pods per plant during first year of planting, whereas, the perennial drumstick gives 40–50 kg green pod per plant. A single tree third year onwards may yield 600 or more pods. A green pod yield of 70–100 q/ha from annual drumstick and about 200–300 q/ha from perennial can be obtained.

POST-HARVEST MANAGEMENT

The properly developed fruits or pods are harvested before they become over mature and fibrous. Pods are graded according to their length and tied in bundled of 10–15 pods, which are packed in crates and sent to the market for sale.

Drumstick pods lose their freshness quickly in transit and storage, thus, are usually consumed on the same day of harvest, however, they can be kept for 4 days at ambient room temperature without much loss in physical and quality parameters. Drumstick leaves dried in partial shade have very less loss of beta-carotene. Cooking of leaves lowers the beta-carotene levels by 8%, while drying in sun lowers the levels by 50%.

Owing to its high nutritional value, attempts are being made to develop various processed products from drumstick. A number of value-added products *viz.*, moringa pickle, dehydrated moringa, moringa powder and moringa flesh mesocarp powder, can be prepared if its availability is in excess.

INSECT PESTS

Pod Fly (*Gitona distigma*)

Pod fly maggots enter into tender fruits by making small bore holes at the terminal end. This causes oozing out of gummy fluid from fruits, which ultimately results in the drying of fruits from tip upwards. A maximum of 20–28 maggots are found in a fruit. Internal contents of the fruits rot. Its maximum activity occurs from April to October and declines thereafter.

Control

- Use attractants like citronella oil, eucalyptus oil, vinegar (acetic acid), dextrose or lactic acid to trap the flies.
- Periodically, collect and destroy all the fallen and damaged fruits by dumping in a pit.
- Frequently, rake up the soil under the trees or plough the infested field to destroy pupae.
- Drench the soil with *neem* seed kernel extract (NSKE) 5% at 2 L per tree at 50% fruit set stage.
- Spray Dichlorvos 76% SC 500 ml or Malathion 50% EC 750 ml in 500–750 L of water per hectare when pods are 20–30 days old and apply *Azadirachtin* 0.03% 1.0 L during 50% fruit set and 35 days later.

Budworm (*Noorda moringae*)

Budworm is a major pest in South India. Its larvae bore into flower buds and cause shedding of buds up to 75%. The damaged buds seldom blossom and fall down prematurely. In early stages, the caterpillars are gregarious and scrape the chlorophyll of leaf lamina giving it a papery white appearance. Later, they become voracious feeders making irregular holes on the leaves. Skeletonization of leaves also occur with the appearance of only veins and petioles. In severe cases, heavy defoliation is also observed. Bored fruits with irregular holes are also seen. Its activity is more during summer months in South India.

Control

- Plough around trees to expose and kill the pupae.
- Collect and destroy the damaged buds along with caterpillars.
- Use light traps 1–2 per hectare to attract and kill the adults.
- Spray Malathion 1.0 L dissolved in 500–750 L of water per hectare.

Leaf Eating Caterpillar (*Noorda blitealis*)

It is a sporadically serious pest of drumstick trees, especially in South India. Caterpillars feed on leaf lamina, turning them into transparent parchment like structures. Peak period of infestation is from March to April and December to January.

Control

- Plough around trees to expose and kill the pupae.
- Use light traps 1–2 per hectare to attract and kill the adults.
- Spray Malathion 1.0 L dissolved in 500–750 L of water per hectare.
- Spray Decis or Cypermethrin at 1 ml/L of water.

HAIRY CATERPILLARS (*Eupterote mollifera*)

It is the most destructive pest of drumstick in South India. Caterpillars feed gregariously by scrapping bark and gnawing foliage. Caterpillars gather in a cluster on stem of the plants during hot hours of day. They are active at night, defoliate the tree quickly and collect on the trunk. The larvae feed on leaves causing complete defoliation of the tree.

Control

- Collect and destroy egg masses and caterpillars.
- Use light traps to attract and kill adults immediately after rains.
- Use burning torch to kill congregating larvae on the trunk.
- Spray Chlorpyriphos 20% EC or Quinalphos 25% EC 1.0 L dissolved in 500–750 L of water per hectare or fish oil resin-soap 25 g/L on trunks and foliage, immediately after rain and 15 days later.

LONG HORN BEETLE (*Batocera rubus*)

This pest is widely distributed all over the Indian subcontinent. Grubs make zigzag burrow beneath the bark, feed on internal tissues, reach sapwood and cause death of affected branch or stem. Adults feed on the bark of young twigs and petioles.

Control

- Clean affected portion of tree by removing all webbed material, excreta etc.
- Insert in each hole the cotton-wool soaked in Monocrotophos 36 WSC 5 ml or any good fumigant like carbon disulfide, carbon tetrachloride, chloroform or even petrol and seal the treated hole with mud.

APHIDS (*Aphis gossypii*)

It is a polyphagous pest. Its nymphs and adults suck vital sap from twigs. As reproduction is mostly parthenogenic, population build-up is very fast.

Control

- Spray dimethoate 30% EC 500 ml or Malathion 1.0 L dissolved in 500–750 L of water per hectare.

THRIPS (*Megalurothrips distalis*)

Thrips damage the crop occasionally by sucking sap from its tender leaves especially at flowering stage.

Control

- Spray dimethoate 30% EC 500 ml or Malathion 1.0 L dissolved in 500–750 L of water per hectare.

DISEASES

Damping-off (*Rhizoctonia solani, Pythium*)

It is a disease of nursery beds and young seedlings, resulting in reduced seed germination and poor stand of seedlings. Very high seedling mortality up to 25–75% is also reported. In case of pre-emergence damping-off, seedlings disintegrate before they come out of soil surface, leading to poor seed germination. Post-emergence damping-off is characterized by development of disease after the seedlings have emerged out of soil but before the stems are lignified. Water soaked lesions form at collar region. Infected areas turn brown and rot. Plants shrivel and collapse as a result of softening of tissues. In *Rhizoctonia solani* attack, the infected stems become hard, thin (wire stem symptoms) and infected seedlings topple. Disease appears in patches both in nursery and in field beds.

Control

- Raise seedlings in light soil with proper drainage.
- Burn the farm trash on surface of the beds.
- Sow seeds on raised beds of 15 cm.
- Use low seed rate of 650 g/40 m² of nursery.
- Cover the beds with polythene sheet of 0.45 mm thickness for 3 weeks before sowing for soil solarization, which will reduce the soil borne diseases.
- Apply bio-control agents in soil *viz., Trichoderma harizianum, Trichoderma. viride, Gliocladium virens* or *Bacillus subtilis* 1 kg/100 kg farmyard manure.
- Drench the soil with carbendazim 0.1% or Tricyclazole 0.06% to control root rot caused by *Rhizoctonia solani*.

Twig Canker (*Fusarium* spp.)

The first symptom of the disease is clearing of the veinlets and chlorosis of the leaves. The younger leaves may die in succession and the entire plant may wilt and die in a course of few days. Soon the petiole and the leaves droop and wilt. In young plants, symptom consists of clearing of veinlets and dropping of petioles. The symptoms continue in subsequent leaves. At later stage, browning of vascular system occurs. Plants become stunted and they die.

Control

- Apply bio-control agents in soil *viz., T. harzianum, T. viride, Gliocladium virens* or *Bacillus subtilis* 1 kg/100 kg farmyard manure.
- Spray carbendazim (0.1%) or Tricyclazole (0.06%) on foliage

■■■

111 Bay Leaf

Pradyumna Tripathy and A.K. Pandav

Botanical Name : *Laurus nobilis* L.
Syn. *Laurus undulata* Mill.

Family : Lauraceae

Chromosome Number : 2n = 48 (tetraploid)

ORIGIN AND DISTRIBUTION

Bay leaf history is very decent and stretched one. The history indicates that in the ancient period, the bay leaf, otherwise known as sweet laurel, was devoted to the god Apollo and its leaves were used to make wreaths ahead of laurel leaf. Also, the wreaths of laurel were crowned to great and proficient individuals such as emperors, kings, conquerors, scholar, victorious athletes or Olympians, rhymesters, etc. in ancient Greece and Rome. In the legends, the wreath laurel is believed to be the tree of the *Sun God* under the celestial sign of *Leo*. It has also been the emblem of decency, festivity and victory as mentioned in *winning your laurels*. Though the exact centre of origin is not known, two primary centres of origin of bay leaf have been suggested such as the eastern Mediterranean region as well as the Asia Minor. Some other reports also suggest that the Asia Minor region was treated as the primary centre of origin, from where, it dispersed to all over the Mediterranean region and subsequently to other parts of the Asian countries with similar climates. The crop bay leaf is extensively cultivated as a commercial crop in Europe, the United States and Arabian countries. It grows usually in scrub land woods in Fujian, Jiangsu, Sichuan, Taiwan, Yunnan Zhejiang, etc. However, it is not cultivated as a commercial crop in India.

INTRODUCTION

The bay leaf or laurel is known in several synonyms such as bay laurel, sweet laurel, wreath laurel, noble laurel, poet's laurel, Roman laurel, Grecian laurel, etc. The typical word *Laurel* is derived from a Greek word *dhafni* named for the myth of nymph Daphne. Apollo made the tree sacred, and *per se*, it became a symbol of honour. The association with honour and glory continues even today. There are poet laureates (Apollo was the God of poets) and baccalaureate, meaning *laurel berries*, from which, the word *bachelor* has been derived and it signifies the completion of a bachelor degree. The term *baccalaureate* originates from this giving of bay leaf crowns to signify success, as does the term *poet laureate*. Doctors were also crowned with laurel, which was considered to cure all. Laurel garlands were awarded to triumphant athletes of ancient Greece and were given to winners at Olympic Games since 776 BC. Today, the Grand Prix winners are decked with a laurel wreath, and the poet of the British Royal Household is given the title of poet laureate. The emperor Tiberius always wore a laurel wreath during thunderstorms.

Bay leaf is also known by many names such as sweet bay, Indian bay, true bay, bay laurel, sweet laurel, wreath laurel, noble laurel, poet's laurel, Roman laurel, Grecian laurel, Apollo's bay leaf, Mediterranean bay, Mediterranean laurel, etc. The name of sweet bay in various languages also differs such as *loureiro* (Portuguese), *dhafni* (Greek), *laur* (Russian), sweet laurel (English), *feuille de laurier* or *laurier franc* (French), *hoja de laurel* (Spanish), *Lorbeerblatt* (German), *Foglia di alloro* or *lauro* (Italian), *Yuch-kuei* (Chinese), *Gekkeiju* (Japanese) and *Ghar* in Arabic. In India, the name of true cinnamon leaves, *i.e.*, *Tamaalpatra* (Sanskrit) or *Tejpatta* (Hindi), is used for noble laurel too. However, botanically, both the crops are different with respect to genus and species as well, *i.e.*, cinnamon leaf (*Cinnamomum tamala*) and bay leaf (*Laurus nobilis* L.). Laurel leaves are commercially packaged as *tejpatta* (Indian bay leaf), which actually creates a confusion between the two herbs such as cinnamon and bay leaf under Indian conditions.

In the world scenario, the trade of dried laurel leaves exceeds more than 2000 tonnes annually. Turkey is the major producer as well as exporter in the world by contributing about two-third of the world trade alone. The import of dried leaves of bay leaf in Western Europe is estimated 800 tonnes annually. The demand for selected dried herbs, including laurel leaves, in the major European markets is expected to grow much faster in the industrial food and institutional catering sectors than the relatively stable growth rate in the retail sector. However, users from most of the countries are quality conscious and new exporters should be able to provide dried herbs of high quality, particularly in terms of cleanliness and absence of pesticide residues. The aroma of the crushed leaves is delicate and fragrant and its taste is aromatic and bitter.

COMPOSITION AND USES

COMPOSITION

In general, the leaves of bay leaf herb are rich in important minerals such as calcium, iron and phosphorus as well as electrolytes such as sodium and potassium. The reports on biochemical analysis of the plant indicate that mature leaves are higher in magnesium, calcium and sodium, whereas the young shoots are higher in copper and iron. Besides these, the leaves are also a good source of carbohydrates, protein, fat and dietary fibre along with an appreciable amount of vitamins A and C and a fair amount of niacin, pyridoxine, pantothenic acid and riboflavin. The average nutritional composition of bay leaves is presented in Table 111.1.

The leaves of bay leaf herb are also rich in essential oil, which vary from 1.5% to 2.5% or up to 3%, primarily as monoterpenoides over 140 components such as α-pinene, *p*-cymene, β-pinene, limonene, etc., with the principal component of 1,8-cineole. The essential oil represents almost 99% compounds. The *in vitro* active bay leaf component is soluble in water. The chemical compositions of the essential oil is analysed through gas chromatography–mass spectrophotometry (GC–MS)

TABLE 111.1
Nutritional Composition of Bay Leaves (per 100 g Edible Portion)

Constituents	Contents	Constituents	Contents
Water (g)	4.50–10.00	Manganese (mg)	8.17
Carbohydrates (g)	63.97	Copper (mg)	0.41
Protein (g)	7.61–10.60	Selenium (mg)	0.28
Fat (g)	4.50–8.36	Vitamin A (IU)	618
Dietary fibre (g)	16.30	Thiamine (mg)	0.10
Calcium (mg)	834.00	Riboflavin (mg)	0.421
Phosphorus (mg)	113.00	Niacin (mg)	2.003
Potassium (mg)	529.00	Pyridoxine (mg)	1.740
Magnesium (mg)	120.00	Folic acid (mg)	0.018
Sodium (mg)	23.00	Vitamin C (mg)	46.5
Iron (mg)	530.00	Ash (g)	3.50–4.20
Zinc (mg)	40.00	Energy (kcal)	313

indicateing the presence of predominant constituents, *i.e.*, 1,8-cineole, which varies from 35.5% to 60.72%. The other important constituents are sesquiterpenes, guaianolides, trypanocidal terpenoids, α-methylene-γ-butyrolactone, costunolide, dehydrocostus lactone, zaluzanin D, reynosin, santamarine, *p*-menthane hydroperoxides (including [1R,4S]-1-hydroperoxy-*p*-menth-2-en-8-ol acetate), 3-α-acetoxyeudesma-1,4(15),11(13)-trien-12,6-α-olide and 3-oxoeudesma-1,4,11(13)-trien-12,6-α-olide. The oleoresin extracted from bay leaf contains 4–8% volatile oil. The component myrcene, an essential oil, used in perfumery, is also extracted from the bay leaf. The young leaves and fresh buds contain 3-octulose (d-gluco-l-glycero-3-octulose), a compound rarely found in nature. Besides leaves, the bay leaf seed also contains 25% fatty oil comprising at least 20 fatty acids with lauric acid (11–52%), palmitic acid (5–17%), oleic acid (15–40%) and linoleic acid (17–23%) as the main constituents. The dried fruits of bay leaf contain 0.6–10% of essential oil, depending on provenance and storage conditions. Like leaves, the aroma is mostly due to terpenes (cineol, terpineol, α- and β-pinene and citral) but also due to cinnamic acid and its methyl ester. A green semi-solid oil (melting point about 30°C) can be extracted from bay leaf fruit, which contains several percentage of essential oil (containing main components two sesquiterpenoids, costunol and dehydrocostuslacton) but is mainly composed of fixed oil, *i.e.*, triglycerides of lauric acid (dodecanoic acid), myristic acid (tetradecanoic acid) and oleic acid (Z-octadec-9-enoic acid). Similarly, the petroleum ether extracted from the bay leaf fruits contains several characteristic compounds such as 10-hydroxyoctacosanyl tetradecanoate, 1-docosanol tetradecanoate and 11-hydroxytriacontan-9-1.

Uses

In general, both the fresh and dried bay leaves are used as spice and imparting an excellent fragrant aromatic flavour in preparation of various food items such as soups, sauces, syrups, jam, stews, stocks, braises, pates, confectionaries, seafood, vegetable dishes and daubes, and also as an appropriate seasoning for fish, meat and poultry dishes. The spice is used in the preparation of *court bouillon*, which is readymade preparation made of water, salt, white wine, vegetable aromatics (onion and celery) and flavoured with *bouquet garni* and black pepper.

The leaves are imparting flavour to many classic French dishes too. Usually, bay leaf has a pleasant odour, while its taste is typically strong, pungent and aromatic as well, that is why the fragrance of bay leaf is considered more noticeable than its taste. Its fresh leaves are normally very mild but they gradually develop their full flavour several weeks after picking and proper drying. The leaves are pungent and have a sharp bitter taste but the bitterness of fresh leaves is easily removed by drying. After drying, the fragrance becomes herbal, slightly floral and somewhat similar to oregano and thyme.

In many European cuisines (mainly Mediterranean and North America), the leaves are used as a fixture in cooking. The bay leaves are also used in special *biryanis* and other rich spicy dishes in Indian and Pakistani cuisine. The leaves are very popular especially in Italian, Spanish and Creole cuisine. Sometimes, the dried bay leaves are commonly placed in bouquet garni, which is essentially a little bundle of culinary herbs used for cooking. In the Philippines, dried bay leaves are also added as a spice in the Filipino dish *Adobo*.

The leaves of bay herb, either after fine crushing or in powdered form, are commercially used as an essential ingredient of mixed pickling spices, vinegar and in bakery products. However, the crushed or ground bay may be utilized in many seasoning blends and other products. It may be substituted for whole leaves and does not need to be removed but it is much stronger due to the increased surface area. In some dishes, the texture may not be desirable. The essential oil extracted from leaves is also used as flavouring agent in various dishes in different parts of the world. The bay oil and oleoresin are used in soluble pickling spices. Dark-green glossy bay leaves can be used as fresh. However, they are at their best after being allowed to wilt under the shade for few days when their bitterness has gone but the leaves still retain their pleasing aroma.

The plants of bay leaf can also be used as an important component in floriculture and landscaping design such as popular container or as an ordinary houseplant in winter by trimming into distinctive shapes, an interesting specimen for shady areas of landscape including patios, herb gardens and woodland gardens. The sweet bay is planted only as an ornamental, mainly at higher elevations in the highlands of Southeast Asia. In some countries, *e.g.,* Italy and Turkey, the leaves of laurel are used for packaging liquorices. Sometimes, its wood is used for posts. The flexible twigs were formerly used to make hoops for small barrels.

The Bay leaf oil has been used in entomology as the active ingredient in killing jars, owing to its lauric acid content, which gives it insecticidal properties. They can also be used as a repellent in pantry to repel meal moths, flies, mice and silverfish. The crushed fresh young leaves are put into the jar under a layer of paper. The vapours they release kill insects slowly but effectively and keep the specimens relaxed and easy to mount. The leaves discourage the growth of moulds. They are not effective for killing large beetles and similar specimens but the insects that have been killed in a cyanide killing jar can be transferred to a laurel jar to await mounting. It is not clear to what extent the effect is due to cyanide released by the crushed leaves, and to what extent other volatile products are responsible. Eucalyptol essential oil present in bay leaves is effective in eliminating bugs or weevils from the kitchen and repelling cockroaches. The Bay leaf essential oil was found to inhibit the growth of *Phytophthora infestans.*

The ripe berries or seeds oil and fat are used in soap manufacture, liqueurs, perfume and in veterinary field. The bay leaf is also known as a good cleanser.

Medicinal Uses

The plants of bay leaf are not cultivated commercially in Indian subcontinent, the leaves, oil, bark and fruits not only have the versatile uses in various food items but also have legendary medicinal properties. It has been used since the ancient times to treat problems associated with the liver, stomach and kidney. The Bay leaf was highly praised by the Greeks and the Romans, who deeply believed that the herb symbolizes wisdom, peace and protection. It has some importance due to its diuretic, diaphoretic, digestive, emetic, stomachic, astringent and carminative properties and a good appetite stimulant. The bay leaves contain compounds called parthenolides, which is useful in the treatment of migraines. It has also been shown to help the body to produce insulin more efficiently, which leads to lower blood sugar levels and reduces the effects of stomach ulcers. The eugenol in bay leaves has several beneficial effects such as anti-inflammatory, antifungal, antibacterial and antioxidant properties. The volatile active compounds in bay leaf such as *a-pinene, β-pinene, myrcene, limonene, linalool, methyl chavicol, neral, a-terpineol, geranyl acetate, eugenol* and *chavicol* are known to have been antiseptic, antioxidant, digestive and thought to have anticancerous properties. Some of the most impressive health benefits of bay leaves include their ability to detoxify the body, slow the ageing process, speed up wound healing, protect the body from bacterial infections, manage diabetes, improve heart health, reduce inflammation, alleviate respiratory issues, optimize digestion and prevent certain types of cancer. The bay leaf berries have astringent, diuretic and appetite stimulant properties. Infusion of herb parts is reputed to soothe stomach ulcers and help relieve flatulence and colic pain.

Bay leaves have a very strong effect on the gastrointestinal system, they not only stimulate urination (as a diuretic), which decreases the toxicity of the body but also stimulate vomiting (as an emetic) when something toxic has been consumed. The phytonutrient parthenolide in bay leaves has the ability to reduce inflammation throughout the body. Furthermore, the organic compounds found in bay leaves are very effective for settling upset stomach, soothing irritable bowel syndrome and even lessening the symptoms of Celiac disease. Some of the more complex proteins in modern diets can be difficult to digest but the unique enzymes found in bay leaves help to facilitate efficient digestion and nutrient intake.

Rutin in bay leaves strengthens capillary walls in the heart and the body's extremities, while caffeic acid helps in eliminating the bad cholesterol from the cardiovascular system, thereby protecting the heart. Bay leaves have very beneficial effects on total cholesterol, low density lipoprotein (LDL) cholesterol, high density lipoprotein (HDL) cholesterol and triglyceride levels in patients with type-2 diabetes. The phytonutrients catechins, linalool and parthenolide in bay leaf help the individuals suffering from anxiety and stress by lowering the stresses. Bay leaves have direct relation with improved insulin receptor function as well as regulated blood sugar levels in diabetic patients.

Bay oil is also used in the treatment of various ailments, *e.g.*, ointments for bruising and sprains. It provides relief from swelling, rheumatic, arthritic pain, muscle pain, headache bronchitis and flu symptoms. The bay leaves essential oil is applied to the chest to help alleviate various respiratory conditions. The same effect can also be achieved with a poultice made of leaves spread on chest and retaining overnight. Inhaling the vapours has a similar effect to aromatherapy and can loosen up phlegm and eliminate dangerous bacteria that may be trapped in respiratory tracts. Rubbing bay leaves on scalp after shampooing improves hair follicles health and eliminates dandruff and dry skin.

PRECAUTIONS

Besides so many benefits of bay leaf in the form of leaves, oils and berries, the bay leaf plant also has some adverse side effects if not used as per the proper recommendation. Dosing adjustment is necessary. Washing of bay leaves before consumption is very essential. This tough leaf may cause choking if consumed whole, especially to little children. Pregnant and breastfeeding women should consume bay leaves under the direct supervision of doctors, as they may have suspected strong chemical compounds, which may cause abortion. Consuming whole bay leaves with food may become lodged in the gastrointestinal tract, causing tears or blockages and may even block breathing. In patients using central nervous system depressants, drowsiness or sedation may occur. Bay leaf may increase the risk of bleeding. Therefore, it is advised to those patients with bleeding disorders or those taking drugs that it may increase the risk of bleeding. Bay leaf or its constituents may cause contact dermatitis including airborne contact dermatitis, hand and face eczema, occupational asthma, perioral dermatitis with eczematous stomatitis and cross-sensitization to *Frullania* in individuals allergic to Lauraceae, Compositae or Asteraceae family plants.

BOTANY

Botanically, the bay leaf belongs to the family Lauraceae and has very capricious chromosomal numbers. The existence of a polyploid series with chromosome numbers 36, 42, 48, 54, 60, 66 and 72 has been demonstrated. It has been postulated that most cultivated forms of *Laurus nobilis* are autotetraploid with basic chromosome number $x = 12$, and that this basic chromosome number may be of masked or ancient polyploid nature with $x = 6$. Apomixis has been observed in tetraploid laurel. The cultivated forms of bay leaf are dicotyledonary, dioecious, large dense evergreen shrub of perennial nature or more rarely a tree for the purpose of tender fresh as well as dried leaf production. It is a tall, conical, hardy, polymorphic, mealy white, erect perennial plant that may extent to a height of 3–9 m and spread up to 1.5–6 m, depending upon availability of moisture and fertility status in soil. It may grow to 12 m in Mediterranean regions, though in 8–10 zones of the United States, it can grow from 1.8 to 7.5 m, if protected from winter winds. It is often pruned to 0.6–1.2 m tall or less for garden purposes; however, the growth rate of the plant is very slow. Branchlets are striate, terete, puberulent on young part, dark brown (young ones) and nearly glabrous or with soft hairs. The colour of bark is almost blackish brown.

Bay leaf or sweet laurel produces simple, alternate, pinnately lobed and oblong to triangular leaves of variable shape. The shape is elliptical and tapering to a point at the base and tip of the leaves. The leaf size is 2.5–7.5 cm long and 1.6–2.5 cm broad and has thin stalks that are about half as long as the leaf blade. The petiole is 0.5–1.5 cm long and purple-red when fresh and the leaf blades (3–15 × 2–5 cm) are oblong elliptical to oblong lanceolate. The leaves are smooth, curved and interconnected at ends near leaf margin. Margins are slightly undulate, coriaceous, entire and foveolate with base cuneate and apex acute to acuminate or obtuse. The leaves are shiny leathery and thick, violet-red (fresh), bright green (young) and darker green (matures) and matte olive green when it dried. The upper surface of the leaf is glabrous or with sparse soft hairs and shiny olive green, and the lower surface is dull olive to brown with 10–12 pairs of lateral veins rib and prominent on both sides.

Inflorescence is axillary, 4-5-flowered umbel, solitary or mostly 2–5 arranged in a fascicle or short raceme, suborbicular, glabrous outside and sericeous inside, peduncle 2–12 mm long, sparsely puberulent or subglabrous, involucral bracts, rounded, inner ones 0.7–1 cm long, outer ones smaller, coriaceous and glabrous outside with silky hairs inside, pedicel 2–5 mm long with soft hairs and four tepals, ovate oblong to rounded, 4.5–6 mm long and obtuse with soft hairs on both sides. Yellow or greenish white star-shaped flowers appear in clusters. In warm areas, bays bear yellowish-white fluffy flowers: male flower bears five in each umbel with 8–12 fertile stamens, in three whorls, outer whorl glandular, second and third whorls with two sessile, reniform glands at middle of filaments, small size, green colour and pilose, perianth tube short and densely pilose outside, perianth four lobed, broadly obovate or sub or bicular and appressed villous on both surfaces, anthers ellipsoidal, two-celled, cells introrse and pistillode present and female flower with four staminodes bearing an appendage at apex of connective, ovary single celled, style short and stigma slightly dilated obtusely trigonous. The fruits are small, globular to ellipsoidal, 1–2 cm long, red-blue and single-seeded berries that later turn dark violet to glossy black at maturity and seed about 12 mm in size. Pot-grown bays seldom flower and fruit. The anthers may dehisce by splitting into a number of valves. The flowers bear in March to May and fruiting occur in June to September.

CLIMATIC REQUIREMENT

Basically, the bay leaf is a temperate evergreen tree but it can be cultivated over a wide range of climates, *i.e.*, drier tropics, subtropics and warm temperate areas. In warm areas, it can grow as high as 18 m. Contemporarily, it is also cultivated as a garden and pot plant all over the world. The bay leaf plant can be grown successfully at an average annual temperature ranging from 8°C to 27°C with an annual precipitation of 300–2200 mm. However, day temperature of 18°C and night temperature of 4–13°C are ideal for its growth and development. The plants can tolerate light frost to some extent but more severe frost may kill the above ground parts, thus, the young plants require protection from both cold winds and frost. Further, low temperature may reduce not only the leaf size but also their oil content. Leaves from trees grown at or near sea level are reported to contain more oil than those from trees grown on inland hills. In general, the bay leaf plants require sunny conditions for better growth and development. As a potted plant, bay leaf grows well with at least 4 hours or more per day of direct sunlight. The size and shape of this plant can easily be organized by pruning.

SOIL REQUIREMENT

The ideal soil for better growth and development of bay leaves should preferably be moist, well drained, deep and fertile with full sun to partially shady conditions. The growth is satisfactory on a wide range of soils with a pH of 4.5–8.3. Although the crop is susceptible to water logging condition, the soil must be slightly moist throughout the lifespan for satisfactory growth and development of bay leaf plants.

CULTIVATED VARIETIES

In general, primarily two types of bay leaf cultivar are under cultivation namely Turkish and Californian type as mentioned below:

- **Turkish type:** The Turkish type cultivars produce 2.5–5 cm long to oval size leaves, which have more delicate flavour as compared to Californian type.

- **Californian type:** The Californian type cultivars produce narrow and 5–7.5-cm-long leaves.

SOWING/PLANTING TIME

The sowing of seeds or planting of cuttings are generally done from second week of April to the last week of September. However, in eastern Mediterranean region, the seeds are sown in June to July. In general, planting should be done immediately after the onset of first shower of rain since at that time, the atmospheric humidity is usually high, which is favourable for better survival of seedlings, cuttings or layering plants as well as the subsequent establishment of plants. However, the planting can also be done at other times, provided assured irrigation facilities are available.

SEED RATE

About 350–650 seedlings of bay leaf are required for planting a hectare land area. If it is grown by cuttings then 200–400 cuttings about 10–15 cm long with 2–3 buds are required for planting a hectare land area.

PROPAGATION

In general, the bay leaf plant can be propagated through sexual (seed) as well as asexual (shoot cuttings and layering) means. In all cases, there is a problem of poor seed germination in bay leaf. Normally, the seeds of bay leaf may take about 3–4 weeks to germinate. The real problem of bay leaf seed is decaying before it germinates when bay leaf is propagated through seeds. However, both direct sown and transplanting method are adopted by the growers.

Cuttings are normally prepared in late summer to early winter. Usually, the cuttings are taken from 6–8-weeks-old shoots. The cuttings must be of about 10–15 cm length. Prepared cuttings can be stored for months together without any significant deterioration in growth. The cutting will form roots after 1 year, thus, it is very difficult to be successful with cuttings. The rate of success of cutting depends on availability of high humidity. However, the propagation of bay leaf plant through layering is better than both cutting and seed methods as well.

SOWING/PLANTING METHOD

In general, the plants raised by seeds, cuttings or layering are planted when they have attained an age not less than 12–14 months. Transplanting is only necessary once every few years. The appropriate planting distance depends on the method of harvesting and availability of water. For harvesting bay leaf manually by the small holders, an initial spacing of 3 m × 3 m, gradually thinned to 6 m × 6 m is recommended. However, under commercial plantations in Israel, a spacing of 2–3 m × 2–3 m is practiced with assured irrigation facilities, whereas, in Russia, a spacing of 0.5 m × 2 m is followed in plantations that are harvested mechanically. The pits of 60 cm³ size, which are filled well with a mixture of good soil and farmyard manure or compost, are used for planting bay leaf plants. Before filling pits with soil mixture, chloropyriphos dust at 50 g per pit should be applied to protect the crop against termite and other soil insects at early stages of planting.

NUTRITIONAL REQUIREMENT

In general, the bay leaf plant grows well in most soils without any fertilizers. Once the plant is established, the extensive and deep root system of bay leaf is efficient in mining nutrients from the soil. However, for optimum growth and yield, 4–6 kg farmyard manure or compost should be added in pits at the time of planting. The requirement of 1–2 years old bay leaf plant for nitrogen, phosphorus and potash is 300, 150 and 175 g per plant.

IRRIGATION REQUIREMENT

Usually, the bay leaf plants do not require much water, thus, regular irrigation is rarely practiced in most of the plantations. However, irrigation during very dry conditions should be done regularly for the first 3–4 months planting for better establishment of the plantations. In general, irrigation should be applied soon after transplanting of seedlings or planting of cutting in order to reduce the rate of mortality for proper establishment of the plants. A well-rooted bay leaf plant can tolerate drought and needs irrigation only when persistent wilting is evident; however, for better growth and yield, the bay leaf plant should be irrigated once a month during winter season and twice a month during summer season.

INTERCULTURAL OPERATIONS

HOEING AND WEEDING

The Bay leaf is a slow growing tree, however, due to its high medicinal properties, weeds pose no serious problem. Only once or twice weeding near the plant at 40–50 days after planting is advised to fend off weeds in early growth stages. When it is grown as commercial scale, mulching can also be practiced for better germination or emergence and to conserve soil moisture, suppress weeds and also to protect from biotic and abiotic stresses. The leaves residues after distillation of sweet bay oil itself may serve as mulch and protection against frost.

PRUNING AND TRAINING

The Bay leaf can be grown in a number of ways. Pruning and training depend on whether the bay plants are trained as topiary or simply grown as a shrub on the ground. It can also be turned into topiary (trees, shrubs cut or trained into specified shapes) specimens, which can be shaped into pyramid, ball or *lollipop* standards, and some have ornately plaited or spirally trained stems. The topiary-trained bay trees are trimmed with secateurs during summer to encourage a dense habit and to maintain a balanced shape. In late spring, light pruning is advised to new shoots to a bud facing in the direction of desired growth and to remove any leaf tips damaged by winter weather. Mature bay trees can tolerate hard pruning but are slow to recover and re-grow. This would be best carried out over two or three seasons in late spring. Bays shrubs can also be trimmed into desired shape by simply cutting back to a lower leaf or bud in spring or summer.

In general, bay trees are very tolerant to light pruning although when they are hard pruned, they may take a year or so to recover. The best time to prune bay tree is late spring through to midsummer. Bay trees often throw up *suckers* or *new shoots* from just below or just above soil surface. They should be pruned off with a sharp pair of secateurs if growing above soil surface. If growing below soil, they are pruned off from as low down as possible after removing some of the soil from around the base of the sucker. The sucker may appear again in a few months, thus, the same practice should again be followed. It does not require many attempts (most of the time, it is done only in one attempt) before the shoot stays pruned. The pruned suckers can be used to propagate new bay trees.

Large bay trees withstand even very hard pruning very well including cutting back the main trunk. It is difficult to permanently damage an established one. However, they are well known for taking 2 or 3 years to get back into shape. Like many trees, they are best pruned in late spring during a dry period. Regular pruning after that annually in early spring will keep it to the desired size.

INTERCROPPING

Sweet bay is planted at a wider distance, thus, it has sufficient space for intercropping. It can be intercropped with annual crops during first 2–4 years. It will not only provide additional income to the grower but also reduce weeding requirement.

HARVESTING

Leaves of laurel are harvested in summer or autumn in temperate regions. The plants established from cuttings in spring can be harvested for the first time in autumn of the same year if it is well irrigated and maintained or in the following year if rainfed. Two to three harvests can be ended annually but two cuts have been shown to give the highest yield. Leaves can be harvested at any time. However, it is usually done manually early in the day. In Russia, systems have been developed to mechanically harvest laurel grown in hedges. When berries are to be harvested, they are mostly picked at the same time as the leaves.

YIELD

The yield of laurel leaves is strongly pretentious by the frequency of harvesting, fertilization and irrigation management. An average yield about 8–9 t/ha of fresh leaves or 4–5 t/ha of dried leaves, as leaves lose about 40% moisture upon drying can be obtained under intensive irrigation and management, even from one harvest per year. In rainfed area, an average yield about 2–3 t/ha of dried leaves is obtained from harvesting twice in a year.

Bay leaf is hardier, when it is planted in the ground. The pruning studies indicate that leave growth is fast during the first 30–40 days, becoming slower in next 15–20 days and generally ceasing after about 60 days. The leaves may persist on the tree intended for 1–3 years, which is helpful for accumulating essential oil in palisade and mesophyll cells of the leaves and it reaches a content of 1–3% on fresh weight basis. It further increases during early summer and reaches its maximum level in midsummer.

POST-HARVEST MANAGEMENT

Laurel leaves are dried in thin layers in trays in a sheltered area for 12–15 days. The leaves are pressed lightly under boards to prevent curling. The leaves of laurel should be whole, dried and olive green since if it is brown, they lose their flavour. Therefore, light should be kept out from airtight containers as the whole leaves retain flavour for over 2 years. An essential oil can also be extracted from the berries through steam distillation.

PHYSIOLOGICAL DISORDERS

YELLOW LEAVES

Yellowing of leaves is more commonly caused by waterlogged roots and cold or wet weather conditions, but sometimes, it can be a problem due to nutrients deficiency. Direct seeded or transplanted plants are more susceptible and plants grown in containers are also very prone to this problem. The affected plants shed their older leaves naturally. The compost has become hoary and exhausted, which indicates that the plants should be repotted into fresh and well-drained compost in spring.

Brown Leaves

Most of the times, it is observed that the bay plant leaves become yellow or brown in colour and gradually dyeing off, thereby reduces not only the yield but also the leaf quality. The studies reveal that a number of causes, either alone or in combination are involved for such physiological disorder. The major causes are over watering of plants, nutrient deficiency, severe cold and wind velocity, lack of general care and maintenance of crop and sometimes infested by sucking pests. Therefore, the brown leaves of bay leaf plant can be corrected after proper diagnosis of the cause(s) followed by adoption of proper remedial measures such as judicious use of water, nutrient management, protection against severe cold and wind, protection against the sucking pests and scientific management of the crops.

Peeling Bark

In general, most of bay trees develop cracking and peeling bark, especially on lower main stems. Although the exact cause of such physiological disorder is unknown, however, winter cold or possibly other stress factors like fluctuating soil moisture levels are likely to be involved to cause the cracking and peeling bark. Though the damage looks alarming, it does not appear to be invariably fatal at all. It is recommended that no remedial measures should be taken if rest of the plant is growing normally or recovering from winter damage (recovery should be apparent by midsummer if it is to happen). However, it is recommended to remove the damaged dead parts by cutting to healthy wood, which means green under the bark or to near soil level, when the growth above the damaged area is dead. Recovery from lower down or soil level often occurs.

Frost Injury

Bay trees can also be affected by hard frosts when they are grown in pots. The best remedial measure to correct the plant is to move them near the house in winter. The house walls will lessen the effect of frost and protect them from harsh winds.

INSECT PESTS

Bay Sucker (*Trioza alacris*)

Bay sucker, also known as jumping plant louse or lice, causes severe damage especially to the leaves and tender portions of the plant. The young nymphs are flattened wingless grey insects, whose bodies are covered with a white fluffy material. They suck sap from the young leaves during summer. The adults are winged, greenish brown and about 2 mm long and they overwinter in sheltered places like corners and crevices, often in the soil surrounding the bay tree. Breaking up the surface of very top soil surrounding the plant to a fine tilth will help in exposing the pests, especially in winter. They lay even smaller eggs on underside of the leaves, almost invisible to the naked eye. These eggs also feed on the sap, increasing the damage. The adults feed the leaves but the nymph causes most of the damage then the adults. They develop curled leaf margins that are abnormally thickened and pale yellow, and later, the discoloured areas or damaged parts of the leaves dry up and turn into brown. The insects may be seen underneath or near the curled leaf margin and they are most active during May to October. Two or three overlapping generations occur between May and September. Bay sucker can affect the appearance of bay but does not usually reduce the vigour of bay plants.

Control

- Pick off and destroy all the infested leaves if the infestation is light.
- Prune out the badly affected shoots in winter.
- Spray the under sides of all leaves with a much diluted washing up liquid solution.
- In severe infestation, spray the crop with Metasystox or Rogor 2 mL/L of water at 14 days interval.

APHIDS (*Myzus persicae* Sulzer)

There are several species of aphids damaging the bay tree. Among them, *Myzus persicae* Sulzer is the most prevalent. The aphid attacks both the leaves and the stem. It causes direct as well as indirect damage. The nymphs and adult females attached to the growing shoots and leaves suck cell sap, and thereby, reduce the plant vigour. The aphids secrete a honeydew like sugary substance, which invites the fungi to grow on it. When the attack is severe, the leaves, stems and small twigs are completely wilted and defoliated.

Control

- Remove the infested plant parts if aphid population is limited.
- Spray simple water with a strong jet pressure to wash the aphids from leaves.
- Use reflective mulch such as silver-coated plastic to deter aphids from feeding on plants.
- Spray the crop with insecticidal soap or oils such as *neem* cake or canola oil to cure the plants.
- Spray the crop with Acephate 2 g or Methamidophos, Pyrethrum, Quark, Cypermethrin, or Ripcord 1 mL/L of water.

SCALE INSECT (*Coccus hesperidum*)

Scale insects, which are generally brown, are easy to overlook and more problematic particularly on indoor plants. They can hide in plain site on the stems and under the leaves. In general, the scale insect looks like the part of plant. The insects have a crawling stage, where they hitch a ride onto plant and crawl around looking for a decent place to settle down. When they find it, they lock down tight on the plant and they suck juice from the tissues. Sometimes, the insects like aphids also exude a sticky shiny substance called honeydew, which adheres to the leaves and surfaces beneath the tree and favours for the development of sooty moulds. There are different types of scale namely soft, armoured scale and horse chestnut scale. Often a host plant, which is always the infected one, will attract just one type of scale. In Italy, scale insects and psyllids cause severe damage.

Control

- Apply or rub alcohol with a cotton swab to the scales at 10 days intervals.
- Spray the plant with horticultural oil, or clean off the scales from the leaves with Q-tips dipped in alcohol.
- In severe infestation, spray diazinon 20 EC 2.5 L/ha.

DISEASES

There is little information on the economic damage in laurel caused by diseases. Diseases are far more important than pests and are often widespread. The most damaging diseases of laurel are laurel wilt, root rot caused by *Phytophthora* spp., leaf spot by *Colletotrichum* spp., powdery mildew and sooty mould.

Laurel Wilt (*Raffaelea lauricola*)

Laurel wilt is caused by a nonnative beetle, which is thought to have been introduced by infected firewood. It is also called ambrosia beetles and it primarily attacks the already stressed bay trees by boring into the wood. The damaged portions of the foliage are wilted or discoloured to shades of red or purple. The foliage eventually turns brown but remains on the branches.

Control

*   Hoe around the plant.
*   Apply systemic granular insecticide such as Phorate or Aldicarb 10G 10–15 kg/ha or Carbofuran 3G 30–33 kg at the time of sowing/planting.
*   Drench the field with fungicide propiconazole to control the wilt.

Root Rot (*Phytophthora cinnamomi*)

Symptoms appear mainly on leaves and tender twigs. It causes stunted growth of the foliage, yellowing and wilting of leaves and dieback of twigs. In severe attack, it may be incurable in both young and mature plants. Sometimes, cankers, stains or streaks may develop on bark of the trunk and darkened spots can also develop on the infected shrub. In extreme infections, it causes stem dieback and portends the life of broadleaf evergreen. The fungus attacks the plants growing in high rainfall, cool temperature and moist soils.

Control

*   Provide good drainage conditions in the field.
*   Provide support to the plant to avoid contact of plant to the soil.
*   Affected plants should be uprooted and burnt.
*   Use healthy root stocks.
*   Spray the crop with Ridomil MZ (0.25%) followed by Indofil M-45 or Blitox (0.25%) at 10–12 days intervals.

Anthracnose/Leaf Spot (*Colletotrichum nobile*)

The disease becomes very severe in wet weather at maturity of the crop. It causes black or brown spots on the leaves and becomes progressively larger. Later, the infected leaves start to wilt, distort, dry and curl and the trunks and stems may develop cankers and swollen or sunken tissues. The low level of this disease infection is very common but it affects the leaf's oil content more severely as compared to damage to the leaves.

Control

*   Uproot and destroy the disease debris.
*   Prune all the affected twigs and foliage from the shrub and dispose of them.
*   Add 5–7.5 cm layer of mulch around the base of bay plant, which helps in preventing fungal spores in soil from splashing onto the shrub.
*   Treat the seeds with Agrosan GN, Thiram, Bavistin or Benomil at 2 g/kg of seed or dip the seed in 0.125% solution of Captan for half an hour.
*   Spray the crop with 0.1–0.2% Bavistin at 15–20 days interval.

Powdery Mildew (*Odium spp.*)

The fungal disease appears as white mycelial patches dotted with the fruiting bodies the abaxial leaf surface and even on stem. The affected plant leaves are wilted, stunted, discoloured and distorted.

The severely infected plants are defoliated and weakened by premature drying up and death of infected leaves. This contagious fungus spreads *via* spores carried from plant to plant by wind.

Control

- Remove and destroy all the affected plant parts.
- Provide proper irrigation and ventilation.
- Apply *neem* oil, potassium bicarbonate or horticultural oil to the pretentious plant parts.
- Spray the plant with 0.025% Benomyl or 0.1% Dinocap.

Sooty Mould (*Alternaria* spp. and *Cladosporium* spp.)

Sooty mould looks worse than it actually is, and generally, it does not cause any serious harm to the bay laurel. Sooty mould grows on honeydew, which is naturally produced by several different species of sap-sucking insects that feed on plant cells, coursing through the bay laurel. Bay laurel is susceptible to aphids and scales and both of these pests secrete honeydew. The black soot-like fungus easily washes off the bay laurel with a water hose. However, if the honeydew-secreting pests are not dealt with, the powdery mildew will return.

Control

- Apply a plant-based insecticide like *neem* oil.
- Introduce lacewings, parasitic wasps or ladybugs to feed on the pests.

Note: If the leaves are to be used for culinary purpose, a minimum of 14 days interval must be left between spraying and using the leaves.

■■■

112 Breadfruit

R.K. Tarai, Dipika Sahoo and Asish Kumar Panda

Botanical Name	: *Artocarpus altilis* (Parkinson) Fosberg
	Artocarpus communis J.R. Forst and G. Forst.; *Artocarpus incisus* L.F
Family	: Moraceae
Chromosome Number	: 2n = 56

ORIGIN AND DISTRIBUTION

Breadfruit is believed to be native to a vast area extending from the Indo-Malayan archipelago to the Philippines and the Moluccas through New Guinea to Western Micronesia. The seeded breadnut (*Artocarpus camansi*), which is considered as an ancestor of breadfruit, originated in Papua New Guinea in the Western Pacific region. Breadfruit has been a staple food and traditional crop in the Pacific for more than 3,000 years and is now being cultivated for food security in the Caribbean and other tropical regions. British and French navigators introduced a few Polynesian seedless varieties to the Caribbean islands during the late eighteenth century, and today, it is grown in some 90 countries throughout South and Southeast Asia, the Pacific Ocean, the Caribbean, Central America and Africa. Breadfruit is widely planted in tropical Asia and has become a popular staple food crop and abundantly planted in garden areas, fallow forest lands and in home gardens throughout Melanesia, Polynesia and Micronesia. These Polynesian varieties were then spread throughout the Caribbean and to Central and South America, Africa, India, Southeast Asia, Madagascar, the Maldives, the Seychelles, Indonesia, Sri Lanka and northern Australia. Breadfruit is also found in south Florida. In some areas, only the seedless type is grown, while seeded types are grown in other areas like Haiti.

INTRODUCTION

Breadfruit has long been an important staple crop and a primary component of traditional agro-forestry systems. It was regarded as a staple crop in the Jamaican diet and now experts believe that it could provide food security on the island, which imports more than half of its food. Though the crop has been distributed widely, it has been scored as an underutilized crop and little attention has been given to its large-scale commercial cultivation owing to limited knowledge on planting material, agronomic practices and production techniques. Recently, the breadfruit has been identified as an alternative source of carbohydrates, and its starch can be processed into many forms that would offer potential opportunities in agro-processing and agro-export industries. When it is just ripe enough to eat, it is more like bread as it is starchy and dry but when it is softer and riper, it tastes sweeter and is moister. In addition to producing abundant, nutritious, tasty fruits, this multipurpose tree provides medicine, construction materials and animal feed. The trees begin bearing in 3–5 years and are productive for many decades. It is easy to propagate, as it requires little attention and input of labour or materials, and it can be grown under a wide range of ecological conditions. Mostly, breadfruit is produced for subsistence purpose and a small quantity is available for sale in town markets as fresh fruit or chips.

COMPOSITION AND USES

COMPOSITION

Breadfruit is a rich source of starch, carbohydrates, sugars, dietary fibre and a fair source of thiamine, riboflavin, niacin, vitamin C, calcium, phosphorus, potassium and copper. The nutritional composition of breadfruit is given below in Table 112.1.

USES

Breadfruit is used more as a vegetable than as a fruit. It is eaten after cooking, boiling, baking, roasting, frying or making soups. It is usually cooked after peeling and slicing and makes a good delicacy in combination with ingredients like coconut cream or grated coconut meat. The entire fruit is often roasted to make a sweet dessert, something like pudding, though not everyone likes its unique taste. Chips and biscuits can be made from it. Breadfruit may be eaten ripe as a fruit or under ripe as a vegetable like banana and plantain. The dried fruit is used for making flour as a substitute for wheat flour in bread making. The combination has been found more nutritious than wheat flour alone. Breadfruit flour is much richer in lysine and other essential amino acids than wheat flour. Soft or over ripe breadfruit is best for making chips. The male and female flower buds are often eaten as a cooked vegetable. Fallen male flower spikes are boiled, peeled and eaten as vegetable. The seeds are boiled, steamed, roasted over a fire or in hot coals and eaten with salt. Its leaves are eagerly eaten by domestic livestock, thus, they are fed to cattle and goats. Under ripe fruits are cooked for feeding to pigs and uncooked soft ripe fruits are used as animal feed.

Breadfruit trees are planted as windbreaks, ornamentals and as shade trees in coffee plantations. Its latex has been used in the past as birdlime on the tips of posts to catch birds. The wood is used for construction and furniture. The inner bark is used to make bark cloth called *tapa*. Material for tape cloth is obtained from the inner bark of young trees and branches. Leaves are used for

TABLE 112.1
Nutritional Composition of Breadfruit (Per 100 g Edible Portion)

Constituent	Contents	Constituent	Contents
Water (g)	70.65	Thiamine (mg)	0.110
Carbohydrates (g)	27.12	Riboflavin (mg)	0.030
Total sugar (g)	11.0	Niacin (mg)	0.900
Protein (g)	1.07	Pyridoxine (mg)	0.329
Total dietary fat (g)	4.9	Vitamin C (mg)	29
Total lipid (g)	0.23	Vitamin E (mg)	0.1
Calcium (mg)	17	Vitamin K (µg)	0.5
Phosphorus (mg)	30	Lysine (g)	0.037
Potassium (mg)	490	Methionine (g)	0.010
Magnesium (mg)	25	Cystine (g)	0.009
Iron (mg)	0.54	Phenylalanine (g)	0.026
Sodium (mg)	2	Tyrosine (g)	0.019
Manganese (mg)	0.060	Valine (g)	0.047
Selenium (µg)	0.6	Lutein-zeazanthin (µg)	22
Zinc (mg)	0.12	Energy (kcal)	103

wrapping food for cooking, for parcelling of fresh food and as plates. The male flower spike used to be blended with the fibre of paper mulberry to make elegant loin cloths.

MEDICINAL USES

Breadfruit is used in many traditional folk medicines. A decoction of its leaf is believed to lower blood pressure and is also said to relieve asthma. Crushed leaves are applied on the tongue as a treatment for thrush. The leaf juice is employed as ear drops. Ash of burned leaves is used on skin infections. A powder of roasted leaves is employed as a remedy for enlarged spleen. The crushed fruit is used as poultice on tumors to ripen them. Toasted flowers are rubbed on the gums around an aching tooth. The latex is used on skin diseases and is bandaged on the spine to relieve backache. Diluted latex is taken internally to overcome diarrhea. The roots have been used as a component for venereal diseases and cancer. The roots and stems having detoxifying and anti-inflammatory effects are used traditionally for the treatment of liver cirrhosis and hypertension.

BOTANY

Breadfruit tree can grow up to a height of 30 m or more at maturity, more commonly around 12–15 m and the trunk may be as large as 2 m in diameter. The leaves, evergreen or deciduous depending on climatic conditions, are more or less deeply cut into 5–11 pointed lobes. Leaf blade is 12–59 cm long and 10–47 cm wide but juvenile leaves often larger with up to 13 lobes cut from one-third to four-fifth of the midrib varying in size and shape on the same tree, below pale green, above dark green and glossy, glabrous to moderately pubescent with pale or colourless rough-walled hairs on the midrib. Leaf size is variable ranging from 15 to 60 cm long, depending on the variety. The tree is monoecious with male and female flowers appearing on the same tree. The mature tree bears both male and female inflorescence in the leaves axils. The male inflorescences appear first and are up to 25 cm long generally and light yellow at first but later become light brown. Male breadfruit inflorescences produce little to no pollen, therefore, the female flowers are not fertilized and no seed is produced by the fruits of traditional cultivars grown in the Caribbean. About 3–4 weeks after the appearance of male inflorescences, round to oval female inflorescences begin to emerge in the leaf axils. These inflorescences consist of hundreds of small female flowers, which fuse together and develop into one large compound fruit, called a syncarp. In breadfruit, the flowers develop into fruitlets without pollination. These are fused for most of their length, therefore, the fruit inside has a spongy texture near the core, and fruit skin has a netted appearance. Young fruits up to four may arise in one leaf axil but at least two fruits fall off usually when they are 4–6 weeks old. It is a cross-pollinated crop. Fruits are variable in shape, size and surface texture. They are usually round, oval or oblong ranging from 9 to 20 cm wide and more than 30 cm long, weighing 0.25–6 kg. The skin texture varies from smoothly to slightly bumpy or spiny. The colour is light green, yellowish-green or yellow when mature. The skin is usually stained with dried latex exudations at maturity. The flesh is creamy white or pale yellow and contains zero to many seeds, depending upon the variety. Fruits are typically mature and ready to harvest in 15–19 weeks. Ripe fruits have a yellow or yellow-brown skin and soft, sweet, creamy flesh that can be eaten raw. Seeds are thin-walled, sub-globose or obovoid, irregularly compressed, 1–2 cm thick and embedded in the pulp. The outer seed coat is usually shiny dark brown with a light brown inner seed coat. Seeds have little or no endosperm and no period of dormancy. They germinate immediately and are unable to withstand desiccation.

CLIMATIC REQUIREMENT

Breadfruit is a tropical fruit crop that can be grown best at temperature between 21°C and 32°C, and its growth is retarded at lower or higher temperature. It thrives best in locations where the annual rainfall receipt is 1500 to over 3000 mm and well-distributed throughout the year with a relative humidity of 70–80%. However, in South India, it is cultivated at sea level and up to an altitude of 1065 m. The trees compete strongly for light by growing quickly to produce a tall trunk with high branches. Therefore, the trees should be exposed to full sunlight to encourage a lower canopy to develop.

SOIL REQUIREMENT

Breadfruit tree can be grown on a wide range of soil types but it prefers sufficiently deep, well-drained and moisture retentive soil. Some varieties are well adapted to shallow sandy soils. Soils that are prone to water logging or seriously eroded and shallow should be avoided for its cultivation. The soils should have fairly high organic matter content, adequate nutrient status and a pH of 6.1 to 7.4.

CULTIVATED VARIETIES

Hundreds of breadfruit varieties are perpetuated clonally by vegetative propagation. The most widely known cultivars of breadfruit are the *Yellow* and the *White*. In some countries, there are no given names but cultivars are distinguished mainly by fruit shape, *i.e.*, round or long.

YELLOW

The fruits are round or oval in shape, and at maturity, the skin is smooth, greenish brown and usually heavily stained with latex. Fruit weight ranges from 1.5 to 3.5 kg. The flesh of the fruit is light yellow and the colour becomes brighter with age and on cooking. The major bearing period extends from June to September and the minor season is from December to February. This is the preferred cultivar in the Caribbean on the basis of its pleasant taste and soft smooth texture. The mature stage of the fruit is considered to be the best for roasting, while the green mature or slightly immature stage is preferred for boiling.

WHITE

This cultivar is similar to *Yellow* cultivar in shape, size and to some extent skin colour, except that there is less brown discoloration and the skin texture may be slightly rougher at maturity. The cultivar can be distinguished from the *Yellow* cultivar mainly by colour and eating quality of the flesh. The flesh colour is off-white to cream and does not change much as the fruit ages or with cooking. The eating quality is good with a mild taste and a smooth but somewhat firmer texture than the *Yellow* cultivar. *White* cultivar is very good for boiling and frying, and its fruits can be roasted at full mature stage.

MAOPO

This cultivar has an almost entire leaf with shallow lobes at the tip. The seedless fruits are oval or broad ovoid with pale white or creamy flesh, 16–26 cm long and 16–18 cm wide, weighing 2–3.5 kg with an average fruit weight of 2.4 kg. The tree reaches a height of 15 m or taller, and the timber is used for house building.

MA'AFALA

This cultivar is common throughout Polynesia. Its tree is generally smaller up to 10 m tall with a spreading canopy. The small leaves are moderately dissected with 3–5 pairs of lobes. The small fruits are oval or oblong with white flesh, 12–16 cm long and 10–13 cm wide, weighing 0.6–1 kg with average fruit weight 0.75 kg and zero to few seeds.

Puou

This cultivar is common throughout Polynesia. Its tree is generally smaller up to 10 m tall with dense spreading canopy. The large leaves are dull and shallowly dissected with 4–6 pairs of lobes. The fruits are round, 12–20 cm long and 11–17 cm wide, weighing 1.2–2.4 kg with average fruit weight 1.5 kg.

Aravei

Fruits of this cultivar are ellipsoidal, large, 10–30 cm long and 15–22 cm wide. Their rind is yellowish-green with brown spots on sunny side and rough with sharp points, which are shed on maturity. Pulp is light-yellow, dry or flaky and of delicious flavour after cooking, which takes very little time. Core is long and slim with many abortive seeds.

Havana

Fruits are oval-round and spiny. Rind is yellowish-green and pulp is golden-yellow, moist and pasty, which separates into loose flakes when cooked and very sweet with excellent flavour. Core is oval and large with a row of abortive seeds. Fruits are highly perishable, thus, must be used within 2 days. Fruits cook quickly over fire. Popularly, it is claimed to be one of the best breadfruits.

Maohi

Fruits are round and 15 cm wide with large core. Rind is rough and spines with bright yellowish green and red-brown patches. Pulp is cream-colored and smooth when cooked and is of very good flavour but slow in cooking. Tree, which is the most common of Tahiti, is a heavy bearer.

Paea

Fruits of this variety are ellipsoidal with length 28 cm and breadth 22.8 cm. Rind colour is spiny with yellowish-green colour. Core is oblong and thick with a row of brown and abortive seeds. Pulp colour is bright yellow, separating into flakes when cooked. Formerly *Paea* was reserved for chiefs only. It needs 1 hour to roast on open fire. The tree is tall, well-formed and elegant.

Pei

Fruit shape is broad ellipsoidal and large. Rind colour is light green, and pulp is light yellow, flaky and aromatic when cooked. The fruit cooks quickly. Its taste is sweet and delicious with fruity flavour. The tree of this variety bears fruit early.

Pucro

Fruit is spherical or elongated and large. Rind is very rough and spiny with yellowish green and small brown spots. Pulp colour is light yellow and smooth of excellent flavour. The fruit cooks quickly. The fruits of this variety are highly ranked.

Rare

Fruit of this variety are broad ovoid and 17.5 cm long. Rind is rough and spiny with bright green colour. The pulp colour is deep creamy, fine-grained, smooth and flaky when cooked. Its taste is very sweet with excellent flavour. Core is small with a great many small abortive seeds. Fruits are borne singly on a tall, open, short-branched tree.

Rare Aumee

Fruit is round with length and breadth of 16.5 cm. Rind is bright green and fairly smooth at the base but rough at the apex. Pulp is deep ivory, firm and smooth when cooked. Its taste is not very sweet but of excellent flavour. The fruit cooks quickly.

Rare Autia

Fruits are roundish with fruit breadth of 15 cm. The rind is dull green with red-brown markings. Pulp is light yellow when cooked and separates into lumps with excellent flavour. Core is large with small abortive seeds.

Tatara

Fruits are broad ellipsoidal and very large weighing up to 4.5 kg. Rind has prominent faces with long green spines. Pulp is light yellow and smooth when cooked with pleasant flavour. Core is oblong. The tree is found only in small coastal valley areas where there is heavy rainfall. The tree has high branching habit with difficulty in harvesting.

Vai Paere

Fruit of this variety is obovoid with length 25–30 cm and fruit breadth 17.5–20 cm. Rind is yellowish green. Pulp is light yellow, firm and smooth, slightly acid but flavour is excellent. Core is oblong with a few abortive seeds. Fruit cooks easily. Tree is very tall and bears fruit in clusters.

PROPAGATION

Breadfruit is easy to propagate from root shoots or root cuttings, by air layering or from seeds. Its plants can also be grafted using various techniques. Stem cuttings are not used for its propagation. Seeds are rarely used for propagation as they do not develop true to type plants. Vegetative propagation is compulsory for seedless varieties, and root shoots or root cuttings are the preferred methods for both seeded and seedless varieties. Breadfruit produces numerous root shoots when roots are cut or damaged. It quickly re-grows new shoots and branches after wind damage or when topped to facilitate harvest. For multiplication, it is better to make root cuttings of about 2.5–6.4 cm thick and 22 cm long. The ends may be dipped into a solution of potassium permanganate to coagulate the latex, and the cuttings are planted close together horizontally in sand. They should be shaded and watered regularly. Callus may form in 6 weeks (though rooting time may vary from 2 to 5 months) and the cuttings are transplanted in pots and watered once or twice a day for several months or until the plants are 60 cm high. In India, breadfruit scions can successfully be grafted or budded onto seedlings of wild jackfruit trees.

Root shoots

The use of root shoots to propagate breadfruit is the traditional method and some varieties like *Puou* produce numerous root shoots. Healthy shoots are collected when they are at least 20–25 cm tall and the stem has become woody and is producing lobed leaves. Shoots up to 1 m tall can be used for propagation. The attached root of 10–15 cm on either side of the shoots is removed by cutting. The shoots and attached root system are then carefully lifted. The percentage of successful rooting and shoot growth ranges from 50% to 90%.

Root Cuttings

Root shoots from a desired variety are not always available, thus, root cuttings can be used for mass propagation of breadfruit. However, the roots should be collected from healthy and vigorous trees just beneath the soil surface. Roots of 1.5–3 cm diameter can be used for propagation. Removing roots larger than 6 cm diameter can be detrimental to the tree because it may damage the root system and the wounded area may heal slowly. Root cuttings should be kept under 60% shade and moist but not wet. The roots should never be allowed to dry out. The percentage of rooting ranges from 75% to 85%. Shoots begin to develop from adventitious buds after 3–4 weeks. When shoots are 20–25 cm tall with their own root system (usually in 4–6 months), they are uprooted and transplanted in pots of 20–30 cm diameter. They may attain a size of 0.6–1.6 m in 6–9 months.

AIR-LAYERING

It is best to air-layer the branches at the beginning of rainy season when the tree is in an active vegetative phase, producing new shoots and leaves and before the fruits appear. Branches of 2–4 cm thickness are prepared for air layering by removing a 2.5–3 cm wide strip of bark around the circumference of branch. Moist sphagnum moss or other organic medium is wrapped around this area. After 2–4 months, new roots will develop and grow through the medium. The air-layer is removed by cutting the branch directly below the roots and placed in a 10–20 cm polyethylene bag containing a well-drained medium until the plant has established a root system within a year.

IN VITRO PROPAGATION

Currently, very limited reports are available on *in vitro* propagation of breadfruit. *In vitro* propagation involves *in vitro* culture initiation, multiplication of shootlets, root initiation and acclimatization of plantlets. Suitable explants like shoot tips, axillary buds, seeds or zygotic embryos and appropriate medium need to be standardized for successful *in vitro* propagation of breadfruit plants.

PLANTING

Trees may be planted at any time provided enough soil moisture or irrigation is available. The best time for planting is during the beginning of rainy season. Planting is done by digging pits of 1.5 m depth and 1 m width. The recommended spacing is 10 m × 10 m that would give a plant population of 100 trees per hectare. Square (8 m × 8 m) or rectangular (8 m × 10 m) system of planting permits easy movement within the field and is suitable for flat areas. A hexagonal system of planting gives 10% more trees per hectare than the square arrangement. The planting is done after removing the plastic bag and slightly roughening the sides of the root ball. The plant should be turned to orient the foliage to the best position for maximum light interception. The stakes are installed either in the pit before it is filled with soil or outside the pit after planting. It may also be necessary to install protection staking where livestock, pests or wild animals such as monkeys have access to the orchard and may damage the young trees.

AFTERCARE

Young plants prefer partial shade. It is best to plant at the onset of rainy season but if the weather is dry, irrigation is applied for the first 1–3 months of plant establishment. Once the breadfruit tree is established, it can withstand a dry season of 3–4 months, although it prefers moist conditions. Mulching of young plants is beneficial by keeping the soil moist and adding a steady supply of nutrients. It also helps in controlling weeds around the root system. Use of herbicides to control weeds around base of the tree can damage the tree if it comes in contact of surface roots or young trunk. Young trees need to be protected from cattle, goats, horses and pigs that will eat the bark and tender shoots.

NUTRITIONAL REQUIREMENT

Proper nutrients management is essential to develop a desired form of root system and canopy. Fertilizer application must be based on soil nutrients analysis, and as the trees develop, analysis of nutrients level in the leaves should be used. Very little or meager information is available so far as application of manure and fertilizers in breadfruit is concerned. However, it has been reported that periodic application of ammonium sulfate 0.5–1.0 kg/tree or 2.0 kg superphosphate increases fruit size and fruit quality significantly. A quantity of 25 kg farmyard manure along with a fertilizer mixture, N:P:K, in the proportion of 7:10:5 should be applied at 1–2 kg/tree. Organic matter as compost or well-decomposed farmyard manure is beneficial to soil with low organic matter, and it should be placed at least 20 cm away from base of the tree trunk to avoid a problem of collar rot. The fertilizers should be applied in a shallow ring around the tree or outside drip line of the canopy and covered with soil to prevent loss through rainfall run-off.

IRRIGATION REQUIREMENT

Breadfruit has a higher water requirement because of its shallow root system and larger canopy. Irrigation during summer helps in checking fruit drop. The young trees perform best under uniformly moist soil conditions throughout the year. Therefore, for rain-fed orchards in locations with short dry spells and with soils that retain moisture, the orchard floor should be mulched with the fallen leaves or other well-decomposed organic materials to conserve soil moisture. Irrigation must be applied immediately at the onset of dry spell to such soils to avoid low moisture stress symptoms, *e.g.* wilted leaves, premature leaf senescence or yellowing, excessive leaf fall and ultimately stunted growth. Drip irrigation may be used for young plants but micro-jet sprinklers are more effective as they encourage growing trees to develop a more extensive root system and they also use water more efficiently than overhead irrigation. Care should be taken to facilitate proper drainage condition because prolonged period of excessive soil moisture may lead to root death and kill young trees.

TRAINING AND PRUNING

Generally, training and pruning are not practiced in breadfruit. However, first lateral branches from the tip can be pruned to restrict the upward growth and to facilitate more canopy spread. Young breadfruit trees should be trained to develop a low branching canopy to facilitate harvesting. Pruning of 1.0–1.2 m tall trees by heading back of main stem encourages branching at less than 1 m. Young trees can be pruned at any time of the year during dry weather. The fruits are borne towards ends of the branches; therefore, training should be completed when the trees are 30 months old to allow them to develop sturdy branches with enough foliage for the mature phase. Excessive leaders and additional shoots that crowd the interior of canopy should be removed because heavily shaded branches may die and those that survive will produce fewer and smaller fruit. Pruning of weak, criss-cross, intermingling, dead and diseased branches should be done at the end of *Kharif* season and after harvesting the fruits for better penetration of sun light to the inner canopy.

INTERCROPPING

In a breadfruit plantation, various short duration fruit crops like banana, papaya and pineapple and vegetable crops like okra, cabbage, cauliflower and legumes can be taken as intercrop for fetching additional income in the initial years.

FLOWERING AND FRUITING

Male flowers and fruits develop at tip of the branches, with the male flowers occurring first. The fruiting season typically coincides with the wet, rainy summer months but a smaller flush may occur about 5 months later for some varieties. New leaves are produced round the year with a heavy flush after a period of rest that follows the end of fruiting season.

HARVESTING

The tree comes to bearing in 3–6 years, depending upon the type of planting material used. The fruits are harvested 60–90 days after emergence of inflorescence. Fruits are harvested both for vegetable and fruit purpose. Fruit maturity can be judged by pattern and colour of the fruit skin, flesh texture and colour and sugar composition. However, changes in skin colour from green to light green or yellow and flattened or wider spines on the skin indicate fruit maturity. Fruits harvested for vegetable purpose should be light green with irregular fruit surface, firm flesh and white pulp, which is hard to squeeze. Fruits with yellow skin colour with brown areas, flattened fruit surface and creamy soft pulp can be used for dessert purpose. The presence of latex stain on fruit surface is a good indicator of fruit maturity. On scratching, if a small drop of latex appears on the fruit surface, it indicates that fruit is ready for harvesting as a starchy vegetable. However, intensification of brown staining on the fruit surface indicates complete ripening of the fruit. Fruits are picked manually by cutting fruit stalk or by using a pruning pole. The harvested fruit should be detained by hand or net before reaching on the ground. Fruits are available during February to March and June to August.

YIELD

With a plant population of 100 trees per hectare, a yield of 160–500 q/ha can be obtained. Generally, 50–150 fruits can be obtained per tree. A fully grown tree can also yield about 500–2,000 fruits each weighing 1–4 kg.

POST-HARVEST MANAGEMENT

Post-harvest management of fruits is necessary for reliably delivering fruits in the condition that the customers want from the retailers. After harvesting, the breadfruit should be washed to remove dirt and field debris in cool water containing 150 ppm sodium hypochlorite and with pH 6.5 to prevent post-harvest disease infection, and then, they are air-dried on a flat surface. Fruits for export should be graded based on size. Well-ventilated cartons should be used and fruits should be packed in a single layer and separated from adjacent fruits by separators to prevent bruising during transport. The fruits have to be stored at a storage temperature of 12.5–13°C since lower temperature will cause chilling injury, while at higher temperature, the fruit will ripen more quickly. The ideal relative humidity to prevent water and weight loss and shrivelling is 90–95%. The storage time between packing and delivery to the market should be as short as possible to ensure that the fruit still has a reasonable shelf life when it reaches the consumer.

INSECT PESTS AND DISEASES

Generally, the tree is relatively pest free. However, there are some specific localized insect pest and disease problems, which are described below.

INSECT PESTS

Fruit Fly (*Bactrocera umbrosa*)

This species causes considerable damage to breadfruit by ovipositing on ripe breadfruit. It also affects younger fruits, causing premature ripening and fruit drop. Its population peaks in December to January. It attacks 30% to 75% of the breadfruits in different parts of the world.

Control

- Install pheromone traps.
- Inspect the host fruits regularly.
- Keep the fruits in bags.
- Collect and destroy the fallen and overripe fruits.
- Harvest the fruits at mature green stage.

MEALY BUG (*Icerya aegyptiaca*)

Mealy bug is responsible for huge crop losses in breadfruit. Heavy infestation of the pest kills young leaves and stems, reduces fruit yield by 50% and may even kill mature trees.

Control

- Plough the orchard around the tree.
- Apply heavy irrigation to destruct the eggs of mealy bugs.
- Introduce predatory ladybird beetle (*Rodolia limbata*).
- Provide a band to tree trunk with a 400 gauge polythene sheet up to 30 cm above ground level to prevent movement of mealy bugs.
- Rake the soil around the tree trunk and mix 2% methyl parathion dust 250 g per tree.

DISEASES

FRUIT ROT (*Glomerella cingulata* and *Phytophthora palmivora*)

Fruit rot is the major problem in breadfruit. Due to the infection of *Glomerela cingulata*, the round brown spots appear on the surface of immature fruits and they may be extended by insects such as bees and by wind. The spots expand to form larger lesions that are grey at the centre and dark brown at the margins. The spores also infect mature fruits, which may fall off the tree. On the other hand, *Phytophthora palmivora* infection appears first as small water-soaked lesions on the surface of mature green fruit that spread and develop light brown centres, with fungal growth and spores at the margins. With both pathogens, the lesions may coalesce to form large discoloured areas on the peel. The entire fruit may become mummified on the tree and if not removed early, it can infect branches also and cause dieback. Both the diseases are spread by rain and water and are more prevalent under high temperature and high relative humidity conditions of rainy season.

Control

- Provide proper drainage in the orchard.
- Prune the tree appropriately for proper air circulation. Maintain proper field sanitation.
- Collect and destroy the fallen fruits from the trees and floor to reduce the inoculums load.
- Spray 1% Bordeaux mixture to control fruit rot, especially during rainy season.
- Apply copper fungicides like copper oxychloride (0.3%) at the onset of rainy season and during fruit development to reduce the incidence.

DIEBACK OR DECLINE (*Fusarium* spp.)

The symptoms are excessive leaf yellowing and leaf fall, followed by dieback of the younger branches. This dieback eventually extends to larger branches and the trunk and can kill the entire tree. The disease occurs mainly on mature trees. Death can occur in 6 months or the tree can decline gradually over a period of 2 years. Several non-aggressive fungi, bacteria and high population of certain nematodes species have been found in association with diseased trees. Certain environmental conditions such as water logging, soil salinity and wind stress due to hurricanes may spread the disease rapidly.

Control

- Maintain proper drainage in the orchard of breadfruits.
- Keep the trees healthy, giving special attention to its root system.
- Maintain proper field sanitation.
- Follow soil solarization to reduce nematode population.
- Spray Bordeaux mixture 1% at a month interval.

ROOT ROT (*Rosellinia* sp.)

The disease spreads rapidly and trees do not usually survive. It is characterized by a mycelial fan that surrounds the stem at ground level, which causes wilting and death of the tree. Identification of the disease can be done by removing the dead tree. Once the fungus is in the soil, it can survive for years. The root rot affected plants are usually smaller, less vigorous, produce fewer leaves, flowers and fruit than healthy plants of equal age. Flowering is delayed. As a result, the crop quality is very irregular.

Control

- Isolate the affected trees.
- Remove the infected material from the soil.
- Provide adequate soil drainage and avoid over irrigation.
- Apply lime to the soil to raise its pH.
- Spray copper oxychloride 0.3% or SAAF 0.2% alternatively at 15 days interval.

■■■

113 Jackfruit

R.K. Tarai, D.K. Dora, S.C. Swain and
Asish Kumar Panda

Botanical Name   : *Artocarpus heterophyllus* Lam.

Family   : Moraceae

Chromosome Number : 2n = 56

ORIGIN AND DISTRIBUTION

No one really knows its place of origin but it is believed to be indigenous to rainforests of the Western Ghats. It is widely cultivated in tropical regions of India, Bangladesh, Nepal, Sri Lanka, Vietnam, Thailand, Malaysia, Indonesia and the Philippines. It is also found across Africa in Cameroon, Uganda, Tanzania and Mauritius as well as throughout Brazil and Caribbean nations like Jamaica. However, jackfruit is considered to be native to India. In India, it is now widely culti-vated throughout the tropical low land in both hemispheres and found distributed in southern states such as Kerala, Tamil Nadu, Karnataka, Goa, coastal Maharashtra and other states such as, Assam, Bihar, Tripura, Uttar Pradesh and foothills of the Himalayas. It is very popular in eastern and south-ern parts of India. In North India, this tropical fruit grows along the foothills of the Himalayas and in South India up to an altitude of 2400 m. It is found growing wild in Western Ghats of India and is a common sight around every home in Kerala. The fleshy carpel, which is botanically the perianth, is the edible portion.

INTRODUCTION

Jackfruit, also known *kathhal* in Hindi, is considered as the poor man's crop in India. It is grown in homestead gardens without any management practices. All parts of the tree contain sticky white latex. Its fruit has a unique compound structure. The exterior of the fruit is green or yellow when ripe and composed of numerous hard cone-like points attached to a thick, rubbery, pale yellow or whitish wall. The interior consists of large fully developed *bulbs* (called perianths) of yellow, banana-flavoured flesh, massed among narrow ribbons of thin, tough undeveloped perianths and a central pithy core. Each bulb encloses a smooth, oval, light-brown seed covered with a thin white membrane. There is one other unique and peculiar aspect about the jackfruit, *i.e.*, when fully ripe, the unopened fruit emits a strong disagreeable odour, resembling that of decayed onions, while the pulp of the opened fruit smells of pineapple and banana. Within the seed, it is thick, white and crisp. One single fruit can have from 100–500 seeds. Ninety per cent of the North Indians consume it as a vegetable.

COMPOSITION AND USES

Composition

The fruit is deliciously sweet in taste. It is rich in energy, dietary fibre, minerals and vitamins and free from saturated fats or cholesterol, making it one of the healthy summer treats to relish. The

TABLE 113.1

Nutritional Composition of Jackfruit (per 100 g Edible Portion)

Constituents	Contents	Constituents	Contents
Water (g)	76.2–85.2	Zinc (mg)	0.42
Carbohydrates (g)	23.5	Vitamin A (IU)	110
Protein (g)	1.72	Beta-carotene (µg)	61
Total fat (g)	0.64	Crypto-xanthin-β (µg)	5.0
Dietary fibre (g)	1.5	Lutein-zeazanthin (µg)	157
Calcium (mg)	34	Thiamine (mg)	0.105
Phosphorus (mg)	36	Pyridoxine (mg)	0.329
Potassium (mg)	303	Riboflavin (mg)	0.055
Magnesium (mg)	37	Niacin (mg)	0.920
Sodium (mg)	3.0	Vitamin C (mg)	13.7
Iron (mg)	0.60	Vitamin E (mg)	0.34
Manganese (mg)	0.197	Energy (Kcal)	95.0
Selenium (mg)	0.6		

ripe fruit is a rich source of β-carotene (a precursor to vitamin A), iron, pectin and contains about 1.9–2.2% protein on a fresh weight basis. The seeds are very rich in carbohydrates and proteins. The nutritional composition of jackfruit is given in Table 113.1.

Uses

Jackfruits are eaten unripe at 25–50% of full size as a vegetable or ripe as a fruit. When used as vegetable, it is peeled, sliced, boiled and then mixed with other food. The seeds are boiled or roasted and used in many culinary preparations. The latex from the bark contains resin. Pickles and dehydrated leather are its preserved delicacies. Canning of flakes can be done. They can be bottled and served after mixing with honey and sugar. Nectar is prepared from its pulp. The rind, rich in pectin, can be used for making jelly.

The skin of the fruit and its leaves are excellent cattle feed. Its timber is valued for making musical instruments, furniture, doors, windows and roof construction since it is rarely attacked by white ants.

Medicinal Uses

Jackfruit is also rich in phytonutrients such as *lignants, lisoflavones* and *saopnin*, which have anti-cancerous and antiaging properties. It is also known to contain anti-ulcer properties, which help in treating ulcers and digestive disorders. Additionally, the presence of high fibre in the jackfruit prevents constipation and helps in smooth bowel movements. Its vitamin A is known to maintain the eyes and skin healthy. Lectine, which is a natural protein found in jackfruit, is used for treatment of cancer. An extract of jackfruit called *jacaline* inhibits the growth of human immunodeficiency virus (HIV) infection in *in-vitro*. It is considered as an energy generating fruit due to the presence of simple sugars like fructose and sucrose, which give an almost immediate energy boost. Potassium contained in jackfruit has been found to be helpful in lowering of blood pressure and thus reducing the risk of heart attack as well as strokes. Its roots have been found to help those who suffer from asthma. It is rich in magnesium, a nutrient that is important in the absorption of calcium and works with calcium to help in strengthening bone and preventing bone-related disorders like osteoporosis. It also contains iron, which helps in preventing anaemia and also helps in proper blood circulation in our body. Extract of leaves has been found to promote glucose tolerance when tested on diabetics. Hot water leaf extract contains flavonoids, anthocyanins, tannins and proanthocyanidin, which increase the glucose tolerance in diabetics.

The leaves of jackfruit tree are useful for curing fever, boils and skin diseases. When heated, they prove useful in curing wounds. The latex of the fruit is helpful in treating dysopia, opthalmities and pharyngitis. The latex can also be mixed with vinegar to heal abscesses, snakebites and glandular swellings. The seed starch is useful in relieving biliousness, while the roasted seeds are regarded as aphrodisiac. To heal ulcers, the ash of its leaves is burnt with corn and coconut shells and used either alone or mixed with coconut oil. Its roots are used to prepare a remedy for skin diseases, fever and diarrhoea. The heartwood of the tree is used by Buddhist forest monastics in Southeast Asia, for dying the robes of the monks to light brown colour. In China, the pulp and seeds of jackfruit are considered as a cooling and nutritious tonic. The fruit is useful in overcoming the influence of alcohol on a person's body system.

BOTANY

The genus *Artocarpus* consists of about 50 species, out of which, only two species are of horticultural significance, *e.g.*, jackfruit (*Artocarpus heterophyllus*) and breadfruit (*Artocarpus altilis*). Jackfruit is a diploid with $2n = 2x = 56$. Jackfruit is a handsome tree that can grow up to a height of 9–21 m, with evergreen, alternate, glossy and leathery leaves of 22.5 cm length. The jackfruit tree is monoecious, bearing male and female flowers on the same tree. Tiny male flowers are borne in oblong clusters, while the female flower clusters are rounded. Large number of flowers is borne on club-shaped rachis. The inflorescence is a spike covered by two spathes. The female spikes are borne on footstalks, while the male spikes appear both on the footstalks as well as on the terminal branchlets. Footstalks bearing female spikes are much more vigorous than carrying only the male or both the male and female spikes. The ratio of male to female inflorescence varies. There is always a higher proportion of male than female inflorescence. The spike can easily be identified when small, as the length and diameter of female inflorescence is larger than those of male inflorescence.

CLIMATIC REQUIREMENT

Jackfruit grows well and gives good yield in warm humid climate of hill slopes, and hot and humid climate of plains. The crop grows successfully from sea level up to an elevation of 1200 m at an optimum temperature range of 22–35°C. It cannot tolerate frost or drought. The yield and quality of fruits are medium under low humidity. The west coast plains with high humidity are found to be highly suitable. Though the jackfruit is essentially a tree of the tropical lowlands, it is adapted to a wider range of conditions. It can tolerate higher altitudes and cold better than the breadfruit.

SOIL REQUIREMENT

Jackfruit can be grown on a variety of soils as long as they are well drained but it thrives best in deep alluvial soils of open texture. A deep rich alluvial or open-textured loamy soil or red laterite soils with slightly acidic condition (pH 6.0–6.5) with good drainage is ideal for jackfruit. However, it can come up in variety of soils. The tree exhibits moderate tolerance to saline soils and poor drought and flood tolerance. The jackfruit flourishes in rich, deep soil of medium or open texture, sometimes on deep gravelly or laterite soil.

CULTIVATED VARIETIES

Being a cross-pollinated and mostly seed propagated tree, its innumerable types of fruits differ widely in density of spines, rind, bearing, size, shape, quality and period of maturity. Local selections are named as *Gulabi* (rose scented), *Champa* (flavour like that of Champak), *Hazari* (bearing large number of fruits). *Rudrakshi* has common pumello-sized fruits with smooth rind and less spines, whereas, *Singapore* or *Ceylon Jack* introduced from Ceylon is highly precocious. Sometimes, it produces light off-season crop between September and December. *Muttam Varikka* is another important variety producing fruits of 7 kg each. Jackfruit *NJT 1, NJT 2, NJT 3* and *NJT 4* collections from Faizabad have large fruits of excellent quality with thin rind and soft flesh. They are suitable for culinary purpose. *Varikka, Koozha* and *Navarikka (Pazam Varikka)* types are available in Kerala, Tamil Nadu and Karnataka.

Based on firmness of the bulb, the cultivated or naturally found jackfruit trees can be broadly classified into two types, *i.e.*, (i) firm fleshed and (ii) soft fleshed. However, there are few important varieties introduced by and released from various organizations.

Konkan Prolific

A selection developed at Dr. BSKKV, Dapoli, Maharashtra has an average plant height of 14.7 m with an average spread of 11.5 m (N-S) and average girth of 1.10 m. Growth habit is semi-spreading and fairly large with leaves dark green and alternate. The tree is monoecious with small male flowers held by pedicel. Female flowers are larger than the male and pedicel is thick. It bears fruit in multiple type and medium size (8–9 kg); the colour of fruit's skin is green when it is immature and when it is ripe, the colour of the skin is greenish yellow to brownish yellow. The fruit contains the edible, sweet, aromatic, crisply bulbs.

Singapore or Ceylon Jack

An introduction from Sri Lanka to Tamil Nadu and it bears medium size fruits, each weighing 7–10 kg. The carpels are crisp, sweet and yellow with strong pleasant aroma. It is a precocious bearer, *i.e.*, even seedling progenies will start bearing 3 years after planting (normally in other types the seedling progenies start bearing 7–8 years after planting). Fruits will be available from March to June and again from September to December.

Hybrid jack

A cultivar developed at Fruit Research Station, Kallar (Tamil Nadu) through a cross between Singapore Jack and Velippala is precocious in bearing. Carpels are bigger in size and sweeter in comparison to their parents.

Burliar-1

A cultivar developed at Fruit Research Station, Burliar, Tamil Nadu has trees of medium height with prolific bearing.

PLR-1 (Palur-1)

A high yielding variety developed at Vegetable Research Station, Palur of Tamil Nadu Agricultural University through a single plant selection isolated in Panikkankuppam village near Panruti of South Arcot District of Tamil Nadu bears fruits twice in a year *viz.*, fruits will be available in the regular jack season from March to June and an off-season crop during October to December is also available. The fully ripe fruits have flat stigmatic surface instead of a spiny surface. Each tree bears about 60–80 fruits. The average fruit weight is 12 kg, containing 115–120 bulbs. The total bulb weight per fruit is 2.36 kg, which accounts for 19.68% of the total fruit weight. Bulbs are pale yellow in colour, crisp and sweet with total soluble solids of 19°Brix.

PPI-1 (PECHIPARAI-1)

A variety developed at Horticultural Research Station, Pechiparai of Tamil Nadu Agricultural University through clonal selection from Mulagumoodu local is suitable for commercial planting as well as for planting in home gardens. Trees are medium tall, maximum bearing in tree trunk. On an average, each tree bears 107 fruits weighing 1818 kg per year in two seasons *viz.*, April to June and November to December. Carpels are sweet, crisp and tasty with pleasant aroma.

PROPAGATION METHOD

The most common and simplest method of raising jackfruit trees is from seeds, though other methods are also known. The trees do not generally breed true from seeds, thus, the practice has led to immense variation in yield, fruit characters and quality. Commercially, inarch grafting on 10 months old jackfruit seedlings is done to produce vegetative progenies. The grafts come to bearing within 4½–5 years when compared with seedlings, which take 7–8 years normally. Soft wood grafting (cleft method) on 2 months old seedlings with scion of 3–4 months old was also found to be successful (70–80%). Since the viability of seeds is very low, seeds have to be sown immediately after extraction to raise rootstocks. Rudrakshi and *Artocarpus hirsuta* are also used as rootstock.

SEED

The use of seeds is generally preferred because vegetative propagation is quite difficult. However, trees may not exhibit the characters of parent plant, take longer time to start flowering, and are generally tall. The seeds should be collected from healthy mature plants, which are prolific producers of fruits with desirable characteristics. Only large seeds are used. Immediately after extraction from the fruit, the seeds are washed in water to remove the slimy coating around the seeds. The horny part of the pericarp is also removed to hasten germination. As a rule, the seeds are sown immediately without drying because they are recalcitrant. The seeds can be stored in airtight plastic containers at 20°C to maintain their viability for about 3 months. While sowing, the seeds are laid flat or with their hilum facing downward. Germination should begin within 10 days. It is expected that 80–100% of the seeds germinate within 35–40 days after sowing. After 1 or 2 years, the seedlings are planted at site. Soaking seeds in 25 ppm Naphthalene Acetic Acid (NAA) for 24 hours improves their germination and seedling growth.

CUTTING

Propagation by cutting is not so common, pre-conditioning (especially etiloation) or stock plant treatment with ethephon before taking cutting and treating the cuttings with Indol-3-Butyric Acid (IBA) (5000 ppm) and *p*-hydroxybenzoic acid (200 ppm) at the time of planting was found effective in inducing rooting of semi-hardwood cutting.

LAYERING

Treatment with IBA markedly improves the root formation in air layering of shoots like girdling, etiolation and treatment with IBA and NAA are more effective in rooting of air layers.
Mound layering was also found to be successful in jackfruit and treatment with IBA improves rooting of layered shoots.

GRAFTING

In South India, inarching of jackfruit is successful using *Artocarpus hirsute* or Rudrakshi as rootstocks. Epicotyl grafting with mature, plump, terminal scion shoot on germinating jackfruit seedling of about 8–10 days by wedge method during April to May could be successful. Epicotyl grafts attain saleable size within a year. The grafts become ready for planting 1 or 2 years after grafting.

BUDDING

Forkert budding, chip budding and patch budding may be used successfully for propagation in jack-fruit among different methods of budding.

MICRO-PROPAGATION

For micro-propagation of jackfruit, the new plantlets can be raised successfully from nodal explants and shoot apices through culturing on Murashige and Skoog (MS) medium supplemented with vari-ous concentration of cytokinin for shoot proliferation and sub-cultured on similar medium supple-mented with different concentrations of auxin for root regeneration.

PLANTING

Square system of planting is generally followed for jackfruit planting. Hexagonal system may be fol-lowed in soil having less fertility. In fertile soil, spacing up to 12 m ×12 m accommodating 70 jack-fruit plants per hectare will be sufficient. For planting jackfruit, 1 m³ pits are dug at least 10 days before planting. About 30 kg well-decomposed farmyard manure and 500 g single superphosphate are mixed with the soil of each pit and the pit is refilled. Chloropyriphos may be applied in the pit to avoid termite attack. *In situ* planting of 3–4 seeds per pit leads to stronger plant but nursing and raising of a large number of plants in this way are difficult. After planting, the soil is pressed firmly to avoid waterlogging in pits during rainy season, as jackfruit cannot withstand water logging.

The best time for planting grafts or seedlings is on the onset of monsoon (June to August). Prolonged dry weather after planting may lead to more mortality of the plants. The taproot should not be disturbed while planting to avoid damage to plants.

AFTER CARE

The leaves of young jackfruit plants being a favourite feed for cattle are frequently damaged by stray goats and cattle unless adequately guarded by providing gabions for about 2 years. Gabions may have to be replaced after a year. Manual watering of young plants during one or two sum-mers is necessary for assured survival and good growth of plants. In cooler regions, protection against frost at least during first few years is safe. Cleaning of basins by spading and ploughing of orchards should be followed as a routine measure. Weeding and mulching are essential to achieve standard plant growth.

NUTRITIONAL REQUIREMENT

A bearing tree requires farmyard manure 10–50 kg, nitrogen 750 g, phosphorus 4 kg, potash 5 kg and ash 5 kg. Tree should be irrigated immediately after manure and fertilizer application. The flower buds appear on trunk, which should be kept free of vegetative growth. The nutrients should be applied as given in Table 113.2.

The fertilizers needed to be applied during rainy season. If irrigation is available, it can be applied in two split doses once in June to July and another in September to October. The manure and fertilizers can be applied in a circular trench about 50–60 cm away from the trunk. It has been noticed that the young fruits suffer from browning, and mature fruits show the symptoms of devel-oping spongy and corky tissues along with whitish pockets in the fruit mesocarp. This problem is

TABLE 113.2
Nutritional Requirement of Jackfruit

Nutrients	1 Year after Planting (kg/plant)	Annual Increase (kg/plant)	5th Year Onwards (kg/plant)
Farmyard manure	10	10	50
Nitrogen	0.150	0.150	0.750
Phosphorus	0.080	0.080	0.400
Potash	0.100	0.100	0.500

caused due to lack of boron in the soil, which can be managed by spraying the trees with 1% solution of Borax at monthly interval starting from January to May or by adding borax at 250 g per tree along with fertilizer application.

IRRIGATION REQUIREMENT

Irrigation is hardly given to jackfruit. The young trees are sensitive to drought, thus, for better growth of the seedlings, irrigation should be done during summer and winter months. The old tree is also sensitive to drought, therefore, irrigation during dry periods is considered essential in arid regions for normal plant growth. It is cultivated in Northeast as a rainfed crop. Generally, ring system of irrigation should be adopted for irrigating jackfruit trees as it also economizes the use of water. For young orchards, manual watering is necessary during first 2–3 years. The frequency of irrigation will depend on the soil moisture conditions.

TRAINING AND PRUNING

Properly trained jackfruit grows with an open centre, which allows better light penetration. The height of jackfruit tree, especially those raised from seed, can be regulated by cutting the main trunk about 2–3 m from the ground. Early pruning of the main trunk can also be done to induce production of branches, allowing 4 or 5 branches to develop, which are evenly distributed when viewed from the top. Weak, dead, diseased and overlapping branches should be removed. This is done to encourage light penetration and air movement and to prevent buildup of insect pests and disease. Branches are also removed if they hinder access to the fruits during harvesting.

INTERCROPPING

Jackfruit starts bearing after 5–7 years of its planting. Suitable intercrops should be grown every year until the trees come into bearing stage. When the soil moisture is not a limiting factor, vegetable crop such as okra, brinjal, chilli, tomato and pulses such as green gram, black gram and Bengal gram, etc., can be conventionally grown. These crops will also improve the fertility status of the soil.

FLOWERING, FRUIT SET AND MATURITY

Flowering generally takes place in the month of December and lasts up to March. Occasionally, though rare, off-season flowering may be noticed in September to October. A single male flower is actually a stamen enclosed in a green leathery tubular perianth. During anthesis, the stamens protrude out of the perianth tube and appear on surface of the spike about 4–6 days. In a female flower, the perianth encloses the ovary and style. Creamy white stigmas protrude out of the surface of the female spike, 4–6 days after opening from the spathes. In another 4–6 days, the whole surface looks wooly. The receptivity of the stigma continues for about 36 hours. The whole surface of the male spike is covered with undehisced yellow anthers or ashy grey dehisced ones held on filaments. After dehiscence, the male spike gradually turns black due to the growth of a mould on it and drops down in another week time. The jackfruit is anemophilous. Under open condition, the fruit set is about 75%, which is improved by hand pollination. The best medium for pollen germination and pollen tube growth was 1% agar and 10% sucrose. Flowering starts at the age of 2–8 years for trees propagated from seed. The vegetatively propagated trees start fruiting within 2–4 years after planting under favourable conditions. The trees bear flowers and fruits year round under favourable conditions, but usually, there are peak harvest periods from April to August or September to December in Malaysia, from January to May in Thailand and during summer in India. Aggregate fruit (Syncarp) is formed by enlargement of the entire female head. Seeds are large without endosperm and germination is hypogeal. Cauliflorous fruiting habit is noticed in jackfruit. In natural condition, the fruit set is about 60–70% and fruit development takes about 120–140 days, depending on the type and climatic conditions. Maturity can be judged by (i) change of skin colour from light green to yellowish or brownish green, (ii) hollow sound produced by fruit when tapped and (iii) well developed and well-spaced fruit spines.

HARVESTING

The immature and tender jackfruit, when used for cooking purpose should be harvested before hardening of seeds. Tender jackfruit is harvested for use as vegetable during early spring and summer until the seeds harden. Its fruits mature in June. The small-sized fruits mature faster than big fruits. The ripe fruits are harvested at 120–140 days. While cutting the fruits, latex is exuded. This can be prevented from sticking by coating the knife and hand with edible oil.

YIELD

Seedling trees start bearing from 7th to 8th year, while the grafted trees from 3rd year onwards. The yield of jackfruit varies widely depending upon the types and climatic conditions. The tree can yield 100–300 fruits annually. The largest of all trees produces fruits that can reach 20–90 cm in length and weighing 0.5–20 kg or as much as 50 kg. On an average, about 40–50 tonnes of fruits per hectare could be obtained.

POST-HARVEST MANAGEMENT

Promotion of scientific means of post-harvest handling, primary processing and preservation of jackfruit is essential so that the raw materials are made available to consumers and entrepreneurs round the year. The market potential of jackfruit can be better exploited if fruits are made available to the consumers in a ready to eat or ready to cook form throughout the year. Some value-added products such as pickle, squash, leather, *pakoda*, jam, *kheer*, custard, wine, etc., can be prepared from jackfruit.

The initial quality and stage of maturity at harvest are important factors, which affect the storage life. In most cases, ripening is not a problem. The fruit ripens when the maximum temperature reaches during the end of the summer season. In colder regions, the fruits mature late. Fruit firmness, dry matter, alcohol, insoluble solids and starch contents decrease, whereas pulp colour (carotenoid pigments), sucrose concentration, total lipids, total fatty acid and total sterols increase during ripening. Jackfruit is not normally stored in cold store. A storage life of about 6 weeks is expected when stored at 11.1°C temperature and 85–90% relative humidity. Spraying of $CaNO_3$ (2%) and $CaCl_2$ 15 days before harvesting reduces the weight loss and maintains shelf life for about

19 days under ambient room conditions with best flavour and firmness. Jackfruit has great potential for value addition. More than 100 items can be prepared from jackfruit right from immature stage to well-ripened stage.

INSECT PESTS

Shoot and Fruit Borer (*Diaphania caesalis*)

The shoot and fruit borer are one of the major insect pests of jackfruit in India and Bangladesh. The insect lays eggs on tender shoots and flower buds. On hatching, the reddish brown larvae bore into shoot, flower buds and fruits, resulting in the wetting of affected parts. Larvae make small holes and enter the fruit. At the initial stage of damage, a small hole with fresh excreta can be seen. Gradually, the hole is extended, and at a later stage, fungal infection occurs. The infested fruits frequently get rotten due to the entrance of rainwater into the fruits. The entrance hole is easily visible and associated with mass of excreta. As a result, the affected parts wilt and dry. Severe infestation may cause dropping of fruits.

Control
- Remove and destruct the affected shoots, buds and fruits along with caterpillars.
- Prune the dead and diseased twigs and small shoots within the canopy regularly soon after harvest or at the end of rainy season to allow sufficient light and air to pass through.
- Cover young, emerging fruits with polythene or alkathene bags to prevent from egg laying.
- Spray *neem* oil 4 mL, or Profenophos 1 mL, or Malathion 2 mL/L of water.
- Fumigate the holes in trees to destroy the borers.

Mealy Bugs (*Drosicha mangiferae*)

These are flat, oval, waxy white-coloured bugs found in clusters on tender shoots, inflorescence and fruits of jackfruit. They suck sap from shoots, inflorescence and fruits, and thereby, the vitality gets reduced.

Control
- Plough the orchard deep immediately after harvest to expose eggs and pupae.
- Apply heavy irrigation to destruct the eggs of mealy bugs.
- Rake the soil around the tree trunk and mix 2% methyl parathion dust at 250 g per tree during November to December.
- Fasten the tree trunk with 400 gauge polythene sheet up to 30 cm above the ground level to prevent the movement of early instar nymphs of mealy bug during December to January.
- Apply *Cryptolaemus montrouzieri* and *Leptomastix danylopi* formulations to bring down the mealy bug nymphal and adult stages of the mealy bug.
- Spray nicotene sulfate in water (1:600) or Metacid 0.1% to control the mealy bug effectively.

Bud Weevil or Leaf Eating Weevil (*Ochyromera artocarpi*)

The bud weevil *is* a specific pest of jackfruit. The adult weevils are greenish brown in colour and are found to eat the leaves. The small whitish grubs bore into tender flower buds and fruits and induce premature drop. Thereafter, the buds and fruits will fall subsequently.

Control
- Remove and destruct the infested shoots, flower buds and fruits to check infestation.
- Collect and kill the grubs and adults followed by a spray of Trizophos 2 mL/L of water.

APHIDS (*Greenidia artocarpi* and *Toxoptera aurantii*)

Greenidia artocarpi is a pale green aphid having long hairy cornicles. Both the species of aphids have been found feeding in clusters on tender shoots and leaves. Due to severe infestation, the leaves are destroyed and covered with sooty mould. Dry weather followed by rains favour the multiplication of aphids, resulting in severe infestation.

Control

* Install yellow fast-coloured sticky traps for monitoring of aphids (one trap for five trees)
* Apply *mahua* oil 2% but harvest the fruits only after 2 weeks of spraying.
* Spray Bio-Magic-F 3 kg/ha and *neem* oil 0.4% or acephate 1.5 g/L of water.

TRUNK BORERS (*Elaphidion mucronaitum, Nyssordrysina haldemani, Leptostylopsdis terraecolor* and *Apriona germani*)

A number of woodborers attack the trunks and branches of jackfruit. These borers bore into the trunks and branches. Affected parts wilt and subsequently dry. In case of severe infestation, the fruits drop down.

Control

* Remove and destruct the affected parts as soon as the attack is first noticed.
* Cover the fruits with polythene bags to protect them from oviposition.
* Fumigate the holes in the trees to destroy the borers.
* Plug all the borer holes except fresh one with mud plastering and plug the fresh hole with 0.5% DDVP.
* 2,2–dichlorovinyl dimethyl phosphate CD
* 2,2–dichlorovinyl dimethyl phosphate (DDVP)

DISEASES

FRUIT OR BLOSSOM ROT (*Rhizopus artocarpi*)

Young fruits and male inflorescences are badly attacked by the fungus and only a small percentage of fruits reach maturity. Female inflorescence and mature fruits are not usually attacked. A large number of affected fruits fall off early. In first stage of attack, the fungus appears as greyish growth with abundant mycelia, which gradually become denser forming a black growth. The fungus gradually advances until the whole fruit or the entire inflorescence rots and falls off. It may cause 15–32% crop loss.

Control

* Prune the tree to encourage good ventilation and to reduce relative humidity in the canopy.
* Remove and destruct the diseased fruits from the tree and the ground.
* Follow proper sanitation and weed management.
* Prevent the orchard from water logging.
* Avoid storage of fruits after harvest in hot and poorly ventilated containers.
* Mix *Trichogamma viride* at 100 g per tree during inter-cultivation or drenching around the tree.
* Spray young fruits with Bordeaux mixture 0.5%, mancozeb 0.25%, or copper oxychloride 0.25% at an interval of 3 weeks during the months of January, February and March.

DIEBACK (*Botryodiplodia theobromae*)

This is a destructive disease of jackfruit. The affected leaves turn brown and their margins roll upwards. At this stage, the twigs or branches die, shrivel and fall, and there may be an exudation of gum from affected branches. Such branches have also been found to be affected by shoot borers and shot hole borers. The infected twigs show internal discolouration when split open. In early stages, epidermal and sub-epidermal cells of twigs are often slightly shriveled. On such twigs, *Colletotrichum gloeosporioides* is also observed. Infected twigs show internal discolouration.

Control

- Shoot the borers by spraying suitable insecticide to reduce the dieback disease.
- Prune the infected twigs followed by a spray of carbendazim 0.1%, chlorothalonil 0.2% or Topsin M 0.1%.

LEAF SPOT (*Phyllosticta artocarpina*)

This is the most common disease occurring on jackfruit. The disease is characterized as dark brown to brick red spot on both the leaf surfaces. These spots later turn to greyish white centres with dark brown boundaries. Dark-coloured acervuli of the fungus are observed on greyish region, which when torn ooze out profuse spore masses. The fungus has been seen to infect the young tender shoots and petioles of the leaves too.

Control

- Spray Bordeaux mixture 0.5%, methyl thiophonate 0.1%, chlorothalonil 0.2% or mancozeb 0.3% alternatively.

PINK DISEASE (*Botryobasidium solmonicolor*)

The disease appears as a pinkish powdery coating on the stem. Pink colour represents profuse conidial production of fungus. Young woody branches of the affected trees lose their leaves and show dieback.

Control

- Prune the affected branches and paste the cut end with Bordeaux paste or copper oxychloride.

■■■

114 Plantain

S.C. Swain, R.K. Tarai, Asish Kumar Panda and D.K. Dora

Botanical Name : *Musa* spp.

Family : Musaceae

Chromosome Number : 2n = 22, 33 or 44

ORIGIN AND DISTRIBUTION

Banana plants were originally classified by Carl Linnaeus into two species *viz.*, *Musa paradisiaca* and *Musa sapientum*. The former is used as cooking banana (plantain) and the later is used as dessert. The edible bananas and plantains (2n = 22, 33 or 44 and the basic number being *n* = *x* = 11) have been derived from two wild species, *i.e.*, *Musa acuminata* (A genome) and *Musa balbisiana* (B genome). The primary centre of diversity of cultivated banana is Southeast Asia. *M. acuminata* is said to have originated from Malaysia and the hardy *M. balbisiana* from Indo-China. The low land areas of West Africa contain the world's largest range of genetic diversity in plantains (*Musa* AAB). All the edible bananas and plantains are indigenous to the warm and moist regions of tropical Asia, comprising the regions of India, Burma, Thailand and Indo-China. The edible plantain is believed to have originated in the hot tropical regions of Southeast Asia. India is believed to be one of the centres of origin of plantain. Its cultivation is distributed throughout the warmer countries and is confined to regions between 30°N and 30°S of the equator. It is also grown in many other countries of the world *viz.*, Bangladesh, the Caribbean Islands, the Canary Islands, Florida, Egypt, Israel, Ghana, Congo, South Africa, Fiji, Hawaii, Taiwan, Indonesia, the Philippines, South China, Queensland and Sri Lanka.

INTRODUCTION

Plantains (*Musa* spp., AAB genome) are plants producing fruits, which remain starchy even at maturity and need processing before consumption. Plantain (*M. paradisiaca*) is an herbaceous perennial plant. Banana and plantain have no formal botanical distinction but they are distinguished based on how their fruits are consumed. The word banana is applied to all table varieties whereas the plantain represents cooking banana and they also have been given the status of separate species. Thus, plantains with persistent bracts and male flowers with large fruits held in a lax manner and eaten after cooking are assigned to *M. paradisiaca*, whereas, banana with small fruits, held compact in a bunch become soft when ripe with non-persistent bracts and male flowers, belongs to *M. sapientum*. Plantains are characterized by orange-yellow colour compound tepals in flowers and orange-yellow colour fruit pulp at ripening. Other morphological features include slender and angular to pointed fruits, incurved petiole margins, slightly hairy peduncles and glabrous fruit skin.

Plantains are firmer and low in sugar content than dessert bananas. When the fruit is removed at raw stage, it is cooked as a vegetable and when removed after ripening, it is eaten as a fruit. Plantains are distinguished from bananas by their fruit, which are although morphologically very similar to banana but they are actually longer, firmer and possess a higher starch content and thicker skin as compared to their sweeter relatives. During ripening, the colour of the plantain changes from green to yellow to black. Plantains usually resemble green bananas. However, ripe plantains may be black in colour. They also have natural brown spots and rough areas.

Banana and plantain are continuously exhibiting a spectacular growth worldwide. Its year round availability, affordability, varietal range, taste and nutritive and medicinal value make it a favourite fruit among all classes of people with good export potential. Banana and plantain are the fourth most important staple crops after rice, wheat and maize in the world and are critical for food security in many tropical countries. Banana and plantain were consumed maximum in African countries with 40–45 kg/year/capita as against 10.5 kg/year/capita in India, and both are the main staple food in East and West African countries. Most of the plantains grown in West Africa are exported to France and European markets. Thus, cultivation of banana and plantain forms the food security of the millions of people and sustenance of small and marginal farmers in African and Asian countries. The plantain averages about 65% moisture and the banana about 83%.

COMPOSITION AND USES

COMPOSITION

Plantains are low in saturated fat, cholesterol and sodium. They are also good sources of minerals, especially potassium and vitamins in particular vitamin C. It is well known that fruits contain various antioxidants such as β-carotene, vitamin C and vitamin E. Both banana and plantain contain complex carbohydrates capable of replacing glycogen and important vitamins, predominantly B_6 and C and minerals such as potassium, calcium, magnesium and iron. Cooked bananas or plantains are rich in energy that the body needs while working and playing. Potassium is a vital mineral for regulating metabolism and maintaining normal blood pressure. The nutritional composition of plantain is given in Table 114.1.

USES

Plantains are eaten as a vegetable and are cooked prior to consumption. They are an important component of many dishes in Western Africa and Caribbean countries. The banana flower buds used as vegetable particularly rich in nutrients. Chips are one of the most popular snacks made out of the plantains (raw or ripe). Plantain flower and bunch ends are used to make dry curry in Tamil Nadu and Kerala and is considered highly nutritious. After harvesting the fruit, the central core of the pseudostem is used for making delicious cuisines. The peeled layers of pseudostems are used by farmers as a binding rope for packaging agricultural produce such as flowers, betel leaves etc. The

TABLE 114.1
Nutritional Composition of Plantain (per 100 g Edible Portion)

Constituents	Contents	Constituents	Contents
Carbohydrates (g)	31.9	Iron (mg)	0.6
Protein (g)	1.3	Vitamin A (IU)	1127.0
Total fat (g)	0.37	Thiamine (mg)	0.052
Dietary fibre (g)	2.30	Riboflavin (mg)	0.054
Calcium (mg)	3.0	Niacin (mg)	0.686
Phosphorus (mg)	34.0	Pyridoxine (mg)	0.299
Potassium (mg)	499.0	Vitamin C (mg)	18.4
Magnesium (mg)	37.0	Vitamin E (mg)	0.14
Sodium (mg)	4.0	Vitamin K (µg)	0.7
Zinc (mg)	0.14	Energy (kcal)	122.0

dried stem peels can be made into fine threads and are used for weaving mats, stringing garlands and packaging wrapper. According to South Indian tradition, it is a ritual to eat food or entire meal served on a plantain leaf. They also have a religious significance in many Hindu rituals. In some parts of East Africa, the plantain is regarded as a staple food and beer-making crop particularly in central and eastern Uganda and Tanzania.

MEDICINAL USES

The high potassium content of plantain helps in regulating the body's electrolyte balance and controls blood pressure. It has been reported that regular intake of the juice taken from the shoot or the shoot consumed as a salad cures various ailments like stomach ulcers and kidney stones. The leaves are used by the tribals of Western Ghats for bandaging cuts, blisters and ulcers. The juice pressed from the pith is given to cure snakebites. The villagers and hill people of Jammu bandage the stomach with tender banana leaves smeared with mustard oil to get relief from loose motion. The tribals of Assam, Meghalaya and Nagaland use a piece of fresh banana leaf as a cool and pleasant shade to eyes in various eye infections. The leaf sheath and leaves are burnt and the ash dissolved in water is given to patients suffering from acidity and indigestion. It has also been reported that an aqueous extract derived from the unripe plantain banana is rich in the flavonoid-leucocyanidin, which has significant anti-ulcerogenic potential. Amino acids in plantain particularly tryptophan help the brain to emit the right levels of serotonin necessary for mood regulation.

BOTANY

Plantains are large perennial herbs with an underground stem called a corm, which is the true stem of the plant. The corm produces aerial shoots that arise from the lateral buds and develop into eyes and later suckers. The aerial shoot is called a pseudostem and it grows to a height of 2–8 m, depending on the variety and the growing conditions. Like banana, the plant is tall and tree-like with a sturdy pseudostem and large broad leaves arranged spirally at the top. The leaves are large blades with a pronounced central midrib and obvious veins. They can reach up to 2.7 m in length and 0.6 m in width. Each pseudostem produces a group of flowers, from which, the fruits develop in a hanging cluster. The pseudostem consists of large overlapping leaf bases that are tightly rolled round each other forming a cylindrical structure. The roots that range from 50 to 100 cm in length are initiated from the corm. The corm also consists of the apical meristem, from which, the leaves and, ultimately, the flowers are initiated. On an average, each plant produces 35–50 leaves in its growth cycle. When the plant has formed on an average of 40 leaves within 8–18 months, the terminal bud of the corm develops directly into the inflorescence, which is carried up on a long smooth unbranched stem through the centre of pseudostem emerging at top in the centre of leaf cluster. The inflorescence is a compound spike of female and male flowers arranged in groups. Each group consists of two rows of flowers, one above the other, and the whole collection is covered by a large subtending bract. The bracts and their axillary groups of flowers are arranged spirally round the axis and the bracts closely overlap each other forming a tight conical inflorescence at the tip. The lower bracts of the axis enclose female flowers, while male flowers occur at the tip of the inflorescence. The female inflorescence develops into fingers that constitute a bunch. About 12–20 flowers are produced per cluster. Individual clusters of fruits are known as hands and individual fruits as fingers. Bunches possess 4–12 hands each with at least 10 fingers. The female flowers have ovaries that develop first by parthenocarpy (without fertilization) to form pulp, which is the edible part of plantain. There are two types of plantain group, *i.e.*, (i) French plantain (male axis persistent) and (ii) horn plantain (male axis absent).

CLIMATIC REQUIREMENT

Plantain is a crop of tropics and subtropics, requiring hot and humid climate. The warm moist climate throughout the year without strong winds is most suitable for its growth. In India, banana is grown in the regions from humid tropics to humid subtropics and semi-arid tropics and from the sea level up to an elevation of 2000 m above mean sea level. It requires an annual rainfall of at least 1000 mm to survive and has a high light requirement. Plantain requires an optimum temperature of 27°C for better growth. Temperature below 10°C is not suitable for the plant since it leads to a condition called *choke* or impeded inflorescence and bunch development. In cooler climate, crop duration is prolonged, sucker production becomes slower and bunches become small. Young plantains are very susceptible to wind damage; thus, it is advisable that they should be planted in sufficient shelter or in a block so that the plants may protect one another. Both banana and plantain are water-loving crop and large areas of banana and plantain are found along the major river basins and in areas where availability of water or rainfall is plenty.

SOIL REQUIREMENT

Plantain requires a deep soil rich in organic matter. Soil depth and drainage are the two most important considerations in selecting the soil for growing plantain. The soil suitable for plantain should be 0.5–1 m in depth, well drained, fertile and moisture retentive, containing plenty of organic matter. The plants grow optimally in soil with a pH range of 6.5 to 7.5. Saline soils with the percentage of salinity exceeding 8.05 are unsuitable. Adding organic amendments is essential in sandy and heavy soils to improve water retention capacity and porosity, respectively.

CULTIVATED VARIETIES

Monthan

A culinary variety widely cultivated for processing is also suitable for cultivation for leaf production in Tamil Nadu. Being hardy, its plant can withstand drought. Bunch weight is 20–25 kg and bears 60 fingers per bunch. Fingers are large, irregularly five sided, with prominent ridges and slightly curved, broad at base and tapering towards apex. Peel is thick, tough and dark green, which changes to yellow when ripe. Pulp is firm and cream coloured. It is fairly tolerant to nematodes, leaf spot and rhizome rot diseases. Male flowers without stamens are also used as a popular dish. The crop matures in 12–14 months. It is highly susceptible to *Fusarium* wilt disease.

Ney Mannan

A vigorous variety with medium plant height bears large and compact bunches, weighing 18 kg with 10 hands. Fingers are small to medium, plump at base and tapering at apex with irregularly five sided and thick beak. Peel is thick, tough, easily removable and dark green, turning orange yellow when ripe. Pulp is firm and not very sweet. Crop matures in 12 months.

Nendran

An important variety for processing is largely grown in Tamil Nadu. Fingers are eaten both as ripe and as vegetable in mature condition. Plant has distinct pink colouration in pseudostem. Bunch is not compact, weighing around 12–15 kg and having four to six hands with three prominent ridges. Fingers are large with thick peel and have a distinct neck with thick green skin, turning yellow on ripening. Fingers remain as starchy even on ripening. Pulp is firm with mild flavour and medium sweet. This variety matures in 11–12 months. Nendran has good keeping quality but is not suitable for ratoon crop. This variety is highly susceptible to Banana Bract Mosaic Virus (BBMV), nematodes and borers.

Sakkai

A variety commercially cultivated under low input conditions owing to its tolerance to drought, salt and nematodes, takes 12–14 months for harvesting. It is dual-purpose variety with medium size plants. Bunch weight is 15–18 kg. Fingers are short and fleshy.

Bontha

Bunch of this variety has 5–6 hands. The fingers are short slightly curved with prominent ridges and rounded apex. The rind is thick and green with whitish pulp. The male bud is also used for culinary purpose.

Savarboni

Savarboni is a cooking type of banana. The well-matured unripe banana fruits are used for culinary purpose. The plants grow to a height of 4 m and girth of 65–70 cm. Initially, they produce one to two side suckers per plant. Bunches of 12–13 kg can be harvested after 425 days of planting under coconut shade, whereas under open conditions, it takes 3–4 weeks less duration. Each bunch contains 6–8 hands, with 10 fingers per hand. Each hand weighs 1.75–2.00 kg. Fruits are of big size, weighing 160–175 g, 19–20 cm long and of 3.7–4.0 cm diameter.

PROPAGATION

Vegetative Method

Commercial plantains are seedless and propagated exclusively by vegetative means. It has a reduced underground stem, called the rhizome, which bears several buds. Each of these buds sprouts and forms its own pseudostem and a new bulbous rhizome. These daughter plants are called suckers. The mother plantain plant produces two types of suckers, *i.e.*, (i) sword suckers and (ii) water suckers. Sword suckers have a well-developed base and conical shape with narrow sword-shaped leaf blades at early stages. Such suckers weighing 1.5–2 kg developed at flowering stage of parental plant having 2–4 months age are best planting material. They produce healthy and fruitful pseudostem when mature. Water suckers have short pseudostems and broad leaves. Water suckers are not strongly attached to the rhizome, thus, they generally produce weaker plants and less fruits. Well-developed rhizomes with dormant lateral buds and dead central bud are also used. Apart from these, cut rhizomes called *Bits* and *Peepers* are also used successfully for propagation. After cutting the parent plant, the rhizomes are removed from the soil, stored in a cool and dry place for about 2 months. During the resting period, the remaining part of pseudostem at the bottom falls off, leaving prominent heart bud. Conical rhizome should be selected, whereas flat rhizomes should be rejected. It should be of 3–4 months age at planting. Very small rhizomes will give bigger size fruits with late flowering, whereas the bigger size rhizomes bear flowers early but small size fruits or bunches. Since banana is highly unstable in genetic constitution, the suckers or rhizomes should be selected from plants, which are healthy, having all the desirable bunch qualities and high yielding ability possessing at least 10 hands in a bunch.

Tissue Culture

Nowadays, plantain plants are also propagated through tissue culture technique. Normally, disease-free plantlets with three to four leaves are supplied in pots for raising secondary nursery. Plants are initially kept in partial shade, and as they harden, shade is reduced gradually. After 6 weeks, the plants do not require any shade. Usually, 2 months of secondary nursery is good enough before the plants are to be planted in the field pits.

PLANTING

In general, planting can be done during June to July–December to February in India. The planting season should be such that there is no check to growth of the plant. The market situation is also important in timing of planting. The land should be ploughed deeply, harrowed and levelled properly. Plant spacing depends on the cultivar being planted. Generally, a spacing of 2 m × 2 m is followed for the planting of plantain. The different methods of planting suckers are given below:

- **Pit method:** Pits of 0.5 m × 0.5 m × 0.5 m are dug for planting the rhizomes. However, this method is very laborious and expensive. The only advantage is that no earthing up is required as planting is done at the required depth. However, this practice is not very popular at present.

- **Furrow method:** This is a very common method, in which, furrows of 20–25 cm depth are opened by a tractor driven ridger at a distance of 1.5 m and rhizomes are planted in the furrows. In this method, earthing up needs to be done frequently to cover the exposed rhizomes.

The pits are filled with 20–30 kg of well-decomposed farmyard manure and 500 g *neem* cake along with top soil. The roots of selected sword suckers are trimmed off and the decayed portion of corm is removed out in order to avoid suckers from nematodes and weevils. To avoid wilt disease in wilt susceptible varieties, the infected portion of corm may be dipped in 0.1% Emisan solution for 5 minutes. The corms, before planting, are soaked for few seconds in 550 mL dibromochloropropane (DBCP) + 40 L of clay for disinfection. This treatment will form a coat on the corm in a persistent nematicidal preparation. In place of DBCP, carbofuran can also be used at 40 g/sucker at planting time and again applied at the 4th month of crop growth. The suckers are planted at the centre of pit leaving 5 cm pseudostem above the soil level. The soil is then pressed around the sucker to avoid hollow air space. Planting is followed by light irrigation. Partial shade, if possible, should be provided immediately after planting.

NUTRITIONAL REQUIREMENT

Plantain is a nutrient loving plant, requiring large quantity of nutrients for growth and development. The nutrients uptake studies in plantain reveal that a crop giving 40–60 t/ha yield removes nitrogen 250–300 kg, phosphorus 25–40 kg and potassium 800–1000 kg. This clearly indicates that the potash and nitrogen requirement of plantain is much higher as compared to phosphorus. For shoot analysis, the petioles of third opened leaves from apex are taken at the bud differentiation stage. Nitrogen deficiency causes slow growth, pale leaves, reduced leaf area, thin roots and reduction in the sucker production. Phosphorus deficiency leads to poor root development, rotting of corm, poor growth and bad colour of leaves. Plantain requires potassium in large quantity, as it stimulates early shooting, increased number of hands, finger size and improves quality and shelf life. It also imparts tolerance to diseases and reduces water uptake. Deficiency of potassium leads to improper bunch filling.

In general, the nutrient requirement of plantain as worked out on all India basis is 10 kg farmyard manure, 200–250 g nitrogen, 60–70 g phosphorus and 300 g potash per plant per year. Full dose of phosphorus and potash is applied at the time of planting, while nitrogen is applied in three equal split doses. Nutrients should be placed 60–75 cm away from the pseudostem in shallow basins made around the plant. Application of nutrients should be completed before 6 months after planting suckers in order to harness higher yield and quality fruits. Zinc sulfate 0.5%, ferrous sulfate 0.2%, copper sulfate 0.2% and boric acid 0.1% sprayed in combination at 3, 5 and 7 months after planting helps in increasing yield and quality of banana. Use of biofertilizers has been warranted considering the soil health and sustainable production. *Azospirillum* and *phosporus solublizing bacteria* (PSB) 25 g each per plant at the time of planting is recommended for increasing production along with organic sources, *i.e.*, 10 kg farmyard manure or 5 kg vermi-compost at the time of planting.

FERTIGATION

In order to avoid loss of nutrients from conventional fertilizers, *i.e.*, loss of nitrogen through leaching, volatilization, evaporation and loss of phosphorus and potash by fixation in the soil, application of water soluble or liquid fertilizers through drip irrigation (fertigation) is adopted. Banana responds very well to fertigation, which can achieve early flowering with high production per unit area. Using fertigation increases the yield 25–30%. Moreover, it saves labour and time and the distribution of nutrients is uniform.

IRRIGATION REQUIREMENT

Plantain, being a mesophite, requires a large amount of water as evident from high moisture content of pseudostem and leaves. Water requirement of plantain has been worked out to be 1800–2000 mm per annum. In winter, irrigation is provided at an interval of 7–8 days, while in summer, it should be given at an interval of 4–5 days. However, during rainy season, irrigation is provided if required, as excess irrigation will lead to root zone congestion due to removal of air from soil pores, thereby affecting plant establishment and growth. In all, about 70–75 irrigations are provided to the crop. In garden lands, furrow and basin system of irrigation is followed.

Nowadays, majority of plantain growers are using a drip irrigation system, which is an innovative method of irrigation for providing precise and measured quantity of water directly to the root zone. The principle involves supply of water drop by drop through plastic emitters or drippers through a low-pressure delivery system. There is a savings of 58% of water and increasing yield by 23–32% under drip as compared to basin system. The drip irrigation schedule for plantain is given in Table 114.2.

INTERCULTURAL OPERATIONS

HOEING AND WEEDING

A period of the first 6 months of growth is most critical for weed growth in plantain orchard, thus, regular weeding is important during the first 6 months. Normally, four hoeings a year are effective in controlling weeds. Integrated weed management by including cover crops, judicious use of herbicides, intercropping and hand weeding, wherever necessary, will contribute to increased

TABLE 114.2
The Drip Irrigation Schedule for Plantain

No	Crop Growth Stage	Duration (weeks)	Quantity of Water (litre/plant)
1	After planting	1–4	4
2	Juvenile phase	5–9	8–10
3	Critical growth stage	10–19	12
4	Flower bud differentiation stage	20–32	16–20
5	Shooting stage	33–37	20 and above
6	Bunch development stage	38–50	20 and above

production. Pre-emergence application of diuron 1 kg a.i./ha or glyphosate 2 kg a.i./ha is effective in controlling grasses and broad-leaved weeds without affecting the yield and quality of plantain. Double cropping of cowpea is equally effective in suppressing the weed growth.

MULCHING

Mulching has major implication for conserving soil moisture with ancillary benefits like suppression of weeds and regulation of plant microclimate. Use of paddy straw, wheat straw and banana straw as a mulch material (12.5 kg/plant) in banana orchards is cost effective apart from addition of organic manure to the soil. It is useful in increasing the bunch weight and conservation of soil moisture. The mulch is applied at the beginning of summer (February). Plastic mulching is now widely applied in commercial cultivation of plantain in India.

INTERCROPPING

Intercropping is a common practice in plantain orchards to check weed growth, improve soil health and to generate additional income to the farmers. Intercropping can easily be raised in plantain field at early stages of growth. Vegetables such as radish, cauliflower, cabbage, spinach, hot pepper, brinjal and okra, and flower crops such as marigold and tuberose can successfully be grown as an intercrop. However, growing of cucurbitaceous vegetables should be avoided, as they are bearer of viruses. In coastal regions of Karnataka, Kerala and Andhra Pradesh, plantain is grown in coconut and areca nut plantations with tall cultivars, as mixed cropping is a widely adopted practice in South India.

DESUCKERING

Plantain is traditionally propagated through suckers. During the life cycle, plantain produces a number of suckers from the underground stem. Sword suckers have a well-developed base with narrow sword-shaped leaf blades at early stages. Water suckers with broad leaves do not produce healthy banana clumps. Planters throughout the world usually plant healthy sword suckers. If all these suckers are allowed to grow, they grow at the expense of main plant growth, and hence, the sucker growth should be discouraged. Removing unwanted suckers, which is known as desuckering, is one of the most critical operations in plantain cultivation. In pure crop of plantain, where only single crop is taken, usually all the suckers, which arise before flowering, are removed, as they will compete with mother plant for nutrients, resulting in reduction of bunch size. Hence, suckers, which are produced by the plant in excess or out of place, are removed periodically to ensure better growth and bunch development of mother plant. In perennial system of plantain culture, the *setting of followers* at proper time will ensure good ratoon crop to the plantain growers.

The suckers are removed either by cutting them off or the heart may be destroyed without detaching the sucker from the parent plant. Removing suckers with a portion of corm at an interval of 5–6 weeks hastens shooting and increases yield. The unwanted suckers are also removed after heading off and killing them by pouring kerosene (5–10 mL per sucker) over the cut surface of the sucker. While using kerosene, it should be avoided to treat suckers, which arise on the corm above ground level and have no root system. These suckers are entirely dependent on the parent plant and the kerosene may move back into the sap stream and injure it. The best method of handling these suckers is to cut off level with the parent corm. Desuckering of plantain by using chemicals like, 2,4-dichlorophenoxyacetic acid (2,4-D) is also found to be in use. A pruning paste consisting of a mixture of 2,4-D, fuel oil and grease may also be applied for desuckering. Removing all suckers up to flowering of the plant and maintaining only one follower afterwards is the best desuckering practice.

EARTHING UP

Earthing up is an important practice followed in plantain cultivation. It provides support to the base of the plant and helps to develop good root system. Earthing up should be done during rainy season to avoid water logging and provide good drainage.

PROPPING

The bearing plants may lodge and uproot due to heavy bunch load. Therefore, the plants are traditionally staked with bamboo, *Eucalyptus*, or *Casuarina* poles. As the supports are placed against the stems on leaning side of the bunch emergence, it has a disadvantage of causing bruising injury to the bunch. These are now replaced with the use of polypropylene strips for supporting the pseudostem. In this method, the strip is fastened to the neck of bearing plant and the two ends of the strip are tied to the base of neighbouring plants on the side opposite to the leaned direction. This reduces the cost of propping as well as the bruising injury to the bunch. Propping operation is carried out in areas with high wind speeds. Tall varieties and cultivars, which produce heavy bunches, need propping.

LEAF REMOVAL

Pruning of surplus leaves helps in reducing the disease from spreading through old leaves. Leaf pruning can change light and temperature factors of microclimate. Pruning of leaves before bunch initiation delays flowering and harvesting cycle. For obtaining maximum yield, a minimum of 12 leaves are to be retained.

BUNCH COVERING

Bagging (bunch covering) is a cultural technique used by planters where export quality plantains are grown. This practice protects bunches against cold, sun scorching, attack of thrips and scarring beetles. It also improves certain visual qualities of the fruits. Bunch covering can be practiced with dry leaves, gunny bags or polyethylene sleeves. It is done after the emergence of last hands of the bunch.

DENAVELLING

Removing male bud after completion of female phase is necessary. Once the process of fruit setting is over, the inflorescence rachis should be cut beyond the last hand, otherwise it grows at the cost of fruit development. This helps in early maturity of the bunch. The removed male buds can be used as vegetable.

BUNCH MANAGEMENT

The male bud is removed after the completion of female phase. It serves a dual purpose, *i.e.*, (i) saving of movement of food into unwanted sink and (ii) avoiding thrips attack on fruits, which remain underneath the male bud. Dehanding at sixth or eighth hands improves the fruit quality parameters to meet the export standards. Banana Research Station, Jalgaon has recommended the two sprays of potassium dihydrogen phosphate 0.5% + urea 1% on plantain bunch 5 and 20 days after last hand opening, which helps in improving the overall quality of bunch, enhancing maturity and yield of banana. The bunches are then covered with perforated (2–6% vent) 100 gauge polyethylene bags, which protect the bunch from thrips, beetles, dust, rains and from low temperature during winter, scorching sun in summer and enhances the maturity with increased yield.

OTHER ORCHARD OPERATIONS

Plantation of plantain should be kept clean by removing the unwanted plant parts such as dried, diseased and decayed leaves and pseudostems after the harvest. During winter months if temperature goes below 10°C, the plant growth is adversely affected. Under such circumstances, irrigation is to be provided at night or smoking is to be done by inducing fire.

FLOWERING AND FRUIT SET

The shoot meristem transforms into an inflorescence at about the time when the eleventh last leaf has been produced, and then, the inflorescence emerges after about a month. The inflorescence, classified as a compound spike, and the peduncle emerges upwards through the centre of pseudostem before bending down under the weight of developing spike. Three types of flower such as hermaphrodite, female and male are observed on the same inflorescence. Female flowers are larger with well-developed trilocular ovary, long style and reduced staminoides. Male flowers have well-developed anthers and abortive ovary with slender style and stigma. Usually, fertile pollen grains are absent in edible plantain. The fruit of triploid plantain cultivars are parthenocarpic (develop without fertilization). The fruit is botanically called as a berry without any seed.

PLANT GROWTH REGULATORS

In order to improve the grade of bunches, 2,4-D 25 ppm may be sprayed after the last hand has opened. This also helps in removing seeds in certain varieties. Applying 2,4-D 30 ppm at 7 days after shooting reduces the number of days required for maturity by 15–20 days and enhances the yield by 43 t/ha. After full development of bunch, potassium dihydrogen phosphate 0.5% and urea 1% or 2,4-D 10 ppm is to be sprayed on the bunches so that finger size may be improved.

HARVESTING

The duration of crop varies from 6 to 16 months, depending on ploidy level, climate and other cultural practices. A bunch usually takes 90–150 days to be mature after shooting. Irrigation of plantain plantation should be stopped well in advance of the harvesting date, preferably a week to facilitate drying of soil for the movement of labour, harvesting, loading, etc. The plantain bunch should be harvested at tender and immature stage for vegetable purpose. For cutting bunches, a cutter and a helper are required. The bunch should be cut in one stroke 20–25 cm above the first band or 7.5–10 cm from tip of the fingers of first hand. The helper should hold the same portion and place it carefully on freshly cut leaves spread on the ground. For carrying bunches to packing shed, it is necessary that after 15 minutes of harvest, when the latex flow ceases, two bunches should be taken at a time on stretchers and should not be allowed to come in contact of soil.

YIELD

Yield of plantain depends on so many factors such as variety, population density, soil fertility and management practices. The average yield of plantain is 200–300 q/ha. Higher yield up to 600–800 q/ha can be obtained under high density planting with proper cultural practices.

MATTOCKING

The practice of cutting pseudostem partially after harvesting of bunch is called mattocking. It is a term which has its origin after the name of an implement used in Jamaica to cut the pseudostem at three stages within 3–4 months of bunch harvesting. The idea of retaining a part of the pseudostem is to supply nutrients contained in the pseudostem to the followers.

PHYSIOLOGICAL DISORDERS

KOTTA VAZHAI

The word kottai means seed. It is a disorder of unknown aetiology. The enlarged ovules are present in fruits. It can be controlled by spraying 2,4-D 120 ppm.

YELLOW PULP

This disorder is characterized by abnormal and pre-mature development of plantain fruits. Excess of potassium in relation to nitrogen and sulfur deficiency is the main reason of this problem. Application of sulfur reduces this incidence in plantain.

DEGRAIN

This physiological disorder is seen in ripe fruits, which become drooped from bunches due to rotting of the pedicels.

GOOSE FLESH

This disorder is observed on plantain peel. Ripe fruits show a wilted and shrivelled appearance and the peel turns to brown. It is mainly noticed during dry winter season when atmospheric humidity falls to a relatively low level.

WHITE LEAF DISORDER

This disorder much relates to unbalanced nitrogen content in soil and plant. The younger leaves of the plant show white discolouration.

POST-HARVEST MANAGEMENT

At the time of harvesting, 15–20-cm-long peduncle is retained, which helps in handling bunches during harvesting and transportation. Care should be taken to avoid direct contact of fingers on any hard surface, which turn to black in colour in later stages. After harvesting, the grading is done according to size, appearance and number of fingers, arrangements of hands and maturity stage of fruits. As per the quality standards, plantain is graded into two grades, *i.e.*, Class I and Class II. Bunches are padded with plantain leaves. Packing of hands or fingers in 100 gauge polyethylene bags with 0.2% holes enhances shelf life. The shelf life of plantains can be enhanced by treating with Waxol (12% wax emulsion). The weight loss and spoilage can be reduced by fungicidal wax emulsion. The fruits can be stored for 3 weeks at 13°C temperature and 85–95% relative humidity. The fruits should not be placed in refrigerator as the fruits may turn black at low temperature. Post-harvest life of plantain can be increased by storing in controlled atmosphere storage.

INSECT PESTS

PSEUDOSTEM WEEVIL (*Odoiporus longicollis*)

This pest affects most of the commercial cultivars of plantain. Exudation of sap is the initial symptom, and blackened mass comes out from the holes made by the weevil larvae. The presence of small holes and jelly exudation on the stem indicates the activity of grub inside the stem. Its feeding makes the pseudostem hollow, and the apical region from this point is broken down due to gush of wind.

As the infestation progresses, the severely affected plants break or topple alone with the bunch, and finally, the whole plant dies. The banana pseudostem weevil attacks the plant during flowering and bunch formation stage and causes severe yield loss by preventing the bunch development.

Control

- Follow clean cultivation.
- Remove and destroy the older and dried leaves.
- Collect the adult weevils by using banana pseudostem traps.
- Put three Celphos tablets per plant inside the pseudostem and plaster the slit with mud for successful control of egg, larva, pupa and adults.
- Apply 3 g carbofuran granules per stool or spray the plants with quinalphos 0.05%.
- Inject monocrotophos 2 mL/plant or dichlorvos 0.25% in opposite direction, one at 60 cm height and the other at 120 cm above ground level at 30° angle on either side of the plant if the infestation is noticed after 7 months of planting.

Rhizome Weevil (*Cosmopolites sordidus*)

Rhizome weevils are nocturnal insects, which feed only at night. Adult insect is a hard-shelled beetle, which is almost black in colour and is commonly found between leaf sheaths. Larvae are creamy-white legless grubs with red-brown head. In advanced stage of infestation, plant shows tapering of stem at crown region, reduction in leaf size, poor bunch formation and choked throat appearance due to grub damage in corms. Plant growth is reduced and fruit production is decreased. Tunnels may be visible in corm as rounded holes up to 8 mm in diameter. Wilting of plants occurs followed by destruction of root system, and ultimately, death of the plant occurs.

Control

- Follow clean cultivation of plantain.
- Use uninfected healthy sucker or rhizomes for planting.
- Treat the cleaned and trimmed suckers with hot water to kill eggs and grubs.
- Wash the suckers properly and dip in 0.25% solution of chlorpyriphos 20 EC before planting.
- Apply phorate 20 g/pit at the time of planting and after 3 months subsequently.
- Dip the suckers in 0.2% monocrotophos for 30 minutes to protect the rhizomes from weevil attack.
- Avoid mattocking (leaving plant after bunch harvest for nutrients recycling) in weevil endemic area.
- Keep banana traps *viz.*, (1) longitudinal cut stem trap of 30 cm size 25–45/ha and place longitudinal split banana traps 100/ha with biocontrol agents like *Beauveria bassiana* or *Heterorhabditis indica* 20 g/trap.

Aphid (*Pentalonia nigronervosa*)

This pest is particularly important, as it is the vector of virus causing bunchy top disease. Banana aphid is soft-bodied and red-brown to almost black in colour. Colonies of aphids are usually found in the crown at the base of pseudostem or between the outer leaf sheaths. Colonies are often inclined by ants and their populations can build rapidly during warm weather. Plants become deformed with curling of leaves. In severe cases, galls may be formed on the leaves.

Control

- Follow clean cultivation of plantain.
- Collect and destroy the plants infected with bunchy top to prevent its spreading.

- Apply 10 g carbofuran 3 G in the leaf axils after 75 and 165 days of planting.
- Spray the plants with malathion 0.2% or oxydemeton-methyl 25 EC 0.2% at first sign of attack.

THRIPS (*Astrothrips parvilimbus* and *Thrips florum*)

Thrips damage the plantain plants by oviposition on young fruits, which causes brown freekling at maturity, and in severe cases, the skin becomes rough. The leaves become yellow and fruits appear rusty. In severe cases, the skin develops longitudinal cracks, and sometimes, the fruit may split. The taste and texture of the fruit within peels remain unaffected but the external appearance of affected fruits reduces their marketability.

Control

- Provide some protection by covering the whole bunch length wise but check the fruits regularly under the bunch cover to ensure that there is no damage.
- Introduce some lacewings, ladybird beetles and predacious mites to control nymph and adult thrips and ants to prey on pupae in the soil.
- Use *Paecilomyces* spp. and *Verticilliumlecanii* fungi to infect the thrips as well.
- Apply dichlorovos at 19% impregnated in 0.5 g plastic cones.

FRUIT AND LEAF SCARRING BEETLE (*Colaspis hypochlora*)

The beetle feeds on young leaves and skin of young fruits. Sometimes, the insect lives in heart of the pseudostem within the roll of central leaf. Occurrence of this pest is usually more during the rainy season. In case of severe scarring of the fruit skin, the infested fruits fetch low prices in the market.

Control

- Remove grassy weeds from the field margins.
- Follow clean cultivation by removing grasses and weeds from the orchard.
- Spray the plants with Aldrex 30 EC 0.25% a.i. or dust with malathion powder if damage is serious.

NEMATODES

Nematodes are now recognized as an important soilborne pathogen causing decline in banana yield. The affected plants do not respond to fertilizer, irrigation or cultural practices. A total of 71 species of nematode belonging to 33 genera have been reported to infest plantain, of which, only six species are pathogenic. The most destructive is burrowing nematode (*Radopholus similis*) followed by lesion nematode (*Pratylenchus coffeae*). The other nematode parasites of plantain are spiral nematode (*Helicotylenchus multicinctus*), root-knot nematode (*Meloidogyne incognita*), cyst nematode (*Heterodera oryzicola*) and reniform nematode (*Rotylenchulus reniformis*). Plant parasitic nematodes have proved to be a limiting factor in plantain production in many countries.

Control

- Follow suitable crop rotation with non-host crops.
- Use certified suckers free from nematodes.
- Grow marigold (*Tagetes* spp.) as an intercrop cum trap crop in plantain field to reduce population of root lesion nematode and to increase yield.
- Grow sun hemp or chrysanthemum 45 days after plantain planting and incorporate into the soil 1 month later.

- Apply *neem* or *Pongamia* cake 200 g per pit before planting.
- Dip the trimmed suckers in clay slurry and sprinkle carbofuran 15 g per sucker.
- Apply carbofuran granules 20 g per pit at the time of planting, 2nd and 4th month after planting or *neem* cake 400 g/plant and carbofuran 20 g/plant at planting along with the bioagent vesicular-arbuscular (VA) mycorrhiza *Glomus fasciculatum* 50 g culture and bacterium *Pasteuria penetrans* 100 g.

DISEASES

FUSARIUM OR PANAMA WILT (*Fusarium oxysporum* f.sp. *cubense*)

This is the most serious disease of plantain. The typical symptoms of this disease are yellowing, bending and breaking at the petiole and cracking of pseudostem, which appears as inverted 'V' shaped. Disease appears at later stages of plant growth, *i.e.*, 4–5 months after planting. The suckers around the infected plants look healthy but they are the carriers of disease. Panama wilt is a soil-borne fungal disease and the fungus gets entry in the plant body through roots. It is most serious in poorly drained soil. Greater incidence of the disease has been noticed in poor soil with continuous cropping of plantain.

Control

- Follow at least 3–4 years crop rotation with paddy, avoiding inclusion of sunflower or sugarcane in rotation.
- Grow resistant cultivar, if available.
- Uproot and burn the severely infected plants.
- Avoid cultivation of plantain in infected soil.
- Use disease-free healthy planting material.
- Apply bioagents such as *Trichoderma viride* or *Pseudomonas* fluorescence in soil.
- Apply *neem* cake (*Azadirachta indica*) 250 kg/ha to control most effectively.
- Dip the suckers in 0.1% carbendazim solution followed by bimonthly drenching starting from 6 months after planting.

LEAF SPOT, LEAF STREAK OR SIGATOKA DISEASE (Yellow sigatoka—*Mycosphaerella musicola* and Black sigatoka or Black leaf streak—*Mycosphaerella fijiensis*)

Yellow sigatoka is of major concern in India, while black sigatoka is a serious disease in plantain growing in the areas of Africa, Central America, South America and Mexico. In India, the disease is most serious in coastal areas where rainfall is high. The first symptom of infection is the presence of light yellowish spots on the leaves. A small number of these yellow spots enlarge, become oval and their colour changes to dark brown. Later, the centre of the spot dies, turning light grey surrounded by a brown ring. In severe cases, numerous spots coalesce, killing large parts of the leaf. Infection occurs through stomata of the young leaves, the lower surface being much more important than the upper. The extensive defoliation results in delayed flowering and reduction in number of hands and fingers. The disease inflicts serious yield loss. Fingers do not fill out properly and often remain immature even after attaining the time of full maturity. The quality of plantain is drastically reduced. Usually, rainfall, dew and temperature determine the production and movement of sigatoka inoculums.

Control

- Follow recommended cultural practices, especially plant spacing.
- Remove the infected leaves regularly once in a month starting from 2nd month of planting and destroy by burning or dumping in manure pit.
- Spray the plants three times with Bordeaux mixture 1% along with linseed oil 2%, carbendazim 0.1%, propicanozole 0.1% or mancozeb 0.25% with sticking agent at 10–15 days interval.

ANTHRACNOSE (*Gloeosporium musae*)

This disease attacks the plantain plants at all stages of growth. The symptoms appear as large brown patches covered with a crimson growth of fungus. The spread of the disease is through airborne conidia and numerous insects visiting banana flowers frequently. The disease is favoured by high temperature and high humidity, wounds and bruises. The diseased fruit turns black and shrivels.

Control

- Follow at least 3–4 years crop rotation with paddy or sugarcane.
- Maintain proper sanitation in plantain field.
- Follow proper post-harvest practices.
- Facilitate proper drainage and weed management in the orchard.
- Remove distal bud when all the hands opened to prevent infection.
- Store plantain at 14°C to minimize the disease in cold storage.
- Spray the plant with 1% Bordeaux mixture when the fruit is still young.
- Dip the fruits after harvesting in Mycostatin 440 ppm, Aureofungin-sol 100 ppm, carbendazim 400 ppm or Benomyl 1000 ppm.

PSEUDOSTEM HEART ROT (*Botryodiplodia* sp., *Gloeosporium* sp. and *Fusarium* sp.)

This is a minor fungal disease. The first indication of heart rot is the presence of heartleaves with part of the lamina missing or decayed. In severe cases, the inner crown leaves first turn yellow, then brown and finally die.

Control

- Maintain proper sanitation in plantain field.
- Facilitate proper drainage and weed management in the orchard.
- Grow the crop at proper spacing to reduce the incidence of disease.
- Spray the plants with 0.2% Captan, Dithane M-45 or Dithane Z-78.

BACTERIAL SOFT ROT OF RHIZOME (*Erwinia chrysanthemi* var. *paradisiaca*)

This is a minor bacterial disease but causing concern in particular localities. It is characterized by a massive soft odorous rot of the centre or a portion of a rhizome. It progresses to upward portion of the pseudostem and destroys the growing point, which in turn causes internal rot and discolouration of the vascular tissues. Yellowing and wilting of leaves are noticed. Sometimes, this disease is confused with *Fusarium* wilt.

Control

- Uproot and destroy the affected plants.
- Drench the field with bleaching powder at an interval of 10–15 days.
- Add biocontrol agents like *Pseudomonas fluorescens* 7 g/plant in soil.

BACTERIAL WILT OR MOKO DISEASE (*Pseudomonas solanacearum*)

The pathogen affects the young plants severely. In initial stages, the bacterial wilt is characterized by yellowish discoloration of the inner leaf lamina close to the petiole. The leaf collapses near junction of the lamina with the petiole, and within a week, most of the leaves exhibit wilting symptoms. The Moko disease induces wilting, which begins with the yellowing and collapse of the youngest leaves and necrosis of the flag leaf. These symptoms progress towards the oldest leaves, and internally, the vascular tissues become necrotic, especially, those located in central area of the pseudostem. Immature fruits of infected plants become yellow and hollow. When infection occurs before flowering, the fruiting bunch may become abnormal. Newly planted seedlings become yellow and when the pseudostem is cut, a few reddish spots or brown lines appear.

Control

- Collect and destroy the infected plants.
- Avoid the use of infected instruments for pruning and cutting operations.
- Disinfect the instruments with 2.6% sodium hypochlorite.
- Grow French marigold around the banana plantation.
- Cover healthy bunches with plastic bags.
- Remove male flowers as soon as the last female hand emerges to avoid the spread of disease.
- Incorporate composted organic matter into the soil and beneficial microorganisms such as *T. viride* and *Bacillus subtilis*.
- Dip the suckers or rhizomes in hot water or 500–1000 ppm solution of tetracycline hydrochloride or 2% solution of benomyl for 48 hours.

BANANA BUNCHY TOP VIRUS

The disease is transmitted to the plant by the aphid *P. nigronervosa*. Dwarf varieties of plantain are more susceptible to this disease. Primary symptoms of the disease are seen when the infected suckers are planted. Such infected suckers put forth narrow leaves, which become chlorotic and exhibit mosaic symptoms. The leaves arise in clusters, giving a rosette appearance. They are brittle with numerous dark green dots or patches with the margins rolled upward. The plants do not usually grow taller than 60–90 cm and they fail to put forth any fruit. The secondary infection is the premature unfurling of leaves and the development of dark green spots and streaks on the blades. These symptoms are common along the secondary veins and on the midrib and petiole. The leaves become pale and much reduced in size and when a few more leaves develop, the characteristic rosette or bunchy top symptom is evident. The symptomless infection or latency is reported to last for two seasons. If late infection occurs, the plant may sometime produce a bunch but the fingers never develop to maturity. Fruits of infected plants are malformed.

Control

- Use disease-free healthy planting material.
- Treat the suckers with malathion 0.2% and mancozeb 0.3% before planting.
- Remove alternate hosts, weeds and unwanted suckers to reduce the aphid population.
- Eradicate the diseased plants, suckers and the clumps or destroy using 4 mL of 2,4-D (50 g in 400 mL of water).
- Spray the plants with monocrotophos 2 mL, phosphamidon 1 mL or imidachloprid 0.50 mL or methyl demeton 1 mL/L of water to control the vectors.
- Inject the diseased trees with 4 mL of Fernoxone solution (50 g dissolved in 400 mL of water) or monocrotophos 1 mL/plant diluted in 4 mL of water.

BANANA STREAK VIRUS

A prominent symptom exhibited by banana streak virus (BSV) is yellow streaking on leaves, which becomes progressively necrotic, producing a black streaked appearance on older leaves. The virus is transmitted mostly through infected planting materials, though citrus mealy bug (*Planococcus citri*) and more probably pink sugarcane mealy bug (*Saccharicoccus sacchari*) are also believed to transmit it. Shoot tip culture does not eliminate it from vegetatively propagated materials. Bunch choking, abortion of bunch and seediness in fruits are seen in infected plants. Diseased plants may be stunted, fruit distorted with thinner peel and bunches small in size.

Control

- Use disease-free healthy planting material.
- Remove the diseased plants from the field and bury them.
- Spray the plants with monocrotophos 2 mL or phosphamidon 1 mL/L of water to control the virus vectors.

BANANA BRACT MOSAIC VIRUS

The symptoms appear as yellow green bands or mottling over an entire area of young leaves. The affected leaves show abnormal thickening of veins. Bunch development is affected severely. The disease is characterized by the presence of spindle-shaped pinkish to reddish streaks on pseudostem, midrib and peduncle. Suckers exhibit unusual reddish brown streaks at emergence and separation of leaf sheath from central axis. The characteristic symptoms are clustering of leaves at crown with appearance of travellers' palm, elongated peduncle and half-filled hands. Virus is transmitted through *Aphis craccivora*, *Aphis gossypii* and *P. nigronervosa* in a non-persistent manner. The virus primarily spreads through infected suckers and tissue culture plants.

Control

- Remove and destroy the affected plants along with rhizomes.
- Use disease-free planting materials for new planting.
- Avoid growing of cucurbits in and around plantain field.
- Spray the plants with Neemkosol 2.5 mL or monocrotophos 2 mL/L of water to control the virus vectors.

MOSAIC VIRUS

The disease is characterized by the presence of typical mosaic-like or discontinuous linear streaking in bands extending from margin to midrib. Rolling of leaf margins, twisting and bunching of leaves are found at the crown. The presence of dead or drying suckers is noticed in advanced stage. The leaves are narrower and smaller than normal and the infected plants are dwarf and lag behind in growth. Such plants do not produce bunches. The causal agent of this disease is cucumber mosaic virus (CMV). The primary transmission is through the use of infected daughter suckers from diseased plants, and the secondary spread of the disease is through melon aphid, *i.e.*, *A. gossypii*.

Control

- Keep the plantain field weed free.
- Use disease-free healthy planting material.
- Avoid growing pumpkin, cucumber and other cucurbits between the rows of plantain.
- Provide dry heat treatment to suckers at 40°C for a day to inactivate the virus.
- Spray the plants with methyl demeton 0.03% at 3–4 weeks interval to control the vectors.

■■■

Index

Printed and bound by CPI Group (UK) Ltd, Croydon, CR0 4YY

24/10/2024

01778310-0005